Lecture Notes in Artificial Intelligence 5212

Edited by R. Goebel, J. Siekmann, and W. Wahlster

Subseries of Lecture Notes in Computer Science

Walter Daelemans Bart Goethals
Katharina Morik (Eds.)

# Machine Learning and Knowledge Discovery in Databases

European Conference, ECML PKDD 2008
Antwerp, Belgium, September 15-19, 2008
Proceedings, Part II

Series Editors

Randy Goebel, University of Alberta, Edmonton, Canada
Jörg Siekmann, University of Saarland, Saarbrücken, Germany
Wolfgang Wahlster, DFKI and University of Saarland, Saarbrücken, Germany

Volume Editors

Walter Daelemans
University of Antwerp, Department of Linguistics
CNTS Language Technology Group
Prinsstraat 13, L-203, 2000 Antwerp, Belgium
E-mail: walter.daelemans@ua.ac.be

Bart Goethals
University of Antwerp
Mathematics and Computer Science Department
Middelheimlaan 1, 2020 Antwerp, Belgium
E-mail: bart.goethals@ua.ac.be

Katharina Morik
Technische Universität Dortmund
Computer Science VIII, Artificial Intelligence Unit
44221 Dortmund, Germany
E-mail: katharina.morik@tu-dortmund.de



Library of Congress Control Number: Applied for

CR Subject Classification (1998): I.2, H.2.8, H.2, H.3, G.3, J.1, I.7, F.2.2, F.4.1

LNCS Sublibrary: SL 7 – Artificial Intelligence

ISSN 0302-9743
ISBN-10 3-540-87480-1 Springer Berlin Heidelberg New York
ISBN-13 978-3-540-87480-5 Springer Berlin Heidelberg New York



Springer is a part of Springer Science+Business Media

springer.com


Printed in Germany

Typesetting: Camera-ready by author, data conversion by Scientific Publishing Services, Chennai, India
Printed on acid-free paper SPIN: 12527915 06/3180 5 4 3 2 1 0

# Preface

When in 1986 Yves Kodratoff started the European Working Session on Learning at Orsay, France, it could not be foreseen that the conference would grow year by year and become the premier European conference of the field, attracting submissions from all over the world. The first European Conference on Principles of Data Mining and Knowledge Discovery was organized by Henryk Jan Komorowski and Jan Zytkow in 1997 in Trondheim, Norway. Since 2001 the two conferences have been collocated, offering participants from both areas the opportunity to listen to each other's talks. This year, the integration has moved even further. Instead of first splitting the field according to ECML or PKDD topics, we flattened the structure of the field to a single set of topics. For each of the topics, experts were invited to the Program Committee. Submitted papers were gathered into one collection and characterized according to their topics. The reviewers were then asked to bid on all papers, regardless of the conference. This allowed us to allocate papers more precisely.

The hierarchical reviewing process as introduced in 2005 was continued. We nominated 30 Area Chairs, each supervising the reviews and discussions of about 17 papers. In addition, 307 reviewers completed the Program Committee. Many thanks to all of them! It was a considerable effort for the reviewers to carefully review the papers, some providing us with additional reviews even at short notice. Based on their reviews and internal discussions, which were concluded by the recommendations of the Area Chairs, we could manage the final selection for the program. We received 521 submissions, of which 100 were presented at the conferences, giving us an acceptance rate of 20%. This high selectivity means, on the one hand, that some good papers could not make it into the conference program. On the other hand, it supports the traditionally high standards of the joint conference. We thank the authors from all over the world for submitting their contributions!

Following the tradition, the first and the last day of the joint conference were dedicated to workshops and tutorials. ECML PKDD 2008 offered 8 tutorials and 11 workshops. We thank the Workshop and Tutorial Chairs Siegfried Nijssen and Arno Siebes for their excellent selection. The discovery challenge is also a tradition of ECML PKDD that we continued. We are grateful to Andreas Hotho and his colleagues from the Bibsonomy project for organizing the discovery challenge of this year. The results were presented at the Web 2.0 Mining Workshop.

One of the pleasures of chairing a conference is the opportunity to invite colleagues whose work we esteem highly. We are grateful to Françoise Fogelman Soulié (KXEN) for opening the industrial track, Yoav Freund (University of California, San Diego), Anil K. Jain (Michigan State University), Ray Mooney (University of Texas at Austin), and Raghu Ramakrishnan (Yahoo! Research) for accepting our invitation to present recent work at the conference.

Some novelties were introduced to the joint conference this year.

First, there was no distinction into long and short papers. Instead, paper length was raised to 16 pages for all submissions.

Second, 14 papers were selected for publication in Springer Journals Seven papers were published in the *Machine Learning Journal* 72:3 (September 2008), and 7 papers were published in the *Data Mining and Knowledge Discovery Journal* 17:1 (August 2008). This LNAI volume includes the abstracts of these papers, each containing a reference to the respective full journal contribution. At the conference, participants received the proceedings, the tutorial notes and workshop proceedings on a USB memory stick.

Third, all papers were additionally allowed to be presented as posters. Since the number of participants has become larger, questions and discussions after a talk are no longer possible for all those interested. Introducing poster presentations for all accepted papers allows for more detailed discussions. Hence, we did not reserve this opportunity for a minority of papers and it was not an alternative to an oral presentation.

Fourth, a special demonstration session was held that is intended to be a forum for showcasing the state of the art in machine learning and knowledge discovery software. The focus lies on innovative prototype implementations in machine learning and data analysis. The demo descriptions are included in the proceedings. We thank Christian Borgelt for reviewing the submitted demos. Finally, for the first time, the conference took place in Belgium!

September 2008

Walter Daelemans
Bart Goethals
Katharina Morik

# Organization

## Program Chairs

| | |
|---|---|
| Walter Daelemans | University of Antwerp, Belgium |
| Bart Goethals | University of Antwerp, Belgium |
| Katharina Morik | Technische Universität Dortmund, Germany |

## Workshop and Tutorial Chairs

| | |
|---|---|
| Siegfried Nijssen | Utrecht University, the Netherlands |
| Arno Siebes | Katholieke Universiteit Leuven, Belgium |

## Demo Chair

| | |
|---|---|
| Christian Borgelt | European Centre for Soft Computing, Spain |

## Publicity Chairs

| | |
|---|---|
| Hendrik Blockeel | Katholieke Universiteit Leuven, Belgium |
| Pierre Dupont | Université catholique de Louvain, Belgium |
| Jef Wijsen | University of Mons-Hainaut, Belgium |

## Steering Committee

Pavel Brazdil
Johannes Fürnkranz
Alípio Jorge
Jacek Koronacki
Stan Matwin
Tobias Scheffer
Myra Spiliopoulou
Rui Camacho
João Gama
Joost N. Kok
Ramon Lopez de Mantaras
Dunja Mladenic
Andrzej Skowron

## Local Organization Team

Boris Cule
Calin Garboni
Iris Hendrickx
Kim Luyckx
Roser Morante
Koen Smets
Koen Van Leemput
Guy De Pauw
Joris Gillis
Wim Le Page
Michael Mampaey
Adriana Prado
Vincent Van Asch

## Area Chairs

Dimitris Achlioptas
Jean-François Boulicaut
Toon Calders
James Cussens
Tapio Elomaa
Johannes Fürnkranz
Szymon Jaroszewicz
Hillol Kargupta
Heikki Mannila
Srinivasan Parthasarathy
Lorenza Saitta
Martin Scholz
John Shawe-Taylor
Maarten van Someren
Stefan Wrobel
Roberto Bayardo
Wray Buntine
Padraig Cunningham
Ian Davidson
Martin Ester
Aris Gionis
Thorsten Joachims
Eamonn Keogh
Taneli Mielikainen
Jian Pei
Tobias Scheffer
Michèle Sébag
Antal van den Bosch
Michalis Vazirgiannis
Osmar Zaiane

## Program Committee

Niall Adams
Alekh Agarwal
Charu Aggarwal
Gagan Agrawal
David Aha
Florence d'Alche-Buc
Enrique Alfonseca
Aris Anagnostopoulos
Annalisa Appice
Thierry Artieres
Yonatan Aumann
Paolo Avesani
Ricardo Baeza-Yates
Jose Balcazar
Aharon Bar Hillel
Ron Bekkerman
Bettina Berendt
Michael Berthold
Steffen Bickel
Ella Bingham
Gilles Blanchard
Hendrik Blockeel
Axel Blumenstock
Francesco Bonchi
Christian Borgelt
Karsten Borgwardt
Henrik Bostrom
Thorsten Brants
Ivan Bratko
Pavel Brazdil
Ulf Brefeld
Bjorn Bringmann
Greg Buehrer
Rui Camacho
Stephane Canu
Carlos Castillo
Deepayan Chakrabarti
Kamalika Chaudhuri
Nitesh Chawla
Sanjay Chawla
David Cheung
Alok Choudhary
Wei Chu
Alexander Clark
Chris Clifton
Ira Cohen
Antoine Cornuejols
Bruno Cremilleux
Michel Crucianu
Juan-Carlos Cubero

Gautam Das
Tijl De Bie
Luc De Raedt
Janez Demsar
Francois Denis
Giuseppe Di Fatta
Laura Dietz
Debora Donato
Dejing Dou
Kurt Driessens
Chris Drummond
Pierre Dupont
Haimonti Dutta
Pinar Duygulu-Sahin
Sašo Dzeroski
Tina Eliassi-Rad
Charles Elkan
Roberto Esposito
Floriana Esposito
Christos Faloutsos
Thomas Finley
Peter Flach
George Forman
Blaz Fortuna
Eibe Frank
Dayne Freitag
Alex Freitas
Elisa Fromont
Thomas Gaertner
Patrick Gallinari
João Gama
Dragan Gamberger
Auroop Ganguly
Gemma Garriga
Eric Gaussier
Floris Geerts
Claudio Gentile
Pierre Geurts
Amol Ghoting
Chris Giannella
Fosca Giannotti
Attilio Giordana
Mark Girolami
Derek Greene
Marko Grobelnik
Robert Grossman
Dimitrios Gunopulos
Maria Halkidi
Lawrence Hall
Jiawei Han
Tom Heskes
Alexander Hinneburg
Frank Höppner
Thomas Hofmann
Jaakko Hollmen
Tamas Horvath
Andreas Hotho
Eyke Hüllermeier
Manfred Jaeger
Sarah Jane Delany
Ruoming Jin
Alipio Jorge
Matti Kaariainen
Alexandros Kalousis
Benjamin Kao
George Karypis
Samuel Kaski
Petteri Kaski
Kristian Kersting
Ralf Klinkenberg
Adam Klivans
Jukka Kohonen
Joost Kok
Aleksander Kolcz
Stefan Kramer
Andreas Krause
Marzena Kryszkiewicz
Jussi Kujala
Pavel Laskov
Dominique Laurent
Nada Lavrac
David Leake
Chris Leckie
Philippe Leray
Carson Leung
Tao Li
Hang Li
Jinyan Li
Chih-Jen Lin
Jessica Lin

Michael Lindenbaum
Giuseppe Lipori
Kun Liu
Xiaohui Liu
Huan Liu
Michael Madden
Donato Malerba
Brad Malin
Giuseppe Manco
Bernard Manderick
Lluis Marquez
Yuji Matsumoto
Stan Matwin
Michael May
Prem Melville
Rosa Meo
Ingo Mierswa
Pericles Mitkas
Dunja Mladenic
Fabian Moerchen
Alessandro Moschitti
Claire Nedellec
John Nerbonne
Jenniver Neville
Joakim Nivre
Richard Nock
Alexandros Ntoulas
Siegfried Nijssen
Arlindo Oliveira
Salvatore Orlando
Miles Osborne
Gerhard Paass
Balaji Padmanabhan
George Paliouras
Themis Palpanas
Spiros Papadimitriou
Mykola Pechenizkiy
Dino Pedreschi
Ruggero Pensa
Jose-Maria Peña
Bernhard Pfahringer
Petra Philips
Enric Plaza
Pascal Poncelet
Doina Precup
Kai Puolamaki
Filip Radlinski
Shyamsundar Rajaram
Jan Ramon
Chotirat Ratanamahatana
Jan Rauch
Christophe Rigotti
Celine Robardet
Marko Robnik-Sikonja
Juho Rousu
Céline Rouveirol
Ulrich Rueckert
Stefan Rüping
Hichem Sahbi
Ansaf Salleb-Aouissi
Taisuke Sato
Yucel Saygin
Bruno Scherrer
Lars Schmidt-Thieme
Matthias Schubert
Marc Sebban
Nicu Sebe
Giovanni Semeraro
Pierre Senellart
Jouni Seppanen
Benyah Shaparenko
Arno Siebes
Tomi Silander
Dan Simovici
Andrzej Skowron
Kate Smith-Miles
Soeren Sonnenburg
Alessandro Sperduti
Myra Spiliopoulou
Ramakrishnan Srikant
Jerzy Stefanowski
Michael Steinbach
Jan Struyf
Gerd Stumme
Tamas Sziranyi
Domenico Talia
Pang-Ning Tan
Zhaohui Tang
Evimaria Terzi
Yannis Theodoridis

## Additional Reviewers

Anne Doucet
Rémi Douence
Jim Dowling
Karel Driesen
Sophia Drossopoulou
Stéphane Ducasse
Natalie Eckel
Marc Evers
Johan Fabry
Leonidas Fegaras
Luca Ferrarini
Rony Flatscher
Jacques Garrigue
Marie-Pierre Gervais
Miguel Goulão
Thomas Gschwind
Pedro Guerreiro
I. Hakki Toroslu
Görel Hedin
Christian Heide Damm
Roger Henriksson
Martin Hitz
David Holmes
James Hoover
Antony Hosking
Cengiz Icdem
Yuuji Ichisugi
Anders Ive
Hannu-Matti Järvinen
Andrew Kennedy
Graham Kirby
Svetlana Kouznetsova
Kresten Krab Thorup
Reino Kurki-Suonio
Thomas Ledoux
Yuri Leontiev
David Lorenz
Steve MacDonald
Ole Lehrmann Madsen
Eva Magnusson
Margarida Mamede
Klaus Marius Hansen
Kim Mens
Tom Mens
Isabella Merlo
Marco Mesiti
Thomas Meurisse
Mattia Monga
Sandro Morasca
M. Murat Ezbiderli
Oner N. Hamali
Hidemoto Nakada
Jacques Noye
Deniz Oguz
José Orlando Pereira
Alessandro Orso
Johan Ovlinger
Marc Pantel
Jean-François Perrot
Patrik Persson
Frédéric Peschanski
Gian Pietro Picco
Birgit Pröll
Christian Queinnec
Osmar R. Zaiane
Barry Redmond
Sigi Reich
Arend Rensink
Werner Retschitzegger
Nicolas Revault
Matthias Rieger
Mario Südholt
Paulo Sérgio Almeida
Ichiro Satoh
Tilman Schaefer
Jean-Guy Schneider
Pierre Sens
Veikko Seppänen
Magnus Steinby
Don Syme
Tarja Systä
Duane Szafron
Yusuf Tambag
Kenjiro Taura
Michael Thomsen
Sander Tichelaar
Mads Torgersen
Tom Tourwé
Arif Tumer
Ozgur Ulusoy

Werner Van Belle
Dries Van Dyck
Vasco Vasconcelos
Karsten Verelst
Cristina Videira Lopes
Juha Vihavainen
John Whaley
Mario Wolzko
Mikal Ziane
Gabi Zodik
Elena Zucca

## Sponsors

We wish to express our gratitude to the sponsors of ECML PKDD 2008 for their essential contribution to the conference.

Fonds Wetenschappelijk Onderzoek
Research Foundation – Flanders

# Table of Contents – Part II

## Regular Papers

## Demo Papers

# Table of Contents – Part I

## Invited Talks (Abstracts)



## Machine Learning Journal Abstracts

## Data Mining and Knowledge Discovery Journal Abstracts



## Regular Papers

# Exceptional Model Mining

Dennis Leman[1], Ad Feelders[1], and Arno Knobbe[1,2]

[1] Utrecht University, P.O. box 80 089,
NL-3508 TB Utrecht, the Netherlands
{dlleman,ad}@cs.uu.nl
[2] Kiminkii, P.O. box 171,
NL-3990 DD, Houten, the Netherlands
a.knobbe@kiminkii.com

**Abstract.** In most databases, it is possible to identify small partitions of the data where the observed distribution is notably different from that of the database as a whole. In classical subgroup discovery, one considers the distribution of a single nominal attribute, and exceptional subgroups show a surprising increase in the occurrence of one of its values. In this paper, we introduce *Exceptional Model Mining* (EMM), a framework that allows for more complicated target concepts. Rather than finding subgroups based on the distribution of a single target attribute, EMM finds subgroups where a model fitted to that subgroup is somehow exceptional. We discuss regression as well as classification models, and define quality measures that determine how exceptional a given model on a subgroup is. Our framework is general enough to be applied to many types of models, even from other paradigms such as association analysis and graphical modeling.

## 1 Introduction

By and large, subgroup discovery has been concerned with finding regions in the input space where the distribution of a single target variable is substantially different from its distribution in the whole database [3,4]. We propose to extend this idea to targets that are models of some sort, rather than just single variables. Hence, in a very general sense, we want to discover subgroups where a model fitted to the subgroup is substantially different from that same model fitted to the entire database.

As an illustrative example, consider the simple linear regression model

$$P_i = a + bS_i + e_i$$

where $P$ is the sales price of a house, $S$ the lot size (measured, say, in square meters), and $e$ the random error term (see Fig. 1 and Section 4 for an actual dataset containing such data). If we think the location of the house might make a difference for the price per square meter, we could consider fitting the same model to the subgroup of houses on a desirable location:

$$P_i = a_D + b_D S_i + e_i,$$

W. Daelemans et al. (Eds.): ECML PKDD 2008, Part II, LNAI 5212, pp. 1–16, 2008.

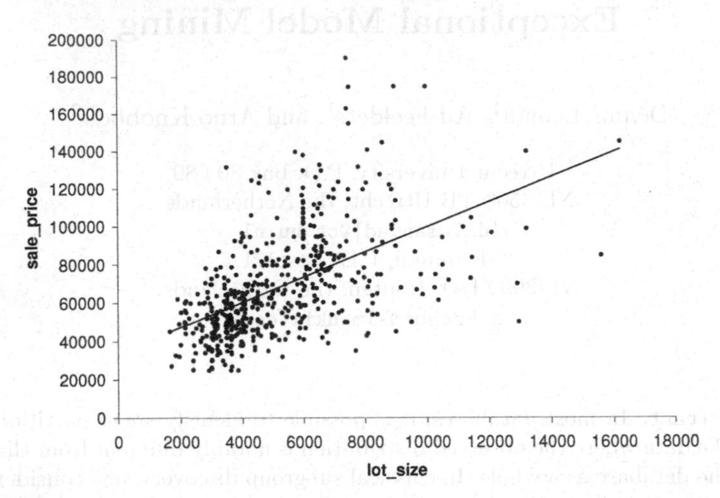


**Fig. 1.** Scatter plot of *lot_size* and *sales_price* for the housing data

where the subscript $D$ indicates we are only considering houses on a desirable location. To test whether the slope for desirable locations is significantly different, we could perform a statistical test of $H_0 : b = b_D$, or more conveniently, $H_0 : b_D = b_{\bar{D}}$, where $\bar{D}$ denotes the complement of $D$.

In the above example, we came up ourselves with the idea that houses on a desirable location might have a different slope in the regression model. The main idea presented in this paper is that we can find such groups automatically by using the subgroup discovery framework. Hence, the subgroups are not limited to simple conditions based on a single variable. Their description may involve conjunctions of conditions, and in case of multi-relational data, existential quantification and aggregation as well. In the general case of simple linear regression, we could be looking for subgroups $G$ where the slope $b_G$ in

$$y_i = a_G + b_G x_i + e_i,$$

is substantially different from the slope $b_{\bar{G}}$. The search process only involves the subgroups; the variables $y$ and $x$ are assumed to be determined by the question of the user, that is, they are fixed.

We have stated that the objective is to find subgroups where a model fitted to the subgroup is substantially different from that same model fitted to the entire database. This statement is deliberately general: we can use different types of models in this scheme, and for each type of model we can consider several measures of difference. In this paper we describe a number of model classes and quality measures that can be useful. All these methods have been implemented in the Multi-Relational Data Mining system Safarii [5].

This paper is organized as follows. In Section 2, we introduce some notation that is used throughout the paper, and define the subgroup discovery and exceptional model mining framework. In Section 3, we give examples of three basic types of models for exceptional model mining: correlation, regression and classification. We also propose appropriate quality measures for the types of models discussed. In Section 4, we present the results of exceptional model mining applied to two real-life datasets. Finally, we draw conclusions in Section 5.

## 2 Exceptional Model Mining

We assume that the database $d$ is a bag of labelled objects $i \in D$, referred to as *individuals*, taken from a domain $D$. We refer to the size of the database as $N = |d|$. At this point, we do not fix the nature of individuals, be it propositional, relational, or graphical, etc. However, each description of an individual includes a number of attributes $x_1, ..., x_k$ and optionally an output attribute $y$. These attributes are used in fitting models to subgroups of the data. In regular subgroup discovery, only the $y$ attribute is used, which is typically binary.

We make no assumptions about the syntax of the pattern language, and treat a pattern simply as a function $p : D \rightarrow \{0, 1\}$. We will say that a pattern $p$ *covers* an individual $i$ *iff* $p(i) = 1$.

**Definition 1 (Subgroup).** *A* subgroup *corresponding to a pattern $p$ is the set of individuals $G_p \subseteq d$ that are covered by $p$: $G_p = \{i \in d | p(i) = 1\}$.*

**Definition 2 (Complement).** *The* complement *of a subgroup $G_p$ is the set of individuals $\bar{G}_p \subseteq d$ that are* not *covered by $p$: $\bar{G}_p = d \backslash G_p$.*

When clear from the context, we will omit the $p$ from now on, and simply refer to a subgroup and its complement as $G$ and $\bar{G}$. We use $n$ and $\bar{n}$ to denote the size of $G$ and $\bar{G}$, respectively. In order to judge the quality of candidate patterns in a given database, a *quality measure* needs to be defined. This measure determines for each pattern in a pattern language $\mathcal{P}$ how interesting (exceptional) a model induced on the associated subgroup is.

**Definition 3 (Quality Measure).** *A* quality measure *for a pattern $p$ is a function $\varphi_d : \mathcal{P} \rightarrow \mathbb{R}$ that computes a unique numeric value for a pattern $p$, given a database $d$.*

Subgroup discovery [3] is a data mining framework aimed at discovering patterns that satisfy a number of user-specified inductive constraints. These constraints typically include an interestingness constraint $\varphi(p) \geq t$, as well as a minimum support threshold $n \geq minsup$ that guarantees the relative frequency of the subgroups in the database. Further constraints may involve properties such as the complexity of the pattern $p$. In most cases, a subgroup discovery algorithm will traverse a search lattice of candidate patterns in a top-down, general-to-specific fashion. The structure of the lattice is determined by a *refinement operator* $\rho : \mathcal{P} \rightarrow 2^{\mathcal{P}}$, a syntactic operation which determines how simple patterns can be extended into more complex ones by atomic additions. In our application (and most others), the refinement operator is assumed to be a *specialisation operator*: $\forall q \in \rho(p) : p \succeq q$ ($p$ is more general than $q$).

The actual search strategy used to consider candidates is a parameter of the algorithm. We have chosen the *beam search* strategy [13], because it nicely balances the benefits of a greedy method with the implicit parallel search resulting from the beam. Beam search effectively performs a level-wise search that is guided by the quality measure $\varphi$. On each level, the best-ranking $w$ patterns are refined to form the candidates for the next level. This means that although the search will be targeted, it is less likely to get stuck in a local optimum, because at each level alternatives are being considered. The search is further bounded by complexity constraints and the *minsup* constraint. The end-result is a ranked list of patterns (subgroups) that satisfy the inductive constraints.

In the case of regular subgroup discovery, with only a single discrete target variable, the quality measure of choice is typically a measure for how different the distribution over the target variable is, compared to that of the whole database (or in fact to that of the complement). As such an unusual distribution is easily produced in small fractions of the database, the deviation is often weighed with the size of the subgroup: a pattern is interesting if it is both exceptional and frequent. Well-known examples of quality measures for binary targets are frequency, confidence, $\chi^2$, and novelty.

The subject of this paper, exceptional model mining (EMM), can now be viewed as an extension of the subgroup discovery framework. The essential difference with standard subgroup discovery is the use of more complex target concepts than the regular single attribute. Our targets are models of some sort, and within each subgroup considered, a model is induced on the attributes $x_1, ..., x_k$, and optionally $y$. We will define quality measures that capture how exceptional the model within the subgroup is in relation to the model induced on its complement. In the next section, we present a number of model types, and propose one or more quality measures for each. When only the subgroup itself is considered, the quality measures tend to focus on the accuracy of the model, such as the fit of a regression line, or the predictive accuracy of a classifier. If the quality measure captures the difference between the subgroup and its complement, it is typically based on a comparison between more structural properties of the two models, such as the slope of the regression lines, or the make-up of the classifiers (e.g. size, attributes used).

*Example 1.* Consider again the housing dataset (Fig. 1). Individuals (houses) are described by a number of attributes such as the number of bathrooms or whether the house is located at a desirable location. An example of a pattern (and associated subgroup $G$) would be:

$$p : nbath \geq 2 \wedge drive = 1$$

which covers 128 houses (about 23% of the data). Its complement (which is often only considered implicitly) is

$$\bar{p} : \neg nbath \geq 2 \vee \neg drive = 1$$

The typical refinement operator will add a single condition on any of the available attributes to the conjunction. In this example, target models are defined over the two attributes $x = lot\_size$ and $y = sales\_price$. Note that these two attributes are therefore not allowed to appear in the subgroup definitions. One possibility is to perform the linear regression of $y$ on $x$. As a quality measure $\varphi_d$, we could consider the absolute difference in slope between the two regression lines fitted to $G$ and $\bar{G}$. In Section 3.2, we propose a more sophisticated quality measure for the difference in slope, that implicitly takes into account the supports $n$ and $\bar{n}$, and thus the significance of the finding.

# 3 Model Classes

In this section, we discuss simple examples of three classes of models, and suggest quality measures for them. As an example of a model without an output attribute, we consider the correlation between two numeric variables. We discuss linear regression for models with a numeric output attribute, and two simple classifiers for models with discrete output attributes.

## 3.1 Correlation Models

As an example of a model without an output attribute, we consider two numeric variables $x_1$ and $x_2$, and their linear association as measured by the correlation coefficient $\rho$. We estimate $\rho$ by the sample correlation coefficient $r$:

$$r = \frac{\sum(x_1^i - \bar{x}_1)(x_2^i - \bar{x}_2)}{\sqrt{\sum(x_1^i - \bar{x}_1)^2 \sum(x_2^i - \bar{x}_2)^2}}$$

where $x^i$ denotes the $i^{th}$ observation on $x$, and $\bar{x}$ denotes its mean.

**Absolute Difference between Correlations ($\varphi_{abs}$).** A logical quality measure is to take the absolute difference of the correlation in the subgroup $G$ and its complement $\bar{G}$, that is

$$\varphi_{abs}(p) = |r_G - r_{\bar{G}}|$$

The disadvantage of this measure is that it does not take into account the size of the groups, and hence does not do anything to prevent overfitting. Intuitively, subgroups with higher support should be preferred.

**Entropy ($\varphi_{ent}$).** As an improvement of $\varphi_{abs}$, the following quality function weighs the absolute difference between the correlations with the *entropy* of the split between the subgroup and its complement. The entropy captures the information content of such a split, and favours balanced splits (1 bit of information for a 50/50 split) over skewed splits (0 bits for the extreme case of either subgroup or complement being empty). The entropy function $H(p)$ is defined (in this context) as:

$$H(p) = -n/N \lg n/N - \bar{n}/N \lg \bar{n}/N$$

The quality measure $\varphi_{ent}$ is now defined as:

$$\varphi_{ent}(p) = H(p) \cdot |r_G - r_{\bar{G}}|$$

**Significance of Correlation Difference ($\varphi_{scd}$).** A more statistically oriented approach to prevent overfitting is to perform a hypothesis test on the difference between the correlation in the subgroup and its complement. Let $\rho_p$ and $\rho_{\bar{p}}$ denote the population coefficients of correlation for $p$ and $\bar{p}$, respectively, and let $r_G$ and $r_{\bar{G}}$ denote their sample estimates. The test to be considered is

$$H_0 : \rho_p = \rho_{\bar{p}} \qquad \text{against} \qquad H_a : \rho_p \neq \rho_{\bar{p}}$$

We would like to use the observed significance ($p$-value) of this test as a quality measure, but the problem is that the sampling distribution of the sample correlation coefficient is not known in general. If $x_1$ and $x_2$ follow a bivariate normal distribution, then application of the Fisher $z$ transformation

$$z' = \frac{1}{2} \ln \left( \frac{1+r}{1-r} \right)$$

makes the sampling distribution of $z'$ approximately normal [11]. Its standard error is given by

$$\frac{1}{\sqrt{m-3}}$$

where $m$ is the size of the sample. As a consequence

$$z^* = \frac{z' - \bar{z}'}{\sqrt{\frac{1}{n-3} + \frac{1}{\bar{n}-3}}}$$

approximately follows a standard normal distribution under $H_0$. Here $z'$ and $\bar{z}'$ are the $z$-scores obtained through the Fisher $z$ transformation for $G$ and $\bar{G}$, respectively. If both $n$ and $\bar{n}$ are greater than 25, then the normal approximation is quite accurate, and can safely be used to compute the $p$-values. Because we have to introduce the normality assumption to be able to compute the $p$-values, they should be viewed as a heuristic measure. Transformation of the original data (for example, taking their logarithm) may make the normality assumption more reasonable. As a quality measure we take 1 minus the computed $p$-value so that $\varphi_{scd} \in [0, 1]$, and higher values indicate a more interesting subgroup.

### 3.2 Regression Model

In this section, we discuss some possibilities of EMM with regression models. For ease of exposition, we only consider the linear regression model

$$y_i = a + bx_i + e_i, \tag{1}$$

but this is in no way essential to the methods we discuss.

**Significance of Slope Difference ($\varphi_{ssd}$).** Consider model (1) fitted to a subgroup $G$ and its complement $\bar{G}$. Of course, there is a choice of distance measures between the fitted models. We propose to look at the difference in the slope $b$ between the two models, because this parameter is usually of primary interest when fitting a regression model: it indicates the change in the expected value of $y$, when $x$ increases with one unit. Another possibility would be to look at the intercept $a$, if it has a sensible interpretation in the application concerned. Like with the correlation coefficient, we use significance testing to measure the distance between the fitted models. Let $b_p$ be the slope for the regression function of $p$ and $b_{\bar{p}}$ the slope for the regression function of $\bar{p}$. The hypothesis to be tested is

$$H_0 : b_p = b_{\bar{p}} \qquad \text{against} \qquad H_a : b_p \neq b_{\bar{p}}$$

We use the least squares estimate

$$\hat{b} = \frac{\sum(x_i - \bar{x})(y_i - \bar{y})}{\sum(x_i - \bar{x})^2}$$

for the slope $b$. An unbiased estimator for the variance of $\hat{b}$ is given by

$$s^2 = \frac{\sum \hat{e}_i^2}{(m-2)\sum(x_i - \bar{x})^2}$$

where $\hat{e}_i$ is the regression residual for individual $i$, and $m$ is the sample size. Finally, we define our test statistic

$$t' = \frac{\hat{b}_G - \hat{b}_{\bar{G}}}{\sqrt{s_G^2 + s_{\bar{G}}^2}}$$

Although $t'$ does not have a $t$ distribution, its distribution can be approximated quite well by one, with degrees of freedom given by (cf. [10]):

$$df = \frac{\left(s_G^2 + s_{\bar{G}}^2\right)^2}{\frac{s_G^4}{n-2} + \frac{s_{\bar{G}}^4}{\bar{n}-2}} \tag{2}$$

Our quality measure $\varphi_{ssd} \in [0, 1]$ is once again defined as one minus the $p$-value computed on the basis of a $t$ distribution with degrees of freedom given in (2). If $n + \bar{n} \geq 40$ the $t$-statistic is quite accurate, so we should be confident to use it unless we are analysing a very small dataset.

### 3.3 Classification Models

In the case of classification, we are dealing with models for which the output attribute $y$ is discrete. In general, the attributes $x_1, ..., x_k$ can be of any type (binary, nominal, numeric, etc). Furthermore, our EMM framework allows for any classification method, as long as some quality measure can be defined in order to judge the models induced. Although we allow arbitrarily complex methods, such as decision trees, support vector machines or even ensembles of classifiers, we only consider two relatively simple classifiers here, for reasons of simplicity and efficiency.

**Logistic Regression.** Analogous to the linear regression case, we consider the logistic regression model

$$\text{logit}(P(y_i = 1|x_i)) = \ln\left(\frac{P(y_i = 1|x_i)}{P(y_i = 0|x_i)}\right) = a + b \cdot x_i,$$

where $y \in \{0,1\}$ is a binary class label. The coefficient $b$ tells us something about the effect of $x$ on the probability that $y$ occurs, and hence may be of interest to subject area experts. A positive value for $b$ indicates that an increase in $x$ leads to an increase of $P(y = 1|x)$ and vice versa. The strength of influence can be quantified in terms of the change in the odds of $y = 1$ when $x$ increases with, say, one unit.

To judge whether the effect of $x$ is substantially different in a particular subgroup $G_p$, we fit the model

$$\text{logit}(P(y_i = 1|x_i)) = a + b \cdot p(i) + c \cdot x_i + d \cdot (p(i) \cdot x_i). \tag{3}$$

Note that

$$\text{logit}(P(y_i = 1|x_i)) = \begin{cases} (a+b) + (c+d) \cdot x_i & \text{if } p(i) = 1 \\ a + c \cdot x_i & \text{if } p(i) = 0 \end{cases}$$

Hence, we allow both the slope and the intercept to be different in the subgroup and its complement. As a quality measure, we propose to use one minus the $p$-value of a test on $d = 0$ against a two-sided alternative in the model of equation (3). This is a standard test in the literature on logistic regression [11]. We refer to this quality measure as $\varphi_{sed}$.

**DTM Classifier.** The second classifier considered is the *Decision Table Majority* (DTM) classifier [7,6], also known as a *simple decision table*. The idea behind this classifier is to compute the relative frequencies of the $y$ values for each possible combination of values for $x_1, \ldots, x_k$. For combinations that do not appear in the dataset, the relative frequency estimates are based on that of the whole dataset. The predicted $y$ value for a new individual is simply the one with the highest probability estimate for the given combination of input values.

*Example 2.* As an example of a DTM classifier, consider a hypothetical dataset of 100 people applying for a mortgage. The dataset contains two attributes describing the age (divided into three suitable categories) and marital status of the applicant. A third attribute indicates whether the application was successful, and is used as the output. Out of the 100 applications, 61 were successful. The following decision table lists the estimated probabilities of success for each combination of *age* and *married?*. The support for each combination is indicated between brackets.

| | *married?* = 'no' | *married?* = 'yes' |
|---|---|---|
| *age* = 'low' | 0.25 (20) | 0.61 (0) |
| *age* = 'medium' | 0.4 (15) | 0.686 (35) |
| *age* = 'high' | 0.733 (15) | 1.0 (15) |

As this table shows, the combination *married?* = 'yes'$\wedge$*age* = 'low' does not appear in this particular dataset, and hence the probability estimate is based on the complete dataset (0.61). This classifier predicts a positive outcome in all cases except when *married?* = 'no' and *age* is either 'low' or 'medium'.

For this instance of the classification model we discuss two different quality measures. The *BDEU* (Bayesian Dirichlet equivalent uniform) score, which is a measure for the performance of the DTM classifier on $G$, and the *Hellinger distance*, which assigns a value to the distance between the conditional probabilities estimated on $G$ and $\bar{G}$.

**BDeu Score ($\varphi_{BDeu}$).** The BDeu score $\varphi_{BDeu}$ is a measure from Bayesian theory [2] and is used to estimate the performance of a classifier on a subgroup, with a penalty for small contingencies that may lead to overfitting. Note that this measure ignores how the classifier performs on the complement. It merely captures how 'predictable' a particular subgroup is.

The BDeu score is defined as

$$\prod_{x_1,\ldots,x_k} \frac{\Gamma(\alpha/q)}{\Gamma(\alpha/q + n(x_1,\ldots,x_k))} \prod_{y} \frac{\Gamma(\alpha/qr + n(x_1,\ldots,x_k,y))}{\Gamma(\alpha/qr)}$$

where $\Gamma$ denotes the gamma function, $q$ denotes the number of value combinations of the input variables, $r$ the number of values of the output variable, and $n(x_1,\ldots,x_k,y)$ denotes the number of cases with that value combination. The parameter $\alpha$ denotes the *equivalent sample size*. Its value can be chosen by the user.

**Hellinger ($\varphi_{Hel}$).** Another possibility is to use the Hellinger distance [12]. It defines the distance between two probability distributions $P(z)$ and $Q(z)$ as follows:

$$H(P,Q) = \sum_{z} \left( \sqrt{P(z)} - \sqrt{Q(z)} \right)^2$$

where the sum is taken over all possible values $z$. In our case, the distributions of interest are

$$P(y \mid x_1,\ldots,x_k)$$

for each possible value combination $x_1,\ldots,x_k$. The overall distance measure becomes

$$\varphi_{Hel}(p) = D(\hat{P}_G, \hat{P}_{\bar{G}}) = \sum_{x_1,\ldots,x_k} \sum_{y} \left( \sqrt{\hat{P}_G(y|x_1,\ldots,x_k)} - \sqrt{\hat{P}_{\bar{G}}(y|x_1,\ldots,x_k)} \right)^2$$

where $\hat{P}_G$ denotes the probability estimates on $G$. Intuitively, we measure the distance between the conditional distribution of $y$ in $G$ and $\bar{G}$ for each possible combination of input values, and add these distances to obtain an overall distance. Clearly, this measure is aimed at producing subgroups for which the conditional distribution of $y$ is substantially different from its conditional distribution in the overall database.

## 4 Experiments

This section illustrates exceptional model mining on two real-life datasets, using different quality measures. Although our implementation in Safarii essentially is multi-relational [5], the two dataset we present are propositional. For each test, Safarii returns a configurable number of subgroups ranked according to the quality measure of choice. The following experiments only present the best ranking subgroup and take a closer look at the interpretation of the results.

### 4.1 Analysis of Housing Data

First, we analyse the Windsor housing data[1] [8]. This dataset contains information on 546 houses that were sold in Windsor, Canada in the summer of 1987. The information for each house includes the two attributes of interest, *lot_size* and *sales_price*, as plotted in Fig. 1. An additional 10 attributes are available to define candidate subgroups, including the number of bedrooms and bathrooms and whether the house is located at a desirable location. The correlation between lot size and sale price is 0.536, which implies that a larger size of the lot coincides with a higher sales price. The fitted regression function is:

$$\hat{y} = 34136 + 6.60 \cdot x$$

As this function shows, on average one extra square meter corresponds to a 6.6 dollar higher sales price. Given this function, one might wonder whether it is possible to find specific subgroups in the data where the price of an additional square meter is significantly less, perhaps even zero. In the next paragraphs, we show how EMM may be used to answer this question.

**Significance of Correlation Difference.** Looking at the restrictions defined in Section 3.1 we see that the support has to be over 25 in order to be confident about the test results for this measure. This number was used as minimum support threshold for a run of Safarii using $\varphi_{scd}$. The following subgroup (and its complement) was found to show the most significant difference in correlation: $\varphi_{scd}(p_1) = 0.9993$.

$$p_1 : drive = 1 \wedge rec\_room = 1 \wedge nbath \geq 2.0$$

This is the group of 35 houses that have a driveway, a recreation room and at least two bathrooms. The scatter plots for the subgroup and its complement are given in Fig. 2. The subgroup shows a correlation of $r_G = -0.090$ compared to $r_{\bar{G}} = 0.549$ for the remaining 511 houses. A tentative interpretation could be that $G$ describes a collection of houses in the higher segments of the markets where the price of a house is mostly determined by its location and facilities. The desirable location may provide a natural limit on the lot size, such that this is not a factor in the pricing. Figure 2 supports this hypothesis: houses in $G$ tend to have a higher price.

[1] Available from the Journal of Applied Econometrics Data Archive at `http://econ.queensu.ca/jae/`

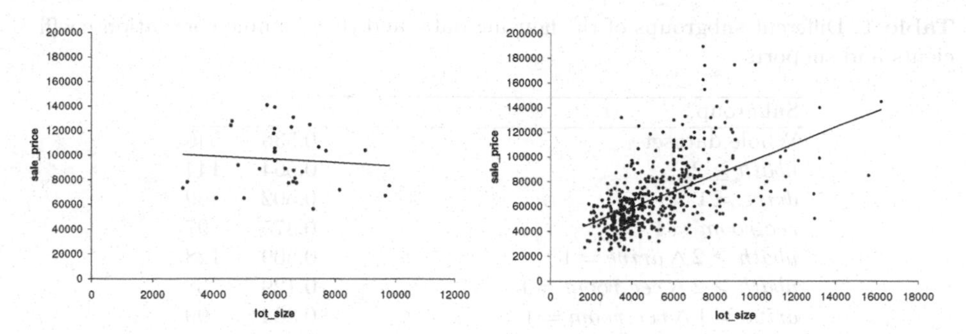


**Fig. 2.** Housing - $\varphi_{scd}$: Scatter plot of *lot_size* and *sales_price* for $drive = 1 \wedge rec\_room = 1 \wedge nbath \geq 2$ (left) and its complement (right)

In general *sales_price* and *lot_size* are positively correlated, but EMM discovers a subgroup with a slightly negative correlation. However, the value in the subgroup is not significantly different from zero: a test of

$$H_0 : b_{p_1} = 0 \qquad \text{against} \qquad H_a : b_{p_1} \neq 0,$$

yields a $p$-value of 0.61. The scatter plot confirms our impression that *sales_price* and *lot_size* are uncorrelated within the subgroup. For purposes of interpretation, it is interesting to perform some post-processing. In Table 1 we give an overview of the correlations within different subgroups whose intersection produces the final result, as given in the last row. It is interesting to see that the condition $nbath \geq 2$ in itself actually leads to a slight increase in correlation compared to the whole database, but the combination with the presence of a recreation room leads to a substantial drop to $r = 0.129$. When we add the condition that the house should also have a driveway we arrive at the final result with $r = -0.090$. Note that adding this condition only eliminates 3 records (the size of the subgroup goes from 38 to 35) and that the correlation between sales price and lot size in these three records (defined by the condition $nbath \geq 2 \wedge \neg drive = 1 \wedge rec\_room = 1$) is $-0.894$. We witness a phenomenon similar to Simpson's paradox: splitting up a subgroup with positive correlation (0.129) produces two subgroups both with a negative correlation ($-0.090$ and $-0.894$, respectively).

**Significance of Slope Difference.** In this section, we perform EMM on the housing data using the Significance of Slope Difference ($\varphi_{ssd}$) as the quality measure. The highest ranking subgroup consists of the 226 houses that have a driveway, no basement and at most one bathroom:

$$p_2 : drive = 1 \wedge basement = 0 \wedge nbath \leq 1$$

The subgroup $G$ and its complement $\bar{G}$ (320 houses) lead to the following two fitted regression functions, respectively:

$$\hat{y} = 41568 + 3.31 \cdot x$$

$$\hat{y} = 30723 + 8.45 \cdot x$$

**Table 1.** Different subgroups of the housing data, and their sample correlation coefficients and supports

| Subgroup | $r$ | $n$ |
|---|---|---|
| Whole dataset | 0.536 | 546 |
| $nbath \geq 2$ | 0.564 | 144 |
| $drive = 1$ | 0.502 | 469 |
| $rec\_room = 1$ | 0.375 | 97 |
| $nbath \geq 2 \wedge drive = 1$ | 0.509 | 128 |
| $nbath \geq 2 \wedge rec\_room = 1$ | 0.129 | 38 |
| $drive = 1 \wedge rec\_room = 1$ | 0.304 | 90 |
| $nbath \geq 2 \wedge rec\_room = 1 \wedge \neg drive = 1$ | −0.894 | 3 |
| $nbath \geq 2 \wedge rec\_room = 1 \wedge drive = 1$ | −0.090 | 35 |

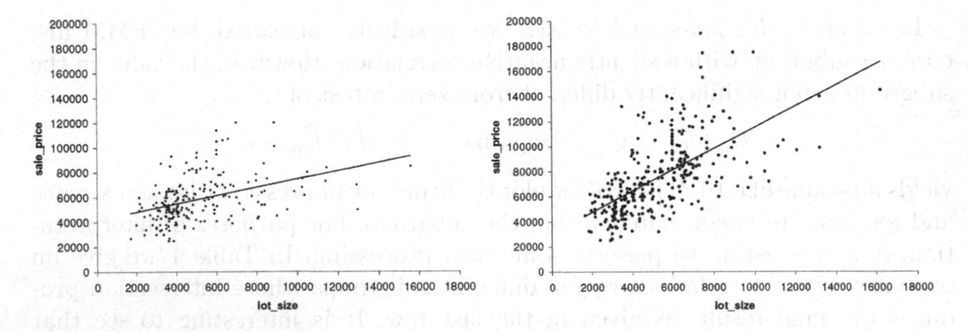


**Fig. 3.** Housing - $\varphi_{ssd}$: Scatter plot of $drive = 1 \wedge basement = 0 \wedge nbath \leq 1$ (left), and its complement (right)

The subgroup quality is $\varphi_{ssd} > 0.9999$, meaning that the $p$-value of the test

$$H_0 : b_{p_2} = b_{\bar{p}_2} \qquad \text{against} \qquad H_a : b_{p_2} \neq b_{\bar{p}_2}$$

is virtually zero. There are subgroups with a larger difference in slope, but the reported subgroup scores higher because it is quite big. Figure 3 shows the scatter plots of *lot_size* and *sales_price* for the subgroup and its complement.

## 4.2 Analysis of Gene Expression Data

The following experiments demonstrate the usefulness of exceptional model mining in the domain of bioinformatics. In genetics, genes are organised in so-called *gene regulatory networks*. This means that the expression (its effective activity) of a gene may be influenced by the expression of other genes. Hence, if one gene is regulated by another, one can expect a linear correlation between the associated expression-levels. In many diseases, specifically cancer, this interaction between genes may be disturbed. The Gene Expression dataset shows the expression-levels of 313 genes as measured by an Affymetrix microarray, for 63 patients that suffer from a cancer known as neuroblastoma [9]. Additionally, the dataset

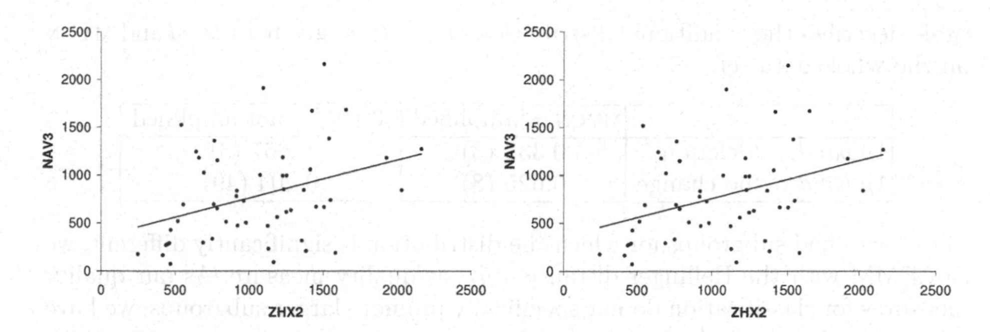


**Fig. 4.** Gene Expression - $\varphi_{abs}$: Scatter plot of *11_band* = 'no deletion' $\wedge$ *survivaltime* $\leq$ 1919 $\wedge$ *XP_498569.1* $\leq$ 57 (left; $r = -0.950$) and its complement (right; $r = 0.363$)

contains clinical information about the patients, including age, sex, stage of the disease, etc.

**Correlation Model Experiment.** As a demonstration of a correlation model, we analyse the correlation between ZHX3 ('Zinc fingers and homeoboxes 2') and NAV3 ('Neuron navigator 3'), in terms of the absolute difference of correlations $\varphi_{abs}$. These genes show a very slight correlation ($r = 0.218$) in the whole dataset. The remaining attributes (both gene expression and clinical information) are available for building subgroups. As the $\varphi_{abs}$ measure does not have any provisions for promoting larger subgroups, we use a minimum support threshold of 10 (15% of the patients). The largest distance ($\varphi_{abs}(p_3) = 1.313$) was found with the following condition:

$$p_3 : \mathit{11\_band} = \text{'no deletion'} \wedge \mathit{survivaltime} \leq 1919 \wedge \mathit{XP\_498569.1} \leq 57$$

Figure 4 shows the plot for this subgroup and its complement with the regression lines drawn in. The correlation in the subgroup is $r_G = -0.95$ and the correlation in the remaining data is $r_{\bar{G}} = 0.363$. Note that the subgroup is very "predictable": all points are quite close to the regression line, with $R^2 \approx 0.9$.

**DTM Experiment.** For the DTM classification experiments on the Gene Expression dataset, we have selected three binary attributes. The first two attributes, which serve as input variables of the decision table, are related to genomic alterations that may be observed within the tumor tissues. The attribute *1p_band* ($x_1$) describes whether the small arm ('p') of the first chromosome has been deleted. The second attribute, MYCN ($x_2$), describes whether one specific gene is amplified or not (multiple copies introduced in the genome). Both attributes are known to potentially influence the genesis and prognosis of neuroblastoma. The output attibute for the classification model is *NBstatus* ($y$), which can be either 'no event' or 'relapse or deceased'. The following decision table describes the conditional distribution of *NBstatus* given *1p_band* and MYCN on the whole data set:

| | MYCN ='amplified' | MYCN = 'not amplified' |
|---|---|---|
| *1p_band* = 'deletion' | 0.333 (3) | 0.667 (3) |
| *1p_band* = 'no change' | 0.625 (8) | 0.204 (49) |

In order to find subgroups for which the distribution is significantly different, we run EMM with the Hellinger distance $\varphi_{Hel}$ as quality measure. As our quality measures for classification do not specifically promote larger subgroups, we have selected a slightly higher minimum support constraint: $minsup = 16$, which corresponds to 25% of the data. The following subgroup of 17 patients was the best found ($\varphi_{Hel} = 3.803$):

$$p_4 : prognosis = \text{'unknown'}$$

| | MYCN ='amplified' | MYCN = 'not amplified' |
|---|---|---|
| *1p_band* = 'deletion' | 1.0 (1) | 0.833 (6) |
| *1p_band* = 'no change' | 1.0 (1) | 0.333 (9) |

Note that for each combination of input values, the probability of 'relapse or deceased' is increased, which makes sense when the prognosis is uncertain. Note furthermore that the overall dataset does not yield a pure classifier: for every combination of input values, there is still some confusion in the predictions. In our second classification experiment, we are interested in "predictable" subgroups. Therefore, we run EMM with the $\varphi_{BDeu}$ measure. All other settings are kept the same. The following subgroup ($n = 16$, $\varphi_{BDeu} = -1.075$) is based on the expression of the gene RIF1 ('RAP1 interacting factor homolog (yeast)')

$$p_5 : \text{RIF1} >= 160.45$$

| | MYCN ='amplified' | MYCN = 'not amplified' |
|---|---|---|
| *1p_band* = 'deletion' | 0.0 (0) | 0.0 (0) |
| *1p_band* = 'no change' | 0.0 (0) | 0.0 (16) |

In this subgroup, the predictiveness is optimal, as all patients turn out to be tumor-free. In fact, the decision table ends up being rather trivial, as all cells indicate the same decision.

**Logistic Regression Experiment.** In the logistic regression experiment, we take *NBstatus* as the output $y$, and *age* (age at diagnosis in days) as the predictor $x$. The subgroups are created using the gene expression level variables. Hence, the model specification is

$$\text{logit}\{P(NBstatus = \text{'relapse or deceased'})\} = a + b \cdot p + c \cdot age + d \cdot (p \cdot age).$$

We find the subgroup

$$p_6 : \text{SMPD1} \geq 840 \wedge \text{HOXB6} \leq 370.75$$

with a coverage of 33, and quality $\varphi_{sed} = 0.994$. We find a positive coefficient of $x$ for the subgroup, and a slightly negative coefficient for its complement. Within the subgroup, the odds of *NBstatus* = 'relapse or deceased' increase with 44% when the age at diagnosis increases with 100 days, whereas in the complement the odds decrease with 8%. More loosely, within the subgroup an increase in age at diagnosis decreases the probability of survival, whereas in the complement an increase in age slightly increases the probability of survival. Such reversals of the direction of influence may be of particular interest to the domain expert.

## 5 Conclusions and Future Research

We have introduced exceptional model mining (EMM) as an extension of the well-known subgroup discovery framework. By focusing on models instead of single target variables, many new interesting analysis possibilities are created. We have proposed a number of model classes that can be used in EMM, and defined several quality measures for them. We illustrated the use of EMM by its application to two real datasets. Like subgroup discovery, EMM is an exploratory method that requires interaction with a user that is knowledgable in the application domain. It can provide useful insights into the subject area, but does not result in ready-to-use predictive models.

We believe there are many possibilities to extend the work presented in this paper. One could look at different models, for example naive Bayes for classification problems or graphical models for modelling the probability distribution of a number of (discrete) variables. Whatever the selected class of models, the user should specify a quality measure that relates to the more fundamental questions a user may have about the data at hand. In the case of our housing example, the choice for the difference in slope is appropriate, as it captures a relevant aspect of the data, namely a significant change in price per square meter. For similar reasons, we used the difference between the coefficients of the explanatory variable (age at diagnosis) in the subgroup and its complement as a quality measure for logistic regression models.

Specifying an appropriate quality measure that is inspired by a relevant question of the user becomes less straightforward when more complex models are considered, although of course one can always focus on some particular aspect (e.g. coefficients) of the models. However, even for sophisticated models such as support vector machines or Bayesian networks, one can think of measures that make sense, such as the linear separability or the edit distance between two networks [14], respectively.

From a computational viewpoint, it is advisable to keep the models to be fitted simple, since many subgroups have to be evaluated in the search process. For example, fitting a naive Bayes model to a large collection of subgroups can be done quite efficiently, but fitting a support vector machine could prove to be too time consuming.

## References

1. Affymetrix (1992), http://www.affymetrix.com/index.affx
2. Heckerman, D., Geiger, D., Chickering, D.: Learning Bayesian Networks: The combination of knowledge and statistical data. Machine Learning 20, 179–243 (1995)
3. Klösgen, W.: Subgroup Discovery. In: Handbook of Data Mining and Knowledge Discovery, ch. 16.3. Oxford University Press, New York (2002)
4. Friedman, J., Fisher, N.: Bump-Hunting in High-Dimensional Data. Statistics and Computing 9(2), 123–143 (1999)
5. Knobbe, A.: Safarii multi-relational data mining environment (2006), http://www.kiminkii.com/safarii.html
6. Knobbe, A., Ho, E.: Pattern Teams. In: Fürnkranz, J., Scheffer, T., Spiliopoulou, M. (eds.) PKDD 2006. LNCS (LNAI), vol. 4213. Springer, Heidelberg (2006)
7. Kohavi, R.: The Power of Decision Tables. In: Lavrač, N., Wrobel, S. (eds.) ECML 1995. LNCS, vol. 912. Springer, Heidelberg (1995)
8. Anglin, P.M., Gençay, R.: Semiparametric Estimation of a Hedonic Price Function. Journal of Applied Econometrics 11(6), 633–648 (1996)
9. van de Koppel, E., et al.: Knowledge Discovery in Neuroblastoma-related Biological Data. In: Data Mining in Functional Genomics and Proteomics workshop at PKDD 2007, Warsaw, Poland (2007)
10. Moore, D., McCabe, G.: Introduction to the Practice of Statistics, New York (1993)
11. Neter, J., Kutner, M., Nachtsheim, C.J., Wasserman, W.: Applied Linear Statistical Models. WCB McGraw-Hill (1996)
12. Yang, G., Le Cam, L.: Asymptotics in Statistics: Some Basic Concepts. Springer, Berlin (2000)
13. Xu, Y., Fern, A.: Learning Linear Ranking Functions for Beam Search. In: Proceedings ICML 2007 (2007)
14. Niculescu-Mizil, A., Caruana, R.: Inductive Transfer for Bayesian Network Structure Learning. In: Proceedings of the 11th International Conference on AI and Statitics (AISTATS 2007) (2007)

# A Joint Topic and Perspective Model for Ideological Discourse

Wei-Hao Lin, Eric Xing, and Alexander Hauptmann*

Language Technologies Institute
School of Computer Science
Carnegie Mellon University
Pittsburgh, PA 15213 U.S.A.
{whlin,epxing,alex}@cs.cmu.edu

**Abstract.** Polarizing discussions on political and social issues are common in mass and user-generated media. However, computer-based understanding of ideological discourse has been considered too difficult to undertake. In this paper we propose a statistical model for ideology discourse. By ideology we mean "a set of general beliefs socially shared by a group of people." For example, Democratic and Republican are two major political ideologies in the United States. The proposed model captures lexical variations due to an ideological text's topic and due to an author or speaker's ideological perspective. To cope with the non-conjugacy of the logistic-normal prior we derive a variational inference algorithm for the model. We evaluate the proposed model on synthetic data as well as a written and a spoken political discourse. Experimental results strongly support that ideological perspectives are reflected in lexical variations.

## 1 Introduction

When people describe a set of ideas as "ideology", the ideas are usually regarded as false beliefs. Marxists associate the dominant class's viewpoints as ideology. Ideology's pejorative connotation is usually used to describe other group's ideas and rarely our own ideas.

In this paper we take a definition of ideology broader than the classic Marxists' definition, but define ideology as "a set of general beliefs socially shared by a group of people" [1]. Groups whose members share similar goals or face similar problems usually share a set of beliefs that define membership, value judgment, and action. These collective beliefs form an ideology. For example, Democratic and Republican are two major political ideologies in the United States.

Written and spoken discourses are critical in the van Dijk's theory of ideology [1]. Ideology is not innate and must be learned through interaction with the

* We would like to thank the anonymous reviewers for their valuable comments for improving this paper, and thank Rong Yan, David Gondek, and Ching-Yung Lin for helpful discussions. This work was supported in part by the National Science Foundation (NSF) under Grants No. IIS-0535056 and CNS-0751185.

W. Daelemans et al. (Eds.): ECML PKDD 2008, Part II, LNAI 5212, pp. 17–32, 2008.

world. Spoken and written texts are major media through which an ideology is understood, transmitted, and reproduced. For example, two presidential candidates, John Kerry and George W. Bush, gave the following answers during a presidential debate in 2004:

*Example 1.* Kerry: What is an article of faith for me is not something that I can legislate on somebody who doesn't share that article of faith. I believe that choice is a woman's choice. It's between a woman, God and her doctor. And that's why I support that.

*Example 2.* Bush: I believe the ideal world is one in which every child is protected in law and welcomed to life. I understand there's great differences on this issue of abortion, but I believe reasonable people can come together and put good law in place that will help reduce the number of abortions.

From their answers we can clearly understand their attitude on the abortion issue.

Interest in computer based understanding of ideology dates back to the sixties in the last century, but the idea of learning ideology *automatically* from texts has been considered almost impossible. Abelson expressed a very pessimistic view on automatic learning approaches in 1965 [2]. We share Abelson's vision but do not subscribe to his view. We believe that ideology can be statistically modeled and learned from a large number of ideological texts.

- In this paper we develop a statistical model for ideological discourse. Based on the empirical observation in Section 2 we hypothesize that ideological perspectives were reflected in lexical variations. Some words were used more frequently because they were highly related to an ideological text's topic (i.e., *topical*), while some words were used more frequently because authors holding a particular ideological perspective chose so (i.e., *ideological*).
- We formalize the hypothesis and proposed a statistical model for ideological discourse in Section 3. Lexical variations in ideological discourse were encoded in a word's topical and ideological weights. The coupled weights and the non-conjugacy of the logistic-normal prior posed a challenging inference problem. We develop an approximate inference algorithm based on the variational method in Section 3.2.
  Such a model can not only uncover topical and ideological weights from data and can predict the ideological perspective of a document. The proposed model will allow news aggregation service to organize and present news by their ideological perspectives.
- We evaluate the proposed model on synthetic data (Section 4.1) as well as on a written text and a spoken text (Section 4.2). In Section 4.3 we show that the proposed model automatically uncovered many discourse structures in ideological discourse.
- In Section 4.4 we show that the proposed model fit ideological corpora better than a model that assumes no lexical variations due to an author or speaker's ideological perspective. Therefore the experimental results strongly suggested that ideological perspectives were reflected in lexical variations.

## 2 Motivation

Lexical variations have been identified as a "major means of ideological expression" [1]. In expressing a particular ideological perspective, word choices can highly reveal an author's ideological perspective on an issue. "One man's terrorist is another man's freedom fighter." Labeling a group as "terrorists" strongly reveal an author's value judgement and ideological stance [3].

We illustrate lexical variations in an ideological text about the Israeli-Palestinian conflict (see Section 4.2). There were two groups of authors holding contrasting ideological perspectives (i.e., Israeli vs. Palestinian). We count the words used by each group of authors and showed the top 50 most frequent words in Figure 1.

abu agreement american arab arafat bank bush conflict disengagement fence gaza government international iraq israel israeli israelis israels jerusalem jewish leadership minister palestine palestinian palestinians peace plan political president prime process public return roadmap security settlement settlements sharon sharons solution state states terrorism time united violence war west world years

american arab arafat authority bank conflict elections end gaza government international israel israeli israelis israels jerusalem land law leadership military minister negotiations occupation palestine palestinian palestinians peace people plan political prime process public rights roadmap security settlement settlements sharon side solution state states territories time united violence wall west world

**Fig. 1.** The top 50 most frequent words used by the Israeli authors (left) and the Palestinian authors (right) in a document collection about the Israeli-Palestinian conflict. A word's size represents its frequency: the larger, the more frequent.

Both sides share many words that are highly related to the corpus's topic (i.e., the Israeli-Palestinian conflict): "Palestinian", "Israeli", "political", "peace", etc. However, each ideological perspective seems to emphasize (i.e., choosing more frequently) different subset of words. The Israeli authors seem to use more "disengagement", "settlement", and "terrorism". On the contrary, the Palestinian authors seem to choose more "occupation", "international", and "land." Some words seem to be chosen because they are about a topic, while some words are chosen because of an author's ideological stance.

We thus hypothesize that lexical variations in ideological discourse are attributed to both an ideological text's topic and an author or speaker's ideological point of view. Word frequency in ideological discourse should be determined by how much a word is related to a text's topic (i.e., *topical*) and how much authors holding a particular ideological perspective emphasize or de-emphasize the word (i.e., *ideological*). A model for ideological discourse should take both topical and ideological aspects into account.

## 3 A Joint Topic and Perspective Model

We propose a statistical model for ideological discourse. The model associates *topical* and *ideological* weights to each word in the vocabulary. Topical weights represent how frequently a word is chosen because of a text's topic independent of an author or speaker's ideological perspective. Ideological weights, on the other hand, modulate topical weights based on an author or speaker's ideological perspective. To emphasize a word (i.e., choosing the word more frequently) we put a larger ideological weight on the word.

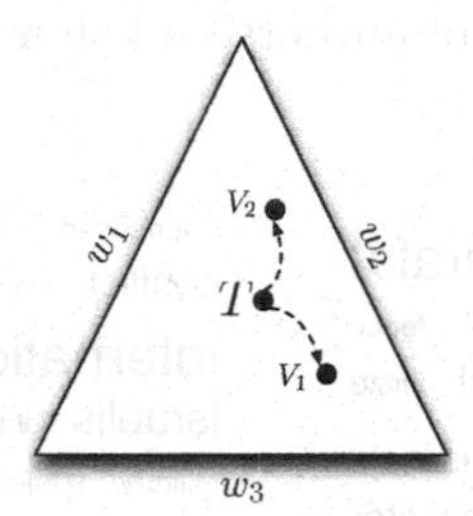


**Fig. 2.** A three-word simplex illustrates how topical weights $T$ are modulated by two differing ideological weights

We illustrate the interaction between topical and ideological weights in a three-word simplex in Figure 2. A point $T$ represents topical weights about a specific topic. Suppose authors holding a particular perspective emphasize the word $w_3$, while authors holding the contrasting perspective emphasize the word $w_1$. Ideological weights associated with the first perspective will move a multinomial distribution's parameter from $T$ to a new position $V_1$, which is more likely to generate $w_3$ than $T$ is. Similarly, ideological weights associated with the second perspective will move the multinomial distribution's parameter from $T$ to $V_2$, which is more likely to generate $w_1$ than $T$ is.

### 3.1 Model Specification

Formally, we combine a word's topical and ideological weights through a logistic function. The complete model specification is listed as follows,

$$\begin{aligned}
P_d &\sim \text{Bernoulli}(\pi), d = 1, \ldots, D \\
W_{d,n} | P_d = v &\sim \text{Multinomial}(\beta_v), n = 1, \ldots, N_d \\
\beta_v^w &= \frac{\exp(\tau^w \times \phi_v^w)}{\sum_{w'} \exp(\tau^{w'} \times \phi_v^{w'})}, v = 1, \ldots, V \\
\tau &\sim \text{N}(\mu_\tau, \Sigma_\tau) \\
\phi_v &\sim \text{N}(\mu_\phi, \Sigma_\phi).
\end{aligned}$$

We assume that there are two contrasting perspectives in an ideological text (i.e., $V = 2$), and model a document's ideological perspective that its author or speaker holds as a Bernoulli variable $P_d$, $d = 1, \ldots, D$, where $D$ is the total number of documents in a collection. Each word in a document, $W_{d,n}$, is sampled from a multinomial distribution conditioned on the document $d$'s perspective, $n = 1, \ldots, N_d$, where $N_d$ is a document's length. The bag-of-words representation has been commonly used and shown to be effective in text classification and topic modeling.

The multinomial distribution's parameter, $\beta_v^w$, indexed by an ideological perspective $v$ and $w$-th word in the vocabulary, consists of two parts: topical weights $\tau$ and ideological weights $\phi$. $\beta$ is an auxiliary variable, and is deterministically determined by (latent) topical $\tau$ and ideological weights $\{\phi_v\}$. The two weights are combined through a logistic function. The relationship between topical and ideological weights is assumed to be multiplicative. Therefore, a word of an ideological weight $\phi = 1$ means that the word is not emphasized or de-emphasized. The prior distributions for topical and ideological weights are normal distributions. The parameters of the joint topic and perspective model, denoted as $\Theta$, include: $\pi, \mu_\tau, \Sigma_\tau, \mu_\phi, \Sigma_\phi$. We call this model a Joint Topic and Perspective Model (jTP). We show the graphical representation of the joint topic and perspective model in Figure 3.

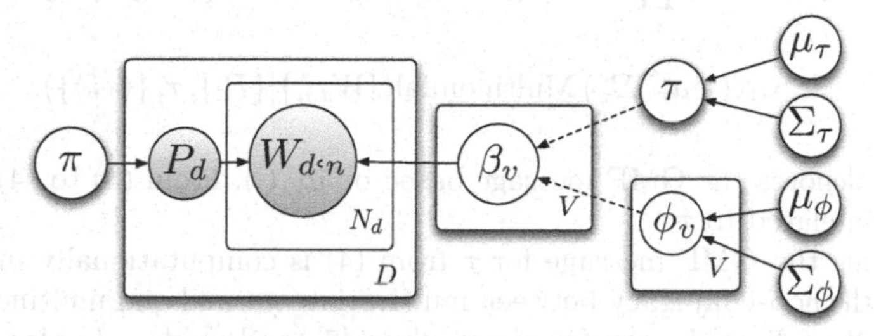


**Fig. 3.** A joint topic and perspective model in a graphical model representation (see Section 3 for details). A dashed line denotes a deterministic relation between parent and children nodes.

### 3.2 Variational Inference

The quantities of most interest in the joint topic and perspective model are (unobserved) topical weights $\tau$ and ideological weights $\{\phi_v\}$. Given a set of $D$ documents on a particular topic from differing ideological perspectives $\{P_d\}$, the joint posterior probability distribution of the topical and ideological weights under the joint topic and perspective model is

$$\begin{aligned}
&P(\tau, \{\phi_v\} | \{W_{d,n}\}, \{P_d\}; \Theta) \\
&\propto P(\tau|\mu_\tau, \Sigma_\tau) \prod_v P(\phi_v|\mu_\phi, \Sigma_\phi) \prod_{d=1}^{D} P(P_d|\pi) \prod_{n=1}^{N_d} P(W_{d,n}|P_d, \tau, \{\phi_v\}) \\
&= \mathrm{N}(\tau|\mu_\tau, \Sigma_\tau) \prod_v \mathrm{N}(\phi_v|\mu_\phi, \Sigma_\phi) \prod_d \mathrm{Bernoulli}(P_d|\pi) \prod_n \mathrm{Multinomial}(W_{d,n}|P_d, \beta),
\end{aligned}$$

where $N(\cdot)$, Bernoulli$(\cdot)$ and Multinomial$(\cdot)$ are the probability density functions of multivariate normal, Bernoulli, and multinomial distributions, respectively.

The joint posterior probability distribution of $\tau$ and $\{\phi_v\}$, however, are computationally intractable because of the non-conjugacy of the logistic-normal prior. We thus approximate the posterior probability distribution using a variational method [4], and estimate the parameters using variational expectation maximization [5]. By the Generalized Mean Field Theorem (GMF) [6], we can approximate the joint posterior probability distribution of $\tau$ and $\{\phi_v\}$ as the product of individual functions of $\tau$ and $\phi_v$:

$$P(\tau, \{\phi_v\}|\{P_d\}, \{W_{d,n}\}; \Theta) \approx q_\tau(\tau) \prod_v q_{\phi_v}(\phi_v), \tag{1}$$

where $q_\tau(\tau)$ and $q_{\phi_v}(\phi_v)$ are the posterior probabilities of the topical and ideological weights conditioned on the random variables on their Markov blanket.

Specifically, $q_\phi$ is defined as follows,

$$q_\tau(\tau) = P(\tau|\{W_{d,n}\}, \{P_d\}, \{\langle\phi_v\rangle\}; \Theta) \tag{2}$$

$$\propto P(\tau|\mu_\tau, \Sigma_\tau) \prod_v P(\langle\phi_v\rangle|\mu_\phi, \Sigma_\phi) P(\{W_{d,n}\}|\tau, \{\langle\phi_v\rangle\}, \{P_d\}) \tag{3}$$

$$\propto \mathrm{N}(\tau|\mu_\tau, \Sigma_\tau)\,\mathrm{Multinomial}(\{W_{d,n}\}|\{P_d\}, \tau, \{\langle\phi_v\rangle\}), \tag{4}$$

where $\langle\phi_v\rangle$ denotes the GMF message based on $q_{\phi_v}(\cdot)$. From (3) to (4) we drop the terms unrelated to $\tau$.

Calculating the GMF message for $\tau$ from (4) is computationally intractable because of the non-conjugacy between multivariate normal and multinomial distributions. We follow the similar approach in [7], and made a Laplace approximation of (4). We first represent the word likelihood $\{W_{d,n}\}$ as the following exponential form:

$$P(\{W_{d,n}\}|\{P_d\}, \tau, \{\langle\phi_v\rangle\}) = \exp\left(\sum_v n_v(\langle\phi_v\rangle \bullet \tau) - \sum_v n_v^T \mathbf{1} C(\langle\phi_v\rangle \bullet \tau)\right) \tag{5}$$

where $\bullet$ is element-wise vector product, $n_v$ is a word count vector under the ideological perspective $v$, $\mathbf{1}$ is a column vector of one, and $C$ function is defined as follows,

$$C(x) = \log\left(1 + \sum_{p=1}^{P} \exp x_p\right), \tag{6}$$

where $P$ is the dimensionality of the vector $x$.

We expand $C$ using Taylor series to the second order around $\hat{x}$ as follows,

$$C(x) \approx C(\hat{x}) + \nabla(x)(x - \hat{x}) + \frac{1}{2}(x - \hat{x})^T H(\hat{x})(x - \hat{x}),$$

where $\nabla$ is the gradient of $C$, and $H$ is the Hessian matrix of $C$. We set $\hat{x}$ as $\langle\tau\rangle^{(t-1)} \bullet \langle\phi_v\rangle$. The superscript denoted the GMF message in the $t-1$ (i.e., previous) iteration.

Finally, we plug the second-order Taylor expansion of $C$ back to (4) and rearranged terms about $\tau$. We obtain the multivariate normal approximation of $q_\tau(\cdot)$ with a mean vector $\mu^*$ and a variance matrix $\Sigma^*$ as follows,

$$\Sigma^* = \left(\Sigma_\tau^{-1} + \sum_v n_v^T \mathbf{1}\langle\phi_v\rangle \downarrow H(\hat{\tau} \bullet \langle\phi_v\rangle) \rightarrow \langle\phi_v\rangle\right)^{-1}$$

$$\mu^* = \Sigma^* \Bigg(\Sigma_\tau^{-1}\mu_\tau + \sum_v n_v \bullet \langle\phi_v\rangle - \sum_v n_v^T \mathbf{1}\nabla C(\hat{\tau} \bullet \langle\phi_v\rangle) \bullet \langle\phi_v\rangle$$

$$+ \sum_v n_v^T \mathbf{1}\langle\phi_v\rangle \bullet (H(\hat{\tau} \bullet \langle\phi_v\rangle)(\hat{\tau} \bullet \langle\phi_v\rangle))\Bigg),$$

where $\downarrow$ is column-wise vector-matrix product, $\rightarrow$ is row-wise vector-matrix product. The Laplace approximation for the logistic-normal prior has been shown to be tight [8].

$q_{\phi_v}$ in (3.2) can be approximated in a similar fashion as a multivariate normal distribution with a mean vector $\mu^\dagger$ and a variance matrix $\Sigma^\dagger$ as follows,

$$\Sigma^\dagger = \left(\Sigma_\phi^{-1} + n_v^T \mathbf{1}\langle\tau\rangle \downarrow H(\langle\tau\rangle \bullet \hat{\phi}_v) \rightarrow \langle\tau\rangle\right)^{-1}$$

$$\mu^\dagger = \Sigma^\dagger \Big(\Sigma_\phi^{-1}\mu_\phi + n_v \bullet \langle\tau\rangle - n_v^T \mathbf{1}\nabla C(\langle\tau\rangle \bullet \hat{\phi}_v) \bullet \langle\tau\rangle$$

$$+ n_v^T \mathbf{1}\langle\tau\rangle \bullet (H(\langle\tau\rangle \bullet \hat{\phi}_v)(\langle\tau\rangle \bullet \hat{\phi}_v))\Big),$$

where we set $\hat{\phi}_v$ as $\langle\phi_v\rangle^{(t-1)}$.

In E-step, we have a message passing loop and iterate over the $q$ functions in (3.2) until converge. We monitor the change in the auxiliary variable $\beta$ and stop when the absolute change is smaller than a threshold. In M-step, $\pi$ can be easily maximized by taking the sample mean of $\{P_d\}$. We monitor the data likelihood and stop the variational EM loop when the change of data likelihood is less than a threshold.

### 3.3 Identifiability

The joint topic and perspective model as specified above is not identifiable. There are multiple assignments of topical and ideological weights that can produce exactly the same data likelihood. Therefore, topic and ideological weights estimated from data may be incomparable.

The first source of un-identifiability is due to the multiplicative relationship between $\tau$ and $\phi_v$. We can easily multiply a constant to $\tau^w$ and divide $\phi_v^w$ by the same constant, and the auxiliary variable $\beta$ stays the same.

The second source of un-identifiability comes from the sum-to-one constraint in the multinomial distribution's parameter $\beta$. Given a vocabulary $\mathcal{W}$, we have only $|\mathcal{W}| - \infty$ number of free parameters for $\tau$ and $\{P_d\}$. Allowing $|\mathcal{W}|$ number of free parameters makes topical and ideological weights unidentifiable.

We fix the following parameters to solve the un-identifiability issue: $\tau^1$, $\{\phi_1^w\}$, and $\phi_v^1$. We fix the values of the $\tau^1$ to be one and $\{\phi_v^1\}$ to be zero, $v = 1, \ldots, V$. We choose the first ideological perspective as a base and fix its ideological weights $\phi_1^w$ to be one for all words, $w = 1, \ldots, |\mathcal{W}|$. By fixing the corner of $\phi$ (i.e., $\{\phi_v^1\}$) we assume that the first word in the vocabulary are not biased by either ideological perspectives, which may not be true. We thus add a dummy word as the first word in the vocabulary, whose frequency is the average word frequency in the whole collection and conveys no ideological information (in the word frequency).

# 4 Experiments

## 4.1 Synthetic Data

We first evaluate the proposed model on synthetic data. We fix the values of the topical and ideological weights, and generated synthetic data according to the generative process in Section 3. We test if the variational inference algorithm for the joint topic and perspective model in Section 3.2 successfully converges. More importantly, we test if the variational inference algorithm can correctly recover the true topical and ideological weights that generated the synthetic data.

Specifically, we generate the synthetic data with a three-word vocabulary and topical weights $\tau = (2, 2, 1)$, shown as $\circ$ in the simplex in Figure 4. We then simulate different degrees to which authors holding two contrasting ideological beliefs emphasized words. We let the first perspective emphasize $w_2$ ($\phi_1 = (1, 1+p, 0)$) and let the second perspective emphasized $w_1$ ($\phi_2 = (1+p, 1, 0)$. $w_3$ is the dummy word in the vocabulary. We vary the value of $p$ ($p = 0.1, 0.3, 0.5$) and plotted the corresponding auxiliary variable $\beta$ in the simplex in Figure 4. We generate the equivalent number of documents for each ideological perspective, and varied the number of documents from 10 to 1000.

We evaluate how closely the variational inference algorithm recovered the true topical and ideological weights by measuring the maximal absolute difference between the true $\beta$ (based on the true topical weights $\tau$ and ideological weights $\{\phi_v\}$) and the estimated $\hat{\beta}$ (using the expected topical weights $\langle\tau\rangle$ and ideological weights $\{\langle\phi_v\rangle\}$ returned by the variational inference algorithm).

The simulation results in Figure 5 suggested that the proposed variational inference algorithm for the joint topic and perspective is valid and effective. Although the variational inference algorithm was based on Laplace approximation, the inference algorithm recovered the true weights very closely. The absolute difference between true $\beta$ and estimated $\hat{\beta}$ was small and close to zero.

## 4.2 Ideological Discourse

We evaluate the joint topic and perspective model on two ideological discourses. The first corpus, bitterlemons, is comprised of editorials written by the Israeli

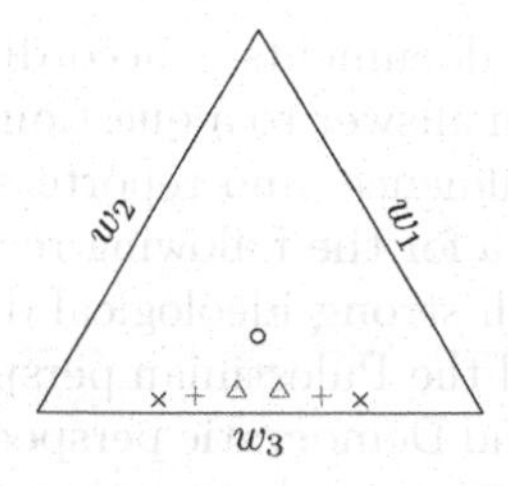


**Fig. 4.** We generate synthetic data with a three-word vocabulary. The ∘ indicates the value of the true topical weight $\tau$. $\triangle$, $+$, and $\times$ are $\beta$ after $\tau$ is modulated by different ideological weights $\{\phi_v\}$.

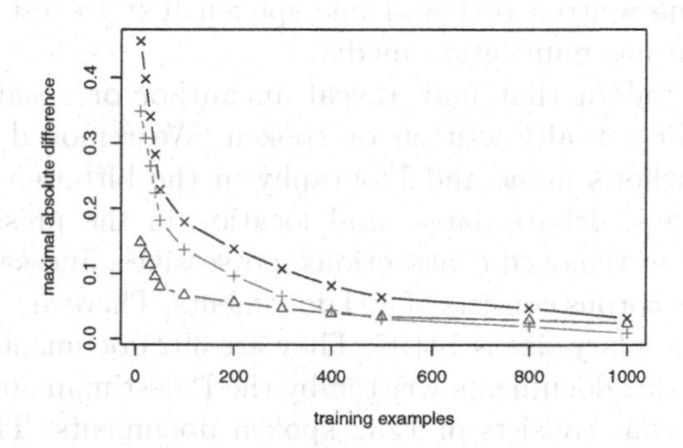


**Fig. 5.** The experimental results of recovering true topical and ideological weights. The x axis is the number of training examples, and the y axis is the maximal absolute difference between true $\beta$ and estimated $\hat{\beta}$. The smaller the difference, the better. The curves in $\triangle$, $+$, and $\times$ correspond to the three different ideological weights in Figure 4.

and Palestinian authors on the Israeli-Palestinian conflict. The second corpus, presidential debates, is comprised of spoken words from the Democratic and Republican presidential candidates in 2000 and 2004.

The bitterlemons corpus consists of the articles published on the website `http://bitterlemons.org/`. The website is set up to "contribute to mutual understanding [between Palestinians and Israelis] through the open exchange of ideas."[1] Every week an issue about the Israeli-Palestinian conflict is selected for discussion (e.g., "Disengagement: unilateral or coordinated?"). The website editors have labeled the ideological perspective of each published article. The bitterlemons corpus has been used to learn individual perspectives [9], but the previous work was based on naive Bayes models and did not simultaneously model topics and perspectives.

The 2000 and 2004 presidential debates corpus consists of the spoken transcripts of six presidential debates and two vice-presidential debates in 2000 and 2004. We downloaded the speech transcripts from the American Presidency Project[2]. The speech transcripts came with speaker tags, and we segmented the

[1] `http://www.bitterlemons.org/about/about.html`

[2] `http://www.presidency.ucsb.edu/debates.php`

transcripts into spoken documents according to speakers. Each spoken document was either an answer to a question or a rebuttal. We discarded the words from moderators, audience, and reporters.

We choose these two corpora for the following reasons. First, the two corpora contain political discourse with strong ideological differences. The bitterlemons corpus contains the Israeli and the Palestinian perspectives; the presidential debates corpus the Republican and Democratic perspectives. Second, they are from multiple authors or speakers. There are more than 200 different authors in the bitterlemons corpus; there are two Republican candidates and four Democratic candidates. We are interested in ideological discourse expressing socially shared beliefs, and less interested in individual authors or candidates' personal beliefs. Third, we select one written text and one spoken text to test how our model behaves on different communication media.

We removed metadata that may reveal an author or speaker's ideological stance but were not actually written or spoken. We removed the publication dates, titles, an author's name and biography in the bitterlemons corpus. We removed speaker tags, debate dates, and location in the presidential debates corpus. Our tokenizer removed contractions, possessives, and cases.

The bitterlemons corpus consists of 594 documents. There are a total of 462308 words, and the vocabulary size is 14197. They are 302 documents written by the Israeli authors and 292 documents written by the Palestinian authors. The presidential debates corpus consists of 1232 spoken documents. There are a total of 122056 words, and the vocabulary size is 16995. There are 235 spoken documents from the Republican candidates, and 214 spoken documents from the Democratic candidates.

### 4.3 Topical and Ideological Weights

We fit the proposed joint topic and perspective model on two text corpora, and the results were shown in Figure 6 and Figure 7 in color text clouds[3]. Text clouds represent a word's frequency in size. The larger a word's size, the more frequently the word appears in a text collection. Text clouds have been a popular method of summarizing tags and topics on the Internet (e.g., bookmark tags on Del.icio.us[4] and photo tags on Flicker [5]. Here we have matched a word's size with its topical weight $\tau$.

To show a word's ideological weight, we paint a word in color shades. We assign each ideological perspective a color (red or blue). A word's color is determined by which perspective uses a word more frequently than the other. Color shades gradually change from pure colors (strong emphasis) to light gray (no emphasis). The degree of emphasis is measured by how extreme a word's ideological weight $\phi$ is from one (i.e., no emphasis). Color text clouds allow us to present three kinds of information at the same time: words, their topical weights, and ideological weights.

[3] We omit the words of low topical and ideological weights due to space limit.

[4] http://del.icio.us/

[5] http://www.flickr.com/

fence terrorism disengagement terrorist jordan leader case bush jews past appears leaders unilateral jewish forces status iraq arafats line egypt green term arafat level approach abu settlers months left territory good arabs idea large syria suicide war strategic arab back democratic year sharons effect settlements decision bank west agreement majority water present mazen gaza pa sharon minister prime withdrawal israels return state israel process american oslo violence support security ariel peace conflict issue president current israeli sides palestinian israelis solution future middle jerusalem settlement world force plan long make issues time leadership public refugees east political administration pressure palestinians camp strip palestine ceasefire roadmap national policy government final order situation military economic hamas elections part states international end community territories negotiations based agreements real side united recent work 1967 party made movement important control authority dont hand violent borders continue change including clear relations problem society resolution parties building people al means move power role refugee ongoing intifada nations major civilians fact occupation areas talks council land struggle efforts hope position compromise rights stop difficult put historic opinion positions give accept reason inside law internal occupied americans years significant result ending things wall resistance

**Fig. 6.** Visualize the topical and ideological weights learned by the joint topic and perspective model from the bitterlemons corpus (see Section 4.3). Red: words emphasized more by the Israeli authors. Blue: words emphasized more by the Palestinian authors.

Let us focus on the words of large topical weights learned from the bitterlemons corpus (i.e., words in large sizes in Figure 6). The word of the largest topical weight is "Palestinian", followed by "Israeli", "Palestinians", "peace", and "political". The topical weights learned by the joint topic and perspective model clearly match our expectation from the discussions about the Israeli-Palestinian conflict. Words in large sizes summarizes well what the bitterlemons corpus is about.

Similarly, a brief glance over words of large topical weights learned from the presidential debates corpus (i.e., words in large sizes in Figure 7) clearly tells us the debates' topic. Words of large topical weights capture what American politics is about (e.g., "people", "president", "America", "government") and specific political and social issues (e.g., "Iraq", "taxes", "Medicare"). Although not every word of large topical weights is attributed to a text's topic, e.g., "im" ("I'm" after contraction is removed) occurred frequently because of the spoken nature of debate speeches, the majority of words of large topical weights appear to convey what the two text collections are about.

Now let us turn our attention to words' ideological weights $\phi$, i.e., color shade in Figure 6. The word "terrorism", followed by "terrorist", is painted pure red, which is highly emphasized by the Israeli authors. "Terrorist" is a word that clearly reveals an author's attitude toward the other group's violent behavior. Many words of large ideological weights can be categorized into the ideology discourse structures previously manually identified by researchers in discourse analysis [1]:

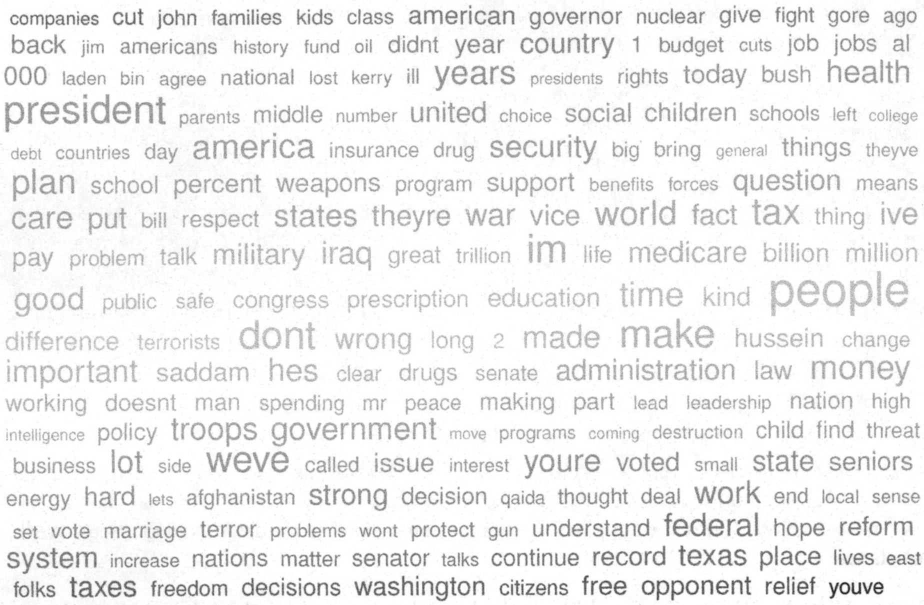

companies cut john families kids class american governor nuclear give fight gore ago
back jim americans history fund oil didnt year country 1 budget cuts job jobs al
000 laden bin agree national lost kerry ill years presidents rights today bush health
president parents middle number united choice social children schools left college
debt countries day america insurance drug security big bring general things theyve
plan school percent weapons program support benefits forces question means
care put bill respect states theyre war vice world fact tax thing ive
pay problem talk military iraq great trillion im life medicare billion million
good public safe congress prescription education time kind people
difference terrorists dont wrong long 2 made make hussein change
important saddam hes clear drugs senate administration law money
working doesnt man spending mr peace making part lead leadership nation high
intelligence policy troops government move programs coming destruction child find threat
business lot side weve called issue interest youre voted small state seniors
energy hard lets afghanistan strong decision qaida thought deal work end local sense
set vote marriage terror problems wont protect gun understand federal hope reform
system increase nations matter senator talks continue record texas place lives east
folks taxes freedom decisions washington citizens free opponent relief youve

**Fig. 7.** Visualize the topical weights and ideological weights learned by the joint topic and perspective model from the presidential debates corpus i(see Section 4.3). Red: words emphasized by the Democratic candidates. Blue: words emphasized by the Republican candidates.

– Membership: Who are we and who belongs to us? "Jews" and "Jewish" are used more frequently by the Israeli authors than the Palestinian authors. "Washington" is used more frequently by the Republican candidates than Democratic candidates.
– Activities: What do we do as a group? "Unilateral", "disengagement", and "withdrawal" are used more frequently by the Israeli authors than the Palestinian authors. "Resistance" is used more frequently by the Palestinian authors than the Israeli authors.
– Goals: What is our group's goal? (Stop confiscating) "land" , "independent", and (opposing settlement) "expansion" are used more frequently by the Palestinian authors than the Israeli authors.
– Values: How do we see ourselves? What do we think is important? "Occupation" and (human) "rights" are used more frequently by Palestinian authors than the Israeli authors. "Schools", "environment", and "middle" "class" are used more frequently by the Democratic candidates than the Republican candidates. "Freedom" and "free" are used more frequently by the Republican candidates.
– Position and Relations: what is our position and our relation to other groups? "Jordan" and "Arafats" (after removing contraction of "Arafat's") are used more frequently by the Israeli authors than by the Palestinian authors.

We do not intend to give a detailed analysis of the political discourse in the Israeli-Palestinian conflict and in American politics. We do, however, want to point out that the joint topic and perspective model seems to "discover" words that play important roles in ideological discourse. The results not only support the hypothesis that ideology is greatly reflected in an author or speaker's lexical choices, but also suggest that the joint topic and perspective model closely captures the lexical variations.

Political scientists and media analysts can formulate research questions based on the uncovered topical and ideological weights, such as: what are the important topics in a text collection? What words are emphasized or de-emphasized by which group? How strongly are they emphasized? In what context are they emphasized? The joint topic and perspective model can thus become a valuable tool to explore ideological discourse.

Our results, however, also point out the model's weaknesses. First, a bag-of-words representation is convenient but fails to capture many linguistic phenomena in political discourse. "Relief" is used to represent tax relief, marriage penalty relief, and humanitarian relief. Proper nouns (e.g., "West Bank" in the bitterlemons corpus and "Al Quida" in the presidential debates corpus) are broken into multiple pieces. N-grams do not solve all the problems. The discourse function of the verb "increase" depends much on the context. A presidential candidate can "increase" legitimacy, profit, or defense, and single words cannot distinguish them.

### 4.4 Prediction

We evaluate how well the joint topic and perspective model predicted words from unseen ideological discourse in terms of perplexity on a held-out set. Perplexity has been a popular metric to assess how well a statistical language model generalizes [10]. A model generalizes well if it achieves lower perplexity. We choose unigram as a baseline. Unigram is a special case of the joint topic and perspective model that assumes *no* lexical variations are due to an author or speaker's ideological perspective (i.e., fixing all $\{\phi_v\}$ to one).

Perplexity is defined as the exponential of the negative log word likelihood with respect to a model normalized by the total number of words:

$$\exp\left(\frac{-\log P(\{W_{d,n}\}|\{P_d\};\Theta)}{\sum_d N_d}\right)$$

We can integrate out topical and ideological weights to calculate the predictive probability $P(\{W_{d,n}\}|\{P_d\};\Theta)$:

$$P(\{W_{d,n}\}|\{P_d\};\Theta) = \int\int\prod_{d=1}^{D}\prod_{n=1}^{N_d} P(W_{d,n}|P_d)d\tau d\phi_v.$$

Instead, we approximate the predictive probability by plugging in the point estimates of $\tau$ and $\phi_v$ from the variational inference algorithm.

For each corpus, we vary the number of training documents from 10% to 90% of the documents, and measured perplexity on the remaining 10% held-out set. The results were shown in Figure 8. We can clearly see that the joint topic and perspective model reduces perplexity on both corpora. The results strongly support the hypothesis that ideological perspectives are reflected in lexical variations. Only when ideology is reflected in lexical variations can we observe the perplexity reduction from the joint topic and perspective model. The results also suggest that the joint topic and perspective model closely captures the lexical variations due to an author or speaker's ideological perspective.

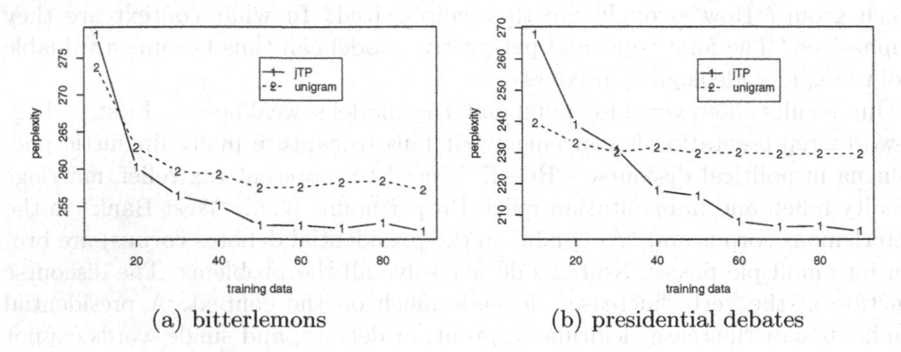


(a) bitterlemons (b) presidential debates

**Fig. 8.** The proposed joint topic and perspective model reduces perplexity on a held-out set

## 5 Related Work

Abelson and Carroll pioneered modeling ideological beliefs in computers in the sixties [2]. Their system modeled the beliefs of a right-wing politician as a set of English sentences (e.g., "Cuba subverts Latin America."). Carbonell proposed a system, POLITICS, that can interpret text from two conflicting ideologies [11]. These early studies model ideology at a more sophisticated level (e.g., goals, actors, and action) than the proposed joint topic and perspective model, but require humans to manually construct a knowledge database. The knowledge-intensive approaches suffer from the "knowledge acquisition bottleneck." We take a completely different approach and aim to *automatically* learn ideology from a large number of documents.

[12] explored a similar problem of identifying media's bias. They found that the sources of news articles can be successfully identified based on word choices using Support Vector Machines. They identified the words that can best discriminate two news sources using Canonical Correlation Analysis. In addition to the clearly different methods between [12] and this paper, there are crucial differences. First, instead of applying two different methods as [12] did, the Joint Topic and Perspective Model (Section 3) is a single unified model that can learn to predict an article's ideological slant and uncover discriminating word choices *simultaneously*. Second, the Joint Topic and Perspective Model makes explicit

the assumption of the underlying generative process on ideological text. In contrast, discriminative classifiers such as SVM do not model the data generation process [13]. However, our methods implicitly assume that documents are about the same news event or issue, which may not be true and could benefit from an extra story alignment step as [12] did.

We borrow statistically modeling and inference techniques heavily from research on topic modeling (e.g., [14], [15] and [16]). They focus mostly on modeling text collections that containing many *different* (latent) topics (e.g., academic conference papers, news articles, etc). In contrast, we are interested in modeling ideology texts that are mostly on the *same* topic but mainly differs in their ideological perspectives. There have been studies going beyond topics (e.g., modeling authors [17]). We are interested in modeling lexical variation collectively from multiple authors sharing similar beliefs, not lexical variations due to individual authors.

## 6 Conclusion

We present a statistical model for ideological discourse. We hypothesized that ideological perspectives were partially reflected in an author or speaker's lexical choices. The experimental results showed that the proposed joint topic and perspective model fit the ideological texts better than a model naively assuming no lexical variations due to an author or speaker's ideological perspectives. We showed that the joint topic and perspective model uncovered words that represent an ideological text's topic as well as words that reveal ideological discourse structures. Lexical variations appeared to be a crucial feature that can enable automatic understanding of ideological perspectives from a large amount of documents.

## References


1. Van Dijk, T.A.: Ideology: A Multidisciplinary Approach. Sage Publications, Thousand Oaks (1998)
2. Abelson, R.P., Carroll, J.D.: Computer simulation of individual belief systems. The American Behavioral Scientist 8, 24–30 (1965)
3. Carruthers, S.L.: The Media At War: Communication and Conict in the Twentieth Century. St. Martin's Press (2000)
4. Jordan, M.I., Ghahramani, Z., Jaakkola, T.S., Saul, L.K.: An introduction to variational methods for graphical models. achine Learning 37, 183–233 (1999)
5. Attias, H.: A variational bayesian framework for graphical models. In: Advances in Neural Information Processing Systems 12 (2000)
6. Xing, E.P., Jordan, M.I., Russell, S.: A generalized mean eld algorithm for variational inference in exponential families. In: Proceedings of the 19th Annual Conference on Uncertainty in AI (2003)
7. Xing, E.P.: On topic evolution. Technical Report CMU-CALD-05-115, Center for Automated Learning & Discovery, Pittsburgh, PA (December 2005)

8. Ahmed, A., Xing, E.P.: On tight approximate inference of the logistic-normal topic admixture model. In: Proceedings of the Eleventh International Conference on Artificial Intelligence and Statistics (2007)
9. Lin, W.H., Wilson, T., Wiebe, J., Hauptmann, A.: Which side are you on? identifying perspectives at the document and sentence levels. In: Proceedings of Tenth Conference on Natural Language Learning (CoNLL) (2006)
10. Manning, C.D., Schütze, H.: Foundations of Statistical Natural Language Processing. The MIT Press, Cambridge (1999)
11. Carbonell, J.G.: POLITICS: Automated ideological reasoning. Cognitive Science 2(1), 27–51 (1978)
12. Fortuna, B., Galleguillos, C., Cristianini, N.: Detecting the bias in media with statistical learning methods. In: Text Mining: Theory and Applications. Taylor and Francis Publisher (2008)
13. Rubinstein, Y.D.: Discriminative vs Informative Learning. PhD thesis, Department of Statistics, Stanford University (January 1998)
14. Hofmann, T.: Probabilistic latent semantic indexing. In: Proceedings of the 22nd Annual International ACM SIGIR Conference on Research and Development in Information Retrieval, pp. 50–57 (1999)
15. Blei, D.M., Ng, A.Y., Jordan, M.: Latent dirichlet allocation. Journal of Machine Learning Research 3, 993–1022 (2003)
16. Griffiths, T.L., Steyvers, M.: Finding scientic topics. Proceedings of the National Academy of Sciences 101, 5228–5235 (2004)
17. Rosen-Zvi, M., Griths, T., Steyvers, M., Smyth, P.: The author-topic model for authors and documents. In: Proceedings of the 20th Conference on Unvertainty in Artificial Intelligence (2004)

# Effective Pruning Techniques for Mining Quasi-Cliques*

Guimei Liu and Limsoon Wong

School of Computing, National University of Singapore, Singapore

**Abstract.** Many real-world datasets, such as biological networks and social networks, can be modeled as graphs. It is interesting to discover densely connected subgraphs from these graphs, as such subgraphs represent groups of objects sharing some common properties. Several algorithms have been proposed to mine quasi-cliques from undirected graphs, but they have not fully utilized the minimum degree constraint for pruning. In this paper, we propose an efficient algorithm called *Quick* to find maximal quasi-cliques from undirected graphs. The *Quick* algorithm uses several effective pruning techniques based on the degree of the vertices to prune unqualified vertices as early as possible, and these pruning techniques can be integrated into existing algorithms to improve their performance as well. Our experiment results show that *Quick* is orders of magnitude faster than previous work on mining quasi-cliques.

## 1 Introduction

Graphs can represent complicated relationships among objects, and they have been used to model many real-world datasets. For example, a protein-protein interaction network can be represented as a graph where each vertex represents a protein and edges represent interactions between proteins. A set of microarray data can be converted to a graph in which each vertex represents a gene and an edge between two vertices represents a strong similarity between the expression data of the two corresponding genes. Highly connected subgraphs in these graphs often have significant biological implications. They can correspond to protein complexes [1] or biologically relevant functional groups [2,3,4].

The discovery of dense subgraphs from one or multiple graphs has attracted increasing attention. Cliques are the densest form of subgraphs. A graph is a clique if there is an edge between every pair of the vertices. However, this requirement is often too restrictive given that real-world datasets are often incomplete and noisy. The concept of quasi-cliques has been proposed to relax the requirement. Different definitions have been given to quasi-cliques. Here we adopt the definition that is based on the degree of individual vertices, that is, a graph is a quasi-clique if every vertex in the graph is adjacent to at least $\lceil\gamma(n-1)\rceil$ other vertices in the graph, where $\gamma$ is a number between 0 and 1 and $n$ is the number of vertices in the graph.

* This work was supported in part by a Singapore A*STAR SERC PSF grant.

W. Daelemans et al. (Eds.): ECML PKDD 2008, Part II, LNAI 5212, pp. 33–49, 2008.

Given a graph, the search space of the quasi-clique mining problem is the power set of its vertex set. How to efficiently and effectively prune the search space is critical to the performance of a quasi-clique mining algorithm. However, the downward closure property no longer holds on quasi-cliques, which makes mining quasi-cliques much more challenging than mining cliques. Existing algorithms for mining quasi-cliques from a single graph all use heuristic or randomized methods and they do not produce the complete set of quasi-cliques [5,6,2]. Although existing algorithms for mining quasi-cliques from a set of graphs generate the complete result, they have not fully exploited the pruning power of the minimum degree constraint [7,8].

In this paper, we propose an efficient algorithm called Quick to mine quasi-cliques, which uses several effective pruning techniques based on the degree of the vertices. These pruning techniques can effectively remove unpromising vertices as early as possible. We conducted a set of experiments to demonstrate the effectiveness of the proposed pruning techniques.

The rest of the paper is organized as follows. Section 2 gives the formal problem definition. Section 3 presents the Quick algorithm, and its performance is studied in Section 4. Related work is described in Section 5. Finally, Section 6 concludes the paper.

## 2 Problem Definition

In this section, we formally define the quasi-clique mining problem. We consider simple graphs only, that is, undirected graphs that have no self-loops and multi-edges. Graph isomorphism test is very complicated and costly. To simplify the problem and the presentation, we restrict our discussion to relational graphs where every vertex has a unique label. In this case, graph isomorphism test can be performed by simply comparing the vertex set and edge set of two graphs. Note that the techniques described in this paper can be applied to non-relational graphs as well. In the rest of this paper, the term "graph" refers to simple relational graphs unless otherwise stated.

A *simple graph* $G$ is defined as a pair $(V, E)$, where $V$ is a set of vertices, and $E$ is a set of edges between the vertices. Two vertices are *adjacent* if there is an edge between them. The *adjacency list* of a vertex $v$ in $G$, denoted as $N^G(v)$, is defined as $\{u|(u, v) \in E\}$. The *degree* of a vertex $v$ in $G$, denoted as $deg^G(v)$, is defined as $|N^G(v)|$. The adjacency list of a vertex set $X$, denoted as $N^G(X)$, is defined as $\{u|\forall v \in X, (u, v) \in E\}$.

The distance between two vertices $u$ and $v$ in a graph $G = (V, E)$, denoted as $dist^G(u, v)$, is defined as the number of edges on the shortest path between $u$ and $v$. Trivially, $dist^G(u, u) = 0$, and $dist^G(u, v) = 1$ if $u \neq v$ and $(u, v) \in E$. We denote the set of vertices that are within a distance of $k$ from vertex $v$ as $N_k^G(v) = \{u|dist^G(u, v) \leq k\}$. The diameter of a graph $G$, denoted as $diam(G)$, is defined as $max_{u,v \in V} dist^G(u, v)$. A graph is called *connected* if $dist^G(u, v) < \infty$ for any $u, v \in V$.

**Definition 1 ($\gamma$-quasi-clique).** *A graph $G = (V, E)$ is a $\gamma$-quasi-clique ($0 \leq \gamma \leq 1$) if $G$ is connected, and for every vertex $v \in V$, $deg^G(v) \geq \lceil \gamma \cdot (|V| - 1) \rceil$.*

According to the definition, a quasi-clique is a graph satisfying a user-specified minimum vertex degree bound, and we call $\gamma$ the *minimum degree threshold*. A clique is a special case of quasi-clique with $\gamma$=1. Figure 1 shows two example quasi-cliques. Graph $G_1$ is a 0.5-quasi-clique, but it is not a 0.6-quasi-clique because the degree of every vertex in $G_1$ is 2, and 2 is smaller than $\lceil 0.6 \cdot (5 - 1) \rceil$. Graph $G_2$ is a 0.6-quasi-clique.

Given a graph $G = (V, E)$, graph $G' = (V', E')$ is a *subgraph* of $G$ if $V' \subseteq V$ and $E' \subseteq E$. Graph $G$ is called a *supergraph* of $G'$. If $V' \subset V$ and $E' \subseteq E$, or $E' \subset E$ and $V' \subseteq V$, then $G'$ is called a *proper subgraph* of $G$, and $G$ is called a *proper supergraph* of $G'$. A subgraph $G'$ of a graph $G$ is called an *induced subgraph* of $G$ if, for any pair of vertices $u$ and $v$ of $G'$, $(u, v)$ is an edge of $G'$ if and only if $(u, v)$ is an edge of $G$. We also use $G(X)$ to denote the subgraph of $G$ induced on a vertex set $X \subseteq V$. Given a minimum degree threshold $\gamma$, if a graph is a $\gamma$-quasi-clique, then its subgraphs usually become uninteresting even if they are also $\gamma$-quasi-cliques. In this paper, we mine only maximal quasi-cliques.

**Definition 2 (Maximal $\gamma$-quasi-clique).** *Given graph $G = (V, E)$ and a vertex set $X \subseteq V$. $G(X)$ is a maximal $\gamma$-quasi-clique of $G$ if $G(X)$ is a $\gamma$-quasi-clique, and there does not exist another vertex set $Y$ such that $Y \supset X$ and $G(Y)$ is a $\gamma$-quasi-clique.*

Cliques have the downward-closure property, that is, if $G$ is a clique, then all of its induced subgraphs must also be cliques. This downward-closure property has been used to mine various frequent patterns in the data mining community. Unfortunately, this property does not hold for quasi-cliques. An induced subgraph of a $\gamma$-quasi-clique may not be a $\gamma$-quasi-clique. For example, graph $G_1$ in Figure 1 is a 0.5-quasi-clique, but one of its induced subgraph is not a 0.5-quasi-clique as shown in Figure 1(c). In fact, none of the induced subgraphs of $G_1$ with four vertices is a 0.5-quasi-clique.

Small quasi-cliques are usually trivial and not interesting. For example, a single vertex itself is a quasi-clique for any $\gamma$. We use a minimum size threshold *min_size* to filter small quasi-cliques.

**Problem statement (Mining maximal quasi-cliques from a single graph).** Given a graph $G = (V, E)$, a minimum degree threshold $\gamma \in [0, 1]$ and a minimum size threshold *min_size*, the problem of mining maximal quasi-cliques from $G$ is to find all the vertex sets $X$ such that $G(X)$ is a maximal $\gamma$-quasi-cliques of $G$ and $X$ contains at least *min_size* vertices.

In some applications, users are interested in finding quasi-cliques that occur frequently in a set of graphs. The techniques proposed in this paper can be applied to mine frequent quasi-cliques (or so-called cross quasi-cliques [7] or coherent quasi-cliques [8]) from a given graph database as well.

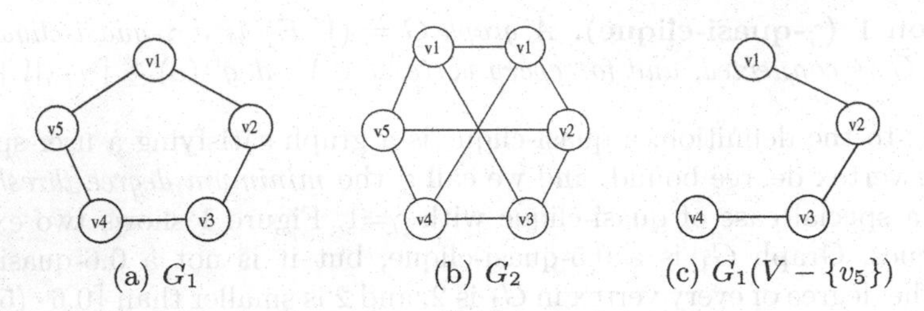


**Fig. 1.** Examples of quasi-cliques

# 3 Efficient Mining of Quasi-Cliques

## 3.1 The Depth-First Search Framework

Given a graph $G = (V, E)$, any subset of $V$ may form a quasi-clique. Therefore, the search space of the maximal quasi-clique mining problem is the power set of $V$, and it can be represented as a set-enumeration tree [9]. Figure 2 shows the search space tree for a graph $G$ with four vertices $\{a, b, c, d\}$. Each node in the tree represents a vertex set. For every vertex set $X$ in the tree, only vertices after the last vertex of $X$ can be used to extend $X$. This set of vertices are called *candidate extensions* of $X$, denoted as $cand\_exts(X)$. For example, in the search space tree shown in Figure 2, vertices are sorted into lexicographic order, so vertex $d$ is in $cand\_exts(\{a, c\})$, but vertex $b$ is not a candidate extension of $\{a, c\}$ because vertex $b$ is before vertex $c$ in lexicographic order.

The Quick algorithm uses the depth-first order to explore the search space. In the example search space tree shown in Figure 2, the Quick algorithm first finds all the quasi-cliques containing vertex $a$, and then finds all the quasi-cliques containing vertex $b$ but not containing vertex $a$, and so on. The size of the search space is exponential to the number of vertices in the graph. The main issue in mining quasi-cliques is how to effectively and efficiently prune the search space. As discussed in Section 2, quasi-cliques do not have the downward-closure property, hence we cannot use the downward-closure property to prune the search space here. According to the definition of quasi-cliques, there is a minimum

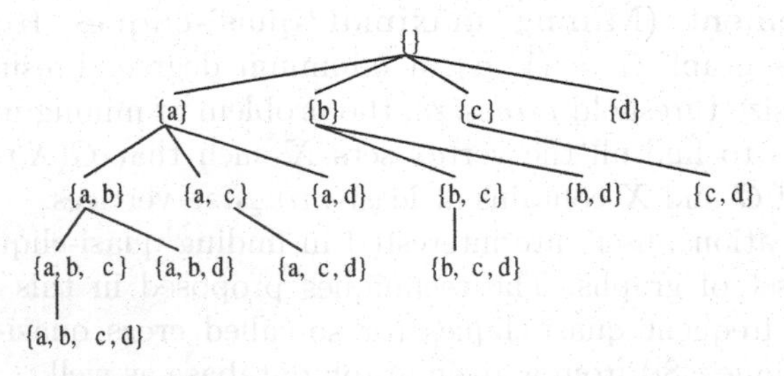


**Fig. 2.** The search space tree ($V = \{a, b, c, d\}$)

requirement on the degree of the vertices in a quasi-clique. We use this constraint to reduce the candidate extensions of each vertex set $X$.

### 3.2 Pruning Techniques Used in Existing Work

Before describing the new pruning techniques used in the Quick algorithm, we first describe the pruning techniques used by existing work. These pruning techniques are adopted in the Quick algorithm.

**Pruning based on graph diameters.** Pei et al. inferred the upper bound of the diameter of a $\gamma$-quasi-clique based on the value of $\gamma$ (Theorem 1 in [7]). In particular, the upper bound of the diameter of a $\gamma$-quasi-clique is 2 when $\gamma \geq 0.5$. They used this upper bound to reduce the candidate extensions of a vertex set as stated in the following lemma.

**Lemma 1.** *Given graph $G = (V, E)$ and two vertex sets $X \subset Y \subseteq V$, if $G(Y)$ is a $\gamma$-quasi-clique, then for every vertex $u \in (Y - X)$, we have $u \in \bigcap_{v \in X} N_k^G(v)$, where $k$ is the upper bound of the diameter of a $\gamma$-quasi-clique.*

Based on the above lemma, those vertices that are not in $\bigcap_{v \in X} N_k^G(v)$ can be removed from $cand\_exts(X)$.

**Pruning based on the minimum size threshold.** The size of a valid $\gamma$-quasi-clique should be no less than $min\_size$. Consequently, the degree of a vertex contained in any valid $\gamma$-quasi-clique should be no less than $\lceil \gamma \cdot (min\_size - 1) \rceil$. Those vertices whose degree is less than $\lceil \gamma \cdot (min\_size - 1) \rceil$ can be removed since no valid $\gamma$-quasi-cliques contain them. Pei et al. [7] used this pruning technique in their algorithm.

**Pruning based on the degree of the vertices.** For a vertex set $X$ in the search space, the Cocain algorithm proposed by Zeng et al. [10] prunes the candidate extensions of $X$ based on the number of their neighbors in $X$ and $cand\_ext(X)$. Given a vertex $u$, we use $indeg^X(u)$ to denote the number of vertices in $X$ that are adjacent to $u$, and $exdeg^X(u)$ to denote the number of vertices in $cand\_exts(X)$ that are adjacent to $u$, that is, $indeg^X(u) = |\{v|(u, v) \in E, v \in X\}|$ and $exdeg^X(u) = |\{v|(u, v) \in E, v \in cand\_ext(X)\}|$. The Cocain algorithm uses the following lemmas to prune the search space and interested readers may refer to [10] for their proof.

**Lemma 2.** *If $m + u < \lceil \gamma \cdot (k + u) \rceil$, where $m, u, k \geq 0$, then $\forall i \in [0, u]$, $m + i < \lceil \gamma \cdot (k + i) \rceil$.*

**Lemma 3.** *Given a vertex set $X$ and a vertex $u \in cand\_exts(X)$, if $indeg^X(u) + exdeg^X(u) < \lceil \gamma \cdot (|X| + exdeg^X(u)) \rceil$, then there does not exist a vertex set $Y$ such that $(X \cup \{u\}) \subseteq Y \subseteq (X \cup cand\_exts(X))$, and $G(Y)$ is a $\gamma$-quasi-clique.*

For a vertex $v \in cand\_exts(X)$, if there does not exist a vertex set $Y$ such that $(X \cup \{u\}) \subseteq Y \subseteq (X \cup cand\_exts(X))$ and $G(Y)$ is a $\gamma$-quasi-clique, then $v$ is called an *invalid candidate extension* of $X$. The Cocain algorithm removes those invalid candidate extensions of $X$ based on Lemma 3. Due to the removal of these invalid candidate extensions, some other candidate extensions of $X$ that appear to be valid originally may become invalid apparently. Cocain does the pruning iteratively until no vertex can be removed from $cand\_exts(X)$. However, not all the invalid candidate extensions can be removed using Lemma 2.

The Cocain algorithm also checks the extensibility of the vertices in $X$ using the following lemma.

**Lemma 4.** *Given vertex set $X$ and a vertex $v \in X$, if $indeg^X(v) < \lceil \gamma \cdot |X| \rceil$ and $exdeg^X(v) = 0$, or $indeg^X(v) + exdeg^X(v) < \lceil \gamma \cdot (|X| - 1 + exdeg^X(v)) \rceil$, then there does not exist a vertex set $Y$ such that $X \subset Y \subseteq (X \cup cand\_exts(X))$ and $G(Y)$ is a $\gamma$-quasi-clique. Vertex $v$ is called a* failed vertex *of $X$.*

If there is a failed vertex in $X$, then there is no need to extend $X$ further.

### 3.3 New Pruning Techniques Used in the Quick Algorithm

The above pruning techniques can effectively prune the search space, but they have not fully utilized the pruning power of the minimum degree constraint yet, and not all the invalid candidate extensions can be detected and removed by them. Next we describe the new pruning techniques used in our Quick algorithm.

**Technique 1: pruning based on the upper bound of the number of vertices that can be added to $X$ concurrently to form a $\gamma$-quasi-clique.** Given vertex set $X$, the maximum number of vertices that can be added to $X$ to form a $\gamma$-quasi-clique is bounded by the minimal degree of the vertices in $X$.

**Lemma 5.** *Let $deg_{min}(X) = \min\{indeg^X(v) + exdeg^X(v) | v \in X\}$, $Y$ be a superset of $X$ such that $Y \subseteq (X \cup cand\_exts(X))$ and $G(Y)$ is a $\gamma$-quasi-clique. We have $|Y| \leq \lfloor deg_{min}(X)/\gamma \rfloor + 1$.*

*Proof.* For every vertex $v \in X$, we have $indeg^X(v) + exdeg^X(v) \geq indeg^Y(v) \geq \lceil \gamma \cdot (|Y| - 1) \rceil$, so we have $deg_{min}(X) \geq \lceil \gamma \cdot (|Y| - 1) \rceil$. Therefore, we have $\lfloor deg_{min}(X)/\gamma \rfloor \geq \lfloor (\lceil \gamma \cdot (|Y| - 1) \rceil)/\gamma \rfloor \geq \lfloor \gamma \cdot (|Y| - 1)/\gamma \rfloor = |Y| - 1$. So we have $|Y| \leq \lfloor deg_{min}(X)/\gamma \rfloor + 1$.

Based on Lemma 5, we derive the following upper bound:

**Definition 3 ($U_X^{min}$).** *The maximal number of vertices in $cand\_exts(X)$ that can be added to $X$ concurrently to form a $\gamma$-quasi-clique should be no larger than $\lfloor deg_{min}(X)/\gamma \rfloor + 1 - |X|$, where $deg_{min}(X) = \min\{indeg^X(v) + exdeg^X(v) | v \in X\}$. We denote this upper bound as $U_X^{min} = \lfloor deg_{min}(X)/\gamma \rfloor + 1 - |X|$.*

We further tighten this lower bound based on the observation that if $G(Y)$ is a $\gamma$-quasi-clique, then for any subset $X$ of $Y$, we have $\sum_{v \in X} indeg^Y(v) \geq |X| \cdot \lceil \gamma \cdot (|Y| - 1) \rceil$.

**Lemma 6.** *Let vertices in $cand\_exts(X)$ be sorted in descending order of their $indeg^X$ value, and the set of sorted vertices be denoted as $\{v_1, v_2, \cdots, v_n\}$. Given an integer $1 \leq k \leq n$, if $\sum_{v\in X} indeg^X(v) + \sum_{1\leq i\leq k} indeg^X(v_k) < |X| \cdot \lceil \gamma \cdot (|X| + k - 1) \rceil$, then for every vertex set $Z$ such that $Z \subseteq cand\_exts(X)$ and $|Z| = k$, $X \cup Z$ is not a $\gamma$-quasi-clique.*

*Proof.* Given a vertex set $Z$ such that $Z \subseteq cand\_exts(X)$ and $|Z| = k$, we have $\sum_{v\in X} indeg^{X\cup Z}(v) = \sum_{v\in X} indeg^X(v) + \sum_{v\in X} indeg^Z(v) = \sum_{v\in X} indeg^X(v) + \sum_{v\in Z} indeg^X(v) \leq \sum_{v\in X} indeg^X(v) + \sum_{1\leq i\leq |Z|} indeg^X(v_i) < |X| \cdot \lceil \gamma \cdot (|X| + |Z| - 1) \rceil$. Therefore, $X \cup Z$ is not a $\gamma$-quasi-clique.

Based on the above lemma, we tighten the upper bound as follows:

**Definition 4 (Upper bound $U_X$).** *Let $U_X = \max\{t | \sum_{v\in X} indeg^X(v) + \sum_{1\leq i\leq t} indeg^X(v_i) \geq |X| \cdot \lceil \gamma \cdot (|X| + t - 1) \rceil, 1 \leq t \leq U_X^{min}\}$ if such $t$ exists, otherwise $U_X = 0$. If $G(Y)$ is a $\gamma$-quasi-clique and $X \subseteq Y \subseteq (X \cup cand\_exts(X))$, then $|Y - X| \leq U_X$.*

**Lemma 7.** *Given a vertex set $X$ and a vertex $u \in cand\_exts(X)$, if $indeg^X(u) + U_X - 1 < \lceil \gamma \cdot (|X| + U_X - 1) \rceil$, then there does not exist a vertex set $Y$ such that $(X \cup \{u\}) \subseteq Y \subseteq (X \cup cand\_exts(X))$, and $G(Y)$ is a $\gamma$-quasi-clique.*

*Proof.* Let $Y$ be a vertex set such that $G(Y)$ is a $\gamma$-quasi-clique and $(X \cup \{u\}) \subseteq Y \subseteq (X \cup cand\_exts(X))$. Since $u \in (Y - X)$, there are at most $|Y| - |X| - 1$ vertices in $Y - X$ that are adjacent to $u$, and $|Y| - |X| - 1 \leq U_X - 1$ based on the definition of $U_X$. Based on Lemma 2 and the fact that $indeg^X(u) + U_X - 1 < \lceil \gamma \cdot (|X| + U_X - 1) \rceil$, we have $indeg^X(u) + |Y| - |X| - 1 < \lceil \gamma \cdot (|X| + |Y| - |X| - 1) \rceil = \lceil \gamma \cdot (|Y| - 1) \rceil$. Therefore, we have $indeg^Y(u) \leq indeg^X(u) + |Y| - |X| - 1 < \lceil \gamma \cdot (|Y| - 1) \rceil$. It contradicts the assumption that $G(Y)$ is a $\gamma$-quasi-clique.

Similarly, we can get the following lemma, and its proof is similar to Lemma 7.

**Lemma 8.** *Given a vertex set $X$ and a vertex $u \in X$, if $indeg^X(u) + U_X < \lceil \gamma \cdot (|X| + U_X - 1) \rceil$, then there does not exist a vertex set $Y$ such that $X \subseteq Y \subseteq (X \cup cand\_exts(X))$, and $G(Y)$ is a $\gamma$-quasi-clique.*

For each vertex set $X$, its candidate extensions are pruned based on the upper bound as follows. We first check whether $U_X = 0$. If it is true, then no vertices in $cand\_exts(X)$ can be added to $X$ to form a $\gamma$-quasi-clique. Next, we check whether there exists some vertex $u \in X$ such that $indeg^X(u) + U_X < \lceil \gamma \cdot (|X| + U_X - 1) \rceil$. If such $u$ exists, then no $\gamma$-quasi-cliques can be generated by extending $X$. Otherwise, we remove the invalid candidate extensions of $X$ identified by Lemma 7 from $cand\_exts(X)$. The removal of these invalid candidate extensions can in turn reduce the degree of other vertices in $cand\_exts(X)$, thus making other invalid vertices identifiable. The pruning is iteratively carried out until no more vertices can be removed from $cand\_exts(X)$.

**Technique 2: pruning based on the lower bound of the number of vertices that can be added to $X$ concurrently to form a $\gamma$-quasi-clique.** Given a vertex set $X$, if there exists a vertex $u \in X$ such that $indeg^X(u) < \lceil\gamma \cdot (|X|-1)\rceil$, then at least a certain number of vertices need to be added to $X$ to increase the degree of $u$ in order to form a $\gamma$-quasi-clique. We denote this lower bound as $L_X^{min}$, and it is defined as follows.

**Definition 5 ($L_X^{min}$).** *Let $indeg_{min}(X) = \min\{indeg^X(v)|v \in X\}$. $L_X^{min}$ is defined as $L_X^{min} = \min\{t|indeg_{min}(X) + t \geq \lceil\gamma \cdot (|X|+t-1)\rceil\}$.*

Again, this lower bound can be further tightened based on Lemma 6. We sort vertices in $cand\_exts(X) = \{v_1, v_2, \cdots, v_n\}$ in descending order of $indeg^X$ value.

**Definition 6 (Lower bound $L_X$).** *Let $L_X = \min\{t| \sum_{v \in X} indeg^X(v) + \sum_{1 \leq i \leq t} indeg^X(v_i) \geq |X| \cdot \lceil\gamma \cdot (|X|+t-1)\rceil, L_X^{min} \leq t \leq n\}$ if such $t$ exists. Otherwise $L_X = |cand\_exts(X)|+1$. If $G(Y)$ is a $\gamma$-quasi-clique and $X \subseteq Y \subseteq (X \cup cand\_exts(X))$, then $|Y - X| \geq L_X$.*

Based on the definition of $L_X$, we can get the following lemmas.

**Lemma 9.** *Let $X$ be a vertex set and $u$ be a vertex in $cand\_exts(X)$. If $indeg^X(u) + exdeg^X(u) < \lceil\gamma \cdot (|X| + L_X - 1)\rceil$, then there does not exist a vertex set $Y$ such that $(X \cup \{u\}) \subseteq Y \subseteq (X \cup cand\_exts(X))$, and $G(Y)$ is a $\gamma$-quasi-clique.*

**Lemma 10.** *Let $X$ be a vertex set and $u$ be a vertex in $X$. If $indeg^X(u) + exdeg^X(u) < \lceil\gamma \cdot (|X| + L_X - 1)\rceil$, then there does not exist a vertex set $Y$ such that $X \subseteq Y \subseteq (X \cup cand\_exts(X))$, and $G(Y)$ is a $\gamma$-quasi-clique.*

For each vertex set $X$, its candidate extensions are pruned based on the lower bound as follows. We first check whether $L_X > U_X$. If it is true, then no need to extend $X$ further. Next, we check whether there exists some vertex $u \in X$ such that $indeg^X(u) + exdeg^X(u) < \lceil\gamma \cdot (|X| + L_X - 1)\rceil$. If such $u$ exists, then no $\gamma$-quasi-cliques can be generated by extending $X$ based on Lemma 10. Otherwise, we remove the invalid candidate extensions of $X$ identified by Lemma 9. Again the removal is carried out iteratively until no more candidate can be removed.

Note that the removal of the invalid candidate extensions based on Lemma 3, 7 and 9 may further tighten the two bounds $L_X$ and $U_X$, which may in turn make more invalid candidate extensions identifiable.

**Technique 3: pruning based on critical vertices.** Let $X$ be a vertex set. If there exists a vertex $v \in X$ such that $indeg^X(v) + exdeg^X(v) = \lceil\gamma \cdot (|X| + L_X - 1)\rceil$, then $v$ is called a *critical vertex* of $X$.

**Lemma 11.** *If $v \in X$ is a critical vertex of $X$, then for any vertex set $Y$ such that $X \subset Y \subseteq (X \cup cand\_exts(X))$ and $G(Y)$ is a $\gamma$-quasi-clique, we have $\{u|(u,v) \in E \wedge u \in cand\_exts(X)\} \subseteq Y$.*

*Proof.* Let $u$ be a vertex such that $u \in cand\_exts(X)$ and $(u,v) \in E$. Suppose that $u \notin Y$, then we have $indeg^Y(v) < indeg^X(v) + exdeg^X(v) = \lceil\gamma \cdot (|X| + L_X - 1)\rceil \leq \lceil\gamma \cdot (|Y| - 1)\rceil$. It contradicts the fact that $Y$ is a $\gamma$-quasi-clique.

Based on the above lemma, we can identify the critical vertices for every vertex set $X$. If such critical vertex exists, let it be $v$, then we add the vertices in $cand\_exts(X)$ that are adjacent to $v$ to $X$. Let $Y$ be the resultant vertex set. The remaining mining is performed on $Y$, and the cost for extending $X \cup \{u\} (u \notin Y)$ is saved.

**Technique 4: pruning based on cover vertices.** This pruning technique is inspired by the technique used in [11] for mining maximal cliques. Tomita et al. use the following lemma to prune non-maximal cliques.

**Lemma 12.** *Let $X$ be a clique and $u$ be a vertex in $N^G(X)$. For any vertex set $Y$ such that $G(Y)$ is a clique and $Y \subseteq (X \cup N^G(X \cup \{u\}))$, $G(Y)$ cannot be a maximal clique.*

*Proof.* Vertex $u$ is adjacent to all the vertices in $X$, and it is also adjacent to all the vertices in $N^G(X \cup \{u\})$. Hence $u$ is adjacent to all the vertices in $Y$. In other words, $Y \cup \{u\}$ is a clique, thus clique $Y$ is not maximal.

Based on the above lemma, Tomita et al. pick a vertex with the maximal degree from $cand\_exts(X)$. Let $u$ be the picked vertex. When $X$ is extended using vertices in $cand\_exts(X)$, the vertex set extended from $X$ must contain at least one vertex that is not in $N^G(X \cup \{u\})$. In this way, the subsets of $(X \cup N^G(X \cup \{u\}))$ are pruned since they are not maximal.

Here we generalize the above lemma to quasi-cliques.

**Lemma 13.** *Let $X$ be a vertex set and $u$ be a vertex in $cand\_exts(X)$ such that $indeg^X(u) \geq \lceil \gamma \cdot |X| \rceil$. If for any vertex $v \in X$ such that $(u,v) \notin E$, we have $indeg^X(v) \geq \lceil \gamma \cdot |X| \rceil$, then for any vertex set $Y$ such that $G(Y)$ is a $\gamma$-quasi-clique and $Y \subseteq (X \cup (cand\_exts(X) \cap N^G(u) \cap (\bigcap_{v \in X \wedge (u,v) \notin E} N^G(v))))$, $G(Y)$ cannot be a maximal $\gamma$-quasi-clique.*

*Proof.* Vertex set $Y$ is a $\gamma$-quasi-clique, then for every vertex $v \in Y$, we have $indeg^Y(v) \geq \lceil \gamma \cdot (|Y| - 1) \rceil$. Let us look at vertex set $Y \cup \{u\}$. (1) Vertex $u$ is adjacent to all the vertices in $cand\_exts(X) \cap N^G(u) \cap (\bigcap_{v \in X \wedge (u,v) \notin E} N^G(v))$ and $indeg^X(u) \geq \lceil \gamma \cdot |X| \rceil$, so we have $indeg^{Y \cup \{u\}}(u) = indeg^X(u) + |Y| - |X| \geq \lceil \gamma \cdot |X| \rceil + |Y| - |X| \geq \lceil \gamma \cdot |Y| \rceil$. (2) Similarly, for every vertex $v \in X$ such that $(u,v) \notin E$, $v$ is adjacent to all the vertices in $cand\_exts(X) \cap N^G(u) \cap (\bigcap_{v \in X \wedge (u,v) \notin E} N^G(v))$ and $indeg^X(u) \geq \lceil \gamma \cdot |X| \rceil$, so we have $indeg^{Y \cup \{u\}}(v) = indeg^X(v) + |Y| - |X| \geq \lceil \gamma \cdot |X| \rceil + |Y| - |X| \geq \lceil \gamma \cdot |Y| \rceil$. (3) For every vertex $v \in X$ such that $(u,v) \in E$, we have $indeg^{Y \cup \{u\}}(v) = indeg^Y(v) + 1 \geq \lceil \gamma \cdot (|Y| - 1) \rceil + 1 \geq \lceil \gamma \cdot |Y| \rceil$. (4) Similarly, for every vertex $v \in (Y - X)$, we have $indeg^{Y \cup \{u\}}(v) = indeg^Y(v) + 1 \geq \lceil \gamma \cdot |Y| \rceil$. In summary, $Y \cup \{u\}$ is a $\gamma$-quasi-clique and $Y$ is not maixmal.

We use $C_X(u)$ to denote the set of vertices that are covered by $u$ with respect to $X$, that is, $C_X(u) = cand\_exts(X) \cap N^G(u) \cap (\bigcap_{v \in X \wedge (u,v) \notin E} N^G(v))$. Based on the above lemma, we find a vertex that maximize the size of $C_X(u)$ from *cand_exts*$(X)$. Let $u$ be the picked vertex. We call $u$ the *cover vertex* of $X$. We prune those vertex sets that are subsets of $X \cup C_X(u)$ by putting the vertices in $C_X(u)$ after all the other vertices in *cand_exts*$(X)$ and then using the vertices in *cand_exts*$(X) - C_X(u)$ to extend $X$.

**Technique 5: the lookahead technique.** This pruning technique has been used in mining maximal frequent itemsets [12]. Its basic idea is that before extending $X$ using any vertex from *cand_exts*$(X)$, we first check whether $X \cup$ *cand_exts*$(X)$ is a $\gamma$-quasi-clique. If it is, then there is no need to extend $X$ further because all the vertex sets extended from $X$ are subsets of $X \cup$ *cand_exts*$(X)$, thus they cannot be maximal except for $X \cup$ *cand_exts*$(X)$ itself.

---

**Algorithm 1.** Quick Algorithm

---

**Input:**
$X$ is a vertex set
*cand_exts* is the valid candidate extensions of $X$
$\gamma$ is the minimum degree threshold
*min_size* is the minimum size threshold
**Output:**
true if some superset of $X$ can form a $\gamma$-quasi-clique, otherwise false;
**Description:**
1. Find the cover vertex $u$ of $X$; Sort vertices in *cand_exts*$(X)$ such that vertices in $C_X(u)$ are after all the other vertices;
2. *bhas_qclq* = false;
3. **for all** vertex $v \in cand_exts(X) - C_X(u)$ **do**
4. **if** $|X| + |cand\_exts(X)| < min\_size$ **then**
5. **return** *bhas_qclq*;
6. **if** $G(X \cup cand\_exts(X))$ is a $\gamma$-quasi-clique **then**
7. Output $X \cup cand\_exts(X)$;
8. **return** t rue;
9. $Y = X \cup \{v\}$;
10. $cand\_exts(X) = cand\_exts(X) - \{v\}$;
11. $cand_Y = cand\_exts(X) \cap N_k^G(v)$, where $k$ is calculated based on Theorem in [7];
12. **repeat**
13. Calculate the upper bound $U_Y$ and lower bound $L_Y$ of the number of vertices that can be added to $Y$ concurrently to form a $\gamma$-quasi-clique;
14. **if** there is a critical vertex $u'$ in $Y$ **then**
15. $Y = Y \cup (cand_Y \cap N^G(u'))$;
16. $cand_Y = cand_Y - (cand_Y \cap N^G(u'))$;
17. update $U_Y$ and $L_Y$;
18. $Z = \{v \mid indeg^Y(v) + exdeg^Y(v) < \lceil \gamma \cdot (|Y| + exdeg^Y(v) - 1) \rceil \bigvee indeg^Y(v) + U_Y < \lceil \gamma \cdot (|Y| + U_Y - 1) \rceil \bigvee indeg^Y(v) + exdeg(Y)(v) < \lceil \gamma \cdot (|Y| + L_Y - 1) \rceil, v \in X\}$;
19. **if** $Z$ is not empty **then**
20. $cand_Y = \{\}$;
21. $Z = \{v \mid indeg^Y(v) + exdeg^Y(v) < \lceil \gamma \cdot (|Y| + exdeg^Y(v)) \rceil \bigvee indeg^Y(v) + U_Y - 1 < \lceil \gamma \cdot (|Y| + U_Y - 1) \rceil \bigvee indeg^Y(v) + exdeg(Y)(v) < \lceil \gamma \cdot (|Y| + L_Y - 1) \rceil, v \in cand_Y\}$;
22. $cand_Y = Cand_Y - Z$;
23. **until** $L_Y > U_Y$ OR $Z = \{\}$ OR $cand_Y = \{\}$
24. **if** $L_Y \leq U_Y$ AND $|cand_Y| > 0$ AND $|Y| + |cand_Y| \geq min\_size$ **then**
25. *bhas_superqclq* = Quick($Y$, $cand_Y$, $\gamma$, *min_size*);
26. *bhas_qclq* = *bhas_qclq* OR *bhas_superqclq*;
27. **if** $|Y| \geq min\_size$ AND $G(Y)$ is a $\gamma$-quasi-clique AND *bhas_superqclq*==false **then**
28. *bhas_qclq* = true;
29. Output $Y$;
30. **return** *bhas_qclq*;

---

### 3.4 The Pseudo-codes of the Quick Algorithm

Algorithm 1 shows the pseudo-codes of the Quick algorithm. When the algorithm is first called on a graph $G = (V, E)$, $X$ is set to the empty set, and $cand\_exts(X)$ is set to $\{v|exdeg^X(v) \geq \lceil\gamma \cdot (min\_size - 1)\rceil, v \in V\}$.

The Quick algorithm explores the search space in depth-first order. For a vertex set $X$ in the search space, Quick first finds its covering vertex $u$, and puts the vertices in $C_X(u)$ after all the other vertices in $cand\_exts(X)$ (line 1). Only the vertices in $cand\_exts(X) - C_X(u)$ are used to extend $X$ to prune the subsets of $X \cup C_X(u)$ based on Lemma 13 (line 3). Before using a vertex $v$ to extend $X$, Quick uses the minimum size constraint (line 4-5) and the lookahead technique to prune search space. Quick checks whether $X \cup cand\_exts(X)$ is a $\gamma$-quasi-clique. If it is, then Quick outputs $X \cup cand\_exts(X)$ and skips the generation of subsets of $X \cup cand\_exts(X)$ (line 6-8).

When using a vertex $v$ to extend $X$, Quick first removes those vertices that are not in $N_k^G(v)$, where $k$ is the upper bound of the diameter of $G(X \cup cand\_exts(X))$ if $G(X \cup cand\_exts(X))$ is a $\gamma$-quasi-clique (line 11). Let $Y = X \cup \{u\}$. Next Quick calculates the upper bound $U_Y$ and lower bound $L_Y$ of the number of vertices that can be added to $Y$ concurrently to form a $\gamma$-quasi-clique, and uses these two bounds to iteratively remove invalid candidate extensions of $Y$ as follows (line 12-23). It first identifies critical vertices in $Y$. If there is a critical vertex $u$ in $Y$, Quick adds all the vertices in $cand\_exts(Y)$ that are adjacent to $u$ to $Y$ based on Lemma 11 (line 14-16). Next, Quick checks the extensibility of the vertices in $Y$ based on Lemma 4, 8 and 10 (line 18-20). Quick then prunes invalid candidate extensions based on Lemma 3, 7 and 9 (line 21-22). If $cand_Y$ is not empty after all the pruning, Quick extends $Y$ recursively using $cand_Y$ (line 25).

The last two pruning techniques described in Section 3.3 can remove some non-maximal quasi-cliques, but they cannot remove all. To further reduce the number of non-maximal quasi-cliques generated, we check whether there is any $\gamma$-quasi-clique generated from some superset of $Y$, and output $Y$ only if there is none (line 26-28). The remaining non-maximal quasi-cliques are removed in a post-processing step. We store all the vertex sets of the $\gamma$-quasi-cliques produced by Algorithm 1 in a prefix-tree. Quasi-cliques represented by internal nodes cannot be maximal. For each quasi-clique represented by a leaf node, we search for its subsets in the tree and mark them as non-maximal. At the end, the quasi-cliques represented by the leaf nodes that are not marked as non-maximal are maximal, and they are put into the final output.

Algorithm 1 prunes the search space based on the lemmas described in Section 3.2 and 3.3, so its correctness and completeness is guaranteed by the correctness of these lemmas.

## 4 A Performance Study

In this section, we study the efficiency of the Quick algorithm and the effectiveness of the pruning techniques used in Quick. Our experiments were conducted on a PC with an Intel Core 2 Duo CPU (2.33GHz) and 3.2GB of RAM. The operating system is Fedora 7. Our algorithm was implemented using C++ and complied using g++.

We used both real datasets and synthetic datasets in our experiments. The real datasets are protein interaction networks downloaded from DIP(`http://dip.doe-mbi.ucla.edu/`). The yeast interaction network contains 4932 vertices (proteins) and 17201 edges (interactions). The E.coli interaction network contains 1846 vertices and 5929 edges. The synthetic graphs are generated using several parameters: $V$ is the number of vertices, $Q$ is the number of quasi-cliques planted in the graph, $\gamma_{min}$ is the minimum degree threshold of the planted quasi-cliques, $MinSize$ and $MaxSize$ are the minimum and maximum size of the planted quasi-cliques, and $d$ is the average degree of the vertices. To generate a synthetic graph, we first generate a value $\gamma$ between $\gamma_{min}$ and 1, and then generate $Q$ $\gamma$-quasi-cliques. The size of the quasi-cliques is uniformly distributed between $MinSize$ and $MaxSize$. If the average degree of the $V$ vertices is less than $d$ after all the $Q$ quasi-cliques are planted, then we randomly add edges into the graph until the average degree reaches $d$.

### 4.1 Comparison with *Cocain*

We compared Quick with Cocain [8] in terms of mining efficiency. The Cocain algorithm is designed for mining coherent closed quasi-cliques from a graph database. A quasi-clique is closed if all of its supergraphs that are quasi-cliques are less frequent than it. When applied to a single graph with minimum support

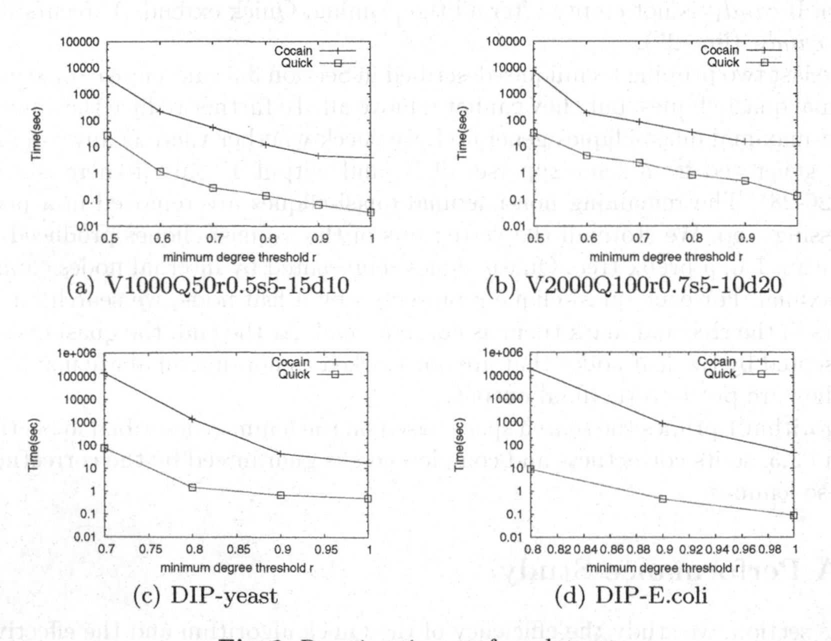


**Fig. 3.** Running time on four datasets

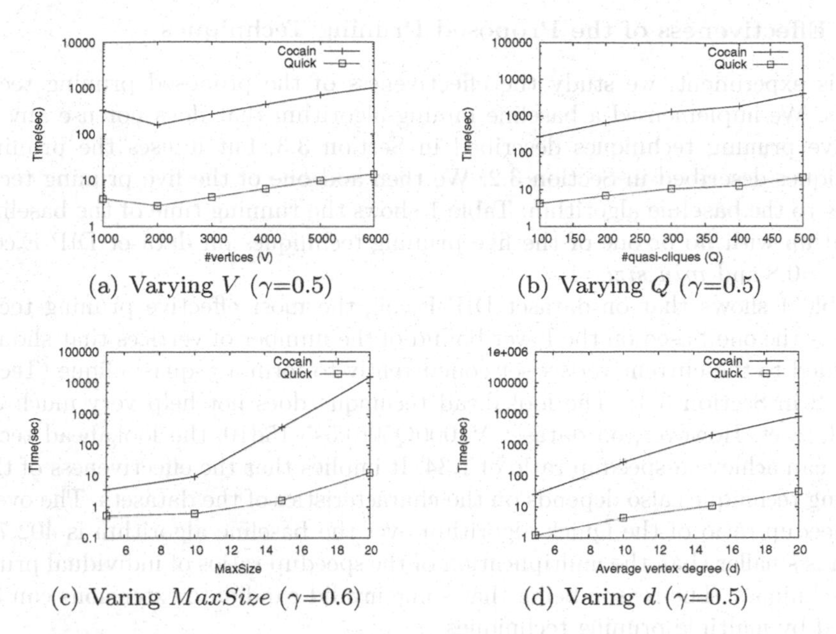


(a) Varying $V$ ($\gamma$=0.5) (b) Varying $Q$ ($\gamma$=0.5)

(c) Varing $MaxSize$ ($\gamma$=0.6) (d) Varing $d$ ($\gamma$=0.5)

**Fig. 4.** Varying dataset generation parameters (default parameters: $V$=3000, $Q$=100 $\gamma_{min}$=0.5, $MinSize$=5, $MaxSize$=10, $d$=10.0)

of 100%, mining coherent closed quasi-cliques is equivalent to mining maximal quasi-cliques. The executable of Cocain was kindly provided by their authors.

Figure 3 shows the running time of the two algorithms with respect to the $\gamma$ threshold on four datasets. The first two datasets are synthetic datasets. Dataset V1000Q50r0.5s5-15d10 was generated with $V$ = 1000, $Q$=50, $\gamma_{min}$=0.5, $MinSize$=5, $MaxSize$=15 and $d$=10. Dataset V2000Q100r0.7s5-10d20 was generated with $V$ = 2000, $Q$ = 100, $\gamma_{min}$=0.7, $MinSize$=5, $MaxSize$=10 and $d$=20. On all four datasets, the $min\_size$ threshold is set to 1. Quick is tens of times or even hundreds of times more efficient than Cocain on all four datasets. It indicates that the pruning techniques used in Quick are very effective in pruning search space. The running time of both algorithms increases with the decrease of the $\gamma$ threshold because more $\gamma$-quasi-cliques are generated and less vertices can be pruned when $\gamma$ decreases.

We studied the scalability of the two algorithms using synthetic datasets. Figure 4 shows the running time of the two algorithms when varying the number of vertices $V$, the number of planted quasi-cliques $Q$, the maximum size the planted quasi-cliques $MaxSize$ and the average degree of the vertices $d$ respectively. The default parameters are set as follows: $V$=3000, $Q$=100, $\gamma_{min}$=0.5, $MinSize$=5, $MaxSize$=10, and $d$=10. The running time of both algorithm increases steadily with the increase of the number of vertices and the number of planted quasi-cliques. They are more sensitive to the increase of the maximum size of the planted quasi-cliques and the average degree of the vertices. We observed the same trend on the number of maximal quasi-cliques generated.

### 4.2 Effectiveness of the Proposed Pruning Techniques

In this experiment, we study the effectiveness of the proposed pruning techniques. We implemented a baseline mining algorithm that does not use any of the five pruning techniques described in Section 3.3, but it uses the pruning techniques described in Section 3.2. We then add one of the five pruning techniques to the baseline algorithm. Table 1 shows the running time of the baseline algorithm with no or one of the five pruning techniques on dataset DIP-E.coli with $\gamma$=0.8 and *min_size*=1.

Table 1 shows that on dataset DIP-E.coli, the most effective pruning technique is the one based on the lower bound of the number of vertices that should be added to the current vertex set concurrently to form a $\gamma$-quasi-clique (Technique 2 in Section 3.3). The lookahead technique does not help very much on this dataset. However, on dataset V1000Q50r0.5s5-15d10, the lookahead technique can achieve a speedup ratio of 1.34. It implies that the effectiveness of the pruning techniques also depends on the characteristics of the datasets. The overall speedup ratio of the Quick algorithm over the baseline algorithm is 402.71, which is smaller than the multiplication of the speedup ratios of individual pruning techniques. The reason being that some invalid candidate extensions can be pruned by multiple pruning techniques.

**Table 1.** The effectiveness of the five pruning techniques on dataset DIP-E.coli

| Algorithms | time | speedup |
|---|---|---|
| Baseline | 3604.67 | - |
| Baseline+UpperBound | 1032.88 | 3.49 |
| Baseline+LowerBound | 18.73 | 192.48 |
| Baseline+CriticalVertex | 3572.09 | 1.01 |
| Baseline+CoverVertex | 2505.75 | 1.44 |
| Baseline+Lookahead | 3601.45 | 1.00 |
| Quick | 8.95 | 402.71 |

# 5 Related Work

The problem of determining whether a graph contains a clique of at least a given size $k$ is a NP-complete problem [13]. It is an even harder problem to enumerate all the maximal cliques or quasi-cliques from a graph. Bron and Kerbosch [14] proposed an efficient algorithm to solve the problem more than 30 years ago, which is still one of the most efficient algorithms for enumerating all maximal cliques today. Their algorithm is recently improved by Tomita et al. [11] by using a tree-like output format.

There is a growing interest in mining quasi-cliques in recent years. Matsuda et al. [5] introduced a graph structure called $p$-quasi complete graph, which is the same as the $\gamma$-quasi-cliques defined in this paper, and they proposed an approximation algorithm to cover all the vertices in a graph with a minimum number of $p$-quasi complete subgraphs. Abello et al. [6] defined a $\gamma$-clique in a graph as a connected induced subgraph with edge density no less than $\gamma$. They proposed a greedy randomized adaptive search algorithm called GRASP to find $\gamma$-cliques. Bu et al. [2] used the spectral clustering method to find quasi-cliques and quasi-bicliques from protein interaction networks.

The above work finds cliques or quasi-cliques from a single graph. Some work mines the clique and quasi-clique from multiple graphs. Pei et al. [7] proposed an algorithm called Crochet to mine cross quasi-cliques from a set of graphs, and they required that a quasi-clique must appear in all the graphs. The pruning techniques used in this paper is mainly based on the co-occurrences of the vertices across all the graphs. Wang et al. [15] studied the problem of mining frequent closed cliques from graph databases. A clique is frequent if it appears in sufficient number of graphs. A clique is closed if all of its super-cliques are less frequent than it. Since cliques still have the downward closure property, so mining cliques is much easier than mining quasi-cliques. Zeng et al. [8] studied a more general problem formulation, that is, mining frequent closed quasi-cliques from graph databases, and proposed an efficient algorithm called Cocain to solve the problem. The same group of authors later extended the algorithm for out-of-core mining of quasi-cliques from very large graph databases [10]. Cocain uses several pruning techniques to prune search space, but it has not fully utilized the pruning power of the minimum degree constraint yet. The pruning techniques proposed in this paper can be integrated into Crochet and Cocain to improve their performance.

There are also some work on finding densely connected subgraphs from one single graph or from a graph database. The connectivity of a subgraph can be measured by the size of its minimum cut [16,17], edge density [3] or by other measures. Hartuv and Shamir [16] proposed an algorithm called HCS which recursively splits the weighted graph into a set of highly connected components along the minimum cut. Each highly connected component is considered as a gene cluster. Yan et al. [17] investigated the problem of mining closed frequent graphs with connectivity constraints in massive relational graphs, and proposed two algorithms, CloseCut and Splat, to solve the problem. Hu et al. [3] proposed an algorithm called Codense to mine frequent coherent dense subgraphs across massive biological networks where all edges in a coherent subgraph should have sufficient support in the whole graph set. Gibson et al. [18] proposed an algorithm to find large dense bipartite subgraphs from massive graphs, and their algorithm is based on a recursive application of fingerprinting via shingles. Ucar et al. [4] used a refinement method based on neighborhoods and the biological importance of hub proteins to find dense subgraphs from protein-protein interaction networks. Bader and Hogue [1] proposed a heuristic algorithm called MCODE which is based on vertex weighting by local neighborhood density and outward traversal from a locally dense seed protein to isolate the dense regions according to given parameters.

## 6 Discussion and Conclusion

In this paper, we proposed several effective pruning techniques for mining quasi-cliques. These techniques can be applied to mining quasi-cliques from a single graph or a graph database. We describe the pruning techniques in the context of relational graphs where each vertex has a unique label. It is not difficult to apply these pruning techniques to non-relational graphs where different vertices may have the same label. Our preliminary experiment results show that by using these pruning techniques, our algorithm can be orders of magnitude faster than existing algorithms for the task of mining quasi-cliques from a single graph. In our future work, we will study the effectiveness of these pruning techniques for mining frequent quasi-cliques from a graph database.

## References

1. Bader, G.D., Hogue, C.W.: An automated method for finding molecular complexes in large protein interaction networks. BMC Bioinformatics 4(2) (2003)
2. Bu, D., Zhao, Y., Cai, L., Xue, H., Zhu, X., Lu, H., Zhang, J., Sun, S., Ling, L., Zhang, N., Li, G., Chen, R.: Topological structure analysis of the protein interaction network in budding yeast. Nucleic Acids Research 31(9), 2443–2450 (2003)
3. Hu, H., Yan, X., Huang, Y., Han, J., Zhou, X.J.: Mining coherent dense subgraphs across massive biological networks for functional discovery. Bioinformatics 21(1), 213–221 (2005)
4. Ucar, D., Asur, S., Çatalyürek, Ü.V., Parthasarathy, S.: Improving functional modularity in protein-protein interactions graphs using hub-induced subgraphs. In: Fürnkranz, J., Scheffer, T., Spiliopoulou, M. (eds.) PKDD 2006. LNCS (LNAI), vol. 4213, pp. 371–382. Springer, Heidelberg (2006)
5. Matsuda, H., Ishihara, T., Hashimoto, A.: Classifying molecular sequences using a linkage graph with their pairwise similarities. Theoretical Computer Science 210(2), 305–325 (1999)
6. Abello, J., Resende, M.G.C., Sudarsky, S.: Massive quasi-clique detection. In: Proc. of the 5th Latin American Symposium on Theoretical Informatics, pp. 598–612 (2002)
7. Pei, J., Jiang, D., Zhang, A.: On mining cross-graph quasi-cliques. In: Proc. of the 11th ACM SIGKDD Conference, pp. 228–238 (2005)
8. Zeng, Z., Wang, J., Zhou, L., Karypis, G.: Coherent closed quasi-clique discovery from large dense graph databases. In: Proc. of the 12th ACM SIGKDD Conference, pp. 797–802 (2006)
9. Rymon, R.: Search through systematic set enumeration. In: Proc. of the Internation Conference on Principles of Knowledge Representation and Reasoning (1992)
10. Zeng, Z., Wang, J., Zhou, L., Karypis, G.: Out-of-core coherent closed quasi-clique mining from large dense graph databases. ACM Transactions on Database Systems (TODS) 32(2), 13 (2007)
11. Tomita, E., Tanaka, A., Takahashi, H.: The worst-case time complexity for generating all maximal cliques and computational experiments. Theoretical Computer Science 363(1), 28–42 (2006)
12. Bayardo Jr., R.J.: Efficiently mining long patterns from databases. In: Proc. of the 1998 ACM SIGMOD International Conference on Management of Data, pp. 85–93 (1998)

13. Karp, R.: Reducibility among combinatorial problems. In: Proc. of a Symposium on the Complexity of Computer Computations, pp. 85–103 (1972)
14. Bron, C., Kerbosch, J.: Algorithm 457: Finding all cliques of an undirected graph. Communications of the ACM 16(9), 575–576 (1973)
15. Wang, J., Zeng, Z., Zhou, L.: Clan: An algorithm for mining closed cliques from large dense graph databases. In: Proc. of the 22nd International Conference on Data Engineering, p. 73 (2006)
16. Hartuv, E., Shamir, R.: A clustering algorithm based on graph connectivity. Information Processing Letters 76(4-6) (2000)
17. Yan, X., Zhou, X.J., Han, J.: Mining closed relational graphs with connectivity constraints. In: Proc. of the 11th ACM SIGKDD conference, pp. 324–333 (2005)
18. Gibson, D., Kumar, R., Tomkins, A.: Discovering large dense subgraphs in massive graphs. In: Proc. of the 31st International Conference on Very Large Data Bases, pp. 721–732 (2005)

# Efficient Pairwise Multilabel Classification for Large-Scale Problems in the Legal Domain

Eneldo Loza Mencía and Johannes Fürnkranz

Knowledge Engineering Group
Technische Universität Darmstadt
{eneldo, juffi}@ke.informatik.tu-darmstadt.de

**Abstract.** In this paper we applied multilabel classification algorithms to the EUR-Lex database of legal documents of the European Union. On this document collection, we studied three different multilabel classification problems, the largest being the categorization into the EUROVOC concept hierarchy with almost 4000 classes. We evaluated three algorithms: (i) the binary relevance approach which independently trains one classifier per label; (ii) the multiclass multilabel perceptron algorithm, which respects dependencies between the base classifiers; and (iii) the multilabel pairwise perceptron algorithm, which trains one classifier for each pair of labels. All algorithms use the simple but very efficient perceptron algorithm as the underlying classifier, which makes them very suitable for large-scale multilabel classification problems. The main challenge we had to face was that the almost 8,000,000 perceptrons that had to be trained in the pairwise setting could no longer be stored in memory. We solve this problem by resorting to the dual representation of the perceptron, which makes the pairwise approach feasible for problems of this size. The results on the EUR-Lex database confirm the good predictive performance of the pairwise approach and demonstrates the feasibility of this approach for large-scale tasks.

## 1 Introduction

The EUR-LEX text collection is a collection of documents about European Union law. It contains many several different types of documents, including treaties, legislation, case-law and legislative proposals, which are indexed according to several orthogonal categorization schemes to allow for multiple search facilities. The most important categorization is provided by the EUROVOC descriptors, which is a topic hierarchy with almost 4000 categories regarding different aspects of European law.

This document collection provides an excellent opportunity to study text classification techniques for several reasons:

- it contains multiple classifications of the same documents, making it possible to analyze the effects of different classification properties using the same underlying reference data without resorting to artificial or manipulated classifications,
- the overwhelming number of produced documents make the legal domain a very attractive field for employing supportive automated solutions and therefore a machine learning scenario in step with actual practice,

W. Daelemans et al. (Eds.): ECML PKDD 2008, Part II, LNAI 5212, pp. 50–65, 2008.

– the documents are available in several European languages and are hence very interesting e.g. for the wide field of multi- and cross-lingual text classification,
– and, finally, the data is freely accessible (at http://eur-lex.europa.eu/)

In this paper, we make a first step towards analyzing this database by applying multilabel classification techniques on three of its categorization schemes. The database is a very challenging multilabel scenario due to the high number of possible labels (up to 4000), which, for example, exceeds the number of labels in the REUTERS databases by one order of magnitude.

We evaluated three methods on this task:

– the conventional binary relevance approach (BR), which trains one binary classifier per label
– the *multilabel multiclass perceptron* (MMP), which also trains one classifier per label but does not treat them independently, instead it tries to minimize a ranking loss function of the entire ensemble [3]
– the *multilabel pairwise perceptron* (MLPP), which trains one classifier for each pair of classes [12]

Previous work on using these algorithms for text categorization [12] has shown that the MLPP algorithm outperforms the other two algorithms, while being slightly more expensive in training (by a factor that corresponds to the average number of labels for each example). However, another key disadvantage of the MLPP algorithm is its need for storing one classifier for each pair of classes. For the EUROVOC categorization, this results in almost 8,000,000 perceptrons, which would make it impossible to solve this task in main memory.

To solve this problem, we introduce and analyze a novel variant that addresses this problem by representing the perceptron in its dual form, i.e. the perceptrons are formulated as a combination of the documents that were used during training instead of explicitly as a linear hyperplane. This reduces the dependence on the number of classes and therefore allows the *Dual MLPP* algorithm to handle the tasks in the EUR-Lex database.

We must note that in this work we do not solve the entire multilabel classification problem, but, following [3], we only provide a ranking of all possible labels. There are three reasons for this: (i) the MMP and the pairwise method naturally provide such a ranking, (ii) the ranking allows to evaluate the performance differences on a finer scale, and (iii) our key motivation is to study the scalability of these approaches which is determined by the rankings. However, there are several methods for finding a delimiter between relevant and irrelevant labels within a provided ranking of the labels, a good overview can be found in [17]. For the pairwise approach, we have recently introduced the idea of using an artificial label that encodes the boundary between relevant and irrelevant labels for each example [2], which has also been successfully applied to the REUTERS text categorization task [7].

The outline of the paper is as follows: We start with a presentation of the EUR-Lex respository and the datasets that we derived from it (Section 2). Section 3 briefly recapitulates the algorithms that we study, followed by the presentation of the dual version of the MLPP classifier (section 4). In Section 5, we compare the computational complexity of all approaches, and present the experimental results in Section 6.

**Title and reference**

Council Directive 91/250/EEC of 14 May 1991 on the legal protection of computer programs

**Classifications**

**EUROVOC descriptor**

- *data-processing law, computer piracy, copyright, software, approximation of laws*

**Directory code**

- 17.20.00.00 *Law relating to undertakings* / Intellectual property law

**Subject matter**

- *Internal market, Industrial and commercial property*

**Text**

COUNCIL DIRECTIVE of 14 May 1991 on the legal protection of computer programs (91/250/EEC)

THE COUNCIL OF THE EUROPEAN COMMUNITIES,

Having regard to the Treaty establishing the European Economic Community and in particular Article 100a thereof, ...

**Fig. 1.** Excerpt of a EUR-Lex sample document with the CELEX ID 31991L0250. The original document contains more meta-information. We trained our classifiers to predict the EUROVOC descriptors, the directory code and the subject matters based on the text of the document.

## 2 The EUR-Lex Repository

The EUR-Lex/CELEX (Communitatis Europeae LEX) Site[1] provides a freely accessible repository for European Union law texts. The documents include the official Journal of the European Union. They are available in most of the languages of the EU. We retrieved the HTML versions with bibliographic notes recursively from all (non empty) documents in the English version of the *Directory of Community legislation in force*[2], in total 19,596 documents. Only documents related to secondary law (in contrast to primary law, the constitutional treaties of the European Union) and international agreements are included. The legal form of the included acts are mostly *decisions* (8,917 documents), *regulations* (5,706), *directives* (1,898) and *agreements* (1,597).

The bibliographic notes of the documents contain information such as dates of effect, authors, etc. and classifications. The classifications include the assignment to several EUROVOC descriptors, directory codes and subject matters, hence all classifications are multilabel ones. EUROVOC is a multilingual thesaurus providing a controlled vocabulary[3]. Documents in the documentation systems of the EU are indexed using this thesaurus.The directory codes are classes of the official classification hierarchy of the

[1] http://eur-lex.europa.eu

[2] http://eur-lex.europa.eu/en/legis/index.htm

[3] http://europa.eu/eurovoc/

*Directory of Community legislation in force*. It contains 20 chapter headings with up to four sub-division levels.

The high number of 3,993 different EUROVOC descriptors were identified in the retrieved documents, each document is associated to 5.37 descriptors on average. In contrast there are only 201 different subject matters appearing in the dataset, with a mean of 2.23 labels per document, and 412 different directory codes, with a label set size of on average 1.29.

Figure 1 shows an excerpt of a sample document with all information that has not been used removed. The full document can be viewed at http://eur-lex.europa.eu/LexUriServ/LexUriServ.do?uri=CELEX:31991L0250:EN:NOT. We extracted the text body from the HTML documents, excluding HTML tags, bibliographic notes or other additional information that could distort the results. The text was tokenized into lower case, stop words were excluded, and the Porter stemmer algorithm was applied. In order to perform cross validation, the instances were randomly distributed into ten folds. The tokens were projected for each fold into the vector space model using the common TF-IDF term weighting.In order to reduce the memory requirements, of the approx. 200,000 resulting features we selected the first 5,000 ordered by their document frequency.

# 3 Preliminaries

We represent an instance or object as a vector $\bar{x} = (x_1, \ldots, x_M)$ in a feature space $\mathcal{X} \subseteq \mathbb{R}^N$. Each instance $\bar{x}_i$ is assigned to a set of relevant labels $\mathcal{y}_i$, a subset of the $K$ possible classes $\mathcal{Y} = \{c_1, \ldots, c_K\}$. For multilabel problems, the cardinality $|\mathcal{y}_i|$ of the label sets is not restricted, whereas for binary problems $|\mathcal{y}_i| = 1$. For the sake of simplicity we use the following notation for the binary case: we define $\mathcal{Y} = \{1, -1\}$ as the set of classes so that each object $\bar{x}_i$ is assigned to a $y_i \in \{1, -1\}$ , $\mathcal{y}_i = \{y_i\}$.

## 3.1 Ranking Loss Functions

In order to evaluate the predicted ranking we use different *ranking losses*. The losses are computed comparing the ranking with the true set of relevant classes, each of them focusing on different aspects. For a given instance $\bar{x}$, a relevant label set $\mathcal{y}$, a negative label set $\overline{\mathcal{y}} = \mathcal{Y} \backslash \mathcal{y}$ and a given predicted ranking $r : \mathcal{Y} \rightarrow \{1 \ldots K\}$, with $r(c)$ returning the position of class $c$ in the ranking, the different loss functions are computed as follows:

- ISERR . The is-error loss determines whether $r(c) < r(c')$ for all relevant classes $c \in \mathcal{y}$ and all irrelevant classes $c' \in \overline{\mathcal{y}}$. It returns 0 for a completely correct, *perfect ranking*, and 1 for an incorrect ranking, irrespective of 'how wrong' the ranking is.
- ONEERR . The one error loss is 1 if the top class in the ranking is not a relevant class, otherwise 0 if the top class is relevant, independently of the positions of the remaining relevant classes.
- RANKLOSS . The ranking loss returns the number of pairs of labels which are not correctly ordered normalized by the total number of possible pairs. As ISERR, it is

0 for a perfect ranking, but it additionally differentiates between different degrees of errors.

$$E \stackrel{\text{def}}{=} \{(c, c') \mid r(c) > r(c')\} \subseteq \mathcal{Y} \times \overline{\mathcal{Y}} \qquad \delta_{\text{RANKLOSS}} \stackrel{\text{def}}{=} \frac{|E|}{|\mathcal{Y}||\overline{\mathcal{Y}}|} \tag{1}$$

MARGIN. The margin loss returns the number of positions between the worst ranked positive and the best ranked negative classes. This is directly related to the number of wrongly ranked classes, i.e. the positive classes that are ordered below a negative class, or vice versa. We denote this set by $F$.

$$F \stackrel{\text{def}}{=} \{c \in \mathcal{Y} \mid r(c) > r(c'), c' \in \overline{\mathcal{Y}}\} \cup \{c' \in \overline{\mathcal{Y}} \mid r(c) > r(c'), c \in \mathcal{Y}\} \tag{2}$$

$$\delta_{\text{MARGIN}} \stackrel{\text{def}}{=} \max(0, \max\{r(c) \mid c \in \mathcal{Y}\} - \min\{r(c') \mid c' \notin \mathcal{Y}\}) \tag{3}$$

AVGP. Average Precision is commonly used in Information Retrieval and computes for each relevant label the percentage of relevant labels among all labels that are ranked before it, and averages these percentages over all relevant labels. In order to bring this loss in line with the others so that an optimal ranking is 0, we revert the measure.

$$\delta_{\text{AVGP}} \stackrel{\text{def}}{=} 1 - \frac{1}{\mathcal{Y}} \sum_{c \in \mathcal{Y}} \frac{|\{c^* \in \mathcal{Y} \mid r(c^*) \leq r(c)\}|}{r(c)}. \tag{4}$$

### 3.2 Perceptrons

We use the simple but fast perceptrons as base classifiers [16]. As Support Vector Machines (SVM), their decision function describes a hyperplane that divides the $N$-dimensional space into two halves corresponding to positive and negative examples. We use a version that works without learning rate and threshold:

$$o'(\bar{\mathrm{x}}) = sgn(\bar{\mathrm{x}} \cdot \bar{\mathrm{w}}) \tag{5}$$

with the internal weight vector $\bar{\mathrm{w}}$ and $sgn(t) = 1$ for $t \geq 0$ and $-1$ otherwise. If there exists a *separating hyperplane* between the two set of points, i.e. they are linearly separable, the following update rule provably finds it (cf., e.g., [1]).

$$\alpha_i = (y_i - o'(\bar{\mathrm{x}}_i)) \qquad \bar{\mathrm{w}}_{i+1} = \bar{\mathrm{w}}_i + \alpha_i \bar{\mathrm{x}}_i \tag{6}$$

It is important to see that the final weight vector can also be represented as linear combination of the training examples:

$$\bar{\mathrm{w}} = \sum_{i=1}^{M} \alpha_i \bar{\mathrm{x}}_i \qquad o'(\bar{\mathrm{x}}) = sgn(\sum_{i=1}^{M} \alpha_i \cdot \bar{\mathrm{x}}_i \bar{\mathrm{x}}) \tag{7}$$

assuming $M$ to be the number of seen training examples and $\alpha_i \in \{-1, 0, 1\}$. The perceptron can hence be coded implicitly as a vector of instance weights $\alpha = (\alpha_1, \ldots, \alpha_M)$ instead of explicitly as a vector of feature weights. This representation is denominated

**Require:** Training example pair $(\bar{x}, y)$, perceptrons $\bar{w}_1, \ldots, \bar{w}_K$
1: calculate $\bar{x}\bar{w}_1, \ldots, \bar{x}\bar{w}_K$, loss $\delta$
2: **if** $\delta > 0$ **then** ▷ only if ranking is not perfect
3: calculate error sets $E, F$
4: **for each** $c \in F$ **do** $\tau_c \leftarrow 0, \sigma \leftarrow 0$ ▷ initialize $\tau$'s, $\sigma$
5: **for each** $(c, c') \in E$ **do**
6: $p \leftarrow$ PENALTY$(\bar{x}\bar{w}_1, \ldots, \bar{x}\bar{w}_K)$
7: $\tau_c \leftarrow \tau_c + p$ ▷ push up pos. classes
8: $\tau_{c'} \leftarrow \tau_{c'} - p$ ▷ push down neg. classes
9: $\sigma \leftarrow \sigma + p$ ▷ for normalization
10: **for each** $c \in F$ **do**
11: $\bar{w}_c \leftarrow \bar{w}_c + \delta \frac{\tau_c}{\sigma} \cdot \bar{x}$ ▷ update perceptrons
12: **return** $\bar{w}_1 \ldots \bar{w}_K$ ▷ return updated perceptrons

**Fig. 2.** Pseudocode of the training method of the MMP algorithm

the dual form and is crucial for developing the memory efficient variant in Section 4. The main reason for choosing the perceptrons as our base classifier is because, contrary to SVMs, they can be trained efficiently in an incremental setting, which makes them particularly well-suited for large-scale classification problems such as the Reuters-RCV1 benchmark [10], without forfeiting too much accuracy though SVMs find the *maximum-margin hyperplane* [5, 3, 18].

### 3.3 Binary Relevance Ranking

In the binary relevance (BR) or one-against-all (OAA) method, a multilabel training set with $K$ possible classes is decomposed into $K$ binary training sets of the same size that are then used to train $K$ binary classifiers. So for each pair $(\bar{x}_i, y_i)$ in the original training set $K$ different pairs of instances and binary class assignments $(\bar{x}_i, y_{i_j})$ with $j = 1 \ldots K$ are generated setting $y_{i_j} = 1$ if $c_j \in y_i$ and $y_{i_j} = -1$ otherwise. Supposing we use perceptrons as base learners, $K$ different $o'_j$ classifiers are trained in order to determine the relevance of $c_j$. In consequence, the combined prediction of the binary relevance classifier for an instance $\bar{x}$ would be the set $\{c_j \mid o'_j(\bar{x}) = 1\}$. If, in contrast, we desire a class ranking, we simply use the inner products and obtain a vector $\bar{o}(\bar{x}) = (\bar{x}\bar{w}_1, \ldots, \bar{x}\bar{w}_K)$. Ties are broken randomly to not favor any particular class.

### 3.4 Multiclass Multilabel Perceptrons

MMPs were proposed as an extension of the one-against-all algorithm with perceptrons as base learners [3]. Just as in binary relevance, one perceptron is trained for each class, and the prediction is calculated via the inner products. The difference lies in the update method: while in the binary relevance method all perceptrons are trained independently to return a value greater or smaller than zero, depending on the relevance of the classes for a certain instance, MMPs are trained to produce a good ranking so that the relevant classes are all ranked above the irrelevant classes. The perceptrons therefore cannot

**Require:** Training example pair $(\bar{x}, y)$,
perceptrons $\{\bar{w}_{u,v} \mid u < v, c_u, c_v \in \mathcal{Y}\}$
1: **for each** $(c_u, c_v) \in y \times \overline{y}$ **do**
2: **if** $u < v$ **then**
3: $\bar{w}_{u,v} \leftarrow$ TRAINPERCEPTRON($\bar{w}_{u,v}$, $(\bar{x}, 1)$) ▷ train as positive example
4: **else**
5: $\bar{w}_{v,u} \leftarrow$ TRAINPERCEPTRON($\bar{w}_{v,u}$, $(\bar{x}, -1)$) ▷ train as negative example
6: **return** $\{\bar{w}_{u,v} \mid u < v, c_u, c_v \in \mathcal{Y}\}$ ▷ updated perceptrons

**Fig. 3.** Pseudocode of the training method of the MLPP algorithm

be trained independently, considering that the target value for each perceptron depends strongly on the values returned by the other perceptrons.

The pseudocode in Fig. 2 describes the MMP training algorithm. In summary, for each new training example the MMP first computes the predicted ranking, and if there is an error according to the chosen loss function $\delta$ (e.g. any of the losses in Sec. 3.1), it computes the set of wrongly ordered class pairs in the ranking and applies to each class in this set a penalty score according to a freely selectable function. We chose the uniform update method, where each pair in $E$ receives the same score [3]. Please refer to [3] and [12] for a more detailed description of the algorithm.

### 3.5 Multilabel Pairwise Perceptrons

In the pairwise binarization method, one classifier is trained for each pair of classes, i.e., a problem with $K$ different classes is decomposed into $\frac{K(K-1)}{2}$ smaller subproblems. For each pair of classes $(c_u, c_v)$, only examples belonging to either $c_u$ or $c_v$ are used to train the corresponding classifier $o'_{u,v}$. All other examples are ignored. In the multilabel case, an example is added to the training set for classifier $o'_{u,v}$ if $u$ is a relevant class and $v$ is an irrelevant class, i.e., $(u, v) \in y \times \overline{y}$ (cf. Figure 4). We will typically assume $u < v$, and training examples of class $u$ will receive a training signal of $+1$, whereas training examples of class $v$ will be classified with $-1$. Figure 3 shows the training algorithm in pseudocode. Of course MLPPs can also be trained incrementally.

In order to return a class ranking we use a simple voting strategy, known as *max-wins*. Given a test instance, each perceptron delivers a prediction for one of its two classes. This prediction is decoded into a vote for this particular class. After the evaluation of all $\frac{K(K-1)}{2}$ perceptrons the classes are ordered according to their sum of votes. Ties are broken randomly in our case.

Figure 5 shows a possible result of classifying the sample instance of Figure 4. Perceptron $o'_{1,5}$ predicts (correctly) the first class, consequently $c_1$ receives one vote and class $c_5$ zero (denoted by $o'_{1,5} = 1$ in the first and $o'_{5,1} = -1$ in the last row). All 10 perceptrons (the values in the upper right corner can be deduced due to the symmetry property of the perceptrons) are evaluated though only six are 'qualified' since they were trained with the original example.

This may be disturbing at first sight since many 'unqualified' perceptrons are involved in the voting process: $o'_{1,2}$ is asked for instance though it cannot know anything relevant in order to determine if $\bar{x}$ belongs to $c_1$ or $c_2$ since it was neither trained on this example

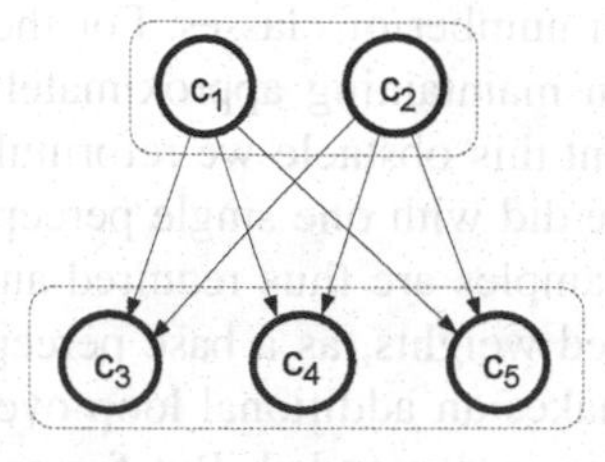


**Fig. 4.** MLPP training: training example $\bar{x}$ belongs to $\mathcal{Y} = \{c_1, c_2\}$, $\overline{\mathcal{Y}} = \{c_3, c_4, c_5\}$ are the irrelevant classes, the arrows represent the trained perceptrons

| | | | | |
|---|---|---|---|---|
| $o'_{1,2} = 1$ | $o'_{2,1} = -1$ | $o'_{3,1} = -1$ | $o'_{4,1} = -1$ | $o'_{5,1} = -1$ |
| $o'_{1,3} = 1$ | $o'_{2,3} = 1$ | $o'_{3,2} = -1$ | $o'_{4,2} = -1$ | $o'_{5,2} = -1$ |
| $o'_{1,4} = 1$ | $o'_{2,4} = 1$ | $o'_{3,4} = 1$ | $o'_{4,3} = -1$ | $o'_{5,3} = -1$ |
| $o'_{1,5} = 1$ | $o'_{2,5} = 1$ | $o'_{3,5} = 1$ | $o'_{4,5} = 1$ | $o'_{5,4} = -1$ |
| $v_1 = 4$ | $v_2 = 3$ | $v_3 = 2$ | $v_4 = 1$ | $v_5 = 0$ |

**Fig. 5.** MLPP voting: an example $\bar{x}$ is classified by all 10 base perceptrons $o'_{u,v}, u \neq v\,,\ c_u, c_v \in \mathcal{Y}$. Note the redundancy given by $o'_{u,v} = -o'_{v,u}$. The last line counts the positive outcomes for each class.

nor on other examples belonging simultaneously to both classes (or to none of both). In the worst case the noisy votes concentrate on a single negative class, which would lead to misclassifications. But note that any class can at most receive $K-1$ votes, so that in the extreme case when the qualified perceptrons all classify correctly and the unqualified ones concentrate on a single class, a positive class would still receive at least $K-|\mathcal{Y}|$ and a negative at most $K-|\mathcal{Y}|-1$ votes. Class $c_3$ in Figure 5 is an example for this: It receives all possible noisy votes but still loses against the positive classes $c_1$ and $c_2$.

The pairwise binarization method is often regarded as superior to binary relevance because it profits from simpler decision boundaries in the subproblems [6, 8]. In the case of an equal class distribution, the subproblems have $\frac{2}{K}$ times the original size whereas binary relevance maintains the size. Typically, this goes hand in hand with an increase of the space where a separating hyperplane can be found. Particularly in the case of text classification the obtained benefit clearly exists. An evaluation of the pairwise approach on the Reuters-RCV1 corpus [10], which contains over 100 classes and 800,000 documents, showed a significant and substantial improvement over the MMP method [12]. This encourages us to apply the pairwise decomposition to the EUR-Lex database, with the main obstacle of the quadratic number of base classifier in relationship to the number of classes. Since this problem can not be coped for the present classifications in EUR-Lex, we propose to reformulate the MLPP algorithm in the way described in the next section.

## 4 Dual Multilabel Pairwise Perceptrons

With an increasing number of classes the required memory by the MLPP algorithm grows quadratically and even on modern computers with a large memory this problem becomes unsolvable for a high number of classes. For the EUROVOC classification, the use of MLPP would mean maintaining approximately 8,000,000 perceptrons in memory. In order to circumvent this obstacle we reformulate the MLPP ensemble of perceptrons in dual form as we did with one single perceptron in Equation 7. In contrast to MLPP, the training examples are thus required and have to be kept in memory in addition to the associated weights, as a base perceptron is now represented as $\bar{w}_{u,v} = \sum_{i=1}^{M} \alpha_{u,v}^{t} \bar{x}_i$. This makes an additional loop over the training examples inevitable every time a prediction is demanded. But fortunately it is not necessary to recompute all $\bar{x}_i\bar{x}$ for each base perceptron since we can reuse them by iterating over the training examples in the outer loop, as can be seen in the following equations:

$$\begin{aligned}
\bar{w}_{1,2}\bar{x} &= \alpha_{1,2}^{1}\bar{x}_1\bar{x} + \alpha_{1,2}^{2}\bar{x}_2\bar{x} + \ldots + \alpha_{1,2}^{M}\bar{x}_M\bar{x} \\
\bar{w}_{1,3}\bar{x} &= \alpha_{1,3}^{1}\bar{x}_1\bar{x} + \alpha_{1,3}^{2}\bar{x}_2\bar{x} + \ldots + \alpha_{1,3}^{M}\bar{x}_M\bar{x} \\
&\vdots \\
\bar{w}_{1,K}\bar{x} &= \alpha_{1,K}^{1}\bar{x}_1\bar{x} + \alpha_{1,K}^{2}\bar{x}_2\bar{x} + \ldots + \alpha_{1,K}^{M}\bar{x}_M\bar{x} \\
\bar{w}_{2,3}\bar{x} &= \alpha_{2,3}^{1}\bar{x}_1\bar{x} + \alpha_{2,3}^{2}\bar{x}_2\bar{x} + \ldots + \alpha_{2,3}^{M}\bar{x}_M\bar{x} \\
&\vdots
\end{aligned} \tag{8}$$

By advancing column by column it is not necessary to repeat the dot products computations, however it is necessary to store the intermediate values, as can also be seen in the pseudocode of the training and prediction phases in Figures 6 and 7. Note also that the algorithm preserves the property of being incrementally trainable. We denote this variant of training the pairwise perceptrons the *dual multilabel pairwise perceptrons* algorithm (DMLPP).

In addition to the savings in memory and run-time, analyzed in detail in Section 5, the dual representation allows for using the kernel trick, i.e. to replace the dot product by a kernel function, in order to be able to solve originally not linearly separable problems. However, this is not necessary in our case since text problems are in general linearly separable.

Note also that the pseudocode needs to be slightly adapted when the DMLPP algorithm is trained in more than one epoch, i.e. the training set is presented to the learning algorithm more than once. It is sufficient to modify the assignment in line 8 in Figure 6 to an additive update $\alpha_{u,v}^{t} = \alpha_{u,v}^{t} + 1$ for a revisited example $\bar{x}_t$. This setting is particularly interesting for the dual variant since, when the training set is not too big, memorizing the inner products can boost the subsequent epochs in a substantial way, making the algorithm interesting even if the number of classes is small.

## 5 Computational Complexity

The notation used in this section is the following: $K$ denotes the number of possible classes, $L$ the average number of relevant classes per instance in the training set, $N$ the number of attributes and $N'$ the average number of attributes not zero (size of the sparse representation of an instance), and $M$ denotes the size of the training set. For each complexity we will give an upper bound $O$ in Landau notation. We will indicate the runtime

**Require:** New training example pair $(\bar{x}_M, y)$,
training examples $\bar{x}_1 \ldots \bar{x}_{M-1}$,
weights $\{\alpha^i_{u,v} \mid c_u, c_v \in \mathcal{Y}, 0 < i < M\}$
1: **for each** $\bar{x}_i \in \bar{x}_1 \ldots \bar{x}_{M-1}$ **do**
2: $p_i \leftarrow \bar{x}_i \cdot \bar{x}_M$
3: **for each** $(c_u, c_v) \in y \times \overline{y}$ **do**
4: **if** $\alpha^i_{u,v} \neq 0$ **then**
5: $s_{u,v} \leftarrow s_{u,v} + \alpha^i_{u,v} \cdot p_t$ ▷ note that $s_{u,v} = -s_{v,u}$
6: **for each** $(c_u, c_v) \in y \times \overline{y}$ **do**
7: **if** $s_{u,v} < 0$ **then**
8: $\alpha^M_{u,v} \leftarrow 1$ ▷ note that $\alpha_{u,v} = -\alpha_{v,u}$
9: **return** $\{\alpha^M_{u,v} \mid (c_u, c_v) \in y \times \overline{y}\}$ ▷ return new weights

**Fig. 6.** Pseudocode of the training method of the DMLPP algorithm

**Require:** example $\bar{x}$ for classification,
training examples $\bar{x}_1 \ldots \bar{x}_{M-1}$,
weights $\{\alpha^i_{u,v} \mid c_u, c_v \in \mathcal{Y}, 0 < i < M\}$
1: **for each** $\bar{x}_i \in \bar{x}_1 \ldots \bar{x}_M$ **do**
2: $p \leftarrow \bar{x}_i \cdot \bar{x}$
3: **for each** $(c_u, c_v) \in y_i \times \overline{y}_t$ **do**
4: **if** $\alpha^i_{u,v} \neq 0$ **then**
5: $s_{u,v} \leftarrow s_{u,v} + \alpha^i_{u,v} \cdot p$
6: **for each** $(c_u, c_v) \in \mathcal{Y} \times \mathcal{Y}$ **do**
7: **if** $u \neq v \vee s_{u,v} > 0$ **then**
8: $v_u \leftarrow v_u + 1$
9: **return** voting $\bar{v} = (v_1, \ldots, v_{|\mathcal{Y}|})$ ▷ return voting

**Fig. 7.** Pseudocode of the prediction phase of the DMLPP algorithm

complexity in terms of real value additions and multiplications ignoring operations that have to be performed by all algorithms such as sorting or internal real value operations. Additionally, we will present the complexities per instance as all algorithms are incrementally trainable. We will also concentrate on the comparison between MLPP and the implicit representation DMLPP.

The MLPP algorithm has to keep $\frac{K(K-1)}{2}$ perceptrons, each with $N$ weights in memory, hence we need $O(K^2N)$ memory. The DMLPP algorithm keeps the whole training set in memory, and additionally requires for each training example $\bar{x}$ access to the weights of all class pairs $y \times \overline{y}$. Furthermore, it has to intermediately store the resulting scores for each base perceptron during prediction, hence the complexity is $O(MLK + MN' + K^2) = O(M(LK + N') + K^2)$.[4] We can see that MLPP is

[4] Note that we do not estimate $L$ as $O(K)$ since both values are not of the same order of magnitude in practice. For the same reason we distinguish between $N$ and $N'$ since particularly in text classification both values are not linked: a text document often turns out to employ around 100 different words whereas the size of the vocabulary of a the whole corpus can easily reach 100,000 words (although this number is normally reduced by feature selection).

**Table 1.** Computational complexity given in expected number of addition and multiplication operations. $K$: *#classes*, $L$: *avg. #labels per instance*, $M$: *#training examples*, $N$: *#attributes*, $N'$: *#attributes*$\neq 0$, $\hat{\delta}$: *avg. Loss*, $\hat{\delta}_{per}, \hat{\delta}_{\text{IsErr}} \leq 1, \hat{\delta}_{\text{Margin}} < K$.

| | training time | testing time | memory requirement |
|---|---|---|---|
| MMP, BR | $O(KN')$ | $O(KN')$ | $O(KN)$ |
| MLPP | $O(LKN')$ | $O(K^2N')$ | $O(K^2N)$ |
| DMLPP | $O(M(LK+N'))$ | $O(M(LK+N'))$ | $O(M(LK+N')+K^2)$ |

applicable especially if the number of classes is low and the number of examples high, whereas DMLPP is suitable when the number of classes is high, however it does not handle huge training sets very well.

For processing one training example, $O(LK)$ dot products have to be computed by MLPP, one for each associated perceptron. Assuming that a dot product computation costs $O(N')$, we obtain a complexity of $O(LKN')$ per training example. Similarly, the DMLPP spends $M$ dot product computations. In addition the summation of the scores costs $O(LK)$ per training instance, leading to $O(M(LK+N'))$ operations. It is obvious that MLPP has a clear advantage over DMLPP in terms of training time, unless $K$ is of the order of magnitude of $M$ or the model is trained over several epochs, as already outlined in the previous Section 4.

During prediction the MLPP evaluates all perceptrons, leading to $O(K^2N')$ computations. The dual variant again iterates over all training examples and associated weights, hence the complexity is $O(M(LK + N'))$. At this phase DMLPP benefits from the linear dependence of the number of classes in contrast to the quadratic relationship of the MLPP. Roughly speaking the breaking point when DMLPP is faster in prediction is approximately when the square of the number of classes is clearly greater than the number of training documents. We can find a similar trade-off for the memory requirements with the difference that the factor between sparse and total number of attributes becomes more important, leading earlier to the breaking point when the sparseness is high. A compilation of the analysis can be found in Table 1, together with the complexities of MMP and BR. A more detailed comparison between MMP and MLPP can be found in [12].

In summary, it can be stated that the dual form of the MLPP balances the relationship between training and prediction time by increasing training and decreasing prediction costs, and especially benefits from a decreased prediction time and memory savings when the number of classes is large. Thus, this technique addresses the main obstacle to applying the pairwise approach to problems with a large number of labels.

## 6 Experiments

For the MMP algorithm we used the IsErr loss function and the uniform penalty function. This setting showed the best results in [3] on the RCV1 data set. The perceptrons of the BR and MMP ensembles were initialized with random values. We performed also tests with a multilabel variant of the multinomial Naive Bayes (MLNB) algorithm in order to provide a baseline.

### 6.1 Ranking Performance

The results for the four algorithms and the three different classifications of EUR-Lex are presented in Table 2. MLPP results are omitted since they are equal to those of DMLPP. The values for ISERR, ONEERR, RANKLOSS and AVGP are shown $\times 100\%$ for better readability, AVGP is also presented in the conventional way (with 100% as the optimal value) and not as a loss function. The number of epochs indicates the number of times that the online-learning algorithms were able to see the training instances. No results are reported for the performance of DMLPP on EUROVOC for more than two epochs due to time restrictions.

The first appreciable characteristic is that DMLPP dominates all other algorithms on all three views of the EUR-Lex data, regardless of the number of epochs or losses. For the directory code DMLPP achieve a result in epoch 2 that is still beyond the reach of the other algorithms in epoch 10, except for MMP's ISERR. Especially on the losses that directly evaluate the ranking performance the improvement is quite pronounced and the results are already unreachable after the first epoch.

It is also interesting to compare the performances of MMP and BR as they have still the advantage of reduced computational costs and memory requirements in comparison to the (dual) pairwise approach and could therefore be more applicable for very complex data sets such as EUROVOC, which is certainly hard to tackle for DMLPP (cf. Section 6.2). Please refer to [13] for a more detailed comparison between MMP and BR.

The fact that in only approximately 4% of the cases a perfect classification is achieved and in only approx. 60% the top class is correctly predicted (MMP) should not lead to an underestimation of the performance of these algorithms. Considering that with almost 4000 possible classes and only 5.3 classes per example the probability of randomly choosing a correct class is less than one percent, namely 0.13%, the performance is indeed substantial.

### 6.2 Computational Costs

In order to allow a comparison independent from external factors such as logging activities and the run-time environment, we ignored minor operations that have to be performed by all algorithms, such as sorting or internal operations. An overview over the amount of real value addition and multiplication computations is given in Table 3 (measured on the first cross validation split, trained for one epoch), together with the CPU-times on an AMD Dual Core Opteron 2000 MHz as additional reference information. We included in this table also the results for the non-dual MLPP, however no values have been received for the EUROVOC problem due to the discussed memory space problem: MLPP requires 539 MB memory for the subject matter and already 1825 MB for the directory code problem, whereas DMLPP consumes 203 MB and resp. 217 MB It is remarkable that MMP uses similar MMP 151 MB resp. 165MB. Also for EUROVOC the usage of 1143 MB is comparable to DMLPP's 2246 MB.

We can observe a clear advantage of the non-pairwise approaches on the subject matter data especially for the prediction phase, however the training costs are in the same order of magnitude. Between MLPP and DMLPP we can see an antisymmetric behavior: while MLPP requires only almost half of the amount of the DMLPP operations

**Table 2.** Average losses for the three views on the data and for the different algorithms

| | | | 1 epoch | | | 2 epochs | | | 5 epochs | | | 10 epochs | | |
|---|---|---|---|---|---|---|---|---|---|---|---|---|---|---|
| | | MLNB | BR | MMP | DMLPP | BR | MMP | DMLPP | BR | MMP | DMLPP | BR | MMP | DMLPP |
| *subject matter* | ISERR ×100 | 99.15 | 61.71 | 53.61 | 52.28 | 57.39 | 48.83 | 45.48 | 52.36 | 43.21 | 39.74 | 49.87 | 40.89 | 37.69 |
| | ONEERR ×100 | 98.63 | 30.67 | 27.95 | 23.62 | 26.44 | 24.4 | 18.64 | 22.17 | 18.82 | 15.01 | 20.41 | 17.08 | 13.7 |
| | RANKLOSS | 8.979 | 16.36 | 2.957 | 1.160 | 14.82 | 2.785 | 0.988 | 11.40 | 2.229 | 0.885 | 10.29 | 2.000 | 0.849 |
| | MARGIN | 25.34 | 59.33 | 13.04 | 4.611 | 54.27 | 11.94 | 4.001 | 44.05 | 9.567 | 3.615 | 40.39 | 8.636 | 3.488 |
| | AVGP | 12.26 | 62.9 | 74.71 | 77.98 | 66.61 | 78.05 | 82.05 | 71.28 | 81.71 | 84.76 | 73.06 | 83.19 | 85.79 |
| *directory code* | ISERR ×100 | 99.25 | 52.46 | 46.08 | 37.58 | 45.89 | 40.78 | 33.31 | 40.97 | 34.27 | 30.41 | 37.67 | 32.52 | 29.58 |
| | ONEERR ×100 | 99.28 | 44.61 | 39.42 | 29.41 | 37.46 | 34.27 | 25.60 | 32.09 | 27.62 | 23.04 | 28.86 | 25.83 | 22.40 |
| | RANKLOSS | 7.785 | 19.30 | 2.749 | 1.109 | 15.12 | 2.294 | 0.999 | 12.10 | 2.199 | 0.961 | 10.17 | 1.783 | 0.953 |
| | MARGIN | 35.89 | 96.16 | 15.47 | 6.271 | 77.31 | 13.30 | 5.690 | 62.98 | 12.30 | 5.478 | 53.82 | 10.20 | 5.436 |
| | AVGP | 6.565 | 57.10 | 70.00 | 77.21 | 63.49 | 74.61 | 80.10 | 68.27 | 78.83 | 81.93 | 71.18 | 80.28 | 82.37 |
| *EUROVOC* | ISERR ×100 | 99.58 | 98.57 | 98.84 | 97.92 | 98.19 | 97.47 | 96.64 | 97.23 | 95.96 | | | | |
| | ONEERR ×100 | 99.99 | 48.69 | 75.89 | 35.50 | 41.50 | 54.41 | 29.50 | 37.30 | 40.15 | | | | |
| | RANKLOSS | 22.88 | 40.35 | 3.906 | 2.779 | 35.46 | 4.350 | 2.504 | 30.96 | 4.701 | | | | |
| | MARGIN | 1644.00 | 3230.68 | 597.59 | 433.89 | 3050.07 | 694.10 | 397.11 | 2842.63 | 761.24 | | | | |
| | AVGP | 1.06 | 26.90 | 29.28 | 46.67 | 31.58 | 39.54 | 52.27 | 35.90 | 47.27 | | | | |

**Table 3.** Computational costs in CPU-time and millions of real value operations (M op.)

| *subject matter* | training | testing |
|---|---|---|
| BR | 35.8 s | 8.36 s |
| | 1,675 M op. | 192 M op. |
| MMP | 31.35 s | 6.28 s |
| | 1,789 M op. | 192 M op. |
| DMLPP | 326.02 s | 145.67 s |
| | 6,089 M op. | 4,628 M op. |
| MLPP | 91.77 s | 196.93 s |
| | 3,870 M op. | 19,210 M op. |

| *directory code* | training | testing |
|---|---|---|
| BR | 49.01 s | 11.99 s |
| | 3,410 M op. | 394 M op. |
| MMP | 49.63 s | 11.03 s |
| | 3,579 M op. | 394 M op. |
| DMLPP | 313.59 s | 192.99 s |
| | 2,986 M op. | 5,438 M op. |
| MLPP | 149.49 s | 775.79 s |
| | 4,712 M op. | 80,938 M op. |

| *EUROVOC* | training | testing |
|---|---|---|
| BR | 405.42 s | 56.71 s |
| | 32,975 M op. | 3,817 M op. |
| MMP | 503.04 s | 53.69 s |
| | 40,510 M op. | 3,817 M op. |
| DMLPP | 11,479.81 s | 7,631.86 s |
| | 17,719 M op. | 127,912 M op. |
| MLPP | – | – |
| | – | – |

for training, DMLPP reduces the amount of prediction operations by a factor of more than 4. For the directory code the rate for MMP and BR more than doubles in correspondence with the increase in number of classes, additionally the MLPP testing time substantially increases due to the quadratic dependency, while DMLPP profits from the decrease in the average number of classes per instance. It even causes less computations in the training phase than MMP/BR. The reason for this is not only the reduced maximum amount of weights per instance (cf. Section 5), but particularly the decreased probability that a training example is relevant for a new training example (and consequently that dot products and scores have to be computed) since it is less probable that both class assignments match, i.e. that both examples have the same pair of positive and negative classes. This becomes particularly clear if we observe the number of non-zero weights and actually used weights during training for each new example. The classifier for subject matter has on average 21 weights set per instance out of 443 ($= L(K - L)$) in the worst case (a ratio of 4.47%), and on average 5.1% of them are required when a new training example arrives. For the directory code with a smaller fraction $L/K$ 35.5 weights are stored (3.96%), of which only 1.11% are used when updating. This also explains the relatively small number of operations for training on EUROVOC, since from the 1,802 weights per instance (8.41%), only 0.55% are relevant to a new training instance. In this context, regarding the disturbing ratio between real value operations and CPU-time for training DMLPP on EUROVOC, we believe that this is caused by a suboptimal storage structure and processing of the weights and we are therefore confident that it is possible to reduce the distance to MMP in terms of actual consumed CPU-time by improving the program code.

Note that MMP and BR compute the same amount of dot products, the computational costs only differ in the number of vector additions, i.e. perceptron updates. A deeper analysis of the contrary behavior of both algorithms when the number of classes increases can be found in [11].

## 7 Conclusions

In this paper, we introduced the EUR-Lex text collection as a promising test bed for studies in text categorization. Among its many interesting characteristics (e.g., multilinguality), our main interest was the large number of categories, which is one order of magnitude above other frequently studied text categorization benchmarks, such as the Reuters-RCV1 collection.

On the EUROVOC classification task, a multilabel classification task with 4000 possible labels, the DMLPP algorithm, which decomposes the problem into training classifiers for each pair of classes, achieves an average precision rate of slightly more than 50%. Roughly speaking, this means that the (on average) five relevant labels of a document will (again, on average) appear within the first 10 ranks in the relevancy ranking of the 4,000 labels. This is a very encouraging result for a possible automated or semi-automated real-world application for categorizing EU legal documents into EUROVOC categories.

This result was only possible by finding an efficient solution for storing the approx. 8,000,000 binary classifiers that have to be trained by this pairwise approach. To this end, we showed that a reformulation of the pairwise decomposition approach into a dual form is capable of handling very complex problems and can therefore compete with the approaches that use only one classifier per class. It was demonstrated that decomposing the initial problem into smaller problems for each pair of classes achieves higher prediction accuracy on the EUR-Lex data, since DMLPP substantially outperforms all other algorithms. This confirms previous results of the non-dual variant on the large Reuters Corpus Volume 1 [12]. The dual form representation allows for handling a much higher number of classes than the explicit representation, albeit with an increased dependence on the training set size. Despite the improved ability to handle large problems, DMLPP is still less efficient than MMP, especially for the EUROVOC data with 4000 classes. However, in our opinion the results show that DMLPP is still competitive for solving large-scale problems in practice, especially considering the trade-off between runtime and prediction performance. Additionally, we are currently investigating hybrid variants to further reduce the computational complexity. The idea is to use a different formulation in training than in the prediction phase depending on the specific memory and runtime requirements of the task. In order e.g. to combine the advantage of MLPP during training and DMLPP during predicting on the subject matter subproblem, we could train the classifier as in the MLPP (with the difference of iterating over the perceptrons first so that only one perceptron has to remain in memory) and than convert it to the dual representation by means of the collected information during training the perceptrons. The use of SVMs during training is also an interesting option.

For future research, on the one hand we see space for improvement for the MMP and pairwise approach for instance by using a calibrated ranking approach [2]. The basic idea of this algorithm is to introduce an artificial label which, for each example, separates the relevant from irrelevant labels in order to return a set of classes instead of only a ranking. On the other hand, we see possible improvements by exploiting advancements in the perceptron algorithm and in the pairwise binarization, e.g. by using one of the several variants of the perceptron algorithm that, similar to SVMs, try to maximize the margin of the separating hyperplane in order to produce more accurate models [4, 9], or by employing a voting technique that takes the prediction weights into account such as the weighted voting technique by Price et al. [15]. Finally, we note that we are currently working on an adaptation of the efficient voting technique introduced in [14] to the multilabel case, of which a further significant reduction in classification time can be expected.

**Acknowledgements.** This work was supported by the EC 6th framework project ALIS (Automated Legal Information System) and by the German Science Foundation (DFG).

## References

[1] Bishop, C.M.: Neural Networks for Pattern Recognition. Oxford University Press, Oxford (1995)

[2] Brinker, K., Fürnkranz, J., Hüllermeier, E.: A Unified Model for Multilabel Classification and Ranking. In: Proceedings of the 17th European Conference on Artificial Intelligence (ECAI 2006) (2006)

[3] Crammer, K., Singer, Y.: A Family of Additive Online Algorithms for Category Ranking. Journal of Machine Learning Research 3(6), 1025–1058 (2003)

[4] Crammer, K., Dekel, O., Keshet, J., Shalev-Shwartz, S., Singer, Y.: Online passive-aggressive algorithms. Journal of Machine Learning Research 7, 551–585 (2006)
[5] Freund, Y., Schapire, R.E.: Large Margin Classification using the Perceptron Algorithm. Machine Learning 37(3), 277–296 (1999)
[6] Fürnkranz, J.: Round Robin Classification. Journal of Machine Learning Research 2, 721–747 (2002)
[7] Fürnkranz, J., Hüllermeier, E., Loza Mencía, E., Brinker, K.: Multilabel classification via calibrated label ranking. Machine Learning (to appear, 2008)
[8] Hsu, C.-W., Lin, C.-J.: A Comparison of Methods for Multi-class Support Vector Machines. IEEE Transactions on Neural Networks 13(2), 415–425 (2002)
[9] Khardon, R., Wachman, G.: Noise tolerant variants of the perceptron algorithm. Journal of Machine Learning Research 8, 227–248 (2007)
[10] Lewis, D.D., Yang, Y., Rose, T.G., Li, F.: RCV1: A New Benchmark Collection for Text Categorization Research. Journal of Machine Learning Research 5, 361–397 (2004)
[11] Loza Mencía, E., Fürnkranz, J.: An evaluation of efficient multilabel classification algorithms for large-scale problems in the legal domain. In: LWA 2007: Lernen -Wissen - Adaption, Workshop Proceedings, pp. 126–132 (2007)
[12] Loza Mencía, E., Fürnkranz, J.: Pairwise learning of multilabel classifications with perceptrons. In: Proceedings of the 2008 IEEE International Joint Conference on Neural Networks (IJCNN 2008), Hong Kong (2008)
[13] Loza Mencía, E., Fürnkranz, J.: Efficient multilabel classification algorithms for large-scale problems in the legal domain. In: Proceedings of the Language Resources and Evaluation Conference (LREC) Workshop on Semantic Processing of Legal Texts, Marrakech, Morocco (2008)
[14] Park, S.-H., Fürnkranz, J.: Efficient pairwise classification. In: Kok, J.N., Koronacki, J., Lopez de Mantaras, R., Matwin, S., Mladenič, D., Skowron, A. (eds.) ECML 2007. LNCS (LNAI), vol. 4701, pp. 658–665. Springer, Heidelberg (2007)
[15] Price, D., Knerr, S., Personnaz, L., Dreyfus, G.: Pairwise Neural Network Classifiers with Probabilistic Outputs. In: Advances in Neural Information Processing Systems, vol. 7, pp. 1109–1116. MIT Press, Cambridge (1995)
[16] Rosenblatt, F.: The perceptron: a probabilistic model for information storage and organization in the brain. Psychological Review 65(6), 386–408 (1958)
[17] Sebastiani, F.: Machine learning in automated text categorization. ACM Computing Surveys 34(1), 1–47 (2002)
[18] Shalev-Shwartz, S., Singer, Y.: A New Perspective on an Old Perceptron Algorithm. In: Auer, P., Meir, R. (eds.) COLT 2005. LNCS (LNAI), vol. 3559, pp. 264–278. Springer, Heidelberg (2005)

# Fitted Natural Actor-Critic: A New Algorithm for Continuous State-Action MDPs*

Francisco S. Melo[1] and Manuel Lopes[2]

[1] Carnegie Mellon University
Pittsburgh, USA
fmelo@isr.ist.utl.pt
[2] Institute for Systems and Robotics
Lisboa, Portugal
macl@isr.ist.utl.pt

**Abstract.** In this paper we address reinforcement learning problems with continuous state-action spaces. We propose a new algorithm, *fitted natural actor-critic* (FNAC), that extends the work in [1] to allow for general function approximation and data reuse. We combine the natural actor-critic architecture [1] with a variant of fitted value iteration using importance sampling. The method thus obtained combines the appealing features of both approaches while overcoming their main weaknesses: the use of a gradient-based actor readily overcomes the difficulties found in regression methods with policy optimization in continuous action-spaces; in turn, the use of a regression-based critic allows for efficient use of data and avoids convergence problems that TD-based critics often exhibit. We establish the convergence of our algorithm and illustrate its application in a simple continuous space, continuous action problem.

## 1 Introduction

In theory, reinforcement learning (RL) can be applied to address any optimal control task, yielding optimal solutions while requiring very little *a priori* information on the system itself. Existing RL methods are able to provide optimal solutions for many real-world control problems featuring discrete state and action spaces and exhibit widely studied convergence properties [2]. However, most such methods do not scale well in problems with large state and/or action spaces.

Many RL works addressing problems with infinite state-spaces combine function approximations with learning methods. Encouraging results were reported, perhaps the most spectacular of which by Tesauro and his learning Gammon player [3]. However, as seen in [4, 5], DP/TD-based methods exhibit unsound

* Work partially supported by the Information and Communications Technologies Institute, the Portuguese Fundação para a Ciência e a Tecnologia, under the Carnegie Mellon-Portugal Program, the Programa Operacional Sociedade de Conhecimento (POS_C) that includes FEDER funds and the project PTDC/EEA-ACR/70174/2006, and by the EU Project (IST-004370) RobotCub.

W. Daelemans et al. (Eds.): ECML PKDD 2008, Part II, LNAI 5212, pp. 66–81, 2008.

convergence behavior when combined with general function approximation. Convergence of methods such as $Q$-learning with general function approximation thus remains an important open issue [6].

If problems with infinite state-spaces pose important challenges when developing efficient RL algorithms, the simultaneous consideration of infinite action-spaces adds significant difficulties. Few RL methods to this day address problems featuring continuous state and action spaces. A fundamental issue in this class of problems is *policy optimization*: many RL methods rely on explicit maximization of a utility function to achieve policy optimization. If the number of available actions is *infinite*, this maximization is generally hard to achieve, especially if we consider that it is not *local* but *global* maximization. This significant difficulty affects many methods with otherwise sound performance guarantees, rendering such performance guarantees unusable [7].

When addressing problems with large/infinite state and/or action spaces, two major approaches have been considered in the RL literature. *Regression-based methods* use sample data collected from the system to estimate some target utility function using regression techniques. This class of methods is particularly suited to address problems with infinite state-spaces, although more general applications have been proposed in the literature [7]. Such algorithms can take advantage of the numerous regression methods available from the machine learning literature while exhibiting solid convergence properties [7, 8] and have been successfully applied in many different problems [9, 10, 11, 12].

Gradient-based methods, on the other hand, are naturally suited to address problems with infinite action-spaces. Such methods consider a parameterized policy and estimate the gradient of the performance with respect to the policy parameters. The parameters are then updated in the direction of this estimated gradient. By construction, gradient-based methods implement an incremental policy optimization and thus avoid the need for explicit maximization; it is no surprise that many RL works addressing problems with continuous action spaces thus rely on a gradient-based architecture [1, 13, 14].

However, gradient-based methods are line-search methods and, therefore, convergence is guaranteed only to local minima. Moreover, "pure" gradient methods usually exhibit large variance and, as argued in [15], make poor use of data. *Actor-critic architectures* [16] provide a suitable extension to pure gradient methods. They have been extensively analyzed in several works (*e.g.*, [1, 15, 17, 18]) and found to exhibit several advantages over pure gradient methods (in particular, in terms of variance and data-usage).

### 1.1 Contributions and Structure of the Paper

In this paper, we combine the appealing properties of actor-critic methods with those of regression-based methods in a new method dubbed as *fitted natural actor-critic* (FNAC). FNAC extends the natural actor-critic (NAC) architecture [1], allowing general function approximation and data reuse. In particular, we modify the TD-based critic in the NAC and implement a variant of fitted value iteration using importance sampling.

FNAC thus combines in a single algorithm the potentially faster convergence of *natural gradients* [19] and the sound convergence properties and efficient use of data of regression algorithms [7]. We also make use of an *importance sampling* strategy that allows the reuse of data, making our algorithm very efficient in terms of data usage and allowing the analysis of convergence of the algorithm.[1]

To this respect, it is also worth mentioning that, in many practical problems, collecting data is costly and time-consuming. In these situations, the efficient use of data in FNAC is a significant advantage over other existing approaches. Finally, it is also important to emphasize that the gradient-based policy updates readily overcome the most obvious difficulties of stand-alone regression-methods with respect to policy optimization. Summarizing, FNAC allows for *general function approximation* in the critic component, while *being able to use all sampled data* in all iterations of the algorithm (unlike most previous methods).

The paper is organized as follows. Section 2 reviews some background material on MDPs and policy gradient methods. Section 3 introduces the fitted natural actor-critic algorithm and its main properties. We evaluate its performance in the continuous mountain-car problem in Section 4 and conclude in Section 5 with some final remarks and directions for future work.

## 2 Background

In this section we review some background material that will be of use in the remainder of the paper. In particular, we briefly review the MDP framework [20], the policy gradient theorem and its application in approximate settings [21] and the use of natural gradients in actor-critic methods [1, 19].

### 2.1 Markov Decision Problems

Given two compact sets $\mathcal{X} \subset \mathbb{R}^p$ and $\mathcal{A} \subset \mathbb{R}^q$, let $\{X_t\}$ be an $\mathcal{X}$-valued controlled Markov chain, with control parameter taking values in $\mathcal{A}$. The transition probabilities for the chain are given by the kernel

$$\mathbb{P}\left[X_{t+1} \in U_X \mid X_t = x, A_t = a\right] = \mathsf{P}_a(x, U_X),$$

for any measurable set $U_X \subset \mathcal{X}$. The $\mathcal{A}$-valued process $\{A_t\}$ represents the control process: $A_t$ is the control action at time instant $t$. A decision-maker must determine the control process $\{A_t\}$ so as to maximize the functional

$$V(\{A_t\}, x) = \mathbb{E}\left[\sum_{t=0}^{\infty} \gamma^t R(X_t, A_t) \mid X_0 = x\right],$$

where $0 \leq \gamma < 1$ is a discount-factor and $R(x, a)$ represents a random "reward" received for taking action $a \in \mathcal{A}$ in state $x \in \mathcal{X}$. To play it safe, we assume

[1] As remarked in [18], such analysis is not immediate in the original NAC algorithm, since the data used in NAC is generated online from the current policy estimate in the algorithm.

throughout this paper that there is a deterministic continuous function $r$ defined on $\mathcal{X} \times \mathcal{A} \times \mathcal{X}$ assigning a reward $r(x, a, y)$ every time a transition from $x$ to $y$ occurs after taking action $a$ and that

$$\mathbb{E}\left[R(x, a)\right] = \int_{\mathcal{X}} r(x, a, y) \mathsf{P}_a(x, dy).$$

This simplifies the notation without introducing a great loss in generality. We further assume that there is a constant $\mathcal{R} \in \mathbb{R}$ such that $|r(x, a, y)| < \mathcal{R}$ for all $x, y \in \mathcal{X}$ and all $a \in \mathcal{A}$. We refer to the 5-tuple $(\mathcal{X}, \mathcal{A}, \mathsf{P}, r, \gamma)$ as a *Markov decision problem* (MDP).

Given an MDP $\mathcal{M} = (\mathcal{X}, \mathcal{A}, \mathsf{P}, r, \gamma)$, the *optimal value function* $V^*$ is defined for each state $x \in \mathcal{X}$ as

$$V^*(x) = \max_{\{A_t\}} \mathbb{E}\left[\sum_{k=0}^{\infty} \gamma^t R(X_t, A_t) \mid X_0 = x\right]$$

and verifies

$$V^*(x) = \max_{a \in \mathcal{A}} \int_{\mathcal{X}} \left[r(x, a, y) + \gamma V^*(y)\right] \mathsf{P}_a(x, dy), \tag{1}$$

which is a form of the Bellman optimality equation.[2] The optimal $Q$-values $Q^*(x, a)$ are defined for each state-action pair $(x, a) \in \mathcal{X} \times \mathcal{A}$ as

$$Q^*(x, a) = \int_{\mathcal{X}} \left[r(x, a, y) + \gamma V^*(y)\right] \mathsf{P}_a(x, dy).$$

From $Q^*$ we can define the mapping $\pi^*(x) = \arg\max_a Q^*(x, a)$ for all $x \in \mathcal{X}$. The control process defined by $A_t = \pi^*(X_t)$ is optimal in the sense that the corresponding value function equals $V^*$. The mapping $\pi^*$ thus defined is an *optimal policy* for the MDP $\mathcal{M}$.

More generally, a *policy* is a (time-dependent) mapping $\pi_t$ defined over $\mathcal{X} \times \mathcal{A}$ that generates a control process $\{A_t\}$ verifying

$$\mathbb{P}\left[A_t \in U_A \mid X_t = x\right] = \int_{U_A} \pi_t(x, a) da, \qquad \forall t,$$

where $U_A \subset \mathcal{A}$ is any measurable set. We write $V^{\pi_t}(x)$ instead of $V(\{A_t\}, x)$ if the control process $\{A_t\}$ is generated by a policy $\pi_t$. A *stationary policy* is a policy $\pi$ that does not depend on $t$.

## 2.2 The Policy Gradient Theorem

Let $\pi^\theta$ be a stationary policy parameterized by some finite-dimensional vector $\theta \in \mathbb{R}^M$. Assume, in particular, that $\pi$ is continuously differentiable with respect

[2] Notice that the maximum in (1) is well defined due to our assumption of compact $\mathcal{A}$ and continuous $r$.

to (w.r.t.) $\theta$. We henceforth write $V^\theta$ instead of $V^{\pi^\theta}$ to denote the corresponding value function. Also, given some probability measure $\mu_0$ over $\mathcal{X}$, we define

$$\rho(\theta) = (\mu_0 V^\theta) = \int_{\mathcal{X}} V^\theta(x)\mu_0(dx).$$

We abusively write $\rho(\theta)$ instead of $\rho(\pi^\theta)$ to simplify the notation. The function $\rho(\theta)$ can be seen as the total discounted reward that an agent expects to receive when following the policy $\pi^\theta$ and the initial state is distributed according to $\mu_0$.

We wish to compute the parameter vector $\theta^*$ such that the corresponding policy $\pi^{\theta^*}$ maximizes the expected income for the agent in the sense of $\rho$. In other words, we wish to compute $\theta^* = \arg\max_\theta \rho(\theta)$. If $\rho$ is differentiable w.r.t. $\theta$, this can be achieved by updating $\theta$ according to

$$\theta_{t+1} = \theta_t + \alpha_t \nabla_\theta \rho(\theta_t),$$

where $\{\alpha_t\}$ is a step-size sequence and $\nabla_\theta$ denotes the gradient w.r.t. $\theta$. We can now introduce the following result from [21, 22].

**Theorem 1.** *Given an MDP $\mathcal{M} = (\mathcal{X}, \mathcal{A}, \mathsf{P}, r, \gamma)$, it holds for every $x \in \mathcal{X}$ that*

$$\nabla_\theta V^\theta(x) = \int_{\mathcal{X}\times\mathcal{A}} \nabla_\theta \pi^\theta(y, a) Q^\theta(y, a) da\ \hat{\mathsf{K}}_\gamma^\theta(x, dy),$$

*where $\hat{\mathsf{K}}_\gamma^\theta$ is the un-normalized $\gamma$-resolvent associated with the Markov chain induced by $\pi^\theta$.*[3]

The fact that $\rho(\theta) = (\mu_0 V^\theta)$ immediately implies that

$$\nabla_\theta \rho(\theta) = \int_{\mathcal{X}} \nabla_\theta V^\theta(x)\mu_0(dx).$$

For simplicity of notation, we henceforth denote by $\mu_\gamma^\theta$ the measure over $\mathcal{X}$ defined by

$$\mu_\gamma^\theta(U_X) = \int_{\mathcal{X}} \left( \int_{U_X} \hat{\mathsf{K}}_\gamma^\theta(x, dy) \right) \mu_0(dx).$$

### 2.3 Policy Gradient with Function Approximation

From Theorem 1 it is evident that, in order to compute the gradient $\nabla\rho$, the function $Q^\theta$ needs to be computed. However, when addressing MDPs with infinite state and/or action spaces (as is the case in this paper), some form of function approximation is needed in order to compute $Q^\theta$.

[3] The $\gamma$-resolvent [23] associated with a Markov chain $(\mathcal{X}, \mathsf{P})$ is the transition kernel $\mathsf{K}_\gamma$ defined as $\mathsf{K}_\gamma(x, U) = (1-\gamma)\sum_{t=0}^{\infty} \gamma^t \mathsf{P}^t(x, U)$ and the un-normalized $\gamma$-resolvent is simply $\hat{\mathsf{K}}_\gamma(x, U) = \sum_{t=0}^{\infty} \gamma^t \mathsf{P}^t(x, U)$.

Let $\{\phi_i, i = 1, \ldots, M\}$ be a set of $M$ linearly independent functions and $\mathcal{L}(\phi)$ its linear span. Let $\hat{Q}^\theta$ be the best approximation of $Q^\theta$ in $\mathcal{L}(\phi)$, taken as the orthogonal projection of $Q^\theta$ on $\mathcal{L}(\phi)$ w.r.t. the inner product

$$\langle f, g \rangle = \int_{\mathcal{X}\times\mathcal{A}} f(x,a) \cdot g(x,a) \pi^\theta(x,a) da \; \mu_\gamma^\theta(dx).$$

As any function in $\mathcal{L}(\phi)$, $\hat{Q}^\theta$ can be written as

$$\hat{Q}^\theta(x,a) = \sum_i \phi_i(x,a) w_i = \phi^\top(x,a) w.$$

This leads to the following result from [21].

**Theorem 2.** *Given an MDP $\mathcal{M} = (\mathcal{X}, \mathcal{A}, \mathsf{P}, r, \gamma)$ and a set of basis functions $\{\phi_i, i = 1, \ldots, M\}$ as defined above, if*

$$\nabla_w \hat{Q}^\theta(x,a) = \nabla_\theta \log(\pi^\theta(x,a)) \tag{2}$$

*then*

$$\nabla_\theta \rho(\theta) = \int_{\mathcal{X}\times\mathcal{A}} \nabla_\theta \pi^\theta(x,a) \hat{Q}^\theta(x,a) da \; \mu_\gamma^\theta(dx).$$

Notice that, in the gradient expression in Theorems 1 and 2, we can add an arbitrary function $b(x)$ to $Q^\theta$ and $\hat{Q}^\theta$. This is clear from noting that

$$\int_{\mathcal{A}} \nabla_\theta \pi^\theta(x,a) b(x) da = 0.$$

Such functions are known as *baseline functions* and, as recently shown in [18], if $b$ is chosen so as to minimize the mean-squared error between $\hat{Q}^\theta$ and $Q^\theta$, the optimal choice of baseline function is $b(x) = V^\theta(x)$. Recalling that the *advantage function* [24] associated with a policy $\pi$ is defined as $A^\pi(x,a) = Q^\pi(x,a) - V^\pi(x)$, we get

$$\nabla_\theta \rho(\theta) = \int_{\mathcal{X}\times\mathcal{A}} \nabla_\theta \pi^\theta(x,a) \hat{A}^\theta(x,a) da \; \mu_\gamma^\theta(dx), \tag{3}$$

where $\hat{A}^\theta(x,a)$ denotes the orthogonal projection of the advantage function associated with $\pi^\theta$, $A^\theta$, into $\mathcal{L}(\phi)$. Finally, recalling that $\hat{A}^\theta(x,a) = \phi^\top(x,a) w$, we can compactly write (3) as $\nabla_\theta \rho(\theta) = \mathbf{G}(\theta) w$, with $\mathbf{G}$ the *all-action matrix* [1].

$$\mathbf{G}(\theta) = \int_{\mathcal{X}\times\mathcal{A}} \nabla_\theta \pi^\theta(x,a) \phi^\top(x,a) da \; \mu_\gamma^\theta(dx). \tag{4}$$

## 2.4 Natural Gradient

Given a general manifold $M$ parameterized by a finite-dimensional vector $\theta \in \mathbb{R}^M$ and a real function $F$ defined over this manifold, the gradient $\nabla_\theta F$ seldom corresponds to the actual steepest descent direction, as it fails to take into account the geometry of the manifold [25]. However, in many practical situations, it is possible to impose a particular structure on the manifold (namely, a Riemannian metric) and compute the steepest descent direction taking into account the geometry of the manifold (in terms of the Riemannian metric). This "natural gradient" is invariant to changes in parameterization of the manifold and can potentially overcome the so-called plateau phenomenon [25].

As seen in [19], the parameterized policy space can be seen as a manifold that can be endowed with an adequate Riemannian metric. One possible metric relies on the *Fisher information matrix*, and is defined by the following matrix [19]

$$\mathbf{F}(\theta) = \int_{\mathcal{X}\times\mathcal{A}} \nabla_\theta \log(\pi^\theta(x,a)) \nabla_\theta \log(\pi^\theta(x,a))^\top \pi^\theta(x,a) da\ \mu_\gamma^\theta(dx).$$

The natural gradient is given, in this case, by $\tilde{\nabla}_\theta \rho(\theta) = \mathbf{F}^{-1}(\theta)\mathbf{G}(\theta)w$. However, as shown in [1], multiplying and dividing the integrand in (4) by $\pi^\theta(x,a)$, we get $\mathbf{G}(\theta) = \mathbf{F}(\theta)$, and the natural gradient comes, simply, $\tilde{\nabla}_\theta \rho(\theta) = w$.[4]

# 3 Fitted Natural Actor-Critic

We now use the ideas from the previous section to derive a new algorithm, *fitted natural actor-critic* (FNAC). This algorithm takes advantage of several appealing properties of fitting methods (namely, the solid convergence guarantees and the effective use of data) while overcoming some of the limitations of this class of methods in problems with continuous action spaces.

## 3.1 The FNAC Architecture

We start by briefly going through the FNAC architecture, illustrated in Figure 1.

The algorithm uses a set $\mathcal{D}$ of samples obtained from the environment, each consisting of a tuple $(x_t, a_t, r_t, y_t)$, where $y_t$ is a sample state distributed according to the measure $\mathsf{P}_{a_t}(x_t, \cdot)$ and $r_t = r(x_t, a_t, y_t)$. For the purposes of the algorithm, it is not important how the samples in $D$ are collected. In particular, they can all be collected before the algorithm is run or they can be collected incrementally, as more iterations of the algorithm are performed. Nevertheless, it is important to remark that, for the purposes of our critic, enough data-samples need to be collected to avoid conditioning problems in the regression algorithms.

At each iteration of the FNAC algorithm, the data in $\mathcal{D}$ is processed by the *critic* component of the algorithm. This component, as detailed below, uses a generic regression algorithm to compute an approximation $\hat{V}^\theta$ of the value function associated with the current policy, $\pi^\theta$. This approximation is then used to estimate an approximation of the advantage function, $\hat{A}^\theta$, using a linear function approximation with compatible basis functions. Finally, the evaluation

[4] Peters et al. [1] actually showed that $\mathbf{F}(\theta)$ is the Fisher information matrix for the probability distribution over possible trajectories associated with a given policy.

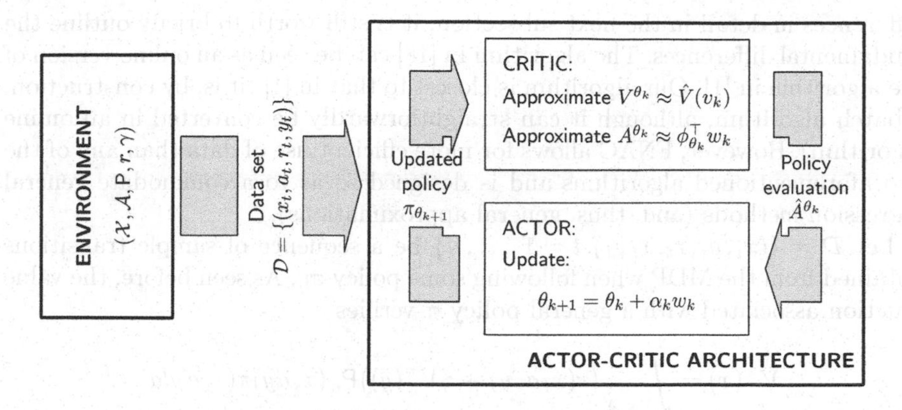


**Fig. 1.** Detailed FNAC architecture

performed by the critic (*i.e.*, the approximation $\hat{A}^\theta$ computed from the data) is used by the actor to update the policy $\pi^\theta$ using a standard policy gradient update.

In the remaining of this section, we detail our FNAC algorithm.

### 3.2 The Actor

The actor component of the FNAC algorithm implements a policy gradient update. As seen in Section 2, this update relies on the *natural gradient.* The fact that the natural gradient manages to take into account the geometry of the policy space may potentially bring significant advantages in terms of performance of the algorithm (namely, in terms of rate of convergence and ability to overcome the plateau phenomenon) [25]. Therefore, given the parameterized policy at iteration $k$, $\pi_{\theta_k}$, the actor component will update the parameter vector as

$$\theta_{k+1} = \theta_k + \alpha_k \tilde{\nabla}_\theta \rho(\theta_k).$$

As seen in Subsection 2.4, the natural gradient is given by $\tilde{\nabla}\rho(\theta_k) = w_k$, where $w_k$ is the linear coefficient vector corresponding to the orthogonal projection of the advantage function $A^{\theta_k}$ in the linear space spanned by the *compatible basis functions*, obtained from 2:

$$\phi_i(x, a) = \frac{\partial \log(\pi_{\theta_k})}{\partial \theta_k(i)}(x, a).$$

Therefore, provided that the critic component yields such an approximation of the advantage function, the update rule for the actor is, simply, $\theta_{k+1} = \theta_k + \alpha_t w_k$.

### 3.3 The Critic

The critic of the FNAC algorithm is the element that distinguishes our algorithm from other gradient-based approaches, such as [1, 18]. Although we discuss these differences in detail in the next subsection, it is still worth to briefly outline the fundamental differences. The algorithm in [18] can be seen as an online version of the algorithm in [1]. Our algorithm is closest to that in [1] (it is, by construction, a batch algorithm, although it can straightforwardly be converted in an online algorithm). However, FNAC allows for more efficient use of data than any of the two aforementioned algorithms and is designed so as to accommodate general regression methods (and, thus, general approximations).

Let $\mathcal{D} = \{(x_t, a_t, r_t, x_{t+1}), t = 1, \ldots, n\}$ be a sequence of sample transitions obtained from the MDP when following some policy $\pi_0$. As seen before, the value function associated with a general policy $\pi$ verifies

$$V^\pi(x) = \int_{\mathcal{X}\times\mathcal{A}} \big(r(x, a, y) + \gamma V^\pi(y)\big)\mathsf{P}_a(x, dy)\pi(x, a)da$$

or, equivalently,

$$\int_{\mathcal{X}\times\mathcal{A}} \big(r(x, a, y) + \gamma V^\pi(y)\big)\mathsf{P}_a(x, dy)\pi(x, a)da - V^\pi(x) = 0.$$

This can be written as

$$\int_{\mathcal{X}\times\mathcal{A}} \big(r(x, a, y) + \gamma V^\pi(y) - V^\pi(x)\big)\mathsf{P}_a(x, dy)\pi(x, a)da = 0.$$

We want to approximate $V^\pi$ by a general parameterized family of functions $\mathcal{V} = \{V_v(x) \mid v \in \mathbb{R}^N\}$. In other words, we want to compute $v^*$ such that $V^\pi(x) \approx V_{v^*}(x)$. For the policy $\pi_0$ used to generate the dataset $\mathcal{D}$, we can use the data in $\mathcal{D}$ and solve the following regression problem:

$$v^* = \arg\min_v \sum_t \frac{1}{\hat{\mu}(x_t)}\big(r_t + \gamma V_v(y_t) - V_v(x_t)\big)^2, \tag{5}$$

where $\hat{\mu}(x_t)$ is the empirical distribution of state $x$ obtained from the dataset $\mathcal{D}$.[5] We remark that the function $V_{v^*}$ thus obtained is the one minimizing the *empirical Bellman residual.* However, in order to adequately perform the above minimization, double sampling is necessary, as pointed out in [4]. In systems where double sampling is not possible, a correction term can be included in the regression to avoid negative correlation effects [26], rendering the regression problem equivalent to the solution of the following fitted-VI iteration:

$$V_{k+1} = \min_V \sum_t \frac{1}{\hat{\mu}(x_t)}\big(r_t + \gamma V_k(y_t) - V(x_t)\big)^2.$$

[5] The inclusion of the term $\hat{\mu}(x_t)$ merely ensures that regions of the state-space that happen to appear more often *in the dataset* do not have "extra weight" in the regression. In fact, since the data in $\mathcal{D}$ can be obtained by any arbitrary sampling process, its distribution will generally be distinct from that induced by the obtained policy. The "normalization" w.r.t. $\hat{\mu}(x_t)$ minimizes any bias that the sampling process may introduce in the regression. Alternative regularizations are possible, however.

Nevertheless, the computation of $V^\pi$ as described above is possible because the data is *distributed according to the policy* $\pi_0$.

Suppose now that we want to approximate the value function associated with some other policy $\pi^\theta \neq \pi_0$. The value function for this policy verifies

$$V^\theta(x) = \int_{\mathcal{X}\times\mathcal{A}} \big(r(x,a,y) + \gamma V^\theta(y)\big)\mathsf{P}_a(x,dy)\pi^\theta(x,a)da.$$

If we want to use the same data to compute $V^\theta$, some modification is needed since the data in $\mathcal{D}$ is not distributed according to $\pi^\theta$. If the policy $\pi^\theta$ is known, the above expression can be modified to yield

$$\int_{\mathcal{X}\times\mathcal{A}} \big(r(x,a,y) + \gamma V^\pi(y) - V^\pi(x)\big)\mathsf{P}_a(x,dy)\cdot \cdot \frac{\pi^\theta(x,a)}{\pi_0(x,a)}\pi_0(x,a)da = 0. \tag{6}$$

This means that we should be able to reuse the data in $\mathcal{D}$ to solve the following regression problem, similar to that in (5):

$$v^* = \arg\min_v \sum_t \frac{1}{\hat{\mu}(x_t)} \frac{\pi^\theta(x_t,a_t)}{\pi_0(x_t,a_t)} \big(r_t + \gamma V_v(y_t) - V_v(x_t)\big)^2.$$

Notice that the above regression makes use of *importance sampling*, by including the term $\frac{\pi^\theta(x,a)}{\pi_0(x,a)}$. Notice also that this importance-sampling term is well-defined for all samples $(x_t, a_t, r_t, y_t)$, since $\pi_0(x_t,a_t) > 0$ necessarily holds. Notice also that, as before, the term $\hat{\mu}(x_t)$ is meant to minimize any bias that the *sampling process* may introduce in the regression, and no change is necessary with the particular policy $\pi^\theta$ considered.

Given an estimate $V_{v^*}$ of the value function associated with a given policy $\pi^\theta$, the corresponding advantage function can now be approximated by solving the following regression problem:

$$w^* = \arg\min_w \sum_t \frac{1}{\hat{\mu}(x_t)} \big(r_t + \gamma V_{v^*}(y_t) - V_{v^*}(x_t) - \phi^\top(x_t,a_t)w\big)^2,$$

where each $\phi_i(x,a)$ is a compatible basis function verifying (2). We remark that no importance sampling is necessary in the above estimation, as can easily be seen by repeating the above computations for the advantage function. The regression problem can now easily be solved by setting

$$\mathbf{M} = \sum_t \frac{1}{\hat{\mu}(x_t)} \phi(x_t,a_t)\phi^\top(x_t,a_t)$$

and

$$\mathbf{b} = \sum_t \frac{\phi(x_t,a_t)}{\hat{\mu}(x_t)} \big(r_t + \gamma V_{v^*}(x_{t+1}) - V_{v^*}(x_{t+1})\big),$$

from where we obtain $w^* = \mathbf{M}^{-1}\mathbf{b}$. We conclude by observing that, with enough samples, the inverse in the expression above is well defined, since we assume the functions $\phi_i$ to be linearly independent.

### 3.4 Analysis and Discussion of the Algorithm

We now discuss several important properties of FNAC and compare it with other related algorithms in the literature.

We start by remarking that the general regression-based critic and the importance sampling "regularization" imply that *the dataset can be made independent of the current learning policy.* The main consequence of this is that, by requiring minimum regularity conditions from the regression algorithm, the following result can easily be established:

**Theorem 3.** *Let $\mathfrak{F}(\theta)$ denote the regression algorithm used in FNAC to produce the estimates $\hat{A}^\theta$ and $\hat{V}^\theta$ for an MDP $\mathcal{M}$, given a $\theta$-dependent policy $\pi^\theta$. Then, if $\mathfrak{F}$ is Lipschitz w.r.t. $\theta$, and the step-size sequence $\{\alpha_t\}$ used in the actor update is such that*

$$\sum_t \alpha_t = \infty \qquad \sum_t \alpha_t^2 < \infty$$

*FNAC converges with probability 1.*

*Proof.* Due to space limitations, we provide only a brief sketch of the proof.

The result arises as a consequence of the convergence results in [18]. In the referred paper, convergence is established by analyzing a two-time-scale update of the general form:

$$\begin{aligned} w_{k+1} &= w_k + \alpha_k F(w_k, \theta_k) \\ \theta_{k+1} &= \theta_k + \beta_k G(w_{k+1}, \theta_k), \end{aligned} \tag{7}$$

where $\alpha = \mathrm{o}(\beta_k)$. By requiring several mild regularity conditions on the underlying process, on the policy space and on the set of basis functions, convergence can be established by an ODE argument. In particular, the proof starts by establishing global asymptotic stability of the "faster" ODE

$$\dot{w}_t = f(w_t, \theta),$$

for every $\theta$, where $f$ is an "averaged" version of $F$. Then, denoting by $w^*(\theta)$ the corresponding limit point, the proof proceeds by establishing the global asymptotic stability of the "slower" ODE

$$\dot{\theta}_t = g(w^*(\theta_t), \theta_t),$$

where, once again, $g$ is an "averaged" version of $G$. Convergence of the coupled iteration (7) is established as long as $w^*(\theta)$ is Lipschitz continuous w.r.t. $\theta$.

In terms of our analysis, and since our critic is a regression algorithm, it holds that at every iteration of the actor the critic *is actually in the corresponding limit $w^*(\theta)$*. Therefore, since our data is policy-independent and we assume our regression algorithm to be Lipschitz on $\theta$, the convergence of our method is a consequence of the corresponding result in [18]. □

Another important difference that distinguished FNAC from other works is the efficient data use in regression methods. To emphasize this point, notice that after each policy update by the actor, the critic must evaluate the updated policy, computing the associated advantage function. In previous actor-critic algorithms [1, 15, 18], this required the acquisition of *new sampled data obtained using the updated policy.* However, in many problems, the acquisition of new data is a costly and time-consuming process. Furthermore, the evaluation of the updated policy makes poor use of previously used data.[6] In our algorithm, *all data collected* can be used in every iteration. Evidently, the importance-sampling term in (6) will weight some samples more than others in the estimation of each $V^\theta$, but this is conducted so as to take full advantage of available data.

Finally, we remark that, from Theorem 2 and subsequent developments, the (natural) gradient is computed from the orthogonal projection of $Q^\theta/A^\theta$ into $\mathcal{L}(\phi)$. However, as all other actor-critic methods currently available [1, 15, 17, 18], our critic cannot compute this projection exactly, since it would require the knowledge of the function to be projected. This will impact the performance of the algorithm in a similar way to that stated in Theorem 1 of [18].

## 4 Experimental Results

We conducted several experiments to evaluate the behavior of our algorithm in a simple continuous state, continuous action problem. We applied FNAC to the well-known mountain-car problem [27]. In this problem, an underpowered car must go up a steep hill, as depicted in Figure 2. As it has not enough acceleration

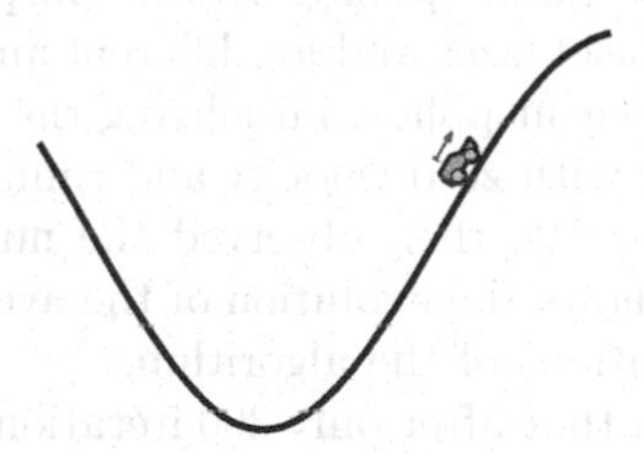

**Fig. 2.** The mountain-car problem: an underpowered car must go up a steep hill

to go all the way up the hill, it must bounce back-and-forth to gain enough velocity to climb up. The car is described by two continuous state-variables, namely position $p$ and velocity $v$, and is controlled by a single continuous action, the acceleration $a$. The range of allowed positions is $[-1.2; +0.5]$ and the velocity ranges between $-0.07$ and $0.07$. The acceleration takes values in the interval $[-1, 1]$. The car can be described by the dynamic equations:

$$v(t+1) = \mathsf{bound}\left[v(t) + 0.001a(t) - 0.0025\cos(3p(t))\right]$$

$$p(t+1) = \mathsf{bound}\left[p(t) + v(t+1)\right]$$

[6] In incremental algorithms [15, 18], the step-size sequence works as a "forgetting" factor.

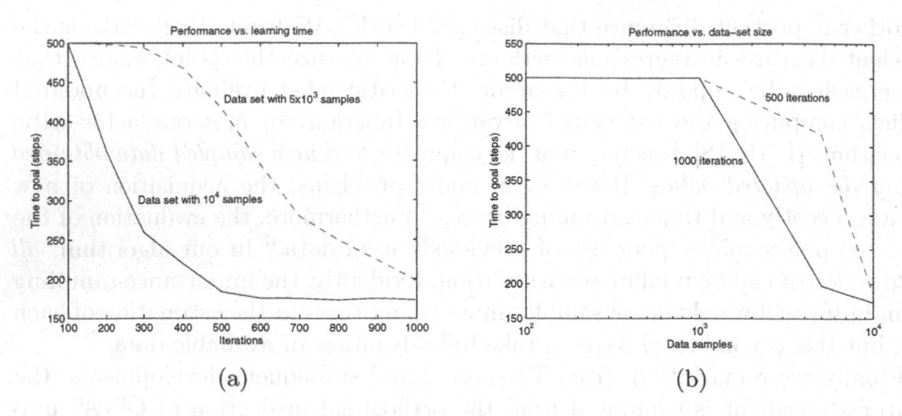


**Fig. 3.** Average performance of FNAC in the mountain-car problem: (a) Average time to reach the goal as the number of iterations is increased. (b) Average time to reach the goal as the size of the dataset is increased. The results depicted correspond to the average over 100 independent Monte-Carlo trials. Average times of 500 steps indicate that the car was not able to reach the goal.

where the function bound maintains the values of $p$ and $v$ within the limits. Whenever the car reaches the position limits, its velocity is set to zero. Whenever the car reaches the top of the hill, its position and velocity are randomly reset and the car gets a reward of 10. Otherwise, it gets a reward of $-1$.

We are interested in analyzing how much data and time are required for the FNAC algorithm to learn a "good" policy. To this purpose, we ran the FNAC algorithm with different dataset sizes and for different amounts of time. For each such run, we evaluated the learnt policy, initializing the car in the bottom-most position of the environment with zero velocity and running the learnt policy up to a maximum of 500 steps. We then observed the number of steps taken to reach the goal. Figure 3.a shows the evolution of the average time to goal as we increase the number of iterations of the algorithm.

It is clear from Figure 3.a that after only 300 iterations the algorithm already learnt a good policy (one that is able to reach the goal) and, for the case of a dataset of $10^4$ points, the policy after 600 iterations is practically as good as the best policy computed. It is important to refer at this point that we used a simple linear approximator with 16 RBFs spread uniformly across the state space. The policy was parameterized using a Boltzmann-like distribution relying on the linear combination of 64 RBF uniformly distributed across the state-action space. Notice also that, by using a linear approximator, the regression in (6) can be computed analytically.

We also present in Figure 3.b the evolution of the average time to goal as we increase the size of the dataset. We tested the performance of the algorithm with datasets containing 100, 500, $10^3$, $5 \times 10^3$ and $10^4$ samples.[7] As clearly

[7] We notice that, due to the deterministic transitions, no double sampling is necessary in this particular example.

seen from the results in Figure 3.b, the size of the dataset greatly influences the performance of the algorithm. In the particular problem considered here, the FNAC algorithm was able to find a "good" policy with $5 \times 10^3$ points and the best performance was attained with a dataset of $10^4$ samples.

## 5 Concluding Remarks

In this paper we presented the fitted natural actor-critic algorithm (FNAC). Our algorithm uses natural policy-gradient updates in the actor. However, unlike other natural actor-critic (NAC) algorithms, the critic component of FNAC relies on regression methods, with several important advantages:

- The use of regression methods allows the estimates of the *value-function* to use *general function approximation*, unlike previous NAC algorithms which, due to their use of TD-based critics, are limited to linear approximators. Notice, however, that Theorem 2 requires $A^\pi$ to be linearly approximated, using a compatible set of basis functions verifying (2). The use of general function approximation for $V^\pi$ will impact (in a positive way) the accuracy with which $A^\pi$ can be represented.
- By adding an importance-sampling component, we allow our critic to *reuse all data* in all iterations of the algorithm, this being a fundamental advantage in problems where collecting data is costly or time consuming;
- The reuse of data allows the algorithm to consider datasets which are policy-independent. This, means that, unlike other NAC methods, the convergence of the algorithm can be conducted by a simple ODE argument;

It is also worth mentioning that FNAC is amenable to a multitude of different implementations, fully exploring the power of general fitting/regression methods.

The work portrayed here also suggests several interesting avenues for future research. First of all, and although not discussed in the paper, an online implementation of our algorithm can easily be obtained by an iterative implementation of the regression routines. Our convergence result holds if the relative weight of each sample in the data-set is stored (in terms of sampling policy).

Also, our initial experimental results illustrate the efficient use of data of our algorithm, since FNAC could attain good performance, reusing the same dataset at every iteration. Currently we are exploring the performance of the algorithm in more demanding tasks (namely, robotic grasping tasks encompassing high-dimensional state and action spaces). It would also be interesting to have some quantitative evaluation of the advantages of FNAC in face of other methods for MDPs with continuous state-action spaces. However, a direct comparison is not possible: in the current implementation of FNAC, the data gathering process is completely decoupled from the actual algorithm, while in most methods both processes occur simultaneously, thus impacting the corresponding learning times.

On a more general perspective, the critic component in FNAC estimates at each iteration the value function $V^\theta$ by minimizing the empirical Bellman residual. Approximations relying on Bellman residual minimization are more stable and more "predictable" than TD-based approaches [28]. It would be interesting to further explore these results to gain a deeper understanding of the advantages of this type of approximation in the setting considered here.

Finally, the simple and elegant results arising from the consideration of natural gradients suggests that it may be possible to further extend this approach and make use of higher-order derivatives (*e.g.*, as in Newton-like methods) to develop policy search methods for RL problems.

## Acknowledgements

The authors would like to acknowledge the useful comments from Jan Peters and the anonymous reviewers.

## References

1. Peters, J., Vijayakumar, S., Schaal, S.: Natural Actor-Critic. In: Proc. European Conf. Machine Learning, pp. 280–291 (2005)
2. Bertsekas, D., Tsitsiklis, J.: Neuro-Dynamic Programming. Athena Scientific (1996)
3. Tesauro, G.: TD-Gammon, a self-teaching backgammon program, achieves master-level play. Neural Computation 6(2), 215–219 (1994)
4. Baird, L.: Residual algorithms: Reinforcement learning with function approximation. In: Proc. Int. Conf. Machine Learning, pp. 30–37 (1995)
5. Tsitsiklis, J., Van Roy, B.: An analysis of temporal-difference learning with function approximation. IEEE Trans. Automatic Control 42(5), 674–690 (1996)
6. Sutton, R.: Open theoretical questions in reinforcement learning. In: Proc. European Conf. Computational Learning Theory, pp. 11–17 (1999)
7. Antos, A., Munos, R., Szepesvári, C.: Fitted $Q$-iteration in continuous action-space MDPs. In: Adv. Neural Information Proc. Systems, vol. 20 (2007)
8. Munos, R., Szepesvári, C.: Finite-time bounds for sampling-based fitted value iteration. J. Machine Learning Research (submitted, 2007)
9. Gordon, G.: Stable fitted reinforcement learning. In: Adv. Neural Information Proc. Systems, vol. 8, pp. 1052–1058 (1996)
10. Ormoneit, D., Sen, S.: Kernel-based reinforcement learning. Machine Learning 49, 161–178 (2002)
11. Ernst, D., Geurts, P., Wehenkel, L.: Tree-based batch mode reinforcement learning. J. Machine Learning Research 6, 503–556 (2005)
12. Riedmiller, M.: Neural fitted $Q$-iteration: First experiences with a data efficient neural reinforcement learning method. In: Gama, J., Camacho, R., Brazdil, P.B., Jorge, A.M., Torgo, L. (eds.) ECML 2005. LNCS (LNAI), vol. 3720, pp. 317–328. Springer, Heidelberg (2005)
13. Kimura, H., Kobayashi, S.: Reinforcement learning for continuous action using stochastic gradient ascent. In: Proc. Int. Conf. Intelligent Autonomous Systems, pp. 288–295 (1998)
14. Lazaric, A., Restelli, M., Bonarini, A.: Reinforcement learning in continuous action spaces through sequential Monte Carlo methods. In: Adv. Neural Information Proc. Systems, vol. 20 (2007)
15. Konda, V., Tsitsiklis, J.: On actor-critic algorithms. SIAM J. Control and Optimization 42(4), 1143–1166 (2003)

16. Barto, A., Sutton, R., Anderson, C.: Neuronlike adaptive elements that can solve difficult learning control problems. IEEE Trans. Systems, Man and Cybernetics 13(5), 834–846 (1983)
17. van Hasselt, H., Wiering, M.: Reinforcement learning in continuous action spaces. In: Proc. 2007 IEEE Symp. Approx. Dynamic Programming and Reinforcement Learning, pp. 272–279 (2007)
18. Bhatnagar, S., Sutton, R., Ghavamzadeh, M., Lee, M.: Incremental natural actor-critic algorithms. In: Adv. Neural Information Proc. Systems, vol. 20 (2007)
19. Kakade, S.: A natural policy gradient. In: Adv. Neural Information Proc. Systems, vol. 14, pp. 1531–1538 (2001)
20. Puterman, M.: Markov Decision Processes: Discrete Stochastic Dynamic Programming. John Wiley & Sons, Inc., Chichester (1994)
21. Sutton, R., McAllester, D., Singh, S., Mansour, Y.: Policy gradient methods for reinforcement learning with function approximation. In: Adv. Neural Information Proc. Systems, vol. 12, pp. 1057–1063 (2000)
22. Marbach, P., Tsitsiklis, J.: Simulation-based optimization of Markov reward processes. IEEE Trans. Automatic Control 46(2), 191–209 (2001)
23. Meyn, S., Tweedie, R.: Markov Chains and Stochastic Stability. Springer, Heidelberg (1993)
24. Baird, L.: Advantage updating. Tech. Rep. WL-TR-93-1146, Wright Laboratory, Wright-Patterson Air Force Base (1993)
25. Amari, S.: Natural gradient works efficiently in learning. Neural Computation 10(2), 251–276 (1998)
26. Antos, A., Szepesvári, C., Munos, R.: Learning near-optimal policies with Bellman-residual minimization based fitted policy iteration and a single sample path. Machine Learning 71, 89–129 (2008)
27. Singh, S., Sutton, R.: Reinforcement learning with replacing eligibility traces. Machine Learning 22, 123–158 (1996)
28. Munos, R.: Error bounds for approximate policy iteration. In: Proc. Int. Conf. Machine Learning, pp. 560–567 (2003)

# A New Natural Policy Gradient by Stationary Distribution Metric

Tetsuro Morimura[1,2], Eiji Uchibe[1], Junichiro Yoshimoto[1,3], and Kenji Doya[1,3,4]

[1] Initial Research Project, Okinawa Institute of Science and Technology
[2] IBM Research, Tokyo Research Laboratory
[3] Graduate School of Information Science, Nara Institute of Science and Technology
[4] ATR Computational Neuroscience Laboratories
{morimura,uchibe,jun-y,doya}@oist.jp

**Abstract.** The parameter space of a statistical learning machine has a Riemannian metric structure in terms of its objective function. Amari [1] proposed the concept of "natural gradient" that takes the Riemannian metric of the parameter space into account. Kakade [2] applied it to policy gradient reinforcement learning, called a natural policy gradient (NPG). Although NPGs evidently depend on the underlying Riemannian metrics, careful attention was not paid to the alternative choice of the metric in previous studies. In this paper, we propose a Riemannian metric for the joint distribution of the state-action, which is directly linked with the average reward, and derive a new NPG named *"Natural State-action Gradient"* (NSG). Then, we prove that NSG can be computed by fitting a certain linear model into the immediate reward function. In numerical experiments, we verify that the NSG learning can handle MDPs with a large number of states, for which the performances of the existing (N)PG methods degrade.

**Keywords:** policy gradient reinforcement learning, natural gradient, Riemannian metric matrix, Markov decision process.

## 1 Introduction

Policy gradient reinforcement learning (PGRL) attempts to find a policy that maximizes the average (or time-discounted) reward, based on the gradient ascent in the policy parameter space [3,4,5]. As long as the policy is represented by a parametric statistical model that satisfies some mild conditions, PGRL can be instantly implemented in the Markov decision process (MDP). Moreover, since it is possible to treat the parameter controlling the randomness of the policy, PGRLs, rather than value-based RLs, can obtain the appropriate stochastic policy and be applied to the partially observable MDP (POMDP). Meanwhile, depending on the tasks, PGRL methods often take a huge number of learning steps. In this paper, we propose a new PGRL method that can improve the slow learning speed by focusing on the metric of the parameter space of the learning model.

W. Daelemans et al. (Eds.): ECML PKDD 2008, Part II, LNAI 5212, pp. 82–97, 2008.

It is easy to imagine that large-scale tasks suffer from a slow learning speed because the dimensionality of the policy parameters increases in conjunction with the task complexity. Besides the problem of dimensionality, the geometric structure of the parameter space also gives rise to slow learning. Ordinary PGRL methods omit the sensitivity of each element of the policy parameter and the correlation between the elements, in terms of the probability distributions of the MDP. However, most probability distributions expressed by the MDP have some manifold structures instead of Euclidean structures. Therefore, the updating direction of the policy parameter by the ordinary gradient method is different from the steepest direction on the manifold; thus, the optimization process occasionally falls into a stagnant state, commonly called a *plateau*. This is mainly due to the regions in which the geometric structure for the objective function with respect to the parameter coordinate system becomes fairly flat and its derivative becomes almost zero [6]. It was reported that a plateau was observed in a very simple MDP with only two states [2]. In order to solve such problem, Amari [1] proposed a "natural gradient" for the steepest gradient method in Riemannian space. Because the direction of the natural gradient is defined on a Riemannian metric, it is an important issue how to design the Riemannian metric. Nevertheless, the metric proposed by Kakade [2] has so far been the only metric in the application of the natural gradient for RL [7,8,9], commonly called *natural policy gradient* (NPG) reinforcement learning.

In this paper, we propose the use of the Fisher information matrix of the state-action joint distribution as the Riemannian metric for RL and derive a new robust NPG learning, *"natural state-action gradient"* (NSG) learning. It is shown that this metric considers the changes in the stationary state-action joint distribution, specifying the average reward as the objective function. In contrast, Kakade's metric takes into account only changes in the action distribution and omits changes in the state distribution, which also depends on the policy in general. A comparison with the Hessian matrix is also given in order to confirm the adequacy of the proposed metric. We also prove that the gradient direction as computed by NSG is equal to the adjustable parameter of the linear regression model with the basis function defined on the policy when it minimizes the mean square error for the rewards. Finally, we demonstrate that the proposed NSG learning improves the performance of conventional (N)PG-based learnings by means of numerical experiments with varying scales of MDP tasks.

## 2 Conventional Natural Policy Gradient Method

We briefly review PGRL in section 2.1 and the natural gradient [1] and the NPG in section 2.2. In section 2.3, we introduce the controversy of NPGs.

### 2.1 Policy Gradient Reinforcement Learning

PGRL is modeled on a discrete-time Markov decision process (MDP) [10,11]. It is defined by the quintuplet $(\mathcal{S}, \mathcal{A}, p, r, \pi_\theta)$, where $\mathcal{S} \ni s$ and $\mathcal{A} \ni a$ are finite sets of states and actions, respectively. Further, $p : \mathcal{S}\times\mathcal{A}\times\mathcal{S} \rightarrow [0,1]$ is a state transition probability function of a state $s_t$, an action $a_t$, and the following state $s_{t+1}$ at a time step $t$, i.e., $p(s_{t+1}|s_t, a_t) \equiv \Pr(s_{t+1}|s_t, a_t)$ [1]. $r : \mathcal{S}\times\mathcal{A}\times\mathcal{S} \rightarrow \mathcal{R}$ is a reward function of $s_t$, $a_t$, and $s_{t+1}$, and is bounded, which defines an immediate reward $r_{t+1} = r(s_t, a_t, s_{t+1})$ observed by a learning agent. $\pi : \mathcal{A}\times\mathcal{S}\times\mathcal{R}^d \rightarrow [0,1]$ is an action probability function of $a_t$, $s_t$, and a policy parameter $\boldsymbol{\theta} \in \mathcal{R}^d$, and is always differentiable with respect to $\boldsymbol{\theta}$ known as a policy, i.e., $\pi(a_t|s_t;\boldsymbol{\theta}) \equiv \Pr(a_t|s_t,\boldsymbol{\theta})$. It defines the decision-making rule of the learning agent and is adjustable by tuning $\boldsymbol{\theta}$. We make an assumption that the Markov chain $\mathrm{M}(\boldsymbol{\theta}) = \{\mathcal{S}, \mathcal{A}, p, \pi_{\boldsymbol{\theta}}\}$ is ergodic for all $\boldsymbol{\theta}$. Then, there exists a unique stationary state distribution $d_{\boldsymbol{\theta}}(s) \equiv \Pr(s|\mathrm{M}(\boldsymbol{\theta}))$, which is equal to the limiting distribution and independent of the initial state, $d_{\boldsymbol{\theta}}(s') = \lim_{t\rightarrow\infty} \Pr(S_t = s'|S_0 = s, \mathrm{M}(\boldsymbol{\theta}))$, $^{\forall} s \in \mathcal{S}$. This distribution satisfies the balance equation:

$$d_{\boldsymbol{\theta}}(s') = \sum_{s\in\mathcal{S}}\sum_{a\in\mathcal{A}} p(s'|s,a)\pi(a|s;\boldsymbol{\theta})d_{\boldsymbol{\theta}}(s). \tag{1}$$

The following equation instantly holds [10]:

$$d_{\boldsymbol{\theta}}(s') = \lim_{T\rightarrow\infty}\frac{1}{T}\sum_{t=1}^{T} \Pr(S_t = s'|S_0 = s, \mathrm{M}(\boldsymbol{\theta})),\ ^{\forall} s \in \mathcal{S}. \tag{2}$$

The goal of PGRL is to find the policy parameter $\boldsymbol{\theta}^*$ that maximizes the average of the immediate rewards called the *average reward*:

$$R(\boldsymbol{\theta}) \equiv \lim_{T\rightarrow\infty}\frac{1}{T}\mathbb{E}\left\{\sum_{t=1}^{T} r_t \middle| s_0, \mathrm{M}(\boldsymbol{\theta})\right\}, \tag{3}$$

where $\mathbb{E}\{\cdot\}$ denotes expectation. It is noted that, under the assumption of ergodicity (eq.2), the average reward is independent of the initial state $s_0$ and can be shown to equal [10]:

$$\begin{aligned} R(\boldsymbol{\theta}) &= \sum_{s\in\mathcal{S}}\sum_{a\in\mathcal{A}}\sum_{s'\in\mathcal{S}} d_{\boldsymbol{\theta}}(s)\pi(a|s;\boldsymbol{\theta})p(s'|s,a)r(s,a,s') \\ &= \sum_{s\in\mathcal{S}}\sum_{a\in\mathcal{A}} d_{\boldsymbol{\theta}}(s)\pi(a|s;\boldsymbol{\theta})\bar{r}(s,a) \\ &\equiv \sum_{s\in\mathcal{S}}\sum_{a\in\mathcal{A}} \Pr(s,a|\mathrm{M}(\boldsymbol{\theta}))\bar{r}(s,a). \end{aligned} \tag{4}$$

where $\bar{r}(s,a) \equiv \sum_{s'\in\mathcal{S}} p(s'|s,a)r(s,a,s')$. The statistical model $\Pr(s,a|\mathrm{M}(\boldsymbol{\theta}))$ is called the stationary state-action (joint) distribution. Since $\bar{r}(s,a)$ is usually independent of the policy parameter, the derivative of the average reward with

[1] Although it should be $\Pr(S_{t+1} = s_{t+1}|S_t = s_t, A_t = a_t)$ for the random variables $S_{t+1}$, $S_t$, and $A_t$ to be precise, we notate $\Pr(s_{t+1}|s_t, a_t)$ for simplicity. The same rule is applied to the other distributions.

respect to the policy parameter, $\nabla_{\boldsymbol{\theta}} R(\boldsymbol{\theta}) \equiv [\partial R(\boldsymbol{\theta})/\partial \theta_1, ..., \partial R(\boldsymbol{\theta})/\partial \theta_d]^\top$ is given by

$$\begin{aligned}\nabla_{\boldsymbol{\theta}} R(\boldsymbol{\theta}) &= \sum_{s\in\mathcal{S}}\sum_{a\in\mathcal{A}} \nabla_{\boldsymbol{\theta}} \{d_{\boldsymbol{\theta}}(s)\pi(a|s;\boldsymbol{\theta})\}\, \bar{r}(s,a) \\ &= \sum_{s\in\mathcal{S}}\sum_{a\in\mathcal{A}} d_{\boldsymbol{\theta}}(s)\pi(a|s;\boldsymbol{\theta})\bar{r}(s,a)\, \{\nabla_{\boldsymbol{\theta}} \ln \pi(a|s;\boldsymbol{\theta}) + \nabla_{\boldsymbol{\theta}} \ln d_{\boldsymbol{\theta}}(s)\}\,, \end{aligned} \tag{5}$$

where $\top$ denotes transpose and $\nabla_{\boldsymbol{\theta}} a_{\boldsymbol{\theta}} b_{\boldsymbol{\theta}} \equiv (\nabla_{\boldsymbol{\theta}} a_{\boldsymbol{\theta}}) b_{\boldsymbol{\theta}}$. Therefore, the average reward $R(\boldsymbol{\theta})$ increases by updating the policy parameter as follows:

$$\boldsymbol{\theta} := \boldsymbol{\theta} + \alpha \nabla_{\boldsymbol{\theta}} R(\boldsymbol{\theta}),$$

where := denotes the right-to-left substitution and $\alpha$ is a sufficiently small learning rate. The above framework is called the PGRL method [5].

### 2.2 Natural Gradient [1]

Natural gradient learning is a gradient method on a Riemannian space. The parameter space being a Riemannian space implies that the parameter $\boldsymbol{\theta} \in \mathcal{R}^d$ is on the Riemannian manifold defined by the Riemannian metric matrix $\boldsymbol{G}(\boldsymbol{\theta}) \in \mathcal{R}^{d\times d}$ (positive definite matrix) and the squared length of a small incremental vector $\Delta\boldsymbol{\theta}$ connecting $\boldsymbol{\theta}$ to $\boldsymbol{\theta} + \Delta\boldsymbol{\theta}$ is given by

$$\|\Delta\boldsymbol{\theta}\|_{\boldsymbol{G}}^2 = \Delta\boldsymbol{\theta}^\top \boldsymbol{G}(\boldsymbol{\theta}) \Delta\boldsymbol{\theta}.$$

Under the constraint $\|\Delta\boldsymbol{\theta}\|_{\boldsymbol{G}}^2 = \varepsilon^2$ for a sufficiently small constant $\varepsilon$, the steepest ascent direction of a function $R(\boldsymbol{\theta})$ is given by

$$\widetilde{\nabla}_{\boldsymbol{G},\boldsymbol{\theta}} R(\boldsymbol{\theta}) = \boldsymbol{G}(\boldsymbol{\theta})^{-1} \nabla_{\boldsymbol{\theta}} R(\boldsymbol{\theta}). \tag{6}$$

It is called the natural gradient of $R(\boldsymbol{\theta})$ in the Riemannian space $\boldsymbol{G}(\boldsymbol{\theta})$. In RL, the parameter $\boldsymbol{\theta}$ is the policy parameter, the function $R(\boldsymbol{\theta})$ is the average reward, and the gradient $\widetilde{\nabla}_{\boldsymbol{G},\boldsymbol{\theta}} R(\boldsymbol{\theta})$ is called the natural policy gradient (NPG) [2]. Accordingly, in order to (locally) maximize $R(\boldsymbol{\theta})$, $\boldsymbol{\theta}$ is incrementally updated by

$$\boldsymbol{\theta} := \boldsymbol{\theta} + \alpha\, \widetilde{\nabla}_{\boldsymbol{G},\boldsymbol{\theta}} R(\boldsymbol{\theta}). \tag{7}$$

When we consider a statistical model of a variable $x$ parameterized by $\boldsymbol{\theta}$, $\Pr(x|\boldsymbol{\theta})$, the Fisher information matrix (FIM) $\boldsymbol{F}_x(\boldsymbol{\theta})$ is often used as the Riemannian metric matrix: [12]

$$\begin{aligned}\boldsymbol{F}_x(\boldsymbol{\theta}) &\equiv \sum_{x\in\mathcal{X}} \Pr(x|\boldsymbol{\theta})\, \nabla_{\boldsymbol{\theta}} \ln \Pr(x|\boldsymbol{\theta})\, \nabla_{\boldsymbol{\theta}} \ln \Pr(x|\boldsymbol{\theta})^\top \\ &= -\sum_{x\in\mathcal{X}} \Pr(x|\boldsymbol{\theta})\, \nabla_{\boldsymbol{\theta}}^2 \ln \Pr(x|\boldsymbol{\theta}), \end{aligned} \tag{8}$$

where $\mathcal{X}$ is a set of possible values taken by $x$. $\nabla^2_{\boldsymbol{\theta}} a_{\boldsymbol{\theta}}$ denotes $\nabla_{\boldsymbol{\theta}}(\nabla_{\boldsymbol{\theta}} a_{\boldsymbol{\theta}})$. The reason for using $\boldsymbol{F}(\boldsymbol{\theta})$ as $\boldsymbol{G}(\boldsymbol{\theta})$ comes from the fact that $\boldsymbol{F}(\boldsymbol{\theta})$ is a unique metric matrix of the second-order Taylor expansion of Kullback-Leibler (KL) divergence[2], which is known as a (pseudo) distance between two probability distributions. That is, the KL divergence of $\Pr(x|\boldsymbol{\theta}+\Delta\boldsymbol{\theta})$ from $\Pr(x|\boldsymbol{\theta})$ is represented by

$$D_{\mathrm{KL}}\{\Pr(x|\boldsymbol{\theta})|\Pr(x|\boldsymbol{\theta}+\Delta\boldsymbol{\theta})\} = \frac{1}{2}\Delta\boldsymbol{\theta}^{\top}\boldsymbol{F}_x(\boldsymbol{\theta})\,\Delta\boldsymbol{\theta} + O(\|\Delta\boldsymbol{\theta}\|^3),$$

where $\|\boldsymbol{a}\|$ denotes the Euclidean norm of a vector $\boldsymbol{a}$.

### 2.3 Controversy of Natural Policy Gradients

PGRL is regarded as an optimizing process of the policy parameter $\boldsymbol{\theta}$ on some statistical models relevant to both a stochastic policy $\pi(a|s;\boldsymbol{\theta})$ and a state transition probability $p(s'|s,a)$. If a Riemannian metric matrix $\boldsymbol{G}(\boldsymbol{\theta})$ can be designed on the basis of the FIM of an apposite statistical model, $\boldsymbol{F}^*(\boldsymbol{\theta})$, an efficient NPG $\widetilde{\nabla}_{\boldsymbol{F}^*,\boldsymbol{\theta}} R(\boldsymbol{\theta})$ is instantly derived by eq.6.

As Kakade [2] pointed out, the choice of the Riemannian metric matrix $\boldsymbol{G}(\boldsymbol{\theta})$ for PGRL is not unique and the question what metric is apposite to $\boldsymbol{G}(\boldsymbol{\theta})$ is still open. Nevertheless, all previous studies on NPG [13,14,8,7,9] did not seriously address the above problem and (naively) used the Riemannian metric matrix proposed by Kakade [2]. In the next section, we will discuss the statistical models and metric spaces for PGRL and propose a new Riemannian metric matrix.

## 3 Riemannian Metric Matrices for PGRL

In section 3.1, we propose a new Riemannian metric matrix for RL and derive its NPG named the NSG. In sections 3.2 and 3.3, we discuss the validity of this Riemannian metric by comparing it with the Riemannian metric proposed by Kakade [2] and the Hessian matrix of the average reward.

### 3.1 A Proposed Riemannian Metric Matrix and NPG Based on State-Action Probability

Since the only adjustable function in PGRL is the policy function $\pi(a|s;\boldsymbol{\theta})$, previous studies on NPG focused on the policy function $\pi(a|s;\boldsymbol{\theta})$, i.e., the statistical models $\Pr(a|s,\mathrm{M}(\boldsymbol{\theta}))$. However, the perturbations in the policy parameter $\boldsymbol{\theta}$ also give rise to the change in the probability of the state $\Pr(s|\mathrm{M}(\boldsymbol{\theta}))$. Because the average reward $R(\boldsymbol{\theta})$ as the objective function of PGRL is specified by the joint probability distribution of the state and the action $(s,a) \in \mathcal{S} \times \mathcal{A}$ (eq.4), it is natural and adequate to focus on the statistical model $\Pr(s,a|\mathrm{M}(\boldsymbol{\theta}))$. For this case, the FIM of $\Pr(s,a|\mathrm{M}(\boldsymbol{\theta}))$ can be used as the Riemannian metric $\boldsymbol{G}(\boldsymbol{\theta})$. Then, its NPG consists with the direction maximizing the average reward under

[2] It is same in the case of all f-divergences in general, except for scale [12].

the constraint that a measure of changes in the KL divergence of the stationary state-action distribution with respect to $\boldsymbol{\theta}$ is fixed by a sufficient small constant $\varepsilon$: $D_{\mathrm{KL}}\{\Pr(s,a|\mathrm{M}(\boldsymbol{\theta}))|\Pr(s,a|\mathrm{M}(\boldsymbol{\theta}+\Delta\boldsymbol{\theta}))\}=\varepsilon^2$. The FIM of this statistical model, $\boldsymbol{F}_{s,a}(\boldsymbol{\theta})$, is calculated with $\Pr(s,a|\mathrm{M}(\boldsymbol{\theta}))=d_{\boldsymbol{\theta}}(s)\pi(a|s;\boldsymbol{\theta})$ and eq.8 to be

$$\begin{aligned}\boldsymbol{F}_{s,a}(\boldsymbol{\theta}) &= \sum_{s\in\mathcal{S}}\sum_{a\in\mathcal{A}}\Pr(s,a|\mathrm{M}(\boldsymbol{\theta}))\nabla_{\boldsymbol{\theta}}\ln\Pr(s,a|\mathrm{M}(\boldsymbol{\theta}))\nabla_{\boldsymbol{\theta}}\ln\Pr(s,a|\mathrm{M}(\boldsymbol{\theta}))^{\top}\\ &= -\sum_{s\in\mathcal{S}}\sum_{a\in\mathcal{A}}d_{\boldsymbol{\theta}}(s)\pi(a|s;\boldsymbol{\theta})\nabla^2_{\boldsymbol{\theta}}\ln\left(d_{\boldsymbol{\theta}}(s)\pi(a|s;\boldsymbol{\theta})\right)\\ &= \boldsymbol{F}_s(\boldsymbol{\theta})+\sum_{s\in\mathcal{S}}d_{\boldsymbol{\theta}}(s)\,\boldsymbol{F}_a(s,\boldsymbol{\theta}),\end{aligned}\tag{9}$$

where

$$\boldsymbol{F}_s(\boldsymbol{\theta})=\sum_{s\in\mathcal{S}}d_{\boldsymbol{\theta}}(s)\nabla_{\boldsymbol{\theta}}\ln d_{\boldsymbol{\theta}}(s)\nabla_{\boldsymbol{\theta}}\ln d_{\boldsymbol{\theta}}(s)^{\top}\tag{10}$$

is the FIM defined from the statistical model comprising the state distribution, $\Pr(s|\mathrm{M}(\boldsymbol{\theta}))=d_{\boldsymbol{\theta}}(s)$, and

$$\boldsymbol{F}_a(s,\boldsymbol{\theta})=\sum_{a\in\mathcal{A}}\pi(a|s;\boldsymbol{\theta})\nabla_{\boldsymbol{\theta}}\ln\pi(a|s;\boldsymbol{\theta})\nabla_{\boldsymbol{\theta}}\ln\pi(a|s;\boldsymbol{\theta})^{\top}\tag{11}$$

is the FIM of the policy comprising the action distribution given the state $s$, $\Pr(a|s,\mathrm{M}(\boldsymbol{\theta}))=\pi(a|s;\boldsymbol{\theta})$. Hence, the new NPG on the FIM of the stationary state-action distribution is

$$\widetilde{\nabla}_{\boldsymbol{F}_{s,a},\boldsymbol{\theta}}R(\boldsymbol{\theta})=\boldsymbol{F}_{s,a}(\boldsymbol{\theta})^{-1}\nabla_{\boldsymbol{\theta}}R(\boldsymbol{\theta}).$$

We term it the "natural state-action gradient" (NSG).

### 3.2 Comparison with Kakade's Riemannian Metric Matrix

The only Riemannian metric matrix for RL that has been proposed so far is the following matrix, which was proposed by Kakade [2] and was the weighted sum of the FIMs of the policy by the stationary state distribution $d_{\boldsymbol{\theta}}(s)$,

$$\overline{\boldsymbol{F}}_a(\boldsymbol{\theta})\equiv\sum_{s\in\mathcal{S}}d_{\boldsymbol{\theta}}(s)\,\boldsymbol{F}_a(s,\boldsymbol{\theta}).\tag{12}$$

This is equal to the second term in eq.9. If it is assumed that the stationary state distribution is not changed by a variation in the policy, i.e., if $\nabla_{\boldsymbol{\theta}}d_{\boldsymbol{\theta}}(s)=\mathbf{0}$ holds, then $\boldsymbol{F}_s(\boldsymbol{\theta})=\mathbf{0}$ holds according to eq.10. While this assumption is not true in general, Kakade's metric $\overline{\boldsymbol{F}}_a(\boldsymbol{\theta})$ is equivalent to $\boldsymbol{F}_{s,a}(\boldsymbol{\theta})$ if it holds. These facts indicate that $\overline{\boldsymbol{F}}_a(\boldsymbol{\theta})$ is the Riemannian metric matrix ignoring the change in the stationary state distribution $d_{\boldsymbol{\theta}}(s)$ caused by the perturbation in the policy parameter $\boldsymbol{\theta}$ in terms of the statistical model of the stationary state-action distribution $\Pr(s,a|\mathrm{M}(\boldsymbol{\theta}))$.

Meanwhile, Bagnell et al. [13] and Peters et al. [14] independently, showed the relationship between the Kakade's metric and the system trajectories $\xi_T = (s_0, a_0, s_1, ..., a_{T-1}, s_T) \in \Xi_T$. When the FIM of the statistical model for the system trajectory $\xi_T$,

$$\Pr(\xi_T|\mathrm{M}(\boldsymbol{\theta})) = \Pr(s_0)\prod_{t=0}^{T-1}\pi(a_t|s_t;\boldsymbol{\theta})p(s_{t+1}|s_t,a_t),$$

is normalized by the time steps $T$ with the limit $T \to \infty$, it is equivalent to the Kakade's Riemannian metric,

$$\begin{aligned}\lim_{T\to\infty}\frac{1}{T}\boldsymbol{F}_{\xi_T}(\boldsymbol{\theta}) &= -\lim_{T\to\infty}\frac{1}{T}\sum_{\xi_T\in\Xi_T}\Pr(\xi_T|\mathrm{M}(\boldsymbol{\theta}))\nabla^2_{\boldsymbol{\theta}}\left\{\sum_{t=0}^{T-1}\ln\pi(a_t|s_t;\boldsymbol{\theta})\right\}\\ &= -\sum_{s\in\mathcal{S}}d_{\boldsymbol{\theta}}(s)\sum_{a\in\mathcal{A}}\pi(a|s;\boldsymbol{\theta})\nabla^2_{\boldsymbol{\theta}}\ln\pi(a|s;\boldsymbol{\theta})\\ &= \overline{\boldsymbol{F}}_a(\boldsymbol{\theta})\end{aligned}$$

Since the PGRL objective, i.e., the maximization of the average reward, is reduced to the optimization of the system trajectory by eq.3 [13,14] suggested that the Kakade's metric $\overline{\boldsymbol{F}}_a(\boldsymbol{\theta})$ could be a good metric. However, being equal to $\overline{\boldsymbol{F}}_a(\boldsymbol{\theta})$, the normalized FIM for the infinite-horizon system trajectory obviously differs with $\boldsymbol{F}_{s,a}(\boldsymbol{\theta})$ and is the metric that ignores the information $\boldsymbol{F}_s(\boldsymbol{\theta})$ about the stationary state distribution $\Pr(s|\mathrm{M}(\boldsymbol{\theta}))$. This is due to the fact that the statistical model of the system trajectory considers not only the state-action joint distribution but also the progress for the (infinite) time steps, as follows.

Here, $s_{+t}$ and $a_{+t}$ are the state and the action, respectively, progressed in $t$ time steps after converging to the stationary distribution. Since the distribution of the system trajectory for $T$ time steps from the stationary distribution, $\xi_{+T} \equiv (s, a_{+0}, s_{+1}, ..., a_{+T-1}, s_{+T}) \in \Xi_T$, is

$$\Pr(\xi_{+T}|\mathrm{M}(\boldsymbol{\theta})) = d_{\boldsymbol{\theta}}(s)\prod_{t=0}^{T-1}\pi(a_{+t}|s_{+t};\boldsymbol{\theta})p(s_{+t+1}|s_{+t},a_{+t}),$$

its FIM is given by

$$\boldsymbol{F}_{\boldsymbol{\xi}_{+T}}(\boldsymbol{\theta}) = \boldsymbol{F}_s(\boldsymbol{\theta}) + T\overline{\boldsymbol{F}}_a(\boldsymbol{\theta}). \tag{13}$$

The derivation of which is shown in appendix A. Because of $\lim_{T\to\infty}\boldsymbol{F}_{\xi_{+T}}/T = \overline{\boldsymbol{F}}_a(\boldsymbol{\theta})$, the Kakade's metric $\overline{\boldsymbol{F}}_a(\boldsymbol{\theta})$ is regarded as the limit $T \to \infty$ of the system trajectory distribution for $T$ time steps from the stationary state distribution. Consequently, $\overline{\boldsymbol{F}}_a(\boldsymbol{\theta})$ omits the FIM of the state distribution, $\boldsymbol{F}_s(\boldsymbol{\theta})$. On the other hand, the FIM of the system trajectory distribution for one time step is obviously equivalent to the FIM of the state-action joint distribution, i.e., $\boldsymbol{F}_{\xi_{+1}}(\boldsymbol{\theta}) = \boldsymbol{F}_{s,a}(\boldsymbol{\theta})$.

Now, we discuss which FIM is adequate for the average reward maximization. As discussed in section 3.1, the average reward in eq.4 is the expectation of $\bar{r}(s, a)$

over the distribution of the state-action (i.e. the +1-time-step system trajectory) and does not depend on the system trajectories after +2 time steps. It indicates that the Kakade's metric $\overline{\boldsymbol{F}}_a(\boldsymbol{\theta})$ supposed a redundant statistical model and the proposed metric for state-action distribution, $\boldsymbol{F}_{s,a}(\boldsymbol{\theta})$, would be more natural and adequate for PGRL. We give comparisons among various metrics such as $\boldsymbol{F}_{s,a}(\boldsymbol{\theta})$, $\overline{\boldsymbol{F}}_a(\boldsymbol{\theta})$, and a unit matrix $\boldsymbol{I}$ through the numerical experiments in section 5.

Similarly, when the reward function is extended a function of $T$ time steps, $r(s_t, a_t, ..., a_{t+T-1}, s_{t+T})$, instead of one time step, $r(s_t, a_t, s_{t+1})$, the FIM of the $T$-time-step system trajectory distribution, $\boldsymbol{F}_{\xi_{+T}}(\boldsymbol{\theta})$, would be a natural metric because the average reward becomes $R(\boldsymbol{\theta}) = \sum_{\xi_{+T}\in\Xi_T} \Pr(\xi_{+T}|\mathrm{M}(\boldsymbol{\theta}))r(\xi_{+T})$.

### 3.3 Analogy with Hessian Matrix

We discuss the analogies between the Fisher information matrices $\boldsymbol{F}_{s,a}(\boldsymbol{\theta})$ and $\overline{\boldsymbol{F}}_a(\boldsymbol{\theta})$ and the Hessian matrix $\boldsymbol{H}(\boldsymbol{\theta})$, which is the second derivative of the average reward with respect to the policy parameter $\boldsymbol{\theta}$,

$$\begin{aligned}
\boldsymbol{H}(\boldsymbol{\theta}) &\equiv \nabla^2_{\boldsymbol{\theta}} R(\boldsymbol{\theta}) \\
&= \sum_{s\in\mathcal{S}}\sum_{a\in\mathcal{A}} \bar{r}(s,a) d_{\boldsymbol{\theta}}(s)\pi(a|s;\boldsymbol{\theta}) \\
&\quad \Big\{\nabla^2_{\boldsymbol{\theta}} \ln\big(d_{\boldsymbol{\theta}}(s)\pi(a|s;\boldsymbol{\theta})\big) + \nabla_{\boldsymbol{\theta}} \ln\big(d_{\boldsymbol{\theta}}(s)\pi(a|s;\boldsymbol{\theta})\big)\nabla_{\boldsymbol{\theta}} \ln\big(d_{\boldsymbol{\theta}}(s)\pi(a|s;\boldsymbol{\theta})\big)^\top\Big\} \qquad (14)
\end{aligned}$$

$$\begin{aligned}
&= \sum_{s\in\mathcal{S}}\sum_{a\in\mathcal{A}} \bar{r}(s,a) d_{\boldsymbol{\theta}}(s)\pi(a|s;\boldsymbol{\theta}) \\
&\quad \Big\{\nabla^2_{\boldsymbol{\theta}} \ln \pi(a|s;\boldsymbol{\theta}) + \nabla_{\boldsymbol{\theta}} \ln \pi(a|s;\boldsymbol{\theta})\nabla_{\boldsymbol{\theta}} \ln \pi(a|s;\boldsymbol{\theta})^\top + \nabla^2_{\boldsymbol{\theta}} \ln d_{\boldsymbol{\theta}}(s) \\
&\quad + \nabla_{\boldsymbol{\theta}} \ln d_{\boldsymbol{\theta}}(s)\nabla_{\boldsymbol{\theta}} \ln d_{\boldsymbol{\theta}}(s)^\top + \nabla_{\boldsymbol{\theta}} \ln d_{\boldsymbol{\theta}}(s)\nabla_{\boldsymbol{\theta}} \ln \pi(a|s;\boldsymbol{\theta})^\top \\
&\quad + \nabla_{\boldsymbol{\theta}} \ln \pi(a|s;\boldsymbol{\theta})\nabla_{\boldsymbol{\theta}} \ln d_{\boldsymbol{\theta}}(s)^\top\Big\}. \qquad (15)
\end{aligned}$$

Comparing eq.12 of the Kakade's metric matrix $\overline{\boldsymbol{F}}_a(\boldsymbol{\theta})$ with eq.15 of the Hessian matrix $\boldsymbol{H}(\boldsymbol{\theta})$, the Kakade's metric does not have any information about the last two terms in braces $\{\cdot\}$ of eq.15, as Kakade [2] pointed out[3]. This is because $\overline{\boldsymbol{F}}_a(\boldsymbol{\theta})$ is derived under $\nabla_{\boldsymbol{\theta}} d_{\boldsymbol{\theta}}(s) = \boldsymbol{0}$. By eq.9 and eq.14, meanwhile, the proposed metric $\boldsymbol{F}_{s,a}(\boldsymbol{\theta})$ obviously has some information about all the terms of $\boldsymbol{H}(\boldsymbol{\theta})$. This comparison with the Hessian matrix suggests that $\boldsymbol{F}_{s,a}(\boldsymbol{\theta})$ should be an appropriate metric for PGRL. Additionally, $\boldsymbol{F}_{s,a}(\boldsymbol{\theta})$ becomes equivalent to the Hessian matrix in the cases using an atypical reward function that depends on $\boldsymbol{\theta}$ (see Appendix B).

It is noted that the average reward would not be a quadratic form with respect to the policy parameter $\boldsymbol{\theta}$ in general. Especially when $\boldsymbol{\theta}$ is far from the optimal parameter $\boldsymbol{\theta}^*$, the Hessian matrix $\boldsymbol{H}(\boldsymbol{\theta})$ occasionally gets into an indefinite matrix. Meanwhile, FIM $\boldsymbol{F}(\boldsymbol{\theta})$ is always positive (semi-)definite, assured

[3] Strictly speaking, $\boldsymbol{H}(\boldsymbol{\theta})$ is sligthtly different from the Hessian matrix used in [2]. However, the essence of argument is the same as in [2].

by its definition in eq.8. Accordingly, the natural gradient method using FIM might be a more versatile covariant gradient ascent for PGRL than the Newton-Raphson method [15], in which the gradient direction is given by $\widetilde{\nabla}_{-\boldsymbol{H},\boldsymbol{\theta}} R(\boldsymbol{\theta})$. Comparison experiments are presented in section 5.

## 4 Computation of Natural State-Action Gradient

In this section, we view the estimation of the NSG. It will be shown that this estimation can be reduced to the regression problem of the immediate rewards.

Consider the following linear regression model

$$f_{\theta}(s,a;\boldsymbol{\omega}) \equiv \boldsymbol{\phi}_{\theta}(s,a)^{\top}\boldsymbol{\omega}, \tag{16}$$

where $\boldsymbol{\omega}$ is the adjustable parameter and $\boldsymbol{\phi}_{\theta}(s,a)$ is the basis function of the state and action, also depending on the policy parameter $\boldsymbol{\theta}$,

$$\begin{aligned}\boldsymbol{\phi}_{\theta}(s,a) &\equiv \nabla_{\boldsymbol{\theta}} \ln \left(d_{\boldsymbol{\theta}}(s)\pi(a|s;\boldsymbol{\theta})\right) \\ &= \nabla_{\boldsymbol{\theta}} \ln d_{\boldsymbol{\theta}}(s) + \nabla_{\boldsymbol{\theta}} \ln \pi(a|s;\boldsymbol{\theta}).\end{aligned} \tag{17}$$

Then, the following theorem holds:

**Theorem 1.** *Let the Markov chain* $\mathrm{M}(\boldsymbol{\theta})$ *have the fixed policy parameter* $\boldsymbol{\theta}$, *if the objective is to minimize the mean square error* $\epsilon(\boldsymbol{\omega})$ *of the linear regression model* $f_{\theta}(s_t,a_t;\boldsymbol{\omega})$ *in eq.16 for the rewards* $r_{t+1}$,

$$\epsilon(\boldsymbol{\omega}) = \lim_{T\to\infty} \frac{1}{2T} \sum_{t=0}^{T-1} \left\{ r_{t+1} - f_{\theta}(s_t,a_t;\boldsymbol{\omega}) \right\}^2, \tag{18}$$

*then the optimal adjustable parameter* $\boldsymbol{\omega}^*$ *is equal to NSG as the natural policy gradient on* $\boldsymbol{F}_{s,a}(\boldsymbol{\theta})$:

$$\widetilde{\nabla}_{\boldsymbol{F}_{s,a},\boldsymbol{\theta}} R(\boldsymbol{\theta}) = \boldsymbol{\omega}^*.$$

**Proof:** By the ergodic property of $\mathrm{M}(\boldsymbol{\theta})$, eq.18 is written as

$$\epsilon(\boldsymbol{\omega}) = \frac{1}{2} \sum_{s\in\mathcal{S}} \sum_{a\in\mathcal{A}} d_{\boldsymbol{\theta}}(s)\pi(a|s;\boldsymbol{\theta}) \left(\bar{r}(s,a) - f_{\theta}(s,a;\boldsymbol{\omega})\right)^2.$$

Since $\boldsymbol{\omega}^*$ satisfies $\nabla_{\boldsymbol{\omega}}\epsilon(\boldsymbol{\omega})|_{\boldsymbol{\omega}=\boldsymbol{\omega}^*} = \mathbf{0}$, we have

$$\sum_{s\in\mathcal{S}} \sum_{a\in\mathcal{A}} d_{\boldsymbol{\theta}}(s)\pi(a|s;\boldsymbol{\theta})\boldsymbol{\phi}_{\theta}(s,a)\boldsymbol{\phi}_{\theta}(s,a)^{\top}\boldsymbol{\omega}^* = \sum_{s\in\mathcal{S}} \sum_{a\in\mathcal{A}} d_{\boldsymbol{\theta}}(s)\pi(a|s;\boldsymbol{\theta})\boldsymbol{\phi}_{\theta}(s,a)\bar{r}(s,a).$$

By the definition of the basis function (eq.17), the following equations hold,

$$\sum_{s,a} d_{\boldsymbol{\theta}}(s)\pi(a|s;\boldsymbol{\theta})\boldsymbol{\phi}_{\theta}(s,a)\boldsymbol{\phi}_{\theta}(s,a)^{\top} = \boldsymbol{F}_{s,a}(\boldsymbol{\theta}),$$

$$\sum_{s,a} d_{\boldsymbol{\theta}}(s)\pi(a|s;\boldsymbol{\theta})\boldsymbol{\phi}_{\theta}(s,a)\bar{r}(s,a) = \nabla_{\boldsymbol{\theta}} R(\boldsymbol{\theta}).$$

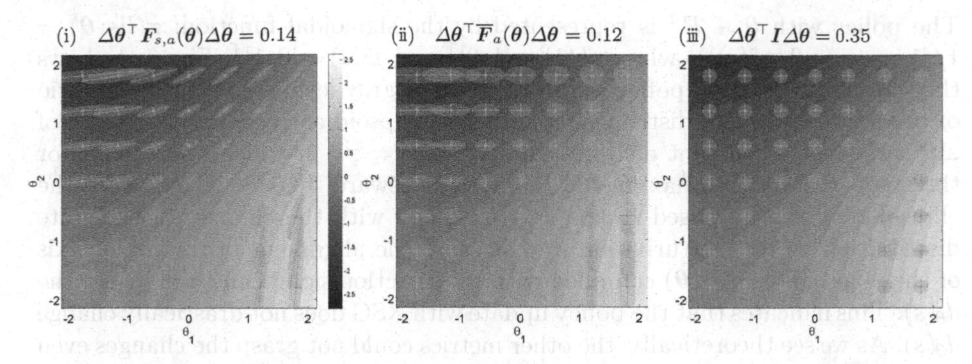


**Fig. 1.** Phase planes of a policy parameter in a two-state MDP: The gray level denotes $\ln d_\theta(1)/d_\theta(2)$. Each ellipsoid denotes the fixed distance spaces by each metric $G(\theta) :=$ (i) $F_{s,a}(\theta)$, (ii) $\overline{F}_a(\theta)$, or (iii) $I$.

Therefore, the following equation holds:

$$\omega^* = F_{s,a}(\theta)^{-1}\nabla_\theta R(\theta) = \widetilde{\nabla}_{F_{s,a},\theta} R(\theta). \qquad \square$$

It is confirmed by theorem 1 that if the least-square regression to the immediate reward $r_{t+1}$ by the linear function approximator $f_\theta(s_t, a_t; \omega)$ with the basis function $\phi_\theta(s,a) \equiv \nabla_\theta \ln (d_\theta(s)\pi(a|s;\theta))$ is performed, the adjustable parameter $\omega$ becomes the unbiased estimate of NSG $\widetilde{\nabla}_{F_{s,a},\theta} R(\theta)$. Therefore, since the NSG estimation problem is reduced to the regression problem of the reward function, NSG would be simply estimated by the least-square technique or by such a gradient descent technique as the method with the eligibility traces proposed by Morimura et al. [7], where the matrix inversion is not required.

It should be noted that, in order to implement this estimation, the computation of both the derivatives, $\nabla_\theta \ln \pi(a|s;\theta)$ and $\nabla_\theta \ln d_\theta(s)$, is required for the basis function $\phi_\theta(s,a)$. While $\nabla_\theta \ln \pi(a|s;\theta)$ can be instantly calculated, $\nabla_\theta \ln d_\theta(s)$ cannot be solved analytically because the state transition probabilities are generally unknown in RL. However, an efficient online estimation manner for $\nabla_\theta \ln d_\theta(s)$, which is similar to the method of estimating the value function, has been established by Morimura et al. [16]. However, we have not discussed the concrete implementations in this paper.

## 5 Numerical Experiments

### 5.1 Comparison of Metrics

We first looked into the differences among the Riemannian metric matrices $G(\theta)$—the proposed metric $F_{s,a}(\theta)$, Kakade's metric $\overline{F}_a(\theta)$, and unit matrix $I$—in a simple two-state MDP [2], where each state $s \in \{1,2\}$ has self- and cross-transition actions $\mathcal{A} = \{l, m\}$ and each state transition is deterministic.

The policy with $\boldsymbol{\theta} \in \mathcal{R}^2$ is represented by the sigmoidal function: $\pi(l|s;\boldsymbol{\theta}) = 1/(1+\exp(-\boldsymbol{\theta}^\top \boldsymbol{\psi}(s)))$, where $\boldsymbol{\psi}(1) = [1,0]^\top$ and $\boldsymbol{\psi}(2) = [0,1]^\top$. Figure 1 shows the phase planes of the policy parameter $\boldsymbol{\theta}$. The gray level denotes the log ratio of the stationary state distribution, and each ellipsoid corresponds to the set of $\Delta\boldsymbol{\theta}$ satisfying a constant distance $\Delta\boldsymbol{\theta}^\top \boldsymbol{G}(\boldsymbol{\theta})\Delta\boldsymbol{\theta} = \varepsilon^2$, in which NPG looks for the steepest direction maximizing the average reward. It is confirmed that the ellipsoids by the proposed metric $\boldsymbol{F}_{s,a}(\boldsymbol{\theta})$ coped with the changes in the state distribution by the perturbation in $\boldsymbol{\theta}$ because the alignment of the minor axis of the ellipsoid on $\boldsymbol{F}_{s,a}(\boldsymbol{\theta})$ complied with the direction significantly changing the $d_{\boldsymbol{\theta}}(s)$. This indicates that the policy update with NSG does not drastically change $d_{\boldsymbol{\theta}}(s)$. As we see theoretically. the other metrics could not grasp the changes even though $\overline{\boldsymbol{F}}_a(\boldsymbol{\theta})$ is the expectation of $\boldsymbol{F}_a(\boldsymbol{\theta})$ over $d_{\boldsymbol{\theta}}(s)$.

### 5.2 Comparison of Learnings

We compared NSG with Kakade's NPG, the ordinary PG, and the (modified) Newton PG learnings in terms of the optimizing performances for $\boldsymbol{\theta}$ through randomly synthesized MDPs with a varying number of states, $|\mathcal{S}| \in \{3, 10, 20, 35, 50, 65, 80, 100\}$. Note that the only difference among these gradients is the definition of the matrix $\boldsymbol{G}(\boldsymbol{\theta})$ in eq.6. The Newton PG uses a modified Hessian matrix $\boldsymbol{H}^\star(\boldsymbol{\theta})$ to assure the negative definiteness:

$$\boldsymbol{H}^\star(\boldsymbol{\theta}) = \boldsymbol{H}(\boldsymbol{\theta}) - \max(0, \lambda_{\max} - \lambda'_{\max})\boldsymbol{I},$$

where $\lambda_{\max}$ and $\lambda'_{\max}$ are the maximum and the largest-negative eigenvalues of $\boldsymbol{H}(\boldsymbol{\theta})$, respectively[4].

It is noted that each gradient was computed analytically because we focussed on the direction of the gradients rather than the sampling issue in this paper.

**Experimental Setup.** We initialized the $|\mathcal{S}|$-state MDP in each episode as follows. The set of the actions was always $|\mathcal{A}| = \{l, m\}$. The state transition probability function was set by using the Dirichlet distribution $\mathtt{Dir}(\boldsymbol{\alpha} \in \mathcal{R}^2)$ and the uniform distribution $\mathtt{U}(|\mathcal{S}|; b)$ generating an integer from 1 to $|\mathcal{S}|$ other than $b$: we first initialized it such that $p(s'|s,a) := 0, \forall(s',s,a)$ and then, with $\boldsymbol{q}(s,a) \sim \mathtt{Dir}(\boldsymbol{\alpha}{=}[.3,.3])$ and $x_{\backslash b} \sim \mathtt{U}(|\mathcal{S}|; b)$,

$$\begin{cases} p(s+1|s,l) := q_1(s,l) \\ p(x_{\backslash s+1}|s,l) := q_2(s,l) \end{cases} \qquad \begin{cases} p(s|s,m) \quad := q_1(s,m) \\ p(x_{\backslash s}|s,m) := q_2(s,m) \end{cases}$$

where $s' = 1$ and $s' = |\mathcal{S}|+1$ are the identical states. The reward function $r(s,a,s')$ was temporarily set for each argument by Gaussian distribution $\mathrm{N}(\mu{=}0, \sigma^2{=}1)$ and was normalized such that $\max_{\boldsymbol{\theta}} R(\boldsymbol{\theta}) = 1$ and $\min_{\boldsymbol{\theta}} R(\boldsymbol{\theta}) = 0$;

$$r(s,a,s') := \frac{r(s,a,s') - \min_{\boldsymbol{\theta}} R(\boldsymbol{\theta})}{\max_{\boldsymbol{\theta}} R(\boldsymbol{\theta}) - \min_{\boldsymbol{\theta}} R(\boldsymbol{\theta})}.$$

[4] We examined various Hessian modifications [15]. The modification adopted here worked best in this task.

The policy parameterization was the same as that for previous experiment. Accordingly, in this MDP setting, there is no local optimum except for the global optimum. Each element of $\boldsymbol{\theta}_0 \in \mathcal{R}^{|\mathcal{S}|}$ and $\boldsymbol{\psi}(s) \in \mathcal{R}^{|\mathcal{S}|}$ for any state $s$ was drawn from N(0, .5) and N(0, 1), respectively. We set the total episode time step at $T = 300$ and the initial learning rate $\alpha_0$ in eq.7 for each (N)PG before each episode at the inverse of RMS,

$$\alpha_0 = \sqrt{|\mathcal{S}|}/\left\|\widetilde{\nabla}_{G,\boldsymbol{\theta}} R(\boldsymbol{\theta})|_{\boldsymbol{\theta}=\boldsymbol{\theta}_0}\right\|.$$

If the learning rate $\alpha$ is decent, $R(\boldsymbol{\theta})$ will always increase by the policy update of eq.7. Hence, when the policy update decreased $R(\boldsymbol{\theta})$, we tuned the learning rate "$\alpha{:=}\alpha/2$" and reattempted the update in the same time step. This tuning was kept until $\Delta R(\theta) \geq 0$. On the other hand, when $\alpha_0 > \alpha$ held true at the following time step, we also tuned "$\alpha := 2\alpha$" to avoid standstills of the learning.

**Results and Discussions.** Figure 2 shows the learning curves for ten individual episodes in 100-state MDPs and reveals that NSG learning was able to succeed in optimizing the policy parameter uniformly and robustly though, compared with the other gradients, NSG was not infrequently slow in improving of performance at a moment. These are consistent with the results about the application of the natural gradient method to the learning of the multilayer perceptron [17].

Figure 3(A) shows the success rate of the learning by 300 episodes at each number of states. Since the maximum of the average reward was set to 1, we regarded the episodes satisfying $R(\boldsymbol{\theta}_T) \geq 0.95$ as "successful" episodes. This suggests that, in the case of the MDPs with a small number of states, NSG and Kakade's NPG methods could avoid falling into the severe plateau phenomena and robustly optimize the policy parameter $\boldsymbol{\theta}$, compared with the other methods. The reason why Kakade's NPG could work as well as NSG would be that the Riemannian metric used in Kakade's method has partial information about the statistical model $\Pr(s, a|\mathrm{M}(\boldsymbol{\theta}))$. Meanwhile, Kakade's method frequently failed to improve the average reward in the cases of the MDPs with a large number of states. This could be due to the fact that Kakade's metric omits the FIM about the state distribution, $\boldsymbol{F}_s(\boldsymbol{\theta})$ unlike the proposed metric, as discussed theoretically in section 3.2. It is also confirmed that Kakade's NPG was inferior to the modified Newton PG in the cases of many states. This could also be a result of whether the gradient has the information about the derivative of $d_{\boldsymbol{\theta}}(s)$ or not.

Finally, we analyzed how severe was the plateau in which these PG learnings were trapped. As this criterion, we utilized the smoothness of the learning curve (approximate curvature),

$$\Delta^2 R(\boldsymbol{\theta}_t) = \Delta R(\boldsymbol{\theta}_{t+1}) - \Delta R(\boldsymbol{\theta}_t),$$

where $\Delta R(\boldsymbol{\theta}_t) \equiv R(\boldsymbol{\theta}_t) - R(\boldsymbol{\theta}_{t-1})$. The criterion for the plateau measure of the episode was defined by

$$\mathtt{PM} = \textstyle\sum_{t=1}^{T-1} \|\Delta^2 R(\boldsymbol{\theta}_t)\|.$$

Figure 3(B) represents the average of PM over all episodes for each PG and shows that NSG learning could learn very smoothly. This result indicates that the

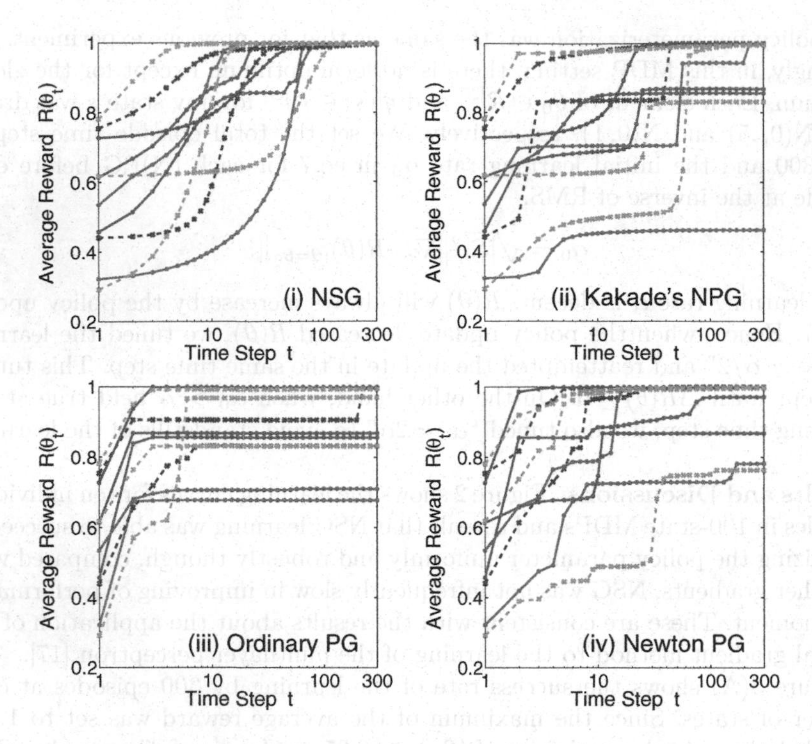


**Fig. 2.** Time courses of $R(\boldsymbol{\theta})$ for ten individual runs by (i) $\widetilde{\nabla}_{\boldsymbol{F_{s,a}},\boldsymbol{\theta}} R(\boldsymbol{\theta})$, (ii) $\widetilde{\nabla}_{\overline{\boldsymbol{F}}_a,\boldsymbol{\theta}} R(\boldsymbol{\theta})$, (iii) $\widetilde{\nabla}_{\boldsymbol{I},\boldsymbol{\theta}} R(\boldsymbol{\theta})$, (iv) $\widetilde{\nabla}_{-\boldsymbol{H}^*,\boldsymbol{\theta}} R(\boldsymbol{\theta})$

(A) (B)

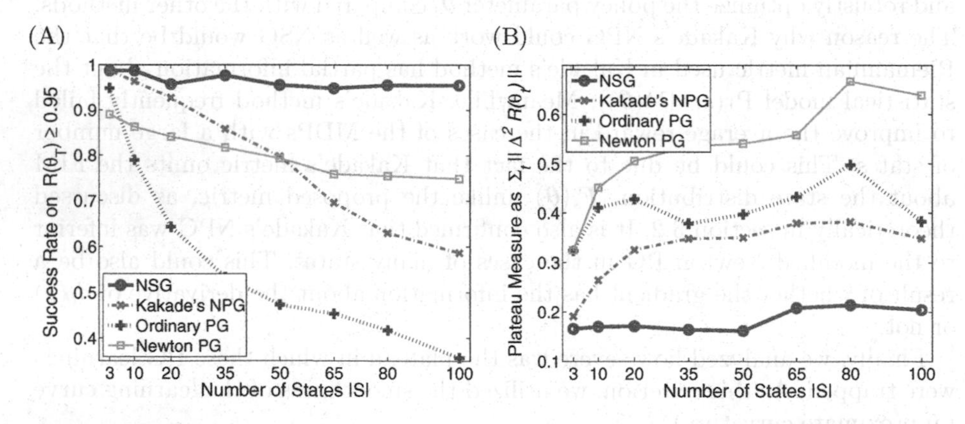


**Fig. 3.** (A) Learning success rates and (B) plateau measures for each number of states

learning by NSG could most successfully escape from a plateau; this is consistent with all other results.

Since NSG could avoid the plateau and robustly optimize $\boldsymbol{\theta}$ without any serious effect of the setting of the MDP and the initial policy parameter, we conclude that NSG could be a more robust and natural NPG than the NPG by Kakade [2].

## 6 Summary and Future Work

This paper proposed a new Riemannian metric matrix for the natural gradient of the average reward, which was the Fisher information matrix of the stationary state-action distribution. We clarified that Kakade's NPG [2], which has been widely used in RL, does not consider the changes in the stationary state distribution caused by the perturbation of the policy, while our proposed NSG does. The difference was confirmed in numerical experiments where NSG learning could dramatically improve the performance and rarely fell into the plateau. Additionally, we proved that, when the immediate rewards were fitted by using the linear regression model with the basis function defined on the policy, its adjustable parameter represented the unbiased NSG estimate.

More algorithmic and experimental studies are necessary to further emphasize the effectiveness of NSG. The significant ones would be to establish an efficient Monte-Carlo estimation way of NSG along with estimating the derivative of the stationary state distribution [16], and then to clarify whether or not the proposed NSG method can still be useful even when the gradient is computed from samples. We will investigate them in future work.

## References


1. Amari, S.: Natural gradient works efficiently in learning. Neural Computation 10(2), 251–276 (1998)
2. Kakade, S.: A natural policy gradient. In: Advances in Neural Information Processing Systems, vol. 14. MIT Press, Cambridge (2002)
3. Williams, R.J.: Simple statistical gradient-following algorithms for connectionist reinforcement learning. Machine Learning 8, 229–256 (1992)
4. Kimura, H., Miyazaki, K., Kobayashi, S.: Reinforcement learning in pomdps with function approximation. In: International Conference on Machine Learning, pp. 152–160 (1997)
5. Baxter, J., Bartlett, P.: Infinite-horizon policy-gradient estimation. Journal of Artificial Intelligence Research 15, 319–350 (2001)
6. Fukumizu, K., Amari, S.: Local minima and plateaus in hierarchical structures of multilayer perceptrons. Neural Networks 13(3), 317–327 (2000)
7. Morimura, T., Uchibe, E., Doya, K.: Utilizing natural gradient in temporal difference reinforcement learning with eligibility traces. In: International Symposium on Information Geometry and its Applications, pp. 256–263 (2005)
8. Peters, J., Vijayakumar, S., Schaal, S.: Natural actor-critic. In: European Conference on Machine Learning (2005)

9. Richter, S., Aberdeen, D., Yu, J.: Natural actor-critic for road traffic optimisation. In: Advances in Neural Information Processing Systems. MIT Press, Cambridge (2007)
10. Bertsekas, D.P.: Dynamic Programming and Optimal Control, vol. 1, 2. Athena Scientific (1995)
11. Sutton, R.S., Barto, A.G.: Reinforcement Learning. MIT Press, Cambridge (1998)
12. Amari, S., Nagaoka, H.: Method of Information Geometry. Oxford University Press, Oxford (2000)
13. Bagnell, D., Schneider, J.: Covariant policy search. In: Proceedings of the International Joint Conference on Artificial Intelligence (July 2003)
14. Peters, J., Vijayakumar, S., Schaal, S.: Reinforcement learning for humanoid robotics. In: IEEE-RAS International Conference on Humanoid Robots (2003)
15. Nocedal, J., Wright, S.J.: Numerical Optimization. Springer, Heidelberg (2006)
16. Morimura, T., Uchibe, E., Yoshimoto, J., Doya, K.: Reinforcement learning with log stationary distribution gradient. Technical report, Nara Institute of Science and Technology (2007)
17. Amari, S., Park, H., Fukumizu, K.: Adaptive method of realizing natural gradient learning for multilayer perceptrons. Neural Computation 12(6), 1399–1409 (2000)

# Appendix

## A Derivation of eq.13

For simplicity, we denote $\pi_{+t} \equiv \pi(a_{+t}|s_{+t};\boldsymbol{\theta})$ and $p_{+t} \equiv p(s_{+t}|s_{+t-1},a_{+t-1})$. Since $\xi_{+T}$ is the system trajectory for $T$ time steps from $d_{\boldsymbol{\theta}}(s)$, $\boldsymbol{F}_{\boldsymbol{\xi}_{+T}}(\boldsymbol{\theta})$ is calculated to be

$$\begin{aligned}
\boldsymbol{F}_{\boldsymbol{\xi}_{+T}}(\boldsymbol{\theta}) &= -\sum_{\xi_{+T}\in\Xi_T} \Pr(\xi_{+T})\,\nabla^2_{\boldsymbol{\theta}}\Big\{\ln d_{\boldsymbol{\theta}}(s) + \sum_{t=0}^{T-1}\ln\pi(a_{+t}|s_{+t};\boldsymbol{\theta})\Big\} \\
&= -\sum_{s\in\mathcal{S}} d_{\boldsymbol{\theta}}(s)\Big(\nabla^2_{\boldsymbol{\theta}}\ln d_{\boldsymbol{\theta}}(s) + \sum_{a_{+0}\in\mathcal{A}}\pi_{+0}\Big(\nabla^2_{\boldsymbol{\theta}}\ln\pi_{+0} + \\
&\qquad \sum_{s_{+1}\in\mathcal{S}} p_{+1}\sum_{a_{+1}\in\mathcal{A}}\pi_{+1}\Big(\nabla^2_{\boldsymbol{\theta}}\ln\pi_{+1} + \cdots + \\
&\qquad\qquad \sum_{s_{+T-1}\in\mathcal{S}} p_{+T-1}\sum_{a_{+T-1}\in\mathcal{A}}\pi_{+T-1}\,\nabla^2_{\boldsymbol{\theta}}\ln\pi_{+T-1}\Big)\cdots\Big)\Big).
\end{aligned}$$

By using the balance equation of the $d_{\boldsymbol{\theta}}(s)$ in eq.1,

$$\begin{aligned}
\boldsymbol{F}_{\boldsymbol{\xi}_{+T}}(\boldsymbol{\theta}) &= \boldsymbol{F}_s(\boldsymbol{\theta}) + \sum_{t=0}^{T-1}\Big(\sum_{s_{+t}\in\mathcal{S}} d_{\boldsymbol{\theta}}(s_{+t})\boldsymbol{F}_a(\boldsymbol{\theta}|s_{+t})\Big) \\
&= \boldsymbol{F}_s(\boldsymbol{\theta}) + T\overline{\boldsymbol{F}}_a(\boldsymbol{\theta}). \qquad \square
\end{aligned}$$

## B Consistency of $\boldsymbol{F}_{s,a}(\boldsymbol{\theta})$ and $\boldsymbol{H}(\boldsymbol{\theta})$

If the immediate reward is dependent on $\boldsymbol{\theta}$

$$r(s,a;\boldsymbol{\theta}) = \frac{\Pr(s,a|\mathrm{M}(\boldsymbol{\theta}^*))}{\Pr(s,a|\mathrm{M}(\boldsymbol{\theta}))}\ln\Pr(s,a|\mathrm{M}(\boldsymbol{\theta})), \tag{19}$$

then the average reward becomes the negative cross entropy,

$$R(\boldsymbol{\theta}) = \sum_{s,a} \Pr(s,a|\mathrm{M}(\boldsymbol{\theta}^*)) \ln \Pr(s,a|\mathrm{M}(\boldsymbol{\theta})).$$

Hence, $\Pr(s,a|\mathrm{M}(\boldsymbol{\theta}^*)) = \Pr(s,a|\mathrm{M}(\boldsymbol{\theta}))$ holds, if the average reward is maximized. The Hessian matrix becomes $\boldsymbol{H}(\boldsymbol{\theta}) = \sum_{s,a} \Pr(s,a|\mathrm{M}(\boldsymbol{\theta}^*)) \nabla_{\boldsymbol{\theta}}^2 \ln \Pr(s,a|\mathrm{M}(\boldsymbol{\theta}))$. If the policy parameter is nearly optimal $\boldsymbol{\theta} \approx \boldsymbol{\theta}^*$, $\Pr(s,a|\mathrm{M}(\boldsymbol{\theta})) \approx \Pr(s,a|\mathrm{M}(\boldsymbol{\theta}^*))$ holds by the assumption of the smoothness of $\pi(a|s;\boldsymbol{\theta})$ with respect to $\boldsymbol{\theta}$. Therefore, at this time, the Hessian matrix approximately equates the negative, proposed FIM:

$$\begin{aligned}\boldsymbol{H}(\boldsymbol{\theta}) &\approx \sum_{s \in \mathcal{S}} \sum_{a \in \mathcal{A}} \Pr(s,a|\mathrm{M}(\boldsymbol{\theta})) \nabla_{\boldsymbol{\theta}}^2 \ln \Pr(s,a|\mathrm{M}(\boldsymbol{\theta})) \\ &= -\boldsymbol{F}_{s,a}(\boldsymbol{\theta}).\end{aligned}$$

$\boldsymbol{H}(\boldsymbol{\theta}^*) = -\boldsymbol{F}_{s,a}(\boldsymbol{\theta}^*)$ obviously holds. Therefore, when the reward function is in eq.19 and the policy parameter is close to the optimal, NSG almost consists with the Newton direction and the NSG learning attains quadratic convergence.

# Towards Machine Learning of Grammars and Compilers of Programming Languages

Keita Imada and Katsuhiko Nakamura

School of Science and Engineering,
Tokyo Denki University, Hatoyama-machi, Saitama-ken,
350-0394 Japan
nakamura@k.dendai.ac.jp

**Abstract.** This paper discusses machine learning of grammars and compilers of programming languages from samples of translation from source programs into object codes. This work is an application of incremental learning of definite clause grammars (DCGs) and syntax directed translation schema (SDTS), which is implemented in the Synapse system. The main experimental result is that Synapse synthesized a set of SDTS rules for translating extended arithmetic expressions with function calls and assignment operators into object codes from positive and negative samples of the translation. The object language is a simple intermediate language based on inverse Polish notation. These rules contain an unambiguous context free grammar for the extended arithmetic expressions, which specifies the precedence and associativity of the operators. This approach can be used for designing and implementing a new programming language by giving the syntax and semantics in the form of the samples of the translation.

## 1 Introduction

This paper discusses machine learning of grammars and compilers of programming languages from positive and negative samples of translation from source programs into object codes. This work is an application of incremental learning of definite clause grammars (DCGs) [16] and syntax-directed translation schema (SDTS) [1]. The grammatical induction of these extended context free grammars (CFGs) is implemented in the *Synapse* system [13,14].

The DCG can be converted to a logic program for parsing and generating strings in the language of the grammar. The DCG is more powerful than the CFG, as the DCG rules can have additional parameters for controlling derivations and for communicating parameters between separate nodes in the derivation tree.

This paper shows our experimental results of incremental learning of DCG rules representing a restricted form of SDTS, which specify the translation by compilers. We intend to apply this approach to design and implement a new programming language by giving the syntax and semantics in the form of the samples of the translation.

W. Daelemans et al. (Eds.): ECML PKDD 2008, Part II, LNAI 5212, pp. 98–112, 2008.

### 1.1 Grammatical Induction of Extended CFG

Our approach to grammatical induction is characterized by rule generation based on bottom-up parsing for positive sample strings, the search for rule sets and incremental learning. In Synapse, the process called *bridging* generates the production rules that bridge, or make up, any lacking parts of an incomplete derivation tree that is the result of bottom-up parsing of a positive string. The system searches for a set of rules that satisfies all the samples by *global search*, in which the system searches for the minimal set of rules that satisfies given sets of positive and negative samples by iterative deepening.

Incremental learning is essential to this approach. In order to learn a grammar from its sample strings, the positive samples are given to the rule generation process in the order of their lengths. This process continues until the system finds a set of rules that derives all the positive samples, but none of negative samples. By incremental learning, a grammar can be synthesized by adding rules to previously learned grammars of either similar languages or a subset of the target language. This is a method to solve the fundamental problem of computational complexity in learning CFG and more complex grammars [7,18].

An important feature of Synapse for the subject of this paper is that the system synthesizes minimal or semi-minimal grammars based on a covering-based approach. Many other grammatical inference systems for CFGs and their extensions are classified into the generalization based approach, in which the systems generate rules by analyzing the samples and generalizing and abstracting the rules. This is a reason that there have been few publications on learning small rule sets of both ambiguous and unambiguous CFGs and/or extended CFGs. Learning SDTS was chosen as the subject of the *Tenjinno* competition [20] at ICGI 2006. Clark [2] solved some of the problems of the competition. Most participants to this competition, as well as the *Omphalos* competition [19] at ICGI 2004 about learning CFG, did not intend to synthesize small grammars.

Learning DCG is closely related to inductive logic programming (ILP) [15]. Several studies dealt with learning grammars based on ILP [3,10]. Cussens and Pulman [4] describes a method of learning missing rules using bottom-up ILP. There have been, however, few publications focusing on learning DCG and its applications based on ILP. Fredouille et. al. [6] describes a method for efficient ILP-based learning of DCGs for biological pattern recognition. Ross [17] presents learning of Definite Clause Translation Grammar, a logical version of attribute grammar, by genetic programming.

### 1.2 Learning Grammars and Compilers

Processing of programming languages has been a key technology in computer science. There has been much work in compiler theory, including reducing the cost of implementing programming languages and optimizing object codes in compilers. On the other hand, there have been few works on applying machine learning of grammars and compilers. Dubey et. al. [5] describes a method of inferring grammar rules of a programming language dialect based on the grammar of the original language for restoring the lost knowledge of the dialect. Other applications of machine learning to compilers are related to optimizing object codes. Monsifrot et. al. [11] and Stephenson et. al. [21] showed methods of improving optimization heuristics in compilers by using machine learning technologies, decision trees and genetic programming, respectively.

The grammars of programming languages need to not only be unambiguous, but also reflect its semantics, i.e. how a program is computed. For example, any grammar for arithmetic expressions should specify the precedence and left, or right, associativity of the operators. Therefore, synthesizing grammars requires some structural or semantic information other than positive and negative samples of source codes. To address this requirement, in our approach the samples for the learning are pairs of source codes and corresponding object codes in an intermediate language.

### 1.3 Organization of the Paper

Section 2 outlines CFG and SDTS, and defines Chomsky normal form of SDTS. Section 3 briefly defines some basics of logic programming and DCG, and shows a method of representing SDTS by DCG. Section 3 shows the bridging rule generation procedure for both CFGs and DCGs. Section 4 describes search strategies for finding minimal or semi-minimal rule sets, and shows recent experimental results of learning CFGs and DCGs. Section 6 describes learning DCG rules that represent SDTS rules for translating extended arithmetic expressions into object codes in a simple intermediate language, which is based on inverse Polish notation. Section 7 presents the concluding remarks.

## 2 CFGs and SDTS

A *context free grammar* (CFG) is a system $(N, T, P, s)$, where: $N$ and $T$ are finite sets of nonterminal and terminal symbols, respectively; $P$ is the set of (production) rules of the form $p \to u$, $p \in N, u \in (N \cup T)^+$; and $s \in N$ is the starting symbol. We write $w \Rightarrow_G x$ for $w, x \in (N \cup T)^+$, if there are a rule $(p \to u) \in P$ and strings $z_1, z_2 \in (N \cup T)^*$ such that $w = z_1 p\, z_2$ and $x = z_1 u\, z_2$. The *language* of $G$ is the set $L(G) = \{w \in T^+ \mid s \Rightarrow_G^* w\}$, where the relation $\Rightarrow_G^*$ is the reflexive transitive closure of $\Rightarrow_G$. Nonterminal symbols are represented by $p, q, r, s, t$, terminal symbols by $a, b, c, d, e$ and either nonterminal or terminal symbols by $\beta, \gamma$.

*Chomsky normal form* (CNF) rules are of the forms $p \to a$ and $p \to qr$. Synapse synthesizes rules of the *extended CNF*, $p \to \beta$ and $p \to \beta\gamma$. A feature of this normal form is that grammars can be made simpler than those in CNF.

A *syntax-directed translation schema* (SDTS) is a system $T = (N, \Sigma, \Delta, R, s)$, where: $\Sigma$ and $\Delta$ are sets of input terminal and output terminal symbols, respectively; and $R$ is a set of rules of the form

$$p \to u, v. \quad p \in N, u \in (N \cup \Sigma)^+, v \in (N \cup \Delta)^+,$$

such that the set of nonterminal symbols that occur in $u$ is equal to that in $v$. If a nonterminal symbol $p$ appears more than once in $u$, we use symbols with the subscripts $p_1, p_2$ to indicate the correspondence of the symbols in $u$ and $v$. We write $w_1, w_2 \Rightarrow_T x_1, x_2$ for $w_1, x_1 \in (N \cup \Sigma)^*$ and $w_2, x_2 \in (N \cup \Delta)^*$, if there is a rule $(p \rightarrow u, v) \in R$ and strings $y_1, y_2 \in (N \cup \Sigma)^*$ and $z_1, z_2 \in (N \cup \Delta)^*$ such that $w_1 = y_1 \, p \, y_2, x_1 = y_1 u \, y_2, w_2 = z_1 \, p \, z_2$ and $x_2 = z_1 \, v \, z_2$. The SDTS $T$ *translates* a string $w \in \Sigma^+$, into $x \in \Delta^+$ and vice versa, if $(s, s) \Rightarrow_T^* (w, x)$.

The SDTS is *regular*, if and only if every rule is of the form of either $p \rightarrow a\,q, b\,r$ or $p \rightarrow a, b$. The regular SDTS is equivalent to a finite sequential transducer.

We restrict the form of SDTS to *Chomsky normal form* (CNF), in which every string in the right hand side of the rule has at most two symbols, e.g. $(p \rightarrow pq, pq)$, $(p \rightarrow pq, qp)$ or $(p \rightarrow a, b)$. Any SDTS $T$ with rule(s) having three symbols in the right hand side can be transformed into a SDTS in CNF, which is equivalent to $T$. The SDTS in this paper is extended so that elements of the strings may be not only constants but also lists of the constants. There are, however, SDTS rules with more than three symbols in the right hand side, that cannot be simply transformed to CNF.

*Example 1: reversing strings* The following SDTS in CNF translates any string into its reversal, for example, *aababb* to *bbabaa*.

$$s \rightarrow a, a \qquad s \rightarrow b, b \qquad s \rightarrow a\,s, s\,a \qquad s \rightarrow b\,s, s\,b$$

This SDTS derives the set of pairs of strings,

$$(a, a), (b, b), (aa, aa), (ab, ba), (ba, ab), (bb, bb), (aaa, aaa), (aab, baa), \cdots .$$

## 3 Definite Clause Grammars (DCG)

We use the notations and syntax of standard Prolog for constants, variables, terms, lists and operators, except that the constants are called atoms in Prolog. A constant (an atom in Prolog) is an identifier that starts with a lower-case character, and a variable starts with an upper-case character, as in standard Prolog. A *subterm* of a term $T$ is either $T$, an argument of a complex term $T$ or recursively a subterm of an argument of a complex term $T$. As any list is a special term, subterms of $[a, b, c]$ are $[a, b, c]$, $a$, $[b, c]$, $b$, $[c]$, $c$, $[\,]$.

A *substitution* $\theta$ is a mapping from a set of variables into a set of terms. For any term $t$, an *instance* $t\theta$ is a term in which each variable $X$ defined in $\theta$ is replaced by its value $\theta(X)$. For any terms $s$ and $t$, we write $s \succeq t$, and say that $s$ is *more general* than $t$, if and only if $t$ is an instance of $s$. A *unifier* for two terms $s$ and $t$ is a substitution $\theta$, such that $s\theta = t\theta$. The unifier $\theta$ is the *most general unifier* (mgu), if there is another unifier $\sigma$ for $s$ and $t$, then $s\theta \succeq s\sigma$ and $t\theta \succeq t\sigma$. We write $s \equiv_\theta t$, if $s$ is unifiable with $t$ by an mgu $\theta$. For any terms $s$ and $t$, a term $u$ is the *lgg: least general generalization*, if and only if $u \succeq s$, $u \succeq t$ and there is no other term $v$ such that $v \succeq s$, $v \succeq t$ and $u \succeq v$.

A *DCG rule*, also called *grammar rule* in the draft of the ISO standard [12], is of the form $P$ `-->` $Q_1, Q_2, \cdots, Q_m$, where:

- $P$ is a nonterminal term, which is either a symbol or a complex term of the form $p(T)$ with a *DCG term* $T$; and
- each of $Q_1, \cdots, Q_m$ is either a constant of the form $[a]$ representing a terminal symbol, or a nonterminal term.

The DCG terms are additional arguments in the Horn clause rules, which are generally used for controlling the derivation and for returning results of the derivation. The deduction by DCG is similar to that of CFG, except that each of the DCG terms is unified with a corresponding term in the deduction. To simplify the synthesis process, we restrict every atom for the nonterminal symbol to have exactly one DCG term.

Most Prolog implementations have a functionality to transform the grammar rules into Horn clauses, such that a string $a_1 a_2 \cdots a_n$ is derived by the rule set from the starting symbol $s$, if and only if the query $s([a_1, a_2, \cdots, a_n], [])$ succeeds for the parsing program composed of the transformed clauses. Note that the two arguments are used for representing strings by the difference lists.

*Example 2: A DCG for non-context-free language* The following set of rules is a DCG for the language $\{a^n b^n c^n \mid n \geq 1\}$.

```
p(1) --> [a].          p(t(N)) --> [a],p(N).
q(1) --> [b].          q(t(N)) --> [b],q(N).
r(1) --> [c].          r(t(N)) --> [c],r(N).
s(N) --> p(N),q(N),r(N).
```

The Horn clauses transformed from these DCG rules are:

```
p(1,[a|X],X).          p(t(N),[a|X],Y) :- p(N,X,Y).
q(1,[b|X],X).          q(t(N),[b|X],Y) :- q(N,X,Y).
r(1,[c|X],X).          r(t(N),[c|X],X) :- r(N,X,Y).
s(N,X0,X3) :- p(N,X0,X1),q(N,X1,X2),r(N,X2,X3).
```

For the query `?-s(N,[a,a,a,b,b,b,c,c,c],[])`, the Prolog system returns the computation result `N = t(t(1))`.

We can transform a SDTS in CNF into a DCG for translating strings by the relations in Table 1. Each pair of the form $X/Y$ represents an output side of a string by a difference list. By this method, the problem of learning SDTS in CNF from pairs of input and output strings can be transformed into that of learning DCG.

*Example 3: Reversal* The following DCG corresponds to the example SDTS for reversing strings in Section 2.

```
s([a|X]/X) --> [a].               s([b|X]/X) --> [b].
s([a|X]/[]) --> s(X/[]),[a].      s([b|X]/[]) --> s(X/[]),[b].
```

For a query `?-s(X,[b,b,a,b,a,a],[])`, the transformed Prolog program returns the value `X = [a,a,b,a,b,b]/[]` of the DCG term, and for a query `?-s([a,a,b,a,b,b]/[],X,[])`, the solution `X = [b,b,a,b,a,a]`.

**Table 1.** Relations between SDTS rules and DCG rules

| SDTS | DCG |
|---|---|
| $p \to q$ | $p(X/Y)$ --> $q$ |
| $p \to a, b$ | $p([b\|Y]/Y)$ --> $[a]$ |
| $p \to ar, ra$ | $p(X/Y)$ --> $[a], r(X/[a\|Y])$ |
| $p \to qr, qr$ | $p(X/Z)$ --> $q(X/Y), r(Y/Z)$ |
| $p \to qr, rq$ | $p(X/Z)$ --> $q(Y/Z), r(X/Y)$ |

## 4 Rule Generation Based on Bottom-Up Parsing

Fig. 1 shows the rule generation procedure, which receives a string $a_1 \cdots a_n$, the starting symbol with a DCG term $s(T)$, and a set of rules from the top-level search procedure from global variable $P$, and returns a set of DCG rules that derives the string from $s(T)$. This nondeterministic procedure is an extension of that for learning CFG in extended CNF [14]. The extension is related to adding DCG terms and generalization for the generated rules.

### 4.1 Rule Generation

The rule generation procedure includes a bottom-up parsing algorithm for parsing an input string $a_1 \cdots a_n$ using the rules in the set $P$. If the parsing does not succeed, the bridging process generates rules in extended CNF, which bridge any lacking parts of the incomplete derivation tree.

The input string $a_1 a_2 \cdots a_n$ is represented by a set of 3-tuples $\{(a_1, 0, 1), (a_2, 1, 2), \cdots, (a_n, n-1, n)\}$, and the resulting derivation tree by a set $D$ of 3-tuples of the form $(p(T), i, j)$, each of which represents that the set of rules derives $a_i \cdots a_j$ from $p(T)$. For the ambiguity check, each time a new term $(p(T), i, j)$ is generated, it is tested whether it has been generated before.

Subprocedure $Bridge(p(T), i, k)$ generates additional rules that bridge missing parts in the incomplete derivation tree represented by the set of terms in $D$. The process nondeterministically chooses six operations, as shown in Fig. 2. In operations 5 and 6, nonterminal symbols $q$ and $r$ are nondeterministically chosen from either previously used symbols or new symbols. The DCG term $U$ and $V$ of these symbols are also nondeterministically chosen from the subterms of the term $T$ of the parent node. Subprocedure $AddRule(R)$ first searches for a rule $R'$ in the set $P$ that has the same form as $R$ in the sense that $R$ and $R'$ differ only in the DCG terms. It then replaces $R'$ with the lgg of $R$ and $R'$, or simply adds $R$ to $P$.

Synapse has a special mode to efficiently generate DCG rules for SDTS. When this mode is selected, each DCG term is restricted to a difference list, and the rules are restricted to forms in Table 1. The system uses these restrictions for generating rules for SDTS in Operation 5 and 6 in Fig. 2 and in generalization in subprocedure *AddRule*.

**Procedure** *RuleGeneration*$(w, s(T_0), K)$ (Comment: $w$ : an input string, $s(T)$ : the starting symbol with a DCG term, $K$ : the bound for the number of rules. Global variable $P$ holds a set of rules.)

**Step 1** (Initialize variables.)
$D \leftarrow \emptyset$. ($D$ is a set of 3-tuples $(\beta(U), i, j)$.)
$k \leftarrow |P|$. ($k$ holds the initial number of rules in $P$.)

**Step 2:** (Parsing by inverse derivation)
For each $a_i, 1 \leq i \leq n = |w|$ in $w$, call *Derive*$(a_i, i-1, i)$ in order.
If $(s(T), 0, n) \in D$ and $T_0 \succeq T$ then terminate (Success).

**Step 3:** (Bridging rule generation)
If $|P| \geq K$ then terminate (Failure).
Call procedure *Bridge*$(s(T), 0, n)$.
Terminate (Success). (Return the set $P$ of rules).

**Procedure** *Derive*$(\beta, i, j)$ ($\beta$ : either a terminal a nonterminal term, $i, j$ : integers for representing the positions of a substring. )

1. Add $(\beta, j, k)$ to $D$. If $p(W) \rightarrow \beta' \in P$ such that $\beta \equiv_\theta \beta'$, then add $(p(W\theta), j, k)$ to $D$.
   To synthesize an unambiguous grammar, check ambiguity.
2. If $p(W) \rightarrow \alpha\beta \in P$ and $(\alpha', i, j) \in D$ with $\alpha \equiv_\theta \alpha'$ and $\beta \equiv_\theta \beta'$, then add $(p(W\theta), i, k)$ to $D$, and call *Derive*$(p(W\theta), i, j)$.

**Procedure** *Bridge*$(p(T), i, k)$ ($p(T)$ : a nonterminal term, $i, k$ : integers.)
Nondetermistically choose one of the operations in Fig. 2.

**Procedure** *AddRule*$(R)$ ($R$: a rule.)
Nondereminsitically choose one of the following process 1 or 2.

1. If $P$ contains a rule $R'$ such that $R$ differs from $R'$ only in DCG terms, delete $R'$ from $P$ and add the lgg of $R$ and $R'$ to $P$. Else add $R$ to $P$.
2. Add $R$ to $P$.

**Fig. 1.** Procedure for Rule Generation by Bridging

# 5 Search for Rule Sets

The inputs to Synapse are ordered sets $S_P$ and $S_N$ of positive and negative samples, respectively, and a set $P_0$ of initial rules for incremental learning of the grammars. Samples for learning a CFG are strings, whereas those for learning DCGs are pairs of strings and atoms of the form $s(T)$ with DCG terms $T$. The system searches for any set $P$ of rules with $P_0 \subseteq P$ such that all the strings in $S_P$ are derived from $P$ but no string in $S_N$ is derived from $P$. Synapse has two search strategies, global and serial search, for finding rule sets.

Fig. 3 shows the top-level procedure for the global search for finding minimal rule sets. The system controls the search by the iterative deepening on the number of rules to be generated. First, the number of initial rules is assigned to the bound $K$ of the number of rules. When the system fails to generate sufficient rules to parse the samples within this bound, it increases the bound by one and iterates the search. By this control, it is assured that the procedure finds a grammar with the minimal number of rules at the expense that the system repeats the same search each time the bound is increased.

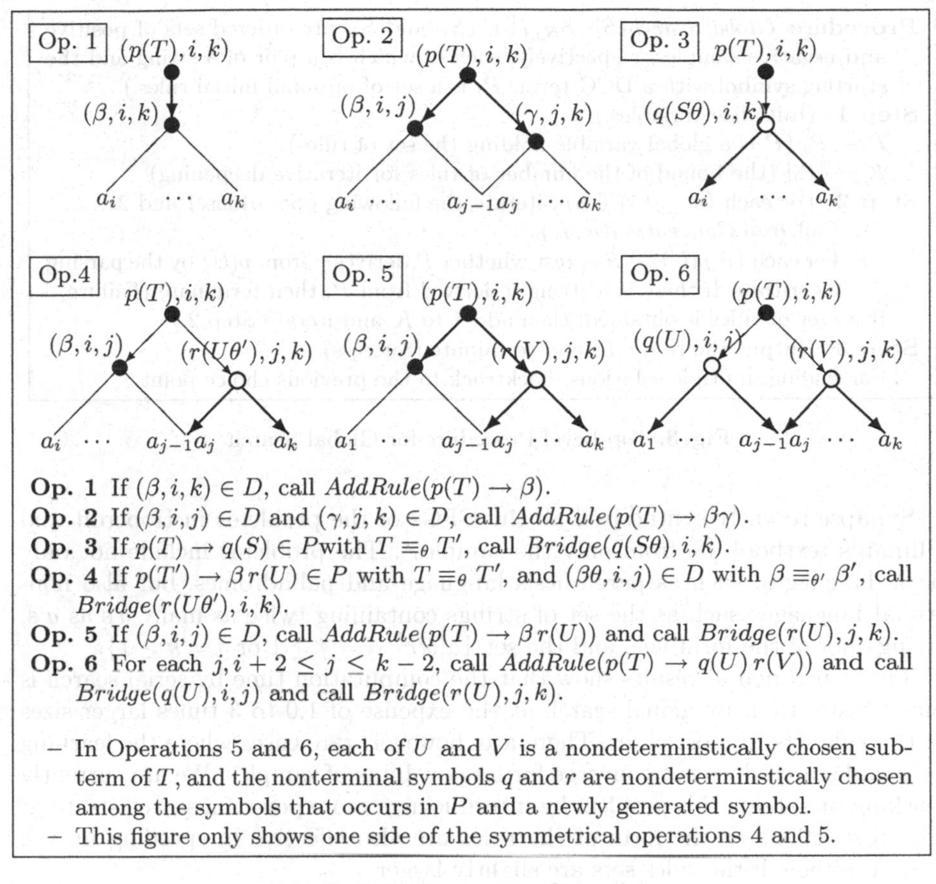


**Fig. 2.** Rule Generation Operations

### 5.1 Serial Search and Search for Semi-minimal Rule Sets

In the serial search, the system generates additional rules for each positive sample by iterative deepening. After the system finds a rule set satisfying a positive sample and no negative samples, the process does not backtrack to redo the search on the previous samples. By this search strategy, the system generally finds semi-minimal rule sets in shorter computation time. Other methods for finding semi-minimal rule sets include using non-minimal nonterminal symbols in rule generation and restricting the form of the generated rules. These methods generally increase the efficiency of searching for rule sets at the cost that the rule sets found may not be minimal.

### 5.2 Experimental Results on Learning CFG and DCG

The experimental results in this paper were obtained with Synapse Version 4, written in Prolog, using a Xeon processor with a 3.6 GHz clock and SWI-Prolog for Linux.

**Procedure** *GlobalSearch*($S_P, S_N, P_0$) ($S_P$ and $S_N$ are ordered sets of positive and negative samples, respectively, each of which is a pair of a string and the starting symbol with a DCG term. $P_0$ is a set of optional initial rules.)

**Step 1** (Initialize variables.)
$P \leftarrow P_0$ ($P$ is a global variable holding the set of rules).
$K \leftarrow |P_0|$ (the bound of the number of rules for iterative deepening).

**Step 2:** For each $(w, s(T)) \in S_P$, iterate the following operations 1 and 2.
1. Call *RuleGeneration*($w, K$).
2. For each $(v, p(U)) \in S_N$, test whether $P$ derives $v$ from $p(U)$ by the parsing algorithm. If there is a string $v$ derived from $P$, then terminate (Failure).

If no set of rules is obtained, then add 1 to $K$ and iterate Step 2.

**Step 3:** Output the result $P$, and terminate (Success).
For finding multiple solutions, backtrack to the previous choice point.

**Fig. 3.** Top-Level Procedure for Global Search

Synapse recently synthesized all the CFGs of the problems in Hopcroft and Ullman's textbook [8] from only the samples[1]. The problems include not only basic languages, such as parenthesis language and palindromes, but also non-trivial languages such as the set of strings containing twice as many $b$'s as $a$'s, strings not of the form $ww$, and the set $\{a^i b^j c^k \mid i = j \geq 1 \text{ or } j = k \geq 1\}$.

The experimental results show that the computation time by serial search is much faster than by global search at the expense of 1.0 to 3 times larger sizes of the rule sets in most cases. There are, however, rare cases where the learning by serial search does not converge for large volume of samples. We are currently working on solving this problem by introducing a more powerful search strategy. The restrictions to the form of the rules are also effective in speeding up the search, although the rules sets are slightly larger.

*Example 4: Reversal* The following pairs are some examples of positive samples for learning a DCG for reversal.

```
[a] - [a].       [b] - [b].         [a,b] - [b,a].
[b,a] - [a,b].   [a,a,b] - [b,a,a]. [a,a,b,a]- [a,b,a,a].
```

Given these samples and no negative samples, Synapse synthesized the following DCG rules after generating 12 rules in 0.13 sec.

```
s([a|X]/X) --> a.                  s([b|X]/X) --> b.
s(X/Y) --> s(Z/Y), s(X/Z).
```

By the serial search, the same rule sets are found after generating 21 rules in less than 0.1 sec. This DCG of three rules is smaller than the DCG for reversing strings in Example 3 in Section 2.

Synapse synthesized the DCG rules for the language $\{a^n b^n c^n \mid n \geq 1\}$ in Example 2 in Section 3 after generating $6.5 \times 10^5$ rules in 3200 sec. by giving

[1] Detailed experimental results at the time 2006 are shown in [14].

three initial rules `p(1) --> [a]`, `q(1) --> [b]`, `r(1) --> [c]` in addition to the positive and negative samples.

## 6 Learning Translation in Simple Compiler

This section shows experimental results of learning translation of arithmetic expressions and assignment statements into an object language called *SIL* (Simple Intermediate Language). This language is similar to P-code and byte-code, which were originally designed for intermediate languages of Pascal and JAVA compilers, respectively. In these languages, object codes for arithmetic and logical expressions are based on the inverse Polish notation. The language SIL includes the following instructions.

- `load(T,V)` pushes the value of variable (symbolic address) `V` of type `T`, which is either `i` or `f` (float), to the top of the stack.
- `push(T,C)` pushes a constant `C` of type `T` to the top of the stack.
- `store(T,V)` stores the value at the top of the stack to variable (symbolic address) `V` of type `T`.
- Arithmetic operations: `fadd, fsubt, fmult, fdivide`. These float type operations are applied to two values on the stack and return the result at the top of the stack instead of the two values.
- `call(n,S)` calls function `S` with `n` parameters, which are placed at the top of the stack. The function returns the value at the top of the stack.

Note that although every instruction is typed, we deal with only object codes of float type in this paper.

### 6.1 Step One: Learning Translation of Arithmetic Expression

For reducing the computation time, we divide the learning of translation into two steps, and use incremental learning. In the first step, we gave Synapse positive and negative samples as shown in Fig. 4 and the initial rules in Fig. 5 (a). Each of the samples is a pair of an arithmetic expression and an object code in SIL. The negative samples containing symbol `X` in the right hand side represent restriction only on the source language. We assume that the rules for `a` and `b` are generated by processing constants, and those for `x` and `y` by declarations of the variables.

Synapse was made to search unambiguous DCGs with the starting symbol `s1`. The system synthesized eight DCG rules in Fig. 6 (a) after generating $3.4 \times 10^6$ rules in approximately 2000 sec by the global search. Among the positive samples, only the first nine samples were directly used for generating the grammar and the other positive samples are used for checking of the translation; the system parsed these samples without generating any additional rules. For the learning, 24 randomly chosen negative samples including Fig. 4 were sufficient.

**Positive samples**

```
a : push(f,a).                            ( a ) : push(f,a).
a + a : push(f,a),push(f,a),fadd.         a * a : push(f,a),push(f,a),fmult.
a / a : push(f,a),push(f,a),fdivide.      a - a : push(f,a),push(f,a),fsubt.
a + a + a : push(f,a),push(f,a),fadd,push(f,a),fadd.
a * a + a : push(f,a),push(f,a),fmult,push(f,a),fadd.
a / a + a : push(f,a),push(f,a),fdivide,push(f,a),fadd.
a - a + a : push(f,a),push(f,a),fsubt,push(f,a),push(f,a),fadd.
( a ) + a : push(f,a),push(f,a),fadd.
a + a * a : push(f,a),push(f,a),push(f,a),fmult,fadd.
a * a * a : push(f,a),push(f,a),fmult,push(f,a),fmult.
```

**Negative samples**

```
( a : X.        a ) : X.        ( + a : X.       ( * a : X.
+ a : X.        * a : X.        a + : X.         a * : X.
a + a : push(f,a),fadd,push(f,a).
a * a : push(f,a),fmult,push(f,a).
a + a + a : push(f,a),push(f,a),push(f,a),fadd,fadd.
a * a + a : push(f,a),push(f,a),push(f,a),fmult,fadd.
a + a * a : push(f,a),push(f,a),fadd,push(f,a),fmult.
```

**Fig. 4.** Samples for translating arithmetic expressions into the object codes in SIL for Step One

(a) Initial rules for Step One

```
n([push(f,a)|Y]/Y) --> a.          n([push(f,b)|Y]/Y) --> b.
n([load(f,x)|Y]/Y) --> x.          v([store(f,x)|Y]/Y) --> x.
n([load(f,y)|Y]/Y) --> y.          v([store(f,y)|Y]/Y) --> y.
op1([fadd|Y]/Y) --> +.             op1([fsubt|Y]/Y) --> -.
op2([fmult|Y]/Y) --> *.            op2([fdivid|Y]/Y) --> /.
lp(Y/Y) --> '('.                   rp(Y/Y) -->')'.
```

(b) Initial rules for Step Two

```
op3(Y/Y) --> =.                    s(X/Y) --> s1.
fn([call(1,sin)|Y]/Y) --> sin.     fn([call(1,cos)|Y]/Y) --> cos.
```

**Fig. 5.** Initial DCG rules for translating arithmetic expressions and assignment statements into object codes in SIL

### 6.2 Step Two: Learning Translation of Function Calls and Assignment Operator

In the second step, we gave Synapse the samples including those in Fig. 7, and made the system search for rules for translating function calls and assignment operator (=), based on the result of the first step. The starting symbol was set

(a) Rules for Arithmetic Expressions Synthesized in Step 1

```
s1(X/Y) --> e(X/Y).                s → e, e
s1(X/Z) --> s1(X/Y), f(Y/Z).       s → s f, f p
f(X/Z) --> op1(Y/Z),e(X/Y).        p → op1 e, e op1
e(X/Y) --> n(X/Y).                 e → n, n
e(X/Z) --> e(X/Y), g(Y/Z).         e → e g, e g
g(X/Z) --> op2(Y/Z), n(X/Y).       r → op2 n, n op2
n(X/Z) --> lp(X/Y), p(Y/Z).        s → lp p, lp p
p(X/Z) --> s1(X/Y), rp(Y/Z).       p → s rp, s rp
```

(b) Rules for Function Calls and "=" Operator Synthesized in Step 2

```
n(X/Z) --> fn(Y/Z), q(X/Y).        n → fn q, q fn
q(X/Z) --> lp(X/Y), p(Y/Z).        n → lp p, lp p
s(X/Z) --> v(Y/Z),r(X/Y).          s → v r, r v
r(X/Z) --> op3(Y/Z), s(X/Y).       q → op3 s, op3 s
```

**Fig. 6.** Synthesized DCG rules and the corresponding SDTS rules for translating extended arithmetic expression into object codes in SIL

to s. The initial rules are composed of the synthesized rules in Step One in Fig. 6 (a) and all the initial rules in Fig. 5, which include `s(X/Y) --> s1`, (s → s1 in SDTS).

By incremental learning, Synapse synthesized the four additional rules in Fig. 6 (b) after generating $1.8 \times 10^7$ rules in 4400 sec. Only the first four positive samples were directly used for generating the grammar and the other positive samples are used for checking of the translation. In Step two, 24 randomly chosen negative samples were also sufficient.

The synthesized DCG contains the CFG for the extended arithmetic expressions, which specifies that:

1. The precedence of the operators in op2 (* and /) is higher than that of op1 (+ and -), which is higher than the assignment operator "="; and
2. All arithmetic operators of op1 and op2 are left associative. The operator "=" is right associative. (Note that this associativity is coincident with that of C language. In many languages, the assignment operator is non-associative.)

The synthesized DCG rules can be converted to usual Prolog rules by adding two arguments for the difference lists representing the input strings. Since the DCG is left recursive, we needed to remove the left recursion by folding some of the clauses. The obtained program is executed as follows.

```
?- s(X/[],[a,*,b,*,b,*,'(',a,+,x,')'],[]).
X = [push(f,a),push(f,b),fmult,push(f,b),fmult,push(f,a),
     load(f,x),fadd,fmult]

?- s(X/[],[x,=,y,=,sin,'(',a,+,b,*,y,')'],[]).
```

**Positive samples**

```
x = a : push(f,a), store(f,x). sin ( a ) : push(a),call(1,sin). x
= a + b : push(f,a), push(f,b),fadd, store(f,x). x = y = a :
push(f,a), store(f,y),store(f,x). sin ( a + b ) :
push(f,a),push(f,b),fadd,call(1,sin). sin ( a ) + b :
push(f,a),call(1,sin),push(f,b),fadd. sin ( a ) * b :
push(f,a),call(1,sin),push(f,b),fmult. sin ( a ) * cos ( b ) :
push(f,a),call(1,sin), push(f,b),
                    call(1,cos),fmul.
sin ( cos ( b ) ) : push(f,b),call(1,cos),call(1,sin). sin ( a + b
) : push(f,a),push(f,b),fadd,call(1,sin).
```

**Negative samples**

```
sin a : X.          sin + a : X.       sin * a : X.      sin a ) : X.
sin ( a ) a : X.    sin ) : X.         sin sin : X.      sin + : X.
sin * : X.          = a a ) : X.       = a : X.          ( = a : X. (
a = a : X.       a = x : X.        = x ( a ) : X.    a = b b : X. x
= a + b : push(f,a),store(f,x),push(f,b),fadd.
```

**Fig. 7.** Samples for learning DCG for translating extended arithmetic expressions with function calls and assignment operators into the object codes in SIL

```
X = [push(f,a),push(f,b),load(f,y),fmult,fadd,call(1,sin),
     store(f,y),store(f,x)] ;
```

The computation is deterministic, and each query has only one solution.

## 7 Concluding Remarks

We showed an approach for machine learning of grammars and compilers of programming languages based on grammatical inference of DCG and SDTS, and showed the experimental results. The results of this paper are summarized as follows.

1. We extended the incremental learning of minimal, or semi-minimal, CFGs in the Synapse system to those of DCG and of SDTS.
2. Synapse synthesized a set of rules in SDTS for translating arithmetic expressions with function calls and assignment operators into object codes from samples of the translation. This set of SDTS rules can be used as a compiler in Prolog that outputs object codes in the intermediate language SIL.
3. The synthesized SDTS rules contain an unambiguous CFG for the extended arithmetic expressions, which specifies the precedence and associativity of the operators.

Although we showed learning of only a portion of compiling process of of the existing language, the learning system synthesized an essential part of the compiler from samples of translation. This approach can be used for produce the grammar of a new language and at the same time implement the language from the samples of source programs and object codes.

We are currently working to improve the methods of learning DCGs and SDTSs, and extending the compiler to include type check and type conversion and to translate other statements and declarations. For type checking, the non-terminal symbols need to have additional parameters for the type. As the control instructions in the object codes generally have labels, we need a non-context-free language for the object codes. Other future subjects include:

- Theoretical analysis of learning DCG and SDTS.
- Clarifying the limitations of our methods in learning grammars and compilers of programming languages.
- Applying our approach for learning DCG to syntactic pattern recognition.
- Applying our approach for learning DCG to general ILP, and inversely applying methods in ILP to learning DCG and SDTS.

## Acknowledgements

The authors would like to thank Akira Oouchi and Tomihiro Yamada for their help in writing and testing the Synapse system. This work is partially supported by KAKENHI 19500131 and the Research Institute for Technology of Tokyo Denki University Q06J-06.

## References

1. Aho, A.V., Ullman, J.E.: The Theory of Parsing, Translation, and Compiling. Prentice-Hall, Englewood Cliffs (1972)
2. Clark, A.: Large Scale Inference of Deterministic Transductions: Tenjinno Problem 1. In: Sakakibara, Y., Kobayashi, S., Sato, K., Nishino, T., Tomita, E. (eds.) ICGI 2006. LNCS (LNAI), vol. 4201, pp. 227–239. Springer, Heidelberg (2006)
3. Bratko, I.: Refining Complete Hypotheses in ILP. In: Džeroski, S., Flach, P.A. (eds.) ILP 1999. LNCS (LNAI), vol. 1634, pp. 44–59. Springer, Heidelberg (1999)
4. Cussens, J., Pulman, S.: Incorporating Linguistic Constraints into Inductive Logic Programming. In: Proc. CoNLL 2000 and LLL 2000, pp. 184–193 (2000)
5. Dubey, A., Jalote, P., Aggarwal, S.K.: Inferring Grammar Rules of Programming Language Dialects. In: Sakakibara, Y., Kobayashi, S., Sato, K., Nishino, T., Tomita, E. (eds.) ICGI 2006. LNCS (LNAI), vol. 4201, pp. 201–213. Springer, Heidelberg (2006)
6. Fredouille, D.C., et al.: An ILP Refinement Operator for Biological Grammar Learning. In: Muggleton, S., Otero, R., Tamaddoni-Nezhad, A. (eds.) ILP 2006. LNCS (LNAI), vol. 4455, pp. 214–228. Springer, Heidelberg (2007)
7. de la Higuera, C., Oncina, J.: Learning Context-Free Languages, PASCAL-Pattern analysis, Static Model and Computational Learning, p. 28 (2004), http://eprints.pascal-network.org/archive/00000061/
8. Hopcroft, J.E., Ullman, J.E.: Introduction to Automata Theory, Languages, and Computation. Addison-Wesley, Reading (1979)
9. Kowalski, R.: Logic for Problem Solving. North-Holland, Amsterdam (1979)

10. Muggleton, S.: Inverse Entailment and Progol. New generation Computing 13, 245–286 (1995)
11. Monsifrot, A., Boodin, F., Quiniou, R.: A Machine Learning Approach to Automatic Production of Compiler Heuristics. In: Scott, D. (ed.) AIMSA 2002. LNCS (LNAI), vol. 2443, pp. 41–50. Springer, Heidelberg (2002)
12. Moura, P.: Definite clause grammar rules, ISO/IEC DTR 13211-3 (2006)
13. Nakamura, K., Matsumoto, M.: Incremental Learning of Context Free Grammars Based on Bottom-up Parsing and Search. Pattern Recognition 38, 1384–1392 (2005)
14. Nakamura, K.: Incremental Learning of Context Free Grammars by Bridging Rule Generation and Semi-Optimal Rule Sets. In: Sakakibara, Y., Kobayashi, S., Sato, K., Nishino, T., Tomita, E. (eds.) ICGI 2006. LNCS (LNAI), vol. 4201, pp. 72–83. Springer, Heidelberg (2006)
15. Nienhuys-Cheng, S.H., de Wolf, R.: Foundations of Inductive Logic Programming. Springer, Heidelberg (1997)
16. Pereira, F., Warren, D.H.D.: Definite Clause Grammars for Language Analysis: A Survey of the Formalism and a Comparison with Augmented Transition Networks. Jour. of Artificial Intelligence 13, 231–278 (1980)
17. Ross, B.J.: Logic-Based Genetic Programming with Definite Clause Translation Grammars. New generation Computing 19, 313–337 (2001)
18. Sakakibara, Y.: Recent advances of grammatical inference. Theoretical Computer Science 185, 15–45 (1997)
19. Starkie, B., Coste, F., van Zaanen, M.: The Omphalos Context-Free Grammar Learning Competition. In: Paliouras, G., Sakakibara, Y. (eds.) ICGI 2004. LNCS (LNAI), vol. 3264, pp. 16–27. Springer, Heidelberg (2004)
20. Starkie, B., van Zaanen, M., Estival, D.: The Tenjinno Machine Translation Competition. In: Sakakibara, Y., Kobayashi, S., Sato, K., Nishino, T., Tomita, E. (eds.) ICGI 2006. LNCS (LNAI), vol. 4201, pp. 214–226. Springer, Heidelberg (2006)
21. Stephenson, M., Martin, M., O'Relly, U.M.: Meta Optimization: Improving Compiler Heuristics with Machine Learning. In: Proc. of PLDI 2003, San Diego, pp. 77–90 (2003)

# Improving Classification with Pairwise Constraints: A Margin-Based Approach

Nam Nguyen and Rich Caruana

Department of Computer Science
Cornell University, USA
{nhnguyen,caruana}@cs.cornell.edu

**Abstract.** In this paper, we address the semi-supervised learning problem when there is a small amount of labeled data augmented with pairwise constraints indicating whether a pair of examples belongs to a same class or different classes. We introduce a discriminative learning approach that incorporates pairwise constraints into the conventional margin-based learning framework. We also present an efficient algorithm, PCSVM, to solve the pairwise constraint learning problem. Experiments with 15 data sets show that pairwise constraint information significantly increases the performance of classification.

**Keywords:** classification, pairwise constraints, margin-based learning.

## 1 Introduction

Learning with partially labeled training data, also known as semi-supervised learning, has received considerable attention, especially for classification and clustering [1,2,3,4,5,6,7,8,9,10,11,12,13,14,15,16,17]. While labeled data is usually expensive, time consuming to collect, and sometimes requires human domain experts to annotate, unlabeled data often is relatively easy to obtain. For this reason, semi-supervised learning has mainly focused on using the large amount of unlabeled data [18], together with a small amount of labeled data, to learn better classifiers. Note that unlabeled data may not always help. For example, [19] showed that unlabeled data can degrade classification performance even in situations where additional labeled data would increase the performance. Hence, partially labeled data is an attractive tradeoff between fully labeled data and unlabeled data.

In this paper, we investigate the usefulness of partially labeled information in the form of pairwise constraints. More specifically, a pairwise constraint between two items indicates whether they belong to the same class or not. Similar to unlabeled data, in many applications pairwise constraints can be collected automatically, e.g. in [1], pairwise constraints are extracted from surveillance video. Pairwise constraints also can be relatively easy to collect from human feedback: unlike labels that would require users to have prior knowledge or experience with a data set, pairwise constraints require often little effort from users. For example, in face recognition, it is far easier for users to determine if two faces

W. Daelemans et al. (Eds.): ECML PKDD 2008, Part II, LNAI 5212, pp. 113–124, 2008.

belong to the same nationality, than it would be for the same users to classify the faces into different nationalities.

In this paper, we propose a discriminative learning approach which incorporates pairwise constraints into the conventional margin-based learning framework. In extensive experiments with a variety of data sets, pairwise constraints significantly increase the performance of classification. The paper is structured as follow: in section 2, we describe in detail our classification algorithm, PCSVM, which incorporates pairwise constraints; in section 3, we review related work on semi-supervised learning with pairwise constraints; the experimental results and conclusion are given in section 4 and 5, respectively.

## 2 Classification with Pairwise Constraints

In the supervised setting, a learning algorithm typically takes a set of labeled training examples, $\mathbf{L} = \{(x_i, y_i)\}_{i=1}^n$ as input, where $x_i \in \mathcal{X}$ and $y_i$ belongs to a finite set of classes called $\mathcal{Y}$. For our learning framework, in addition to the labeled data, there is additional partially labeled data in the form of pairwise constraints $\mathbf{C} = \{(x_i^\alpha, x_j^\beta, \widetilde{y}_i)\}_{i=1}^m$ where $x_i^\alpha, x_j^\beta \in \mathcal{X}$ and $\widetilde{y}_i \in \{+1, -1\}$ is the indicator of whether $x_i^\alpha$ and $x_j^\beta$ belong to the same class ($\widetilde{y}_i = +1$), or not ($\widetilde{y}_i = -1$). Ultimately, the goal of classification is to form a hypothesis $h : \mathcal{X} \mapsto \mathcal{Y}$.

First, we review the margin-based multiclass classification, also known as the multiclass-SVM proposed by [20]. Consider a mapping $\Phi : \mathcal{X} \times \mathcal{Y} \mapsto \mathcal{F}$ which projects each item-label pair $(x, y) \in \mathcal{X} \times \mathcal{Y}$ to $\Phi(x, y)$ in a new space $\mathcal{F}$,

$$\Phi(x, y) = \begin{bmatrix} x \cdot \mathcal{I}(y = 1) \\ \cdots \\ x \cdot \mathcal{I}(y = |\mathcal{Y}|) \end{bmatrix},$$

where $\mathcal{I}(\cdot)$ is the indicator function. The multiclass-SVM learns a weight vector $w$ and slack variables $\xi$ via the following quadratic optimization problem:

Optimization Problem I: Multiclass-SVM

$$\min_{w, \xi \geq 0} : \frac{\lambda}{2} \|w\|^2 + \frac{1}{n} \sum_{i=1}^{n} \xi_i \tag{1}$$

subject to:

$$\forall (x_i, y_i) \in \mathbf{L}, \overline{y}_i \in \mathcal{Y} \backslash y_i : \; w^T \left[ \Phi(x_i, y_i) - \Phi(x_i, \overline{y}_i) \right] \geq 1 - \xi_i.$$

After we have learned $w$ and $\xi$, the classification of a new example,$x$, is done by

$$h(x) = \operatorname*{argmax}_{y \in \mathcal{Y}} w^T \Phi(x, y).$$

In this margin-based learning framework, we observed that for a training example $(x_i, y_i) \in \mathbf{L}$ the score associated with the correct label $y_i$, $w^T \Phi(x_i, y_i)$,

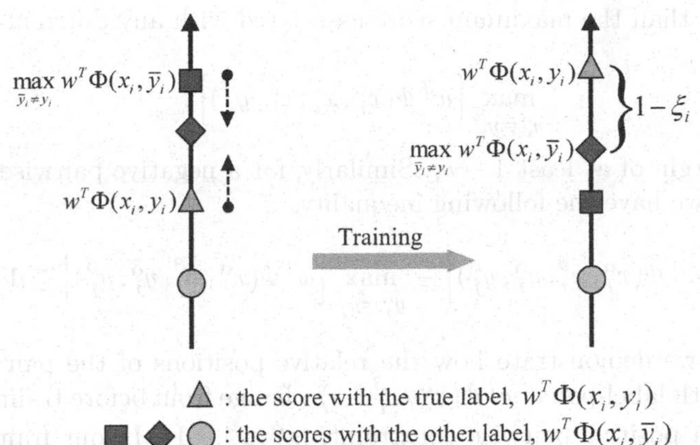


**Fig. 1.** Illustration of how the relative positions of the scores associated with different labels, $w^T \Phi(x_i, \cdot)$, change from before training to after training for a fully labeled example

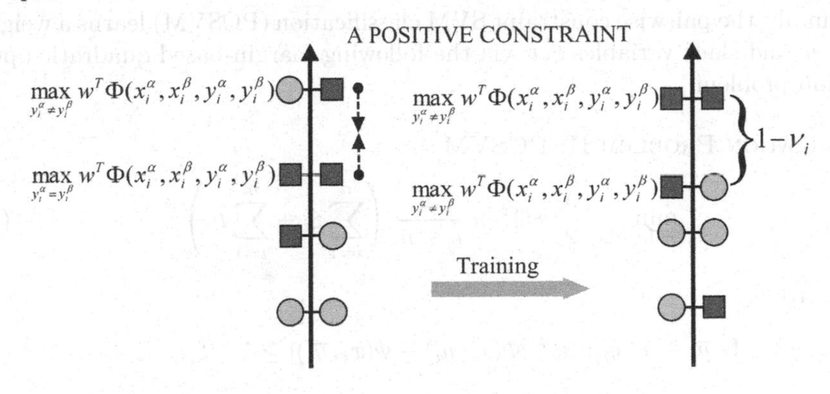


: the pairwise scores with the same-label pair, $w^T \Phi(x_i^\alpha, x_i^\beta, y_i^\alpha = y_i^\beta)$

: the pairwise scores with the different-label pair, $w^T \Phi(x_i^\alpha, x_i^\beta, y_i^\alpha \neq y_i^\beta)$

**Fig. 2.** Illustration of how the relative positions of the pairwise scores associated with label-pairs, $w^T \Phi(x_i^\alpha, x_i^\beta, \cdot, \cdot)$, change from before training to after training for a positive pairwise constraint

is greater than the scores associated with any other labels $\bar{y}_i \neq y_i$, $w^T \Phi(x_i, \bar{y}_i)$, by at least the amount, $1 - \xi_i$. In Figure 1, we demonstrate how the relative positions of the scores associated with different labels, $w^T \Phi(x_i, \cdot)$, change from before training to after training for a fully labeled example, $(x_i, y_i)$.

In a similar manner, we will incorporate the pairwise constraint information into the margin-based learning framework. Specifically, given a positive pairwise constraint $(x_i^\alpha, x_i^\beta, +1)$, we want the maximum score associated with the same-label pairs $y_i^\alpha = y_i^\beta$,

$$\max_{y_i^\alpha = y_i^\beta} \left[ w^T \Phi(x_i^\alpha, x_i^\beta, y_i^\alpha, y_i^\beta) \right],$$

to be greater than the maximum score associated with any different-label pairs $y_i^\alpha \neq y_i^\beta$,

$$\max_{y_i^\alpha \neq y_i^\beta} \left[ w^T \Phi(x_i^\alpha, x_i^\beta, y_i^\alpha, y_i^\beta) \right],$$

by a soft margin of at least $1 - \nu_i$. Similarly, for a negative pairwise constraint $(x_j^\alpha, x_j^\beta, -1)$ we have the following inequality,

$$\max_{y_j^\alpha \neq y_j^\beta} \left[ w^T \Phi(x_j^\alpha, x_j^\beta, y_j^\alpha, y_j^\beta) \right] - \max_{y_j^\alpha = y_j^\beta} \left[ w^T \Phi(x_j^\alpha, x_j^\beta, y_j^\alpha, y_j^\beta) \right] \geq 1 - \nu_j.$$

In Figure 2, we demonstrate how the relative positions of the pairwise scores associated with label-pairs, $w^T \Phi(x_i^\alpha, x_i^\beta, \cdot, \cdot)$, change from before training to after training for a positive pairwise constraint, $(x_i^\alpha, x_i^\beta, +1)$. In our framework, we define the mapping of a pairwise constraint as the sum of the individual example-label scores,

$$\Phi(x_i^\alpha, x_i^\beta, y_i^\alpha, y_i^\beta) = \Phi(x_i^\alpha, y_i^\alpha) + \Phi(x_i^\beta, y_i^\beta).$$

Formally, the pairwise constraint SVM classification (PCSVM) learns a weight vector $w$ and slack variables $\xi, \nu$ via the following margin-based quadratic optimization problem:

OPTIMIZATION PROBLEM II: PCSVM

$$\min_{w, \xi \geq 0, \nu \geq 0} : \frac{\lambda}{2} \|w\|^2 + \frac{1}{n+m} \left( \sum_{i=1}^{n} \xi_i + \sum_{i=1}^{m} \nu_i \right) \tag{2}$$

subject to:

$$\forall (x_i, y_i) \in \mathbf{L}, \overline{y}_i \in \mathcal{Y} \backslash y_i : \; w^T \left[ \Phi(x_i, y_i) - \Phi(x_i, \overline{y}_i) \right] \geq 1 - \xi_i,$$

$$\forall (x_i^\alpha, x_j^\beta, \widetilde{y}_i) \in \mathbf{C}^+ :$$

$$\max_{y_i^\alpha = y_i^\beta} \left[ w^T \Phi(x_i^\alpha, x_i^\beta, y_i^\alpha, y_i^\beta) \right] - \max_{y_i^\alpha \neq y_i^\beta} \left[ w^T \Phi(x_i^\alpha, x_i^\beta, y_i^\alpha, y_i^\beta) \right] \geq 1 - \nu_i,$$

$$\forall (x_i^\alpha, x_j^\beta, \widetilde{y}_i) \in \mathbf{C}^- :$$

$$\max_{y_i^\alpha \neq y_i^\beta} \left[ w^T \Phi(x_i^\alpha, x_i^\beta, y_i^\alpha, y_i^\beta) \right] - \max_{y_i^\alpha = y_i^\beta} \left[ w^T \Phi(x_i^\alpha, x_i^\beta, y_i^\alpha, y_i^\beta) \right] \geq 1 - \nu_i,$$

where $\mathbf{C}^+ = \{(x_i^\alpha, x_j^\beta, \widetilde{y}_i) \in \mathbf{C} \mid \widetilde{y}_i = +1\}$ and $\mathbf{C}^- = \{(x_i^\alpha, x_j^\beta, \widetilde{y}_i) \in \mathbf{C} \mid \widetilde{y}_i = -1\}$ are the set of same/positive constraints and different/negative constraints respectively. The classification of test examples is done in the same manner as for the multiclass SVM classification.

In order to solve the pairwise constraint SVM classification, we extend the Primal QP solver by [21]. The PCSVM is a simple and effective iterative algorithm for solving the above QP and does not require transforming to the dual

**Algorithm 1.** Pairwise Constraint SVM Classification (PCSVM)

**Input:** $\mathbf{L}$ - the labeled data, $\mathbf{C}$ - the pairwise constraint data
$\lambda$ and $T$ - parameters of the QP

Initialize: Choose $w_1$ such that $\|w_1\| \leq 1/\sqrt{\lambda}$
**for** $t = 1$ **to** $T$ **do**

Set $\mathbf{A} = \left\{ (x_i, y_i) \in \mathbf{L} \mid w_t^T \Phi(x_i, y_i) - \max_{\overline{y}_i \neq y_i} w_t^T \Phi(x_i, \overline{y}_i) < 1 \right\}$

$$\text{Set } \mathbf{A}^+ = \left\{ (x_i^\alpha, x_i^\beta, \widetilde{y}_i) \in \mathbf{C}^+ \mid \max_{y_i^\alpha = y_i^\beta} \left[ w_t^T \Phi(x_i^\alpha, x_i^\beta, y_i^\alpha, y_i^\beta) \right] - \max_{y_i^\alpha \neq y_i^\beta} \left[ w_t^T \Phi(x_i^\alpha, x_i^\beta, y_i^\alpha, y_i^\beta) \right] < 1 \right\}$$

$$\text{Set } \mathbf{A}^- = \left\{ (x_i^\alpha, x_i^\beta, \widetilde{y}_i) \in \mathbf{C}^- \mid \max_{y_i^\alpha \neq y_i^\beta} \left[ w_t^T \Phi(x_i^\alpha, x_i^\beta, y_i^\alpha, y_i^\beta) \right] - \max_{y_i^\alpha = y_i^\beta} \left[ w_t^T \Phi(x_i^\alpha, x_i^\beta, y_i^\alpha, y_i^\beta) \right] < 1 \right\}$$

Set $\eta_t = \dfrac{1}{\lambda t}$

$$\text{Set } w_{t+\frac{1}{2}} = (1 - \eta_t \lambda) w_t + \frac{\eta_t}{n+m} \left\{ \sum_{(x_i, y_i) \in \mathbf{A}} [\Phi(x_i, y_i) - \Phi(x_i, \overline{y}_i)] + \sum_{(x_i^\alpha, x_i^\beta, \widetilde{y}_i) \in \mathbf{A}^+} \left[ \Phi(x_i^\alpha, x_i^\beta, y_+^\alpha, y_+^\beta) - \Phi(x_i^\alpha, x_i^\beta, y_-^\alpha, y_-^\beta) \right] + \sum_{(x_i^\alpha, x_i^\beta, \widetilde{y}_i) \in \mathbf{A}^-} \left[ \Phi(x_i^\alpha, x_i^\beta, y_-^\alpha, y_-^\beta) - \Phi(x_i^\alpha, x_i^\beta, y_+^\alpha, y_+^\beta) \right] \right\}$$

where $\overline{y} = \operatorname{argmax}_{\overline{y}_i \neq y} w_t^T \Phi(x_i, \overline{y}_i)$,

$(y_+^\alpha, y_+^\beta) = \operatorname{argmax}_{y^\alpha = y^\beta} w_t^T \Phi(x_i^\alpha, x_i^\beta, y^\alpha, y^\beta)$,

$(y_-^\alpha, y_-^\beta) = \operatorname{argmax}_{y^\alpha \neq y^\beta} w_t^T \Phi(x_i^\alpha, x_i^\beta, y^\alpha, y^\beta)$

Set $w_{t+1} = \min \left\{ 1, \dfrac{1/\sqrt{\lambda}}{\|w_{t+\frac{1}{2}}\|} \right\} w_{t+\frac{1}{2}}$

**end for**

**Output:** $w_{T+1}$

formulation. The algorithm alternates between gradient descent steps and projection steps. In each iteration, the algorithm first computes a set of labeled examples $\mathbf{A} \subset \mathbf{L}$, a set of positive pairwise constraints $\mathbf{A}^+ \subset \mathbf{C}^+$, and a set of negative pairwise constraints $\mathbf{A}^- \subset \mathbf{C}^-$ that contain violated examples and pairwise constraints. Then the weight vector $w$ is updated according to the violated sets $\mathbf{A}$, $\mathbf{A}^+$, and $\mathbf{A}^-$. In the projection step, the weight vector $w$ is projected to the sphere of radius $1/\sqrt{\lambda}$. The details of the PCSVM are given in Algorithm 1.

We observed that if $w_1 = 0$ then $w_t$ can be written as

$$w_t = \sum_{x,y} \varphi_{xy} \Phi(x,y).$$

Hence, we can incorporate the usage of kernel when computing inner product operations, i.e.:

$$\langle w, \Phi(x',y') \rangle = \sum_{x,y} \varphi_{xy} \mathbf{K}(x,y,x',y')$$
$$\|w\|^2 = \sum_{x,y} \sum_{x',y'} \varphi_{xy} \varphi_{x'y'} \mathbf{K}(x,y,x',y')$$

In our experiments, we use the polynomial kernel,

$$\mathbf{K}(x,y,x',y') = \langle \Phi(x,y), \Phi(x',y') \rangle^d,$$

where polynomial kernel degree $d$ is chosen from the set $\{1,2,3,4,5\}$.

The efficiency and guaranteed performance of PCSVM in solving the quadratic optimization problem is shown by the following theorem:

**Theorem 1.** *Let*

$$R = 2\max \left\{ \begin{array}{c} \max\limits_{x,y} \|\Phi(x,y)\|, \\ \max\limits_{x^\alpha, x^\beta, y^\alpha, y^\beta} \|\Phi(x^\alpha, x^\beta, y^\alpha, y^\beta)\| \end{array} \right\}$$

*then the number of iterations for Algorithm 1 to achieving a solution of accuracy* $\delta > 0$ *is* $\tilde{O}(R^2/(\lambda\delta))$.[1]

## 3 Related Work

For classification, pairwise constraints have been shown to improve the performance of classifiers. In [3,4,5,6,7,8,9], pairwise constraints is used to learn a Mahalanobis metric and then apply distance-based classifier such as KNN to the transformed data. Unlike our proposed method, most metric learning algorithms deal with labeled data indirectly by converting into pairwise constraints. In addition, the work of [1,2] is most related to our proposed algorithm. In [1], the authors also presented a discriminative learning framework which can learn the decision boundary with labeled data as well as additional pairwise constraints. However, in the binary algorithm, PKLR proposed by [1], a logistic regression loss function is used for binary classification instead of the hinge loss. In [2], the authors proposed a binary classifier which also utilizes pairwise constraint information. The proposed classifier, Linear-PC, is a sign-insensitive estimator of the optimal linear decision boundary.

[1] The proof of Theorem 1 is omitted since it is similar to the one given in [21].

Similarly, pairwise constraints have also shown to be successful in the semi-supervised clustering [10,11,12,13,14,15,16,17]. In particular, COPKmeans [11] is a semi-supervised variant of Kmeans. COPKmeans follows the same clustering procedure of Kmeans while avoiding violations of pairwise constraints. In addition, MPCKmeans [17] utilized both metric learning and pairwise constraints in the clustering process. In MPCKmeans, a separate weight matrix for each cluster is learned to minimize the distance between must-linked instances and maximize the distance between cannot-link instances. Hence, the objective function of MPCKmeans minimizes cluster dispersion under the learned metrics while reducing constraint violations. However, most existing algorithms can only find a local-optimal solution for the clustering problem with pairwise constraints as users' feedback.

## 4 Experiments

We evaluate our proposed algorithms on fifteen data sets from the UCI repository [22] and the LIBSVM data [23]. A summary of the data sets is given in Table 1. For the PCSVM algorithm, we set the parameters used in the experiments as follows: (i) the SVM $\lambda$ parameter is chosen from $\{10^i\}_{i=-3}^{3}$; (ii) the kernel degree, $d$, is selected from the set $\{1, 2, 3, 4, 5\}$; (iii) the number of pairwise constraints is from the set $\{10, 20, 40, 80, 160\}$; (iv) the number of label examples is chosen from the set $\{1, \ldots, 5\}$[2]. The parameters, $\lambda$ and $d$, are selected using two fold cross validation on the training pairwise constraints.

**Table 1.** A summary of the data sets

| DATA SETS | CLASSES | SIZE | FEATURES |
|---|---|---|---|
| AUSTRALIAN | 2 | 690 | 14 |
| SPAMBASE | 2 | 2300 | 57 |
| IONOSPHERE | 2 | 351 | 34 |
| GERMAN | 2 | 1000 | 24 |
| HEART | 2 | 270 | 13 |
| DIABETES | 2 | 768 | 8 |
| LIVER-DISORDER | 2 | 345 | 6 |
| SPLICE | 2 | 3175 | 60 |
| MUSHROOM | 2 | 8124 | 112 |
| SVMGUIDE2 | 3 | 391 | 20 |
| VEHICLE | 4 | 846 | 18 |
| DERMATOLOGY | 6 | 179 | 34 |
| SATIMAGE | 6 | 6435 | 36 |
| SEGMENT | 7 | 2310 | 19 |
| VOWEL | 11 | 990 | 10 |

[2] Both the pairwise constraints and label examples are randomly generated.

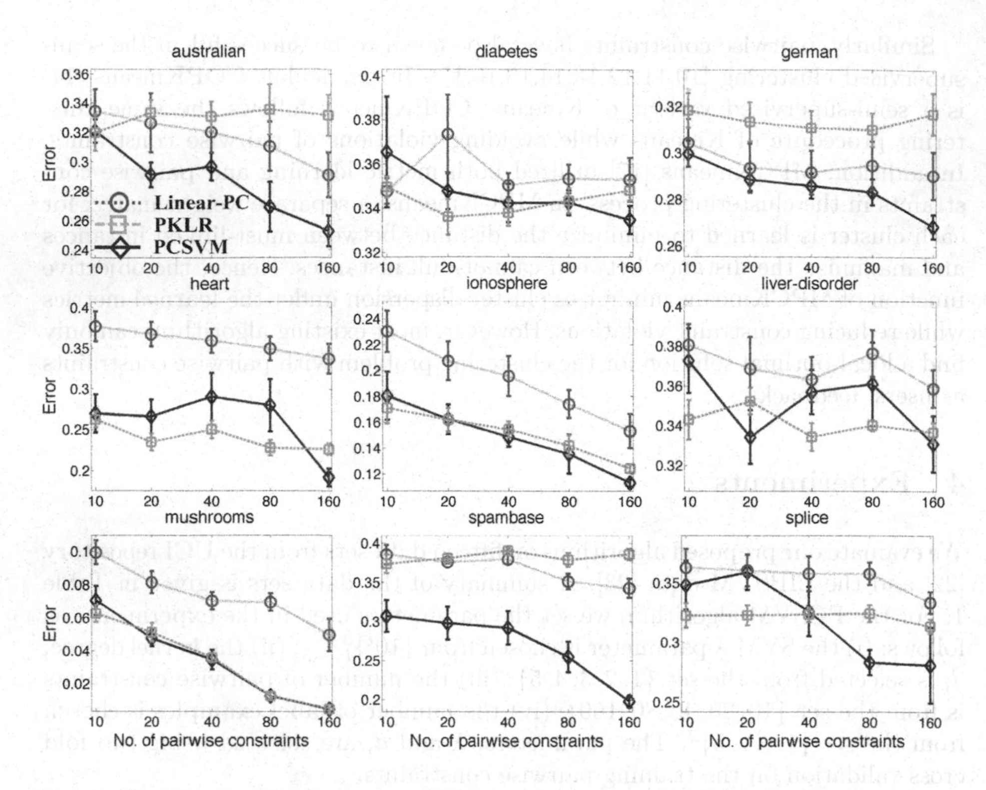


**Fig. 3.** Classification Performance of 9 binary data sets using 5 label points per class: PCSVM, Linear-PC, and PKLR

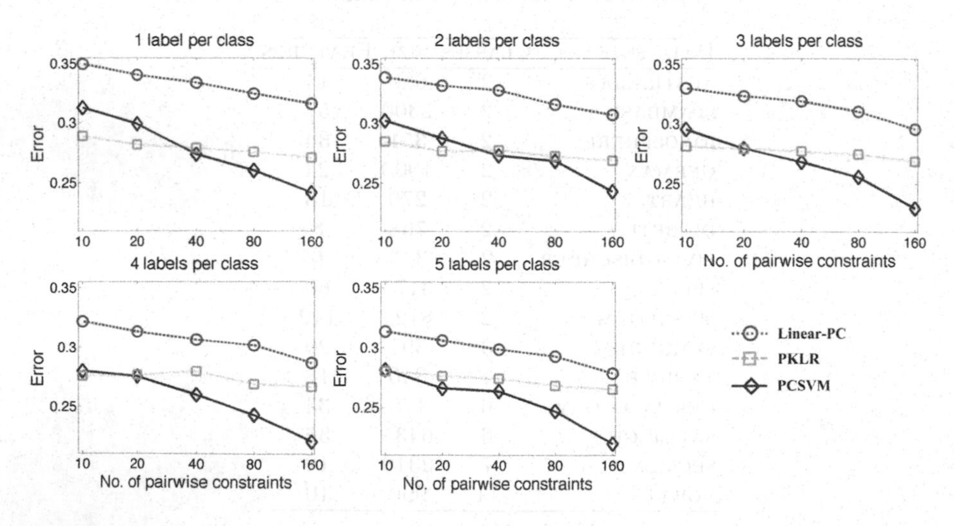


**Fig. 4.** Average classification performance of 9 binary data sets: PCSVM, Linear-PC, and PKLR

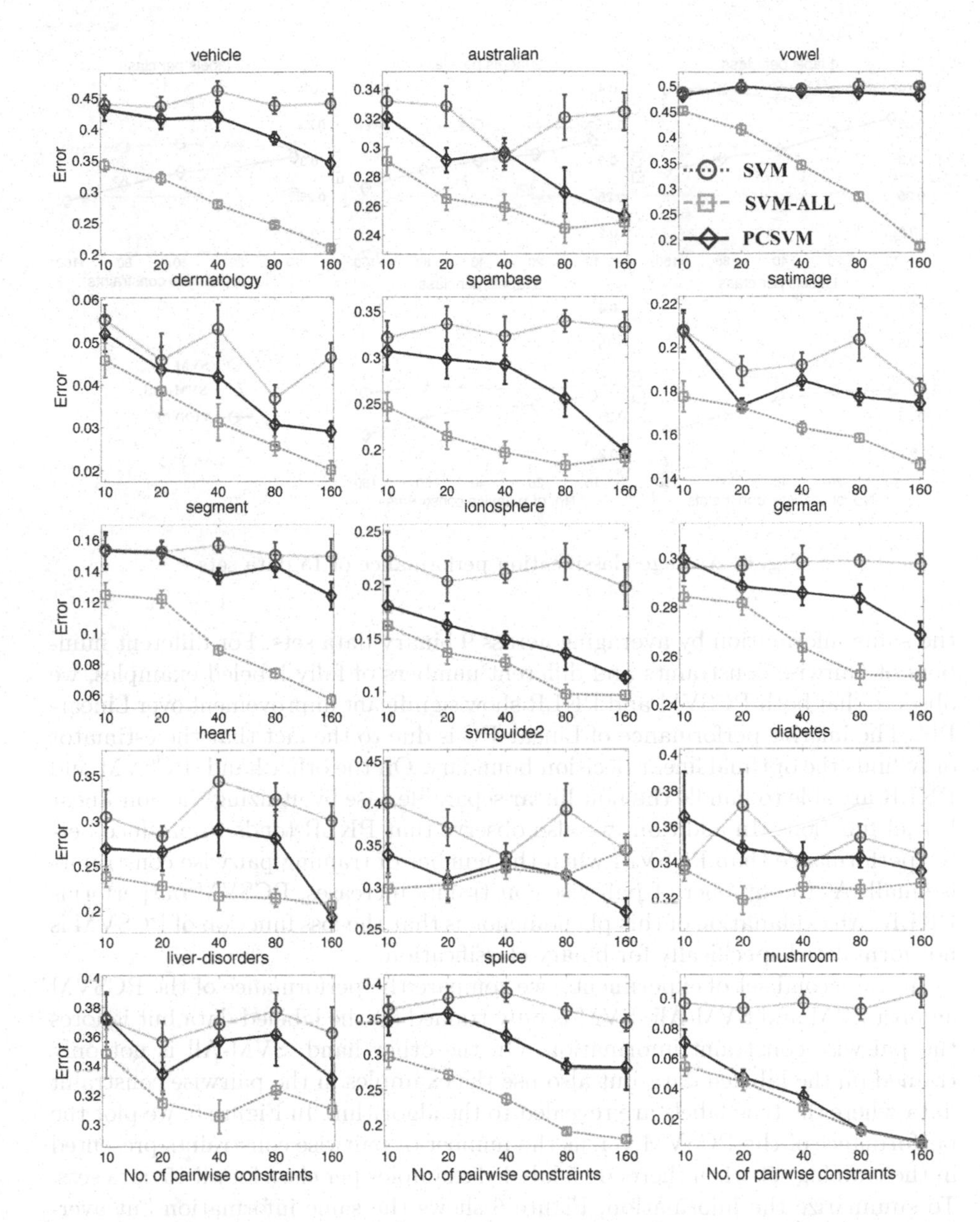


**Fig. 5.** Classification Performance of 15 data sets using 5 label points per class: PCSVM, SVM, and SVM-All

In the first set of experiments, we compare the performance of the PCSVM against the two other methods proposed by [1,2], called PKLR and Linear-PC respectively, on 9 binary data sets. In Figure 3, we plot the performance of PCSVM, Linear-PC, and PKLR versus the number of pairwise constraints when there are 5 fully labeled examples per class. To summarize the information, Figure 4 presents

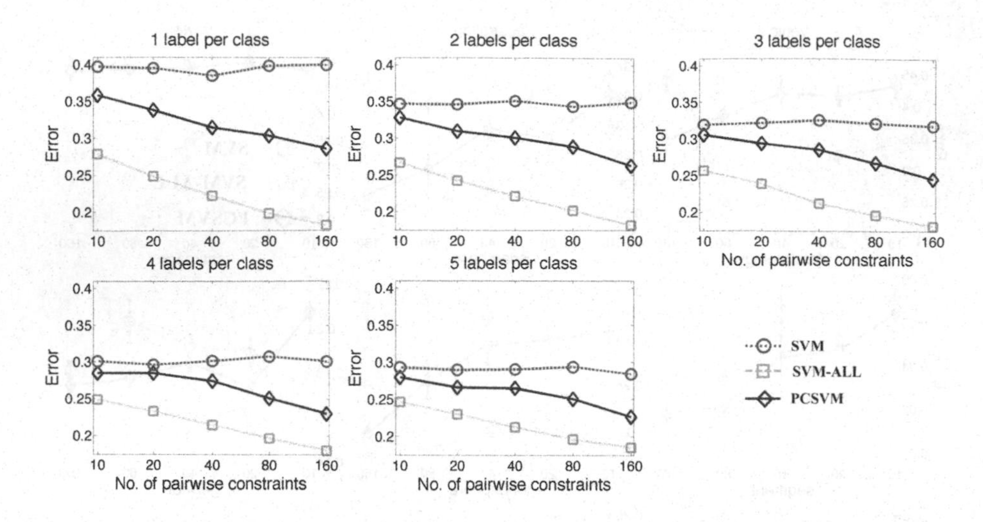


**Fig. 6.** Average classification performance of 15 data sets

the same information by averaging across 9 binary data sets. For different numbers of pairwise constraints and different numbers of fully labeled examples, we observe that both PCSVM and PKLR show significant improvement over Linear-PC. The inferior performance of Linear-PC is due to the fact that the estimator only finds the optimal linear decision boundary. On the other hand, PCSVM and PKLR are able to handle the non-linear separable case by utilizing the non-linear kernel functions. In addition, we also observe that PKLR tends to produce better performance than PCSVM when the number of training pairwise constraints is small. As the number of pairwise constraints increases, PCSVM outperforms PKLR. An explanation of this phenomenon is that the loss function of PCSVM is not formulated specifically for binary classification.

In the second set of experiments, we compare the performance of the PCSVM against SVM and SVM-All. SVM is only trained on the labeled data but ignores the pairwise constraint information. On the other hand, SVM-All is not only trained on the labeled data but also use the examples in the pairwise constraint data where the true labels are revealed to the algorithm. In Figure 5, we plot the performance of the PCSVM versus the number of pairwise constraints presented in the training set when there are 5 labeled examples per class for all 15 data sets. To summarize the information, Figure 6 shows the same information but averaging across 15 data sets. Across all data sets, we observe that the performance of the PCSVM is between that of SVM and SVM-All. This behavior is what we should expect since pairwise constraint information helps to improve the performance of PCSVM over SVM which does not use this information; and labeled data should still provide more discriminative information to the SVM-All than pairwise constraint information could do to the PCSVM. Note that PCSVM, by learning from the pairwise constraints, on average yields half or more of the error reduction that could be achieved by learning with labels. Hence, SVM and

SVM-All can be viewed as the lower and upper bound on the performance of PCSVM.

## 5 Conclusion

In this paper, we study the problem of classification in the presence of pairwise constraints. We propose a discriminative learning approach which incorporates pairwise constraints into the margin-based learning framework. We also present an efficient algorithm, PCSVM, that integrates pairwise constraints into the multiclass-SVM classification. In experiments with 15 data sets, pairwise constraints not only improves the performance of the binary classification in comparison with two other methods (Linear-PC and PKLR) but also significantly increase the performance of the multiclass classification.

**Acknowledgments.** This work was supported by NSF CAREER Grant # 0347318.

## References

1. Yan, R., Zhang, J., Yang, J., Hauptmann, A.G.: A discriminative learning framework with pairwise constraints for video object classification. IEEE Transactions on Pattern Analysis and Machine Intelligence 28(4), 578–5932 (2006)
2. Zhang, J., Yan, R.: On the value of pairwise constraints in classification and consistency. In: Proceedings of the 24th International Conference on Machine Learning, pp. 1111–1118 (2007)
3. Davis, J.V., Kulis, B., Jain, P., Sra, S., Dhillon, I.S.: Information-theoretic metric learning. In: Proceedings of the 24th International Conference on Machine Learning (2007)
4. Globerson, A., Roweis, S.: Metric learning by collapsing classes. In: Advances in Neural Information Processing Systems (NIPS) (2005)
5. Goldberger, J., Roweis, S., Hinton, G., Salakhutdinov, R.: Neighbourhood components analysis. In: Advances in Neural Information Processing Systems (NIPS) (2004)
6. Schultz, M., Joachims, T.: Learning a distance metric from relative comparisons. In: Advances in Neural Information Processing Systems (NIPS) (2004)
7. Shalev-Shwartz, S., Singer, Y., Ng, A.Y.: Online and batch learning of pseudo-metrics. In: Proceedings of the 21st International Conference on Machine Learning (2004)
8. Weinberger, K.Q., Blitzer, J., Saul, L.K.: Distance metric learning for large margin nearest neighbor classification. In: Advances ing Neural Information Processing Systems (NIPS) (2006)
9. Chopra, S., Hadsell, R., LeCun, Y.: Learning a similarity metric discriminatively, with application to face verification. In: Proceedings of the IEEE Conference on Computer Vision and Pattern Recognition (CVPR) (2005)
10. Bar-Hillel, A., Hertz, T., Shental, N., Weinshall, D.: Learning distance functions using equivalence relations. In: Proceedings of the 20th International Conference on Machine Learning (2003)

11. Wagstaff, K., Cardie, C., Rogers, S., Schroedl, S.: Constrained k-means clustering with background knowledge. In: Proceedings of 18th International Conference on Machine Learning (2001)
12. Bilenko, M., Basu, S., Mooney, R.J.: Semi-supervised clustering by seeding. In: Proceedings of 19th International Conference on Machine Learning (2002)
13. Klein, D., Kamvar, S.D., Manning, C.D.: From instance-level constraints to spacelevel constraints: Making the most of prior knowledge in data clustering. In: Proceedings of 19th International Conference on Machine Learning (2002)
14. Xing, E.P., Ng, A.Y., Jordan, M.I., Russell, S.: Distance metric learning, with application to clustering with side-information. Advances in Neural Information Processing Systems 15 (2003)
15. Cohn, D., Caruana, R., McCallum, A.: Semi-supervised clustering with user feedback. In: Cornell University Technical Report TR2003-1892 (2003)
16. Bar-Hillel, A., Hertz, T., Shental, N., Weinshall, D.: Learning distance functions using equivalence relations. In: Proceedings of 20th International Conference on Machine Learning (2003)
17. Bilenko, M., Basu, S., Mooney, R.J.: Integrating constraints and metric learning in semi-supervised clustering. In: Proceedings of 21th International Conference on Machine Learning (2004)
18. Zhu, X.: Semi-supervised learning literature survey. Technical Report 1530, Computer Sciences, University of Wisconsin-Madison (2005)
19. Cozman, F., Cohen, I., Cirelo, M.: Semi-supervised learning of mixture models and bayesian networks. In: Proceedings of the Twentieth International Conference of Machine Learning (2003)
20. Crammer, K., Singer, Y.: On the algorithmic implementation of multiclass kernel-based vector machines. Journal of Machine Learning Research 2, 265–292 (2001)
21. Shalev-Shwartz, S., Singer, Y., Srebro, N.: Pegasos: Primal estimated sub-gradient solver for svm. In: Proceedings of the 24th International Conference on Machine Learning, pp. 807–814. ACM, New York (2007)
22. Asuncion, A., Newman, D.: UCI machine learning repository (2007), `http://www.ics.uci.edu/~mlearn/MLRepository.html`
23. Chang, C.C., Lin, C.J.: Libsvm data (2001), `http://www.csie.ntu.edu.tw/~cjlin/libsvmtools/datasets/`

# Metric Learning: A Support Vector Approach

Nam Nguyen and Yunsong Guo

Department of Computer Science
Cornell University, USA
{nhnguyen,guoys}@cs.cornell.edu

**Abstract.** In this paper, we address the metric learning problem utilizing a margin-based approach. Our metric learning problem is formulated as a quadratic semi-definite programming problem (QSDP) with local neighborhood constraints, which is based on the Support Vector Machine (SVM) framework. The local neighborhood constraints ensure that examples of the same class are separated from examples of different classes by a margin. In addition to providing an efficient algorithm to solve the metric learning problem, extensive experiments on various data sets show that our algorithm is able to produce a new distance metric to improve the performance of the classical K-nearest neighbor (KNN) algorithm on the classification task. Our performance is always competitive and often significantly better than other state-of-the-art metric learning algorithms.

**Keywords:** metric learning, K-nearest neighbor classification, SVM.

## 1 Introduction

The distance metric learning problem, i.e. learning a distance measure over an input space, has received much attention in the machine learning community recently [1,2,3,4,5,6,7,8,9,10]. This is because designing a good distance metric is essential to many distance-based learning algorithms. For example, the K-nearest neighbor (KNN) [11], which is a classical classification method and requires no training effort, critically depends on the quality of the distance measures among examples. The classification of a new example is determined by the class labels of the $K$ training examples with shortest distances. Traditionally the distance measure between two examples is defined to be the Euclidean distance between the features of the examples.

Although KNN is a simple method by nowadays standard, it is still an active research area [12,13], widely used in real-world applications [14], and serves as a basic component in more complex learning models [15,16]. Hence, improving KNN prediction accuracy would have a significant impact on this part of the machine learning community.

The performance of the KNN algorithm is influenced by three main factors: (i) the distance function or distance metric used to determine the nearest neighbors; (ii) the decision rule used to derive a classification from the K-nearest neighbors;

W. Daelemans et al. (Eds.): ECML PKDD 2008, Part II, LNAI 5212, pp. 125–136, 2008.

and (iii) the number of neighbors used to classify the new example. Ideally the distance function should bring examples with the same labels closer to one another and push examples with different labels further apart. However, the ideal distance metric is hard to achieve. On the other hand, it is clear that with the right distance metric, KNN can perform exceptionally well using a simple majority-voting decision rule and a fixed number of nearest neighbors. Hence, in this paper we focus on the first main factor of KNN to derive a good distance function utilizing the maximum-margin approach [17].

Our approach presented in this paper is to learn a Mahalanobis distance function by minimizing a quadratic objective function subject to local neighborhood constraints. The goal of this optimization is, for every example, to bring examples of the same class close and to separate examples from other classes by a large margin. This goal is very similar to the one proposed in [7]. However, in [7] the authors solved a different optimization problem via semidefinite programming. In our framework, similar to most other support vector machine approaches our proposed algorithm is able to handle the non-linear separable cases by utilize the kernel trick. Finally, our extensive experiments with various data sets from the UCI repository show that our proposed metric learning algorithm is able to produce a better distance function that helps improve the performance of KNN classification in comparison to other state-of-the-art metric learning algorithms.

The paper is organized as follows: in section 2, we describe our metric learning algorithm via the support vector machine approach in detail; in section 3, we review other state-of-the-art metric learning algorithms; the experimental results and conclusions are presented in section 4 and 5.

## 2 Metric Learning: A Margin-Based Approach

In this section, we consider the following supervised learning framework. The learning algorithm takes a set of labeled training examples, $\mathbf{L} = \{(\mathbf{x}_1, y_1), (\mathbf{x}_2, y_2), ..., (\mathbf{x}_n, y_n)\}$ as input, where $\mathbf{x}_i \in \mathcal{X}$ the input space and $y_i$ belongs to a finite set of classes denoted by $\mathcal{Y}$. The goal of the metric learning problem is to produce a Mahalanobis distance function which can be used in computing nearest neighbors for the KNN classifier. In most metric learning algorithms, a positive semidefinite matrix $\mathbf{A} \succeq 0$, known as the Mahalanobis distance matrix, is learned as a transformation matrix in computing the (squared) distance between examples $\mathbf{x}_i$ and $\mathbf{x}_j$:

$$d_{\mathbf{A}}(\mathbf{x}_i, \mathbf{x}_j) = (\mathbf{x}_i - \mathbf{x}_j)^T \mathbf{A} (\mathbf{x}_i - \mathbf{x}_j).$$

Alternatively, we rewrite the distance function as,

$$d_{\mathbf{A}}(\mathbf{x}_i, \mathbf{x}_j) = \langle \mathbf{A}, (\mathbf{x}_i - \mathbf{x}_j)(\mathbf{x}_i - \mathbf{x}_j)^T \rangle,$$

where $\langle \cdot, \cdot \rangle$ represents the matrix inner-product.

A desirable distance function preserves the local neighborhood property that examples within the same class are separated from examples of different classes by a large margin in the new distance space. This translates to the following local neighborhood constraint that, for each example its distance to all neighboring examples of the same class is smaller than its distance to any neighboring examples of different classes by at least a predefined margin, such as 1. Figure 1 demonstrates the idea how the neighborhood of a training example is transformed after applying the new distance function. Formally, the above constraint can be represented by the following inequalities,

$$\forall(\mathbf{x}_i, y_i) \in \mathbf{L}:$$

$$\min_{\substack{\mathbf{x}_j \in \mathcal{N}(\mathbf{x}_i) \\ y_j \neq y_i}} d_{\mathbf{A}}(\mathbf{x}_i, \mathbf{x}_j) \geq \max_{\substack{\mathbf{x}_j \in \mathcal{N}(\mathbf{x}_i) \\ y_j = y_i}} d_{\mathbf{A}}(\mathbf{x}_i, \mathbf{x}_j) + 1, \tag{1}$$

where $\mathcal{N}(\mathbf{x}_i)$ represents the set of neighbor examples of $\mathbf{x}_i$. In the absence of prior knowledge, the neighborhood set for each example is determined by the Euclidean distance. Along with the labeled training data $\mathbf{L}$, our metric learning algorithm also takes an integer $K$ in the input which servers as the size of the constructed neighborhood $\mathcal{N}(\cdot)$, as well as the input to the KNN classification algorithm.

In our learning framework, we seek the positive semidefinite matrix that minimizes a quadratic objective function subject to the above constraints presented in inequality (1). Rather than using the hard margin constraints, we incorporate slack variables, $\xi$, to obtain the quadratic optimization problem with the soft margin constraints. Formally, the metric learning algorithm we propose

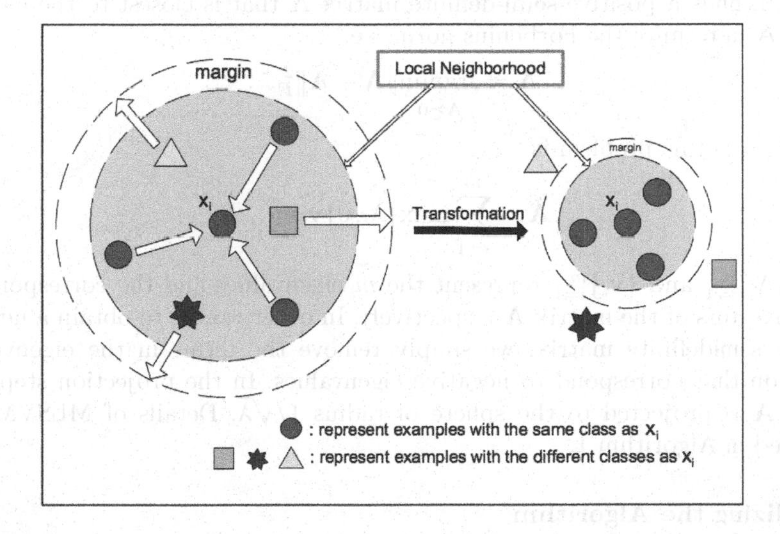


**Fig. 1.** Illustration of the local neighborhood property for an example $\mathbf{x}_i$ where before applying the transformation (on the left) the neighborhood of $\mathbf{x}_i$ contains both examples from the same and different classes; and after applying the transformation (on the right), examples within the same class of $\mathbf{x}_i$ cluster together and are separated from examples from different classes by a large margin

(MLSVM) learns a positive semidefinite matrix $\mathbf{A}$ and slack variables $\xi$ via the following quadratic semi-definite programming problem (QSDP):

OPTIMIZATION PROBLEM: MLSVM

$$\min_{\mathbf{A}\succeq 0, \xi \geq 0} : \frac{\lambda}{2}\|\mathbf{A}\|_F^2 + \frac{1}{n}\sum_{i=1}^{n} \xi_i \tag{2}$$

subject to:
$\forall(\mathbf{x}_i, y_i) \in \mathbf{L}$ :

$$\min_{\substack{\mathbf{x}_j \in \mathcal{N}(\mathbf{x}_i) \\ y_j \neq y_i}} d_{\mathbf{A}}(\mathbf{x}_i, \mathbf{x}_j) \geq \max_{\substack{\mathbf{x}_j \in \mathcal{N}(\mathbf{x}_i) \\ y_j = y_i}} d_{\mathbf{A}}(\mathbf{x}_i, \mathbf{x}_j) + 1 - \xi_i,$$

where $\|\cdot\|_F$ represents the Frobenius norm for matrices, $\|\mathbf{A}\|_F^2 = \sum_i \sum_j \mathbf{A}_{ij}^2$.

In order to solve MLSVM, we extend the Pegasos method from [18], which is an effective iterative algorithm for solving the SVM QP via gradient descent without the need to transform the formulation into its dual form as many other SVM based methods do [19,20]. Our algorithm is called MLSVM, as from the optimization problem. In each iteration, MLSVM follows three steps: gradient descent, positive semidefinite approximation, and projection steps. In the first step, the algorithm computes a set $\mathbf{V} \subset \mathbf{L}$ that contains examples violating the local neighborhood constraints. Then the matrix $\mathbf{A}$ is updated according to the violated constraint set $\mathbf{V}$. In the positive semidefinite approximation step, the algorithm finds a positive semi-definite matrix $\hat{\mathbf{A}}$ that is closest to the current matrix $\mathbf{A}$ in term of the Forbenius norm, i.e.

$$\hat{\mathbf{A}} = \underset{\tilde{\mathbf{A}}\succeq 0}{\operatorname{argmin}} \|\mathbf{A} - \tilde{\mathbf{A}}\|_F^2. \tag{3}$$

A solution to this problem is

$$\hat{\mathbf{A}} = \sum_{i=1}^{m} \max\{\lambda_i, 0\} \mathbf{v}_i \mathbf{v}_i^T,$$

where $\{\lambda_i\}_{i=1}^m$ and $\{\mathbf{v}_i\}_{i=1}^m$ represent the $m$ eigenvalues and the corresponding $m$ eigenvectors of the matrix $\mathbf{A}$ respectively. In other words, to obtain a nearest positive semidefinite matrix, we simply remove the terms in the eigenvector expansion that correspond to negative eigenvalues. In the projection step, the matrix $\mathbf{A}$ is projected to the sphere of radius $1/\sqrt{\lambda}$. Details of MLSVM are presented in Algorithm 1.

### Kernelizing the Algorithm

We now consider kernelizing our metric learning algorithm. In Algorithm 1, if we initialize $\mathbf{A}_1 = 0$ then $\mathbf{A}_t$ can be expressed as

$$\mathbf{A}_t = \sum_{i=1}^{n} \sum_{j=1}^{n} \varphi_{ij} \mathbf{x}_i \mathbf{x}_j^T.$$

**Algorithm 1.** Metric Learning SVM (MLSVM)

**Input:** $\mathbf{L}$ - the labeled data
$\lambda$ and $T$ - parameters of the QP

Initialization: Choose $\mathbf{A}_1$ such that $\|\mathbf{A}_1\|_F \leq 1/\sqrt{\lambda}$
**for** $t = 1$ **to** $T$ **do**

Set $\mathbf{V} = \{(\mathbf{x}_i, y_i) \in \mathbf{L} \mid \min_{\substack{\mathbf{x}_j \in \mathcal{N}(\mathbf{x}_i) \\ y_j \neq y_i}} d_{\mathbf{A}}(\mathbf{x}_i, \mathbf{x}_j) - \max_{\substack{\mathbf{x}_j \in \mathcal{N}(\mathbf{x}_i) \\ y_j = y_i}} d_{\mathbf{A}}(\mathbf{x}_i, \mathbf{x}_j) < 1\}$

Set $\eta_t = \frac{1}{\lambda t}$

Set $\mathbf{A}_{t+\frac{1}{3}} = (1 - \eta_t \lambda)\mathbf{A}_t + \frac{\eta_t}{n} \sum_{(\mathbf{x}_i, y_i) \in \mathbf{V}} \Big[ (\mathbf{x}_i - \mathbf{x}_i^-)(\mathbf{x}_i - \mathbf{x}_i^-)^T - (\mathbf{x}_i - \mathbf{x}_i^+)(\mathbf{x}_i - \mathbf{x}_i^+)^T \Big]$

where $\mathbf{x}_i^- = \operatorname{argmin}_{\substack{\mathbf{x}_i^- \in \mathcal{N}(\mathbf{x}_i) \\ y_i^- \neq y_i}} d_{\mathbf{A}}(\mathbf{x}_i, \mathbf{x}_i^-)$,
$\mathbf{x}_i^+ = \operatorname{argmax}_{\substack{\mathbf{x}_i^+ \in \mathcal{N}(\mathbf{x}_i) \\ y_i^+ = y_i}} d_{\mathbf{A}}(\mathbf{x}_i, \mathbf{x}_i^+)$

Set $\mathbf{A}_{t+\frac{2}{3}} = \operatorname{argmin}_{\mathbf{A} \succeq 0} \|\mathbf{A}_{t+\frac{1}{3}} - \mathbf{A}\|_F^2$

Set $\mathbf{A}_{t+1} = \min\left\{1, \frac{1/\sqrt{\lambda}}{\|\mathbf{A}_{t+\frac{2}{3}}\|_F}\right\} \mathbf{A}_{t+\frac{2}{3}}$

**end for**

**Output:** $\mathbf{A}_{T+1}$

Hence, we can incorporate the use of kernels when computing the matrix inner-product operation and the Frobenius norm:

$$\langle \mathbf{A}, \mathbf{x}\mathbf{x}'^T \rangle = \sum_{i=1}^{n} \sum_{j=1}^{n} \varphi_{ij} \kappa(\mathbf{x}, \mathbf{x}_i) \kappa(\mathbf{x}', \mathbf{x}_j)$$
$$\|\mathbf{A}\|_F^2 = \sum_{i,j} \sum_{i',j'} \varphi_{ij} \varphi_{i'j'} \kappa(\mathbf{x}_i, \mathbf{x}_{i'}) \kappa(\mathbf{x}_j, \mathbf{x}_{j'}),$$

where $\kappa(\mathbf{x}_i, \mathbf{x}_j) = \Phi(\mathbf{x}_i)^T \Phi(\mathbf{x}_j)$ is the kernel function and $\Phi : \mathcal{X} \mapsto \mathcal{F}$ projects examples to a new space. In our experiment, we consider the polynomial kernel function, i.e. $\kappa(\mathbf{x}_i, \mathbf{x}_j) = \langle \mathbf{x}_i, \mathbf{x}_j \rangle^d$.

In the kernel space, the matrix $\mathbf{A}$ does not have an explicit form since $\mathbf{A}$ is only expressed as $\mathbf{A} = \sum_{i=1}^n \sum_{j=1}^n \varphi_{ij} \Phi(\mathbf{x}_i) \Phi(\mathbf{x}_j)^T$ where $\Phi(\mathbf{x})$ may have infinite dimensions. Similar to the problem of Kernel Principal Component Analysis [21], in order to carry out the semidefinite matrix approximation step, we now have to find eigenvalues $\lambda \geq 0$ and corresponding eigenvectors $\mathbf{v} \in \mathcal{F} - \{\mathbf{0}\}$ satisfying

$$\lambda \mathbf{v} = \mathbf{A}\mathbf{v}. \tag{4}$$

By taking the inner-product of each $\Phi(\mathbf{x}_k)$ with both sides of equation 4, we obtain the set of equations

$$\lambda\langle\Phi(\mathbf{x}_k),\mathbf{v}\rangle = \langle\Phi(\mathbf{x}_k),\mathbf{A}\mathbf{v}\rangle,\ \forall k \in \{1,\ldots,n\}. \tag{5}$$

In addition, all eigenvectors $\mathbf{v}$ that correspond with non-zero eigenvalues must lie in the span of $\{\Phi(\mathbf{x}_1),\Phi(\mathbf{x}_2),...,\Phi(\mathbf{x}_n)\}$. Hence there exists a set of coefficients $\{\alpha_i\}_{i=1}^n$ such that

$$\mathbf{v} = \sum_{i=1}^{n} \alpha_i \Phi(\mathbf{x}_i) \tag{6}$$

By substituting $\mathbf{A} = \sum_{ij} \varphi_{ij}\Phi(\mathbf{x}_i)\Phi(\mathbf{x}_j)^T$ and equation (6) into the set of equations (5), we obtain the following set of equations

$$\lambda \sum_{i=1}^{n} \alpha_i \langle\Phi(\mathbf{x}_k),\Phi(\mathbf{x}_i)\rangle = \sum_{i,j,l=1}^{n} \varphi_{ij}\alpha_l \langle\Phi(\mathbf{x}_k),\Phi(\mathbf{x}_i)\rangle\langle\Phi(\mathbf{x}_l),\Phi(\mathbf{x}_j)\rangle,\ \forall k \in \{1,\ldots,n\} \tag{7}$$

By stacking the set of $n$ equations (7), we get the simplified matrix equation,

$$\lambda\mathbf{K}\overline{\alpha} = \mathbf{K}\overline{\varphi}\mathbf{K}\overline{\alpha}, \tag{8}$$

where $\mathbf{K}$ is an $n \times n$ matrix whose elements $\mathbf{K}_{ij} = \langle\Phi(\mathbf{x}_i),\Phi(\mathbf{x}_j)\rangle$; $\overline{\alpha}$ is the column vector $\overline{\alpha} = [\alpha_1,...,\alpha_n]^T$; and $\overline{\varphi}$ is an $n \times n$ matrix with $\overline{\varphi}_{ij} = \varphi_{ij}$. To find a solution of equation (8), we solve the eigenvalue problem

$$\lambda\overline{\alpha} = (\overline{\varphi}\mathbf{K})\overline{\alpha} \tag{9}$$

for nonzero eigenvalues. This will give us all solutions of equation (8) that we are interested in.

Let $\lambda_1 \leq \lambda_2 \leq ... \leq \lambda_n$ and $\{\overline{\alpha}^1,\overline{\alpha}^2,...,\overline{\alpha}^n\}$ denote the eigenvalues and the corresponding eigenvectors of $\overline{\varphi}\mathbf{K}$ (i.e. the solutions of equation (9)), with $\lambda_p$ being the first positive eigenvalue. Since the solution eigenvectors of equation (4) must be unit vectors, i.e $\langle\mathbf{v}_k,\mathbf{v}_k\rangle = 1$ for all $k \in \{p,\ldots,n\}$, we normalize $\mathbf{v}_k$:

$$\mathbf{v}_k = \frac{\sum_{i=1}^{n} \overline{\alpha}_i^k \Phi(\mathbf{x}_i)}{\sum_{i,j=1}^{n} \overline{\alpha}_i^k \overline{\alpha}_j^k \langle\Phi(\mathbf{x}_i),\Phi(\mathbf{x}_j)\rangle}. \tag{10}$$

The nearest positive semidefinite matrix of $\mathbf{A}$ can be computed as

$$\hat{\mathbf{A}} = \sum_{i=p}^{n} \lambda_i \mathbf{v}_k \mathbf{v}_k^T. \tag{11}$$

In summary, to compute the nearest positive semidefinite matrix of $\mathbf{A}$ we follow the three steps: first, compute the product matrix $\overline{\varphi}\mathbf{K}$; second, compute its eigenvalues and corresponding eigenvectors, after which normalize the eigenvectors of $\mathbf{A}$; finally, compute the nearest positive semidefinite matrix as shown in equation (11).

## 3 Related Work

Metric learning has been an active research topic in the machine learning community. Here we briefly review some representative recent works on this topic. The method proposed in [7] is the most related prior work to our learning method since both methods aim to bring examples of the same class closer together and to separate examples of different classes. The main difference between the two methods is that we formulate our objective function as a quadratic semi-definite programming problem (QSDP) instead of a semidefinite program described in [7]. In addition, our proposed algorithm is able to incorporate many different kernel functions easily for non-linearly separable cases.

An earlier work by [6] also used semidefinite programming based on similarity and dissimilarity constraints to learn a Mahalanobis distance metric for clustering. The authors proposed a convex optimization model to minimize the sum of squared distances between example pairs of similar labels subject to a lower bound constraint on the sum of distances between example pairs of different labels. While [6] focuses on a global constraint which brings all similar examples closer together and separates dissimilar examples, our proposed method enforces local neighborhood constraints to improve the KNN classification performance.

Most recently, [1] proposed an information-theoretic approach to learning a Mahalanobis distance function. The authors formulate the metric learning problem as minimizing the LogDet divergence, subject to similarity and dissimilarity constraints. In addition to the difference in the objective function formulation, the number of constraints for our proposed optimization is only linear with respect to the number of training labeled examples. In contrast, the number of similarity and dissimilarity constraints of the optimization formulation in [1] is potentially quadratic in the number of labeled training examples.

In [5], the authors proposed an online learning algorithm for learning a Mahalanobis distance function which is also based on similarity and dissimilarity constraints on example pairs. The online metric learning algorithm is based on successive projections onto the positive semi-definite cone subject to the constraint that all similarly labeled examples have small pairwise distances and all differently labeled examples have large pairwise distances. Instead of using global similarity and dissimilarity constraints, our proposed method enforces local neighborhood constraints for each labeled training example.

In [2], Maximally Collapsing Metric Learning (MCML) algorithm is proposed to learn a Mahalanobis distance metric based on similarity and dissimilarity constraints. The authors construct a convex optimization problem which aims to collapse all examples in the same class into a single point and push examples in other classes infinitely far apart. The main difference between our proposed method and MCML is also the usage of the local neighborhood constraints versus the pairwise constraints between similarly labeled and differently labeled examples.

In [3], the authors proposed Neighborhood Component Analysis (NCA) to learn a distance metric specifically for the KNN classifier. The optimization aims at improving the leave-one-out performance of the KNN algorithm on the training data. The algorithm minimizes the probability of error under stochastic neighborhood assignments using gradient descent. Although both NCA and our method try to derive a distance function in order to improve the performance of the KNN algorithm, we employed different optimization problem formulations.

In [8], the authors proposed a framework for similarity metric learning by the energy-based model (EBM) from a pair of convolutional neural networks that shared the same learning parameters. The cost function incorporated into the EBM aimed at penalizing large distances among examples with the same label, and small distance among example pairs of different classes. In our work, the distance function is parameterized by a transformation matrix instead of a convolutional neural network.

## 4 Experiments

In this section, we evaluate our metric learning algorithm (MLSVM) on a number of data sets from the UCI repository [22] and the LIBSVM data [23]. A summary of the data sets is given in Table 1.

In our experiments, we compare MLSVM against recent proposed metric learning methods for KNN classification, namely the Large Margin Nearest Neighbor [7] (LMNN) and Information-Theoretic Metric Learning (ITML) [1]. In addition, we also compare our method against a benchmark margin-based multiclass classification method, i.e SVM [24]. All the compared methods learn a new distance metric before applying KNN except SVM, which is directly used as a classification method. The parameters of the MLSVM and SVM, $\lambda$ and the degree of the polynomial kernel, are selected from the sets $\{10^i\}_{i=-3}^{3}$ and $\{1, 2, 3, 4, 5\}$, respectively. For ITML, the trade-off parameter $\gamma$ is also chosen from the set $\{10^i\}_{i=-3}^{3}$. For all metric learning algorithms, the K-nearest neighbor value is set to 4 which is also used in [1]. The parameters are selected using two fold cross validation on the training data. For all data sets, 50% of data is randomly selected for the training phase and the rest is used for test. The experiment is repeated for 10 random trials. The mean performance and standard error of various methods are reported for each data set.

In Figure 2, we present the performance of KNN, LMNN, ITML, and SVM against the new metric learning method MLSVM for the 9 data sets. From the

**Table 1.** A summary of the data sets

| DATA SETS | CLASSES | SIZE | FEATURES |
|---|---|---|---|
| AUSTRALIAN | 2 | 690 | 14 |
| BALANCE | 3 | 625 | 4 |
| DERMATOLOGY | 6 | 179 | 34 |
| GERMAN | 2 | 1000 | 24 |
| HEART | 2 | 270 | 13 |
| IONOSPHERE | 2 | 351 | 34 |
| IRIS | 3 | 150 | 4 |
| SPAMBASE | 2 | 2300 | 57 |
| WINE | 3 | 178 | 13 |

**Table 2.** Performance summary by the three metric learning algorithms and SVM

| DATA SETS | BEST PERFORMANCE | BEST MODELS |
|---|---|---|
| AUSTRALIAN | 0.1683 | MLSVM, ITML |
| BALANCE | 0.0047 | MLSVM, SVM |
| DERMATOLOGY | 0.0244 | MLSVM, SVM |
| GERMAN | 0.2383 | MLSVM, SVM |
| HEART | 0.1770 | MLSVM, LMNN |
| IONOSPHERE | 0.0835 | MLSVM |
| IRIS | 0.0192 | MLSVM, SVM |
| SPAMBASE | 0.1023 | MLSVM, LMNN, ITML |
| WINE | 0.0383 | MLSVM, ITML |

figure we clearly see that all three metric learning methods (MLSVM, ITML and LMNN) significantly improved the KNN classification accuracy by comparing with the first column, which is the raw KNN performance, with the only exception for LMNN on the "german" dataset. SVM, as a non-metric-distance based method also performed competitively. MLSVM is consistently among the best performing methods across all datasets. We also present the significance test results in Table 2. In Table 2, the second column is the best error rate achieved by any method for a dataset, and column 3 shows all the methods whose results are not statistically significantly different from the best result by pair-wise t-test at the 95% level. It is interesting to notice that MLSVM is the only model that always perform the best even comparing with the non-metric-learning SVM classifier. This is an evidence that KNN, as a simple model, can achieve competitive or superior classification performance compared with SVM, if a proper distance metric can be learned. In comparison with the other two metric-distance learning methods, namely LMNN and ITML, in our conducted experiments MLSVM is more reliable in improving KNN's performance, as LMNN and ITML are only within the best models for 2 and 4 times respectively among the 9 datasets.

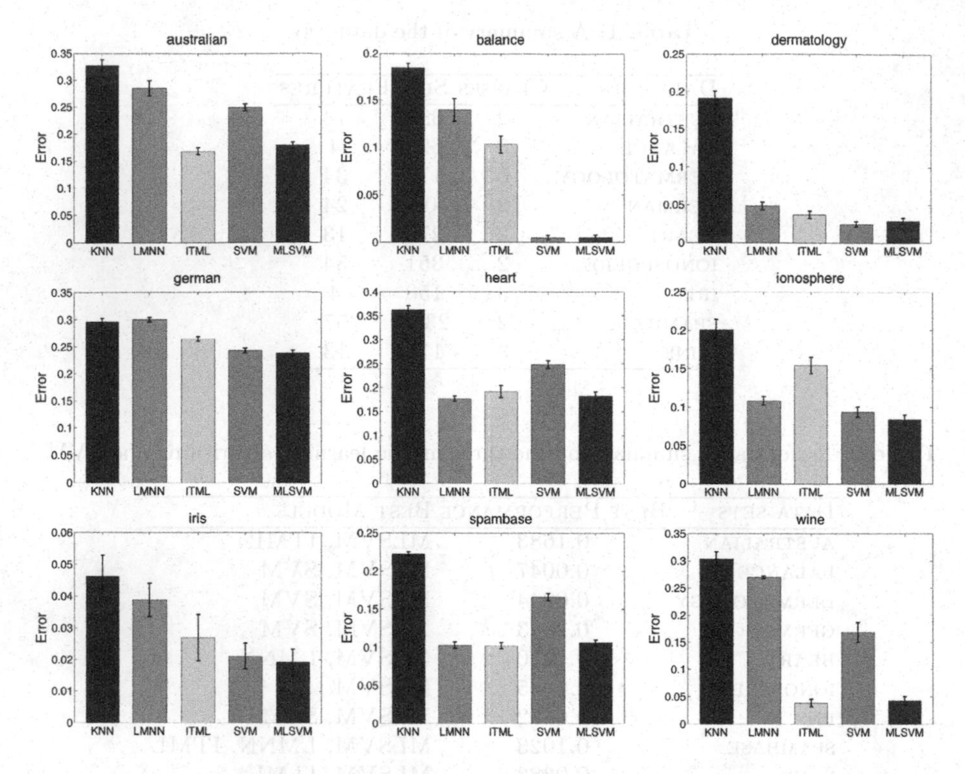


**Fig. 2.** Classification Error Rates of 6 different learning algorithms: KNN, LMNN, ITML, SVM, MLSVM for 9 data sets: wine, heart, dermatology, ionosphere, balance, iris, australian, german, and spambase

## 5 Conclusion

In this paper, we approached the metric learning problem utilizing the margin-based framework with the proposed method MLSVM. The problem is formulated as a quadratic semi-definite programming problem (QSDP) subject to local neighborhood constraints. We solve the optimization problem by iteratively relaxing the semidefiniteness constraint of the distance matrix and only approximate the closest semidefinite matrix after the gradient descent step. The SVM-based formulation allows MLSVM to incorporate various kernels conveniently. From the conducted experiments, MLSVM is able to produce a distance metric that helps improve this performance of the KNN classifier significantly. Our experiments also show that our metric learning algorithm always performs competitively and often better than the two other state-of-the-art metric learning approaches. Although KNN is arguably the most obvious application where metric distance learning plays an important role, we plan to adapt MLSVM to other interesting applications where a good distance metric is essential in obtaining competitive performances.

## References

1. Davis, J.V., Kulis, B., Jain, P., Sra, S., Dhillon, I.S.: Information-theoretic metric learning. In: Proceedings of the 24th International Conference on Machine learning, pp. 209–216 (2007)
2. Globerson, A., Roweis, S.: Metric learning by collapsing classes. In: Advances in Neural Information Processing Systems (NIPS) (2005)
3. Goldberger, J., Roweis, S., Hinton, G., Salakhutdinov, R.: Neighbourhood components analysis. In: Advances in Neural Information Processing Systems (NIPS) (2004)
4. Schultz, M., Joachims, T.: Learning a distance metric from relative comparisons. In: Advances in Neural Information Processing Systems (NIPS) (2004)
5. Shalev-Shwartz, S., Singer, Y., Ng, A.Y.: Online and batch learning of pseudo-metrics. In: Proceedings of the 21st International Conference on Machine Learning (2004)
6. Xing, E.P., Ng, A.Y., Jordan, M.I., Russell, S.: Distance metric learning, with application to clustering with side-information. In: Advances in Neural Information Processing Systems (NIPS) (2003)
7. Weinberger, K.Q., Blitzer, J., Saul, L.K.: Distance metric learning for large margin nearest neighbor classification. In: Advances ing Neural Information Processing Systems (NIPS) (2006)
8. Chopra, S., Hadsell, R., LeCun, Y.: Learning a similarity metric discriminatively, with application to face verification. In: Proceedings of the IEEE Conference on Computer Vision and Pattern Recognition (CVPR) (2005)
9. Tsang, I.W., Kwok, J.T.: Distance metric learning with kernels. In: Proceedings of the International Conferences on Artificial Neural Networks (2003)
10. Shental, N., Hertz, T., Weinshall, D., Pavel, M.: Adjusting learning and relevant component analysis. In: Proceedings of the Seventh European Conference on Computer Vision (ECCV), pp. 776–792 (2002)
11. Cover, T., Hart, P.: Nearest neighbor pattern classification. IEEE Transactions on Information Theory 13(1), 21–27 (1967)
12. Domeniconi, C., Gunopulos, D., Peng, J.: Large margin nearest neighbor classifiers. IEEE Transactions on Neural Networks 16(4), 899–909 (2005)
13. Hastie, T., Tibshirani, R.: Discriminant adaptive nearest neighbor classification and regression. In: Advances in Neural Information Processing Systems (NIPS), vol. 8, pp. 409–415. MIT Press, Cambridge (1996)
14. He, H., Hawkins, S., Graco, W., Yao, X.: Application of genetic algorithm and k-nearest neighbor method in real world medical fraud detection problem. Journal of Advanced Computational Intelligence and Intelligent Informatics 4(2), 130–137 (2000)
15. Tenenbaum, J.B., de Silva, V., Langford, J.C.: A global geometric framework for nonlinear dimensionality reduction. Science 290, 2319–2322 (2000)
16. Roweis, S.T., Saul, L.K.: Nonlinear dimensionality reduction by locally linear embedding. Science 290 (2000)
17. Vapnik, V.N.: Statistical Learning Theory. John Wiley & Sons, Chichester (1998)
18. Shalev-Shwartz, S., Singer, Y., Srebro, N.: Primal estimated sub-gradient solver for SVM. In: Proceedings of the 24th International Conference on Machine Learning, pp. 807–814. ACM, New York (2007)
19. Joachims, T.: Making large-scale support vector machine learning practical. In: Schölkopf, B., Burges, C., Smola, A.J. (eds.) Advances in Kernel Methods: Support Vector Machines, MIT Press, Cambridge (1998)

20. Platt, J.C.: Fast training of support vector machines using sequential minimal optimization. In: Schölkopf, B., Burges, C., Smola, A.J. (eds.) Advances in Kernel Methods: Support Vector Machines, MIT Press, Cambridge (1999)
21. Schölkopf, B., Smola, A.J., Müller, K.R.: Kernel principal component analysis. Neural Computation 10(5), 1299–1319 (1998)
22. Asuncion, A., Newman, D.: UCI machine learning repository (2007), `http://www.ics.uci.edu/~mlearn/MLRepository.html`
23. Chang, C.C., Lin, C.J.: Libsvm data (2001), `http://www.csie.ntu.edu.tw/~cjlin/libsvmtools/datasets/`
24. Crammer, K., Singer, Y.: On the algorithmic implementation of multiclass kernel-based vector machines. Journal of Machine Learning Research 2, 265–292 (2001)

# Support Vector Machines, Data Reduction, and Approximate Kernel Matrices

XuanLong Nguyen[1], Ling Huang[2], and Anthony D. Joseph[3]

[1] SAMSI & Duke University
xuanlong.nguyen@stat.duke.edu
[2] Intel Research
ling.huang@intel.com
[3] UC Berkeley
adj@eecs.berkeley.edu

**Abstract.** The computational and/or communication constraints associated with processing large-scale data sets using support vector machines (SVM) in contexts such as distributed networking systems are often prohibitively high, resulting in practitioners of SVM learning algorithms having to apply the algorithm on approximate versions of the kernel matrix induced by a certain degree of data reduction. In this paper, we study the tradeoffs between data reduction and the loss in an algorithm's classification performance. We introduce and analyze a consistent estimator of the SVM's achieved classification error, and then derive approximate upper bounds on the perturbation on our estimator. The bound is shown to be empirically tight in a wide range of domains, making it practical for the practitioner to determine the amount of data reduction given a permissible loss in the classification performance.[1]

**Keywords:** Support vector machines, kernel methods, approximate kernel matrices, matrix perturbation, classification.

## 1 Introduction

The popularity of using support vector machines (SVM) for classification has led to their application in a growing number of problem domains and to increasingly larger data sets [1,2,3,4]. An appealing key feature of the SVM is that the only interface of the learning algorithm to the data is through its kernel matrix. In many applications, the communication-theoretic constraints imposed by limitations in the underlying distributed data collection infrastructure, or the computational bottleneck associated with a large-scale kernel matrix, naturally requires some degree of data reduction. This means that practitioners usually do not have the resources to train the SVM algorithm on the original kernel matrix. Instead, they must rely on an approximate, often simplified, version of the kernel matrix induced by data reduction.

Consider, for instance, the application of an SVM to a detection task in a distributed networking system. Each dimension of the covariate $X$ represents the data captured by a

[1] The authors would like to thank Michael I. Jordan, Noureddine El Karoui and Ali Rahimi for helpful discussions.

W. Daelemans et al. (Eds.): ECML PKDD 2008, Part II, LNAI 5212, pp. 137–153, 2008.

monitoring device (e.g., network node or sensor), which continuously ships its data to a coordinator for an aggregation analysis using the SVM algorithm. Due to the communication constraints between nodes within the network and the power constraints of each node (e.g., for battery-powered sensors), the monitoring devices do not ship all of their observations to the coordinator; rather, they must appropriately down-sample the data. From the coordinator's point of view, the data analysis (via the SVM or any other algorithm) is not applied to the original data collected by the monitoring devices, but rather to an approximate version. This type of in-network distributed processing protocol has become increasingly popular in various fields, including systems and databases [5,6], as well as in signal processing and machine learning [7,8,9]. In the case where the coordinator uses an SVM for classification analysis, the SVM has access not to the original data set, but rather to only an approximate version, which thus yields an approximate kernel matrix. The amount of kernel approximation is dictated by the amount of data reduction applied by the monitoring devices.

Within the machine learning field, the need for training with an approximate kernel matrix has long been recognized, primarily due to the computational constraints associated with large kernel matrices. As such, there are various methods that have been developed for replacing an original kernel matrix $K$ with a simplified version $\tilde{K}$: matrices with favorable properties such as sparsity, low-rank, etc [10,11,12,13,14].

To our knowledge, there has been very little work focusing on the tradeoffs between the amount of data reduction and the classification accuracy. This issue has only been recently explored in the machine learning community; see [15] for a general theoretical framework. Understanding this issue is important for learning algorithms in general, and especially for SVM algorithms, as it will enable their application in distributed systems, where large streams of data are generated in distributed devices, but not all data can be centrally collected. Furthermore, the tradeoff analysis has to be achieved in simple terms if it is to have impact on practitioners in applied fields.

The primary contribution of this paper is an analysis of the tradeoff between data reduction and the SVM classification error. In particular, we aim to produce simple and practically useful upper bounds that specify the amount of loss of classification accuracy for a given amount of data reduction (to be defined formally). To this end, the contributions are two-fold: (i) First, we introduce a novel estimate, called the *classification error coefficient* $C$, for the classification error produced by the SVM, and prove that it is a consistent estimate under appropriate conditions. The derivation of this estimator is drawn from the relationship between the hinge loss (used by the SVM) and the 0-1 loss [16]. (ii) Second, using the classification error coefficient $C$ as a surrogate for the classification accuracy, we introduce upper bounds on the change in $C$ given an amount of data reduction. Specifically, let $K$ be the kernel matrix on the original data that we don't have access to, $\tilde{K}$ the kernel matrix induced by data reduction, and suppose that each element of $\Delta = \tilde{K} - K$ has variance bounded by $\sigma^2$. Let $\tilde{C}$ be the classification error coefficient associate to $\tilde{K}$. We express an upper bound of $\tilde{C} - C$ in terms of $\sigma$ and matrix $\tilde{K}$. The bound is empirically shown to be remarkably tight for a wide range of data domains, making it practical for the practitioner of the SVM to determine the amount of data reduction given a permissible loss in the classification performance.

The remainder of the paper is organized as follows: in Section 2, we provide background information about the SVM algorithm, and describe the contexts that motivate the need for data reduction and approximate kernel matrices; in Section 3, we describe the main results of this paper, starting with a derivation and consistency analysis of the classification error coefficient $C$, and then presenting upper bounds on the change of $C$ due to kernel approximation; in Section 4, we present an empirical evaluation of our analyses; and in Section 5, we discuss our conclusions.

## 2 SVM, Data Reduction and Kernel Matrix Approximation

### 2.1 SVM Background

In a classification algorithm, we are given as our training data $m$ i.i.d. samples $(x_i, y_i)_{i=1}^m$ in $\mathcal{X} \times \{\pm 1\}$, where $\mathcal{X}$ denotes a bounded subset of $\mathrm{R}^d$. A classification algorithm involves finding a discriminant function $y = \mathrm{sign}(f(x))$ that minimizes the classification error $P(Y \neq \mathrm{sign}(f(X)))$.

Central to a kernel-based SVM classification algorithm is the notion of a kernel function $K(x, x')$ that provides a measure of similarity between two data points $x$ and $x'$ in $\mathcal{X}$. Technically, $K$ is required to be a symmetric positive semidefinite kernel. For such a function, Mercer's theorem implies that there must exist a reproducing kernel Hilbert space $\mathcal{H} = \mathrm{span}\{\Phi(x) | x \in \mathcal{X}\}$ in which $K$ acts as an inner product, i.e., $K(x, x') = \langle \Phi(x), \Phi(x') \rangle$. The SVM algorithm chooses a linear function in this feature space $f(x) = \langle \mathrm{w}, \Phi(x) \rangle$ for some w that minimizes the regularized training error:

$$\min_{\mathrm{w} \in \mathcal{H}} \frac{1}{m} \sum_{i=1}^{m} \phi(y_i f(x_i)) + \lambda_m \|\mathrm{w}\|^2 / 2. \tag{1}$$

Here $\lambda_m$ denotes a regularization parameter, and $\phi$ denotes an appropriate loss function that is a convex surrogate to the 0-1 loss $\mathbb{I}(y \neq \mathrm{sign}(f(x)))$. In particular, the SVM uses hinge loss $\phi(yf(x)) = (1 - yf(x))_+$ [3]. It turns out that the above optimization has the following dual formulation in quadratic programming:

$$\max_{0 \leq \alpha \leq 1} \frac{1}{m} \sum_i \alpha_i - \frac{1}{2m^2 \lambda_m} \sum_{i,j} \alpha_i \alpha_j y_i y_j K(x_i, x_j). \tag{2}$$

For notational convenience, we define matrix $Q$ such that $Q_{ij} = K(x_i, x_j) y_i y_j$. The solution $\alpha$ of the above dual formulation defines the optimal $f$ and w of the primal formulation via the following:

$$\mathrm{w} = \frac{1}{m\lambda_m} \sum_{i=1}^{m} \alpha_i \Phi(x_i) \tag{3}$$

$$f(x) = \frac{1}{m\lambda_m} \sum_{i=1}^{m} \alpha_i K(x_i, x). \tag{4}$$

## 2.2 In-Network Data Reduction and Approximate Kernel Matrices

As seen from the dual formulation (2), the kernel matrix $K = \{K(x_i, x_j)_{i,j}\}$ and the label vector $y = [y_1, \ldots, y_m]$ form sufficient statistics of the SVM. However, there is substantial previous work that focuses on the application of an SVM to an approximate version $\tilde{K}$ of the kernel matrix from data reduction. We extend this work to focus in particular on the application of SVM in distributed system environments.

*Suppression of data streams and data quantization in distributed systems.* A primary motivation regarding this work is the application of SVM-based classification analysis to distributed settings in a number of fields, including databases, distributed systems, and sensor networks [5,6,9]. In a distributed system setting, there are $d$ monitoring devices which receive streams of raw data represented by a $d$-dimensional covariate $X$ and send the data to a central coordinator for classification analysis. Because of communication constraints, each monitoring devices cannot send all its received data; instead, they must send as little data as possible. An $\epsilon$-suppression algorithm is frequently used: each monitoring devices $j, j = 1, \ldots, d$, send the $i$-th data point to the coordinator only if: $|X_i^j - X_{i-1}^j| > \epsilon$. Using these values, the coordinator reconstructs an approximate view $\tilde{X}$ of the true data $X$, such that $\|X - \tilde{X}\|_\infty \leq \epsilon$. A key question in the design of such systems is how to determine the data reduction parameter $\epsilon$, given a permissible level of loss in the classification accuracy.

In signal processing, data reduction is achieved by quantization or binning: each dimension of $X$ is discretized into a given number of bins before being sent to the central coordinator [7,8]. The bin size is determined by the number of bits available for transmission: for bins of equal size $\epsilon$, the number of bins is proportional to $1/\epsilon$, corresponding to using $\log(1/\epsilon)$ number of bits. As before, the coordinator receives an approximate version $\tilde{X}$, such that $\|X - \tilde{X}\|_\infty \leq \epsilon$. Once $\tilde{X}$ is received by the coordinator, one obtains an approximate kernel matrix by applying the kernel function $K$ to $\tilde{X}$. Suppose that a Gaussian kernel with width parameter $\omega > 0$ is used, then we obtain the approximate kernel $\tilde{K}$ as $\tilde{K}(\tilde{X}_i, \tilde{X}_j) = \exp\left(-\frac{\|\tilde{X}_i - \tilde{X}_j\|^2}{2\omega^2}\right)$.

*Kernel matrix sparsification and approximation.* Beside applications in in-network and distributed data processing, a variety of methods have been devised to approximate a large kernel matrix by a more simplified version with desirable properties, such as sparsity and low-rank (e.g., [10,11,12,13,14]). For instance, [17] proposes a simple method to approximate $K$ by randomly zeroing out its entries:

$$\tilde{K}_{ij} = \tilde{K}_{ji} = \begin{cases} 0 & \text{with probability } 1 - 1/\delta, \\ \delta K_{ij} & \text{with probability } 1/\delta, \end{cases}$$

where $\delta \geq 1$ controls the degree of sparsification on the kernel.[2] This sparsification was shown to greatly speed up the construction and significantly reduce the space required

[2] This method may not retain the positive definiteness of the kernel matrix, in which case positive values have to be added to the matrix diagonal.

to store the matrix. Our analysis can also be applied to analyze the tradeoff of kernel approximation error and the change in classification error.

# 3 Classification Error Coefficient and Effects of Data Reduction

We begin by describing the set-up of our analysis. Let $\tilde{K}$ be a (random) kernel matrix that is an approximate version of kernel matrix $K$ induced by a data reduction scheme described above (e.g., quantization or suppression). Let $C_0$ and $\tilde{C}_0$ be the (population) classification error associated with the SVM classifier trained with kernel matrix $K$ and $\tilde{K}$, respectively. We wish to bound $|\tilde{C}_0 - C_0|$ in terms of the "magnitude" of the error matrix $\Delta = \tilde{K} - K$, which we now define. For a simplified analysis, we make the following assumption about the error matrix $\Delta$:

A0. Conditioned on $\tilde{K}$ and $y$, all elements $e_{ij}$ $(i,j = 1,\ldots,m; i \neq j)$ of $\Delta$ are uncorrelated, have zero mean, and the variance bounded by $\sigma^2$.

We use $\sigma$ to control the degree of our kernel matrix approximation scheme, abstracting away from further detail. It is worth noting that certain kernel matrix approximation schemes may not satisfy the independence assumption. On the one hand, it is possible to incorporate the correlation of elements of $\Delta$ into our analysis. On the other hand, we find that the correlation is typically small, such that elaboration does not significantly improve our bounds in most cases.

Our ultimate goal is to produce practically useful bounds on $\tilde{C}_0 - C_0$ in terms of $\sigma$ and kernel matrix $\tilde{K}$. This is a highly nontrivial task, especially since we have access only to approximate data (through $\tilde{K}$, but not $K$).

## 3.1 Classification Error Coefficient

In order to quantify the effect on the population SVM classification error $C_0$, we first introduce a simple estimate of $C_0$ from empirical data. In a nutshell, our estimator relies on the following intuitions:

1. The SVM algorithm involves minimizing over a surrogate loss (the hinge loss), while we are interested in the performance in terms of 0-1 loss. Thus, we need to be able to compare between these two losses.
2. We are given only empirical data, and we replace the risk (population expectation of a loss function) by its empirical version.
3. We avoid terms that are "nonstable" for the choice of learning parameters, which is important for our subsequent perturbation analysis.

The first key observation comes from the fact that the optimal expected $\phi$-risk using the hinge loss is shown to be twice the optimal Bayes error (i.e., using 0-1 loss) (cf. [16], Sec. 2.1):

$$\min_{f \in \mathcal{F}} P(Y \neq f(X)) = \frac{1}{2} \min_{f \in \mathcal{F}} \mathrm{E}\phi(Yf(X)), \tag{5}$$

where $\mathcal{F}$ denotes an arbitrary class of measurable functions that contains the optimal Bayes classifier.

Note that we can estimate the optimal expected $\phi$-risk by its empirical version defined in Eqn. (1), which equals its dual formulation (2). Let $\hat{\mathrm{w}}$ be the solution of (1). As shown in the proof of Theorem 1, if $\lambda_m \to 0$ sufficiently slowly as $m \to \infty$, the penalty term $\lambda_m \|\hat{\mathrm{w}}\|^2$ vanishes as $m \to \infty$. Due to (3), the second quantity in the dual formulation (2) satisfies

$$\frac{1}{2m^2\lambda_m}\sum_{i,j}\alpha_i\alpha_j y_i y_j K(x_i, x_j) = \lambda_m\|\hat{\mathrm{w}}\|^2/2 \to 0.$$

As a result, we have:

$$\inf_{\mathrm{w}\in\mathcal{H}} \hat{\mathrm{E}}\phi(Yf(X)) + \lambda_m\|\mathrm{w}\|^2/2 = \frac{1}{m}\sum_{i=1}^{m}\alpha_i - \lambda_m\|\hat{\mathrm{w}}\|^2/2. \tag{6}$$

Approximating the optimal $\phi$-risk in (5) by its empirical version over $\mathcal{H}$, and dropping off the vanishing term $\lambda_m\|\hat{\mathrm{w}}\|^2$ from Eqn. (6), we obtain the following estimate:

**Definition 1.** *Let $\alpha$ be the solution of the SVM's dual formulation* (2)*, the following quantity is called the* **classification error coefficient:**

$$C = \frac{1}{2m}\sum_{i=1}^{m}\alpha_i. \tag{7}$$

An appealing feature of $C$ is that $C \in [0, 1/2]$. Furthermore, it is a simple function of $\alpha$. As we show in the next section, this simplicity significantly facilitates our analysis of the effect of kernel approximation error. Applying consistency results of SVM classifiers (e.g., [18]) we can show that $C$ is also a universally consistent estimate for the optimal classification error under appropriate assumptions. These assumptions are:

A1. $K$ is a universal kernel on $\mathcal{X}$, i.e., the function class $\{\langle \mathrm{w}, \Phi(\cdot)|\mathrm{w} \in \mathcal{H}\rangle\}$ is dense in the space of continuous functions on $\mathcal{X}$ with respect to the sup-norm (see [18] for more details). Examples of such kernels include the Gaussian kernel $K(x, x') = \exp\left(-\frac{\|x-x'\|^2}{2\omega^2}\right)$, among others.

A2. $\lambda_m \to 0$ such that $m\lambda_m \to \infty$.

**Theorem 1.** *Suppose that $(X_i, Y_i)_{i=1}^m$ are drawn i.i.d. from a Borel probability measure $P$. Under assumptions* A1 *and* A2*, there holds as $m \to \infty$:*

$$C - \inf_{f\in\mathcal{F}} P(Y \neq f(X)) \to 0 \textit{ in probability.}$$

See the appendix for a proof. It is worth noting that this result is kernel-independent.

Let $\tilde{K}, \tilde{\alpha}, \tilde{f}, \tilde{C}$ denote the corresponding counterparts for kernel matrix $K$, the dual formulation's solutions $\alpha$, classifier $f$, and the classification coefficient $C$, respectively. For the data suppression and quantization setting described in Section 2, suppose that a universal kernel (such as Gaussian kernel) is applied to both original and approximate data. By Theorem 1, both $C$ and $\tilde{C}$ are consistent estimates of the classification error applied on original and approximate data, respectively. Thus, the difference $\tilde{C} - C$ can be used to evaluate the loss of classification accuracy of the SVM. This is the focus of the next section. [3]

### 3.2 Effects of Data Reduction on Classification Error Coefficient

In this section, we analyze the effects of the approximation of the kernel matrix $K - \tilde{K}$ on the classification error coefficient difference $C - \tilde{C}$.

Let $r = \#\{i : \alpha_i \neq \tilde{\alpha}_i\}$. From Eqn. (7), the difference of the classification coefficients is bounded via Cauchy-Schwarz inequality:

$$|\tilde{C} - C| \leq \frac{1}{2m}\|\tilde{\alpha} - \alpha\|_1 \leq \frac{1}{2m}\sqrt{r}\|\tilde{\alpha} - \alpha\|, \tag{8}$$

from which we can see the key point lies in deriving a tight bound on the $L_2$ norm $\|\tilde{\alpha} - \alpha\|$. Define two quantities:

$$R_1 = \frac{\|\tilde{\alpha} - \alpha\|^2}{(\tilde{\alpha} - \alpha)^T Q (\tilde{\alpha} - \alpha)}, \quad R_2 = \frac{(\tilde{\alpha} - \alpha)^T (Q - \tilde{Q})\tilde{\alpha}}{\|\tilde{\alpha} - \alpha\|}.$$

**Proposition 1.** *If $\alpha$ and $\tilde{\alpha}$ are the optimal solution of the program* (2) *using kernel matrix $K$ and $\tilde{K}$ respectively, then:*

$$|\tilde{C} - C| \leq \frac{\sqrt{r}}{2m}\|\tilde{\alpha} - \alpha\| \leq \frac{\sqrt{r}}{2m} R_1 R_2.$$

For a proof, see the Appendix. Although it is simple to derive rigorous absolute bounds on $R_1$ and $R_2$, such bounds are not practically useful. Indeed, $R_1$ is upper bounded by the inverse of the smallest eigenvalue of $Q$, which tends to be very large. An alternative solution is to obtain probabilistic bounds that hold with high probability, using Prop. 1 as a starting point. Note that given a data reduction scheme, there is an induced joint distribution generating kernel matrix $K$, its approximate version $\tilde{K}$, as well as the label vector $y$. Matrix $Q = K \circ yy^T$ determines the value of vector $\alpha$ through an optimization problem (2). Likewise, $\tilde{Q} = \tilde{K} \circ yy^T$ determines $\tilde{\alpha}$. Thus, $\alpha$ and $\tilde{\alpha}$ are random under the distribution that marginalizes over random matrices $Q$ and $\tilde{Q}$, respectively.

The difficult aspect of our analysis lies in the fact that we do not have closed forms of either $\alpha$ or $\tilde{\alpha}$, which are solutions of quadratic programs parameterized by $Q$ and

[3] We make several remarks: (i) The rates at which $C$ and $\tilde{C}$ converge to the respective misclassification rate may not be the same. To understand this issue one has to take into account additional assumptions on both the kernel function, and the underlying distribution $P$. (ii) Although quantization of data does not affect the consistency of the classification error coefficient since one can apply the same universal kernel function to quantized data, quantizing/approximating *directly* the kernel matrix (such as those proposed in [17] and described in Sec. 2) may affect both consistency and convergence rates in a nontrivial manner. An investigation of approximation rates of the *quantized/sparsified* kernel function class is an interesting open direction.

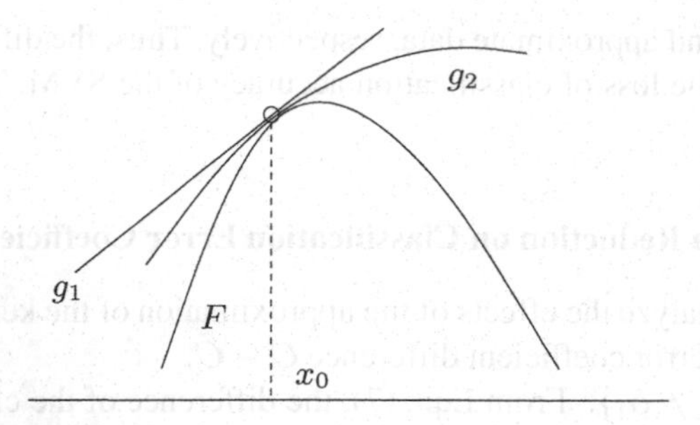


**Fig. 1.** Illustration of upper bounds via perturbation analysis: Linear approximation $g_1$ and upper bound $g_2$ via second-order perturbation analysis of a concave function $F$ around a point $x_0$ in the domain. The bounds continue to hold for large perturbation around $x_0$.

$\tilde{Q}$, respectively. We know a useful fact, however, regarding the distributions of vector $\alpha$ and $\tilde{\alpha}$. Since the training data are i.i.d., the roles of $\alpha_i$ and $\tilde{\alpha}_i$ for $i = 1, \ldots, m$ are equivalent. Thus $(\alpha_i, \tilde{\alpha}_i)$ have marginally identical distributions for $i = 1, \ldots, m$.

We first motivate our subsequent perturbation analysis by an observation that the optimal classification error defined by Eq. (5), for which $C$ is an estimate, is a concave function with respect jointly to the class probability distributions $(P(X|Y = 1), P(X|Y = -1))$ (cf. [16], Sec. 2). When the data is perturbed (e.g., via quantization/suppression) the joint distribution $(P(X|Y = 1), P(X|Y = -1))$ is also perturbed. Intuitively, upper bounds for a concave functional via either linear or second-order approximation under a small perturbation on its variables should also hold under larger perturbations, even if such bounds tend to be less tight in the latter situation. See Fig. 3.2 for an illustration. Thus, to obtain useful probabilistic bounds on $\tilde{C} - C$, we restrict our analysis to the situation where $\tilde{K}$ is a small perturbation from the original matrix $K$. Under a small perturbation, the following assumptions can be made:

B1. The random variables $\tilde{\alpha}_i - \alpha_i$ for $i = 1, \ldots, n$ are non-correlated.
B2. The random variables $\tilde{\alpha}_i - \alpha_i$ have zero means.

Given Assumption B1, coupled with the fact that $(\tilde{\alpha}_i - \alpha_i)$ have identical distributions for $i = 1, \ldots, m$, by the central limit theorem, as $m$ gets large, a rescaled version of $\tilde{C} - C$ behaves like a standard normal distribution. Using a result for standard normal random variables, for any constant $t > 0$, we obtain that with probability at least $1 - \frac{1}{\sqrt{2\pi}t} e^{-t^2/2}$:

$$\tilde{C} - C \lesssim t\sqrt{\mathrm{Var}(\tilde{C} - C)} + \mathrm{E}(\tilde{C} - C) \overset{Ass.(B1)}{=} \frac{t}{2m}\sqrt{\sum_{i=1}^{m} \mathrm{Var}(\tilde{\alpha}_i - \alpha_i)} + \mathrm{E}(\tilde{C} - C)$$
$$\leq \frac{t}{2m}\sqrt{\mathrm{E}\|\alpha - \tilde{\alpha}\|^2} + \mathrm{E}(\tilde{C} - C) \overset{Prop.\,1}{\leq} \frac{t}{2m}\sqrt{\mathrm{E}R_1^2 R_2^2} + \mathrm{E}(\tilde{C} - C).$$

Our next step involves an observation that under certain assumptions to be described below, random variable $R_1$ is tightly concentrated around a constant, and that $\mathrm{E}R_2^2$ can be easily bounded.

$$R_1 \approx \frac{m}{\text{tr}(K)} \quad \text{by Lemma 2} \tag{9}$$

$$\text{E}R_2^2 \leq \sigma^2 m \text{E}\|\tilde{\alpha}\|^2 \quad \text{by Lemma 1.} \tag{10}$$

As a result, we obtain the following approximate bound:

$$\tilde{C} - C \lesssim t\frac{\sigma\sqrt{m\text{E}\|\tilde{\alpha}\|^2}}{2\text{tr}(K)} + \text{E}(\tilde{C} - C) \overset{Ass.B2}{=} t\frac{\sigma\sqrt{m\text{E}\|\tilde{\alpha}\|^2}}{2\text{tr}(K)}, \tag{11}$$

where Eqn. (11) is obtained by invoking Assumption B2.

Suppose that in practice we do not have access to $K$, then $\text{tr}(K)$ can be approximated by $\text{tr}(\tilde{K})$. In fact, for a Gaussian kernel $\text{tr}(K) = \text{tr}(\tilde{K}) = m$. One slight complication is estimating $\text{E}\|\tilde{\alpha}\|^2$. Since we have only one training sample for $\tilde{K}$, which induces a single sample for $\tilde{\alpha}$, this expectation is simply estimated by $\|\tilde{\alpha}\|^2$.

When we choose $t = 1$ in bound (11), the probability that the bound is correct is approximately $1 - e^{-1/2}/\sqrt{2\pi} = 75\%$. For $t = 2$, the probability improves to $1 - e^{-2}/2\sqrt{2\pi} = 97\%$. While $t = 1$ yields relatively tighter bound, we choose $t = 2$ in practice. In summary, we have obtained an approximate bound:

$$\text{classif. coeff. (approx. data)} \leq \text{classif. coeff. (original data)} + \frac{\sigma\sqrt{m\|\tilde{\alpha}\|^2}}{\text{tr}(\tilde{K})} \tag{12}$$

*Remark.* (i) Even though our analysis is motivated by the context of small perturbations to the kernel matrix, bound (12) appears to hold up well in practice when $\sigma$ is large. This agrees with our intuition on the concavity of (5) discussed earlier. (ii) Our analysis is essentially that of second-order matrix perturbation which requires the perturbation be small so that both Assumptions B1 and B2 hold. Regarding B1, for $i = 1, \ldots, m$, each pair $(\alpha_i, \tilde{\alpha}_i)$ corresponds to the $i$-th training data point, which is drawn i.i.d. As a result, $(\alpha_i - \tilde{\alpha}_i)$ are very weakly correlated with each other. We show that this is empirically true through a large number of simulations. (ii) Assumption B2 is much more stringent by comparison. When $\tilde{K}$ is only a small perturbation of matrix $K$, we have also found through simulations that this assumption is very reasonable, especially in the contexts of the data quantization and kernel sparsification methods described earlier.

### 3.3 Technical Issues

**Probabilistic bounds of $R_1$ and $R_2$.** Here we elaborate on the assumptions under which the probabilistic bounds for $R_1$ and $R_2$ are obtained, which motivate the approximation method given above. Starting with $R_2$, it is simple to obtain:

**Lemma 1.** *Under Assumption* A0, $\text{E}R_2 \leq \sqrt{\text{E}R_2^2} \leq \sigma\sqrt{m\text{E}[\|\tilde{\alpha}\|^2]}$.

See the Appendix for a proof. Turning to the inverse of Raleigh quotient term $R_1$, our approximation is motivated by the following fact, which is a direct consequence of Thm 2.2. of [19]:

**Lemma 2.** *Let $A$ be fixed $m \times m$ symmetric positive definite matrix with* bounded *eigenvalues $\lambda_1, \ldots, \lambda_m$, and $z$ be an $m$-dim random vector drawn from any spherically symmetric distribution,*

$$\mathrm{E}[z^T Az/\|z\|^2] = \mathrm{tr}(A)/m$$

$$\mathrm{Var}[z^T Az/\|z\|^2] = \frac{2}{m+2}\Big(\sum_{i=1}^{m} \lambda_i^2/m - (\sum_{i=1}^{m} \lambda_i/m)^2\Big).$$

By this result, $z^T Az/\|z\|^2$ has vanishing variance as $m \to \infty$. Thus, this quantity is tightly concentrated around its mean. Note that if $\mathrm{tr}(A)/m$ is bounded away from 0, we can also approximate $1/z^T Az$ by $m/\mathrm{tr}(A)$. This is indeed the situation with most kernels in practice: As $m$ becomes large, $\mathrm{tr}(\tilde{K})/m \to \mathrm{E}\tilde{K}(X,X) > 0$. As a result, we obtain approximation (9).

It is worth noting that the "goodness" of this heuristic approximation relies on the assumption that $\alpha - \tilde{\alpha}$ follows an approximately spectrally symmetric distribution. On the other hand, the concentration of the Raleigh quotient term also holds under more general conditions (cf. [20]). An in-depth analysis of such conditions on $\alpha$ and $\tilde{\alpha}$ is beyond the scope of this paper.

### 3.4 Practical Issues

The bound we derived in Eqn. (12) is readily applicable to practical applications. Recall from Section 2 the example of the detection task in a distributed networking system using a SVM. Each monitoring device independently applies a quantization scheme on their data before sending to the coordinator. The size of the quantized bin is $\epsilon$. Equivalently, one could use an $\epsilon$-suppression scheme similar to [9]. The coordinator (e.g., network operation center) has access only to approximate data $\tilde{X}$, based on which it can compute $\tilde{C}, \tilde{K}, \tilde{\alpha}$ by applying a SVM on $\tilde{X}$. Given $\epsilon$, one can estimate the amount of kernel matrix approximation error $\sigma$ and vice versa (see, e.g., [9]). Thus, Eqn. (12) gives the maximum possible loss in the classification accuracy due to data reduction. The tightness of bound (12) is crucial: it allows the practitioner to tune the data reduction with good a confidence on the detection performance of the system. Conversely, suppose that the practitioner is willing to incur a loss of classification accuracy due to data reduction by an amount at most $\delta$. Then, the appropriate amount of kernel approximation due to data reduction is:

$$\sigma^* = \frac{\delta \cdot \mathrm{tr}(\tilde{K})}{\sqrt{m\|\tilde{\alpha}\|^2}}. \tag{13}$$

## 4 Evaluation

In this section, we present an empirical evaluation of our analysis on both synthetic and real-life data sets. For exhaustive evaluation of the behavior of the classification error coefficient $C$ and the tradeoff analysis captured by bound (12), we replicate our experiments on a large number of of synthetic data sets of different types in moderate dimensions; for illustration in two dimensions, see Fig 2. To demonstrate the practical

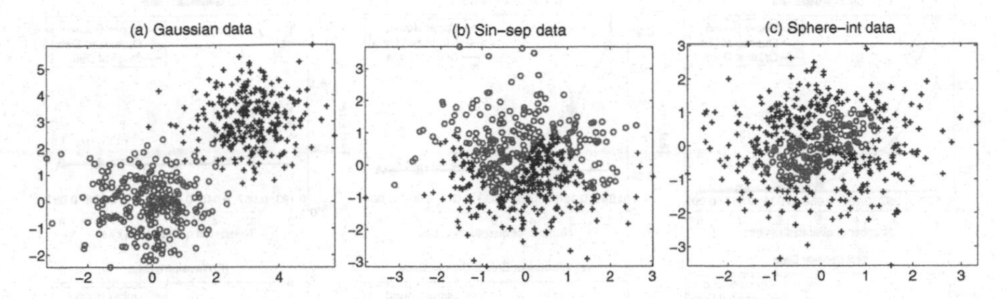


**Fig. 2.** Synthetic data sets illustrated in two dimensions

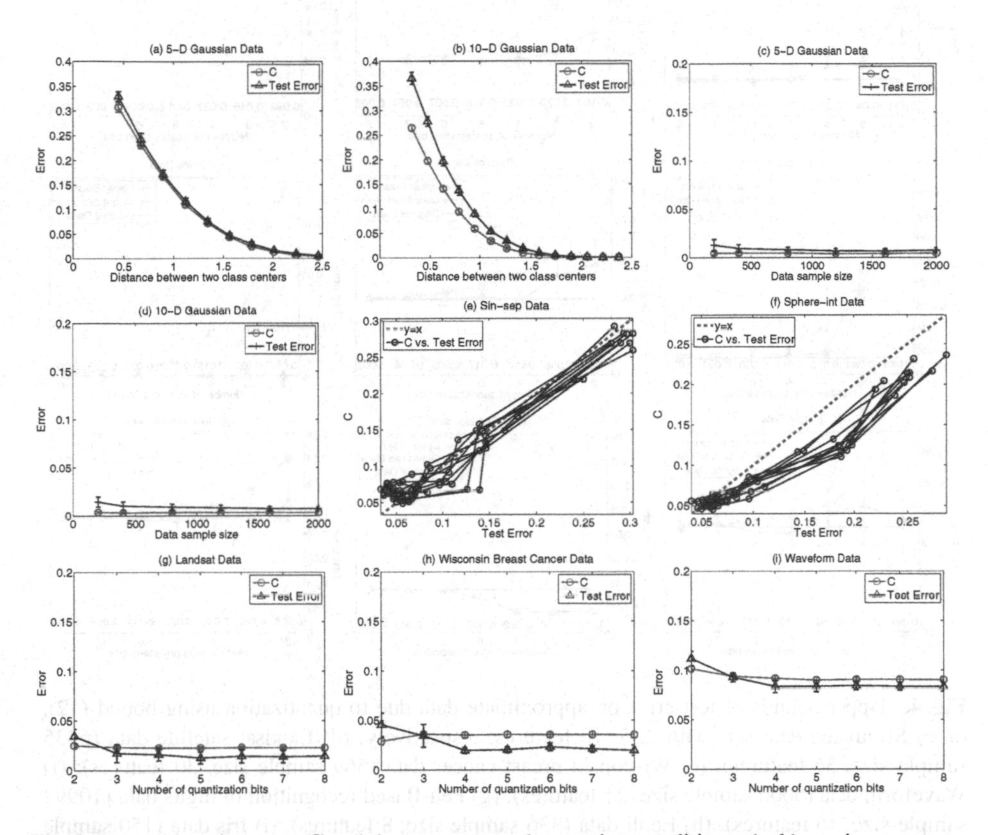


**Fig. 3.** Comparison between $C$ and the test error under varying conditions: (a–b) varying amount of overlap between two classes (both training and test data sets have 2,000 sample points. Error bars are derived from 25 replications); (c–d) varying sample sizes; (e–f) varying amount of data reduction via scatter plots (each path in the scatter plots connects points corresponding to varying number of quantization bits ranging from 8 in low-left corner to 2 bits in upper-right corner); (g–i) varying amount of data reduction via error bar plots. All plots show $C$ remains a good estimate of the test error even with data reduction. We use Gaussian kernels for all experiments.

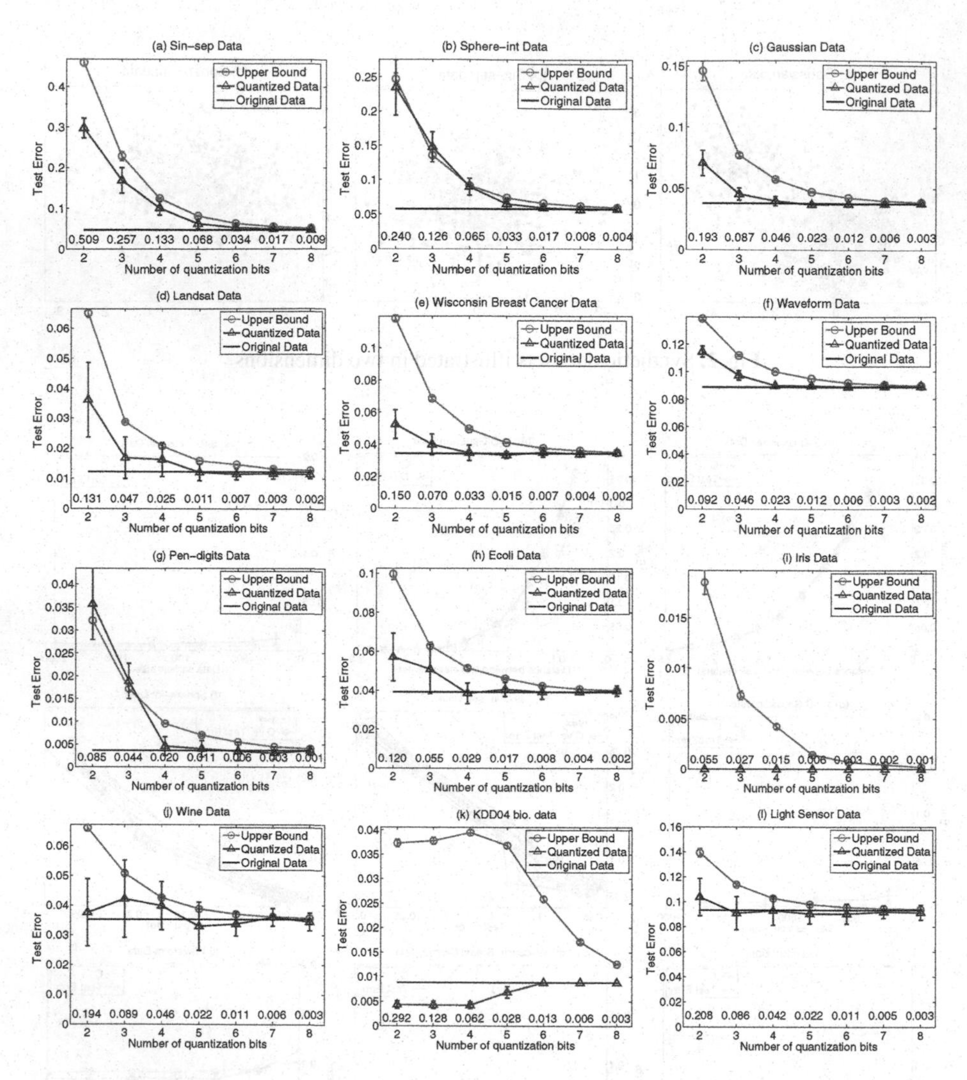


**Fig. 4.** Upper bounds of test error on approximate data due to quantization using bound (12). (a–c) Simulated data sets with 2, 5, 10 features, respectively; (d) Landsat satellite data (6435 sample size, 36 features); (e) Wisconsin breast cancer data (569 sample size, 30 features); (f) Waveform data (5000 sample size, 21 features); (g) Pen-Based recognition of digits data (10992 sample size, 16 features); (h) Ecoli data (336 sample size, 8 features). (i) Iris data (150 sample size, 4 features); (j) Wine data (178 sample size, 13 features); (k) KDD04 Bio data (145K sample size, 74 features); (l) Intel Lab light sensor data (81 sample size, 25 features). We use Gaussian kernels for (a–i), and linear kernels for (j–l). The x-axis shows increased bit numbers and the correspondingly decreasing matrix error $\sigma$.

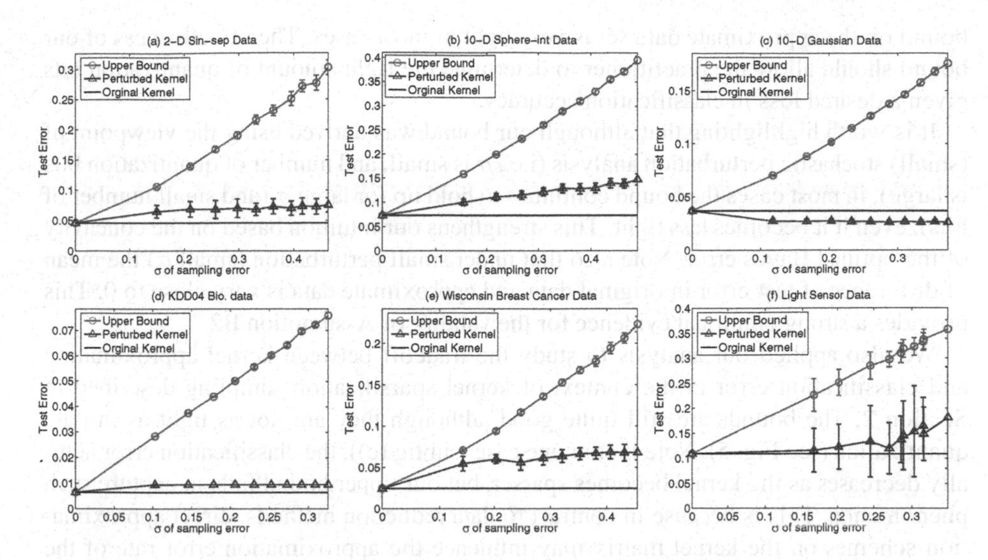


**Fig. 5.** Upper bounds of test error based on bound (12) using approximate Gaussian kernel matrices obtained from kernel sparsification sampling. (a–c) Simulated data sets; (d) KDD04 Bio data; (e) Wisconsin breast cancer data; (f) Intel Lab light sensor data. We use Gaussian kernels for all experiments. The x-axis shows the increasing matrix error $\sigma$ due to down sampling on the kernel matrix.

usefulness of our analysis, we have tested (12) on nine real-life data sets (from the UCI repository [21] and one light sensor data set from the IRB Laboratory [7]), which are subject to varying degrees of data reduction (quantization bits). The data domains are diverse, including satellite images, medicine, biology, agriculture, handwritten digits, and sensor network data, demonstrating the wide applicability of our analysis.

*Evaluation of estimate $C$.* The first set of results in Fig. 3 verify the relationship between the classification error coefficient $C$ and test error on held-out data under varying conditions on: (i) overlap between classification classes (subfigs (a–b)), (ii) sample sizes (subfigs (c–d)) and (iii) amount of data reduction (subfigs (e–i)). It is observed that $C$ estimates the test error very well in all such situations for both simulated and real data sets, and even when the misclassification rate is high (i.e. noisy data). In particular, Fig. 3 (e)(f) show scatter plots comparing C against test error. Each path connects points corresponding to varying amount of data reduction on the same data set. They are very closely parallel to the $y = x$ line, with the points in the upper-right corner corresponding to the most severe data reduction.

*Effects of data reduction on test error.* Next, we evaluate the effect of data reduction via quantization (suppression). Fig. 4 plots the misclassification rate for data sets subject to varying degree of quantization, and the upper bound developed in this paper. Our bound is defined as a sum of test error on original (non-quantized) data set plus the upper bound of $\tilde{C} - C$ provided by (12). As expected, the misclassification rate increases as one decreases the number of quantization bits. What is remarkable is that our upper bound on the approximate data set is very tight in most cases. The effectiveness of our bound should allow the practitioner to determine the right amount of quantization bits given a desired loss in classification accuracy.

It is worth highlighting that although our bound was derived using the viewpoint of (small) stochastic perturbation analysis (i.e., $\sigma$ is small, and number of quantization bits is large), in most cases the bound continues to hold up for large $\sigma$ (and small number of bits), even if it becomes less tight. This strengthens our intuition based on the concavity of the optimal Bayes error. Note also that under small perturbation (small $\sigma$) the mean of difference of test error in original data and approximate data is very close to 0. This provides a strong empirical evidence for the validity of Assumption B2.

We also applied our analysis to study the tradeoff between kernel approximation and classification error in the context of kernel sparsification sampling described in Section 2. The bounds are still quite good, although they are not as tight as in data quantization (see Fig. 5). Note that in one case (subfig (c)), the classification error actually decreases as the kernel becomes sparser, but our upper bound fails to capture such phenomenon. This is because in contrast to data reduction methods, direct approximation schemes on the kernel matrix may influence the approximation error rate of the induced kernel function class in a nontrivial manner. This aspect is not accounted for by our classification error coefficient $C$ (see remarks following Theorem 1).

## 5 Conclusion

In this paper, we studied the tradeoff of data reduction and classification error in the context of the SVM algorithm. We introduced and analyzed an estimate of the test error for the SVM, and by adopting a viewpoint of stochastic matrix perturbation theory, we derived approximate upper bounds on the test error for the SVM in the presence of data reduction. The bound's effectiveness is demonstrated in a large number of synthetic and real-world data sets, and thus can be used to determine the right amount of data reduction given a permissible loss in classification accuracy in applications. Our present analysis focuses mainly on the effect of data reduction on the classification error estimate $C$ while ignoring the its effect on approximability and the approximation rate of the quantized (or sparsified) kernel function class. Accounting for the latter is likely to improve the analysis further, and is an interesting open research direction.

## References


1. Cortes, C., Vapnik, V.: Support-vector networks. Machine Learning 20(3), 273–297 (1995)
2. Joachims, T.: Making large-scale SVM learning practical. In: Schölkopf, B., Burges, C., Smola, A. (eds.) Advances in Kernel Methods – Support Vector Learning. MIT Press, Cambridge (1999)
3. Schölkopf, B., Smola, A.: Learning with Kernels. MIT Press, Cambridge (2002)
4. Shawe-Taylor, J., Cristianini, N.: Kernel methods for pattern analysis. Cambridge University Press, Cambridge (2004)
5. Keralapura, R., Cormode, G., Ramamirtham, J.: Communication-efficient distributed monitoring of thresholded counts. In: Proceedings of ACM SIGMOD (2006)

6. Silberstein, A., Gelfand, G.P.A., Munagala, K., Yang, J.: Suppression and failures in sensor networks: A bayesian approach. In: Proceedings of VLDB (2007)
7. Nguyen, X., Wainwright, M.J., Jordan, M.I.: Nonparametric decentralized detection using kernel methods. IEEE Transactions on Signal Processing 53(11), 4053–4066 (2005)
8. Shi, T., Yu, B.: Binning in Gaussian kernel regularization. Statist. Sinic. 16, 541–567 (2005)
9. Huang, L., Nguyen, X., Garofalakis, M., Joseph, A.D., Jordan, M.I., Taft, N.: In-network PCA and anomaly detection. In: Advances in Neural Information Processing Systems (NIPS) (2006)
10. Fine, S., Scheinberg, K.: Efficient SVM training using low-rank kernel representations. Journal of Machine Learning Research 2, 243–264 (2001)
11. Smola, A., Schölkopf, B.: Sparse greedy matrix approximation for machine learning. In: Proceedings of the 17th International Conference on Machine Learning (2000)
12. Williams, C., Seeger, M.: Using the Nyström method to speed up kernel machines. In: Advances in Neural Information Processing Systems (NIPS) (2001)
13. Yang, C., Duraiswami, R., Davis, L.: Efficient kernel machines using the improved fast gauss transform. In: Advances in Neural Information Processing Systems (NIPS) (2004)
14. Rahimi, A., Recht, B.: Random features for large-scale kernel machines. In: Advances Neural Information Processing Systems (NIPS) (2007)
15. Bottou, L., Bousquet, O.: The tradeoffs of large scale learning. In: Advances in Neural Information Processing Systems (NIPS) (2007)
16. Nguyen, X., Wainwright, M.J., Jordan, M.I.: On surrogate loss functions and $f$-divergences. Annals of Statistics (to appear, 2008)
17. Achlioptas, D., McSherry, F., Schölkopf, B.: Sampling techniques for kernel methods. In: Advances in Neural Information Processing Systems (NIPS) (2001)
18. Steinwart, I.: Consistency of support vector machines and other regularized kernel classifiers. IEEE Trans. on Information Theory 51, 128–142 (2005)
19. Böttcher, A., Grudsky, S.: The norm of the product of a large matrix and a random vector. Electronic Journal of Probability 8, 1–29 (2003)
20. Ledoux, M.: The concentration of measure phenomenon. AMS Society (2001)
21. Asuncion, A., Newman, D.: UCI Machine Learning Repository, Department of Information and Computer Science (2007), http://www.ics.uci.edu/ mlearn/MLRepository.html
22. Bartlett, P., Mendelson, S.: Gaussian and Rademacher complexities: Risk bounds and structural results. Journal of Machine Learning Research 3, 463–482 (2002)

# Appendix

*Proof sketch of Theorem 1:* Let $R_m(f) := \frac{1}{m}\sum_{i=1}^{m}\phi(Yf(X))$, $R(f) := \mathrm{E}\phi(Yf(X))$, and let $I(f) = \|f\|_{\mathcal{H}}$ for any $f \in \mathcal{H}$. To signify the dependence on sample size $m$ we shall use $f_m$ in this proof to denote the SVM classifier defined by (4). The primal form (1) can be re-written as

$$f_m = \mathrm{argmin}_{f\in\mathcal{H}} R_m(f) + \lambda_m I(f)^2/2.$$

The classification error coefficient can be expressed by:

$$C = \frac{1}{2}(R_m(f) + \lambda_m I(f_m)^2).$$

$K$ being a universal kernel implies that $\inf_{w \in \mathcal{H}} R(f) = \min_{f \in \mathcal{F}} R(f)$ (cf. [18], Prop. 3.2). For arbitrary $\epsilon > 0$, let $f_0 \in \mathcal{H}$ such that $R(f_0) \le \min_{f \in \mathcal{F}} R(f) + \epsilon$.

By the definition of $f_m$, we obtain that $R_m(f_m) + \lambda_m I(f_m)^2/2 \le R_m(0) + \lambda_m I(0)^2/2 = R_m(0) = \phi(0)$, implying that $I(f_m) = O(1/\sqrt{\lambda_m})$. We also have:

$$R_m(f_m) + \lambda_m I(f_m)^2/2 \le R_m(f_0) + \lambda_m I(f_0)^2/2.$$

Rearranging gives:

$$R(f_m) - R(f_0) \le (R(f_m) - R_m(f_m)) + (R_m(f_0) - R(f_0)) + \lambda_m (I(f_0)^2 - I(f_m)^2/2.$$

Now, note that for any $B > 0$, if $I(f) \le B$, then $f(x) = \langle \mathrm{w}, \Phi(x) \rangle \le \|w\| \sqrt{K(x,x)} \le M \cdot B$, where $M := \sup_{x \in \mathcal{X}} \sqrt{K(x,x)}$. Note also that the hinge loss $\phi$ is a Lipschitz function with unit constant. We can now apply a result on the concentration of the supremum of empirical processes to bound $R(\cdot) - R_m(\cdot)$. Indeed, applying Thm. 8 of [22] to function class $\{\frac{1}{MB} f | I(f) \le B\}$ (using their Thm. 12 to bound the Rademacher complexity of the kernel function class), we obtain that for any $\delta > 0$, with probability at least $1 - \delta$:

$$R(f_m) - R_m(f_m) \le \frac{4MI(f_m)}{\sqrt{m}} + MI(f_m)\sqrt{\frac{8\ln(2/\delta)}{m}}.$$

We obtain with probability at least $1 - 2\delta$:

$$R(f_m) + \lambda_m I(f_m)^2/2 \le R(f_0) + \frac{4M(I(f_m) + I(f_0))}{\sqrt{m}} + M(I(f_m) + I(f_0))\sqrt{\frac{8\ln(2/\delta)}{m}} +$$

$$\lambda_m I(f_0)^2/2 \le \min_{f \in \mathcal{F}} R(f) + \epsilon + \frac{4M(I(f_m) + I(f_0))}{\sqrt{m}} + M(I(f_m) + I(f_0))\sqrt{\frac{8\ln(2/\delta)}{m}} + \lambda_m (I(f_0))^2/2.$$

Combining Assumption A2 with the fact that $I(f_m) = O(1/\sqrt{\lambda_m})$, the RHS tends to $\min_{f \in \mathcal{F}} R(f) + \epsilon$ as $m \to \infty$. But $R(f_m) \ge \min_{f \in \mathcal{F}} R(f)$ by definition, so $R(f_m) - \min_{f \in \mathcal{F}} R(f) \to 0$ and $\lambda_m I(f_m)^2/2 \to 0$ in probability.

Thus we obtain that

$$C = \frac{1}{2}(R_m(f_m) + \lambda_m I(f_m)^2) \xrightarrow{p} \frac{1}{2} \min_{f \in \mathcal{F}} R(f) = \min_{f \in \mathcal{F}} P(Y \neq f(X)),$$

where the last equality is due to (5).

Before completing the proof, it is worth noting that to the rate of convergence also depends on the rate that $\lambda_m I(f_0)^2 \to 0$ as $\epsilon \to 0$. This requires additional knowledge of the approximating kernel class $\mathcal{H}$ driven by kernel function $K$, and additional properties of the optimal Bayes classifier that $f_0$ tends to.

*Proof sketch of Proposition 1:* If $x_0$ is a minimizer of a differentiable function $F : \mathrm{R}^d \to \mathrm{R}$ over a convex domain, then for any $z$ in the domain, $(z - x_0)^T \nabla F(x_0) \ge 0$.

Applying this fact to both $\alpha$ and $\tilde{\alpha}$ which are the optimizers of Eqn. (2) using $Q$ and $\tilde{Q}$, respectively:

$$(\tilde{\alpha}-\alpha)^T\left(\frac{1}{2m^2\lambda_m}Q\alpha-\frac{1}{m}\mathbb{I}_m\right)\geq 0$$
$$(\alpha-\tilde{\alpha})^T\left(\frac{1}{2m^2\lambda_m}\tilde{Q}\tilde{\alpha}-\frac{1}{m}\mathbb{I}_m\right)\geq 0,$$

where $\mathbb{I}_m=[1\ldots 1]^T$. Adding up the two inequalities yields

$$(\alpha-\tilde{\alpha})^T(\tilde{Q}\tilde{\alpha}-Q\alpha)\geq 0.$$

A minor rearrangement yields

$$(\tilde{\alpha}-\alpha)^T(Q-\tilde{Q})\tilde{\alpha}\geq(\tilde{\alpha}-\alpha)^T Q(\tilde{\alpha}-\alpha),$$

from which the proposition follows immediately.

*Proof of Lemma 1:* By Cauchy-Schwarz, $R_2\leq\|(\tilde{Q}-Q)\tilde{\alpha}\|$. The $i$-th element of the vector inside $\|.\|$ in the RHS is $a_i=y_i\sum_{j=1}^m e_{ij}y_j\tilde{\alpha}_j$. Note that $\tilde{K},y$ determines the value of $\tilde{\alpha}$. Thus, by Assumption A0, we have:

$$\mathrm{E}[a_i^2|\tilde{K},y]=\sum_{j=1}^m\mathrm{E}[e_{ij}^2|\tilde{K},y]\mathrm{E}[\tilde{\alpha}_j^2|\tilde{K},y]\leq\sigma^2\mathrm{E}[\|\tilde{\alpha}\|^2|\tilde{K},y].$$

Marginalizing over $(\tilde{K},y)$ gives $\mathrm{E}a_i^2\leq\sigma^2\mathrm{E}\|\tilde{\alpha}\|^2$. Thus, $\mathrm{E}R_2\leq(\mathrm{E}R_2^2)^{1/2}\leq\sigma\sqrt{m\mathrm{E}\|\tilde{\alpha}\|^2}$.

# Mixed Bregman Clustering with Approximation Guarantees

Richard Nock[1], Panu Luosto[2], and Jyrki Kivinen[2]

[1] CEREGMIA — Université Antilles-Guyane, Schoelcher, France
rnock@martinique.univ-ag.fr
[2] Department of Computer Science, University of Helsinki, Finland
{panu.luosto,jyrki.kivinen}@cs.helsinki.fi

**Abstract.** Two recent breakthroughs have dramatically improved the scope and performance of $k$-means clustering: squared Euclidean seeding for the initialization step, and Bregman clustering for the iterative step. In this paper, we first unite the two frameworks by generalizing the former improvement to *Bregman seeding* — a biased randomized seeding technique using Bregman divergences — while generalizing its important theoretical approximation guarantees as well. We end up with a complete Bregman hard clustering algorithm integrating the distortion at hand in both the initialization and iterative steps. Our second contribution is to further generalize this algorithm to handle *mixed Bregman distortions*, which smooth out the asymetricity of Bregman divergences. In contrast to some other symmetrization approaches, our approach keeps the algorithm simple and allows us to generalize theoretical guarantees from regular Bregman clustering. Preliminary experiments show that using the proposed seeding with a suitable Bregman divergence can help us discover the underlying structure of the data.

## 1 Introduction

Intuitively, the goal of clustering is to partition a set of data points into *clusters* so that similar points end up in the same cluster while points in different clusters are dissimilar. (This is sometimes called *hard clustering*, since each data point is assigned to a unique cluster. In this paper we do not consider so-called soft clustering.) One of the most influential contributions to the field has been Lloyd's $k$-means algorithm [Llo82]. It is beyond our scope to survey the vast literature on the theory and applications of the $k$-means algorithm. For our purposes, it is sufficient to note three key features of the basic algorithm that can serve as starting points for further development: (i) Each cluster is represented by its *centroid*. (ii) Initial *seeding* chooses random data points as centroids. (iii) Subsequently the algorithm improves the quality of the clustering by locally optimizing its *potential*, defined as the sum of the squared Euclidean distances between each data point and its nearest centroid.

Our starting points are two recent major improvements that address points (ii) and (iii) above. First, Banerjee et al. [BMDG05] have generalized the $k$-means

W. Daelemans et al. (Eds.): ECML PKDD 2008, Part II, LNAI 5212, pp. 154–169, 2008.

algorithm to allow, instead of just squared Euclidean distance, any *Bregman divergence* [Bre67] as a distortion measure in computing the potential. Bregman divergences are closely associated with exponential families of distributions and include such popular distortion measures as Kullback-Leibler divergence and Itakura-Saito divergence. As these divergences are in general not symmetrical, they introduce nontrivial technical problems. On the other hand, they give us a lot of freedom in fitting the performance measure of our algorithm to the nature of the data (say, an exponential family of distributions we feel might be appropriate) which should lead to qualitatively better clusterings. Bregman divergences have found many applications in other types of machine learning (see *e.g.* [AW01]) and in other fields such as computational geometry [NBN07], as well.

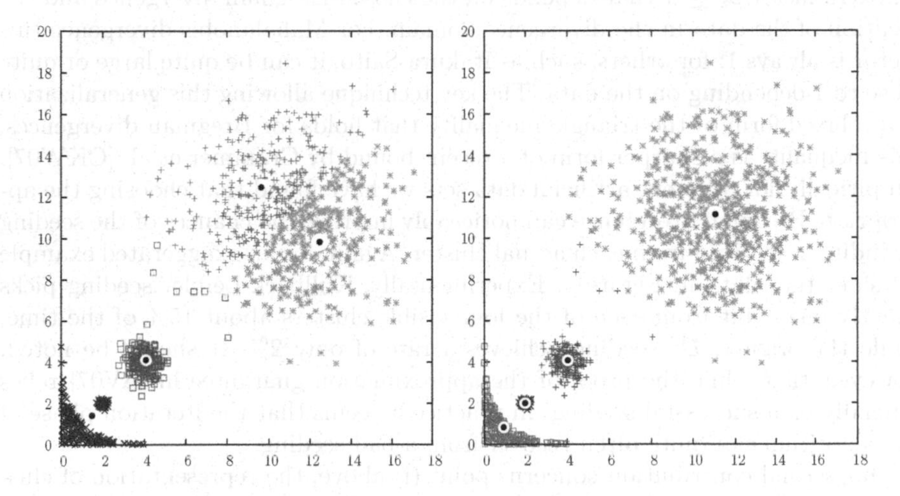

**Fig. 1.** Clusterings obtained by minimizing Euclidean (left) and Kullback-Leibler (right) potential. The centroids are shown as black dots.

To appreciate the effect of the distortion measure on clustering results, consider the exaggerated toy example in Figure 1. Visually, the data consists of four clusters. The first one is centered around the origin and spreads along the $x$ and $y$ axes. It can be seen as an approximate Itakura-Saito ball [NN05]. The other three clusters come from isotropic Gaussian distributions with different variances and centers at $(2, 2)$, $(4, 4)$ and $(11, 11)$. The cluster around $(11, 11)$ has 600 data points, while the other three clusters have 200 data points each. For $k = 4$, a Euclidean distortion measure favors clusterings that use a single centroid to cover the two small clusters close to the origin and uses two centroids to cover the one big cluster. In contrast, Kullback-Leibler divergence gives a better score to solutions that correspond to the visually distinguishable clusters.

A second recent improvement to $k$-means clustering is $D^2$ seeding by Arthur and Vassilvitskii [AV07]. Instead of choosing the $k$ initial cluster centroids uniformly from all the data points, we choose them in sequence so that in choosing the next initial centroid we give a higher probability to points that are not close to any of the already chosen centroids. Intuitively, this helps to get centroids that cover the data set better; in particular, if the data does consist of several clearly separated clusters, we are more likely to get at least one representative from each cluster. Surprisingly, this simple change in the seeding guarantees that the squared Euclidean potential of the resulting clustering is in expectation within $O(\log k)$ of optimal. For another recent result that obtains approximation guarantees by modifying the seeding of the $k$-means algorithm, see [ORSS06].

The first contribution in this paper is to combine the previous two advances by replacing squared Euclidean distance in the $D^2$ seeding by an arbitrary Bregman divergence. The resulting *Bregman seeding* gives a similar approximation guarantee as the original $D^2$ seeding, except that the approximation factor contains an extra factor $\rho_\psi \geq 1$ that depends on the chosen Bregman divergence and the location of the data in the divergence domain. For Mahalanobis divergence this factor is always 1; for others, such as Itakura-Saito, it can be quite large or quite close to 1 depending on the data. The key technique allowing this generalization is a relaxed form of the triangle inequality that holds for Bregman divergences; this inequality is a sharper form of a recent bound by Crammer et al. [CKW07]. Empirically, for different artificial data sets we have found that choosing the appropriate Bregman divergence can noticeably improve the chances of the seeding including a centroid from each actual cluster. Again, for an exaggerated example consider that data in Figure 1. Experimentally, Kullback-Leibler seeding picks exactly one point from each of the four visible clusters about 15% of the time, while the original $D^2$ seeding achieves a rate of only 2%. It should be noted, however, that while the proof of the approximation guarantee in [AV07] relies crucially on a successful seeding, in practice it seems that the iteration phase of the algorithm can quite often recover from a bad seeding.

Our second contribution concerns point (i) above, the representation of clusters by a centroid. Since Bregman divergences are asymmetric, it is very significant whether our potential function considers divergence from a data point to the centroid, or from the centroid to a data point. One choice keeps the arithmetic mean of the data points in a cluster as its optimal centroid, the other does not [BMDG05, BGW05]. The strong asymmetricity of Bregman divergences may seem undesirable in some situations, so a natural thought is to symmetrize the divergence by considering the average of the divergences in the two different directions. However, this makes finding the optimal centroid quite a nontrivial optimization problem [Vel02] and makes the statistical interpretation of the centroid less clear. As a solution, we suggest using *two* centroids per cluster, one for each direction of the Bregman divergence. This makes the centroid computations easy and allows a nice statistical interpretation. We call this symmetrized version with two centroids a *mixed Bregman divergence*.

Previously, approximation bounds for Bregman clustering algorithms have been given by [CM08] and [ABS08]. Chadhuri and McGregor [CM08] consider the KL divergence, which is a particularly interesting case as the KL divergence between two members of the same exponential family is a Bregman divergence between their natural parameters [BMDG05]. Ackermann *et al.* [ABS08] consider a statistically defined class of distortion measures which includes the KL divergence and other Bregman divergences. In both of these cases, the algorithms achieve $(1+\varepsilon)$-approximation for arbitrary $\varepsilon > 0$. This is a much stronger guarantee than the logarithmic factor achieved here using the technique of [AV07]. On the other hand, the $(1+\varepsilon)$-approximation algorithms are fairly complex, whereas our algorithm based on [AV07] is quite easy to implement and runs in time $O(nkd)$.

Section 2 presents definitions; Section 3 presents our seeding and clustering algorithm. Section 4 discusses some results. Section 5 provides experiments, and Section 6 concludes the paper with open problems.

## 2 Definitions

*Divergences.* Let $\psi : \mathbb{X} \rightarrow \mathbb{R}$ be a strictly convex function defined on a convex set $\mathbb{X} \subseteq \mathbb{R}^d$, with the gradient $\nabla\psi$ defined in the interior of $\mathbb{X}$. (Hereafter, for the sake of simplicity, we do not make the difference between a set and its interior.) We denote by $\psi^\star(\boldsymbol{x}) \doteq \langle \boldsymbol{x}, (\boldsymbol{\nabla}\psi)^{-1}(\boldsymbol{x})\rangle - \psi((\boldsymbol{\nabla}\psi)^{-1}(\boldsymbol{x}))$ its convex conjugate. The *Bregman divergence* $\Delta_\psi(\boldsymbol{x}\|\boldsymbol{y})$ between any two point $\boldsymbol{x}$ and $\boldsymbol{y}$ of $\mathbb{X}$ is [Bre67]:

$$\Delta_\psi(\boldsymbol{x}\|\boldsymbol{y}) \doteq \psi(\boldsymbol{x}) - \psi(\boldsymbol{y}) - \langle \boldsymbol{x} - \boldsymbol{y}, \boldsymbol{\nabla}\psi(\boldsymbol{y})\rangle \ .$$

Popular examples of Bregman divergences include Mahalanobis divergence with $D_{\mathrm{M}}(\boldsymbol{x}\|\boldsymbol{y}) \doteq (\boldsymbol{x}-\boldsymbol{y})^\top \mathrm{M}(\boldsymbol{x}-\boldsymbol{y})$ ($\mathbb{X} = \mathbb{R}^d$, M symmetric positive definite), Kullback-Leibler divergence, $D_{\mathrm{KL}}(\boldsymbol{x}\|\boldsymbol{y}) \doteq \sum_{i=1}^d (x_i \log(x_i/y_i) - x_i + y_i)$ ($\mathbb{X} = \mathbb{R}^d_{+*}$), Itakura-Saito divergence, $D_{\mathrm{IS}}(\boldsymbol{x}\|\boldsymbol{y}) \doteq \sum_{i=1}^d ((x_i/y_i) - \log(x_i/y_i) - 1)$ ($\mathbb{X} = \mathbb{R}^d_{+*}$), and many others [NBN07, BMDG05]. It is not hard to prove that Mahalanobis divergence is the *only* symmetric Bregman divergence. This general asymmetry, which arises naturally from the links with the exponential families of distributions [BMDG05], is not really convenient for clustering. Thus, we let:

$$\triangle_{\psi,\alpha}(\boldsymbol{x}\|\boldsymbol{y}\|\boldsymbol{z}) \doteq (1-\alpha)\Delta_\psi(\boldsymbol{x}\|\boldsymbol{y}) + \alpha\Delta_\psi(\boldsymbol{y}\|\boldsymbol{z}) \quad (1)$$

denote the *mixed* Bregman divergence of parameters $(\psi, \alpha)$, with $0 \leq \alpha \leq 1$. When $\alpha = 0, 1$, this is just a regular Bregman divergence. The special case $\alpha = 1/2$ and $\boldsymbol{x} = \boldsymbol{z}$ is known as a symmetric Bregman divergence [Vel02].

*Clustering.* We are given a set $\mathcal{S} \subseteq \mathbb{X}$. For some $\mathcal{A} \subseteq \mathcal{S}$ and $\boldsymbol{y} \in \mathbb{X}$, let

$$\psi_\alpha(\mathcal{A}, \boldsymbol{y}) \doteq \sum_{\boldsymbol{x}\in\mathcal{A}} \triangle_{\psi,\alpha}(\boldsymbol{y}\|\boldsymbol{x}\|\boldsymbol{y}) \ ,$$

$$\psi^\star_\alpha(\mathcal{A}, \boldsymbol{y}) \doteq \psi_{1-\alpha}(\mathcal{A}, \boldsymbol{y}) = \sum_{\boldsymbol{x}\in\mathcal{A}} \triangle_{\psi,\alpha}(\boldsymbol{x}\|\boldsymbol{y}\|\boldsymbol{x}) \ . \quad (2)$$

Let $\mathcal{C} \subset \mathbb{X}^2$. The *potential* for Bregman clustering with the centroids of $\mathcal{C}$ is:

$$\psi_\alpha(\mathcal{C}) \doteq \sum_{\boldsymbol{x}\in\mathcal{S}} \min_{(\boldsymbol{c},\boldsymbol{c}^\star)\in\mathcal{C}} \triangle_{\psi,\alpha}(\boldsymbol{c}^\star\|\boldsymbol{x}\|\boldsymbol{c}) \ . \quad (3)$$

When $\alpha = 0$ or $\alpha = 1$, we can pick $\mathcal{C} \subset \mathbb{X}$, and we return to regular Bregman clustering [BMDG05]. The contribution to this potential of some subset $\mathcal{A}$, not necessarily defining a cluster, is noted $\psi_\alpha(\mathcal{A})$, omitting the clustering that shall be implicit and clear from context. An optimal clustering, $\mathcal{C}_{\mathrm{opt}}$, can be defined either as its set of centroids, or the partition of $\mathcal{S}$ induced. It achieves:

$$\psi_{\mathrm{opt},\alpha} \doteq \min_{\mathcal{C} \subset \mathbb{X}^2, |\mathcal{C}|=k} \psi_\alpha(\mathcal{C}) \ . \quad (4)$$

In this clustering, the contribution of some cluster $\mathcal{A}$ is:

$$\psi_{\mathrm{opt},\alpha}(\mathcal{A}) \doteq \sum_{\boldsymbol{x} \in \mathcal{A}} \Delta_{\psi,\alpha}(\boldsymbol{c}^\star_{\mathcal{A}} \| \boldsymbol{x} \| \boldsymbol{c}_{\mathcal{A}}) \ ,$$

where $(\boldsymbol{c}_{\mathcal{A}}, \boldsymbol{c}^\star_{\mathcal{A}}) \in \mathcal{C}_{\mathrm{opt}}$ is the pair of centroids which minimizes $\psi_\alpha(\mathcal{A})$ over all possible choices of $(\boldsymbol{c}, \boldsymbol{c}^\star)$ in (3). It turns out that these two centroids are always respectively the arithmetic and *Bregman* averages of $\mathcal{A}$:

$$\boldsymbol{c}_{\mathcal{A}} \doteq \frac{1}{|\mathcal{A}|} \sum_{\boldsymbol{x} \in \mathcal{A}} \boldsymbol{x} \ , \quad (5)$$

$$\boldsymbol{c}^\star_{\mathcal{A}} \doteq (\boldsymbol{\nabla}\psi)^{-1} \left( \frac{1}{|\mathcal{A}|} \sum_{\boldsymbol{x} \in \mathcal{A}} \boldsymbol{\nabla}\psi(\boldsymbol{x}) \right) \ . \quad (6)$$

To see that it holds for the arithmetic average, we may write:

$$\forall \boldsymbol{c} \in \mathcal{A}, \sum_{\boldsymbol{x} \in \mathcal{A}} \Delta_\psi(\boldsymbol{x} \| \boldsymbol{c}) - \sum_{\boldsymbol{x} \in \mathcal{A}} \Delta_\psi(\boldsymbol{x} \| \boldsymbol{c}_{\mathcal{A}}) = |\mathcal{A}| \Delta_\psi(\boldsymbol{c}_{\mathcal{A}} \| \boldsymbol{c}) \ . \quad (7)$$

Since the right hand side is not negative and zero only when $\boldsymbol{c} = \boldsymbol{c}_{\mathcal{A}}$, (5) is the best choice for $\boldsymbol{c}$. On the other hand, if we compute (7) on $\psi^\star$ and then use the following well-known dual symmetry relationship which holds for any Bregman divergence,

$$\Delta_\psi(\boldsymbol{x} \| \boldsymbol{y}) = \Delta_{\psi^\star}(\boldsymbol{\nabla}\psi(\boldsymbol{y}) \| \boldsymbol{\nabla}\psi(\boldsymbol{x})) \ ,$$

then we obtain:

$$\forall \boldsymbol{c} \in \mathcal{A}, \sum_{\boldsymbol{x} \in \mathcal{A}} \Delta_\psi(\boldsymbol{c} \| \boldsymbol{x}) - \sum_{\boldsymbol{x} \in \mathcal{A}} \Delta_\psi(\boldsymbol{c}^\star_{\mathcal{A}} \| \boldsymbol{x}) = |\mathcal{A}| \Delta_\psi(\boldsymbol{c} \| \boldsymbol{c}^\star_{\mathcal{A}}) \ , \quad (8)$$

and we conclude that (6) is the best choice for $\boldsymbol{c}^\star$. Since $\boldsymbol{c}^\star_{\mathcal{A}} \neq \boldsymbol{c}_{\mathcal{A}}$ except when $\Delta_{\psi,\alpha}$ is proportional to Mahalanobis divergence, the mixed divergence (1) is only a partial symmetrization of the Bregman divergence with respect to approaches like *e.g.* [Vel02] that enforce $\boldsymbol{c}^\star_{\mathcal{A}} = \boldsymbol{c}_{\mathcal{A}}$. There are at least two good reasons for this symmetrization to remain partial for Bregman divergences. The first is statistical: up to additive and multiplicative factors that would play no role in its optimization, (1) is an exponential family's log-likelihood in which $\alpha$ tempers the probability to fit in the expectation parameter's space versus the natural

**Algorithm 1.** MBS($\mathcal{S}$, $k$, $\alpha$, $\psi$)

**Input**: Dataset $\mathcal{S}$, integer $k > 0$, real $\alpha \in [0, 1]$, strictly convex $\psi$;
Let $\mathcal{C} \leftarrow \{(\boldsymbol{x}, \boldsymbol{x})\}$;
//where $\boldsymbol{x}$ is chosen uniformly at random in $\mathcal{S}$;
**for** $i = 1, 2, ..., k - 1$ **do**
Pick at random point $\boldsymbol{x} \in \mathcal{S}$ with probability:

$$\pi_{\mathcal{S}}(\boldsymbol{x}) \doteq \frac{\Delta_{\psi,\alpha}(\boldsymbol{c_x}\|\boldsymbol{x}\|\boldsymbol{c_x})}{\sum_{\boldsymbol{y}\in\mathcal{S}} \Delta_{\psi,\alpha}(\boldsymbol{c_y}\|\boldsymbol{y}\|\boldsymbol{c_y})} , \quad (9)$$

//where $(\boldsymbol{c_x}, \boldsymbol{c_x}) \doteq \arg\min_{(\boldsymbol{z},\boldsymbol{z})\in\mathcal{C}} \Delta_{\psi,\alpha}(\boldsymbol{z}\|\boldsymbol{x}\|\boldsymbol{z})$;
$\mathcal{C} \leftarrow \mathcal{C} \cup \{(\boldsymbol{x}, \boldsymbol{x})\}$;

**Output**: Set of initial centroids $\mathcal{C}$;

parameter's space [BMDG05]. This adds a twist in the likelihood for the uncertainty of the data to model which is, in the context of clustering, desirable even against regular Bregman divergences. However, it does not hold for approaches like [Vel02]. The second reason is algorithmic: mixed divergences incur no complexity counterpart if we except the computation of the inverse gradient for $\boldsymbol{c}^\star_{\mathcal{A}}$; in the complete symmetric approaches, there is no known general expression for the centroid, and it may be time consuming to get approximations even when it is trivial to compute $\boldsymbol{c}^\star_{\mathcal{A}}$ [Vel02]. Finally, we define a dual potential for the optimal clustering, obtained by permuting the parameters of the divergences:

$$\psi^\star_{\mathrm{opt},\alpha}(\mathcal{A}) \doteq \sum_{\boldsymbol{x}\in\mathcal{A}} \Delta_{\psi,1-\alpha}(\boldsymbol{c}_{\mathcal{A}}\|\boldsymbol{x}\|\boldsymbol{c}^\star_{\mathcal{A}}) = \sum_{\boldsymbol{x}\in\mathcal{A}} (\alpha \Delta_\psi(\boldsymbol{c}_{\mathcal{A}}\|\boldsymbol{x}) + (1-\alpha)\Delta_\psi(\boldsymbol{x}\|\boldsymbol{c}^\star_{\mathcal{A}})) .$$

# 3 Mixed Bregman Clustering

## 3.1 Mixed Bregman Seeding

Algorithm 1 (Mixed Bregman Seeding) shows how we seed the initial cluster centroids. It is generalizes the approach of [AV07] and gives their $D^2$ seeding as a special case when using the squared Euclidean distance as distortion measure. Since the Bregman divergence between two points can usually be computed in the same $O(d)$ time as the Euclidean distance, our algorithm has the same $O(nkd)$ running time as the original one by [AV07]. The main result of [AV07] is an approximation bound for the squared Euclidean case:

**Theorem 1.** *[AV07] The average initial potential resulting from $D^2$ seeding satisfies $\boldsymbol{E}[\psi] \leq 8(2 + \log k)\psi_{\mathrm{opt}}$, where $\psi_{\mathrm{opt}}$ is the smallest squared Euclidean potential possible by partitioning $\mathcal{S}$ in $k$ clusters.*

We prove a generalization of Theorem 1 by generalizing each of the lemmas used by [AV07] in their proof.

**Lemma 1.** *Let $\mathcal{A}$ be an arbitrary cluster of $\mathcal{C}_{\mathrm{opt}}$. Then:*

$$\boldsymbol{E}_{\boldsymbol{c}\sim U_{\mathcal{A}}}[\psi_\alpha(\mathcal{A},\boldsymbol{c})] = \psi_{\mathrm{opt},\alpha}(\mathcal{A}) + \psi^\star_{\mathrm{opt},\alpha}(\mathcal{A})) \ , \tag{10}$$

$$\boldsymbol{E}_{\boldsymbol{c}\sim U_{\mathcal{A}}}[\psi^\star_\alpha(\mathcal{A},\boldsymbol{c})] = \psi_{\mathrm{opt},1-\alpha}(\mathcal{A}) + \psi^\star_{\mathrm{opt},1-\alpha}(\mathcal{A})) \ , \tag{11}$$

*where $U_{\mathcal{A}}$ is the uniform distribution over $\mathcal{A}$.*

**Proof.** We use (7) and (8) in (13) below and obtain:

$$\begin{aligned}
\mathbf{E}_{\boldsymbol{c}\sim U_{\mathcal{A}}}[\psi_\alpha(\mathcal{A},\boldsymbol{c})] &= \frac{1}{|\mathcal{A}|}\sum_{\boldsymbol{c}\in\mathcal{A}}\sum_{\boldsymbol{x}\in\mathcal{A}}\{\alpha\Delta_\psi(\boldsymbol{x}\|\boldsymbol{c}) + (1-\alpha)\Delta_\psi(\boldsymbol{c}\|\boldsymbol{x})\} &(12)\\
&= \frac{1}{|\mathcal{A}|}\sum_{\boldsymbol{c}\in\mathcal{A}}\left\{\alpha\left(\sum_{\boldsymbol{x}\in\mathcal{A}}\Delta_\psi(\boldsymbol{x}\|\boldsymbol{c}_{\mathcal{A}}) + |\mathcal{A}|\Delta_\psi(\boldsymbol{c}_{\mathcal{A}}\|\boldsymbol{c})\right)\right.\\
&\quad \left. +(1-\alpha)\left(\sum_{\boldsymbol{x}\in\mathcal{A}}\Delta_\psi(\boldsymbol{c}^\star_{\mathcal{A}}\|\boldsymbol{x}) + |\mathcal{A}|\Delta_\psi(\boldsymbol{c}\|\boldsymbol{c}^\star_{\mathcal{A}})\right)\right\} &(13)\\
&= \alpha\sum_{\boldsymbol{x}\in\mathcal{A}}\Delta_\psi(\boldsymbol{x}\|\boldsymbol{c}_{\mathcal{A}}) + \alpha\sum_{\boldsymbol{x}\in\mathcal{A}}\Delta_\psi(\boldsymbol{c}_{\mathcal{A}}\|\boldsymbol{x})\\
&\quad +(1-\alpha)\sum_{\boldsymbol{x}\in\mathcal{A}}\Delta_\psi(\boldsymbol{c}^\star_{\mathcal{A}}\|\boldsymbol{x}) + (1-\alpha)\sum_{\boldsymbol{x}\in\mathcal{A}}\Delta_\psi(\boldsymbol{x}\|\boldsymbol{c}^\star_{\mathcal{A}})\\
&= \alpha\psi_{\mathrm{opt},1}(\mathcal{A}) + (1-\alpha)\psi_{\mathrm{opt},0}(\mathcal{A}) + \alpha\psi^\star_{\mathrm{opt},1}(\mathcal{A}) &(14)\\
&\quad +(1-\alpha)\psi^\star_{\mathrm{opt},0}(\mathcal{A})\\
&= \psi_{\mathrm{opt},\alpha}(\mathcal{A}) + \psi^\star_{\mathrm{opt},\alpha}(\mathcal{A}) \ .
\end{aligned}$$

This gives (10). Applying (2) to (10) gives (11). □

Analyzing the biased distribution case requires a triangle inequality for Bregman divergences, stated below. For any positive semidefinite matrix M, $\mathrm{M}^{1/2}$ denotes the positive semidefinite matrix such that $\mathrm{M}^{1/2}\mathrm{M}^{1/2} = \mathrm{M}$.

**Lemma 2.** *For any three points $\boldsymbol{x},\boldsymbol{y},\boldsymbol{z}$ of $\mathrm{co}(\mathcal{S})$, the convex closure of $\mathcal{S}$,*

$$\Delta_\psi(\boldsymbol{x},\boldsymbol{z}) \leq 2\rho^2_\psi(\Delta_\psi(\boldsymbol{x},\boldsymbol{y}) + \Delta_\psi(\boldsymbol{y},\boldsymbol{z})) \ , \tag{15}$$

*where $\rho_\psi$ is defined as:*

$$\rho_\psi \doteq \sup_{\boldsymbol{s},\boldsymbol{t},\boldsymbol{u},\boldsymbol{v}\in\mathrm{co}(\mathcal{S})}\frac{\|\mathrm{H}^{1/2}_{\boldsymbol{s}}(\boldsymbol{u}-\boldsymbol{v})\|_2}{\|\mathrm{H}^{1/2}_{\boldsymbol{t}}(\boldsymbol{u}-\boldsymbol{v})\|_2} \ , \tag{16}$$

*where $\mathrm{H}_{\boldsymbol{s}}$ denotes the Hessian of $\psi$ in $\boldsymbol{s}$.*

**Proof.** The key to the proof is the Bregman triangle equality:

$$\Delta_\psi(\boldsymbol{x}\|\boldsymbol{z}) = \Delta_\psi(\boldsymbol{x}\|\boldsymbol{y}) + \Delta_\psi(\boldsymbol{y}\|\boldsymbol{z}) + (\boldsymbol{\nabla}\psi(\boldsymbol{z}) - \boldsymbol{\nabla}\psi(\boldsymbol{y}))^\top(\boldsymbol{y}-\boldsymbol{x}) \ . \tag{17}$$

A Taylor-Lagrange expansion on Bregman divergence $\Delta_\psi$ yields:

$$\Delta_\psi(\boldsymbol{a}\|\boldsymbol{b}) = \frac{1}{2}(\boldsymbol{a}-\boldsymbol{b})^\top\mathrm{H}_{\boldsymbol{ab}}(\boldsymbol{a}-\boldsymbol{b}) = \frac{1}{2}\|\mathrm{H}^{1/2}_{\boldsymbol{ab}}(\boldsymbol{a}-\boldsymbol{b})\|_2^2 \ , \tag{18}$$

for some value $\mathrm{H}_{\boldsymbol{ab}}$ of the Hessian of $\psi$ in the segment $\boldsymbol{ab} \subseteq \mathrm{co}(\mathcal{S})$. Another expansion on the gradient part of (17) yields:

$$\nabla\psi(\boldsymbol{z}) - \nabla\psi(\boldsymbol{y}) = \mathrm{H}_{\boldsymbol{zy}}(\boldsymbol{z} - \boldsymbol{y}) \ . \tag{19}$$

Putting this altogether, (17) becomes:

$$\begin{aligned} \Delta_\psi(\boldsymbol{x}\|\boldsymbol{z}) &\overset{(19)}{=} \Delta_\psi(\boldsymbol{x}\|\boldsymbol{y}) + \Delta_\psi(\boldsymbol{y}\|\boldsymbol{z}) + (\mathrm{H}^{1/2}_{\boldsymbol{zy}}(\boldsymbol{z}-\boldsymbol{y}))^\top(\mathrm{H}^{1/2}_{\boldsymbol{zy}}(\boldsymbol{y}-\boldsymbol{x})) \\ &\leq \Delta_\psi(\boldsymbol{x}\|\boldsymbol{y}) + \Delta_\psi(\boldsymbol{y}\|\boldsymbol{z}) + \|\mathrm{H}^{1/2}_{\boldsymbol{zy}}(\boldsymbol{z}-\boldsymbol{y})\|_2\|\mathrm{H}^{1/2}_{\boldsymbol{zy}}(\boldsymbol{y}-\boldsymbol{x})\|_2 \qquad (20) \\ &\leq \Delta_\psi(\boldsymbol{x}\|\boldsymbol{y}) + \Delta_\psi(\boldsymbol{y}\|\boldsymbol{z}) + \rho_\psi^2\left(\|\mathrm{H}^{1/2}_{\boldsymbol{yz}}(\boldsymbol{z}-\boldsymbol{y})\|_2\|\mathrm{H}^{1/2}_{\boldsymbol{xy}}(\boldsymbol{y}-\boldsymbol{x})\|_2\right) \\ &\overset{(18)}{=} \Delta_\psi(\boldsymbol{x}\|\boldsymbol{y}) + \Delta_\psi(\boldsymbol{y}\|\boldsymbol{z}) + 2\rho_\psi^2\sqrt{\Delta_\psi(\boldsymbol{x}\|\boldsymbol{y})\Delta_\psi(\boldsymbol{y}\|\boldsymbol{z})} \end{aligned}$$

where (20) makes use of Cauchy-Schwartz inequality. Since $\rho_\psi \geq 1$, the right-hand side of the last inequality is of the form $a+b+2\rho_\psi^2\sqrt{ab} \leq \rho_\psi^2(a+b+2\sqrt{ab}) \leq \rho_\psi^2(2a+2b) = 2\rho_\psi^2(a+b)$. Since $a = \Delta_\psi(\boldsymbol{x}\|\boldsymbol{y})$ and $b = \Delta_\psi(\boldsymbol{y}\|\boldsymbol{z})$, we obtain the statement of the Lemma. □

Lemma 2 is a sharper version of the bound used by [CKW07]. The improvement is basically that we use the same vector $\boldsymbol{u}-\boldsymbol{v}$ in the numerator and denominator in (16), so we are not automatically hurt by anisotropy in the divergence. In particular, we have $\rho_\psi = 1$ for any Mahalanobis distance.

The following lemma generalizes [AV07, Lemma 3.2]. We use Lemmas 1 and 2 instead of special properties of the squared Euclidean distance. Otherwise the proof is essentially the same.

**Lemma 3.** *Let $\mathcal{A}$ be an arbitrary cluster of $\mathcal{C}_{\mathrm{opt}}$, and $\mathcal{C}$ an arbitrary clustering. If we add a random pair $(\boldsymbol{y}, \boldsymbol{y})$ from $\mathcal{A}^2$ to $\mathcal{C}$ in Algorithm 1, then*

$$\boldsymbol{E}_{\boldsymbol{y}\sim\pi_\mathcal{S}}[\psi_\alpha(\mathcal{A},\boldsymbol{y})|\boldsymbol{y}\in\mathcal{A}] = \boldsymbol{E}_{\boldsymbol{y}\sim\pi_\mathcal{A}}[\psi_\alpha(\mathcal{A},\boldsymbol{y})] \leq 4\rho_\psi^2(\psi_{\mathrm{opt},\alpha}(\mathcal{A}) + \psi^\star_{\mathrm{opt},\alpha}(\mathcal{A})) \ .$$

**Proof.** The equality comes from the fact that the expectation is constrained to the choice of $\boldsymbol{y}$ in $\mathcal{A}$. The contribution of $\mathcal{A}$ to the potential is thus:

$$\begin{aligned} &\mathbf{E}_{\boldsymbol{y}\sim\pi_\mathcal{A}}[\psi_\alpha(\mathcal{A},\boldsymbol{y})] \\ &= \sum_{\boldsymbol{y}\in\mathcal{A}}\left\{\frac{\Delta_{\psi,\alpha}(\boldsymbol{c_y}\|\boldsymbol{y}\|\boldsymbol{c_y})}{\sum_{\boldsymbol{x}\in\mathcal{A}}\Delta_{\psi,\alpha}(\boldsymbol{c_x}\|\boldsymbol{x}\|\boldsymbol{c_x})}\sum_{\boldsymbol{x}\in\mathcal{A}}\min\{\Delta_{\psi,\alpha}(\boldsymbol{c_x}\|\boldsymbol{x}\|\boldsymbol{c_x}),\Delta_{\psi,\alpha}(\boldsymbol{y}\|\boldsymbol{x}\|\boldsymbol{y})\}\right\} \qquad (21) \end{aligned}$$

We also have:

$$\begin{aligned} &\Delta_{\psi,\alpha}(\boldsymbol{c_y}\|\boldsymbol{y}\|\boldsymbol{c_y}) \\ &= \alpha\Delta_\psi(\boldsymbol{y}\|\boldsymbol{c_y}) + (1-\alpha)\Delta_\psi(\boldsymbol{c_y}\|\boldsymbol{y}) \\ &\leq \alpha\Delta_\psi(\boldsymbol{y}\|\boldsymbol{c_x}) + (1-\alpha)\Delta_\psi(\boldsymbol{c_x}\|\boldsymbol{y}) \\ &\leq 2\rho_\psi^2(\alpha\Delta_\psi(\boldsymbol{y}\|\boldsymbol{x}) + \alpha\Delta_\psi(\boldsymbol{x}\|\boldsymbol{c_x}) + (1-\alpha)\Delta_\psi(\boldsymbol{c_x}\|\boldsymbol{x}) + (1-\alpha)\Delta_\psi(\boldsymbol{x}\|\boldsymbol{y})) \\ &= 2\rho_\psi^2(\Delta_{\psi,\alpha}(\boldsymbol{c_x}\|\boldsymbol{x}\|\boldsymbol{c_x}) + \Delta_{\psi,\alpha}(\boldsymbol{x}\|\boldsymbol{y}\|\boldsymbol{x})) \ , \end{aligned}$$

where we have used Lemma 2 on the last inequality. Summing over $\boldsymbol{x} \in \mathcal{A}$ yields:

$$\triangle_{\psi,\alpha}(\boldsymbol{c_y}\|\boldsymbol{y}\|\boldsymbol{c_y}) \leq 2\rho_\psi^2 \left( \frac{1}{|\mathcal{A}|} \sum_{\boldsymbol{x}\in\mathcal{A}} \triangle_{\psi,\alpha}(\boldsymbol{c_x}\|\boldsymbol{x}\|\boldsymbol{c_x}) + \frac{1}{|\mathcal{A}|} \sum_{\boldsymbol{x}\in\mathcal{A}} \triangle_{\psi,\alpha}(\boldsymbol{x}\|\boldsymbol{y}\|\boldsymbol{x}) \right) ;$$

plugging this into (21) and replacing the min by its left or right member in the two sums yields:

$$\begin{aligned}\mathbf{E}_{\boldsymbol{y}\sim\pi_{\mathcal{A}}}[\psi_\alpha(\mathcal{A},\boldsymbol{y})] &\leq 4\rho_\psi^2 \frac{1}{|\mathcal{A}|} \sum_{\boldsymbol{y}\in\mathcal{A}} \sum_{\boldsymbol{x}\in\mathcal{A}} \triangle_{\psi,\alpha}(\boldsymbol{x}\|\boldsymbol{y}\|\boldsymbol{x}) \\ &= 4\rho_\psi^2 (\psi_{\mathrm{opt},\alpha}(\mathcal{A}) + \psi^\star_{\mathrm{opt},\alpha}(\mathcal{A})) ,\end{aligned}$$

where we have used (12). □

For any subset of clusters $\mathcal{A}$ of some optimal clustering $\mathcal{C}_{\mathrm{opt}}$, let $\tilde{\psi}_{\mathrm{opt},\alpha}(\mathcal{A}) \doteq (1/2)(\psi_{\mathrm{opt},\alpha}(\mathcal{A}) + \psi^\star_{\mathrm{opt},\alpha}(\mathcal{A}))$. We remark that:

$$\mathbf{E}_{\boldsymbol{y}\sim\pi_{\mathcal{A}}}[\psi_\alpha(\mathcal{A},\boldsymbol{y})] \leq 8\rho_\psi^2 \tilde{\psi}_{\mathrm{opt},\alpha}(\mathcal{A}) , \quad (22)$$

$$\forall \mathcal{A},\mathcal{B} : \mathcal{A}\cap\mathcal{B} = \emptyset, \tilde{\psi}_{\mathrm{opt},\alpha}(\mathcal{A}\cup\mathcal{B}) = \tilde{\psi}_{\mathrm{opt},\alpha}(\mathcal{A}) + \tilde{\psi}_{\mathrm{opt},\alpha}(\mathcal{B}) . \quad (23)$$

**Lemma 4.** *Let $\mathcal{C}$ be an arbitrary clustering. Choose $u > 0$ clusters from $\mathcal{C}_{\mathrm{opt}}$ that are still not covered by $\mathcal{C}$, and let $\mathcal{S}_u$ denote the set of points in these clusters. Also, let $\mathcal{S}_c \doteq \mathcal{S} - \mathcal{S}_u$. Now suppose that we add $t \leq u$ random pairs of centroids, chosen according to $\pi_{\mathcal{S}}$ as in Algorithm 1. Let $\mathcal{C}'$ denote the resulting clustering. Define $H_t \doteq 1 + (1/2) + ... + (1/t)$. Then*

$$\boldsymbol{E}_{\boldsymbol{c}\sim\pi_{\mathcal{S}}}[\psi_\alpha(\mathcal{C}')] \leq (1+H_t)\left(\psi_\alpha(\mathcal{S}_c) + 8\rho_\psi^2\tilde{\psi}_{\mathrm{opt},\alpha}(\mathcal{S}_u)\right) + \left(\frac{u-t}{u}\right)\psi_\alpha(\mathcal{S}_u) .$$

Again, the proof is obtained from the proof of [AV07, Lemma 3.3] by just applying (22) and (23) to handle $\tilde{\psi}$. We omit the details. As in [AV07] we now obtain the main approximation bound as a special case of Lemma 4.

**Theorem 2.** *The average initial potential obtained by Mixed Bregman Seeding (Algorithm 1) satisfies $\boldsymbol{E}[\psi_\alpha] \leq 8\rho_\psi^2(2+\log k)\tilde{\psi}_{\mathrm{opt},\alpha}$, where $\tilde{\psi}_{\mathrm{opt},\alpha}$ is the minimal mixed Bregman divergence possible by partitioning $\mathcal{S}$ into $k$ clusters as defined in (4).*

When the Hessian of $\psi$ satisfies $\mathrm{H}_. = \sigma\mathrm{I}$ for $\sigma > 0$, we return to regular $k$-means and the bound of Theorem 1 [AV07]. Interestingly, the bound remains the same for general Mahalanobis divergence ($\rho_\psi = 1$, $\psi_{\mathrm{opt},\alpha}(\mathcal{S}) = \psi^\star_{\mathrm{opt},\alpha}(\mathcal{S}) = \tilde{\psi}_{\mathrm{opt}}(\mathcal{S})$).

### 3.2 Integrating Mixed Bregman Seeding into Clustering

Bregman seeding in the special case $\alpha = 1$ can be integrated with Bregman clustering [BMDG05] to provide a complete clustering algorithm in which the

**Algorithm 2.** MBC($\mathcal{S}$, $k$, $\alpha$, $\psi$)

**Input**: Dataset $\mathcal{S}$, integer $k > 0$, real $\alpha \in [0, 1]$, strictly convex $\psi$;
Let $\mathcal{C} = \{(c_{\mathcal{A}_i}, c^\star_{\mathcal{A}_i})\}_{i=1}^k \leftarrow \text{MBS}(\mathcal{S}, k, \alpha, \psi)$;
**repeat**
  //Assignment
  **for** $i = 1, 2, ..., k$ **do**
    $\mathcal{A}_i \leftarrow \{s \in \mathcal{S} : i = \arg\min_j \Delta_{\psi,\alpha}(c^\star_{\mathcal{A}_j} \| s \| c_{\mathcal{A}_j})\}$;
  //Re-estimation
  **for** $i = 1, 2, ..., k$ **do**
    $c_{\mathcal{A}_i} \leftarrow \frac{1}{|\mathcal{A}_i|}\sum_{s\in\mathcal{A}_i} s$;
    $c^\star_{\mathcal{A}_i} \leftarrow (\nabla\psi)^{-1}\left(\frac{1}{|\mathcal{A}_i|}\sum_{s\in\mathcal{A}_i} \nabla\psi(s)\right)$;
**until** *convergence* ;
**Output**: Partition of $\mathcal{S}$ in $k$ clusters following $\mathcal{C}$;

divergence at hand is integrated in all steps of clustering. What remains to do is take this algorithm as a whole and lift it further to handle mixed Bregman divergences, that is, generalize the Bregman clustering of [BMDG05] to hold for any $0 \leq \alpha \leq 1$. This is presented in Algorithm 2 (Mixed Bregman Clustering). This algorithm is conceptually as simple as Bregman clustering [BMDG05], and departs from the complexity of approaches that would be inspired by fully symmetrized Bregman divergences [Vel02]. However, for this algorithm to be a suitable generalization of Bregman clustering, we have to ensure that it monotonously achieves a local minimum of the mixed potential in finite time. This is done in the following Lemma, whose proof, omitted to save space, follows similar steps as in [BMDG05] while making use of (7) and (8).

**Lemma 5.** *Algorithm 2 monotonically decreases the function in (3). Furthermore, it terminates in a finite number of steps at a locally optimal partition.*

## 4 Discussion

One question arises on such a scheme, namely how the choice of the main free parameter, the generator $\psi$, impacts on the final output. This question is less relevant to the clustering phase, where the optimization is local and all that may be required is explicitly given in Lemma 5, independently of $\psi$. It is more relevant to the seeding phase, and all the more interesting as the upper bound in Theorem 2 exhibits two additional penalties that depend on $\psi$: one relies on the way we measure the potential and seed centroids ($\tilde{\psi}$), the other relies on convexity ($\rho_\psi$). The analysis of $D^2$ seeding by [AV07] is tight on average, as they show that for some clusterings the upper bound of 1 is within a constant factor of the actual performance of the algorithm.

Beyond $D^2$ seeding, it is not hard to show that the analysis of [AV07] is in fact tight for Mahalanobis divergence. To see this, we only have to make a variable change, and set $\tilde{\boldsymbol{x}} \doteq \mathrm{M}^{-1/2}\boldsymbol{x}$ for any point $\boldsymbol{x}$ in the lower bound proof of [AV07].

Mahalanobis divergence on the new points equals the $k$-means potential on the initial points, the optimal centroids do not change, and the proof remains as is. For arbitrary divergences, the upper bound of Theorem 2 gets unfastened in stronger convex regimes of the generator, that is when $\rho_\psi$ increases. Some non-metric analysis of seeding that would avoid the use of a triangle inequality might keep it tighter, as stronger convex regimes do not necessarily penalize that much seeding. Sometimes, artificial improvements are even possible. The following Lemma, whose proofsketch is available in an appendix at the end of the paper, gives a lower bound for the uniform approximation that seeding achieves in some cluster.

**Lemma 6.** *Let $\mathcal{A}$ be an arbitrary cluster. Then:*

$$\boldsymbol{E}_{\mathbf{c}\sim U_{\mathcal{A}}}[\psi_\alpha(\mathcal{A}, \mathbf{c})] \geq \frac{2\rho_\psi^2}{2\rho_\psi^2 - 1}\psi_\alpha(\mathcal{A}) \ . \tag{24}$$

(24) matches ratio 2 that follows from Lemma 1 for Mahalanobis divergence. The average participation of the seeds in (24) hides large discrepancies, as there do exist seeds whose clustering potential come arbitrarily close to the lower bound (24) as $\rho_\psi$ increases. In other words, since this lower bound is decreasing with $\rho_\psi$, increasing $\rho_\psi$ may make seeding artificially more efficient if we manage to catch these seeds, a fact that Theorem 2 cannot show. A toy example shows that we can indeed catch such seeds with high probability: we consider $k = 2$ clusters on $n > 2$ points with $\alpha = 1$. The first cluster contains two points $p$ and $q$ as in Figure 2, with $p$ located at abscissa 0 and $q$ at abscissa $\delta$ (for the sake of simplicity, $\psi$ is assumed defined on $[0, +\infty)$). Add $n-2$ points $x_1, x_2, ..., x_{n-2}$, all at abscissa $\Delta > \delta$, and pick $\Delta$ sufficiently large to ensure that these $n-2$ points define a single cluster, while $p$ and $q$ are grouped altogether in cluster $\mathcal{A}$. It follows that $\psi_{\mathrm{opt},1} = 2\mathrm{BR}_\psi(\{0, \delta\}) = \psi_{\mathrm{opt},1}(\mathcal{A})$. The probability to seed one of the $x_i$ in the two centers is at least $(n-2)/n + (2/n)\cdot(n-2)/(n-1) > 1 - 4/n^2$, which makes that the expected potential is driven by the event that we seed exactly one of the $x_i$. The associated potential is then either $\Delta_\psi(\delta\|0)$ (we seed $p$ with $x_i$) or $\Delta_\psi(0\|\delta)$ (we seed $q$ with $x_i$). Take $\psi(x) = (x+1)^K$ for $K \notin [0, 1]$. Then the ratio between these seeding potentials and $\psi_{\mathrm{opt},1}$ respectively satisfy $\rho_p \leq 2^{K-1}/(2^{K-1} - 1)$ and $\rho_q = \theta(K)$, while $\rho_\psi^2 = (1 + \Delta)^K$. When $K \to +\infty$, we have (i) $\rho_p \to 1$, and so seeding $p$ rapidly approaches the lower bound (24); (ii) $\rho_q \to +\infty$, and so seeding $q$ drives $\mathbf{E}[\psi_1]$; (iii) ratio $\rho_\psi^2$ is extremely large compared to $\rho_q$.

## 5 Experiments

Empirical tests were made with synthetic point sets which had a distinctive structure with high probability. The number of clusters was always 20 and every cluster had 100 points in $\mathbb{R}_{+*}^{50}$. Each of the 20 distributions for the clusters was generated in two phases. In the first phase, for each coordinate lots were drawn independently with a fixed probability $p$ whether the coordinate value is a

**Table 1.** Percentage of seeding runs in which one center was picked from every original cluster. Numbers in the labels KL-0, KL-0.25 etc. refer to different values of $\alpha$.

| $p$ | unif. | $D^2$ | KL-0 | KL-0.25 | KL-0.5 | KL-0.75 | KL-1 | IS-0 | IS-0.25 | IS-0.5 | IS-0.75 | IS 1 |
|---|---|---|---|---|---|---|---|---|---|---|---|---|
| 0.1 | 0 | 9.70 | 56.8 | 75.5 | 77.1 | 76.1 | 57.2 | 34.4 | 94.3 | 95.4 | **96.0** | 48.4 |
| 0.5 | 0 | 24.0 | 77.8 | 83.1 | 81.8 | 80.7 | 79.3 | 94.9 | 95.9 | **96.5** | 95.8 | 94.4 |
| 0.9 | 0 | 7.10 | 34.6 | 38.8 | 42.2 | 39.4 | 29.5 | 28.1 | 72.6 | **75.8** | 68.6 | 20.5 |
| 1.0 | 0 | 4.10 | 7.20 | **10.0** | 7.90 | 9.30 | 5.90 | 0 | 0 | 0 | 0 | 0.100 |

**Table 2.** Percentage of original clusters from which no point was picked in the seeding phase. Average over 1000 runs.

| $p$ | unif. | $D^2$ | KL-0 | KL-0.25 | KL-0.5 | KL-0.75 | KL-1 | IS-0 | IS-0.25 | IS-0.5 | IS-0.75 | IS-1 |
|---|---|---|---|---|---|---|---|---|---|---|---|---|
| 0.1 | 35.8 | 7.60 | 2.48 | 1.32 | 1.19 | 1.24 | 2.32 | 5.50 | 0.285 | 0.230 | **0.200** | 3.39 |
| 0.5 | 35.5 | 5.47 | 1.18 | 0.880 | 0.945 | 0.975 | 1.09 | 0.260 | 0.205 | **0.180** | 0.210 | 0.285 |
| 0.9 | 35.8 | 8.54 | 4.15 | 3.75 | 3.45 | 3.74 | 4.68 | 4.42 | 1.46 | **1.31** | 1.67 | 6.45 |
| 1.0 | 35.9 | 9.81 | 8.27 | **7.86** | 8.27 | 8.27 | 9.00 | 27.8 | 23.1 | 23.4 | 25.1 | 21.7 |

**Table 3.** Bregman clustering potentials with Kullback-Leibler divergence ($\alpha = 0.25$)

| $p$ | unif. | $D^2$ | KL-0 | KL-0.25 | KL-0.5 | KL-0.75 | KL-1 | IS-0 | IS-0.25 | IS-0.5 | IS-0.75 | IS-1 |
|---|---|---|---|---|---|---|---|---|---|---|---|---|
| 0.1 | 31.2 | 4.59 | 2.64 | 1.59 | 1.46 | 1.46 | 1.85 | 6.45 | 1.10 | 1.09 | **1.06** | 2.33 |
| 0.5 | 19.5 | 4.55 | 1.74 | 1.59 | 1.50 | 1.66 | 1.74 | 1.14 | 1.14 | **1.08** | 1.10 | 1.15 |
| 0.9 | 7.29 | 2.92 | 1.97 | 1.82 | 1.73 | 1.92 | 2.05 | 1.90 | 1.34 | **1.29** | 1.39 | 2.57 |
| 1.0 | 4.13 | 2.21 | **1.97** | 1.99 | 2.01 | 2.05 | 2.11 | 3.93 | 3.64 | 3.69 | 3.89 | 3.54 |

Poisson distributed random variable or the constant 0. Then in the second phase the expectations of the Poisson random variables were chosen independently and uniformly from range $]0, 100[$. 100 points were generated from each distribution and after that the value $\epsilon = 10^{-6}$ was added to every coordinate of every point in order to move the points to the domain of Kullback-Leibler and Itakura-Saito divergences. Because not only the seeding methods but also the datasets were random, 10 datasets were generated for each value of $p$. The seeding and clustering test were repeated 100 times for each dataset.

When $p$ was less than 1, each cluster was characterized with high probability by the position of coordinates whose value was not $\epsilon$. That made the mixed Itakura-Saito divergences between two points belonging to different clusters very high, and picking one point from every original cluster was strikingly easy using those distortion measures (Tables 1 and 2). However, when all the coordinates were Poisson distributed, the task of finding a center candidate from every cluster was far more difficult. In that case the Kullback-Leibler seeding performed best.

In the clustering tests uniform, $D^2$, Kullback-Leibler and Itakura-Saito seeding ($\alpha \in \{0, 0.25, \ldots, 1\}$) were used with KL-divergences (same set of values for $\alpha$ as in the seeding) in the iterative phase and by evaluation of the clustering potential. Table 3 illustrates the situation when value $\alpha = 0.25$ is used in the iterative phase. All the values in the table were normalized using the clustering potentials which were achieved by refining the known centers of the distributions with Bregman clustering. When $p$ was 0.1 uniform seeding brought an over twenty-eight times and $D^2$ seeding over four times larger average potential than the mixed versions of Itakura-Saito seeding.

In general, there was a clear correlation between the quality of the seeding and the final clustering potential, even if the relative differences in the final potentials tended to diminish gradually when $p$ increased. That means the mixed versions of Bregman seeding algorithms led to low clustering potentials also when a regular Bregman divergence was used in the clustering phase. Additional tests were run with Poisson random variables replaced by Binomial($n$, $r$) distributed variables, so that $n = 100$ and $r$ was taken uniformly at random from range $]0, 1[$. The results were quite similar to those shown here.

## 6 Conclusions and Open Problems

We have seen that the $D^2$ seeding of [AV07] can be generalized for Bregman clustering while maintaining some form of approximation guarantee. Our other main contribution was symmetrization of Bregman clustering by using pairs of centroids. Experiments suggest that the resulting new algorithm can significantly improve the quality of both the seeding and the final clustering. The experiments are somewhat preliminary, though, and should be extended to cover more realistic data sets. We also need a better understanding of how much the seeding affects the end result in practice.

On theoretical side, it is not clear if the factor $\rho_\psi$ really is necessary in the bounds. Conceivably, some proof technique not relying on the triangle inequality could give a sharper bound. Alternatively, one could perhaps prove a lower bound that shows the $\rho_\psi$ factor necessary. It would also be interesting to consider other divergences. One possibility would be the $p$-norm divergences [Gen03] which in some other learning context give results similar to Kullback-Leibler divergence but do not have similar extreme behavior at the boundary of the domain.

## Acknowledgments

R. Nock gratefully acknowledges support from the University of Helsinki for a stay during which this work was done. R. Nock was supported by ANR "Blanc" project ANR-07-BLAN-0328-01 "Computational Information Geometry and Applications." P. Luosto and J. Kivinen were supported by Academy of Finland grants 118653 (ALGODAN) and 210796 (ALEA) and the PASCAL Network of Excellence. We thank the anonymous referees for helpful comments.

## References

[ABS08] Ackermann, M.R., Blömer, J., Sohler, C.: Clustering for metric and nonmetric distance measures. In: Proc. of the 19th ACM-SIAM Symposium on Discrete Algorithms, pp. 799–808 (2008)

[AV07] Arthur, D., Vassilvitskii, S.: $k$-means++: the advantages of careful seeding. In: Proc. of the 18th ACM-SIAM Symposium on Discrete Algorithms, pp. 1027–1035 (2007)

[AW01] Azoury, K.S., Warmuth, M.K.: Relative loss bounds for on-line density estimation with the exponential family of distributions. Machine Learning Journal 43(3), 211–246 (2001)

[BGW05] Banerjee, A., Guo, X., Wang, H.: On the optimality of conditional expectation as a Bregman predictor. IEEE Trans. on Information Theory 51, 2664–2669 (2005)

[BMDG05] Banerjee, A., Merugu, S., Dhillon, I., Ghosh, J.: Clustering with Bregman divergences. Journal of Machine Learning Research 6, 1705–1749 (2005)

[Bre67] Bregman, L.M.: The relaxation method of finding the common point of convex sets and its application to the solution of problems in convex programming. USSR Comp. Math. and Math. Phys. 7, 200–217 (1967)

[CKW07] Crammer, K., Kearns, M., Wortman, J.: Learning from multiple sources. In: Advances in Neural Information Processing Systems 19, pp. 321–328. MIT Press, Cambridge (2007)

[CM08] Chaudhuri, K., McGregor, A.: Finding metric structure in information-theoretic clustering. In: Proc. of the 21st Conference on Learning Theory (2008)

[DD06] Deza, E., Deza, M.-M.: Dictionary of distances. Elsevier, Amsterdam (2006)

[Gen03] Gentile, C.: The robustness of the $p$-norm algorithms. Machine Learning Journal 53(3), 265–299 (2003)

[Llo82] Lloyd, S.: Least squares quantization in PCM. IEEE Trans. on Information Theory 28, 129–136 (1982)

[NBN07] Nielsen, F., Boissonnat, J.-D., Nock, R.: On Bregman Voronoi diagrams. In: Proc. of the 18th ACM-SIAM Symposium on Discrete Algorithms, pp. 746–755 (2007)

[NN05] Nock, R., Nielsen, F.: Fitting the smallest enclosing Bregman ball. In: Gama, J., Camacho, R., Brazdil, P.B., Jorge, A.M., Torgo, L. (eds.) ECML 2005. LNCS (LNAI), vol. 3720, pp. 649–656. Springer, Heidelberg (2005)

[ORSS06] Ostrovsky, R., Rabani, Y., Schulman, L.J., Swamy, C.: The effectiveness of Lloyd-type methods for the $k$-means problem. In: Proc. of the 47th IEEE Symposium on the Foundations of Computer Science, pp. 165–176. IEEE Computer Society Press, Los Alamitos (2006)

[Vel02] Veldhuis, R.: The centroid of the symmetrical Kullback-Leibler distance. IEEE Signal Processing Letters 9, 96–99 (2002)


## Appendix: Proofsketch of Lemma 6

Fix $\mathcal{A} = \{\boldsymbol{x}_1, \boldsymbol{x}_2, ..., \boldsymbol{x}_K\}$ and let

$$\mathrm{BR}_\psi(\mathcal{A}) \doteq \frac{\sum_{i=1}^K \psi(\boldsymbol{x}_i)}{K} - \psi\left(\frac{\sum_{i=1}^K \boldsymbol{x}_i}{K}\right)$$

be the Burbea-Rao divergence generated by $\psi$ on $\mathcal{A}$, that is, the non negative remainder of Jensen's inequality [DD06]. It shall be convenient to abbreviate the

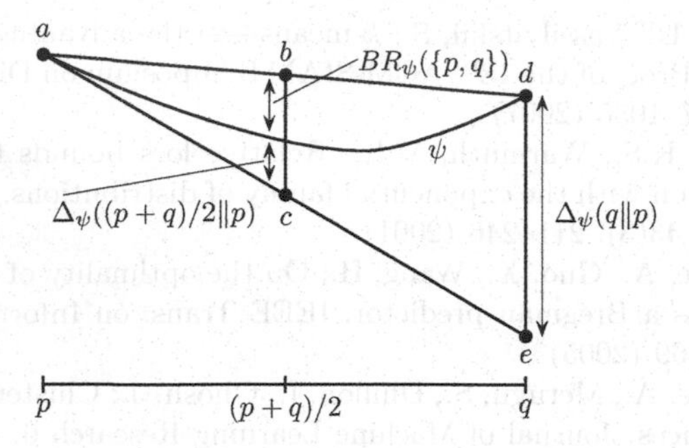


**Fig. 2.** Plot of some convex function $\psi$ defined on some segment $[p, q]$. Here, $a = (p, \psi(p))$, $b = (q, \psi(q))$ and segment $ae$ is tangent to $\psi$ in $a$. Thales theorem in triangles $(a, b, c)$ and $(a, d, e)$ proves (29), as it gives indeed $|de|/|bc| = |ad|/|ab| = |ae|/|ac|$ (here, $|.|$ denotes the Euclidean length).

arithmetic and Bregman averages of $\mathcal{A}$ as $\boldsymbol{c}$ and $\boldsymbol{c}^\star$ respectively. Then we want to estimate the ratio between the uniform seeding potential and the optimal potential for $\mathcal{A}$:

$$\rho_{\mathcal{A}} \doteq \frac{1}{K}\sum_{i=1}^{K} \frac{\sum_{\boldsymbol{x}\in\mathcal{A}} \Delta_{\psi,\alpha}(\boldsymbol{x}_i\|\boldsymbol{x}\|\boldsymbol{x}_i)}{\sum_{\boldsymbol{x}\in\mathcal{A}} \Delta_{\psi,\alpha}(\boldsymbol{c}^\star\|\boldsymbol{x}\|\boldsymbol{c})} \ . \tag{25}$$

Fix $i \in \{1, 2, ..., K\}$. First, it comes from (7) that $\sum_{\boldsymbol{x}\in\mathcal{A}} \Delta_\psi(\boldsymbol{x}\|\boldsymbol{x}_i) - \sum_{\boldsymbol{x}\in\mathcal{A}} \Delta_\psi(\boldsymbol{x}\|\boldsymbol{c}) = K\Delta_\psi(\boldsymbol{c}\|\boldsymbol{x}_i)$, and we also get from (8) that $\sum_{\boldsymbol{x}\in\mathcal{A}} \Delta_\psi(\boldsymbol{x}_i\|\boldsymbol{x}) - \sum_{\boldsymbol{x}\in\mathcal{A}} \Delta_\psi(\boldsymbol{c}^\star\|\boldsymbol{x}) = K\Delta_\psi(\boldsymbol{x}_i\|\boldsymbol{c}^\star)$. The numerator of (25) becomes:

$$\sum_{\boldsymbol{x}\in\mathcal{A}} \Delta_{\psi,\alpha}(\boldsymbol{x}_i\|\boldsymbol{x}\|\boldsymbol{x}_i) = \alpha\left(\sum_{\boldsymbol{x}\in\mathcal{A}} \Delta_\psi(\boldsymbol{x}\|\boldsymbol{c}) + K\Delta_\psi(\boldsymbol{c}\|\boldsymbol{x}_i)\right)$$

$$+(1-\alpha)\left(\sum_{\boldsymbol{x}\in\mathcal{A}} \Delta_\psi(\boldsymbol{c}^\star\|\boldsymbol{x}_i) + K\Delta_\psi(\boldsymbol{x}_i\|\boldsymbol{c}^\star)\right) \tag{26}$$

$$= \sum_{\boldsymbol{x}\in\mathcal{A}} \Delta_{\psi,\alpha}(\boldsymbol{c}^\star\|\boldsymbol{x}\|\boldsymbol{c}) + K\Delta_{\psi,1-\alpha}(\boldsymbol{c}\|\boldsymbol{x}_i\|\boldsymbol{c}^\star) \ . \tag{27}$$

The left summand in (27) is the optimal potential for the cluster. Finally, the denominator of (25) can be rewritten as $\sum_{i=1}^{K} \Delta_{\psi,\alpha}(\boldsymbol{c}^\star\|\boldsymbol{x}_i\|\boldsymbol{c}) = K(\alpha\mathrm{BR}_\psi(\mathcal{A}) + (1-\alpha)\mathrm{BR}_{\psi^\star}(\boldsymbol{\nabla}\psi\mathcal{A}))$, where $\boldsymbol{\nabla}\psi\mathcal{A}$ is the set of gradient images of the elements of $\mathcal{A}$. We get:

$$\rho_{\mathcal{A}} = 1 + \frac{1}{K}\sum_{i=1}^{K} \frac{\Delta_{\psi,1-\alpha}(\boldsymbol{c}\|\boldsymbol{x}_i\|\boldsymbol{c}^\star)}{\alpha\mathrm{BR}_\psi(\mathcal{A}) + (1-\alpha)\mathrm{BR}_{\psi^\star}(\boldsymbol{\nabla}\psi\mathcal{A})} \ . \tag{28}$$

For the sake of simplicity, let $\boldsymbol{c}_{i,j}$ denote the weighted arithmetic average of $\mathcal{A}$ in which the weight of each $\boldsymbol{x}_k$ is $\frac{1}{2^j K}$ for $k \neq i$, and the weight of $\boldsymbol{x}_i$ is $\frac{1}{2^j K} + 1 - \frac{1}{2^j}$

($\forall j \geq 0$). We also let $\boldsymbol{c}^{\star}_{i,j}$ denote the weighted Bregman average of $\mathcal{A}$ under this same distribution. Thus, as $j$ increases, the averages get progressively close to $\boldsymbol{x}_i$. Then, we have $\forall j \geq 0$:

$$\begin{aligned} \Delta_{\psi}(\boldsymbol{c}_{i,j}\|\boldsymbol{x}_i) &= 2(\mathrm{BR}_{\psi}(\{\boldsymbol{c}_{i,j},\boldsymbol{x}_i\}) + \Delta_{\psi}((\boldsymbol{c}_{i,j}+\boldsymbol{x}_i)/2\|\boldsymbol{x}_i)) && (29)\\ &= 2(\mathrm{BR}_{\psi}(\{\boldsymbol{c}_{i,j},\boldsymbol{x}_i\}) + \Delta_{\psi}(\boldsymbol{c}_{i,j+1}\|\boldsymbol{x}_i)) \ . \end{aligned}$$

$$\begin{aligned} \Delta_{\psi}(\boldsymbol{x}_i\|\boldsymbol{c}^{\star}_{i,j}) &= \Delta_{\psi^{\star}}(\boldsymbol{\nabla}\psi(\boldsymbol{c}^{\star}_{i,j})\|\boldsymbol{\nabla}\psi(\boldsymbol{x}_i)) && (30)\\ &= 2(\mathrm{BR}_{\psi^{\star}}(\{\boldsymbol{\nabla}\psi(\boldsymbol{c}^{\star}_{i,j}),\boldsymbol{\nabla}\psi(\boldsymbol{x}_i)\}) && (31)\\ &\quad +\Delta_{\psi^{\star}}((\boldsymbol{\nabla}\psi(\boldsymbol{c}^{\star}_{i,j})+\boldsymbol{\nabla}\psi(\boldsymbol{x}_i))/2\|\boldsymbol{\nabla}\psi(\boldsymbol{x}_i))) && (32)\\ &= 2(\mathrm{BR}_{\psi^{\star}}(\{\boldsymbol{\nabla}\psi(\boldsymbol{c}^{\star}_{i,j}),\boldsymbol{\nabla}\psi(\boldsymbol{x}_i)\}) + \Delta_{\psi}(\boldsymbol{x}_i\|\boldsymbol{c}^{\star}_{i,j+1})) \ . \end{aligned}$$

While (30) is just stating the convex conjugate, Thales Theorem proves both (29) and (32). Figure 2 presents a simple graphical view to state this result in the context of Bregman and Burbea-Rao divergences. We get:

$$\begin{aligned} \Delta_{\psi,1-\alpha}(\boldsymbol{c}_{i,j}\|\boldsymbol{x}_i\|\boldsymbol{c}^{\star}_{i,j}) &= \alpha\Delta_{\psi}(\boldsymbol{c}_{i,j}\|\boldsymbol{x}_i) + (1-\alpha)\Delta_{\psi}(\boldsymbol{x}_i\|\boldsymbol{c}^{\star}_{i,j})\\ &= 2(\alpha\mathrm{BR}_{\psi}(\{\boldsymbol{c}_{i,j},\boldsymbol{x}_i\}) && (33)\\ &\quad +(1-\alpha)\mathrm{BR}_{\psi^{\star}}(\{\boldsymbol{\nabla}\psi(\boldsymbol{c}^{\star}_{i,j}),\boldsymbol{\nabla}\psi(\boldsymbol{x}_i)\}))\\ &\quad +\Delta_{\psi,1-\alpha}(\boldsymbol{c}_{i,j+1}\|\boldsymbol{x}_i\|\boldsymbol{c}^{\star}_{i,j+1}) \ . && (34) \end{aligned}$$

Note that $\boldsymbol{c} = \boldsymbol{c}_{i,0}$ and $\boldsymbol{c}^{\star} = \boldsymbol{c}^{\star}_{i,0}$, $\forall i = 1, 2, ..., K$. We let:

$$b_0 \doteq \frac{2}{K}\cdot\frac{\sum_{i=1}^{K}\left\{\alpha\mathrm{BR}_{\psi}(\{\boldsymbol{c}_{i,0},\boldsymbol{x}_i\}) + (1-\alpha)\mathrm{BR}_{\psi^{\star}}(\{\boldsymbol{\nabla}\psi(\boldsymbol{c}^{\star}_{i,0}),\boldsymbol{\nabla}\psi(\boldsymbol{x}_i)\})\right\}}{\alpha\mathrm{BR}_{\psi}(\mathcal{A}) + (1-\alpha)\mathrm{BR}_{\psi^{\star}}(\boldsymbol{\nabla}\psi\mathcal{A})}$$

and for all $j > 0$

$$b_j \doteq 2\cdot\frac{\sum_{i=1}^{K}\left\{\alpha\mathrm{BR}_{\psi}(\{\boldsymbol{c}_{i,j},\boldsymbol{x}_i\}) + (1-\alpha)\mathrm{BR}_{\psi^{\star}}(\{\boldsymbol{\nabla}\psi(\boldsymbol{c}^{\star}_{i,j}),\boldsymbol{\nabla}\psi(\boldsymbol{x}_i)\})\right\}}{\sum_{i=1}^{K}\left\{\alpha\mathrm{BR}_{\psi}(\{\boldsymbol{c}_{i,j-1},\boldsymbol{x}_i\}) + (1-\alpha)\mathrm{BR}_{\psi^{\star}}(\{\boldsymbol{\nabla}\psi(\boldsymbol{c}^{\star}_{i,j-1}),\boldsymbol{\nabla}\psi(\boldsymbol{x}_i)\})\right\}} \ .$$

Furthermore, $\forall j \geq 0$, we let:

$$r_j \doteq \frac{\sum_{i=1}^{K}\Delta_{\psi,1-\alpha}(\boldsymbol{c}_{i,j}\|\boldsymbol{x}_i\|\boldsymbol{c}^{\star}_{i,j})}{2\sum_{i=1}^{K}\left\{\alpha\mathrm{BR}_{\psi}(\{\boldsymbol{c}_{i,j},\boldsymbol{x}_i\}) + (1-\alpha)\mathrm{BR}_{\psi^{\star}}(\{\boldsymbol{\nabla}\psi(\boldsymbol{c}^{\star}_{i,j}),\boldsymbol{\nabla}\psi(\boldsymbol{x}_i)\})\right\}} \ .$$

Plugging (34) into (28) and using the last notations yields:

$$\begin{aligned} \rho_{\mathcal{A}} &= 1 + b_0(1 + b_1(...(1 + b_J r_J)))\\ &\geq 1 + b_0(1 + b_1(...(1 + b_{J-1})))\\ &\geq \sum_{j=0}^{J-1}\left(\frac{1}{2\rho_{\psi}^2}\right)^j = \frac{(2\rho_{\psi}^2)^J - 1}{(2\rho_{\psi}^2)^J}\cdot\frac{2\rho_{\psi}^2}{2\rho_{\psi}^2 - 1}, \forall J \geq 0 \ . \end{aligned}$$

The last inequality is obtained after various suitable Taylor expansions of $\psi$ are used for $b_i, i \geq 0$, which gives $b_i \geq 1/(2\rho_{\psi}^2)$ (not shown to save space).

# Hierarchical, Parameter-Free Community Discovery

Spiros Papadimitriou[1], Jimeng Sun[1], Christos Faloutsos[2], and Philip S. Yu[3]

[1] IBM T.J. Watson Research Center, Hawthorne, NY, USA
spapadim,jimeng@us.ibm.com
[2] Carnegie Mellon University, Pittsburgh, PA, USA
christos@cs.cmu.edu
[3] University of Illinois, Chicago, IL, USA
psyu@cs.uic.edu

**Abstract.** Given a large bipartite graph (like document-term, or user-product graph), how can we find meaningful communities, quickly, and automatically? We propose to look for community hierarchies, with communities-within-communities. Our proposed method, the *Context-specific Cluster Tree (CCT)* finds such communities at multiple levels, with no user intervention, based on information theoretic principles (MDL). More specifically, it partitions the graph into progressively more refined subgraphs, allowing users to quickly navigate from the global, coarse structure of a graph to more focused and local patterns. As a fringe benefit, and also as an additional indication of its quality, it also achieves better compression than typical, non-hierarchical methods. We demonstrate its scalability and effectiveness on real, large graphs.

## 1 Introduction

Bipartite graphs (or, equivalently, sparse binary matrices) are natural representations of relations between two sets of nodes, namely source and destination nodes. Such large bipartite graphs arise naturally in many applications, like information retrieval (document-term graphs), collaborative filtering and recommendation systems (person-product graphs), social networks, and many more.

Graph mining aims at discovering the useful patterns hidden in the graphs. Various tools geared towards large graphs have been proposed in the literature. All of those techniques usually examine the graph at two extreme levels: 1) global, i.e., patterns present in the entire graph such as power law distribution on graphs [9], graph partitioning [4, 8, 16], community evolution [25, 27]; or, 2) local, i.e, patterns related to a subgraph such as center-piece graph [28], neighborhood formation [26], quasi-cliques [21].

In this paper, we aim to fill the gap between global and local patterns, by proposing a technique that allows users to effectively discover and explore communities in large graphs at multiple levels, starting from a global view and narrowing down to more local information. More specifically, we study ways to quickly and *automatically* construct a recursive community structure of a

W. Daelemans et al. (Eds.): ECML PKDD 2008, Part II, LNAI 5212, pp. 170–187, 2008.

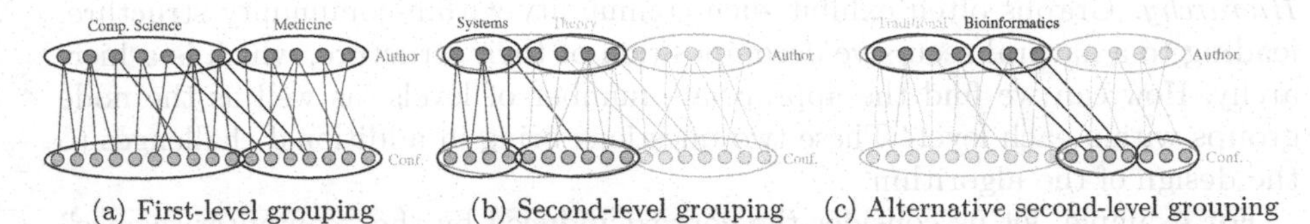


(a) First-level grouping (b) Second-level grouping (c) Alternative second-level grouping

**Fig. 1.** Hierarchy and context

large bipartite graph at multiple levels, namely, a Context-specific Cluster Tree (CCT). The resulting CCT can identify relevant context-specific clusters. It also provides an efficient data summarization scheme and facilitates visualization of large graphs, which is a difficult, open problem itself [13, 14]. Intuitively, a *context* is a subgraph which is implicitly defined by a pair of source and destination node groups (and, thus, includes exactly those edges that connect nodes of those groups)—see Definition 8. The entire graph and a single edge are the two extreme contexts, at the global and local level, respectively.

Our approach allows users to start from groups of nodes and edges at the global, coarse level and quickly focus on the appropriate context to discover more focused and fine-grained patterns. We shall illustrate the insight and intuition behind our proposed framework with an example. Consider a set of authors (blue nodes, top of Figure 1) and a set of conferences (red nodes, bottom of Figure 1), with edges indicating that the author published in that conference.

At first blush, one might discover a natural partitioning of the graph at the global level as follows:

- Node groups: Assume there are predominantly two groups of authors, computer scientists and medical researchers. Further assume there are two corresponding predominant conference groups. In matrix form, they correspond to the two row and column partitions, respectively, shown in Figure 2a.
- Contexts: The above node grouping leads to four contexts (i.e., edge groups, or subgraphs), one for each possible combination of the two node groups of each type (authors and conferences). In matrix form, they correspond to the four submatrices in Figure 1b. The dominant contexts are the two submatrices on the diagonal of Figure 2a corresponding to computer science (intersection of computer scientists and CS conferences) and medicine (intersection of of doctors and medical conferences), respectively.

This first-level decomposition already reveals much information about the structure of the data and answers the question: "given the mutual associations between *all* authors and *all* conferences, which are the groups of nodes that are most closely associated." Each of those group associations reveals a new context, which is a subgraph of the original graph. Thus, it can be likewise analyzed recursively, to reveal further contexts of finer granularity. In fact, if we stop at the first level and consider the computer science and medicine contexts, we may miss bioinformatics which will likely appear as associations between a subset of computer scientists and medical conferences (see Figure 2). To realize this intuition, we proceed to explain the two key concepts of hierarchy and context.

*Hierarchy.* Graphs often exhibit such community-within-community structure, leading to a natural recursive decomposition of their structure, which is a hierarchy. How can we find the appropriate number of levels, as well as the node groups within each level? These two questions bring in additional challenges to the design of the algorithm.

For example, let us consider the context induced by the "computer science" author and the "computer science" conference group (see Figure 1b, or the top-left part of Figure 2b). Performing a similar analysis as before, we may discover additional structure in terms of node groups and contexts in the second level. The computer science field may be further subdivided into systems and theory authors, with a corresponding division in computer science conferences.

*Context.* In our example, the dominant contexts are those of "computer science" and "medicine," as explained above. However, there is nothing special about those "diagonal" contexts. In fact, we argue that one need also examine "off-diagonal" contexts. For example, the context defined by the intersection of "computer science" authors and "medical" conferences (see Figure 1c) may also be further partitioned into multiple lower-level contexts, with one of them corresponding to "bioinformatics".

In general, a particular choice of subgraph during the recursive decomposition consists of a pair of node groups and the context provided by the edges that associate them. Different contexts may reveal different aspects of the data. Taking this idea to its logical conclusion, the overall result is a rich hierarchy, CCT, that captures the graph structure at multiple levels.

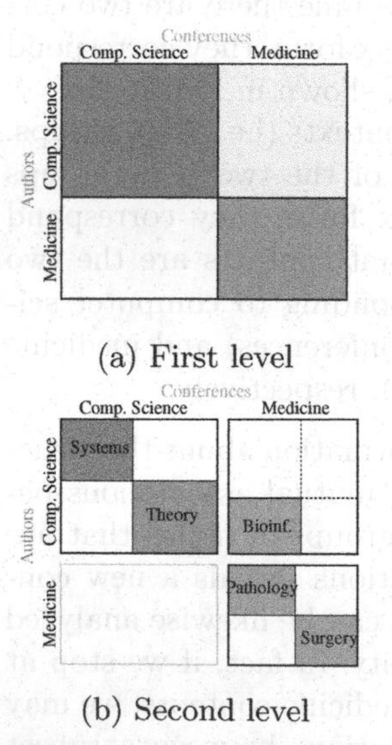


(a) First level

(b) Second level

**Fig. 2.** Adjacency matrix view (cf. Figure 1)

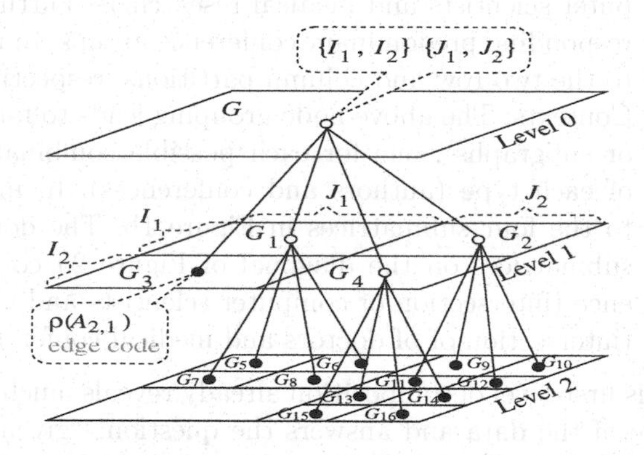


**Fig. 3.** Cluster tree (CCT) corresponding to Figure 2. Unfilled nodes correspond to non-leaves (subgraphs for which the partitioned model was best) and filled nodes correspond to leaves (subgraphs for which the random graph model was best). The two popups show examples of information represented at each type of node, with dark blue representing parts of the model and light green parts of the code, given the model.

Our goal is to automatically find this hierarchy and allow users to quickly navigate it. For example, given a theoretician, there are multiple relevant contexts at different levels. Depending on the conferences, the most relevant context for her could be "computer theory" if the relevant conference is FOCS, SODA, or maybe "bioinformatics" if relevant conference is RECOMB or ISMB (see experiments for details). In general, the most relevant context can be automatically identified given the set of query nodes through a simple tree reversal.

*Contributions.* The main contributions of this paper are the following:

- We employ a parameter-free scheme based on minimum description language (MDL) that automatically finds the best context-specific cluster tree (CCT) to summarize the graph.
- The method is linear on the number of edges, and thus scalable for large, possibly disk-resident, graphs.
- We provide a scheme for users to navigate from global, coarse structure to more focused and local patterns.

Because our method is based on sound, information theoretic principles, it also leads to better compression as a fringe benefit. Moreover, we develop a GUI prototype that allow users to visualize and explore large graphs in an intuitive manner. We demonstrate the efficiency and effectiveness of our framework on a number of datasets. In particular, a number of interesting clusters are identified in different levels.

The rest of the paper is organized as follows: Section 2 presents the necessary background and Section 3 introduces the fundamental definitions. Section 4 presents our proposed method and Section 5 evaluates it on a number of datasets. Finally, Section 6 briefly discusses related work and Section 7 concludes.

## 2 Background

In this section we give a brief overview of a practical formulation of the minimum description length (MDL) principle. For further information see, e.g., [7, 10]. Intuitively, the main idea behind MDL is the following: Let us assume that we have a family $\mathcal{M}$ of *models* with varying degrees of complexity. More complex models $M \in \mathcal{M}$ involve more parameters but, *given* these parameters (i.e., the model $M \in \mathcal{M}$), we can describe the observed data more concisely.

As a simple, concrete example, consider a binary sequence $A := [a(1), a(2), \ldots, a(n)]$ of $n$ coin tosses. A simple model $M^{(1)}$ might consist of specifying the number $h$ of heads. Given this model $M^{(1)} \equiv \{h/n\}$, we can encode the dataset $A$ using $C(A|M^{(1)}) := nH(h/n)$ bits [22], where $H(\cdot)$ is the Shannon entropy function. However, in order to be fair, we should also include the number $C(M^{(1)})$ of bits to transmit the fraction $h/n$, which can be done using $\log^{\star} n$ bits for the denominator and $\lceil \log(n+1) \rceil$ bits for the numerator $h \in \{0, 1, \ldots, n\}$, for a total of $C(M^{(1)}) := \log^{\star} n + \lceil \log(n+1) \rceil$ bits.

**Table 1.** Symbols and definitions

| Symbol | Definition | Symbol | Definition |
|---|---|---|---|
| $A$ | Binary adjacency matrix. | $m_p, n_q$ | Dimensions of $A_{p,q}$. |
| $m, n$ | Dimensions of $A$. | $\|A\|$ | Number of elements $\|A\| := mn$. |
| $k, \ell$ | No. of source and dest. partitions. | $\rho(A)$ | Edge density in $\rho(A) = e(A)/\|A\|$. |
| $A_{p,q}$ | Submatrix for intersection of $p$-th source and $q$-th dest. partitions. | $H(\cdot)$ | Shannon entropy function. |
| | | $C(A)$ | Codelength for $A$. |

**Definition 1.** (Code length and description complexity) *$C(A|M^{(1)})$ is* code length *for $A$, given the model $M^{(1)}$. $C(M^{(1)})$ is the* model description complexity *and $C(A, M^{(1)}) := C(A|M^{(1)}) + C(M^{(1)})$ is the* total code length.

A slightly more complex model might consist of segmenting the sequence in two pieces of length $n_1 \geq 1$ and $n_2 = n - n_1$ and describing each one independently. Let $h_1$ and $h_2$ be the number of heads in each segment. Then, to describe the model $M^{(2)} \equiv \{h_1/n_1, h_2/n_2\}$, we need $C(M^{(2)}) := \log^\star n + \lceil \log n \rceil + \lceil \log(n - n_1) \rceil + \lceil \log(n_1 + 1) \rceil + \lceil \log(n_2 + 1) \rceil$ bits. Given this information, we can describe the sequence using $C(A|M^{(2)}) := n_1 H(h_1/n_1) + n_2 H(h_2/n_2)$ bits.

Now, assume that our family of models is $\mathcal{M} := \{M^{(1)}, M^{(2)}\}$ and we wish to choose the "best" one for a particular sequence $A$. We will examine two sequences of length $n = 16$, both with 8 zeros and 8 ones, to illustrate the intuition.

Let $A_1 := \{0, 1, 0, 1, \cdots, 0, 1\}$, with alternating values. We have $C(A_1|M_1^{(1)}) = 16H(1/2) = 16$ and $C(M_1^{(1)}) = \log^\star 16 + \lceil \log(16 + 1) \rceil = 10 + 5 = 15$. However, for $M_1^{(2)}$ the best choice is $n_1 = 15$, with $C(A_1|M_1^{(2)}) \approx 15$ and $C(M_1^{(2)}) \approx 19$. The *total* code lengths are $C(A_1, M_1^{(1)}) \approx 16 + 15 = 31$ and $C(A_1, M_1^{(2)}) \approx 15 + 19 = 34$. Thus, based on total code length, the simpler model is better[1]. The more complex model may give us a lower code length, but that benefit is not enough to overcome the increase in description complexity: $A_1$ does not exhibit a pattern that can be exploited *by a two-segment model* to describe the data.

Let $A_2 := \{0, \cdots, 0, 1, \cdots, 1\}$ with all similar values contiguous. We have again $C(A_2|M_2^{(1)}) = 16$ and $C(M_2^{(1)}) = 15$. But, for $M_2^{(2)}$ the best choice is $n_1 = n_2 = 8$ so that $C(A_2|M_2^{(2)}) = 8H(0) + 8H(1) = 0$ and $C(M_2^{(2)}) \approx 24$. The *total* code lengths are $C(A_2, M_2^{(1)}) \approx 16 + 15 = 31$ and $C(A_2, M_2^{(2)}) \approx 0 + 24 = 24$. Thus, based on total code length, the two-segment model is better. Intuitively, it is clear that $A_2$ exhibits a pattern that can help reduce the total code length. This intuitive fact is precisely captured by the total code length.

## 3 CCT: Encoding and Partitioning

We want to subdivide the adjacency matrix in tiles (or "contexts"), with possible reordering of rows and columns, and compress them, either as-is (if they are

[1] The absolute codelengths are not important; the bit overhead compared to the straight transmission of $A$ tends to zero, as $n$ grows to infinity.

homogeneous enough) or by further subdividing. First, we formalize the problem and set the stage, by defining a lossless hierarchical encoding. As we shall see, the cluster-tree structure corresponds to the model, whereas the code for the data given the model is associated only with leaf nodes. This encoding allows us to apply MDL for automatically finding the desired progressive decomposition. In this section we define the codelength, assuming that a tree structure is given. Next, in Section 4, we present a practical algorithm to find the structure.

### 3.1 Problem Definition

Assume we are given a set of $m$ source nodes, $\mathcal{I} := \{1, 2, \ldots, m\}$ and a set of $n$ destination nodes, $\mathcal{J} := \{1, 2, \ldots, n\}$. Each node pair $(i, j)$, for $1 \leq i \leq m$ and $1 \leq j \leq n$, may be linked with an edge. Let $A = [a(i, j)]$ denote the corresponding $m \times n$ $(m, n \geq 1)$ binary adjacency matrix.

**Definition 2 (Bipartite graph and subgraph).** *The bipartite graph $\mathcal{G}$ is the triple $\mathcal{G} \equiv (\mathcal{I}, \mathcal{J}, A)$. A subgraph of this graph is a triple $\mathcal{G}' \equiv (\mathcal{I}', \mathcal{J}', A')$, where $\mathcal{I}' \subseteq \mathcal{I}$, $\mathcal{J}' \subseteq \mathcal{J}$ and $A' := [a(i', j')]$ for all $i' \in \mathcal{I}'$ and $j' \in \mathcal{J}'$.*

Our goal is to discover groups of edges that closely link groups of source nodes and destination nodes.

**Definition 3 (Subgraph partitioning).** *Given a graph $\mathcal{G} \equiv (\mathcal{I}, \mathcal{J}, A)$, we will partition it into a set of subgraphs $\{\mathcal{G}_1, \mathcal{G}_2, \ldots, \mathcal{G}_T\}$ such that their union equals the original graph $\mathcal{G}$.*

More specifically, we seek to decompose the original graph into a set of subgraphs, which should have the following properties:

- Connectedness: Each of the subgraphs should ideally be either fully connected or fully disconnected, i.e., it should be as homogeneous as possible.
- Flexible: The structure of the decomposition into subgraphs should be rich enough, without imposing too many constraints. On the other hand, it should lead to tractable and practical algorithms to find the decomposition.
- Progressive: The decomposition should allow users to navigate from global, coarse structure to more focused and local patterns, in the form of progressively more dense subgraphs.

Furthermore, we seek to automatically find such a decomposition, without requiring any parameters from the user. To that end, we employ MDL on an encoding of the bipartite adjacency matrix. The encoding we choose is hierarchical, so as to satisfy the last two properties.

### 3.2 Hierarchical Encoding

In order to achieve the previously stated goals, we employ a top-down approach. Consider an $m \times n$ adjacency matrix $A$, which may correspond to any bipartite subgraph (including the entire graph). We proceed to explain how we can build a code for a *given* hierarchical partitioning.

*Base case.* Our first and simplest option is that no patterns are present in the data. In this case, we may safely model the data by simply assuming that each edge is independently drawn with probability $\rho(A)$, where $\rho(A)$ is the density of edges (or, of ones in the adjacency matrix $A$).

**Definition 4 (Random graph model).** *In this case, we may encode the entire matrix using*

$$C_0(A) := \lceil \log(|A|+1) \rceil + \lceil |A| H(\rho(A)) \rceil \text{ bits.} \quad (1)$$

More specifically, we use $\lceil \log(|A|+1) \rceil$ bits to transmit $\rho(A)$ and finally $\lceil |A| H(\rho(A)) \rceil$ bits to transmit the individual edges. This assumes that we already know the graph size (i.e., $m$ and $n$). For the initial graph $\mathcal{G}$, we may safely assume this. For its subgraphs, this information is provided by our model, as will be explained shortly.

*Recursive case.* The second option is to try to find joint groups of nodes and edges, as described before, which partition the original graph into subgraphs. Note that the partitioning (see Definition 3) is equivalent to a tiling of the adjacency matrix with $T$ tiles, allowing for row and column reordering and, in the most general case, possibly overlapping tiles. Although we can allow arbitrary tilings, this leads to significant penalty in terms of complexity. Therefore, we impose certain constraints on the structure of the partitioning, so as to make the problem more tractable, while still allowing enough flexibility in the model to capture interesting patterns.

First, we require that the tiling is *exclusive* (i.e., no two tiles overlap) and *complete* (i.e., the tiles completely cover the entire adjacency matrix, without "gaps"). Next, we proceed to construct the tiling in a hierarchical fashion. We constrain the tiling to follow a checkerboard structure only within a single level of the hierarchy. The first-level decomposition of Figures 2a and 3 follows such a structure, consisting of $\mathcal{G}_1 = (\mathcal{I}_1, \mathcal{J}_1, A_{1,1})$, $\mathcal{G}_2 = (\mathcal{I}_1, \mathcal{J}_2, A_{1,2})$, $\mathcal{G}_3 = (\mathcal{I}_2, \mathcal{J}_1, A_{2,1})$, and $\mathcal{G}_4 = (\mathcal{I}_2, \mathcal{J}_2, A_{2,2})$, where $\mathcal{I}_1$ is the set of "computer science researchers" and $\mathcal{I}_2$ the set of "medical researchers" and similarly for the conference sets $\mathcal{J}_1$ and $\mathcal{J}_2$. Formally, the checkerboard structure means that set of source-destination group pairs, $\{(\mathcal{I}_1, \mathcal{J}_1), (\mathcal{I}_1, \mathcal{J}_2), (\mathcal{I}_2, \mathcal{J}_1), (\mathcal{I}_2, \mathcal{J}_2)\}$, can be written as a Cartesian product of individual sets of source and destination groups, $\{\mathcal{I}_1, \mathcal{I}_2\} \times \{\mathcal{J}_1, \mathcal{J}_2\}$.

In general, we can define a checkerboard decomposition into $T = k \cdot \ell$ subgraph tiles, using $k$ source groups $\mathcal{I}_p$, for $1 \le p \le k$, and $\ell$ destination groups $\mathcal{J}_q$, for $1 \le q \le \ell$. We denote the sizes of $\mathcal{I}_p$ and $\mathcal{J}_q$ by $m_p := |\mathcal{I}_p|$ and $n_q := |\mathcal{J}_q|$, respectively, and the corresponding adjacency submatrices by $A_{p,q} := [a(\mathcal{I}_p, \mathcal{J}_q)]$, for $1 \le p \le k$ and $1 \le q \le \ell$.

**Definition 5 (Partitioned graph model).** *The cost of encoding the partitioned graph is*

$$C_1(A) := \lceil \log m \rceil + \lceil \log n \rceil + \left\lceil \log \binom{m}{m_1 \cdots m_k} \right\rceil + \\ + \left\lceil \log \binom{n}{n_1 \cdots n_\ell} \right\rceil + \sum_{p=1}^{k} \sum_{q=1}^{\ell} C(A_{p,q}). \quad (2)$$

We need $\lceil \log m \rceil$ bits to transmit $k$ and $\lceil \log n \rceil$ bits to transmit $\ell$. Furthermore, if we assume each mapping of $m$ source nodes into $k$ source groups equally likely, then we need $\lceil \log \binom{m}{m_1 \cdots m_k} \rceil$ bits to transmit the source partitioning $\{\mathcal{I}_1, \ldots, \mathcal{I}_k\}$, and similarly for the destination partitioning. Note that this partitioning implicitly determines the size $m_p \times n_q$ of each subgraph (which is assumed known by the random graph model). Finally, we need to recursively encode each of the $k \cdot \ell$ adjacency submatrices, one for each subgraph, which is represented by the last term in Equation (2).

Using Stirling's approximation $\ln n! \approx n \ln n - n$ and the fact that $\sum_i m_i = m$, we can easily derive that

$$\log \binom{m}{m_1 \cdots m_k} \approx m H\left(\tfrac{m_1}{m}, \ldots, \tfrac{m_k}{m}\right),$$

where $H(\cdot)$ denotes the Shannon entropy. Now we are ready to define the overall codelength cost $C(A)$.

**Definition 6 (Total hierarchical codelength).** *Given a hierarchical decomposition, the total codelength cost for transmitting the graph $(\mathcal{I}, \mathcal{J}, A)$ is*

$$C(A) := 1 + \min\{C_0(A), C_1(A)\}. \tag{3}$$

We choose the best of the two options, (i) pure random graph, or (ii) partitioned graph. Additionally, we need one bit to transmit which of these two options was the best. Ties are broken in favor of the simpler, random graph model. Note that the definition of the total cost is recursive, since $C(A_{p,q})$ appears in Equation (2).

*Final result.* To summarize, we have recursively built a hierarchical encoding based on a tiling of the adjacency matrix, each tile uniquely corresponding to a subgraph of the original bipartite graph. At each level of the hierarchy, we use checkerboard tiles. Each of those may be further subdivided in the same manner (such as all three tiles except the bottom left one from Figure 2a).

**Definition 7 (Context-specific Cluster Tree).** *The set of all subgraphs in the progressive, hierarchical decomposition consists the* context-specific cluster tree (CCT). *The leaf nodes correspond to subgraphs for which the best choice is the random graph model. These subgraphs comprise the* leaf-level partitioning. *The code for the data given the model consists of the information for individual edges within subgraphs only at the leaf level.*

For example, in Figure 3, the root node would encode the partitioning $\{\mathcal{I}_1, \mathcal{I}_2\}$ and $\{\mathcal{J}_1, \mathcal{J}_2\}$; this is part of the model. The node corresponding to $\mathcal{G}_3 \equiv (\mathcal{I}_2, \mathcal{J}_1, A_{2,1})$ would encode the density $\rho(A_{2,1})$—which is also part of the model—and subsequently, the individual edges of $\mathcal{G}_3$ using entropy coding—which is part of the code given the model. In addition to the root $\mathcal{G}$, the CCT consists of all 16 nodes corresponding to subgraphs $\mathcal{G}_1$ through $\mathcal{G}_{16}$. The leaf-level partition consists of 13 graphs $\{\mathcal{G}_3, \mathcal{G}_5, \mathcal{G}_6, \ldots, \mathcal{G}_{16}\}$, which are represented by filled nodes.

It is clear from the construction that the leaf-level partitioning is also an exclusive and complete tiling, but with a richer structure than the per-level checkerboard tiles.

Finally, we can define the *context* for a set of nodes.

**Definition 8 (Context).** *Given as input a pair of source and destination node sets* $(\mathcal{I}_i, \mathcal{J}_i)$ *of interest, a* context *of* $(\mathcal{I}_i, \mathcal{J}_i)$ *is any pair* $(\mathcal{I}_c, \mathcal{J}_c)$ *such that* $\mathcal{I}_i \subseteq \mathcal{I}_c$ *and* $\mathcal{J}_i \subseteq \mathcal{J}_c$.

In other words, a context for $(\mathcal{I}_i, \mathcal{I}_j)$ is any subgraph of the original graph that fully includes $\mathcal{I}_i$ and $\mathcal{J}_i$. We will typically constrain $(\mathcal{I}_c, \mathcal{J}_c)$ to be only those pairs that appear in some node of our hierarchical decomposition. Given that constraint, we can give the next definition.

**Definition 9 (Minimal hierarchical context).** *The* minimal hierarchical context *among a set of contexts is the context* $(\mathcal{I}_{mc}, J_{mc})$ *such that no other context* $(\mathcal{I}_c, \mathcal{J}_c)$ *exists with* $\mathcal{I}_{mc} \subseteq \mathcal{I}_c$ *and* $\mathcal{J}_{mc} \subseteq \mathcal{J}_c$

Intuitively, the minimal hierarchical context is the deepest-level tree node in CCT that fully contains the input context. Note that, if $\mathcal{I}_i \neq \emptyset \neq \mathcal{J}_i$, then the minimal hierarchical context of $(\mathcal{I}_i, \mathcal{J}_i)$ is unique. If one of $\mathcal{I}_i$ or $\mathcal{J}_i$ is empty, then there may be multiple overlapping minimal contexts and, if both are empty, then all leaf nodes are trivially minimal contexts.

## 4 Finding the CCT

In the previous section we described the cost objective function that determines how "good" a given CCT partitioning is. However, if we don't know the partitioning we need a way to find it. The total codelength provides a yardstick to compare different hierarchical encodings with different number of partitions at each level of the hierarchy, but we need a practical search strategy to find such an encoding given only the initial graph.

In order to build a scalable and practical algorithm, we choose to employ a top-down strategy for building the hierarchy, rather than a bottom-up approach. Starting with the original graph, we try to find a good "checkerboard" tiling for the first level of the decomposition. Then, we fix this tiling and we recursively attempt the same procedure on each of the tiles.

However, there are two problems that need to be addressed. First, the recursive definition of Equation (2) is too expensive to evaluate for each possible assignment of nodes into partitions, so we use the following equation instead,

$$C_1'(A) := \lceil \log m \rceil + \lceil \log n \rceil + \lceil \log \binom{m}{m_1 \cdots m_k} \rceil + \\ + \lceil \log \binom{n}{n_1 \cdots n_\ell} \rceil + \textstyle\sum_{p=1}^{k} \sum_{q=1}^{\ell} C_0(A_{p,q}), \quad (4)$$

where we have substituted $C_0$ for $C$ in the summation at the end. This *surrogate cost* heuristic is fairly effective in practice, as we shall also see in the experiments.

Even with this simplification, finding the optimal checkerboard tiling (i.e., assignment of nodes into partitions) is NP-hard [4], even if the number of tiles (or, equivalently, source and destination node partitions) is known. Additionally, we also seek the number of tiles. Therefore, we will employ an alternating minimization [4] scheme that converges towards a local minimum.

Algorithm SHUFFLE:

Start with an arbitrary partitioning of the matrix $A$ into $k$ source partitions $\mathcal{I}_p^{(0)}$ and $\ell$ column partitions $\mathcal{J}_q^{(0)}$. Subsequently, at each iteration $t$ perform the following steps:

1. For this step, we will hold destination partitions, i.e., $\mathcal{J}_q^{(t)}$, for all $1 \leq q \leq \ell$, fixed. We start with $\mathcal{I}_p^{(t+1)} := \mathcal{I}_p^{(t)}$ for all $1 \leq p \leq k$. Then, we consider each source node $i, 1 \leq i \leq n$ and move it into the $p^*$-th partition $\mathcal{I}_{p^*}^{(t+1)}$ so that the choice maximizes the "surrogate cost gain" $C_1'(A)$ of Equation (4).
2. Similar to step 1, but swapping destination nodes instead to find new partitions $\mathcal{J}_q^{(t+2)}$ for $1 \leq q \leq \ell$.
3. If there is no decrease in surrogate cost $C_1'(A)$, stop. Otherwise, set $t \leftarrow t+2$, go to step 1, and iterate.

**Fig. 4.** Source and destination node partitioning, given the number of partitions

Algorithm SPLIT:

Start with $k^0 = \ell^0 = 1$ and at each iteration $\tau$:

1. Try to increase the number of source partitions, holding the number of destination partitions fixed. We choose to split the source partition $p^*$ with maximum per-node entropy, i.e.,
$$p^* := \arg\max_{1 \leq p \leq k} \sum_{1 \leq q \leq \ell} |A_{p,q}| H(\rho(A_{p,q}))/m_p.$$
Increase the number of row partitions, $k^{\tau+1} = k^\tau + 1$ and construct a partitioning $\{\mathcal{I}_1^{(\tau+1)}, \ldots, \mathcal{I}_{k^{T+1}}^{(\tau+1)}\}$ by moving each node $i$ of the partition $\mathcal{I}_{p^*}^{(\tau)}$ that will be split into the new source partition $\mathcal{I}_{k^{T+1}}^{(\tau+1)}$, if and only if this decreases the per-node entropy of the $p^*$-th partition.
2. Apply algorithm SHUFFLE with initial state $\{\mathcal{I}_p^{(\tau+1)} \mid 1 \leq p \leq k^{\tau+1}\}$ and $\{\mathcal{J}_p^{(\tau)} \mid 1 \leq p \leq \ell^\tau\}$, to find better assignments of nodes into partitions.
3. If there is no decrease in total cost, stop and return $(k, \ell) = (k^\tau, \ell^\tau)$ with corresponding partitions. Otherwise, set $\tau \leftarrow \tau + 1$ and continue.

4–6. Similar to steps 1–3, but trying to increase destination partitions instead.

**Fig. 5.** Algorithm to find number of source and destination partitions

We recursively search for the best checkerboard partitions and stop when partitioned graph model is worse than the random graph model, which indicates the subgraph is sufficiently homogeneous. The search algorithm then proceeds in two steps: (i) an outer step, SPLIT, that attempts to progressively increase the number of source and destination partitions; and (ii) an inner step, SHUFFLE, that, given a fixed number of partitions, tries to find the best assignment of nodes into partitions. The pseudocode in Figures 4, 5, and 6 shows the steps of the overall process in detail.

*Complexity.* Algorithm SHUFFLE is linear with respect to the number of edges and the number of iterations. Algorithm SPLIT invokes SHUFFLE for each split, for a worst-case total of $2(k + \ell + 1)$ splits. For each level of the recursion in HIERARCHICAL, the total number of edges among all partitions of one level is at most equal to the number of edges in the original graph. Thus, the total time

Algorithm HIERARCHICAL:

1. Try SPLIT to find the best partitioned graph model.
2. Compare its codelength $C_1'(A)$ with that of the random graph model, $C_0(A)$.
3. If the partitioned graph model is better then, for each subgraph $(\mathcal{I}_p, \mathcal{J}_q, A_{p,q})$, for all $1 \le p \le k$ and $1 \le q \le \ell$, apply HIERARCHICAL recursively.

**Fig. 6.** Algorithm to find the cluster tree

is proportional to the total number of edges, as well as the average leaf depth and number of partitions. Section 5 presents wall-clock time comparisons of the single-level and hierarchical algorithms.

## 5 Experimental Evaluation

In this section we demonstrate our method on a number of real datasets. We compare against previous non-hierarchical/flat node partitioning schemes [4] and demonstrate that our proposed hierarchical graph decomposition provides significant benefits—in terms of revealing meaningful, interesting contexts and facilitating data exploration, summarization and compression—while still being scalable to large graphs and suitable for interactive visualization.

We implemented our algorithms in Matlab 7, with certain crucial parts (in particular, the main loop of SHUFFLE which iterates over all nodes) written in C as Matlab extensions (MEX). We have also developed a Matlab GUI[2] to facilitate navigation of the results and allow easy exploration of the clustering results.

The goal of the experiments is to show that CCT discovers intuitively meaningful subgraphs of various degrees of coarseness. We provide the evaluation from three aspects: navigation case-study, cluster quality, and method scalability. In particular we show that

- The hierarchical decomposition improves subgraph uniformity, with a reasonable number of subgraphs and an easy-to-navigate structure.
- The hierarchical decomposition achieves significantly lower total codelengths.
- Our approach is scalable to large datasets, with results progressively reported within very reasonable time.

### 5.1 Datasets

The first dataset, DBLP, consists of 8,539 source nodes representing authors that have published at least 20 papers and 3,092 destination nodes representing conferences. An edge represents that an author has published in the corresponding conference. The graph has 157,262 edges, or 0.60% non-zero entries in the adjacency matrix. The second dataset, CLASSIC, is from an information retrieval

[2] http://www.cs.cmu.edu/~spapadim/dist/hcc/

**Table 2.** All authors in the most dense "theory" context (see text). The conferences of this context are SODA, STOC, FOCS, and ICALP.

| Theory authors | | | | |
|---|---|---|---|---|
| Dhalia Malkhi | Nancy M. Amato | Yossi Matias | Monika R. Henzinger | Ronald L. Rivest |
| Joan Feigenbaum | **Robert Endre Tarjan** | Moni Naor | Michael T. Goodrich | David P. Dobkin |
| Thomas Lengauer | **Frank T. Leighton** | Jon M. Kleinberg | Ravi Kumar | Madhu Sudan |
| Nikhil Bansal | Eli Upfal | Lars Arge | Edith Cohen | Noga Alon |
| Richard Cole | Yishay Mansour | Randeep Bhatia | Sanjeev Khanna | Rajeev Motwani |
| Yonatan Aumann | Amit Kumar | Avi Widgerson | Arne Andersson | Vijaya Ramachandran |
| Micah Adler | Stefano Leonardi | Arnold Rosenberg | Gianfranco Bilardi | Ivan Hal Sudborough |
| **Haim Kaplan** | Jeffrey Scott Vitter | Cynthia Dwork | **Bhaskar Dasgupta** | Avrim Blum |
| **Michael Mitzenmacher** | **Mihalis Yannakakis** | **Anne Condon** | David R. Karger | Vwani P. Roychowdhuri |
| Richard E. Ladner | Wojiciech Szpankowski | Amihood Amir | Sampath Kannan | **Tandy Warnow** |

setting, with 3,893 source nodes representing documents and 4,303 destination nodes representing terms. The collection consists of papers from three different disciplines, medicine (MEDLINE), information retrieval (CISI) and aerodynamics (CRANFIELD), and has 176,347 edges, or 1.05% non-zero entries. The last dataset, `ENRON`, is from a social network setting, with 37,335 email addresses. Source nodes correspond to senders and destination nodes to recipients, with an edge representing the fact that the corresponding parties exchanged an email at some point in time. The graph has 367,660 edges, or 0.03% non-zero entries.

### 5.2 Navigating the Results—Case Study

We focus our intuitive explanation of results on the `DBLP` dataset, due to space constraints and also better familiarity with the domain. The dataset consists of author-conference associations from DBLP. We have kept authors with at least 20 publications. In this setting, homogeneous subgraphs consist of authors that publish in similar venues. One should also keep in mind that most researchers have worked in more than one areas over the years and thus the venues they publish in differ over time.

We start our navigation by seeking the most specific theory context. For this purpose, we choose SODA, STOC and FOCS as representative of "theory" and we seek the densest leaf subgraph that contains at least these three conferences. This leaf is three levels deep in our decomposition. Table 2 shows the complete list of 50 authors in that subgraph, which indeed consists mostly of well-known theory researchers. Additionally, our method automatically included ICALP (Intl. Colloq. on Automata, Lang. and Programming) in the set of conferences for that subgraph. The density of this cluster is approximately 90%.

**Table 3.** Theory authors in most specific context w.r.t. RECOMB. Authors common with Table 2 are highlighted in bold.

| Theory/bioinformatics authors | | | | |
|---|---|---|---|---|
| **Robert Endre Tarjan** | Xin He | Francis Y.L. Chin | **Frank T. Leighton** | **Haim Kaplan** |
| Frank Hoffmann | **Bhaskar Dasgupta** | **Tandy Warnow** | **Mihalis Yannakakis** | **Anne Condon** |
| Tao Jiang | Christos H. Papadimitriou | **Michael Mitzenmacher** | Richard M. Karp | Piotr Berman |

**Table 4.** Pair contexts

| (Rakesh Agrawal, SIGMOD) – 3rd level, 100% | | | | |
|---|---|---|---|---|
| Authors (35) | | | | Conf. (3) |
| Rakesh Agrawal | Joseph M. Hellerstein | Bruce G. Lindsay | Daniel M. Dias | SIGMOD |
| Peter J. Haas | Soumen Chakrabarti | Kenneth C. Sevcik | Luis Gravano | VLDB |
| Johannes Gehrke | Jim Gray | S. Muthukrishnan | Bettina Kemme | ICDE |
| Anastassia Ailamaki | Samuel Madden | S. Sheshadri | ... | |
| (Hari Balakrishnan, SIGCOMM) – 4th level, 58% | | | | |
| Authors (17) | | | | Conf. (3) |
| Hari Balakrishnan | Ion Stoica | Srinivasan Sheshan | Leonard Kleinrock | SIGCOMM |
| M. Frans Kaashoek | Ramesh Govindan | Mario Gerla | J.J. Garcia-Luna-Aceves | MOBICOM |
| Christophe Diot | ... | | | ICNP |

Next, we navigate up to the top-level subgraph that contains the same theory authors and seek the most specific context with respect to bioinformatics conferences. First, we chose RECOMB, which is more theory-oriented. Table 3 shows those 15 authors from that subgraph, which is also at the third level and has a density of about 40%. This leaf subgraph also includes the IEEE Conf. on Comp. Complexity, the Structure in Compl. Th. Conf. and CPM, as well as an additional 18 authors that have published in these conferences only. In general, the subgraph of theory people publishing in bioinformatics is far less dense than those subgraphs in their core areas, but still exhibits structure. Note that, since the ("theory", RECOMB) node is a sibling of the SODA-STOC-FOCS node, the list of authors in Table 3 is not a subset of the list in Table 2; authors common in both are highlighted in bold. Certain authors who have recently focused on bioinformatics are now included.

We also chose ISMB, another bioinformatics conference with less theoretical focus—hence it is in a different top-level partition than RECOMB. By navigating to the most dense leaf subgraph for the same set of theory authors, we quickly find that its density is merely 5%, with almost no theory people publishing there.

Finally, in Table 4 we show the most specific contexts for two author-conference pairs from different disciplines. We show a partial list of authors in the most specific context, due to space. The headings list the total number of author and conference nodes, as well as the level and density of the most specific subgraph.

First, from data management and mining we chose (Rakesh Agrawal, SIGMOD). Note that the the conference list automatically includes VLDB and ICDE, which are the other two main database conferences. The author list includes many previous collaborators or coworkers of Rakesh Agrawal, mostly senior, who have (co-)published in similar venues over time. Some junior people with similar publishing records are also included.

Next, from networking we chose (Hari Balakrishnan, SIGCOMM). The conference list automatically includes MOBICOM and ICNP (Intl. Conf. on Net. Protocols), as well as well-known networking researchers that have published in similar venues. Interestingly, INFOCOM is placed in a different subgraph from the highest level of the decomposition, since it has a much broader set of authors, who have also published in other areas.

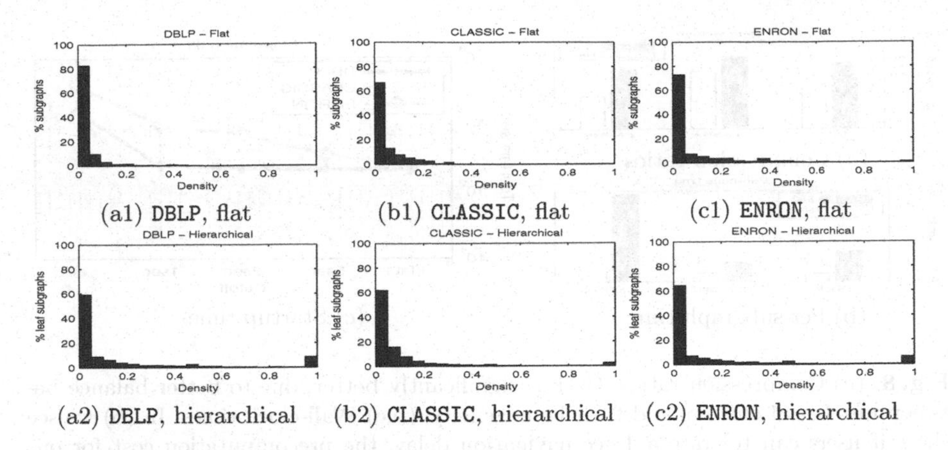


(a1) DBLP, flat (b1) CLASSIC, flat (c1) ENRON, flat

(a2) DBLP, hierarchical (b2) CLASSIC, hierarchical (c2) ENRON, hierarchical

**Fig. 7.** Subgraph density distribution: CCT gives more high-density clusters close to 1

## 5.3 Quantitative Evaluation

In this section we compare "Flat," which is the non-hierarchical, context-free approach in [4], and "Hier," which is our proposed CCT method, in terms of (i) subgraph uniformity, (ii) compression ratios, and (iii) computation time.

*Subgraph uniformity.* Figures 7(a1–c1) shows the distribution of subgraph edge densities $\rho$ of the non-hierarchical approach, whereas Figures 7(a2–c2) show the same for the leaf-level subgraphs of our CCT decomposition. As expected, the leaf-level decomposition consists of more homogeneous subgraphs. The distribution of densities is clearly more concentrated towards either zero or one. The flat decomposition occasionally fails to produce any fully connected (or even close to fully connected) subgraphs. At the same time, the number of subgraphs in CCT is still fairly reasonable, typically at most 6–7% of the number of individual edges. The increased homogeneity of the subgraphs in conjunction with the appropriate complexity of the overall decomposition also leads to good compression, as we shall see later. Finally, as the level of the decomposition increases, subgraph sizes become small in comparison to the size of the original graph, at a rate that is roughly exponential with respect to the depth. In almost all cases, the average depth is close to three, with the maximum ranging from eight to nine. As we shall see later, this is significant to allow progressive reporting of results in interactive graph exploration.

*Compression.* Figure 8a shows the compression achieved by the non-hierarchical approach and by CCT. We estimate the number of bits needed to store the original, "raw" matrix as $\lceil mnH(\rho(A)) \rceil$, i.e., we use the random graph model for the entire matrix. One can get very close to this estimate in practice by using techniques such as arithmetic coding [22]. Similarly, for the hierarchical decomposition we use the cost $C(A)$, from Equation (3). For the non-hierarchical approach, we use the cost formula from [4]. The figure shows the compression ratios for each method. It is clear from Figure 8a that CCT achieves significantly

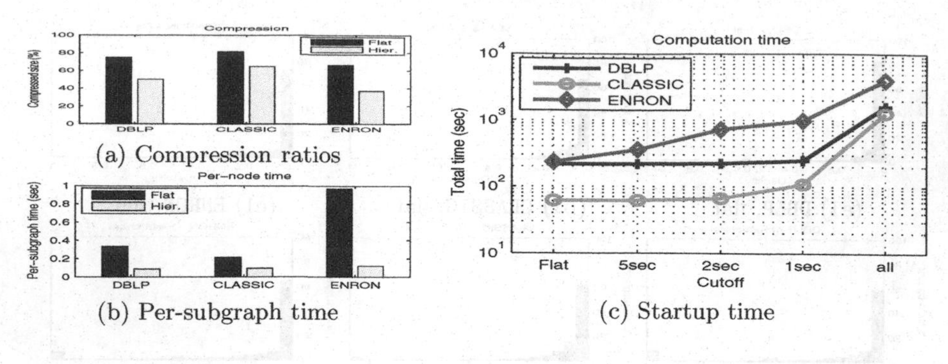


(a) Compression ratios

(b) Per-subgraph time

(c) Startup time

**Fig. 8.** (a) Compression ratios. CCT is significantly better, due to better balance between number of clusters and their homogeneity. (b,c) Wall-clock times. In (b) we see that, if users can tolerate a 1 sec navigation delay, the precomputation cost for our hierarchical approach (CCT) is almost as cheap as the context-free approach.

better compression, since it produces much more homogeneous subgraphs (see also Figure 7), while maintaining an appropriate number of partitions.

*Scalability and progressive result reporting.* Figures 8(b,c) shows measurements of wall-clock time for our prototype implementation in Matlab 7. The experiments were measured on a Pentium 4 running at 3GHz, with 2GB of memory. Figure 8b shows the total wall-clock time divided by the number of subgraphs produced by each method (previous non-hierarchical approach and CCT). As discussed before, subgraph sizes decrease quickly with respect to node depth. Thus, the processing time required to further decompose them decreases dramatically. Figure 8c shows the total wall-clock time if we were to compute: (i) just the first level of the decomposition, (ii) all nodes in the decomposition that require at least 5, 2, or 1 seconds, and (iii) all nodes at all levels of the decomposition. In an interactive graph exploration scheme, this effectively means that if we are willing to tolerate a delay of at most 5, 2 or 1 seconds when we wish to drill down a subgraph context, then the time required to pre-compute all other results would be equal to the corresponding y-axis value in Figure 8c. For example, if we are willing to tolerate at most 2 seconds "click lag," then `DBLP` requires 221 seconds (3.5 minutes) of pre-computation, `ENRON` 717 seconds (or, 12 minutes), versus 1609 seconds (27 minutes) and 3996 seconds (1 hour 7 minutes), respectively, for pre-computing everything.

## 6 Related Work

We now survey related work beyond graph mining mentioned in Section 1 [9, 16, 21, 25, 26, 27, 28].

*Biclustering.* Biclustering/co-clustering [19] simultaneously clusters both rows and columns into coherent submatrices (biclusters). Cheng and Church [5] proposed a biclustering algorithm for gene expression data analysis, using a greedy algorithm that identifies one bicluster at a time by minimizing the sum squared residue. Cho et al. [6] use a similar measure of coherence but find all the biclusters simultaneously by alternating $K$-means. Information-theoretic Co-clustering [8] uses an alternating minimization algorithm for KL-divergence between the biclusters and the original matrix. The work in [4] formulates the biclustering problem as a binary matrix compression problem and also employs a local search scheme based on alternating minimization. However, it does not study hierarchical decomposition schemes or context-specific graph analysis and exploration. More recently, streaming extensions of biclustering has been proposed in [1, 25].

There are two main distinctions between our proposed method and existing work: 1) Most existing methods require a number of parameters to be set, such as number of biclusters or minimum support. 2) Existing methods are context-free approaches, which find biclusters global to the entire dataset, while our approach is context-specific and finds communities at multiple levels. Liu et al. [18] have leveraged the existing ontology for biclustering, which assumes the hierarchy is given, while our method automatically learns hierarchy from the data.

*Hierarchical clustering.* Hierarchical clustering builds a cluster hierarchy over the data points. The basic methods are agglomerative (bottom-up) or divisive (top-down) approaches through linkage metrics [12]. Following that spirit, a number of more sophisticated methods are developed, such as CURE that takes special care of outliers [11], CHAMELEON that relies on partitioning the k-NN graph of the data [15], and BIRCH that constructs a Cluster-Feature (CF) tree to achieve good scalability [30]. All these methods are one-dimensional, in the sense that all records (rows) are clustered based on all features (columns), while our proposed method clusters both records (rows) and features (columns) into coherent and context-specific groups. Another difference is that all these methods require a number of ad-hoc parameters, while our method is completely parameter-free.

Multilevel or multigrid methods for graph clustering [2, 16] and, more remotely related, local mesh refinement techniques pioneered in [3] also employ hierarchical schemes, but still require a few parameters (e.g., density thresholds). Finally, Yang et al develop techniques for 2D image compression via hierarchical, quadtree-like partitionings, which employ heuristics that are more powerful but still require a few parameters (e.g., for deciding when to stop recursion) and, more importantly, are less scalable [29].

*Parameter-free mining.* Recently, "parameter-free" as a desirable property has received more and more attention in many places. Keogh et al. [17] developed a simple and effective scheme for mining time-series data through compression. Actually, compression or Minimum Description Language (MDL) have become the workhorse of many parameter-free algorithms: frequent itemsets [24], biclustering [4, 23], time-evolving graph clustering [25], and spatial-clustering [20].

## 7 Conclusion

In this paper we develop the Context-specific Cluster Tree (CCT) for community exploration on large bipartite graphs. CCT has the following desirable properties: **(1) Parameter-free:** CCT is automatically constructed without any user intervention, using the MDL principle. **(2) Context-Specific:** Communities are detected at multiple levels and presented depending upon what contexts are being examined. **(3) Efficiency:** CCT construction is scalable to large graphs, and the resulting CCT can provide a compressed representation of the graph and facilitate visualization. Experiments showed that both space and computational efficiency are achieved in several large real graphs. Additionally, interesting context-specific clusters are identified in the DBLP graph. Future work could focus on parallelizing the CCT computation in order to speed up the construction.

## References


[1] Aggarwal, C.C., Han, J., Wang, J., Yu, P.S.: A framework for projected clustering of high dimensional data streams. In: VLDB (2004)

[2] Bekkerman, R., El-Yaniv, R., McCallum, A.: Multi-way distributional clustering via pairwise interactions. In: ICML (2005)

[3] Berger, M.J., Oliger, J.: Adaptive mesh refinement for hyperbolic partial differential equations. J. Comp. Phys. 53 (1984)

[4] Chakrabarti, D., Papadimitriou, S., Modha, D.S., Faloutsos, C.: Fully automatic cross-assocaiations. In: KDD (2004)

[5] Cheng, Y., Church, G.M.: Biclustering of expression data. In: ISMB (2000)

[6] Cho, H., Dhillon, I.S., Guan, Y., Sra, S.: Minimum sum squared residue co-clustering of gene expression data. In: SDM (2004)

[7] Cover, T.M., Thomas, J.A.: Elements of Information Theory. Wiley, Chichester (1991)

[8] Dhillon, I.S., Mallela, S., Modha, D.S.: Information-theoretic co-clustering. In: KDD (2003)

[9] Faloutsos, M., Faloutsos, P., Faloutsos, C.: On power-law relationships of the Internet topology. In: SIGCOMM, pp. 251–262 (1999)

[10] Grünwald, P.: A tutorial introduction to the minimum description length principle. In: Advances in Min. Desc. Length: Theory and Applications. MIT Press, Cambridge (2005)

[11] Guha, S., Rastogi, R., Shim, K.: CURE: an efficient clustering algorithm for large databases. In: SIGMOD (1998)

[12] Han, J., Kamber, M.: Data Mining. Morgan Kaufmann, San Francisco (2000)

[13] Henry, N., Fekete, J.-D.: MatrixExplorer: a dual-representation system to explore social networks. In: InfoVis (2006)

[14] Holten, D.: Hierarchical edge bundles: Visualization of adjacency relations in hierarchical data. In: InfoVis (2006)

[15] Karypis, G., Han, E.-H.S., Kumar, V.: Chameleon: Hierarchical clustering using dynamic modeling. IEEE Computer 32(8) (1999)

[16] Karypis, G., Kumar, V.: Multilevel k-way partitioning scheme for irregular graphs. Journal of Parallel and Distributed Computing 48(1), 96–129 (1998)

[17] Keogh, E., Lonardi, S., Ratanamahatana, C.A.: Towards parameter-free data mining. In: KDD, pp. 206–215. ACM Press, New York (2004)
[18] Liu, J., Wang, W., Yang, J.: A framework for ontology-driven subspace clustering. In: KDD (2004)
[19] Madeira, S.C., Oliveira, A.L.: Biclustering algorithms for biological data analysis: a survey. IEEE/ACM TCBB 1, 24–45 (2004)
[20] Papadimitriou, S., Gionis, A., Tsaparas, P., Väisänen, R.A., Faloutsos, C., Mannila, H.: Parameter-free spatial data mining using MDL. In: ICDM (2005)
[21] Pei, J., Jiang, D., Zhang, A.: On mining cross-graph quasi-cliques. In: KDD (2005)
[22] Rissanen, J., Langdon Jr., G.G.: Arithmetic coding. IBM J. Res. 23(2) (1979)
[23] Rosvall, M., Bergstrom, C.T.: An information-theoretic framework for resolving community structure in complex networks. PNAS 104(18) (2007)
[24] Siebes, A., Vreeken, J., van Leeuwen, M.: Item sets that compress. In: SDM (2006)
[25] Sun, J., Papadimitriou, S., Yu, P.S., Faloutsos, C.: Graphscope: Parameter-free mining of large time-evolving graphs. In: KDD (2007)
[26] Sun, J., Qu, H., Chakrabarti, D., Faloutsos, C.: Relevance search and anomaly detection in bipartite graphs. SIGKDD Explorations 7 (2005)
[27] Sun, J., Tao, D., Faloutsos, C.: Beyond streams and graphs: Dynamic tensor analysis. In: KDD (2006)
[28] Tong, H., Faloutsos, C.: Center-piece subgraphs: problem definition and fast solutions. In: KDD (2006)
[29] Yang, Q., Lonardi, S., Melkman, A.: A compression-boosting transform for two-dimensional data. In: AAIM (2006)
[30] Zhang, T., Ramakrishnan, R., Livny, M.: Birch: An efficient data clustering method for very large databases. In: SIGMOD (1996)

# A Genetic Algorithm for Text Classification Rule Induction

Adriana Pietramala[1], Veronica L. Policicchio[1], Pasquale Rullo[1,2], and Inderbir Sidhu[3]

[1] University of Calabria Rende - Italy
{a.pietramala,policicchio,rullo}@mat.unical.it
[2] Exeura s.r.l. Rende, Italy
[3] Kenetica Ltd Chicago, IL-USA
isidhu@computer.org

**Abstract.** This paper presents a Genetic Algorithm, called Olex-GA, for the induction of rule-based text classifiers of the form "classify document $d$ under category $c$ if $t_1 \in d$ or ... or $t_n \in d$ and not ($t_{n+1} \in d$ or ... or $t_{n+m} \in d$) holds", where each $t_i$ is a term. Olex-GA relies on an efficient *several-rules-per-individual* binary representation and uses the $F$-measure as the fitness function. The proposed approach is tested over the standard test sets REUTERS-21578 and OHSUMED and compared against several classification algorithms (namely, Naive Bayes, Ripper, C4.5, SVM). Experimental results demonstrate that it achieves very good performance on both data collections, showing to be competitive with (and indeed outperforming in some cases) the evaluated classifiers.

## 1 Introduction

Text Classification is the task of assigning natural language texts to one or more thematic categories on the basis of their contents.

A number of machine learning methods have been proposed in the last few years, including $k$-nearest neighbors ($k$-NN), probabilistic Bayesian, neural networks and SVMs. Overviews of these techniques can be found in [13, 20].

In a different line, rule learning algorithms, such as [2, 3, 18], have become a successful strategy for classifier induction. Rule-based classifiers provide the desirable property of being readable and, thus, easy for people to understand (and, possibly, modify).

Genetic Algorithms (GA's) are stochastic search methods inspired to the biological evolution [7, 14]. Their capability to provide good solutions for classical optimization tasks has been demonstrated by various applications, including TSP [9, 17] and Knapsack [10]. Rule induction is also one of the application fields of GA's [1, 5, 15, 16]. The basic idea is that each *individual* encodes a candidate solution (i.e., a classification rule or a classifier), and that its fitness is evaluated in terms of predictive accuracy.

W. Daelemans et al. (Eds.): ECML PKDD 2008, Part II, LNAI 5212, pp. 188–203, 2008.

In this paper we address the problem of inducing propositional text classifiers of the form

$$c \leftarrow (t_1 \in d \vee \cdots \vee t_n \in d) \wedge \neg(t_{n+1} \in d \vee \cdots \vee t_{n+m} \in d)$$

where $c$ is a category, $d$ a document and each $t_i$ a term (n-gram) taken from a given vocabulary. We denote a classifier for $c$ as above by $\mathcal{H}_c(Pos, Neg)$, where $Pos = \{t_1, \cdots, t_n\}$ and $Neg = \{t_{n+1} \cdots t_{n+m}\}$. Positive terms in $Pos$ are used to cover the training set of $c$, while negative terms in $Neg$ are used to take precision under control.

The problem of learning $\mathcal{H}_c(Pos, Neg)$ is formulated as an optimization task (hereafter referred to as MAX-F) aimed at finding the sets $Pos$ and $Neg$ which maximize the $F$-measure when $\mathcal{H}_c(Pos, Neg)$ is applied to the training set. MAX-F can be represented as a 0-1 combinatorial problem and, thus, the GA approach turns out to be a natural candidate resolution method. We call Olex-GA the genetic algorithm designed for the task of solving problem MAX-F.

Thanks to the simplicity of the hypothesis language, in Olex-GA an individual represents a candidate classifier (instead of a single rule). The fitness of an individual is expressed in terms of the $F$-measure attained by the corresponding classifier when applied to the training set. This *several-rules-per-individual* approach (as opposed to the *single-rule-per-individual* approach) provides the advantage that the fitness of an individual reliably indicates its quality, as it is a measure of the predictive accuracy of the encoded classifier rather than of a single rule.

Once the population of individuals has been suitably initialized, evolution takes place by iterating elitism, selection, crossover and mutation, until a predefined number of generations is created.

Unlike other rule-based systems (such as Ripper) or decision tree systems (like C4.5) the proposed method is a one-step learning algorithm which does not need any post-induction optimization to refine the induced rule set. This is clearly a notable advantage, as rule set-refinement algorithms are rather complex and time-consuming tasks.

The experimentation over the standard test sets REUTERS-21578 and OHSUMED confirms the goodness of the proposed approach: on both data collections, Olex-GA showed to be highly competitive with some of the top-performing learning algorithms for text categorization, notably, Naive Bayes, C4.5, Ripper and SVM (both polynomial and rbf). Furthermore, it consistently defeated the greedy approach to problem MAX-F we reported in [19]. In addition, Olex-GA turned out to be an efficient rule induction method (e.g., faster than both C4.5 and Ripper).

In the rest of the paper, after providing a formulation of the learning optimization problem (Section 2) and proving its NP-hardness, we describe the GA approach proposed to solve it (Section 3). Then, we present the experimental results (Section 5) and provide a performance comparison with some of the top-performing learning approaches (Section 6). Finally, we briefly discuss the relation to other rule induction systems (Sections 7 and 8) and give the conclusions.

## 2 Basic Definitions and Problem Statement

In the following we assume the existence of:

1. a finite set $\mathcal{C}$ of categories, called *classification scheme*;
2. a finite set $\mathcal{D}$ of documents, called *corpus*; $\mathcal{D}$ is partitioned into a *training set* $TS$, a *validation set* and a *test set*; the training set, along with the validation set, represents the so-called *seen data*, used to induce the model, while the test set represents the *unseen data*, used to asses performance of the induced model;
3. a binary relationship which assigns each document $d \in D$ to a number of categories in $\mathcal{C}$ (*ideal classification*). We denote by $TS_c$ the subset of $TS$ whose documents belong to category $c$ (the *training set of* $c$);
4. a set $\Phi = \{f_1, \cdots, f_k\}$ of scoring functions (or feature selection functions), such as Information Gain, Chi Square, etc. [4, 22], used for vocabulary reduction (we will hereafter often refer to them simply as "functions").

A *vocabulary* $V(k, f)$ over the training set $TS$ is a set of terms (n-grams) defined as follows: let $V_c(k, f)$ be the set of the $k$ terms occurring in the documents of $TS_c$ that score highest according to the scoring function $f \in \Phi$; then, $V(k, f) = \cup_{c \in \mathcal{C}} V_c(k, f)$, that is, $V(k, f)$ is the union of the $k$ best terms, according to $f$, of each category of the classification scheme.

We consider a hypothesis language of the form

$$c \leftarrow (t_1 \in d \vee \cdots \vee t_n \in d) \wedge \neg(t_{n+1} \in d \vee \cdots \vee t_{n+m} \in d) \tag{1}$$

where each $t_i$ is a term taken from a given vocabulary. In particular, each term in $Pos = \{t_1, \cdots, t_n\}$ is a *positive* term, while each term in $Neg = \{t_{n+1} \cdots t_{n+m}\}$ is a *negative* term. A classifier as above, denoted $\mathcal{H}_c(Pos, Neg)$, states the condition "if *any* of the terms $t_1, \cdots, t_n$ occurs in $d$ and *none* of the terms $t_{n+1}, \cdots, t_{n+m}$ occurs in $d$ then classify $d$ under category $c$"[1]. That is, the occurrence of a positive term in a document $d$ requires the contextual *absence* of the (possibly empty) set of negative terms in order for $d$ be classified under $c$[2].

To assess the accuracy of a classifier for $c$, we use the classical notions of Precision $Pr_c$, Recall $Re_c$ and $F$-measure $F_{c,\alpha}$, defined as follows:

$$Pr_c = \frac{a}{a+b}; \quad Re_c = \frac{a}{a+e}; \quad F_{c,\alpha} = \frac{Pr_c \cdot Re_c}{(1-\alpha)Pr_c + \alpha Re_c} \tag{2}$$

Here $a$ is the number of *true positive* documents w.r.t. $c$ (i.e., the number of documents of the test set that have correctly been classified under $c$), $b$ the number of *false positive* documents w.r.t. $c$ and $e$ the number of *false negative* documents w.r.t. $c$. Further, in the definition of $F_{c,\alpha}$, the parameter $\alpha \in [0 \cdots 1]$

[1] It is immediate to recognize that $c \leftarrow (t_1 \in d \vee \cdots \vee t_n \in d) \wedge \neg(t_{n+1} \in d \vee \cdots \vee t_{n+m} \in d)$ is equivalent to the following set of classification rules: $\{c \leftarrow t_1 \in d \wedge \neg(t_{n+1} \in d) \wedge \cdots \wedge \neg(t_{n+m} \in d), \cdots, c \leftarrow t_n \in d \wedge \neg(t_{n+1} \in d) \wedge \cdots \wedge \neg(t_{n+m} \in d)\}$.

[2] $d$ may "satisfy" more then one; thus, it may be assigned to multiple categories.

is the relative degree of importance given to Precision and Recall; notably, if $\alpha = 1$ then $F_{c,\alpha}$ coincides with $Pr_c$, and if $\alpha = 0$ then $F_{c,\alpha}$ coincides with $Re_c$ (a value of $\alpha = 0.5$ attributes the same importance to $Pr_c$ and $Re_c$)[3].

Now we are in a position to state the following optimization problem:

**PROBLEM MAX-F.** Let a category $c \in \mathcal{C}$ and a vocabulary $V(k, f)$ over the training set $TS$ be given. Then, find two subsets of $V(k, f)$, namely, $Pos = \{t_1, \cdots, t_n\}$ and $Neg = \{t_{n+1}, \cdots, t_{n+m}\}$, with $Pos \neq \emptyset$, such that $\mathcal{H}_c(Pos, Neg)$ applied to $TS$ yields a maximum value of $F_{c,\alpha}$ (over $TS$), for a given $\alpha$.

**Proposition 1.** Problem *MAX-F* is NP-hard.

*Proof.* We next show that the decision version of problem *MAX-F* is NP-complete (i.e., it requires time exponential in the size of the vocabulary, unless $P = NP$). To this end, we restrict our attention to the subset of the hypothesis language where each rule consists of only positive terms, i.e, $c \leftarrow t_1 \in d \vee \cdots \vee t_n \in d$ (that is, classifiers are of the form $\mathcal{H}_c(Pos, \emptyset)$). Let us call *MAX-F+* this special case of problem *MAX-F*.

Given a generic set $Pos \subseteq V(k, f)$, we denote by $S \subseteq \{1, \cdots, p\}$ the set of indices of the elements of $V(k, f) = \{t_1, \cdots, t_q\}$ that are in $Pos$, and by $\Delta(t_i)$ the set of documents of the training set $TS$ where $t_i$ occurs. Thus, the set of documents classifiable under $c$ by $\mathcal{H}_c(Pos, \emptyset)$ is $D(Pos) = \cup_{i \in S} \Delta(t_i)$. The $F$-measure of this classification is

$$F_{c,\alpha}(Pos) = \frac{|D(Pos) \cap TS_c|}{(1 - \alpha)|TS_c| + \alpha|D(Pos)|}. \tag{3}$$

Now, problem *MAX-F+*, in its decision version, can be formulated as follows: "$\exists\ Pos \subseteq V(k, f)$ such that $F_{c,\alpha}(Pos) \geq K$?", where $K$ is a constant (let us call it *MAX-F+(D)*).

*Membership.* Since $Pos \subseteq V(k, f)$, we can clearly verify a YES answer of *MAX-F+(D)*, using equation 3, in time polynomial in the size $|V(k, f)|$ of the vocabulary.

*Hardness.* We define a partition of $D(Pos)$ into the following subsets: (1) $\Psi(Pos) = D(Pos) \cap TS_c$, i.e., the set of documents classifiable under $c$ by $\mathcal{H}_c(Pos, \emptyset)$ that belong to the training set $TS_c$ (true classifiable); (2) $\Omega(Pos) = D(Pos) \setminus TS_c$, i.e., the set of documents classifiable under $c$ by $\mathcal{H}_c(Pos, \emptyset)$ that do not belong to $TS_c$ (false classifiable).

Now, it is straightforward to see that the $F$-measure, given by equation 3, is proportional to the size $\psi(Pos)$ of $\Psi(Pos)$ and inversely proportional to the size $\omega(Pos)$ of $\Omega(Pos)$ (just replace in equation 3 the quantities $|D(Pos) \cap TS_c|$ by $\psi(Pos)$ and $|D(Pos)|$ by $\psi(Pos) + \omega(Pos)$). Therefore, problem *MAX-F+(D)* can equivalently be stated as follows: "$\exists\ Pos \subseteq V(k, f)$ such that $\psi(Pos) \geq V$ and $\omega(Pos) \leq C$?", where $V$ and $C$ are constants.

[3] Since $\alpha \in [0 \cdots 1]$ is a parameter of our model, we find convenient using $F_{c,\alpha}$ rather than the equivalent $F_\beta = \frac{(\beta^2+1)Pr_c \cdot Re_c}{\beta^2 Pr_c + Re_c}$, where $\beta \in [0 \cdots \infty]$ has no upper bound.

Now, let us restrict *MAX-F+(D)* to the simpler case in which terms in *Pos* are pairwise disjoint, i.e., $\Delta(t_i) \cap \Delta(t_j) = \emptyset$ for all $t_i, t_j \in V(k, f)$ (let us call $\mathcal{DSJ}$ this assumption). Next we show that, under $\mathcal{DSJ}$, *MAX-F+(D)* coincides with the Knapsack problem. To this end, we associate with each $t_i \in V(k, f)$ two constants, $v_i$ (the *value* of $t_i$) and a $c_i$ (the *cost* of $t_i$) as follows:

$$v_i = |\Delta(t_i) \cap TS_c|, \quad c_i = |\Delta(t_i) \setminus TS_c|.$$

That is, the value $v_i$ (resp. cost $c_i$) of a term $t_i$ is the number of documents containing $t_i$ that are true classifiable (resp., false classifiable).

Now we prove that, under $\mathcal{DSJ}$, the equality $\Sigma_{i \in S} v_i = \psi(Pos)$ holds, for any $Pos \subseteq V(k, f)$. Indeed:

$$\Sigma_{i \in S} v_i = \Sigma_{i \in S} |\Delta(t_i) \cap TS_c| = |(\cup_{i \in S} \Delta(t_i)) \cap TS_c| = |(D(Pos) \cap TS_c| = \psi(Pos).$$

To get the first equality above we apply the definition of $v_i$; for the second, we exploit assumption $\mathcal{DSJ}$; for the third and the fourth we apply the definitions of $D(Pos)$ and $\psi(Pos)$, respectively.

In the same way as for $v_i$, the equality $\Sigma_{i \in S} c_i = \omega(Pos)$ can be easily drawn.

Therefore, by replacing $\psi(Pos)$ and $\omega(Pos)$ in our decision problem, we get the following new formulation, valid under $\mathcal{DSJ}$: "$\exists$ $Pos \subseteq V(k, f)$ such that $\Sigma_{i \in S} v_i \geq V$ and $\Sigma_{i \in S} c_i \leq C$?". That is, under $\mathcal{DSJ}$, *MAX-F+(D)* is the Knapsack problem – a well known NP-complete problem. Therefore, *MAX-F+(D)* under $\mathcal{DSJ}$ is NP-complete. It turns out that (the general case of) *MAX-F+(D)* (which is at least as complex as *MAX-F+(D)* under $\mathcal{DSJ}$) is NP-hard and, thus, the decision version of problem *MAX-F+* is NP-hard as well. It turns out that the decision version of problem *MAX-F* is NP-hard.

Having proved both membership (in NP) and hardness, we conclude that the decision version of problem *MAX-F* is NP-complete.

A solution of problem MAX-F is a best classifier for $c$ *over the training set* $TS$, for a given vocabulary $V(f, k)$. We assume that categories in $\mathcal{C}$ are mutually independent, i.e., the classification results of a given category are not affected by the results of any other. It turns out that we can solve problem MAX-F for that category independently on the others. For this reason, in the following, we will concentrate on a single category $c \in \mathcal{C}$.

## 3 Olex-GA: A Genetic Algorithm to Solve MAX-F

Problem MAX-F is a combinatorial optimization problem aimed at finding a best combination of terms taken from a given vocabulary. That is, MAX-F is a typical problem for which GA's are known to be a good candidate resolution method.

A GA can be regarded as composed of three basic elements: (1) A *population*, i.e., a set of candidate solutions, called *individuals* or *chromosomes*, that will evolve during a number of iterations (*generations*); (2) a *fitness function* used to assign a score to each individual of the population; (3) an evolution mechanism based on operators such as *selection*, *crossover* and *mutation*. A comprehensive description of GA can be found in [7].

Next we describe our choices concerning the above points.

### 3.1 Population Encoding

In the various GA-based approaches to rule induction used in the literature (e.g., [5, 15, 16]), an individual of the population may either represent a single rule or a rule set. The former approach (single-rule-per-individual) makes the individual encoding simpler, but the fitness of an individual may not be a meaningful indicator of the quality of the rule. On the other hand, the several-rules-per-individual approach, where an individual may represent an entire classifier, requires a more sophisticated encoding of individuals, but the fitness provides a reliable indicator. So, in general, there is a trade-off between simplicity of encoding and effectiveness of the fitness function.

Thanks to the structural simplicity of the hypothesis language, in our model an individual encodes a classifier in a very natural way, thus combining the advantages of the two aforementioned approaches. In fact, an individual is simply a binary representation of the sets $Pos$ and $Neg$ of a classifier $\mathcal{H}_c(Pos, Neg)$.

Our initial approach was that of representing $\mathcal{H}_c(Pos, Neg)$ through an individual consisting of $2 \cdot |V(k, f)|$ bits, half for the terms in $Pos$, and half for the terms in $Neg$ (recall that both $Pos$ and $Neg$ are subsets of $V(k, f)$). This however proved to be very ineffective (and inefficient), as local solutions representing classifiers with hundreds of terms were generated.

More effective classifiers were obtained by restricting the search of both positive and negative terms, based on the following simple observations: (1) scoring functions are used to assess goodness of terms w.r.t. a given category; hence, it is quite reasonable searching positive terms for $c$ within $V_c(k, f)$, i.e., among the terms that score highest for $c$ according to a given function $f$; (2) the role played by negative terms in the training phase is that of avoiding classification of documents (of the training set) containing some positive term, but falling outside the training set $TS_c$ of $c$.

Based on the above remarks, we define the following subsets $Pos*$ and $Neg*$ of $V(k, f)$:

- $Pos* = V_c(k, f)$, i.e., $Pos*$ is the subset of $V(k, f)$ consisting of the $k$ terms occurring in the documents of the training set $TS_c$ of $c$ that score highest according to scoring function $f$; we say that $t_i \in Pos*$ is a *candidate positive term* of $c$ over $V(k, f)$;
- $Neg* = \{t \in V(k, f) \mid (\cup_{t_j \in Pos*} \Delta(t_j) \setminus TS_c) \cap \Delta(t) \neq \emptyset\}$, where $\Delta(t_j) \subseteq TS$ (resp. $\Delta(t) \subseteq TS$) is the set of documents of $TS$ containing term $t_j$ (resp. $t$); we say that $t_i \in Neg*$ is a *candidate negative term* of $c$ over $V(k, f)$. Intuitively, a candidate negative term is one which occurs in a document containing some candidate positive term and not belonging to the training set $TS_c$ of $c$.

Now, we represent a classifier $\mathcal{H}_c(Pos, Neg)$ over $V(k, f)$ as a chromosome $K$ made of $|Pos*| + |Neg*|$ bits. The first $|Pos*|$ bits of $K$, denoted $K^+$, are used to represent the terms in $Pos$, and the remaining $|Neg*|$, denoted $K^-$, to represent the terms in $Neg$. We denote the bit in $K^+$ (resp. $K^-$) representing the candidate positive (resp. negative) term $t_i$ as $K^+[t_i]$ (resp. $K^-[t_i]$). Thus, the value 0-1 of $K^+[t_i]$ (resp. $K^-[t_i]$) denotes whether $t_i \in Pos*$ (resp. $t_i \in Neg*$) is included in $Pos$ (resp. $Neg$) or not.

A chromosome $K$ is *legal* if there is no term $t_i \in Pos* \cap Neg*$ such that $K^+[t_i] = K^-[t_i] = 1$ (that is, $t_i$ cannot be at the same time a positive and a negative term)[4].

### 3.2 Fitness Function

The fitness of a chromosome $K$, representing $\mathcal{H}_c(Pos, Neg)$, is the value of the $F$-measure resulting from applying $\mathcal{H}_c(Pos, Neg)$ to the training set $TS$. This choice naturally follows from the formulation of problem MAX-F. Now, if we denote by $D(K) \subseteq TS$ the set of all documents containing any positive term in $Pos$ and no negative term in $Neg$, i.e.,

$$D(K) = \cup_{t \in Pos} \Delta(t) \setminus \cup_{t \in Neg} \Delta(t),$$

starting from the definition of $F_{c,\alpha}$, after some algebra, we obtain the following formula for $F_{c,\alpha}$:

$$F_{c,\alpha}(K) = \frac{|D(K) \cap TS_c|}{(1-\alpha)|TS_c| + \alpha|D(K)|}.$$

### 3.3 Evolutionary Operators

We perform selection via the roulette-wheel method, and crossover by the uniform crossover scheme. Mutation consists in the flipping of each single bit with a given (low) probability. In order not to lose good chromosomes, we apply elitism, thus ensuring that the best individuals of the current generation are passed to the next one without being altered by a genetic operator. All the above operators are described in [7].

### 3.4 The Genetic Algorithm

A description of Olex-GA is sketched in Figure 1. First, the sets $Pos*$ and $Neg*$ of candidate positive and negative terms, respectively, are computed from the input vocabulary $V(k, f)$. After population has been initialized, evolution takes place by iterating elitism, selection, crossover and mutation, until a pre-defined number $n$ of generations is created. At each step, a repair operator $\rho$, aimed at correcting possible illegal individuals generated by crossover/mutation, is applied. In particular, if $K$ is an illegal individual with $K^+[t] = K^-[t] = 1$,

[4] Note that the classifier encoded by an illegal individual is simply redundant, as it contains a "dummy" rule of the form $c \leftarrow t \in d, \cdots \neg(t \in d), \cdots$.

*Algorithm Olex-GA*

**Input**: vocabulary $V(f,k)$ over the training set $TS$; number $n$ of generations;
**Output**: "best" classifier $\mathcal{H}_c(Pos, Neg)$ of $c$ over $TS$;

- **begin**
- Evaluate the sets of candidate positive and negative terms from $V(f,k)$;
- Create the population *oldPop* and initialize each chromosome;
- **Repeat** $n$ times
- Evaluate the fitness of each chromosome in *oldPop*;
- $newPop = \emptyset$;
- Copy in *NewPop* the best $r$ chromosomes of *oldPop* (elitism - $r$ is determined on the basis of the elitism percentage)
- **While** *size*(*newPop*) < *size*(*oldPop*)
- select *parent1* and *parent2* in *oldPop* via roulette wheel
- generate *kid1*, *kid2* through *crossover*(*parent*1,*parent*2)
- apply mutation, i.e., $kid1 = mut(kid1)$ and $kid2 = mut(kid2)$
- apply the repair operator $\rho$ to both *kid*1 and *kid*2;
- add *kid1* and *kid2* to *newPop*;
- **end-while**
- $oldPop = newPop$;
- **end-repeat**;
- Select the best chromosome $K$ in *oldPop*;
- Eliminate redundancies from $K$;
- **return** the classifier $\mathcal{H}_c(Pos, Neg)$ associated with $K$.

**Fig. 1.** Evolutionary Process for category $c$

the application of $\rho$ to $K$ simply set $K^+[t] = 0$ (thus, the redundant rule $c \leftarrow t \in d, \cdots \neg(t \in d), \cdots$ is removed from the encoded classifier). After iteration completes, the elimination of *redundant* bits/terms from the best chromosome/classifier $K$ is performed. A bit $K[t] = 1$ is redundant in a chromosome $K$, representing $\mathcal{H}_c(Pos, Neg)$, if the chromosome $K'$ obtained from $K$ by setting $K[t] = 0$ is such that $F_{c,\alpha}(K) = F_{c,\alpha}(K')$. This may happen if either $t \in Pos$ and $\Delta(t) \subseteq \cup_{t' \in (Pos - \{t\})} \Delta(t')$ (i.e., $t$ does not provide any contribution to the classification of new documents) or $t \in Neg$ and $\Delta(t) \cap \cup_{t' \in Pos} \Delta(t') = \emptyset$ (i.e., $t$ does not contribute to remove any misclassified document).

## 4 Learning Process

While Olex-GA is the search of a "best" classifier for category $c$ *over the training set*, for *a given* input vocabulary, the learning process is the search of a "best" classifier for $c$ *over the validation set*, for *all* input vocabularies. More precisely, the learning process proceeds as follows:

- *Learning*: for each input vocabulary
  - *Evolutionary Process*: execute a predefined number of runs of Olex-GA over the *training set*, according to Figure 1;
  - *Validation*: run over the *validation set* the best chromosome/classifier generated by Olex-GA;
- *Testing*: After all runs have been executed, pick up the best classifier (over the validation set), and assess its accuracy over the *test set* (unseen data).

The output of the learning process is assumed to be the "best classifier" of $c$.

# 5 Experimentation

## 5.1 Benchmark Corpora

We have experimentally evaluated our method using both the REUTERS-21578 [12] and and the OHSUMED [8] test collections.

The REUTERS-21578 consists of 12,902 documents. We have used the Mod-Apté split, in which 9,603 documents are selected for training (*seen data*) and the other 3,299 form the test set (*unseen data*). Of the 135 categories of the TOPICS group, we have considered the 10 with the highest number of positive training examples (in the following, we will refer to this subset as R10). We remark that we used *all* 9603 documents of the training corpus for the learning phase, and performed the test using *all* 3299 documents of the test set (including those not belonging to any category in R10).

The second data set we considered is OHSUMED, in particular, the collection consisting of the first 20,000 documents from the 50,216 medical abstracts of the year 1991. Of the 20,000 documents of the corpus, the first 10,000 were used as *seen data* and the second 10,000 as *unseen data*. The classification scheme consisted of the 23 MeSH *diseases*.

## 5.2 Document Pre-processing

Preliminarily, documents were subjected to the following pre-processing steps: (1) First, we removed all words occurring in a list of common stopwords, as well as punctuation marks and numbers; (2) then, we extracted all n-grams, defined as sequences of maximum three words consecutively occurring within a document (after stopword removal)[5]; (3) at this point we have randomly split the set of *seen data* into a *training set* (70%), on which to run the GA, and a *validation set* (30%), on which tuning the model parameters. We performed the split in such a way that each category was proportionally represented in both sets (stratified holdout); (4) finally, for each category $c \in \mathcal{C}$, we scored all n-grams occurring in the documents of the training set $TS_c$ by each scoring function $f \in \{CHI, IG\}$, where IG stands for Information Gain and CHI for Chi Square [4, 22].

[5] Preliminary experiments showed that n-grams of length ranging between 1 and 3 perform slightly better than single words.

**Table 1.** Best classifiers for categories in R10 and micro-averaged performance

| Cat. | Input ($\alpha = 0.5$) | | Training (GA) | Test | | Classifier | |
|---|---|---|---|---|---|---|---|
| | $f$ | $k$ | $F$-meas | $F$-meas | BEP | #pos | #neg |
| earn | CHI | 80 | 93.91 | 95.34 | 95.34 | 36 | 9 |
| acq | CHI | 60 | 83.44 | 87.15 | 87.49 | 22 | 35 |
| mon-fx | CHI | 150 | 79.62 | 66.47 | 66.66 | 33 | 26 |
| grain | IG | 60 | 91.05 | 91.61 | 91.75 | 9 | 8 |
| crude | CHI | 70 | 85.08 | 77.01 | 77.18 | 26 | 14 |
| trade | IG | 80 | 76.77 | 61.80 | 61.81 | 20 | 11 |
| interest | CHI | 150 | 73.77 | 64.20 | 64.59 | 44 | 23 |
| wheat | IG | 50 | 92.46 | 89.47 | 89.86 | 8 | 6 |
| ship | CHI | 30 | 87.88 | 74.07 | 74.81 | 17 | 33 |
| corn | IG | 30 | 93.94 | 90.76 | 91.07 | 4 | 12 |
| $\mu$-BEP | | | | 86.35 | 86.40 | | |

### 5.3 Experimental Setting

We conducted a number of experiments for inducing the best classifiers for both REUTERS-21578 and OHSUMED according to the learning process described in Section 4. To this end, we used different input vocabularies $V(k, f)$, with $k \in \{10, 20, \cdots, 90, 100, 150, 200\}$ and $f \in \{CHI, IG\}$. For each input vocabulary, we ran Olex-GA three times. The parameter $\alpha$ of the fitness function $F_{c,\alpha}$ was set to 0.5 for all the experiments (thus attributing equal importance to Precision and Recall).

In all cases, the population size has been held at 500 individuals, the number of generations at 200, the crossover rate at 1.0, the mutation rate at 0.001 and the elitism probability at 0.2.

For each chromosome $K$ in the population, we initialized $K^+$ at random, while we set $K^-[t] = 0$, for each $t \in Neg*$ (thus, $K$ initially encodes a classifier $\mathcal{H}_c(Pos, \emptyset)$ with no negative terms).

Initialization, as well as all the above mentioned GA parameters, have been empirically set based on a preliminary experimentation.

### 5.4 Experimental Results

The first data set we considered is REUTERS-21578 . Table 1 shows the performance of the best classifiers for the categories of R10. Here, for each classifier, we report: (1) the values of $f$ and $k$ of the input vocabulary $V(f, k)$; (2) the $F$-measure over the training set (i.e., the fitness of the best individual); (3) F-measure and Break Even Point (BEP) over the test set ("Test"); (4) the number of positive and negative terms occurring in the classifier ("Classifier").

From results of Table 1 it clearly appears that the evolutionary model is sensitive to the input parameters, as each category achieves its maximum performance with different values of $k$ and $f$. For an instance, the best classifier of category "earn" is learned from a vocabulary $V(k,f)$ where $k = 80$ and $f = CHI$; the respective $F$-measure and breakeven point are both 95.34. Looking at the last row, we see that the micro-averaged values of $F$-measure and BEP over the test set are equal to 86.35 and 86.40, respectively. Finally, looking at the last two columns labelled "classifier", we see that the induced classifiers are rather compact: the maximum number of positive terms is 44 ("interest"), and the minimum is 4 ("corn"); likewise, the maximum number of negative terms is 35 ("acq") and the minimum is 6 ("wheat").

As an example, the classifier $\mathcal{H}_c(Pos, Neg)$ for "corn" has

$$Pos = \{\text{"corn"}, \text{"maize"}, \text{"tonnes maize"}, \text{"tonnes corn"}\}$$

$$Neg = \{\text{"jan"}, \text{"qtr"}, \text{"central bank"}, \text{"profit"}, \text{"4th"}, \text{"bonds"}, \text{"pact"}, \text{"offering"}, \text{"monetary"}, \text{"international"}, \text{"money"}, \text{"petroleum"}\}.$$

Thus, $\mathcal{H}_c(Pos, Neg)$ is of the following form:

$$c \leftarrow \text{"corn"} \vee \cdots \vee \text{"tonnes corn"} \;\; \wedge \neg \, (\text{"jan"} \vee \cdots \vee \text{"petroleum"}).$$

To show the sensitivity of the model to parameter settings, in Table 2 we report the F-measure values (on the validation set) for category "corn" obtained by using different values for $\phi$ and $\nu$. As we can see, the performance of the induced classifiers varies from a minimum of 84.91 to a maximum of 93.58 (note that, for both functions, the $F$-measure starts decreasing for $\nu > 50$, i.e., a reduction of the vocabulary size provides a benefit in terms of performance [22]).

**Table 2.** Effect of varying $\phi$ and $\nu$ on the F-measure for category "corn"

| | $\nu$ | | | | |
|---|---|---|---|---|---|
| $\phi$ | 10 | 50 | 100 | 150 | 200 |
| $CHI$ | 90.74 | 91.00 | 88.7 | 88.68 | 84.91 |
| $IG$ | 92.04 | 93.58 | 91.59 | 91.59 | 89.09 |

The second data set we considered is OHSUMED . In Table 3 we provide the best performance for the ten most frequent Mesh categories and micro-averaged performance over all 23. In particular, micro-averaged $F$-measure and BEP (over the test set) are both equal to 62.30. Also in this case, with a few exceptions, classifiers are rather compact.

## 5.5 Time Efficiency

Experiments have been run on a 2.33 GHz Xeon 4 Gb RAM. The average execution time of the evolutionary process of Figure 1 is around 10 seconds per category for both data sets (recall that in both cases the population is made of 500 individuals which evolve for 200 generations).

**Table 3.** Best classifiers for the ten most frequent MeSH "diseases" categories of Ohsumed and micro-averaged performance over all 23

| Cat. | Input ($\alpha = 0.5$) | | Training (GA) | Test | | Classifier | |
|---|---|---|---|---|---|---|---|
| name | $f$ | $k$ | $F$-meas | $F$-meas | BEP | #pos | #neg |
| C23 | CHI | 100 | 54.93 | 50.67 | 51.39 | 25 | 13 |
| C14 | CHI | 70 | 79.67 | 75.46 | 75.61 | 32 | 7 |
| C04 | CHI | 100 | 80.79 | 80.18 | 80.36 | 33 | 7 |
| C10 | CHI | 70 | 61.63 | 54.60 | 54.65 | 24 | 25 |
| C06 | IG | 100 | 73.51 | 68.69 | 68.80 | 95 | 3 |
| C21 | CHI | 200 | 79.17 | 62.11 | 62.22 | 113 | 6 |
| C20 | IG | 80 | 75.64 | 70.98 | 71.12 | 35 | 4 |
| C12 | CHI | 100 | 81.59 | 73.60 | 74.28 | 55 | 7 |
| C08 | IG | 30 | 69.63 | 63.99 | 64.07 | 18 | 24 |
| C01 | CHI | 100 | 76.14 | 64.32 | 64.46 | 93 | 3 |
| avg-BEP (top 10) | | | | 66.46 | 66.96 | | |
| $\mu$-BEP (all 23) | | | | 62.30 | 62.30 | | |

## 6 Performance Comparisons

In this section we compare Olex-GA with four classical learning algorithms: SVM (both, polynomial and radial basis function - rbf), Ripper (with two optimization steps), C4.5 and Naive Bayes (NB). Further, we make a comparison with the greedy approach to problem MAX-F we presented in our previous work [19] (hereafter called Olex-Greedy).

To determine the performance of the above algorithms (apart from Olex-Greedy) we utilized the Weka library of machine learning algorithms, version 3.5.6 [21].

In order to make results really comparable, documents of both corpora were preliminarily pre-processed exactly as described in Subsection 5.2. Then, for each of the above learners, we carried out a number of experiments over the training set, using the validation set for tuning the model parameters. In particular, we ran all methods with vocabularies consisting of 500, 1000, 2000, 5000 terms. For each vocabulary, polynomial SVM was run with degree $d$ ranging from 1 to 5, while rbf SVM was executed with variance $\gamma \in \{0.2, 0.4, 0.6, 0.8, 1.0, 1.2\}$. Once completed all experiments, we selected the best classifier (over the validation set) of *each* category, i.e., the one with the maximum value of the $F$-measure. Finally, we used the test set to evaluate the performance of the best classifiers and the micro-averaged accuracy measures[6].

REUTERS-21578 . The performance results of the tested algorithms are reported in Table 4. In particular, for each algorithm, we provide the BEP over

[6] To attain more reliable estimations of the differences in predictive accuracy of different methods, n-fold cross-validation along with statistical significance tests should be performed. We envision this task as part of our future work.

**Table 4.** Recall/Precision breakeven points on R10

| Category | NB | C4.5 | Ripper | SVM | | Olex | |
|---|---|---|---|---|---|---|---|
| | | | | poly | rbf | Greedy | GA |
| earn | 96.61 | 95.77 | 95.31 | 97.32 | 96.57 | 93.13 | **95.34** |
| acq | 90.29 | 85.59 | 86.63 | 90.37 | 90.83 | 84.32 | **87.49** |
| money | 56.67 | 63.08 | 62.94 | 72.89 | 68.22 | 68.01 | **66.66** |
| grain | 77.82 | 89.69 | 89.93 | 92.47 | 88.94 | 91.28 | **91.75** |
| crude | 78.84 | 82.43 | 81.07 | 87.82 | 86.17 | 80.84 | **77.18** |
| trade | 57.90 | 70.04 | 75.82 | 77.77 | 74.14 | 64.28 | **61.81** |
| interest | 61.71 | 52.93 | 63.15 | 68.16 | 58.71 | 55.96 | **64.59** |
| wheat | 71.77 | 91.46 | 90.66 | 86.13 | 89.25 | 91.46 | **89.86** |
| ship | 68.65 | 71.92 | 75.91 | 82.66 | 80.40 | 78.49 | **74.81** |
| corn | 59.41 | 86.73 | 91.79 | 87.16 | 84.74 | 89.38 | **91.07** |
| $\mu$-BEP | 82.52 | 85.82 | 86.71 | 89.91 | 88.80 | 84.80 | **86.40** |
| learning times (min) | 0.02 | 425 | 800 | 46 | 696 | 2.30 | **116** |

the test set of each category in R10, the micro-avg BEP and the overall learning time. Concerning predictive accuracy, with a $\mu$-BEP of 86.40, our method surpasses Naive Bayes (82.52), which shows the worst behavior, and Olex-Greedy (84.80); is competitive with C4.5 (85.82) and Ripper (86.71), while performs worse than both SVM's (poly = 89.91, rbf = 88.80). Concerning efficiency, NB (0.02 min) and Olex-Greedy (2.30 min) are by far the fastest methods. Then poly SVM (46 min) and Olex-GA (116 min), followed at some distance by C4.5 (425 min), rbf SVM (696 min) and Ripper (800).

OHSUMED . As we can see from Table 5, with a $\mu$-BEP = 62.30, the proposed method is the top-performer. On the other side, C4.5 shows to be the worst

**Table 5.** Recall/Precision breakeven points on the ten most frequent MeSH diseases categories of OHSUMED and micro-averaged performance over all 23

| Category | NB | C4.5 | Ripper | SVM | | Olex | |
|---|---|---|---|---|---|---|---|
| | | | | poly | rbf | Greedy | GA |
| C23 | 47.59 | 41.93 | 35.01 | 45.00 | 44.21 | 47.32 | **51.39** |
| C14 | 77.15 | 73.79 | 74.16 | 73.81 | 75.34 | 74.52 | **75.61** |
| C04 | 75.71 | 76.22 | 80.05 | 78.18 | 76.65 | 77.78 | **80.36** |
| C10 | 45.96 | 44.88 | 49.73 | 52.22 | 51.54 | 54.72 | **54.65** |
| C06 | 65.19 | 57.47 | 64.99 | 63.18 | 65.10 | 63.25 | **68.80** |
| C21 | 54.92 | 61.68 | 61.42 | 64.95 | 62.59 | 61.62 | **62.22** |
| C20 | 68.09 | 64.72 | 71.99 | 70.23 | 66.39 | 67.81 | **71.12** |
| C12 | 63.04 | 65.42 | 70.06 | 72.29 | 64.78 | 67.82 | **74.28** |
| C08 | 57.70 | 54.29 | 63.86 | 60.40 | 55.33 | 61.57 | **64.07** |
| C01 | 58.36 | 48.89 | 56.05 | 43.05 | 52.09 | 55.59 | **64.46** |
| avg-BEP (top 10) | 61.37 | 58.92 | 62.73 | 62.33 | 61.40 | 62.08 | **66.69** |
| $\mu$-BEP (all 23) | 57.75 | 55.14 | 59.65 | 60.24 | 59.57 | 59.38 | **62.30** |
| learning times (min) | 0.04 | 805 | 1615 | 89 | 1100 | 6 | **249** |

performer (55.14) (so confirming the findings of [11]). Then, in the order, Naive Bayes (57.75), rbf SVM (59.57), Ripper (59.65), polynomial SVM (60.24) and Olex-Greedy (61.57). As for time efficiency, the OHSUMED results essentially confirm the hierarchy coming out from the REUTERS-21578 .

## 7 Olex-GA vs. Olex-Greedy

One point that is noteworthy is the relationship between Olex-Greedy and Olex-GA, in terms of both predictive accuracy and time efficiency.

Concerning the former, we have seen that Olex-GA consistently beats Olex-Greedy on both data sets. This confirms Freitas' findings [6], according to which effectiveness of GA's in rule induction is a consequence of their inherent ability to cope with *attribute interaction* as, thanks to their global search approach, more attributes at a time are modified and evaluated as a whole. This in contrast with the local, one-condition-at-a-time greedy rule generation approach.

On the other hand, concerning time efficiency, Olex-Greedy showed to be much faster than Olex-GA. This should not be surprising, as the greedy approach, unlike GA's, provides a search strategy which straight leads to a suboptimal solution.

## 8 Relation to Other Inductive Rule Learners

Because of the computational complexity of the learning problem, all real systems employ heuristic search strategies which prunes vast parts of the hypothesis space. Conventional inductive rule learners (e.g, RIPPER [3]) usually adopt, as their general search method, a covering approach based on a separate-and-conquer strategy. Starting from an empty rule set, they learn a set of rules, one by one. Different learners essentially differ in how they find a single rule. In RIPPER, the construction of a single rule is a two-stage process: a greedy heuristics constructs an initial rule set (IREP*) and, then, one or more optimization phases improve compactness and accuracy of the rule set.

Also decision tree techniques, e.g., C4.5 [18], rely on a two-stage process. After the decision tree has been transformed into a rule set, C4.5 implements a pruning stage which requires more steps to produce the final rule set - a rather complex and time consuming task.

In contrast, Olex-GA, like Olex-Greedy and other GA-based approaches (e.g., [5]), relies on a single-step process whereby an "optimal" classifier, i.e., one consisting of few high-quality rules, is learned. Thus, no pruning strategy is needed, with a great advantage in terms of efficiency. This may actually account for the time results of Tables 4 and 5, which show the superiority of Olex-GA w.r.t. both C4.5 and Ripper.

## 9 Conclusions

We have presented a Genetic Algorithm, Olex-GA, for inducing rule-based text classifiers of the form "if a document $d$ includes either term $t_1$ or ... or term $t_n$, but not term $t_{n+1}$ and ... and not term $t_{n+m}$, then classify $d$ under category $c$".

Olex-GA relies on a simple binary representation of classifiers (several-rules-per-individual approach) and uses the $F$-measure as the fitness function. One design aspect related to the encoding of a classifier $\mathcal{H}_c(Pos, Neg)$ was concerned with the choice of the length of individuals (i.e., the size of the search space). Based on preliminary experiments, we have restricted the search of $Pos$ and $Neg$ to suitable subsets of the vocabulary (instead of taking it entirely), thus getting more effective and compact classifiers.

The experimental results obtained on the standard data collections REUTERS-21578 and OHSUMED show that Olex-GA quickly converges to very accurate classifiers. In particular, in the case of OHSUMED , it defeats all the other evaluated algorithms. Further, on both data sets, Olex-GA consistently beats Olex-Greedy. As for time efficiency, Olex-GA is slower than Olex-Greedy but faster than the other rule learning methods (i.e., Ripper and C4.5).

We conclude by remarking that we consider the experiments reported in this paper somewhat preliminary, and feel that performance can further be improved through a fine-tuning of the GA parameters.

## References

1. Alvarez, J.L., Mata, J., Riquelme, J.C.: Cg03: An oblique classification system using an evolutionary algorithm and c4.5. International Journal of Computer, Systems and Signals 2(1), 1–15 (2001)
2. Apté, C., Damerau, F.J., Weiss, S.M.: Automated learning of decision rules for text categorization. ACM Transactions on Information Systems 12(3), 233–251 (1994)
3. Cohen, W.W., Singer, Y.: Context-sensitive learning methods for text categorization. ACM Transactions on Information Systems 17(2), 141–173 (1999)
4. Forman, G.: An extensive empirical study of feature selection metrics for text classification. Journal of Machine Learning Research 3, 1289–1305 (2003)
5. Freitas, A.A.: A genetic algorithm for generalized rule induction. In: Advances in Soft Computing-Engineering Design and Manufacturing, pp. 340–353. Springer, Heidelberg (1999)
6. Freitas, A.A.: In: Klosgen, W., Zytkow, J. (eds.) Handbook of Data Mining and Knowledge Discovery, ch. 32, pp. 698–706. Oxford University Press, Oxford (2002)
7. Goldberg, D.E.: Genetic Algorithms in Search, Optimization and Machine Learning. Addison-Wesley, Reading (1989)
8. Hersh, W., Buckley, C., Leone, T., Hickman, D.: Ohsumed: an interactive retrieval evaluation and new large text collection for research. In: Croft, W.B., van Rijsbergen, C.J. (eds.) Proceedings of SIGIR-1994, 17th ACM International Conference on Research and Development in Information Retrieval, Dublin, IE, pp. 192–201. Springer, Heidelberg (1994)
9. Homaifar, A., Guan, S., Liepins, G.E.: Schema analysis of the traveling salesman problem using genetic algorithms. Complex Systems 6(2), 183–217 (1992)

10. Hristakeva, M., Shrestha, D.: Solving the 0/1 Knapsack Problem with Genetic Algorithms. In: Midwest Instruction and Computing Symposium 2004 Proceedings (2004)
11. Joachims, T.: Text categorization with support vector machines: learning with many relevant features. In: Nédellec, C., Rouveirol, C. (eds.) ECML 1998. LNCS, vol. 1398, pp. 137–142. Springer, Heidelberg (1998)
12. Lewis, D.D.: Reuters-21578 text categorization test collection. Distribution 1.0 (1997)
13. Lewis, D.D., Hayes, P.J.: Guest editors' introduction to the special issue on text categorization. ACM Transactions on Information Systems 12(3), 231 (1994)
14. Michalewicz, Z.: Genetic Algorithms+Data Structures=Evolution Programs, 3rd edn. Springer, Heidelberg (1999)
15. Noda, E., Freitas, A.A., Lopes, H.S.: Discovering interesting prediction rules with a genetic algorithm. In: Proc. Congress on Evolutionary Computation (CEC-1999), July 1999. IEEE, Washington (1999)
16. Pei, M., Goodman, E.D., Punch, W.F.: Pattern discovery from data using genetic algorithms. In: Proc. 1st Pacific-Asia Conf. Knowledge Discovery and Data Mining (PAKDD-1997), Febuary 1997. World Scientific, Singapore (1997)
17. Jung, Y., Jog, S.P., van Gucht, D.: Parallel genetic algorithms applied to the traveling salesman problem. SIAM Journal of Optimization 1(4), 515–529 (1991)
18. Quinlan, J.R.: Generating production rules from decision trees. In: Proc. of IJCAI-1987, pp. 304–307 (1987)
19. Rullo, P., Cumbo, C., Policicchio, V.L.: Learning rules with negation for text categorization. In: Proc. of SAC - Symposium on Applied Computing, Seoul, Korea, March 11-15 2007, pp. 409–416. ACM, New York (2007)
20. Sebastiani, F.: Machine learning in automated text categorization. ACM Computing Surveys 34(1), 1–47 (2002)
21. Witten, I.H., Frank, E.: Data Mining: Practical machine learning tools and techniques, 2nd edn. Morgan Kaufmann, San Francisco (2005)
22. Yang, Y., Pedersen, J.O.: A comparative study on feature selection in text categorization. In: Fisher, D.H. (ed.) Proceedings of ICML-97, 14th International Conference on Machine Learning, Nashville, US, pp. 412–420. Morgan Kaufmann Publishers, San Francisco (1997)

# Nonstationary Gaussian Process Regression Using Point Estimates of Local Smoothness

Christian Plagemann[1], Kristian Kersting[2], and Wolfram Burgard[1]

[1] University of Freiburg, Georges-Koehler-Allee 79, 79110 Freiburg, Germany
{plagem,burgard}@informatik.uni-freiburg.de
[2] Fraunhofer Institute IAIS, Sankt Augustin, Germany
kristian.kersting@iais.fraunhofer.de

**Abstract.** Gaussian processes using nonstationary covariance functions are a powerful tool for Bayesian regression with input-dependent smoothness. A common approach is to model the local smoothness by a latent process that is integrated over using Markov chain Monte Carlo approaches. In this paper, we demonstrate that an approximation that uses the estimated mean of the local smoothness yields good results and allows one to employ efficient gradient-based optimization techniques for jointly learning the parameters of the latent and the observed processes. Extensive experiments on both synthetic and real-world data, including challenging problems in robotics, show the relevance and feasibility of our approach.

## 1 Introduction

Gaussian processes (GPs) have emerged as a powerful yet practical tool for solving various machine learning problems such as nonlinear regression or multi-class classification [16]. As opposed to making parametric assumptions about the underlying functions, GP models are directly expressed in terms of the training data and thus belong to the so called nonparametric methods. Their increasing popularity is due to the fact that nonlinear problems can be solved in a principled Bayesian framework for learning, model selection, and density estimation while the basic model just requires relatively simple linear algebra. A common assumption when specifying a GP prior is stationarity, i.e., that the covariance between function values $f(\mathbf{x})$ and $f(\mathbf{x}')$ only depends on the distance $\|\mathbf{x}-\mathbf{x}'\|$ between the indexes and not on their locations directly. Consequently, standard GPs lack the ability to adapt to variable smoothness in the function of interest.

Modeling input-dependent smoothness, however, is essential in many fundamental problems in the geo-sciences, mobility mining, activity recognition, and robotics, among other areas. Consider, for example, the problem of modeling terrain surfaces given sets of noisy elevation measurements. Accurately modeling such data requires the ability to deal with a varying data density and to balance smoothing against the preservation of discontinuities. Discontinuities arise for instance at steps, stairs, curbs, or at walls. These features are important for planning paths of mobile robots, for estimating traversability, and in terrain

W. Daelemans et al. (Eds.): ECML PKDD 2008, Part II, LNAI 5212, pp. 204–219, 2008.

segmentation tasks. Accordingly, the ability to flexibly adapt a regression model to the local properties of the data may greatly enhance the applicability of such techniques.

In the past, several approaches for specifying nonstationary GP models haven been proposed in the literature [12,13]. A particularly promising approach is due to Paciorek and Schervish [8] who proposed to explicitly model input-depending smoothness using additional, latent GPs. This approach (a) provides the user with a continuous latent space of local kernels, (b) allows the user to analyze the estimated covariance function yielding important insights into the problem domain, and (c) fully stays in the GP framework so that computational methods for speeding up GP inference and fitting can be adapted.

Paciorek and Schervish provide a flexible and general framework based on MCMC integration, which unfortunately – as also noted by the authors – is computationally demanding for large data sets and which is thus not feasible in the real world situations that are typically encountered in robotics and engineering tasks. In this paper, we present a simple approximation that does not integrate over all latent values but uses the predicted mean values only. Specifically, we parameterize the nonstationary covariances using a second GP with $m$ latent length-scales. Assuming $m \ll n$, where $n$ is the number of training points, this results in a nonstationary GP regression method with practically no overhead compared to standard GPs. More importantly, using point estimates naturally leads to gradient-based techniques for efficiently learning the parameters of both processes jointly, which is the main contribution of this paper.

We present experiments carried out on on synthetic and real-world data sets from challenging application domains such as robotics and embedded systems showing the relevance and feasibility of our approach. More specifically, our nonstationary GP approach significantly outperforms standard GPs in terms of prediction accuracy, while it is significantly faster then [8]. We regard these empirical results as an additional substantial contribution of this paper as they tighten the link between advanced regression techniques based on GPs and application domains such as robotics and embedded systems. To the best of our knowledge, it is the first time that nonstationary GPs have been learned in a principled way in these challenging domains.

This paper is organized as follows. After reviewing related work, we introduce nonstationary Gaussian processes regression and how to make predictions in Section 3. In Section 4, we then show how to learn the hyperparameters using gradient-based methods. Before concluding, we demonstrate the feasibility and relevance of our approach in an extensive set of experiments.

## 2 Related Work

Gaussian process models [11] have the advantage of providing predictive uncertainties for their estimates while not requiring, for instance, a fixed discretization of the space. This has led to their successful application in a wide range of application areas including robotics and ubiquitous computing. For example, Schwaighofer *et al.* [14] applied the model for realizing positioning systems using cellular networks. GPs have been proposed as measurement models [1] and for model-based failure detection [10] in robotics because they naturally deal with noisy measurements, unevenly distributed observations, and fill small gaps in the data with high confidence while assigning higher predictive uncertainty in sparsely sampled areas. Many robotics applications, however, call for non-standard GP models. Kersting *et al.* [5], for example, have shown that heteroscedastic GP regression, i.e., regression with input-dependent noise outperforms state-of-the-art approaches in mobile robot localization. Whereas they also use a GP prior to model local noise rates, they do not estimate the hyperparameters jointly using gradient-based optimization but alternate each process in a sampling-based EM fashion. Lang *et al.* [6] modeled 3D terrain data using nonstationary GPs by following the approach of Paciorek and Schervish [8]. They derived a specific adaptation procedure that also does not require MCMC integration as originally proposed by Paciorek and Schervish, but that is not derived from first principles. Another approach to modeling nonstationarity is to use ensembles of GPs, where every GP is assigned to a specific region, an idea akin to GP mixture models such as presented by Williams' [16]. A related idea has also been proposed by Pfingsten *et al.* [9]. Cornford *et al.* [3] model straight discontinuities in wind fields by placing auxiliary GPs along the edge on both sides of the discontinuity. They are then used to learn GPs representing the process on either side of the discontinuity.

Apart from Paciorek and Schervish's [8] (see also the references in there) approach of directly modeling the covariance function using additional latent GPs, several other approaches for specifying nonstationary GP models can be found in the literature. For instance, Sampson and Guttorp [12] map a nonstationary spatial process (not based on GPs) into a latent space, in which the problem becomes approximately stationary. Schmidt and O'Hagan [13] followed this idea and used GPs to implement the mapping. Similar in spirit, Pfingsten *et al.* [9] proposed to augment the input space by a latent extra input to tear apart regions of the input space that are separated by abrupt changes of the function values. All GP approaches proposed so far, however, followed a Markov chain Monte Carlo approach to inference and learning. Instead, we present a novel maximum-a-posterior treatment of Paciorek and Schervish's approach that fully stays in the GP framework, explicitly models the covariance function, provides continuous estimates of the local kernels, and that naturally allows for gradient-based joint optimization of its parameters.

## 3 Nonstationary Gaussian Process Regression

The nonlinear regression problem is to recover a functional dependency $y_i = f(\mathbf{x}_i) + \epsilon_i$ from $n$ observed data points $\mathcal{D} = \{(\mathbf{x}_i, y_i)\}_{i=1}^n$, where $y_i \in \mathbb{R}$ are the noisy observed outputs at input locations $\mathbf{x}_i \in \mathbb{R}^d$. Throughout this paper we will also use $\mathbf{X} \in \mathbb{R}^{n \times d}$ to refer to all input locations. For the sake of simplicity, we will concentrate on one-dimensional outputs, but all formulas naturally

**Table 1.** Notation used to derive the gradient of the model selection criterion w.r.t. the joint hyperparameters $\boldsymbol{\theta}$ of the nonstationary GP

| | |
|---|---|
| Observed GP | $\mathcal{GP}_y$ |
| Hyperparameters of $\mathcal{GP}_y$ | $\boldsymbol{\theta}_y = \langle \sigma_f, \sigma_n \rangle$ |
| Training set | $\mathcal{D} = \langle \mathbf{X}, \mathbf{y} \rangle, \mathbf{X} \in \mathbb{R}^{n \times d}, \mathbf{y} \in \mathbb{R}^n$ |
| Prediction | $y^* \in \mathbb{R}$ at location $\mathbf{X}^* \in \mathbb{R}^{1 \times d}$ |
| Latent length-scale process | $\mathcal{GP}_\ell$ |
| Latent length-scale support values | $\overline{\boldsymbol{\ell}} \in \mathbb{R}^m$ at locations $\overline{\mathbf{X}} \in \mathbb{R}^{m \times d}$ |
| Latent length-scales at training points of $\mathcal{GP}_y$ | $\boldsymbol{\ell} \in \mathbb{R}^n$ at locations $\mathbf{X}$ |
| Latent length-scale at test point | $\ell^* \in \mathbb{R}$ at location $\mathbf{X}^*$ |
| Hyperparameters of $\mathcal{GP}_\ell$ | $\boldsymbol{\theta}_\ell = \langle \overline{\sigma}_f, \overline{\sigma}_\ell, \overline{\sigma}_n \rangle$ |
| Joint hyperparameters | $\boldsymbol{\theta} = \langle \boldsymbol{\theta}_y, \boldsymbol{\theta}_\ell, \overline{\boldsymbol{\ell}} \rangle = \langle \sigma_f, \sigma_n, \overline{\sigma}_f, \overline{\sigma}_\ell, \overline{\sigma}_n, \overline{\boldsymbol{\ell}} \rangle$ |

generalize to the multidimensional case. The regression task is to learn a model for $p(y^*|\mathbf{x}^*, \mathcal{D})$, i.e., the predictive distribution of new target values $y^*$ at $\mathbf{x}^*$ given $\mathcal{D}$. The notation we will use is listed in Table 1.

**Stationary Gaussian Process Regression:** In the standard Gaussian process model for regression (STD-GP), we assume independent, normally distributed noise terms $\epsilon_i \sim \mathcal{N}(0, \sigma_n^2)$ with a constant noise variance parameter $\sigma_n^2$. The central idea is to model every finite set of samples $y_i$ from the process as *jointly* Gaussian distributed, such that the predictive distribution $p(y^*|\mathbf{x}^*, \mathcal{D})$ at arbitrary query points $\mathbf{x}^*$ is a Gaussian distribution $\mathcal{N}(\mu, \sigma^2)$ with mean

$$\mu = \mathbf{k}_{\mathbf{x}^*,\mathbf{x}}^T (\mathbf{K}_{\mathbf{x},\mathbf{x}} + \sigma_n^2 \mathbf{I})^{-1} \mathbf{y} \tag{1}$$

and variance

$$\sigma^2 = k(\mathbf{x}^*, \mathbf{x}^*) - \mathbf{k}_{\mathbf{x}^*,\mathbf{x}}^T (\mathbf{K}_{\mathbf{x},\mathbf{x}} + \sigma_n^2 \mathbf{I})^{-1} \mathbf{k}_{\mathbf{x}^*,\mathbf{x}} + \sigma_n^2 \,. \tag{2}$$

Here, we have $\mathbf{K}_{\mathbf{x},\mathbf{x}} \in \mathbb{R}^{n \times n}$ with $\mathbf{K}_{\mathbf{x},\mathbf{x}}(i,j) = k(\mathbf{x}_i, \mathbf{x}_j)$, $\mathbf{k}_{\mathbf{x}^*,\mathbf{x}} \in \mathbb{R}^n$ with $\mathbf{k}_{\mathbf{x}^*,\mathbf{x}}(i) = k(\mathbf{x}^*, \mathbf{x}_i)$, $\mathbf{y} = (y_1, \ldots, y_n)^T$, and $\mathbf{I}$ the identity matrix.

An integral part of GP regression is the covariance function $k(\cdot,\cdot)$, which specifies the covariance of the corresponding targets (see [11] for more details). A common choice is the squared exponential covariance function $k_{se}(\mathbf{x}, \mathbf{x}') = \sigma_f^2 \exp\left(-1/2 \cdot (s(\mathbf{x}, \mathbf{x}')/\sigma_\ell)^2\right)$ with $s(\mathbf{x}, \mathbf{x}') = \|\mathbf{x} - \mathbf{x}'\|$. The term $\sigma_f$ denotes the amplitude (or signal variance) and $\sigma_\ell$ is the characteristic length-scale. The parameters $\boldsymbol{\theta}_y = (\sigma_f, \sigma_\ell, \sigma_n)$ are called the *hyperparameters* of the process. Note that we – as opposed to some other authors – also treat the global noise rate $\sigma_n^2$ as a hyperparameter for ease of notation.

**Modeling Input-Dependent Smoothness:** A limitation of the standard GP framework as described above is the assumption of constant length-scales $\sigma_\ell$ over the whole input space. Intuitively, length-scales define the extent of the area in which observations strongly influence each other. For 3D terrain modeling,

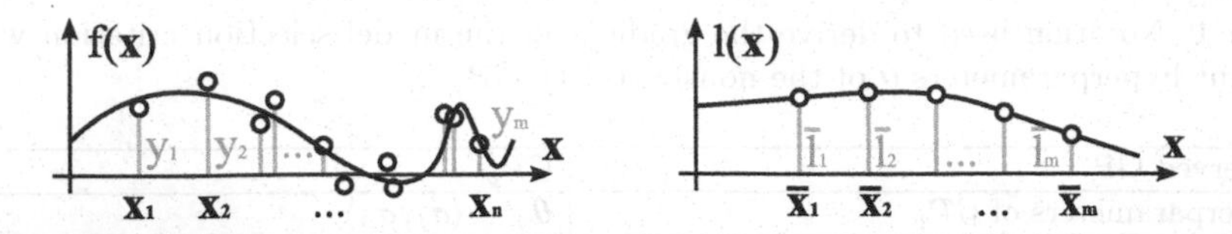


**Fig. 1.** Placing a GP prior over the latent length-scales for nonstationary GP regression. An observed Gaussian process $\mathcal{GP}_y$ is sketched on left-hand side and the latent $\mathcal{GP}_\ell$ governing the local length-scales is shown on the right-hand side.

for instance, within the context of mobile robot localization, one would like to use locally varying length-scales to account for the different situations. For example in flat plains, the terrain elevations are strongly correlated over long distances. In high-variance, "wiggly" terrain, on the other hand and at strong discontinuities, the terrain elevations are correlated over very short distances only, if at all. To address this problem of varying correlation scale, an extension of the squared exponential (SE) covariance function was proposed by Paciorek and Schervish [8], which takes the form

$$k(\mathbf{x}_i, \mathbf{x}_j) = \sigma_f^2 \, |\Sigma_i|^{\frac{1}{4}} \, |\Sigma_j|^{\frac{1}{4}} \left| \frac{\Sigma_i + \Sigma_j}{2} \right|^{-\frac{1}{2}} \cdot \exp\left[ -\mathbf{d}_{ij}^T \left( \frac{\Sigma_i + \Sigma_j}{2} \right)^{-1} \mathbf{d}_{ij} \right] , \quad (3)$$

where $\mathbf{d}_{ij} = (\mathbf{x}_i - \mathbf{x}_j)$. The intuition is that each input location $\mathbf{x}'$ is assigned a local Gaussian kernel matrix $\Sigma'$ and the covariance between two targets $y_i$ and $y_j$ is calculated by averaging between the two local kernels at the input locations $\mathbf{x}_i$ and $\mathbf{x}_j$. In this way, the local characteristics at both locations influence the modeled covariance of the corresponding target values. For the sake of simplicity, we consider the isotropic case only in this paper. The general case can be treated in the same way. In the isotropic case, where the eigenvectors of the local kernels are aligned to the coordinate axes and their eigenvalues are equal, the matrices $\mathbf{\Sigma}_i$ simplify to $\ell_i^2 \cdot \mathbf{I}_n$ with a real-valued length-scale parameter $\ell_i$. In the one-dimensional case, for instance, Eq. (3) then simplifies to

$$k(x_i, x_j) = \sigma_f^2 \cdot (\ell_i^2)^{\frac{1}{4}} \cdot (\ell_j^2)^{\frac{1}{4}} \cdot \left( \frac{1}{2}\ell_i^2 + \frac{1}{2}\ell_j^2 \right)^{-\frac{1}{2}} \cdot \exp\left[ -\frac{(x_i - x_j)^2}{\frac{1}{2}\ell_i^2 + \frac{1}{2}\ell_j^2} \right] . \quad (4)$$

We do not specify a functional form for the length-scale $\ell(x)$ at location $x$ but place a GP prior over them. More precisely, an independent GP is used to model the logarithms $\log(\ell(x))$ of the length-scales, to avoid negative values. This process, denoted as $\mathcal{GP}_\ell$ is governed by a different covariance function specified by the hyperparameters $\boldsymbol{\theta}_\ell = \langle \overline{\sigma}_f, \overline{\sigma}_\ell, \overline{\sigma}_n \rangle$. Additionally, we have to maintain the set of $m$ support values $\overline{\boldsymbol{\ell}}$ as part of $\boldsymbol{\theta}$ as depicted in Figure 1.

**Making Predictions:** In the extended model, we now have to integrate over all possible latent length-scales to get the predictive distribution

$$p(y^*|\mathbf{X}^*, \mathcal{D}, \boldsymbol{\theta}) = \iint p(y^*|\mathbf{X}^*, \mathcal{D}, \exp(\boldsymbol{\ell}^*), \exp(\boldsymbol{\ell}), \boldsymbol{\theta}_y) \cdot p(\boldsymbol{\ell}^*, \boldsymbol{\ell}|\mathbf{X}^*, \mathbf{X}, \overline{\boldsymbol{\ell}}, \overline{\mathbf{X}}, \boldsymbol{\theta}_\ell) \, d\boldsymbol{\ell} \, d\boldsymbol{\ell}^*$$

of a regressand $y^*$ at location $\mathbf{X}^*$ given a dataset $\mathcal{D}$ and hyperparameters $\boldsymbol{\theta}$ (Note that we explicitly highlight here that $\mathcal{GP}_\ell$ is defined over the log length-scales). Because this marginalization is intractable, [8] apply MCMC to approximate it. Instead, we seek for the solution using the most probable length-scale estimates, i.e., $p(y^*|\mathbf{X}^*, \mathcal{D}, \boldsymbol{\theta}) \approx p(y^*|\mathbf{X}^*, \exp(\ell^*), \exp(\boldsymbol{\ell}), \mathcal{D}, \boldsymbol{\theta}_y)$ where $(\ell^*, \boldsymbol{\ell})$ are the mean predictions of the length-scale process at $\mathbf{X}^*$ and the locations in $\mathcal{D}$. Since the length-scales are independent latent variables in the combined regression model, making predictions amounts to making two standard GP predictions using Eqs. (1) and (2), one using $\mathcal{GP}_\ell$ to get $(\ell^*, \boldsymbol{\ell})$ and one using $\mathcal{GP}_y$ with $(\ell^*, \boldsymbol{\ell})$ treated as fixed parameters.

# 4 Learning Hyperparameters

So far, we have described our model assuming that we have the joint hyperparameters $\boldsymbol{\theta}$ of the overall process. In practice, we are unlikely to have these parameters a-priori and, instead, we wish to estimate them from observations $\mathbf{y}$.

Assume a given set of $n$ observations $\mathbf{y}$ at locations $\mathbf{X}$. We seek to find those hyperparameters that maximize the probability of observing $\mathbf{y}$ at $\mathbf{X}$, i.e., we seek to maximize $p(\mathbf{y}|\mathbf{X}, \boldsymbol{\theta}) = \int p(\mathbf{y}|\mathbf{X}, \boldsymbol{\ell}, \boldsymbol{\theta}_y) \cdot p(\boldsymbol{\ell}|\mathbf{X}, \overline{\boldsymbol{\ell}}, \overline{\mathbf{X}}, \boldsymbol{\theta}_\ell)\, d\boldsymbol{\ell}$ . As for making predictions, such a marginalization is intractable. Instead, we seek to make progress by seeking a solution that maximizes the a-posteriori probability of the latent length-scales

$$\log p(\boldsymbol{\ell}|\mathbf{y}, \mathbf{X}, \boldsymbol{\theta}) = \log p(\mathbf{y}|\mathbf{X}, \exp(\boldsymbol{\ell}), \boldsymbol{\theta}_y) + \log p(\boldsymbol{\ell}|\mathbf{X}, \overline{\boldsymbol{\ell}}, \overline{\mathbf{X}}, \boldsymbol{\theta}_\ell) + \text{const.}\,, \qquad (5)$$

where, again, the $\boldsymbol{\ell}$ are the mean predictions of $\mathcal{GP}_\ell$. The gradient of this objective function w.r.t. to hyperparameters $\boldsymbol{\theta}$ or a subset of them can be employed within gradient-based optimization to find the corresponding solution. In our experiments, we optimized $\overline{\sigma}_f$, $\overline{\sigma}_n$, and $\overline{\sigma}_\ell$ of the latent kernel width process in an outer cross-validation loop on an independent validation set and assumed $\partial L(\boldsymbol{\theta})/\partial \bullet = 0$, where $\bullet$ denotes one of them, within the inner gradient optimization. The locations $\overline{\mathbf{X}}$ of the latent kernel width variables were sampled uniformly on the bounding rectangle given by $\mathbf{X}$.

In the following, we will detail the objective function and the gradient of it with respect to the hyperparameter.

## 4.1 The Objective Function

We maximize the marginal likelihood (5) of the data with respect to the joint hyperparameters as well as the support values $\overline{\boldsymbol{\ell}}$ of the length-scale process. The first term in this equation is the standard objective function for Gaussian processes

$$\log p(\mathbf{y}|\mathbf{X}, \exp(\boldsymbol{\ell}), \boldsymbol{\theta}_y) = -\frac{1}{2}\mathbf{y}^T(\mathbf{K}_{\mathbf{x},\mathbf{x}} + \sigma_n^2\mathbf{I})^{-1}\mathbf{y} - \frac{1}{2}\log|\mathbf{K}_{\mathbf{x},\mathbf{x}} + \sigma_n^2\mathbf{I}| - \frac{n}{2}\log(2\pi)\,,$$

where $|\mathbf{M}|$ denotes the determinant of a matrix $\mathbf{M}$ and $\mathbf{K}_{\mathbf{x},\mathbf{x}}$ stands for the noise-free nonstationary covariance matrix for the training locations $\mathbf{X}$ that will be detailed below. Our point estimate approach considers the most likely latent length-scales $\boldsymbol{\ell}$, i.e. the mean predictions of $\mathcal{GP}_\ell$ at locations $\mathbf{X}$. Thus, the second term of Eq. (5) has the form

$$\log p(\boldsymbol{\ell}|\mathbf{X}, \overline{\boldsymbol{\ell}}, \overline{\mathbf{X}}, \boldsymbol{\theta}_\ell) = -\frac{1}{2}\log|\mathbf{K}_{\overline{\mathbf{x}},\overline{\mathbf{x}}} + \overline{\sigma}_n^2\mathbf{I}| - \frac{n}{2}\log(2\pi)\,.$$

Putting both together, we get the objective function

$$L(\boldsymbol{\theta}) = \log p(\boldsymbol{\ell}|\mathbf{y}, X, \boldsymbol{\theta}) = c_1 + c_2 \cdot \left[\mathbf{y}^T\mathbf{A}^{-1}\mathbf{y} + \log|\mathbf{A}| + \log|\mathbf{B}|\right]\,, \tag{6}$$

where $c_1$ and $c_2$ are real-valued constants, and $\mathbf{A} := \mathbf{K}_{\mathbf{x},\mathbf{x}} + \sigma_n^2\mathbf{I}$ and $\mathbf{B} := \mathbf{K}_{\overline{\mathbf{x}},\overline{\mathbf{x}}} + \overline{\sigma}_n^2\mathbf{I}$ are covariance matrices. The noise-free part of the nonstationary covariance matrix $\mathbf{K}_{\mathbf{x},\mathbf{x}}$ is calculated according to Eq. (3). As mentioned above, we consider the isotropic case only for the sake of simplicity. We express Eq. (4) for the case of multivariate inputs $\mathbf{x}_i$ using the compact matrix-vector notation suggested in [2]. Recalling that $\boldsymbol{\ell}$ represents the local length-scales at the training locations $\mathbf{X}$, we get

$$\mathbf{K}_{\mathbf{x},\mathbf{x}} = \sigma_f^2 \cdot \mathbf{P_r}^{\frac{1}{4}} \circ \mathbf{P_c}^{\frac{1}{4}} \circ (1/2)^{-\frac{1}{2}}\,\mathbf{P_s}^{-\frac{1}{2}} \circ \mathbf{E} \tag{7}$$

with

$$\mathbf{P_r} = \mathbf{p}\cdot\mathbf{1}_n^T\,, \quad \mathbf{P_c} = \mathbf{1}_n^T\cdot\mathbf{p}^T\,, \quad \mathbf{p} = \boldsymbol{\ell}^T\boldsymbol{\ell}\,,$$
$$\mathbf{P_s} = \mathbf{P_r} + \mathbf{P_c}\,, \quad \mathbf{E} = \exp[-s(\mathbf{X}) \div \mathbf{P_s}]\,, \quad \boldsymbol{\ell} = \exp\left[\overline{\mathbf{K}}_{\mathbf{x},\overline{\mathbf{x}}}^T\left[\mathbf{K}_{\overline{\mathbf{x}},\overline{\mathbf{x}}} + \overline{\sigma}_n^2\mathbf{I}\right]^{-1}\overline{\boldsymbol{\ell}}\right].$$

Note that $\mathbf{p} \in \mathbb{R}^n$ and, thus, $\mathbf{P_r}$ and $\mathbf{P_c}$ are matrices built using the outer vector product. Here, $s(\mathbf{X})$ calculates the $n \times n$ matrix of squared distances between the input vectors $\mathbf{x}$ contained in $\mathbf{X}$. $\circ$ and $\div$ denote element-wise multiplication and division respectively and matrix exponentiation $\mathbf{M}^\alpha$ is also defined element-wise for $\alpha \neq -1$. In the same notation, the covariance function for the latent length-scale process $\mathcal{GP}_\ell$ becomes (in the stationary squared exponential form)

$$\mathbf{K}_{\overline{\mathbf{x}},\overline{\mathbf{x}}} = \overline{\sigma}_f{}^2 \cdot \exp\left[-\frac{1}{2}s(\overline{\sigma_\ell}^{-2}\overline{\mathbf{X}})\right]$$

and, analogously, for making predictions within $\mathcal{GP}_\ell$

$$\mathbf{K}_{\mathbf{x},\overline{\mathbf{x}}} = \overline{\sigma}_f{}^2 \cdot \exp\left[-\frac{1}{2}s(\overline{\sigma_\ell}^{-2}\mathbf{X}, \overline{\sigma_\ell}^{-2}\overline{\mathbf{X}})\right]\,.$$

### 4.2 The Gradient

Using standard results from matrix calculus, the partial derivative of the objective (6) w.r.t. an element $\bullet$ of $\boldsymbol{\theta}$ turns out to be

$$\frac{\partial L(\boldsymbol{\theta})}{\partial\bullet} = -\mathbf{y}^T\mathbf{A}^{-1}\frac{\partial\mathbf{A}}{\partial\bullet}\mathbf{A}^{-1}\mathbf{y} + \mathrm{tr}(\mathbf{A}^{-1}\frac{\partial\mathbf{A}}{\partial\bullet}) + \mathrm{tr}(\mathbf{B}^{-1}\frac{\partial\mathbf{B}}{\partial\bullet})\,, \tag{8}$$

where $\mathrm{tr}(\mathbf{M})$ is the trace of a matrix $\mathbf{M}$. For the two hyperparameters of $\mathcal{GP}_y$ we get the straight-forward results

$$\frac{\partial \mathbf{A}}{\partial \sigma_n} = 2\sigma_n \mathbf{I}\,, \qquad \frac{\partial \mathbf{B}}{\partial \sigma_n} = 0\,, \qquad \frac{\partial \mathbf{A}}{\partial \sigma_f} = 2\sigma_f \mathbf{K}_{\mathbf{x},\mathbf{x}}\ , \qquad \frac{\partial \mathbf{B}}{\partial \sigma_f} = 0\,.$$

The case $\bullet = \overline{\boldsymbol{\ell}}$ yields $(\partial \mathbf{B}/\partial \overline{\boldsymbol{\ell}}) = 0$ and $(\partial \mathbf{A})/(\partial \overline{\boldsymbol{\ell}}) = (\partial \mathbf{K}_{\mathbf{x},\mathbf{x}})/(\partial \overline{\boldsymbol{\ell}}) =$

$$\sigma_f^2\ (1/2)^{-\frac{1}{2}} \cdot \left[ \left( \frac{\partial(\mathbf{P_r}^{\frac{1}{4}})}{\partial \overline{\boldsymbol{\ell}}} \circ \mathbf{P_c}^{\frac{1}{4}} \circ \mathbf{P_s}^{-\frac{1}{2}} \circ \mathbf{E} \right) + \left( \mathbf{P_r}^{\frac{1}{4}} \circ \frac{\partial(\mathbf{P_c}^{\frac{1}{4}})}{\partial \overline{\boldsymbol{\ell}}} \circ \mathbf{P_s}^{-\frac{1}{2}} \circ \mathbf{E} \right) + \right.$$

$$\left. \left( \mathbf{P_r}^{\frac{1}{4}} \circ \mathbf{P_c}^{\frac{1}{4}} \circ \frac{\partial(\mathbf{P_s}^{-\frac{1}{2}})}{\partial \overline{\boldsymbol{\ell}}} \circ \mathbf{E} \right) + \left( \mathbf{P_r}^{\frac{1}{4}} \circ \mathbf{P_c}^{\frac{1}{4}} \circ \mathbf{P_s}^{-\frac{1}{2}} \circ \frac{\partial(\mathbf{E})}{\partial \overline{\boldsymbol{\ell}}} \right) \right] .$$

The remaining simplifications can be achieved by substitution with the definitions given after Eq. (7) and by applying general rules for differentiation such as the chain rule

$$\frac{\partial f(g(\mathbf{X}))}{\partial \mathbf{x}} = \frac{\partial (f(\mathbf{U}):)}{\partial \mathbf{U}} \cdot \frac{\partial g(\mathbf{x})}{\partial \mathbf{x}} \Bigg|_{\mathbf{U}=g(\mathbf{X})}$$

where $\mathbf{X}:$ denotes the vectorization of a matrix by stacking its columns, e.g., as applied to a term containing element-wise division

$$\frac{\partial (\mathbf{A} \div \mathbf{B})}{\partial \mathbf{x}} = \mathbf{A} \circ \frac{\partial\ \mathrm{inv}(\mathbf{U}):}{\partial \mathbf{U}:} \cdot \frac{\partial \mathbf{B}:}{\partial \mathbf{x}} \Bigg|_{\mathbf{U}=\mathbf{B}}$$

for a matrix $\mathbf{A}$ that does not depend on $\mathbf{x}$. Substituting the resulting partial derivatives in Eq. (8) yields the gradient $\partial L(\boldsymbol{\theta})/\partial \boldsymbol{\theta}$, which can be used in gradient-based optimization techniques, such as Møller's [7] scaled conjugate gradients (SCG), to jointly optimize the hyperparameters of $\mathcal{GP}_y$ and $\mathcal{GP}_\ell$.

## 5 Experiments

The goal of our experimental evaluation was to investigate to which extent the point estimate approach to nonstationary GP regression is able to handle input-dependent smoothness and to quantify the gains relative to the stationary model. Specifically, we designed several experiments to investigate whether the approach can solve standard regression problems from the literature. We also applied it to two hard and relevant regression problems from embedded systems and robotics. On the two standard test sets, we demonstrate that the prediction accuracy of our approach is comparable to the one achieved by the MCMC-based method by Paciorek and Schervish [8], which, compared to our algorithm, is substantially more demanding regarding the computational resources.

We have implemented and evaluated our approach in Matlab[1]. Using the compact matrix notation for all derivations, the core algorithm is implemented in less than 150 lines of code and, more importantly, advanced optimization strategies like sparse matrix approximations or parallelization can be realized with virtually no additional implementation efforts. As optimization procedure,

[1] The source code and data sets are available at `http://robreg.org/projects/lagp`

we applied Møller's scaled conjugate gradient (SCG) [7] approach. In all our experiments, the SCG converged after at most 20 iterations. To quantitatively evaluate the performance of our nonstationary regression technique, we ran 30 to 50 independent test runs for each of the following test cases. Each run consisted of (a) randomly selecting or generating training data, (b) fitting the nonstationary model, and (c) evaluating the predictive distribution of the learned model at independent test locations. The latter was done either using the known ground truth function values or by assessing the likelihood of independent observations in the cases in which the ground truth was not known (e.g., for the RFID and terrain mapping experiments).

In all test scenarios, we evaluate the accuracy of the mean predictions and also the fit of the whole predictive distribution using the **standardized mean squared error**

$$\text{sMSE} := n^{-1} \sum\nolimits_{i=1}^{n} \text{var}(y)^{-1} (y_i - m_i)^2$$

and the **negative log predictive density**

$$\text{NLPD} := n^{-1} \sum\nolimits_{i=1}^{n} \log p_{\text{model}}(y_i | \mathbf{x}_i)$$

respectively. Here, $\{(\mathbf{x}_i, y_i)\}_{i=1}^{n}$ denotes the test data set, $p_{\text{model}}(\cdot|\mathbf{x}_i)$ stands for the predictive distribution at location $\mathbf{x}_i$, and $m_i := \mathbb{E}[p_{\text{model}}(\cdot|\mathbf{x}_i)]$ denotes the predicted mean. Statistical significance was assessed using two-sample t-tests with 95% confidence intervals.

All experiments were conducted using Matlab on a Linux desktop PC with a single 2 GHz CPU. The typical runtime for fitting the full nonstationary model to 100 training points was in the order of 50 seconds. The runtime requirements of the MCMC-based approach [8] which does not employ any gradient information were reported to be in the order of hours for a C-implementation on standard hardware in year 2004. In the following, we term our nonstationary approach as LA-GP (Locally Adaptive GP), the standard model employing the isotropic, squared exponential covariance function as STD-GP and Paciorek and Schervish's MCMC-based approach as NS-GP (Nonstationary GP).

### 5.1 Simulation Results in 1D and 2D

First, we verified that our approach accurately solves standard regression problems described in the literature. To this aim, we considered the two simulated functions shown in Figure 2. Both functions were also used for evaluation purposes by Dimatteo *et al.* [4] and in [8]. In the remainder, these test scenarios will be referred to as SMOOTH-1D and JUMP-1D. Whereas SMOOTH-1D is a smoothly varying function with a substantial "bump" close to 0, JUMP-1D has a sharp jump at 0.4. For SMOOTH-1D, we sampled 101 training points and 400 test points from the interval $(-2, 2)$. In the case of JUMP-1D, we sampled 111 training points and 111 for testing from $(0, 1)$. Table 2 gives the results for theses experiments (averaged over 50 independent runs). Additionally, this table

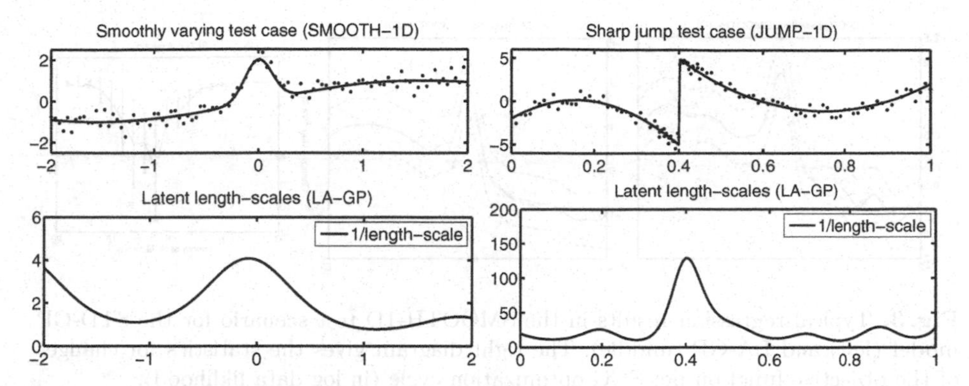


**Fig. 2.** Two standard nonstationary test cases SMOOTH-1D (top left) and JUMP-1D (top right) that were used for evaluation purposes in previous work [4] and [8]. The lower two plots give the inverse latent length-scales as optimized by our approach. Higher values in these plots indicate a larger local frequency.

contains results for a two-dimensional simulated function NONSTAT-2D, which is described further below in this sub-section.

The results can be summarized as follows: with respect to the sMSE, the accuracy of our approach is comparable to the MCMC-based method of Paciorek and Schervish. Note that values given here were taken from their publication [8]. Both approaches significantly ($\alpha$=0.05) outperform standard GPs. Our approach also provides a significantly better performance compared to standard GPs with respect to the NLPD. For a visual comparison of the regression results, consider the left two diagrams in Figure 3. Whereas the standard GP (left plot) – having a constant length-scale for the whole domain – cannot adapt to all local properties well, our LA-GP accurately fits the bump and also the smoother parts (center plot). It should be noted that LA-GP tends to assign higher frequencies to the border regions of the training set, since there is less constraining data there compared to the center regions (see also the lower left plot in Figure 2).

**Table 2.** Quantitative evaluation of the proposed nonstationary approach (LA-GP) and the standard Gaussian process (STD-GP) as well as the MCMC-based approach of [8] (NS-GP). We compare the prediction accuracies using the negative log predictive density (NLPD) and the standardized mean squared errors (sMSE), see text. Results marked by • differ significantly ($\alpha = 0.05$) from the others in their category.

| Test Scenario | NLPD LA-GP | NLPD STD-GP | sMSE LA-GP | sMSE STD-GP | sMSE NS-GP [8] |
|---|---|---|---|---|---|
| SMOOTH-1D | -1.100 | -1.026 (•) | 0.0156 | 0.021 (•) | 0.015 |
| JUMP-1D | -0.375 | -0.440 (•) | 0.0268 | 0.123 (•) | 0.026 |
| NONSTAT-2D | -3.405 | -3.315 (•) | 0.0429 | 0.0572 (•) | - |

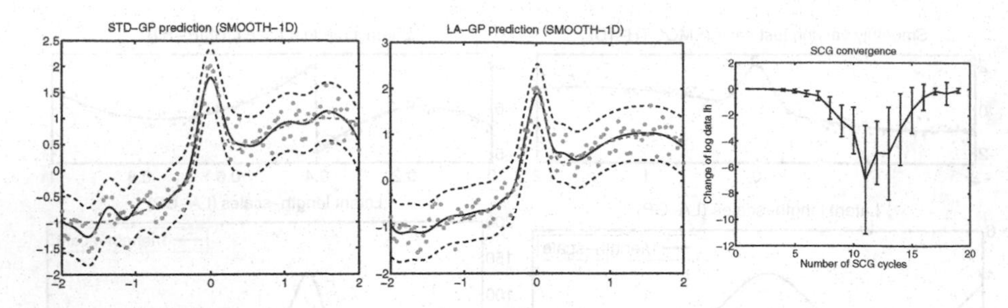


**Fig. 3.** Typical regression results in the SMOOTH-1D test scenario for the STD-GP model (left) and LA-GP (middle). The right diagram gives the statistics for changes of the objective function per SCG optimization cycle (in log data liklihood).

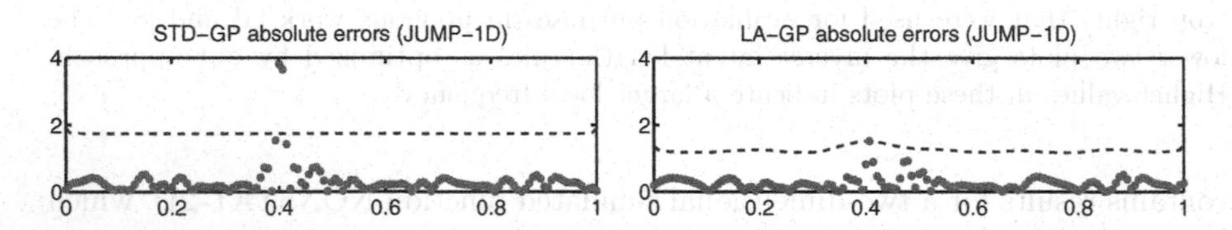


**Fig. 4.** Absolute distances of the test points from the predicted means in one run of the JUMP-1D scenario using the STD-GP model (left) and LA-GP (right). The model confidence bounds (2 standard deviations of the predictive distribution) are given by dashed lines.

The right diagram of Figure 3 provides statistics about the individual gains during the SCG cycles for 50 independent test runs. As can be seen from this plot, after about 20 cycles the objective function, which corresponds to the negative log data likelihood, does not change notably any more. ‘ Figure 4 compares the confidence bounds of the different regression models to the actual prediction errors made. It can be seen that the LA-GP model has more accurate bounds. It should be noted that the predictive variance of the STD-GP model depends only on the local data density and not on the target values and, thus, it is constant in the non-border regions.

We give the absolute average errors of the mean predictions in the different test cases in Figure 5. To highlight the more accurate confidence bounds of the LA-GP model, we also give the statistics for the 50% most confident predictions.

In addition to the two one-dimensional standard test cases, we evaluated the performance or our approach on a bivariate function (NONSTAT-2D). In particular, we simulated observations $y(x_1, x_2) \sim f(x_1, x_2) + \mathcal{N}(0, 0.025)$ using the noise-free bivariate function $f(x_1, x_2) = 1/10 \cdot (\sin(x_1\, b(x_1, x_2)) + \sin(x_2\, b(x_1, x_2))$ and the underlying bandwidth function $b(x_1, x_2) = \pi\ (2x_1 + 0.5x_2 + 1)$. This function and typical observations are depicted in the left diagram of Figure 6. During training, we sampled $11 \cdot 11 = 121$ points from a uniform distribution over $[-0.5, 1] \times [-0.5, 1]$ and corresponding simulated observations (the latter were drawn independently for each run). For testing, we uniformly sampled

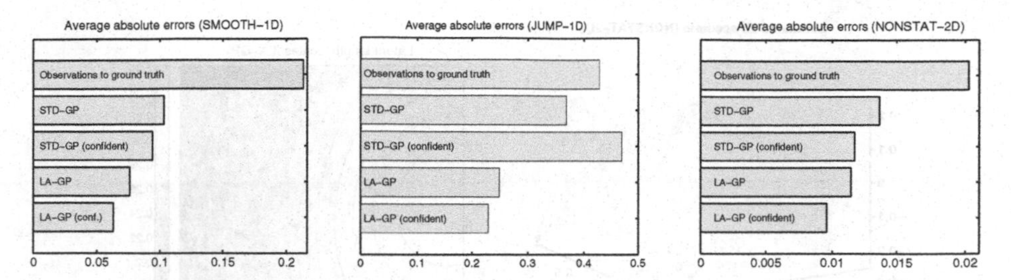


**Fig. 5.** Absolute average errors of the mean predictions in the SMOOTH-1D test scenario (left), JUMP-1D (middle), and NONSTAT-2D (right). We give the absolute distances of the simulated observations to the true function values, the overall average errors for the different models, and the average errors of the 50% most confidently predicted means respectively.

$31 \cdot 31 = 961$ points from $[-0.5, 1] \times [-0.5, 1]$ including their true function values. A typical example of the resulting optimized length-scales are visualized in the upper right contour plot of Figure 6. It can be seen that larger length-scales (which correspond to stronger smoothing) are assigned to the flat part of the surface around $(-0.5, 0)^T$ and smaller ones towards $(1, 1)^T$.

The quantitative results in terms of NLPD and sMSE for 30 independent test runs are given in Table 2. The absolute errors of the mean predictions are given in the right bar chart of Figure 5. The two lower plots of Figure 6 give a visual impression about the accuracy of the two regression models. We give the NLPD errors at equidistantly sampled test locations overlayed by contour plots of the predictive uncertainties. Note that the LA-GP model assigns higher confidence to the flat part of the function, which – given the uniform sampling of training points – can be reconstructed more accurately than the higher-frequency parts.

## 5.2 Modeling RFID Signal Strength

We have applied our nonstationary regression approach to the problem of learning the signal strength distribution of RFID (Radio Frequency Identification) tags. For this experiment, 21.794 signal strength measurements (logarithmic to the base of 10) have been recorded in a test setup at the University of Freiburg using a static antenna and a mobile, externally localized RFID tag. For efficiency reasons, only the left half-space of the antenna was sampled with real measurements and then mirrored along the respective axis. We randomly sampled 121 training points for learning the regression models and 500 different ones for evaluation. Note that although larger training sets lead to better models, we learn from this comparably small number of observations only to achieve faster evaluation runs. Table 3 gives the quantitative comparison to the standard GP model (STD-GP). As can be seen from the results, the standard model is outperformed by our nonstationary extension both in terms of sMSE and NLPD. Figure 7 shows predicted mean log signal strengths of the two models as color maps overlayed with contour plots of the corresponding predictive uncertainties.

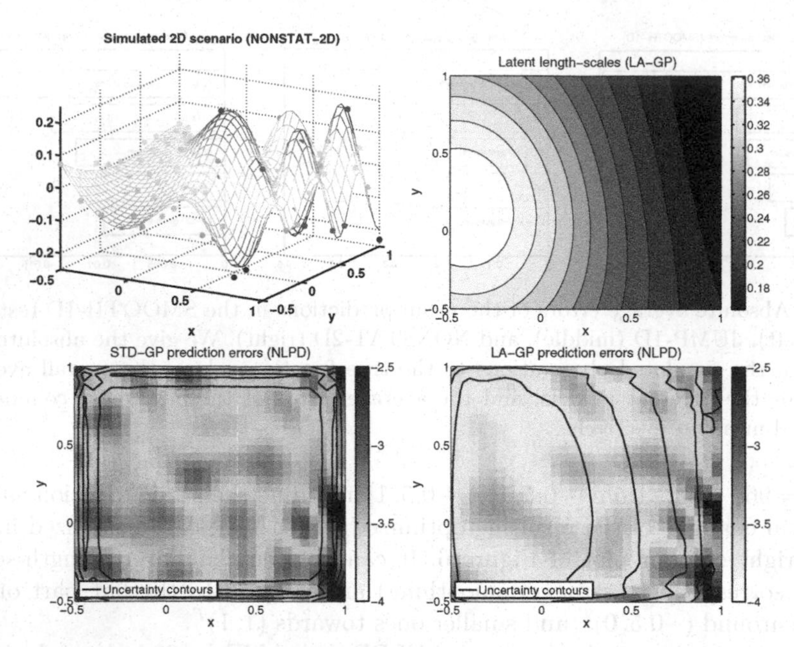


**Fig. 6.** The true function and noisy observations in the NONSTAT-2D test case (top left). Note the spatially varying oscillation frequency. The top right plot depicts the contours of the latent length-scales as estimated by our LA-GP model. In the two lower diagrams, we give the individual prediction errors (NLPD) of the Standard GP model (bottom left) and LA-GP (bottom right). The predictive uncertainty of the models is visualized using overlayed contours.

**Table 3.** Quantitative results for the RFID-2D experiment. Results marked by • differ significantly ($\alpha = 0.05$) from the others in their category.

| **Test Scenario** | NLPD | | sMSE | |
|---|---|---|---|---|
| | **LA-GP** | **STD-GP** | **LA-GP** | **STD-GP** |
| RFID-2D | -0.0101 (•) | 0.1475 | 0.3352 (•) | 0.4602 |

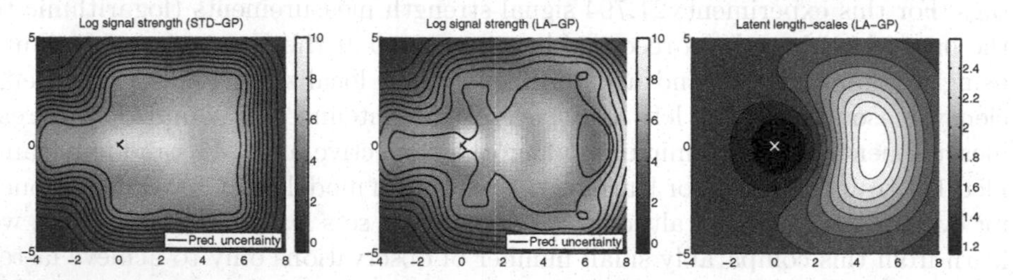


**Fig. 7.** Predicted mean log signal strengths of RFID tags using the standard GP (left) and the locally adapted GP (middle). The sensor location (0,0) is marked by a cross and the predictive uncertainties of the models are visualized by overlayed contours. The right plot visualizes the adapted latent length-scales of the LA-GP model. Coordinates are given in Meters.

**Table 4.** Quantitative results for the simulated (TERSIM-2D) and the real (TERREAL-2D) terrain mapping experiment. Results marked by • differ significantly ($\alpha = 0.05$) from the others in their category.

| | NLPD | | sMSE | |
|---|---|---|---|---|
| **Test Scenario** | **LA-GP** | **STD-GP** | **LA-GP** | **STD-GP** |
| TERSIM-2D | -4.261 (•) | -4.198 | 0.127 | 0.126 |
| TERREAL-2D | -3.652 | -3.626 | 0.441 (•) | 0.475 |

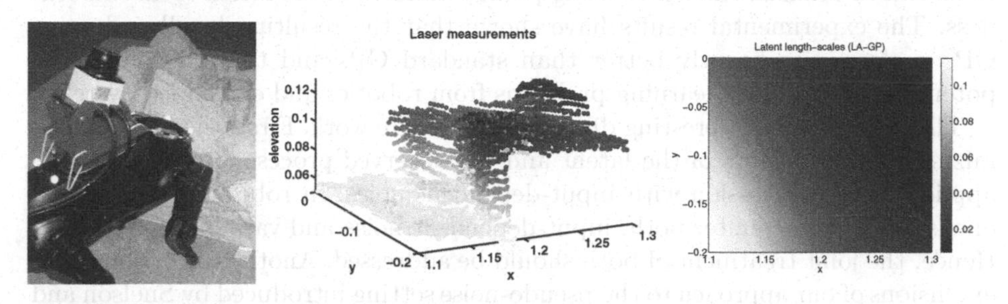


**Fig. 8.** A quadruped robot equipped with a laser sensor (left) acquires elevation measurements of a rough terrain surface (middle) by executing a 'pushup' motion. From a subset of elevation samples, our LA-GP approach learns a predictive model that captures the nonstationary nature of the data set (right).

We also visualize the contours of the latent length-scales modeling higher frequencies in the proximity of the sensor location and lower ones in front of it.

### 5.3 Laser-Based Terrain Mapping

We also applied our model to the particularly hard robotics problem of learning probabilistic terrain models from laser range measurements. In a joint project with the Massachusetts Institute of Technology, we have equipped a quadruped robot with a Hokuyo URG laser range sensor (see the left picture in Figure 8). The robot was programmed to perform a 'pushup' motion sequence in order to acquire a 3D scan of the local environment. For evaluation, we selected a $20 \times 20cm$ part of a rough terrain (with a maximum height of around 9 cm) including its front edge (see the middle plot of Figure 8). 4.282 laser end points of the 3D scan fall into this area.

We have trained the standard GP model and our nonstationary variant on 80 randomly selected training points from a noise-free simulation of the real terrain (TERSIM-2D) and evaluated the prediction accuracy for 500 test points (averaged over 30 independent runs). We repeated the same procedure on the real data (TERREAL-2D) and evaluated the prediction accuracy for other, independently selected test points from the real scan. Thus, the latter evaluation quantifies how well the models are able to predict other samples from the same distribution while the former gives the prediction errors relative to a known ground truth function. Table 4 gives the quantitative results for these two experiments. The right colormap in Figure 8 depicts the optimized length-scales of the LA-GP model. It can be seen that the flat part of the terrain is assigned larger local kernels compared to the rougher parts.

## 6 Conclusions

This paper has shown that GP regression with nonstationary covariance functions can be realized efficiently using point estimates of the latent local smoothness. The experimental results have shown that the resulting locally adaptive GPs perform significantly better than standard GPs and that they have the potential to solve hard learning problems from robotics and embedded systems.

There are several interesting directions for future work. First, the idea of optimizing the parameters of the latent and the observed process jointly should be applied to GP regression with input-dependent noise. In robotics applications, one is likely to encounter both, input-dependent noise and variable smoothness. Hence, the joint treatment of both should be addressed. Another direction is the extensions of our approach to the pseudo-noise setting introduced by Snelson and Ghahramani, see e.g. [15], so that the locations of the length-scale support values are learned from data, too. Finally, one should investigate multi-task learning e.g. along the lines of Yu *et al.* [17] to generalize e.g. across different types of terrains.

**Acknowledgments.** The authors would like to thank the anonymous reviewers for their helpful comments. Many thanks to Christopher Paciorek for helpful discussions on nonstationary GP regression. The RFID data set was acquired by Dominik Joho at the University of Freiburg. Many thanks to Nicholas Roy, Sam Prentice, and Russ Tedrake at MIT CSAIL and to the DARPA Learning Locomotion program for making the robot terrain mapping experiment possible. Kristian Kersting was supported by the Fraunhofer Attract fellowship STREAM. Christian Plagemann and Wolfram Burgard were supported by the EC under contract number FP6-004250-CoSy and by the German Ministry for Education and Research (BMBF) through the DESIRE project.

## References

1. Brooks, A., Makarenko, A., Upcroft, B.: Gaussian process models for sensor-centric robot localisation. In: Proc. of ICRA (2006)
2. Brooks, M.: The matrix reference manual, `http://www.ee.ic.ac.uk/hp/staff/dmb/matrix/intro.html`
3. Cornford, D., Nabney, I., Williams, C.: Adding constrained discontinuities to gaussian process models of wind fields. In: Advances in Neural Information Processing Systems 11 (NIPS), Cambridge, MA (1999)
4. Dimatteo, I., Genovese, C.R., Kass, R.E.: Bayesian curve-fitting with free-knot splines. Biometrika 88(4), 1055–1071 (2001)
5. Kersting, K., Plagemann, C., Pfaff, P., Burgard, W.: Most likely heteroscedastic gaussian processes regression. In: Gharahmani, Z. (ed.) ICML 2007 (2007)

6. Lang, T., Plagemann, C., Burgard, W.: Adaptive non-stationay kernel regression for terrain modelling. In: Proc. of Robotics: Science and Systems (RSS) (2007)
7. Møller, M.: A Scaled Conjugate Gradient Algoritm for Fast Supervised Learning. Neural Networks 6, 525–533 (1993)
8. Paciorek, C., Schervish, M.: Nonstationary covariance functions for Gaussian process regression. In: Thrun, S., Saul, L., Schoelkopf, B. (eds.) Advances in Neural Information Processing Systems 16. MIT Press, Cambridge (2004)
9. Pfingsten, T., Kuss, M., Rasmussen, C.E.: Nonstationary gaussian process regression using a latent extension of the input space. In: Proc. of ISBA Eighth World Meeting on Bayesian Statistics, Valencia, Spain (Extended Abstract, 2006)
10. Plagemann, C., Fox, D., Burgard, W.: Efficient failure detection on mobile robots using particle filters with gaussian process proposals. In: Proc. of the International Joint Conference on Artificial Intelligence (IJCAI), Hyderabad, India (2007)
11. Rasmussen, C.E., Williams, C.K.I.: Gaussian Processes for Machine Learning. Adaptive Computation and Machine Learning. MIT Press, Cambridge (2006)
12. Sampson, P.D., Guttorp, P.: Nonparametric estimation of nonstationary spatial covariance structure. Journal of the American Stat. Association 87, 108–119 (1992)
13. Schmidt, A.M., OHagan, A.: Bayesian inference for nonstationary spatial covariance structure via spatial deformations. JRSS, series B 65, 745–758 (2003)
14. Schwaighofer, A., Grigoras, M., Tresp, V., Hoffmann, C.: A Gaussian process positioning system for cellular networks. In: Advances in Neural Information Processing Systems 16. MIT Press, Cambridge (2004)
15. Snelson, E., Ghahramani, Z.: Variable noise and dimensionality reduction for sparse gaussian processes. In: UAI (2006)
16. Williams, O.: A switched Gaussian process for estimating disparity and segmentation in binocular stereo. In: Neural Info. Proc. Systems (NIPS) (2006)
17. Yu, K., Tresp, V., Schwaighofer, A.: Learning gaussian processes from multiple tasks. In: ICML 2007 (2007)

# Kernel-Based Inductive Transfer

Ulrich Rückert and Stefan Kramer

Institut für Informatik/I12, Technische Universität München, Boltzmannstr. 3,
D-85748 Garching b. München, Germany
{rueckert,kramer}@in.tum.de

**Abstract.** Methods for inductive transfer take advantage of knowledge from previous learning tasks to solve a newly given task. In the context of supervised learning, the task is to find a suitable bias for a new dataset, given a set of known datasets. In this paper, we take a kernel-based approach to inductive transfer, that is, we aim at finding a suitable kernel for the new data. In our setup, the kernel is taken from the linear span of a set of predefined kernels. To find such a kernel, we apply convex optimization on two levels. On the base level, we propose an iterative procedure to generate kernels that generalize well on the known datasets. On the meta level, we combine those kernels in a minimization criterion to predict a suitable kernel for the new data. The criterion is based on a meta kernel capturing the similarity of two datasets. In experiments on small molecule and text data, kernel-based inductive transfer showed a statistically significant improvement over the best individual kernel in almost all cases.

**Keywords:** kernels, inductive transfer, transfer learning, regularized risk minimization.

## 1 Introduction

It is well known that the success or failure of a supervised learning method depends on its bias. If the bias matches well with the underlying learning problem, the system will be able to construct predictive models. If the bias does not match very well, the generated classifier will perform poorly. One of the great advantages of kernel-based methods is the fact that the learning bias can be flexibly adjusted by choosing a customized kernel for the data at hand. However, building custom kernels from scratch for individual applications can be a tedious task. Recent research has dealt with the problem of learning kernels automatically from data, see e.g. the work by Ong *et al.* [17] and Lanckriet *et al.* [14]. In practice actual training data is often rare and in most cases it is better to invest it for the actual learning task than for kernel selection. Even though data from the same source may be rare, it is sometimes the case that data on related or similar learning problems is available. As an example, for text classification problems, plenty of related text data might be available on the internet. Similarly, for some problems from computational chemistry, research on related endpoints might lead to related datasets.

W. Daelemans et al. (Eds.): ECML PKDD 2008, Part II, LNAI 5212, pp. 220–233, 2008.

The task of using such related data to improve accuracy for the the learning problem at hand is known as *inductive transfer*. Here, the main idea is that a kernel (i.e., a bias) that has worked well on the related *transfer datasets* should also work well on the new data. One particularly pragmatic approach to inductive transfer is to build a range of classifiers with varying kernels and settings on the transfer data and to evaluate the predictive accuracy of those classifiers. The kernel that performed best on the transfer datasets could then be selected for the new data. While conceptually simple, this method has three disadvantages. First, classifier evaluation takes quite a lot of time, because evaluation methods like cross validation require the generation of many classifiers. Second, the method evaluates only single kernels and does not take into account the case where a combination of many kernels might perform better than each individual kernel. Third, it does not consider the fact that some transfer datasets are more similar to the learning problem at hand than others.

In this paper we would like to address these issues. As a first contribution we present a method that finds kernels, which generalize well on the existing transfer data without the need to resort to expensive evaluation methods. Having these "known good" kernels for the transfer data, we frame the problem of finding a good kernel for the new data at hand as a *meta learning* problem. Roughly, this learning problem can be stated as follows: given a set of transfer datasets together with the corresponding good kernels, what would a good kernel for the data at hand look like? We propose to solve this meta learning task using a strategy based on regularized convex loss minimization with a meta-kernel. For the case where the design of domain-specific meta-kernels is too tedious or impossible (perhaps due to lack of suitable background knowledge), we propose the *histogram kernel*, a generic meta-kernel that is applicable for standard propositional datasets.

The paper is organized as follows. After a brief review of some related work in Section 2, we introduce the setting and describe our approach in Section 3. We start with the problem of finding good kernels for the transfer data in Section 3.1, present the histogram kernel in section 3.2 and discuss our approach to the meta learning problem in Section 3.3. Finally, we report on experiments in Section 4 and conclude in Section 5.

## 2 Related Work

The presented work is related to research in three areas. On one side, there has been considerable progress in learning kernels from data. The original formulation as a semi-definite optimization problem by Lanckriet *et al.* [14] has been extended in various directions [16,19]. Other techniques for kernel learning include hyper kernels [17] and boosting [4,12]. All these approaches aim at learning a good kernel from training data rather than from transfer data. On the other side, there is a long history of work on inductive transfer, see e.g. Baxter [2]. Of course, inductive transfer is related to multi-task learning [3], where the goal is to induce classifiers for many tasks at once. Multi-task learning with kernels

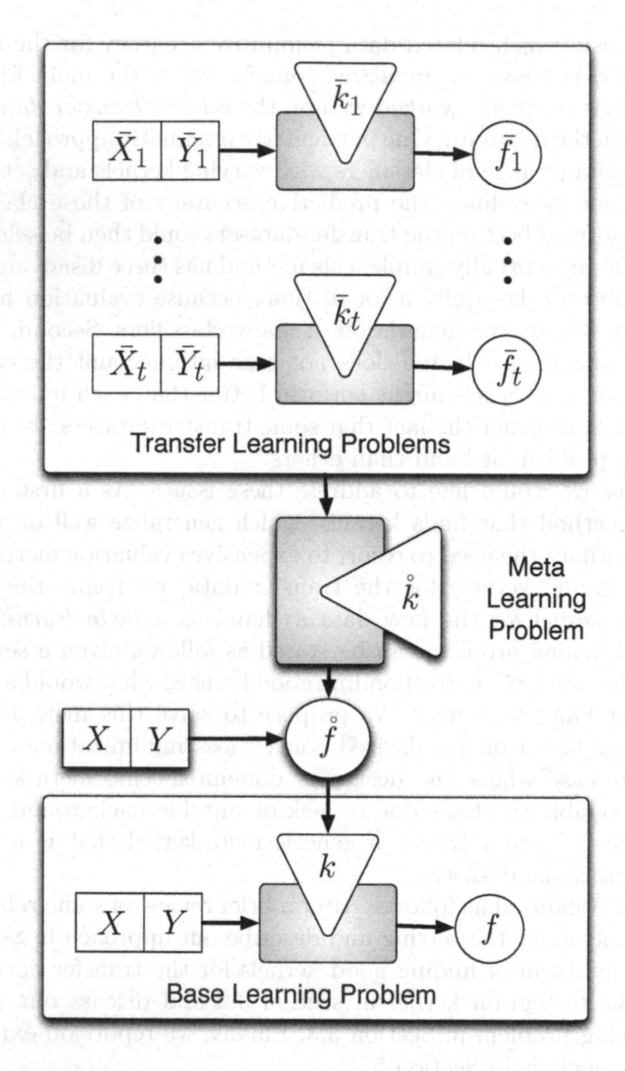


**Fig. 1.** We aim at finding a good kernel $k$ for the base learning problem. To do so, we search for good kernels $\bar{k}_1, \ldots, \bar{k}_n$ for the $n$ transfer learning problems. The meta learning problem deals with learning a predictor $\mathring{f}$, which outputs a good base kernel $k$ for the base data.

has been the subject of research by Evgeniou *et al.* [6] and Argyriou *et al.* [1]. It is often approached using Bayesian methods [10,24,22]. This paper deals with a more asymmetric setting, where we use the older datasets only to increase predictive performance on the new data at hand. Similar settings have been the subject of work by Kaski and Peltonen [13] and Erhan *et al.* [5]. While Kaski and Peltonen consider the case where only a few instances in the transfer datasets are relevant, the study by Erhan *et al.* aims at generalizing from the transfer data only, so that no base data is necessary.

Finally, the presented work is also loosely related to research on *meta learning*. Here, the goal is to induce meta-classifiers, which, given a dataset and its characteristics, propose a learning algorithm that is supposed to perform well on the data at hand. We refer to Pfahringer *et al.* [18] for a meta learning study based on landmarking and a short overview of related work.

## 3 Kernel-Based Inductive Transfer

For the following discussion, we assume a setting, where a user is interested in finding a good classifier for some new data. Additionally, she has access to a range of older datasets, which are assumed to be similar to the new data in some regard. In the following, we call the older datasets *transfer datasets* and the new data *base dataset*. The main question we are dealing with is how to find a good kernel (i.e. bias) for the base learning problem, given the old transfer datasets. As illustrated in Figure 1, we frame this problem as a *meta learning* task. We proceed in three steps. First, for each transfer dataset, we generate a kernel that leads to a highly predictive classifier on this data. Then, from the transfer datasets and kernels, we learn a *meta classifier* $\mathring{f}$, which predicts a new base kernel when given a base dataset. The meta learning algorithm makes use of the *meta kernel* $\mathring{k}$ to compare two datasets. Finally, in the last step, the meta classifier is applied to the base data at hand, leading to a kernel $k$ for the base learning problem. This kernel is then plugged into a standard SVM to construct a suitable classifier for the base data.

Let us introduce some notation before we delve into the details. As usual, we assume that the training data is given as a set of labeled examples $(X, Y)$, where the rows $x_1, \ldots, x_n$ of the training matrix $X \in \mathbb{R}^{n \times m}$ constitute the examples, and the $y_1, \ldots, y_n \in \{-1, 1\}$ represent the class labels. The goal is to induce a classifier from $(X, Y)$ that predicts well on new, previously unseen test instances. Since we are dealing with the *inductive transfer* setting, we are given an additional set of $t$ *transfer datasets* $(\bar{X}_1, \bar{Y}_1), \ldots, (\bar{X}_t, \bar{Y}_t)$, which are supposed to pose similar learning problems as the one given by $(X, Y)$. To allow for a concise description, we mark all parameters and variables referring to the transfer problems with a bar. For instance, given a particular transfer dataset $(\bar{X}, \bar{Y})$, we let $(\bar{x}_1, \bar{y}_1), \ldots, (\bar{x}_n, \bar{y}_n)$ denote its examples. Similarly, if an SVM is applied on this data, we denote the coefficient vector and threshold output by the SVM with $\bar{\alpha} := (\bar{\alpha}_1, \ldots, \bar{\alpha}_n)^T$ and $\bar{b}$.

As explained above, we proceed in three steps to find a kernel, which is likely to perform well on the base data. First of all, we compute for each transfer dataset $(\bar{X}_i, \bar{Y}_i)$ a kernel $\bar{k}_i^*$ that generalizes well from (transfer) training data to (transfer) test data. Then, in the second step, we tackle the problem of finding a good kernel $k$ for the base learning data $(X, Y)$. We frame this as a meta learning problem. In particular, we make use of a *meta kernel* $\mathring{k} : ((\bar{X}, \bar{Y}), (\bar{X}', \bar{Y}')) \mapsto r \in \mathbb{R}$, defined on the space of (transfer) datasets, to induce a meta model, represented by a coefficient vector $\mathring{\alpha}$ and a threshold $\mathring{b}$. Given a (base) dataset, the meta model outputs a kernel that is likely to work well for learning on the base data. For notational clarity, we again mark all parameters and variables that deal with the meta learning task (as opposed to the transfer and base learning tasks) with a circle (e.g., we write $\mathring{k}$ to represent the meta kernel). Finally, in the last step, we apply the meta model to the training data $(X, Y)$ at hand, yielding a base kernel $k$. This kernel is then employed in the base SVM to build a final classifier. In the following sections we describe the three steps in more detail.

### 3.1 Learning Kernels from Transfer Data

In the first step, we would like to discover which bias works well for each of the $k$ transfer data sets. Since we are working with SVMs, this essentially boils down to the question of what kernel should be chosen in each case. Recall that the soft-margin SVM optimizes the regularized hinge loss of classifier $\bar{f}$ on the training set. More formally, let $\bar{k}$ be some positive semi-definite kernel, $l_h(\bar{y}, \bar{f}(\bar{x})) = \max(0, 1 - \bar{y}\bar{f}(\bar{x}))$ denote the *hinge loss*, and let $C > 0$ be a tuning parameter. Then, the SVM minimizes the following functional over all linear classifiers $\bar{f} \in \mathcal{H}_{\bar{k}}$ in the Hilbert space produced by kernel $\bar{k}$, where $\|\cdot\|_{\bar{k}}$ denotes the norm in this space.

$$S(\bar{X}, \bar{Y}, \bar{f}, \bar{k}) := C \sum_{(\bar{x},\bar{y})\in(\bar{X},\bar{Y})} l_h(\bar{y}, \bar{f}(\bar{x})) + \|\bar{f}\|_{\bar{k}} \tag{1}$$

The standard SVM computes the optimal classifier $\bar{f}^*$ by minimizing (1) over $\bar{f}$, while keeping the kernel $\bar{k}$ fixed: $\bar{f}^* := \text{argmin}_{\bar{f}\in\mathcal{H}_{\bar{k}}} S(\bar{X}, \bar{Y}, \bar{f}, \bar{k})$. Since the hinge loss can be seen as a robust upper bound of the zero-one loss, it is a sensible strategy to select not only the classifier $\bar{f}$, but also the kernel $\bar{k}$ by minimizing $S(\bar{X}, \bar{Y}, \bar{f}, \bar{k})$. In other words, to find a good kernel for a given dataset, one can solve $(\bar{f}^*, \bar{k}^*) := \arg\min_{\bar{f}\in\mathcal{H}_{\bar{k}}, \bar{k}\in\mathcal{K}} S(\bar{X}, \bar{Y}, \bar{f}, \bar{k})$, where $\mathcal{K}$ denotes some pre-defined space of possible kernels. If $\mathcal{K}$ is a convex set, this enlarged optimization problem is still convex and can thus be solved efficiently [14].

Unfortunately, minimizing (1) over a transfer data set $(\bar{X}, \bar{Y})$ does not necessarily lead to a kernel that generalizes well on new data. This is for two reasons. First, by optimizing $\bar{k}$ and $\bar{f}$ at the same time, one finds a kernel $\bar{k}^*$ that works well together with the optimal $\bar{f}^*$. However, when one applies the kernel later on new data, the SVM might induce a $\bar{f}$, which might differ from $\bar{f}^*$ and does therefore not match well with the $\bar{k}^*$. In other words, we are not looking for a $\bar{k}^*$ that works well with $\bar{f}^*$, but a kernel that works well with an $\bar{f}$ that is typically chosen by an SVM on new data. Second, for some classes of kernels the kernel matrix has always full rank. This means that there is always a subspace $H_0 \subseteq \mathcal{H}_k$ whose classifiers $\bar{f} \in H_0$ achieve $\sum_{i=1}^n l_h(\bar{y}_i, \bar{f}(\bar{x}_i)) = 0$. This is also true for practically relevant classes of kernels, for instance, radial basis kernels. In those cases, minimizing $S(\bar{X}, \bar{Y}, \bar{f}, \bar{k})$ focuses almost exclusively on the regularization term $\|\bar{f}\|_{\bar{k}}$ and the data-dependent term is largely ignored (because it is zero in most cases). In other words, a kernel matrix of full rank leads to overfitting in the sense that the optimization procedure prefers kernels that match well with the regularization criterion rather than kernels that catch the bias inherent in the data.

**Algorithm 1.** An iterative procedure to find a kernel that generalizes well on the dataset $(\bar{X}, \bar{Y})$

**procedure** FINDGOODKERNEL$(\bar{X}, \bar{Y}, C)$
  select subset $(\bar{X}', \bar{Y}')$ from $(\bar{X}, \bar{Y})$
  select initial $\bar{k}^{(0)} \in \mathcal{K}$
  $i \leftarrow 0$
  **repeat**
    $i \leftarrow i+1$
    $\bar{f}^{(i)} \leftarrow \operatorname{argmin}_{\bar{f} \in \mathcal{H}_{\bar{k}}} S(\bar{X}', \bar{Y}', \bar{f}, \bar{k}^{(i-1)})$
    $\bar{k}^{(i)} \leftarrow \operatorname{argmin}_{\bar{k} \in \mathcal{K}} S(\bar{X}, \bar{Y}, \bar{f}^{(i)}, \bar{k})$
  **until** $S(\bar{X}, \bar{Y}, \bar{f}^{(i)}, \bar{k}^{(i)}) \geq S(\bar{X}, \bar{Y}, \bar{f}^{(i-1)}, \bar{k}^{(i-1)})$
  **return** $(\bar{k}^{(i)}, \bar{f}^{(i)})$
**end procedure**

To avoid the two problems, we split $(\bar{X}, \bar{Y})$ into two parts and modify the optimization criterion, so that $\bar{f}$ depends only on the first part of the data, whereas the kernel $\bar{k} \in \mathcal{K}$ is evaluated on the whole dataset. In this way, $f^*$ is chosen from a much smaller and more restricted space of classifiers. Consequently, the optimization procedure needs to better adjust $\bar{k}^*$ to ensure that even a $\bar{f}^*$ from the rather restricted subspace generalizes well to the remaining instances. This setup is similar to classifier evaluation with a hold-out set, where a classifier is induced from the training set and then evaluated on a separate test set. More formally, let $(\bar{X}', \bar{Y}')$ be some subset of $(\bar{X}, \bar{Y})$. We use the standard SVM regularized risk functional (1) to rate a classifier $\bar{f}$ for a fixed kernel $\bar{k}$ on this subset:

$$\bar{f}^*_{\bar{k}} := \operatorname*{argmin}_{\bar{f} \in \mathcal{H}_{\bar{k}}} S(\bar{X}', \bar{Y}', \bar{f}, \bar{k}) \tag{2}$$

Then, we choose the optimal kernel so that it performs best with $\bar{f}^*_{\bar{k}}$ on all examples. More precisely, the desired optimal kernel $\bar{k}^*$ is

$$\bar{k}^* = \operatorname*{argmin}_{\bar{k} \in \mathcal{K}} S(\bar{X}, \bar{Y}, \operatorname*{argmin}_{\bar{f} \in \mathcal{H}_{\bar{k}}} S(\bar{X}', \bar{Y}', \bar{f}, \bar{k}), \bar{k}) \tag{3}$$

This criterion contains two nested optimization problems, one to determine the best classifier, the other, which depends on the first one, to compute the optimal kernel. Generally, the functional is not convex when optimizing over $\bar{k}$ and $\bar{f}^*_{\bar{k}}$ at the same time. However, on their own, the two single parts are convex optimization problems: The first part $\operatorname{argmin}_{\bar{f} \in \mathcal{H}_{\bar{k}}} S(\bar{X}', \bar{Y}', \bar{f}, \bar{k})$ is the standard SVM criterion, that is, a quadratic program. The second part $\operatorname{argmin}_{\bar{k} \in \mathcal{K}} S(\bar{X}, \bar{Y}, \bar{f}^*_{\bar{k}}, \bar{k})$ is a convex criterion, if one sets $\bar{f}^*_{\bar{k}}$ to a fixed value and optimizes only over $\bar{k}$. This naturally leads to an optimization procedure that alternates between optimizing (2) with a fixed $\bar{k}$ and optimizing (3) with a fixed $\bar{f}^*_{\bar{k}}$. The loop is terminated as soon as neither of the two steps improves on the score anymore. The approach is outlined in Algorithm 1.

Of course, it is impractical to deal with the explicit representations of $\bar{f}^*_{\bar{k}}$ in a high-dimensional Hilbert space. Fortunately, the representer theorem ensures that the optimal $\bar{f}_{\bar{k}}^*$ is always contained in the linear span of $\{k(\bar{x}', \cdot)|\bar{x}' \in \bar{X}'\}$. Therefore, a classifier $\bar{f}$ can be conveniently represented by a coefficient vector $\bar{\alpha}$ and a threshold $\bar{b}$, so that $\bar{f}(x) = \sum_{i=1}^{n'} \bar{y}_i' \bar{\alpha}_i \bar{k}(x, \bar{x}_i') + \bar{b}$. With this, the first criterion (2) can be equivalently stated as follows:

$$\operatorname*{argmin}_{\bar{\alpha} \in \mathbb{R}_+^n, \bar{b} \in \mathbb{R}} C \sum_{i=1}^{n'} l_h(\bar{y}_i, [\bar{K}'\bar{D}'\bar{\alpha}]_i + \bar{b}) + \bar{\alpha}^T \bar{D}'\bar{K}'\bar{D}'\bar{\alpha} \tag{4}$$

Here, $\bar{\alpha}$ and $\bar{b}$ are the coefficient vector and threshold of the linear classifier to be found, $n'$ is the number of instances in $(\bar{X}', \bar{Y}')$, $\bar{K}'$ denotes the $n' \times n'$ kernel matrix with $\bar{K}_{ij}' = \bar{k}(\bar{x}_i, \bar{x}_j)$, $\bar{D}'$ is a $n' \times n'$ diagonal matrix whose diagonal contains the class labels $\bar{D}_{ii}' = \bar{y}_i'$, and $[x]_i$ denotes the $i$th component of vector $x$.

The exact form of the second criterion depends on the structure of $\mathcal{K}$. For the experiments in section 4 we set $\mathcal{K}$ to the space of positive linear combinations of $l$ main kernels $\bar{k}_1, \ldots, \bar{k}_l$. This means that a kernel $\bar{k} \in \mathcal{K}$ can be represented by a vector $\bar{\mu} \in \mathbb{R}_+^l, \|\bar{\mu}\| \leq 1$ of linear coefficients, because $\bar{k}(\bar{x}_1, \bar{x}_2) = \sum_{i=1}^{l} \bar{\mu}_i \bar{k}_i(\bar{x}_1, \bar{x}_2)$. With this, criterion (3) becomes:

$$\operatorname*{argmin}_{\bar{\mu} \in \mathbb{R}_+^n, \|\bar{\mu}\| \leq 1} C \sum_{i=1}^{n} l_h(\bar{y}_i, [\bar{M}\bar{\mu}]_i + \bar{b}) + \bar{r}^T \bar{\mu} \tag{5}$$

Here, $\bar{M} \in \mathbb{R}^{n \times l}$ with $\bar{M}_{ij} = \bar{y}_i \sum_{k=1}^{n'} \bar{y}_k \bar{\alpha}_k \bar{k}_j(\bar{x}_i, \bar{x}_k)$, and $\bar{r} \in \mathbb{R}^l$ with $\bar{r}_k = \sum_{i=1}^{n'} \sum_{j=1}^{n'} \bar{y}_i \bar{y}_j \bar{\alpha}_i \bar{\alpha}_j \bar{k}_k(\bar{x}_i, \bar{x}_j)$. Of course, the overall transfer learning scheme works also with other kernel spaces $\mathcal{K}$. For instance, one could choose $\mathcal{K} = \{\bar{k}_p | p \in \mathbb{R}\}$, where $\bar{k}_p$ is a kernel function parameterized with $p$. If the resulting kernel space $\mathcal{K}$ is a convex set, the corresponding transfer optimization criterion (3) and the meta learning problem (6) (see section 3.3) can be cast as convex optimization problems and are thus solvable with standard methods.

### 3.2 The Histogram Kernel for the Meta Learning Task

In the preceding section we described a method to find a kernel $\bar{k}_i^*$ that encodes the bias inherent in each transfer dataset $(\bar{X}_i, \bar{Y}_i)$. Now, we address the main question of the paper: How can we make use of this transfer information when dealing with the base learning problem, where we wish to learn a classifier for the base dataset $(X, Y)$? In particular, given the $\bar{k}_i^*$, what should a "good" base kernel $k$ for the base data look like? Since we assume that the transfer learning problems are similar to the base learning problem, choosing the average over the $\bar{k}_i^*$ appears to be a good option. On the second sight, though, it is clear that some transfer learning problems are more similar to the base setup than others. Thus, it makes sense to frame the task of finding a good base kernel $k$ as a *meta learning* problem. Formally, this problem can be stated as follows: Given a set of $t$ transfer datasets $(\bar{X}_1, \bar{Y}_1), \ldots, (\bar{X}_t, \bar{Y}_t)$ and the corresponding $t$ known good kernels $\bar{k}_1^*, \ldots, \bar{k}_t^*$, we would like to predict a new kernel $k$ that performs as good as possible on the base data $(X, Y)$.

We tackle this meta learning problem using a kernel-based approach. As a first step, we need a meta kernel $\mathring{k} : (\mathcal{X}, \mathcal{Y}) \times (\mathcal{X}, \mathcal{Y}) \to \mathbb{R}$. As it is the case with all kernel-based classification systems, it is best to apply a kernel that incorporates domain specific knowledge about the problem at hand. In our setting one could use information about the features of the transfer and base datasets to construct informative kernels. For instance, if the learning task at hand deals with the classification of text documents, a meta kernel could compare two datasets by counting the tokens that are shared in the two documents. As a default kernel for the case where no usable meta-information is available for the construction of a custom kernel, we propose the *histogram kernel*.

Given two datasets $(\bar{X}, \bar{Y})$ and $(\bar{X}', \bar{Y}')$, where $\bar{X} \in \mathbb{R}^{n \times m}$ and $\bar{X}' \in \mathbb{R}^{n' \times m'}$ are training matrices, the histogram kernel is the sum of the radial basis kernel applied on the difference of the two histograms of feature averages and the radial basis kernel on the difference of the histograms of instance averages. More precisely, the kernel is computed as follows. Assume one is given the two training datasets $X$ and $X'$. First, we compute the averages $h_c(\bar{X}) = \frac{1}{n} e_n^T X$ and $h_r(\bar{X}) = \frac{1}{m} X e_m$ for $\bar{X}$ and $\bar{X}'$ ($e_n$ denotes the vector of $n$ ones) over the rows and columns of those two datasets. Let $h_c, h'_c, h_r$ and $h'_r$ denote the average vectors for $\bar{X}$ and $\bar{X}'$ respectively. In the next step we sort the components in the four vectors by descending size. With that, each vector represents the distribution of feature and instance averages in the datasets, similar to a histogram. Unfortunately, we can not compare the corresponding vectors directly, because the number of columns and rows may differ between $X$ and $X'$ so that the histogram vectors differ in the number of components. Thus, we normalize the histograms $h_c, h'_c$ and $h_r, h'_r$ by duplicating components evenly in the smaller of the two vectors until the two histograms have the same number of components. Finally, we compute the absolute differences between the two corresponding histograms $d_c := \frac{1}{m} \sum_{j=1}^{m} |[h_c - h'_c]_j|$ and $d_r := \frac{1}{n} \sum_{j=1}^{n} |[h_r - h'_r]_j|$. The final kernel value is then $\mathring{k}((\bar{X}, \bar{Y}), (\bar{X}', \bar{Y}')) := \frac{1}{2}(\exp(-d_c) + \exp(-d_r))$. It is easy to see that $\mathring{k}$ is positive semi-definite, because $d_c$ and $d_r$ constitute the one-norm of the histogram differences. While the histogram kernel is designed as a generic meta kernel for the case where no application-specific choice is available, it appears to work quite well in the experiments in section 4.1. It is an open question, whether it could also be applied in other transfer or meta learning schemes, and which other application-specific meta kernels could provide successful alternatives.

### 3.3 The Meta Learning Procedure

With this, we have the building blocks to address the meta learning problem. Recall that we wish to learn a meta-model that relates the transfer datasets $(\bar{X}_1, \bar{Y}_1), \ldots, (\bar{X}_t, \bar{Y}_t)$ and the corresponding "known good" kernels $\bar{k}_1^*, \ldots, \bar{k}_t^*$. The model is later used to predict a suitable base kernel for the base dataset. Also recall from section 3.1 that each kernel $\bar{k}_i^*$ is a linear combination $\bar{k}_i^* = \sum_{j=1}^{l} \bar{\mu}_j \bar{k}_j$ of some main kernels $\bar{k}_j$. Thus, the transfer kernels $\bar{k}_1^*, \ldots, \bar{k}_t^*$ are actually represented by the corresponding coefficient vectors $\bar{\mu}_1, \ldots, \bar{\mu}_t$.

For the meta learning problem, we chose a regularized loss minimization approach that resembles a Leave-One-Out-SVM [21]. First of all, it is clear, that the zero-one or hinge losses are not applicable in this setting, because the quantity to predict is a whole vector rather than a binary variable. As a natural replacement, we select the 2-norm to measure the loss between the predicted and true coefficient vector: $l_2(\bar{\mu}, \mathring{f}(\bar{X}, \bar{Y})) := \|\bar{\mu} - \mathring{f}(\bar{X}, \bar{Y})\|_2$. Following the approach of the SVM, we now aim at finding a coefficient vector $\mathring{\alpha} \in \mathbb{R}^t$ and a threshold $\mathring{b} \in \mathbb{R}^l$, which minimize the loss of the kernel classifier $\mathring{f}(X, Y) := \sum_{i=1}^{t} \mathring{k}((\bar{X}_i, \bar{Y}_i), (X, Y))\mathring{\alpha}_i \bar{\mu}_i + \mathring{b}$. For the training of this classifier, though, we follow the philosophy of the Leave-One-Out-SVM and do not consider the contribution of a training example for its own classification. More precisely, when evaluating the classifier given by $(\mathring{\alpha}, \mathring{b})$ on training instance $(\bar{X}_i, \bar{Y}_i)$ during the learning step, we measure the 2-norm loss of the modified classifier $\mathring{f}^{\setminus i}(X, Y) := \sum_{j \neq i} \mathring{k}((\bar{X}_j, \bar{Y}_j), (X, Y))\mathring{\alpha}_j \bar{\mu}_j + \mathring{b}$, which does not incorporate the contribution of the instance it is evaluated on. This ensures that the induced classifier $(\mathring{\alpha}, \mathring{b})$ focuses more on generalization from similar training instances, rather than the rote-learning of dataset-weight associations. In other words, the approach encourages stronger generalization, which is helpful for the typically quite small meta datasets. Altogether, the optimization procedure for the meta-learning task is given by:

$$\underset{\mathring{\alpha} \geq 0, \mathring{b}}{\operatorname{argmin}}\, C \sum_{i=1}^{t} l_2(\bar{\mu}, \mathring{f}^{\setminus i}(\bar{X}, \bar{Y})) + \mathring{\alpha}^T \mathring{D} \mathring{\alpha} \tag{6}$$

Here, $\mathring{D}$ is the meta kernel matrix normalized with the main kernel weight vectors: $\mathring{D}_{ij} = \mathring{k}((\bar{X}_i, \bar{Y}_i), (\bar{X}_j, \bar{Y}_j))\bar{\mu}_i^T \bar{\mu}_j$.

Finally, after the meta classifier is induced from the transfer datasets, we can apply it to the base data $(X, Y)$ to obtain a kernel $k$. This kernel, in turn, can be expected to work well for the data at hand, because it was derived from the related transfer datasets. In the next section we report results on experiments with the described transfer learning scheme.

## 4 Experiments

We evaluated the described approach to inductive transfer in two experiments. The first one deals with the prediction of biological activity of molecules, the second with text categorization.

### 4.1 Structure-Activity Relationship Data

In this experiment, we evaluated kernel-based inductive transfer on the task of learning the biological activity of compounds given their molecular structure as a graph. Learning challenges of this sort are known as structure-activity relationships (SARs), and are common in medicinal chemistry and pharmaceutical research. For the transfer and base data, we chose six datasets from the literature.

**Table 1.** Results: estimated predictive accuracy of the classifiers on the structure-activity relationship (SAR) data

| | Induct. Trans. | Kernel Learn. | Lin 100 | Quad 100 | RBF 100 | Lin 500 | Quad. 500 | RBF 500 | Lin all | Quad all | RBF all |
|---|---|---|---|---|---|---|---|---|---|---|---|
| bloodbarr | 75.6 | 70.6• | 75.9 | 75.7 | 72.8• | 72.7• | 74.1• | 71.3• | 71.7• | 73.6• | 71.3• |
| factorxa | 95.7 | 93.9• | 90.9• | 94.9• | 78.9• | 94.0• | 95.0• | 71.7• | 94.7• | 95.5 | 68.1• |
| hia | 81.8 | 77.0• | 76.7• | 77.4• | 70.6• | 79.1• | 79.4• | 67.2• | 77.3• | 79.2• | 66.3• |
| mutagenesis | 77.1 | 71.9• | 72.0• | 73.7• | 67.7• | 72.0• | 73.6• | 57.8• | 71.5• | 72.8• | 56.3• |
| nctrer | 81.6 | 77.6• | 79.3• | 78.2• | 65.0• | 80.0• | 79.9• | 62.8• | 78.5• | 80.1• | 61.1• |
| yoshida | 72.5 | 64.9• | 63.6• | 65.9• | 68.8• | 67.4• | 67.5• | 61.0• | 67.7• | 67.7• | 61.0• |

The bloodbarr dataset classifies 415 molecules according to the degree to which they can cross the blood-brain barrier [15]. The HIA (Human Intestinal Absorption) dataset contains 164 molecules from various sources, rated according to their oral bio-availability [20]. For the FactorXa set, the task is to discriminate between factor Xa inhibitors of high and low affinity [8]. The NCTRER dataset [7] deals with the prediction of binding activity of small molecules at the estrogen receptor, while the mutagenesis dataset [11] deals with the mutagenicity of compounds. Finally, the Yoshida dataset [23] consists of 265 molecules classified according to their bio-availability.

To transform the graph data into an attribute-value representation, we determined for each dataset the subgraphs that occur in more than a fixed fraction $p_{min}$ of graphs in the set. We lowered the threshold $p_{min}$ for each dataset until around 1,000 non-redundant subgraphs[1] were found and used these subgraphs as binary features ("subgraph occurs/does not occur"). For the resulting datasets, we applied the method described in section 3.1. We randomly chose half of the instances for the estimation of $f^*_{\bar{k}}$ and ran Algorithm 1 with $C = 1$ to compute good transfer kernels $\bar{k}^*$. Each such kernel is a linear combination of nine main kernels. For the main kernels, we chose a linear kernel, a quadratic kernel, and a Gaussian kernel with $\sigma = 1$. Since the subgraph features are sorted by size (i.e. number of edges), and since it is known that smaller subgraphs tend to be more informative, we applied the three kernels first on the first 100 features, then on the first 500 features, and finally on all features. The $\bar{k}^*_i$ for each transfer dataset are then linear combinations of these nine main kernels.

Finally, we set one dataset as base learning problem aside and kept the remaining datasets as transfer data. We generated the meta model from this transfer data using the histogram kernel outlined in section 3.2 and the optimization criterion described in section 3.3. The application of the meta model on the base data yields the desired base kernel, which was used in an SVM with $C = 1$ on the base data to learn a classifier. An evaluation of this classifier is given in

[1] Here, "non-redundant" means that we omitted features whose occurrence vector agreed exactly with an already exiting subgraph feature. This step is necessary to avoid the combinatorial explosion, which takes place when a large amount of slight subgraph variants occur frequently in a dataset.

**Table 2.** Results: estimated predictive accuracy of the classifiers induced on text categorization data

| Dataset | Induct. Trans. | Kernel Learn. | Lin 25% | Quad 25% | RBF 25% | Lin 50% | Quad 50% | RBF 50% | Lin all | Quad all | RBF all |
|---|---|---|---|---|---|---|---|---|---|---|---|
| 10341-14525 | 55.9 | 52.2• | 49.9• | 49.2• | 56.6 | 50.2• | 49.2• | 57.2 | 50.1• | 49.2• | 53.4• |
| 1092-1110 | 78.5 | 63.3• | 74.6• | 51.7• | 61.3• | 74.6• | 51.8• | 61.7• | 74.9• | 51.8• | 61.9• |
| 114202-190888 | 69.3 | 55.3• | 64.6• | 52.6• | 50.2• | 64.6• | 52.6• | 48.8• | 64.9• | 52.8• | 50.8• |
| 1155181-138526 | 85.2 | 64.8• | 85.2 | 57.5• | 57.0• | 85.2 | 58.4• | 56.1• | 85.1 | 58.3• | 55.9• |
| 123412-17899 | 68.7 | 66.7 | 66.3• | 64.6• | 63.5• | 66.3• | 64.5• | 63.7• | 66.3• | 64.2• | 64.4• |
| 14271-194927 | 67.1 | 56.9• | 65.3 | 52.7• | 49.4• | 65.1 | 52.4• | 49.8• | 64.9 | 52.0• | 49.1• |
| 14630-20186 | 73.4 | 56.1• | 64.7• | 44.6• | 59.7• | 64.6• | 45.2• | 59.9• | 67.5• | 45.8• | 54.8• |
| 173089-524208 | 79.8 | 68.0• | 73.5• | 63.4• | 62.5• | 73.7• | 63.9• | 64.5• | 73.8• | 63.9• | 64.2• |
| 17360-20186 | 69.5 | 56.0• | 62.8• | 51.6• | 53.2• | 62.6• | 51.7• | 53.9• | 63.1• | 51.9• | 51.2• |
| 186330-314499 | 59.3 | 56.0 | 52.3• | 49.0• | 57.4• | 52.5• | 49.0• | 57.3• | 52.7• | 49.0• | 55.9• |

Table 1. For comparison, we state the accuracy of a classifier that was induced by kernel learning without the use of the transfer data. To obtain this classifier, we split the training data into two parts of equal size, learn a kernel from the first part of the data and a classifier from the second part. We also give the accuracies of the classifiers induced by the nine main kernels. The estimates are averages over ten ten-fold cross-validation runs. An accuracy estimate for the comparison classifiers is marked with a bullet ("•"), if it is significantly worse than the one for the inductive transfer method according to a paired Wilcoxon rank test at the 5% significance level. The classifiers generated by the presented inductive transfer method were never significantly worse than those induced by the main kernels and by kernel learning. Inductive transfer outperforms kernel learning on all six datasets and it is better than all main kernels on five of them (significantly so on four datasets). Only on the bloodbarr dataset, the classifier constructed by the linear kernel proved to be competitive with the inductive transfer approach.

## 4.2 Text Data

For the second experiment, we evaluated the proposed method on text categorization data. This domain is well suited for our setting, because the internet provides a wealth of labeled text documents that can be used for inductive transfer. The experiments are based on the datasets from TechTC, the Technion repository of text categorization datasets [9]. Each dataset contains between 150 and 200 web documents, which were obtained by crawling the english language section of a web directory. The binary classification task posed by a dataset consists in telling which category of the web directory a document was taken from. An instance is represented by a vector whose components state the number of occurrences of a word. We sorted the features in descending order from frequently to infrequently occurring words. For the experiments, we randomly chose fifty datasets as transfer data, and selected ten base datasets for evaluation.

As a first step, we applied Algorithm 1 to construct a predictive kernel for each transfer dataset. As before, we used half of the data for classifier estimation and set $C = 1$. The kernels were again represented as linear combinations of nine main kernels $\bar{k}_1, \ldots, \bar{k}_9$. The first three main kernels were applied on only the first quarter of the features (that is, the 25% of most frequently occurring words), the second three main kernels used the 50% most frequently occurring words, and the last three kernels were computed on all features. As in the previous experiment, we had a linear, a quadratic and a Gauss main kernel for each feature (sub-)set. In the next step, we computed the meta kernel matrix for the fifty transfer datasets. The meta kernel was computed based on the overlap of word features between two datasets. More precisely, we chose $\mathring{k}(\bar{X}_1, \bar{X}_2) := |\bar{W}_1 \cap \bar{W}_2| / \max(|\bar{W}_1|, |\bar{W}_2|)$, where $\bar{W}_1$ and $\bar{W}_2$ denote the set of words in $\bar{X}_1$ and $\bar{X}_2$, respectively.

Plugging this meta kernel into the optimization criterion (6), we obtain a meta classifier that predicts a kernel for each base dataset. Table 2 gives an evaluation of the predictive accuracy of the classifiers that were induced with those base kernels (setting $C = 1$ for the base SVM). As before, we give also the predictive performance of a kernel induced from the training data and the single main kernels. All estimates are averaged over ten runs of tenfold-cross validation. An estimate is marked with a bullet ("•"), if it is significantly worse than the estimate for the inductive transfer method according to a paired Wilcoxon rank test on the 5% significance level, and with a circle ("◦"), if it is better. The inductive transfer method outperformed the kernel learning on all datasets (significantly so on eight out of the ten) and it was better than all main kernels in nine cases (in seven of them significantly), a clear indication that the predicted base kernels provide indeed a good bias for text categorization tasks on web documents.

## 5 Conclusion

In this paper, we described a kernel-based approach to inductive transfer. In this setting, the main goal is to make use of existing datasets to increase the predictive performance of classifiers induced on a new dataset. We framed the problem of finding a good bias for the new data as a meta learning problem: Given the transfer data, what should a kernel for the new data look like? We proposed a kernel-based approach to this problem. First, we presented an iterative procedure to find transfer kernels, which encode the bias necessary to perform well on the transfer datasets. Second, we introduced a convex optimization criterion for the meta learning problem to predict a suitable base kernel for the new data. Third, we described the histogram kernel, a general purpose kernel for the meta learning task. Of course, the work can be extended in various directions. For example, it would be interesting to investigate other meta kernels that use more (possibly domain specific) knowledge about the datasets.

## References

1. Argyriou, A., Evgeniou, T., Pontil, M.: Multi-task feature learning. In: Schölkopf, B., Platt, J.C., Hoffman, T. (eds.) Proceedings of the Twentieth Annual Conference on Neural Information Processing Systems 19. Advances in Neural Information Processing Systems, pp. 41–48. MIT Press, Cambridge (2006)
2. Baxter, J.: A model of inductive bias learning. Journal of Artificial Intelligence Research 12, 149–198 (2000)
3. Caruana, R.: Multitask learning. Machine Learning 28(1), 41–75 (1997)
4. Crammer, K., Keshet, J., Singer, Y.: Kernel design using boosting. In: Becker, S., Thrun, S., Obermayer, K. (eds.) Advances in Neural Information Processing Systems 15, pp. 537–544. MIT Press, Cambridge (2002)
5. Erhan, D., Bengio, Y., L'Heureux, P.-J., Yue, S.Y.: Generalizing to a zero-data task: a computational chemistry case study. Technical Report 1286, Département d'informatique et recherche opérationnelle, Université de Montréal (2006)
6. Evgeniou, T., Micchelli, C.A., Pontil, M.: Learning multiple tasks with kernel methods. J. Mach. Learn. Res. 6, 615–637 (2005)
7. Fang, H., Tong, W., Shi, L.M., Blair, R., Perkins, R., Branham, W., Hass, B.S., Xie, Q., Dial, S.L., Moland, C.L., Sheehan, D.M.: Structure-activity relationships for a large diverse set of natural, synthetic, and environmental estrogens. Chemical Research in Toxicology 14(3), 280–294 (2001)
8. Fontaine, F., Pastor, M., Zamora, I., Sanz, F.: Anchor-GRIND: Filling the gap between standard 3D QSAR and the grid-independent descriptors. Journal of Medicinal Chemistry 48(7), 2687–2694 (2005)
9. Gabrilovich, E., Markovitch, S.: Parameterized generation of labeled datasets for text categorization based on a hierarchical directory. In: Proceedings of The 27th Annual International ACM SIGIR Conference, Sheffield, UK, pp. 250–257. ACM Press, New York (2004)
10. Girolami, M., Rogers, S.: Hierarchic Bayesian models for kernel learning. In: ICML 2005: Proceedings of the 22nd international conference on Machine learning, pp. 241–248. ACM Press, New York (2005)
11. Helma, C., Cramer, T., Kramer, S., De Raedt, L.: Data mining and machine learning techniques for the identification of mutagenicity inducing substructures and structure activity relationships of noncongeneric compounds. Journal of Chemical Information and Computer Sciences 44(4), 1402–1411 (2004)
12. Hertz, T., Hillel, A.B., Weinshall, D.: Learning a kernel function for classification with small training samples. In: ICML 2006: Proceedings of the 23rd international conference on Machine learning, pp. 401–408. ACM, New York (2006)
13. Kaski, S., Peltonen, J.: Learning from relevant tasks only. In: Kok, J.N., Koronacki, J., de Mántaras, R.L., Matwin, S., Mladenic, D., Skowron, A. (eds.) ECML 2007. LNCS (LNAI), vol. 4701, pp. 608–615. Springer, Heidelberg (2007)
14. Lanckriet, G.R.G., Cristianini, N., Bartlett, P., El Ghaoui, L., Jordan, M.I.: Learning the kernel matrix with semidefinite programming. J. Mach. Learn. Res. 5, 27–72 (2004)
15. Li, H., Yap, C.W., Ung, C.Y., Xue, Y., Cao, Z.W., Chen, Y.Z.: Effect of selection of molecular descriptors on the prediction of blood-brain barrier penetrating and nonpenetrating agents by statistical learning methods. Journal of Chemical Information and Modeling 45(5), 1376–1384 (2005)
16. Micchelli, C.A., Pontil, M.: Learning the kernel function via regularization. J. Mach. Learn. Res. 6, 1099–1125 (2005)

17. Ong, C.S., Smola, A.J., Williamson, R.C.: Learning the kernel with hyperkernels. J. Mach. Learn. Res. 6, 1043–1071 (2005)
18. Pfahringer, B., Bensusan, H., Giraud-Carrier, C.G.: Meta-learning by landmarking various learning algorithms. In: ICML 2000: Proceedings of the Seventeenth International Conference on Machine Learning, pp. 743–750. Morgan Kaufmann, San Francisco (2000)
19. Sonnenburg, S., Rätsch, G., Schäfer, C., Schölkopf, B.: Large scale multiple kernel learning. J. Mach. Learn. Res. 7, 1531–1565 (2006)
20. Wegner, J.K., Fröhlich, H., Zell, A.: Feature selection for descriptor based classification models. 1. theory and ga-sec algorithm. Journal of Chemical Information and Modeling 44(3), 921–930 (2004)
21. Weston, J., Herbrich, R.: Adaptive margin support vector machines. In: Advances in Large-Margin Classifiers, pp. 281–295. MIT Press, Cambridge (2000)
22. Xue, Y., Liao, X., Carin, L., Krishnapuram, B.: Multi-task learning for classification with dirichlet process priors. J. Mach. Learn. Res. 8, 35–63 (2007)
23. Yoshida, F., Topliss, J.: QSAR model for drug human oral bioavailability. J. Med. Chem. 43, 2575–2585 (2000)
24. Yu, K., Tresp, V., Schwaighofer, A.: Learning gaussian processes from multiple tasks. In: ICML 2005: Proceedings of the 22nd international conference on Machine learning, pp. 1012–1019. ACM Press, New York (2005)

# State-Dependent Exploration for Policy Gradient Methods

Thomas Rückstieß, Martin Felder, and Jürgen Schmidhuber

Technische Universität München, 85748 Garching, Germany
{ruecksti,felder,schmidhu}@in.tum.de

**Abstract.** Policy Gradient methods are model-free reinforcement learning algorithms which in recent years have been successfully applied to many real-world problems. Typically, Likelihood Ratio (LR) methods are used to estimate the gradient, but they suffer from high variance due to random exploration at every time step of each training episode. Our solution to this problem is to introduce a state-dependent exploration function (SDE) which during an episode returns the same action for any given state. This results in less variance per episode and faster convergence. SDE also finds solutions overlooked by other methods, and even improves upon state-of-the-art gradient estimators such as Natural Actor-Critic. We systematically derive SDE and apply it to several illustrative toy problems and a challenging robotics simulation task, where SDE greatly outperforms random exploration.

## 1 Introduction

Reinforcement Learning (RL) is a powerful concept for dealing with semi-supervised control tasks. There is no teacher to tell the agent the correct action for a given situation, but it does receive feedback (the *reinforcement signal*) about how well it is doing. While exploring the space of possible actions, the reinforcement signal can be used to adapt the parameters governing the agent's behavior. Classical RL algorithms [1,2] are designed for problems with a limited, discrete number of states. For these scenarios, sophisticated exploration strategies can be found in the literature [3,4].

In contrast, Policy Gradient (PG) methods as pioneered by Williams [5] can deal with continuous states and actions, as they appear in many real-life settings. They can handle function approximation, avoid sudden discontinuities in the action policy during learning, and were shown to converge at least locally [6]. Successful applications are found e.g. in robotics [7], financial data prediction [8] or network routing [9].

However, a major problem in RL remains that feedback is rarely available at every time step. Imagine a robot trying to exit a labyrinth within a set time, with a default policy of driving straight. Feedback is given at the end of an *episode*, based on whether it was successful or not. PG methods most commonly use a random exploration strategy [5,7], where the deterministic action ("if wall ahead, go straight") at each time step is perturbed by Gaussian noise. This way,

W. Daelemans et al. (Eds.): ECML PKDD 2008, Part II, LNAI 5212, pp. 234–249, 2008.

the robot may wiggle free from time to time, but it is very hard to improve the policy based on this success, due to the high variance in the gradient estimation. Obviously, a lot of research has gone into devising smarter, more robust ways of estimating the gradient, as detailed in the excellent survey by Peters [7].

Our novel approach is much simpler and targets the exploration strategy instead: In the example, the robot would use a deterministic function providing an exploration offset consistent throughout the episode, but still depending on the state. This might easily change the policy into something like "if wall ahead, veer a little left", which is much more likely to lead out of the labyrinth, and thus can be identified easily as a policy improvement. Hence, our method, which we call *state-dependent exploration* (SDE), causes considerable variance reduction and therefore faster convergence. Because it only affects exploration and does not depend on a particular gradient estimation technique, SDE can be enhanced with any episodic likelihood ratio (LR) method, like REINFORCE [5], GPOMDP [10], or ENAC [11], to reduce the variance even further.

Our exploration strategy is in a sense related to Finite Difference (FD) methods like SPSA [12] as both create policy deltas (or strategy variations) rather than perturbing single actions. However, direct parameter perturbation has to be handled with care, since small changes in the policy can easily lead to unforseen and unstable behavior and a fair amount of system knowledge is therefore necessary. Furthermore, FD are very sensitive to noise and hence not suited for many real-world tasks. SDE does not suffer from these drawbacks—it embeds the power of FD exploration into the stable LR framework.

The remainder of this paper is structured as follows: Section 2 introduces the policy gradient framework together with a thorough derivation of the equations and applications to function approximation. Our novel exploration strategy SDE will be explained in detail in section 3. Experiments and their results are described in section 4. The paper concludes with a short discussion in section 5.

## 2 Policy Gradient Framework

An advantage of policy gradient methods is that they don't require the environment to be Markovian, i.e. each controller action may depend on the whole history encountered. So we will introduce our policy gradient framework for general non-Markovian environments but later assume a Markov Decission Process (MDP) for ease of argument.

### 2.1 General Assumptions

A policy $\pi(u|h, \theta)$ is the probability of taking action $u$ when encountering history $h$ under the policy parameters $\theta$. Since we use parameterized policies throughout this paper, we usually ommit $\theta$ and just write $\pi(u|h)$. We will use $h^\pi$ for the history of all the observations $x$, actions $u$, and rewards $r$ encountered when following policy $\pi$. The history at time $t = 0$ is defined as the sequence $h_0^\pi = \{x_0\}$, consisting only of the start state $x_0$. The history at time $t$ consists of all the observations, actions and rewards encountered so far and is defined as $h_t^\pi = \{x_0, u_0, r_0, x_1, \ldots, u_{t-1}, r_{t-1}, x_t\}$.

The return for the controller whose interaction with the environment produces history $h^\pi$ is written as $R(h^\pi)$, which is defined as $R(h^\pi) = a_\Sigma \sum_{t=0}^{T} a_D\, r_t$ with $a_\Sigma = (1-\gamma)$, $a_D = \gamma^t$ for discounted (possibly continuous) tasks and $a_\Sigma = 1/T$, $a_D = 1$ for undiscounted (and thus necessarily episodic) tasks. In this paper, we deal with episodic learning and therefore will use the latter definition. The expectation operator is written as $E\{\cdot\}$.

The overall performance measure of policy $\pi$, independent from any history $h$, is denoted $J(\pi)$. It is defined as $J(\pi) = E\{R(h^\pi)\} = \int p(h^\pi)R(h^\pi)\, dh^\pi$. Instead of $J(\pi)$ for policy $\pi$ parameterized with $\theta$, we will also write $J(\theta)$.

To optimize policy $\pi$, we want to move the parameters $\theta$ along the gradient of $J$ to an optimum with a certain learning rate $\alpha$:

$$\theta_{t+1} = \theta_t + \alpha \nabla_\theta J(\pi). \tag{1}$$

The gradient $\nabla_\theta J(\pi)$ is

$$\nabla_\theta J(\pi) = \nabla_\theta \int_{h^\pi} p(h^\pi)R(h^\pi)\, dh^\pi = \int_{h^\pi} \nabla_\theta p(h^\pi)R(h^\pi)\, dh^\pi. \tag{2}$$

## 2.2 Likelihood Ratio Methods

Rather than perturbing the policy directly, as it is the case with FD methods [12,7], LR methods [5] perturb the resulting action instead, leading to a stochastic policy (which we assume to be differentiable with respect to its parameters $\theta$), such as

$$u = f(h, \theta) + \epsilon, \ \epsilon \sim \mathcal{N}(0, \sigma^2) \tag{3}$$

where $f$ is the controller and $\epsilon$ the exploration noise. Unlike FD methods, the new policy that leads to this behavior is not known and consequently the difference quotient

$$\frac{\partial J(\theta)}{\partial \theta_i} \approx \frac{J(\theta + \delta\theta) - J(\theta)}{\delta \theta_i} \tag{4}$$

can not be calculated. Thus, LR methods use a different approach in estimating $\nabla_\theta J(\theta)$.

Following the derivation of e.g. Wierstra et al. [13], we start with the probability of observing history $h^\pi$ under policy $\pi$, which is given by the probability of starting with an initial observation $x_0$, multiplied by the probability of taking action $u_0$ under $h_0$, multiplied by the probability of receiving the next observation $x_1$, and so on. Thus, (5) gives the probability of encountering a certain history $h^\pi$.

$$p(h^\pi) = p(x_0) \prod_{t=0}^{T-1} \pi(u_t | h_t^\pi)\, p(x_{t+1} | h_t^\pi, u_t) \tag{5}$$

Inserting this into (2), we can rewrite the equation by multiplying with $1 = {p(h^\pi)}/{p(h^\pi)}$ and using $\frac{1}{x}\nabla x = \nabla \log(x)$ to get

$$\nabla_\theta J(\pi) = \int \frac{p(h^\pi)}{p(h^\pi)} \nabla_\theta p(h^\pi) R(h^\pi)\, dh^\pi \tag{6}$$

$$= \int p(h^\pi) \nabla_\theta \log p(h^\pi) R(h^\pi)\, dh^\pi . \tag{7}$$

For now, let us consider the gradient $\nabla_\theta \log p(h^\pi)$. Substituting the probability $p(h^\pi)$ according to (5) gives

$$\nabla_\theta \log p(h^\pi) = \nabla_\theta \log \Big[ p(x_0) \prod_{t=0}^{T-1} \pi(u_t|h_t^\pi)\, p(x_{t+1}|h_t^\pi, u_t) \Big]$$

$$= \nabla_\theta \Big[ \log p(x_0) + \sum_{t=0}^{T-1} \log \pi(u_t|h_t^\pi) + \sum_{t=0}^{T-1} \log p(x_{t+1}|h_t^\pi, u_t) \Big] . \tag{8}$$

On the right side of (8), only the policy $\pi$ is dependent on $\theta$, so the gradient can be simplified to

$$\nabla_\theta \log p(h^\pi) = \sum_{t=0}^{T-1} \nabla_\theta \log \pi(u_t|h_t^\pi). \tag{9}$$

We can now resubstitute this term into (7) and get

$$\nabla_\theta J(\pi) = \int p(h^\pi) \sum_{t=0}^{T-1} \nabla_\theta \log \pi(u_t|h_t^\pi) R(h^\pi)\, dh^\pi$$

$$= E\left\{ \sum_{t=0}^{T-1} \nabla_\theta \log \pi(u_t|h_t^\pi) R(h^\pi) \right\} . \tag{10}$$

Unfortunately, the probability distribution $p(h^\pi)$ over the histories produced by $\pi$ is not known in general. Thus we need to approximate the expectation, e.g. by *Monte-Carlo sampling.* To this end, we collect $N$ samples through world interaction, where a single sample comprises a complete history $h^\pi$ (one episode or rollout) to which a return $R(h^\pi)$ can be assigned, and sum over all samples which basically yields Williams' [5] episodic REINFORCE gradient estimation:

$$\nabla_\theta J(\pi) \approx \frac{1}{N} \sum_{h^\pi} \sum_{t=0}^{T-1} \nabla_\theta \log \pi(u_t|h_t^\pi) R(h^\pi) \tag{11}$$

At this point there are several approaches to improve gradient estimates, as mentioned in the introduction. Neither these nor ideas like baselines [5], the PEGASUS trick [14] or other variance reduction techniques [15] are treated here. They are complementary to our approach, and their combination with SDE will be covered by a future paper.

## 2.3 Application to Function Approximation

Here we describe how the results above, in particular (11), can be applied to general parametric function approximation. Because we are dealing with multi-dimensional states $\boldsymbol{x}$ and multi-dimensional actions $\boldsymbol{u}$, we will now use bold font for (column) vectors in our notation for clarification.

To avoid the issue of a growing history length and to simplify the equations, we will assume the world to be Markovian for the remainder of the paper, i.e. the current action only depends on the last state encountered, so that $\pi(u_t|h_t^\pi) = \pi(u_t|x_t)$. But due to its general derivation, the idea of SDE is still applicable to non-Markovian environments.

The most general case would include a multi-variate normal distribution function with a covariance matrix $\boldsymbol{\Sigma}$, but this would square the number of parameters and required samples. Also, differentiating this distribution requires calculation of $\boldsymbol{\Sigma}^{-1}$, which is time-consuming. We will instead use a simplification here and add independent uni-variate normal noise to each element of the output vector seperately. This corresponds to a covariance matrix $\boldsymbol{\Sigma} = \mathrm{diag}(\sigma_1, \ldots, \sigma_n)$.[1] The action $\boldsymbol{u}$ can thus be computed as

$$\boldsymbol{u} = f(\boldsymbol{x}, \boldsymbol{\theta}) + \boldsymbol{e} = \begin{bmatrix} f_1(\boldsymbol{x}, \boldsymbol{\theta}) \\ \vdots \\ f_n(\boldsymbol{x}, \boldsymbol{\theta}) \end{bmatrix} + \begin{bmatrix} e_1 \\ \vdots \\ e_n \end{bmatrix} \tag{12}$$

with $\boldsymbol{\theta} = [\theta_1, \theta_2, \ldots]$ being the parameter vector and $f_j$ the $j$th controller output element. The exploration values $e_j$ are each drawn from a normal distribution $e_j \sim \mathcal{N}(0, \sigma_j^2)$. The policy $\boldsymbol{\pi}(\boldsymbol{u}|\boldsymbol{x})$ is the probability of executing action $\boldsymbol{u}$ when in state $\boldsymbol{x}$. Because of the independence of the elements, it can be decomposed into $\boldsymbol{\pi}(\boldsymbol{u}|\boldsymbol{x}) = \prod_{k \in \mathbb{O}} \pi_k(u_k|\boldsymbol{x})$ with $\mathbb{O}$ as the set of indices over all outputs, and therefore $\log \boldsymbol{\pi}(\boldsymbol{u}|\boldsymbol{x}) = \sum_{k \in \mathbb{O}} \log \pi_k(u_k|\boldsymbol{x})$. The element-wise policy $\pi_k(u_k|\boldsymbol{x})$ is the probability of receiving value $u_k$ as $k$th element of action vector $\boldsymbol{u}$ when encountering state $\boldsymbol{x}$ and is given by

$$\pi_k(u_k|\boldsymbol{x}) = \frac{1}{\sqrt{2\pi}\sigma_k} \exp\left(-\frac{(u_k - \mu_k)^2}{2\sigma_k^2}\right), \tag{13}$$

where we substituted $\mu_k := f_k(\boldsymbol{x}, \boldsymbol{\theta})$. We differentiate with respect to the parameters $\theta_j$ and $\sigma_j$:

$$\begin{aligned} \frac{\partial \log \boldsymbol{\pi}(\boldsymbol{u}|\boldsymbol{x})}{\partial \theta_j} &= \sum_{k \in \mathbb{O}} \frac{\partial \log \pi_k(u_k|\boldsymbol{x})}{\partial \mu_k} \frac{\partial \mu_k}{\partial \theta_j} \\ &= \sum_{k \in \mathbb{O}} \frac{(u_k - \mu_k)}{\sigma_k^2} \frac{\partial \mu_k}{\partial \theta_j} \end{aligned} \tag{14}$$

[1] A further simplification would use $\boldsymbol{\Sigma} = \sigma \boldsymbol{I}$ with $\boldsymbol{I}$ being the unity matrix. This is advisable if the optimal solution for all parameters is expected to lay in similar value ranges.

$$\frac{\partial \log \boldsymbol{\pi}(\boldsymbol{u}|\boldsymbol{x})}{\partial \sigma_j} = \sum_{k \in \mathbb{O}} \frac{\partial \log \pi_k(u_k|\boldsymbol{x})}{\partial \sigma_j} = \frac{(u_j - \mu_j)^2 - \sigma_j^2}{\sigma_j^3} \tag{15}$$

For the linear case, where $\boldsymbol{f}(\boldsymbol{x}, \boldsymbol{\theta}) = \boldsymbol{\Theta x}$ with the parameter matrix $\boldsymbol{\Theta} = [\theta_{ji}]$ mapping states to actions, (14) becomes

$$\frac{\partial \log \boldsymbol{\pi}(\boldsymbol{u}|\boldsymbol{x})}{\partial \theta_{ji}} = \frac{(u_j - \sum_i \theta_{ji} x_i)}{\sigma_j^2} x_i \tag{16}$$

An issue with nonlinear function approximation (NLFA) is a parameter dimensionality typically much higher than their output dimensionality, constituting a huge search space for FD methods. However, in combination with LR methods, they are interesting because LR methods only perturb the resulting outputs and not the parameters directly. Assuming the NLFA is differentiable with respect to its parameters, one can easily calculate the log likelihood values for each single parameter.

The factor $\frac{\partial \mu_k}{\partial \theta_j}$ in (14) describes the differentiation through the function approximator. It is convenient to use existing implementations, where instead of an error, the log likelihood derivative with respect to the mean, i.e. the first factor of the sum in (14), can be injected. The usual backward pass through the NLFA—known from supervised learning settings—then results in the log likelihood derivatives for each parameter [5].

## 3 State-Dependent Exploration

As indicated in the introduction, adding noise to the action $u$ of a stochastic policy (3) at each step enables random exploration, but also aggravates the credit assignment problem: The overall reward for an episode (also called *return*) cannot be properly assigned to individual actions because information about which actions (if any) had a positive effect on the return value is not accessible.[2]

Our alternative approach adds a *state-dependent offset* to the action at each timestep, which can still carry the necessary exploratory randomness through variation between episodes, but will always return the same value in the same state within an episode. We define a function $\hat{\boldsymbol{\epsilon}}(\boldsymbol{x}; \hat{\boldsymbol{\theta}})$ on the states, which will act as a pseudo-random function that takes the state $\boldsymbol{x}$ as input. Randomness originates from parameters $\hat{\boldsymbol{\theta}}$ being drawn from a normal distribution $\hat{\theta}_j \sim \mathcal{N}(0, \hat{\sigma}_j^2)$. As discussed in section 2.3, simplifications to reduce the number of variance parameters can be applied. The action is then calculated by

$$\boldsymbol{u} = \boldsymbol{f}(\boldsymbol{x}; \boldsymbol{\theta}) + \hat{\boldsymbol{\epsilon}}(\boldsymbol{x}; \hat{\boldsymbol{\theta}}), \quad \hat{\theta}_j \sim \mathcal{N}(0, \hat{\sigma}_j^2). \tag{17}$$

[2] GPOMDP [10], also known as the Policy Gradient Theorem [6], does consider single step rewards. However, it still introduces a significant amount of variance to a rollout with traditional random exploration.

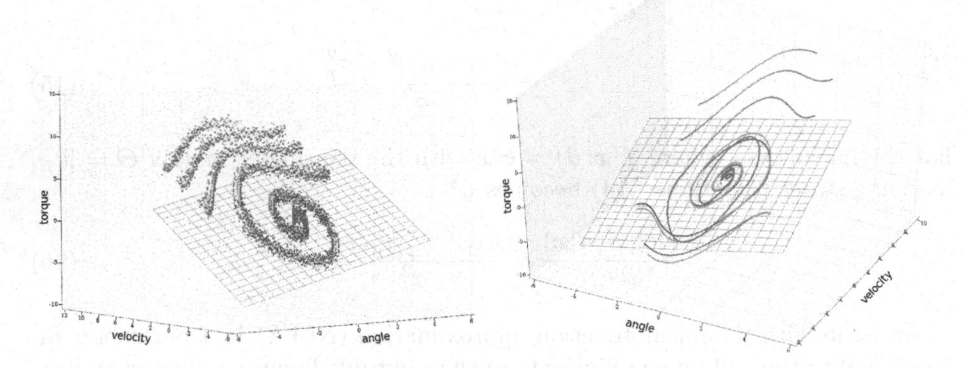


**Fig. 1.** Illustration of the main difference between random (left) and state-dependent (right) exploration. Several rollouts in state-action space of a task with state $\boldsymbol{x} \in \mathbb{R}^2$ (x- and y-axis) and action $u \in \mathbb{R}$ (z-axis) are plotted. While random exploration follows the same trajectory over and over again (with added noise), SDE instead tries different *strategies* and can quickly find solutions that would take a long time to discover with random exploration.

If the parameters $\hat{\boldsymbol{\theta}}$ are drawn at each timestep, we have an LR algorithm as in (3) and (12), although with a different exploration variance. However, if we keep $\hat{\boldsymbol{\theta}}$ constant for a full episode, then our action will have the same exploration added whenever we encounter the same state (Figure 1). Depending on the choice of $\hat{\boldsymbol{\epsilon}}(\boldsymbol{x})$, the randomness can further be "continuous", resulting in similar offsets for similar states. Effectively, by drawing $\hat{\boldsymbol{\theta}}$, we actually create a *policy delta*, similar to FD methods. In fact, if both $\boldsymbol{f}(\boldsymbol{x};\boldsymbol{\Theta})$ with $\boldsymbol{\Theta} = [\theta_{ji}]$ and $\hat{\boldsymbol{\epsilon}}(\boldsymbol{x},\hat{\boldsymbol{\Theta}})$ with $\hat{\boldsymbol{\Theta}} = [\hat{\theta}_{ji}]$ are linear functions, we see that

$$\begin{aligned} \boldsymbol{u} &= \boldsymbol{f}(\boldsymbol{x};\boldsymbol{\Theta}) + \hat{\boldsymbol{\epsilon}}(\boldsymbol{x};\hat{\boldsymbol{\Theta}}) \\ &= \boldsymbol{\Theta}\boldsymbol{x} + \hat{\boldsymbol{\Theta}}\boldsymbol{x} \\ &= (\boldsymbol{\Theta} + \hat{\boldsymbol{\Theta}})\boldsymbol{x}, \end{aligned} \tag{18}$$

which shows that direct parameter perturbation methods (cf. (4)) are a special case of SDE and can be expressed in this more general reinforcement framework.

### 3.1 Updates of Exploration Variances

For a linear exploration function $\hat{\boldsymbol{\epsilon}}(\boldsymbol{x};\hat{\boldsymbol{\Theta}}) = \hat{\boldsymbol{\Theta}}\boldsymbol{x}$ it is possible to calculate the derivative of the log likelihood with respect to the variance. We will derive the adaptation for general $\hat{\sigma}_{ji}$, any parameter reduction techniques from 2.3 can be applied accordingly.

First, we need the distribution of the action vector elements $u_j$:

$$u_j = f_j(\boldsymbol{x},\boldsymbol{\Theta}) + \hat{\boldsymbol{\Theta}}_j\boldsymbol{x} = f_j(\boldsymbol{x},\boldsymbol{\Theta}) + \sum_i \hat{\theta}_{ji}x_i \tag{19}$$

with $f_j(\boldsymbol{x}, \boldsymbol{\Theta})$ as the $j$th element of the return vector of the deterministic controller $f$ and $\hat{\theta}_{ji} \sim \mathcal{N}(0, \hat{\sigma}_{ji}^2)$. We now use two well-known properties of normal distributions: First, if $X$ and $Y$ are two independent random variables with $X \sim \mathcal{N}(\mu_a, \sigma_a^2)$ and $Y \sim \mathcal{N}(\mu_b, \sigma_b^2)$ then $U = X + Y$ is normally distributed with $U \sim \mathcal{N}(\mu_a + \mu_b, \sigma_a^2 + \sigma_b^2)$. Second, if $X \sim \mathcal{N}(\mu, \sigma^2)$ and $a, b \in \mathbb{R}$, then $aX + b \sim \mathcal{N}(a\mu + b, (a\sigma)^2)$.

Applied to (19), we see that $\hat{\theta}_{ji} x_i \sim \mathcal{N}(0, (x_i \hat{\sigma}_{ji})^2)$, that the sum is distributed as $\sum_i \hat{\theta}_{ji} x_i \sim \mathcal{N}(0, \sum_i (x_i \hat{\sigma}_{ji})^2)$, and that the action element $u_j$ is therefore distributed as

$$u_j \sim \mathcal{N}(f_j(\boldsymbol{x}, \boldsymbol{\Theta}), \sum_i (x_i \hat{\sigma}_{ji})^2), \tag{20}$$

where we will substitute $\mu_j := f_j(\boldsymbol{x}, \boldsymbol{\Theta})$ and $\sigma_j^2 := \sum_i (x_i \hat{\sigma}_{ji})^2$ to obtain expression (13) for the policy components again. Differentiation of the policy with respect to the free parameters $\hat{\sigma}_{ji}$ yields:

$$\begin{aligned} \frac{\partial \log \boldsymbol{\pi}(\boldsymbol{u}|\boldsymbol{x})}{\partial \hat{\sigma}_{ji}} &= \sum_k \frac{\partial \log \pi_k(u_k|\boldsymbol{x})}{\partial \sigma_j} \frac{\partial \sigma_j}{\partial \hat{\sigma}_{ji}} \\ &= \frac{(u_j - \mu_j)^2 - \sigma_j^2}{\sigma_j^4} x_i^2 \hat{\sigma}_{ji} \end{aligned} \tag{21}$$

For more complex exploration functions, calculating the exact derivative for the sigma adaptation might not be possible and heuristic or manual adaptation (e.g. with slowly decreasing $\hat{\sigma}$) is required.

### 3.2 Stochastic Policies

The original policy gradient setup as presented in e.g. [5] conveniently unifies the two stochastic features of the algorithm: the stochastic exploration and the stochasticity of the policy itself. Both were represented by the Gaussian noise added on top of the controller. While elegant on the one hand, it also conceals the fact that there are two different stochastic processes. With SDE, randomness has been taken out of the controller completely and is represented by the seperate exploration function. So if learning is switched off, the controller only returns deterministic actions. But in many scenarios the best policy is necessarily of stochastic nature.

It is possible and straight-forward to implement SDE with stochastic policies, by combining both random and state-dependent exploration in one controller, as in

$$\boldsymbol{u} = f(\boldsymbol{x}; \boldsymbol{\theta}) + \boldsymbol{\epsilon} + \hat{\boldsymbol{\epsilon}}(\boldsymbol{x}; \hat{\boldsymbol{\theta}}), \tag{22}$$

where $\epsilon_j \sim N(0, \sigma_j)$ and $\hat{\theta}_j \sim N(0, \hat{\sigma}_j)$. Since the respective noises are simply added together, none of them affects the derivative of the log-likelihood of the other and $\sigma$ and $\hat{\sigma}$ can be updated independently. In this case, the trajectories through state-action space would look like a noisy version of Figure 1, righthand side.

### 3.3 Negative Variances

For practical applications, we also have to deal with the issue of negative variances. Obviously, we must prevent $\sigma$ from falling below zero, which can happen since the right side of (15) can become negative. We therefore designed the following smooth, continuous function and its first derivative:

$$\text{expln}(\sigma) = \begin{cases} \exp(\sigma) & \sigma \leq 0 \\ \ln(\sigma + 1) + 1 & \text{else} \end{cases} \tag{23}$$

$$\text{expln}'(\sigma) = \begin{cases} \exp(\sigma) & \sigma \leq 0 \\ \frac{1}{\sigma+1} & \text{else} \end{cases} \tag{24}$$

Substitution of $\sigma^* := \text{expln}(\sigma)$ will keep the variance above zero (exponential part) and also prevent it from growing too fast (logarithmic part). For this, the derivatives in (15) and (21) have to be multiplied by $\text{expln}'(\sigma)$. In the experiments in section 4, this factor is included.

## 4 Experiments

Two different sets of experiments are conducted to investigate both the theoretical properties and the practical application of SDE. The first looks at plain function minimization and analyses the properties of SDE compared to REX. The second demonstrates SDE's usefulness for real-world problems with a simulated robot hand trying to catch a ball.

### 4.1 Function Minimization

The following sections compare SDE and random exploration (REX) with regard to sensitivity to noise, episode length, and parameter dimensionality. We chose a very basic setup where the task was to minimize $g(x) = x^2$. This is sufficient for first convergence evaluations since policy gradients are known to only converge locally. The agent's state $x$ lies on the abscissa, its action is multiplied with a step-size factor $s$ and the result is interpreted as a step along the abscissa in either direction. To make the task more challenging, we always added random noise to the agent's action. Each experiment was repeated 30 times, averaging the results. Our experiments were all episodic, with the return $R$ for each episode being the average reward as stated in section 2.1. The reward per timestep was defined as $r_t = -g(x_t)$, thus a controller reducing the costs (negative reward) minimizes $g(x)$.

For a clean comparison of SDE and REX, we used the SDE algorithm in both cases, and emulated REX by drawing new exploration function parameters $\hat{\boldsymbol{\theta}}$ after each step (see Section 3). Unless stated otherwise, all experiments were conducted with a REINFORCE gradient estimator with optimal baseline [7] and the following parameters: learning rate $\alpha = 0.1$, step-size factor $s = 0.1$, initial parameter $\theta_0 = -2.0$, episode length EL $= 15$ and starting exploration noise $\hat{\sigma} = e^{-2} \approx 0.135$.

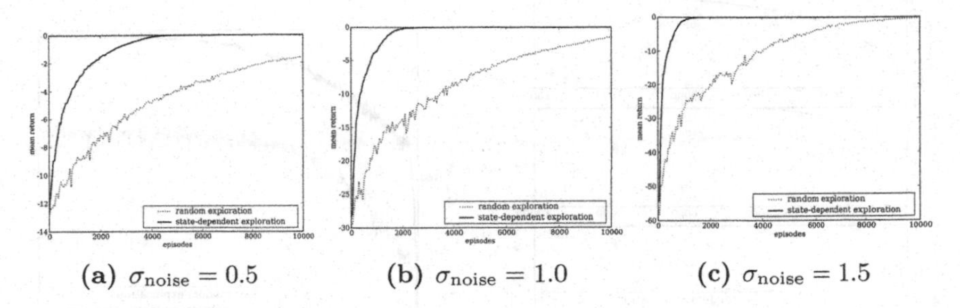


(a) $\sigma_{\text{noise}} = 0.5$ (b) $\sigma_{\text{noise}} = 1.0$ (c) $\sigma_{\text{noise}} = 1.5$

**Fig. 2.** Convergence for different levels of noise, averaged over 30 runs per curve. The upper solid curve shows SDE, the lower dotted curve REX.

**Noise Level.** First, we investigated how both SDE and REX deal with noise in the setting. We added normally distributed noise with variance $\sigma^2_{\text{noise}}$ to each new state after the agent's action was executed: $x_{t+1} = x_t + su_t + \mathcal{N}(0, \sigma^2_{\text{noise}})$, where $s$ is the step-size factor and $u_t$ is the action at time $t$. The results of experiments with three different noise levels are given in Figure 2 and the right part of Table 1.

The results show that SDE is much more robust to noise, since its advantage over REX grows with the noise level. This is a direct effect of the credit assignment problem, which is more severe as the randomness of actions increases.

An interesting side-effect can also be found when comparing the convergence times for different noise levels. Both methods, REX and SDE, ran at better convergence rates with higher noise. The reason for this behavior can be shown best for a one-dimensional linear controller. In the absence of (environmental) noise, we then have:

$$x_t = x_{t-1} + su_{t-1}$$
$$u_t = \theta x_t + \epsilon_{\text{explore}}$$

**Table 1.** Noise and episode length (EL) sensitivity of REX and SDE. $\sigma_{\text{noise}}$ is the standard deviation of the environmental noise. The steps designate the number of episodes until convergence, which was defined as $R_t > R_{\text{lim}}$ (a value that all controllers reached). The quotient REX/SDE is given as a speed-up factor.

| | | | # steps | | | | | | # steps | | |
|---|---|---|---|---|---|---|---|---|---|---|---|
| $\sigma_{\text{noise}}$ | EL | $R_{\text{lim}}$ | REX | SDE | speed-up | $\sigma_{\text{noise}}$ | EL | $R_{\text{lim}}$ | REX | SDE | speed-up |
| | 5 | | 9450 | 3350 | 2.82 | 0.5 | | | 9950 | 2000 | 4.98 |
| | 15 | | 9150 | 1850 | 4.95 | 1.0 | 15 | -1.6 | 9850 | 1400 | 7.04 |
| 0.5 | 30 | -1.8 | 9700 | 1050 | 9.24 | 1.5 | | | 7700 | 900 | 8.56 |
| | 45 | | 8800 | 650 | 13.54 | | | | | | |
| | 60 | | 8050 | 500 | 16.10 | | | | | | |

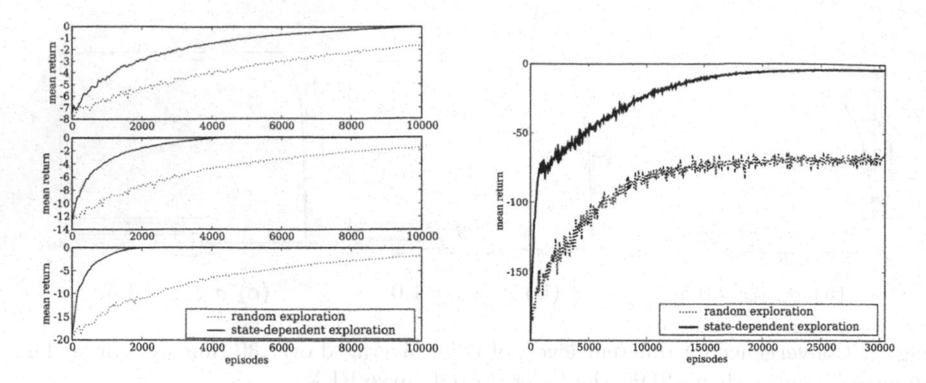


**Fig. 3.** Left: Results for different episode lengths, from top to bottom: 5, 15, 30. Right: Nonlinear controller with 18 free parameters on a 2-dimensional task. With REX, the agent became stuck in a local optimum, while SDE found the same optimum about 15 times faster and then converged to a better solution.

Adding noise to the state update results in

$$\begin{aligned} x'_t &= x_t + \epsilon_{\text{noise}} = x_{t-1} + su_{t-1} + \epsilon_{\text{noise}} \\ u'_t &= \theta(x_{t-1} + su_{t-1} + \epsilon_{\text{noise}}) + \epsilon_{\text{explore}} \\ &= \theta(x_{t-1} + su_{t-1}) + \theta\epsilon_{\text{noise}} + \epsilon_{\text{explore}} \\ &= \theta x_t + \epsilon'_{\text{explore}} \end{aligned}$$

with $\epsilon'_{\text{explore}} = \theta\epsilon_{\text{noise}} + \epsilon_{\text{explore}}$. In our example, increasing the environmental noise was equivalent to increasing the exploratory noise of the agent, which obviously accelerated convergence.

**Episode Length.** In this series of experiments, we varied only the episode length and otherwise used the default settings with $\sigma_{\text{noise}} = 0.5$ for all runs. The results are shown in Figure 3 on the left side and Table 1, left part. Convergence speed with REX only improved marginally with longer episodes. The increased variance introduced by longer episodes almost completely outweighed the higher number of samples for a better gradient estimate. Since SDE does not introduce more noise with longer episodes during a single rollout, it could profit from longer episodes enormously. The speed-up factor rose almost proportionally with the episode length.

**Parameter Dimensionality.** Here, we increased the dimensionality of the problem in two ways: Instead of minimizing a scalar function, we minimized $g(x, y) = [x^2, y^2]^T$. Further, we used a nonlinear function approximator, namely a multilayer perceptron with 3 hidden units with sigmoid activation and a bias unit connected to hidden and output layer. We chose a single parameter $\hat{\sigma}$ for exploration variance adaptation. Including $\hat{\sigma}$ the system consisted of 18 adjustable parameters, which made it a highly challenging task for policy gradient methods.

The exploration variance was initially set to $\hat{\sigma} = -2$ which corresponds to an effective variance of $\sim 0.135$. The parameters were initialized with $\theta_i \in [-1, 0[$ because positive actions quickly lead to high negative rewards and destabilized learning. For smooth convergence, the learning rate $\alpha = 0.01$ needed to be smaller than in the one-dimensional task.

As the righthand side of Figure 3 shows, the agent with REX became stuck after 15,000 episodes at a local optimum around $R = -70$ from which it could not recover. SDE on the other hand found the same local optimum after a mere 1,000 episodes, and subsequently was able to converge to a much better solution.

## 4.2 Catching a Ball

This series of experiments is based on a simulated robot hand with realistically modelled physics. We chose this experiment to show the predominance of SDE over random exploration, especially in a realistic robot task. We used the Open Dynamics Engine[3] to model the hand, arm, body, and object. The arm has 3 degrees of freedom: shoulder, elbow, and wrist, where each joint is assumed to be a 1D hinge joint, which limits the arm movements to forward-backward and up-down. The hand itself consists of 4 fingers with 2 joints each, but for simplicity we only use a single actor to move all finger joints together, which gives the system the possibility to open and close the hand, but it cannot control individual fingers. These limitations to hand and arm movement reduce the overall complexity of the task while giving the system enough freedom to catch the ball. A 3D visualization of the robot attempting a catch is shown in Fig. 4. First, we used REINFORCE gradient estimation with optimal baseline and a learning rate of $\alpha = 0.0001$. We then repeated the experiment with Episodic Natural Actor-Critic (ENAC), to see if SDE can be used for different gradient estimation techniques as well.

**Experiment Setup.** The information given to the system are the three coordinates of the ball position, so the robot "sees" where the ball is. It has four degrees of freedom to act, and in each timestep it can add a positive or negative torque to the joints. The controller therefore has 3 inputs and 4 outputs. We map inputs directly to outputs, but squash the outgoing signal with a tanh-function to ensure output between -1 and 1.

The reward function is defined as follows: upon release of the ball, in each time step the reward can either be $-3$ if the ball hits the ground (in which case the episode is considered a failure, because the system cannot recover from it) or else the negative distance between ball center and palm center, which can be any value between $-3$ (we capped the distance at 3 units) and $-0.5$ (the closest possible distance considering the palm heights and ball radius). The return for a whole episode is the mean over the episode: $R = \frac{1}{N}\sum_{n=1}^{N} r_t$. In practice, we found an overall episodic return of $-1$ or better to represent nearly optimal

[3] The Open Dynamics Engine (ODE) is an open source physics engine, see `http://www.ode.org/` for more details.

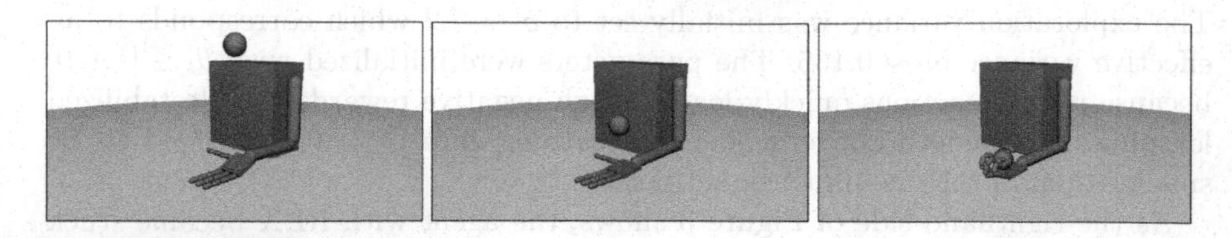

**Fig. 4.** Visualization of the simulated robot hand while catching a ball. The ball is released 5 units above the palm, where the palm dimensions are 1 x 0.1 x 1 units. When the fingers grasp the ball and do not release it throughout the episode, the best possible return (close to $-1.0$) is achieved.

catching behavior, considering the time from ball release to impact on palm, which is penalized with the capped distance to the palm center.

One attempt at catching the ball was considered to be one episode, which lasted for 500 timesteps. One simulation step corresponded to 0.01 seconds, giving the system a simulated time of 5 seconds to catch and hold the ball.

For the policy updates, we first executed 20 episodes with exploration and stored the complete history of states, actions, and rewards in an episode queue. After executing one learning step with the stored episodes, the first episode was discarded and one new rollout was executed and added to the front of the queue, followed by another learning step. With this "online" procedure, a policy update could be executed after each single step, resulting in smoother policy changes. However, we did not evaluate each single policy but ran every twentieth a few times without exploration. This yields a return estimate for the deterministic policy. Training was stopped after 500 policy updates.

**Results.** We will first describe the results with REINFORCE. The whole experiment was repeated 100 times. The left side of Figure 5 shows the learning curves over 500 episodes. Please note that the curves are not perfectly smooth because we only evaluated every twentieth policy. As can be seen, SDE finds a near-perfect solution in almost every case, resulting in a very low variance. The mean of the REX experiments indicate a semi-optimal solution, but in fact some of the runs found a good solution while others failed, which explains the high variance throughout the learning process.

The best controller found by SDE yielded a return of $-0.95$, REX reached $-0.97$. While these values do not differ much, the chances of producing a good controller are much higher with SDE. The right plot in Figure 5 shows the percentage of runs where a solution was found that was better than a certain value. Out of 100 runs, REX only found a mere 7 policies that qualified as "good catches", where SDE found 68. Almost all SDE runs, 98%, produced rewards $R \geq -1.1$, corresponding to behavior that would be considered a "catch" (closing the hand and holding the ball), although not all policies were as precise and quick as the "good catches". A typical behavior that returns $R \simeq -1.5$ can be described as one that keeps the ball on the fingers throughout the episode but hasn't learned to close the hand. $R \simeq -2.0$ corresponds to a behavior where

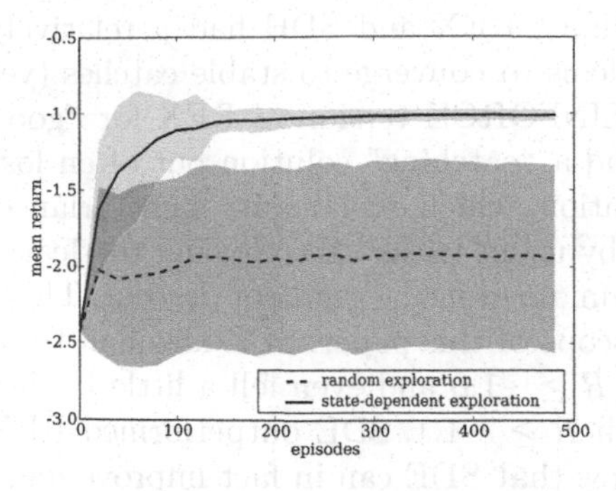

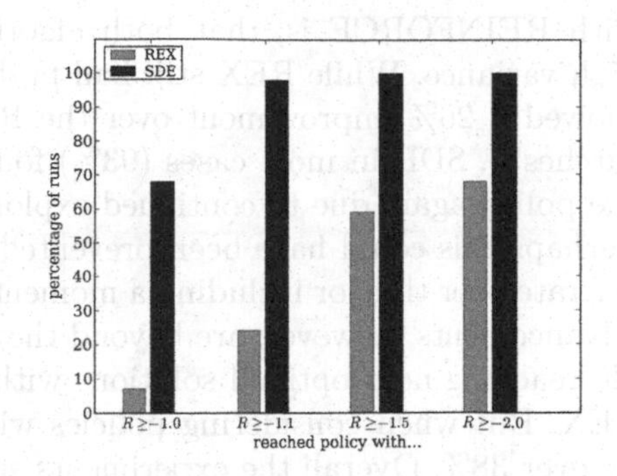


**Fig. 5.** Results after 100 runs with REINFORCE. Left: The solid and dashed curves show the mean over all runs, the filled envelopes represent the standard deviation. While SDE (solid line) managed to learn to catch the ball quickly in every single case, REX occasionally found a good solution but in most cases did not learn to catch the ball. Right: Cumulative number of runs (out of 100) that achieved a certain level. $R \geq -1$ means "good catch", $R \geq -1.1$ corresponds to all "catches" (closing the hand and holding the ball). $R \geq -1.5$ describes all policies managing to keep the ball on the hand throughout the episode. $R \geq -2$ results from policies that at least slowed down ball contact to the ground. The remaining policies dropped the ball right away.

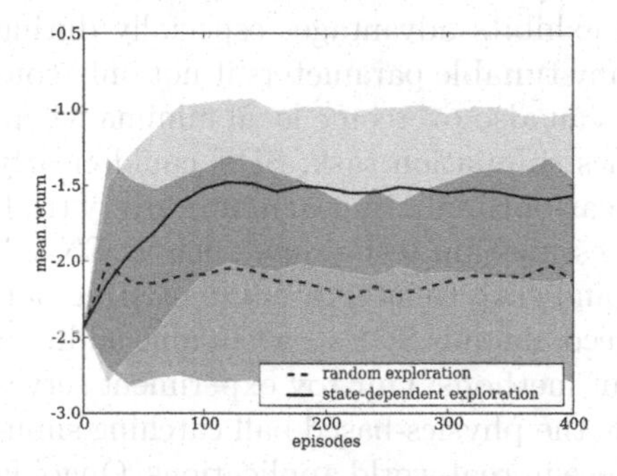

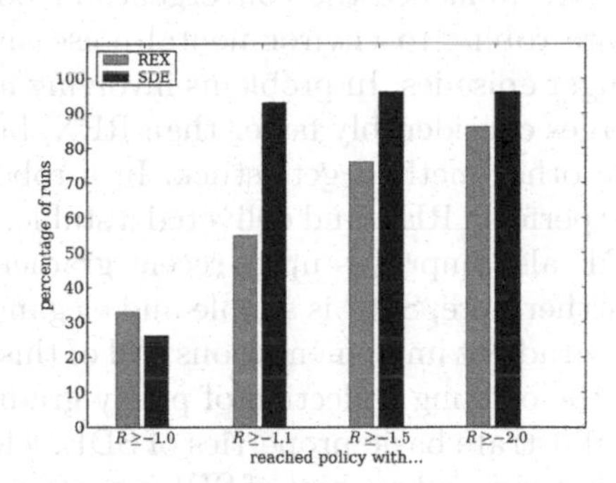


**Fig. 6.** Results after 100 runs with ENAC. Both learning curves had relatively high variances. While REX often didn't find a good solution, SDE found a catching behavior in almost every case, but many times lost it again due to continued exploration. REX also found slightly more "good catches" but fell far behind SDE considering both "good" and "average" catches.

the hand is held open and the ball falls onto the palm, rolls over the fingers and is then dropped to the ground. Some of the REX trials weren't even able to reach the $-2.0$ mark. A typical worst-case behavior is pulling back the hand and letting the ball drop to the ground immediately.

To investigate if SDE can be used with different gradient estimation techniques, we ran the same experiments with ENAC [11] instead of REINFORCE. We used a learning rate of 0.01 here, which lead to similar convergence speed. The results are presented in Figure 6. The difference compared to the results with REINFORCE is, that both algorithms, REX and SDE had a relatively high variance. While REX still had problems to converge to stable catches (yet showed a 26% improvement over the REINFORCE version of REX for "good catches"), SDE in most cases (93%) found a "catching" solution but often lost the policy again due to continued exploration, which explains its high variance. Perhaps this could have been prevented by using tricks like reducing the learning rate over time or including a momentum term in the gradient descent. These advancements, however, are beyond the scope of this paper. SDE also had trouble reaching near-optimal solutions with $R \geq -1.0$ and even fell a little behind REX. But when considering policies with $R \geq -1.1$, SDE outperformed REX by over 38%. Overall the experiments show that SDE can in fact improve more advanced gradient estimation techniques like ENAC.

## 5 Conclusion

We introduced state-dependent exploration as an alternative to random exploration for policy gradient methods. By creating strategy variations similar to those of finite differences but without their disadvantages, SDE inserts considerably less variance into each rollout or episode. We demonstrated how various factors influence the convergence of both exploration strategies. SDE is much more robust to environmental noise and exhibits advantages especially during longer episodes. In problems involving many tunable parameters it not only converges considerably faster than REX, but can also overcome local minima where the other method gets stuck. In a robotics simulation task, SDE could clearly outperform REX and delivered a stable, near-optimal result in almost every trial. SDE also improves upon recent gradient estimation techniques such as ENAC. Furthermore, SDE is simple and elegant, and easy to integrate into existing policy gradient implementations. All of this recommends SDE as a valuable addition to the existing collection of policy gradient methods. Our toy experiment serves to illustrate basic properties of SDE, while the physics-based ball catching simulation gives a first hint of SDE's performance in real-world applications. Ongoing work is focusing on realistic robot domains.

## Acknowledgements

This work was funded within the Excellence Cluster *Cognition for Technical Systems* (CoTeSys) by the German Research Foundation (DFG).

## References

1. Watkins, C., Dayan, P.: Q-learning. Machine Learning 8(3), 279–292 (1992)
2. Sutton, R., Barto, A.: Reinforcement Learning: An Introduction. MIT Press, Cambridge (1998)
3. Kaelbling, L.P., Littman, M.L., Moore, A.W.: Reinforcement learning: a survey. Journal of AI research 4, 237–285 (1996)

4. Wiering, M.A.: Explorations in Efficient Reinforcement Learning. PhD thesis, University of Amsterdam / IDSIA (February 1999)
5. Williams, R.J.: Simple statistical gradient-following algorithms for connectionist reinforcement learning. Machine Learning 8, 229–256 (1992)
6. Sutton, R.S., McAllester, D., Singh, S., Mansour, Y.: Policy gradient methods for reinforcement learning with function approximation. In: Advances in Neural Information Processing Systems (2000)
7. Peters, J., Schaal, S.: Policy gradient methods for robotics. In: Proc. 2006 IEEE/RSJ Intl. Conf. on Intelligent Robots and Systems (2006)
8. Moody, J., Saffell, M.: Learning to trade via direct reinforcement. IEEE Transactions on Neural Networks 12(4), 875–889 (2001)
9. Peshkin, L., Savova, V.: Reinforcement learning for adaptive routing. In: Proc. 2002 Intl. Joint Conf. on Neural Networks (IJCNN 2002) (2002)
10. Baxter, J., Bartlett, P.: Reinforcement learning in POMDP's via direct gradient ascent. In: Proc. 17th Intl. Conf. on Machine Learning, pp. 41–48 (2000)
11. Peters, J., Vijayakumar, S., Schaal, S.: Natural actor-critic. In: Proceedings of the Sixteenth European Conference on Machine Learning (2005)
12. Spall, J.C.: Implementation of the simultaneous perturbation algorithm for stochastic optimization. IEEE Transactions on Aerospace and Electronic Systems 34(3), 817–823 (1998)
13. Wierstra, D., Foerster, A., Peters, J., Schmidhuber, J.: Solving deep memory POMDPs with recurrent policy gradients. In: de Sá, J.M., Alexandre, L.A., Duch, W., Mandic, D.P. (eds.) ICANN 2007. LNCS, vol. 4668, pp. 697–706. Springer, Heidelberg (2007)
14. Ng, A., Jordan, M.: PEGASUS: A policy search method for large MDPs and POMDPs. In: Proceedings of the Sixteenth Conference on Uncertainty in Artificial Intelligence, pp. 406–415 (2000)
15. Aberdeen, D.: Policy-gradient Algorithms for Partially Observable Markov Decision Processes. Australian National University (2003)

# Client-Friendly Classification over Random Hyperplane Hashes

Shyamsundar Rajaram and Martin Scholz

Hewlett Packard Laboratories
1501 Page Mill Road
Palo Alto, CA
shyam.rajaram@hp.com, scholz@hp.com

**Abstract.** In this work, we introduce a powerful and general feature representation based on a locality sensitive hash scheme called random hyperplane hashing. We are addressing the problem of centrally learning (linear) classification models from data that is distributed on a number of clients, and subsequently deploying these models on the same clients. Our main goal is to balance the accuracy of individual classifiers and different kinds of costs related to their deployment, including communication costs and computational complexity. We hence systematically study how well schemes for sparse high-dimensional data adapt to the much denser representations gained by random hyperplane hashing, how much data has to be transmitted to preserve enough of the semantics of each document, and how the representations affect the overall computational complexity. This paper provides theoretical results in the form of error bounds and margin based bounds to analyze the performance of classifiers learnt over the hash-based representation. We also present empirical evidence to illustrate the attractive properties of random hyperplane hashing over the conventional baseline representation of bag of words with and without feature selection.

## 1 Introduction

In times of increasingly web-oriented information architectures, it becomes more and more natural to push analytical software down to clients, and to have them report back critical and prototypical events that require additional attention or indicate specific business opportunities. Examples of analytical software running on user PCs include spam filtering, malware detection, and diagnostic tools for different kinds of system failures.

We are concerned with a family of classification problems where high-dimensional, sparse training data is available on a large number of clients. We want to centrally collect data to train classifiers for deployment on the clients. Although the techniques we are studying apply to a broader set of problems, like eDiscovery, for the sake of this paper, we will exemplarily only consider the problem of classifying (reasonably large) text documents, like web pages a user visits. We aim to classify with respect to a quickly changing taxonomy of relevant concepts.

W. Daelemans et al. (Eds.): ECML PKDD 2008, Part II, LNAI 5212, pp. 250–265, 2008.

Our constraints in this setting stem from the natural goal of minimizing resource consumption on clients. This includes network bandwidth, but also memory and CPU footprint of classifiers and related software. During a cycle of training and deploying classifiers we go through the following phases: Data is preprocessed on clients before it is uploaded to a server. The preprocessing is required to reduce data volumes, and sometimes also to preserve privacy. The classifiers are learnt on the server. Clients download a potentially large number of classifiers, so we would like to minimize the required bandwidth. Finally, the models are deployed on the clients and triggered for each document under consideration, so we are concerned with the associated costs of preprocessing each document and of applying a linear classifier on top of that representation.

The questions we are addressing in this paper are therefore, which representations of sparse and high-dimensional data are **compact enough** to be transmitted over the web, **general enough** to be used for all kinds of upcoming multi-class classification problems, **cheap enough** to be applicable at deployment time, and are **close enough** to the performance of the models that are not narrowed down by operational costs.

The novelty of this work is in exploiting a locality sensitive hashing technique called random hyperplane hashing (cf. Algorithm 1) towards building a compact, general, cheap and powerful representation scheme. This technique has the additional benefit of being content-agnostic, which implies that it does not require any deep domain understanding in the preprocessing phase. Further, it can be used in the text domain, without any feature selection and for any language. The contributions of this paper are two fold: (i) we present key theoretical results in terms of error bounds and margin based bounds to quantify the loss of information due to random hyperplane hashing and (ii) we present experimental results to bring out the above mentioned attractive properties of the hashing-based representation scheme.

The closely related random projection technique [3,4,9,19,10] has been used successfully as a tool for dimensionality reduction because of its simplicity and nice theoretical Euclidean distance preserving properties. Though the random hyperplane hash method has a striking resemblance to the random projection method (as seen in Algorithm 1), there is a key difference as random hyperplane hashing preserves the angular distance while random projection preserves Euclidean distance, leading to a completely different representation scheme. More importantly, from a practical standpoint, the bit-based representation based on random hyperplane hashing turns out to be significantly more compact, which we elaborate later in the paper.

The rest of the paper is organized as follows. Sec. 2 provides the required technical preliminaries and reviews locality sensitive hashing with emphasis on the angular distance preserving random hyperplane representation scheme. In this paper, we confine ourselves to the problem of learning linear classification models on top of this representation; in Sec. 3 we discuss this setting and present theoretical results based on error bounds and margin based bounds. Sec. 3 includes a comparative study of our representation scheme with that of random projection

**Require:**
- Input document $d$
- Number $k$ of output dimensions
- Type of transformation, either RP or RHH

**Ensure:**
$k$-dimensional boolean (for RHH) or integer (for RP) vector representing $d$

```
Computation:
Create a k dimensional vector v with v[i] = 0 for 1 ≤ i ≤ k
for all terms w in document d do
   Set random seed to w          // cast w to integer or use hash value
   for all i in (1,...,k) do
      b ← sample random bit uniformly from {−1,+1}
      v[i] ← v[i] + b
for all i in (1,...,k) do
   v[i] ← sign(v[i])             // only for RHH, skip this step for RP
return v
```

**Algorithm 1.** A random projection (RP) / hyperplane (RHH) algorithm

based representation. We complement our theoretical findings with a number of empirical results on benchmark datasets in Sec. 4. Finally, we summarize and conclude in Sec. 5.

## 2 Random Hyperplane Hashing

Locality Sensitive Hash functions are invaluable tools for approximate near neighbor problems in high dimensional spaces. Traditionally, nearest neighbor search in high dimensional spaces has been expensive, because with increasing dimensionality, indexing schemes such as KD Trees very quickly deteriorate to a linear scan of all the items. Locality Sensitive Hash Functions [12] were introduced to solve the approximate nearest neighbor problem in high dimensional spaces, and several advancements [6,2,5] have been done in this area.

**Definition 1 (Locality Sensitive Hashing [5,12]).** *A locality sensitive hashing scheme is a distribution on a family $\mathcal{F}$ of hash functions on a set of instances, such that for two instances $x$ and $y$,*

$$\mathbf{Pr}_{h \in \mathcal{F}}\left[h(x) = h(y)\right] = f(\mathrm{sim}(x, y)) \tag{1}$$

*where,* $\mathrm{sim}(x, y)$ *is some similarity function defined on the instances and $f$ is a monotonically increasing function i.e., more the similarity, higher the probability.*

Simply put, a locality sensitive hash function is designed in such a way that if two vectors are close in the intended distance measure, the probability that they hash to the same value is high; if they are far in the intended distance measure, the probability that they hash to the same value is low. Next, we provide a formal review of locality sensitive hash functions. In particular, we focus on the cosine similarity as it is a popular one for a variety of applications such as in document retrieval [17], natural language processing [16] and image retrieval [18].

**Definition 2 (Cosine Similarity).** *The cosine of two vectors* $u \in \mathbb{R}^m$ *and* $v \in \mathbb{R}^m$ *is defined as* $cos(u,v) = \frac{u.v}{|u||v|}$.

As we will discuss, the random hyperplane hashing algorithm provides a locality sensitive hash function that corresponds to the cosine similarity measure. Let us first introduce the notion of a cosine hash family.

**Definition 3 (Cosine Hash Family [7]).** *A set of functions* $H = \{h_1, h_2, \ldots\}$ *constitute a cosine hash family over* $\mathbb{R}^m$ *iff for some finite* $\mathbf{U} \subset \mathbb{N}$,

- $h_k \in H : \mathbb{R}^m \rightarrow \mathbf{U}$
- *For any four vectors* $u, u', v, v'$ *in* $\mathbb{R}^m$ *where* $\cos(u,u') = \cos(v,v')$, *the following is true:*

$$\mathbf{Pr}_{h \in H}(h(u) = h(u')) = \mathbf{Pr}_{h \in H}(h(v) = h(v'))$$

- *For any four vectors* $u, u', v, v'$ *in* $\mathbb{R}^m$ *where* $\cos(u,u') > \cos(v,v')$, *the following is true:*

$$\mathbf{Pr}_{h \in H}(h(u) = h(u')) > \mathbf{Pr}_{h \in H}(h(v) = h(v'))$$

The following definition is similar to Algorithm 1, but more convenient for analytical purposes. We will compare the two at a later point.

**Definition 4 (Random Hyperplane Hash (RHH) algorithm [5]).** *The random hyperplane hashing algorithm works as follows: for a desired hash length of* $k$, *generate an* $m \times k$ *projection matrix* $M$, *where each element is chosen i.i.d from a* $\mathcal{N}(0,1)$ *distribution. The* $k$ *dimensional hash of a vector* $u \in \mathbb{R}^m$ *is computed in two steps,*

- *Compute the vector* $P = uM$
- *The* $i$*th entry of the hash of* $u$ *is* $-1$ *if* $P_i < 0$, *and* $1$ *otherwise.*

We will conclude this section with a number of important properties of random projections that will be used throughout the paper.

**Lemma 1.** *Let* $u$, $v$ *be vectors in* $\mathbb{R}^m$ *such that* $cos(u,v) = \rho$ *and their corresponding* $k$*-dimensional random hyperplane hash are denoted as* $u'$ *and* $v'$ *respectively. We denote the distance between* $u'$ *and* $v'$ *as* $d'(u',v')$ *which is defined below. Then,*

1. $P(u'_i = v'_i) = 1 - \frac{\arccos(\rho)}{\pi}$
2. $\frac{1}{4}(u'_i - v'_i)^2 \sim Bernoulli(\frac{\arccos(\rho)}{\pi})$
3. $d'(u',v') := \frac{1}{4}\sum_{i=1}^{k}(u'_i - v'_i)^2 \sim Binomial(k, \frac{\arccos(\rho)}{\pi})$

4. *Additive bound: For any* $\epsilon > 0$,

$$\mathbf{Pr}\left(\frac{1}{k}d'(u',v') - \frac{\arccos(\rho)}{\pi} > \epsilon\right) \leq e^{-2\epsilon^2 k} \tag{2}$$

$$\mathbf{Pr}\left(\frac{1}{k}d'(u',v') - \frac{\arccos(\rho)}{\pi} < -\epsilon\right) \leq e^{-2\epsilon^2 k} \tag{3}$$

5. *Multiplicative bound: For any* $\lambda \in [0,1]$,

$$\mathbf{Pr}\left(\frac{1}{k}d'(u',v') \geq (1+\lambda)\frac{\arccos(\rho)}{\pi}\right) \leq e^{-\frac{\lambda^2 k \arccos(\rho)}{3\pi}} \tag{4}$$

$$\mathbf{Pr}\left(\frac{1}{k}d'(u',v') \leq (1-\lambda)\frac{\arccos(\rho)}{\pi}\right) \leq e^{-\frac{\lambda^2 k \arccos(\rho)}{2\pi}} \tag{5}$$

*Proof.* The first result is a fundamental property of random hyperplanes exploited in [11] and was later used to develop the random hyperplane hashing algorithm in [5]. This key property results in random hyperplane hashes satisfying the properties of a cosine hash family. The second and third results follow from 1. The fourth and fifth results are obtained by applying Hoeffding and Chernoff bounds respectively on $d'(u',v')$ by expressing it as a sum of independent Bernoulli distributed random variables $(u'_1 - v'_1)^2, (u'_2 - v'_2)^2, \ldots, (u'_k - v'_k)^2$. □

The key contribution of this work is the use of random hyperplane hashing as a representation scheme over which classifiers can be learnt. The efficient random hyperplane hashing algorithm leads to a general and compact angular distance preserving representation. In the next section, we exploit the properties of random hyperplane hashing provided in Lemma 1 to obtain bounds on the performance of linear classifiers learnt over such a representation and further, compare it with the closely related random projection algorithm.

## 3 Classifier over Random Hyperplane Hashes

We start with a formal definition of a linear classifier in both spaces, Euclidean space and random hyperplane hash space.

**Definition 5 (Linear classifier and its hash representation).** *Let* $T$ *represent a concept that maps instances* $x_i$ *from an* $m$*-dimensional space of reals* $\mathbb{R}^m$ *to labels* $y_i$ *that belongs to* $\{-1,+1\}$*. The concept* $T$ *allows for a linear classifier* $h \in \mathbb{R}^m$*, if there exists a* $h$ *satisfying,* $y_i(h^T x_i) \geq 0$ *which can alternately be written as* $y_i\left(\frac{1}{2} - \frac{1}{\pi}\arccos(\frac{h^T x_i}{|h||x_i|})\right) \geq 0$*. Since,* $h \in \mathbb{R}^m$*, it allows for a random hyperplane hash representation* $h'$*. In the hash space, we consider linear classifiers of the form* $\frac{y_i}{k}(h'^T x'_i) \geq 0$ [1].

[1] The $\frac{1}{k}$ is for convenience and results in the classifier output to be in the range $[-1, 1]$.

Our first important error bound requires a stronger notion of separability, however. In the next subsection, we will show that a slightly less restrictive constraint is not sufficient.

**Definition 6 ($\epsilon$-robust concept).** *For any real $0 < \epsilon < 0.5$, a concept $T$ along with a distribution $\mathcal{D}$ on $\mathbb{R}^m$ is $\epsilon$-robust, if it allows for a linear classifier $h$ that satisfies $\frac{1}{2} - \frac{1}{\pi}\arccos(\frac{h^T x}{|h||x|}) > \epsilon$ for positive and $\frac{1}{2} - \frac{1}{\pi}\arccos(\frac{h^T x}{|h||x|}) < -\epsilon$ for negative instances.*

**Theorem 1.** *Let $h$ be a linear classifier that can correctly classify instances in $\mathbb{R}^m$ according to some $\epsilon$-robust target concept $T$ and let $h'$ denote its hash representation. Consider $x \in \mathbb{R}^m$ and its $k$ dimensional random hyperplane hash $x'$, for $k \geq \frac{\alpha}{2\epsilon^2}$ and projection matrix $M$. Then,*

$$\forall x : \mathcal{D}(x) > 0, \mathbf{Pr}_{M \sim [N(0,1)]^{m \times k}} \left[ label_h(x) \neq label_{h'}(x') \right] \leq e^{-\alpha} \tag{6}$$

*Proof.* Consider a positive instance $x$. Let the random hyperplane hash of $x$ be $x'$. By definition of $\epsilon$ robustness, $h$ satisfies

$$\frac{1}{\pi}\arccos(\frac{h^T x_i}{|h||x_i|}) \leq \frac{1}{2} - \epsilon. \tag{7}$$

Now for $0 < \epsilon < 0.5$, let us compute the following probability,

$$\mathbf{Pr}\left(\frac{1}{k}d'(h', x') > \frac{1}{2} + \epsilon\right) \tag{8}$$

The Hoeffding bound from Lemma 1,

$$\mathbf{Pr}\left(\frac{1}{k}d'(h', x') - \frac{1}{\pi}\arccos(\frac{h^T x_i}{|h||x_i|}) > \epsilon\right) \leq e^{-2k\epsilon^2} \tag{9}$$

and Eqn. 7, with the definition of $d'$, can be combined to obtain,

$$\mathbf{Pr}\left(\frac{1}{4k}\left(|h'|^2 + |x'|^2 - 2h'^T x'\right) - (\frac{1}{2} - \epsilon) > \epsilon\right) \leq e^{-2k\epsilon^2} \tag{10}$$

and using the property that $|h'| = |x'| = \sqrt{k}$ (by construction[2]) leads to,

$$\mathbf{Pr}\left(\frac{1}{k}h'^T x' < 0\right) \leq e^{-2k\epsilon^2} \tag{11}$$

which corresponds to the probability that $x'$ is mislabeled as negative by $h'$. A similar result can be shown for negative instances which leads to the error bound in Eqn. 6. □

Using Theorem 6, Fig. 1 illustrates the lower bounds on hash length $k$ required for different label error rate constraints. It is important to note that the bound is obtained based on the hash of the classifier. A classifier explicitly learnt in the hash space can only lead to a better margin.

[2] The conventional random hyperplane algorithm results in a $k$ dimensional vector of 0s and 1s, but we construct a vector of $-1$s and 1s to allow for a constant norm of $\sqrt{k}$ for hash length $k$.

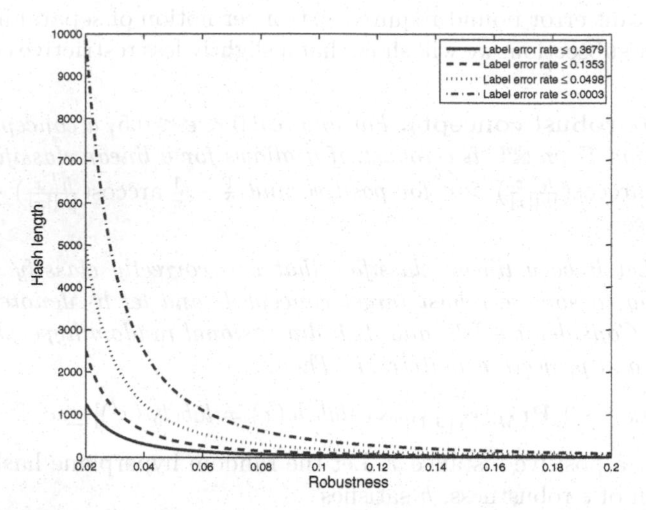


**Fig. 1.** Plot showing lower bounds on hash length for the random hyperplane hashing algorithm for different upper bounds on label error rates (set by varying $\alpha$). The x-axis represents the level of robustness of the target concept in terms of $\epsilon$.

### 3.1 Comparison with Random Projections

The method of random projections (RP) has been used extensively to perform dimensionality reduction [19] in a distance preserving manner. The random projection method is the same as the random hyperplane method described in Def. 1 without the thresholding performed in the second step. The thresholding step results in a key difference: Euclidean distance is preserved with high probability in the case of random projections [13] whereas the angular distance (Lemma 1) is preserved in the random hyperplane case. As mentioned before, cosine similarity is extensively used in domains like text and image retrieval. The work reported in [3] provides error bounds for classifiers learnt over random projections. The error bound (based on Lemma 3 from [3]), indicates the number of projections, $k_{RP} \geq \frac{8\alpha}{\epsilon^2}$ compared to our case where the number of hash elements $k_{RHH} \geq \frac{\alpha}{2\epsilon^2}$ while ensuring the same error bound $e^{-\alpha}$ for different notions of robustness. In [19], the authors define the notion of $l$-robustness which simply put, disallows instances with different labels to be less than $l$ close to each other. Our notion of $\epsilon$ robustness is much stronger than the $l$-robustness. Next, we present empirical results to show the need for such a strong notion of robustness.

In this synthetic experiment, we demonstrate the label errors introduced by the random projection algorithm and random hyperplane hashing algorithm. Generate $p$ different random hyperplanes $h_1, h_2, \ldots, h_p$. For every random hyperplane, we use the $l$ robustness assumption used in [3] and generate $N$ instances at random from $\mathbb{R}^m$ in such a way that no two instances with different labels are more than $l$ close to each other. The labeling $y_{ij}$ of each instance $x_i$ is obtained

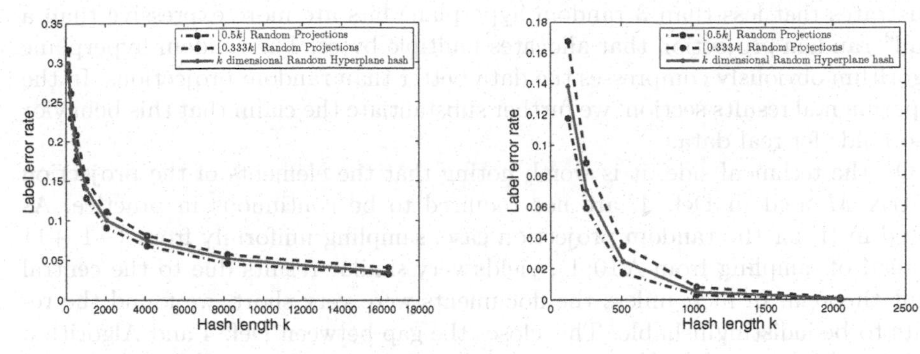


**Fig. 2.** Plot illustrating the drop in label error rate with increasing projections/hashlength where the classification obeys $l$-robustness (left plot) and $\epsilon$-robustness (right plot)

based on $y_{ij}(h_j^T x_i) \geq 0$ [3]. We generate $K$ different projection matrices of size $k \times m$ where each element is chosen i.i.d from $\mathcal{N}(0, 1)$. The random projection and random hyperplane hashing algorithms are performed using all $l$ projection matrices, on all $x_{il}$, $h_{jl}$, to obtain $\hat{x}_{il}$,$\hat{h}_{jl}$ and $x'_{il}$, $h'_{jl}$ respectively. The labels in the hash space are obtained according to $\hat{y}_{ijl}(\hat{h}_{jl}^T \hat{x}_{il}) \geq 0$ and $y'_{ijl}({h'}_{jl}^T x'_{il}) \geq 0$. Now, we evaluate the label error rate given by, $\frac{1}{K \times N \times p} \sum_{i,j,l} I\left(y_{ij} \neq \hat{y}_{ijl}\right)$ and $\frac{1}{K \times N \times p} \sum_{i,j,l} I\left(y_{ij} \neq y'_{ijl}\right)$ for the random projection method and the random hyperplane case for different values of $k$. Fig. 2 (left) shows that, even with hashlength 16384, the label error rate is still around 4% which is clearly unexpected based on the lower bounds on hash length. The same inconsistency arises with the random projection case as well. The reason for this effect is that the notion of $l$-robustness does not account for disallowing instances that are very close to the hyperplane, which leads to high label error rates. In the following simulation, we enforce the $\epsilon$-robustness constraint, which simply put, disallows instances to be too close to the hyperplane. We set $\epsilon$ to be 0.0319. The rest of the experimental setup is the same as the first experiment. Results presented in Fig. 2 (right) show a more desirable fast drop-off in label error rate and at the same time, comfortably satisfying the lower bounds presented in Theorem 6 and illustrated through Fig. 1.

It is important to note that one projection obtained through random projection requires a significantly larger chunk of memory compared to one bit required to represent one element of a random hyperplane hash; each hyperplane hash is just one bit long, whereas random projections are of continuous nature, and would usually be represented by 4 or even 8 bytes. The results from Fig. 2 (right) lead to another very interesting observation. The label error rate of the $k$-dimensional random hyperplane hashing is empirically bound by the label error rates of $\lfloor \frac{k}{2} \rfloor$ random projections and $\lfloor \frac{k}{3} \rfloor$ random projections. It also

[3] The random projection method and the random hyperplane algorithm are comparable only in the case of hyperplanes passing through the origin.

illustrates that less than 3 random hyperplane bits are more expressive than a "full" random projection that allocates multiple bytes. The random hyperplane algorithm obviously compresses the data better than random projections. In the experimental results section, we further substantiate the claim that this behavior also holds for real data.

On the technical side, it is worth noting that the elements of the projection matrix $M$ used in Def. 4, are not required to be continuous in practice. As noted in [1] for the random projection case, sampling uniformly from $\{-1, +1\}$ instead of sampling from $\mathcal{N}(0, 1)$, yields very similar results due to the central limit theorem. In fact, unless the documents were very short, we found the results to be indistinguishable. This closes the gap between Def. 4 and Algorithm 1. Regarding preprocessing, random projections are cheaper. Even if we need only about 2.5 times more random hyperplane "bits" than random projections to get similarity approximations of the same quality, this still means that the algorithm has to generate more than twice as many random bits in the first place. This is required only once for each document, however, and whenever we expect to apply a large number of classifiers to that representation we may in return benefit from the cheaper complexity of the models: random hyperplane representations are just bits, so linear models just have to add associated weights. There is a trivial way of simplifying such classifiers: After transforming the decision function of the classifier from $\{-1, +1\}$ feature space to $\{0, +1\}$ space (which just requires to adapt the threshold) it only has to do an addition for every second bit on average. In contrast, random projections yield integer or real-valued vectors, which even requires additional multiplications. In practice, random hyperplanes should hence be cheaper, even if they operate on twice as many dimensions. When it comes to uploading models to clients, which happens only once per model-client pair, the higher dimensionality of the random hyperplanes algorithm is a disadvantage however, because models that get the same quality of results will usually have a larger number of associated weights.

Whether random hyperplanes or random projections are favorable depends on an application-dependent weighting of CPU and memory footprint, the expected number of examples to be classified per model, the number of examples to be uploaded, and the expected frequency of model upgrades.

### 3.2 Large Margin Classifiers over Random Hyperplane Hashes

The results previously derived in this section show that the preservation of labels in the random hyperplane hash space is closely related to the notion of a margin. In this subsection, we consequently analyze how margins of SVM classifiers in the hash space are affected by the hash length. Further, we determine the hash length which will preserve the margin with high probability. In [14], the authors present such a theorem on the effect of learning large margin classifiers over random projections. We first require a few more definitions.

**Definition 7 (SVM classifiers, margin and angular margin).** *Consider a linearly separable data set $S = \{(x_i, y_i), i = 1 \dots n\}, x_i \in \mathbb{R}^m, y_i \in \{+1, -1\}$, the SVM optimization problem is set up as follows:*

$$\max_h \frac{1}{|h|} \text{ s.t. } y_i(h^T x_i) \geq 1, i = 1, \dots, n \tag{12}$$

*where, $\frac{1}{|h|}$ is defined as the margin and hence, the solution to the above optimization problem is termed the maximum margin solution. Traditionally, an offset is included in the constraint and we address this issue at the end of this section.*

*For the sake of convenience, we define the angular margin $l_a$, as*

$$l_a := \min_i y_i \left( \frac{1}{2} - \frac{1}{\pi} \arccos \frac{h^T x_i}{|h||x_i|} \right) \tag{13}$$

*where we have exploited the fact that a classifier satisfying $y_i(h^T x_i) \geq 0$ can be rewritten as, $y_i \left( \frac{1}{2} - \frac{1}{\pi} \arccos( \frac{h^T x_i}{|h||x_i|}) \right) \geq 0$.*

These definitions allow us to formulate the desired probabilistic bound on the margin:

**Theorem 2.** *Given a linearly separable data set $S$ ($|S| = n$), let the optimal maximum margin solution on $S$ be given by $h$ with angular margin $l_a$. Let $S'_k$ represent the set of random hyperplane hashes of instances in the set $S$. If $k \geq \frac{\frac{1}{2}+l_a}{\gamma^2 l_a^2} \log \frac{n}{\delta}$, $0 < \gamma < 1$, $0 < \delta < 1$, then the maximum margin solution on $S'_k$ satisfies the following bound,*

$$\mathbf{Pr}(l_p > (1-\gamma) l_a) \geq 1 - \delta \tag{14}$$

*where $l_p$ corresponds to the margin in the hash space.*

*Proof.* Consider a positive example $x$, and in particular we consider the worst case where the instance lies on the margin i.e., $\frac{1}{2} - \frac{1}{\pi} \arccos \frac{h^T x}{|h||x|} = l_a$. Let $x'$ and $h'$ represent the $k$-dimensional random hyperplane hash of $x$ and the optimal solution $h$ respectively. Now, let us compute the following probability:

$$\mathbf{Pr}\left( \frac{1}{k} d'(h', x') < \frac{1}{2} - (1-\gamma) l_a \right) \tag{15}$$

which can be reformulated in a more amenable multiplicative form:

$$\frac{1}{k} d'(h', x') < \frac{1}{2} - (1-\gamma) l_a \equiv \frac{1}{k} d'(h', x') < (1+\lambda) \left( \frac{1}{2} - l_a \right) \tag{16}$$

Solving for $\lambda$, we obtain $\lambda = \frac{\gamma l_a}{\frac{1}{2} - l_a}$. Eqn. 15, can be rewritten as:

$$\mathbf{Pr}\left( \frac{1}{2k} h'^T x' > (1-\gamma) l_a \right) = \mathbf{Pr}\left( \frac{h'^T x'}{\sqrt{k}} > 2\sqrt{k}(1-\gamma) l_a \right). \tag{17}$$

Using the multiplicative Chernoff bounds from Lemma 1, we obtain the following bound:

$$\mathbf{Pr}\left(\frac{h'^T x'}{\sqrt{k}} > 2\sqrt{k}(1-\gamma)l_a\right) = 1 - \mathbf{Pr}\left(\frac{1}{k}d'(h',x') > \frac{1}{2} - (1-\gamma)l_a\right)$$

$$> 1 - e^{-\frac{\gamma^2 l_a^2}{\frac{1}{2}-l_a}\frac{k}{3}}$$

$$\Rightarrow \mathbf{Pr}\left(\frac{h'^T x'}{\sqrt{k}} > (1-\gamma)l_a\right) > 1 - e^{-\frac{\gamma^2 l_a^2}{\frac{1}{2}-l_a}\frac{k}{3}} \tag{18}$$

Similarly for any negative instance $x'$, we obtain the bound:

$$\mathbf{Pr}\left(\frac{h'^T x'}{\sqrt{k}} < -(1-\gamma)l_a\right) = 1 - \mathbf{Pr}\left(\frac{1}{k}d'(h',x') < \frac{1}{2} + (1-\gamma)l_a\right)$$

$$> 1 - e^{-\frac{\gamma^2 l_a^2}{\frac{1}{2}+l_a}\frac{k}{2}} \tag{19}$$

By definition, the margin in the hash space expressed as $l_p$ is given by $\min y' \frac{h'^T x'}{\sqrt{k}}$. The union bound can be used on Eqns. 18 and 19, to obtain guarantees of a hyperplane in the hash space with a margin $l_p$ that is at least $(1-\gamma)l_a$ where

$$\gamma \leq \frac{\epsilon\sqrt{l_a + \frac{1}{2}}}{l_a} \tag{20}$$

with probability at least $1 - ne^{-\epsilon^2 \frac{k}{3}}$. The corresponding value of $k$ is given by:

$$ne^{-\frac{\gamma^2 l_a^2}{\frac{1}{2}+l_a}\frac{k}{3}} \leq \delta \Rightarrow k \geq \frac{\frac{1}{2}+l_a}{\gamma^2 l_a^2}\log\frac{n}{\delta} \tag{21}$$

Notice that the sub-optimal $h'$, which corresponds to the random hyperplane hash of the optimal classifier $h$ achieves the bound. So, the optimal classifier on $S'_k$ can only achieve a better margin than $h'$. □

Finally, we want to justify for not having included an offset term into Eqn. (12) for mathematical simplicity. The common SVM constraint is given by

$$y_i(h^T x_i - b) \geq 1, i = 1, \ldots, n. \tag{22}$$

It is slightly more expressive in theory, but we found the difference in practice to be negligible. However, in doubt we can always include a "constant" component to each example vector, e.g. a new first component which is always equal to 1. In this case, the vector $h$ effectively also provides the offset $b$. The only remaining difference to the regular SVM is then that having the offset $b$ inside $h$ makes it part of the margin. In the case of high-dimensional data we are discussing in this paper this obviously does not change matters too much. In the next section we will complement this argument with empirical evidence that the hashing-based representation is well suited for building text classifiers, even without spending any particular attention to the matter of offsets.

**Table 1.** LibSVM results for a number of multi-class classification problems previously used in the context of feature selection. Reported are 10fold cross-validated accuracies on different representations and corresponding standard deviations.

| dataset | classes | size | BOW | 8192 RHH | 2048 RHH | 2048 RP |
|---|---|---|---|---|---|---|
| Cora | 36 | 1800 | $49.50 \pm 2.24$ | $66.56 \pm 2.51$ | $58.00 \pm 3.45$ | $52.22 \pm 3.42$ |
| FBIS | 17 | 2463 | $85.87 \pm 1.61$ | $84.94 \pm 2.08$ | $83.92 \pm 1.29$ | $84.65 \pm 1.91$ |
| LA1 | 6 | 3204 | $90.20 \pm 1.34$ | $89.95 \pm 1.32$ | $85.55 \pm 1.96$ | $87.42 \pm 0.60$ |
| LA2 | 6 | 3075 | $90.51 \pm 2.28$ | $90.41 \pm 1.92$ | $86.86 \pm 1.70$ | $87.51 \pm 1.98$ |
| OH0 | 10 | 1003 | $89.53 \pm 3.75$ | $88.83 \pm 2.69$ | $87.13 \pm 2.68$ | $87.63 \pm 2.65$ |
| OH10 | 10 | 1050 | $81.43 \pm 3.79$ | $81.71 \pm 2.44$ | $76.86 \pm 2.37$ | $78.19 \pm 2.71$ |
| OH15 | 10 | 913 | $81.49 \pm 2.69$ | $81.71 \pm 3.54$ | $79.62 \pm 5.09$ | $80.83 \pm 5.12$ |
| OH5 | 10 | 918 | $87.69 \pm 3.76$ | $88.57 \pm 3.43$ | $83.33 \pm 3.80$ | $85.08 \pm 3.88$ |
| RE0 | 13 | 1504 | $84.17 \pm 2.69$ | $84.31 \pm 2.56$ | $83.04 \pm 2.86$ | $82.44 \pm 2.45$ |
| RE1 | 25 | 1657 | $84.55 \pm 2.32$ | $80.87 \pm 2.80$ | $79.85 \pm 3.01$ | $82.62 \pm 2.77$ |
| WAP | 20 | 1560 | $85.38 \pm 1.34$ | $82.95 \pm 2.84$ | $80.06 \pm 2.49$ | $82.50 \pm 2.48$ |
| OHSCAL | 10 | 11162 | $78.07 \pm 0.93$ | $78.31 \pm 1.14$ | $71.58 \pm 1.19$ | $74.82 \pm 0.97$ |

# 4 Experimental Results

The main goal of the subsequently described experiments is to complement the theoretical guarantees by empirical evidence that for typical text classification tasks hashing-based representations are in fact able to capture enough of the similarities between documents to allow for competitive predictive classifier performance. More detailed questions of interest include how much information is lost when switching to less exact representations and how this loss qualitatively changes with the number of bits we allow for each of the representations. We stress that the experiments do *not* aim at reaching the best individual classifier performances reported for each dataset. We are hence not applying any sophisticated preprocessing methods and avoid the issue of parameter tuning by fixing the experimental setup. We just vary the representations of input data sets.

The representations we want to evaluate are random hyperplane hashing (RHH) and random projections (RP). We compare the classifiers learnt over these representations to the classifiers learnt over the regular bag of words vector space representation of text (BOW). In order to raise the bar of the baseline method, we also report results when classifying on top of a BOW representation after feature selection. We do not want to narrow down the setting to the somewhat unnatural case of strictly binary classifiers, so all of our experiments involve multi-class problems. For this kind of data, Information Gain feature selection is a widely used option.

All experimental results reported in this section are the result of tenfold cross-validation. We chose the LibSVM operator that is part of RapidMiner [15] as an inner learner, because it is capable of building competitive logistic models on top of SVM classifiers for multi-class problems. We used a linear kernel with default settings, and fixed the $C$ parameter to 0.1 for all experiments to establish a common baseline. Please note that this low value of $C$ introduces a slight

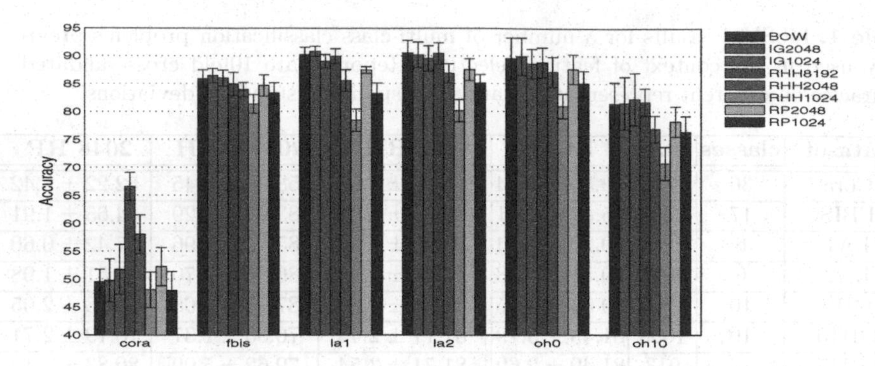


**Fig. 3.** Accuracies and standard deviations of different methods on benchmark datasets

disadvantage for hashing-based representations, since the original vector space contains up to 30,000+ dimensions and for the most part much fewer data points in our experiments; the data can be expected to be linearly separable in the original space, but after the noisy projections there might be a considerable number of examples on the wrong side of the hyperplane, which requires a larger value of $C$. In our experimental setup we allow for the baseline information gain based feature selection to select features on the complete data set (including the test set) which results in a minor disadvantage for the hashing representation.

### 4.1 Medium-Sized Corpora

For brevity, we first report experiments on a *set* of publicly available benchmark data sets, namely corpora that were used in [8] to evaluate feature selection techniques[4]. Compression is not a concern for data sets at this scale, but they allow for estimates of how many projections or RHH bits are required to preserve similarities in hash space sufficiently well for classification purposes.

Table 1 lists the data sets and summarizes the results for the plain bag of words representation, random hyperplane hashing with 8192 and 2048 bits (1 KByte and 256 Bytes), and random projections with 2048 projections. We report results for random projections where we encoded each projection by 2 bytes, which means that the representation using only 2048 projections is still four times longer than the 8192 RHH bit representation. Figures 3 and 4 show the same results and in addition compare to Information Gain with 2048 and 1024 dimensions and 1024 random projections.

The results of our experiments suggest that 8192 RHH bits are enough to represent documents based on hashes with a negligible loss in classification accuracy (see bar plots). Only in rare cases, the RHH representation and the vector space model give significantly different results. Examples include the Cora data

[4] It is available at the JMLR website:
`http://www.jmlr.org/papers/volume3/forman03a/forman03a_data.tgz`.

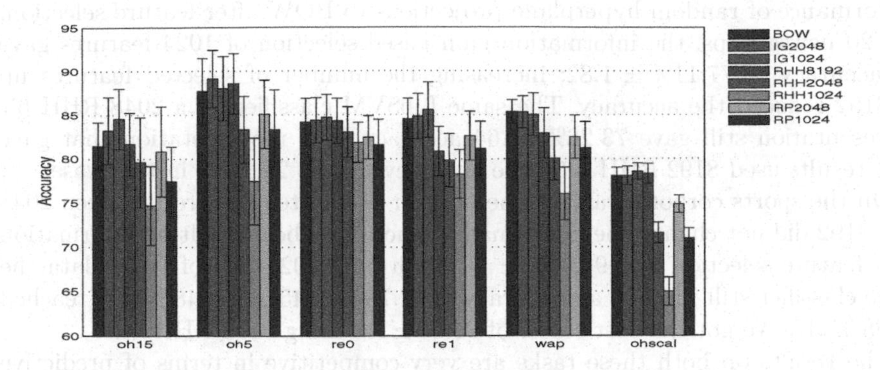


**Fig. 4.** Accuracies and standard deviations of different methods on benchmark datasets

set, where RHH performs much better, and RE1, where the original representations and feature selection perform better. Using only 2048 RHH bits reduces the accuracy typically by about 3 to 4%. Keeping the number of classes and the generality of the representation in mind, this is still a decent result. Interestingly, reducing the number of bits even further to 1024 (factor of 2) the performance drops more than when we reduced it by a factor of 4 before. This reflects the exponentially decreasing quality of approximations.

Using random projections in the order of 200 dimensions, a case that has been studied in the literature, clearly does not serve the purpose. It is interesting to note though that even if we assume just 4 bytes per projection (float, single precision) we end up with 8192 bits representing only 256 dimensions. In our experiments we generally found that – depending on whether accuracy or shorter representations are more important – choices between 100 bytes and 1KByte seem to work reasonably well. A pleasant property of the hashing-based representation we found is that the accuracy always increased monotonically with an increasing number of features. This does not hold for feature selection.

## 4.2 Experiments with Larger Corpora

The practically more relevant task is to scale up the sketched method to data sets of very high dimensionality and size. To study the effects with larger corpora we hence ran experiments on the widely used 20 newsgroups data set and on a hand-crafted corpus that contained examples of categorized web pages. The former contains about 20K examples, 40K terms, and 20 different categories. For the latter we crawled pages from the different sports subcategories of the Open Directory (DMOZ). We removed all non-textual parts from the web pages, and tokenized and Porter stemmed the plain text. To reduce noise, we deleted pages with less than 2 KBytes. Finally, we removed all categories with less than 20 remaining documents, leaving about 30K documents, 200K terms, and 52 skewed classes for the evaluation. The largest category is "Soccer" with 4858 documents. Due to the high dimensionality of the original corpus, we only compared the performance of random hyperplane projections to BOW after feature selection. For 20 newsgroups, the information gain based selection of 1024 features gave an accuracy of 77.11% ± 1.82; increasing the number of selected features up to 8192 reduced the accuracy. The same LibSVM classifier on a 2048 RHH bit representation still gave 73.26% ± .64 accuracy. The representation that gave best results used 8192 RHH bits; the accuracy was 83.28 ± .62 in that case.

On the sports corpora, varying the number of selected features between 2048 and 8192 did not change the performance much. The best result on information gain feature selection was 90.99% ± .5. With only 1024 bits of RHH data the same classifier still reached an accuracy of 84.75% ± .45, on 2048 bits it reached 88.68 ± .37. We got the best result of 91.66 ± .41 using 8192 RHH bits.

The results on both these tasks are very competitive in terms of predictive performance. They illustrate that global feature selection can be avoided without compromising accuracy, e.g., if transmitting raw data to a central server for learning is prohibitively expensive in client-server settings.

## 5 Conclusion

We studied a specific locality sensitive hashing scheme called random hyperplane hashing as a representation for building linear classifiers. It differs from random projections in that it preserves angular rather than Euclidean distances. A margin-based robustness criterion allowed us to both upper-bound the error rate and to lower-bound the margin of the resulting classifier in the hash-space in a strong probabilistic sense. We illustrated that (i) a certain pre-defined weaker notion of robustness fails to achieve desirably low label error rates, but that (ii) label errors when enforcing our margin-based robustness criterion decrease exponentially fast with increasing hash length (dimensionality). Moreover, we showed that less than 3 random hyperplane hash bits are as expressive in terms of preserving labels as a full, real-valued random projection.

The error and margin bounds coupled with the high density and expressiveness motivated an evaluation of the hash representation for learning text classifiers in a client-centered setting where the representation length of documents and models as well as the computational costs of classifiers are crucial. We demonstrated significant gains in representation length over bag of words and the random projection method, and a classification performance that is competitive to standard feature selection, despite using a generic, content-agnostic preprocessing scheme.

## Acknowledgments

We would like to thank Kave Eshghi, Rajan Lukose, George Forman, Ram Swaminathan, and Jaap Suermondt for several fruitful discussions regarding this work. We would also like to thank an anonymous reviewer for helpful comments.

## References


1. Achlioptas, D.: Database-friendly random projections. In: Symposium on Principles of Database Systems (PODS 2001), pp. 274–281. ACM Press, New York (2001)
2. Andoni, A., Indyk, P.: Near-optimal hashing algorithms for approximate nearest neighbor in high dimensions. In: Symposium on Foundations of Computer Science (FOCS 2006), pp. 459–468. IEEE Computer Society, Los Alamitos (2006)
3. Arriaga, R.I., Vempala, S.: An algorithmic theory of learning: Robust concepts and random projection. In: IEEE Symposium on Foundations of Computer Science, pp. 616–623 (1999)
4. Bingham, E., Mannila, H.: Random projection in dimensionality reduction: applications to image and text data. In: Int. Conf. on Knowledge Discovery and Data Mining (KDD 2001), pp. 245–250. ACM Press, New York (2001)
5. Charikar, M.S.: Similarity estimation techniques from rounding algorithms. In: Symposium on Theory of computing (STOC 2002), pp. 380–388. ACM Press, New York (2002)
6. Datar, M., Immorlica, N., Indyk, P., Mirrokni, V.S.: Locality-sensitive hashing scheme based on p-stable distributions. In: Symposium on Computational geometry (SCG 2004), pp. 253–262. ACM Press, New York (2004)
7. Eshghi, K., Rajaram, S.: Locality-sensitive hash functions based on concommitant rank order statistics. In: Int. Conf. on Knowledge discovery and data mining (KDD 2008). ACM Press, New York (2008)
8. Forman, G.: An extensive empirical study of feature selection metrics for text classification. Journal of Machine Learning Research (JMLR) (3), 1289–1305 (2003)
9. Fradkin, D., Madigan, D.: Experiments with random projections for machine learning. In: Int. Conf. on Knowledge discovery and data mining (KDD 2003), pp. 517–522. ACM Press, New York (2003)
10. Goel, N., Bebis, G., Nefian, A.: Face recognition experiments with random projection. In: SPIE, Bellingham, WA, pp. 426–437 (2005)
11. Goemans, M.X., Williamson, D.P.: Improved approximation algorithms for maximum cut and satisfiability problems using semidefinite programming. J. ACM 42(6), 1115–1145 (1995)
12. Indyk, P., Motwani, R.: Approximate nearest neighbors: towards removing the curse of dimensionality. In: Symposium on Theory of computing (STOC 1998), pp. 604–613 (1998)
13. Johnson, W., Lindenstrauss, J.: Extensions of lipschitz maps into a hilbert space. Contemporary Mathematics 26, 189–206 (1984)
14. Kumar, K., Bhattacharya, C., Hariharan, R.: A randomized algorithm for large scale support vector learning. In: Advances in Neural Information Processing Systems (NIPS 2007), pp. 793–800. MIT Press, Cambridge (2008)
15. Mierswa, I., Wurst, M., Klinkenberg, R., Scholz, M., Euler, T.: YALE: Rapid prototyping for complex data mining tasks. In: Int. Conf. on Knowledge discovery and data mining (KDD 2006). ACM Press, New York (2006)
16. Ravichandran, D., Pantel, P., Hovy, E.: Randomized algorithms and NLP: using locality sensitive hash function for high speed noun clustering. In: Association for Computational Linguistics (ACL 2005), pp. 622–629 (2005)
17. Salton, G., Wong, A., Yang, C.S.: A vector space model for automatic indexing. Commun. ACM 18(11), 613–620 (1975)
18. Singh, K., Ma, M., Park, D.W.: A content-based image retrieval using FFT & cosine similarity coefficient. Signal and Image Processing (2003)
19. Vempala, S.: The Random Projection Method. American Mathematical Society (2004)

# Large-Scale Clustering through Functional Embedding

Frédéric Ratle[1], Jason Weston[2], and Matthew L. Miller[2]

[1] IGAR, University of Lausanne, Amphipôle, 1015 Lausanne, Switzerland
frederic.ratle@unil.ch
[2] NEC Labs America, 4 Independence Way, Princeton NJ, USA

**Abstract.** We present a new framework for large-scale data clustering. The main idea is to modify functional dimensionality reduction techniques to directly optimize over discrete labels using stochastic gradient descent. Compared to methods like spectral clustering our approach solves a single optimization problem, rather than an ad-hoc two-stage optimization approach, does not require a matrix inversion, can easily encode prior knowledge in the set of implementable functions, and does not have an "out-of-sample" problem. Experimental results on both artificial and real-world datasets show the usefulness of our approach.

## 1 Introduction

Clustering, which aims at identifying groups in the data in an unsupervised manner, is one of the main tasks in the field of machine learning, and different approaches to tackle this problem have been popularized over the last decades, among them: $k$-means and hierarchical clustering, spectral clustering and its variants (e.g., [20,24]), support vector and maximum margin clustering [5,30,31].

Clustering and dimensionality reduction algorithms have, in recent years, grown to become a single field of study through spectral embedding methods. Indeed, both tasks aim at producing a compact and visually meaningful representation of data, and strong links between methods to achieve both tasks have been highlighted. For example, methods based on the analysis of the graph Laplacian of a similarity matrix have provided a common framework to perform clustering and embedding using the top eigenvectors of the Laplacian [3]. Typically methods like spectral clustering, however, require a two-stage approach of first embedding and then using $k$-means to cluster data points in the found embedding space.

In this paper we aim at providing a framework for data clustering based on large-scale direct optimization by stochastic gradient descent. We describe a general class of objective functions that are the analogue of nonlinear embedding pairwise loss functions, but using loss functions that are directly related to the task of clustering. After a review of existing embedding algorithms for clustering, we present an instance of our approach, the Ncut embedding (NCutEmb thereafter) method. We first describe a comparison of explicit and functional embedding methods in order to highlight the benefits of using a function-based approach. Experiments with NCutEmb clustering clearly show that we can achieve

W. Daelemans et al. (Eds.): ECML PKDD 2008, Part II, LNAI 5212, pp. 266–281, 2008.

error rates inferior or *at most* similar to those obtained with deterministic batch methods using stochastic gradient descent, while obtaining at the same time a model for new examples as they become available, thus avoiding the so-called "out-of-sample" problem.

## 2 Existing Methods: Embedding and Clustering

The problem of learning the underlying data structure from unlabeled examples has been intensively studied with both functional and explicit (point-based) methods. This problem is usually referred to as manifold learning or dimensionality reduction in the machine learning literature. Self-organizing maps and autoassociator networks are classical examples of function-based embedding methods. While these methods have traditionally been based on various heuristics, recent work [4,13,14] has shown that functional methods can benefit from the theoretical understanding of dimensionality reduction made since the nineties, and offer several practical advantages over explicit methods.

### 2.1 Point-Based Methods

Point-based, or explicit, methods are most commonly rooted either in principal component analysis or multidimensional scaling, two well-known linear embedding methods, which provide them a sound mathematical background. They provide a point-to-point correspondence between the input space and the intrinsic space in which the data lie. Explicit methods have been multiplied in recent years, and include the following methods and their variants: kernel PCA [23], Isomap [25], Locally Linear Embedding [21], Maximum Variance Unfolding [28] and Laplacian Eigenmaps [3]. Thorough reviews of this family of methods can be found in [19,22].

**Laplacian Eigenmaps and Spectral Clustering.** We will briefly describe here Laplacian Eigenmaps (LE), as it is central to the remainder of the paper. The idea behind LE is to map nearby inputs to nearby outputs, hence preserving the neighborhood relations between data points. This preservation of neighborhood structure renders the method insensitive to outliers, thus making it appropriate for clustering. LE finds the embedding that minimizes the following loss function:

$$\sum_{ij} L(f_i, f_j, W_{ij}) = \sum_{ij} W_{ij} ||f_i - f_j||^2 \quad (1)$$

where $f_k \in \mathbb{R}^d$ is the embedding of training example $k$. The algorithm finds an embedding of examples given a distance metric between the examples encoded in the graph Laplacian $L = D - W$, where $W$ is a similarity ("affinity") matrix between examples and $D_{ii} = \sum_j W_{ij}$ is the (diagonal) degree matrix. The basis vectors of the embedding are given by the top eigenvectors of the graph Laplacian. LE can be summarized as follows:

1. Construct the **neighborhood graph**.
2. Choose **weights** on the edges (e.g., $W_{ij} = 1$ if $i$ and $j$ are neighbors, or $W_{ij} = e^{\frac{||x_i - x_j||^2}{2\sigma^2}}$).
3. Solve the **eigenvalue problem** $L\mathbf{v} = \lambda D\mathbf{v}$, where $D_{ii} = \sum_j W_{ij}$ and $L = D - W$).

There is a direct connection between LE and spectral clustering (SC). Both methods are based on the graph Laplacian of a similarity matrix. Spectral clustering only involves a supplementary step, that is, clustering the spectral images using a conventional clustering algorithm such as $k$-means.

Explicit methods have a unique solution. However, their cost becomes prohibitive when confronted with large datasets. Moreover, a vector $f_i$ has to be learnt in order to embed an example $x_i$ into a $K$-dimensional space; each example is embedded separately, and after training, given a new example $x^*$ there is no straightforward way to embed it. This is referred to as the "out-of-sample" problem, which several authors have tried to address with special techniques (e.g., [6,16,26]). Learning a function-based embedding might prove useful to solve the above problems.

### 2.2 Functional Methods

While explicit methods learn directly the underlying low-dimensional representation, functional methods learn a mapping from the input space to the low-dimensional space. In this paper, we show that with the latter approach, learning can often be faster, while being easily generalizable to new data points, as we obtain a function and thus avoid the so-called out-of-sample problem.

As proposed in [14], one can instead of learning vectors learn a function-based embedding, i.e., $y_i = f(x_i)$, using some family of functions to choose $f$. Adopting this approach can provide several gains:

- Training is faster since if two examples $x_i$ and $x_j$ are highly correlated, the embedding for $f(x_j)$ will be good before we have even seen it during training if we have trained well on $f(x_i)$ already.
- Prior knowledge can be expressed via the set of possible embedding functions, as noted in [14]. In their work they showed how choosing a set of functions based on convolutions exploiting prior knowledge of images they achieve improved embedding results. In the clustering algorithms we develop, we will be able to encode this type of prior in exactly the same way.
- There is no out-of-sample problem, as we have the embedding $y^* = f(x^*)$.
- By choosing $f(x)$ of sufficiently high capacity, we can find the solution provided by a point-based embedding. By capacity we mean the complexity of the class of functions $f(x)$, measured by, e.g., the VC dimension.

Several methods have been devised to provide a model that is based on an embedding criterion. In [4], the authors incorporate a LE-type of regularization (see Eq. 1) in the cost function of the SVM, which provides a functional version of this embedding method, where the function used is a kernel-based classifier. This method is known as the Laplacian SVM (LapSVM).

The above mentioned algorithms are functional embedding algorithms. Of course there are many functional clustering algorithms as well, e.g. classical $k$-means. Several authors also showed how to implement functional clustering within the context of a support vector machine by proposing objectives whereby good clusterings have large margins [5,30,31].

**DrLIM Embedding and Siamese Networks.** Hadsell, Chopra and LeCun [14] recently suggested to minimize the following loss function for embedding:

$$\sum_{ij} L(f_i, f_j, W_{ij}) = \sum_{ij} W_{ij}||f_i - f_j||^2 + \sum_{ij}(1 - W_{ij})\max(0, m - ||f_i - f_j||)^2 \quad (2)$$

where $W_{ij} = 1$ if $i$ and $j$ are deemed similar and 0 otherwise, and $m$ is the size of the margin. DrLIM encourages similar examples to be mapped closely, and dissimilar ones to be separated by at least the distance $m$. This is related to energy-based models in general, such as Ising models or Hopfield networks, but it is based on neighborhood relations rather than actual labels.

Another similar approach is the Normalized Embedding [9]. The loss function, independent of magnitude, encourages similar examples to be positively correlated:

$$\sum_{ij} L(f_i, f_j, W_{ij}) = \sum_{ij} -W_{ij} \frac{f_i \cdot f_j}{||f_i||||f_j||} + \sum_{ij}(1 - W_{ij}) \frac{f_i \cdot f_j}{||f_i||||f_j||} \quad (3)$$

These two loss functions were implemented as instantiations of a general class of neural networks called *siamese networks*. However, potentially any class of functions can be used to implement $f$.

## 3 Functional Embeddings for Clustering

In the above we considered the task of embedding data in a $K$-dimensional space $K < p$, where $p$ is the dimensionality of the input. This can be viewed as soft clustering; for instance, PCA and soft $k$-means are guaranteed to provide, under certain conditions, the same result (see e.g. [12]).

We now propose a general objective for using functional embeddings for clustering. In our framework, one optimizes the following loss:

$$\sum_c \sum_{ij} L(f(x_i), c)\, Y_c(f(x_i), f(x_j), W_{ij}) \quad (4)$$

- $W$ is a pairwise similarity matrix defined *a priori* as in previous algorithms.
- $L(\cdot)$ is a classification based loss function such as the hinge loss:

$$L(f(x), y) = H(yf(x)) = \max(0, 1 - yf(x))) \quad (5)$$

For the multiclass case we can use $L(f(x), y) = \sum_{c=1}^{K} H(y(c)f_c(x))$. Here, $y(c) = 1$ if $y = c$ and -1 otherwise.

– $Y_c(f(x_i), f(x_j), W_{ij})$ encodes the weight to assign to point $i$ being in cluster $c$. In the general case it can be a function of $f(x_i)$, $f(x_j)$ and the similarity score $W_{ij}$. Different choices of $Y_c(\cdot)$ implement different algorithms.

This class of algorithms learns a clustering $c(x) = \text{argmax } f(x) \in \mathbb{R}^K$ (or $c(x) = \text{sign}(f(x))$ in the two-class case). Our objective differs from usual clustering algorithms in that we use *pairs of examples* to learn the clustering. This objective is similar to embedding algorithms in that we essentially embed the data *at the same time as performing a clustering* by viewing the $K$-dimensional output as $K$ clusters.

Intuitively, we directly encode into our clustering algorithm that neighbors with $W_{ij} > 0$ should have the same cluster assignment. For example, one choice for $Y_c(\cdot)$ that we explore in this paper, which we call NCutEmb, is as follows (two-class case, $c \in \{\pm 1\}$ ):

$$Y_c(f_i, f_j, W_{ij}) = \begin{cases} \eta^{(+)} & \text{if } \text{sign}(f_i + f_j) = c \text{ and } W_{ij} = 1 \\ -\eta^{(-)} & \text{if } \text{sign}(f_j) = c \text{ and } W_{ij} = 0 \\ 0 & \text{otherwise.} \end{cases} \quad (6)$$

Equation (6) assigns a pair of neighbors to the cluster with the most confident label from the pair. Examples $x_j$ that are not neighbors of $x_i$, i.e. when $W_{ij} = 0$, are encouraged to fall into different clusters. Multiclass versions of NCutEmb are described in Section 3.3.

In the following we describe in detail some particular specializations of this objective function, and our particular implementation of it by training online by stochastic gradient descent using a set of functions implementable by a neural network.

### 3.1 Balancing Constraint

Many unconstrained clustering objective functions lead to a trivial solution, e.g., all the points end up in one single cluster, therefore it is often necessary to enforce some constraints. Even though we do not have access to label information, we can use the predicted labels to implement such constraints.

Unsupervised and semi-supervised learning (SSL) both usually make use of balancing constraints, albeit different since in SSL we have partial label information. In an unsupervised setting, the usual assumption is that each cluster should be reasonably large. For instance, RatioCut [15] and NormalizedCut [24] weigh, respectively, the cut by the number of vertices $|c_i|$ in cluster $c_i$ and its volume $vol(c_i)$, i.e., the sum of the weights on the edges. In [2], the authors simply impose that each cluster has at least $\frac{N}{K}$ examples, where $N$ is the size of the training set and $K$, the number of clusters.

In a semi-supervised setting, the problem is somehow easier, as the labeled points provide an estimate of the class balance (if the examples are i.i.d.). In [17], the author imposes that the fraction of each class in the unlabeled set is the same as in the labeled set. A similar constraint is used with low-density

**Algorithm 1.** Two-class NCutEmb algorithm with hard constraints

**for** each iteration **do**
  Pick $x_i$ and $x_j$ such that $x_j \in neighb\,(x_i)$
  Find $g = \text{argmax}_{i,j}(|f(x_i)|, |f(x_j)|)$ and $h = \text{sign}\,(f(x_g))$
  **if** $seen(h) \leq \frac{N_t}{K} + \zeta$ **then**
    Update the network w.r.t. to $\eta^{(+)}\nabla L\,(f(x_i), h)$ and $\eta^{(+)}\nabla L\,(f(x_j), h)$
    Update $seen(h)$, we assigned an example to cluster $h$
  **end if**
  Pick $x_m$ and $x_n$ such that $x_n \notin neighb\,(x_m)$
  Find $p = \text{sign}\,(x_m)$ and $q = \text{sign}\,(x_n)$
  Update the network w.r.t. $\eta^{(-)}\nabla L\,(f(x_m), -q)$ and $\eta^{(-)}\nabla L\,(f(x_n), -p)$
**end for**

separation [10]. In [18], the authors ignore examples of a class that has been seen more than its expected frequency, as provided by the labeled set.

Generally speaking, we can classify the constraints used in clustering simply as "hard" or "soft". We have tried both types, in the following way:

1. At each iteration, control the number of examples assigned to cluster $c$, denoted as $seen(c)$. We then define $Y_c(\cdot)$ as

$$Y_c(f_i, f_j, W_{ij}) = \begin{cases} \eta^{(+)} & \text{if } \text{sign}(f_i + f_j) = c \text{ and } W_{ij} = 1 \text{ and} \\ & seen(c) \leq \frac{N_t}{K} + \zeta \\ -\eta^{(-)} & \text{if } \text{sign}(f_j) = c \text{ and } W_{ij} = 0 \\ 0 & \text{otherwise.} \end{cases} \tag{7}$$

   where $N_t$ is the number of examples seen at time $t$, $K$ is the number of clusters and $\zeta$ is a negative or positive factor allowing for some variation in cluster size. One can apply this constraint on a fixed-sized window of the last, e.g., 100 examples seen, which corresponds to a particular choice of the function $seen(\cdot)$.

2. Diminish the "learning rate" proportionally to the number of examples seen of each class. $Y_c(\cdot)$ then becomes

$$Y_c(f_i, f_j, W_{ij}) = \begin{cases} \frac{\eta^{(+)}}{seen(c)+1} & \text{if } \text{sign}(f_i + f_j) = c \text{ and } W_{ij} = 1 \\ -\eta^{(-)} & \text{if } \text{sign}(f_j) = c \text{ and } W_{ij} = 0 \\ 0 & \text{otherwise.} \end{cases} \tag{8}$$

   where again $seen(c)$ is the number of examples assigned to cluster $c$ seen in the last 100 examples (+1 to avoid zeros).

**Algorithm 2.** Two-class NCutEmb algorithm with soft constraints

**for** each iteration **do**
  Pick $x_i$ and $x_j$ such that $x_j \in neighb\,(x_i)$
  Find $g = \text{argmax}_{i,j}(|f(x_i)|, |f(x_j)|)$ and $h = \text{sign}\,(f(x_g))$
  Update the network w.r.t. to $\frac{\eta^{(+)}}{seen(h)+1}\nabla L\,(f(x_i), h)$ and $\frac{\eta^{(+)}}{seen(c)+1}\nabla L\,(f(x_j), h)$
  Update $seen(h)$, we assigned an example to cluster $h$
  Pick $x_m$ and $x_n$ such that $x_n \notin neighb\,(x_m)$
  Find $p = \text{sign}\,(x_m)$ and $q = \text{sign}\,(x_n)$
  Update the network w.r.t. $\eta^{(-)}\nabla L\,(f(x_m), -q)$ and $\eta^{(-)}\nabla L\,(f(x_n), -p)$
**end for**

The hard constraint is similar to the approaches described in [2,18], while the soft constraint is very similar to the optimal learning rate for stochastic $k$-means, as reported in [8]. These approaches will be referred to as NCutEmb$^h$ and NCutEmb$^s$, respectively.

### 3.2 Two-Class Clustering

In a two-class setting, we employ a neural network $f(x)$ with one output $y \in \{\pm 1\}$, trained online via stochastic gradient descent. We employ either a hard constraint (Algorithm 1) or a soft constraint (Algorithm 2). In these algorithms, $neighb\,(x_i)$ refers to the neighborhood of $x_i$.

Simply put, similar points are pushed towards the same label. In the case of dissimilar points, the points are pulled in opposite classes. We use two different learning rates $\eta^{(+)}$ and $\eta^{(-)}$. The latter is used to pull non-neighbors in different classes and is typically smaller. The idea behind the method is illustrated in Figure 1.

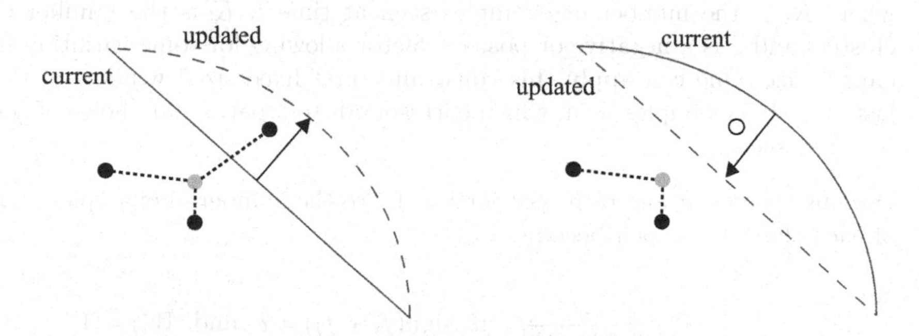


**Fig. 1.** Clustering using NCutEmb. A hyperplane that separates neighbors is pushed to classify them in the same class (**left**; the classifier cuts through an edge of the graph and is pushed upwards), whilst one that classifies non-neighbors in the same class is modified to separate them (**right**; the hyperplane is pushed to separate the unconnected points).

### 3.3 The Multiclass Case

In the multiclass setting, we also tried two alternative approaches.

1. The same strategy as binary case: push neighbors towards the **most confident label**, i.e., if we define the most confident example of a pair as

$$m(f_i, f_j) = \text{argmax}_{i,j} \left(\max(f_i), \max(f_j)\right))$$

then we can define the function $Y_c(\cdot)$ using:

$$Y_c(f_i, f_j, W_{ij}) = \begin{cases} \eta^{(+)} & \text{if argmax } f_{m(f_i,f_j)} = c \text{ and } W_{ij} = 1 \\ -\eta^{(-)} & \text{if argmax } f_j = c \text{ and } W_{ij} = 0 \\ 0 & \text{otherwise.} \end{cases} \quad (9)$$

The balancing constraint can then be enforced in a hard way by a fixed-sized window, as in the binary case.

2. Push towards the $K$ clusters simultaneously, with one learning rate $\eta$ per class, weighted by the outputs of the point with the most confident label. That is,

$$Y_c(f_i, f_j, W_{ij}) = \begin{cases} \eta_c & \text{if } W_{ij} = 1 \\ 0 & \text{otherwise} \end{cases} \quad (10)$$

where

$$\eta_c \leftarrow \eta^{(+)} f_c(x_i)$$

The balancing constraint is soft, i.e, the learning rate decreases with the size of the cluster.

These approaches will be referred to as NCutEmb$^{max}$ and NCutEmb$^{all}$. In the latter approach, the softmax$(\cdot)$ function is used at the output of $f(\cdot)$ in order to have $f(x_i) \in [0, 1]$:

$$\text{softmax}\,(y_1, ..., y_k) = \frac{\exp(y_i)}{\sum_i \exp(y_i)}$$

This is a common technique in Neural Networks [7].

### 3.4 Large-Scale Extension

Building a graph *a priori* is not feasible when dealing with very large datasets (i.e., millions of examples). One can easily avoid this step by computing neighbors on the fly with epsilon-balls or local weighted distances. Indeed, whenever two points are picked, it is straightforward to verify a condition such as

$$||x_i - x_j|| < \epsilon$$

where $\epsilon$ is a neighborhood parameter. Note that any similarity measure can be used. For efficiency purposes, a table with the neighbors found up to date can be kept in memory.

# 4 Experiments

Table 1 summarizes the small-scale datasets that have been used throughout Section 4. `g50c`, `text` and `usps` are datasets that are often used in the semi-supervised learning literature (e.g., [10,11]). Numerous clustering techniques were applied to the `usps` digit dataset in [29]. The authors, however, studied two subsets of 2 and 4 digits. In this paper we consider the 10 `usps` classes.

The datasets `bcw` (Breast Cancer Wisconsin) and `glass` (Glass Identification) are taken from the UCI repository [1]. The `ellips` dataset consists of four 50-dimensional ellipses connected by one extremity.

In order to have a relative measure of class imbalance, which will be useful for the clustering section, we define the degree of class imbalance (DCI):

$$DCI = \frac{1}{N}\left(\frac{1}{K-1}\sum_{k}\left(|c_k| - \frac{N}{K}\right)^2\right)^{\frac{1}{2}} \tag{11}$$

This is the standard deviation of the clusters with respect to the "expected value" of the size of clusters $c_k$ in a balanced problem, divided by $N$, the number of examples. A DCI of 0 thus represents a perfectly balanced problem. Simply taking the ratio between the biggest and smallest classes would not take into account the variability across all the classes in a multiclass setting.

## 4.1 Functional vs. Explicit Embedding

We first performed some experiments that aimed at comparing embedding methods as a justification of the use of functional embedding for clustering. To this end, we have used the `g50c`, `text` and `usps` datasets. We provide results for explicit and functional DrLIM, and these are compared with Laplacian Eigenmaps. The goal is to reach a value as close as possible to LE, as DrLIM is an approximate method. However, the latter has the advantage of being useful on much larger datasets as it does not require costly operations such as matrix inversion. The performance criterion that has been used is the fraction of embedded points

**Table 1.** Small-scale datasets used throughout the experiments. `g50c`, `text` and `usps` are used in Section 4.1, while all datasets are used in Section 4.2.

| data set | classes | dims | points | DCI ×100 |
|---|---|---|---|---|
| g50c | 2 | 50 | 550 | 0 |
| text | 2 | 7511 | 1946 | 0.87 |
| bcw | 2 | 9 | 569 | 21.19 |
| ellips | 4 | 50 | 1064 | 12.21 |
| glass | 6 | 10 | 214 | 13.89 |
| usps | 10 | 256 | 2007 | 2.83 |

that do not share the same label with their nearest neighbor, i.e.,

$$L_{embed}\left(f_1, ..., f_n, \alpha\right) = \frac{1}{N} \sum_{i=1}^{n} \delta\left(f_i, f_{nn(i)}\right) \qquad (12)$$

where $f_{nn(i)}$ is $f_i$'s nearest neighbor in the embedding (the relation needs not be mutual), $\alpha$ are the parameters, $N$ is the size of the training set, and

$$\delta\left(f_i, f_{nn(i)}\right) = \begin{cases} 0 \text{ if } y_i = y_{nn(i)} \\ 1 \text{ if } y_i \neq y_{nn(i)} \end{cases}$$

Here, $y_i$ designates the label of $f_i$. We have chosen this criterion because unlike Eq. 1, it does not depend on the specific weights $W_{ij}$ that are used for each dataset.

Table 2 reports the best values of the embedding criterion obtained. The number of iterations required to achieve that result is given in brackets. The results are compared against Laplacian Eigenmaps. Both LE and DrLIM are assessed using the criterion of Eq. 12. We have tried both a linear network ("Linear") and a one-hidden layer network ("NN"). The linear network performed best for the g50c dataset, while the NN performed better for the `text` and `usps` datasets. For all the methods, the number of neighbors was optimized. An appropriate learning rate was selected for nf-DrLIM and f-DrLIM.

It can be observed from Table 2 that functional DrLIM provides a gain of accuracy compared to its non-functional counterpart, and requires a smaller number of iterations. Using a functional model rather than a point-based one appears more important than the choice of the class of functions used within such a model. Indeed, even though there are differences specific to each dataset between a linear network and a one hidden layer network, the differences between functional and point-based methods are larger, in terms of both convergence speed and value of the embedding criterion.

**Table 2.** Lowest obtained value of the embedding criterion (Eq. 12) for LE, functional DrLIM and non-functional DrLIM, on an average of 10 runs. Low values indicate that points sharing the same label are mapped closely. The number of iterations (in iterations $\times 10^3$) to reach a minimum value for f-DrLIM and nf-DrLIM is given in brackets. Convergence is considered achieved when the value of the embedding criterion stops decreasing. The neural network has one hidden layer and 5, 40 and 40 hidden units for the three datasets, respectively. For every dataset, the appropriate learning rate was selected.

| | g50c | text | usps |
|---|---|---|---|
| LE | 7.45 [-] | 10.69 [-] | 7.28 [-] |
| nf-DrLIM | 12.82 [100] | 29.58 [70] | 11.55 [80] |
| f-DrLIM (Linear) | **7.14** [50] | 14.18 [50] | 10.02 [70] |
| f-DrLIM (NN) | 7.55 [80] | **12.00** [40] | **8.99** [70] |

The network that has been trained using functional DrLIM can be applied to any other data point; we thus avoid the out-of-sample problem, as we have a functional relationship between the high-dimensional input and the output. The fact that functional DrLIM is also more accurate and faster than the point-based method further encourages the use of functional methods for embedding. Indeed, the difference with the values obtained with LE, towards which we wish to converge, is small. Note that LE optimizes the criterion of Eq. 1, but we make the assumption that a good embedding will minimize both Eq. 1 and 12.

### 4.2 Small-Scale Clustering Experiments

Here, all the datasets from Table 1 are used. We compare the NCutEmb algorithm to $k$-means and spectral clustering (the algorithm presented in [20]). For spectral clustering, a $k$-nearest neighbor graph (sc-knn) and a full radial basis function graph (sc-rbf) have been used for the calculation of the weights, and the corresponding parameters ($k$ and $\sigma$) have been selected. For the NCutEmb method, we have selected for each experiment the appropriate number of neighbors for $W$ (typically 10, 50 or 100) and learning rate (a value between 0.001 and 1).

Tables 3 and 4 report the cluster assignment error for two-class and multiclass problems, respectively. The cluster assignment error is simply the fraction of points for which $c_i \neq y_i$, where $c_i$ is the cluster in which we assign $x_i$. This has to be calculated considering all permutations from clusters to indices. When confronted to a large number of classes, one can use optimization techniques to solve this problem, as underlined in [27]. For the small-scale experiments, training times ranged approximately from 5 seconds (`bcw`) to a minute (`usps`) on a standard PC.

The results show that the NCutEmb method performs either similarly or provides superior performance than conventional clustering algorithms. Since we have other benefits when using NCutEmb, such as the availability of a functional model and the scalability to large datasets, the approach is quite attractive. For

**Table 3.** Cluster assignment error (%) on two-class small-scale datasets with the NCutEmb method, compared to $k$-means and spectral clustering. The spectral clustering algorithms have been optimized over the choice of $\sigma$ and $k$. Each value is averaged over 10 random replications of the initial dataset. A linear network was used for `bcw` and `text`, while a one-hidden layer NN with 5 hidden units was used for `g50c`.

| | bcw | g50c | text |
|---|---|---|---|
| $k$-means | 3.89 | 4.64 | 7.26 |
| sc-rbf | 3.94 | 5.56 | 6.73 |
| sc-knn | 3.60 | 6.02 | 12.9 |
| NCutEmb$^h$ | 3.63 | 4.59 | **7.03** |
| NCutEmb$^s$ | **3.15** | **4.41** | 7.89 |

**Table 4.** Cluster assignment error (%) on multiclass small-scale datasets with the NCutEmb method, compared to $k$-means and spectral clustering. The spectral clustering algorithms have been optimized over the choice of $\sigma$ and $k$. Each value is averaged over 10 random replications of the initial dataset. A linear network was used for `ellips` while a one-hidden layer network with 3 hidden units was used for `glass` and 40 hidden units for `usps`.

| | ellips | glass | usps |
|---|---|---|---|
| $k$-means | 20.29 | 25.71 | 30.34 |
| sc-rbf | 10.16 | 39.30 | 32.93 |
| sc-knn | **2.51** | 40.64 | 33.82 |
| NCutEmb$^{max}$ | 4.76 | **24.58** | 19.36 |
| NCutEmb$^{all}$ | 2.75 | 24.91 | **19.05** |

the two-class datasets, the difference between the type of constraint used appears minor.

Regarding the multiclass experiments, it is interesting to study the performance of the methods with respect to the class imbalance of each dataset. NCutEmb$^{all}$ outperforms the other methods (except sc-knn) for the `ellips` dataset, which is very imbalanced. However, the `glass` dataset is also imbalanced, and both NCutEmb$^{all}$ and NCutEmb$^{max}$ perform well. Consequently, it is hard to draw any conclusion regarding the superior ability of one method or another to handle clusters of different sizes; the two methods can handle them well given the right parameters.

On average, NCutEmb appears superior to both $k$-means and spectral clustering on the datasets that we have used. This is a surprising result, as we could expect spectral clustering to be the "ground truth" result, as with LE in Section 4.1. However, this could be the effect of the use of a one-stage approach for the optimization. Indeed, even if a good embedding is obtained with spectral clustering, the clustering stage might be poor (and vice versa). Performing the embedding and clustering tasks simultaneously seems to overcome this problem. We note, however, that similarly to $k$-means, the method performs better when clusters are close to Gaussian objects. Since NCutEmb is neighborhood based, we cannot expect good performance with highly complex clusters. If $||x_i - x_j|| < \epsilon$, we must have a high probability that $x_i$ and $x_j$ truly belong to the same cluster. In other words, the smoothness assumption must hold.

We then conducted a series of experiments where we split the datasets in order to obtain a test error, as an estimation of the out-of-sample error. In the following experiments, 70% of the data is used for training, and the error is calculated on the remaining 30%, which has not been seen. Tables 5 and 6 report this error for two-class and multiclass methods.

We note that the difference between clustering error and out-of-sample error is small for each dataset. A network that has been trained on a large enough set of examples can cluster new points accurately.

**Table 5.** Test error (i.e., out-of-sample error) for the two-class datasets (%)

| | bcw | g50c | text |
|---|---|---|---|
| $k$-means | 4.22 | **6.06** | 8.75 |
| NCutEmb$^h$ | **3.21** | **6.06** | 7.68 |
| NCutEmb$^s$ | 3.64 | 6.36 | **7.38** |

**Table 6.** Test error (i.e., out-of-sample error) for the multiclass datasets (%)

| | ellips | glass | usps |
|---|---|---|---|
| $k$-means | 20.85 | 28.52 | 29.44 |
| NCutEmb$^{max}$ | 5.11 | 25.16 | 20.80 |
| NCutEmb$^{all}$ | **2.88** | **24.96** | **17.31** |

### 4.3 Large-Scale Clustering Experiments

Here, we repeat the experiments of the previous section with the MNIST digits database, consisting of 60,000 training examples and 10,000 test examples of digits 0 to 9. Each image has 28 × 28 pixels. MNIST is a fairly well balanced dataset (DCI×100 of 0.57 using Eq. 11). NCutEmb requires about 60 minutes to converge to an optimum on a database such as MNIST. For visualization purposes, projections of MNIST data are shown in Fig. 2 using Laplacian Eigenmaps. Only a subset of 3000 examples is projected in Fig. 2 because of computational limitations, as LE requires the eigenvectors of the similarity matrix of the data.

**Table 7.** Clustering the MNIST database with the NCutEmb$^{max}$ and NCutEmb$^{all}$ methods, compared to $k$-means, for different numbers of clusters. The training error (i.e., clustering error) and the test error (out-of-sample error) are provided. A one-hidden layer network with 40 units has been used for the 10 and 20 clusters problems, and 80 units for the 50 clusters problem.

| # clusters | method | train | test |
|---|---|---|---|
| 50 | $k$-means | 18.46 | 17.70 |
| | NCutEmb$^{max}$ | **13.82** | **14.23** |
| | NCutEmb$^{all}$ | 18.67 | 18.37 |
| 20 | $k$-means | 29.00 | 28.03 |
| | NCutEmb$^{max}$ | 20.12 | 23.43 |
| | NCutEmb$^{all}$ | **17.64** | **21.90** |
| 10 | $k$-means | 40.98 | 39.89 |
| | NCutEmb$^{max}$ | **21.93** | **24.37** |
| | NCutEmb$^{all}$ | 24.10 | 24.90 |

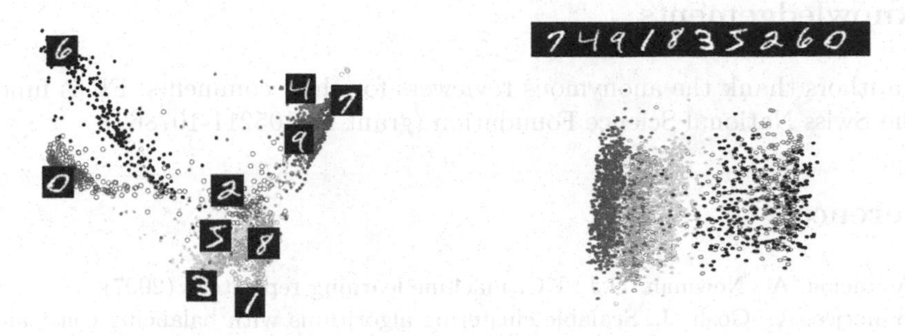

**Fig. 2.** Projections of an MNIST subset using Laplacian Eigenmaps with a nearest-neighbor graph

Table 7 presents the training and test error of the network used to cluster MNIST. We compare our method to $k$-means. For 10 and 20 clusters, NCutEmb$^{max}$ and NCutEmb$^{all}$ provide equivalent error rates. We seem to achieve a much better performance than $k$-means in most cases, except with NCutEmb$^{all}$ using 50 clusters. This may be due to the fact that within this approach, we push neighbors towards all classes at each iteration. It is interesting to note that the gap with $k$-means is reduced when increasing the number of clusters. This could be expected; the more groups we use to divide the dataset, the more simple these groups become in their structure. With a very large number of clusters, NCutEmb and $k$-means should be expected to provide the same results.

## 5 Conclusion

We have presented a framework for data clustering based on functional models for dimensionality reduction optimized by stochastic gradient descent. Our framework suggests direct algorithms for clustering using an embedding approach rather than the two-stage approach of spectral clustering. The experiments have shown that using this approach brings about many benefits: (i) the algorithms scale well as we do not need to compute eigenvectors, (ii) a model is available and can be applied to new data points, and (iii) the method provides superior results on average than spectral clustering or $k$-means.

Some limitations were highlighted, however, in the way it can handle highly non-linear datasets. The more we violate the smoothness assumption, the more chance that pairs of neighbors do not belong to the same cluster and the less accurate the method is expected to be. However, this limitation is dependent on the method for choosing neighbors: if one has a way of choosing pairs of neighbors which can guarantee they belong to the same cluster, then the method can perform well.

We believe that this approach could be of great use in many important applications areas where new examples are continuously available.

## Acknowledgements

The authors thank the anonymous reviewers for their comments. FR is funded by the Swiss National Science Foundation (grant no.105211-107862).

## References

1. Asuncion, A., Newman, D.J.: UCI machine learning repository (2007)
2. Banerjee, A., Gosh, J.: Scalable clustering algorithms with balancing constraints. Data Mining and Knowledge Discovery 13(3), 365–395 (2006)
3. Belkin, M., Niyogi, P.: Laplacian eigenmaps for dimensionality reduction and data representation. Neural Computation 15(6), 1373–1396 (2003)
4. Belkin, M., Niyogi, P., Sindhwani, V.: Manifold regularization: a geometric framework for learning from Labeled and Unlabeled Examples. Journal of Machine Learning Research 7, 2399–2434 (2006)
5. Ben-Hur, A., Horn, D., Siegelmann, H.T., Vapnik, V.: Support vector clustering. Journal of Machine Learning Research 2, 125–137 (2001)
6. Bengio, Y., Delalleau, O., Le Roux, N., Paiement, J.-F., Vincent, P., Ouimet, M.: Learning eigenfunctions links spectral embedding and kernel PCA. Neural Computation 16(10), 2197–2219 (2004)
7. Bishop, C.M.: Neural Networks for Pattern Recognition. Oxford University Press, USA (1995)
8. Bottou, L.: Stochastic learning. In: Bousquet, O., von Luxburg, U., Rätsch, G. (eds.) Machine Learning 2003. LNCS (LNAI), vol. 3176, pp. 146–168. Springer, Heidelberg (2004)
9. Bromley, J., Bentz, J.W., Bottou, L., Guyon, I., LeCun, Y., Moore, C., Sackinger, E., Shah, R.: Signature verification using a siamese time delay neural network. International Journal of Pattern Recognition and Artificial Intelligence 7(4) (August 1993)
10. Chapelle, O., Zien, A.: Semi-supervised classification by low density separation. In: AISTATS, pp. 57–64 (January 2005)
11. Collobert, R., Sinz, F., Weston, J., Bottou, L.: Large scale transductive SVMS. Journal of Machine Learning Research 7, 1687–1712 (2006)
12. Ding, C., He, X.: K-means clustering via principal component analysis. In: Proc. of the Int. Conference on Machine Learning (ICML 2004) (2004)
13. Gong, H.F., Pan, C., Yang, Q., Lu, H.Q., Ma, S.: Neural network modeling of spectral embedding. In: BMVC 2006, p. I–227 (2006)
14. Hadsell, R., Chopra, S., LeCun, Y.: Dimensionality reduction by learning an invariant mapping. In: Proc. Computer Vision and Pattern Recognition Conference (CVPR 2006). IEEE Press, Los Alamitos (2006)
15. Hagen, L., Kahng, A.: New spectral methods for ratio cut partitioning and clustering. IEEE Trans. on Computer Aided-Design 11(9), 1074–1085 (1992)
16. He, X., Yan, S.C., Hu, Y., Niyogi, P., Zhang, H.J.: Face recognition using laplacianfaces. IEEE Trans. PAMI 27(3), 328
17. Joachims, T.: Transductive inference for text classification using support vector machines. In: International Conference on Machine Learning, ICML (1999)
18. Karlen, M., Weston, J., Erken, A., Collobert, R.: Large scale manifold transduction. In: Proc. of the Int. Conference on Machine Learning (ICML 2008) (2008)

19. Lee, J.A., Verleysen, M.: Nonlinear Dimensionality Reduction. Springer, New York (2007)
20. Ng, A.Y., Jordan, M., Weiss, Y.: On spectral clustering: analysis and an algorithm. In: Advances in Neural Information Processing Systems (NIPS 13) (2001)
21. Roweis, S.T., Saul, L.K.: Nonlinear dimensionality reduction by locally linear embedding. Science 290(5500), 2323–2326 (2000)
22. Saul, L.K., Weinberger, K.Q., Ham, J.H., Sha, F., Lee, D.D.: Spectral methods for dimensionality reduction. In: Semi-Supervised Learning. MIT Press, Cambridge (2006)
23. Schölkopf, B., Smola, A.J., Müller, K.R.: Nonlinear component analysis as a kernel eigenvalue problem. Neural Computation 10, 1299–1319 (1998)
24. Shi, J., Malik, J.: Normalized cuts and image segmentation. IEEE Trans. on Pattern Analysis and Machine Intelligence 22(8) (2000)
25. Tenenbaum, J.B., de Silva, V., Langford, J.C.: A global geometric framework for nonlinear dimensionality reduction. Science 290(5500), 2319–2323 (2000)
26. Trosset, M.W., Priebe, C.E.: The out-of-sample problem for multidimensional scaling. Technical Report 06-04, Dept. of Statistics, Indiana University (2006)
27. Verma, D., Meila, M.: Comparison of spectral clustering methods. In: Advances in Neural Information Processing Systems (NIPS 15) (2003)
28. Weinberger, K.Q., Saul, L.K.: Nonlinear dimensionality reduction by semidefinite programming and kernel matrix factorization. In: Proc. of the Tenth International Workshop on AI and Statistics (AISTATS 2005) (2005)
29. Wu, M., Schölkopf, B.: A local learning approach for clustering. In: Advances in Neural Information Processing Systems (NIPS 19) (2006)
30. Xu, L., Neufeld, J., Larson, B., Schuurmans, D.: Maximum margin clustering. In: Advances in Neural Information Processing Systems (NIPS 16) (2004)
31. Zhang, K., Tsang, I., Kwok, J.T.: Maximum margin clustering made practical. In: Proc. of the Int. Conference on Machine Learning (ICML 2007) (2007)

# Clustering Distributed Sensor Data Streams

Pedro Pereira Rodrigues[1,2], João Gama[1,3], and Luís Lopes[2,4]

[1] LIAAD - INESC Porto L.A.
Rua de Ceuta, 118 - 6 andar, 4050-190 Porto, Portugal
[2] Faculty of Sciences of the University of Porto, Portugal
pprodrigues@fc.up.pt
[3] Faculty of Economics of the University of Porto, Portugal
jgama@fep.up.pt
[4] CRACS - INESC Porto L.A.
lblopes@dcc.fc.up.pt

**Abstract.** Nowadays applications produce infinite streams of data distributed across wide sensor networks. In this work we study the problem of continuously maintain a cluster structure over the data points generated by the entire network. Usual techniques operate by forwarding and concentrating the entire data in a central server, processing it as a multivariate stream. In this paper, we propose *DGClust*, a new distributed algorithm which reduces both the dimensionality and the communication burdens, by allowing each local sensor to keep an online discretization of its data stream, which operates with constant update time and (almost) fixed space. Each new data point triggers a cell in this univariate grid, reflecting the current state of the data stream at the local site. Whenever a local site changes its state, it notifies the central server about the new state it is in. This way, at each point in time, the central site has the global multivariate state of the entire network. To avoid monitoring all possible states, which is exponential in the number of sensors, the central site keeps a small list of counters of the most frequent global states. Finally, a simple adaptive partitional clustering algorithm is applied to the frequent states central points in order to provide an anytime definition of the clusters centers. The approach is evaluated in the context of distributed sensor networks, presenting both empirical and theoretical evidence of its advantages.

**Keywords:** online adaptive clustering, distributed data streams, sensor networks, incremental discretization, frequent items monitoring.

## 1 Introduction

Data gathering and analysis have become ubiquitous, in the sense that our world is evolving into a setting where all devices, as small as they may be, will be able to include sensing and processing ability. Nowadays applications produce infinite streams of data distributed across wide sensor networks. Therefore, data should also be processed by sensor networks in a distributed fashion. Usual techniques operate by forwarding and concentrating the entire data in a central server,

W. Daelemans et al. (Eds.): ECML PKDD 2008, Part II, LNAI 5212, pp. 282–297, 2008.

processing it as a multivariate stream. In this work we study the problem of continuously maintain a cluster structure over the data points generated by the entire network, where each sensor produces a univariate stream of data. The main problem in applying clustering to data streams is that systems should consider data evolution, being able to compress old information and adapt to new concepts. In this paper, we propose a new distributed algorithm which reduces both the dimensionality and the communication burdens.

In the next section we present some related work on the research areas addressed by this paper. In section 3 our method is explained, with relevant analysis of the overall processing. Section 4 focus on the advantages presented by our proposal in terms of memory and communication resources, especially important in distributed sensor networks. Validation of the system and experimental results on real-world scenarios are presented in section 5. Final section concludes the paper, with thorough discussion and conclusions, including foreseen future work.

## 2 Related Work

The method we present in this paper is related with three other areas of research in data streams: sensor network processing, frequent items monitoring, and distributed clustering.

### 2.1 Sensor Data and Networks

Sensors are usually small, low-cost devices capable of sensing some attribute of a physical phenomenon. In terms of hardware development, the state-of-the-art is well represented by a class of multi-purpose sensor nodes called *motes* [8], which were originally developed at UC Berkeley and are being deployed and tested by several research groups and start-up companies. Although common applications are traditionally developed on low-level programming of the *motes*, recent programming languages and environments such as Regiment [21] and EnviroSuite [18], provide high level programming abstractions, allowing more complex programming and usage of sensor devices as processing units for knowledge extraction scenarios. Sensor networks are composed of a variable number of sensors (depending on the application), which have several distinguishing features: (a) the number of nodes is potentially very large and thus scalability is a problem, (b) the individual sensors are prone to failure given the often challenging conditions they experiment in the field, (c) the network topology changes dynamically, (d) broadcast protocols are used to route messages in the network, (e) limited power, computational, and memory capacity, and (f) lack of global identifiers [3]. In sensor network applications, data routing is usually based on two main approaches: (a) sensors broadcast advertisements for the availability of the data and wait for interested nodes, or; (b) sinks broadcast interest in the data and wait for replies from the sensors. Common problems with these strategies are *implosion* and *overlap* [3].

### 2.2 Frequent Items in Data Streams

The problem of finding the most frequent items in a data stream $S$ of size $N$ is, roughly put, the problem to find the elements $e_i$ whose relative frequency $f_i$ is higher than a user specified support $\phi N$, with $0 \leq \phi \leq 1$. Given the space requirements that exact algorithms addressing this problem would need [5], several algorithms were already proposed to find the top-$k$ frequent elements, being roughly classified into *counter-based* and *sketch-based* [20]. *Counter-based* techniques keep counters for each individual element in the monitored set, which is usually a lot smaller than the entire set of elements. When an element is seen which is not currently being monitored, different algorithms take different actions in order to adapt the monitored set accordingly. *Sketch-based* techniques provide less rigid guarantees, but they do not monitor a subset of elements, providing frequency estimators for the entire set.

Simple *counter-based* algorithms such as *Sticky Sampling* and *Lossy Counting* were proposed in [19], which process the stream in reduced size. Yet, they suffer from keeping a lot of irrelevant counters. *Frequent* [10] keeps only $k$ counters for monitoring $k$ elements, incrementing each element counter when it is observed, and decrementing all counters when a unmonitored element is observed. Zeroed-counted elements are replaced by new unmonitored element. This strategy is similar to the one applied by *Space-Saving* [20], which gives guarantees for the top-$m$ most frequent elements. *Sketch-based* algorithms usually focus on families of hash functions which project the counters into a new space, keeping frequency estimators for all elements. The guarantees are less strict but all elements are monitored. The *CountSketch* algorithm [5] solves the problem with a given success probability, estimating the frequency of the element by finding the median of its representative counters, which implies sorting the counters. Also, *GroupTest* method [6] employs expensive probabilistic calculations to keep the majority elements within a given probability of error. Although generally accurate, its space requirements are large and no information is given about frequencies or ranking.

### 2.3 Clustering Data Streams

The main problem in applying clustering to data streams is that systems should consider data evolution, being able to compress old information and adapt to new concepts. The range of clustering algorithms that operate online over data streams is wide, including *partitional*, *hierarchical*, *density-based* and *grid-based* methods. A common connecting feature is the definition of unit cells or representative points, from which clustering can be obtained with less computational costs. In this paper we address two types of clustering procedures: *point-based* clustering and *grid-based* clustering.

**Point-Based Clustering.** Several algorithms operate over summaries or samples of the original stream. Bradley et al. [4] proposed the *Single Pass K-Means*, increasing the capabilities of *k-means* for large datasets, by using a buffer where points of the dataset are kept in a compressed way. The *STREAM* [22] system can be seen as an extension of [4] which keeps the same goal but has as restriction the use of available memory. After filling the buffer, STREAM clusters the buffer into $k$ clusters, retaining only the $k$ centroids weighted by the number of examples in each cluster. The process is iteratively repeated with new points. The BIRCH hierarchical method [25] uses *Clustering Features* to keep sufficient statistics for each cluster at the nodes of a balanced tree, the *CF-tree*. Given its hierarchical structure, each non-leaf node in the tree aggregates the information gathered in the descendant nodes. This algorithm tries to find the best groups with respect to the available memory, while minimizing the amount of input and output. Another use of the *CF-tree* appears in [1]. A different strategy is used in another hierarchical method, the *CURE* system [17], where each cluster is represented by a constant number of points well distributed within the cluster, which capture the extension and shape of the cluster. This process allows the identification of clusters with arbitrary shapes on a random sample of the dataset, using Chernoff bounds in order to obtain the minimum number of required examples. The same principle of error-bounded results was recently used in VFKM to apply consecutive runs of k-means, with increasing number of examples, until the error bounds were satisfied [11]. This strategy supports itself on the idea of guaranteeing that the clustering definition does not differ significantly from the one gather with infinite data. Hence, it does not consider data evolution.

**Grid-Based Clustering.** The main focus of *grid-based* algorithms is the so called *spatial data*, which model the geometric structure of objects in space. These algorithms divide the data space in small units, defining a grid, and assigning each object to one of those units, proceeding to divisive and aggregative operations hierarchically. These features make this type of methods similar to hierarchical algorithms, with the main difference of applying operations based on a parameter rather than the dissimilarities between objects. A sophisticated example of this type of algorithms is *STING* [24], where the space area is divided in cells with different levels of resolution, creating a layered structure. The main features and advantages of this algorithm include being incremental and able of parallel execution. Also, the idea of dense units, usually present in *density-based* methods [12], has been successfully introduced in *grid-based* systems. The *CLIQUE* algorithm tries to identify sub-spaces of a large dimensional space which can allow a better clustering of the original data [2]. It divides each dimension on the same number of equally ranged intervals, resulting in exclusive units. One unit is accepted as dense if the fraction of the total number of points within the unit is higher than a parameter value. A cluster is the largest set of contiguous dense units within a subspace. This technique's main advantage is the fact that it automatically finds subspaces of maximum dimensionality in a way that high density clusters exist in those subspaces. The inclusion of the notion of dense units in simpler *grid-based* methods presents several benefits. However, in distributed systems, the increase in communication given the need to keep sufficient statistics may be prejudicial.

**Distributed Clustering.** Since current applications generate many pervasive distributed computing environments, data mining systems must nowadays be designed to work not as a monolithic centralized application but as a distributed collaborative process. The centralization of information yields problems not only with resources such as communication and memory, but also with privacy of sensitive information. Instead of centralizing relevant data in a single server and afterwards perform the data mining operations, the entire process should be distributed and, therefore, paralleled throughout the entire network of processing units. A recent example of such techniques was proposed by Subramaniam et al., where an online system uses density distribution estimation to detect outliers in distributed sensor data [23]. Recent research developments in clustering are directed towards distributed algorithms for continuous clustering of examples over distributed data streams. In [9] the authors present a distributed majority vote algorithm which can be seen as a primitive to monitor a k-means clustering over peer-to-peer networks. The k-means monitoring algorithm has two major parts: monitoring the data distribution in order to trigger a new run of k-means algorithm and computing the centroids actually using the k-means algorithm. The monitoring part is carried out by an exact local algorithm, while the centroid computation is carried out by a centralization approach. The local algorithm raises an alert if the centroids need to be updated. At this point data is centralized, a new run of k-means is executed, and the new centroids are shipped back to all peers. Cormode et al. [7] proposed different strategies to achieve the same goal, with local and global computations, in order to balance the communication costs. They considered techniques based on the *furthest point* algorithm [16], which gives a approximation for the radius and diameter of clusters with guaranteed cost of two times the cost of the optimal clustering. They also present the *parallel guessing* strategy, which gives a slightly worse approximation but requires only a single pass over the data. They conclude that, in actual distributed settings, it is frequently preferable to track each site locally and combine the results at the coordinator site. These methods of combining local and central processing are paradigmatic examples of the path that distributed data mining algorithms should traverse.

## 3 *DGClust* – Distributed Grid Clustering

In this section we present *DGClust*, a distributed grid clustering system for sensor data streams. Each local sensor receives data from a given source, producing a univariate data stream, which is potentially infinite. Therefore, each sensor's data is processed locally, being incrementally discretized into a univariate adaptive grid. Each new data point triggers a cell in this grid, reflecting the current state of the data stream at the local site. Whenever a local site changes its state, that is, the triggered cell changes, the new state is communicated to a central site. Furthermore, the central site keeps the global state of the entire network where each local site's state is the cell number of each local site's grid. Nowadays, sensor networks may include thousands of sensors. This scenario yields an exponential number of cell combinations to be monitored by the central site. However, it is expected that only a small number of this combinations are frequently triggered by the whole network, so, parallel to the aggregation, the central site keeps a small list of counters of the most frequent global states. Finally, the current clustering definition is defined and maintained by a simple adaptive partitional clustering algorithm applied on the frequent states central points.

### 3.1 Notation and Formal Setup

The goal is to find $k$ cluster centers of the data produced by a network of $d$ local sensors. Let $X = \{X_1, X_2, ..., X_d\}$ be the set of $d$ univariate data streams, each of which is produced by one sensor in one local site. Each local site $i$ keeps a two-layered discretization of the univariate stream, with $p_i$ intervals in the first layer and $w_i$ intervals in the second layer, where $k < w_i << p_i$ but $w_i \in O(k)$. At each time $t$, each local sensor produces a value $X_i(t)$ and defines its local discretized state $s_i(t)$, drawn from the set of possible states $S_i$, the unit cells in the univariate grid ($|S_i| = w_i$). If no value is read, or $s_i(t) = s_i(t-1)$, no information is sent to the central site. The central site monitors the global state of the network at time $t$, by combining each local discretized state $s(t) = \langle s_1(t), s_2(t), ..., s_d(t)\rangle$. Each $s(t)$ is drawn from a finite set of cell combinations $E = \{e_1, e_2, ..., e_{|E|}\}$, with $|E| = \prod_{i=1}^{d} w_i$. Given the exponential size of $E$, the central site monitors only a subset $F$ of the top-$m$ most frequent elements of $E$, with $k < |F| << |E|$. Relevant focus is given to size requirements, as $|E| \in O(k^d)$ but $|F| \in O(dk^\beta)$, with small $\beta$. Finally, the top-$m$ frequent states central points are used in an online adaptive partitional clustering algorithm, which defines the current $k$ cluster centers, being afterwards continuously adapted.

### 3.2 Local Adaptive Grid

Discretization of continuous attributes is an important task for certain types of machine learning algorithms. Although discretization is a well-known topic in data analysis and machine learning, most of the works refer to a batch discretization where all the examples are available for discretization. Few works refer to incremental discretization. However, it is commonly seen as an essential tool for high-speed processing of data streams with limited memory resources [14]. For example, grid clustering algorithms operate on discrete cells to define dense regions of points [24]. In our approach, the main reason to perform discretization of each univariate data stream in sensor networks is to considerably reduce the communication with the central server, which tries to capture the most frequent regions of the entire input space.

**Partitional Incremental Discretization.** In our approach, we apply incremental discretization at each sensor univariate stream $X_i$ using the *Partition Incremental Discretization (PiD)* algorithm [13], which is composed by two layers. The first layer simplifies and summarizes the data, while the second layer constructs the final grid. For the scope of this work, we only consider equal-width discretization. The first layer is initialized based on the number of intervals $p_i$ (that should be much larger than the desired final number of intervals $w_i$) and the range of the variable. The range of the variable is only indicative, as it is used to define the intervals in the first layer using a equal-width strategy, but it is not rigid as layer-one bins can be split if necessary, monitoring the actual range of the values. Each time a new value $X_i(t)$ is read by the sensor $i$, we increment the counter in the corresponding interval. After a given number of examples, the second layer can be defined. The process of updating the first layer works online, doing a single scan over the data stream, hence being able to process infinite sequences of data, processing each example in constant time and (almost) constant space. We consider a split operator to be triggered in a given layer-one bin as soon as the number of hits in it is above a user-defined threshold $\alpha$ (a percentage of the total number of points considered so far). This split operator generates new intervals in layer one, adapting the initial intervals with the actual distribution of values. This is especially important to correct badly initialized range values and to adapt to slight changes in the distribution of values. The second layer merges the set of intervals defined by the first layer, and defines the final univariate discretization grid $\langle b_1, b_2, ..., b_{w_i} \rangle$, with size $w_i > k$ linear to the number of final clusters to find. The grid this way defined represents an approximated histogram of the variable produced at the local site. The update process of the second layer works online along with the first layer. For each new example $X_i(t)$, the system increments the counter in the second-layer cell where the triggered first-layer cell is included, defining the discretized state $s_i(t)$. Algorithm 1 presents the adaptive discretization procedure executed at each local site.

---

**Algorithm 1.** LocalAdaptiveGrid($X_i$, $\alpha$)

---

**Input:** univariate stream $X_i$ and threshold value $\alpha$
**Output:** *void* (sends discretized states $s_i$ to central site)
1: let $n \leftarrow 0$
2: **while** $X_i(t) \leftarrow read\_sensor(i)$ **do**
3: $n \leftarrow n + 1$
4: let $c$ be the layer-one cell triggered by $X_i(t)$
5: $count_c \leftarrow count_c + 1$
6: **if** $count_c/n > \alpha$ **then**
7: split cell $c$, dividing $count_c$ evenly by the two cells
8: update second-layer intervals $\langle b_1, b_2, ..., b_{w_i} \rangle$
9: send $\langle b_1, b_2, ..., b_{w_i} \rangle$ to central site
10: **end if**
11: let $b$ be the layer-two cell triggered by $c$
12: $count_b \leftarrow count_b + 1$
13: send $s_i \leftarrow \langle t, b \rangle$ to the central site
14: **end while**

---

### 3.3 Centralized Frequent State Monitoring

In this work we consider synchronous processing of sensor data. The global state is updated at each time stamp as a combination of each local site's state, where each value is the cell number of each local site's grid, $s(t) = \langle s_1(t), s_2(t), ..., s_d(t)\rangle$. If in that period no information arrives from a given local site $i$, the central site assumes that site $i$ stays in the previous local state ($s_i(t) \leftarrow s_i(t-1)$). The number $|E|$ of cell combinations to be monitored by the central site is exponential to the number of sensors, $|E| = O(w^d)$. However, only a small number of this combinations represent states which are frequently visited by the whole network. This way the central site keeps a small list, $F$, of counters of the most frequent global states, whose central points will afterwards be used in the final clustering algorithm, with $|F| = O(dk^{\beta})$, for small $\beta$.

**Space-Saving Top-$m$ Elements Monitoring.** Each seen global state $e \in E$ is a frequent element $f_r$ whose counter $count_r$ currently estimates that it is the $r^{th}$ most frequent state. The system applies the *Space-Saving* algorithm [20], to monitor only the top-$m$ elements. If we observe a state $s(t)$ that is monitored in rank $r$ ($f_r = s(t)$), we just increment its counter, $count_r$. If $s(t)$ is not monitored, replace $f_m$, the element that currently has the least estimated hits, $count_m$, with $s(t)$, and increment $count_m$. For each monitored element $f_i$, we keep track of its over-estimation, $\varepsilon_i$, resulting from the initialization of its counter when it was inserted into the list. That is, when starting to monitor $f_i$, set $\varepsilon_i$ to the value of the evicted counter. An element $f_i$ is *guaranteed* to be among the top-$m$ elements if its guaranteed number of hits, $count_i - \varepsilon_i$, exceeds $count_{m+1}$. The authors report that, even if it is not possible to guarantee top-$m$ elements, the algorithm can guarantee top-$m'$ elements, with $m' \approx m$. Hence, suitable values for $m$ should be considered. Furthermore, due to errors in estimating the frequencies of the elements, the order of the elements in the data structure might not reflect their exact ranks. Thus, when performing clustering on the top-$m$ elements, we should be careful not to directly weight each point by its rank. The goal of this strategy is to monitor top-$m$ states, using only the *guaranteed* top-$m$ elements as points for the final clustering algorithm. One important characteristic of this algorithm is that it tends to give more importance to recent examples, enhancing the adaptation of the system to data evolution.

### 3.4 Centralized Online Clustering

The goal of *DGClust* is to find and continuously keep a cluster definition, reporting the $k$ cluster centers. Each frequent state $f_i$ represents a multivariate point, defined by the central points of the corresponding unit cells $s_i$ for each local site $X_i$. As soon as the central site has a top-$m$ set of states, with $m > k$, a simple partitional algorithm can be applied.

**Initial Centers.** In the general task of finding $k$ centers given $m$ points, there are two major objectives: minimize the *radius* (maximum distance between a point and its closest cluster center) or minimize the *diameter* (maximum distance between two points assigned to the same cluster) [7]. The *Furthest Point* algorithm [16] gives a guaranteed 2-approximation for both the *radius* and *diameter* measures. It begins by picking an arbitrary point as the first center, $c_1$, then finding the remaining centers $c_i$ iteratively as the point that maximizes its distance from the previously chosen centers $\{c_1, ..., c_{i-1}\}$. After $k$ iterations, one can show that the chosen centers $\{c_1, c_2, ..., c_k\}$ represent a factor 2 approximation to the optimal clustering [16]. See [7] for a proof. This strategy gives a good initialization of the cluster centers, computed by finding the center $k_i$ of each cluster after attracting remaining points to the closest center $c_i$. This algorithm is applied as soon as the system finds a set of $m' > k$ guaranteed top-$m$ states.

**Continuous Adaptive Clustering.** It is known that a single iteration is not enough to converge to the actual centers in simple *k-means* strategies. Hence we consider two different states on the overall system operation: *converged* and *non-converged*. At every new state $s(t)$ that is gathered by the central site, if the system has not yet converged, it adapts the clusters centers using the $m'$ guaranteed top-$m$ states. On the other hand, if the system has converged, two different scenarios may occur. If the current state is being monitored as one of the $m'$ top-$m$ states, then the set of points actually used in the final clustering is the same, so the clustering centers remain the same. No update is performed. On the other hand, if the current state has just become *guaranteed* top-$m$, then the clusters may have change so we move into a non-converged state of the system, updating the cluster centers. Another scenario where the clusters centers require adaptation is when one or more local sites transmit their new grid intervals, which are used to define the central points of each state. In this case we also update and move to non-converged state. A different scenario is created when a new state enters the top-$m$, replacing the least frequent one. In this case, some of the previously *guaranteed* top-$m$ may loose their *guarantee*. However, if the list of frequent items is small (imposed by resources restrictions) this will happen very frequently so we disregard this scenarios to prevent excessive computation when cluster centers have already converged. Future work will focus on these scenarios for concept drift and cluster evolution purposes. In scenarios where clusters centers adaptation is needed, our system updates the clustering definition by applying a single iteration of point-to-cluster assignment and cluster centers computation. This process assures a smooth evolution of the cluster centers, while it nevertheless adapts them to the most recent data, as old data points tend to be less frequent. Algorithm 2 presents the central adaptive procedure executed at the server site. The algorithm for ClusterCentersUpdate($K$, $F$) is omitted for simplicity and space saving. Figure 1 presents a final grid, frequent cells and cluster centers for a specific case with $d = 2, k = 5$, for different values of $w$ and $m$. The flexibility of the system is exposed, as different parameter values yield different levels of results. Moreover, the continuous update keeps track of the most frequent cells keeping the gathered centers within acceptable bounds. A good characteristic of the system is this ability to adapt to resource restricted

**Algorithm 2.** CentralAdaptiveProcessing($L$, $k$, $m$)

**Input:** list of local sites $L = \{l_1, l_2, ..., l_d\}$, number of clusters $k$ and frequent states $m$
**Output:** set $K$ of $k$ cluster centers

1: $F \leftarrow \{\}$ (set of frequent global states)
2: $K \leftarrow \{\}$ (set of $k$ cluster centers)
3: $conv \leftarrow false$ (are the centers stable?)
4: **for** each timestamp $t$ **do**
5:   **for** each local site $i$ **do**
6:     **if** $s_i(t)$ has not been received **then** $s_i(t) \leftarrow s_i(t-1)$
7:   **end for**
8:   $s(t) \leftarrow \langle s_1(t), s_2(t), ..., s_d(t) \rangle$
9:   $\langle F, m', f_{s(t)}, \varepsilon_{s(t)} \rangle \leftarrow$ SpaceSavingUpdate($F$, $m$, $s(t)$) (as in [20])
10:   **if** $m' > k$ **and** $K = \{\}$ **then**
11:     $K \leftarrow$ FurthestPoint ($F$, $k$) (as in [16])
12:   **else**
13:     **if not** $conv$ **or** $f_{s(t)} - \varepsilon_{s(t)} = f_{m+1} + 1$ **then**
14:       $\langle K, conv \rangle \leftarrow$ ClusterCentersUpdate($K$, $F$)
15:     **end if**
16:   **end if**
17: **end for**
18: **return** K

environments: system granularity can be defined given the resources available in the network's processing sites.

## 4 Algorithm Analysis

Each univariate data stream is discretized, with only the discretized state being forwarded to the central site. At this point, the granularity of each sensor's grid will directly influence the error in that dimension. Since the construction of the second layer is directly restricted to the intervals defined in layer one, the final histograms will also be an approximation of the exact histograms that would be defined directly if all data was considered. Nevertheless, with this two-layer strategy the update of the final grid is straightforward. The layer-two intervals just need to be recomputed when the split operator in layer one is triggered. Moreover, the number of intervals in the second layer can be adjusted individually for each sensor, in order to address different needs of data granularity and resources requirements. In this proposal we address univariate sensor readings. The data stream model we consider in sensor networks assumes that a sensor value represents its state in a given moment in time. If the readings of a local sensor fall consecutively in the same layer-two interval, no sound information would be given to the central site. Thus, local sites only centralize information when a new value triggers an interval different from the previously sent to the central server. The central site only monitors the top-$m$ most frequent global states, disregarding infrequent states which could influence the final clusters. Finally, the system performs partitional clustering over the $m'$ guaranteed top-$m$ frequent

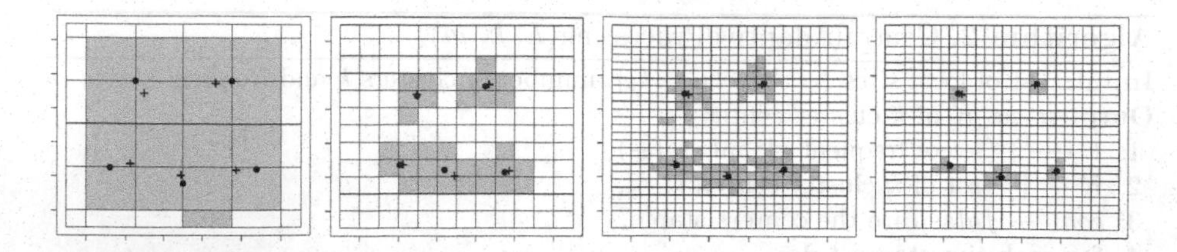

**Fig. 1.** Example of final definition for 2 sensors data, with 5 clusters. Each coordinate shows the actual grid for each sensor, with top-$m$ frequent states (shaded cells), gathered (circles) and real (crosses) centers, run with: $(w = 6, m = 20)$, $(w = 12, m = 60)$, $(w = 24, m = 180)$, and $(w = 24, m = 180)$ presenting only *guaranteed* top-$m$.

states which is a sample of the actual states, being biased to dense cells. Moreover, although the furthest point algorithm may give guarantees on the initial centers for the clustering of the frequent states, the adaptive update is biased towards small changes in the concept generating the data streams. Each sensor $X_i$ produces a univariate adaptive grid. This process uses the *PiD* algorithm which, after the initial definition of the two layers based on $n_i$ examples, in $O(n_i \log p_i)$ time and $O(p_i)$ space, is continuously updated in constant $O(\log p_i)$ time and (almost) constant space. Since this is done parallel across the network, the time complexity of the discretization of one example in the entire network is $O(\log p)$ where $p = \max(p_i)$, $\forall i \in \{1, 2, ..., d\}$. The central site aggregates the state of each of the $d$ local sites. The focus is on monitoring the top-$m$ frequent global states, which are kept in $O(md)$ space (the actual $m$ frequent states) and continuously updated in $O(m)$ time (linear search for the current state). The initial clustering of frequent states, and its subsequent adaptation is made in $O(km)$ time. Data communication occurs only in one direction, between the local sites and the central site. All queries are answered by the central server. Also, this communication does not include sending the original data, rather informing the central server of the current discrete state of the univariate stream of each local site. This feature of only communicating the state when and if it has changed reduces the network's communication requirements. The main reason of this is that, in usual sensor data, sensor readings tend to be highly correlated with previously read value [15], hence tending to stay in the same discretized state. In the worst case scenario, where all $d$ local sites need to communicate their state to the central site, the system processes $d$ messages of one discrete value. However, every time the local site $X_i$ changes its univariate grid, the central site must be informed on the change so that it can correctly compute the points used to find the cluster centers, which imply sending $w_i$ values. In the worst case scenario, the central site may have to receive $O(wd)$ data, where $w = \max(w_i)$.

## 5 Experimental Evaluation

Evaluation of streaming algorithms is a *hot-topic* in research as no standard exists for evaluating models over streams of data. Hence the difficulty to define exact evaluation processes. Nevertheless, we conducted a series of experiments on synthetic scenarios, to assess the quality of our proposal. All scenarios were generated by applying the data generator used in [11], considering each dimension separated across sensors, in univariate streams. The global view of the network scenarios is created by mixtures of $k$ spherical Gaussians, with means $\mu_i$ in the unit hypercube. Each scenario was generated according to three parameters: dimensionality $d$ of the network, the number of mixture components $k$, and the standard deviation of each sensor stream in each component $\sigma$. Each scenario $(d, k, \sigma)$ is created with 100000 examples. Given the scope of this validation, the system's quality is measured by assigning each of the found cluster centers to the closest real cluster center, using a greedy strategy. The *loss* of the chosen assignment is given by

$$L_K = \sum_{i=1}^{k} \sum_{j=1}^{d} (\hat{c}_{ij} - c_{ij})^2 \qquad (1)$$

where $\hat{c}_{ij}$ and $c_{ij}$ are the gathered and real values, respectively, for center $i$ in dimension $j$.

## 5.1 Parameter Sensitivity

For each scenario, the system's sensitivity to parameters is evaluated. Studied parameters are the granularity of the univariate adaptive grid, $w_i, \forall i \in \{1, 2, ..., d\}$, and the number of frequent states to monitor, $m$. We fixed $p_i = 1000, \forall i \in \{1, 2, ..., d\}$. We look for a good relation between the scenario ($k$ and $d$) and parameters ($w$ and $m$). Our assumption is that parameters should follow $w \in O(k)$ and $m \in O(dk^{\beta})$, for small $\beta$ (possibly $\beta = 1$). To study this possibility, we define two factors $\omega$ and $\phi$, where $w = \omega k + 1$ (allows extra cell which will mostly keep outliers) and $m = \phi w d$. As first layers in the univariate grids have size >> 20, we set $\alpha = 0.05$ stating that a first-layer cell of the univariate grid will be split if it contains more that 5% of the total number of points. The initial range is set to $[0, 1]$. We set $d \in \{2, 3, 4, 5\}$ and $\sigma = 0.1$, varying $k \in \{3, 4, 5\}$, $\omega \in \{1, 2, 3, 4, 5\}$ and $\phi \in \{1, 2, 3, 4, 5\}$. Each scenario was evaluated with results averaged over 10 datasets. All $w_i$ are set with the same value $w = \omega k + 1$. We vary $k$ within small values to inspect the ability of the system to find well-separable clusters. After aggregating all experiments, we computed Pearson's correlation between the parameters ($\omega$ and $\phi$) and the resulting loss. The $\omega$ parameter reported (as expected) negative correlation with the loss ($\rho = -0.7524$), as better granularity diminishes the error implied by performing clustering on a grid instead of the real values. However, higher granularity also implies higher values for $m$, so a compromise should be found to optimize computational costs. After running some empirical tests (which will be subject of thorough evaluation in the future), we found that $\omega$ should be larger than 1, in order to allow the existence of infrequent cells between frequent ones, hence improving separability. For the $\phi$ parameter, the study reported a positive correlation with loss ($\rho = 0.3832$). Since this goes against the empirical intuition, we decided to try a different approach. Let $m = \frac{dw^2}{\gamma}$, with $\gamma \in \{10, 8, 6\}$. Tables 1 and 2 present the results gathered

**Table 1.** Averaged loss over $\omega \in \{2, 4\}$ and $\gamma \in \{6, 8, 10\}$, and 10 different data sets for each combination of parameters and scenarios, for *DGCluster* and *Continuous K-Means*

| | $k = 2$ | | $k = 3$ | | $k = 4$ | | $k = 5$ | | $k = 10$ | |
|---|---|---|---|---|---|---|---|---|---|---|
| *d* | *DGC* | *CKM* | *DGC* | *CKM* | *DGC* | *CKM* | *DGC* | *CKM* | *DGC* | *CKM* |
| **2** | 0.015305 | 0.000003 | 0.152415 | 0.000048 | 0.056467 | 0.032453 | 0.011277 | 0.000062 | **0.201048** | 0.238827 |
| **3** | 0.018003 | 0.000010 | 0.181453 | 0.069749 | **0.102557** | 0.157503 | 0.279721 | 0.209871 | **0.194634** | 0.423646 |
| **4** | 0.006005 | 0.000012 | 0.004045 | 0.000022 | **0.178782** | 0.185916 | 1.010598 | 0.436916 | 0.717055 | 0.386985 |
| **5** | 0.007152 | 0.000015 | **0.007816** | 0.130807 | **0.200100** | 0.256851 | **0.678498** | 0.687560 | **0.544533** | 0.908984 |
| **10** | 0.020770 | 0.000034 | **0.146219** | 0.285631 | **0.388414** | 0.939210 | **0.910776** | 1.895545 | 2.188842 | 1.990172 |
| **20** | 0.094645 | 0.000065 | **1.200420** | 1.386870 | **0.961394** | 1.211905 | **1.610627** | 2.015545 | 7.027379 | 5.083694 |

**Table 2.** Averaged loss over all $k \in \{2, 3, 4, 5, 10\}$ for each fixed parameter over 10 different data sets

| *d* | $\omega = 2$ | $\omega = 4$ | $\gamma = 10$ | $\gamma = 8$ | $\gamma = 6$ | *Continuous KM* |
|---|---|---|---|---|---|---|
| **2** | 0.066647 | 0.107959 | 0.072887 | 0.096263 | 0.092757 | 0.054279 |
| **3** | **0.162751** | **0.147796** | **0.151097** | **0.166548** | **0.148176** | 0.172156 |
| **4** | 0.377053 | 0.389541 | 0.381646 | 0.366113 | 0.402133 | 0.201970 |
| **5** | **0.242538** | **0.332701** | **0.295861** | **0.281109** | **0.285889** | 0.396843 |
| **10** | **0.702328** | **0.759680** | **0.748789** | **0.711933** | **0.732290** | 1.022118 |
| **20** | 2.137683 | 2.220103 | 2.188506 | 2.117922 | 2.230250 | 1.939616 |

for different scenarios, comparing with a simple centralized online *k-means* strategy, to which we refer as *Continuous K-Means*, which is a simplification of the STREAM algorithm [22]. This strategy performs a *K-Means* at each chunk of examples, keeping only the centers gathered with the last chunk of data weighted by the amount of points that were assigned to each center. Once again we note how hard it becomes to define a clear relation between $w$, $d$ and $m$, although for higher values of $k$ and $d$ we could see some possible progressive paths towards an improvement in the competitiveness of our proposal. However, the identification of general parameters is always discussable. Although we plan to perform more exhaustive sensitivity tests, in order to achieve at least acceptable ranges for the parameters, we should stress that the flexibility included in the system allows for better deployment in sensor networks and resource restricted environments.

## 5.2 Application to Physiological Sensor Data

The Physiological Data Modeling Contest Workshop (PDMC) was held at the ICML 2004 and aimed at information extraction from streaming sensors data. The training data set for the competition consists of approximately 10,000 hours of this data, containing several variables: userID, sessionID, sessionTime, characteristic[1..2], annotation, gender and sensor[1..9]. We have concentrated on sensors 2 to 9, extracted by userID, resulting in several experimental scenarios of eight sensors, one scenario per userID. For each scenario, we run the system with different values for the parameters, and compare the results both with the *Continuous K-Means* and full data *K-Means*, the latter serving as "real" centers definition. Since different sensors produce readings in different scales, we inspect the distribution of each sensor on an initial chunk of data, defining the initial ranges to percentiles 25% and 75%. This process is acceptable in the sensor networks framework as expert knowledge about the range is usually available. Hence, we are also allowing the system to adapt the local grids accordingly. The system ran with $k \in \{2, 3, 4, 5\}$, $\omega \in \{2, 4, 6, 8, 10\}$ and $\gamma \in \{10, 8, 6, 4, 2\}$. Also

**Table 3.** Performance in terms of communicated values (% of total examples × dimensions), evicted states (% of total examples), cluster centers adaptations (% of total examples) and number of guaranteed top-$m$ (% of total $m$) with $k \in \{2,3,4,5\}$, $\omega \in \{2,4,6,8,10\}$ and $\gamma \in \{10,8,6,4,2\}$

| | $k=2$ | $k=3$ | $k=4$ | $k=5$ |
|---|---|---|---|---|
| Communication (%) | 20.4±08.3 | 26.8±09.09 | 31.6±10.5 | 36.0±11.4 |
| Evicted (%) | 18.0±12.6 | 28.5±16.6 | 36.5±18.5 | 42.7±18.6 |
| Updated (%) | 03.2±01.6 | 04.3±02.8 | 05.9±04.5 | 07.8±06.6 |
| Guaranteed (%) | 24.7±17.0 | 17.4±14.1 | 12.2±10.2 | 10.1±09.0 |
| Best Loss | 1329.67 | 5222.17 | 9529.40 | 34131.75 |
| Worst Loss | 114645.7 | 669330.5 | 393919.9 | 554944.9 |
| CKM Loss | 1785.97 | 112830.4 | 105091.0 | 276932.5 |

monitored was the amount of communication and cluster adaptation in each run. Table 3 presents performance results for the physiological sensor data. Given the characteristics of sensor data, subsequent readings tend to stay in the same interval. Hence the advantage of local discretization: the system transmits only around 30% of the total amount of values, including transmissions of recalculated second layer intervals. Also, we should note that only a small part of the points require cluster center adaptation (less than 10%). Overall we should stress that, given the flexibility of the system, a suitable combination of parameters can yield better results than centralizing the data, while preventing excessive communication in the sensor network.

## 6 Conclusions and Future Work

In this paper, we have proposed a new algorithm to perform clustering of distributed data streams, which reduces both the dimensionality and the communication burdens. Each sensor's data is processed locally, being incrementally discretized into a univariate adaptive grid, with each new value triggering a specific cell of this grid, defining the local state of the sensor. Whenever a local site change its state, that is, the triggered cell changes, the new state is communicated to a central site. This highly reduces the network's communication requirements with the central site which keeps a small list of counters of the most frequent global states. This is extremely helpful as sensor networks may include thousands of sensors, leading to an exponential number of cell combinations that should be monitored by the central site. The final clustering structure is defined and maintained by a simple partitional clustering algorithm, applied on the frequent states, which is capable of online update and adaptation to the evolution of the data. Experiments are presented in terms of sensitivity tests and application to a real-world data set. Major characteristics of the system include: compact representation of possibly infinite streams produced in distributed sensors, with possibly different granularities for each sensor; possibility to adapt discretization and central processing to resources available in the network; centralized monitoring of the global state of the network, with lightweight maintenance of frequent states; and fast and adaptable clustering of frequent states to estimate actual clusters centers. Current work is concentrated on determining acceptable ranges for the parameters of the system and application to more real-world data. Future work will focus on techniques to monitor the evolution of the clusters, taking advantage from the adaptive update of clusters already implemented in the system. Furthermore, we are preparing the deployment of the system in real wireless sensor networks, in order to better assess the sensitivity and advantages of the system with respect to restricted resources requirements.

**Acknowledgment.** The work of P.P. Rodrigues is supported by the Portuguese Foundation for Science and Technology (FCT) under the PhD Grant SFRH/BD/29219/2006. The authors thank FCT's Plurianual financial support attributed to LIAAD and CRACS and the participation of Project ALES II under FCT's Contract POSC/EIA/55340/2004 and Project CALLAS under FCT's Contract PTDC/EIA/71462/2006.

## References

1. Aggarwal, C.C., Han, J., Wang, J., Yu, P.S.: A framework for clustering evolving data streams. In: VLDB 2003, pp. 81–92. Morgan Kaufmann, San Francisco (2003)
2. Agrawal, R., Gehrke, J., Gunopulos, D., Raghavan, P.: Automatic subspace clustering of high dimensional data for data mining applications. In: Proceedings of the ACM-SIGMOD International Conference on Management of Data, Seattle, Washington, June 1998, pp. 94–105. ACM Press, New York (1998)
3. Akyildiz, I., Su, W., Sankarasubramaniam, Y., Cayirci, E.: A Survey on Sensor Networks. IEEE Communications Magazine 40(8), 102–114 (2002)
4. Bradley, P., Fayyad, U., Reina, C.: Scaling clustering algorithms to large databases. In: Proceedings of the Fourth International Conference on Knowledge Discovery and Data Mining, pp. 9–15. AAAI Press, Menlo Park (1998)
5. Charikar, M., Chen, K., Farach-Colton, M.: Finding frequent items in data streams. In: Widmayer, P., Triguero, F., Morales, R., Hennessy, M., Eidenbenz, S., Conejo, R. (eds.) ICALP 2002. LNCS, vol. 2380, pp. 693–703. Springer, Heidelberg (2002)
6. Cormode, G., Muthukrishnan, S.: What's hot and what's not: Tracking most frequent items dynamically. In: Proceedings of the 22nd Symposium on Principles of Database Systems, pp. 296–306 (2003)
7. Cormode, G., Muthukrishnan, S., Zhuang, W.: Conquering the divide: Continuous clustering of distributed data streams. In: Proceedings of the 23nd International Conference on Data Engineering (ICDE 2007), pp. 1036–1045 (2007)
8. Culler, D.E., Mulder, H.: Smart Sensors to Network the World.Scientific American (2004)
9. Datta, S., Bhaduri, K., Giannella, C., Wolff, R., Kargupta, H.: Distributed data mining in peer-to-peer networks. IEEE Internet Computing 10(4), 18–26 (2006)
10. Demaine, E.D., Lopez-Ortiz, A., Munro, J.I.: Frequency estimation of internet packet streams with limited space. In: Proceedings of the 10th Annual European Symposium on Algorithms, pp. 348–360 (2002)
11. Domingos, P., Hulten, G.: A general method for scaling up machine learning algorithms and its application to clustering. In: Proceedings of the Eighteenth International Conference on Machine Learning (ICML 2001), pp. 106–113 (2001)

12. Ester, M., Kriegel, H.-P., Sander, J., Xu, X.: A density-based algorithm for discovering clusters in large spatial databases with noise. In: Second International Conference on Knowledge Discovery and Data Mining, Portland, Oregon, pp. 226–231. AAAI Press, Menlo Park (1996)
13. Gama, J., Pinto, C.: Discretization from data streams: applications to histograms and data mining. In: Proceedings of the, ACM Symposium on Applied Computing (SAC 2006), pp. 662–667. ACM Press, New York (2006)
14. Gama, J., Rodrigues, P.P.: Data stream processing. In: Learning from Data Streams - Processing Techniques in Sensor Networks, ch. 3, pp. 25–39. Springer, Heidelberg (2007)
15. Gama, J., Rodrigues, P.P.: Stream-based electricity load forecast. In: Kok, J.N., Koronacki, J., López de Mántaras, R., Matwin, S., Mladenič, D., Skowron, A. (eds.) PKDD 2007. LNCS (LNAI), vol. 4702, pp. 446–453. Springer, Heidelberg (2007)
16. Gonzalez, T.F.: Clustering to minimize the maximum inter-cluster distance. Theoretical Computer Science 38, 293–306 (1985)
17. Guha, S., Rastogi, R., Shim, K.: CURE: An efficient clustering algorithm for large databases. In: Proceedings of the 1998 ACM-SIGMOD International Conference on Management of Data, pp. 73–84. ACM Press, New York (1998)
18. Luo, L., Abdelzaher, T., He, T., Stankovic, J.: EnviroSuite: An Environmentally Immersive Programming Framework for Sensor Networks. ACM TECS 5(3), 543–576 (2006)
19. Manku, G., Motwani, R.: Approximate frequency counts over data streams. In: Proceedings of the 28th International Conference on Very Large Data Bases, pp. 346–357 (2002)
20. Metwally, A., Agrawal, D., Abbadi, A.E.: Efficient computation of frequent and top-k elements in data streams. In: Proceedings of the 10th International Conference on Database Theory, pp. 398–412. Springer, Heidelberg (2005)
21. Newton, R., Welsh, M.: Region Streams: Functional Macroprogramming for Sensor Networks. In: DMSN 2004 Workshop (2004)
22. O'Callaghan, L., Meyerson, A., Motwani, R., Mishra, N., Guha, S.: Streaming-data algorithms for high-quality clustering. In: Proceedings of the Eighteenth Annual IEEE International Conference on Data Engineering, pp. 685–696. IEEE Computer Society Press, Los Alamitos (2002)
23. Subramaniam, S., Palpanas, T., Papadopoulos, D., Kalogeraki, V., Gunopulos, D.: Online outlier detection in sensor data using non-parametric models. In: VLDB, pp. 187–198 (2006)
24. Wang, W., Yang, J., Muntz, R.R.: STING: A statistical information grid approach to spatial data mining. In: Proceedings of the Twenty-Third International Conference on Very Large Data Bases, Athens, Greece, pp. 186–195. Morgan Kaufmann, San Francisco (1997)
25. Zhang, T., Ramakrishnan, R., Livny, M.: BIRCH: An efficient data clustering method for very large databases. In: Proceedings of the 1996 ACM SIGMOD International Conference on Management of Data, pp. 103–114. ACM Press, New York (1996)

# A Novel Scalable and Data Efficient Feature Subset Selection Algorithm

Sergio Rodrigues de Morais[1] and Alex Aussem[2]

[1] INSA-Lyon, LIESP, F-69622 Villeurbanne, France
sergio.rodrigues-de-morais@insa-lyon.fr
[2] Université de Lyon 1, LIESP, F-69622 Villeurbanne, France
aaussem@univ-lyon1.fr

**Abstract.** In this paper, we aim to identify the minimal subset of discrete random variables that is relevant for probabilistic classification in data sets with many variables but few instances. A principled solution to this problem is to determine the *Markov boundary* of the class variable. Also, we present a novel scalable, data efficient and correct Markov boundary learning algorithm under the so-called *faithfulness* condition. We report extensive empiric experiments on synthetic and real data sets scaling up to 139,351 variables.

## 1 Introduction

The identification of relevant subsets of random variables among thousands of potentially irrelevant and redundant variables with comparably smaller sample sizes is a challenging topic of pattern recognition research that has attracted much attention over the last few years [1, 2, 3]. By relevant subsets of variable, we mean the variables that conjunctly prove useful to construct an efficient classifier from data. This contrasts with the suboptimal problem of ranking the variables individually. Our specific aim is to solve the feature subset selection problem with thousands of variables but few instances using Markov boundary learning techniques. The Markov boundary of a variable $T$, denoted by $\mathbf{MB}_T$, is the minimal subset of $\mathbf{U}$ (the full set) that renders the rest of $\mathbf{U}$ independent of $T$.

Having to learn a Bayesian network $G$ in order to learn a Markov boundary of $T$ can be very time consuming for high-dimensional databases. This is particularly true for those algorithms that are asymptotically correct under faithfulness condition, which are the ones we are interested in. Fortunately, there exist algorithms that search the Markov boundary of a target without having to construct the whole Bayesian network first [3]. Hence their ability to scale up to thousands of variables. Unfortunately, they miss many variables due to the unreliability of the conditional independence tests when the conditioning set is large. Hence the need to increase data-efficiency of these algorithms, that is, the ability of the algorithm to keep the size of the conditional set as small as possible during the search.

W. Daelemans et al. (Eds.): ECML PKDD 2008, Part II, LNAI 5212, pp. 298–312, 2008.

In this paper, we discuss a divide-and-conquer method in order to increase the data-efficiency and the robustness of the Markov boundary (MB for short) discovery while still being scalable and correct under the faithfulness condition. The proposed method aims at producing an accurate MB discovery algorithm by combining fast, rough and moderately inaccurate (but correct) MB learners. The proposed method is compared against two recent powerful constraint-based algorithms PCMB [3], IAMB [4] and Inter-IAMB [5]. We call our algorithm MBOR, it stands for "Markov Boundary search using the OR condition". MBOR was designed with a view to keep the conditional test sizes of the tests as small as possible.

The experiments on the synthetic databases focus on the accuracy and the data efficiency of MBOR, whereas the experiments on real data also addresses its scalability. The benchmarks used for the empiric test are: ASIA, INSURANCE, INSULINE, ALARM, HAILFINDER and CARPO. We report the average number of missing and extra variables in the output of MBOR with various sample sizes. The method is proved by extensive empirical simulations to be an excellent trade-off between time and quality of reconstruction. To show that MBOR is scalable, experiments are conducted on the THROMBIN database which contains 139,351 features [6].

The paper is organized as follows. In Section 2, we briefly discuss the lack of reliability of the conditional independence tests. In Section 3, we present and discuss our proposed algorithm called MBOR. Synthetic and real data sets from benchmarks are used in section 4 to evaluate MBOR against PCMB and InterIAMB.

## 2 Preliminaries

For the paper to be accessible to those outside the domain, we recall first the principle of Bayesian networks (BN), Markov boundaries and constraint-based learning BN methods. A BN is a tuple $< \mathcal{G}, P >$, where $\mathcal{G} =< \mathcal{V}, \mathcal{E} >$ is a directed acyclic graph (DAG) with nodes representing the random variables $\mathcal{V}$ and $P$ a joint probability distribution on $\mathcal{V}$. In addition, $\mathcal{G}$ and $P$ must satisfy the Markov condition: every variable, $X \in \mathcal{V}$, is independent of any subset of its non-descendant variables conditioned on the set of its parents. We denote the conditional independence of the variable $X$ and $Y$ given $\mathbf{Z}$, in some distribution $P$ with $X \perp_P Y|\mathbf{Z}$. The independence constraints implied by the Markov condition necessarily hold in the joint distribution represented by any Bayesian network with structure $\mathcal{G}$. They can be identified by the d-separation criterion (See Pearl 1988). We use $X \perp_\mathcal{G} Y|\mathbf{Z}$ to denote the assertion that the DAG $\mathcal{G}$ imposes the constraint, via d-separation, that for all values $z$ of the set $Z$, $X$ is independent of $Y$ given $Z = z$. We say that $P$ is *faithful* with respect to $\mathcal{G}$ iff the d-separations in the DAG identify *all and only* the conditional independencies in $P$.

A Markov blanket $\mathbf{M}_T$ of the $T$ is any set of variables such that $T$ is conditionally independent of all the remaining variables given $\mathbf{M}_T$. A Markov boundary, $\mathbf{MB}_T$, of $T$ is any Markov blanket such that none of its proper subsets is a

Markov blanket of $T$. Suppose $< \mathcal{G}, P >$, satisfies the faithfulness condition, then for all variable $T$, the set of parents of $T$, the children of $T$, and parents of children of $T$, is the **unique** Markov boundary of $T$. We denote by $\mathbf{PC}_T$, the set of parents and children of $T$ in $\mathcal{G}$, and by $\mathbf{SP}_T$, the set of *spouses* of $T$ in $\mathcal{G}$. The *spouses* of $T$ are the parents of the children of $T$. We denote by $\mathbf{dSep}(X)$, the set that d-separates $X$ from the (implicit) target $T$.

The identification of variable Markov boundary is a challenging topic of pattern recognition research. In recent years, there has been great interest in automatically inducing the Markov boundary from data using constraint-based (CB for short) learning procedures. The correctness, scalability and data efficiency of these methods have been proved and also illustrated by extensive experiments [3]. By correct (or sound), we mean that, under the assumptions that independence test are reliable and that the learning database is a sample from a distribution $P$ faithful to a DAG $\mathcal{G}$, the algorithm returns the correct Markov boundary. The (ideal) assumption that the independence tests are reliable means that they decide (in)dependence iff the (in)dependence holds in $P$. Despite their great efficiency and scalability, these CB methods suffer from several drawbacks as we will see in the next section.

# 3 Problems with Constraint-Based MB Discovery

CB methods have the advantage of possessing clear stopping criteria and deterministic search procedures. On the other hand, they are prone to several instabilities: namely if a mistake is made early on in the search, it can lead to incorrect edges which may in turn lead to bad decisions in the future, which can lead to even more incorrect edges. This instability has the potential to cascade, creating many errors in the final graph [7]. We discuss next two well-known sources of test failure.

## 3.1 Conditional Independence Test Failures with Sparse Data

CB procedures systematically check the data for independence relationships to infer the structure. The association between two variables $X$ and $Y$ given a conditioning set $\mathbf{Z}$ is a measure of the strength of the dependence with respect to D. It is usually implemented with a statistical measure of association. Typically, the algorithms run a $\chi^2$ independence test in order to decide on dependence or independence, that is, upon the acceptance or rejection of the null hypothesis of conditional independence. If the p-value of the test is smaller than a specific significance level, we reject the null hypothesis and consider that the two variables in the test are dependent. Insufficient data presents a lot of problems when working with statistical inference techniques like the independence test mentioned earlier. This occurs typically when the expected counts in the contingency table are small. The decision of accepting or rejecting the null hypothesis depends implicitly upon the degree of freedom which increases exponentially with the number of variables in the conditional set. So the larger the size of the conditioning test, the less accurate are the estimates of conditional probabilities and hence the less reliable are the independence tests.

### 3.2 Almost Deterministic Relationships

Another difficulty arises when true- or almost-deterministic relationships (ADR) are observed among the variables. Loosely speaking, a relationship is said to be almost deterministic (and denoted by $\mathbf{X} \Rightarrow Y$) when the fraction of tuples that violate the deterministic dependency is at most equal to some threshold. True DR are source of unfaithfulness but the existence of ADR among variables doesn't invalidate the faithfulness assumption. The existence of ADR in the data may arise incidentally in smaller data samples. To remedy the problem, the variables that are almost-deterministically related to others may simply be excluded from the discovery process. However, if they are to be excluded, they first need to be identified before the DAG construction. This yields two problems. First, the identification is already exponentially complex. Second, a variable may have both deterministic and probabilistic relationships with other variables. On the other hand if we neither exclude deterministic variables nor handle appropriately the problem, then the unfaithful nature of deterministic nodes brings missing or extra edges to the acquired structure.

### 3.3 Practical Alternatives

Several proposals have been discussed in the literature in order to reduce the cascading effect of early errors that causes many errors to be present in the final graph. The general idea is to keep the size of the conditional sets as small as possible in the course of the learning process. For instance, Fast-IAMB [4] conducts significantly fewer conditional tests compared to IAMB [5] while MMMB [8] and PCMB [3] are more data efficient because $\mathbf{MB}_T$ can be identified by conditioning on sets much smaller than those used by IAMB. Another solution is to determine if enough data is available for the test to be deemed reliable. Following the approach in [9] Inter-IAMB considers a test to be reliable when the number of (complete) instances in D is at least five time the number of degrees of freedom *df* and skips it otherwise. This means that the number of instances required by the test is at least exponential in the size of the conditional set, because *df* is exponential in the size of the conditional set. In [10], they consider that 80% of the cells should have expected values greater than five. If the test is not considered as reliable, the variables are assumed to be independent without actually performing the test. This rule is rather arbitrary and errors may occur well before the lack of data is detected.

As a heuristic, the idea is generally to reduce the degree of freedom of the statistical conditional independence test by some ways. The aim is twofold: to improve the data efficiency and to allow an early detection of ADR. Various strategies exist to reduce the degrees of freedom [11]. In [12] for instance, if the reduced degree of freedom is small, then an ADR $\mathbf{Z} \Rightarrow X$ is suspected, a "safe choice" is taken : dependence is assumed $X \perp_P Y | \mathbf{Z}$ for all $Y$. Similarly, in

[13, 14], association rules miners are used to detect ADR and in [15, 16], the ADR are detected during the MB discovery. Once the ADR are detected, any CB algorithm can be used to construct a DAG such that, for every pair $X$ and $Y$ in $\mathbf{V}$, $(X, Y)$ is connected in $\mathcal{G}$ if $X$ and $Y$ remains dependent conditionally on every set $\mathbf{S} \subseteq \mathbf{V} \setminus \{X, Y\}$ such that $\mathbf{S} \not\Rightarrow X$ and $\mathbf{S} \not\Rightarrow Y$.

# 4 New Method

In this section, we present in detail our learning algorithm called MBOR. We recall that MBOR was designed in order to endow the search procedure with the ability to: 1) handle efficiently data sets with thousands of variables but very few instances, 2) be correct under faithfulness condition, 3) handle implicitly some approximate deterministic relationships (ADR) without detecting them. We discuss next how we tackle each problem.

First of all, MBOR scales up to hundreds of thousands of variables in reasonable time because it searches the Markov boundary of the target without having to construct the whole Bayesian network first. Like PCMB [3] and MMMB [8], MBOR takes a divide-and-conquer approach that breaks the problem of identifying $\mathbf{MB}_T$ into two subproblems : first, identifying $\mathbf{PC}_T$ and, second, identifying the parents of the children (the spouses $\mathbf{SP}_T$) of $T$. According to Peña et al., this divide-and-conquer approach is supposed to be more data efficient than IAMB [5] and its variants, e.g., Fast-IAMB [10] and Interleaved-IAMB [4], because $\mathbf{MB}_T$ can be identified by conditioning on sets much smaller than those used by IAMB. Indeed, IAMB and its variants seek directly the minimal subset of $\mathbf{U}$ (the full set) that renders the rest of $\mathbf{U}$ independent of $T$, given $\mathbf{MB}_T$. Moreover, MBOR keeps the size of the conditional sets to the minimum possible without sacrificing the performance as discussed next.

The advantage of the divide-and-conquer strategy in terms of data efficiency does not come without some cost. MMMB [8] and PCMB [3] apply the "AND condition" to prove correctness under faithfulness condition. In other words, two variables $X$ and $Y$ are considered as neighbors if $Y \in PC_X$ AND $X \in PC_Y$. We believe this condition is far too severe and yields too many false negatives in the output. Instead, MBOR stands for "Markov Boundary search using the OR condition". This "OR condition" is a major difference between MBOR and all the above mentioned correct divide-and-conquer algorithms: two variables $X$ and $Y$ are considered as neighbors with MBOR if $Y \in PC_X$ OR $X \in PC_Y$. Clearly, the OR condition makes it easier for true positive nodes to enter the Markov boundary, hence the name and the practical efficiency of our algorithm. Moreover, the OR condition is a simple way to handle some ADR. For illustration, consider the sub-graph $X \Rightarrow T \rightarrow Y$, since $X \Rightarrow T$ is an ADR, $T \perp Y|X$ so Y will not be considered as a neighbor of $T$. As $Y$ still sees $T$ in its neighborhood, $Y$ and $T$ will be considered as adjacent. The main difficulty was to demonstrate the correctness under the faithfulness condition despite the OR condition. The proof is provided in the next section.

MBOR (Algorithm 1) works in three steps and it is based on four subroutines called *PCSuperset*, *SPSuperset* and *MBtoPC* (Algorithms 2-4). Before we describe the algorithm step by step, we recall that the general idea underlying MBOR is to use a weak MB learner to create a stronger MB learner. By weak learner, we mean a simple and fast method that may produce many mistakes due to its data inefficiency. In other words, the proposed method aims at producing an accurate MB discovery algorithm by combining fast and moderately inaccurate (but correct) MB learners. The weak MB learner is used in *MBtoPC* (Algorithm 4) to implement a correct Parents and Children learning procedure. It works in two steps. First, the weak MB learner called *CorrectMB* is used at line 1 to output a candidate MB. *CorrectMB* may be implemented by any algorithm of the IAMB family. In our implementation, we use Inter-IAMB for its simplicity and performance [5]. The key difference between IAMB and Inter-IAMB is that the shrinking phase is interleaved into the growing phase in Inter-IAMB. The second step (lines 3-6) of *MBtoPC* removes the spouses of the target.

In phase I, MBOR calls *PCSuperset* to extract **PCS**, a superset for the parents and children, and then calls *SPSuperset* to extract **SPS**, a superset for the target spouses (parents of children). Filtering reduces as much as possible the number of variables before proceeding to the MB discovery. In *PCSuperset* and *SPSuperset*, the size of the conditioning set **Z** in the tests is severely restricted: $card(\mathbf{Z}) \leq 1$ in *PCSuperset* (lines 3 and 10) and $card(\mathbf{Z}) \leq 2$ in *SPSuperset* (lines 5 and 11). As discussed before, conditioning on larger sets of variables would increase the risk of missing variables that are weakly associated to the target. It would also lessen the reliability of the independence tests. So the MB superset, **MBS** (line 3), is computed based on a scalable and highly data-efficient procedure. Moreover, the filtering phase is also a way to handle some ADR. For illustration, consider the sub-graph $Z \Rightarrow Y \rightarrow T \Leftarrow X$, since $X \Rightarrow T$ and $Z \Rightarrow Y$ are ADRs, $T \perp Y|X$ and $Y \perp T|Z$, Y would not be considered as a neighbor of $T$ and vice-versa. The OR-condition in Phase II would not help in this particular case. Fortunately, as Phase I filters out variable $Z$, $Y$ and $T$ will be considered as adjacent

Phase II finds the parents and children in the restricted set of variables using the OR condition. Therefore, all variables that have $T$ in their vicinity are included in $\mathbf{PC}_T$ (lines 7-8).

Phase III identifies the target's spouses in **MBS** in exactly the same way PCMB does [3]. Note however that the OR condition is not applied in this last phase because it would not be possible to prove its correctness anymore.

## 5 Proof of Correctness Under Faithfulness Condition

Several intermediate theorems are required before we demonstrate MBOR's correctness under faithfulness condition. Indeed, as **MBS** is a subset of **U**, a difficulty arises: a marginal distribution $P^{\mathbf{V}}$ of $\mathbf{V} \subset \mathbf{U}$ may not satisfy the faithfulness condition with any DAG even if $P^{\mathbf{U}}$ does. This is an example of embedded faithfulness, which is defined as follow:

**Algorithm 1.** *MBOR*

**Require:** $T$: target; $D$: data set ($\mathbf{U}$ is the set of variables)
**Ensure:** [**PC**,**SP**]: Markov boundary of $T$

**Phase I:** *Find MB superset* (**MBS**)
1: $[\mathbf{PCS}, \mathbf{dSep}] = PCSuperSet(T, D)$
2: $\mathbf{SPS} = SPSuperSet(T, D, \mathbf{PCS}, \mathbf{dSep})$
3: $\mathbf{MBS} = \mathbf{PCS} \cup \mathbf{SPS}$
4: $\mathcal{D} = \mathcal{D}(\mathbf{MBS} \cup T)$ *i.e., remove from data set all variables in* $\mathbf{U}/\{\mathbf{MBS} \cup T\}$

**Phase II:** *Find parents and children of the target*
5: $\mathbf{PC} = MBtoPC(T, \mathcal{D})$
6: **for all** $X \in \mathbf{PCS} \setminus \mathbf{PC}$ **do**
7: **if** $T \in MBtoPC(X, \mathcal{D})$ **then**
8: $\mathbf{PC} = \mathbf{PC} \cup X$
9: **end if**
10: **end for**

**Phase III:** *Find spouses of the target*
11: $\mathbf{SP} = \emptyset$
12: **for all** $X \in \mathbf{PC}$ **do**
13: **for all** $Y \in MBtoPC(X, D) \setminus \{\mathbf{PC} \cup T\}$ **do**
14: Find minimal $\mathbf{Z} \subset \mathbf{MBS} \setminus \{T \cup Y\}$ such that $T \perp Y | \mathbf{Z}$
15: **if** $(T \not\perp Y | \mathbf{Z} \cup X)$ **then**
16: $\mathbf{SP} = \mathbf{SP} \cup Y$
17: **end if**
18: **end for**
19: **end for**

**Definition 1.** *Let $P$ be a distribution of the variables in $\mathbf{V}$ where $\mathbf{V} \subset \mathbf{U}$ and let $\mathcal{G} =< \mathbf{U}, \mathbf{E} >$ be a DAG. $< \mathcal{G}, P >$ satisfies the embedded faithfulness condition if $\mathcal{G}$ entails all and only the conditional independencies in $P$, for subsets including only elements of $\mathbf{V}$.*

We obtain embedded faithfulness by taking the marginal of a faithful distribution as shown by the next theorem:

**Theorem 1.** *Let $P$ be a joint probability of the variables in $\mathbf{U}$ with $\mathbf{V} \subseteq \mathbf{U}$ and $\mathcal{G} =< \mathbf{U}, \mathbf{E} >$. If $< \mathcal{G}, P >$ satisfies the faithfulness condition and $P^{\mathbf{V}}$ is the marginal distribution of $\mathbf{V}$, then $< \mathcal{G}, P^{\mathbf{V}} >$ satisfies the* embedded *faithful condition.*

The proof can be found in [17]. Note that every distribution doesn't admit an embedded faithful representation. This property is useful to prove the correctness of our MBOR under the faithfulness condition. Let $\mathbf{PC}_X^{\mathbf{U}}$ denote the variables $Y \in \mathbf{U}$ such that there is no set $\mathbf{Z} \in \mathbf{U} \setminus \{X, Y\}$ such that $X \perp_P Y | \mathbf{Z}$. If $< \mathcal{G}, P >$ satisfies the faithfulness condition, $\mathbf{PC}_X^{\mathbf{U}}$ are the parents and children of $X$ in $\mathbf{U}$. Otherwise, $\mathbf{PC}_X^{\mathbf{U}}$ is the unique set of the variables that remains dependent on $X$ conditioned on any set $\mathbf{Z} \in \mathbf{U} \setminus \{X, Y\}$.

**Algorithm 2.** *PCSuperSet*

**Require:** $T$: target; $D$: data set (**U** is the set of variables)
**Ensure:** **PCS**: PC superset of $T$; **dSep**: d-separation set;

**Phase I:** *Remove* $X$ *if* $T \perp X$
1: $\mathbf{PCS} = \mathbf{U} \setminus T$
2: **for all** $X \in \mathbf{PCS}$ **do**
3: **if** $(T \perp X)$ **then**
4: $\mathbf{PCS} = \mathbf{PCS} \setminus X$
5: $\mathbf{dSep}(X) = \emptyset$
6: **end if**
7: **end for**

**Phase II:** *Remove* $X$ *if* $T \perp X|Y$
8: **for all** $X \in \mathbf{PCS}$ **do**
9: **for all** $Y \in \mathbf{PCS} \setminus X$ **do**
10: **if** $(T \perp X \mid Y)$ **then**
11: $\mathbf{PCS} = \mathbf{PCS} \setminus X$
12: $\mathbf{dSep}(X) = Y$
13: **end if**
14: **end for**
15: **end for**

**Algorithm 3.** *SPSuperSet*

**Require:** $T$: target; $D$: data set (**U** is the set of variables); **PCS**: PC superset of $T$; **dSep**: d-separation set;
**Ensure:** **SPS**: SP superset of $T$;

1: $\mathbf{SPS} = \emptyset$
2: **for all** $X \in \mathbf{PCS}$ **do**
3: $\mathbf{SPS}_X = \emptyset$
4: **for all** $Y \in \mathbf{U} \setminus \{T \cup \mathbf{PCS}\}$ **do**
5: **if** $(T \not\perp Y | \mathbf{dSep}(Y) \cup X)$ **then**
6: $\mathbf{SPS}_X = \mathbf{SPS}_X \cup Y$
7: **end if**
8: **end for**
9: **for all** $Y \in \mathbf{SPS}_X$ **do**
10: **for all** $Z \in \mathbf{SPS}_X \setminus Y$ **do**
11: **if** $(T \perp Y | X \cup Z)$ **then**
12: $\mathbf{SPS}_X = \mathbf{SPS}_X \setminus Y$
13: **end if**
14: **end for**
15: **end for**
16: $\mathbf{SPS} = \mathbf{SPS} \cup \mathbf{SPS}_X$
17: **end for**

**Algorithm 4.** *MBtoPC*

**Require:** $T$: target; $D$: data set
**Ensure:** **PC**: Parents and children of $T$;

1: $\mathbf{MB} = CorrectMB(T, D)$
2: $\mathbf{PC} = \mathbf{MB}$
3: **for all** $X \in \mathbf{MB}$ **do**
4:   **if** $\exists \mathbf{Z} \subset (\mathbf{MB} \setminus X)$ such that $T \perp X \mid \mathbf{Z}$ **then**
5:     $\mathbf{PC} = \mathbf{PC} \setminus X$
6:   **end if**
7: **end for**

**Theorem 2.** *Let* $\boldsymbol{U}$ *be a set of random variables and* $\mathcal{G} =< \mathbf{U}, \mathbf{E} >$. *If* $< \mathcal{G}, P >$ *satisfies the faithfulness condition, then every target* $T$ *admits a unique Markov boundary* $\boldsymbol{MB}_T^{\mathbf{U}}$. *Moreover, for all* $\boldsymbol{V}$ *such that* $\boldsymbol{MB}_T^{\boldsymbol{U}} \subseteq \boldsymbol{V} \subseteq \boldsymbol{U}$, $T$ *admits a unique Markov boundary over* $\boldsymbol{V}$ *and* $\boldsymbol{MB}_T^{\boldsymbol{V}} = \boldsymbol{MB}_T^{\boldsymbol{U}}$.

*Proof.* If $\mathbf{MB}_T^{\mathbf{U}}$ is the Markov boundary of $T$ in $\mathbf{U}$, then $T$ is independent of $\mathbf{V} \setminus \{\mathbf{MB}_T^{\mathbf{U}} \cup T\}$ conditionally on $\mathbf{MB}_T^{\mathbf{U}}$ so $\mathbf{MB}_T^{\mathbf{U}}$ is a Markov blanket in $\mathbf{V}$. Moreover, none of the proper subsets of $\mathbf{MB}_T^{\mathbf{U}}$ is a Markov blanket of $T$ in $\mathbf{V}$, so $\mathbf{MB}_T^{\mathbf{U}}$ is also a Markov boundary of $T$ in $\mathbf{V}$. So if it is not the unique MB for $T$ in $\mathbf{V}$ there exists some other set $\mathbf{S}_T$ not equal to $\mathbf{MB}_T^{\mathbf{U}}$, which is a MB of $T$ in $\mathbf{V}$. Since $\mathbf{MB}_T^{\mathbf{U}} \neq \mathbf{S}_T$ and $\mathbf{MB}_T^{\mathbf{U}}$ cannot be a subset of $\mathbf{S}_T$, there is some $X \in \mathbf{MB}_T^{\mathbf{U}}$ such that $X \notin \mathbf{S}_T$. Since $\mathbf{S}_T$ is a MB for $T$, we would have $T \perp_P X|\mathbf{S}_T$. If $X$ is a parent or child of $T$, we would not have $T \perp_{\mathcal{G}} X|\mathbf{S}_T$ which means we would have a conditional independence which is not entailed by d-separation in $\mathcal{G}$ which contradicts the faithfulness condition. If $X$ is a parent of a child of $T$ in $\mathcal{G}$, let $Y$ be their common child in $\mathbf{U}$. If $Y \in \mathbf{S}_T$ we again would not have $T \perp_{\mathcal{G}} X|\mathbf{S}_T$. If $Y \notin \mathbf{S}_T$ we would have $T \perp_P Y|\mathbf{S}_T$ because $\mathbf{S}_T$ is a MB of $T$ in $\mathbf{V}$ but we do not have $T \perp_{\mathcal{G}} Y|\mathbf{S}_T$ because $T$ is a parent of $Y$ in G. So again we would have a conditional independence which is not a d-separation in $\mathcal{G}$. This proves that there can not be such set $\mathbf{S}_T$. □

**Theorem 3.** *Let* $\boldsymbol{U}$ *be a set of random variables and* $T$ *a target variable. Let* $\mathcal{G} =< \mathbf{U}, \mathbf{E} >$ *be a DAG such that* $< \mathcal{G}, P >$ *satisfies the faithfulness condition. Let* $\boldsymbol{V}$ *be such that* $\boldsymbol{MB}_T^{\boldsymbol{U}} \subseteq \boldsymbol{V} \subseteq \boldsymbol{U}$ *then,* $\boldsymbol{PC}_T^{\boldsymbol{V}} = \boldsymbol{PC}_T^{\boldsymbol{U}}$.

*Proof.* Clearly $\mathbf{PC}_T^{\mathbf{U}} \subseteq \mathbf{PC}_T^{\mathbf{V}}$ as $\mathbf{MB}_T^{\mathbf{U}} \subseteq \mathbf{V} \subseteq \mathbf{U}$. If $X \in \mathbf{PC}_T^{\mathbf{V}}$ and $X \notin \mathbf{PC}_T^{\mathbf{U}}$, $\exists \mathbf{Z} \subset \mathbf{MB}_T^{\mathbf{U}} \setminus X$ such that $T \perp_P X|\mathbf{Z}$ because all non adjacent nodes may be d-separated in $\mathcal{G}$ by a subset of its Markov boundary. As $\mathbf{MB}_T^{\mathbf{U}} = \mathbf{MB}_T^{\mathbf{V}}$ owing to Theorem 2, so $X$ and $T$ can be d-separated in $\mathbf{V} \setminus \{X, T\}$. Therefore, $X$ cannot be adjacent to $T$ in $\mathbf{V}$. □

**Theorem 4.** *Let* $\boldsymbol{U}$ *be a set of random variables and* $T$ *a target variable. Let* $\mathcal{G} =< \mathbf{U}, \mathbf{E} >$ *be a DAG such that* $< \mathcal{G}, P >$ *satisfies the faithfulness condition. Let* $\boldsymbol{V}$ *be such that* $\boldsymbol{MB}_T^{\boldsymbol{U}} \subseteq \boldsymbol{V} \subseteq \boldsymbol{U}$. *Under the assumption that the independence* *tests are reliable, MBtoPC(T,* $\mathbf{V}$*) returns* $\mathbf{PC}_T^{\mathbf{U}}$*. Moreover, let* $X \in \mathbf{V} \setminus T$*, then* $T$ *is in the output of MBtoPC(X,* $\mathbf{V}, \mathcal{D}$*) iff* $X \in \mathbf{PC}_T^{\mathbf{U}}$*.*

*Proof.* We prove first that MBtoPC(T,**V**) returns $\mathbf{PC}_T^{\mathbf{U}}$. In the first stage of MBtoPC, CorrectMB(T,**V**) seeks a minimal set $\mathbf{S}_T \in \mathbf{V} \setminus T$ that renders $\mathbf{V} \setminus \mathbf{S}_T$ independent of $T$ conditionally on $\mathbf{S}_T$. This set is unique owing to Theorem 2, therefore $\mathbf{S}_T = \mathbf{MB}_T^{\mathbf{V}} = \mathbf{MB}_T^{\mathbf{U}}$. In the backward phase, MBtoPC removes the variables $X \in \mathbf{MB}_T^{\mathbf{V}}$ such that $\exists \mathbf{Z} \subset (\mathbf{MB}_T^{\mathbf{V}} \setminus X)$ for which $T \perp X \mid \mathbf{Z}$. These variables are the spouses of $T$ in $\mathcal{G}$, so MBtoPC(T,**V**) returns $\mathbf{PC}_T^{\mathbf{U}}$. Now, if $X \notin \mathbf{PC}_T^{\mathbf{U}}$ then $X \notin \mathbf{PC}_T^{\mathbf{V}}$ owing to Theorem 3. So there is a set $\mathbf{Z} \subset \mathbf{V} \setminus \{X, Y\}$ such that $T \perp X \mid \mathbf{Z}$. Therefore, $X$ cannot be in the output of MBtoPC(T,**V**). □

**Theorem 5.** *Under the assumptions that the independence tests are reliable and that the database is a sample from a probability distribution P faithful to a DAG* $\mathcal{G}$*, MBOR(T) returns* $\mathbf{MB}_T^{\mathbf{U}}$*.*

*Proof.* Let **MBS** be the MB superset constructed at line 3 of *MBOR*. It is straightforward to show that $\mathbf{MB}_T^{\mathbf{U}} \subset \mathbf{MBS}$. So the Markov boundary of $T$ in **MBS** is that of **U** owing to Theorem 2 so the problem is well defined. In Phase II at line 7, if $T$ is in the output of MBtoPC($X, \mathbf{V}, \mathcal{D}$) then $X$ should be in the output of MBtoPC($T, \mathbf{V}, \mathcal{D}$) owing to Theorem 4. So phase II ends up with the $\mathbf{PC}_T^{\mathbf{U}}$. In Phase III, lines 11-19 identify all and only the spouse of $T$ in $\mathcal{G}$ when the faithfulness condition is asssumed as shown in [3]. When the assumption doesn't hold anymore for $< \mathcal{G}, P'\mathbf{V} >$, we need to show that a fake spouse will not enter the set **SP**. In phase III line 12, it is easy to see that MBtoPC($X, \mathbf{V}, \mathcal{D}$) returns a set $\mathbf{PC}_X^{\mathbf{V}}$ that may differ from $\mathbf{PC}_X^{\mathbf{U}}$. Suppose $Y \notin \mathbf{PC}_X^{\mathbf{U}}$ and $Y$ is in the output of MBtoPC($X, \mathbf{V}, \mathcal{D}$). This means that there exists at least one active path between $X$ and $Y$ in $\mathcal{G}$ that contains a node in $\mathbf{U} \setminus \mathbf{V}$. At lines 13-14, $Y$ is considered as spouse of $T$ if there is a set $\mathbf{Z} \subset \mathbf{MBS} \setminus \{T \cup Y\}$ so that $T \perp Y | \mathbf{Z}$ and $T \not\perp Y | \mathbf{Z} \cup X$. Therefore, this path in $\mathcal{G}$ should necessarily be of the type $T \rightarrow X \leftarrow A \leftrightsquigarrow B \leftrightsquigarrow Y$ where $\leftrightsquigarrow$ denotes an active path otherwise we would not have $T \not\perp Y | \mathbf{Z} \cup X$. As $A$ is a spouse of $T$, $A \in \mathbf{MB}_T^{\mathbf{U}}$ and so $A$ is in **V**. Suppose $B$ is not in **V**, then $A$ still d-separates $X$ and $Y$ so $Y$ cannot be in the output of MBtoPC($X, \mathbf{V}, \mathcal{D}$) since we found a set $\mathbf{Z} \subseteq \mathbf{V}$ such that $X \not\perp_P Y | \mathbf{Z}$. So $Y$ is included in **SP** at line 16 iff $Y$ is a spouse of $T$ in **U**. □.

## 6 Experimental Validation

In this section, we compare the performance of InterIAMB, PCMB and MBOR through experiments on synthetic and real databases with very few instances compared to the number of variables. They are written in MATLAB and all the experiments are run on a Intel Core 2 Duo T77500 with 2Gb RAM running Windows Vista. To implement the conditional independence test, we calculate the $G^2$ statistic as in [11], under the null hypothesis of the conditional independence. The significance level of the test in all compared algorithms is 0.05 except on the high-dimensional THROMBIN data where it is 0.0001. All three algorithms are correct under the faithfulness condition and are also scalable. We do not consider MMMB and HITON-MB because we are not interested in any algorithm that does not guarantee the correctness under faithfulness assumption. We do not consider GS because IAMB outperforms it [5]. Even if PCMB was also shown experimentally in [3] to be more accurate than IAMB and its variants, we consider InterIAMB because it is used as a subroutine in MBOR.

It might very well happen that several variables have the same association value with the target in data sets with very few instances. In this particular case, somewhat arbitrary (in)dependence decisions are taken. This can be seen as a source of randomness inherent to all CB procedures. To handle this problem, our implementation breaks ties at random: a random permutation of the variables is carried out before MBOR is run. This explains the variability of MBOR with very few instances and/or extremely large number of variables (e.g., THROMBIN).

### 6.1 Synthetic Data

Figure 1 illustrates the results of our experiments on six common BN benchmarks : BREAST-CANCER or ASIA (8 nodes/8 arcs), INSURANCE (27/52), INSULINE (35/52), ALARM (37/46), HAILFINDER (56/66) and CARPO (61/74). These benchmarks are available from the UCI Machine Learning Repository. All three algorithms have been run on each variable for all data sets. Figure 1 (upper part) summarizes graphically the results in terms of missing and extra nodes in the output of the MB averaged over 10 runs for 200, 500 and 1000 i.i.d. samples. The upper part shows the average false positive and and lower part shows the false negative rates. The overall accuracy is very similar for nodes with Markov boundaries with less than 4 variables. For larger MBs, however, the advantages of MBOR against the other two algorithms are far more noticeable. For instance, MBOR consistently outperforms the other algorithms on variable *IPA* in the INSULINE benchmark as may be seen in Table 2. Figure 2 (lower part) show the performance for nodes with more than 4 variables. Results are averaged over all the above mentioned benchmarks. As observed, MBOR reduces drastically the average number of false negatives compared to PCMB and InterIAMB (up to 40% on INSULINE). This benefit comes at very little expense: the false positive rate is slightly higher. This is not a surprise as PCMB makes it harder for true positives to enter the output.

**Table 1.** INSULINE benchmark : number of extra and missing variables for PCMB, Inter-IAMB and MBOR for variable *IPA* run on 1000 instances. Results are averaged over 10 runs.

| *Algorithm* | false positive | false negative |
|---|---|---|
| PCMB | 0.4 | 11.8 |
| InterIAMB | 0 | 12.6 |
| MBOR | 2.1 | 2.1 |

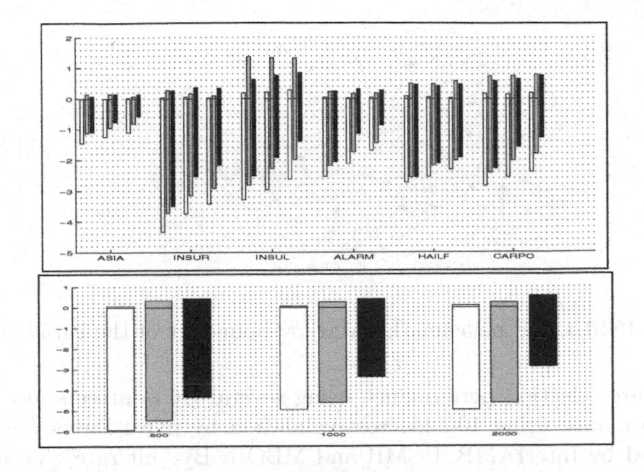


**Fig. 1.** Upper plot : average missing (lower part) and extra (upper part) variables for learning Markov boundaries of all variables of ASIA, INSURANCE, INSULINE, ALARM, HAILFINDER and CARPO networks. The results of PCMB, InterIAMB and MBOR are shaded in white, gray and black respectively. For each benchmark the bars show the results on 200, 500 and 1000 instances respectively. All results are averaged over 10 runs. Lower plot : results are averaged over all benchmarks for nodes that have a MB with more than 4 variables in the MB, for 500, 1000 and 2000 i.i.d. samples.

### 6.2 Real Data

In this section, we assess the performance of the probabilistic classification using the feature subset output by MBOR. To this purpose, we consider several categorical data bases from the UCI Machine Learning Repository in order to evaluate the accuracy of MBOR against InterIAMB and PCMB. The database description and the results of the experiments with the Car Evaluation, Chess, Molecular Biology, SPECT heart, Tic-Tac-Toe, Wine and Waveform are shown in Table 1. Performance is assessed by hit rate (correct classification rate), relative absolute error (R.A.E.), and Kappa Statistics obtained by 10-fold cross-validation. Kappa can be thought of as the chance-corrected proportional agreement, and possible values range from +1 (perfect agreement) via 0 (no agreement above that expected by chance) to -1 (complete disagreement). As may be seen, the classification performance by naive Bayes classifier on the features selected by MBOR has always outperformed that of InterIAMB and PCMB by a noticeable margin, especially on the Molecular Biology database.

### 6.3 Real Data : Thrombin Database

Our last experiments demonstrate the ability of MBOR to solve a real world FSS problem involving thousands of features. We consider the THROMBIN database which was provided by DuPont Pharmaceuticals for KDD Cup 2001. It is exemplary of a real drug design [6]. The training set contains 1909 instances characterized by 139,351 binary features. The accuracy of a Naive Bayesian classifier

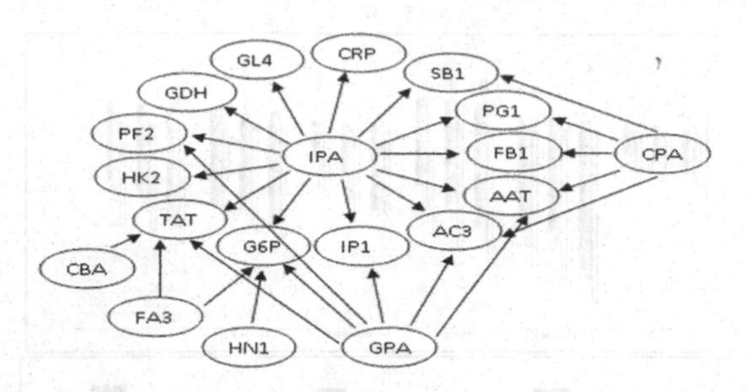


**Fig. 2.** INSULINE benchmark : Markov boundary of the variable *IPA*

**Table 2.** Feature selection performance using several UCI datasets in terms of classification performance using 10-fold cross-validation by naive Bayes classifier on the features selected by InterIAMB, PCMB and MBOR. By "hit rate", we mean correct classification rate.

| *Data Sets* | *Inst.* | *Attr.* | *Accuracy* | *Algorithms* | | |
|---|---|---|---|---|---|---|
| | | | | InterIAMB | PCMB | MBOR |
| Car Evaluation | 1728 | 6 | Hit Rate | 79.11% | 79.11% | 85.36% |
| | | | Kappa | 0.5204 | 0.5204 | 0.6665 |
| | | | Rel. abs. error. | 56.59% | 56.59% | 49.88% |
| Chess (King-Rook vs. King-Pawn) | 3196 | 36 | Hit Rate | 94.37% | 92.43% | 93.15% |
| | | | Kappa | 0.8871 | 0.8479 | 0.8624 |
| | | | Rel. abs. error | 45.09% | 47.23% | 44.51% |
| Molecular Biology (Splice-junction Gene Sequences) | 3190 | 61 | Hit Rate | 74.01% | 74.01% | 95.61% |
| | | | Kappa | 0.5941 | 0.5941 | 0.9290 |
| | | | Rel. abs. error | 56.94% | 56.94% | 10.27% |
| SPECT Heart | 267 | 22 | Hit Rate | 76.40% | 79.40% | 84.27% |
| | | | Kappa | 0.2738 | 0 | 0.4989 |
| | | | Rel. abs. error | 77.83% | 93.92% | 71.71% |
| Tic-Tac-Toe Endgame | 958 | 9 | Hit Rate | 67.95% | 67.95% | 72.44% |
| | | | Kappa | 0.1925 | 0.1951 | 0.3183 |
| | | | Rel. abs. error | 85.62% | 85.75% | 82.53% |
| Wine | 178 | 13 | Hit Rate | 94.38% | 94.38% | 98.88% |
| | | | Kappa | 0.9148 | 0.9148 | 0.9830 |
| | | | Rel. abs. error | 19.08% | 19.08% | 2.67% |
| Waveform - Version 1 | 5000 | 21 | Hit Rate | 76.42% | 76.42% | 81.32% |
| | | | Kappa | 0.6462 | 0.6462 | 0.7196 |
| | | | Rel. abs. error | 44.50% | 44.50% | 29.43% |

was computed as the average of the accuracy on true binding compounds and the accuracy on true non-binding compounds on the 634 compounds of the test set. As the data is unbalanced, the accuracy is calculated as the average of true positive rate and the true negative rate. A significance level of 0.0001 avoids better than 0.01 the spurious dependencies that may exist in the data due to the large number of features. MBS with $\alpha = 0.0001$ returns a set of 21 variables, SPS select 39 variables and MBOR outputs 5 variables on average in about 3h running time.

Note shown here, the 10 runs return each time a different MB, all of them containing 5 features. They mostly differ by one or two variables. MBOR scores between 36% (really bad) to 66% with an average 53% on average which seems really deceiving compared to PCMB and IAMB that achieves respectively 63% and 54% as shown in [3]. Nonetheless, MBOR is highly variable and was able to identify 3 different MBs that outperform those found by IAMB and 90% of those by PCMB. For instance, the MB which scores 66% contains the two variables 20973, 63855. These two variables, when used conjunctly, score 66,9% which is impressive according to [6, 3] for such a small feature set. Note that a MB with the four features obtained by the winner of KDD cup 2001 scores 67% accuracy.

The execution time was not reported as it is too dependent on the specific implementation. We were unable to run PCMB on the Thrombin database in reasonable time with *our* MATLAB implementation. On synthetic data, MBOR runs (say) 30% faster than PCMB.

## 7 Discussion and Conclusion

We discussed simple solutions to improve the data efficiency of current constraint-based Markov boundary discovery algorithms. We proposed a novel approach called MBOR that combines the main advantages of PCMB and IAMB while still being correct under faithfulness condition. Our experimental results show a clear benefit in several situations: densely connected DAGs, weak associations or approximate functional dependencies among the variables. Though not discussed here, a topic of considerable interest would be to ascertain the data distributions for which MBOR, PCMB or the stochastic variant of IAMB termed KIAMB proposed in [3], is most suited. This needs further substantiation through more experiments and analysis.

## References


[1] Guyon, I., Elisseeff, A.: An introduction to variable and feature selection. Journal of Machine Learning Research 3, 1157–1182 (2003)

[2] Nilsson, R., Peña, J., Bjrkegren, J., Tegnr, J.: Consistent feature selection for pattern recognition in polynomial time. Journal of Machine Learning Research 8, 589–612 (2007)

[3] Peña, J., Nilsson, R., Bjrkegren, J., Tegnr, J.: Towards scalable and data efficient learning of markov boundaries. International Journal of Approximate Reasoning 45(2), 211–232 (2007)

[4] Yaramakala, S., Margaritis, D.: Speculative markov blanket discovery for optimal feature selection. In: ICDM, pp. 809–812 (2005)

[5] Tsamardinos, I., Aliferis, C.F., Statnikov, A.R.: Algorithms for large scale markov blanket discovery. In: FLAIRS Conference, pp. 376–381 (2003)

[6] Cheng, J., Hatzis, C., Hayashi, H., Krogel, M., Morishita, S., Page, D., Sese, J.: KDD Cup 2001 Report. In: ACM SIGKDD Explorations, pp. 1–18 (2002)
[7] Dash, D., Druzdzel, M.J.: Robust independence testing for constraint-based learning of causal structure. In: UAI, pp. 167–174 (2003)
[8] Tsamardinos, I., Brown, L.E., Aliferis, C.F.: The max-min hill-climbing bayesian network structure learning algorithm. Machine Learning 65(1), 31–78 (2006)
[9] Spirtes, P., Glymour, C., Scheines, R.: Causation, prediction, and search. Springer, Heidelberg (1993)
[10] Yaramakala, S.: Fast markov blanket discovery. In MS-Thesis. Iowa State University (2004)
[11] Spirtes, P., Glymour, C., Scheines, R.: Causation, Prediction, and Search, 2nd edn. MIT Press, Cambridge (2000)
[12] Yilmaz, Y.K., Alpaydin, E., Akin, H.L., Bilgiç, T.: Handling of deterministic relationships in constraint-based causal discovery. In: Probabilistic Graphical Models (2002)
[13] Luo, W.: Learning bayesian networks in semi-deterministic systems. In: Canadian Conference on AI, pp. 230–241 (2006)
[14] Kebaili, Z., Aussem, A.: A novel bayesian network structure learning algorithm based on minimal correlated itemset mining techniques. In: IEEE Int. Conference on Digital Information Management ICDIM 2007, pp. 121–126 (2007)
[15] Aussem, A., de Morais, S.R., Corbex, M.: Nasopharyngeal carcinoma data analysis with a novel bayesian network skeleton learning. In: Bellazzi, R., Abu-Hanna, A., Hunter, J. (eds.) AIME 2007. LNCS (LNAI), vol. 4594, pp. 326–330. Springer, Heidelberg (2007)
[16] Rodrigues de Morais, S., Aussem, A., Corbex, M.: Handling almost-deterministic relationships in constraint-based bayesian network discovery: Application to cancer risk factor identification. In: 16th European Symposium on Artificial Neural Networks ESANN 2008, pp. 101–106 (2008)
[17] Neapolitan, R.E.: Learning Bayesian Networks. Prentice-Hall, Englewood Cliffs (2004)

# Robust Feature Selection Using Ensemble Feature Selection Techniques

Yvan Saeys, Thomas Abeel, and Yves Van de Peer

Department of Plant Systems Biology, VIB, Technologiepark 927, 9052 Gent, Belgium and Department of Molecular Genetics, Ghent University, Gent, Belgium
{yvan.saeys,thomas.abeel,yves.vandepeer}@psb.ugent.be

**Abstract.** Robustness or stability of feature selection techniques is a topic of recent interest, and is an important issue when selected feature subsets are subsequently analysed by domain experts to gain more insight into the problem modelled. In this work, we investigate the use of ensemble feature selection techniques, where multiple feature selection methods are combined to yield more robust results. We show that these techniques show great promise for high-dimensional domains with small sample sizes, and provide more robust feature subsets than a single feature selection technique. In addition, we also investigate the effect of ensemble feature selection techniques on classification performance, giving rise to a new model selection strategy.

## 1 Introduction

Feature selection is an important preprocessing step in many machine learning applications, where it is often used to find the smallest subset of features that maximally increases the performance of the model. Besides maximizing model performance, other benefits of applying feature selection include the ability to build simpler and faster models using only a subset of all features, as well as gaining a better understanding of the processes described by the data, by focusing on a selected subset of features [1].

Feature selection techniques can be divided into three categories, depending on how they interact with the classifier. Filter methods directly operate on the dataset, and provide a feature weighting, ranking or subset as output. These methods have the advantage of being fast and independent of the classification model, but at the cost of inferior results. Wrapper methods perform a search in the space of feature subsets, guided by the outcome of the model (e.g. classification performance on a cross-validation of the training set). They often report better results than filter methods, but at the price of an increased computational cost [2]. Finally, embedded methods use internal information of the classification model to perform feature selection (e.g. use of the weight vector in support vector machines). They often provide a good trade-off between performance and computational cost [1,3].

During the past decade, the use of feature selection for knowledge discovery has become increasingly important in many domains that are characterized by

W. Daelemans et al. (Eds.): ECML PKDD 2008, Part II, LNAI 5212, pp. 313–325, 2008.

a large number of features, but a small number of samples. Typical examples of such domains include text mining, computational chemistry and the bioinformatics and biomedical field, where the number of features (problem dimensionality) often exceeds the number of samples by orders of magnitude [3]. When using feature selection in these domains, not only model performance but also robustness of the feature selection process is important, as domain experts would prefer a stable feature selection algorithm over an unstable one when only small changes are made to the dataset. Robust feature selection techniques would allow domain experts to have more confidence in the selected features, as in most cases these features are subsequently analyzed further, requiring much time and effort, especially in biomedical applications.

Surprisingly, the robustness (stability) of feature selection techniques is an important aspect that received only relatively little attention during the past. Recent work in this area mainly focuses on the stability indices to be used for feature selection, introducing measures based on Hamming distance [4], correlation coefficients [5], consistency [6] and information theory [7]. Kalousis and coworkers also present an extensive comparative evaluation of feature selection stability over a number of high-dimensional datasets [5]. However, most of this work only focuses on the stability of single feature selection techniques, an exception being the work of [4] which describes an example combining multiple feature selection runs.

In this work, we investigate whether the use of ensemble feature selection techniques can be used to yield more robust feature selection techniques, and whether combining multiple methods has any effect on the classification performance. The rationale for this idea stems from the field of ensemble learning, where multiple (unstable) classifiers are combined to yield a more stable, and better performing ensemble classifier. Similarly, one could think of more robust feature selection techniques by combining single, less stable feature selectors. As this issue is especially critical in large feature/small sample size domains, the current work focuses on ensemble feature selection techniques in this area.

The rest of this paper is structured as follows. Section 2 introduces the methodology used to assess robustness of the algorithms we evaluated. Subsequently, we introduce ensemble feature selection techniques in section 3 and present the results of our experiments in section 4. We conclude with some final remarks and ideas for future work.

## 2 Robustness of Feature Selection Techniques

The robustness of feature selection techniques can be defined as the variation in feature selection results due to small changes in the dataset. When applying feature selection for knowledge discovery, robustness of the feature selection result is a desirable characteristic, especially if subsequent analyses or validations of selected feature subsets are costly. Modification of the dataset can be considered at different levels: perturbation at the instance level (e.g. by removing or adding samples), at the feature level (e.g. by adding noise to features), or a combination of both. In the current work, we focus on perturbations at the instance level.

### 2.1 Estimating Stability with Instance Perturbation

To measure the effect of instance perturbation on the feature selection results, we adopt a subsampling based strategy. Consider a dataset $\mathcal{X} = \{x_1, \ldots, x_M\}$ with $M$ instances and $N$ features. Then $k$ subsamples of size $\lceil xM \rceil$ $(0 < x < 1)$ are drawn randomly from $\mathcal{X}$, where the parameters $k$ and $x$ can be varied. Subsequently, feature selection is performed on each of the $k$ subsamples, and a measure of stability or robustness is calculated. Here, following [5], we take a similarity based approach where feature stability is measured by comparing the outputs of the feature selectors on the $k$ subsamples. The more similar all outputs are, the higher the stability measure will be. The overall stability can then be defined as the average over all pairwise similarity comparisons between the different feature selectors:

$$S_{\text{tot}} = \frac{2 \sum_{i=1}^{k} \sum_{j=i+1}^{k} S(\mathbf{f}_i, \mathbf{f}_j)}{k(k-1)}$$

where $\mathbf{f}_i$ represents the outcome of the feature selection method applied to subsample $i$ $(1 \leq i \leq k)$, and $S(\mathbf{f}_i, \mathbf{f}_j)$ represents a similarity measure between $\mathbf{f}_i$ and $\mathbf{f}_j$.

To generalize this approach to all feature selection methods, it has to be noted that not all feature selection techniques report their result in the same form, and we can distinguish between feature weighting, feature ranking and feature subset selection. Evidently, a feature weighting can be converted to a feature ranking by sorting the weights, and a ranking can be converted to a feature subset by choosing an appropriate threshold. For the remainder of the paper, we choose the similarity function $S(.,.)$ to compare only results of the same type. Thus, $\mathbf{f}_i$ can be considered a vector of length $N$, where $\mathbf{f}_i^j$ represents (a) the weight for feature $j$ in the case of comparing feature weightings, (b) the rank of feature $j$ in the case of feature ranking (the worst feature is assigned rank 1, the best one rank $N$) and (c) $\mathbf{f}_i^j = 1$ if the feature is present in the subset, and zero otherwise in the case of feature subset selection.

### 2.2 Similarity Measures

Appropriate similarity measures for feature weighting, ranking and subset selection can be derived from different correlation coefficients. For feature weighting, the Pearson correlation coefficient can be used:

$$S(\mathbf{f}_i, \mathbf{f}_j) = \frac{\sum_l (\mathbf{f}_i^l - \mu_{\mathbf{f}_i})(\mathbf{f}_j^l - \mu_{\mathbf{f}_j})}{\sqrt{\sum_l (\mathbf{f}_i^l - \mu_{\mathbf{f}_i})^2 \sum_l (\mathbf{f}_j^l - \mu_{\mathbf{f}_j})^2}}$$

For feature ranking, the Spearman rank correlation coefficient can be used:

$$S(\mathbf{f}_i, \mathbf{f}_j) = 1 - 6 \sum_l \frac{(\mathbf{f}_i^l - \mathbf{f}_j^l)^2}{N(N^2 - 1)}$$

For feature subsets, we use the Jaccard index:

$$S(\mathbf{f}_i, \mathbf{f}_j) = \frac{|\mathbf{f}_i \cap \mathbf{f}_j|}{|\mathbf{f}_i \cup \mathbf{f}_j|} = \frac{\sum_l I(\mathbf{f}_i^l = \mathbf{f}_j^l = 1))}{\sum_l I(\mathbf{f}_i^l + \mathbf{f}_j^l > 0)}$$

where the indicator function $I(.)$ returns 1 if its argument is true, and zero otherwise.

Finally, it is important to note that robustness of feature selection results should not be considered per se, but always in combination with classification performance, as domain experts are not interested in a strategy that yields very robust feature sets, but returns a badly performing model. Hence, these two aspects need always be investigated together.

## 3 Ensemble Feature Selection Techniques

In ensemble learning, a collection of single classification or regression models is trained, and the output of the ensemble is obtained by aggregating the outputs of the single models, e.g. by majority voting in the case of classification, or averaging in the case of regression. Dietterich [8] shows that the result of the ensemble might outperform the single models when weak (unstable) models are combined, mainly because of three reasons: a) several different but equally optimal hypotheses can exist and the ensemble reduces the risk of choosing a wrong hypothesis, b) learning algorithms may end up in different local optima, and the ensemble may give a better approximation of the true function, and c) the true function cannot be represented by any of the hypotheses in the hypothesis space of the learner and by aggregating the outputs of the single models, the hypothesis space may be expanded.

### 3.1 The Ensemble Idea for Feature Selection

Similar to the case of supervised learning, ensemble techniques might be used to improve the robustness of feature selection techniques. Indeed, in large feature/small sample size domains it is often reported that several different feature subsets may yield equally optimal results [3], and ensemble feature selection may reduce the risk of choosing an unstable subset. Furthermore, different feature selection algorithms may yield feature subsets that can be considered local optima in the space of feature subsets, and ensemble feature selection might give a better approximation to the optimal subset or ranking of features. Finally, the representational power of a particular feature selector might constrain its search space such that optimal subsets cannot be reached. Ensemble feature selection could help in alleviating this problem by aggregating the outputs of several feature selectors.

### 3.2 Components of Ensemble Feature Selection

Similar to the construction of ensemble models for supervised learning, there are two essential steps in creating a feature selection ensemble. The first step involves creating a set of different feature selectors, each providing their output, while the second step aggregates the results of the single models. Variation in the feature selectors can be achieved by various methods: choosing different feature selection techniques, instance level perturbation, feature level perturbation, stochasticity in the feature selector, Bayesian model averaging, or combinations of these techniques [8,9]. Aggregating the different feature selection results can be done by weighted voting, e.g. in the case of deriving a consensus feature ranking, or by counting the most frequently selected features in the case of deriving a consensus feature subset.

In this work, we focus on ensemble feature selection techniques that work by aggregating the feature rankings provided by the single feature selectors into a final consensus ranking. Consider an ensemble $E$ consisting of $s$ feature selectors, $E = \{F_1, F_2, \ldots, F_s\}$, then we assume each $F_i$ provides a feature ranking $\mathbf{f}_i = (f_i^1, \ldots, f_i^N)$, which are aggregated into a consensus feature ranking $\mathbf{f}$ by weighted voting:

$$\mathbf{f}^l = \sum_{i=1}^{s} w(\mathbf{f}_i^l)$$

where $w(.)$ denotes a weighting function. If a *linear aggregation* is performed using $w(\mathbf{f}_i^l) = \mathbf{f}_i^l$, this results in a sum where features contribute in a linear way with respect to their rank. By modifying $w(\mathbf{f}_i^l)$, more or less weight can be put to the rank of each feature. This can be e.g. used to accommodate for rankings where top features can be forced to influence the ranking significantly more than lower ranked features.

# 4 Experiments

In this section, we present the results of our analysis of ensemble feature selection techniques on large feature/small sample size domains. First, the data sets and the feature selection techniques used in this analysis are briefly described. Subsequently, we analyze two aspects of ensemble feature selection techniques: robustness and classification performance. All experiments were run using Java-ML[1], an open source machine learning library.

## 4.1 Data Sets

Datasets were taken from the bioinformatics and biomedical domain, and can be divided into two parts: microarray datasets (MA) and mass spectrometry (MS) datasets (Table 1). For each domain, three datasets were included, typically consisting of several thousands of features and tens of instances in the case of microarray datasets, and up to about 15,000 features and a few hundred of instances in the case of mass spectrometry datasets. Due to their high dimensionality and low sample size, these datasets pose a great challenge for both classification and feature selection algorithms. Another important aspect

[1] Available at `http://java-ml.sourceforge.net`

**Table 1.** Data set characteristics. Sample to dimension rate (SDR) is calculated as $100M/N$.

| | Name | # Class 1 | # Class 2 | # Features | SDR | Reference |
|---|---|---|---|---|---|---|
| MA | Colon | 40 | 22 | 2000 | 3.1 | [15] |
| MA | Leukemia | 47 | 25 | 7129 | 1.0 | [16] |
| MA | Lymphoma | 22 | 23 | 4026 | 1.1 | [17] |
| MS | Ovarian | 162 | 91 | 15154 | 1.7 | [18] |
| MS | Prostate | 69 | 253 | 15154 | 2.1 | [19] |
| MS | Pancreatic | 80 | 101 | 6771 | 2.7 | [20] |

of this data is the fact that the outcome of feature selection techniques is an essential prerequisite for further validation, such as verifying links between particular genes and diseases. Therefore, domain experts require the combination of feature selection and classification algorithm to yield both a high accuracy as well as robustness of the selected features.

### 4.2 Feature Selection Techniques

In this work, we focus on the application of filter and embedded feature selection techniques. We discarded wrapper approaches because they commonly require on the order of $N^2$ classification models being built if a complete ranking of $N$ features is desired. Filter methods require no model being built, and embedded models only build a small amount of models. Thus, the wrapper approach, certainly when used in the ensemble setting is computationally not feasible for the large feature sizes we are dealing with. We choose a benchmark of four feature selection techniques: two filter methods and two embedded methods. For the filter methods, we selected one univariate and one multivariate method. Univariate methods consider each feature separately, while multivariate methods take into account feature dependencies, which might yield better results. The univariate method we choose was the Symmetrical Uncertainty (SU, [10]):

$$SU(F, C) = 2\frac{H(F) - H(F|C)}{H(F) + H(C)}$$

where $F$ and $C$ are random variables representing a feature and the class respectively, and the function $H$ calculates the entropy. As a multivariate method, we choose the RELIEF algorithm [11], which estimates the relevance of features according to how well their values distinguish between the instances of the same and different classes that are near each other. Furthermore, the computational complexity of RELIEF $\mathcal{O}(MN)$ scales well to large feature/small sample size data sets, compared to other multivariate methods which are often quadratic in the number of features. In our experiments, five neighboring instances were chosen. When using real-valued features, equal frequency binning was used to discretize the features.

As embedded methods we used the feature importance measures of Random Forests [12] and linear support vector machines (SVM). In a Random Forest (RF), feature importance is measured by randomly permuting the feature in the out-of-bag samples and calculating the percent increase in misclassification rate as compared to the out-of-bag rate with all variables intact. In our feature selection experiments we used forests consisting of 10 trees.

For a linear SVM, the feature importance can be derived from the weight vector of the hyperplane [13], a procedure known as recursive feature elimination (SVM_RFE). In this work, we use SVM_RFE as a feature ranker: first a linear SVM is trained on the full feature set, and the $C$-parameter is tuned using an internal cross-validation of the training set. Next, features are ranked according to the absolute value of their weight in the weight vector of the hyperplane, and the 10% worst performing features are discarded. The above procedure is then repeated until the empty feature set is reached.

### 4.3 Ensemble Feature Selection Techniques

For each of the four feature selection techniques described above, an ensemble version was created by instance perturbation. We used bootstrap aggregation (bagging, [14]) to generate 40 bags from the data. For each of the bags, a separate feature ranking was performed, and the ensemble was formed by aggregating the single rankings by weighted voting, using linear aggregation.

### 4.4 Robustness of Feature Selection

To assess the robustness of feature selection techniques, we focus here on comparing feature rankings and feature subsets, as these are most often used by domain experts. Feature weightings are almost never used, and instead converted to a ranking or subset. Furthermore, directly comparing feature weights may be problematic as different methods may use different scales and intervals for the weights.

To compare feature rankings, the Spearman rank correlation coefficient was used, while for feature subsets the Jaccard index was used. The last one was analyzed for different subset sizes: the top 1% and top 5% best features of the rankings were chosen.

To estimate the robustness of feature selection techniques, the strategy explained in section 2.1 was used with $k = 10$ subsamples of size $0.9M$ (i.e. each subsample contains 90% of the data). This percentage was chosen because we use small sample datasets and thus cannot discard too much data when building models, and further because we want to assess robustness with respect to relatively small changes in the dataset. Then, each feature selection algorithm (both the single and the ensemble version) was run on each subsample, and the results were averaged over all pairwise comparisons.

Table 2 summarizes the results of the robustness analysis across the different datasets, using the linear aggregation method for ensemble feature selection. For each feature selection algorithm, the Spearman correlation coefficient (Sp)

**Table 2.** Robustness of the different feature selectors across the different datasets. Spearman correlation coefficient, Jaccard index on the subset of 1% and 5% best features are denoted respectively by Sp, JC1 and JC5.

| Dataset | | SU | | Relief | | SVM_RFE | | RF | |
|---|---|---|---|---|---|---|---|---|---|
| | | Single | Ensemble | Single | Ensemble | Single | Ensemble | Single | Ensemble |
| Colon | Sp | 0.61 | 0.76 | 0.62 | 0.85 | 0.7 | 0.81 | 0.91 | 0.99 |
| | JC5 | 0.33 | 0.49 | 0.44 | 0.64 | 0.47 | 0.45 | 0.44 | 0.79 |
| | JC1 | 0.3 | 0.55 | 0.45 | 0.56 | 0.44 | 0.5 | 0.01 | 0.64 |
| Leukemia | Sp | 0.68 | 0.76 | 0.58 | 0.79 | 0.73 | 0.79 | 0.97 | 0.99 |
| | JC5 | 0.48 | 0.57 | 0.39 | 0.54 | 0.53 | 0.58 | 0.8 | 0.91 |
| | JC1 | 0.54 | 0.6 | 0.44 | 0.55 | 0.49 | 0.57 | 0.36 | 0.8 |
| Lymphoma | Sp | 0.59 | 0.74 | 0.49 | 0.76 | 0.77 | 0.81 | 0.96 | 0.99 |
| | JC5 | 0.31 | 0.49 | 0.35 | 0.53 | 0.54 | 0.54 | 0.74 | 0.9 |
| | JC1 | 0.37 | 0.55 | 0.42 | 0.56 | 0.43 | 0.46 | 0.22 | 0.73 |
| Ovarian | Sp | 0.93 | 0.95 | 0.91 | 0.97 | 0.91 | 0.95 | 0.96 | 0.99 |
| | JC5 | 0.76 | 0.79 | 0.66 | 0.78 | 0.75 | 0.79 | 0.7 | 0.93 |
| | JC1 | 0.84 | 0.85 | 0.85 | 0.88 | 0.8 | 0.84 | 0.1 | 0.83 |
| Pancreatic | Sp | 0.57 | 0.65 | 0.46 | 0.73 | 0.69 | 0.77 | 0.9 | 0.99 |
| | JC5 | 0.2 | 0.24 | 0.16 | 0.3 | 0.43 | 0.41 | 0.52 | 0.76 |
| | JC1 | 0.13 | 0.15 | 0.09 | 0.19 | 0.41 | 0.36 | 0.01 | 0.48 |
| Prostate | Sp | 0.88 | 0.91 | 0.9 | 0.96 | 0.81 | 0.92 | 0.96 | 0.99 |
| | JC5 | 0.68 | 0.7 | 0.61 | 0.71 | 0.6 | 0.63 | 0.72 | 0.88 |
| | JC1 | 0.67 | 0.7 | 0.52 | 0.64 | 0.6 | 0.6 | 0.13 | 0.78 |
| **Average** | **Sp** | **0.71** | **0.8** | **0.66** | **0.84** | **0.77** | **0.84** | **0.94** | **0.99** |
| | **JC5** | **0.46** | **0.55** | **0.44** | **0.58** | **0.55** | **0.57** | **0.65** | **0.86** |
| | **JC1** | **0.47** | **0.57** | **0.46** | **0.57** | **0.53** | **0.56** | **0.14** | **0.71** |

and Jaccard index on the subset of 1% (JC1) and 5% best features (JC5) are shown. In general, it can be observed that ensemble feature selection provides more robust results than a single feature selection algorithm, the difference in robustness being dependent on the dataset and the algorithm.

RELIEF is one of the less stable algorithms, but clearly benefits from an ensemble version, as well as the Symmetrical Uncertainty filter method. SVM_RFE on the other hand proves to be a more stable feature selection method, and creating an ensemble version of this method only slightly improves robustness. For Random Forests, the picture is a bit more complicated. While for Sp and JC5, a single Random Forest seems to outperform the other methods, results are much worse on the JC1 measure. This means that the very top performing features vary a lot with regard to different data subsamples. Especially for knowledge discovery, the high variance in the top selected features by Random Forests may be a problem. However, also Random Forests clearly benefit from an ensemble version, the most drastic improvement being made on the JC1 measure. Thus, it seems that ensembles of Random Forests clearly outperform other feature selection methods regarding robustness.

The effect of the number of feature selectors on the robustness of the ensemble is shown in Figure 1. In general, robustness is mostly increased in the first steps, and slows down after about 20 selectors in the ensemble, an exception being the Random Forest. In essence, a single Random Forest can already be seen as an ensemble feature selection technique, averaging over the different trees in the forest, which can explain the earlier convergence of ensembles of Random Forests.

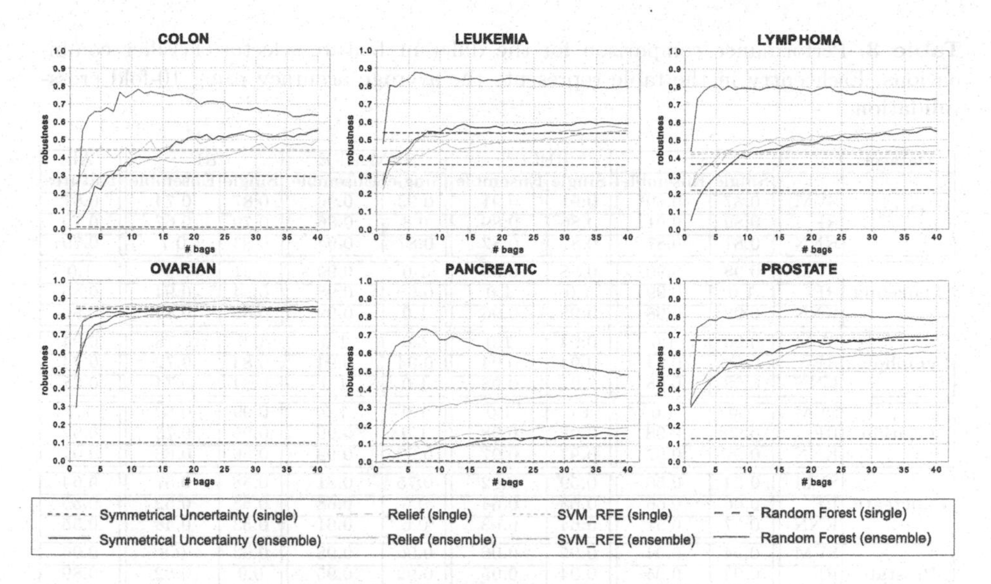


**Fig. 1.** Robustness in function of the ensemble size. Robustness is measured using the Jaccard index on the 1% top ranked features.

One could wonder to what extent an ensemble of Random Forests would be comparable to just one single Random Forest consisting of more trees. Preliminary experiments on the datasets analysed in this work suggest that larger Random Forests often lead to less robust results than smaller ones. Hence, if robust feature selection results are desired, it would be computationally cheaper to average over a number of small Random Forests in an ensemble way, rather than creating one larger Random Forest.

### 4.5 Robustness Versus Classification Performance

Considering only robustness of a feature selection technique is not an appropriate strategy to find good feature rankings or subsets, and also model performance should be taken into account to decide which features to select. Therefore, feature selection needs to be combined with a classification model in order to get an estimate of the performance of the feature selector-classifier combination. Embedded feature selection methods like Random Forests and SVM_RFE have an important computational advantage in this respect, as they combine model construction with feature selection.

To analyze the effect on classification performance using single versus ensemble feature selection, we thus set up a benchmark on the same datasets as used to assess robustness. Due to their capacity to provide a feature ranking, as well as their state-of-the-art performance, Random Forests and linear SVMs were included as classifiers, as well as the distance based k-nearest neighbor algorithm (KNN). The number of trees in the Random Forest classifier was set to 50, and the number of nearest neighbors for KNN was set to 5.

**Table 3.** Performance comparison for the different feature selector-classifier combinations. Each entry in the table represents the average accuracy using 10-fold cross-validation.

| Dataset | | SU | | RELIEF | | SVM_RFE | | RF | | All |
|---|---|---|---|---|---|---|---|---|---|---|
| | | Single | Ensemble | Single | Ensemble | Single | Ensemble | Single | Ensemble | features |
| Colon | SVM | 0.87 | 0.89 | 0.91 | 0.91 | 0.93 | 0.96 | 0.87 | 0.74 | 0.87 |
| | RF | 0.89 | 0.91 | 0.86 | 0.89 | 0.8 | 0.86 | 0.79 | 0.67 | 0.79 |
| | KNN | 0.81 | 0.87 | 0.87 | 0.87 | 0.87 | 0.94 | 0.83 | 0.7 | 0.79 |
| Leukemia | SVM | 0.98 | 0.96 | 0.98 | 0.99 | 1.0 | 0.99 | 0.91 | 0.88 | 1.0 |
| | RF | 1.0 | 0.99 | 1.0 | 1.0 | 0.98 | 0.98 | 0.94 | 0.94 | 0.85 |
| | KNN | 0.99 | 0.98 | 0.98 | 0.96 | 1.0 | 0.99 | 0.88 | 0.83 | 0.86 |
| Lymphoma | SVM | 0.96 | 1.0 | 0.94 | 1.0 | 1.0 | 1.0 | 0.9 | 0.78 | 0.94 |
| | RF | 0.94 | 0.94 | 0.94 | 0.94 | 0.94 | 0.94 | 0.84 | 0.72 | 0.74 |
| | KNN | 0.98 | 0.98 | 0.92 | 0.98 | 1.0 | 1.0 | 0.84 | 0.74 | 0.68 |
| Ovarian | SVM | 1.0 | 1.0 | 1.0 | 1.0 | 1.0 | 1.0 | 0.99 | 0.82 | 1.0 |
| | RF | 0.99 | 0.98 | 0.98 | 0.98 | 1.0 | 0.99 | 0.95 | 0.73 | 0.92 |
| | KNN | 0.97 | 0.97 | 0.97 | 0.97 | 0.99 | 0.99 | 0.96 | 0.66 | 0.92 |
| Pancreatic | SVM | 0.54 | 0.56 | 0.59 | 0.62 | 0.75 | 0.81 | 0.58 | 0.57 | 0.64 |
| | RF | 0.66 | 0.66 | 0.63 | 0.64 | 0.6 | 0.68 | 0.53 | 0.52 | 0.55 |
| | KNN | 0.57 | 0.57 | 0.64 | 0.63 | 0.6 | 0.61 | 0.53 | 0.48 | 0.55 |
| Prostate | SVM | 0.94 | 0.94 | 0.95 | 0.96 | 0.96 | 0.98 | 0.93 | 0.8 | 0.96 |
| | RF | 0.94 | 0.95 | 0.94 | 0.95 | 0.92 | 0.95 | 0.9 | 0.82 | 0.89 |
| | KNN | 0.96 | 0.95 | 0.94 | 0.94 | 0.97 | 0.97 | 0.87 | 0.82 | 0.89 |
| **Average** | **SVM** | **0.88** | **0.89** | **0.90** | **0.91** | **0.94** | **0.96** | **0.86** | **0.77** | **0.90** |
| | **RF** | **0.90** | **0.91** | **0.89** | **0.9** | **0.87** | **0.9** | **0.83** | **0.73** | **0.79** |
| | **KNN** | **0.88** | **0.89** | **0.89** | **0.89** | **0.91** | **0.92** | **0.82** | **0.71** | **0.78** |

For each classifier, we analyzed all combinations with the four feature selection algorithms explained in section 4.4. Classification performance was assessed using a 10-fold cross-validation setting, using accuracy as the performance measure. For each fold, feature selection was performed using only the training part of the data, and a classifier was built using the 1% best features returned by the feature selector, as it was often observed in these domains that only such a small amount of features was relevant [3]. For this experiment, $k = 40$ bootstraps of each training part of the fold were used to create the ensemble versions of the feature selectors. This model was then evaluated on the test part of the data for each fold, and results were averaged over all 10 folds. The results of this experiment are displayed in Table 3.

Averaged over all datasets, we can see that the best classification results are obtained using the SVM classifier, using the ensemble version of RFE as feature selection mechanism. Also for the other classifiers, the combination with the ensemble version of RFE performs well over all datasets. Given that the ensemble version of RFE was also more robust than the single version (Table 2, JC1 rows), this method can thus be used to achieve both robust feature subsets and good classification performance.

In general, it can be observed that the performance of ensemble feature selection techniques is about the same (or slightly better) than the version using a single feature selector, an exception being the Random Forest feature selection technique. Comparing the performance of the Random Forest ensemble feature selection version to the single version, it is clear that the substantial increase in robustness (see Table 2, JC1 rows) comes at a price, and results in lower accuracies for all datasets.

Comparing the results of ensemble feature selection to a classifier using the full feature set (last column in Table 3), it can be observed that in most cases performance is increased, an exception again being the Random Forest feature selector. However, this performance is now obtained at the great advantage of using only 1% of the features. Furthermore, the selected features are robust, greatly improving knowledge discovery and giving more confidence to domain experts, who generally work by iteratively investigating the ranked features in a top-down fashion.

If robustness of the feature selection results is of high importance, then a combined analysis of classification performance and robustness, like the one we presented here, would be advisable. In the case of single and ensemble methods performing equally well, the generally more robust ensemble method can then be chosen to yield both good performance and robustness. In other cases, an appropriate trade-off between robustness and classification performance should be chosen, possibly taking into account the preference of domain experts.

### 4.6 Automatically Balancing Robustness and Classification Performance

In order to provide a formal and automatic way of jointly evaluating the trade-off between robustness and classification performance, we use an adaptation of the F-measure [21]. The F-measure is a well known evaluation performance in data mining, and represents the harmonic mean of precision and recall.

In a similar way, we propose the robustness-performance trade-off (RPT) as being the harmonic mean of the robustness and classification performance.

$$\mathrm{RPT}_\beta = \frac{(\beta^2+1)\ \text{robustness performance}}{\beta^2\ \text{robustness} + \ \text{performance}}$$

The parameter $\beta$ controls the relative importance of robustness versus classification performance, and can be used to either put more influence on robustness or on classification performance. A value of $\beta = 1$ is the standard formulation, treating robustness and classification performance equally important.

Table 4 summarizes the results for the different feature selector-classifier combinations when only 1% of the features is used ($\mathrm{RPT}_1$). For the robustness measure, the Jaccard index was used, while for classification performance the accuracy was used. It can be observed that in almost all cases, the ensemble feature selection version results in a better RPT measure. The best RPT values were obtained using the ensemble version of the Random Forest feature

**Table 4.** Harmonic mean of robustness and classification performance ($RPT_1$) for the different feature selector-classifier combinations using 1% of the features

| Dataset | | SU | | RELIEF | | SVM_RFE | | RF | |
|---|---|---|---|---|---|---|---|---|---|
| | | Single | Ensemble | Single | Ensemble | Single | Ensemble | Single | Ensemble |
| Colon | SVM | 0.45 | 0.68 | 0.6 | 0.69 | 0.6 | 0.66 | 0.02 | 0.69 |
| | RF | 0.45 | 0.69 | 0.59 | 0.69 | 0.57 | 0.63 | 0.02 | 0.65 |
| | KNN | 0.44 | 0.67 | 0.59 | 0.68 | 0.58 | 0.65 | 0.02 | 0.67 |
| Leukemia | SVM | 0.7 | 0.74 | 0.61 | 0.71 | 0.66 | 0.72 | 0.52 | 0.84 |
| | RF | 0.7 | 0.75 | 0.61 | 0.71 | 0.65 | 0.72 | 0.52 | 0.86 |
| | KNN | 0.7 | 0.74 | 0.61 | 0.7 | 0.66 | 0.72 | 0.51 | 0.81 |
| Lymphoma | SVM | 0.53 | 0.71 | 0.58 | 0.72 | 0.6 | 0.63 | 0.35 | 0.75 |
| | RF | 0.53 | 0.69 | 0.58 | 0.7 | 0.6 | 0.62 | 0.35 | 0.72 |
| | KNN | 0.54 | 0.7 | 0.58 | 0.71 | 0.6 | 0.63 | 0.35 | 0.73 |
| Ovarian | SVM | 0.91 | 0.92 | 0.92 | 0.94 | 0.89 | 0.91 | 0.18 | 0.82 |
| | RF | 0.91 | 0.91 | 0.91 | 0.93 | 0.89 | 0.91 | 0.18 | 0.78 |
| | KNN | 0.9 | 0.91 | 0.91 | 0.92 | 0.88 | 0.91 | 0.18 | 0.74 |
| Pancreatic | SVM | 0.21 | 0.24 | 0.16 | 0.29 | 0.53 | 0.5 | 0.02 | 0.52 |
| | RF | 0.22 | 0.24 | 0.16 | 0.29 | 0.49 | 0.47 | 0.02 | 0.5 |
| | KNN | 0.21 | 0.24 | 0.16 | 0.29 | 0.49 | 0.45 | 0.02 | 0.48 |
| Prostate | SVM | 0.78 | 0.8 | 0.67 | 0.77 | 0.74 | 0.74 | 0.23 | 0.79 |
| | RF | 0.78 | 0.81 | 0.67 | 0.76 | 0.73 | 0.74 | 0.23 | 0.8 |
| | KNN | 0.79 | 0.81 | 0.67 | 0.76 | 0.74 | 0.74 | 0.23 | 0.8 |
| **Average** | **SVM** | **0.6** | **0.68** | **0.59** | **0.69** | **0.67** | **0.69** | **0.22** | **0.74** |
| | **RF** | **0.6** | **0.68** | **0.59** | **0.68** | **0.65** | **0.68** | **0.22** | **0.72** |
| | **KNN** | **0.6** | **0.68** | **0.59** | **0.68** | **0.66** | **0.68** | **0.22** | **0.71** |

selector, which can be explained by the very high robustness values (see Table 2), compared to the other feature selectors.

## 5 Conclusions and Future Work

In this work we introduced the use of ensemble methods for feature selection. We showed that by constructing ensemble feature selection techniques, robustness of feature ranking and feature subset selection could be improved, using similar techniques as in ensemble methods for supervised learning. When analyzing robustness versus classification performance, ensemble methods show great promise for large feature/small sample size domains. It turns out that the best trade-off between robustness and classification performance depends on the dataset at hand, giving rise to a new model selection strategy, incorporating both classification performance as well as robustness in the evaluation strategy. We believe that robustness of feature selection techniques will gain importance in the future, and the topic of ensemble feature selection techniques might open many new avenues for further research. Important questions to be addressed include the development of stability measures for feature ranking and feature subset selection, methods for generating diversity in feature selection ensembles, aggregation methods to find a consensus ranking or subset from single feature selection models and the design of classifiers that jointly optimize model performance and feature robustness.

## References

1. Guyon, I., Elisseeff, A.: An Introduction to Variable and Feature Selection. Journal of Machine Learning Research 3, 1157–1182 (2003)
2. Kohavi, R., John, G.: Wrappers for feature subset selection. Artif. Intell. 97(1-2), 273–324 (1997)
3. Saeys, Y., Inza, I., Larrañaga, P.: A review of feature selection techniques in bioinformatics. Bioinformatics 23(19), 2507–2517 (2007)
4. Dunne, K., Cunningham, P., Azuaje, F.: Solutions to instability problems with sequential wrapper-based approaches to feature selection. Technical report TCD-2002-28. Dept. of Computer Science, Trinity College, Dublin, Ireland (2002)
5. Kalousis, A., Prados, J., Hilario, M.: Stability of feature selection algorithms: a study on high-dimensional spaces. Knowl. Inf. Syst. 12(1), 95–116 (2007)
6. Kuncheva, L.: A stability index for feature selection. In: Proceedings of the 25th International Multi-Conference on Artificial Intelligence and Applications, pp. 390–395 (2007)
7. Krízek, P., Kittler, J., Hlavác, V.: Improving Stability of Feature Selection Methods. In: Proceedings of the 12th International Conference on Computer Analysis of Images and Patterns, pp. 929–936 (2007)
8. Dietterich, T.: Ensemble methods in machine learning. In: Proceedings of the 1st International Workshop on Multiple Classifier Systems, pp. 1–15 (2000)
9. Hoeting, J., Madigan, D., Raftery, A., Volinsky, C.: Bayesian model averaging. Statistical Science 14, 382–401 (1999)
10. Press, W.H., Flannery, B.P., Teukolsky, S.A., Vetterling, W.T.: Numerical Recipes in C (1988)
11. Kononenko, I.: Estimating Attributes: Analysis and Extensions of RELIEF. In: Proceedings of the 7th European Conference on Machine Learning, pp. 171–182 (1994)
12. Breiman, L.: Random Forests. Machine Learning 45(1), 5–32 (2001)
13. Guyon, I., Weston, J., Barnhill, S., Vapnik, V.: Gene Selection for Cancer Classification using Support Vector Machines. Machine Learning 46(1-3), 389–422 (2002)
14. Breiman, L.: Bagging Predictors: Machine Learning 24(2), 123–140 (1996)
15. Alon, U., Barkai, N., Notterman, D.A., Gish, K., Ybarra, S., Mack, D., Levine, A.J.: Broad patterns of gene expression revealed by clustering of tumor and normal colon tissues probed by oligonucleotide arrays. Proc. Natl. Acad. Sci. USA 96(12), 6745–6750 (1999)
16. Golub, T.R., Slonim, D.K., Tamayo, P., Huard, C., Gaasenbeek, M., Mesirov, J.P., Coller, H.: Molecular Classification of Cancer: Class Discovery and Class Prediction by Gene Expression Monitoring. Science 286, 531–537 (1999)
17. Alizadeh, A.A., Eisen, M.B., Davis, R.E., Ma, C., Lossos, I.S., Rosenwald, A., Boldrick, J.C., Sabet, H.: Distinct types of diffuse large B-cell lymphoma identified by gene expression profiling. Nature 403(3), 503–511 (2000)
18. Petricoin, E.F., Ardekani, A.M., Hitt, B.A., Levine, P.J., Fusaro, V.A., Steinberg, S.M., Mills, G.B.: Use of proteomics patterns in serum to identify ovarian cancer. The Lancet 359(9306), 572–577 (2002)
19. Petricoin, E.F., Ornstein, D.K., Paweletz, C.P., Ardekani, A., Hackett, P.S., Hitt, B.A., Velassco, A., Trucco, C.: Serum proteomic patterns for detection of prostate cancer. J. Natl. Cancer Inst. 94(20), 1576–1578 (2002)
20. Hingorani, S.R., Petricoin, E.F., Maitra, A., Rajapakse, V., King, C., Jacobetz, M.A., Ross, S.: Preinvasive and invasive ductal pancreatic cancer and its early detection in the mouse. Cancer Cell. 4(6), 437–450 (2003)
21. van Rijsbergen, C.J.: Information Retrieval, 2nd edn. Butterworths, London (1979)

# Effective Visualization of Information Diffusion Process over Complex Networks

Kazumi Saito[1], Masahiro Kimura[2], and Hiroshi Motoda[3]

[1] School of Administration and Informatics, University of Shizuoka
52-1 Yada, Suruga-ku, Shizuoka 422-8526, Japan
k-saito@u-shizuoka-ken.ac.jp

[2] Department of Electronics and Informatics, Ryukoku University
Otsu, Shiga 520-2194, Japan
kimura@rins.ryukoku.ac.jp

[3] Institute of Scientific and Industrial Research, Osaka University
8-1 Mihogaoka, Ibaraki, Osaka 567-0047, Japan
motoda@ar.sanken.osaka-u.ac.jp

**Abstract.** Effective visualization is vital for understanding a complex network, in particular its dynamical aspect such as information diffusion process. Existing node embedding methods are all based solely on the network topology and sometimes produce counter-intuitive visualization. A new node embedding method based on conditional probability is proposed that explicitly addresses diffusion process using either the IC or LT models as a cross-entropy minimization problem, together with two label assignment strategies that can be simultaneously adopted. Numerical experiments were performed on two large real networks, one represented by a directed graph and the other by an undirected graph. The results clearly demonstrate the advantage of the proposed methods over conventional spring model and topology-based cross-entropy methods, especially for the case of directed networks.

## 1 Introduction

Analysis of the structure and function of complex networks, such as social, computer and biochemical networks, has been a hot research subject with considerable attention [10]. A network can play an important role as a medium for the spread of various information. For example, innovation, hot topics and even malicious rumors can propagate through social networks among individuals, and computer viruses can diffuse through email networks. Previous work addressed the problem of tracking the propagation patterns of topics through network spaces [1, 5], and studied effective "vaccination" strategies for preventing the spread of computer viruses through networks [2, 11]. Widely-used fundamental probabilistic models of information diffusion through networks are the *independent cascade (IC) model* and the *linear threshold (LT) model* [5, 8]. Researchers have recently investigated the problem of finding a limited number of influential nodes that are effective for the spread of information through a network under these models [8, 9]. In these studies, understanding the flow of information through networks is an important research issue.

W. Daelemans et al. (Eds.): ECML PKDD 2008, Part II, LNAI 5212, pp. 326–341, 2008.

This paper focuses on the problem of visualizing the information diffusion process, which is vital for understanding its characteristic over a complex network. Existing node embedding methods such as spring model method [7] and cross entropy method [14] are solely based on the network topology. They do not take account how information diffuses across the network. Thus, it often happens that the visualized information flow do not match our intuitive understanding, *e.g.*, abrupt information flow gaps, inconsistency between the nodes distance and the reachability of information, irregular pattern of information spread, etc. This sometimes happens when visualizing the diffusion process for a network represented by a directed graph.

Thus, it is important that node embedding explicitly reflects the diffusion process to produce more natural visualization. We have devised a new node embedding method that incorporates conditional probability of information diffusion between two nodes, a target source node where the information is initially issued and a non-target influenced node where the information has been received via intermediate nodes. Our postulation is that good visualization should satisfy the two conditions: path continuity, *i.e.* any information diffusion path is continuous and path separability, *i.e.* each different information diffusion path is clearly separated from each other. To this end, the above node embedding is coupled with two label assignment strategies, one with emphasis on influence of initially activated nodes, and the other on degree of information reachability.

Extensive numerical experiments were performed on two large real networks, one generated from a large connected trackback network of blog data, resulting in a directed graph of $12,047$ nodes and $53,315$ links, and the other, a network of people, generated from a list of people within a Japanese Wikipedia, resulting in an undirected graph of $9,481$ nodes and $245,044$ links. The results clearly indicate that the proposed probabilistic visualization method satisfies the above two conditions and demonstrate its advantage over the well-known conventional methods: spring model and topology-based cross-entropy methods, especially for the case of a directed network. The method appeals well to our intuitive understanding of information diffusion process.

## 2 Information Diffusion Models

We mathematically model the spread of information through a directed network $G = (V, E)$ under the IC or LT model, where $V$ and $E$ $(\subset V \times V)$ stands for the sets of all the nodes and links, respectively. We call nodes *active* if they have been influenced with the information. In these models, the diffusion process unfolds in discrete time-steps $t \geq 0$, and it is assumed that nodes can switch their states only from inactive to active, but not from active to inactive. Given an initial set $S$ of active nodes, we assume that the nodes in $S$ have first become active at time-step 0, and all the other nodes are inactive at time-step 0.

### 2.1 Independent Cascade Model

We define the IC model. In this model, for each directed link $(u, v)$, we specify a real value $\beta_{u,v}$ with $0 < \beta_{u,v} < 1$ in advance. Here $\beta_{u,v}$ is referred to as the *propagation probability* through link $(u, v)$. The diffusion process proceeds from a given initial active set $S$ in the following way. When a node $u$ first becomes active at time-step $t$, it is given a single chance to activate each currently inactive child node $v$, and succeeds with probability $\beta_{u,v}$. If $u$ succeeds, then $v$ will become active at time-step $t+1$. If multiple parent nodes of $v$ first become active at time-step $t$, then their activation attempts are sequenced in an arbitrary order, but all performed at time-step $t$. Whether or not $u$ succeeds, it cannot make any further attempts to activate $v$ in subsequent rounds. The process terminates if no more activations are possible.

For an initial active set $S$, let $\varphi(S)$ denote the number of active nodes at the end of the random process for the IC model. Note that $\varphi(S)$ is a random variable. Let $\sigma(S)$ denote the expected value of $\varphi(S)$. We call $\sigma(S)$ the *influence degree* of $S$.

### 2.2 Linear Threshold Model

We define the LT model. In this model, for every node $v \in V$, we specify, in advance, a *weight* $\omega_{u,v}$ ($> 0$) from its parent node $u$ such that

$$\sum_{u \in \Gamma(v)} \omega_{u,v} \leq 1,$$

where $\Gamma(v) = \{u \in V; (u,v) \in E\}$. The diffusion process from a given initial active set $S$ proceeds according to the following randomized rule. First, for any node $v \in V$, a *threshold* $\theta_v$ is chosen uniformly at random from the interval $[0,1]$. At time-step $t$, an inactive node $v$ is influenced by each of its active parent nodes, $u$, according to weight $\omega_{u,v}$. If the total weight from active parent nodes of $v$ is at least threshold $\theta_v$, that is,

$$\sum_{u \in \Gamma_t(v)} \omega_{u,v} \geq \theta_v,$$

then $v$ will become active at time-step $t+1$. Here, $\Gamma_t(v)$ stands for the set of all the parent nodes of $v$ that are active at time-step $t$. The process terminates if no more activations are possible.

The LT model is also a probabilistic model associated with the uniform distribution on $[0,1]^{|V|}$. Similarly to the IC model, we define a random variable $\varphi(S)$ and its expected value $\sigma(S)$ for the LT model.

### 2.3 Influence Maximization Problem

Let $K$ be a given positive integer with $K < |V|$. We consider the problem of finding a set of $K$ nodes to target for initial activation such that it yields the largest expected spread of information through network $G$ under the IC or LT model. The problem is referred to as the *influence maximization problem*, and mathematically defined as follows: Find a subset $S^*$ of $V$ with $|S^*| = K$ such that $\sigma(S^*) \geq \sigma(S)$ for every $S \subset V$ with $|S| = K$.

For a large network, any straightforward method for exactly solving the influence maximization problem suffers from combinatorial explosion. Therefore, we approximately solve this problem. Here, $U_K = \{u_1, \cdots, u_K\}$ is the set of $K$ nodes to target for initial activation, and represents the approximate solution obtained by this algorithm. We refer to $U_K$ as the *greedy solution*.

Using large collaboration networks, Kempe et al. [8] experimentally demonstrated that the greedy algorithm significantly outperforms node-selection heuristics that rely on the well-studied notions of degree centrality and distance centrality in the sociology literature. Moreover, the quality of $U_K$ is guaranteed:

$$\sigma(U_K) \geq \left(1 - \frac{1}{e}\right)\sigma(S_K^*),$$

where $S_K^*$ stands for the exact solution to this problem.

To implement the greedy algorithm, we need a method for calculating $\{\sigma(U_k \cup \{v\}); v \in V \setminus U_k\}$ for $1 \leq k \leq K$. However, it is an open question to exactly calculate influence degrees by an efficient method for the IC or LT model [8]. Kimura et al. [9] presented the bond percolation method that efficiently estimates influence degrees $\{\sigma(U_k \cup \{v\}); v \in V \setminus U_k\}$. Therefore, we estimate the greedy solution $U_K$ using their method.

# 3 Visualization Method

We especially focus on visualizing the information diffusion process from the target nodes selected to be a solution of the influence maximization problem. To this end, we propose a visualization method that has the following characteristics: 1) utilizing the target nodes as a set of pivot objects for visualization, 2) applying a probabilistic algorithm for embedding all the nodes in the networks into an Euclidean space, and 3) varying appearance of the embedded nodes on the basis of two label assignment strategies. In what follows, we describe some details of the probabilistic embedding algorithm and the label assignment strategies.

## 3.1 Probabilistic Embedding Algorithm

Let $U_K = \{u_k : 1 \leq k \leq K\} \subset V$ be a set of target nodes, which maximizes an expected number of influenced nodes in the network based on an information diffusion model such as IC or LT. Let $v_n \notin U_K$ be a non-target node in the network, then we can consider the conditional probability $p_{k,n} = p(v_n|u_k)$ that a node $v_n$ is influenced when one target node $u_k$ alone is set to an initial information source. Here note that we can regard $p_{k,n}$ as a binomial probability with respect to a pair of nodes $u_k$ and $v_n$. In our visualization approach, we attempt to produce embedding of the nodes so as to preserve the relationships expressed as the conditional probabilities for all pairs of target and non-target nodes in the network. We refer to this visualization strategy as the *conditional probability embedding (CE) algorithm.*

**Objective Function.** Let $\{\mathbf{x}_k : 1 \leq k \leq K\}$ and $\{\mathbf{y}_n : 1 \leq n \leq N\}$ be the embedding positions of the corresponding $K$ target nodes and $N = |V| - K$ non-target nodes in an $M$ dimensional Euclidean space. Hereafter, the $\mathbf{x}_k$ and $\mathbf{y}_n$ are called target and non-target vectors, respectively. As usual, we define the Euclidean distance between $\mathbf{x}_k$ and $\mathbf{y}_n$ as follows:

$$d_{k,n} = \|\mathbf{x}_k - \mathbf{y}_n\|^2 = \sum_{m=1}^{M}(x_{k,m} - y_{n,m})^2.$$

Here, we introduce a monotonic decreasing function $\rho(s) \in [0, 1]$ with respect to $s \geq 0$, where $\rho(0) = 1$ and $\rho(\infty) = 0$.

Since $\rho(d_{k,n})$ can also be regarded as a binomial probability with respect to $\mathbf{x}_k$ and $\mathbf{y}_n$, we can introduce a cross-entropy (cost) function between $p_{k,n}$ and $\rho(d_{k,n})$ as follows:

$$\mathcal{E}_{k,n} = -p_{k,n} \ln \rho(d_{k,n}) - (1 - p_{k,n}) \ln(1 - \rho(d_{k,n})).$$

Since $\mathcal{E}_{k,n}$ is minimized when $\rho(d_{k,n}) = p_{k,n}$, this minimization with respect to $\mathbf{x}_k$ and $\mathbf{y}_n$ is consistent with our problem setting. In this paper, we employ a function of the form

$$\rho(s) = \exp\left(-\frac{s}{2}\right)$$

as the monotonic decreasing function, but note that our approach is not restricted to this form. Then, the total cost function (objective function) can be defined as follows:

$$\mathcal{E} = \frac{1}{2} \sum_{n=1}^{N} \sum_{k=1}^{K} p_{k,n} d_{k,n} - \sum_{n=1}^{N} \sum_{k=1}^{K} (1 - p_{k,n}) \ln(1 - \rho(d_{k,n})). \tag{1}$$

Namely, our approach is formalized as a minimization problem of the objective function defined in (1) with respect to $\{\mathbf{x}_k : 1 \leq k \leq K\}$ and $\{\mathbf{y}_n : 1 \leq n \leq N\}$.

**Learning Algorithm.** As the basic structure of our learning algorithms, we adopt a coordinate strategy just like the *EM* (Expectation-Maximization) algorithm. First, we adjust the target vectors, so as to minimize the objective function by freezing the non-target vectors, and then, we adjust the non-target vectors by freezing the target vectors. These two steps are repeated until convergence is obtained.

In the former minimization step for the *CE* algorithm, we need to calculate the derivative of the objective function with respect to $\mathbf{x}_n$ as follows:

$$\mathcal{E}_{\mathbf{x}_k} = \frac{\partial \mathcal{E}}{\partial \mathbf{x}_k} = \sum_{n=1}^{N} \frac{p_{k,n} - \rho(d_{k,n})}{1 - \rho(d_{k,n})} (\mathbf{x}_k - \mathbf{y}_n). \tag{2}$$

Since $\mathbf{x}_{k'}$ $(k' \neq k)$ disappears in (2), we can update $\mathbf{x}_k$ without considering the other target vectors. In the latter minimization step for the *CE* algorithm, we need to calculate the following derivative,

$$\mathcal{E}_{\mathbf{y}_n} = \frac{\partial \mathcal{E}}{\partial \mathbf{y}_n} = \sum_{k=1}^{K} \frac{p_{k,n} - \rho(d_{k,n})}{1 - \rho(d_{k,n})} (\mathbf{y}_n - \mathbf{x}_k).$$

In this case, we update $\mathbf{y}_n$ by freezing the other non-target vectors. Overall, our algorithm can be summarized as follows:

**1.** Initialize vectors $\mathbf{x}_1, \cdots, \mathbf{x}_K$ and $\mathbf{y}_1, \cdots, \mathbf{y}_N$.
**2.** Calculate gradient vectors $\mathcal{E}_{\mathbf{x}_1}, \cdots, \mathcal{E}_{\mathbf{x}_K}$.
**3.** Update target vectors $\mathbf{x}_1, \cdots, \mathbf{x}_K$.
**4.** Calculate gradient vectors $\mathcal{E}_{\mathbf{y}_1}, \cdots, \mathcal{E}_{\mathbf{y}_N}$.
**5.** Update non-target vectors $\mathbf{y}_1, \cdots, \mathbf{y}_N$.
**6.** Stop if $\max_{k,n}\{\|\mathcal{E}_{\mathbf{x}_k}\|, \|\mathcal{E}_{\mathbf{y}_n}\|\} < \epsilon$.
**7.** Return to **2.**

Here, a small positive value $\epsilon$ controls the termination condition.

### 3.2 Label Assignment Strategies

In an attempt to effectively understand information diffusion process, we propose two label assignment strategies, on which the appearance of the embedded target and non-target nodes depends. The first strategy assigns labels to non-target nodes according to the standard Bayes decision rule.

$$l_1(v_n) = \arg \max_{1 \le k \le K} \{p_{k,n}\}$$

It is obvious that this decision naturally reflects influence of the target nodes. Note that the target node identification number $k$ corresponds to the order determined by the greedy method, *i.e.*, $l_1(u_k) = k$.

In the second strategy, we introduce the following probability quantization by noting $0 \le \max_{1 \le k \le K}\{p_{k,n}\} \le 1$,

$$l_2(v_n) = \left[ -\log_b \max_{1 \le k \le K} \{p_{k,n}\} \right] + 1,$$

where $[x]$ returns the greatest integer not greater than $x$, and $b$ stands for the base of logarithm. To each node belonging to $Z = \{v_n : \max_{1 \le k \le K}\{p_{k,n}\} = 0\}$, we assign as the label the maximum number determined by the nodes not belonging to $Z$. We believe that this quantization reasonably reflects the degree of information reachability. Here note that $l_2(u_k) = 1$ because it always becomes active at time step $t = 0$. These labels are further mapped to colors scales according to some monotonic mapping functions.

## 4 Experimental Evaluation

### 4.1 Network Data

In our experiments, we employed two sets of real networks used in [9], which exhibit many of the key features of social networks. We describe the details of these network data.

The first one is a trackback network of blogs. Blogs are personal on-line diaries managed by easy-to-use software packages, and have rapidly spread through the World Wide Web [5]. Bloggers (*i.e.*, blog authors) discuss various topics by using trackbacks. Thus, a piece of information can propagate from one blogger to another blogger through a trackback. We exploited the blog "Theme salon of blogs" in the site "goo" [2], where a blogger can recruit trackbacks of other bloggers by registering an interesting theme. By tracing up to ten steps back in the trackbacks from the blog of the theme "JR Fukuchiyama Line Derailment Collision", we collected a large connected trackback network in May, 2005. The resulting network had 12,047 nodes and 53,315 directed links, which features the so-called "power-law" distributions for the out-degree and in-degree that most real large networks exhibit. We refer to this network data as the blog network.

[2] `http://blog.goo.ne.jp/usertheme/`

The second one is a network of people that was derived from the "list of people" within Japanese Wikipedia. Specifically, we extracted the maximal connected component of the undirected graph obtained by linking two people in the "list of people" if they co-occur in six or more Wikipedia pages. The undirected graph is represented by an equivalent directed graph by regarding undirected links as bidirectional ones[3]. The resulting network had 9,481 nodes and 245,044 directed links. We refer to this network data as the Wikipedia network.

Newman and Park [12] observed that social networks represented as undirected graphs generally have the following two statistical properties that are different from non-social networks. First, they show positive correlations between the degrees of adjacent nodes. Second, they have much higher values of the *clustering coefficient* $C$ than the corresponding *configuration models* (*i.e.*, random network models). For the undirected graph of the Wikipedia network, the value of $C$ of the corresponding configuration model was 0.046, while the actual measured value of $C$ was 0.39, and the degrees of adjacent nodes were positively correlated. Therefore, the Wikipedia network has the key features of social networks.

## 4.2 Experimental Settings

In the IC model, we assigned a uniform probability $\beta$ to the propagation probability $\beta_{u,v}$ for any directed link $(u, v)$ of a network, that is, $\beta_{u,v} = \beta$. We, first, determine the typical value of $\beta$ for the blog network, and use it in the experiments. It is known that the IC model is equivalent to the bond percolation process that independently declares every link of the network to be "occupied" with probability $\beta$ [10]. Let $J$ denote the expected fraction of the maximal strongly connected component (SCC) in the network constructed by the occupied links. Note that $J$ is an increasing function of $\beta$. We focus on the point $\beta_*$ at which the average rate of change of $J$, $dJ/d\beta$, attains the maximum, and regard it as the typical value of $\beta$ for the network. Note that $\beta_*$ is a critical point of $dJ/d\beta$, and defines one of the features intrinsic to the network. Figure 1 plots $J$ as a function of $\beta$. Here, we estimated $J$ using the bond percolation method with the same parameter value as below [9]. From this figure we experimentally estimated $\beta_*$ to be 0.2 for the blog network. In the same way, we experimentally estimated $\beta_*$ to be 0.05 for the Wikipedia network.

In the LT model, we uniformly set weights as follows. For any node $v$ of a network, the weight $\omega_{u,v}$ from a parent node $u \in \Gamma(v)$ is given by $\omega_{u,v} = 1/|\Gamma(v)|$.

Once these parameters were set, we estimated the greedy solution $U_K = \{u_1, \cdots, u_K\}$ of targets and the conditional probabilities $\{p_{k,n};\ 1 \leq k \leq K,\ 1 \leq n \leq N\}$ using the bond percolation method with the parameter value 10,000 [9]. Here, the parameter represents the number of bond percolation processes for estimating the influence degree $\sigma(S)$ of a given initial active set $S$.

## 4.3 Brief Description of Other Visualization Methods Used for Comparison

We have compared the proposed method with the two well known methods: spring model method [7] and standard cross-entropy method [14].

[3] For simplicity, we call a graph with bi-directional links an undirected graph.

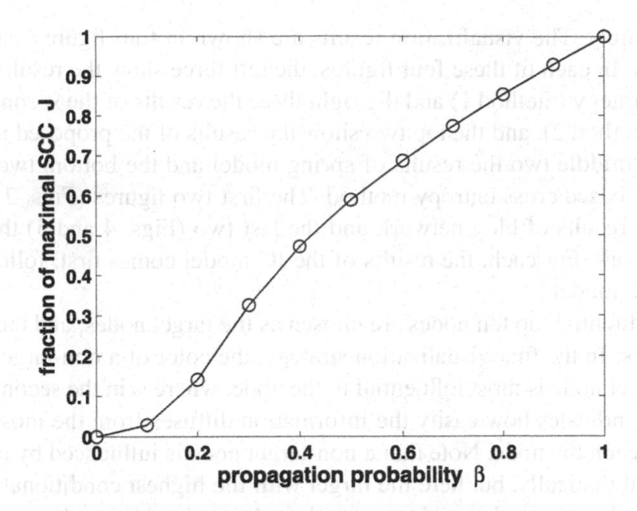


**Fig. 1.** The fraction $J$ of the maximal SCC as a function of the propagation probability $\beta$

Spring model method assumes that there is a hypothetical spring between each connected node pair and locates nodes such that the distance of each node pair is closest to its minimum path length at equilibrium. Mathematically it is formulated as minimizing (3).

$$\mathcal{K}(\mathbf{x}) = \sum_{u=1}^{|V|-1} \sum_{v=u+1}^{|V|} \alpha_{u,v}(g_{u,v} - \|\mathbf{x}_u - \mathbf{x}_v\|)^2, \quad (3)$$

where $g_{u,v}$ is the minimum path length between node $u$ and node $v$, and $\alpha_{u,v}$ is a spring constant which is normally set to $1/(2g_{u,v}^2)$. Standard cross-entropy method first defines a similarity $\rho(\|\mathbf{x}_u - \mathbf{x}_v\|^2) = \exp(-\|\mathbf{x}_u - \mathbf{x}_v\|^2/2)$ between the embedding coordinates $x_u$ and $x_v$ and uses the corresponding element $a_{u,v}$ of the adjacency matrix as a measure of distance between the node pair, and tries to minimize the total cross entropy between these two. Mathematically it is formulated as minimizing (4).

$$C(\mathbf{x}) = \sum_{u=1}^{|V|-1} \sum_{v=u+1}^{|V|} \left\{-a_{u,v} \log \rho(\|\mathbf{x}_u - \mathbf{x}_v\|^2) - (1 - a_{u,v}) \log(1 - \rho(\|\mathbf{x}_u - \mathbf{x}_v\|^2))\right\}, \quad (4)$$

Here, note that we used the same function $\rho$ as before.

As is clear from the above formulation, both methods are completely based on graph topology. They are both non-linear optimization problem and easily solved by a standard coordinate descent method. Here note that the applicability of the spring model method and cross-entropy method is basically limitted to undirected networks. Thus, in order to obtain the embedding results by using these methods we neglected the direction in the directed blog network and regarded it as undirected one.

### 4.4 Experimental Results

Two label assignment strategies are independent to each other. They can be used either separately or simultaneously. Here, we used a color mapping to both, and thus, use them separately. The visualization results are shown in four figures, each with six network figures. In each of these four figures, the left three show the results of the first visualization strategy (method 1) and the right three the results of the second visualization strategy (method 2), and the top two show the results of the proposed method (*CE* algorithm), the middle two the results of spring model and the bottom two the results of the topology-based cross entropy method. The first two figures (Figs. 2 and 3) corresponds to the results of blog network and the last two (Figs. 4 and 5) the results of Wikipedia network. For each, the results of the IC model comes first, followed by the results of the LT model.

The most influential top ten nodes are chosen as the target nodes, and the rest are all non-target nodes. In the first visualization strategy, the color of a non-target node indicates which target node is most influential to the node, whereas in the second visualization strategy, it indicates how easily the information diffuses from the most influential target node to reach the node. Note that a non-target node is influenced by multiple target nodes probabilistically, but here the target with the highest conditional probability is chosen. Thus, the most influential target node is determined for each non-target node.

Observation of the results of the proposed method (Figs. 2a, 2b, 3a, 3b, 4a, 4b, 5a, and 5b) indicates that the proposed method has the following desirable characteristics: 1) the target nodes tend to be allocated separately from each other, and from each target node, 2) the non-target nodes that are most affected by the same target node are laid out forming a band and 3) the reachability changes continuously from the highest at the target node to the lowest at the other end of the band. From this observation, it is confirmed that the two conditions we postulated are satisfied for the both diffusion models. Observation 2) above, however, needs further clarification. Note that our visualization does not necessarily cause the information diffusion to neighboring nodes to be in the form of a line in the embedded space. For example, if there is only one source (K=1), the information would diffuse concentrically. A node in general receives information from multiple sources. The fact that the embedding result forms a line, on the contrary, reveals an important characteristic that little information is coming from the other sources for the networks we analyzed.

In the proposed method, non-target nodes that are readily influenced are easily identified, whereas those that are rarely influenced are placed together. Overlapping of the color well explains the relationship between each target and a non-target node. For example, in Figures 3a and 3b it is easily observed that the effect of the target nodes 5, 2 on non-target nodes interferes with the three bands that are spread from the target nodes 8, 3, 10, and non-target nodes overlap as they move away from the target nodes, demonstrating that a simple two-dimensional visualization facilitates how different node groups overlap and how the information flows from different target nodes interfere each other. The same observation applies for the target nodes 6, 1, 9, 7. On the contrary, the target node 4 has its own effect separately. A similar argument is possible for relationship within target nodes. For example, in Figures 2a target nodes 4, 5, 6, 8 are located in relatively near positions compared with the other target nodes. It is crucial to abstract and visualize the essence of information diffusion by deleting the unnecessary details (node to node diffusion). A good explanation for the overlap like the above is not possible by other visualization methods. Further, the visualization results of both IC and LT models are spatially well balanced. In addition, there are no significant differences on the results of visualization between the directed network and undirected network. Both are equally good.

Observation of the results of the spring model (Figs. 2c, 2d, 3c, 3d, 4c, 4d, 5c, and 5d) and the topology-based cross entropy method (Figs. 2e, 2f, 3e, 3f, 4e, 4f, 5e, and 5f) reveals the followings. The clear difference of these from the proposed method is that it is not that easy to locate the target nodes. This is true, in particular, for the spring model. It is slightly easier for the standard cross-entropy method because the target nodes are placed in the cluster centers, but clusters often overlap, which makes visualization less understandable. It is also noted that those nodes with high reachability, *i.e.*, nodes with red, which should be placed separately due to the influence of different target nodes are placed in mixture. Further, unlike the proposed method, there is clear difference between the IC model and the LT model. In the IC model, we can easily recognize non-target nodes with high reachability, which cover a large portion of the network, whereas in the LT model, such nodes covering only a small portion are almost invisible in the network. In contrast, we can easily pick up such non-target nodes with high reachability even for the LT model in the proposed method.

We observe that the standard cross-entropy method is in general better than the spring model method in terms of the clarity of separability. The standard cross-entropy method does better for the IC model than for the LT model, and is comparable to the proposed method in terms of the clarity of reachability. However, the results of the standard cross-entropy method (e.g., Fig. 2f) are unintuitive, where the high reachability non-target nodes are placed away from the target nodes, and some target node forms several isolated clusters. We believe that this point is an intrinsic limitation of the standard cross-entropy method.

The concept of our visualization is based on the notion that how the information diffuses should primarily determine how the visualization is made, irrespective of the graph topology. We observe that the visualization which is based solely on the topology has intrinsic limitation when we deal with a huge network from the point of both computational complexity (*e.g.*, the spring model does not work for a network with millions nodes) and understandability. Overall, we can conclude that the proposed method provides better visualization which is more intuitive and easily understandable.

# 5 Related Work and Discussion

As defined earlier, let $K$ and $N$ be the numbers of target and non-target nodes in a network. Then the computational complexity of our embedding method amounts to $O(NK)$, where we assume the number of learning iterations and the embedding dimension to be constants. This reduced complexity greatly expands the applicability of our method over the other representative network embedding methods, *e.g.*, the spring model method [7] and the standard cross-entropy method [14], both of which require the computational complexity of $O(N^2)$ under the setting that $K \ll N$.

In view of computational complexity, our visualization method is closely related to those conventional methods, such as FastMap or Landmark Multidimensional Scaling (LMDS), which are based on the Nyström approximation [13]. Typically, these

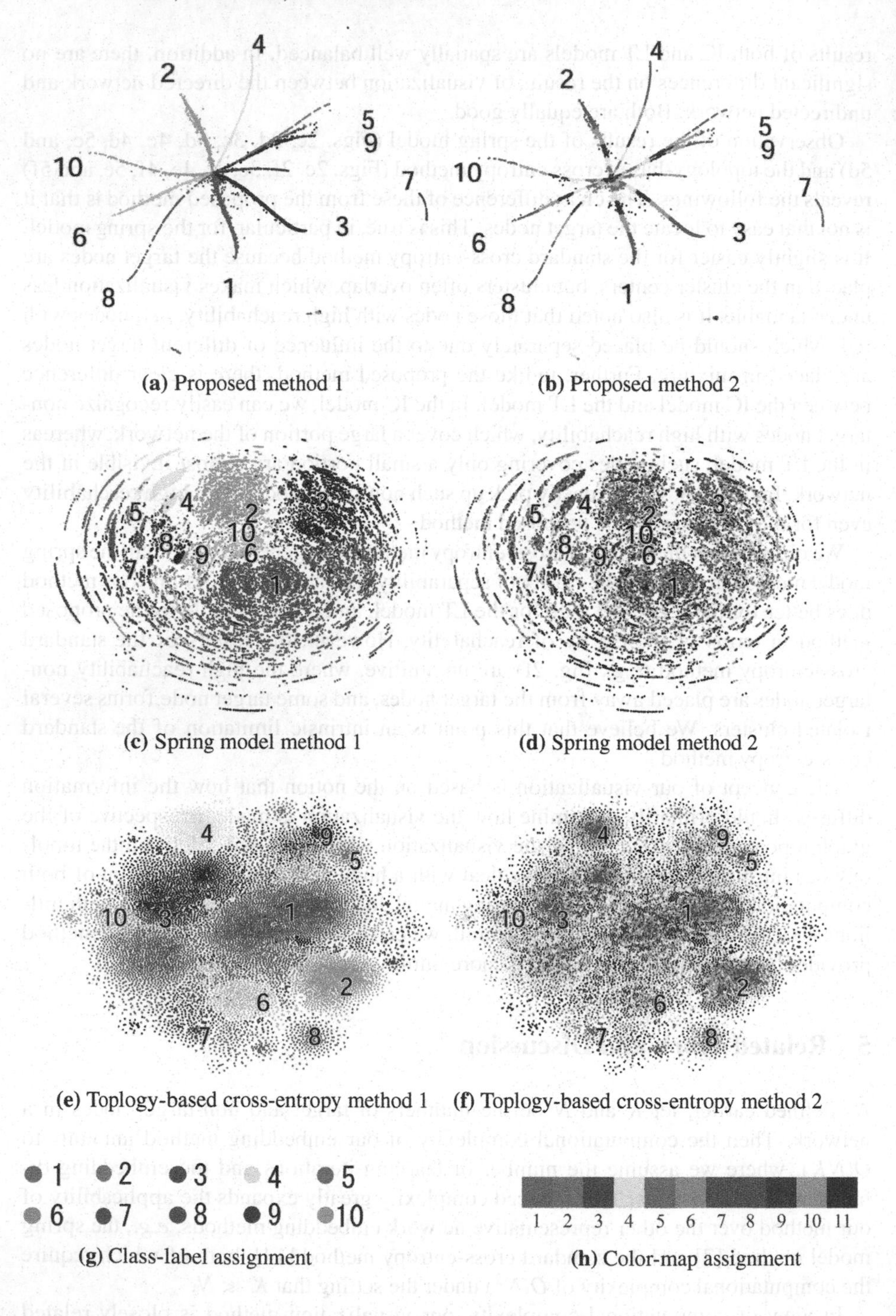


(a) Proposed method 1 (b) Proposed method 2

(c) Spring model method 1 (d) Spring model method 2

(e) Toplogy-based cross-entropy method 1 (f) Toplogy-based cross-entropy method 2

(g) Class-label assignment (h) Color-map assignment

**Fig. 2.** Visualization of IC model for blog network

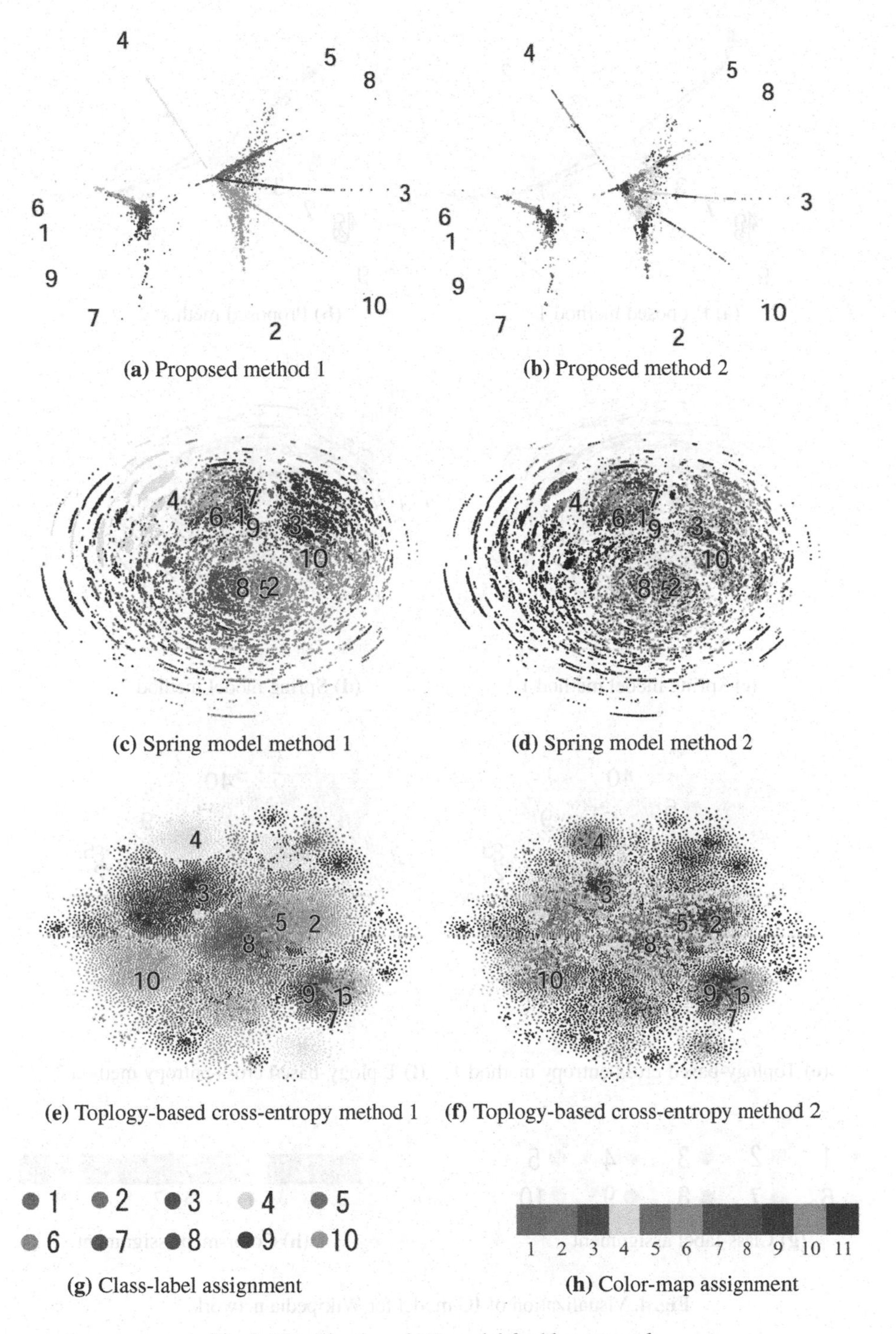


**Fig. 3.** Visualization of LT model for blog network

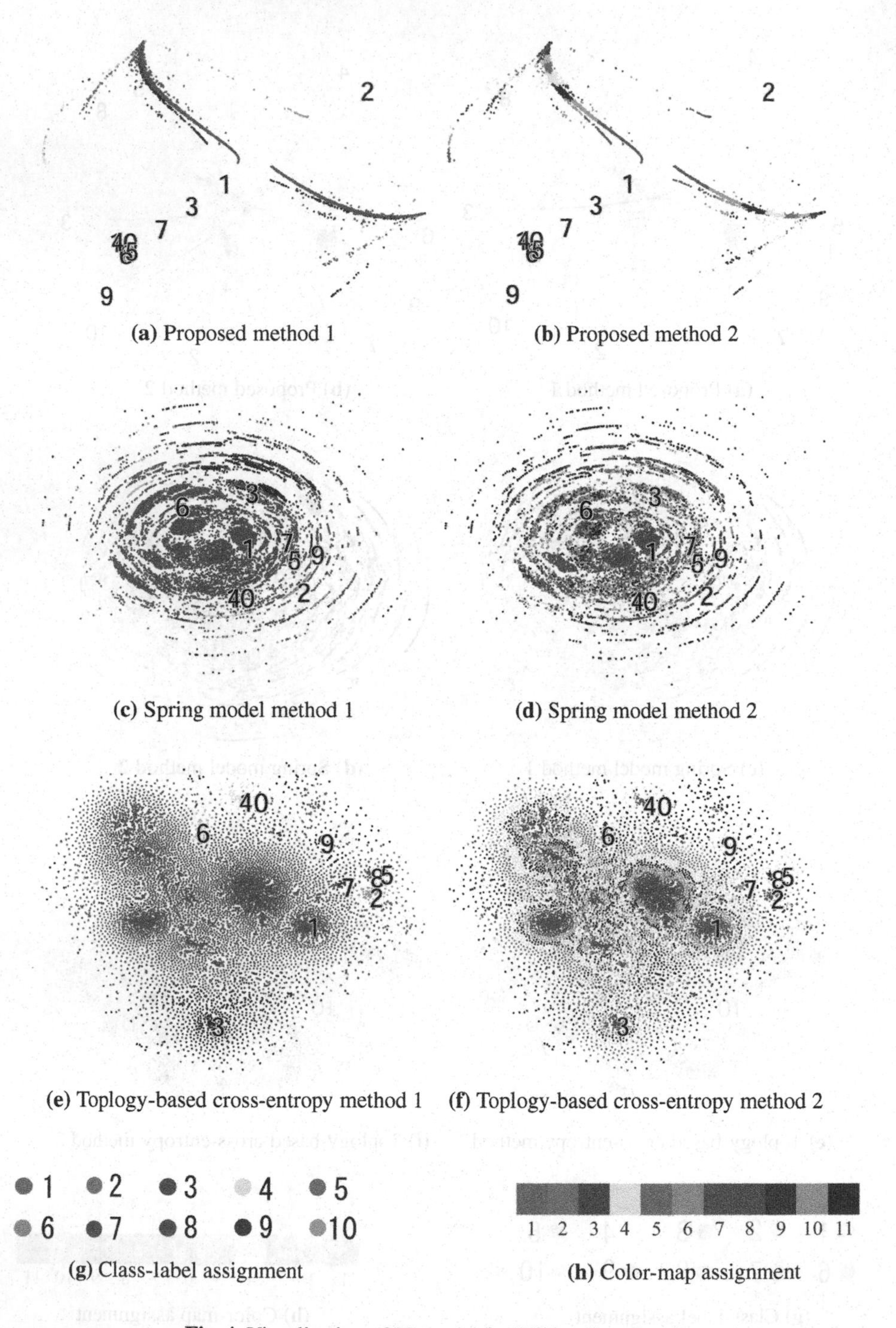


(a) Proposed method 1 (b) Proposed method 2

(c) Spring model method 1 (d) Spring model method 2

(e) Toplogy-based cross-entropy method 1 (f) Toplogy-based cross-entropy method 2

(g) Class-label assignment (h) Color-map assignment

**Fig. 4.** Visualization of IC model for Wikipedia network

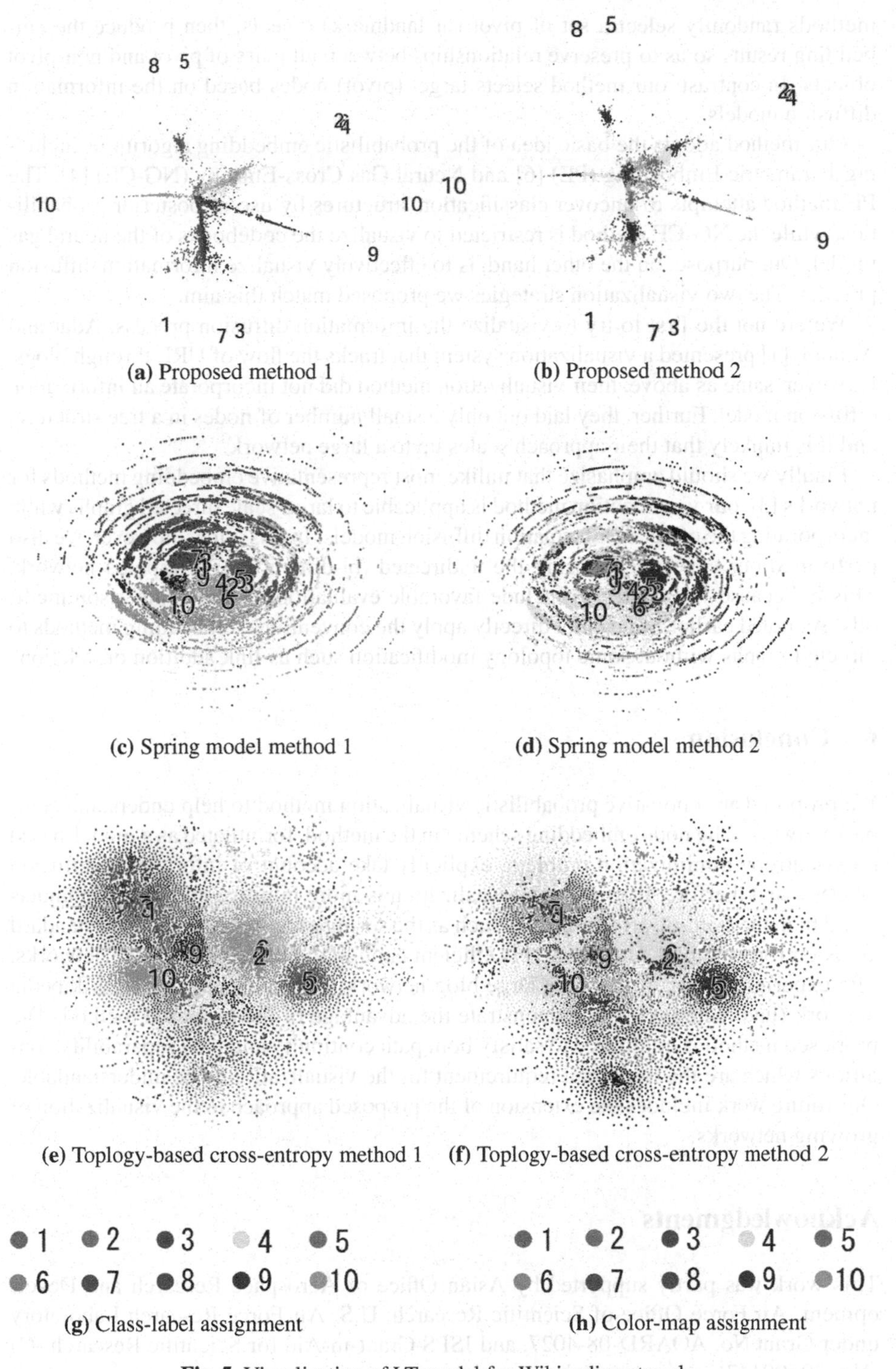


**Fig. 5.** Visualization of LT model for Wikipedia network

methods randomly select a set of pivot (or landmark) objects, then produce the embedding results so as to preserve relationships between all pairs of pivot and non-pivot objects. In contrast, our method selects target (pivot) nodes based on the information diffusion models.

Our method adopts the basic idea of the probabilistic embedding algorithms including Parametric Embedding (PE) [6] and Neural Gas Cross-Entropy (NG-CE) [4]. The PE method attempts to uncover classification structures by use of posterior probabilities, while the NG-CE method is restricted to visualize the codebooks of the neural gas model. Our purpose, on the other hand, is to effectively visualize information diffusion process. The two visualization strategies we proposed match this aim.

We are not the first to try to visualize the information diffusion process. Adar and Adamic [1] presented a visualization system that tracks the flow of URL through blogs. However, same as above, their visualization method did not incorporate an information diffusion model. Further, they laid out only a small number of nodes in a tree structure, and it is unlikely that their approach scales up to a large network.

Finally we should emphasize that unlike most representative embedding methods for networks [3], our visualization method is applicable to large-scale directed graphs while incorporating the effect of information diffusion models. In this paper, however, we also performed our experiments using the undirected (bi-directional) Wikipedia network. This is because we wanted to include favorable evaluation for the comparison methods. As noted earlier, we cannot directly apply the conventional embedding methods to directed graphs without some topology modification such as link addition or deletion.

## 6 Conclusion

We proposed an innovative probabilistic visualization method to help understand complex network. The node embedding scheme in the method, formulated as a model-based cross-entropy minimization problem, explicitly take account of information diffusion process, and therfore, the resulting visualization is more intuitive and easier to understand than the state-of-art approaches such as the spring model method and the standard cross-entropy method. Our method is efficient enough to be applied to large networks. The experiments performed on a large blog network (directed) and a large Wikipedia network (undirected) clearly demonstrate the advantage of the proposed method. The proposed method is confirmed to satisfy both path continuity and path separability conditions which are the important requirement for the visualization to be understandable. Our future work includes the extension of the proposed approach to the visualization of growing networks.

## Acknowledgments

This work was partly supported by Asian Office of Aerospace Research and Development, Air Force Office of Scientific Research, U.S. Air Force Research Laboratory under Grant No. AOARD-08-4027, and JSPS Grant-in-Aid for Scientific Research (C) (No. 20500147).

## References


1. Adar, E., Adamic, L.: Tracking information epidemics in blogspace. In: Proceedings of the 2005 IEEE/WIC/ACM International Conference on Web Intelligence, pp. 207–214 (2005)
2. Balthrop, J., Forrest, S., Newman, M.E.J., Williampson, M.W.: Technological networks and the spread of computer viruses. Science 304, 527–529 (2004)
3. Battista, G., Eades, P., Tamassia, R., Tollis, I.: Graph drawing: An annotated bibliography. Prentice-Hall, New Jersy (1999)
4. Estévez, P.A., Figueroa, C.J., Saito, K.: Cross-entropy embedding of high-dimensional data using the neural gas model. Neural Networks 18, 727–737 (2005)
5. Gruhl, D., Guha, R., Liben-Nowell, D., Tomkins, A.: Information diffusion through blogspace. In: Proceedings of the 13th International World Wide Web Conference, pp. 107–117 (2004)
6. Iwata, T., Saito, K., Ueda, N., Stromsten, S., Griffiths, T.L., Tenenbaum, J.B.: Parametric embedding for class visualization. Neural Computation 19, 2536–2556 (2007)
7. Kamada, K., Kawai, S.: An algorithm for drawing general undirected graph. Information Processing Letters 31, 7–15 (1989)
8. Kempe, D., Kleinberg, J., Tardos, E.: Maximizing the spread of influence through a social network. In: Proceedings of the 9th ACM SIGKDD International Conference on Knowledge Discovery and Data Mining, pp. 137–146 (2003)
9. Kimura, M., Saito, K., Nakano, R.: Extracting influential nodes for information diffusion on a social network. In: Proceedings of the 22nd AAAI Conference on Artificial Intelligence, pp. 1371–1376 (2007)
10. Newman, M.E.J.: The structure and function of complex networks. SIAM Review 45, 167–256 (2003)
11. Newman, M.E.J., Forrest, S., Balthrop, J.: Email networks and the spread of computer viruses. Physical Review E 66, 035101 (2002)
12. Newman, M.E.J., Park, J.: Why social networks are different from other types of networks. Physical Review E 68, 036122 (2003)
13. Platt, J.C.: Fastmap, metricmap, and landmark MDS are all nyström algorithms. In: Proceedings of the 10th International Workshop on Artificial Intelligence and Statistics, pp. 261–268 (2005)
14. Yamada, T., Saito, K., Ueda, N.: Cross-entropy directed embedding of network data. In: Proceedings of the 20th International Conference on Machine Learning, pp. 832–839 (2003)

# Actively Transfer Domain Knowledge

Xiaoxiao Shi[1], Wei Fan[2], and Jiangtao Ren[1,⋆]

[1] Department of Computer Science
Sun Yat-sen University, Guangzhou, China
{isshxx,issrjt}@mail.sysu.edu.cn
[2] IBM T.J.Watson Research, USA
weifan@us.ibm.com

**Abstract.** When labeled examples are not readily available, active learning and transfer learning are separate efforts to obtain labeled examples for inductive learning. Active learning asks domain experts to label a small set of examples, but there is a cost incurred for each answer. While transfer learning could borrow labeled examples from a different domain without incurring any labeling cost, there is no guarantee that the transferred examples will actually help improve the learning accuracy. To solve both problems, we propose a framework to actively transfer the knowledge across domains, and the key intuition is to use the knowledge transferred from other domain as often as possible to help learn the current domain, and query experts only when necessary. To do so, labeled examples from the other domain (out-of-domain) are examined on the basis of their likelihood to correctly label the examples of the current domain (in-domain). When this likelihood is low, these out-of-domain examples will not be used to label the in-domain example, but domain experts are consulted to provide class label. We derive a sampling error bound and a querying bound to demonstrate that the proposed method can effectively mitigate risk of domain difference by transferring domain knowledge only when they are useful, and query domain experts only when necessary. Experimental studies have employed synthetic datasets and two types of real world datasets, including remote sensing and text classification problems. The proposed method is compared with previously proposed transfer learning and active learning methods. Across all comparisons, the proposed approach can evidently outperform the transfer learning model in classification accuracy given different out-of-domain datasets. For example, upon the remote sensing dataset, the proposed approach achieves an accuracy around 94.5%, while the comparable transfer learning model drops to less than 89% in most cases. The software and datasets are available from the authors.

## 1 Introduction

Supervised learning methods require sufficient labeled examples in order to construct accurate models. However, in real world applications, one may easily

⋆ The author is supported by the National Natural Science Foundation of China under Grant No. 60703110.

W. Daelemans et al. (Eds.): ECML PKDD 2008, Part II, LNAI 5212, pp. 342–357, 2008.

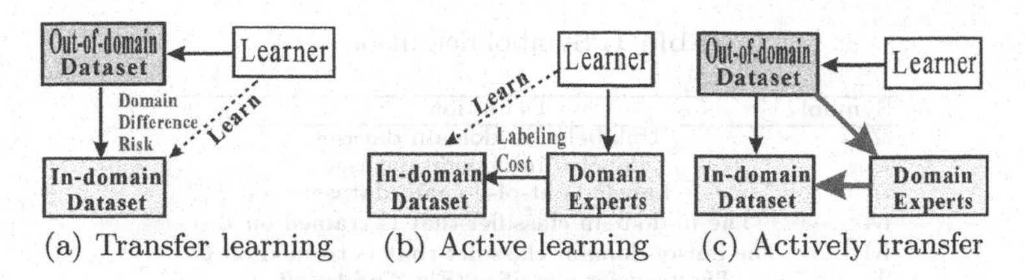

(a) Transfer learning (b) Active learning (c) Actively transfer

**Fig. 1.** Different models to resolve label deficiency

encounter those situations in which labeled examples are deficient, such as data streams, biological sequence annotation or web searching, etc. To alleviate this problem, two separate approaches, transfer learning and active learning, have been proposed and studied. Transfer learning mainly focuses on how to utilize the data from a related domain called out-of-domain, to help learn the current domain called in-domain, as depicted in Fig 1(a). It can be quite effective when the out-of-domain dataset is very similar to the in-domain dataset. As a different solution, active learning does not gain knowledge from other domains, but mainly focuses on selecting a small set of essential in-domain instances for which it requests labels from the domain experts, as depicted in Fig 1(b). However, both transfer learning and active learning have some practical constraints. For transfer learning, if the knowledge from the out-of-domain is essentially different from the in-domain, the learning accuracy might be reduced, and this is called the "domain difference risk". For active learning, the obvious issue is the cost associated with the answer from domain experts.

*Our Method.* To mitigate domain difference risk and reduce labeling cost, we propose a framework that can actively transfer the knowledge from out-of-domain to in-domain, as depicted in Fig 1(c). Intuitively, in daily life, we normally first try to use our related knowledge in learning, but if the related knowledge is unable to guide, we turn to teachers. For example, when learning a foreign language, one normally associates it with the mother tongue. This transfer is easy between, for example, Spanish and Portuguese. But it is not so obvious between Chinese and English. In this situation, one normally pays a teacher instead of picking up himself. The proposed framework is based on these intuitions. We first select an instance that is supposed to be essential to construct an inductive model from the new or in-domain dataset, and then a transfer classifier, trained with labeled data from in-domain and out-of-domain dataset, is used to predict this unlabeled in-domain example. According to defined transfer confidence measure, this instance is either directly labeled by the transfer classifier or labeled by the domain experts if needed. In order to guarantee the performance when "importing" out-of-domain knowledge, the proposed transfer classifier is bounded to be no worse than an instance-based ensemble method in error rate (Section 3 and Theorem 1).

*Contributions.* The most important contributions of the proposed approach can be summarized as follows:

**Table 1.** Symbol definition

| Symbol | Definition |
|---|---|
| $\mathcal{U}$ | Unlabeled in-domain dataset |
| $\mathcal{L}$ | Labeled in-domain dataset |
| $\mathcal{O}$ | Labeled out-of-domain dataset |
| $\mathbf{M}^{\mathcal{L}}$ | The in-domain classifier that is trained on $\mathcal{L}$ |
| $\mathbf{M}^{\mathcal{O}}$ | The out-of-domain classifier that is trained on $\mathcal{O}$ |
| $\mathbf{T}$ | The transfer classifier (Fig 3 and Equ. 1) |
| $\mathbb{F}(\mathbf{x})$ | A decision function (Equ. 3 and Equ. 4) |
| $\ell$ | The actively transfer learner (Algorithm 1) |
| | The following are some symbols only used in Fig 3 |
| $\mathcal{L}^+$ | $\mathcal{L}^+ = \{\mathbf{x} \mid \mathbf{x} \in L \wedge \mathbf{M}^{\mathcal{O}}(\mathbf{x}) = \text{“}+\text{''}\}$ |
| $\mathcal{L}^-$ | $\mathcal{L}^- = \{\mathbf{x} \mid \mathbf{x} \in L \wedge \mathbf{M}^{\mathcal{O}}(\mathbf{x}) = \text{“}-\text{''}\}$ |
| $\mathbf{M}^{\mathcal{L}^+}$ | The classifier that is trained on $\mathcal{L}^+$ |
| $\mathbf{M}^{\mathcal{L}^-}$ | The classifier that is trained on $\mathcal{L}^-$ |

- We propose a framework that can transfer the knowledge across domains actively. We derive the bounds in Theorem 2 and Theorem 3 to prove that the proposed framework not only can mitigate the domain difference risk by transferring out-of-domain knowledge only when they are useful, but also reduce labeling cost by querying fewer examples labeled by experts as compared with traditional active learners.
- We also propose a transfer classifier whose error is bounded.
- Most of the previous active learners can be directly adopted in the proposed framework without changing their original preferences and strategies to select essential examples.

## 2 Actively Transfer

The main flow of proposed approach AcTraK (**Ac**tively **Tra**nsfer **K**nowledge) is summarized in Fig 2 and Algorithm 1, and the most important symbols are in Table 1. The key idea is to use the out-of-domain data to predict in-domain data as often as possible. But when the prediction confidence is too low, in-domain experts are consulted to provide the label. To do so, the algorithm first applies a traditional active learner to select a critical instance $\mathbf{x}$ from the in-domain dataset, then a transfer classifier is trained and used to classify this selected example. According to the prediction confidence of the transfer classifier, the algorithm decides how to label the instance, either using the predicted label given by the transfer classifier or asking domain experts to label. Then, the process is iteratively performed to select and label important examples. Shown as Fig 2, the essential elements of the proposed algorithm are the "transfer classifier" and the "decision function", as described below.

*Transfer Classifier.* Given an unlabeled in-domain dataset $\mathcal{U}$, a small set of labeled in-domain examples $\mathcal{L}$, as well as a labeled out-of-domain dataset $\mathcal{O}$, a transfer classifier is constructed from both $\mathcal{O}$ and $\mathcal{L}$ to classify unlabeled examples in $\mathcal{U}$. In previous work on transfer learning, out-of-domain dataset $\mathcal{O}$ is assumed to share similar distributions with in-domain dataset $\mathcal{U} \cup \mathcal{L}$ ([3][16]).

**Input**: Unlabeled in-domain dataset: $\mathcal{U}$;
Labeled in-domain dataset: $\mathcal{L}$;
Labeled out-of-domain dataset: $\mathcal{O}$;
Maximum number of examples labeled by experts: N.
**Output**: The actively transfer learner: $\ell$

```
1  Initial the number of examples that have
   been labeled by experts: n ← 0;
2  repeat
3  |  x ← select an instance from U by a
   |  traditional active learner;
4  |  Train the transfer classifier T (Fig 3);
5  |  Predict x by T(x) (Fig 3 and Equ. 1);
6  |  Calculate the decision function F(x)
   |  (Details in Equ. 3 and Equ. 4);
7  |  if F(x) = 0 then
8  |  |  y ← label by T(x);
9  |  else
10 |  |  y ← label by the experts;
11 |  |  n ← n + 1;
12 |  end
13 |  L ← L ∪ {(x, y)};
14 until n ≥ N
15 Train the learner ℓ with L
16 Return the learner ℓ.
```

**Algorithm 1.** Framework

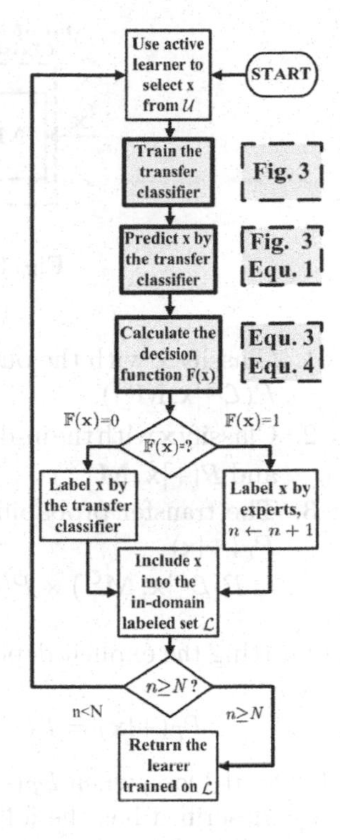


**Fig. 2.** Algorithm flow

Thus, exploring the similarities and exploiting them is expected to improve accuracy. However, it is unclear on how to formally determine whether the out-of-domain dataset shares sufficient similarity with the in-domain dataset, and how to guarantee transfer learning improves accuracy. Thus, in this paper, we propose a transfer learning model whose expected error is bounded.

Intuitively, if one uses the knowledge in out-of-domain dataset $\mathcal{O}$ to make prediction for an in-domain example $\mathbf{x}$, and then double check the predicted label with an in-domain classifier to see if the prediction is the same, it is more likely that the predicted label is correct. Before discussing in detail, we define some common notations. Let $\mathbf{M}^{\mathcal{O}}$ denote the out-of-domain classifier trained on the out-of-domain dataset $\mathcal{O}$. Also, let $\mathcal{L}_t$ denote a set of labeled data from in-domain, but they are labeled as $y_t$ by the out-of-domain classifier $\mathbf{M}^{\mathcal{O}}$ ($y_t$ is the label of the $t$th class). Formally, $\mathcal{L}_t = \{\mathbf{x}|\mathbf{x} \in \mathcal{L} \wedge \mathbf{M}^{\mathcal{O}}(\mathbf{x}) = y_t\}$. Note that the true labels of examples in $\mathcal{L}_t$ are not necessarily $y_t$, but they just happen to be labeled as class $y_t$ by the out-of-domain classifier. We illustrate the transfer classifier for a binary-class problem in Fig 3. There are two classes, "+" and "-", and $\mathcal{L}^+ = \{\mathbf{x}|\mathbf{x} \in \mathcal{L} \wedge \mathbf{M}^{\mathcal{O}}(\mathbf{x}) = \text{"+"}\}$, and $\mathcal{L}^- = \{\mathbf{x}|\mathbf{x} \in \mathcal{L} \wedge \mathbf{M}^{\mathcal{O}}(\mathbf{x}) = \text{"-"}\}$. The transfer classifier $\mathbf{T}(\mathbf{x})$ executes the following steps to label an instance $\mathbf{x}$:

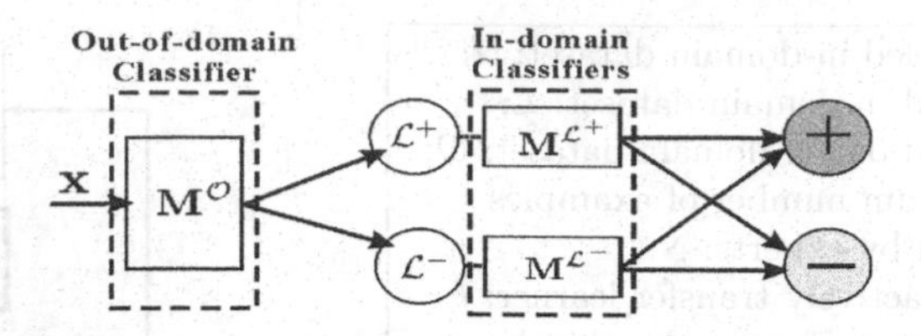


**Fig. 3.** Transfer classifier in Algorithm 1

1. Classify $\mathbf{x}$ with the out-of-domain classifier $\mathbf{M}^{\mathcal{O}}$ to obtain $P(\mathcal{L}^+|\mathbf{x},\mathbf{M}^{\mathcal{O}})$ and $P(\mathcal{L}^-|\mathbf{x},\mathbf{M}^{\mathcal{O}})$.
2. Classify $\mathbf{x}$ with the in-domain classifiers $\mathbf{M}^{\mathcal{L}+}$ and $\mathbf{M}^{\mathcal{L}-}$ to obtain $P(+|\mathbf{x},\mathbf{M}^{\mathcal{L}^+})$ and $P(+|\mathbf{x},\mathbf{M}^{\mathcal{L}^-})$.
3. The transfer probability for $\mathbf{x}$ being "+" is
$P_T(+|\mathbf{x})$
$= P(\mathcal{L}^+|\mathbf{x},\mathbf{M}^{\mathcal{O}}) \times P(+|\mathbf{x},\mathbf{M}^{\mathcal{L}^+}) + P(\mathcal{L}^-|\mathbf{x},\mathbf{M}^{\mathcal{O}}) \times P(+|\mathbf{x},\mathbf{M}^{\mathcal{L}^-})$

Omitting the explicit dependency on models the above formula can be simplified as:

$$P_T(+|\mathbf{x}) = P(+|\mathcal{L}^+,\mathbf{x}) \times P(\mathcal{L}^+|\mathbf{x}) + P(+|\mathcal{L}^-,\mathbf{x}) \times P(\mathcal{L}^-|\mathbf{x}) \quad (1)$$

Under 0-1 loss, when $P_T(+|\mathbf{x}) > 0.5$, $\mathbf{x}$ is classified as "+". This transfer classifier just described has the following important property.

**Theorem 1.** *Let $\varepsilon_1$ and $\varepsilon_2$ denote the expected error of the in-domain classifiers $\mathbf{M}^{\mathcal{L}+}$ and $\mathbf{M}^{\mathcal{L}-}$ respectively, and let $\varepsilon_t$ denote the expected error of the transfer classifier $\mathbf{T}(\mathbf{x})$. Then,*

$$\min(\varepsilon_1,\varepsilon_2) \le \varepsilon_t \le \frac{1}{2}(\varepsilon_1+\varepsilon_2) \quad (2)$$

*Proof.* $\forall \mathbf{x} \in \mathcal{U}$, we consider the situations in which the transfer classifier $\mathbf{T}(\mathbf{x})$ assigns it the wrong label. Let the true label of $\mathbf{x}$ be "+". Further assume that "+" examples are more likely classified into $\mathcal{L}^+$ or $P(\mathcal{L}^+|\mathbf{x}) \ge P(\mathcal{L}^-|\mathbf{x})$. Thus, the probability that $\mathbf{x}$ is mistakenly labeled as "-" is:
$\varepsilon_t(\mathbf{x}) = P(-|\mathbf{x})$
$= P(\mathcal{L}^-|\mathbf{x}) \times P(-|\mathcal{L}^-,\mathbf{x}) + (1 - P(\mathcal{L}^-|\mathbf{x})) \times P(-|\mathcal{L}^+,\mathbf{x})$
$= P(-|\mathcal{L}^+,\mathbf{x}) + P(\mathcal{L}^-|\mathbf{x}) \times (P(-|\mathcal{L}^-,\mathbf{x}) - P(-|\mathcal{L}^+,\mathbf{x}))$

Since $P(-|\mathcal{L}^-,\mathbf{x}) = \frac{P(\mathbf{x}|\mathcal{L}^-,-)P(\mathcal{L}^-,-)}{P(\mathcal{L}^-|\mathbf{x})P(\mathbf{x})} > \frac{P(\mathbf{x}|\mathcal{L}^+,-)P(\mathcal{L}^+,-)}{P(\mathcal{L}^+|\mathbf{x})P(\mathbf{x})} = P(-|\mathcal{L}^+,\mathbf{x})$, then $P(-|\mathbf{x}) \ge P(-|\mathcal{L}^+,\mathbf{x}) = \min(P(-|\mathbf{x},\mathbf{M}^{\mathcal{L}^+}), P(-|\mathbf{x},\mathbf{M}^{\mathcal{L}^-}))$. In addition, since $P(\mathcal{L}^+|\mathbf{x}) \ge P(\mathcal{L}^-|\mathbf{x})$, we have $0 \le P(\mathcal{L}^-|\mathbf{x}) \le \frac{1}{2}$. Then, $P(-|\mathbf{x}) \le P(-|\mathbf{x},\mathbf{M}^{\mathcal{L}^+}) + \frac{1}{2}(P(-|\mathbf{x},\mathbf{M}^{\mathcal{L}^-}) - P(-|\mathbf{x},\mathbf{M}^{\mathcal{L}^+})) = \frac{1}{2}(P(-|\mathbf{x},\mathbf{M}^{\mathcal{L}^+}) + P(-|\mathbf{x},\mathbf{M}^{\mathcal{L}^-}))$. Thus, we have $\min(\varepsilon_1,\varepsilon_2) \le \varepsilon_t \le \frac{1}{2}(\varepsilon_1+\varepsilon_2)$. □

Hence, Theorem 1 indicates that if the out-of-domain knowledge is similar to the current domain, or $P(\mathcal{L}^-|\mathbf{x})$ is small, the model obtains the expected error close to $\varepsilon_t = \min(\varepsilon_1, \varepsilon_2)$. When the out-of-domain knowledge is different, the expected error is bounded by $\frac{1}{2}(\varepsilon_1 + \varepsilon_2)$. In other words, in the worst case, the performance of the transfer classifier is no worse than the average performances of the two in-domain classifiers $\mathbf{M}^{\mathcal{L}^+}$ and $\mathbf{M}^{\mathcal{L}^-}$.

*Decision Function.* After the transfer classifier $\mathbf{T}(\mathbf{x})$ predicts the selected example $\mathbf{x}$, a decision function $\mathbb{F}(\mathbf{x})$ is calculated and further decides how to label the example. In the following situations, one should query the experts for the class label in case of mislabeling.

- When the transfer classifier assigns $\mathbf{x}$ with a class label that is different from that given by an in-domain classifier.
- When the transfer classifier's classification is low in confidence.
- When the size of the labeled in-domain dataset $\mathcal{L}$ is very small.

Recall that $\mathbf{M}^{\mathcal{L}}$ is the in-domain classifier trained on labeled in-domain dataset $\mathcal{L}$. According to the above considerations, we design a "querying indicator" function $\theta(\mathbf{x})$ to reflect the necessity to query experts.

$$\begin{aligned} \theta(\mathbf{x}) &= \Big(1 + \alpha(\mathbf{x})\Big)^{-1} \\ \alpha(\mathbf{x}) &= \Big(1 - [\![\mathbf{M}^{\mathcal{L}}(\mathbf{x}) \neq \mathbf{T}(\mathbf{x})]\!]\Big) \cdot \mathbf{P}_{\mathcal{T}}(\mathbf{T}(\mathbf{x}) = y|\mathbf{x}) \cdot \exp(-\frac{1}{|\mathcal{L}|}) \end{aligned} \tag{3}$$

where $[\![\pi]\!] = 1$ if $\pi$ is true. And $\mathbf{P}_{\mathcal{T}}(\mathbf{T}(\mathbf{x}) = y|\mathbf{x})$ is the transfer probability given by the transfer classifier $\mathbf{T}(\mathbf{x})$. Thus, $0 \leq \alpha(\mathbf{x}) \leq 1$ and it reflects the confidence that the transfer classifier has correctly labeled the example $\mathbf{x}$: it increases with the transfer probability $\mathbf{P}_{\mathcal{T}}(\mathbf{T}(\mathbf{x}) = y|\mathbf{x})$, and $\alpha(\mathbf{x}) = 0$ if the two classifiers $\mathbf{M}^{\mathcal{L}}(\mathbf{x})$ and $\mathbf{T}(\mathbf{x})$ have assigned different labels to $\mathbf{x}$. Hence, the larger of $\alpha(\mathbf{x})$, the less we need to query the experts to label the example. Formally, the "querying indicator" function $\theta(\mathbf{x})$ requires $\theta(\mathbf{x}) \propto \alpha(\mathbf{x})^{-1}$. Moreover, because mislabeling of the first few selected examples can exert significant negative effect on accuracy, we further set $\theta(\mathbf{x}) = \Big(1+\alpha(\mathbf{x})\Big)^{-1}$ so as to guarantee the possibility (necessity) to query experts is greater than 50%. In other words, labeling by the experts is the priority and we trust the label given by the transfer classifier only when its confidence reflected by $\alpha(\mathbf{x})$ is very high. Thus, the proposed algorithm asks the experts to label the example with probability $\theta(\mathbf{x})$. Accordingly, with the value of $\theta(\mathbf{x})$, we randomly generate a real number $R$ within 0 to 1, and then the decision function $\mathbb{F}(\mathbf{x})$ is defined as

$$\mathbb{F}(\mathbf{x}) = \begin{cases} 0 & if\ R > \theta(\mathbf{x}) \\ 1 & otherwise \end{cases} \tag{4}$$

According to Eq. 4, if the randomly selected real number $R > \theta(\mathbf{x})$, $\mathbb{F}(\mathbf{x}) = 0$, and it means Algorithm 1 labels the example by the transfer classifier; otherwise, the example is labeled by the domain experts. In other words, AcTraK labels the example $\mathbf{x}$ by transfer classifier with probability $1 - \theta(\mathbf{x})$.

### 2.1 Formal Analysis of AcTraK

We have proposed the approach AcTraK to transfer knowledge across domains actively. Now, we formally derive its sampling error bound to demonstrate its ability to mitigate domain difference risk, which guarantees that out-of-domain examples are transferred to label in-domain data only when they are useful. Additionally, we prove its querying bound to validate the claim that AcTraK can reduce labeling cost by querying fewer examples labeled by experts by any based level active learner incorporated into the framework.

**Theorem 2.** *In the algorithm* **AcTraK** *(Algorithm 1), let $\varepsilon_t$ denote the expected error of the transfer classifier, and let $N$ denote the maximum number of examples labeled by experts, then the sampling error $\varepsilon$ for* **AcTraK** *satisfies*

$$\varepsilon \leq \frac{\varepsilon_t^2}{1 + (1 - \varepsilon_t) \times \exp(-|N|^{-1})} \tag{5}$$

*Proof.* The proof for Theorem 2 is straightforward. According to the transfer classifier $\mathbf{T}(\mathbf{x})$ and the decision function $\mathbb{F}(\mathbf{x})$ described above, **AcTraK** makes wrong decision only when both the transfer classifier $\mathbf{T}(\mathbf{x})$ and the in-domain classifier $\mathbf{M}^{\mathcal{L}}$ agree on the wrong label. And in this case, **AcTraK** has probability $1 - \theta(\mathbf{x})$ to trust the classification result given by $\mathbf{T}(\mathbf{x})$, where $\theta(\mathbf{x})$ is defined in Eq. 3. Thus, the sampling error for **AcTraK** can be written as $\varepsilon \leq (\varepsilon_t)^2(1-\theta(\mathbf{x}))$. Moreover, in this situation, $\theta(\mathbf{x}) = \frac{1}{1+(1-\varepsilon_t)e^{-\frac{1}{|\mathcal{L}|}}} \geq \frac{1}{1+(1-\varepsilon_t)e^{-\frac{1}{N}}}$. Thus, $\varepsilon \leq \varepsilon_t^2(1 - \theta(\mathbf{x})) \leq \frac{\varepsilon_t^2 \times (1-\varepsilon_t) \times \exp(-|N|^{-1})}{1+(1-\varepsilon_t)\times\exp(-|N|^{-1})} \leq \frac{\varepsilon_t^2}{1+(1-\varepsilon_t)\times\exp(-|N|^{-1})}$. □

**Theorem 3.** *In the algorithm* **AcTraK** *(Algorithm 1), let $\varepsilon_t$ and $\varepsilon_i$ denote the expected error of the transfer classifier and in-domain classifier respectively. And let $\alpha = \varepsilon_t + \varepsilon_i$. Then for an in-domain instance, the probability that* **AcTraK** *queries the label from the experts (with cost) satisfies:*

$$P[Query] \leq \alpha + \frac{1 - \alpha}{1 + (1 - \varepsilon_t) \times \exp(-\frac{1}{|N|})} \tag{6}$$

*Proof.* According to the labeling-decision function $\mathbb{F}(\mathbf{x})$, **AcTraK** will query the experts to label the instance when $\mathbf{T}(\mathbf{x})$ and $\mathbf{M}^{\mathcal{L}}$ hold different predictions on the classification result. Even when the two classifiers agree on the result, it still has probability $\theta(\mathbf{x})$ to query the experts. Thus, $P[Query] = \varepsilon_i(1 - \varepsilon_t) + \varepsilon_t(1 - \varepsilon_i) + [\varepsilon_t\varepsilon_i + (1 - \varepsilon_t)(1 - \varepsilon_i)]\theta(\mathbf{x}) = \theta(\mathbf{x}) + (\varepsilon_t + \varepsilon_i - 2\varepsilon_t\varepsilon_i)(1 - \theta(\mathbf{x})) \leq \alpha + (1 - \alpha)\theta(\mathbf{x}) \leq \alpha + \frac{1-\alpha}{1+(1-\varepsilon_t)\times\exp(-\frac{1}{|N|})}$. □

From Theorem 2, we can find that the sampling error of the proposed approach AcTraK is bounded by $O(\frac{\varepsilon_t^2}{1-\varepsilon_t})$, where $\varepsilon_t$ is the expected error of the transfer classifier, and $\varepsilon_t$ is also bounded according to Theorem 1. Thus, the proposed

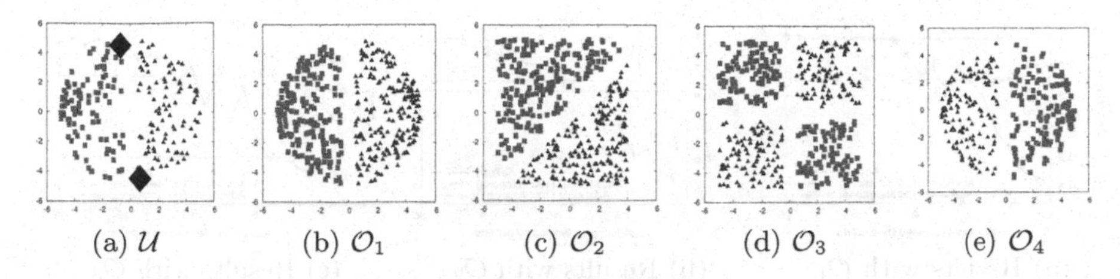

(a) $\mathcal{U}$ (b) $\mathcal{O}_1$ (c) $\mathcal{O}_2$ (d) $\mathcal{O}_3$ (e) $\mathcal{O}_4$

**Fig. 4.** Synthetic dataset

method AcTraK can effectively bound the sampling error to reduce the domain difference risk. In addition, we derive Theorem 3 to understand why AcTraK can query fewer examples labeled by experts as compared with traditional active learners. From Theorem 3, we can see that the querying probability of AcTraK is bounded, and the querying bound decreases with the decreasing $\varepsilon_t$. In other words, the more accurate of the transfer classifier, the less likely will AcTraK query the experts to label the instance. Hence, one can perceive that AcTraK can actively decide how to gain its knowledge.

## 3 Experiments

We first design synthetic datasets to demonstrate how AcTraK mitigates the domain difference risk that can make transfer learning fail, and then study how out-of-domain knowledge can help AcTraK to query fewer examples labeled by experts, as compared with traditional active learner. Then several transfer learning problems composed from remote sensing and text classification datasets are used for evaluation. We use SVM as the basic learners, and logistic regression to simulate the classification probabilities. Furthermore, for active learner employed in AcTraK, we adopt ERS (Error Reduction Sampling method [11]). AcTraK is compared with both a transfer learning model TrAdaBoost ([4]) and the active learning model ERS ([11]). These are some of the most obvious choices, commonly adopted in the research community.

### 3.1 Synthetic Datasets

We design synthetic datasets to empirically address the following questions:

1. Domain difference risk: can AcTraK overcome domain difference if the out-of-domain knowledge is significantly different from the current domain?
2. Number of examples labeled by experts: do experts label fewer examples in AcTraK under the help of out-of-domain knowledge?

We generate five two-dimensional datasets shown in Fig 4 (electronic copy of this paper contains color figures). Fig 4(a) draws the in-domain dataset $\mathcal{U}$, which has two labeled examples highlighted by "♦". Importantly, four out-of-domain datasets with different distributions are plotted in Fig 4(b)~Fig 4(e).

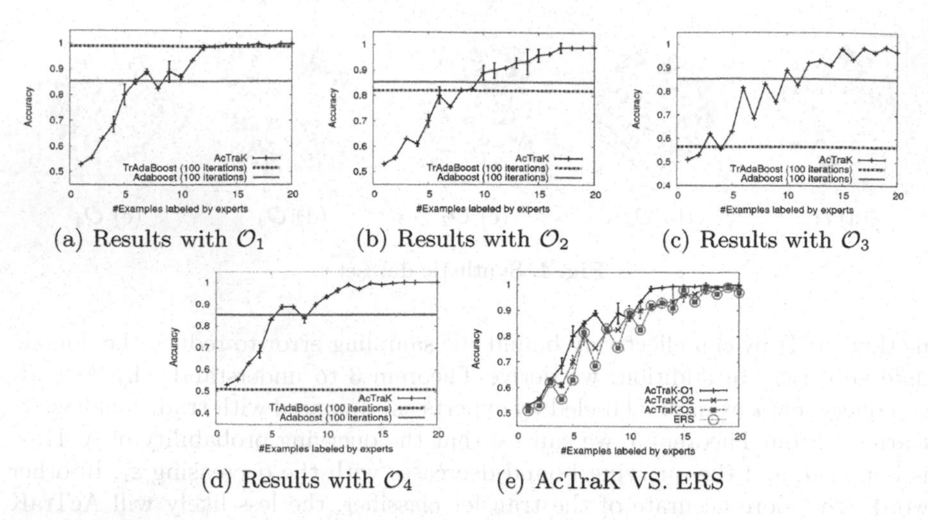


(a) Results with $\mathcal{O}_1$ (b) Results with $\mathcal{O}_2$ (c) Results with $\mathcal{O}_3$

(d) Results with $\mathcal{O}_4$ (e) AcTraK VS. ERS

**Fig. 5.** Results on synthetic dataset

Note: To resolve label deficiency, TrAdaBoost does not query experts but passively trains with the in-domain and all the labeled out-of-domain data. Thus, the its learning curve is straight line.

Fig 4(b) presents a dataset $\mathcal{O}_1$, similarly distributed as the in-domain dataset $\mathcal{U}$. But the dataset $\mathcal{O}_2$ is clearly different from $\mathcal{U}$ though they may still share some similarity. Fig 4(d) plots an "XOR" dataset $\mathcal{O}_3$. In addition, we include the dataset $\mathcal{O}_4$ depicted as Fig 4(e), which has a similar "shape" but totally reversed class labels with the in-domain dataset $\mathcal{U}$. Thus, four experiments are conduced by using the same in-domain dataset $\mathcal{U}$ but different out-of-domain datasets $\mathcal{O}_1 \sim \mathcal{O}_4$. We vary the number of examples labeled by experts up to 50. Moreover, to reveal how domain difference affects transfer learning, we also run on AdaBoost, or boosting without transferring knowledge from other domains. Each experiment is repeated 10 times and average results are reported. For both TrAdaBoost and AdaBoost, the iteration is set to be 100. For the sake of clarity, we plot the most distinctive results in Fig 5.

*Can AcTraK Overcome Domain Difference?* Fig 5(a) to Fig 5(d) plot the performance comparison of AcTraK vs. TrAdaBoost as they are given the four out-of-domain datasets. The result given by AdaBoost is to compare with TrAdaBoost to study effect of domain difference. It is important to note that, to resolve label deficiency, TrAdaBoost does not query the experts but trains on in-domain and all labeled out-of-domain data for many iterations (100 in our experiment). Thus, its result is a straight line. From Fig 5, TrAdaBoost is effective when the out-of-domain dataset is $\mathcal{O}_1$ or Fig 4(b), which shares similar distribution with the in-domain dataset. In this case, TrAdaBoost obviously outperforms the original AdaBoost. However, when the out-of-domain dataset distributes differently from the in-domain dataset, the classification accuracy given by TrAdaBoost significantly drops: 0.82 when the out-of-domain dataset is $\mathcal{O}_2$; 0.57 when the

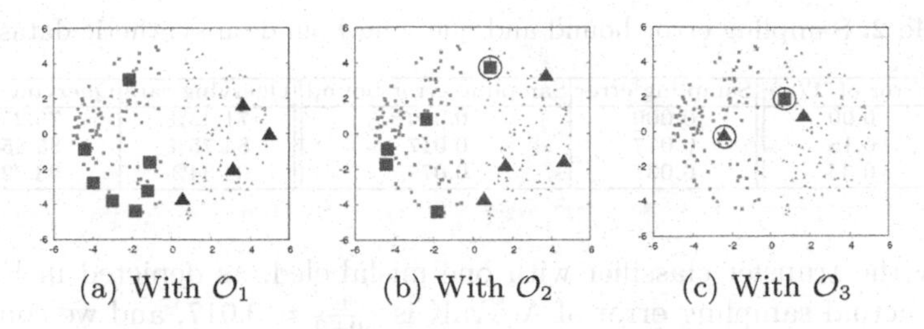

(a) With $\mathcal{O}_1$ (b) With $\mathcal{O}_2$ (c) With $\mathcal{O}_3$

**Fig. 6.** Examples (in $\mathcal{U}$) labeled by the transfer classifier

out-of-domain dataset is $\mathcal{O}_3$ and only 0.51 when with $\mathcal{O}_4$ (the last two results are just slightly better than random guessing). Importantly, these numbers are even worse than the original AdaBoost that achieves an accuracy of 0.85 without using the knowledge from $\mathcal{O}_2$, $\mathcal{O}_3$ or $\mathcal{O}_4$. It is obvious that the culprit is the domain differences from these datasets.

Importantly under these same challenging situations, from Fig 5(b) to Fig 5(d), this domain difference does not significantly affect the proposed algorithm AcTraK. The classification accuracies of AcTraK with different out-of-domain datasets are over 0.9 when the number of examples labeled by experts is 12, demonstrating its ability to overcome domain difference risk. It is interesting to notice that when the out-of-domain dataset is $\mathcal{O}_4$, AcTraK acts similar as that with $\mathcal{O}_1$. This is because that the transfer classifier described in Fig 3 is not sensitive to the actual "name" of the labels of the out-of-domain dataset. For example, if $\mathcal{L}^+$ in Fig 3 actually includes most examples with class label $-$, the term $P(-|\mathbf{x}, \mathcal{L}^+)$ will be likely large and make the final classification result more likely to be $-$. Thus, with respect to their similar structure, $\mathcal{O}_4$ is homogeneous with $\mathcal{O}_1$ to some extend. Hence, we do not consider $\mathcal{O}_4$ but $\mathcal{O}_3$ as the most different distributed dataset with the in-domain dataset $\mathcal{U}$ in this experiment. And owing to the limited space and the homogeneity of $\mathcal{O}_1$ and $\mathcal{O}_4$ to AcTraK, the result of $\mathcal{O}_4$ is omitted in the following of the paper.

Importantly, Fig 5 shows that, even with the dataset $\mathcal{O}_2$ and $\mathcal{O}_3$, AcTraK can ensure the accuracy. It is mainly due to the actively transfer strategy: it does not all depend on the out-of-domain dataset passively. To further reveal this active strategy in AcTraK, we plot the examples labeled by the transfer classifier in Fig 6. The examples mislabeled by the transfer classifier are marked with circles. Shown in Fig 6, when the out-of-domain dataset is $\mathcal{O}_1$, the most similar to the in-domain dataset, the transfer classifier help label more examples than those with $\mathcal{O}_2$ or $\mathcal{O}_3$. Especially when the out-of-domain dataset is $\mathcal{O}_3$, the transfer classifier help label only 3 examples and this is due to domain difference.

The sampling error bound of AcTraK under domain difference is derived in Theorem 2. We calculate these bounds and compare them with the actual sampling errors in Table 2. It is important to mention that the actual sampling error or sampling error bound discussed here is the labeling error for the selected examples, but not the accuracy results given in Fig 5, which is the accuracy on the whole in-domain dataset. To calculate the actual sampling error of AcTraK, for example, when the out-of-domain dataset is $\mathcal{O}_2$, there are a total of 9 examples

**Table 2.** Sampling error bound and querying bound on synthetic dataset

| Datasets | Error of $\mathbf{T}(\mathbf{x})$ | Sampling error | Sampling error bound | Querying rate | Querying bound |
|---|---|---|---|---|---|
| $\mathcal{O}_1$ | 0.00 | 0.000 | 0.000 | 71.43% | 75.25% |
| $\mathcal{O}_2$ | 0.18 | 0.017 | 0.017 | 84.75% | 85.45% |
| $\mathcal{O}_3$ | 0.34 | 0.037 | 0.070 | 94.34% | 93.72% |

labeled by the transfer classifier with one mislabeled, as depicted in Fig 6(b). Thus, the actual sampling error of AcTraK is $\frac{1}{50+9} = 0.017$, and we compare it with the bound calculated according to Theorem 2. The results are summarized in Table 2. The sampling error bound given in Theorem 2 is obviously tight for the synthetic dataset. Moreover, it is evident that AcTraK can effectively reduce sampling error. For instance, when the true error of the transfer classifier $\mathbf{T}(\mathbf{x})$ is 0.34 with the dataset $\mathcal{O}_3$, AcTraK bounds its sampling error as 0.07 and gets the actual error 0.04. Thus, it can be concluded that AcTraK can effectively resolve domain difference by bounding the sampling error shown as Theorem 2.

*Do Experts Label Fewer Examples in AcTraK?* In the proposed approach AcTraK, knowledge transferred from other domain is used to help label the examples. In other words, it saves the number of examples to ask the experts. Thus, compared with traditional active learner, AcTraK is expected to reduce the number of examples labeled by experts, thus to reduce labeling cost. We present Fig 5(e) to demonstrate this claim by comparing AcTraK with the traditional active learner ERS. It is important to note that the original ERS only works on the in-domain dataset. Thus, there is only one result plotted for ERS in Fig 5(e) but three for AcTraK with different out-of-domain datasets. From Fig 5(e), we can see that in most cases, "AcTraK-$\mathcal{O}_1$" and "AcTraK-$\mathcal{O}_2$" can evidently outperform ERS by reaching the same accuracy but with fewer examples labeled by experts. And this is because that some of the examples have been labeled by the transfer classifier under the help of the out-of-domain datasets. Additionally, the out-of-domain dataset $\mathcal{O}_1$ seems more helpful than $\mathcal{O}_2$ to AcTraK, due to the similar distribution between $\mathcal{O}_1$ and the in-domain dataset.

When we use the XOR dataset $\mathcal{O}_3$ to be the out-of-domain dataset, the learning curve of AcTraK overlaps with that of ERS depicted as Fig 5(e). It implies that the transfer learning process is unable to help label examples in this case, and both AcTraK and ERS select the same examples and label them all by experts. Depicted as Fig 4(a) and Fig 4(d), the distribution of $\mathcal{O}_3$ is significantly different from the in-domain dataset $\mathcal{U}$, and thus AcTraK judiciously drops the knowledge transferred from $\mathcal{O}_3$ but queries the experts instead in order to avoid mislabeling. This is formally discussed in Theorem 3, in which we have shown that the bound of the probability to query experts increases when the transfer classifier can not confidently label the examples. We also calculate these querying bounds and the actual querying rates in Table 2. The querying bound given in Theorem 3 is tight. Moreover, we can clearly see that AcTraK queries the experts with probability 94% when the out-of-domain dataset is $\mathcal{O}_3$. It explains why AcTraK can not outperform ERS with $\mathcal{O}_3$ in Fig 5(e): the transfer classifier has too little chance $(1 - 94\% = 6\%)$ to label the examples. Additionally, the querying bound of $\mathcal{O}_1$ is less than that of $\mathcal{O}_2$. In other words, AcTraK may label more examples by the transfer classifier when the out-of-domain dataset is $\mathcal{O}_1$. It explains why $\mathcal{O}_1$ is more helpful than $\mathcal{O}_2$ to AcTraK in Fig 5(e).

### 3.2 Real World Dataset

We use two sets of real world datasets, remote sensing problem as well as text classification problem, to empirically evaluate the proposed method. But we first employ an evaluation metric to compare two active learners. One mainly cares about the performance with the increasing number of examples labeled by experts, shown by the learning curves of $\mathcal{A}_1$ and $\mathcal{A}_2$ in Fig 7. A superior active learner is supposed to gain a higher accuracy under the same number of queried examples, or reach the same classification accuracy with fewer labeled examples. This is shown by $n_1$ vs. $n_2$ in Fig 7. Thus, the superiority of $\mathcal{A}_1$ compared with $\mathcal{A}_2$ can be reflected by the area surrounded by the two learning curves, highlighted by dotted lines in Fig 7. In order to qualify this difference, we employ an evaluation metric IEA(*) (Integral Evaluation on Accuracy), and apply it to evaluate the proposed method in the following experiments.

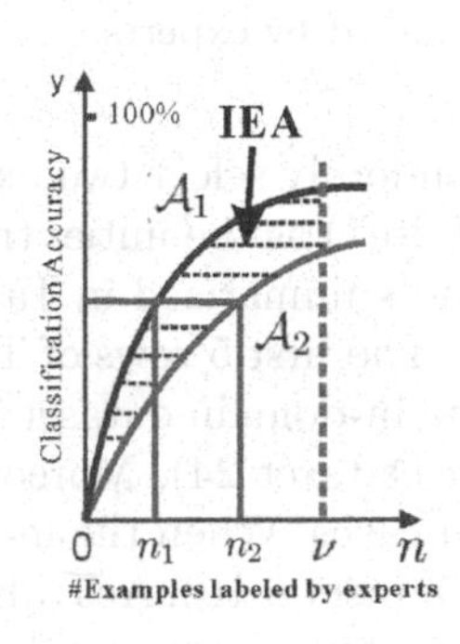


**Fig. 7.** IEA

**Definition 1.** *Given a classifier* **M**, *two active learners* $\mathcal{A}_1$ *and* $\mathcal{A}_2$, *let* $\mathcal{A}(n)$ *denote the classification accuracy of* **M** *trained on the dataset selected by the active learner* $\mathcal{A}$ *when the number of examples labeled by experts is* $n$. *Then,*

$$\mathbf{IEA}(\mathcal{A}_1, \mathcal{A}_2, \nu) = \int_0^{\nu} (\mathcal{A}_1(n) - \mathcal{A}_2(n))d\mathbf{n} = \sum_{n=0}^{\nu} (\mathcal{A}_1(n) - \mathcal{A}_2(n))\Delta n \tag{7}$$

*Remote Sensing Problem.* The remote sensing problem is based on data collected from real landmines[1]. In this problem, there are a total of 29 sets of data, collected from different landmine fields. Each data is represented as a 9-dimensional feature vector extracted from radar images, and the class label is true mine or false mine. Since each of the 29 datasets are collected from different regions that may have different types of ground surface conditions, these datasets are considered to be dominated by different distributions. According to [17], datasets 1 to 10 are collected at foliated regions while datasets 20 to 24 are collected from regions that are bare earth or desert. Then, we combine the datasets 1 to 5 as the unlabeled in-domain dataset, while the datasets 20 to 24 as the labeled out-of-domain dataset respectively. Furthermore, we also combine datasets 6 to 10 as another out-of-domain dataset that is assumed to have a very similar distribution with the in-domain dataset. We conduct the experiment 10 times and

[1] http://www.ee.duke.edu/~lcarin/LandmineData.zip

**Table 3.** Accuracy comparisons on remote sensing (landmine) dataset

| Out-of-domain Dataset | SVM | TrAdaBoost(100 iterations) | AcTraK | IEA(AcTraK, ERS, 100) |
|---|---|---|---|---|
| Dataset 20 | 57% | 89.76% | **94.49%** | +0.101 |
| Dataset 21 | 57% | 86.04% | **94.48%** | +0.108 |
| Dataset 22 | 57% | 90.5% | **94.49%** | +0.103 |
| Dataset 23 | 57% | 88.42% | **94.49%** | +0.123 |
| Dataset 24 | 57% | 90.7% | **94.49%** | +0.12 |
| Dataset 6-10 | 57% | **94.76%** | 94.49% | +0.134 |

Note: The results of AcTraK under the 4th column is when only one example is labeled by experts.

randomly select two examples (one with label "true" while the other with label "false") as the initial training set each time. The experiment results are averaged and summarized in Table 3.

The first 5 rows of Table 3 show that AcTraK outperforms TrAdaBoost when the in-domain dataset is not so similar to the out-of-domain dataset (Dataset 20 to Dataset 24). Moreover, AcTraK also outperforms the active learner ERS in all cases. When the in-domain dataset is similar with the out-of-domain dataset (Dataset 6-10), AcTraK achieves the highest gain on ERS, demonstrating domain transfer can improve learning and reduce number of examples labeled by experts.

*Text Classification Problem.* Another set of experiment on text classification problem uses the 20 Newsgroups. It contains approximately 20,000 newsgroup documents, partitioned across 20 different newsgroups. We generate 6 cross-domain learning tasks. 20 Newsgroups has a two-level hierarchy so that each learning task involves a top category classification problem but the training and test data are drawn from different sub categories. For example, the goal is to distinguish documents from two top newsgroup categories: rec and talk. So a training set involves documents from "rec.autos", and "talk.politics.misc" whereas the test set includes sub-categories "rec.sport.baseball" "talk.religions.misc", etc. The strategy is to split the sub-categories among the training and the test sets so that the distributions of the two sets are similar but not exactly the same. The tasks are generated in the same way as in [4] and more details can be found there. Similar to other experiments, each of the in-domain datasets has only 2 randomly selected labeled examples, one positive and another negative. Reported results in Table 4 are averaged over 10 runs. The results of the first two datasets are also plotted in Fig 8.

It is important to note that the classification results of AcTraK shown in Fig 4 is when the number of examples labeled by experts is 250. It is relatively small in size since each dataset in our experiment has 3500 to 3965 unlabeled documents ([4]). As summarized in Table 4, TrAdaBoost can increase the learning accuracy in some cases, such as with the dataset "comp vs. talk". However, one can hardly guarantee that the exclusive use of transfer learning is enough to learn the current task. For example, when the dataset is "comp vs. sci", TrAdaBoost does not increase the accuracy significantly. But the proposed algorithm AcTraK can achieve an accuracy 78% compared with 57.3% given by TrAdaBoost. It implies the efficiency of AcTraK to actively gain its knowledge both from transfer

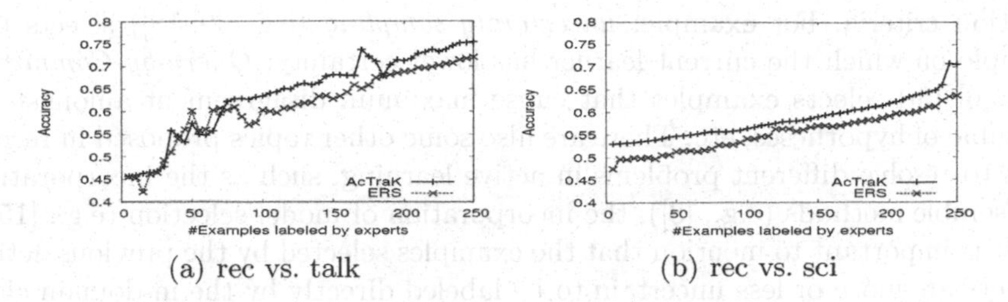


**Fig. 8.** Comparisons with ERS on 20 Newsgroups dataset

**Table 4.** Accuracy comparisons on 20 Newsgroups dataset

| Dataset | SVM | TrAdaBoost(100 iterations) | AcTraK | IEA(AcTraK, ERS, 250) |
|---|---|---|---|---|
| rec vs. talk | 60.2% | 72.3% | **75.4%** | +0.91 |
| rec vs. sci | 59.1% | 67.4% | **70.6%** | +1.83 |
| comp vs. talk | 53.6% | 74.4% | **80.9%** | +0.21 |
| comp vs. sci | 52.7% | 57.3% | **78.0%** | +0.88 |
| comp vs. rec | 49.1% | 77.2% | **82.1%** | +0.35 |
| sci vs. talk | 57.6% | 71.3% | **75.1%** | +0.84 |

learning and domain experts, while TrAdaBoost adopts the passive strategy to thoroughly depend on transfer learning. In addition, from Table 4 and Fig 8, we find that AcTraK can effectively reduce the number of examples labeled by experts as compared with ERS. For example, upon the dataset "rec vs. talk", to reach the accuracy 70%, AcTraK is with 160 examples labeled by experts while ERS needs over 230 such examples.

## 4 Related Work

Transfer learning utilizes the knowledge from other domain(s) to help learn the current domain so as to resolve label deficiency. One of the main issues in this area is how to resolve the different data distributions. One general approach proposed to solve the problem with different data distributions is based on instance weighting (e.g., [2][4][5][10][7]). The motivation of these methods are to "emphasize" those "similar" and discriminated instances. Another line of work tries to change the representation of the observation **x** by projecting them into another space in which the projected instances from different domains are similar to each other (e.g., [1][12]). Most of the previous work assume that the knowledge transferred from other domain(s) can finally help the learning. However, this assumption can be easily violated in practice. The knowledge transferred from other domains may reduce the learning accuracy due to implicit domain differences. We call it the domain difference risk, and we effectively solve the problem by actively transfer the knowledge across domains to help the learning only when they are useful.

Active learning is another way to solve label deficiency. It mainly focuses on carefully selecting a few additional examples for which it requests labels, so as to increase the learning accuracy. Thus, different active learners use different selection criteria. For example, *uncertainty sampling* (e.g., [9][18]) selects the example on which the current learner has lower certainty; *Query-by-Committee* (e.g., [6][13]) selects examples that cause maximum disagreement amongst an ensemble of hypotheses, etc. There are also some other topics proposed in recent years to resolve different problems in active learning, such as the incorporation of ensemble methods (e.g., [8]), the incorporation of model selection (e.g., [15]), etc. It is important to mention that the examples selected by the previous active learners are more or less uncertain to be labeled directly by the in-domain classifier. Then in this paper, we use the knowledge transferred from other domain to help label these selected examples, so as to reduce labeling cost.

## 5 Conclusion

We propose a new framework to actively transfer the knowledge from other domain to help label the instances from the current domain. To do so, we first select an essential example and then apply a transfer classifier to label it. But if the classification given by the transfer classifier is of low confidence, we ask domain experts instead to label the example. We develop Theorem 2 to demonstrate that this strategy can effectively resolve domain difference risk by transferring domain knowledge only when they are useful. Furthermore, we also derive Theorem 3 to prove that the proposed framework can reduce labeling cost by querying fewer examples labeled by experts, as compared with traditional active learners. In addition, we also propose a new transfer learning model adopted in the framework, and this transfer learning model is bounded to be no worse than an instance-based ensemble method in error rate, proven in Theorem 1. There are at least two important advantages of the proposed approach. First, it effectively solves the domain difference risk problem that can easily make transfer learning fail. Second, most of previous active learning models can be directly adopted in the framework to reduce the number of examples labeled by experts.

We design a few synthetic datasets to study how the proposed framework resolves domain difference and reduce the number of examples labeled by experts. The proposed sampling error bound in Theorem 2 and querying bound in Theorem 3 are also empirically demonstrated to be tight bounds in this experiment. Furthermore, two categories of real world datasets, including remote sensing and text classification datasets have been used to generate several transfer learning problems. Experiment shows that the proposed method can significantly outperform the comparable transfer learning model by resolving domain difference. For example, with one of the text classification datasets, the proposed method achieves an accuracy 78.0%, while the comparable transfer learning model drops to 57.3%, due to domain difference. Moreover, the proposed method can also effectively reduce labeling cost by querying fewer examples labeled by experts as compared with the traditional active learner. For instance, in an experiment on the text classification problem, the comparable active learner requires over 230 examples labeled by experts to gain the accuracy 70%, while the proposed method is with at most 160 such examples to reach the same accuracy.

## References


1. Ben-David, S., Blitzer, J., Crammer, K., Pereira, F.: Analysis of representations for domain adaptation. In: Proc. of NIPS 2006 (2007)
2. Bickel, S., Brückner, M., Scheffer, T.: Discriminative learning for differing training and test distributions. In: Proc. of ICML 2007 (2007)
3. Daumé III, H., Marcu, D.: Domain adaptation for statistical classifiers. Journal of Artificial Intelligence Research 26, 101–126 (2006)
4. Dai, W., Yang, Q., Xue, G., Yu, Y.: Boosting for transfer learning. In: Proc. of ICML 2007 (2007)
5. Fan, W., Davidson, I.: On Sample Selection Bias and Its Efficient Correction via Model Averaging and Unlabeled Examples. In: Proc. of SDM 2007 (2007)
6. Freund, Y., Seung, H., Shamir, E., Tishby, N.: Selective sampling using the Query By Committee algorithm. Machine Learning Journal 28, 133–168 (1997)
7. Huang, J., Smola, A.J., Gretton, A., Borgwardt, K.M., Schölkopf, B.: Correcting sample selection bias by unlabeled data. In: Proc. of NIPS 2006 (2007)
8. Körner, C., Wrobel, S.: Multi-class Ensemble-Based Active Learning. In: Fürnkranz, J., Scheffer, T., Spiliopoulou, M. (eds.) ECML 2006. LNCS (LNAI), vol. 4212. Springer, Heidelberg (2006)
9. Lewis, D., Gale, W.: A sequential algorithm for training text classifiers. In: Proc. of SIGIR 1994 (1994)
10. Ren, J., Shi, X., Fan, W., Yu, P.: Type-Independent Correction of Sample Selection Bias via Structural Discovery and Re-balancing. In: Proc. of SDM 2008 (2008)
11. Roy, N., McCallum, A.: Toward optimal active learning through sampling estimation of error reduction. In: Proc. of ICML 2001 (2001)
12. Satpal, S., Sarawagi, S.: Domain adaptation of conditional probability models via feature subsetting. In: Kok, J.N., Koronacki, J., López de Mántaras, R., Matwin, S., Mladenič, D., Skowron, A. (eds.) PKDD 2007. LNCS (LNAI), vol. 4702. Springer, Heidelberg (2007)
13. Senung, H.S., Opper, M., Sompolinsky, H.: Query by committee. In: Proc. 5th Annual ACM Workshop on Computational Learning Theory (1992)
14. Shimodaira, H.: Inproving predictive inference under covariate shift by weighting the log-likehood function. Journal of Statistical Planning and Inference 90, 227–244 (2000)
15. Sugiyama, M., Rubens, N.: Active Learning with Model Selection in Linear Regression. In: Proc. of SDM 2008 (2008)
16. Xing, D., Dai, W., Xue, G., Yu, Y.: Bridged refinement for transfer learning. In: Kok, J.N., Koronacki, J., López de Mántaras, R., Matwin, S., Mladenič, D., Skowron, A. (eds.) PKDD 2007. LNCS (LNAI), vol. 4702. Springer, Heidelberg (2007)
17. Xue, Y., Liao, X., Carin, L., Krishnapuram, B.: Multi-task learning for classification with dirichlet process priors. Journal of Machine Learning Research 8, 35–63 (2007)
18. Xu, Z., Tresp, Y.K., Xu, V., Wang, X.: Representative sampling for text classification using support vector machines. In: Sebastiani, F. (ed.) ECIR 2003. LNCS, vol. 2633. Springer, Heidelberg (2003)

# A Unified View of Matrix Factorization Models

Ajit P. Singh and Geoffrey J. Gordon

Machine Learning Department
Carnegie Mellon University
Pittsburgh PA 15213, USA
{ajit,ggordon}@cs.cmu.edu

**Abstract.** We present a unified view of matrix factorization that frames the differences among popular methods, such as NMF, Weighted SVD, E-PCA, MMMF, pLSI, pLSI-pHITS, Bregman co-clustering, and many others, in terms of a small number of modeling choices. Many of these approaches can be viewed as minimizing a generalized Bregman divergence, and we show that (i) a straightforward alternating projection algorithm can be applied to almost any model in our unified view; (ii) the Hessian for each projection has special structure that makes a Newton projection feasible, even when there are equality constraints on the factors, which allows for matrix co-clustering; and (iii) alternating projections can be generalized to simultaneously factor a set of matrices that share dimensions. These observations immediately yield new optimization algorithms for the above factorization methods, and suggest novel generalizations of these methods such as incorporating row and column biases, and adding or relaxing clustering constraints.

## 1 Introduction

Low-rank matrix factorization is a fundamental building block of machine learning, underlying many popular regression, factor analysis, dimensionality reduction, and clustering algorithms. We shall show that the differences between many of these algorithms can be viewed in terms of a small number of modeling choices. In particular, our unified view places dimensionality reduction methods, such as singular value decomposition [1], into the same framework as matrix co-clustering algorithms like probabilistic latent semantic indexing [2]. Moreover, recently-studied problems, such as relational learning [3] and supervised/semi-supervised matrix factorization [4], can be viewed as the simultaneous factorization of several matrices, where the low-rank representations share parameters. The modeling choices and optimization in the multiple-matrix models are very similar to the single-matrix case.

The first contribution of this paper is descriptive: our view of matrix factorization subsumes many single- and multiple-matrix models in the literature, using only a small set of modeling choices. Our basic single-matrix factorization model can be written $X \approx f(UV^T)$; choices include the prediction link $f$, the definition of $\approx$, and the constraints we place on the factors $U$ and $V$. Different combinations of these choices also yield several new matrix factorization models.

W. Daelemans et al. (Eds.): ECML PKDD 2008, Part II, LNAI 5212, pp. 358–373, 2008.

The second contribution of this paper is computational: we generalize the alternating projections technique for matrix factorization to handle constraints on the factors—*e.g.*, clustering or co-clustering, or the use of margin or bias terms. For most common choices of $\approx$, the loss has a special property, *decomposability*, which allows for an efficient Newton update for each factor. Furthermore, many constraints, such as non-negativity of the factors and clustering constraints, can be distributed across decomposable losses, and are easily incorporated into the per-factor update. This insight yields a common algorithm for most factorization models in our framework (including both dimensionality reduction and co-clustering models), as well as new algorithms for existing single-matrix models.

A parallel contribution [3] considers matrix factorization as a framework for relational learning, with a focus on multiple relations (matrices) and large-scale optimization using stochastic approximations. This paper, in contrast, focuses on modeling choices such as constraints, regularization, bias terms, and more elaborate link and loss functions in the single-matrix case.

# 2 Preliminaries

## 2.1 Notation

Matrices are denoted by capital letters, $X$, $Y$, $Z$. Elements, rows and columns of a matrix are denoted $X_{ij}$, $X_{i\cdot}$, and $X_{\cdot j}$. Vectors are denoted by lower case letters, and are assumed to be column vectors—*e.g.*, the columns of factor $U$ are $(u_1, \dots, u_k)$. Given a vector $x$, the corresponding diagonal matrix is $\mathrm{diag}(x)$. $A \odot B$ is the element-wise (Hadamard) product. $A \circ B$ is the matrix inner product $tr(A^T B) = \sum_{ij} A_{ij} B_{ij}$, which reduces to the dot product when the arguments are vectors. The operator $[A\ B]$ appends the columns of $B$ to $A$, requiring that both matrices have the same number of rows. Non-negative and strictly positive restrictions of a set $\mathbb{F}$ are denoted $\mathbb{F}_+$ and $\mathbb{F}_{++}$. We denote matrices of natural parameters as $\Theta$, and a single natural parameter as $\theta$.

## 2.2 Bregman Divergences

A large class of matrix factorization algorithms minimize a Bregman divergence: *e.g.*, singular value decomposition [1], non-negative matrix factorization [5], exponential family PCA [6]. We generalize our presentation of Bregman divergences to include non-differentiable losses:

**Definition 1.** *For a closed, proper, convex function $F : \mathbb{R}^{m \times n} \to \mathbb{R}$ the generalized Bregman divergence [7,8] between matrices $\Theta$ and $X$ is*

$$\mathbb{D}_F(\Theta \,||\, X) = F(\Theta) + F^*(X) - X \circ \Theta$$

*where $F^*$ is the convex conjugate, $F^*(\mu) = \sup_{\Theta \in \mathrm{dom}\, F} [\Theta \circ \mu - F(\Theta)]$. We overload the symbol $F$ to denote an element-wise function over matrices. If*

*$F : \mathbb{R} \to \mathbb{R}$ is an element-wise function, and $W \in \mathbb{R}_+^{m \times n}$ is a constant weight matrix, then the weighted decomposable generalized Bregman divergence is*

$$\mathbb{D}_F(\Theta \,||\, X, W) = \sum_{ij} W_{ij} (F(\Theta_{ij}) + F^*(X_{ij}) - X_{ij}\Theta_{ij}) .$$

*If $F : \mathbb{R} \to \mathbb{R}$ is additionally differentiable, $\nabla F = f$, and $w_{ij} = 1$, the decomposable divergence is equivalent to the standard definition [9,10]:*

$$\mathbb{D}_F(\Theta \,||\, X, W) = \sum_{ij} F^*(X_{ij}) - F^*(f(\Theta_{ij})) - \nabla F^*(f(\Theta_{ij}))(X_{ij} - f(\Theta_{ij}))$$
$$= D_{F^*}(X \,||\, f(\Theta))$$

Generalized Bregman divergences are important because (i) they include many common separable divergences, such as squared loss, $F(x) = \frac{1}{2}x^2$, and KL-divergence, $F(x) = x \log_2 x$; (ii) there is a close relationship between Bregman divergences and maximum likelihood estimation in regular exponential families:

**Definition 2.** *A parametric family of distributions $\psi_F = \{p_F(x|\theta) : \theta\}$ is a regular exponential family if each density has the form*

$$\log p_F(x|\theta) = \log p_0(x) + \theta \cdot x - F(\theta)$$

*where $\theta$ is the vector of natural parameters for the distribution, $x$ is the vector of minimal sufficient statistics, and $F(\theta) = \log \int p_0(x) \cdot \exp(\theta \cdot x)\, dx$ is the log-partition function.*

A distribution in $\psi_F$ is uniquely identified by its natural parameters. It has been shown that for regular exponential families

$$\log p_F(x|\theta) = \log p_0(x) + F^*(x) - D_{F^*}(x \,||\, f(\theta)),$$

where the prediction link $f(\theta) = \nabla F(\theta)$ is known as the matching link for $F$ [11,12,6,13]. Generalized Bregman divergences assume that the link $f$ and the loss match, though alternating projections generalizes to non-matching links.

The relationship between matrix factorization and exponential families is made clear by viewing the data matrix as a collection of samples $\{X_{11}, \ldots, X_{mn}\}$. Let $\Theta = UV^T$ be the parameters. For a decomposable regular Bregman divergence, minimizing $D_{F^*}(X \,||\, f(\Theta))$ is equivalent to maximizing the log-likelihood of the data under the assumption that $X_{ij}$ is drawn from the distribution in $\psi_F$ with natural parameter $\Theta_{ij}$.

## 3 Unified View of Single-Matrix Factorization

Matrix factorization is both more principled and more flexible than is commonly assumed. Our arguments fall into the following categories:

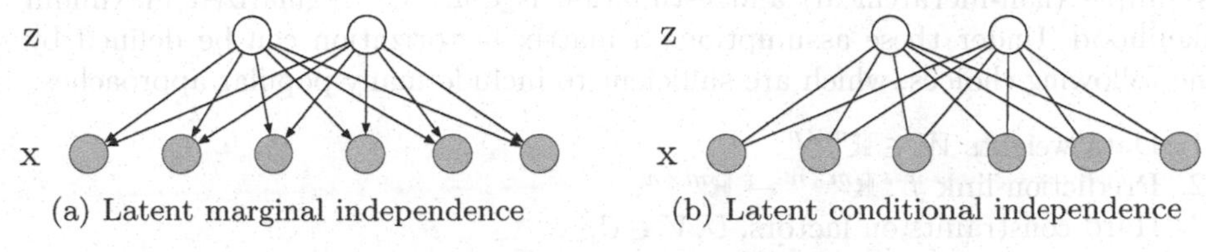


(a) Latent marginal independence (b) Latent conditional independence

**Fig. 1.** Two layer model of matrix factorization

**Decomposability:** Matrix factorization losses tend to be decomposable, expressible as the sum of losses for each element in the matrix, which has both computational and statistical motivations. In many matrices the ordering of rows and columns is arbitrary, permuting the rows and columns separately would not change the distribution of the entries in the matrix. Formally, this idea is known as row-column exchangeability [14,15]. Moreover, for such matrices, there exists a function $\varphi$ such that $X_{ij} = \varphi(\mu, \mu_i, \mu_j, \epsilon_{ij})$ where $\mu$ represents behaviour shared by the entire matrix (*e.g.*, a global mean), $\mu_i$ and $\mu_j$ per-row and per-column effects, and $\epsilon_{ij}$ per-entry effects. The $\epsilon_{ij}$ terms lead naturally to decomposable losses. The computational benefits of decomposability are discussed in Sect. 5.

**Latent Independence:** Matrix factorization can be viewed as maximum likelihood parameter estimation in a two layer graphical model, Fig. 1. If the rows of $X$ are exchangeable, then each training datum corresponds to $x = X_{i\cdot}$, whose latent representation is $z = U_{i\cdot}$, and where $\Theta_{i\cdot} = U_{i\cdot}V^T$ are the parameters of $p(x|z)$. Most matrix factorizations assume that the latents are marginally independent of the observations, Fig. 1(a); an alternate style of matrix factorizations assumes that the latents are conditionally independent given the observations, Fig. 1(b), notably exponential family harmoniums [16].

**Parameters vs. Predictions:** Matrix factorizations can be framed as minimizing the loss with respect to model parameters; or minimizing the loss with respect to reconstruction error. Since many common losses are regular Bregman divergences, and there is a duality between expectation and natural parameters—$D_{F^*}(x \,||\, f(\theta)) = D_F(\theta \,||\, f^{-1}(x))$, the two views are usually equivalent. This allows one to view many plate models, such as pLSI, as matrix factorization.

**Priors and Regularizers:** Matrix factorizations allow for a wide variety of priors and regularizers, which can both address overfitting and the need for pooling information across different rows and columns. Standard regression regularizers, such as the $\ell_p$ norm of the factors, can be adapted. Hierarchical priors can be used to produce a fully generative model over rows and columns, without resorting to folding-in, which can easily lead to optimistic estimates of test errors [17].

**Bayesian stance:** The simplest distinction between the Bayesian and maximum a posteriori/maximum likelihood approaches is that the former computes a distribution over $U$ and $V$, while the latter generates a point estimate. Latent Dirichlet allocation [18] is an example of Bayesian matrix factorization.

Collective matrix factorization assumes that the loss is decomposable and that the latents are marginally independent. Our presentation assumes that the prior is simple (non-hierarchical) and estimation is done via regularized maximum likelihood. Under these assumptions a matrix factorization can be defined by the following choices, which are sufficient to include many popular approaches:

1. Data weights $W \in \mathbb{R}_{+}^{m \times n}$.
2. Prediction link $f : \mathbb{R}^{m \times n} \to \mathbb{R}^{m \times n}$.
3. Hard constraints on factors, $U, V \in \mathcal{C}$.
4. Weighted loss between $X$ and $\hat{X} = f(UV^T)$, $\mathcal{D}(X \,||\, \hat{X}, W) \geq 0$.
5. Regularization penalty, $\mathcal{R}(U, V) \geq 0$.

Given these choices the optimization for the model $X \approx f(UV^T)$ is

$$\underset{(U,V)\in\mathcal{C}}{\operatorname{argmin}} \mathcal{D}(X \,||\, f(UV^T), W) + \mathcal{R}(U, V)$$

Prediction links allow nonlinear relationships between $\Theta = UV^T$ and the data $X$. We focus on the case where $\mathcal{D}$ is a generalized Bregman divergences and $f$ is the matching link. Constraints, weights, and regularizers, along with the ensuing optimization issues, are discussed in Sect. 5.

### 3.1 Models Subsumed by the Unified View

To justify our unified view we discuss a representative sample of single matrix factorizations (Table 1) and how they can be described by their choice of loss, link, constraints, weights, and regularization. Notice that all the losses in Table 1 are decomposable, and that many of the factorizations are closely related to linear models for independent, identically distributed (iid) data points: SVD is the matrix analogue of linear regression; E-PCA/G$^2$L$^2$M are the matrix analogues of generalized linear models; MMMF is the matrix analogue of ordinal regression under hinge loss[1] $h$; $k$-medians is the matrix analogue of quantile regression [19], where the quantile is the median; and $\ell_1$-SVD is the matrix analogue of the LASSO [20]. The key difference between the regression/clustering algorithm and its matrix analogue is changing the assumption from iid observations, where each row of $X$ is drawn from a single distribution, to row exchangeability.

Many of the models in Table 1 differ in the loss, the constraints, and the optimization. In many cases the loss and link do not match, and the optimization is non-convex in $\Theta$, which is usually far harder than minimizing a similar convex problem. We speculate that replacing the non-matching link in pLSI with its matching link may yield an alternative that is easier to optimize.

Similarities between matrix factorizations have been noted elsewhere, such as the equivalence of pLSI and NMF with additional constraints [21]. pLSI requires that the matrix be parameters of a discrete distribution, $1 \circ X = 1$. Adding an orthogonality constraint to $\ell_2$-NMF yields a relaxed form of $k$-means [22]. Orthogonality of a column factors $V^T V = I$ along with integrality of $V_{ij}$ corresponds to hard clustering the columns of $X$, at most one entry in $V_{i\cdot}$ can be non-zero. Even without the integrality constraint, orthogonality plus non-negativity still implies

[1] In Fast-MMMF a smoothed, differentiable version of hinge loss is used, $h_\gamma$.

a stochastic (clustering) constraint: $\forall i \sum_\ell V_{i\ell} = 1, V_{i\ell} \geq 0$. In the $k$-means algorithm, $U$ acts as the cluster centroids and $V$ as the clustering indicators, where the rank of the decomposition and the number of clusters is $k$. In alternating projections, each row update for a factor with the clustering constraint reduces to assigning hard or soft cluster membership to each point/column of $X$.

A closely related characterization of matrix factorization models, which relates NMF, pLSI, as well as Bayesian methods like Latent Dirichlet Allocation, is Discrete PCA [23]. Working from a Bayesian perspective the regularizer and divergence are replaced with a prior and likelihood. This restricts one to representing models where the link and loss match, but affords the possibility of Bayesian averaging where we are restricted to regularized maximum likelihood. We speculate that a relationship exists between variational approximations to these models and Bregman information [12], which averages a Bregman divergence across the posterior distribution of predictions.

Our unified view of matrix factorization is heavily indebted to earlier work on exponential family PCA and $G^2L^2M$. Our approach on a single matrix differs in several ways from $G^2L^2$Ms: we consider extensions involving bias/margin terms, data weights, and constraints on the factors, which allows us to place matrix factorizations for dimensionality reduction and co-clustering into the same alternating Newton-projections approach. Even when the loss is a regular Bregman divergence, which corresponds to a regular exponential family, placing constraints on $U$ and $V$, and thus on $\Theta$, leads to models which do not correspond to regular exponential families. For the specific case of log-loss and its matching link, Logistic PCA proposes alternating projections on a lower bound of the log-likelihood. Max-margin matrix factorization is one of the more elaborate models: ordinal ratings $\{1, \ldots, R\}$[2] are modeled using $R-1$ parallel separating hyperplanes, corresponding to the binary decisions $X_{ij} \leq 1, X_{ij} \leq 2, X_{ij} \leq 3, \ldots, X_{ij} \leq R-1$. The per-row bias term $B_{ir}$ allows the distance between hyperplanes to differ for each row. Since this technique was conceived for user-item matrices, the biases capture differences in each user. Predictions are made by choosing the value which minimizes the loss of the $R-1$ decision boundaries, which yields a number in $\{1, \ldots, R\}$ instead of $\mathbb{R}$.

## 4 Collective Matrix Factorization

A set of related matrices involves entity types $\mathcal{E}_1, \ldots, \mathcal{E}_t$, where the elements of each type are indexed by a row or column in at least one of the matrices. The number of entities of type $\mathcal{E}_i$ is denoted $n_i$. The matrices themselves are denoted $X^{(ij)}$ where each row corresponds to an element of type $\mathcal{E}_i$ and each column to an element of type $\mathcal{E}_j$. If there are multiple matrices on the same types we disambiguate them with the notation $X^{(ij,u)}, u \in \mathbb{N}$. Each data matrix is factored under the model $X^{(ij)} \approx f^{(ij)}(\Theta^{(ij)})$ where $\Theta^{(ij)} = U^{(i)}(U^{(j)})^T$ for low rank factors $U^{(i)} \in \mathbb{R}^{n_i \times k_{ij}}$. The embedding dimensions are $k_{ij} \in \mathbb{N}$. Let $k$ be the largest embedding dimension. In low-rank factorization $k \ll \min(n_1, \ldots, n_t)$.

[2] Zeros in the matrix are considered missing values, and are assigned zero weight.

**Table 1.** Single matrix factorization models. dom $X_{ij}$ describes the types of values allowed in the data matrix. Unweighted matrix factorizations are denoted $W_{ij} = 1$. If constraints or regularizers are not used, the entry is marked with a em-dash.

| Method | dom $X_{ij}$ | Link $f(\theta)$ | Loss $\mathcal{D}(X\|\hat{X} = f(\Theta), W)$ | $W_{ij}$ |
|---|---|---|---|---|
| SVD [1] | $\mathbb{R}$ | $\theta$ | $\|\|W \odot (X - \hat{X})\|\|^2_{Fro}$ | 1 |
| W-SVD [24,25] | $\mathbb{R}$ | $\theta$ | $\|\|W \odot (X - \hat{X})\|\|^2_{Fro}$ | $\geq 0$ |
| $k$-means [26] | $\mathbb{R}$ | $\theta$ | $\|\|W \odot (X - \hat{X})\|\|^2_{Fro}$ | 1 |
| $k$-medians | $\mathbb{R}$ | $\theta$ | $\sum_{ij} \|W_{ij} \left(X_{ij} - \hat{X}_{ij}\right)\|$ | 1 |
| $\ell_1$-SVD [27] | $\mathbb{R}$ | $\theta$ | $\sum_{ij} \|W_{ij} \left(X_{ij} - \hat{X}_{ij}\right)\|$ | $\geq 0$ |
| pLSI [2] | $\mathbf{1} \circ X = 1$ | $\theta$ | $\sum_{ij} W_{ij} \left(X_{ij} \log \frac{X_{ij}}{\hat{X}_{ij}}\right)$ | 1 |
| NMF [5] | $\mathbb{R}_+$ | $\theta$ | $\sum_{ij} W_{ij} \left(X_{ij} \log \frac{X_{ij}}{\Theta_{ij}} + \Theta_{ij} - X_{ij}\right)$ | 1 |
| $\ell_2$-NMF [28,5] | $\mathbb{R}_+$ | $\theta$ | $\|\|W \odot (X - \hat{X})\|\|^2_{Fro}$ | 1 |
| Logistic PCA [29] | $\{0,1\}$ | $(1+e^{-\theta})^{-1}$ | $\sum_{ij} W_{ij} \left(X_{ij} \log \frac{X_{ij}}{\hat{X}_{ij}} + (1 - X_{ij}) \log \frac{1-X_{ij}}{1-\hat{X}_{ij}}\right)$ | 1 |
| E-PCA [6] | many | many | decomposable Bregman ($D_F$) | 1 |
| $\text{G}^2\text{L}^2\text{M}$ [8] | many | many | decomposable Bregman ($\mathbb{D}_F$) | 1 |
| MMMF [30] | $\{0, \ldots, R\}$ | min-loss | $\sum_{r=1}^{R-1} \sum_{ij:X_{ij}\neq 0} W_{ij} \cdot h(\Theta_{ij} - B_{ir})$ | 1 |
| Fast-MMMF [31] | $\{0, \ldots, R\}$ | min-loss | $\sum_{r=1}^{R-1} \sum_{ij:X_{ij}\neq 0} W_{ij} \cdot h_\gamma(\Theta_{ij} - B_{ir})$ | 1 |

| Method | Constraints $U$ | Constraints $V$ | Regularizer $\mathcal{R}(U,V)$ | Algorithm(s) |
|---|---|---|---|---|
| SVD | $U^TU = I$ | $V^TV = \Lambda$ | — | Gaussian Elimination, Power Method |
| W-SVD | — | — | — | Gradient, EM |
| $k$-means | — | $V^TV = I$, $V_{ij} \in \{0,1\}$ | — | EM |
| $k$-medians | — | $V^TV = I$, $V_{ij} \in \{0,1\}$ | — | Alternating |
| $\ell_1$-SVD | — | — | — | Alternating |
| pLSI | $\mathbf{1}^TU\mathbf{1} = 1$, $U_{ij} \geq 0$ | $\mathbf{1}^TV = \mathbf{1}$, $V_{ij} \geq 0$ | — | EM |
| NMF | $U_{ij} \geq 0$ | $V_{ij} \geq 0$ | — | Multiplicative |
| $\ell_2$-NMF | $U_{ij} \geq 0$ | $V_{ij} \geq 0$ | — | Multiplicative, Alternating |
| Logistic PCA | — | — | — | EM |
| E-PCA | — | — | — | Alternating |
| $\text{G}^2\text{L}^2\text{M}$ | — | — | $\|\|U\|\|^2_{Fro} + \|\|V\|\|^2_{Fro}$ | Alternating (Subgradient, Newton) |
| MMMF | — | — | $tr(UV^T)$ | Semidefinite Program |
| Fast-MMMF | — | — | $\frac{1}{2}(\|\|U\|\|^2_{Fro} + \|\|V\|\|^2_{Fro})$ | Conjugate Gradient |

Collective matrix factorization addresses the problem of simultaneously factoring a set of matrices that are related, where the rows or columns of one matrix index the same type as the row or column of another matrix. An example of such data is joint document-citation models where one matrix consists of the word counts in each documents, and another matrix consists of hyperlinks or citations between documents. The types are documents ($\mathcal{E}_1$), words ($\mathcal{E}_2$), and cited documents ($\mathcal{E}_3$), which can include documents not in the set $\mathcal{E}_1$. The two relations (matrices) are denoted $\mathcal{E}_1 \sim \mathcal{E}_2$ and $\mathcal{E}_2 \sim \mathcal{E}_3$. If a matrix is unrelated to the others, it can be factored independently, and so we consider the case where the schema, the links between types $\{\mathcal{E}_i\}_i$, is fully connected. Denote the schema $E = \{(i,j) : \mathcal{E}_i \sim \mathcal{E}_j \wedge i < j\}$.

We assume that each matrix in the set $\{X^{(ij)}\}_{(i,j)\in E}$ is reconstructed under a weighted generalized Bregman divergence with factors $\{U^{(i)}\}_{i=1}^t$ and constant data weight matrices $\{W^{(ij)}\}_{(i,j)\in E}$. Our approach is trivially generalizable to any twice-differentiable and decomposable loss. The total reconstruction loss on all the matrices is the weighted sum of the losses for each reconstruction:

$$\mathcal{L}_u = \sum_{(i,j)\in E} \alpha^{(ij)} \mathbb{D}_F(\Theta^{(ij)} \,||\, X^{(ij)}, W^{(ij)})$$

where $\alpha^{(ij)} \geq 0$. We regularize on a per-factor basis to mitigate overfitting:

$$\mathcal{L} = \sum_{(i,j)\in E} \alpha^{(ij)} \mathbb{D}_F(\Theta^{(ij)} \,||\, X^{(ij)}, W^{(ij)}) + \sum_{i=1}^{t} \mathcal{R}(U^{(i)})$$

Learning consists of finding factors $U^{(i)}$ that minimize $\mathcal{L}$.

## 5 Parameter Estimation

The parameter space for a collective matrix factorization is large, $O(k \sum_{i=1}^t n_i)$, and $\mathcal{L}$ is non-convex, even in the single matrix case. One typically resorts to techniques that converge to a local optimum, like conjugate gradient or EM. A direct Newton step is infeasible due to the number of parameters in the Hessian. Another approximate approach is alternating projection, or block coordinate descent: iteratively optimize one factor $U^{(r)}$ at a time, fixing all the other factors. Decomposability of the loss implies that the Hessian is block diagonal, which allows a Newton coordinate update on $U^{(r)}$ to be reduced to a sequence of independent update over the rows $U^{(r)}_{i\cdot}$.

Ignoring terms that are constant with respect to the factors, the gradient of the objective with respect to one factor, $\nabla_r \mathcal{L} = \frac{\partial \mathcal{L}}{\partial U^{(r)}}$, is

$$\nabla_r \mathcal{L} = \sum_{(r,s)\in E} \alpha^{(rs)} \left( W^{(rs)} \odot \left( f^{(rs)}\left(\Theta^{(rs)}\right) - X^{(rs)} \right) \right) U^{(s)} + \nabla \mathcal{R}(U^{(r)}). \quad (1)$$

The gradient of a collective matrix factorization is the weighted sum of the gradients for each individual matrix reconstruction. If all the per-matrix losses, as well as the regularizers $\mathcal{R}(\cdot)$, are decomposable, then the Hessian of $\mathcal{L}$ with respect to $U^{(r)}$ is block-diagonal, with each block corresponding to a row of $U^{(r)}$. For a single matrix the result is proven by noting that a decomposable loss implies that the estimate of $X_{i\cdot}$ is determined entirely by $U_{i\cdot}$ and $V$. If $V$ is fixed then the Hessian is block diagonal. An analogous argument applies when $U$ is fixed and $V$ is optimized. For a set of related matrices the result follows immediately by noting that Equation 1 is a linear function (sum) of per-matrix losses and the differential is a linear operator. Differentiating the gradient of the loss with respect to $U_{i\cdot}^{(r)}$,

$$\nabla_{r,i}\mathcal{L} = \sum_{(r,s)\in E} \alpha^{(rs)} \left(W_{i\cdot}^{(rs)} \odot \left(f^{(rs)}\left(\Theta_{i\cdot}^{(rs)}\right) - X^{(rs)}\right)\right) U^{(s)} + \nabla\mathcal{R}(U_{i\cdot}^{(r)}),$$

yields the Hessian for the row:

$$\nabla_{r,i}^2\mathcal{L} = \sum_{(r,s)\in E} \alpha^{(rs)} \left(U^{(s)}\right)^T \operatorname{diag}\left(W_{i\cdot}^{(rs)} \odot f^{(rs)}\left(\Theta_{i\cdot}^{(rs)}\right)\right) U^{(s)} + \nabla^2\mathcal{R}(U_{i\cdot}^{(r)}).$$

Newton's rule yields the step direction $\nabla_{r,i}\mathcal{L} \cdot [\nabla_{r,i}^2\mathcal{L}]^{-1}$. We suggest using the Armijo criterion [32] to set the step length. While each projection can be computed fully, we found it suffices to take a single Newton step. For the single matrix case, with no constraints, weights, or bias terms, this approach is known as $G^2L^2M$ [8]. The Hessian is a $k \times k$ matrix, regardless of how large the data matrix is. The cost of a gradient update for $U_{i\cdot}^{(r)}$ is $O(k\sum_{j:\mathcal{E}_i\sim\mathcal{E}_j} n_j)$. The cost of Newton update for the same row is $O(k^3 + k^2\sum_{j:\mathcal{E}_i\sim\mathcal{E}_j} n_j)$. If the matrix is sparse, $n_j$ can be replaced with the number of entries with non-zero weight. The incremental cost of a Newton update over the gradient is essentially only a factor of $k$ more expensive when $k \ll \min(n_1, \ldots, n_t)$.

If $\mathcal{E}_j$ participates in more than one relationship, we allow our model to use only a subset of the columns of $U^{(j)}$ for each relationship. This flexibility allows us to have more than one relation between $\mathcal{E}_i$ and $\mathcal{E}_j$ without forcing ourselves to predict the same value for each one. In an implementation, we would store a list of participating column indices from each factor for each relation; but for simplicity, we ignore this possibility in our notation.

The advantages to our alternating-Newton approach include:

- **Memory Usage:** A solver that optimizes over all the factors simultaneously needs to compute residual errors to compute the gradient. Even if the data is sparse, the residuals rarely are. In contrast, our approach requires only that we store one row or column of a matrix in memory, plus $O(k^2)$ memory to perform the update. This make out-of-core factorizations, where the matrix cannot be stored in RAM, relatively easy.
- **Flexibility of Representation:** Alternating projections works for any link and loss, and can be applied to any of the models in Table 1[3]. In the following sections, we show that the alternating Newton step can also be used

[3] For integrally constrained factors, like $V$ in $k$-means, the Newton projection, a continuous optimization, is replaced by an integer optimization, such as hard clustering.

with stochastic constraints, allowing one to handle matrix co-clustering algorithms. Additionally, the form of the gradient and Hessian make it easy to replace the per-matrix prediction link with different links for each element of a matrix.

### 5.1 Relationship to Linear Models

Single-matrix factorization $X \approx f(UV^T)$ is a bilinear form, *i.e.*, linear when one of the factors is fixed. An appealing property of alternating projection for collective matrix factorization is that the projection reduces an optimization over matrices $U^{(r)}$ into an optimization over data vectors $U_{i\cdot}^{(r)}$—essentially linear models where the "features" are defined by the fixed factors. Since the projection is a linear form, we can exploit many techniques from regression and clustering for iid data. Below, we use this relationship to adapt optimization techniques for $\ell_1$-regularized regression to matrix factorization.

### 5.2 Weights

Weight $W_{ij} = 0$ implies that the corresponding entry in the data matrix has no influence on the reconstruction, which allows us to handle missing values. Moreover, weights can be used to scale terms so that the loss of each matrix reconstruction in $\mathcal{L}$ is a per-element loss, which prevents larger data matrices from exerting disproportionate influence. If the Bregman divergences correspond to exponential families, then we can use log-likelihood as a common scale. Even when the divergences are not regular, computing the average value of $\mathbb{D}_{F^{(ij)}}/\mathbb{D}_{F^{(rs)}}$, given uniform random natural parameters, provides an adequate estimate of the relative scale of the two divergences, which can be accounted for in the weights.

### 5.3 Regularization

The most common regularizers used in matrix factorization are $\ell_p$ regularizers: $\mathcal{R}(U) \propto \lambda \sum_{ij} |u_{ij}|^p$, where $\lambda$ controls the strength of the regularizer, are decomposable. In our experiments we use $\ell_2$-regularization:

$$\mathcal{R}(U) = \lambda ||U||_{Fro}^2/2, \quad \nabla\mathcal{R}(U_{i\cdot}) = U_{i\cdot}/\lambda, \quad \nabla^2\mathcal{R}(U_{i\cdot}) = \text{diag}(\lambda^{-1}\mathbf{1}).$$

The $\ell_1$-regularization constraint can be reduced to an inequality constraint on each row of the factors: $|U_{i\cdot}^{(r)}| \leq t/\lambda, \exists t > 0$. One can exploit a variety of approaches for $\ell_1$-regularized regression (see [33] for survey) in the projection step. For example, using the sequential quadratic programming approach (ibid), the step direction $d$ is found by solving the following quadratic program: Let $x = U_{i\cdot}^{(r)}$, $x^+ = \max(0, x)$, $x^- = -\min(0, x)$, so $x = x^+ - x^-$:

$$\begin{aligned}
&\min_d \, (\nabla_{r,i}\mathcal{L}_u + \lambda\mathbf{1}) \circ d + \frac{1}{2} d^T \cdot \nabla^2_{r,i}\mathcal{L}_u \cdot d \\
&\text{s.t. } \forall i \;\; x_i^+ + d_i^+ \geq 0 \\
&\qquad \forall i \;\; x_i^- + d_i^- \geq 0
\end{aligned}$$

### 5.4 Clustering and Other Equality Constraints

Inequality constraints turn the projection into a constrained optimization; but equality constraints can be incorporated into an unconstrained optimization, such as our Newton projection. Equality constraints can be used to place matrix co-clustering into our framework. With no constraints on the factors, each matrix factorization can be viewed as dimensionality reduction or factor analysis: an increase in the influence of one latent variable does not require a decrease in the influence of other latents. In clustering the stochastic constraint, $\forall i \ \sum_j U_{ij}^{(r)} = 1, U_{ij}^{(r)} \geq 0$, implies that entities of $\mathcal{E}_i$ must belong to one of $k$ latent clusters, and that $U_{i\cdot}^{(r)}$ is a distribution over cluster membership. In matrix co-clustering stochastic constraints are placed on both factors. Since the Newton step is based on a quadratic approximation to the objective, a null space argument ([34] Chap. 10) can be used to show that with a stochastic constraint the step direction $d$ for row $U_{i\cdot}^{(r)}$ is the solution to

$$\begin{bmatrix} \nabla_{r,i}^2 \mathcal{L} & \mathbf{1} \\ \mathbf{1}^T & 0 \end{bmatrix} \begin{bmatrix} d \\ \nu \end{bmatrix} = \begin{bmatrix} -\nabla_{r,i}\mathcal{L} \\ 1 - \mathbf{1}^T U_{i\cdot}^{(r)} \end{bmatrix} \tag{2}$$

where $\nu$ is the Lagrange multiplier for the stochastic constraint. The above technique is easily generalized to $p > 1$ linear constraints, yielding a $k+p$ Hessian.

### 5.5 Bias Terms

Under our assumption of decomposable $\mathcal{L}$ we have that $X_{ij}^{(rs)} \approx f(\Theta_{ij}^{(rs)})$, but matrix exchangeability suggests there may be an advantage to modeling per-row and per-column behaviour. For example, in collaborative filtering, bias terms can calibrate for a user's mean rating. A straightforward way to account for bias is to append an extra column of parameters to $U$ paired with a constant column in $V$: $\tilde{U} = [U \ u_{k+1}]$ and $\tilde{V} = [V \ \mathbf{1}]$. We do not regularize the bias. It is equally straightforward to allow for bias terms on both rows and columns: $\tilde{U} = [U \ u_{k+1} \ \mathbf{1}]$ and $\tilde{V} = [V \ \mathbf{1} \ v_{k+1}]$, and so $\tilde{U}\tilde{V}^T = (UV^T) + u_{k+1}\mathbf{1}^T + \mathbf{1}v_{k+1}^T$. Note that these are biases in the space of natural parameters, a special case being a margin in the hinge or logistic loss—*e.g.*, the per-row (per-user, per-rating) margins in MMMF are just row biases. The above biases maintain the decomposability of $\mathcal{L}$, but there are cases where this is not true. For example, a version of MMMF that shares the same bias for all users for a given rating—all the elements in $u_{k+1}$ must share the same value.

### 5.6 Stochastic Newton

The cost of a full Hessian update for $U_{i\cdot}^{(r)}$ is essentially $k$ times more expensive than the gradient update, which is independent of the size of the data. However, the cost of computing the gradient depends linearly on the number of observations in any row or column whose reconstruction depends on $U_{i\cdot}^{(r)}$. If the data matrices are dense, the computational concern is the cost of the gradient. We refer readers to a parallel contribution [3], which describes a provably convergent stochastic approximation to the Newton step.

## 6 Related Work

The three-factor schema $\mathcal{E}_1 \sim \mathcal{E}_2 \sim \mathcal{E}_3$ includes supervised and semi-supervised matrix factorization, where $X^{(12)}$ contains one or more types of labellings of the rows of $X^{(23)}$. An example of supervised matrix factorization is SVDM [35], where $X^{(23)}$ is factored under squared loss and $X^{(12)}$ is factored under hinge loss. A similar model was proposed by Zhu et al. [36], using a smooth variant of the hinge loss. Supervised matrix factorization has been recently generalized to regular Bregman divergences [4]. Another example is supervised LSI [37], which factors both the data and label matrices under squared loss, with an orthogonality constraint on the shared factors. Principal components analysis, which factors a doubly centered matrix under squared loss, has also been extended to the three-factor schema [38]. An extension of pLSI to two related matrices, pLSI-pHITS, consists of two pLSI models that share latent variables [39].

While our choice of a bilinear form $UV^T$ is common, it is not the only option. Matrix co-clustering often uses a trilinear form $X \approx C_1 A C_2^T$ where $C_1 \in \{0,1\}^{n_1 \times k}$ and $C_2 \in \{0,1\}^{n_2 \times k}$ are cluster indicator matrices, and $A \in \mathbb{R}^{k \times k}$ contains the predicted output for each combination of clusters. This trilinear form is used in $k$-partite clustering [40], an alternative to collective matrix factorization which assumes that rows and columns are hard-clustered under a Bregman divergence, minimized under alternating projections. The projection step requires solving a clustering problem, for example, using $k$-means. Under squared loss, the trilinear form can also be approximately minimized using a spectral clustering relaxation [41]. Or, for general Bregman divergences, the trilinear form can be minimized by alternating projections with iterative majorization for the projection [42]. A similar formulation in terms of log-likelihoods uses EM [43]. Banerjee et al. propose a model for Bregman clustering in matrices and tensors [12,44], which is based on Bregman information instead of divergence. While the above approaches generalize matrix co-clustering to the collective case, they make a clustering assumption. We show that both dimensionality reduction and clustering can be placed into the same framework. Additionally, we show that the difference in optimizing a dimensionality reduction and soft co-clustering model is small, an equality constraint in the Newton projection.

## 7 Experiments

We have argued that our alternating Newton-projection algorithm is a viable approach for training a wide variety of matrix factorization models. Two questions naturally arise: is it worthwhile to compute and invert the Hessian, or is gradient descent sufficient for projection? And, how does alternating Newton-projection compare to techniques currently used for specific factorization models?

While the Newton step is more expensive than the gradient step, our experiments indicate that it is definitely beneficial. To illustrate the point, we use an example of a three-factor model: $X^{(12)}$ corresponds to a user-movie matrix containing ratings, on a scale of 1–5 stars, from the Netflix competition [45]. There are $n_1 = 500$ users and $n_2 = 3000$ movies. Zeros in the ratings matrix correspond to unobserved entries, and are assigned zero weight. $X^{(23)}$ contains movie-genre information from IMDB [46], with $n_3 = 22$ genres. We reconstruct $X^{(12)}$ under I-divergence, and use its matching link—*i.e.*, $X_{ij}^{(12)}$ is Poisson distributed. We reconstruct $X^{(23)}$ under log-loss, and use its matching link—*i.e.*, $X_{js}^{(23)}$ is Bernoulli, a logistic model. From the same starting point, we measure the training loss of alternating projections using either a Newton step or a gradient step for each projection. The results in Fig. 2 are averaged over five trials, and clearly favour the Newton step.

The optimization over $\mathcal{L}$ is a non-convex problem, and the inherent complexity of the objective can vary dramatically from problem to problem. In some problems, our alternating Newton-projection approach appears to perform better than standard alternatives; however, we have found other problems where existing algorithms typically lead to better scores.

Logistic PCA is an example of where alternating projections can outperform an EM algorithm specifically designed for this model [29]. We use a binarized version of the rating matrix described above, whose entries indicate whether a user rated a movie. For the same settings, $k = 25$ and no regularization, we compare the test set error of the model learned using EM[4] vs. the same model learned using alternating projections.[5] Each method is run to convergence, a change of less than one percent in the objective between iterations, and the experiment is repeated ten times. The test error metric is balanced error rate, the average of the error rates on held-out positive and negative entries, so lower is better. Using the EM optimizer, the test error is $0.1813 \pm 0.020$; using alternating projections, the test error is $0.1253 \pm 0.0061$ (errors bars are 1-standard deviation).

Logistic Fast-MMMF is a variant of Fast-MMMF which uses log-loss and its matching link instead of smoothed Hinge loss, following [47]. Alternating Newton-projection does not outperform the recommended optimizer, conjugate gradients[6]. To compare the behaviour on multiple trials run to convergence, we use a small sample of the Netflix ratings data (250 users and 700 movies). Our evaluation metric is prediction accuracy on the held-out ratings under mean absolute error. On five repetitions with a rank $k = 20$ factorization, moderate $\ell_2$-regularization ($\lambda = 10^5$), and for the Newton step the Armijo procedure described above, the conjugate gradient solver yielded a model with zero error; the alternating-Newton method converged to models with test error $> 0.015$.

The performance of alternating Newton-projections suffers when $k$ is large. On a larger Netflix instance (30000 users, 2000 movies, 1.1M ratings) an iteration

[4] We use Schein et al.'s implementation of EM for Logistic PCA.

[5] We use an Armijo line search in the Newton projection, considering step lengths as small as $\eta = 2^{-4}$.

[6] We use Rennie et al.'s conjugate gradient code for Logistic Fast-MMMF.

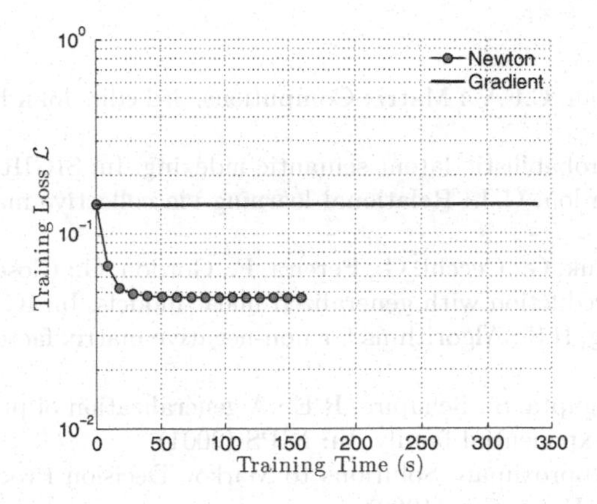


**Fig. 2.** Gradient vs. Newton steps in alternating projection

of our approach takes almost 40 minutes when $k = 100$; an iteration of the conjugate gradient implementation takes 80–120 seconds.

# 8 Conclusion

The vast majority of matrix factorization algorithms differ only in a small number of modeling choices: the prediction link, loss, constraints, data weights, and regularization. We have shown that a wide variety of popular matrix factorization approaches, such as weighted SVD, NMF, and MMMF, and pLSI can be distinguished by these modeling choices. We note that this unified view subsumes both dimensionality reduction and clustering in matrices using the bilinear model $X \approx f(UV^T)$, and that there is no conceptual difference between single- and multiple-matrix factorizations.

Exploiting a common property in matrix factorizations, decomposability of the loss, we extended a well-understood alternating projection algorithm to handle weights, bias/margin terms, $\ell_1$-regularization, and clustering constraints. In each projection, we recommended using Newton's method: while the Hessian is large, it is also block diagonal, which allows the update for a factor to be performed independently on each of its rows. We tested the relative merits of alternating Newton-projections against plain gradient descent, an existing EM approach for logistic PCA, and a conjugate gradient solver for MMMF.

## Acknowledgments

The authors thank Jon Ostlund for his assistance in merging the Netflix and IMDB data. This research was funded in part by a grant from DARPA's RADAR program. The opinions and conclusions are the authors' alone.

## References


1. Golub, G.H., Loan, C.F.V.: Matrix Computions, 3rd edn. John Hopkins University Press (1996)
2. Hofmann, T.: Probabilistic latent semantic indexing. In: SIGIR, pp. 50–57 (1999)
3. Singh, A.P., Gordon, G.J.: Relational learning via collective matrix factorization. In: KDD (2008)
4. Rish, I., Grabarnik, G., Cecchi, G., Pereira, F., Gordon, G.: Closed-form supervised dimensionality reduction with generalized linear models. In: ICML (2008)
5. Lee, D.D., Seung, H.S.: Algorithms for non-negative matrix factorization. In: NIPS (2001)
6. Collins, M., Dasgupta, S., Schapire, R.E.: A generalization of principal component analysis to the exponential family. In: NIPS (2001)
7. Gordon, G.J.: Approximate Solutions to Markov Decision Processes. PhD thesis. Carnegie Mellon University (1999)
8. Gordon, G.J.: Generalized$^2$ linear$^2$ models. In: NIPS (2002)
9. Bregman, L.: The relaxation method of finding the common points of convex sets and its application to the solution of problems in convex programming. USSR Comp. Math and Math. Phys. 7, 200–217 (1967)
10. Censor, Y., Zenios, S.A.: Parallel Optimization: Theory, Algorithms, and Applications. Oxford University Press, Oxford (1997)
11. Azoury, K.S., Warmuth, M.K.: Relative loss bounds for on-line density estimation with the exponential family of distributions. Mach. Learn. 43, 211–246 (2001)
12. Banerjee, A., Merugu, S., Dhillon, I.S., Ghosh, J.: Clustering with Bregman divergences. J. Mach. Learn. Res. 6, 1705–1749 (2005)
13. Forster, J., Warmuth, M.K.: Relative expected instantaneous loss bounds. In: COLT, pp. 90–99 (2000)
14. Aldous, D.J.: Representations for partially exchangeable arrays of random variables. J. Multivariate Analysis 11(4), 581–598 (1981)
15. Aldous, D.J.: 1. In: Exchangeability and related topics, pp. 1–198. Springer, Heidelberg (1985)
16. Welling, M., Rosen-Zvi, M., Hinton, G.: Exponential family harmoniums with an application to information retrieval. In: NIPS (2005)
17. Welling, M., Chemudugunta, C., Sutter, N.: Deterministic latent variable models and their pitfalls. In: SDM (2008)
18. Blei, D.M., Ng, A.Y., Jordan, M.I.: Latent Dirichlet allocation. J. Mach. Learn. Res. 3, 993–1022 (2003)
19. Koenker, R., Bassett, G.J.: Regression quantiles. Econometrica 46(1), 33–50 (1978)
20. Tibshirani, R.: Regression shrinkage and selection via the lasso. J. Royal. Statist. Soc. B. 58(1), 267–288 (1996)
21. Ding, C.H.Q., Li, T., Peng, W.: Nonnegative matrix factorization and probabilistic latent semantic indexing: Equivalence chi-square statistic, and a hybrid method. In: AAAI (2006)
22. Ding, C.H.Q., He, X., Simon, H.D.: Nonnegative Lagrangian relaxation of -means and spectral clustering. In: Gama, J., Camacho, R., Brazdil, P.B., Jorge, A.M., Torgo, L. (eds.) ECML 2005. LNCS (LNAI), vol. 3720, pp. 530–538. Springer, Heidelberg (2005)
23. Buntine, W.L., Jakulin, A.: Discrete component analysis. In: Saunders, C., Grobelnik, M., Gunn, S., Shawe-Taylor, J. (eds.) SLSFS 2005. LNCS, vol. 3940, pp. 1–33. Springer, Heidelberg (2006)

24. Gabriel, K.R., Zamir, S.: Lower rank approximation of matrices by least squares with any choice of weights. Technometrics 21(4), 489–498 (1979)
25. Srebro, N., Jaakola, T.: Weighted low-rank approximations. In: ICML (2003)
26. Hartigan, J.: Clustering Algorithms. Wiley, Chichester (1975)
27. Ke, Q., Kanade, T.: Robust $l_1$ norm factorization in the presence of outliers and missing data by alternative convex programming. In: CVPR, pp. 739–746 (2005)
28. Paatero, P., Tapper, U.: Positive matrix factorization: A non-negative factor model with optimal utilization of error estimates of data values. Environmetrics 5, 111–126 (1994)
29. Schein, A.I., Saul, L.K., Ungar, L.H.: A generalized linear model for principal component analysis of binary data. In: AISTATS (2003)
30. Srebro, N., Rennie, J.D.M., Jaakkola, T.S.: Maximum-margin matrix factorization. In: NIPS (2004)
31. Rennie, J.D.M., Srebro, N.: Fast maximum margin matrix factorization for collaborative prediction. In: ICML, pp. 713–719. ACM Press, New York (2005)
32. Nocedal, J., Wright, S.J.: Numerical Optimization. Series in Operations Research. Springer, Heidelberg (1999)
33. Schmidt, M., Fung, G., Rosales, R.: Fast optimization methods for L1 regularization: A comparative study and two new approaches. In: Kok, J.N., Koronacki, J., Lopez de Mantaras, R., Matwin, S., Mladenič, D., Skowron, A. (eds.) ECML 2007. LNCS (LNAI), vol. 4701, pp. 286–297. Springer, Heidelberg (2007)
34. Boyd, S., Vandenberghe, L.: Convex Optimization. Cambridge University Press, Cambridge (2004)
35. Pereira, F., Gordon, G.: The support vector decomposition machine. In: ICML, pp. 689–696. ACM Press, New York (2006)
36. Zhu, S., Yu, K., Chi, Y., Gong, Y.: Combining content and link for classification using matrix factorization. In: SIGIR, pp. 487–494. ACM Press, New York (2007)
37. Yu, K., Yu, S., Tresp, V.: Multi-label informed latent semantic indexing. In: SIGIR, pp. 258–265. ACM Press, New York (2005)
38. Yu, S., Yu, K., Tresp, V., Kriegel, H.P., Wu, M.: Supervised probabilistic principal component analysis. In: KDD, pp. 464–473 (2006)
39. Cohn, D., Hofmann, T.: The missing link–a probabilistic model of document content and hypertext connectivity. In: NIPS (2000)
40. Long, B., Wu, X., Zhang, Z.M., Yu, P.S.: Unsupervised learning on k-partite graphs. In: KDD, pp. 317–326. ACM Press, New York (2006)
41. Long, B., Zhang, Z.M., Wú, X., Yu, P.S.: Spectral clustering for multi-type relational data. In: ICML, pp. 585–592. ACM Press, New York (2006)
42. Long, B., Zhang, Z.M., Wu, X., Yu, P.S.: Relational clustering by symmetric convex coding. In: ICML, pp. 569–576. ACM Press, New York (2007)
43. Long, B., Zhang, Z.M., Yu, P.S.: A probabilistic framework for relational clustering. In: KDD, pp. 470–479. ACM Press, New York (2007)
44. Banerjee, A., Basu, S., Merugu, S.: Multi-way clustering on relation graphs. In: SDM (2007)
45. Netflix: Netflix prize dataset (January 2007), `http://www.netflixprize.com`
46. Internet Movie Database Inc.: IMDB alternate interfaces (January 2007), `http://www.imdb.com/interfaces`
47. Rennie, J.D.: Extracting Information from Informal Communication. PhD thesis, Massachusetts Institute of Technology (2007)

# Parallel Spectral Clustering

Yangqiu Song[1,4], Wen-Yen Chen[2,4], Hongjie Bai[4],
Chih-Jen Lin[3,4], and Edward Y. Chang[4]

[1]Department of Automation, Tsinghua University, Beijing, China
[2]Department of Computer Science, University of California, Santa Barbara, USA
[3]Department of Computer Science, National Taiwan University, Taipie, Taiwan
[4]Google Research, USA/China
songyq99@mails.tsinghua.edu.cn, wychen@cs.ucsb.edu, hjbai@google.com,
cjlin@csie.ntu.edu.tw, edchang@google.com

**Abstract.** Spectral clustering algorithm has been shown to be more effective in finding clusters than most traditional algorithms. However, spectral clustering suffers from a scalability problem in both memory use and computational time when a dataset size is large. To perform clustering on large datasets, we propose to parallelize both memory use and computation on distributed computers. Through an empirical study on a large document dataset of 193,844 data instances and a large photo dataset of 637,137, we demonstrate that our parallel algorithm can effectively alleviate the scalability problem.

**Keywords:** Parallel spectral clustering, distributed computing.

## 1 Introduction

Clustering is one of the most important subroutine in tasks of machine learning and data mining. Recently, spectral clustering methods, which exploit pairwise similarity of data instances, have been shown to be more effective than traditional methods such as $k$-means, which considers only the similarity to $k$ centers. (We denote $k$ as the number of desired clusters.) Because of its effectiveness in finding clusters, spectral clustering has been widely used in several areas such as information retrieval and computer vision. Unfortunately, when the number of data instances (denoted as $n$) is large, spectral clustering encounters a quadratic resource bottleneck in computing pairwise similarity between $n$ data instances and storing that large matrix. Moreover, the algorithm requires considerable computational time to find the smallest $k$ eigenvalues of a Laplacian matrix.

Several efforts have attempted to address aforementioned issues. Fowlkes et al. propose using the Nyström approximation to avoid calculating the whole similarity matrix [8]. That is, they trade accurate similarity values for shortened computational time. Dhillon et al. [4] assume the availability of the similarity matrix and propose a method that does not use eigenvectors. Although these methods can reduce computational time, they trade clustering accuracy for computational speed gain, or they do not address the bottleneck of memory use. In

W. Daelemans et al. (Eds.): ECML PKDD 2008, Part II, LNAI 5212, pp. 374–389, 2008.

**Table 1.** Notations. The following notations are used in the paper.

| | |
|---|---|
| $n$ | number of data |
| $d$ | dimensionality of data |
| $k$ | number of desired clusters |
| $p$ | number of nodes (distributed computers) |
| $t$ | number of nearest neighbors |
| $m$ | Arnoldi length in using an eigensolver |
| $\boldsymbol{x}_1, \ldots, \boldsymbol{x}_n \in R^d$ | data points |
| $S \in R^{n \times n}$ | similarity matrix |
| $L \in R^{n \times n}$ | Laplician matrix |
| $\boldsymbol{v}_1, \ldots, \boldsymbol{v}_k \in R^n$ | first $k$ eigenvectors of $L$ |
| $V \in R^{n \times k}$ | eigenvector matrix |
| $\boldsymbol{e}_1, \ldots, \boldsymbol{e}_k \in R^n$ | cluster indicator vectors |
| $E \in R^{n \times k}$ | cluster indicator matrix |
| $\boldsymbol{c}_1, \ldots, \boldsymbol{c}_k \in R^n$ | cluster centers of $k$-means |

this paper, we parallelize spectral clustering on distributed computers to address resource bottlenecks of both memory use and computation time. Parallelizing spectral clustering is much more challenging than parallelizing $k$-means, which was performed by e.g., [2,5,25].

Our parallelization approach first distributes $n$ data instances onto $p$ distributed machine nodes. On each node, we then compute the similarities between local data and the whole set in a way that uses minimal disk I/O. These two steps, together with parallel eigensolver and distributed tuning of parameters (including $\sigma$ of the Gaussian function and the initial $k$ centers of $k$-means), speed up clustering time substantially. Our empirical study validates that our parallel spectral clustering outperforms $k$-means in finding quality clusters and that it scales well with large datasets.

The remainder of this paper is organized as follows: In Section 2, we present spectral clustering and analyze its memory and computation bottlenecks. In Section 3, we show some obstacles for parallelization and propose our solutions to work around the challenges. Experimental results in Section 4 show that our parallel spectral clustering algorithm achieves substantial speedup on 128 machines. The resulting cluster quality is better than that of $k$-means. Section 5 offers our concluding remarks.

## 2 Spectral Clustering

We present the spectral clustering algorithm in this section so as to understand the bottlenecks of its resources. To assist readers, Table 1 defines terms and notations used throughout this paper.

### 2.1 Basic Concepts

Given $n$ data points $\boldsymbol{x}_1, \ldots, \boldsymbol{x}_n$, the spectral clustering algorithm constructs a similarity matrix $S \in R^{n \times n}$, where $S_{ij} \geq 0$ reflects the relationships between $\boldsymbol{x}_i$

and $\boldsymbol{x}_j$. It then uses the similarity information to group $\boldsymbol{x}_1, \ldots, \boldsymbol{x}_n$ into $k$ clusters. There are many variants of spectral clustering. Here we consider a commonly used *normalized* spectral clustering [19]. (For a complete account of all variants, please see [17].) An example similarity function is the Gaussian:

$$S_{ij} = \exp\left(-\frac{\|\boldsymbol{x}_i - \boldsymbol{x}_j\|^2}{2\sigma^2}\right). \tag{1}$$

In our implementation, we use an adaptive approach to decide the parameter $\sigma^2$ (details are presented in Section 3.4). For conserving computational time, one often reduces the matrix $S$ to a sparse one by considering only significant relationship between data instances. For example, we may retain $S_{ij}$ satisfying that $j$ (or $i$) is among the $t$-nearest neighbors of $i$ (or $j$). Typically $t$ is a small number (e.g., $t$ a small fraction of $n$ or $t = \log n$)[1].

Consider the normalized Laplacian matrix [3]:

$$L = I - D^{-1/2} S D^{-1/2}, \tag{2}$$

where $D$ is a diagonal matrix with

$$D_{ii} = \sum_{j=1}^{n} S_{ij}.$$

In the ideal case, where data in one cluster are not related to those in others, non-zero elements of $S$ (and hence $L$) only occur in a block diagonal form:

$$L = \begin{bmatrix} L_1 & & \\ & \ddots & \\ & & L_k \end{bmatrix}.$$

It is known that $L$ has $k$ zero eigenvalues, which are also the $k$ smallest ones [17, Proposition 4]. Their corresponding eigenvectors, written as an $R^{n \times k}$ matrix, are

$$V = [\boldsymbol{v}_1, \boldsymbol{v}_2, \ldots, \boldsymbol{v}_k] = D^{1/2} E, \tag{3}$$

where $\boldsymbol{v}_i \in R^{n \times 1}, i = 1, \ldots, k$.

$$E = \begin{bmatrix} \boldsymbol{e}_1 & & \\ & \ddots & \\ & & \boldsymbol{e}_k \end{bmatrix}, \tag{4}$$

where $\boldsymbol{e}_i, i = 1, \ldots, k$ (in different length) are vectors of all ones. As $D^{1/2}E$ has the same structure as $E$, simple clustering algorithms such as $k$-means can easily

[1] Another simple strategy for making $S$ a sparse matrix is to zero out those $S_{ij}$ larger than a pre-specified threshold. Since the focus of this paper is on speeding up spectral clustering, we do not compare different methods to make a matrix sparse. Nevertheless, our empirical study shows that the $t$-nearest-neighbor approach yields good results.

**Algorithm 1.** Spectral Clustering

Input: Data points $\boldsymbol{x}_1, \ldots, \boldsymbol{x}_n$; $k$: number of clusters to construct.

1. Construct similarity matrix $S \in R^{n \times n}$.
2. Modify $S$ to be a sparse matrix.
3. Compute the Laplacian matrix $L$ by Eq. (2).
4. Compute the first $k$ eigenvectors of $L$; and construct $V \in R^{n \times k}$, which columns are the $k$ eigenvectors.
5. Compute the normalized matrix $U$ of $V$ by Eq. (5).
6. Use $k$-means algorithm to cluster $n$ rows of $U$ into $k$ groups.

cluster the $n$ rows of $V$ into $k$ groups. Thus, what one needs is to find the first $k$ eigenvectors of $L$ (i.e., eigenvectors corresponding to the $k$ smallest eigenvalues). However, practically eigenvectors we obtained are in the form of

$$V = D^{1/2}EQ,$$

where $Q$ is an orthogonal matrix. Ng et al. [19] propose normalizing $V$ so that

$$U_{ij} = \frac{V_{ij}}{\sqrt{\sum_{r=1}^{k} V_{ir}^2}}, i = 1, \ldots, n, j = 1, \ldots, k. \tag{5}$$

The row sum of $U$ is one. Due to the orthogonality of $Q$, (5) is equivalent to

$$U = EQ = \begin{bmatrix} Q_{1,1:k} \\ \vdots \\ Q_{1,1:k} \\ Q_{2,1:k} \\ \vdots \end{bmatrix}, \tag{6}$$

where $Q_{i,1:k}$ indicates the $i^{th}$ row of $Q$. Then $U$'s $n$ rows correspond to $k$ orthogonal points on the unit sphere. The $n$ rows of $U$ can thus be easily clustered by $k$-means or other simple clustering algorithms. A summary of the method is presented in Algorithm 1.

Instead of analyzing properties of the Laplacian matrix, spectral clustering algorithms can be derived from the graph cut point of view. That is, we partition the matrix according to the relationship between points. Some representative graph-cut methods are Normalized Cut [20], Min-Max Cut [7] and Radio Cut [9].

## 2.2 Computational Complexity and Memory Usage

Let us examine computational cost and the memory use of Algorithm 1. We omit discussing some inexpensive steps.

**Construct the similarity matrix.** Assume each $S_{ij}$ involves at least an inner product between $\boldsymbol{x}_i$ and $\boldsymbol{x}_j$. The cost of obtaining an $S_{ij}$ is $O(d)$, where $d$ is the dimensionality of data. Constructing similarity matrix $S$ requires

$$O(n^2 d) \text{ time and } O(n^2) \text{ memory.} \tag{7}$$

To make $S$ a sparse matrix, we employ the approach of $t$-nearest neighbors and retain only $S_{ij}$ where $i$ (or $j$) is among the $t$-nearest neighbors of $j$ (or $i$). By scanning once of $S_{ij}$ for $j = 1, \ldots, n$ and keeping a max heap with size $t$, we sequentially insert the similarity that is smaller than the maximal value of the heap and then restructure the heap. Thus, the complexity for one point $\boldsymbol{x}_i$ is $O(n \log t)$ since restructuring a max heap is in the order of $\log t$. The overall complexity of making the matrix $S$ to sparse is

$$O(n^2 \log t) \text{ time and } O(nt) \text{ memory.} \tag{8}$$

**Compute the first $k$ eigenvectors.** Once that $S$ is sparse, we can use sparse eigensolvers. In particular, we desire a solver that can quickly obtain the first $k$ eigenvectors of $L$. Some example solvers are [11,13] (see [10] for a comprehensive survey). Most existing approaches are variants of the Lanczos/Arnoldi factorization. We employ a popular eigensolver ARPACK [13] and its parallel version PARPACK [18]. ARPACK implements an implicitly restarted Arnoldi method. We briefly describe its basic concepts hereafter; more details can be found in the user guide of ARPACK. The $m$-step Arnoldi factorization gives that

$$LV = VH + (\text{a matrix of small values}), \tag{9}$$

where $V \in R^{n \times m}$ and $H \in R^{m \times m}$ satisfy certain properties. If the "matrix of small values" in (9) is indeed zero, then $V$'s $m$ columns are $L$'s first $m$ eigenvectors. Therefore, (9) provides a way to check how good we approximate eigenvectors of $L$. To perform this check, one needs all eigenvalues of the dense matrix $H$, a procedure taking $O(m^3)$ operations. For quickly finding the first $k$ eigenvectors, ARPACK employs an iterative procedure called "implicitly restarted" Arnoldi. Users specify an Arnoldi length $m > k$. Then at each iteration (restarted Arnoldi) one uses $V$ and $H$ of the previous iteration to conduct the eigendecomposition of $H$, and finds a new Arnoldi factorization. Each Arnoldi factorization involves at most $(m - k)$ steps, where each step's main computational complexity is $O(nm)$ for a few dense matrix-vector products and $O(nt)$ for a sparse matrix-vector product. In particular, $O(nt)$ is for

$$L\boldsymbol{v}, \tag{10}$$

where $\boldsymbol{v}$ is an $n \times 1$ vector. As on average the number of nonzeros per row of $L$ is $O(t)$, the cost of this sparse matrix multiply is $O(nt)$.

Based on the above analysis, the overall cost of ARPACK is

$$(O(m^3) + (O(nm) + O(nt)) \times O(m - k)) \times (\# \text{ restarted Arnoldi}), \tag{11}$$

where $O(m-k)$ is a value no more than $m-k$. Obviously, the value $m$ selected by users affects the computational time. One often sets $m$ to be several times larger than $k$. The memory requirement of ARPACK is $O(nt)+O(nm)$.

**$k$-means to cluster the normalized matrix $U$.** Algorithm $k$-means aims at minimizing the total intra-cluster variance, which is the squared error function in the spectral space:

$$J = \sum_{i=1}^{k} \sum_{\boldsymbol{u}_j \in C_i} ||\boldsymbol{u}_j - \boldsymbol{c}_i||^2. \tag{12}$$

We assume that data are in $k$ clusters $C_i, \{i = 1, 2, \ldots, k\}$, and $\boldsymbol{c}_i \in R^{k \times 1}$ is the centroid of all the points $\boldsymbol{u}_j \in C_i$. Similar to Step 5 in Algorithm 1, we also normalize centers $\boldsymbol{c}_i$ to be of unit length.

The traditional $k$-means algorithm employs an iterative procedure. At each iteration, we assign each data point to the cluster of its nearest center, and recalculate cluster centers. The procedure stops after reaching a stable error function value. Since the algorithm evaluates the distance between any point and the current $k$ cluster centers, the time complexity of $k$-means is

$$O(nk^2) \times \# \text{ } k\text{-means iterations}. \tag{13}$$

**Overall analysis.** The step that consumes the most memory is constructing the similarity matrix. For instance, $n = 600,000$ data instances, assuming double precision storage, requires 2.8 Tera Bytes of memory, which is not available on a general-purpose machine. Since we make $S$ sparse, $O(nt)$ memory space may suffice. However, if $n$ is huge, say in billions, no single general-purpose machine can handle such a large memory requirement. Moreover, the $O(n^2d)$ computational time in (7) is a bottleneck. This bottleneck has been discussed in earlier work. For example, the authors of [16] state that "The majority of the time is actually spent on constructing the pairwise distance and affinity matrices. Comparatively, the actually clustering is almost *negligible*."

# 3 Parallel Spectral Clustering

Based on the analysis performed in Section 2.2, it is essential to conduct spectral clustering in a distributed environment to alleviate both memory and computational bottlenecks. In this section, we discuss these challenges and then propose our solutions. We implement our system on a distributed environment using Message Passing Interface (MPI) [22].

## 3.1 Similarity Matrix and Nearest Neighbors

Suppose $p$ machines (or nodes) are allocated in a distributed environment for our target clustering task. Figure 1 shows that we first let each node construct $n/p$ rows of the similarity matrix $S$. We illustrate our procedure using the first node, which is responsible for rows 1 to $n/p$. To obtain the $i^{th}$ row, we use Eq. (1) to

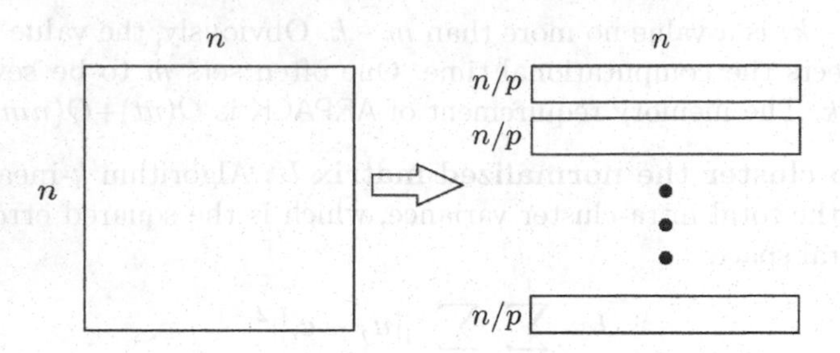


**Fig. 1.** The similarity matrix is distributedly stored in multiple machines

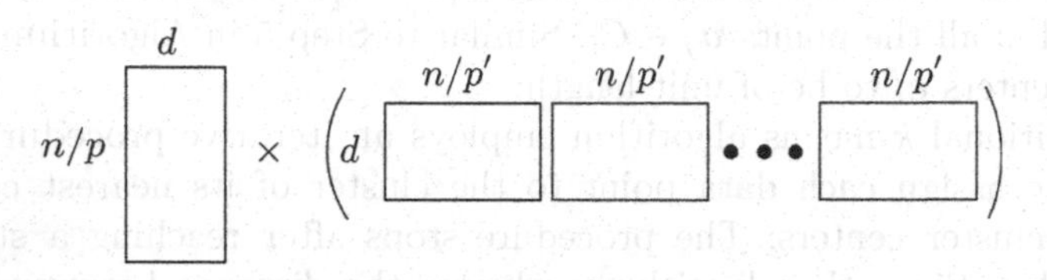


**Fig. 2.** Calculating $n/p$ rows of the similarity at a node. We use matrix-matrix products for inner products between $n/p$ points and all data $\boldsymbol{x}_1, \ldots, \boldsymbol{x}_n$. As data cannot be loaded into memory, we separate $\boldsymbol{x}_1, \ldots, \boldsymbol{x}_n$ into $p'$ blocks.

calculate the similarity between $\boldsymbol{x}_i$ and all the data points, respectively. Using $\|\boldsymbol{x}_i - \boldsymbol{x}_j\|^2 = \|\boldsymbol{x}_i\|^2 + \|\boldsymbol{x}_j\|^2 - 2\boldsymbol{x}_i^T \boldsymbol{x}_j$ to compute similarity between instances $\boldsymbol{x}_i$ and $\boldsymbol{x}_j$, we can precompute $\|\boldsymbol{x}_i\|^2$ for all instances and cache on all nodes to conserve time.

Let $X = [\boldsymbol{x}_1, \ldots, \boldsymbol{x}_n] \in R^{d \times n}$ and $X_{1:n/p} = [\boldsymbol{x}_1, \ldots, \boldsymbol{x}_{n/p}]$. One can perform a matrix-matrix product to obtain $X_{1:n/p}^T X$. If the memory of a node cannot store the entire $X$, we can split $X$ into $p'$ blocks as shown in Figure 2. When each of the $p'$ blocks is memory resident, we multiply it and $X_{1:n/p}^T$.

When data are densely stored, even if $X$ can fit into the main memory, splitting $X$ into small blocks takes advantage of optimized BLAS (Basic Linear Algebra Subroutines) [1]. BLAS places the inner-loop data instances in CPU cache and ensures their cache residency. Table 2 compares the computational time with and without BLAS. It shows that blocking operation can reduce the computational time significantly.

### 3.2 Parallel Eigensolver

After we have calculated and stored the similarity matrix, it is important to parallelize the eigensolver. Section 3.1 shows that each node now stores $n/p$ rows of $L$. For the eigenvector matrix $V$ (see (3)) generated during the call to ARPACK, we also split it into $p$ partitions, each of which possesses $n/p$ rows. As mentioned in Section 2.2, major operations at each step of the Arnoldi factorization include a few dense and a sparse matrix-vector multiplications, which cost $O(mn)$ and

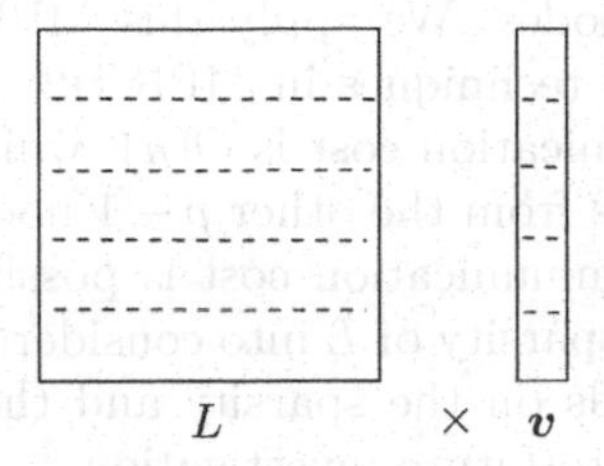


**Fig. 3.** Sparse matrix-vector multiplication. We assume $p = 5$ here. $L$ and $\boldsymbol{v}$ are respectively separated to five block partitions.

**Table 2.** Computational time (in seconds) for the similarity matrix ($n = 637, 137$ and number of features $d = 144$)

| 1 machine without BLAS | 1 machine with BLAS | 16 machines with BLAS |
|---|---|---|
| $3.14 \times 10^5$ | $6.40 \times 10^4$ | $4.00 \times 10^3$ |

$O(nt)$, respectively. We parallelize these computations so that the complexity of finding eigenvectors becomes:

$$\left(O(m^3) + (O(\frac{nm}{p}) + O(\frac{nt}{p})) \times O(m-k)\right) \times (\text{\# restarted Arnoldi}). \quad (14)$$

Note that communication overhead between nodes occurs in the following three situations:

1. Sum $p$ values and broadcast the result to $p$ nodes.
2. Parallel sparse matrix-vector product (10).
3. Dense matrix-vector product: Sum $p$ vectors of length $m$ and broadcast the resulting vector to all $p$ nodes.

The first and the third cases transfer only short vectors, but the sparse matrix vector product may move a larger vector $\boldsymbol{v} \in R^n$ to several nodes. We next discuss how to conduct the parallel sparse matrix-vector product to reduce communication cost.

Figure 3 shows matrix $L$ and vector $\boldsymbol{v}$. Suppose $p = 5$. The figure shows that both $L$ and $\boldsymbol{v}$ are horizontally split into 5 parts and each part is stored on one computer node. Take node 1 as an example. It is responsible to perform

$$L_{1:n/p,1:n} \times \boldsymbol{v}, \quad (15)$$

where $\boldsymbol{v} = [v_1, \ldots, v_n]^T \in R^n$. $L_{1:n/p,1:n}$, the first $n/p$ rows of $L$, is stored at node 1, but only $v_1, \ldots, v_{n/p}$ are available at node 1. Hence other nodes must send to node 1 the elements $v_{n/p+1}, \ldots, v_n$. Similarly, node 1 should dispatch its $v_1, \ldots, v_{n/p}$ to other nodes. This task is a gather operation in MPI: data at each node are gathered on all nodes. We apply this MPI operation on distributed computers by following the techniques in MPICH2[2] [24], a popular implementation of MPI. The communication cost is $O(n)$, which cannot be reduced as a node must get $n/p$ elements from the other $p-1$ nodes.

Further reducing the communication cost is possible only we reduce $n$ to a fraction of $n$ by taking the sparsity of $L$ into consideration. The reduction of the communication cost depends on the sparsity and the structure of the matrix. We defer this optimization to future investigation.

### 3.3 Parallel $k$-Means

After the eigensolver computes the first $k$ eigenvectors of Laplacian, the matrix $V$ is distributedly stored. Thus the normalized matrix $U$ can be computed in parallel and stored on $p$ local machines. Each row of the matrix $U$ is regarded as one data point in the $k$-means algorithm. To start the $k$-means procedure, the master machine chooses a set of initial cluster centers and broadcasts them to all machines. (See next section for our distributed initialization procedure.) At each node, new labels of its data are assigned and local sums of clusters are calculated without any inter-machine communication. The master machine then obtains the sum of all points in each cluster to calculate new centers. The loss function (12) can also be computed in parallel in a similar way. Therefore, the computational time of parallel $k$-means is reduced to $1/p$ of that in (13). The communication cost per iteration is on broadcasting $k$ centers to all machines. If $k$ is not large, the total communication cost is usually smaller than that involved in finding the first $k$ eigenvectors.

### 3.4 Other Implementation Details

We discuss two implementation issues of the parallel spectral clustering algorithm. The first issue is that of assigning parameters in Gaussian function (1), and the second is initializing the centers for $k$-means.

**Parameters in Gaussian function.** We adopt the self-tuning technique [27] to adaptively assign the parameter $\sigma$ in (1). The original method used in [27] is

$$S_{ij} = \exp\left(-\frac{\|\boldsymbol{x}_i - \boldsymbol{x}_j\|^2}{2\sigma_i\sigma_j}\right). \tag{16}$$

Suppose $\boldsymbol{x}_i$ has $t$ nearest neighbors. If we sort these neighbors in ascending order, $\sigma_i$ is defined as the distance between $\boldsymbol{x}_i$ and $\boldsymbol{x}_{i_t}$, the $\lfloor t/2 \rfloor$th neighbor of $\boldsymbol{x}_i$: $\sigma_i = \|\boldsymbol{x}_i - \boldsymbol{x}_{i_t}\|$. Alternatively, we can consider the average distance between $\boldsymbol{x}_i$ and its $t$ nearest neighbors[3]. In a parallel environment, each local machine first computes $\sigma_i$'s of its local data points. Then $\sigma_i$'s are gathered on all machines.

**Initialization of $k$-means.** Revisit (6). In the ideal case, the centers of data instances calculated based on the matrix $U$ are orthogonal to each other. Thus, an

[2] http://www.mcs.anl.gov/research/projects/mpich2

[3] In the experiments, we use the average distance as our self-tuning parameters.

intuitive initialization of centers can be done by selecting a subset of $\{\boldsymbol{x}_1, \ldots, \boldsymbol{x}_n\}$ whose elements are almost orthogonal [26]. To begin, we use the master machine to randomly choose a point as the first cluster center. Then it broadcasts the center to all machines. Each machine identifies the most orthogonal point to this center by finding the minimal cosine distance between its points and the center. By gathering the information of different machines, we choose the most orthogonal point to the first center as the second center. This procedure is repeated to obtain $k$ centers. The communication involved in the initialization includes broadcasting $k$ cluster centers and gathering $k \times p$ minimal cosine distances.

# 4 Experiments

We designed our experiments to validate the quality of parallel spectral clustering and its scalability. Our experiments used two large datasets: 1) RCV1 (Reuters Corpus Volume I), a filtered collection of 193,844 documents, and 2) 637,137 photos collected from PicasaWeb, a Google photo sharing product. We ran experiments on up to 256 machines at our distributed data centers. While not all machines are identical, each machine was configured with a CPU faster than 2GHz and memory larger than 4GBytes. All reported results are the average of nine runs.

## 4.1 Clustering Quality

To check the performance of spectral clustering algorithm, we compare it with traditional $k$-means. We looked for a dataset with ground truth. RCV1 is an archive of 804,414 manually categorized newswire stories from Reuters Ltd [14]. The news documents are categorized with respect to three controlled vocabularies: *industries*, *topics* and *regions*. Data were split into 23,149 training documents and 781,256 test documents. In this experiment, we used the test set and category codes based on the *industries* vocabulary. There are originally 350 categories in the test set. For comparing clustering results, data which are multi-labeled were not considered, and categories which contain less than 500 documents were removed. We obtained 193,844 documents and 103 categories. Each document is represented by a cosine normalization of a log transformed TF-IDF (term frequency, inverse document frequency) feature vector.

For both spectral and $k$-means, we set the number of clusters to be 103, and Arnoldi space dimension $m$ to be two times the number of clusters. We used the *document categories* in the RCV1 dataset as the ground truth for evaluating cluster quality. We measured quality via using the Normalized Mutual Information (NMI) between the produced clusters and the ground-truth categories.

NMI between two random variables CAT (category label) and CLS (cluster label) is defined as $\text{NMI(CAT; CLS)} = \frac{\text{I(CAT; CLS)}}{\sqrt{\text{H(CAT)H(CLS)}}}$, where I(CAT; CLS) is the mutual information between CAT and CLS. The entropies H(CAT) and H(CLS) are used for normalizing the mutual information to be in the range of

**Table 3.** NMI comparisons for $k$-means, spectral clustering with 100 nearest neighbors

| Algorithms | E-$k$-means | S-$k$-means | Spectral Clustering |
|---|---|---|---|
| NMI | 0.2586(±0.0086) | 0.2702(±0.0059) | 0.2875(±0.0011) |

$[0, 1]$. In practice, we made use of the following formulation to estimate the NMI score [23]:

$$\mathrm{NMI} = \frac{\sum_{i=1}^{k} \sum_{j=1}^{k} n_{i,j} \log\left(\frac{n \cdot n_{i,j}}{n_i \cdot n_j}\right)}{\sqrt{\left(\sum_i n_i \log \frac{n_i}{n}\right)\left(\sum_j n_j \log \frac{n_j}{n}\right)}}, \tag{17}$$

where $n$ is the number of documents, $n_i$ and $n_j$ denote the number of documents in category $i$ and cluster $j$, respectively, and $n_{i,j}$ denotes the number of documents in category $i$ as well as in cluster $j$. The NMI score is 1 if the clustering results perfectly match the category labels, and the score is 0 if data are randomly partitioned. The higher this NMI score, the better the clustering quality.

We compared $k$-means algorithm based on Euclidean distance (E-$k$-means), spherical $k$-means based on cosine distance (S-$k$-means) [6], and our parallel spectral clustering algorithm using 100 nearest neighbors. Table 3 reports that parallel spectral clustering algorithm outperforms E-$k$-means and S-$k$-means. This result confirms parallel spectral clustering to be effective in finding clusters.

### 4.2 Scalability: Runtime Speedup

We used both the RCV1 dataset and a PicasaWeb dataset to conduct a scalability experiment. The RCV1 can fit into main memory of one machine, whereas the PicasaWeb dataset cannot. PicasaWeb is an online platform for users to upload, share and manage images. The PicasaWeb dataset we collected consists of 637, 137 images accompanied with 110, 342 tags.

For each image, we extracted 144 features including color, texture, and shape as its representation [15]. In the color channel, we divided color into 12 color bins including 11 bins for culture colors and one bin for outliers [12]. For each color bin, we recorded nine features to capture color information at finer resolution. The nine features are color histogram, color means (in H, S, and V channels), color variances (in H, S, and V channels), and two shape characteristics: elongation and spreadness. Color elongation defines the shape of color, and spreadness defines how the color scatters within the image. In the texture channel, we employed a discrete wavelet transformation (DWT) using quadrature mirror filters [21] due to its computational efficiency. Each DWT on a image yielded four subimages including scale-down image and its wavelets in three orientations. We then obtained nine texture combinations from subimages of three scales (coarse, medium, fine) and three orientations (horizontal, vertical, diagonal). For each texture, we recorded four features: energy mean, energy variance, texture elongation and texture spreadness.

**Table 4.** RCV1 data set. Runtime comparisons for different number of machines. $n$=193,844, $k$=103, $m$=206.

| | Eigensolver | | $k$-means | |
|---|---|---|---|---|
| Machines | Time (sec.) | Speedup | Time (sec.) | Speedup |
| 1 | $9.90 \times 10^2$ | 1.00 | $4.96 \times 10^1$ | 1.00 |
| 2 | $4.92 \times 10^2$ | 2.01 | $2.64 \times 10^1$ | 1.88 |
| 4 | $2.83 \times 10^2$ | 3.50 | $1.53 \times 10^1$ | 3.24 |
| 8 | $1.89 \times 10^2$ | 5.24 | $1.10 \times 10^1$ | 4.51 |
| 16 | $1.47 \times 10^2$ | 6.73 | $9.90 \times 10^0$ | 5.01 |
| 32 | $1.29 \times 10^2$ | 7.67 | $1.05 \times 10^1$ | 4.72 |
| 64 | $1.30 \times 10^2$ | 7.62 | $1.34 \times 10^1$ | 3.70 |

We first report the speedup on the RCV1 dataset in Table 4. As discussed in Section 3.1, the computation of similarity matrix can achieve linear speedup. In this experiment, we focus on the time of finding the first $k$ eigenvectors and conducting $k$-means. Here the $k$-means is referred to Step 6 in Algorithm 1. It is important to notice that we could not ensure the quiesce of the allocated machines at Google's distributed data centers. There were almost always other jobs running simultaneously with ours on each machine. Therefore, the runtime is partially dependent on the slowest machine being allocated for the task. (We consider an empirical setting like this to be reasonable, since no modern machine is designed or expected to be single task.) When 32 machines were used, the parallel version of eigensolver achieved 7.67 times speedup. When more machines were used, the speedup actually decreased. Similarly, we can see that parallelization sped up $k$-means more than five times when 16 machines were used. The speedup is encouraging. For a not-so-large dataset like RCV1, the Amdahl's law kicks in around $p = 16$. Since the similarity matrix in this case is not huge, the communication cost dominates computation time, and hence further increasing machines does not help. (We will see next that the larger a dataset, the higher speedup our parallel implementation can achieve.)

Next, we looked into the speedup on the PicasaWeb dataset. We grouped the data into $1,000$ clusters, where the corresponding Arnoldi space is set to be $2,000$. Note that storing the eigenvectors in Arnoldi space with $2,000$ dimensions requires 10GB of memory. This memory configuration is not available on off-the-shelf machines. We had to use at least two machines to perform clustering. We thus used two machines as the baseline and assumed the speedup of two machines is 2. This assumption is reasonable since we will see shortly that our parallelization can achieve linear speedup on up to 32 machines.

Table 5 reports the speedups of eigensovler and $k$-means. We can see in the table that both eigensolver and $k$-means enjoy near-linear speedups when the number of machine is up to 32. For more than 32 machines, the speedups of $k$-means are better than that of eigensolver. However both speedups became sublinear as the synchronization and communication overheads started to slow down the speedups. The "saturation" point on the PicasaWeb dataset is $p = 128$ machines. Using more than 128 machines is counter-productive to both steps.

**Table 5.** Picasa data set. Runtime comparisons for different numbers of machines. $n$=637,137, $k$=1,000, $m$=2,000.

| | Eigensolver | | $k$-means | |
|---|---|---|---|---|
| Machines | Time (sec.) | Speedup | Time (sec.) | Speedup |
| 1 | – | – | – | – |
| 2 | $8.074 \times 10^4$ | 2.00 | $3.609 \times 10^4$ | 2.00 |
| 4 | $4.427 \times 10^4$ | 3.65 | $1.806 \times 10^4$ | 4.00 |
| 8 | $2.184 \times 10^4$ | 7.39 | $8.469 \times 10^3$ | 8.52 |
| 16 | $9.867 \times 10^3$ | 16.37 | $4.620 \times 10^3$ | 15.62 |
| 32 | $4.886 \times 10^3$ | 33.05 | $2.021 \times 10^3$ | 35.72 |
| 64 | $4.067 \times 10^3$ | 39.71 | $1.433 \times 10^3$ | 50.37 |
| 128 | $3.471 \times 10^3$ | 46.52 | $1.090 \times 10^3$ | 66.22 |
| 256 | $4.021 \times 10^3$ | 40.16 | $1.077 \times 10^3$ | 67.02 |

(a) Sample images of $k$-means.

(b) Sample images of spectral clustering.

**Fig. 4.** Clustering results of $k$-means and spectral clustering

From the experiments with RCV1 and PicasaWeb, we can observe that the larger a dataset, the more machines can be employed to achieve higher speedup. Since several computation intensive steps grow faster than the communication cost, the larger the dataset is, the more opportunity is available for parallelization to gain speedup.

Figure 4 shows sample clusters generated by $k$-means and spectral clustering. The top two rows are clusters generated by $k$-means, the bottom two rows are by spectral clustering. First, spectral clustering finds lions and leopards more effectively. Second, in the flower cluster, spectral clustering can find flowers of different colors, whereas $k$-means is less effective in doing that. Figure 5 provides

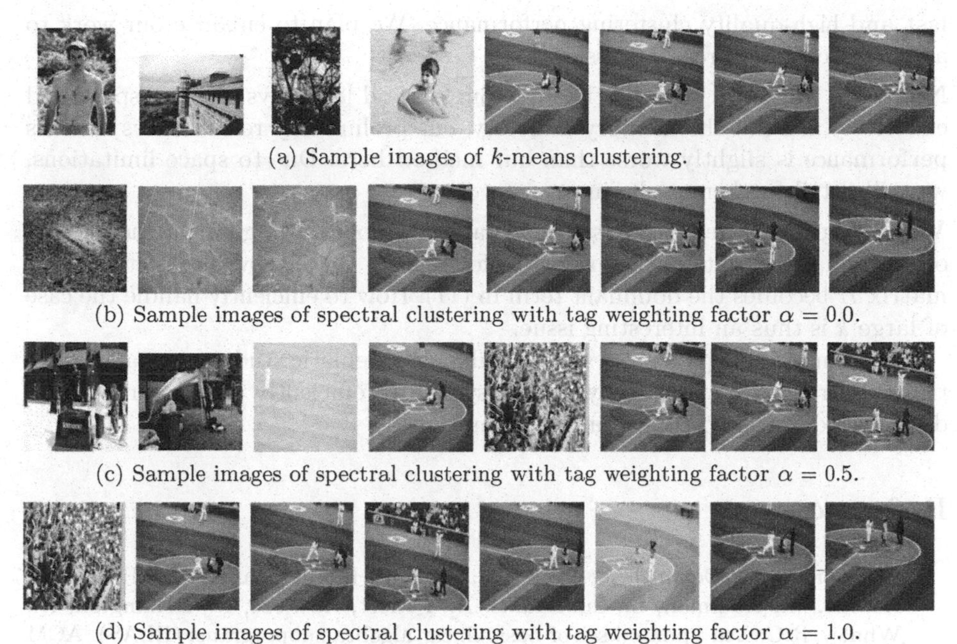

(a) Sample images of $k$-means clustering.

(b) Sample images of spectral clustering with tag weighting factor $\alpha = 0.0$.

(c) Sample images of spectral clustering with tag weighting factor $\alpha = 0.5$.

(d) Sample images of spectral clustering with tag weighting factor $\alpha = 1.0$.

**Fig. 5.** Clustering results of $k$-means and spectral clustering. The cluster topic is "baseball game."

a visual comparison of the clustering results produced by four different clustering schemes (of ours). On the top is our parallel $k$-means. Rows 2 to 4 display results of using parallel spectral clustering with different tag weighting settings ($\alpha$). In addition to perceptual features, tags are useful for image searching and clustering. We use the tag weighting factor $\alpha$ to incorporate tag overlapping information in constructing the similarity matrix. The more tags are overlapped between images, the larger the similarity between the images. When the tag weighting factor is set to zero, spectral clustering considers only the 144 perceptual features depicted in the beginning of this section. When tag information is incorporated, we can see that the clustering performance improves. Though we cannot use one example in Figure 5 to prove that the spectral clustering algorithm is always superior to $k$-means, thanks to the kernel, spectral clustering seems to be more effective in identifying clusters of non-linear boundaries (such as photo clusters).

## 5 Conclusions

In this paper, we have shown our parallel implementation of the spectral clustering algorithm to be both correct and scalable. No parallel algorithm can escape from the Amdahl's law, but we showed that the larger a dataset, the more machines can be employed to use parallel spectral clustering algorithm to enjoy fast and high-quality clustering performance. We plan to enhance our work to address a couple of research issues.

**Nyström method.** Though the Nyström method [8] enjoys a better speed and effectively handles the memory difficulty, our preliminary result shows that its performance is slightly worse than our method here. Due to space limitations, we will detail further results in future work.

**Very large number of clusters.** A large $k$ implies a large $m$ in the process of Arnoldi factorization. Then $O(m^3)$ for finding the eigenvalues of the dense matrix $H$ becomes the dominant term in (11). How to efficiently handle the case of large $k$ is thus an interesting issue.

In summary, this paper gives a general and systematic study on parallel spectral clustering. We successfully built a system to efficiently cluster large image data on a distributed computing environment.

## References


1. Blackford, L.S., Demmel, J., Dongarra, J., Duff, I., Hammarling, S., Henry, G., Heroux, M., Kaufman, L., Lumsdaine, A., Petitet, A., Pozo, R., Remington, K., Whaley, R.C.: An updated set of basic linear algebra subprograms (BLAS). ACM Trans. Math. Software 28(2), 135–151 (2002)
2. Chu, C.-T., Kim, S.K., Lin, Y.-A., Yu, Y., Bradski, G., Ng, A.Y., Olukotun, K.: Map-reduce for machine learning on multicore. In: Proceedings of NIPS, pp. 281–288 (2007)
3. Chung, F.: Spectral Graph Theory. CBMS Regional Conference Series in Mathematics, vol. 92. American Mathematical Society (1997)
4. Dhillon, I.S., Guan, Y., Kulis, B.: Weighted graph cuts without eigenvectors: A multilevel approach. IEEE Trans. on Pattern Analysis and Machine Intelligence 29(11), 1944–1957 (2007)
5. Dhillon, I.S., Modha, D.S.: A data-clustering algorithm on distributed memory multiprocessors. In: Large-Scale Parallel Data Mining, pp. 245–260 (1999)
6. Dhillon, I.S., Modha, D.S.: Concept decompositions for large sparse text data using clustering. Machine Learning 42(1-2), 143–175 (2001)
7. Ding, C.H.Q., He, X., Zha, H., Gu, M., Simon, H.D.: A min-max cut algorithm for graph partitioning and data clustering. In: Proceedings of ICDM (2001)
8. Fowlkes, C., Belongie, S., Chung, F., Malik, J.: Spectral grouping using the Nyström method. IEEE Trans. on Pattern Analysis and Machine Intelligence 26(2), 214–225 (2004)
9. Hagen, L., Kahng, A.: New spectral methods for ratio cut partitioning and clustering. IEEE Trans. on Computer-Aided Design of Integrated Circuits and Systems 11(9), 1074–1085 (1992)
10. Hernandez, V., Roman, J., Tomas, A., Vidal, V.: A Survey of Software for Sparse Eigenvalue Problems. Technical report, Universidad Politecnica de Valencia (2005)
11. Hernandez, V., Roman, J., Vidal, V.: SLEPc: A scalable and exible toolkit for the solution of eigenvalue problems. ACM Trans. Math. Software 31, 351–362 (2005)
12. Hua, K.A., Vu, K., Oh, J.-H.: Sammatch: a exible and efficient sampling-based image retrieval technique for large image databases. In: Proceedings of ACM MM, pp. 225–234. ACM, New York (1999)

13. Lehoucg, R.B., Sorensen, D.C., Yang, C.: ARPACK User's Guide. SIAM, Philadelphia (1998)
14. Lewis, D.D., Yang, Y., Rose, T.G., Li, F.: RCV1: A new benchmark collection for text categorization research. J. Mach. Learn. Res. 5, 361–397 (2004)
15. Li, B., Chang, E.Y., Wu, Y.-L.: Discovery of a perceptual distance function for measuring image similarity. Multimedia Syst. 8(6), 512–522 (2003)
16. Liu, R., Zhang, H.: Segmentation of 3D meshes through spectral clustering. In: Proceedings of Pacifc Conference on Computer Graphics and Applications (2004)
17. Luxburg, U.: A tutorial on spectral clustering. Statistics and Computing 17(4), 395–416 (2007)
18. Maschho, K., Sorensen, D.: A portable implementation of ARPACK for distributed memory parallel architectures. In: Proceedings of Copper Mountain Conference on Iterative Methods (1996)
19. Ng, A.Y., Jordan, M.I., Weiss, Y.: On spectral clustering: Analysis and an algorithm. In: Proceedings of NIPS, pp. 849–856 (2001)
20. Shi, J., Malik, J.: Normalized cuts and image segmentation. IEEE Tran. on Pattern Analysis and Machine Intelligence 22(8), 888–905 (2000)
21. Smith, J.R., Chang, S.-F.: Automated image retrieval using color and texture. IEEE Trans. on Pattern Analysis and Machine Intelligence (1996)
22. Snir, M., Otto, S.: MPI-The Complete Reference: The MPI Core. MIT Press, Cambridge (1998)
23. Strehl, A., Ghosh, J.: Cluster ensembles - a knowledge reuse framework for combining multiple partitions. J. Mach. Learn. Res. 3, 583–617 (2002)
24. Thakur, R., Gropp, W.: Improving the performance of collective operations in MPICH. In: Dongarra, J., Laforenza, D., Orlando, S. (eds.) EuroPVM/MPI 2003. LNCS, vol. 2840, pp. 257–267. Springer, Heidelberg (2003)
25. Xu, S., Zhang, J.: A hybrid parallel web document clustering algorithm and its performance study. Journal of Supercomputing 30(2), 117–131
26. Yu, S.X., Shi, J.: Multiclass spectral clustering. In: Proceedings of ICCV, Washington, DC, USA, p. 313. IEEE Computer Society, Los Alamitos (2003)
27. Zelnik-Manor, L., Perona, P.: Self-tuning spectral clustering. In: Proceeding of NIPS, pp. 1601–1608 (2005)

# Classification of Multi-labeled Data: A Generative Approach

Andreas P. Streich and Joachim M. Buhmann

Institute of Computational Science
ETH Zurich
8092 Zurich, Switzerland
{andreas.streich,jbuhmann}@inf.ethz.ch

**Abstract.** Multi-label classification assigns a data item to one or several classes. This problem of multiple labels arises in fields like acoustic and visual scene analysis, news reports and medical diagnosis. In a generative framework, data with multiple labels can be interpreted as additive mixtures of emissions of the individual sources. We propose a deconvolution approach to estimate the individual contributions of each source to a given data item. Similarly, the distributions of multi-label data are computed based on the source distributions. In experiments with synthetic data, the novel approach is compared to existing models and yields more accurate parameter estimates, higher classification accuracy and ameliorated generalization to previously unseen label sets. These improvements are most pronounced on small training data sets. Also on real world acoustic data, the algorithm outperforms other generative models, in particular on small training data sets.

## 1 Introduction

Data classification, the problem of assigning each data point to a set of categories or classes, is the presumably best studied but still challenging machine learning problem. Dichotomies or binary classifications distinguish between two classes, whereas multi-class classification denotes the case of several class choices.

Multi-label classification characterizes pattern recognition settings where each data point may belong to more than one category. Typical situations where multi-labeled data are encountered are classification of acoustic and visual scenes, text categorization and medical diagnosis. Examples are the well-known Cocktail-Party problem [1], where several signals are mixed together and the objective is to detect the original signal, or a news report about Sir Edmund Hillary, which would probably belong to the categories *Sports* as well as to *New Zealand*. The label set for such an article would thus be $\{Sports, NewZealand\}$.

In this paper, we restrict ourselves to generative models, where each data item is assumed to be generated by one (in the single-label case) or several (in the multilabel case) sources. In a probabilistic setting, the goal is to determine which set of sources is most likely to have produced the given data.

Despite its significance for a large number of application areas, multi-label classification has received comparatively little attention. All current approaches

W. Daelemans et al. (Eds.): ECML PKDD 2008, Part II, LNAI 5212, pp. 390–405, 2008.

we are aware of reduce the problem to a single-label classification task. The trivial approach for this conceptual simplification either ignores data with multiple labels or considers those items with multiple labels as a new class [2]. Such a modeling strategy generates models which we will denote by $\mathcal{M}_{New}$. More advanced approaches decompose the task into a series of independent binary classification problems, deciding for each of the $K$ classes whether the data at hand belongs to it, and then combine the $K$ classifier outputs to a solution of the original problem. We review these approaches in Sect. 2.

All approaches have significant drawbacks. The trivial approach mainly suffers from data sparsity, as the number of possible label sets is in $\mathcal{O}(K^{d_{\max}})$, where $d_{\max}$ is the maximal size of the sets. Even for moderate $K$ or $d_{\max}$, this is typically intractable. Furthermore, these methods can only assign label sets that already were present in the training data.

The main criticism on the reduction of the multi-label task to a series of binary decision tasks is the confusion between frequent co-occurrence and similar source statistics – in all approaches we are aware of, the more often two sources occur together, the more similar their statistics will be. In this way, these methods neglect the information which multi-labeled data contains about all classes in its label set, which deteriorates the source estimates and leads to poor recall rates. Dependencies between multi-labels are considered in [3], but the approach remains limited to the correlation of sources in the label sets.

In this paper, we propose a novel approach for multi-labeled data, which is inspired by the fundamental physical principle of superposition. We assume a source for each class and consider data with multiple labels as an additive mixture of independent samples of the respective classes. A deconvolution enables us to estimate the contributions of each source to the observed data point and thus to use multi-labeled data for inference of the class distributions. Similarly, the distributions of multi-label data are computed based on the source distributions. Doing so, this approach allows us to consistently model jointly occurring single- and multi-label data with a small amount of parameters. Such a deconvolutive learning technique is only possible for generative models. We therefore exclude purely discriminative classifiers from the further analysis.

In the following, we assume that data is generated by a set $\mathcal{S}$ of $K$ sources. For convenience, we assume $\mathcal{S} = \{1, \ldots, K\}$. For each data item $x_i \in \mathbb{R}^D$, $\mathcal{L}_i = \{\lambda_i^{(1)}, \ldots, \lambda_i^{(d_i)}\}$ denotes the set of sources involved in the generation of $x_i$. $d_i = \deg \mathcal{L}_i = |\mathcal{L}_i|$ will be called the *degree* of the label set $\mathcal{L}_i$, and $\lambda_i^{(j)} \in \{1, \ldots, K\}$ for all $j = 1, \ldots, d_i$. Label sets with $d_i = 1$ will be called single labels. We denote by $\mathbb{L}$ the set of all admissible label sets – in the most general case, this is simply the power set of the classes except the empty set $\emptyset$, i.e. $\mathbb{L} = 2^{\mathcal{S}} \setminus \{\emptyset\}$, but restrictions to simplify the learning task often are available from the application area. Finally, $\mathcal{X} = (x_1, \ldots, x_N)$ will denote a tuple of data items, and $\mathcal{L} = (\mathcal{L}_1, \ldots, \mathcal{L}_N)$ the corresponding label sets.

The remainder of this paper is structured as follows: Sect. 2 reviews related word, and Sect. 3 then presents the underlying assumption of our method. In Sect. 4, we present the training and classification phase of the proposed algorithm, both in general and for the special case of Gaussian distributions. Sect. 5 reports results of both synthetic and real world data. A summary and outlook in Sect. 6 concludes the paper.

## 2 Related Work

Multi-label classification has attracted an increasing research interest in the last decade. It was originally proposed in [4] when introducing error-correcting output codes for solving multiclass learning problems. Later on, a modified entropy formula was employed in [5] to adapt the C4.5-algorithm for knowledge discovery in multi-label phenotype data. Given the modified entropy formula, frequently co-occurring classes are distinguished only on the bottom of the decision tree. Support vector machines as a further type of discriminative classifiers are employed to solve multi-label problems in [6].

An important application area for the problem at hand is text mining. Support vector machines were introduced for this task in [7] and were shown to outperform competing algorithms such as nearest neighbor and C4.5 algorithms. A mixture model for text classification has been presented a year later in [8], where the word distribution of a document is represented as a mixture of the source distributions of the categories the document belongs to.

More recent work includes an application of several multi-label learning techniques to scene classification [2], where it was shown that taking data items with multiple labels as samples of each of the classes yields to discriminative classifiers with higher performance. This approach is called *cross-training* ($\mathcal{M}_{Cross}$).

A similar idea is used in probabilistic learning: Each data item has the same weight, which is then equally distributed among all classes in the label set of the data. $\mathcal{M}_{Prob}$ denotes models which are generated by this technique, which are comprehensively reviewed in [9].

The vaguely related topic of multitask learning is treated in [10]. We understand multitask learning mainly as a method to classify with respect to several different criteria (e.g. street direction and type of the markings, an example in the mentioned paper), the multilabel classification task can be formulated as multitask problem when class membership is coded with binary indicator variables. Each task is then to decide whether a given data item belongs to a class or not. Insofar, we confirm the result that joint training increases the performance. Additionally, our model provides a clearly interpretable generative model for the data at hand, which is often not or only partially true for neural networks.

## 3 Generative Models for Multi-label Data

The nature of multi-labeled data is best understood by studying how such data are generated. In the following, we contrast our view of multi-labeled data with the standard parametric model of classification where data are generated by one unique source, i.e., data of one specific source do not contain any information on parameters of any other source.

### 3.1 Standard Generative Model for Classification

In a standard generative model for classification, each data item $x_i$ is a sample of a single source. The source of $x_i$ is identified by the label $\lambda_i \in \mathcal{S}$, and the source distribution will be denoted by $P_{\lambda_i}$. Formally, we thus have $x_i \sim P_{\lambda_i}$.

In the learning phase, a set of data points along with corresponding labels is given. Based on this training sample, the class distributions are usually learned such that the likelihood of the observed data, given the class labels, is maximized. Class priors $\Pi = (\pi_1, \ldots, \pi_K)$ can also be learned based on the labels of the training set.

When classifying a new data item $x_{new}$, the estimated label $\hat{\lambda}_{new}$ is the one with maximal likelihood:

$$\hat{\lambda}_{new} = \arg\max_{\lambda \in \mathcal{S}} L(\lambda | x_{new}, P_\lambda) = \arg\max_{\lambda \in \mathcal{S}} \pi_\lambda L(x_{new} | P_\lambda) \tag{1}$$

This corresponds to a search over the set of possible labels.

### 3.2 A Generative Model of Multi-labeled Data

We propose an approach to classification of multi-labeled data which extends the generative model for single-label data by interpreting multi-labeled data as a superposition of the emissions of the individual sources. A data item $x_j$ with label set $\mathcal{L}_j = \{\lambda_j^{(1)}, \ldots, \lambda_j^{(d_j)}\}$ of degree $d_j$ is assumed to be the sum of one sample from each of the contributing sources, i.e.

$$x_j = \sum_{s=1}^{d_j} \chi_{\lambda_j^{(s)}} \quad \text{with} \quad \chi_{\lambda_j^{(s)}} \sim P_{\lambda_j^{(s)}} \tag{2}$$

The distribution of $x_j$ is thus given by the convolution of all contributing sources:

$$x_j \sim P_{\lambda_j^{(1)}} * \ldots * P_{\lambda_j^{(d_j)}} =: P_{\mathcal{L}_j} \tag{3}$$

Thus, unlike in the single-label model, the distribution of data with multiple labels is traced back to the distribution of the contributing sources. We therefore propose the name *Additive-Generative Multi-Label* Model ($\mathcal{M}_{AdGen}$).

Note that it is possible to explicitly give the distribution $P_{\mathcal{L}_j}$ for data with label set $\mathcal{L}_j$. In contrast to $\mathcal{M}_{New}$, which would estimate $P_{\mathcal{L}_j}$ based solely on the data with this label set, we propose to compute $P_{\mathcal{L}_j}$ based on the distribution of all sources contained in $\mathcal{L}_j$. On the other hand, the estimation of each source distribution is based on all data items which contain the respective source in their label sets.

## 4 Learning a Generative Model for Multi-labeled Data

In the following, we will first describe the learning and classification steps in general and then we give explicit formula for the special case of Gaussian distributions. In order to simplify the notation, we will limit ourselves to the case of data generated by at most two sources. The generalization to label sets of higher degree is straightforward.

### 4.1 Learning and Classification in the General Case

The probability distribution of multi-labeled data is given by (3). The likelihood of a data item $x_i$ given a label set $\mathcal{L}_i = \{\lambda_i^{(1)}, \lambda_i^{(2)}\}$, is

$$P_{\{\lambda_1^{(1)},\lambda_1^{(2)}\}}(x_i) = \left(P_{\lambda_i^{(1)}} * P_{\lambda_i^{(2)}}\right)(x_i)$$

$$= \int P_{\lambda_i^{(1)}}(\chi) P_{\lambda_i^{(2)}}(x_i - \chi)\,\mathrm{d}\chi \tag{4}$$

$$= \mathbb{E}_{\chi \sim P_{\lambda_i^{(1)}}}\left[P_{\lambda_i^{(2)}}(x_i - \chi)\right]. \tag{5}$$

In general, it may not be possible to solve this convolution integral analytically. In such cases, the formulation as an expected value renders Monte Carlo sampling possible to compute a numerical estimate of the data likelihood.

In the training phase, the optimal parameters $\theta_s$ of the distribution $P_s$ are chosen such that they fulfil the condition

$$\frac{\partial}{\partial \theta_s}\left\{\prod_{\mathcal{L}\in\mathbb{L}} \prod_{i:\mathcal{L}_i=\mathcal{L}} P_{\mathcal{L}}(x)\right\} \stackrel{!}{=} 0 \quad \text{for } s = 1, \ldots, K \tag{6}$$

If the convolution integral (4) can not be expressed analytically, the formulation as expected value ((5), and similar terms for superpositions of three and more sources) can be used to estimate the optimal parameter set $(\theta_1, \ldots, \theta_K)$.

When classifying new data, label sets are assigned according to (1). Again, if the probability distribution of a data item $x_i$ with label sets $\{\lambda_i^{(1)}, \lambda_i^{(2)}\}$ of degree 2 can not be expressed in closed form, (5) might be used to get an estimate of $P_{(\lambda_i^{(1)},\lambda_i^{(2)})}(x_i)$ by sampling $\chi$ from $P_{\lambda_i^{(1)}}$. The generalization to label sets of degree larger than 2 is straight forward.

The methods presented here are very general and they are applicable to all parametric distributions. For specific distributions, a closed form expression for the convolution integral and then analytically solving (6) for optimal parameter values will lead to much faster training and classification. The following subsection exemplifies this claim for the Gaussian distributions. Similar explicit convolution formulae can also be computed for, e.g., the chi-square or the Poisson distribution, and approximations exist for many distributions and convolutions of different distributions.

### 4.2 Gaussian Distributions

Let us assume for the remainder of this section that all source distributions are Gaussians, i.e. $P_s = \mathcal{N}(\mu_s, \Sigma_s)$ for $s = 1, \ldots, K$. The convolution of Gaussian distributions is again a Gaussian distribution, where the mean vectors and the covariance matrices are added:

$$\mathcal{N}(\mu_1, \Sigma_1) * \mathcal{N}(\mu_2, \Sigma_2) = \mathcal{N}(\mu_1 + \mu_2, \Sigma_1 + \Sigma_2). \tag{7}$$

A corresponding rule holds for convolutions of more than two Gaussians. This property drastically simplifies the algebraic expressions in our model.

**Training for Gaussian Distributions.** To find the optimal values for the means and the covariance matrices, we have to solve the ML equations

$$\frac{\partial}{\partial \mu_s}\left\{\prod_{\mathcal{L}\in\mathbb{L}}\prod_{i:\mathcal{L}_i=\mathcal{L}} P_{\mathcal{L}}(x)\right\} \stackrel{!}{=} 0 \qquad \frac{\partial}{\partial \Sigma_s}\left\{\prod_{\mathcal{L}\in\mathbb{L}}\prod_{i:\mathcal{L}_i=\mathcal{L}} P_{\mathcal{L}}(x)\right\} \stackrel{!}{=} 0 \tag{8}$$

for $s = 1, \ldots, K$. These conditions yield a set of coupled nonlinear equations, which can be decoupled by proceeding iteratively. As initial values, we choose the sample mean and variance of the single-labeled training data:

$$\mu_s^{(0)} = \frac{\sum_{i:\mathcal{L}_i=\{s\}} x_i}{|\{i : \mathcal{L}_i = \{s\}\}|} \qquad \Sigma_s^{(0)} = \frac{\sum_{i:\mathcal{L}_i=\{s\}}(x_i - \mu_s^{(0)})(x_i - \mu_s^{(0)})^T}{|\{i : \mathcal{L}_i = \{s\}\}|}. \tag{9}$$

For simpler notation, we define the following intermediate values:

$$m_{\mathcal{L}_i\setminus\{s\}}^{(t)} = \sum_{\substack{j\in\mathcal{L}_i\\ j\neq s}} \mu_j^{(t)} \qquad S_{\mathcal{L}_i\setminus\{s\}}^{(t)} = \sum_{\substack{j\in\mathcal{L}_i\\ j\neq s}} \Sigma_j^{(t)} \qquad V_{i\mathcal{L}_i}^{(t)} = (x_i - \mu_{\mathcal{L}_i}^{(t)})(x_i - \mu_{\mathcal{L}_i}^{(t)})^T,$$

where upper indices indicate the iteration steps. Using an iterative approach, the condition for the mean values yields the following update formula for $\mu_s$, $s = 1, \ldots, K$:

$$\mu_s^{(t)} = \left(\sum_{i:\mathcal{L}_i\ni s} (x_i - m_{\mathcal{L}_i\setminus\{s\}}^{(t-1)})\left(\Sigma_{\mathcal{L}_i}^{(t-1)}\right)^{-1}\right)\left(\sum_{i:\mathcal{L}_i\ni s}\left(\Sigma_{\mathcal{L}_i}^{(t-1)}\right)^{-1}\right)^{-1}.$$

Deriving the data likelihood with respect to the covariance matrix $\Sigma_s$ yields the following condition:

$$\frac{1}{2}\sum_{i:\mathcal{L}_i\ni s}\left(\left(\mathbb{I}_d - (x_i - \mu_{\mathcal{L}_i})(x_i - \mu_{\mathcal{L}_i})^T \Sigma_{\mathcal{L}_i}^{-1}\right)\Sigma_{\mathcal{L}_i}^{-1}\right) \stackrel{!}{=} 0,$$

where $\mathbb{I}_d$ denotes the identity matrix in $d$ dimensions. With $\Sigma_{\mathcal{L}_i} = \Sigma_s + S_{\mathcal{L}_i\setminus\{s\}}$, the left hand side of the condition can be rewritten as

$$\sum_{i:\mathcal{L}_i=\{s\}}\left(\left(\mathbb{I}_d - V_{i\mathcal{L}_i}\Sigma_s^{-1}\right)\Sigma_s^{-1}\right) + \sum_{\substack{i:\mathcal{L}_i\ni s\\ |\mathcal{L}_i|>1}}\left(\left(\mathbb{I}_d - V_{i\mathcal{L}_i}(S_{\mathcal{L}_i\setminus s} + \Sigma_s)^{-1}\right)(S_{\mathcal{L}_i\setminus s} + \Sigma_s)^{-1}\right)$$

Note that for a training set containing only single label data, the second sum vanishes, and the condition implies estimating $\Sigma_s$ by the sample variance. If the training set does contain data with multiple labels, the optimality condition can – in general – not be solved analytically, as the condition for $\Sigma_s$ corresponds to a polynomial which degree is twice the number of allowed label sets in $\mathbb{L}$

containing $s$. In this case, the optimal value of $\Sigma_s^{(t)}$ can either be determined numerically, or the Taylor approximation

$$(S_{\mathcal{L}_i\setminus s} + \Sigma_s)^{-1} = \Sigma_s^{-1}(S_{\mathcal{L}_i\setminus s}\Sigma_s^{-1} + \mathbb{I}_d)^{-1} \approx \Sigma_s^{-1}(\mathbb{I}_d - \Sigma_s S_{\mathcal{L}_i\setminus s}^{-1}) = \Sigma_s^{-1} - S_{\mathcal{L}_i\setminus s}^{-1}$$

can be used. The approximation is typically quite crude, we therefore prefer using a numerical solver to determine $\Sigma_s^{(t)}$ for all sources $s$ after having determined the mean values $\mu_s^{(t)}$. Whenever a sufficient number of data is available, the covariance matrix of a source $s$ might also be estimated purely based on the data with $s$ as single label.

In spite of the rather complicated optimization procedure for the covariance matrices, we observed that the estimator for the mean values is quite robust with respect to changes in the covariance matrix. Furthermore, the relative importance per data item for the estimation of $\mu_s^{(t)}$ decreases as the degree of its label increases. If enough data with low degree label sets is available in the training phase, the convergence of the training step can be increased by discarding data items with high label degrees with only minor changes in the accuracy of the parameter estimates.

**Classification for Gaussian Distributions.** Recall the explicit formula for the convolution of two Gaussians (7). This relation yields a simple expression for the likelihood of the data $x_{new}$ given a particular candidate label set $\mathcal{L}_{new} = \{\lambda_{new}^{(1)}, \lambda_{new}^{(2)}\}$:

$$P_{\mathcal{L}_{new}}(x_{new}) = \mathcal{N}(x_{new}; \mu_{\lambda_{new}^{(1)}} + \mu_{\lambda_{new}^{(2)}}, \Sigma_{\lambda_{new}^{(1)}} + \Sigma_{\lambda_{new}^{(2)}})$$

Again, the assignment of the label set for the new data item is done according to (1). As the density functions for data with multiple labels are computed based on the single source densities, this yields more accurate density estimates namely for data with medium to large label degree. This is the second major advantage of the proposed algorithm.

The task of finding the most likely label set, given the data and the source parameters, may become prohibitively expensive if a large number of sources are observed, or if the allowed label degree is large. In the following section, we present an approximation technique that leads to drastically reduced computation costs, while incurring a computable error probability.

### 4.3 Efficient Classification

In the proposed model, the classification tasks consist of choosing a subset of given sources such that the observed data item has maximal likelihood. The classification task thus comprises a combinatorial optimization problem. While this type of problem is NP-hard in most cases, good approximations are possible in the present case, as we exemplify in the following for Gaussian sources.

For Gaussian distributions with equal spherical covariance matrix $\Sigma_s = \sigma^2 I_D$ for all sources $s = 1, \ldots, K$, maximum likelihood classification of a new data item $x_{new} \in \mathbb{R}^D$ can be reduced to

$$\hat{\mathcal{L}}_{new} = \arg\max_{\mathcal{L}\in\mathbb{L}} \left\{ \frac{\pi_{\mathcal{L}}}{\sigma^D (2\pi d_{\mathcal{L}})^{D/2}} \exp\left( -\frac{||x_{new} - \mu_{\mathcal{L}}||_2^2}{2 d_{\mathcal{L}} \sigma^2} \right) \right\}$$
$$= \arg\min_{\mathcal{L}\in\mathbb{L}} \left\{ ||x_{new} - \mu_{\mathcal{L}}||_2^2 + d_{\mathcal{L}}\sigma^2 (D \log(d_{\mathcal{L}}) - 2\log(\pi_{\mathcal{L}})) \right\}, \quad (10)$$

where $\pi_{\mathcal{L}}$ is the prior probability of the label set $\mathcal{L}$, and $d_{\mathcal{L}} = |\mathcal{L}|$ is its degree.

In cases where the set $\mathbb{L}$ of admissible label sets is relatively small, label set $\hat{\mathcal{L}}_{new}$ with maximal likelihood can be found directly within reasonable computation time. Such a case e.g. arises when the new data can only be assigned to a label set that was also present in the training set, i.e. if $\mathbb{L}$ is the set of all label sets contained in the training sample.

However, in a more general setting, there are no such constraints, and the classifier should also be able to assign a label set that was not seen during the training phase. In this case, $\mathbb{L}$ contains $|2^{\mathcal{S}}| - 1 = 2^K - 1$ possible label sets. The time for direct search thus grows exponentially with the number of sources $K$.

Our goal is therefore to determine a subset of sources $\mathcal{S}^- \subset \mathcal{S}$ which – with high probability – have not contributed to $x_{new}$. This constraint to $\mathcal{S}^-$ will limit the search space for the arg min operation and consequently will speed up data processing.

Note that all terms in (10) are positive. The label set prior typically decreases as the degree increases, and the second term grows logarithmically in the size of the label set. The later term thus tends to privilege smaller label sets, and neglecting these two terms might thus yield larger label sets. This is a type of regularization which we omit in the following, as we approximate (10) by the following subset selection problem:

$$\hat{\mathcal{L}}_{new} = \arg\min_{\mathcal{L}\in\mathbb{L}} \left\{ ||x_{new} - \mu_{\mathcal{L}}||_2^2 \right\}, \quad (11)$$

Defining the indicator vector $\hat{\beta}_{new} \in \{0,1\}^K$, with $\hat{\beta}_{new}^{(s)} = 1$ if $s \in \hat{\mathcal{L}}_{new}$ and $\hat{\beta}_{new}^{(s)} = 0$ otherwise, for all sources $s$, this can be written as

$$\hat{\beta}_{new} = \arg\min_{\beta\in\{0,1\}^K} \left\{ \sum_{s=1}^{K} \beta^{(s)} \mu_s - x_{new} \right\}.$$

Relaxing the constraints on $\hat{\beta}_{new}$, we get the following regression problem:

$$\tilde{\beta}_{new} = \arg\min_{\beta\in\mathbb{R}^K} \left\{ \sum_{s=1}^{K} \tilde{\beta}^{(s)} \mu_s - x_{new} \right\}.$$

Defining the matrix $M$ of mean vectors as $M = [\mu_1, \ldots, \mu_K]$, we obtain the least-squares solution for the regression problem:

$$\tilde{\beta}_{new} = (M^T M)^{-1} M^T x_{new} \quad (12)$$

In order to reduce the size of the search space for the label set, we propose to compute a threshold $\tau$ for the components of $\tilde{\beta}_{new}$. Only sources $s$ with $\tilde{\beta}_{new}^{(s)} > \tau$ will be considered further as potential members of the label set $\tilde{\mathcal{L}}_{new}$.

As we have omitted the constraints favoring small label sets, it may happen that single sources with mean close to $x_{new}$ are discarded. This effect can be compensated by adding label sets of small degree (up to 2 is mostly sufficient) containing only discarded classes to the reduced label set. Formally, we have $\mathcal{S}^+ = \{s \in \mathcal{S} | \tilde{\beta}^{(s)}_{new} > \tau\}$, $\mathcal{S}^- = \mathcal{S} \setminus \mathcal{S}^+$, $\mathbb{L}^+ = \left(2^{\mathcal{S}^+} \setminus \{\emptyset\}\right) \cup \mathcal{S}^- \cup \mathcal{S}^- \times \mathcal{S}^-$, and $\mathbb{L}$ replaced by $\mathbb{L}^+$ in (10).

In our experiments, we found that this heuristic can drastically reduce computation times in the classification task. The error probability introduced by this technique is discussed in the following.

We assume the true label set of $x_{new}$ is $\mathcal{L}_{new}$, with the corresponding indicator vector $\beta_{new}$ and degree $d_{new} = |\mathcal{L}_{new}|$. The heuristic introduces an error whenever $\tilde{\beta}^{(s)}_{new} < \tau$ but $\beta^{(s)}_{new} = 1$ for any $s \in \mathcal{L}_{new}$. Thus,

$$P[\text{error}] = 1 - \prod_{s \in \mathcal{L}_{new}} P[(\tilde{\beta}^{(s)}_{new} > \tau) \wedge (\beta^{(s)}_{new} = 1)].$$

For the analysis, we assume that all source distributions have the same variance $\sigma^2 \cdot \mathbb{I}_d$. Then, we have $x_{new} = M\beta_{new} + \epsilon$, with $\epsilon \sim \mathcal{N}(0, d_{new} \cdot \sigma^2 \mathbb{I}_d)$. Inserting this into (12), we derive

$$\tilde{\beta}_{new} = \beta_{new} + (M^T M)^{-1} M^T \epsilon =: \beta_{new} + \epsilon',$$

where we have defined $\epsilon' = (M^T M)^{-1} M^T \epsilon$, with $\epsilon' \sim \mathcal{N}(0, d_{new}\sigma^2 (M^T M)^{-1})$. Using the eigendecomposition of the symmetric matrix $M^T M$, $M^T M = U \Lambda U^T$, the distribution of $\epsilon'$ can be rewritten as $\epsilon' \sim U\mathcal{N}(0, d_{new}\sigma^2 \Lambda^{-1})$. Note that $\Lambda$ scales with the squared 2-norm of the mean vectors $\mu$, which typically scales with the number of dimensions $D$.

For the special case when $U = \mathbb{I}_D$, we then have

$$P[\text{error}] = 1 - \prod_{s \in \mathcal{L}_{new}} \left(1 - \Phi\left(\frac{\tau - 1}{\sigma\sqrt{d_{new}\Lambda_{ss}^{-1}}}\right)\right)$$

where $\Phi(\cdot)$ is the cumulative distribution function of the standardized Gaussian. Summing up, the probability of an error due to the heuristic decreases whenever the dimensionality grows ($\Lambda_{ss}$ grows), sources become more concentrated ($\sigma$ gets smaller), or the degree of the true label set decreases ($d_{new}$ grows).

For a given classification task, $\mathcal{L}_{new}$ will not be known. In our experiments, we derived an upper limit $d_{\max}$ for the label degree from the distribution of the label set degrees in the training set. For $\Lambda_{ss}$, we used the average eigenvalue $\bar{\lambda}$ of the eigendecomposition of $M^T M$. Finally, $\sigma$ can be estimated from the variance of the single labeled data.

With these estimates, we finally get

$$P[\text{error}] \leq 1 - \left(1 - \Phi\left(\frac{\tau - 1}{\sigma\sqrt{d_{\max}\bar{\lambda}^{-1}}}\right)\right)^{d_{\max}} \tag{13}$$

Given an acceptable error probability, this allows us to choose an appropriate value for the threshold $\tau$. Note that the bound is typically quite pessimistic, as most of the real-world data samples have a large number of data with label sets of small degree. For these data items, the effective error probability is much lower than indicated by (13). Keeping this in mind, (13) provides a reasonable error bound also in the general case where $U \neq \mathbb{I}_D$.

# 5 Experimental Evaluation

The experiments include artificial and real-world data with multiple labels. In the following, we first introduce a series of quality measures and then present the results.

## 5.1 Performance Measures

Precision and recall are common quality measures in information retrieval and multi-label classification. These measures are defined on each source. For a source $s$ and a data set $\mathcal{X} = (x_1, \ldots, x_N)$, let $tp_s$, $fn_s$, $fp_s$ and $tn_s$ denote the number of true positives, true negatives, false positives and false negatives as defined in Table 1. Then, *precision* and *recall* on source $s$ are defined as follows:

$$Precision_s = \frac{tp_s}{tp_s + fp_s} \qquad Recall_s = \frac{tp_s}{tp_s + fn_s}$$

Intuitively speaking, recall is the fraction of true instances of a base class correctly recognized as such, while precision is the fraction of classified instances that are correct. The F-score is the harmonic mean of the two:

$$F_s = \frac{2 \cdot Recall_s \cdot Precision_s}{Recall_s + Precision_s}$$

All these measures take values between 0 (worst) and 1 (best).

Furthermore, we define the *Balanced Error Rate* (BER) as the average of the ratio of incorrectly classified samples per label set over all label sets:

$$BER = \frac{1}{|\mathbb{L}|} \sum_{\mathcal{L} \in \mathbb{L}} \frac{|\{i|\hat{\mathcal{L}}_i \neq \mathcal{L}_i = \mathcal{L}\}|}{|\{i|\mathcal{L}_i = \mathcal{L}\}|}$$

**Table 1.** Definition of true positives, true negatives, false positives and false negatives for a base class $s$

| true classification | estimated classification $x_i \in s$ | estimated classification $x_i \notin s$ |
|---|---|---|
| $x_i \in s$ | $tp_C$ (true positive) | $fn_s$ (false negative) |
| $x_i \notin s$ | $fp_s$ (false positive) | $tn_s$ (true negative) |

Note that the BER is a quality measure computed on an entire data set, while $Precision_s$, $Recall_s$ and the $F_s$-score are determined for each source $s$.

## 5.2 Artificial Data

We use artificial data sampled from multivariate Gaussian distributions to compute the accuracy of the source parameter estimates of different models.

The artificial data scenario consisted of 10 sources denoted by $\{1, \ldots, 10\}$. In order to avoid hidden assumptions or effects of hand-chosen parameters, the mean values of the sources were uniformly chosen in the 10-dimensional hypercube $[-2; 2]^{10}$. The covariance matrix was diagonal with diagonal elements uniformly sampled from $[0; 1]$. 25 different subsets of $\{1, \ldots, 10\}$ were randomly chosen and used as label sets. Training sets of different sizes as well as a test set were sampled based on the label sets and the additivity assumption (2). This procedure was repeated 10 times for cross-validation.

Figure 1 shows the average deviation of the mean vectors and the average deviation of the largest eigenvalue from the corresponding true values. For the estimates of the source means, it can be clearly seen that the proposed model is the most accurate. The deviation of the parameters of $\mathcal{M}_{New}$ is explained by the small effective sample size available to estimate each of the mean vectors: As $\mathcal{M}_{New}$ learns a separate source for each label set, there are only two samples per

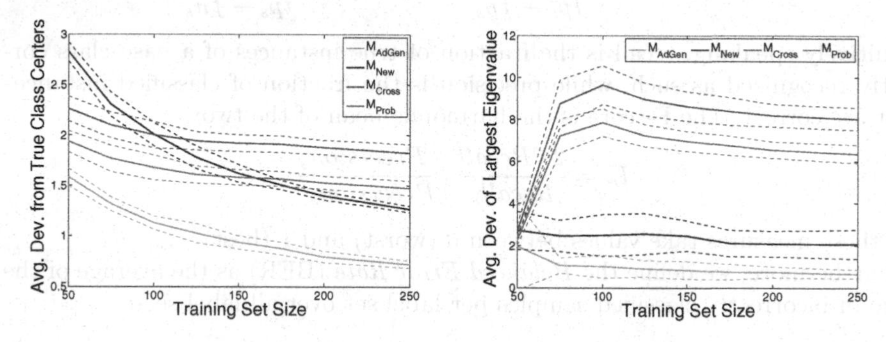


**Fig. 1.** Accuracy of the parameter estimation of different models. The left panel shows the deviation of the mean estimate, the right one shows the relative deviation between the true and the estimated value of the largest eigenvalue of the covariance matrix. For each model, the average (continuous bold line) over all classes and the standard deviation based on 10-fold cross-validation (dashed lines) is plotted.

We used a setting with 10 sources in 10 dimensions. The mean of each source was chosen uniformly in $[-1, 1]^{10}$. The sources were randomly combined to 25 label sets. Training data sets of different sizes were then sampled according to the generative model.

The generative multi-label model clearly yields the most accurate parameter estimates. The $\mathcal{M}_{New}$ suffers from the small sample size problem, while $\mathcal{M}_{Cross}$ and $\mathcal{M}_{Prob}$ can not clearly improve the estimates of the source parameters.

For the training on sample size 50, we used a default starting value for the covariance matrices. All models have therefore covariance estimates of roughly the same quality.

source when the training set size is 50. $\mathcal{M}_{AdGen}$, on the other hand, decomposes the contributions of each source to every data item. On the average, $\mathcal{M}_{AdGen}$ has thus 2.5 times more training samples per parameter than $\mathcal{M}_{New}$. Furthermore, and the samples used by $\mathcal{M}_{New}$ to estimate the density distribution of multi-labeled data have higher variance than the single label data.

For the estimation of the covariance, $\mathcal{M}_{AdGen}$ still yields distinctly more precise values, but the difference to $\mathcal{M}_{New}$ is not as large as in the estimation of the mean values. This is due to the more complicated optimization problem that has to be solved to estimate the covariance matrix.

The estimates of $\mathcal{M}_{New}$ and $\mathcal{M}_{Prob}$ for both the mean and the covariance are clearly less accurate. Using a data item with multiple label as a training sample independently for each class brings the source parameters closer to each other – and away from their true values. As multi-labeled data have a reduced weight for the estimation of the single sources, this effect is less pronounced in $\mathcal{M}_{Prob}$ than in $\mathcal{M}_{Cross}$.

As in many other machine learning problems, the estimation of the covariance matrix is a hard problem. As no analytic solution of the optimality condition exists and numerical methods have to be used, the computational effort to estimate the covariance grows linearly or even quadratically in the number of dimensions (depending on whether a diagonal or a full covariance matrix is assumed).

Only for spherical covariances, the conditions can be solved to get a coupled set of equations, which can be used for an iterative solution scheme. A possible remedy is to estimate the source covariances based on single label data only, and to use the deconvolution approach only for estimating the mean values. For classification, the proposed method yields considerably more accurate parameter estimates for the distributions of all label sets and therefore performs clearly better.

The estimation of the source means is much more stable and it performs independently of the dimensionality of the data. As expected, the amelioration due to $\mathcal{M}_{AdGen}$ is larger if the covariance matrix does not have to be estimated, and also the improvements in the classification are more pronounced.

### 5.3 Acoustic Data

For the experiments on real data, we used the research database provided by a collaborating hearing instrument company. This challenging data set serves as benchmark for next generation hearing instruments and captures the large variety of acoustic environments that are typically encountered by a hearing aid user. It contains audio streams of every day acoustic scenes recorded with state of the art hearing instruments. Given the typically difficult acoustic situations in day to day scenes, the recordings have significant artefacts.

Each sound clip is assigned to one of the four classes *Speech* ($SP$), *Speech in Noise* ($SN$), *Noise* ($NO$) and *Music* ($MU$). While $\mathcal{M}_{New}$ learns a separate source for each of the four label sets, $\mathcal{M}_{Cross}$, $\mathcal{M}_{Prob}$ and $\mathcal{M}_{AdGen}$ interpret $SN$ as a mixture of $SP$ and $NO$. $SN$ is the only multi-label in our real data setting.

It should be noted that intra-class variance is very high – just consider various genres of music, or different sources of noise! Additionally, mixtures arise in different proportions, i.e. the noise level in the mixture class varies strongly between different sound clips. All these factors render the classification problem a difficult challenge: Even with specially designed features and a large training data set, we have been unable to train a classifier that is able to reach an accuracy of more than 0.75. Precision, recall and the F-score are around 0.80 for all three sources.

Mel Frequency Cepstral Coefficients (MFCCs) [11] have been extracted from the sound clips at a rate of about 100Hz, yielding a 12-dimensional feature vector per time window. As classification is expected to be independent of the signal volume, we have used normalized coefficients. Thus, the additivity assumption (2) has been changed to

$$x_{SP,NO} = \frac{x_{SP} + x_{NO}}{2} \tag{14}$$

Since the extraction of MFCCs is nonlinear, this modified additivity property in the signal space has been transformed into the feature space. A sequence of 10 MFCC feature sets is used as feature vector, describing also the short-time evolution of the signal. Features for the training and test sets have been extracted from different sound clips.

Hidden Markov models (HMM) are widely used in signal processing and speech recognition [12]. We use a HMM with Gaussian output and two states per sound source a simple generative model. In the training phase, we use the approximations

$$\mathbb{E}_{\chi \sim P_{NO}}\left[P_{SP}(x_i - \chi)\right] \approx P_{SP}(x_i - \mathbb{E}_{\chi \sim P_{NO}}[\chi])$$
$$\mathbb{E}_{\chi \sim P_{SP}}\left[P_{NO}(x_i - \chi)\right] \approx P_{NO}(x_i - \mathbb{E}_{\chi \sim P_{SP}}[\chi])$$

to get a rough estimate of the individual source contributions to a data item $x_i$ with label $\mathcal{L}_i = SN = \{SP, NO\}$. In the classification phase, the formulation of the convolution as expected value (5) is used to estimate the probability of the binary label by sampling from one of the two contributing sources.

Experiments are cross-validated 10 times. In every cross validation set, the number of training samples has been gradually increased from 4 (i.e. one per label set) to 60. The differences in F-score and BER are depicted in Fig. 2. The test sets consist of 255 data items.

Comparing the results of the four algorithms on the test data set, we observe only minor differences in the precision, with $\mathcal{M}_{AdGen}$ tending to yield slightly less precise results. The recall rate of $\mathcal{M}_{AdGen}$, however, is consistently higher than the corresponding results of its three competitors. The F-score obtained by the generic multi-label algorithm is consistently above the F-scores obtained by $\mathcal{M}_{New}$, $\mathcal{M}_{Cross}$ and $\mathcal{M}_{Prob}$. As can be observed in the plots, $\mathcal{M}_{New}$ approaches $\mathcal{M}_{AdGen}$ as the size of the training set increases. The difference between $\mathcal{M}_{AdGen}$ and the two other models does not shows a clear dependency on the size of the training set.

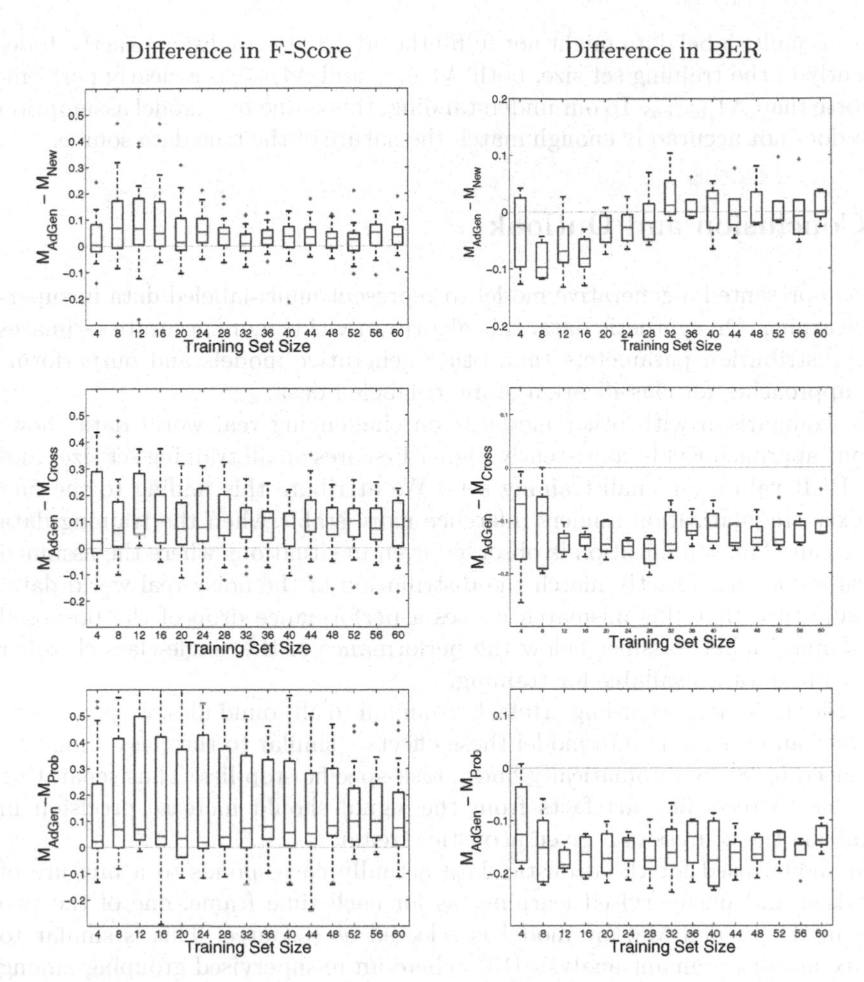


**Fig. 2.** Difference of quality measures between the proposed method and the three mentioned competing methods. The left column shows the differences in F-Score (higher is better), the right one the differences in BER (lower is better). The differences in F-score are less pronounced and have higher variance than the differences in BER. $\mathcal{M}_{New}$ is the strongest competitor to $\mathcal{M}_{AdGen}$. As the training set grows, $\mathcal{M}_{New}$ comes very close to $\mathcal{M}_{AdGen}$ in terms of F-score and occasionally gets slightly lower BER values. $\mathcal{M}_{Cross}$ and $\mathcal{M}_{Prob}$ are clearly lagging behind $\mathcal{M}_{AdGen}$ in terms of BER and also yield consistently lower F-scores. The absolute values are around 0.6 for the F-score and around 0.4 for the BER at the very small sample sizes.

In all plots, the green horizontal line at 0 indicates equal performance of the two compared algorithms. Note the difference in scale between the two columns.

Differences are more pronounced in terms of the BER. $\mathcal{M}_{New}$ is clearly outperformed on small training sets, but it is able to perform competitively as more training data are available. For larger training sets, learning a separate, independent class for the multi-labeled data as $\mathcal{M}_{New}$ does, sometimes even performs slightly better, as multi-label data might not fulfill the additivity condition exactly. Independently of the training set size, both $\mathcal{M}_{Cross}$ and $\mathcal{M}_{Prob}$ are clearly performing worse than $\mathcal{M}_{AdGen}$. To our understanding, this is due to a model assumption which does not accurately enough match the nature of the true data source.

## 6 Conclusion and Outlook

We have presented a generative model to represent multi-labeled data in supervised learning. On synthetic data, this algorithm yields more accurate estimates of the distribution parameters than other generative models and outperforms other approaches for classification of multi-labeled data.

The comparison with other methods on challenging real world data shows that our approach yields consistently higher F-scores on all training set sizes and lower BER values on small training sets. We attribute this finding to the fact that extra regularization renders inference more stable when the training data set is small. This stabilization is observed even in situations where the assumed structure does not exactly match the distribution of the noisy real world data. We conjecture that this mismatch causes a performance drop of the proposed generic multi-label classifier below the performance of the single-class classifier when more data is available for training.

In order to handle recording artefacts common to all sound files, we propose to introduce an extra source to model these effects – similar to the class "English" introduced in [8] to automatically find a task-specific stop list. Thus separating noise due to recording artefacts from the signal should increase precision in recognizing the sources of a given acoustic stream.

Our model used for the acoustic data actually corresponds to a mixture of supervised and unsupervised learning, as for each time frame, one of the two states in the hidden Markov model is selected as a source. This is similar to the mixture discriminant analysis [13], where an unsupervised grouping among all data of one class yield several prototypes within each class. Tracking down these prototypes for multi-label data might yield the distinction between, say, different types of background noise in a mixture with speech.

Furthermore, we have not yet taken into account the fact that several noise sources might be mixed together at different intensities. For example, we might have 70% speech and 30% noise in a conversation situation with moderate background noise, or the opposite of only 30% speech and 70% noise in a very loud environment. In the presented model, both situations are treated equally and lead to a difficult learning and classification task. Modeling mixtures at different intensities is subject to future work.

Finally, the proposed model for data generation is also applicable to unsupervised learning. We expect more precise parameter estimations also in this scenario and thus more stable clustering. The binary assignments in the learning phase of data to its generating classes would be replaced by an estimated responsibility for each (data, class) pair, and the model could then be learned by expectation maximization.

**Acknowledgement.** This work was in part funded by CTI grant Nr. 8539.2;2 EPSS-ES.

## References

1. Arons, B.: A review of the cocktail party effect. Journal of the American Voice I/O Society 12, 35–50 (1992)
2. Boutell, M., Luo, J., Shen, X., Brown, C.: Learning multi-label scene classification. Pattern Recognition, 1757–1771 (2004)
3. Zhu, S., Ji, X., Xu, W., Gong, Y.: Multi-labelled classification using maximum entropy method. In: Proceedings of SIGIR 2005 (2005)
4. Dietterich, T.G., Bakiri, G.: Solving multiclass learning problems via error-correcting output codes. J. of Artificial Intelligence Research 2, 263–286 (1995)
5. Clare, A., King, R.D.: Knowledge discovery in multi-label phenotype data. In: Siebes, A., De Raedt, L. (eds.) PKDD 2001. LNCS (LNAI), vol. 2168, pp. 42–53. Springer, Heidelberg (2001)
6. Elisseeff, A., Weston, J.: Kernel methods for multi-labelled classification and categorical regression problems. In: Proceedings of NIPS 2002 (2002)
7. Joachims, T.: Text categorization with support vector machines: learning with many relevant features. In: Nédellec, C., Rouveirol, C. (eds.) ECML 1998. LNCS, vol. 1398. Springer, Heidelberg (1998)
8. McCallum, A.K.: Multi-label text classification with a mixture model trained by EM. In: Proceedings of NIPS 1999 (1999)
9. Tsoumakas, G., Katakis, I.: Multi label classification: An Overview. Int. J. of Data Warehousing and Mining 3(3), 1–13 (2007)
10. Caruana, R.: Multitask learning. Machine Learning 28(1), 41–75 (1997)
11. Pols, L.: Spectral analysis and identification of Dutch vowels in monosyllabic words. PhD thesis, Free University of Amsterdam (1966)
12. Rabiner, L.R.: A tutorial on hidden markov models and selected applications in speech recognition. In: Readings in speech recognition, pp. 267–296 (1990)
13. Hastie, T., Tibshirani, R.: Discriminant analysis by Gaussian Mixtures. J. of the Royal Statist. Soc. B 58, 155–176 (1996)
14. Dempster, A., Laird, N., Rubin, D.: Maximum likelihood from incomplete data via the EM algorithm. J. of the Royal Statist. Soc. B 39(1), 138 (1977)

# Pool-Based Agnostic Experiment Design in Linear Regression

Masashi Sugiyama[1] and Shinichi Nakajima[2]

[1] Department of Computer Science, Tokyo Institute of Technology,
2-12-1 O-okayama, Meguro-ku, Tokyo 152-8552, Japan
sugi@cs.titech.ac.jp
http://sugiyama-www.cs.titech.ac.jp/~sugi/
[2] Nikon Corporation,
201-9 Oaza-Miizugahara, Kumagaya-shi, Saitama 360-8559, Japan
nakajima.s@nikon.co.jp

**Abstract.** We address the problem of batch active learning (or experiment design) in regression scenarios, where the best input points to label is chosen from a 'pool' of unlabeled input samples. Existing active learning methods often assume that the model is correctly specified, i.e., the unknown learning target function is included in the model at hand. However, this assumption may not be fulfilled in practice (i.e., agnostic) and then the existing methods do not work well. In this paper, we propose a new active learning method that is robust against model misspecification. Simulations with various benchmark datasets as well as a real application to wafer alignment in semiconductor exposure apparatus illustrate the usefulness of the proposed method.

## 1 Introduction

Active learning (AL) is a problem of optimally designing the location of training input points in supervised learning scenarios [1]. Choice of training input location is particularly important when the sampling cost of output values is very high, e.g., in the analysis of, medical data, biological data, or chemical data. In this paper, we address *batch AL* (a.k.a. experiment dasign), where the location of all training input points are designed in the beginning (cf. *on-line AL* where input points are chosen sequentially).

**Population-based vs. Pool-based AL:** Depending on the situations, AL can be categorized into two types: *population-based* and *pool-based.*

Population-based AL indicates the situation where we know the distribution of test input points and we are allowed to locate training input points at any desired positions [2,3,4]. The goal of population-based AL is to find the optimal training input distribution from which we generate training input points.

On the other hand, in pool-based AL, the test input distribution is unknown but samples from the test input distribution are given [5,6]. The goal of pool-based AL is to choose the best input samples to label from the pool of test input

W. Daelemans et al. (Eds.): ECML PKDD 2008, Part II, LNAI 5212, pp. 406–422, 2008.

samples. If we have infinitely many test input samples, the pool-based problem is reduced to the population-based problem. In this paper, we address the problem of pool-based AL and propose a new algorithm.

**AL for Misspecified Models:** In traditional AL research [1,7,8], it is often assumed that the model used for function learning is *correctly specified*, i.e., it can exactly realize the learning target function. However, such an assumption may not be satisfied in reality (i.e., agnostic) and the violation of this assumption can cause significant performance degradation [2,3,4,6]. For this reason, we do not assume from the beginning that our model is correct in this paper. This highly enlarges the range of application of AL techniques.

In the AL scenarios, the distribution of training input points is generally different from that of test input points since the location of training input points is designed by users. Such a situation is referred to as *covariate shift* in statistics [9]. When we deal with misspecified models, covariate shift has a significant influence—for example, *Ordinary Least-Squares (OLS)* is no longer unbiased even asymptotically. Therefore, we need to explicitly take into account the bias caused by covariate shift. A standard approach to alleviating the influence of covariate shift is to use an *importance-weighting* technique [10], where the term 'importance' refers to the ratio of test and training input densities. For example, in parameter learning, OLS is biased, but *Importance-Weighted Least-Squares (IWLS)* is asymptotically unbiased [9].

**Importance Estimation in Pool-based AL:** In population-based AL, importance-weighting techniques can be employed for bias reduction in a straightforward manner since the test input distribution is accessible by assumption (and the training input distribution is also known since it is designed by ourselves) [2,3,4]. However, in pool-based AL, the test and training input distributions may both be unknown and therefore the importance weights cannot be directly computed. A naive approach to coping with this problem is to estimate the training and test input distributions from training and test input samples. However, density estimation is known to be a hard problem particularly in high dimensional problems. Therefore, such a naive approach may not be useful in practice. This difficulty could be eased by employing recently developed methods of *direct importance estimation* [11,12,13], which allow us to obtain the importance weight without going through density estimation. However, these methods still contain some estimation error.

A key observation in pool-based AL is that we choose training input points from the pool of test input points. This implies that our training input distribution is defined *over* the test input distribution, i.e., the training input distribution can be expressed as a product of the test input distribution and a *resampling bias function*. This decomposition allows us to directly compute the importance weight based on the resampling bias function, which is more accurate and computationally more efficient than the naive density estimation approach and the direct importance estimation approaches.

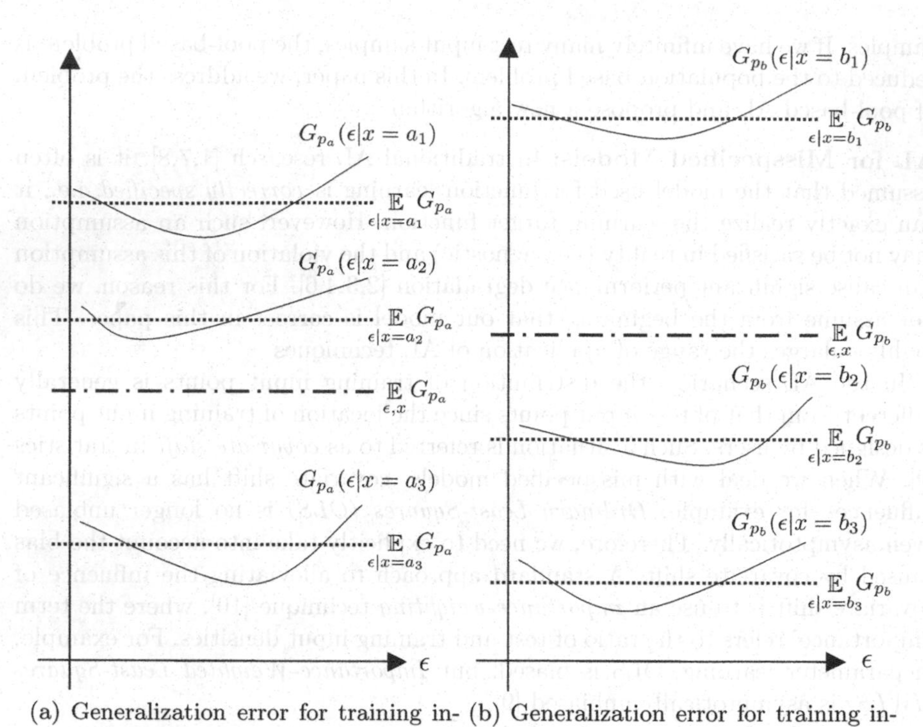


(a) Generalization error for training input density $p_a$

(b) Generalization error for training input density $p_b$

**Fig. 1.** Schematic illustrations of the conditional-expectation and full-expectation of the generalization error

**Single-trial Analysis of Generalization Error:** In practice, we are only given a single realization of training samples. Therefore, ideally, we want to have an estimator of the generalization error that is accurate in each *single trial.* However, we may not be able to avoid taking the expectation over the training output noise since it is not generally possible to know the realized value of noise. On the other hand, the location of the training input points is accessible by nature. Motivated by this fact, we propose to estimate the generalization error *without* taking the expectation over training input points. That is, we evaluate the unbiasedness of the generalization error in terms of the *conditional* expectation of training output noise given training input points.

To illustrate a possible advantage of this conditional expectation approach, let us consider a simple population-based active learning scenario where only one training sample $(x, y)$ is gathered (see Figure 1). Suppose thatthe input $x$ is drawn from a user-chosen training input distribution and $y$ is contaminated by additive noise $\epsilon$. The solid curves in Figure 1(a) depict $G_{p_a}(\epsilon|x)$, the generalization error for a training input density $p_a$ as a function of the training output noise $\epsilon$ given a training input point $x$. The three solid curves correspond to the cases where the realizations of the training input point $x$ are $a_1$, $a_2$, and $a_3$, respectively. The value of the generalization error for the training input density $p_a$ in the full-expectation approach is depicted by the dash-dotted line, where the generalization error is expected over both the training output noise $\epsilon$ and the training input points $x$ (i.e., the mean of the three solid curves). The values of the generalization error in the conditional-expectation approach are depicted by the dotted lines, where the generalization errors are expected only over the training output noise $\epsilon$, given $x = a_1, a_2, a_3$, respectively (i.e., the mean of each solid curve). The graph in Figure 1(b) depicts the generalization errors for another training input density $p_b$ in the same manner.

In the full-expectation framework, the density $p_a$ is judged to be better than $p_b$ regardless of the realization of the training input point since the dash-dotted line Figure 1(a) is lower than that in Figure 1(b). However, as the solid curves show, $p_a$ is often worse than $p_b$ in single trials. On the other hand, in the conditional-expectation framework, the goodness of the density is adaptively judged depending on the realizations of the training input point $x$. For example, $p_b$ is judged to be better than $p_a$ if $a_2$ and $b_3$ are realized, or $p_a$ is judged to be better than $p_b$ if $a_3$ and $b_1$ are realized. That is, the conditional-expectation framework may provide a finer choice of the training input density (and the training input points) than the full-expectation framework.

**Contributions of This Paper:** We extend two population-based AL methods proposed by [2] and [4] to pool-based scenarios. The pool-based extension of the method proposed in [2] allows us to obtain a closed-form solution of the best resampling bias function; thus it is computationally very efficient. However, this method is based on the full-expectation analysis of the generalization error, so the obtained solution is not necessarily optimal in terms of the single-trial generalization error. On the other hand, the pool-based extension of the method proposed in [4] can give a better solution since it is based on the conditional-expectation analysis of the generalization error. However, it does not have a closed-form solution and therefore some additional search strategy is needed.

To cope with this problem, we propose a practical procedure by combining the above two methods—we use the analytic optimal solution of the full-expectation method for efficiently searching for a better solution in the conditional-expectation method. Extensive simulations show that the proposed AL method consistently outperforms the baseline passive learning scheme and compares favorably with other active learning methods. Finally, we apply the proposed AL method to a real-world wafer alignment problem in semiconductor exposure apparatus and show that the alignment accuracy can be improved.

## 2 A New Pool-Based AL Method

In this section, we formulate the pool-based AL problem in regression scenarios and describe our new algorithm. Derivation and justification of the proposed algorithm are given in the next section.

### 2.1 Formulation of Pool-Based AL in Regression

We address a regression problem of learning a real-valued function $f(\boldsymbol{x})$ defined on $\mathcal{D} \subset \mathbb{R}^d$. We are given a 'pool' of test *input* points $\{\boldsymbol{x}_j^{\mathrm{te}}\}_{j=1}^{n_{\mathrm{te}}}$, which are drawn independently from an *unknown* test input distribution with strictly positive density $p_{\mathrm{te}}(\boldsymbol{x})$. From the pool, we are allowed to choose $n_{\mathrm{tr}}$ ($\ll n_{\mathrm{te}}$) input points for observing output values. Let $\{\boldsymbol{x}_i^{\mathrm{tr}}\}_{i=1}^{n_{\mathrm{tr}}}$ be input points selected from the pool and $\{y_i^{\mathrm{tr}}\}_{i=1}^{n_{\mathrm{tr}}}$ be corresponding output values, which we call *training samples*:

$$\{(\boldsymbol{x}_i^{\mathrm{tr}}, y_i^{\mathrm{tr}}) \mid y_i^{\mathrm{tr}} = f(\boldsymbol{x}_i^{\mathrm{tr}}) + \epsilon_i^{\mathrm{tr}}\}_{i=1}^{n_{\mathrm{tr}}},$$

where $\{\epsilon_i^{\mathrm{tr}}\}_{i=1}^{n_{\mathrm{tr}}}$ are i.i.d. noise with mean zero and unknown variance $\sigma^2$.

The goal of the regression task is to accurately predict the output values $\{f(\boldsymbol{x}_j^{\mathrm{te}})\}_{j=1}^{n_{\mathrm{te}}}$ at all test input points[1] $\{\boldsymbol{x}_j^{\mathrm{te}}\}_{j=1}^{n_{\mathrm{te}}}$. We adopt the squared loss as our error metric:

$$\frac{1}{n_{\mathrm{te}}}\sum_{j=1}^{n_{\mathrm{te}}}\left(\widehat{f}(\boldsymbol{x}_j^{\mathrm{te}}) - f(\boldsymbol{x}_j^{\mathrm{te}})\right)^2, \tag{1}$$

where $\widehat{f}(\boldsymbol{x})$ is a function learned from the training samples $\{(\boldsymbol{x}_i^{\mathrm{tr}}, y_i^{\mathrm{tr}})\}_{i=1}^{n_{\mathrm{tr}}}$.

### 2.2 Weighted Least-Squares for Linear Regression Models

We use the following linear regression model for learning:

$$\widehat{f}(\boldsymbol{x}) = \sum_{\ell=1}^{t}\theta_\ell\varphi_\ell(\boldsymbol{x}), \tag{2}$$

where $\{\varphi_\ell(\boldsymbol{x})\}_{\ell=1}^{t}$ are fixed linearly independent basis functions. $\boldsymbol{\theta}=(\theta_1, \theta_2, \ldots, \theta_t)^\top$ are parameters to be learned, where $^\top$ denotes the transpose of a vector or a matrix.

We learn the parameter $\boldsymbol{\theta}$ of the regression model by *Weighted Least-Squares (WLS)* with a weight function $w(\boldsymbol{x})$ ($> 0$ for all $\boldsymbol{x} \in \mathcal{D}$), i.e.,

$$\widehat{\boldsymbol{\theta}}_{\mathrm{W}} = \underset{\boldsymbol{\theta}}{\mathrm{argmin}}\left[\sum_{i=1}^{n_{\mathrm{tr}}} w(\boldsymbol{x}_i^{\mathrm{tr}})\left(\widehat{f}(\boldsymbol{x}_i^{\mathrm{tr}}) - y_i^{\mathrm{tr}}\right)^2\right], \tag{3}$$

where the subscript 'W' denotes 'Weighted'. Let $\boldsymbol{X}$ be the $n_{\mathrm{tr}} \times t$ matrix with $X_{i,\ell} = \varphi_\ell(\boldsymbol{x}_i^{\mathrm{tr}})$, and let $\boldsymbol{W}$ be the $n_{\mathrm{tr}} \times n_{\mathrm{tr}}$ diagonal matrix with $W_{i,i} = w(\boldsymbol{x}_i^{\mathrm{tr}})$. Then $\widehat{\boldsymbol{\theta}}_{\mathrm{W}}$ is given in a closed-form as

$$\widehat{\boldsymbol{\theta}}_{\mathrm{W}} = \boldsymbol{L}_{\mathrm{W}}\boldsymbol{y}^{\mathrm{tr}}, \tag{4}$$

[1] Under the assumption that $n_{\mathrm{tr}} \ll n_{\mathrm{te}}$, the difference between the prediction error at all test input points $\{\boldsymbol{x}_j^{\mathrm{te}}\}_{j=1}^{n_{\mathrm{te}}}$ and the remaining test input points $\{\boldsymbol{x}_j^{\mathrm{te}}\}_{j=1}^{n_{\mathrm{te}}} \backslash \{\boldsymbol{x}_i^{\mathrm{tr}}\}_{i=1}^{n_{\mathrm{tr}}}$ is negligibly small. More specifically, if $n_{\mathrm{tr}} = o(\sqrt{n_{\mathrm{te}}})$, all the discussions in this paper is still valid even when the prediction error is evaluated only at the remaining test input points.

where

$$\boldsymbol{L}_{\mathrm{W}} = (\boldsymbol{X}^\top \boldsymbol{W}\boldsymbol{X})^{-1}\boldsymbol{X}^\top \boldsymbol{W},$$
$$\boldsymbol{y}^{\mathrm{tr}} = (y_1^{\mathrm{tr}}, y_2^{\mathrm{tr}}, \ldots, y_{n_{\mathrm{tr}}}^{\mathrm{tr}})^\top.$$

### 2.3 Proposed AL Algorithm: P-CV$_\mathrm{W}$

Here we describe our AL algorithm for choosing the training input points from the pool of test input points; its derivation and justification are provided in the next section.

First, we prepare a candidate set of training input points $\{\boldsymbol{x}_i^{\mathrm{tr}}\}_{i=1}^{n_{\mathrm{tr}}}$, which is a subset of $\{\boldsymbol{x}_j^{\mathrm{te}}\}_{j=1}^{n_{\mathrm{te}}}$. More specifically, we prepare a *resampling bias function* $b(\boldsymbol{x})$ ($> 0$ for all $\boldsymbol{x} \in \mathcal{D}$) and choose $n_{\mathrm{tr}}$ training input points from the pool of test input points $\{\boldsymbol{x}_j^{\mathrm{te}}\}_{j=1}^{n_{\mathrm{te}}}$ with probability proportional to

$$\{b(\boldsymbol{x}_j^{\mathrm{te}})\}_{j=1}^{n_{\mathrm{te}}}.$$

Later, we explain how we prepare a family of useful resampling bias functions. We evaluate the 'quality' of the candidate training input points $\{\boldsymbol{x}_i^{\mathrm{tr}}\}_{i=1}^{n_{\mathrm{tr}}}$ by

$$\text{P-CV}_{\mathrm{W}} = \mathrm{tr}(\widehat{\boldsymbol{U}}\boldsymbol{L}_{\mathrm{W}}\boldsymbol{L}_{\mathrm{W}}^\top), \tag{5}$$

where the weight function $w(\boldsymbol{x})$ included in $\boldsymbol{L}_{\mathrm{W}}$ is defined as

$$w(\boldsymbol{x}_j^{\mathrm{te}}) = b(\boldsymbol{x}_j^{\mathrm{te}})^{-1}.$$

$\widehat{\boldsymbol{U}}$ is the $t \times t$ matrix with

$$\widehat{U}_{\ell,\ell'} = \frac{1}{n_{\mathrm{te}}}\sum_{j=1}^{n_{\mathrm{te}}} \varphi_\ell(\boldsymbol{x}_j^{\mathrm{te}})\varphi_{\ell'}(\boldsymbol{x}_j^{\mathrm{te}}).$$

We call the above criterion *pool-based CV$_W$ (P-CV$_W$)*, which is a pool-based extension of a population-based AL criterion *CV$_W$ (Conditional Variance of WLS)* [4]; we will explain the meaning and derivation of P-CV$_\mathrm{W}$ in Section 3.

We repeat the above evaluation for each resampling bias function in our candidate set and choose the best one with the smallest P-CV$_\mathrm{W}$ score. Once the resampling bias function and the training input points are chosen, we gather training output values $\{y_i^{\mathrm{tr}}\}_{i=1}^{n_{\mathrm{tr}}}$ at the chosen location and train a linear regression model (2) using WLS with the chosen weight function.

In the above procedure, the choice of the candidates of the resampling bias function $b(\boldsymbol{x})$ is arbitrary. As a heuristic, we propose using the following family of resampling bias functions parameterized by a scalar $\gamma$:

$$b_\gamma(\boldsymbol{x}) = \left(\sum_{\ell,\ell'=1}^{t} [\widehat{\boldsymbol{U}}^{-1}]_{\ell,\ell'}\varphi_\ell(\boldsymbol{x})\varphi_{\ell'}(\boldsymbol{x})\right)^\gamma. \tag{6}$$

**Input:** Test input points $\{\boldsymbol{x}_j^{\mathrm{te}}\}_{j=1}^{n_{\mathrm{te}}}$ and basis functions $\{\varphi_\ell(\boldsymbol{x})\}_{\ell=1}^{t}$
**Output:** Learned parameter $\widehat{\boldsymbol{\theta}}_{\mathrm{W}}$

Compute the $t \times t$ matrix $\widehat{\boldsymbol{U}}$ with $\widehat{U}_{\ell,\ell'} = \frac{1}{n_{\mathrm{te}}} \sum_{j=1}^{n_{\mathrm{te}}} \varphi_\ell(\boldsymbol{x}_j^{\mathrm{te}}) \varphi_{\ell'}(\boldsymbol{x}_j^{\mathrm{te}})$;
**For** several different values of $\gamma$ (intensively around $\gamma = 1/2$)
  Compute $\{b_\gamma(\boldsymbol{x}_j^{\mathrm{te}})\}_{j=1}^{n_{\mathrm{te}}}$ with $b_\gamma(\boldsymbol{x}) = \left(\sum_{\ell,\ell'=1}^{t} [\widehat{\boldsymbol{U}}^{-1}]_{\ell,\ell'} \varphi_\ell(\boldsymbol{x}) \varphi_{\ell'}(\boldsymbol{x})\right)^{\gamma}$;
  Choose $\mathcal{X}_\gamma^{\mathrm{tr}} = \{\boldsymbol{x}_i^{\mathrm{tr}}\}_{i=1}^{n_{\mathrm{tr}}}$ from $\{\boldsymbol{x}_j^{\mathrm{te}}\}_{j=1}^{n_{\mathrm{te}}}$ with probability proportional to $\{b_\gamma(\boldsymbol{x}_j^{\mathrm{te}})\}_{j=1}^{n_{\mathrm{te}}}$;
  Compute the $n_{\mathrm{tr}} \times t$ matrix $\boldsymbol{X}_\gamma$ with $[X_\gamma]_{i,\ell} = \varphi_\ell(\boldsymbol{x}_i^{\mathrm{tr}})$;
  Compute the $n_{\mathrm{tr}} \times n_{\mathrm{tr}}$ diagonal matrix $\boldsymbol{W}_\gamma$ with $[W_\gamma]_{i,i} = b_\gamma(\boldsymbol{x}_i^{\mathrm{tr}})^{-1}$;
  Compute $\boldsymbol{L}_\gamma = (\boldsymbol{X}_\gamma^\top \boldsymbol{W}_\gamma \boldsymbol{X}_\gamma)^{-1} \boldsymbol{X}_\gamma^\top \boldsymbol{W}_\gamma$;
  Compute $\text{P-CV}_{\mathrm{W}}(\gamma) = \mathrm{tr}(\widehat{\boldsymbol{U}} \boldsymbol{L}_\gamma \boldsymbol{L}_\gamma^\top)$;
**End**
Compute $\widehat{\gamma} = \mathrm{argmin}_\gamma \, \text{P-CV}_{\mathrm{W}}(\gamma)$;
Gather training output values $\{y_i^{\mathrm{tr}}\}_{i=1}^{n_{\mathrm{tr}}}$ at $\mathcal{X}_{\widehat{\gamma}}^{\mathrm{tr}}$;
Compute $\widehat{\boldsymbol{\theta}}_{\mathrm{W}} = \boldsymbol{L}_{\widehat{\gamma}} (y_1^{\mathrm{tr}}, y_2^{\mathrm{tr}}, \ldots, y_{n_{\mathrm{tr}}}^{\mathrm{tr}})^\top$;

**Fig. 2.** Pseudo code of proposed pool-based AL algorithm. In practice, the best $\gamma$ may be intensively searched around $\gamma = 1/2$.

The parameter $\gamma$ controls the 'shape' of the training input distribution—when $\gamma = 0$, the resampling weight is uniform over all test input samples. Thus the above choice includes *passive learning* (the training and test distributions are equivalent) as a special case. We seek the best $\gamma$ by simple multi-point search, i.e., we compute the value of P-CV$_{\mathrm{W}}$ for several different values of $\gamma$ and choose the minimizer. In practice, we propose performing the search intensively around $\gamma = 1/2$, e.g., Eq.(13); the reason for this will be explained in the next section.

A pseudo code of the proposed pool-based AL algorithm is described in Figure 2.

## 3 Derivation and Justification of Proposed AL Algorithm

The proposed P-CV$_{\mathrm{W}}$ criterion (5) and our choice of candidates of the training input distribution (6) are motivated by population-based AL criteria called CV$_{\mathrm{W}}$ (Conditional Variance of WLS; [4]) and FV$_{\mathrm{W}}$ (Full Variance of WLS; [2]). In this section, we explain how we came up with the pool-based AL algorithm given in Section 2.

### 3.1 Population-Based AL Criterion: CV$_{\mathrm{W}}$

Here we review a population-based AL criterion CV$_{\mathrm{W}}$.

In the population-based framework, we are given the test input density $p_{\mathrm{te}}(\boldsymbol{x})$, and the goal is to determine the best training input density $p_{\mathrm{tr}}(\boldsymbol{x})$ from which we draw training input points $\{\boldsymbol{x}_i^{\mathrm{tr}}\}_{i=1}^{n_{\mathrm{tr}}}$ [8,2,3,4].

The aim of the regression task in the population-based framework is to accurately predict the output values for all test input samples drawn from $p_{\mathrm{te}}(\boldsymbol{x})$.

Thus the error metric (often called the *generalization error*) is

$$G' = \int \left(\widehat{f}(\boldsymbol{x}^{\mathrm{te}}) - f(\boldsymbol{x}^{\mathrm{te}})\right)^2 p_{\mathrm{te}}(\boldsymbol{x}^{\mathrm{te}}) d\boldsymbol{x}^{\mathrm{te}} \equiv \|\widehat{f} - f\|_{p_{\mathrm{te}}}^2.$$

Suppose the regression model (2) *approximately* includes the learning target function $f(\boldsymbol{x})$, i.e., for a scalar $\delta$ such that $|\delta|$ is small, $f(\boldsymbol{x})$ is expressed as

$$f(\boldsymbol{x}) = g(\boldsymbol{x}) + \delta r(\boldsymbol{x}). \tag{7}$$

In the above, $g(\boldsymbol{x})$ is the optimal approximation to $f(\boldsymbol{x})$ by the model (2):

$$g(\boldsymbol{x}) = \textstyle\sum_{\ell=1}^{t} \theta_\ell^* \varphi_\ell(\boldsymbol{x}),$$

where $\boldsymbol{\theta}^* = (\theta_1^*, \theta_2^*, \ldots, \theta_t^*)^\top = \mathrm{argmin}_{\boldsymbol{\theta}} G'$ is the unknown optimal parameter. $\delta r(\boldsymbol{x})$ in Eq.(7) is the residual function, which is orthogonal to $\{\varphi_\ell(\boldsymbol{x})\}_{\ell=1}^{t}$ under $p_{\mathrm{te}}(\boldsymbol{x})$, i.e., $\langle r, \varphi_\ell \rangle_{p_{\mathrm{te}}} = 0$ for $\ell = 1, 2, \ldots, t$. The function $r(\boldsymbol{x})$ governs the nature of the model error, while $\delta$ is the possible magnitude of this error. In order to separate these two factors, we further impose $\|r\|_{p_{\mathrm{te}}} = 1$.

Let $\mathbb{E}_{\boldsymbol{\epsilon}}$ be the expectation over the noise $\{\epsilon_i^{\mathrm{tr}}\}_{i=1}^{n_{\mathrm{tr}}}$. Then, the generalization error expected over the training output noise can be decomposed into the (squared) *bias* term $B$, the *variance* term $V$, and the model error $\delta^2$:

$$\mathbb{E}_{\boldsymbol{\epsilon}} G' = B + V + \delta^2,$$

where

$$B = \|\mathbb{E}_{\boldsymbol{\epsilon}} \widehat{f} - g\|_{p_{\mathrm{te}}}^2, \quad V = \mathbb{E}_{\boldsymbol{\epsilon}} \|\widehat{f} - \mathbb{E}_{\boldsymbol{\epsilon}} \widehat{f}\|_{p_{\mathrm{te}}}^2.$$

Since $\delta$ is constant which depends neither on $p_{\mathrm{tr}}(\boldsymbol{x})$ nor $\{\boldsymbol{x}_i^{\mathrm{tr}}\}_{i=1}^{n_{\mathrm{tr}}}$, we subtract $\delta^2$ from $G'$ and define it by $G$.

$$G = G' - \delta^2.$$

Here we use *Importance-Weighted Least-Squares (IWLS)* for parameter learning [9], i.e., Eq.(3) with weight function $w(\boldsymbol{x})$ being the ratio of densities called the *importance ratio*:

$$w(\boldsymbol{x}) = \frac{p_{\mathrm{te}}(\boldsymbol{x})}{p_{\mathrm{tr}}(\boldsymbol{x})}. \tag{8}$$

The solution $\widehat{\boldsymbol{\theta}}_{\mathrm{W}}$ is given by Eq.(4).

Let $G_{\mathrm{W}}$, $B_{\mathrm{W}}$, and $V_{\mathrm{W}}$ be $G$, $B$, and $V$ for the learned function obtained by IWLS, respectively. Let $\boldsymbol{U}$ be the $t \times t$ matrix with

$$U_{\ell,\ell'} = \int \varphi_\ell(\boldsymbol{x}^{\mathrm{te}}) \varphi_{\ell'}(\boldsymbol{x}^{\mathrm{te}}) p_{\mathrm{te}}(\boldsymbol{x}^{\mathrm{te}}) d\boldsymbol{x}^{\mathrm{te}}.$$

Then, for IWLS with an approximately correct model, $B_{\mathrm{W}}$ and $V_{\mathrm{W}}$ are expressed as follows [4]:

$$B_{\mathrm{W}} = \mathcal{O}_p(\delta^2 n_{\mathrm{tr}}^{-1}), \quad V_{\mathrm{W}} = \sigma^2 \mathrm{tr}(\boldsymbol{U} \boldsymbol{L}_{\mathrm{W}} \boldsymbol{L}_{\mathrm{W}}^\top) = \mathcal{O}_p(n_{\mathrm{tr}}^{-1}).$$

The above equations imply that if $\delta = o_p(1)$,

$$\mathbb{E}_{\boldsymbol{\epsilon}} G_{\mathrm{W}} = \sigma^2 \mathrm{tr}(\boldsymbol{U}\boldsymbol{L}_{\mathrm{W}}\boldsymbol{L}_{\mathrm{W}}^\top) + o_p(n_{\mathrm{tr}}^{-1}).$$

The AL criterion $\mathrm{CV_W}$ is motivated by this asymptotic form, i.e., $\mathrm{CV_W}$ chooses the training input density $p_{\mathrm{tr}}(\boldsymbol{x})$ from the set $\mathcal{P}$ of all strictly positive probability densities as

$$p_{\mathrm{tr}}^{\mathrm{CV_W}} = \mathop{\mathrm{argmin}}_{p_{\mathrm{tr}} \in \mathcal{P}} \mathrm{CV_W}, \quad \mathrm{CV_W} = \mathrm{tr}(\boldsymbol{U}\boldsymbol{L}_{\mathrm{W}}\boldsymbol{L}_{\mathrm{W}}^\top).$$

Practically, $\mathcal{P}$ may be replaced by a finite set $\widehat{\mathcal{P}}$ of strictly positive probability densities and choose the one that minimizes $\mathrm{CV_W}$ from the set $\widehat{\mathcal{P}}$.

### 3.2 Extension of $\mathrm{CV_W}$ to Pool-Based Scenarios: P-$\mathrm{CV_W}$

Our basic idea of P-$\mathrm{CV_W}$ is to extend $\mathrm{CV_W}$ to the pool-based scenario, where we do not know $p_{\mathrm{te}}(\boldsymbol{x})$, but we are given a pool of test input samples $\{\boldsymbol{x}_i^{\mathrm{te}}\}_{i=1}^{n_{\mathrm{te}}}$ drawn independently from $p_{\mathrm{te}}(\boldsymbol{x})$. Under the pool-based setting, the following two quantities included in $\mathrm{CV_W}$ are not accessible:

**(A)** The expectation over $p_{\mathrm{te}}(\boldsymbol{x})$ in $\boldsymbol{U}$,
**(B)** The importance ratio $p_{\mathrm{te}}(\boldsymbol{x})/p_{\mathrm{tr}}(\boldsymbol{x})$ at training input points $\{\boldsymbol{x}_i^{\mathrm{tr}}\}_{i=1}^{n_{\mathrm{tr}}}$ in $\boldsymbol{L}_{\mathrm{W}}$.

Regarding (A), we may simply approximate the expectation over $p_{\mathrm{te}}(\boldsymbol{x})$ by the empirical average over the test input samples $\{\boldsymbol{x}_i^{\mathrm{te}}\}_{i=1}^{n_{\mathrm{te}}}$, which is known to be *consistent*.

On the other hand, approximation regarding (B) can be addressed as follows. In pool-based AL, we choose training input points from the pool of test input points following a resampling bias function $b(\boldsymbol{x})$. This implies that our training input distribution is defined *over* the test input distribution, i.e., the training input distribution is expressed as a product of the test input distribution and a resampling bias function $b(\boldsymbol{x})$:

$$p_{\mathrm{tr}}(\boldsymbol{x}_j^{\mathrm{te}}) \propto p_{\mathrm{te}}(\boldsymbol{x}_j^{\mathrm{te}}) b(\boldsymbol{x}_j^{\mathrm{te}}). \tag{9}$$

This immediately shows that the importance weight $w(\boldsymbol{x}_j^{\mathrm{te}})$ is given by

$$w(\boldsymbol{x}_j^{\mathrm{te}}) \propto b(\boldsymbol{x}_j^{\mathrm{te}})^{-1}. \tag{10}$$

Note that the scaling factor of $w(\boldsymbol{x})$ is irrelevant in IWLS (cf. Eq.(3)), so the above proportional form is sufficient here. By this, we can avoid density estimation which is known to be very hard.

Summarizing the above results, we obtain the P-$\mathrm{CV_W}$ criterion (5).

### 3.3 Population-Based AL Criterion: $\mathrm{FV_W}$

Next, we show how we came up with the candidate set of resampling bias functions given in Eq.(6). Our choice is based on a population-based AL method proposed by [2]. First, we consider the population-based setting and briefly review this method.

For IWLS, [3] proved that the generalization error expected over training input points $\{\boldsymbol{x}_i^{\rm tr}\}_{i=1}^{n_{\rm tr}}$ and training output noise $\{\epsilon_i^{\rm tr}\}_{i=1}^{n_{\rm tr}}$ is asymptotically expressed as

$$\mathbb{E}_{\boldsymbol{x}}\mathbb{E}_{\boldsymbol{\epsilon}}G_{\rm W} = \frac{\mathrm{tr}(\boldsymbol{U}^{-1}(\boldsymbol{S}+\sigma^2\boldsymbol{T}))}{n_{\rm tr}} + \mathcal{O}({n_{\rm tr}}^{-\frac{3}{2}}), \tag{11}$$

where $\mathbb{E}_{\boldsymbol{x}}$ is the expectation over training input points $\{\boldsymbol{x}_i^{\rm tr}\}_{i=1}^{n_{\rm tr}}$. $\boldsymbol{S}$ and $\boldsymbol{T}$ are the $t \times t$ matrices with

$$S_{\ell,\ell'} = \delta^2 \int \varphi_\ell(\boldsymbol{x})\varphi_{\ell'}(\boldsymbol{x})\,(r(\boldsymbol{x}))^2 \frac{p_{\rm te}(\boldsymbol{x})^2}{p_{\rm tr}(\boldsymbol{x})} d\boldsymbol{x},$$

$$T_{\ell,\ell'} = \int \varphi_\ell(\boldsymbol{x})\varphi_{\ell'}(\boldsymbol{x}) \frac{p_{\rm te}(\boldsymbol{x})^2}{p_{\rm tr}(\boldsymbol{x})} d\boldsymbol{x}.$$

Note that $\frac{1}{n_{\rm tr}}\mathrm{tr}(\boldsymbol{U}^{-1}\boldsymbol{S})$ corresponds to the squared bias while $\frac{\sigma^2}{n_{\rm tr}}\mathrm{tr}(\boldsymbol{U}^{-1}\boldsymbol{T})$ corresponds to the variance.

It can be shown [3,4] that if $\delta = o(1)$,

$$\mathbb{E}_{\boldsymbol{x}}\mathbb{E}_{\boldsymbol{\epsilon}}G_{\rm W} = \frac{\sigma^2}{n_{\rm tr}}\mathrm{tr}(\boldsymbol{U}^{-1}\boldsymbol{T}) + o({n_{\rm tr}}^{-1}).$$

Based on this asymptotic form, a population-based AL criterion, which we refer to as *$FV_W$ (Full Variance of WLS)*, is given as follows [2]:

$$p_{\rm tr}^{\rm FV_W} = \operatorname*{argmin}_{p_{\rm tr}\in\mathcal{P}} \mathrm{FV_W}, \quad \mathrm{FV_W} = \frac{1}{n_{\rm tr}}\mathrm{tr}(\boldsymbol{U}^{-1}\boldsymbol{T}).$$

A notable feature of $\mathrm{FV_W}$ is that the optimal training input density $p_{\rm tr}^{\rm FV_W}(\boldsymbol{x})$ can be obtained in a closed-form [2]:

$$p_{\rm tr}^{\rm FV_W}(\boldsymbol{x}) \propto p_{\rm te}(\boldsymbol{x}) b_{\rm FV_W}(\boldsymbol{x}), \quad b_{\rm FV_W}(\boldsymbol{x}) = \sqrt{\sum_{\ell,\ell'=1}^{t} [\boldsymbol{U}^{-1}]_{\ell,\ell'}\varphi_\ell(\boldsymbol{x})\varphi_{\ell'}(\boldsymbol{x})}.$$

Note that the importance ratio for the optimal training input density $p_{\rm tr}^{\rm FV_W}(\boldsymbol{x})$ is given by

$$w_{\rm FV_W}(\boldsymbol{x}) \propto b_{\rm FV_W}(\boldsymbol{x})^{-1}.$$

### 3.4 Extension of $\mathrm{FV_W}$ to Pool-Based Scenarios: P-$\mathrm{FV_W}$

If the values of the function $b_{\rm FV_W}(\boldsymbol{x})$ at the test input points $\{\boldsymbol{x}_j^{\rm te}\}_{j=1}^{n_{\rm te}}$ are available, they can be used as a resampling bias function in pool-based AL. However, since $\boldsymbol{U}$ is unknown in the pool-based scenario, it is not possible to directly compute the values of $b_{\mathrm{FV_W}}(\boldsymbol{x})$ at the test input points $\{\boldsymbol{x}_j^{\mathrm{te}}\}_{j=1}^{n_{\mathrm{te}}}$. To cope with this problem, we propose simply replacing $\boldsymbol{U}$ with an empirical estimate $\widehat{\boldsymbol{U}}$. Then, the resampling bias function $\{b_{\mathrm{P\text{-}FV_W}}(\boldsymbol{x}_j^{\mathrm{te}})\}_{j=1}^{n_{\mathrm{te}}}$ is given by

$$b_{\mathrm{P\text{-}FV_W}}(\boldsymbol{x}_j^{\mathrm{te}}) = \sqrt{\sum_{\ell,\ell'=1}^{t} [\widehat{\boldsymbol{U}}^{-1}]_{\ell,\ell'}\varphi_\ell(\boldsymbol{x}_j^{\mathrm{te}})\varphi_{\ell'}(\boldsymbol{x}_j^{\mathrm{te}})}. \tag{12}$$

The importance weight is simply given by

$$w_{\mathrm{P\text{-}FV_W}}(\boldsymbol{x}_j^{\mathrm{te}}) \propto b_{\mathrm{P\text{-}FV_W}}(\boldsymbol{x}_j^{\mathrm{te}})^{-1}.$$

### 3.5 Combining P-CV$_\mathrm{W}$ and P-FV$_\mathrm{W}$

It was shown that P-FV$_\mathrm{W}$ has a closed-form solution of the optimal resampling bias function. This simply suggests using $b_{\mathrm{P\text{-}FV_W}}(\boldsymbol{x}_j^{\mathrm{te}})$ for AL. Nevertheless, we argue that it is possible to further improve the solution.

The point of our argument is the way the generalization error is analyzed—the optimality of FV$_\mathrm{W}$ is in terms of the expectation over *both* training input points $\{\boldsymbol{x}_i^{\mathrm{tr}}\}_{i=1}^{n_{\mathrm{tr}}}$ and training output noise $\{\epsilon_i^{\mathrm{tr}}\}_{i=1}^{n_{\mathrm{tr}}}$, while CV$_\mathrm{W}$ is optimal in terms of the *conditional* expectation over training output noise $\{\epsilon_i^{\mathrm{tr}}\}_{i=1}^{n_{\mathrm{tr}}}$ given $\{\boldsymbol{x}_i^{\mathrm{tr}}\}_{i=1}^{n_{\mathrm{tr}}}$. However, in reality, what we really want to evaluate is the *single-trial* generalization error (i.e., without any expectation; both $\{\boldsymbol{x}_i^{\mathrm{tr}}\}_{i=1}^{n_{\mathrm{tr}}}$ and $\{\epsilon_i^{\mathrm{tr}}\}_{i=1}^{n_{\mathrm{tr}}}$ are given and fixed). Unfortunately, it is not possible to directly evaluate the single-trial generalization error since the training output noise $\{\epsilon_i^{\mathrm{tr}}\}_{i=1}^{n_{\mathrm{tr}}}$ cannot be observed directly; on the other hand, the training input points $\{\boldsymbol{x}_i^{\mathrm{tr}}\}_{i=1}^{n_{\mathrm{tr}}}$ are available. It was shown that the conditional expectation approach is provably more accurate in the single-trial analysis than the full expectation approach: if $\delta = o_p(n_{\mathrm{tr}}^{-1/4})$ and terms of $o_p(n_{\mathrm{tr}}^{-3})$ are ignored, the following inequality holds [4]:

$$\mathbb{E}_{\boldsymbol{\epsilon}}(\sigma^2\mathrm{FV_W} - G_\mathrm{W})^2 \geq \mathbb{E}_{\boldsymbol{\epsilon}}(\sigma^2\mathrm{CV_W} - G_\mathrm{W})^2.$$

This implies that $\sigma^2\mathrm{CV_W}$ is asymptotically a more accurate estimator of the single-trial generalization error $G_\mathrm{W}$ than $\sigma^2\mathrm{FV_W}$.

This analysis suggests that using P-CV$_\mathrm{W}$ is more suitable than P-FV$_\mathrm{W}$. However, a drawback of P-CV$_\mathrm{W}$ is that a closed-form solution is not available—thus, we may practically need to prepare candidates of training input samples and search for the best solution from the candidates. To ease this problem, our heuristic is to use the closed-form solution of P-FV$_\mathrm{W}$ as a 'base' candidate and search around its vicinity. More specifically, we consider a family of resampling bias functions (6), which is parameterized by $\gamma$. This family consists of the optimal solution of P-FV$_\mathrm{W}$ ($\gamma = 1/2$) and its variants ($\gamma \neq 1/2$); passive learning is also included as a special case ($\gamma = 0$) in this family.

The experimental results in Section 4 show that an additional search using P-CV$_\mathrm{W}$ tends to significantly improve the AL performance over P-FV$_\mathrm{W}$.

# 4 Simulations

In this section, we quantitatively compare the proposed and existing AL methods through numerical experiments.

## 4.1 Toy Dataset

We first illustrate how the proposed and existing AL methods behave under a controlled setting.

Let the input dimension be $d = 1$ and let the learning target function be

$$f(x) = 1 - x + x^2 + \delta r(x),$$

where $r(x) = (z^3 - 3z)/\sqrt{6}$ with $z = (x - 0.2)/0.4$. $r(x)$ defined here is a third order polynomial and is chosen to satisfy $\langle r, \varphi_\ell \rangle_{p_{\mathrm{te}}} = 0$ and $\|r\|_{p_{\mathrm{te}}} = 1$. Let us consider three cases $\delta = 0, 0.03, 0.06$.

Let the number of training examples to gather be $n_{\mathrm{tr}} = 100$ and let $\{\epsilon_i^{\mathrm{tr}}\}_{i=1}^{n_{\mathrm{tr}}}$ be i.i.d. Gaussian noise with mean zero and standard deviation $\sigma = 0.3$, where $\sigma$ is treated as unknown here. Let the test input density $p_{\mathrm{te}}(x)$ be Gaussian with mean 0.2 and standard deviation 0.4; $p_{\mathrm{te}}(x)$ is also treated as unknown here. We draw $n_{\mathrm{te}} = 1000$ test input points independently from the test input distribution.

We use a polynomial model of order 2 for learning:

$$\widehat{f}(x) = \theta_1 + \theta_2 x + \theta_3 x^2.$$

We compare the performance of the following sampling strategies:

**(A) P-CV$_{\mathrm{W}}$:** We draw training input points following the resampling bias function (6) with

$$\gamma \in \{0, 0.1, 0.2, \ldots, 1\} \cup \{0.4, 0.41, 0.42, \ldots, 0.6\}. \tag{13}$$

Then we choose the best $\gamma$ from the above candidates based on P-CV$_{\mathrm{W}}$ (5). IWLS is used for parameter learning.

**(B) P-FV$_{\mathrm{W}}$:** We draw training input points following the resampling bias function (12). IWLS is used for parameter learning.

**(C) Q-OPT [1,7,8]:** We draw training input points following the resampling bias function (6) with Eq.(13), and choose the best $\gamma$ based on

$$\text{Q-OPT} = \mathrm{tr}(\widehat{\boldsymbol{U}} \boldsymbol{L}_{\mathrm{O}} \boldsymbol{L}_{\mathrm{O}}^\top),$$

where $\boldsymbol{L}_{\mathrm{O}} = (\boldsymbol{X}^\top \boldsymbol{X})^{-1} \boldsymbol{X}^\top$. OLS is used for parameter learning.

**(D) Passive:** We draw training input points uniformly from the pool of test input samples. OLS is used for parameter learning.

In Table 1, the mean squared test error (1) obtained by each method is described. The numbers in the table are means and standard deviations over 100 trials.

When $\delta = 0$, Q-OPT and P-CV$_\text{W}$ are comparable to each other and are better than P-FV$_\text{W}$ and Passive. When $\delta = 0.03$, the performance of P-CV$_\text{W}$ and P-FV$_\text{W}$ is almost unchanged, while the performance of Q-OPT is degraded significantly. Consequently, P-CV$_\text{W}$ gives the best performance among all. When $\delta = 0.06$, the performance of P-CV$_\text{W}$ and P-FV$_\text{W}$ are still almost unchanged, while Q-OPT performs very poorly and is outperformed even by the baseline Passive method.

The above results show that P-CV$_\text{W}$ and P-FV$_\text{W}$ are highly robust against model misspecification, while Q-OPT is very sensitive to the violation of the model correctness assumption. P-CV$_\text{W}$ tends to outperform P-FV$_\text{W}$, which would be caused by the fact that CV$_\text{W}$ is a more accurate estimator of the single-trial generalization error than FV$_\text{W}$.

## 4.2 Benchmark Datasets

Here we use the *Bank*, *Kin*, and *Pumadyn* regression benchmark data families provided by DELVE [14]. Each data family consists of 8 different datasets: The input dimension is either $d = 8$ or 32, the target function is either 'fairly linear' or 'non-linear' ('f' or 'n'), and the unpredictability/noise level is either 'medium' or 'high' ('m' or 'h'). Thus we use 24 datasets in total. Each dataset includes 8192 samples, consisting of $d$-dimensional input and 1-dimensional output data. For convenience, we normalize every attribute into $[0, 1]$.

We use all 8192 input samples as the pool of test input points (i.e., $n_{\mathrm{te}} = 8192$), and choose $n_{\mathrm{tr}} = 100$ training input points from the pool when $d = 8$ and $n_{\mathrm{tr}} = 300$ training input points when $d = 32$. We use the following linear regression model:

$$\widehat{f}(\boldsymbol{x}) = \sum_{\ell=1}^{50} \theta_\ell \exp\left(-\frac{\|\boldsymbol{x} - \boldsymbol{c}_\ell\|^2}{2}\right),$$

where $\{\boldsymbol{c}_\ell\}_{\ell=1}^{50}$ are template points randomly chosen from the pool of test input points. Other settings are the same as the toy experiments in Section 4.1.

Table 2 summarizes the mean squared test error (1) over 1000 trials, where all the values are normalized by the mean error of the Passive method.

When $d = 8$, all 3 AL methods tend to be better than the Passive method. Among them, P-CV$_\text{W}$ significantly outperforms P-FV$_\text{W}$ and Q-OPT. When $d = 32$, Q-OPT outperforms P-CV$_\text{W}$ and P-FV$_\text{W}$ for several datasets. However, the performance of Q-OPT is highly unstable and is very poor for the *kin-32fm*, *kin-32fh*, and *pumadyn-32fm* datasets. Consequently, the average error of Q-OPT over all 12 datasets is worse than the baseline Passive method. On the other hand, P-CV$_\text{W}$ and P-FV$_\text{W}$ are still stable and consistently outperform the Passive method. Among these two methods, P-CV$_\text{W}$ significantly outperforms P-FV$_\text{W}$.

From the above experiments, we conclude that P-CV$_\text{W}$ and P-FV$_\text{W}$ are more reliable than Q-OPT, and P-CV$_\text{W}$ tends to outperform P-FV$_\text{W}$.

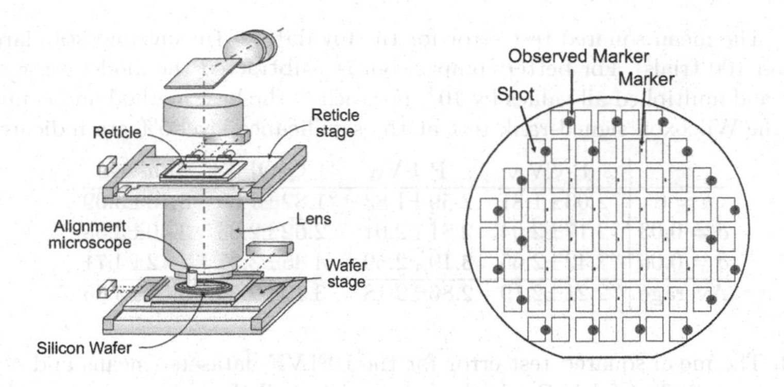


**Fig. 3.** Exposure apparatus (left) and a wafer (right)

## 5 Real-World Applications

Finally, we apply the proposed AL method to a wafer alignment problem in semiconductor exposure apparatus (see the left picture of Figure 3).

Recent semiconductors have the layered circuit structure, which are built by exposing circuit patterns multiple times. In this process, it is extremely important to align the wafer at the same position with very high accuracy. To this end, the location of markers are measured to adjust the shift and rotation of wafers. However, measuring the location of markers is time-consuming and therefore there is a strong need to reduce the number of markers to measure for speeding up the semiconductor production process.

The right picture of Figure 3 illustrates a wafer, where markers are printed uniformly over the wafer. Our goal here is to choose the most 'informative' markers to measure for better alignment of the wafer. A conventional choice is to measure markers far from the center in a symmetric way (see the right picture of Figure 3 again), which would provide robust estimation of the rotation angle. However, this naive approach is not necessarily the best since misalignment is not only caused by affine transformation, but also by several other non-linear factors such as a warp, a biased characteristic of measurement apparatus, and different temperature conditions. In practice, it is not easy to model such non-linear factors accurately, so the linear affine model or the second-order model is often used in wafer alignment. However, this causes model misspecification and therefore our proposed AL method would be useful in this application.

Let us consider the functions whose input $\boldsymbol{x} = (u, v)^\top$ is the location on the wafer and whose output is the horizontal discrepancy $\Delta u$ or the vertical discrepancy $\Delta v$. We learn these functions by the following second-order model.

$$\Delta u \text{ or } \Delta v = \theta_0 + \theta_1 u + \theta_2 v + \theta_3 uv + \theta_4 u^2 + \theta_5 v^2.$$

We totally have 220 wafer samples and our experiment is carried out as follows. For each wafer, we choose $n_{\mathrm{tr}} = 20$ points from $n_{\mathrm{te}} = 38$ markers and observe the horizontal and the vertical discrepancies. Then the above model is trained and its prediction performance is tested using all 38 markers in the 220 wafers. This

**Table 1.** The mean squared test error for the toy dataset (means and standard deviations over 100 trials). For better comparison, we subtracted the model error $\delta^2$ from the error and multiplied all values by $10^3$. For each $\delta$, the best method and comparable ones by the Wilcoxon signed-rank test at the significance level 5% are indicated with '∘'.

| | P-CV$_W$ | P-FV$_W$ | Q-OPT | Passive |
|---|---|---|---|---|
| $\delta = 0$ | ∘2.03±1.81 | 2.59±1.83 | ∘1.82±1.69 | 3.10±3.09 |
| $\delta = 0.03$ | ∘2.17±2.04 | 2.81±2.01 | 2.62±2.05 | 3.40±3.55 |
| $\delta = 0.06$ | ∘2.42±2.65 | 3.19±2.59 | 4.85±3.37 | 4.12±4.71 |
| Average | ∘2.21±2.19 | 2.86±2.18 | 3.10±2.78 | 3.54±3.85 |

**Table 2.** The mean squared test error for the DELVE datasets (means and standard deviations over 1000 trials). For better comparison, all the values are normalized by the mean error of the Passive method.

| | P-CV$_W$ | P-FV$_W$ | Q-OPT | Passive |
|---|---|---|---|---|
| bank-8fm | ∘0.89±0.14 | 0.95±0.16 | 0.91±0.14 | 1.00±0.19 |
| bank-8fh | 0.86±0.14 | 0.94±0.17 | ∘0.85±0.14 | 1.00±0.20 |
| bank-8nm | ∘0.89±0.16 | 0.95±0.20 | 0.91±0.18 | 1.00±0.21 |
| bank-8nh | 0.88±0.16 | 0.95±0.20 | ∘0.87±0.16 | 1.00±0.21 |
| kin-8fm | ∘0.78±0.22 | 0.87±0.24 | 0.87±0.22 | 1.00±0.25 |
| kin-8fh | ∘0.80±0.17 | 0.88±0.21 | 0.85±0.17 | 1.00±0.23 |
| kin-8nm | ∘0.91±0.14 | 0.97±0.16 | 0.92±0.14 | 1.00±0.17 |
| kin-8nh | ∘0.90±0.13 | 0.96±0.16 | 0.90±0.13 | 1.00±0.17 |
| pumadyn-8fm | ∘0.89±0.13 | 0.95±0.16 | ∘0.89±0.12 | 1.00±0.18 |
| pumadyn-8fh | 0.89±0.13 | 0.98±0.16 | ∘0.88±0.12 | 1.00±0.17 |
| pumadyn-8nm | ∘0.91±0.13 | 0.98±0.17 | 0.92±0.13 | 1.00±0.18 |
| pumadyn-8nh | ∘0.91±0.13 | 0.97±0.14 | 0.91±0.13 | 1.00±0.17 |
| Average | ∘0.87±0.16 | 0.95±0.18 | 0.89±0.15 | 1.00±0.20 |

| | P-CV$_W$ | P-FV$_W$ | Q-OPT | Passive |
|---|---|---|---|---|
| bank-32fm | 0.97±0.05 | 0.99±0.05 | ∘0.96±0.04 | 1.00±0.06 |
| bank-32fh | 0.98±0.05 | 0.99±0.05 | ∘0.96±0.04 | 1.00±0.05 |
| bank-32nm | 0.98±0.06 | 0.99±0.07 | ∘0.96±0.06 | 1.00±0.07 |
| bank-32nh | 0.97±0.05 | 0.99±0.06 | ∘0.96±0.05 | 1.00±0.06 |
| kin-32fm | ∘0.79±0.07 | 0.93±0.09 | 1.53±0.14 | 1.00±0.11 |
| kin-32fh | ∘0.79±0.07 | 0.92±0.08 | 1.40±0.12 | 1.00±0.10 |
| kin-32nm | 0.95±0.04 | 0.97±0.04 | ∘0.93±0.04 | 1.00±0.05 |
| kin-32nh | 0.95±0.04 | 0.97±0.04 | ∘0.92±0.03 | 1.00±0.05 |
| pumadyn-32fm | ∘0.98±0.12 | 0.99±0.13 | 1.15±0.15 | 1.00±0.13 |
| pumadyn-32fh | 0.96±0.04 | 0.98±0.05 | ∘0.95±0.04 | 1.00±0.05 |
| pumadyn-32nm | 0.96±0.04 | 0.98±0.04 | ∘0.93±0.03 | 1.00±0.05 |
| pumadyn-32nh | 0.96±0.03 | 0.98±0.04 | ∘0.92±0.03 | 1.00±0.04 |
| Average | ∘0.94±0.09 | 0.97±0.07 | 1.05±0.21 | 1.00±0.07 |

**Table 3.** The mean squared test error for the wafer alignment problem (means and standard deviations over 220 wafers). 'Conv.' indicates the conventional heuristic of choosing the outer markers.

| P-CV$_W$ | P-FV$_W$ | Q-OPT | Passive | Conv. |
|---|---|---|---|---|
| ∘1.93±0.89 | 2.09±0.98 | 1.96±0.91 | 2.32±1.15 | 2.13±1.08 |

process is repeated for all 220 wafers. Since the choice of the sampling location by AL methods is stochastic, we repeat the above experiment for 100 times with different random seeds and take the mean value.

The mean and standard deviation of the squared test error over 220 wafers are summarized in Table 3. This shows that the proposed P-CV$_W$ works significantly better than other sampling strategies and it provides about 10-percent reduction in the squared error from the conventional heuristic of choosing the outer markers. We also conducted similar experiments with the first-order or the third-order models and confirmed that P-CV$_W$ still works the best. However, the errors were larger than the second-order model and therefore we omit the detail.

## 6 Conclusions

We extended a population-based AL method (FV$_W$) to a pool-based scenario (P-FV$_W$) and derived a closed-form 'optimal' resampling bias function. This closed-form solution is optimal within the full-expectation framework, but is not necessarily optimal in the single-trial analysis. To further improve the performance, we extended another population-based method (CV$_W$) to a pool-based scenario (P-CV$_W$), which is input-dependent and therefore more accurate. However, P-CV$_W$ does not allow us to obtain a closed-form solution. To cope with this problem, we proposed a practical procedure which efficiently searches for a better solution around the P-FV$_W$ optimal solution. Simulations showed that the proposed method consistently outperforms the baseline passive learning scheme and compares favorably with other AL methods.

## References


1. Fedorov, V.V.: Theory of Optimal Experiments. Academic Press, New York (1972)
2. Wiens, D.P.: Robust weights and designs for biased regression models: Least squares and generalized M-estimation. Journal of Statistical Planning and Inference 83(2), 395–412 (2000)
3. Kanamori, T., Shimodaira, H.: Active learning algorithm using the maximum weighted log-likelihood estimator. Journal of Statistical Planning and Inference 116(1), 149–162 (2003)
4. Sugiyama, M.: Active learning in approximately linear regression based on conditional expectation of generalization error. Journal of Machine Learning Research 7, 141–166 (2006)
5. McCallum, A., Nigam, K.: Employing EM in pool-based active learning for text classification. In: Proceedings of the 15th International Conference on Machine Learning (1998)
6. Bach, F.: Active learning for misspecified generalized linear models. In: Schölkopf, B., Platt, J., Hoffman, T. (eds.) Advances in Neural Information Processing Systems 19. MIT Press, Cambridge (2007)
7. Cohn, D.A., Ghahramani, Z., Jordan, M.I.: Active learning with statistical models. Journal of Artificial Intelligence Research 4, 129–145 (1996)
8. Fukumizu, K.: Statistical active learning in multilayer perceptrons. IEEE Transactions on Neural Networks 11(1), 17–26 (2000)

9. Shimodaira, H.: Improving predictive inference under covariate shift by weighting the log-likelihood function. Journal of Statistical Planning and Inference 90(2), 227–244 (2000)
10. Fishman, G.S.: Monte Carlo: Concepts, Algorithms, and Applications. Springer, Berlin (1996)
11. Huang, J., Smola, A., Gretton, A., Borgwardt, K.M., Schölkopf, B.: Correcting sample selection bias by unlabeled data. In: Schölkopf, B., Platt, J., Hoffman, T. (eds.) Advances in Neural Information Processing Systems 19, pp. 601–608. MIT Press, Cambridge (2007)
12. Bickel, S., Brückner, M., Scheffer, T.: Discriminative learning for differing training and test distributions. In: Proceedings of the 24th International Conference on Machine Learning (2007)
13. Sugiyama, M., Nakajima, S., Kashima, H., von Bünau, P., Kawanabe, M.: Direct importance estimation with model selection and its application to covariate shift adaptation. In: Advances in Neural Information Processing Systems 20. MIT Press, Cambridge (2008)
14. Rasmussen, C.E., Neal, R.M., Hinton, G.E., van Camp, D., Revow, M., Ghahramani, Z., Kustra, R., Tibshirani, R.: The DELVE manual (1996)

# Distribution-Free Learning of Bayesian Network Structure

Xiaohai Sun

Max Planck Institute for Biological Cybernetics
Spemannstr. 38, 72076 Tübingen, Germany
xiaohai.sun@tuebingen.mpg.de

**Abstract.** We present an independence-based method for learning Bayesian network (BN) structure without making any assumptions on the probability distribution of the domain. This is mainly useful for continuous domains. Even mixed continuous-categorical domains and structures containing vectorial variables can be handled. We address the problem by developing a non-parametric conditional independence test based on the so-called kernel dependence measure, which can be readily used by any existing independence-based BN structure learning algorithm. We demonstrate the structure learning of graphical models in continuous and mixed domains from real-world data without distributional assumptions. We also experimentally show that our test is a good alternative, in particular in case of small sample sizes, compared to existing tests, which can only be used in purely categorical or continuous domains.

**Keywords:** graphical models, independence tests, kernel methods.

## 1 Introduction

There are two basic classes of BN learning algorithms, namely constraint-based (or independence-based) and model-based approaches. The model-based approaches often employ a search in the space of all possible structures guided by a heuristic function, usually penalized log-likelihood or Bayesian metric [1,2,3]. They are also called score-based approach in many literature. One of the challenges of applying score-based methods is the assessment of informative priors on possible structures and on parameters for those structures. Constraint-based approaches carry out conditional independence (CI) tests on the database and build a BN structure in agreement with the obtained independence restrictions. They make weak commitments as to the nature of relationships between variables, i.e., faithfulness/stability [4,5]. The best-known example of this kind of approaches is the so-called PC algorithm [4]. The output of PC is the Markov equivalence class of BN structures, which can often be interpreted as causal graph under some additional technical assumptions, e.g., causal sufficiency. Such independence-based algorithms can essentially be broken into an adjacency phase and an orientation phase. The rule of learning adjacency is to test whether a set variables $Z$

W. Daelemans et al. (Eds.): ECML PKDD 2008, Part II, LNAI 5212, pp. 423–439, 2008.

**Fig. 1.** The "$\wedge$"-structure (left) represents the Markov equivalence class of "$X \leftarrow Y \rightarrow Z$", "$X \rightarrow Y \rightarrow Z$", and "$X \leftarrow Y \leftarrow Z$", where constraints $X \not\perp\!\!\!\perp Y$ and $X \perp\!\!\!\perp Y|Z$ hold by construction. In the "$\vee$"-structure (right), $X \perp\!\!\!\perp Y$ and $X \not\perp\!\!\!\perp Y|Z$ hold by construction

exists which makes marginally dependent $X$ and $Y$ conditionally independent, denoted by $X \not\perp\!\!\!\perp Y$ and $X \perp\!\!\!\perp Y|Z$ (the so-called "$\wedge$"-structure in Fig. 1, left). The orientation strategy bases on the so-called collider ("$\vee$"-structure) identification, namely if some subset $S$ exists such that $X \perp\!\!\!\perp Y|S$ and $X \not\perp\!\!\!\perp Y|(S, Z)$, a $\vee$-structure (Fig. 1, right) can be inferred. Consequently, the accuracy and reliability of CI tests play a central role in independence-based algorithms.

One of the crucial problems of learning BN structure from data is making a choice on the kind of probability distributions. Commonly, a local probability distribution function (PDF) needs to be defined between every variable and its parents. Such distributions are used to model local interactions among subsets of variables in the model. CI tests that do not assume any particular family of distributions are called non-parametric. Although such tests exist for categorical variables, e.g., $\chi^2$ test or test via mutual information (MI) is perhaps the most common one, the problem in continuous or mixed continuous-categorical domains is considerably harder. The common assumption is for the local PDFs between parents and children to be linear relations with additive Gaussian errors, as the correlation analysis in PC algorithm. However, there are many situations where this assumption fails and the underlying interactions are far from linear. In some situation, interactions cannot be captured by correlation analysis at all. To make this apparent, we demonstrate the following toy data example. We sampled dataset $(X_0, Y_0)$ as shown in the leftmost plot of Fig. 2 and then transformed the original dataset by a rotation of angle $\omega$ (in degree) into a new dataset $(X_\omega, Y_\omega)$. The second and the third plot (from left) in Fig. 2 visualize the transformed data $(X_{45}, Y_{45})$ and $(X_{90}, Y_{90})$ respectively. According to the generating model with $P(X_0|Y_0 < 0) = P(X_0|Y_0 \geq 0)$ (see Fig. 2 for the description of the underlying model), $X_0$ and $Y_0$ are independent. It is obvious that $X_\omega$ and $Y_\omega$ are only independent for $\omega = 0, 90, 180, \ldots$.

It is easy to see that the correlation matrix $\rho_0$ of the data matrix

$$\mathcal{D}_0 := \begin{pmatrix} X_0 \\ Y_0 \end{pmatrix} \text{ is a unit matrix, namely } \rho_0 := \begin{pmatrix} \rho_{X_0X_0} & \rho_{X_0Y_0} \\ \rho_{Y_0X_0} & \rho_{Y_0Y_0} \end{pmatrix} = \begin{pmatrix} 1 & 0 \\ 0 & 1 \end{pmatrix}.$$

Further, it is well known that the rotation matrix $R_\omega$ of angle $\omega$ in an anticlockwise direction has the form of

$$R_\omega = \begin{pmatrix} \cos(\frac{\pi}{180}\omega) & \sin(\frac{\pi}{180}\omega) \\ -\sin(\frac{\pi}{180}\omega) & \cos(\frac{\pi}{180}\omega) \end{pmatrix}. \text{ Hence, } \begin{pmatrix} X_\omega \\ Y_\omega \end{pmatrix} =: \mathcal{D}_\omega = R_\omega \mathcal{D}_0 = R_\omega \begin{pmatrix} X_0 \\ Y_0 \end{pmatrix},$$

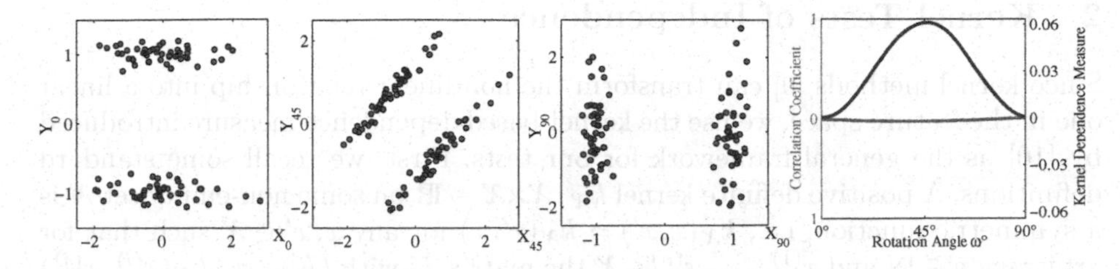


**Fig. 2.** 100 data points of $Y_0$ is sampled from $\frac{1}{2}\mathcal{N}(1,0.01)+\frac{1}{2}\mathcal{N}(-1,0.01)$, where $\mathcal{N}(\mu,\sigma^2)$ denotes a Gaussian distribution with expectation $\mu$ and variance $\sigma^2$. Data points of $X_0$ are sampled from $P(X_0|Y_0<0)\propto\mathcal{N}(0,1)$ and $P(X_0|Y_0\geq 0)\propto\mathcal{N}(0,1)$. $(X_\omega, Y_\omega)$ is transformed dataset with rotation angle $\omega$ in an anticlockwise direction. $X_0$ and $Y_0$ (leftmost plot) as well as $X_{90}$ and $Y_{90}$ (second plot from right) are mutually independent by construction, whereas $X_{45}$ and $Y_{45}$ (second plot from left) are strongly dependent due to the rotation angle 45. The rightmost plot visualizes the typical curve of the empirical correlation coefficients (red line) for $\omega\in[0,90]$, and a typical curve of the empirical kernel dependence measures (blue line), which will be defined in the next section. Correlation coefficient vanishes for any $\omega$. Kernel dependence measures seem to capture the magnitude of dependences reasonably.

depicts the dataset transformed by a rotation angle $\omega$. The corresponding correlation matrix $\rho_\omega$ is given by

$$\rho_\omega = \mathrm{E}\left[R_\omega\,(\mathcal{D}_0-\mathrm{E}[\mathcal{D}_0])(\mathcal{D}_0-\mathrm{E}[\mathcal{D}_0])^{\mathrm{T}}\,R_\omega^{\mathrm{T}}\right] = R_\omega\,\rho_0\,R_\omega^{\mathrm{T}} = \rho_0\,,$$

where "E[·]" depicts the expectation. This means that the correlation coefficient vanishes for an arbitrary $\omega$, while the dependence indeed vanishes only for few specific rotation angles $\omega=0,90,180,\ldots$. This is the reason why it is not surprising that the performance of the conventional BN learning algorithms, e.g. PC, is sometimes unsatisfactory, since it takes only linear interactions into account.

A non-parametric Bayesian method of testing independence on continuous domains is recently proposed by Margaritis et al. [6,7]. Their method is not based on a conventional hypothesis test but on the calculation of probability of independence given data by the Bayesian approach. To determine whether two variables are (conditionally) independent, they discretized the domains by maximizing the posterior probability of dependence given data. If the probability of independence $P(\text{Independence})$ larger than $\frac{1}{2}$, the independence is verified, otherwise dependence. More precisely, the method determines the probability of dependence by calculating the likelihoods of modeling the data as dependent with a joint multinomial distribution or as independent with two marginal multinomial distribution. Margaritis' Bayesian method is impressive because it is the first practicable distribution-free learning of BN structure in purely continuous domains, although it involves a sophisticated process of domain discretization. We will compare our test method with this method later.

## 2 Kernel Tests of Independence

Since kernel methods [9] can transform the non-linear relationship into a linear one in the feature space, we use the kernel-based dependence measure introduced by [10] as the general framework for our tests. First, we recall some standard definitions. A positive definite kernel $k_{\mathcal{X}}:\mathcal{X}\times\mathcal{X}\rightarrow\mathbb{R}$ on some non-empty set $\mathcal{X}$ is a symmetric function, i.e., $k_{\mathcal{X}}(x,x') = k_{\mathcal{X}}(x',x)$ for any $x,x' \in \mathcal{X}$ such that for arbitrary $n \in \mathbb{N}$ and $x^{(1)},\ldots,x^{(n)} \in \mathcal{X}$ the matrix $K$ with $(K)_{ij} := k_{\mathcal{X}}(x^{(i)},x^{(j)})$ is positive definite, i.e., $\sum_{i,j=1}^{n} c_i c_j k_{\mathcal{X}}(x^{(i)},x^{(j)}) \geq 0$ for all $c_1,\ldots,c_n \in \mathbb{R}$. A reproducing kernel Hilbert space (RKHS) $\mathcal{H}_{\mathcal{X}}$ is a Hilbert space defined by the completion of an inner product space of functions $k_{\mathcal{X}}(x,\cdot)$ with $x \in \mathcal{X}$ and the inner product defined by $\langle k_{\mathcal{X}}(x,\cdot), k_{\mathcal{X}}(x',\cdot)\rangle = k_{\mathcal{X}}(x,x')$ for all $x,x' \in \mathcal{X}$. In other words, $\Phi(x)(\cdot) = k_{\mathcal{X}}(x,\cdot)$ defines a map from $\mathcal{X}$ into a feature space $\mathcal{H}_{\mathcal{X}}$. With the so-called "kernel trick", a linear algorithm can easily be transformed into a non-linear algorithm, which is equivalent to the linear algorithm operating in the space of $\Phi$. However, because the kernel function is used for calculating the inner product, the mapping $\Phi$ is never explicitly computed. This is desirable, because the high-dimensional space may be infinite-dimensional, as is the case when the kernel, e.g., is a Gaussian: $k_{\mathcal{X}}:\mathbb{R}^m\times\mathbb{R}^m\rightarrow\mathbb{R}$, $k_{\mathcal{X}}(x,x') = \exp(-\|x-x'\|^2/2\sigma^2)$.

Within the kernel framework, the conditional cross-covariance operator expressing correlations between variables in the RKHS can be introduced. Let $(\mathcal{H}_{\mathcal{X}},k_{\mathcal{X}})$, $(\mathcal{H}_{\mathcal{Y}},k_{\mathcal{Y}})$, $(\mathcal{H}_{\mathcal{Z}},k_{\mathcal{Z}})$ be RKHSs on measurable spaces $\mathcal{X},\mathcal{Y},\mathcal{Z}$ respectively, $(X,Y,Z)$ be a random vector on $\mathcal{X}\times\mathcal{Y}\times\mathcal{Z}$. For all $f\in\mathcal{H}_{\mathcal{X}}$ and $g\in\mathcal{H}_{\mathcal{Y}}$, there exists (under some technical assumptions [11]) a unique conditional cross-covariance operator $\Sigma_{YX|Z}$ such that $\langle g,\Sigma_{YX|Z}f\rangle_{\mathcal{H}_{\mathcal{Y}}} = \mathrm{E}_Z[\mathrm{Cov}[f(X),g(Y)\mid Z]]$. If the RKHSs are induced by characteristic kernels [11], e.g., Gaussian kernels, vanishing of the operator is equivalent to the conditional independence [10]: $\Sigma_{\ddot{Y}\ddot{X}|Z} = O \Leftrightarrow X \perp\!\!\!\perp Y \mid Z$ where $\ddot{X} := (X,Z)$ and $\ddot{Y} := (Y,Z)$. It should be stressed that $X,Y,Z$ can be continuous, categorical, vectorial or mixed. For strictly nominal-categorical variables, the natural way to represent the $d$ nominal alternatives, namely $d$ unit vectors in a $d$-dimensional Cartesian coordinate system: $\{(1,0,\ldots)^{\mathrm{T}}, (0,1,\ldots)^{\mathrm{T}}, \ldots, (\ldots,0,1)^{\mathrm{T}}\} \subset \mathbb{R}^d$.

Following the work of [12,10], we evaluate the operator with Hilbert-Schmidt (HS) norm and denote the empirical estimator by $\widehat{\mathbb{H}}_{YX|Z} := \|\Sigma_{\ddot{Y}\ddot{X}|Z}\|^2_{\mathrm{HS}}$. All these statements hold also for the unconditional operator and marginal independence $\Sigma_{YX} = O \Leftrightarrow X \perp\!\!\!\perp Y$. Given RKHS induced by Gaussian kernels, where functions being less smooth correspond to larger RKHS-norms, large values will then indicate correlations between smooth functions of $X$ and $Y$. A finite cut-off value for the kernel dependence measure corresponds to neglecting correlations if they are small or if they occur only on complex (not sufficiently smooth) functions. Instead of thresholding the kernel measure directly as in our previous work [10], we employ permutation tests [13] for judging whether dependencies are significant or not. Although permutation tests require a large number of calculations of HS-norms, the computation is feasible if we use the incomplete Cholesky decomposition $\widehat{K} = LL^{\mathrm{T}}$ [14] where $L$ is a lower triangular matrix determined uniquely by this equation. This may lead to considerably fewer columns than the original matrix. If $k$ columns are returned, the storage requirements are $O(kn)$ instead of $O(n^2)$, and the running time of many matrix operations reduces to $O(nk^2)$ instead of $O(n^3)$. An alternative kernel statistical test based on moment matching is currently proposed by Gretton et al. [15]. Instead of computing the HS-norm of the operator directly, they designed a test statistics based on entries of kernel matrices. But, for one thing, this alternative test is designed for unconditional cases, because the null distribution only in such cases is known. For the other thing, they claimed that the permutation test may outperform their alternative in some situation, particularly if the sample size is small (e.g., less than 200), since the estimation of second moments of entries of kernel matrices tends to be unreliable (see [15] for experiments with text data).

Given a data matrix $(X, Y, Z)$ with $X := (x^{(1)}, \dots, x^{(n)})^{\mathrm{T}}$ and so on, testing unconditional independence $X \perp\!\!\!\perp Y$ via random permutations is straightforward. But how to approximate the null distribution of the kernel dependence measure $\widehat{\mathbb{H}}_{YX|Z}$ under conditional independency is non-trivial. The idea is that random permutations should release the connection between $X$ and $Y$ to simulate the independency. On the other hand, the mutual relations between $X$ and $Z$ and between $Y$ and $Z$ have to be kept, since $Z$ is tied to a specific value. In order to yield this effect, we propose to restrict the random permutations $\pi$ to those that satisfy the condition $z^{(\pi(i))} = z^{(i)}$ for $i = 1, \dots, n$. In other words, if $Z$ is categorical, $\pi$ is restricted to random permutations within the same category of $Z$. If $Z$ is real-valued or vectorial, the condition could be said to hold if $z^{(\pi_j(i))}$ and $z^{(i)}$ are "similar" in some sense. This suggests the use of clustering techniques to search for an appropriate partition of $Z$. Then, the data points within the same partition are similar w.r.t. values of $Z$ (see Tab. 1). This way, we designed a kernel independence test providing a general non-parametric tool for verifying CI constraints in both conditional and unconditional cases. In our following experiments, we employed the standard k-means clustering to construct the partition of data points of conditioning variables and computed the approximate null distribution under conditional independency. The significance level is set to be 0.05 throughout this paper.

**Table 1.** Set of random permutations that are used for the test of conditional independence $X \perp\!\!\!\perp Y|Z$. $Z$ data are clustered into $n$ partition. Within each partition $z^{(i)}$, generate simulated conditionally independent data $(X^{\pi_j}, Y)$.

| | partition $z^{(1)}$ | partition $z^{(2)}$ | ... | ... | ... | partition $z^{(n)}$ |
|---|---|---|---|---|---|---|
| $Z$ | $z^{(1_1)}\ z^{(1_2)} \dots z^{(1_m)}$ | $z^{(2_1)}\ z^{(2_2)} \dots z^{(2_m)}$ | ... | ... | ... | $z^{(n_1)}\ z^{(n_2)} \dots z^{(n_m)}$ |
| $Y$ | $y^{(1_1)}\ y^{(1_2)} \dots y^{(1_m)}$ | $y^{(2_1)}\ y^{(2_2)} \dots y^{(2_m)}$ | ... | ... | ... | $y^{(n_1)}\ y^{(n_2)} \dots y^{(n_m)}$ |
| $X^{\pi_j}$ | $x^{(1_1)}\ x^{(1_2)} \dots x^{(1_m)}$ | $x^{(2_1)}\ x^{(2_2)} \dots x^{(2_m)}$ | ... | ... | ... | $x^{(n_1)}\ x^{(n_2)} \dots x^{(n_m)}$ |
| $\pi_j$ | permute | permute | permute | ··· | permute | permute |

# 3 Experiments

Learning BN structure in real applications is challenging, because real-world data are often on mixed domains and the relationships are often nonlinear. Sometimes, the real data even do not necessarily have the representation of a BN structure. The conventional approaches either can not handle them or ignore possible problems. By means of some benchmark real-world data, we will show some respects, in which some improvements can be achieved by our method in comparison to conventional ones, in particular in relation to causal structure learning, to which the independence-based approaches are supposed to be often applied.

## 3.1 Real-World BN Structure Learning

**Digoxin Clearance.** The study of the passage of drugs through the body is of essential interest in medical science. A real-world dataset on 35 consecutive patients under treatment for heart failure with the drug digoxin is analyzed in [16] (see also [17] p. 323 and [18] p. 42). The renal clearances of digoxin, creatinine, and urine flow were determined simultaneously in each of the patients receiving digoxin, in most of whom there was prerenal azotemia. Halkin et al. [16] and Edwards [18] based their analysis on the (partial) correlation coefficient. In comparison to the correlation analysis and Margaritis' Bayesian method, we conducted the kernel tests (see Tab. 2 for results). A visual inspection of the data indicates that the linearity assumption appears to be reasonable for the dependence between the creatinine and digoxin clearances (Fig. 3, leftmost). A linear relation between them was first suggested by Jelliffe et al. [19] and later confirmed by various clearance studies, which revealed a close relationship between creatinine and digoxin clearance in many patients. The ready explanation is that both creatinine and digoxin are mainly eliminated by the kidneys. In agreement with this explanation, all three tests found the unconditional and conditional dependence (Tab. 2, first and second row).

As one can see from Fig. 3, the relations between creatinine clearance and urine flow (second plot) and between digoxin clearances and urine flow (third plot) are less linear than the relation between creatinine and digoxin clearance

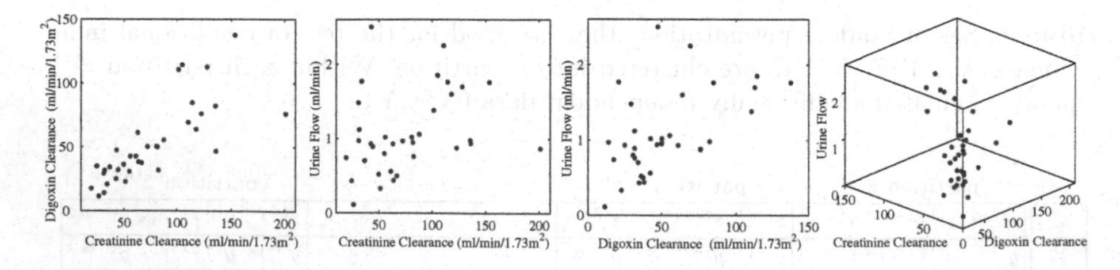


**Fig. 3.** Data on 35 consecutive patients under treatment for heart failure with the drug digoxin. Clearances are given in ml/min/1.73m$^2$, urine flow in ml/min.

**Table 2.** Comparison of independence tests on digoxin clearance data. Significance level $\alpha = 0.05$ is chosen for analysis via correlations and kernel dependence measures.

| | Correlation | | Margaritis' Bayesian | | Kernel Dependence | |
|---|---|---|---|---|---|---|
| Independence Hypothesis | p-value | Test | $P$(independence) | Test | p-value | Test |
| Creatinine $\perp\!\!\!\perp$ Digoxin | 0.00 | Reject | 0.0030 | Reject | 0.00 | Reject |
| Creatinine $\perp\!\!\!\perp$ Digoxin \| Urine Flow | 0.00 | Reject | 0.1401 | Reject | 0.00 | Reject |
| Creatinine $\perp\!\!\!\perp$ Urine Flow | 0.07 | **Accept** | 0.7018 | **Accept** | 0.01 | **Reject** |
| Creatinine $\perp\!\!\!\perp$ Urine Flow \| Digoxin | 0.40 | Accept | 0.8538 | Accept | 0.58 | Accept |
| Digoxin $\perp\!\!\!\perp$ Urine Flow | 0.00 | Reject | 0.0474 | Reject | 0.00 | Reject |
| Digoxin $\perp\!\!\!\perp$ Urine Flow \| Creatinine | 0.02 | **Reject** | 0.7851 | **Accept** | 0.17 | **Accept** |

(first plot). The correlation analysis (see also [18] p. 43) did not reveal the dependence between creatinine clearance and urine flow, whereas kernel test did (third row of Tab. 2). We conjecture that Margaritis' Bayesian method failed to detect dependence in this case because of the small sample size, which gives evidence for better performance of kernel tests in case of small datasets. All three tests found that, given digoxin clearance, creatinine clearance was not significantly related to urine flow rate (Tab. 2, fourth row). All methods found that in these patients digoxin clearance was significantly related to urine flow rate (Tab. 2, fifth row). This finding is consistent with the opinion of Halkin et al. [16], who suspected that the elimination of digoxin might be subject to reabsorption, which might give rise to a correlation with urine flow. However, if the linear dependence model is wrong, a biased estimate of the partial correlation and a biased test for independence via linear model may result. Both non-parametric tests, i.e., kernel tests and Margaritis' Bayesian method accepted the hypothesis that, given creatinine clearance, digoxin clearance is independent of urine flow, whereas the partial correlation did not confirm this hypothesis (last row of Tab. 2). The finding that digoxin clearance is independent of urine flow controlling for creatinine clearance is particularly of medical interest. In summary, the test results revealed that the non-parametric tests are superior to correlation analysis. This example makes also clear that, in practice, independence by kernel measures does not necessarily require the independence by correlation analysis, although it is theoretically apparent that non-vanishing of correlation implies non-vanishing of kernel dependence measure.

Even though all CI constraints are correctly detected by some test, we can have problem to learn BN structures. The problem is that the CI constraints could be inconsistent w.r.t. representation by BN structures. In this example, we have "Creatinine $\perp\!\!\!\perp$ Urine Flow|Digoxin" and "Digoxin $\perp\!\!\!\perp$ Urine Flow|Creatinine identified by both non-parametric tests. Due to the rule of learning adjacency, direct links between Urine Flow and both clearances are excluded. However, the kernel test confirmed "Urine Flow $\not\!\perp\!\!\!\perp$ (Creatinine, Digoxin)" with a p-value of 0.007, i.e., urine flow is dependent of clearances, which is indeed plausible from the medical viewpoint. Such inconsistence violates the so-called intersection property [20] of a BN structure, which states $(Y_1 \perp\!\!\!\perp X|Y_2) \wedge (Y_2 \perp\!\!\!\perp X|Y_1) \Rightarrow X \perp\!\!\!\perp (Y_1, Y_2)$. In fact, the intersection property does not hold in general. A trivial example for such violation is that $Y_1$ and $Y_2$ are related deterministically with each other (see [21,22] for more theoretical discussions about the problem in learning BN with deterministic nodes), i.e., $Y_1$ and $Y_2$ contain entire information about each other. The uncertainty of $Y_1$ (or $Y_2$) vanishes due to the knowledge of $Y_2$ (or $Y_1$), then testing conditional dependences between $X$ and $Y_1$ given $Y_2$ and between $X$ and $Y_2$ given $Y_1$ cannot provide any evaluable information about the dependence between $X$ and $(Y_1, Y_2)$. Martín [23] showed that this property only holds, when $Y_1$ and $Y_2$ are measurably separated. The so-called "measurable separability" concept is introduced by Florens et al. [24] and provides a sufficient assumption to make the intersection property valid [23]. Since such violation essentially reveals some symmetry in CI constraints between the measurably inseparable $Y_1$ and $Y_2$ w.r.t. $X$, we propose to consider $(Y_1, Y_2)$ as one factor in the resulting structure, which makes such conflicting information of CI constraints irrelevant. By means of kernel independence tests, structures containing vectorial variables can be straightforward handled. The following examples will show that such construction of nodes gains advantage in the structure learning, in particular, in relation to the potential causal interpretation of the structure.

**Montana Outlook Poll.** The data contain the outcome in the Montana Economic Outlook Poll in May 1992. This benchmark dataset is listed at `http://lib.stat.cmu.edu/DASL/Stories/montana.html`. After removing records with missing values, the dataset contains 163 entries. The Montana poll asked a random sample of Montana residents whether the respondent feels his/her personal financial status is worse, the same, or better than a year ago, and whether they view the state economic outlook as better over the next year. Respondents are classified by sex, age, income, political orientation, and area of residence in the state. In the triple {Financial,Political,Outlook}, we observe (Outl. $\perp\!\!\!\perp$ Fin. | Pol.) and (Outl. $\perp\!\!\!\perp$ Pol. | Fin.). According to the rule of learning adjacency, the first two constraints excluded a direct link between Outl. and Fin., and a direct link between Outl. and Pol. (Fig. 4, left), which seems to be implausible, due to the constraint (Outl. $\not\!\perp\!\!\!\perp$ (Fin., Pol.)) identified by kernel test. If we merge Fin. and Pol. together to a new node, we will obtain indeed a structure (Fig. 4, right) which can be plausibly interpreted causally.

**US Crime Data.** The US crime data [25] are crime-related and demographic statistics for 47 US states in 1960. This dataset is listed as benchmark for causal structure learning on the homepage of TETRAD project. The dataset consists of

**Fig. 4.** Learning BN structure on Montana data. The left plot represents the structure learned with inconsistent CI constraints. The right plot illustrates a structure without representing the violation of the intersection property of a BN structure.

14 variable: CRIME rate; the number of YOUNG MALE; STATES: binary indicator variable for Southern states; EDUCATION; EX59, EX60: 1959, 1960 per capita expenditure on police by state and local government; YOUNG LABOR force participation rate; the number of MALE; POPULATION; the number of NON-WHITES; U1: unemployment rate of urban males of age 14 – 24; U2: unemployment rate of urban males of age 35 – 39; ASSETS: value of transferable goods and assets or family income; POVERTY: the number of families earning below 1/2 the median income. It is remarkable that the output of PC is not a BN contains 4 bi-directed edges (Fig. 5, left), which are traced back to the inconsistence of CI constraints. If we wish to find out the causal relationships between crime rate and other factors, the result of PC is unsatisfactory, although they provide some plausible connections between the expenditure on police and crime rate, some relationships among demographic statistics. Regarding the meaning of variables, it is obvious that some variables must be strongly related. In order to better understand the phenomenon of crime rate, we propose to reconstruct a demographic and geographic factor, called DEMO-GEOGRAPHIC (comprising POPULATION, NON-WHITE, MALE, YOUNG MALE, and STATES), a factor called EXPENDITURE (containing EX59 and EX60), a factor called EMPLOYMENT (containing YOUNG LABOR, U1, and U2), and a factor WEALTH (containing ASSETS and POVERTY). The variable CRIME remains unchanged. We conducted the so-called KCL algorithm [10] to learn the causal ordering of variables on the complete connected graph. After that, we used the kernel independence tests to remove the unnecessary edges. The output is shown in the right plot of Fig. 5. The variable CRIME is reasonably detected as the effect of other factors. Interestingly, CRIME is conditional independent of EMPLOYMENT given EXPENDITURE. This example demonstrates the main advantage of kernel tests, i.e., the possibility to analyze the relationship between vectorial factors on continuous, categorical, vectorial or mixed domains. Nonetheless, it is actually difficult to judge the performance of a structure learning algorithm by experiments with real-world data, where the ground truth is not completely known, although the outputs of previous examples seem to be plausible in relation to intuitive causal interpretation. For this reason, we conduct experiments with the kernel tests in comparison to other

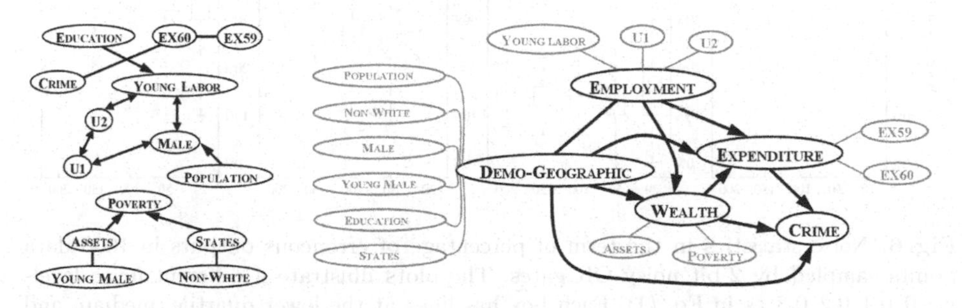


**Fig. 5.** Outputs learned by PC algorithm (left), KCL and kernel independence tests (right) from US crime data

methods on purely categorical or continuous domains to get more evidence of performance improvements.

### 3.2 Comparison of Performance on Categorical Domains

First, we use toy data on categorical domains. The data are sampled from logically linked models, namely noisy OR gates. Such Boolean functions are simplified models for relations in many real-world situations. In general, an n-bit $X_1, \ldots, X_n \in \{0,1\}$ noisy OR gate can be characterized by the conditional probabilities $P(X_{n+1}=1 \mid x_1, \ldots, x_n) = (1-r_2)(1-r_1^{x_1+\ldots+x_n}) + r_2$ with parameters $r_1, r_2 \in [0,1]$. $r_1$ can be interpreted as the probability of suppressing the input 1; $r_2$ can be interpreted as the probability for a spontaneous inversion of the output. If $r_1$ and $r_2$ vanish, the OR gate is deterministic. For the sake of notational simplicity, we chose $r_1 = r_2 =: r$ in this paper, i.e.,

$$P\left(X_{n+1}=1 \mid x_1, \ldots, x_n\right) = (1-r)\left(1-r^{x_1+\ldots+x_n}\right) + r\,. \tag{1}$$

Shorthand $\mathrm{OR}_r\{X_1, \ldots, X_n\}$ is used to depict a noisy OR gate with noise level $r \in [0,1]$. Data are sampled from a 2-bit noisy OR (Fig. 1, right) with $P(X=1)=0.6$ and $P(Y=1)=0.5$ as inputs and $Z=\mathrm{OR}_r\{X,Y\}$ as output. The underlying model implies $X \perp\!\!\!\perp Y$ and $X \not\perp\!\!\!\perp Y|Z$. We sampled 1000 datasets with different noise levels $r = 0.0, 0.1, 0.2, 0.3$ and different sample sizes $20, 50, 100, 150, 200$. Fig. 6 shows the real noise statistics in term of percentage of erroneous outputs in 1000 data points. We perform the permutation test via kernel dependence measure (KD) and two popular independence tests on categorical domains, i.e., likelihood ratio $\chi^2$ test, permutation test via MI. Since all these three CI tests are non-parametric, their performance are expected to be similar, in the sense

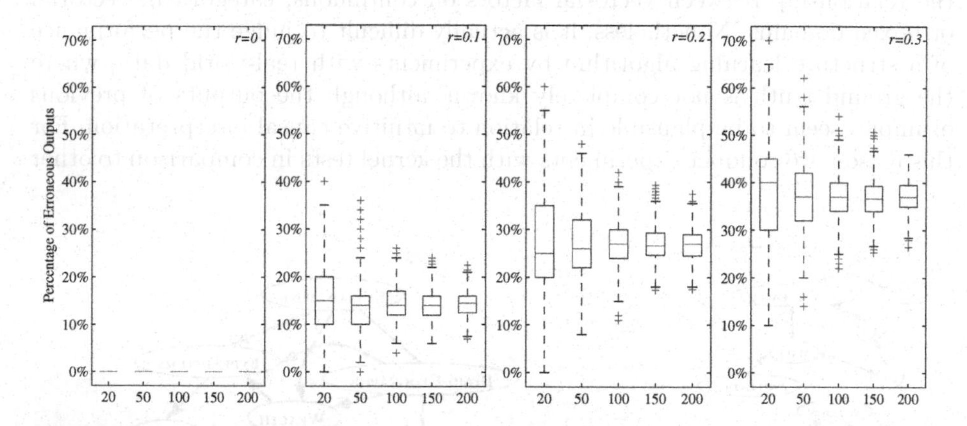


**Fig. 6.** Noise statistics in the term of percentage of erroneous outputs in 1000 data points sampled by 2-bit noisy OR gates. The plots illustrate 4 different noise levels $r = 0, 0.1, 0.2, 0.3$ as in Eq. (1). Each box has lines at the lower quartile, median, and upper quartile values of the percentage of erroneous outputs. The whiskers are lines extending from each end of the box to show the extent of the rest of the percentage. Outliers are the percentage beyond the ends of the whiskers.

**Table 3.** Numerical comparison of three different independence tests on categorical domain, i.e., likelihood ratio $\chi^2$ test, permutation test via mutual information (MI), and permutation test via kernel dependence (KD) measure. The generating models are noisy OR gates with 4 different noise levels $r = 0.0, 0.1, 0.2, 0.3$ as shown in Eq. (1). The experiments are conducted with 1000 replications. The entries show how often (in percentage) the constraint $X \perp\!\!\!\perp Y$ or $X \not\perp\!\!\!\perp Y|Z$ is identified.

| | Accepting $X \perp\!\!\!\perp Y$ | | | | | | | | | | | | | | |
|---|---|---|---|---|---|---|---|---|---|---|---|---|---|---|---|
| Sample Size | 20 | | | 50 | | | 100 | | | 150 | | | 200 | | |
| Noise Level | $\chi^2$ | MI | KD | $\chi^2$ | MI | KD | $\chi^2$ | MI | KD | $\chi^2$ | MI | KD | $\chi^2$ | MI | KD |
| $r = 0.0$ | 94.0 | 97.4 | 88.0 | 94.7 | 96.3 | 90.8 | 95.6 | 95.9 | 92.5 | 94.3 | 93.7 | 92.4 | 94.1 | 94.3 | 91.8 |
| $r = 0.1$ | 93.1 | 96.4 | 86.5 | 94.6 | 96.0 | 90.7 | 94.2 | 94.0 | 91.1 | 95.8 | 96.3 | 94.4 | 94.2 | 94.8 | 92.3 |
| $r = 0.2$ | 93.6 | 96.9 | 86.9 | 94.9 | 96.1 | 91.3 | 96.3 | 96.1 | 93.1 | 95.7 | 95.7 | 93.5 | 93.6 | 94.0 | 91.4 |
| $r = 0.3$ | 94.5 | 97.1 | 87.3 | 95.9 | 97.0 | 93.0 | 93.5 | 93.6 | 90.7 | 93.6 | 94.1 | 91.6 | 94.4 | 94.8 | 93.2 |
| Noise Level | Rejecting $X \perp\!\!\!\perp Y\|Z$ | | | | | | | | | | | | | | |
| $r = 0.0$ | 24.8 | 54.8 | 23.5 | 94.5 | 97.7 | 91.9 | 100 | 100 | 98.0 | 100 | 100 | 100 | 100 | 100 | 100 |
| $r = 0.1$ | 23.5 | 33.7 | 16.9 | 57.6 | 57.0 | 54.8 | 85.9 | 84.7 | 89.2 | 97.6 | 97.3 | 98.0 | 99.3 | 99.2 | 99.7 |
| $r = 0.2$ | 14.9 | 18.9 | 8.9 | 25.1 | 22.7 | 22.5 | 39.8 | 40.7 | 40.5 | 56.2 | 57.2 | 60.0 | 71.6 | 72.5 | 74.8 |
| $r = 0.3$ | 9.9 | 10.9 | 7.0 | 10.4 | 10.3 | 9.6 | 16.3 | 16.5 | 16.3 | 19.1 | 21.5 | 19.2 | 23.1 | 23.9 | 23.9 |

that the levels of type I and II errors are almost the same. The larger the sample size and the less noisy the model, the better the performance (see Tab. 3 for results). The kernel test is slightly worse than the other two in the case of 20 data points. The actual benefit of kernel tests does not lie in the tests on categorical domains, but in tests on continuous or hybrid domains. Nonetheless, the kernel independence test provides a good alternative to the popular tests on categorical domains.

### 3.3 Comparison of Performance on Continuous Domains

Toy data are generated by functional models on continuous domains. Fisher's Z test, Margaritis' Bayesian method and permutation test via kernel dependence measures are evaluated. Although the estimation of MI on continuous domains is an unsolved problem in its generality, there are some attempt to do that. We performed permutation test via MI using the estimation method proposed by Kraskov et al. [8]. First, we sampled the so-called Meander data. The generating model of Meander data is proposed by Margaritis [26]. It resembles a spiral (Fig. 7). This dataset is challenging because the joint distribution of $X$ and $Y$ given $Z$ changes dramatically with the given value of $Z$. According to the functional relation, $X$ and $Y$ are conditionally independent given $Z$, however, unconditionally dependent, in fact strongly correlated (rightmost plot of Fig. 7). We generated 1000 datasets and ran tests. Tab. 4 shows the results for samples sizes ranging from 20 to 200. The dependence between $X$ and $Y$ can already be captured by the linear relation. All methods achieved already very good performance at testing $X \not\perp\!\!\!\perp Y$ from merely 20 data points. Testing

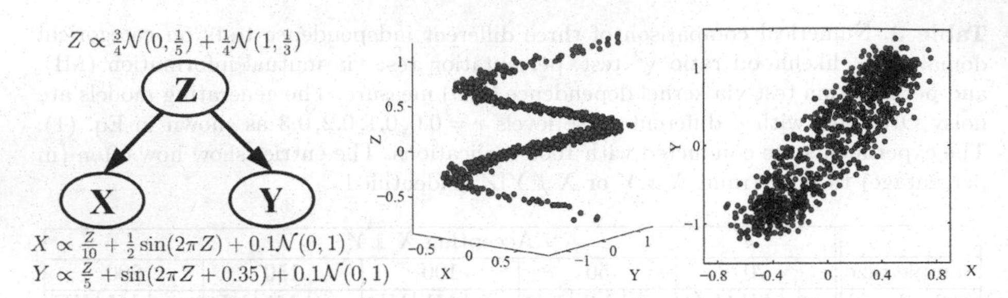


**Fig. 7.** Generating model of meander dataset (leftmost), which implies $X \not\perp\!\!\!\perp Y$ and $X \perp\!\!\!\perp Y|Z$. 3-dimensional plot of the Meander dataset (middle) and projection of data along $Z$ axis (rightmost)

**Table 4.** Comparison of independence tests on continuous domains, i.e., Fisher's Z test, Margaritis' Bayesian method, and permutation test via mutual information and kernel dependence measures. The underlying model Meander is given by Fig. 7. The experiments are conducted with 1000 replications. The entries show how often (in percentage) the constraint $X \not\perp\!\!\!\perp Y$ and $X \perp\!\!\!\perp Y|Z$ are identified.

| | Rejecting $X \perp\!\!\!\perp Y$ | | | | | Accepting $X \perp\!\!\!\perp Y\|Z$ | | | | |
|---|---|---|---|---|---|---|---|---|---|---|
| Sample Size | 20 | 50 | 100 | 150 | 200 | 20 | 50 | 100 | 150 | 200 |
| Fisher's Z | 100 | 100 | 100 | 100 | 100 | 0 | 0 | 0 | 0 | 0 |
| Margaritis' Bayesian | 94.3 | 100 | 100 | 100 | 100 | 4.8 | 15.1 | 21.2 | 23.2 | 33.2 |
| Mutual Information | 99.6 | 100 | 100 | 100 | 100 | 8.6 | 19.4 | 33.8 | 38.9 | 43.7 |
| Kernel Dependence | 99.9 | 100 | 100 | 100 | 100 | 35.1 | 49.7 | 67.0 | 75.3 | 79.9 |

| $M_k := (f_i, f_j)$ | $f_2 := 2\sin(x)$ | $f_3 := \ln(\|x\|)$ | $f_4 := \frac{1}{\frac{x}{5}+1}$ | $f_5 := \exp(x)$ |
|---|---|---|---|---|
| $f_1 := x$ | $M_1$ | $M_2$ | $M_3$ | $M_4$ |
| $f_2 := 2\sin(x)$ | – | $M_5$ | $M_6$ | $M_7$ |
| $f_3 := \ln(\|x\|)$ | – | – | $M_8$ | $M_9$ |
| $f_4 := \frac{1}{\frac{x}{5}+1}$ | – | – | – | $M_{10}$ |

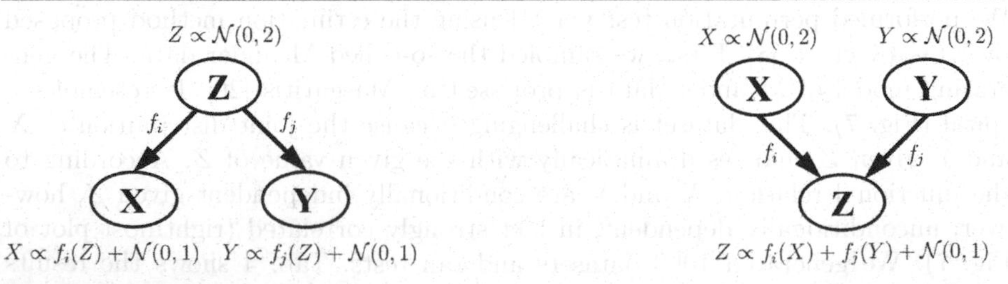


**Fig. 8.** Functional models with a $\wedge$-structure (left plot) and a $\vee$-structure (right plot). The pairs of functions $M_k = (f_i, f_j)$ with $i, j = 1, \ldots, 5$ and $k = 1, \ldots, 10$ for both structures are defined in the table.

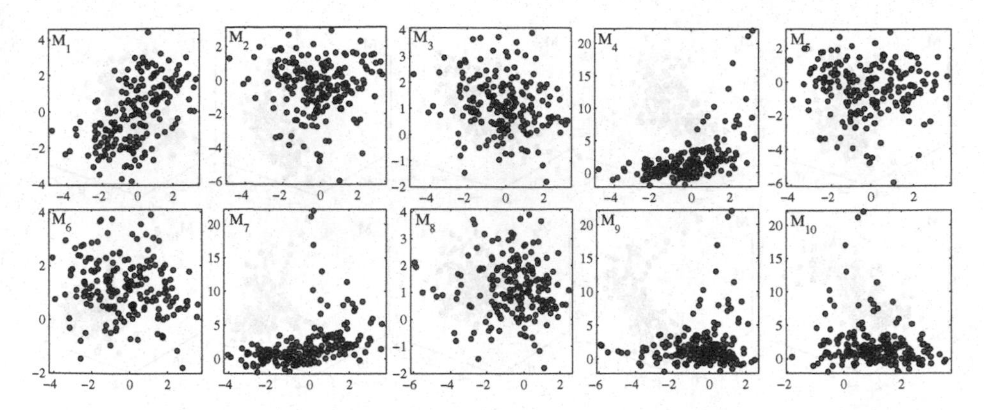


**Fig. 9.** 2-dimensional plots of data sampled from a $\wedge$-structure (Fig. 8, first plot), where $X$ and $Y$ have a functional relation $M_k = (f_i, f_j)$ (see table in Fig. 8) with $Z$, respectively. The illustrated sample contains 200 data points.

**Table 5.** Comparison of various independence tests on continuous domains sampled by a $\wedge$-structure (Fig. 8, first plot). The parameter $M_k = (f_i, f_j)$ of models is defined in the table in Fig. 8. The entries show how often (in percentage) $X \not\perp\!\!\!\perp Y$ and $X \perp\!\!\!\perp Y|Z$ are identified after 1000 replications of simulations.

| | Rejecting $X \perp\!\!\!\perp Y$ | | | | | | | | | |
|---|---|---|---|---|---|---|---|---|---|---|
| $M_k = (f_i, f_j)$ | $M_1$ | $M_2$ | $M_3$ | $M_4$ | $M_5$ | $M_6$ | $M_7$ | $M_8$ | $M_9$ | $M_{10}$ |
| Fisher's Z | 100 | 4.2 | 93.6 | 100 | 1.8 | 57.6 | 71.5 | 17.5 | 41.0 | 58.6 |
| Margaritis' Bayesian | 100 | 2.0 | 42.5 | 100 | 2.0 | 27.6 | 100 | 1.7 | 4.0 | 18.6 |
| Mutual Information | 91.0 | 43.2 | 31.9 | 100 | 33.3 | 60.2 | 100 | 5.2 | 65.5 | 9.2 |
| Kernel Dependence | 100 | 95.6 | 63.9 | 100 | 63.9 | 56.1 | 100 | 11.5 | 97.8 | 68.7 |
| | Accepting $X \perp\!\!\!\perp Y|Z$ | | | | | | | | | |
| Fisher's Z | 94.0 | 95.6 | 94.1 | 95.6 | 95.5 | 72.2 | 10.6 | 81.5 | 1.2 | 64.3 |
| Margaritis' Bayesian | 97.0 | 97.6 | 97.9 | 98.7 | 97.0 | 97.9 | 98.9 | 98.3 | 98.7 | 98.8 |
| Mutual Information | 80.1 | 92.5 | 93.6 | 88.4 | 85.7 | 92.9 | 88.2 | 97.5 | 93.6 | 89.4 |
| Kernel Dependence | 93.8 | 93.8 | 92.5 | 93.4 | 93.3 | 93.5 | 93.4 | 94.5 | 94.2 | 92.9 |

conditional independence $X \perp\!\!\!\perp Y|Z$ is more challenging. Here, the kernel test clearly outperforms other three methods. Fisher's Z test fails completely due to the incorrect multivariate Gaussian assumption. The kernel independence test made significantly less errors than Margaritis' Bayesian method and test via MI. In order to gain more evidence of performance in learning BN structure, we sampled datasets of 200 data points by different functional models with the $\wedge$- and $\vee$-structures (Fig. 8). We define the functional relations $f_{1,\ldots,5}$ in the same way as proposed in [26] and use all pairs of them $M_k = (f_i, f_j)$, i.e., 10 different combinations $M_1, \ldots, M_{10}$ as shown in the table in Fig. 8, added by a Gaussian noise as underlying ground-truth for the sampling. One sample of 200 data points for the $\wedge$-structure (left plot in Fig. 8) with $M_1, \ldots, M_{10}$ is visualized in Fig. 9 and Fig. 10. The performance of various independence tests after 1000

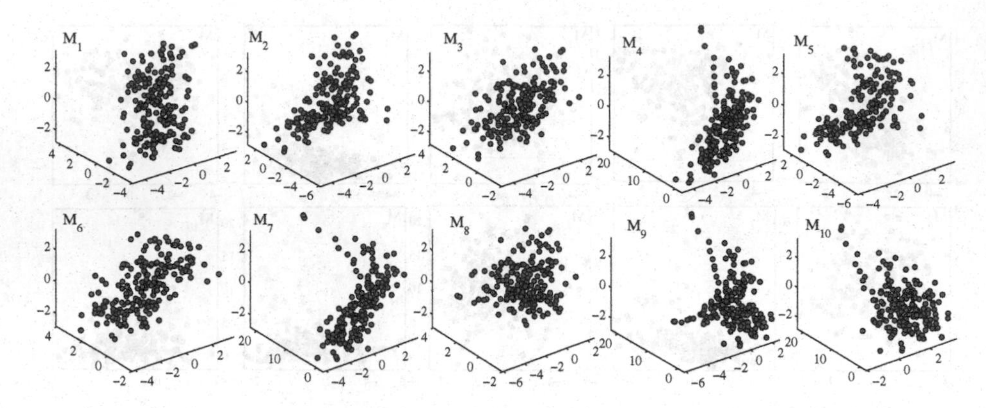


**Fig. 10.** 2-dimensional plots of data sampled from a $\wedge$-structure (Fig. 8, first plot), where $X$ and $Y$ have a functional relation $M_k = (f_i, f_j)$ (see table in Fig. 8) with $Z$, respectively. The illustrated sample contains 200 data points.

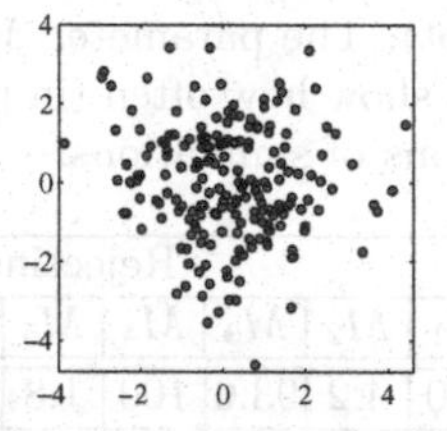

**Fig. 11.** 2-dimensional plots of data sampled from a $\vee$-structure (Fig. 8, second plot), where $Z$ has a functional relation $M_k = (f_i, f_j)$ (see table in Fig. 8) with $X$ and $Y$. The illustrated sample contains 200 data points.

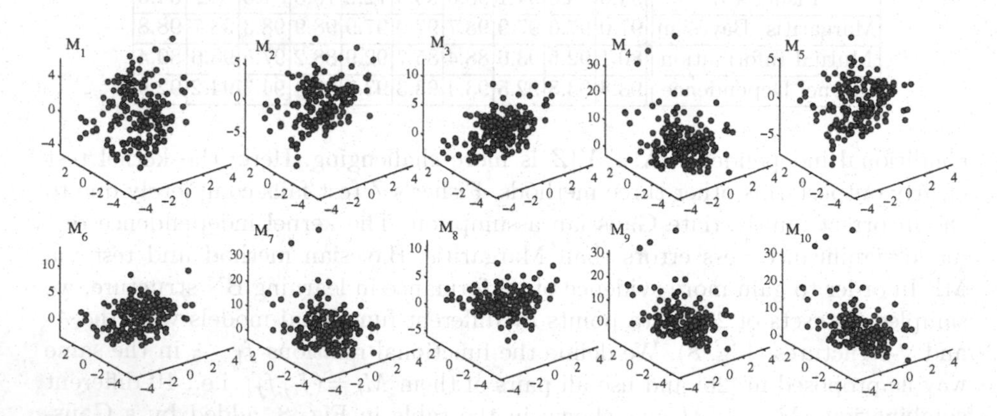


**Fig. 12.** 3-dimensional plots of data from a $\vee$-structure (Fig. 8, second plot), where $Z$ has a functional relation $M_k = (f_i, f_j)$ (see table in Fig. 8) with $X$ and $Y$. The illustrated sample contains 200 data points.

**Table 6.** Comparison of various independence tests on continuous domains sampled by a $\wedge$-structure (Fig. 8, first plot). The parameter $M_k = (f_i, f_j)$ of models is defined in the table in Fig. 8. The entries show how often (in percentage) $X \perp\!\!\!\perp Y$ and $X \not\perp\!\!\!\perp Y|Z$ are identified after 1000 replications of simulations.

| | Accepting $X \perp\!\!\!\perp Y$ | Rejecting $X \perp\!\!\!\perp Y\|Z$ | | | | | | | | | |
|---|---|---|---|---|---|---|---|---|---|---|---|
| $M_k = (f_i, f_j)$ | $M_{1,\ldots,10}$ | $M_1$ | $M_2$ | $M_3$ | $M_4$ | $M_5$ | $M_6$ | $M_7$ | $M_8$ | $M_9$ | $M_{10}$ |
| Fisher's Z | 94.6 | 100 | 4.1 | 92.1 | 77.1 | 4.7 | 58.8 | 61.2 | 5.1 | 3.9 | 20.8 |
| Margaritis' Bayesian | 98.1 | 91.4 | 3.9 | 10.9 | 84.8 | 3.1 | 9.1 | 75.0 | 2.1 | 3.7 | 6.7 |
| Mutual Information | 92.3 | 92.3 | 66.8 | 17.5 | 76.9 | 65.6 | 16.9 | 76.3 | 1.2 | 21.9 | 8.0 |
| Kernel dependence | 94.1 | 100 | 92.0 | 60.7 | 100 | 96.0 | 51.7 | 100 | 18.7 | 93.6 | 46.9 |

replications on these datasets is summarized in Tab. 5. One sample of 200 data points for the $\vee$-structure (right plot in Fig. 8) with $M_1, \ldots, M_{10}$ is visualized in Fig. 11 and Fig. 12. The performance of various independence tests of 1000 replications on these datasets is summarized in Tab. 6.

One can see that all methods make relatively few errors at discovering independence, i.e., $X \perp\!\!\!\perp Y$ in the $\vee$-structure (left half of Tab. 6) and $X \perp\!\!\!\perp Y|Z$ in the $\wedge$-structure (under half of Tab. 5). Fisher's Z test performed very bad in the case of testing conditional independence on data sampled by models $M_7$ and $M_9$ (first row of the under half of Tab. 5). In summary, the kernel tests outperformed the other non-parametric approaches, i.e., Margaritis' Bayesian method and test via mutual information, in cases of testing dependence, i.e., $X \not\perp\!\!\!\perp Y$ in the $\wedge$-structure (upper half of Tab. 5) and $X \not\perp\!\!\!\perp Y|Z$ in the $\vee$-structure (see the right half of Tab. 6). In addition, the results indicate that the fluctuation of the kernel tests within different models is significantly smaller than that of the correlation analysis based on the linear model. For this reason, we can reasonably expect more accuracy and reliability in testing CI constraints, and consequently better performance in identification of $\wedge$- and $\vee$-structures, thus structure learning in general by using kernel tests in existing independence-based algorithms.

## 4 Conclusion

A new method for structure learning of Bayesian networks with arbitrary distributions on arbitrary domains is demonstrated. This was made possible by the use of the probabilistic non-parametric conditional independence tests presented in the paper. Our evaluation on both real and artificial data shows that our method performs well against existing alternatives. Using kernel independence tests for learning Bayesian network structure is expected to make less errors than other state-of-the-art methods used in the independence-based algorithms.

## References


1. Heckerman, D., Meek, C., Cooper, G.: A Bayesian approach to causal discovery. In: Glymour, C., Cooper, G. (eds.) Computation, Causation, and Discovery, pp. 141–165. MIT Press, Cambridge (1999)
2. Cooper, G.: The computational complexity of probabilistic inference using Bayesian belief networks. Journal of Artificial Intelligence 42(3–4), 393–405 (1990)
3. Chickering, D., Heckerman, D., Meek, C.: Large-sample learning of Bayesian networks is NP-hard. Journal of Machine Learning Research 5, 1287–1330 (2004)
4. Spirtes, P., Glymour, C., Scheines, R.: Causation, prediction, and search. Lecture notes in statistics. Springer, New York (1993)
5. Pearl, J.: Causality: Models, reasoning, and inference. Cambridge University Press, Cambridge (2000)
6. Margaritis, D.: A Bayesian multiresolution independence test for continuous variables. In: Proceedings of the 17th conference on uncertainty in artificial intelligence, Pittsburgh, PA, pp. 346–353 (2001)
7. Margaritis, D.: Distribution-free learning of Bayesian network structure in continuous domains. In: Proceedings of the 20th National Conference on Artificial Intelligence, Seattle, WA, pp. 825–830 (2005)
8. Kraskov, A., Stögbauer, H., Grassberger, P.: Estimating mutual information. Physical Review E 69(6), 66138 (2000)
9. Schölkopf, B., Smola, A.: Learning with kernels. MIT Press, Cambridge (2002)
10. Sun, X., Janzing, D., Schölkopf, B., Fukumizu, K.: A kernel-based causal learning algorithm. In: Ghahramani, Z. (ed.) Proceedings of the 24th International Conference on Machine Learning, Corvallis, OR, pp. 855–862 (2007)
11. Fukumizu, K., Gretton, A., Sun, X., Schölkopf, B.: Kernel measures of conditional dependence. In: Platt, J., Koller, D., Singer, Y., Roweis, S. (eds.) Proceedings of the 21th Neural Information Processing Systems Conference, pp. 489–496. MIT Press, Cambridge (2007)
12. Gretton, A., Bousquet, O., Smola, A., Schölkopf, B.: Measuring statistical dependence with Hilbert-Schmidt norms. In: Jain, S., Simon, H.U., Tomita, E. (eds.) ALT 2005. LNCS (LNAI), vol. 3734, pp. 63–77. Springer, Heidelberg (2005)
13. Good, P.: Permutation tests: A practical guide to resampling methods for testing hypothesis. Birkhäuer, Boston (1994)
14. Fine, S., Scheinberg, K.: Efficient SVM training using low-rank kernel representations. Journal of Machine Learning Research 2, 243–264 (2001)
15. Gretton, A., Fukumizu, K., Teo, C., Song, L., et al.: A kernel statistical test of independence. In: Platt, J., Koller, D., Singer, Y., Roweis, S. (eds.) Proceedings of the 21th Neural Information Processing Systems Conference, pp. 585–592. MIT Press, Cambridge (2007)
16. Halkin, H., Sheiner, L., Peck, C., Melmon, K.: Determinants of the renal clearance of digoxin. Clinical Pharmacology and Therapeutics 17(4), 385–394 (1975)
17. Altman, D.G.: Practical statistics for medical research. Chapman and Hall, London (1991)
18. Edwards, D.: Introduction to graphical modelling. Springer, New York (2000)
19. Jelliffe, R., Blankenhorn, D.: Improved method of digitalis therapy in patients with reduced renal function. Circulation 35, 11–150 (1967)
20. Pearl, J.: Probabilistic reasoning in intelligent systems: Networks of plausible inference. Morgan Kaufmann, San Mateo (1988)

21. Shachter, R.: Probabilistic inference and influence diagrams. Operations Research 36(4), 589–604 (1988)
22. Geiger, D.: Graphoids: A qualitative framework for probabilistic inference. PhD thesis, Cognitive Systems Laboratory, Department of Computer Science, University of California, Los Angeles, CA (1990)
23. Martín, E.: Ignorable common information, null sets and Basu's first theorem. The Indian Journal of Statistics 67(4), 674–698 (2005)
24. Florens, J., Mouchart, M., Rolin, J.: Elements of Bayesian Statistics. Marcel Dekker, New York (1990)
25. Vandaele, W.: Participation in illegitimate activites: Erlich revisited. In: Blumstein, A., Cohen, J., Nagin, D. (eds.) Deterrence and incapacitation, pp. 270–335. National Academy of Sciences, Washington (1978)
26. Margaritis, D.: Distribution-free learning of graphical model structure in continuous domains. Technical Report TR-ISU-CS-04-06, Computer Science, Iowa State University (2004)

# Assessing Nonlinear Granger Causality from Multivariate Time Series

Xiaohai Sun

Max Planck Institute for Biological Cybernetics
Spemannstr. 38, 72076 Tübingen, Germany
xiaohai.sun@tuebingen.mpg.de

**Abstract.** A straightforward nonlinear extension of Granger's concept of causality in the kernel framework is suggested. The kernel-based approach to assessing nonlinear Granger causality in multivariate time series enables us to determine, in a model-free way, whether the causal relation between two time series is present or not and whether it is direct or mediated by other processes. The trace norm of the so-called covariance operator in feature space is used to measure the prediction error. Relying on this measure, we test the improvement of predictability between time series by subsampling-based multiple testing. The distributional properties of the resulting p-values reveal the direction of Granger causality. Experiments with simulated and real-world data show that our method provides encouraging results.

**Keywords:** time series, Granger causality, kernel methods.

## 1 Introduction

In this paper, a time series $\mathrm{X} := (\ldots, x_{t-1}, x_t, x_{t+1}, \ldots)$ is a discrete time, continuous state process where $t \in \mathbb{Z}$ is a certain discrete time point. Time points are usually taken at equally spaced intervals. We consider, without loss of generality, a triviariate time series $(\mathrm{X}, \mathrm{Y}, \mathrm{Z})$ measured simultaneously, where $\mathrm{X}, \mathrm{Y}, \mathrm{Z}$ can, in general, again be multivariate time series. The problem that we focus on is that whether the underlying process of X is causal to the underlying process of Y and/or the other way around, when the underlying process Z is known. The well-known concept of causality in analysis of times series is the so-called Granger causality: The process X Granger causes another process Y given a third process Z (subsequently denoted as $\mathrm{X} \Rightarrow \mathrm{Y} \mid \mathrm{Z}$), if future values of Y can be better predicted using the past values of $(\mathrm{X}, \mathrm{Y}, \mathrm{Z})$ compared to using the past values of $(\mathrm{Y}, \mathrm{Z})$ alone.

For the sake of simplicity, we first start with the bivariate case $(\mathrm{X}, \mathrm{Y})$. The time-delayed embedding vector reconstructing the state (or phase) space of times series X is expressed as

$$X_{t,r,m} = (x_{t-(m-1)r}, \ldots, x_{t-2r}, x_{t-r}, x_t)^T \,, \tag{1}$$

W. Daelemans et al. (Eds.): ECML PKDD 2008, Part II, LNAI 5212, pp. 440–455, 2008.

where $m$ is the embedding dimension and $r$ is the time delay (or lag) between successive elements of the state vector [1,2]. The choice of $m, r$ depends on the dynamics of underlying process. If not otherwise stated in this paper, $r = 1$, $m = 100$ and the expression $X_t = (x_{t-99}, \ldots, x_{t-2}, x_{t-1}, x_t)^T$ is used. $Y_t$ can be defined analogously. For a more principled choice of the embedding dimension and the time lag we refer to [3,4]. Like many of the existing causality measures, the test described in this paper is also based on the notion of the embedding vector. To assess the causal hypothesis $\mathrm{X} \Rightarrow \mathrm{Y}$, Granger causality considers the following autoregressive predictions:

$$Y_{t+1} = a^T X_t + \epsilon_t^{(\mathrm{Y})} \quad \text{and} \quad Y_{t+1} = b_1^T Y_t + b_2^T X_t + \epsilon_t^{(\mathrm{Y}|\mathrm{X})} \tag{2}$$

where $\epsilon_t^{(\mathrm{Y})}$ and $\epsilon_t^{(\mathrm{Y}|\mathrm{X})}$ represent the prediction errors, and $a, b_1, b_2$ denote regression coefficient vectors. The coefficient vectors and constants are determined so as to minimize the variance of $\epsilon_t^{(\mathrm{Y})}$ and $\epsilon_t^{(\mathrm{Y}|\mathrm{X})}$. Once the coefficient vectors have been calculated, the causal inference of X on Y can be verified if the variance $\mathrm{Var}[\epsilon_t^{(\mathrm{Y}|\mathrm{X})}]$ is significantly smaller than $\mathrm{Var}[\epsilon_t^{(\mathrm{Y})}]$. The opposite direction $\mathrm{X} \Leftarrow \mathrm{Y}$ can be analogously tested. Such standard test of Granger causality developed by Granger [5] assumes linear regression models, and its application to nonlinear systems may not be appropriate in the general case. A nonlinear extension of Granger causality, called the extended Granger causality index (EGCI) was proposed in [6]. The main idea of this technique is to divide the phase space into a set of small neighborhoods and approximate the globally nonlinear dynamics by local linear regression models. However, the local linearity is obviously a restrictive assumption. Another recently introduced nonlinear extensions of Granger causality (NLGC) [7] is based on kernel autoregression scheme, instead of a linear autoregression. Although this method takes no restrictive assumption of linearity, the method is fixed to a certain class of functions, i.e., nonlinear radial based functions (RBFs) and assume the additivity of the interaction between these functions. In this paper, we generalize the Granger causality in the kernel framework and present a straightforward nonlinear extension.

## 2 Kernel Framework

A positive definite kernel $k_{\mathcal{X}} : \mathcal{X} \times \mathcal{X} \to \mathbb{R}$ on a non-empty set $\mathcal{X}$ is a symmetric function, i.e., $k_{\mathcal{X}}(x, x') = k_{\mathcal{X}}(x', x)$ for any $x, x' \in \mathcal{X}$ such that for arbitrary $n \in \mathbb{N}$ and $x^{(1)}, \ldots, x^{(n)} \in \mathcal{X}$ the matrix $K$ with $(K)_{ij} := k_{\mathcal{X}}(x^{(i)}, x^{(j)})$ is positive definite, i.e., $\sum_{i,j=1}^{n} c_i c_j k_{\mathcal{X}}(x^{(i)}, x^{(j)}) \geq 0$ for all $c_1, \ldots, c_n \in \mathbb{R}$. A reproducing kernel Hilbert space (RKHS) [8] $\mathcal{H}_{\mathcal{X}}$ is a Hilbert space defined by the completion of an inner product space of functions $k_{\mathcal{X}}(x, \cdot)$ with $x \in \mathcal{X}$ and the inner product defined by $\langle k_{\mathcal{X}}(x, \cdot), k_{\mathcal{X}}(x', \cdot) \rangle = k_{\mathcal{X}}(x, x')$ for all $x, x' \in \mathcal{X}$. In other words, $\Phi(x)(\cdot) = k_{\mathcal{X}}(x, \cdot)$ defines a map from $\mathcal{X}$ into a feature space $\mathcal{H}_{\mathcal{X}}$. With the so-called "kernel trick", a linear algorithm can easily be transformed into a non-linear algorithm, which is equivalent to the linear algorithm operating in the space of $\Phi$. However, because the kernel function is used for calculating the inner product, the mapping $\Phi$ is never explicitly computed. This is desirable, because the high-dimensional space may be infinite-dimensional, as is the case when the kernel is ,e.g., a Gaussian or Laplacian.

In kernel framework, the autoregression models of Eq. 2 can be replaced by

$$a^T\Phi(Y_{t+1}) = b_1^T\Psi(Y_t) + \epsilon_t^{(Y)} \quad \text{and} \quad a^T\Phi(Y_{t+1}) = b_2^T\Psi(Y_t, X_t) + \epsilon_t^{(Y|X)} \quad (3)$$

where $\Phi, \Psi$ are nonlinear maps into feature spaces $\mathcal{H}_\mathcal{Y}, \mathcal{H}_{\mathcal{Y}\otimes\mathcal{X}}$ respectively. In the conditional case (conditioning on Z), we will have

$$a^T\Phi(Y_{t+1}) = b_1^T\Psi(Y_t, Z_t) + \epsilon_t^{(Y)} \text{ and } a^T\Phi(Y_{t+1}) = b_2^T\Psi(Y_t, Z_t, X_t) + \epsilon_t^{(Y|X)} \quad (4)$$

where $\Phi, \Psi$ are nonlinear maps into $\mathcal{H}_\mathcal{Y}, \mathcal{H}_{\mathcal{Y}\otimes\mathcal{Z}\otimes\mathcal{X}}$ respectively. Then, if $\mathrm{Var}[\epsilon_t^{(Y|X)}]$ is significant smaller than $\mathrm{Var}[\epsilon_t^{(Y)}]$, $X \Rightarrow Y \mid Z$ can be verified, otherwise no evidence for Granger causality. To test this significance, we define the covariance operator [9] expressing variance of variables in the feature space.

## 2.1 Covariance Operator

Suppose we have random vector $(X, Y)$ taking values on $\mathcal{X}\times\mathcal{Y}$. The base spaces $\mathcal{X}$ and $\mathcal{Y}$ are topological spaces. Measurability of these spaces is defined with respect to the Borel $\sigma$-field. The joint distribution of $(X, Y)$ is denoted by $P_{XY}$ and the marginal distributions by $P_X$ and $P_Y$. Let $(\mathcal{X}, \mathcal{B}_\mathcal{X})$ and $(\mathcal{Y}, \mathcal{B}_\mathcal{Y})$ be measurable spaces and $(\mathcal{H}_\mathcal{X}, k_\mathcal{X}), (\mathcal{H}_\mathcal{Y}, k_\mathcal{Y})$ be RKHSs of functions on $\mathcal{X}$ and $\mathcal{Y}$ with positive definite kernels $k_\mathcal{X}, k_\mathcal{Y}$. We consider only random vectors $(X, Y)$ on $\mathcal{X}\times\mathcal{Y}$ such that the expectations $\mathrm{E}_X[k_\mathcal{X}(X, X)], \mathrm{E}_Y[k_\mathcal{Y}(Y, Y)]$ are finite, which guarantees $\mathcal{H}_\mathcal{X}$ and $\mathcal{H}_\mathcal{Y}$ are included in $L^2(P_X)$ and $L^2(P_Y)$ respectively, where $L^2(\mu)$ denotes the Hilbert space of square integrable functions with respect to a measure $\mu$. It is known that there exists a unique operator $\Sigma_{YX}$, called cross-covariance operator, from $\mathcal{H}_\mathcal{X}$ to $\mathcal{H}_\mathcal{Y}$ such that

$$\langle g, \Sigma_{YX} f\rangle_{\mathcal{H}_\mathcal{Y}} = \mathrm{E}_{XY}[f(X)g(Y)] - \mathrm{E}_X[f(X)]\mathrm{E}_Y[g(Y)] = \mathrm{Cov}[f(X), g(Y)]$$

for all $f \in \mathcal{H}_\mathcal{X}, g \in \mathcal{H}_\mathcal{Y}$. Here, $\mathrm{Cov}[\cdot]$ denotes the covariance. $\mathrm{E}_X[\cdot]$, $\mathrm{E}_Y[\cdot]$ and $\mathrm{E}_{XY}[\cdot]$ denote the expectation over $P_X$, $P_Y$ and $P_{XY}$, respectively. Baker [9] showed that $\Sigma_{YX}$ has a representation of the form $\Sigma_{YX} = \Sigma_{YY}^{1/2} V_{YX} \Sigma_{XX}^{1/2}$ with a unique bounded operator $V_{YX} : \mathcal{H}_\mathcal{X} \to \mathcal{H}_\mathcal{Y}$ such that $\|V_{YX}\| \leq 1$, where $\|\cdot\|$ is used for the operator norm of a bounded operator, i.e., $\|V\| = \sup_{\|f\|=1} \|Vf\|$. Moreover, it is obvious that $\Sigma_{XY} = \Sigma_{YX}^*$, where $\Sigma^*$ denotes the adjoint of an operator $\Sigma$. If $X$ is equal to $Y$, the positive self-adjoint operator $\Sigma_{YY}$ is called the covariance operator.

Based on the cross-covariance operator, we introduce the conditional covariance operator. Let $(\mathcal{H}_\mathcal{S}, k_\mathcal{S})$, $(\mathcal{H}_\mathcal{T}, k_\mathcal{T})$, $(\mathcal{H}_\mathcal{Y}, k_\mathcal{Y})$ be RKHSs on measurable spaces $\mathcal{S}, \mathcal{T}, \mathcal{Y}$ respectively. Let $(X, Y) = (S, T, Y)$ be a random vector on $\mathcal{S}\times\mathcal{T}\times\mathcal{Y}$, where $X = (S, T)$ and $\mathcal{H}_\mathcal{X} = \mathcal{H}_\mathcal{S} \otimes \mathcal{H}_\mathcal{T}$. The positive self-adjoint operator $\Sigma_{YY|X} := \Sigma_{YY} - \Sigma_{YY}^{1/2} V_{YX} V_{XY} \Sigma_{YY}^{1/2}$ is called the conditional covariance operator, where $V_{YX}$ and $V_{XY}$ are the bounded operators derived from $\Sigma_{YX}$ and $\Sigma_{XY}$. If $\Sigma_{XX}^{-1}$ exists, we can rewrite $\Sigma_{YY|X}$ as $\Sigma_{YY|X} = \Sigma_{YY} - \Sigma_{YX}\Sigma_{XX}^{-1}\Sigma_{XY}$. Fukumizu et al. [10, Proposition 2] showed that $\langle g, \Sigma_{YY|X} g\rangle_{\mathcal{H}_\mathcal{Y}} = \inf_{f\in\mathcal{H}_\mathcal{X}} \mathrm{E}_{XY}\big[\big|(g(Y) - \mathrm{E}_Y[g(Y)]) - (f(X) - \mathrm{E}_X[f(X)])\big|^2\big]$ for any $g \in \mathcal{H}_\mathcal{Y}$. This is an analogous to the well-known results on covariance matrices and linear regression: The conditional covariance matrix $C_{YY|X} = C_{YY} - C_{YX}C_{XX}^{-1}C_{XY}$ expresses the residual error of the least square regression problem as $b^\mathrm{T} C_{YY|X} b = \min_a \mathrm{E}_{XY} \|b^\mathrm{T} Y - a^\mathrm{T} X\|^2$. To relate this residual error to the conditional variance of $g(Y)$ given $X$, the following assumption for RKHSs is made.

**Assumption 1.** *Let $(\mathcal{H}_\mathcal{X}, k_\mathcal{X})$ be an RKHS on measurable space $\mathcal{X}$, then $\mathcal{H}_\mathcal{X} + \mathbb{R}$ is dense in $L^2(P_X)$ for every probability measure $P_X$ on $\mathcal{X}$, where "+" means the direct sum of Hilbert spaces.*

This assumption is the necessary and sufficient condition that the kernels are "characteristic". The notation of the characteristic kernels is a generalization of the characteristic function $\mathrm{E}_X[\exp(\sqrt{-1}u^\mathrm{T} X)]$, which is the expectation of the (complex-valued) positive definite kernel $k(x,u) = \exp(\sqrt{-1}u^\mathrm{T} x)$ (see [11] for details). One popular class of characteristic kernels is the universal kernels [12] on a compact metric space, e.g., the Gaussian RBF kernel or Laplacian kernel on a compact subset of $\mathbb{R}^m$, because the Banach space of bounded continuous functions on a compact subset $\mathcal{X}$ of $\mathbb{R}^m$ is dense in $L^2(P_X)$ for any $P_X$ on $\mathcal{X}$. Another example is the Gaussian or Laplacian kernel on the entire Euclidean space, since many random variables are defined on non-compact spaces. One can prove that, Assumption 1 holds for these kernels (see Appendix A1 in [13] for the proof). Under Assumption 1, one can show that $\langle g, \Sigma_{YY|X} g\rangle_{\mathcal{H}_\mathcal{Y}} = \mathrm{E}_X\big[\mathrm{Var}_{Y|X}[g(Y)|X]\big]$ for all $g \in \mathcal{H}_\mathcal{Y}$. Thus, the conditional covariance operator expresses the conditional variance of $g(Y)$ given $X$ in the feature space. As a side note, Ancona et al. [14] claimed that not all kernels are suitable for their nonlinear prediction schemes. They presented only sufficient conditions, which hold for Gaussian kernels. The kernel framework allows wider class of kernels and the kernel functions for $k_\mathcal{X}$ and $k_\mathcal{Y}$ (or $k_{\mathcal{Y}\otimes\mathcal{Z}}$), consequently maps $\Phi, \Psi$ in Eq. 3 (or in Eq. 4), are not necessarily to belong to the same function class. In particular, even though one can overall use Gaussian kernels $k_\mathcal{X} : \mathbb{R}^m \times \mathbb{R}^m \to \mathbb{R}$, $k_\mathcal{X}(x, x') = \exp(-\|x - x'\|^2/2\sigma^2)$ (as we did in all experiments), the parameter $\sigma^2$ of kernels can be adapted to different variables independently. We set the parameter such that $2\sigma^2$ equals the variance of the variable. For these reasons, our framework is more general and flexible than the framework of the state-of-the-art approaches, as described in [6,7,15].

To evaluate the conditional covariance operator, we use the trace norm, because it is not difficult to see that the trace norm of the operator is directly linked with the sum of residual errors, namely

$$\mathrm{Tr}(\Sigma_{YY|X}) = \sum_i \min_{f\in\mathcal{H}_\mathcal{X}} \mathrm{E}_{XY}\left[\left|(\phi_i(Y) - \mathrm{E}_Y[\phi_i(Y)]) - (f(X) - \mathrm{E}_X[f(X)])\right|^2\right],$$

where $\{\phi_i\}_{i=1}^{\infty}$ is the complete orthonormal system of the separable RKHS $\mathcal{H}_\mathcal{Y}$. An RKHS $(\mathcal{H}_\mathcal{Y}, k_\mathcal{Y})$ is separable, when the topological space $\mathcal{Y}$ is separable and $k_{\mathcal{Y}}$ is continuous on $\mathcal{Y}\times\mathcal{Y}$ [16]. It can be shown that $\Sigma_{YY|X} \leq \Sigma_{YY|S}$ with $X=(S,T)$, where the inequality refers to the usual order of self-adjoint operators, namely if $A \leq B \Leftrightarrow \langle Ag, g\rangle \leq \langle Bg, g\rangle$ for all $g \in \mathcal{H}_{\mathcal{Y}}$. Further, if $\mathcal{H}_{\mathcal{X}}$, $\mathcal{H}_{\mathcal{Y}}$, $\mathcal{H}_{\mathcal{S}}$ are given by characteristic kernels, $\Sigma_{YY|X} = \Sigma_{YY|S} \Leftrightarrow Y \perp\!\!\!\perp T \mid S$, which denotes that $Y$ and $T$ are conditionally independent, given $S$ (denoted by $Y \perp\!\!\!\perp T|S$). In terms of the trace norm, we have the following property:

**Property 1.** *Let $\mathbb{T}_{YY|X}$ denote the trace norm of the conditional covariance operator* $\mathrm{Tr}(\Sigma_{YY|X})$ *with* $X := (S,T)$*. Then we have* $\mathbb{T}_{YY|S,T} < \mathbb{T}_{YY|S} \Leftrightarrow Y \not\perp\!\!\!\perp T \mid S$ *and* $\mathbb{T}_{YY|S,T} = \mathbb{T}_{YY|S} \Leftrightarrow Y \perp\!\!\!\perp T \mid S$.

All technical details are omitted due to space limitations. Property 1 generalizes the (P1)-property, required by Ancona et al. [7, Section II.A] for any nonlinear extension of Granger causality, since (P1)-property describes merely the bivariate case, while Property 1 holds for multivariate cases.

## 2.2 Empirical Estimation of Operators

In analogy to [11], we introduce the estimations of $\mathbb{T}_{YY}$ and $\mathbb{T}_{YY|X}$ based on sample $(x^{(1)}, y^{(1)}), \ldots, (x^{(n)}, y^{(n)})$ from the joint distribution. Using the empirical mean elements $\widehat{m}_X^{(n)} = \frac{1}{n}\sum_{i=1}^n k_{\mathcal{X}}(x^{(i)}, \cdot)$ and $\widehat{m}_Y^{(n)} = \frac{1}{n}\sum_{i=1}^n k_{\mathcal{Y}}(y^{(i)}, \cdot)$, an estimator of $\Sigma_{YX}$ is $\widehat{\Sigma}_{YX}^{(n)} = \frac{1}{n}\sum_{i=1}^n \left(k_{\mathcal{Y}}(y^{(i)}, \cdot) - \widehat{m}_Y^{(n)}\right)\langle k_{\mathcal{X}}(x^{(i)}, \cdot) - \widehat{m}_X^{(n)}, \cdot\rangle_{\mathcal{H}_{\mathcal{X}}}$. $\widehat{\Sigma}_{YY}^{(n)}$ and $\widehat{\Sigma}_{XX}^{(n)}$ can be defined accordingly. An estimator of $\Sigma_{YY|X}$ is then defined by $\widehat{\Sigma}_{YY|X}^{(n,\epsilon)} = \widehat{\Sigma}_{YY}^{(n)} - \widehat{\Sigma}_{YX}^{(n)}(\widehat{\Sigma}_{XX}^{(n)} + \epsilon I)^{-1}\widehat{\Sigma}_{XY}^{(n)}$, where $\epsilon > 0$ is a regularization constant that enables inversion.[1] It can be shown that $\widehat{\mathbb{T}}_{YY}^{(n)} = \mathrm{Tr}(\widehat{\Sigma}_{YY}^{(n)})$ is a consistent estimator of $\mathbb{T}_{YY}$, which guarantees to converge in Hilbert-Schmidt norm at rate $n^{-1/2}$. Moreover, $\widehat{\mathbb{T}}_{YY|X}^{(n,\epsilon)} = \mathrm{Tr}(\widehat{\Sigma}_{YY|X}^{(n,\epsilon)})$ is a consistent estimator of $\mathbb{T}_{YY|X}$. If $\epsilon$ converges to zero more slowly than $n^{-1/2}$, this estimator converges to $\mathbb{T}_{YY|Z}$ (see [13] for proofs). For notational convenience, we will henceforth omit the upper index and use $\widehat{\mathbb{T}}_{YY}$ and $\widehat{\mathbb{T}}_{YY|X}$ to denote the empirical estimators. The computation with kernel matrices of $n$ data points becomes infeasible for very large $n$. In our practical implementation, we use the incomplete Cholesky decomposition $\widehat{K} = LL^{\mathrm{T}}$ [19] where $L$ is a lower triangular matrix determined uniquely by this equation. This may lead to considerably fewer columns than the original matrix. If $k$ columns are returned, the storage requirements are $O(kn)$ instead of $O(n^2)$, and the running time of many matrix operations reduces to $O(nk^2)$ instead of $O(n^3)$.

[1] The regularizer is required as the number of observed data points is finite, whereas the feature space could be infinite-dimensional. The regularization may be understood as a smoothness assumption on the eigenfunctions of $\mathcal{H}_{\mathcal{X}}$. It is analogous to Tikhonov regularization [17] or ridge regression [18]. Many simulated experiments showed that the empirical measures are insensitive to $\epsilon$, if it is chosen in some appropriate interval, e.g., $[10^{-10}, 10^{-2}]$. We chose $\epsilon = 10^{-5}$ in all our experiments.

## 3 Subsampling-Based Testing of Granger Causality

Essentially, Granger causality is expressed in terms of predictability: If data vector $X$ causally influences data vector $Y$, then $Y$ would be predictable by $X$, we denote $X \rightarrow Y$. Shorthand $X \not\rightarrow Y$ denotes the fact that the predictability on $Y$ by $X$ cannot be verified. Note that, in this paper, we distinguish between simple arrow "$\rightarrow$" expressing predictability and the double arrow "$\Rightarrow$" expressing Granger causality. To capture the predictability in a nonlinear setting by a significance test, we take the empirical estimation of the trace norm $\widehat{\mathbb{T}}_{YY|X}$ and conduct permutation tests to simulate this measure under the null hypothesis, i.e., unpredictable. Given a set of random permutations $\{\pi_1, \ldots, \pi_m\}$, we shuffle the data vector $X = (x^{(1)}, \ldots, x^{(n)})^T$ and calculate the empirical null distribution $\{\widehat{\mathbb{T}}_{YY|X^{\pi_1}}, \ldots, \widehat{\mathbb{T}}_{YY|X^{\pi_m}}\}$. Then, we determine the p-value $p$, which is the percentage of values in $\{\widehat{\mathbb{T}}_{YY|X^{\pi_1}}, \ldots, \widehat{\mathbb{T}}_{YY|X^{\pi_m}}\}$, which are properly less than $\widehat{\mathbb{T}}_{YY|X}$. Smaller p-values suggest stronger evidence against the null hypothesis (unpredictable), and thus stronger evidence favoring the alternative hypothesis (predictable). As a cut-off point for p-values, a significance level $\alpha$ is pre-specified. A typical choice is $\alpha = 0.05$. If $p < \alpha$, the null hypothesis is rejected, which means the predictability $X \rightarrow Y$ is significantly verified; Otherwise, we have $X \not\rightarrow Y$. This procedure can be straightforwardly extended to testing conditional predictability, i.e., $X \rightarrow Y \mid Z$. The measure and the corresponding empirical null distribution will be $\widehat{\mathbb{T}}_{YY|Z,X}$ and $\{\widehat{\mathbb{T}}_{YY|Z,X^{\pi_1}}, \ldots, \widehat{\mathbb{T}}_{YY|Z,X^{\pi_m}}\}$. It is remarkable that the test of predictability can be straightforwardly applied to multi-dimensional conditioning variables. Such cases are usually difficult by partitioning and Parzen-window approaches, which are common practice for non-parametric conditional independence tests. $X, Y, Z$ in "$X \rightarrow Y \mid Z$" could be usual random variables, and are not necessarily time series. The test of predictability alone does not use the temporal information. Granger's concept bases the asymmetry of causality on the time flow. Consequently, to test whether a time series X Granger causes another time series Y, we should first filter out the prediction power of past values of Y. For this purpose, we expand X with embedding vectors $\{\ldots, X_{t-1}, X_t, X_{t+1}, \ldots\}$ and the system Y to $\{\ldots, Y_{t-1}, Y_t, Y_{t+1}, \ldots\}$. In doing so, we can test predictability in a temporal context.

So far, we have shown the test of predictability on a single sample. The statistical power of such single tests is limited if the number of data points involved is not very large. Although long time series are often available in real applications, working with kernel matrices is only tractable, if the number of data points is not too large. For these reasons, we propose to utilize subsampling-based multiple testing. If the subsample size is moderate, the multiple testing remains feasible in spite of replications of the single test. In particular, such a multiple test can significantly increase the power of statistical tests in comparison to a single test [20]. In addition,the subsampling-based procedure characterizes the predictability through local segments of time series, i.e., sub-time-series, which makes the test somewhat robust with respect to e.g., global non-stationarity of the time series. Fig. 1 summarizes the multiple testing procedure for detecting Granger causality. Step 1 runs single hypothesis tests on $N$ random subsamples and obtains $N$ p-values, one for each subsample. Based on the distributional properties of these p-values, Step 2 makes the decision on the predictability. In details, Step 1.1 searches for the smallest integer $l_Y \in \{0, 1, 2, \ldots\}$, which states that $\{Y_{t-l_Y}, \ldots, Y_{t-1}\}$ (empty set for $l_Y = 0$) achieves the maximum knowledge that would be useful to predict $Y_t$. Step 1.2 yields a p-value which contains the information about whether $X_{t-1}$ can additionally improve the prediction of $Y_t$, given the maximum past knowledge of Y. Due to the strict time order requirement of Granger's concept, the instantaneous influence within one time slice is excluded. We test the one-step-ahead prediction, because an influence from $X_{t-j}$ ($j = 2, 3, \ldots$) to $Y_t$ can be usually be captured by $X_{t-1}$ to $Y_t$, if $X_{t-j}$ is dependent of $X_{t-1}$.

To make the second step apparent, let us take a closer look at the distribution of p-values. According to Property 1, we have $\mathbb{T}_{YY|X} \leq \mathbb{T}_{YY}$ and $\mathbb{T}_{YY|X^{\pi_i}} \leq$

**Hypothesis:** X$\Rightarrow$Y|Z.

**Step 1:** For each sub-time series $(X^{(i)}, Y^{(i)}, Z^{(i)})$ $(i = 1, \ldots, N)$ of time series (X, Y, Z):

**Step 1.1:** If $(Y_{t-1}^{(i)}, Z_{t-1}^{(i)}) \not\rightarrow Y_t^{(i)}$, set time lag $l_Y^{(i)} = 0$, else find the smallest lag $l_Y^{(i)} \in \{1, 2, \ldots\}$ such that $(Y_{t-l_Y-1}^{(i)}, Z_{t-l_Y-1}^{(i)}) \not\rightarrow Y_t^{(i)} \mid Y_{t-1}^{(i)}, \ldots, Y_{t-l_Y}^{(i)}, Z_{t-1}^{(i)}, \ldots, Z_{t-l_Y}^{(i)}$.

**Step 1.2:** If $l_Y^{(i)} = 0$, test $X_{t-1}^{(i)} \rightarrow Y_t^{(i)}$, else test $X_{t-1}^{(i)} \rightarrow Y_t^{(i)} \mid Y_{t-1}^{(i)}, \ldots, Y_{t-l_Y}^{(i)}, Z_{t-1}^{(i)}, \ldots, Z_{t-l_Y}^{(i)}$. A p-value $p_i$ is obtained.

**Step 2:** Calculate the skewness of p-values $p_1, \ldots, p_N$ and the probability $P^{\mathrm{Unif}}$ that $p_1, \ldots, p_N$ are from a uniform distribution over $[0, 1]$. If $P^{\mathrm{Unif}} < \frac{1}{2}$ and the p-values are positively skewed, accept the hypothesis, i.e., X$\Rightarrow$Y|Z; Otherwise, the hypothesis cannot be verified.

**Fig. 1.** Subsampling-based multiple testing of Granger causality

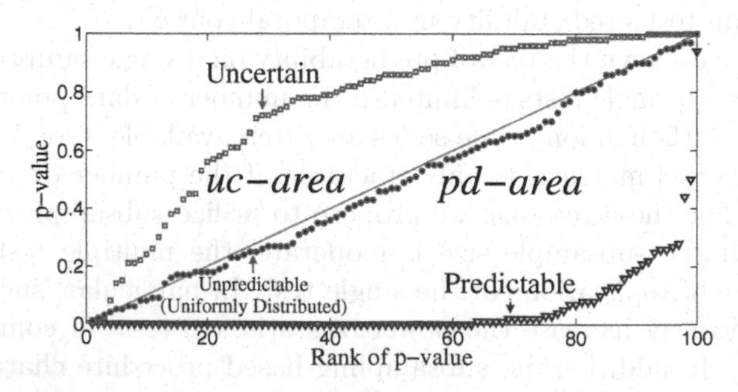


**Fig. 2.** Q-Q plots of p-values of three predictability tests. The field above and under the diagonal line are called uncertain (uc) area and predictable (pd) area respectively.

$\mathbb{T}_{YY}$ for any permutation $\pi_i$. If $Y$ is predictable by $X$, we expect $\mathbb{T}_{YY|X} < \mathbb{T}_{YY|X^{\pi_i}}$ for almost all random permutations $\pi_i$. Intuitively, the predictability on $Y$ by $X$ is reflected in a significant reduction of the sum of residual errors, captured by $\mathbb{T}_{YY|X}$, in comparison to $\mathbb{T}_{YY|X^{\pi_i}}$. Consequently, the majority of p-values are closer to 0 and the distribution of p-values is rather positively skewed (right-skewed). If $Y$ is unpredictable by $X$, then $\mathbb{T}_{YY|X} < \mathbb{T}_{YY|X^{\pi_i}}$ is as likely as $\mathbb{T}_{YY|X^{\pi_i}} \leq \mathbb{T}_{YY|X}$, p-values are uniformly distributed over $[0,1]$ and the skewness of p-values vanishes. If $\mathbb{T}_{YY|X^{\pi_i}} \leq \mathbb{T}_{YY|X}$ is true for the majority of random permutations $\pi_i$, more p-values are closer to 1 and the distribution of p-values is negatively skewed (left-skewed). This case, called uncertain situation, can occur due to various reasons. A very special case is, e.g., $\mathbb{T}_{YY|X} = \mathbb{T}_{YY|X^{\pi_i}}$, the null distribution is degenerate (all its probability mass is concentrated on one point). For instance, if $Z$ is high-dimensional or the statistical fluctuation of $Y$ is very small, it could happen that $\mathbb{T}_{YY|Z,X} \approx \mathbb{T}_{YY|Z,X^{\pi_i}} \approx \mathbb{T}_{YY|Z} \approx 0$. Then, the test of predictability cannot provide reliable information about the relation between $X$ and $Y$. In such cases, an uncertain situation can be interpreted as non-evidence for the hypothesis $X \to Y \mid Z$. Another interpretation of uncertain situations will be discussed later.

Inspired by [21,20], we visualize these observations with an intuitive graphical tool. First, we sort the set of p-values $\{p_{(1)}, \ldots, p_{(N)}\}$ in an increasing order, i.e., $p_1 \leq p_2 \leq \ldots \leq p_N$. If $p_i$ behaves as an ordered sample from the uniform distribution over $[0,1]$, the expected value of $p_i$ is approximately $\frac{i}{N}$. The slope of $p_i$ versus $i$, also called Q-Q plot ("Q" stands for quantile), should exhibit a linear relationship, along a line of slope $\frac{1}{N}$ passing through the origin (diagonal line in the Q-Q plot as shown in Fig. 2). If p-values are positively skewed, the reordered p-values are located in the subfield under the diagonal line, called pd-area ("pd": predictable). If p-values are negatively skewed, the reordered p-values are located in the so-called uc-area ("uc": uncertain). For a reliable decision based on the resulting p-values, the subsamples should be independent to some degree, since independent statistics on p-values are needed. This is the case, when the given sample size is much larger than the subsample size. In all our experiments, we fixed the subsample size at 100, since time series in our experiments contain at least 5000 data points. The other parameter of the multiple testing is the number of replications of the single test: $N$. In principle, $N$ should be large enough to enable a reliable identification of uniform distribution from $N$ p-values. In our experiments, we chose $N \geq 100$. For large sample sizes, we chose $N = 1000$.

The remaining problem is how to judge, in spite of the fluctuation of one specific set of p-values, whether the $N$ resulting p-values are uniformly distributed or not. We transform this problem to a two-sample-problem. More precisely, we simulate 1000 samples of $N$ values from the uniform distribution over $[0,1]$. For each of the 1000 simulated samples, we test whether the $N$ resulting p-values are identically distributed with the $N$ values from truly uniform distribution. The percentage of the positive results, i.e., the resulting p-values and the simulated values come from the same distribution, can be considered as the probability that the resulting p-values are uniformly distributed: $P^{\text{Unif}}$. If $P^{\text{Unif}} < \frac{1}{2}$, the p-values are less likely from a uniform distribution than from a non-uniform distribution. In our experiments, we employ the kernel-based test for the two-sample-problem proposed by Gretton et al. [22]. After all, the decision of predictability relies on whether $P^{\text{Unif}} < \frac{1}{2}$ and whether the p-values are positively skewed.

# 4 Experiments

To demonstrate the effectiveness of the proposed approach, we test our method on simulated data generated by chaotic maps and benchmark data of real-life systems of different scientific fields.

## 4.1 Hénon Maps

As the first simulated example, we consider the following two noisy Hénon maps:

$$\begin{aligned} x_{t+1} &= a + c_1\, x_{t-1} - d_1\, x_t^2 + \mu\, \xi_1 \\ y_{t+1} &= a + c_2\, y_{t-1} - b\, x_t y_t - (1-b)\, d_2\, y_t^2 + \mu\, \xi_2 \end{aligned}$$

represented as systems X and Y, respectively. Here, system X drives system Y with coupling strength $b \in [0, 1]$. If $b = 0$, X and Y are uncoupled; If $b > 0$, we have a uni-directed coupling X$\Rightarrow$Y. This example is also discussed by Bhattacharya et al. [23], who proposed to choose $b < 0.7$ to avoid strong synchronization. Similar to [23], we fixed the parameters at $a = 1.4$, $c_1 = 0.3$, $c_2 = 0.1$, $d_1 = 1$, $d_2 = 0.4$, $\mu = 0.01$. $\xi_{1,2}$ are unit variance Gaussian distributed noise terms. Note that $X$ and $Y$ are different systems even in the case of $b = 0$, because $c_1 \neq c_2$ and $d_1 \neq d_2$. Therefore, identical synchronization is impossible. We start with points $x_{1,2} = y_{1,2} = 0$. The upper line of Fig. 3 shows the time series of 10000 data points. We ran our test procedure on uncoupled time series ($b = 0$) and weakly unidirectionally coupled time series ($b = 0.25$). The reordered p-values obtained in both cases are visualized in the lower line of Fig. 3. In the case of $b = 0$, our test rejected the Granger causality in both directions. In the case of $b = 0.25$, our test revealed X$\Rightarrow$Y and gained no evidence for X$\Leftarrow$Y. Actually, the reordered p-values of testing $Y_{t-1} \rightarrow X_t \mid X_{t-1}, \ldots, X_{t-l_X}$ are located in the uc-area. Based on the underlying model, the past values of X can, in fact, improve the prediction of the future of Y. At the same time, values of Y in the future (random permutation of $Y_{t-1}$ might yield that) could be helpful for guessing the past of X. If this effect is present, we might have $\widehat{\mathbb{T}}_{X_t X_t | X_{t-1}, \ldots, X_{t-l_X}, Y_{t-1}^{\pi_i}} < \widehat{\mathbb{T}}_{X_t X_t | X_{t-1}, \ldots, X_{t-l_X}, Y_{t-1}}$. Thus, an uncertain situation might be interpreted not only as non-evidence for the direction X$\Leftarrow$Y that is just tested, but also as indirect evidence for the reversed direction X$\Rightarrow$Y.

## 4.2 Logistic Maps

As a second example, we consider the following pair of noisy logistic maps:

$$\begin{aligned} x_{t+1} &= (1-b_1)\, a\, x_t(1-x_t) + b_1\, a\, y_t(1-y_t) + \mu\, \xi_1 \\ y_{t+1} &= (1-b_2)\, a\, y_t(1-y_t) + b_2\, a\, x_t(1-x_t) + \mu\, \xi_2 \end{aligned} \qquad (5)$$

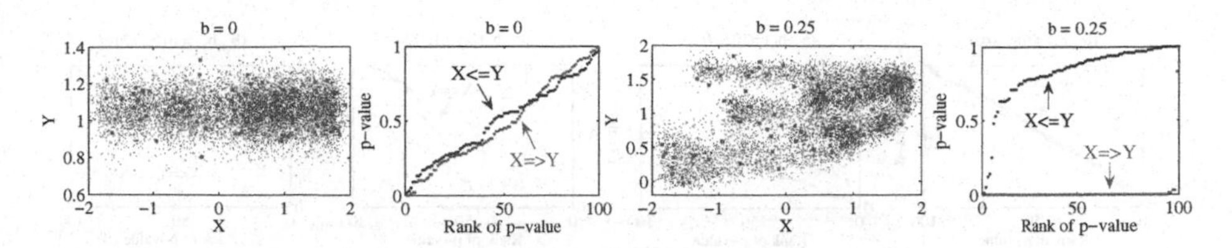


**Fig. 3.** The first and third plot from left show bivariate time series of 10000 observations (fine points) generated by Hénon maps and one random sub-time-series of 100 observations (bold points). The second and forth plot from left shows the corresponding Q-Q plots of p-values obtained by tests of predictability on 100 sub-time-series.

represented as systems X and Y. $b_{1,2} \in [0,1]$ describe the coupling strengths between X and Y. If $b_{1,2} = 0$, X and Y are uncoupled; If $b_{1,2} > 0$, we have a bi-directed coupling X $\Leftrightarrow$ Y; If $b_1 = 0$ and $b_2 > 0$, we have a uni-directed coupling X $\Rightarrow$ Y. The last case is also studied by Ancona et al. [7]. They claimed that in the noise-free case, i.e., $\mu = 0$, a transition to synchronization occurs at $b_2 = 0.37$ based on the calculation of the Lyapunov exponents. For this reason, we chose $b_{1,2} < 0.37$. As proposed in [7], parameter $a$ is fixed to 3.8; $\xi_{1,2}$ are unit variance Gaussian distributed noise terms; and $\mu$ is set at 0.01. We chose the start point $(x_1, y_1)$ randomly from $[0,1] \times [0,1]$ and generated time series of length 10000 with $(b_1, b_2) \in \{(0,0), (0,0.3), (0.1,0.3), (0.3,0.3)\}$. We repeated the same experiment with 100 different start points that are randomly chosen in $[0,1] \times [0,1]$. All these time series were not observed to diverge. The resulting directions of Granger causality were always the same. Fig. 4 shows Q-Q plots of p-values after 100 subsamples. In all 4 cases, our method identified the directions of coupling correctly: If $b_{1,2} = 0$, both X$\Rightarrow$Y and Y$\Rightarrow$X are rejected due to uniform distributions; If $(b_1, b_2) = (0, 0.3)$, X$\Rightarrow$Y is accepted and Y$\Rightarrow$X gained non-evidence. In testing Y$\Rightarrow$X, we have due to the uc-area indirect evidence for the reversed direction, which is consistent with the generating model. In the case of $(b_1, b_2) \in \{(0.1, 0.3), (0.3, 0.3)\}$, both X $\Rightarrow$ Y and Y $\Rightarrow$ X are accepted. Interestingly, by means of kernel-based two-sample-test, we can additionally confirm that when $b_1 = b_2 = 0.3$ the resulting p-values corresponding to testing X $\Rightarrow$ Y and to testing Y$\Rightarrow$X are identically distributed, while in the case of $b_1 \neq b_2$ the resulting p-values corresponding to testing X$\Rightarrow$Y and to testing Y$\Rightarrow$X come from different distributions. This is reasonable, since the coupling is absolutely symmetric in X and Y, if $b_1 = b_2$. It seems plausible that the more right-skewed, the stronger the coupling (see the case $0.1 = b_1 < b_2 = 0.3$). But, we do not speculate on that.

To explore the limitation of the method, we conduct our test on data generated by highly noisy relations. Suppose X causes Y mediated by some hidden process Z, i.e., X$\Rightarrow$Z$\Rightarrow$Y. The uni-directed coupling between $X$ and $Z$, and between $Z$ and $Y$ are given by noisy logistic maps as in Eq. 5 with $b_1 = 0$. The coupling strength $b_2$ varies from 0.05 to 0.30. The causal relation from $X$ to $Y$ is highly noisy and more sophisticated due to the indirect relation. We sampled time series of 10000 points from these models and tested the causal relation between $X$ and

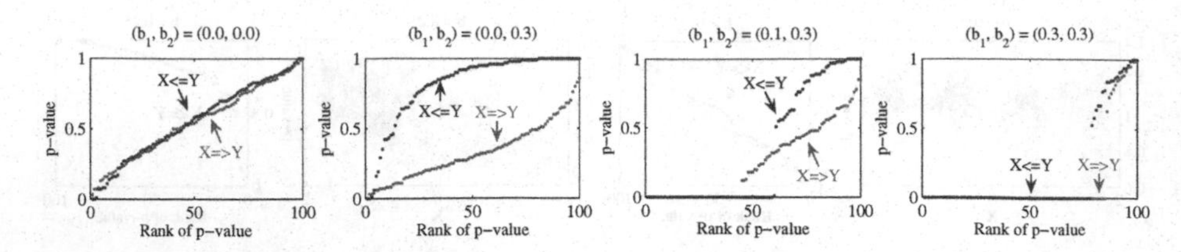


**Fig. 4.** Q-Q plots of p-values obtained by tests of predictability on 100 random sub-time-series from time series generated by noisy logistic maps

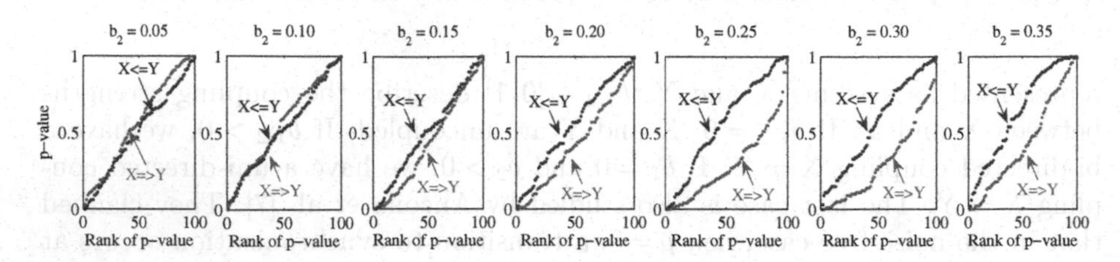


**Fig. 5.** Q-Q plots of p-values obtained by tests of predictability on 100 random sub-time-series from time series of indirect relations generated by noisy logistic maps.

**Table 1.** Comparison of tests based on various causal measures (see text). The generating model is X$\Rightarrow$Y.

| | Rejecting X$\Leftarrow$Y | | | | | | | Accepting X$\Rightarrow$Y | | | | | | |
|---|---|---|---|---|---|---|---|---|---|---|---|---|---|---|
| Coupling Strength $r$ | 0.05 | 0.10 | 0.15 | 0.20 | 0.25 | 0.30 | 0.35 | 0.05 | 0.10 | 0.15 | 0.20 | 0.25 | 0.30 | 0.35 |
| Kernel Test | 98 | 99 | 97 | 100 | 99 | 100 | 100 | 3 | 28 | 47 | 78 | 100 | 97 | 100 |
| EGCI | 97 | 95 | 94 | 91 | 92 | 93 | 92 | 0 | 3 | 7 | 23 | 34 | 48 | 84 |
| NLGC | 99 | 98 | 100 | 100 | 100 | 99 | 100 | 0 | 0 | 0 | 0 | 2 | 10 | 24 |
| TE | 100 | 99 | 98 | 97 | 90 | 95 | 99 | 0 | 0 | 0 | 0 | 0 | 0 | 0 |

$Y$ using different causality measures. Results for testing causal relation between X and Y on one sample is visualized in Fig. 5. In this sample, X$\Leftarrow$Y is correctly rejected at different coupling strengths $b_2$. X$\Rightarrow$Y is correctly accepted when the coupling strengths $b_2$ larger than 0.15. Tab. 1 summarizes the results after 100 replications. In comparison, we conducted permutation tests based on the other causality measures, i.e., extended Granger causality index (EGCI) [6], nonlinear extensions of Granger causality (NLGC) [7], and Transfer entropy (TE) [24] (Significance level 0.05). The results show that all measures performance well in rejecting X$\Leftarrow$Y. In accepting X$\Rightarrow$Y, our kernel test performs better than EGCI. NLGC works bad, while TE fails completely. We conjecture that the reason that TE can not identify the relationship significantly is that the subsample size of 100 is too small for TE. After all, we expect that our kernel-based approach provides also a reasonable measure of Granger causality even in the case of small sample sizes.

### 4.3 Stock Price-Volume Relation

We undertake an investigation of price and volume co-movement using the daily closing value DJ industrial average index and the daily volume of shares traded on NYSE from October 1928 to January 2008. The raw data are transformed to a series of day-to-day differences to obtain the daily movements (DM). The final dataset is a bivariate time series of 19920 observations (Fig. 6, left). Due to the large number of data points, we increase the number of subsamples to $N = 1000$. The probability of the uniform distribution for the p-values corresponding to the hypothesis that the daily movement of price causes that of volume is $P^{\text{Unif}} = 0 < \frac{1}{2}$ (p-values are positively skewed); For the reserved causal direction, we have $P^{\text{Unif}} = 0.940 \geq \frac{1}{2}$. Thus, we found strong evidence of asymmetry in the relationship between the daily movement of stock prices and the daily movement of trading volume. The daily movement of prices is much more important in predicting the movement of the trading volume, than the other way around.

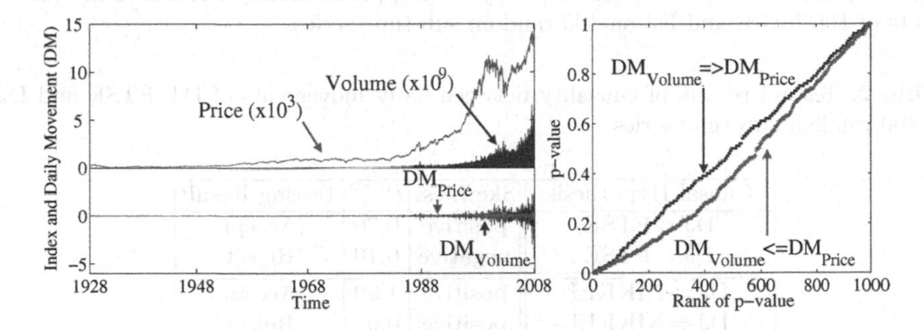


**Fig. 6.** DJ daily index, volume and movements (left). Q-Q plots of p-values (right) obtained by testing predictability between "DM$_{\text{Volume}}$" and "DM$_{\text{Price}}$" on 1000 random sub-time-series.

### 4.4 Co-movement of Stock Indexes

The analyzed raw dataset consists of daily closing values (adjusted for dividends and splits) of Dow Jones (DJ) industrial average index, Financial Times Stock Exchange (FTSE 100) and NIKKEI 225 stock average index during the time between April 1984 and January 2008. Only days with trading activity in both stock exchanges were considered. The time increment in the raw data is not always exactly one day due to weekends and moving holidays. We transform it into a series of day-to-day differences to describe the daily movements of indexes. After all, we have a trivariate time series with 5602 observations. We ran our test procedure on this dataset. The p-values obtained for all causal hypotheses are visualized in Fig. 7. The final testing results are summarized in Tab. 2. After all, our testing procedure showed evidence of following causality: DJ ⇒ FTSE,

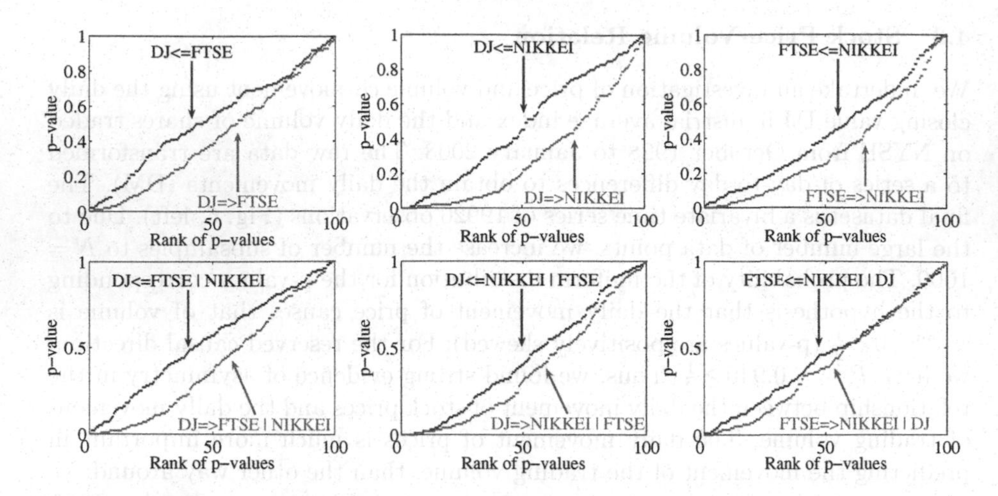


**Fig. 7.** Q-Q plots of p-values obtained by testing predictability between daily movements of DJ, FTSE and DJ on 100 random sub-time-series

**Table 2.** Testing results of causality between daily movements of DJ, FTSE and DJ on 100 random sub-time-series

| Causal Hypothesis | Skewness | $P^{\text{Unif}}$ | Testing Result |
|---|---|---|---|
| DJ $\Rightarrow$ FTSE | positive | 0.70 | Accept |
| DJ $\Leftarrow$ FTSE | negative | 0.10 | Reject |
| DJ $\Rightarrow$ NIKKEI | positive | 1.00 | Accept |
| DJ $\Leftarrow$ NIKKEI | positive | 0.03 | Reject |
| FTSE $\Rightarrow$ NIKKEI | positive | 1.00 | Accept |
| FTSE $\Leftarrow$ NIKKEI | positive | 0.65 | Accept |
| DJ $\Rightarrow$ FTSE $\mid$ NIKKEI | positive | 0.98 | Accept |
| DJ $\Leftarrow$ FTSE $\mid$ NIKKEI | negative | 0.13 | Reject |
| DJ $\Rightarrow$ NIKKEI $\mid$ FTSE | positive | 1.00 | Accept |
| DJ $\Leftarrow$ NIKKEI $\mid$ FTSE | positive | 0.49 | Reject |
| FTSE $\Rightarrow$ NIKKEI $\mid$ DJ | positive | 0.78 | Accept |
| FTSE $\Leftarrow$ NIKKEI $\mid$ DJ | positive | 0.08 | Reject |

DJ $\Rightarrow$ NIKKEI and FTSE $\Leftrightarrow$ NIKKEI, whereas the causality running from the daily movement of NIKKEI to the daily movement of FTSE is spurious, since FTSE $\Leftarrow$ NIKKEI|DJ is rejected. The knowledge of the dynamics of DJ can significantly improve a prediction of the dynamics of FTSE and NIKKEI, but the dynamics of FTSE and NIKKEI has a very limited, yet non-significant impact on the future dynamics of DJ. The finding that the movement of DJ influences the movement of FTSE and NIKKEI not vice versa, which may seem trivial as a purely economical fact, but actually confirms in an independent way the validity of the kernel test formalism.

### 4.5 Cardiorespiratory Interaction

As the last example of real-life systems, we consider the benchmark bivariate time series of heart rate and respiration force of a sleeping human suffering from sleep apnea (data set B of the Santa Fe Institute time series competition [25]) recorded in the sleep laboratory of the Beth Israel Hospital in Boston, MA. The magnitudes considered are heart rate and respiration force. The data are plotted in Fig. 8 (left). The time interval between measurements is 0.5 seconds (sampling frequency 2 Hz). As described in [26,27], under normal, physiological conditions, the heart rate is modulated by respiration through a process known as Respiratory Sinus Arrhythmia (RSA). It is the natural cycle of arrhythmia that occurs through the influence of breathing on the flow of sympathetic and vagus impulses to the sinoatrial node of the heart. When we inhale, vagus nerve activity is impeded and the heart rate begins to increase. When we exhale, this pattern is reversed. This quasi-periodic modulation of heart rate by respiration is most notable in young, healthy subjects and decreases with age, which means Heart $\Leftarrow$ Respiration. However, this dataset corresponds to a patient suffering from sleep apnea, which is a breathing disorder characterized by brief interruptions of breathing during sleep. Sleep apnea affects the normal process of RSA, disturbing the usual patterns of interaction between the heart rate and respiration. As a result, the control of the heart rate by respiration becomes unclear. It may well be blocked, in accordance with the change in dynamics, that is characteristic of the so-called "dynamical diseases". Some studies [7,23,24] claimed a coupling in the reversed direction: Heart $\Rightarrow$ Respiration. In summary, the bidirected causation Heart $\Leftrightarrow$ Respiration might be likely the ground truth in this example. The result of our test procedure is consistent with this prior knowledge, since for both directions we have $P^{\mathrm{Unif}} = 0 < \frac{1}{2}$ (p-values are positively skewed). Since the dynamics of this times series are rhythmic over time, we used embedding vectors with different $m, r$ as defined in Eq. 1. Only when $m \leq 20$ and $r \geq 12$, our test could not detect any significant causal relations between "Heart" and "Respiration", in other cases, a bidirected causality is always verified. The results

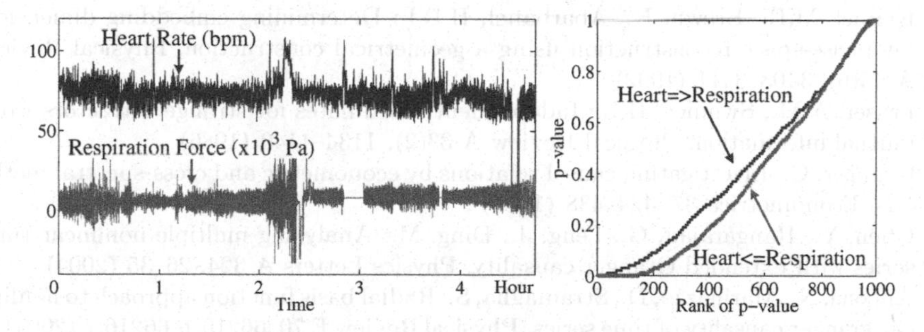


**Fig. 8.** Time series of the heart rate and respiration force of a patient suffering sleep apnea (left). Q-Q plots of p-values (right) obtained by testing predictability between "Heart" and "Respiration" on 1000 random sub-time-series.

are plausible, because it is well known that the high-frequency band (0.25 – 0.45 Hz) is characteristic of the respiratory rhythm. The identified bi-directed causation between heart rate and respiration suggests a probably causal link between sleep apnea and cardiovascular disease [28], although the exact mechanisms that underlie this relationship remain unresolved [29].

## 5 Conclusion

We have presented the kernel framework for detecting nonlinear Granger causality and have proposed to measure the improvement of the predictability by the trace norm of the conditional covariance operator and test whether the prediction error of the first time series can be significantly reduced by including knowledge from the second time series, given a third one. In comparison to other nonlinear extensions of Granger causality [6,7,15,24], our approach is designed in the general kernel framework and straightforwardly applicable. It is also appealing for users that the kernelized analysis of covariance is easy to implement and its statistical properties are well understood. Although kernel methods are rather popular nonlinear techniques in statistical pattern recognition, this is the first time, to the best of our knowledge, that the nonlinear extension of Granger's concept of causality in such a straightforward way is demonstrated.

**Acknowledgement.** Valuable discussions with Kenji Fukumizu are warmly acknowledged.

## References

1. Packard, N.H., Crutchfield, J.P., Farmer, J.D., Shaw, R.S.: Geometry from a time series. Physical Review Letters 45, 712–716 (1980)
2. Takens, M.: Detecting strange attractors in turbulence. In: Rand, D.A., Young, L.S. (eds.) Dynamical Systems and Turbulence. Lecture Notes in Mathematics, pp. 366–381. Springer, Berlin (1982)
3. Kennel, M.B., Brown, R., Abarbanel, H.D.I.: Determining embedding dimension for phase-space reconstruction using a geometrical construction. Physical Review A 45(6), 3403–3411 (1992)
4. Fraser, A.M., Swinney, H.L.: Independent coordinates for strange attractors from mutual information. Physical Review A 33(2), 1134–1140 (1986)
5. Granger, C.: Investigating causal relations by econometric and cross-spectral methods. Econometrica 37, 424–438 (1969)
6. Chen, Y., Rangarajan, G., Feng, J., Ding, M.: Analyzing multiple nonlinear time series with extended Granger causality. Physics Letters A 324, 26–35 (2004)
7. Ancona, N., Marinazzo, D., Stramaglia, S.: Radial basis function approach to nonlinear granger causality of time series. Physical Review E 70, 66216.1–66216.7 (2004)
8. Aronszajn, N.: Theory of reproducing kernels. Transactions of the American Mathematical Society 68(3), 337–404 (1950)
9. Baker, C.: Joint measures and cross-covariance operators. Transactions of the American Mathematical Society 186, 273–289 (1973)

10. Fukumizu, K., Bach, F., Jordan, M.: Kernel dimension reduction in regression. Technical Report 715, University of California, Berkeley, CA (2006)
11. Fukumizu, K., Gretton, A., Sun, X., Schölkopf, B.: Kernel measures of conditional dependence. In: Platt, J., Koller, D., Singer, Y., Roweis, S. (eds.) Proceedings of the 21th Neural Information Processing Systems Conference, pp. 489–496. MIT Press, Cambridge (2007)
12. Steinwart, I.: On the influence of the kernel on the consistency of support vector machines. Journal of Machine Learning Research 2, 67–93 (2001)
13. Sun, X.: Causal inference from statistical data. PhD thesis, Computer Science Faculty, University of Karlsruhe (TH), Germany (2008)
14. Ancona, N., Stramaglia, S.: An invariance property of predictors in kernel-induced hypothesis spaces. Neural Computation 18, 749–759 (2006)
15. Marinazzo, D., Pellicoro, M., Stramaglia, S.: Nonlinear parametric model for Granger causality of time series. Physical Review E 73(6), 66216.1–66216.6 (2006)
16. Lukić, M., Beder, J.: Stochastic processes with sample paths in reproducing kernel Hilbert spaces. Transactions of the American Mathematical Society 353(10), 3945–3969 (2001)
17. Groetsch, C.: The theory of Tikhonov regularization for Fredholm equations of the first kind. Pitman Publishing Program, Boston (1984)
18. Hoerl, A., Kennard, R.: Ridge regression: Biased estimation for nonorthogonal problems. Technometrics 12(1), 55–67 (1970)
19. Fine, S., Scheinberg, K.: Efficient SVM training using low-rank kernel representations. Journal of Machine Learning Research 2, 243–264 (2001)
20. Hochberg, Y.: More powerful procedures for multiple significance testing. Statistics in Medicine 9(7), 811–818 (1990)
21. Schweder, T.: Plots of p-values to evaluate many tests simultaneously. Biometrika 69(3), 493–502 (1982)
22. Gretton, A., Borgwardt, K., Rasch, M., Schölkopf, B., et al.: A kernel method for the two-sample-problem. In: Schölkopf, B., Platt, J., Hoffman, T. (eds.) Proceedings of the 20th Neural Information Processing Systems Conference, pp. 513–520. MIT Press, Cambridge (2006)
23. Bhattacharya, J., Pereda, E., Petsche, H.: Effective detection of coupling in short and noisy bivariate data. IEEE Transactions on Systems, Man, and Cybernetics 33(1), 85–95 (2003)
24. Schreiber, T.: Measuring information transfer. Physical Review Letters 85(2), 461–464 (2000)
25. Rigney, D., Goldberger, A., Ocasio, W., Ichimaru, Y., et al.: Multi-channel physiological data: description and analysis (data set B). In: Weigend, A., Gershenfeld, N. (eds.) Time series prediction: Forecasting the future and understanding the past, pp. 105–129. Addison-Wesley, Reading (1993)
26. Ichimaru, Y., Clark, K., Ringler, J., Weiss, W.: Effect of sleep stage on the relationship between respiration and heart rate variability. In: Proceedings of Computers in Cardiology 1990, pp. 657–660. IEEE Computer Society Press, Los Alamitos (1990)
27. Verdes, P.: Assessing causality from multivariate time series. Physical Review E 72(2), 066222.1–066222.9 (2005)
28. Roux, F., D'Ambrosio, C., Mohsenin, V.: Sleep-related breathing disorders and cardiovascular disease. The American Journal of Medicine 108(5), 396–402 (2000)
29. Duchna, H., Grote, L., Andreas, S., Schulz, R., et al.: Sleep-disordered breathing and cardio- and cerebrovascular diseases: 2003 update of clinical significance and future perspectives. Somnologie 7(3), 101–121 (2003)

# Clustering Via Local Regression

Jun Sun[1,2], Zhiyong Shen[1,2], Hui Li[2], and Yidong Shen[1]

[1] State Key Laboratory of Computer Science,
Institute of Software, Chinese Academy of Sciences, Beijing 100190, China
[2] Graduate University, Chinese Academy of Sciences, Beijing 100049, China
{junsun,zyshen}@ios.ac.cn, lihui@is.iscas.ac.cn, ydshen@ios.ac.cn

**Abstract.** This paper deals with the local learning approach for clustering, which is based on the idea that in a good clustering, the cluster label of each data point can be well predicted based on its neighbors and their cluster labels. We propose a novel local learning based clustering algorithm using kernel regression as the local label predictor. Although sum of absolute error is used instead of sum of squared error, we still obtain an algorithm that clusters the data by exploiting the eigen-structure of a sparse matrix. Experimental results on many data sets demonstrate the effectiveness and potential of the proposed method.

**Keywords:** Clustering, Local Learning, Sum of Absolute Error.

## 1 Introduction

Data clustering has been extensively studied and practiced across multiple disciplines for several decades [8]. It aims to group objects, usually represented as data points in $\mathbb{R}^d$, into several clusters in a meaningful way. Generally, the clustering objective is formulated to maximize intra-cluster cohesion and inter-cluster separability. Clustering techniques have been applied to many tasks, such as image segmentation [11], unsupervised document organization, grouping genes and proteins with similar functionality, and so on.

Many clustering algorithms have emerged over the years. One of the most popular clustering methods in recent years is the spectral clustering approach which exploits the eigen-structure of a specially constructed matrix. Generally, spectral clustering can be motivated from a graph partitioning perspective. Various graph clustering objectives, including ratio cut [6], normalized cut [11], and min-max cut [7], can be solved effectively by the spectral clustering method.

In this paper, we propose a clustering method that's also based on eigen-decomposing a matrix as is done in spectral clustering. However, we motivate it from the local learning idea, namely, in a good clustering, the cluster label of each data point can be well predicted based on its neighbors and their cluster labels. First, our method constructs local label predictors for each data point using its neighbors and their cluster labels as the training data. Then it minimizes the discrepancy between the data points' cluster labels and the prediction results of all the local predictors to get a final clustering.

W. Daelemans et al. (Eds.): ECML PKDD 2008, Part II, LNAI 5212, pp. 456–471, 2008.

The local learning idea has already been used in supervised learning [3], transductive classification [20] and dimensionality reduction [21]. In supervised learning, for each test data point, data points in the vicinity of the test point are selected for learning an output function. Then this function is used to predict the label of the test point. In practice, this method can achieve better performance than global learning machines since only data points relevant to the test point are used for training [3].

When the local learning idea is used in an unsupervised manner, a clustering objective is formulated in [19] and achieves good performance. This objective is also combined with a global label smoothness regularizer to obtain a method called "clustering with local and global regularization" in [16,17]. However, their local clustering objective uses sum of squared error to measure the discrepancy between the data points' cluster labels and the prediction results of all the local predictors. In regression analysis, a model using the sum of squared error measure can be sensitive to noise and outliers, since large error entries are overemphasized. To obtain a better model, we use sum of absolute error which is more robust and reliable. Traditionally, a regression model using absolute error will lead to a linear programming optimization problem [26]. However, by combining sum of absolute error and kernel regression as the local label predictors, we still obtain an algorithm that clusters the data by exploiting the eigen-structure of a specially constructed matrix, thus inheriting the advantages of the powerful spectral clustering approach. Experimental results on many data sets from real-world domains demonstrate the superiority of our proposed approach in obtaining high quality clusterings.

The rest of the paper is organized as follows. In Section 2, we begin with an introduction to the notations and representation of cluster labels, then formulate the model in detail. The algorithm is derived in Section 3. In Section 4, we evaluate the proposed method on many data sets. We make concluding remarks in Section 5.

## 2 Model Formulation

### 2.1 Notations

In this section we introduce the notations adopted in this paper. Boldface lowercase letters, such as $\mathbf{x}$ and $\mathbf{y}$, denote column vectors. Boldface uppercase letters, such as $\mathbf{M}$ and $\mathbf{A}$, denote matrices. The superscript $T$ is used to denote the transpose of a vector or matrix. $\mathbf{M} \geq 0$ means that every entry in $\mathbf{M}$ is nonnegative. For $\mathbf{x} \in \mathbb{R}^d$, $\|\mathbf{x}\|$ denotes the $L_2$ norm, and $\|\mathbf{x}\|_1$ denotes the $L_1$ norm. Specifically, $\|\mathbf{x}\| = \sqrt{\sum_{i=1}^d x_i^2}$ and $\|\mathbf{x}\|_1 = \sum_{i=1}^d |x_i|$. For $\mathbf{M} \in \mathbb{R}^{s \times t}$, $\|\mathbf{M}\|_{\mathrm{F}}$ denotes the Frobenius norm, and $\|\mathbf{M}\|_{\mathrm{SAV}}$ denotes the Sum-Absolute-Value norm [4]. Specifically,

$$\|\mathbf{M}\|_{\mathrm{F}} = \sqrt{\mathrm{trace}(\mathbf{M}^T\mathbf{M})} = \left(\sum_{i=1}^{s}\sum_{j=1}^{t} m_{ij}^2\right)^{\frac{1}{2}} \quad \text{and} \quad \|\mathbf{M}\|_{\mathrm{SAV}} = \sum_{i=1}^{s}\sum_{j=1}^{t} |m_{ij}|$$

**Table 1.** Summary of notations

| Symbols | Description |
|---|---|
| $n$ | total number of input data points |
| $d$ | data dimensionality |
| $\mathbf{X}$ | total data matrix, $\mathbf{X} \in \mathbb{R}^{n \times d}$ |
| $\mathcal{X}$ | input space from which data is drawn, $\mathcal{X} \subseteq \mathbb{R}^d$ |
| $c$ | number of output clusters |
| $\pi_l$ | the set of points in the $l$-th cluster where $1 \leq l \leq c$ |
| $\|\pi_l\|$ | the number of points in the $l$-th cluster where $1 \leq l \leq c$ |
| $\mathcal{N}_i$ | the set of "neighbors" of $\mathbf{x}_i$, here $\mathbf{x}_i \notin \mathcal{N}_i$ |
| $\mathbf{1}_m$ | $m$-dimensional column vector whose entries are all 1's |
| $\mathbf{I}_m$ | the identity matrix of order $m$ |
| Diag($\mathbf{M}$) | the diagonal matrix whose size and diagonal elements are the same as the square matrix $\mathbf{M}$ |
| Deg($\mathbf{M}$) | the degree matrix of $\mathbf{M}$, i.e., the diagonal matrix whose diagonal elements are the sums of rows of $\mathbf{M}$ |
| Trace($\mathbf{M}$) | the trace of the square matrix $\mathbf{M}$ |

Other important notations are summarized in Table 1. In addition, neighborhood $\mathcal{N}_i$ simply denotes a set of nearest neighbors (measured by some distance metric) of point $\mathbf{x}_i$. Typically, given $k \ll n$, we define each neighborhood $\mathcal{N}_i$ as the set of $k$-nearest neighbors of $\mathbf{x}_i$, not including $\mathbf{x}_i$.

### 2.2 Cluster Labels

Given a set of data points $\{\mathbf{x}_i\}_{i=1}^n \subseteq \mathcal{X} \subseteq \mathbb{R}^d$, the goal of clustering is to find a disjoint partitioning $\{\pi_l\}_{l=1}^c$ of the data where $\pi_l$ is the $l$-th cluster.

We represent a clustering of the data points by a *partition matrix* $\mathbf{P} = [\mathbf{p}_1, \cdots, \mathbf{p}_c] = [p_{il}] \in \{0,1\}^{n \times c}$ and $\mathbf{P}\mathbf{1}_c = \mathbf{1}_n$. Thus, exactly one element in each row of $\mathbf{P}$ is 1. Specifically,

$$p_{il} = \begin{cases} 1 & \text{if } \mathbf{x}_i \in \pi_l, \\ 0 & \text{if } \mathbf{x}_i \notin \pi_l. \end{cases} \tag{1}$$

Instead of directly using the entries of partition matrix $\mathbf{P}$ as the cluster labels, we use a *Scaled Partition Matrix* $\mathbf{Y} = \mathbf{P}(\mathbf{P}^T\mathbf{P})^{-1}$ where $\mathbf{Y} = [\mathbf{y}_1, \ldots, \mathbf{y}_c] = [y_{il}] \in \mathbb{R}^{n \times c}$. Specifically,

$$y_{il} = \begin{cases} \frac{1}{|\pi_l|} & \text{if } \mathbf{x}_i \in \pi_l, \\ 0 & \text{if } \mathbf{x}_i \notin \pi_l. \end{cases} \tag{2}$$

Thereby each data point $\mathbf{x}_i$ is associated with a $c$-dimensional cluster label $[y_{i1}, y_{i2}, \cdots, y_{ic}]$. The scaling is used for obtaining balanced clusters and balanced clusters usually lead to better performance in practice.

### 2.3 Clustering Objective

In a typical local learning approach for supervised learning [3], for each test data point, data points in the vicinity of the test point is selected for learning an output function. Then this function is used to predict the label of the test point. Although this method looks "simple and stupid" [3], it can achieve good performance in practice since only data points relevant to the test point are used for training. This approach can be also adapted to the clustering problem [19] and the idea is translated into:

> In a good clustering, the cluster label of a data point can be well estimated based on its neighbors and their cluster labels.

Based on this idea, a clustering objective can be formulated to obtain a clustering that satisfies the above property. For each point $\mathbf{x}_i$, if we construct a local label predictor $o_{il}(\cdot)$ based on its neighborhood information $\{(\mathbf{x}_j, y_{jl}) \,|\, \mathbf{x}_j \in \mathcal{N}_i\}$, then the prediction result $o_{il}(\mathbf{x}_i)$ should be similar to the cluster label $y_{il}$ of $\mathbf{x}_i$. Hence, a brute force approach to select the best final clustering would be:

1. Enumerate all possible labelings of $\{\mathbf{x}_i\}_{i=1}^n$.
2. For each labeling,
   (a) For each neighborhood $\mathcal{N}_i$ $(1 \leq i \leq n)$ and the corresponding labels, build local learners $[o_{i1}(\cdot), o_{i2}(\cdot), \ldots, o_{ic}(\cdot)]$. Here, $o_{il}(\cdot)$ is the output function learned using the training data $\{(\mathbf{x}_j, y_{jl}) \,|\, \mathbf{x}_j \in \mathcal{N}_i\}$ where $1 \leq l \leq c$.
   (b) Predict each data point $\mathbf{x}_i$ using the above local label predictors $[o_{i1}(\cdot), o_{i2}(\cdot), \ldots, o_{ic}(\cdot)]$, obtaining $[o_{i1}(\mathbf{x}_i), o_{i2}(\mathbf{x}_i), \ldots, o_{ic}(\mathbf{x}_i)]$. Calculate the error $\sum_{l=1}^c |y_{il} - o_{il}(\mathbf{x}_i)|$.
   (c) Calculate the total error sum $\sum_{i=1}^n \sum_{l=1}^c |y_{il} - o_{il}(\mathbf{x}_i)|$.
3. Pick the labeling with the smallest total error sum.

Obviously, we won't directly use the above exhaustive search approach since the number of possible clusterings is too large. In fact, for $n$ data points and $c$ clusters, there're $\frac{1}{c!}\sum_{l=1}^c \binom{c}{l}(-1)^{c-l}l^n$ different clusterings [8].

Suppose $\mathbf{o}_l = [o_{1l}(\mathbf{x}_1), o_{2l}(\mathbf{x}_2), \ldots, o_{nl}(\mathbf{x}_n)]^T \in \mathbb{R}^{n\times 1}$ and $\mathbf{O} = [\mathbf{o}_1, \ldots, \mathbf{o}_c] \in \mathbb{R}^{n\times c}$, which is the combination of all the predictions made by the local label predictors. Then, the clustering objective can be written as follows:

$$\min_{\mathbf{Y}\in\mathbb{R}^{n\times c}} \quad \mathcal{J}(\mathbf{Y}) = \|\mathbf{Y} - \mathbf{O}\|_{\text{SAV}} = \sum_{l=1}^{c} \|\mathbf{y}_l - \mathbf{o}_l\|_1 \tag{3}$$

$$\text{subject to} \quad \mathbf{Y} \text{ is a scaled partition matrix defined in (2)} \tag{4}$$

Note that instead of using the Frobenius norm $\|\cdot\|_{\text{F}}$ as the error measure, we use the Sum-Absolute-Value norm $\|\cdot\|_{\text{SAV}}$ which is more robust. The problem of what kind of local predictor we will use will be addressed in the next subsection.

### 2.4 Local Label Predictor

For the clustering objective (3) to be tractable, we want to select a local label predictor that's easy to deal with and yet sufficiently powerful for learning the local structure. So, we choose *kernel regression* model as our local label predictor. Kernel regression [1,13,18] is a widely used nonparametric technique for nonlinear regression. The prediction of a kernel regression model for a test point takes the form of a weighted average of the target values observed at training points. The weighting coefficients are related to the *kernel function* $K(\mathbf{x}_i, \mathbf{x}_j)$ for $\mathbf{x}_i, \mathbf{x}_j \in \mathcal{X}$ where $\mathcal{X}$ is the data space. With $\mathbf{x}_i$ fixed, $K(\mathbf{x}_i, \mathbf{x})$ can be interpreted as an unnormalized probability density function centered around $\mathbf{x}_i$. Therefore, two key properties of the kernel function are

$$K(\mathbf{x}_i, \mathbf{x}_j) \geq 0 \qquad \text{for all } \mathbf{x}_i, \mathbf{x}_j \in \mathcal{X}. \tag{5}$$

$$\int_{\mathbf{x}\in\mathcal{X}} K(\mathbf{x}_i, \mathbf{x})\, d\mathbf{x} \in (0, \infty) \qquad \text{for all } \mathbf{x}_i \in \mathcal{X} \tag{6}$$

Given a training set $\{(\mathbf{x}_i, y_i)\}_{i=1}^n$, we want to construct a learner to predict the target value $y$ for a test point $\mathbf{x}$. Motivated from kernel density estimation [2], the target value for a test point $\mathbf{x}$ can be estimated by

$$y = \frac{\sum_{i=1}^n y_i K(\mathbf{x}_i, \mathbf{x})}{\sum_{i=1}^n K(\mathbf{x}_i, \mathbf{x})} \tag{7}$$

which is called the kernel regression formula, also known as Nadaraya-Watson model [10]. This model can also be motivated from the interpolation problem when the input variables are noisy [2].

Generally, a distance-based kernel function [18] can be written as

$$K(\mathbf{x}_i, \mathbf{x}_j) = \varphi(\mathcal{D}(\mathbf{x}_i, \mathbf{x}_j)) \tag{8}$$

where $\mathcal{D}(\cdot, \cdot)$ is a distance metric. $\varphi(\cdot)$ is a nonnegative function and monotonically decreases with increasing $\mathcal{D}(\mathbf{x}_i, \mathbf{x}_j)$. In addition, $\varphi(\cdot)$ often have parameters pertaining to the rate of decay.

Various kernel functions have been studied in the literature, such as Gaussian, Epanechnikov, rectangular, triangular, and so on. The following two kernels will be used in our experiments as in [19]:

- The Gaussian kernel is defined as

$$K(\mathbf{x}_i, \mathbf{x}_j) = \exp\left(-\frac{\mathcal{D}(\mathbf{x}_i, \mathbf{x}_j)^2}{\gamma}\right) \tag{9}$$

  where $\varphi(t) = \exp\left(-\frac{t^2}{\gamma}\right)$ for $\gamma > 0$.

- The cosine kernel defined over nonnegative data points on the unit hypersphere is defined as follows

$$K(\mathbf{x}_i, \mathbf{x}_j) = \mathbf{x}_i^T \mathbf{x}_j = 1 - \frac{\|\mathbf{x}_i - \mathbf{x}_j\|^2}{2} \tag{10}$$

  where $\mathbf{x}_i, \mathbf{x}_j \in \mathcal{X} = \{\mathbf{x} \,|\, \mathbf{x}^T\mathbf{x} = 1, \mathbf{x} \geq 0\}$. Here, $\varphi(t) = 1 - \frac{t^2}{2}$ for $0 \leq t \leq \sqrt{2}$ and $\varphi(t) = 0$ for $t \geq \sqrt{2}$. This kernel generally leads to good performance when used for document data sets [19].

### 2.5 Constructing and Combining Local Predictors

In our clustering problem, for each data point $\mathbf{x}_i$ and $\{(\mathbf{x}_j, y_{jl}) \mid \mathbf{x}_j \in \mathcal{N}_i\}$, we want to construct a local label predictor $o_{il}(\cdot)$ to estimate the cluster label of $\mathbf{x}_i$. Here we choose kernel regression model introduced in the previous subsection as our local label predictor. According to equation (7), the solution of local label prediction for $\mathbf{x}_i$ and $l$ is given by

$$o_{il}(\mathbf{x}_i) = \frac{\sum_{\mathbf{x}_j \in \mathcal{N}_i} K(\mathbf{x}_i, \mathbf{x}_j) y_{jl}}{\sum_{\mathbf{x}_j \in \mathcal{N}_i} K(\mathbf{x}_i, \mathbf{x}_j)} \tag{11}$$

We can construct a matrix $\mathbf{A} = [a_{ij}] \in \mathbb{R}^{n \times n}$ as follows

$$a_{ij} = \begin{cases} \frac{K(\mathbf{x}_i, \mathbf{x}_j)}{\sum_{\mathbf{x}_j \in \mathcal{N}_i} K(\mathbf{x}_i, \mathbf{x}_j)} & \text{if } \mathbf{x}_j \in \mathcal{N}_i \\ 0 & \text{if } \mathbf{x}_j \notin \mathcal{N}_i \end{cases} \tag{12}$$

Two key properties that $\mathbf{A}$ satisfies are

$$\mathbf{A} \geq 0 \tag{13}$$

$$\mathbf{A}\mathbf{1}_n = \mathbf{1}_n \tag{14}$$

These two properties will be used in the next section for the derivation of our main clustering algorithm.

Recall from section 2.3 that

$$\mathbf{o}_l = [o_{1l}(\mathbf{x}_1), o_{2l}(\mathbf{x}_2), \ldots, o_{nl}(\mathbf{x}_n)]^T \in \mathbb{R}^{n \times 1} \tag{15}$$

$$\mathbf{O} = [\mathbf{o}_1, \mathbf{o}_2, \ldots, \mathbf{o}_c] \in \mathbb{R}^{n \times c} \tag{16}$$

where $1 \leq l \leq c$. By combining equations (11) and (12), it's not difficult to see that

$$\mathbf{o}_l = \mathbf{A}\mathbf{y}_l \quad \text{and} \quad \mathbf{O} = \mathbf{A}\mathbf{Y} \tag{17}$$

Therefore, the clustering objective in (3) can be rewritten as follows:

$$\min_{\mathbf{Y} \in \mathbb{R}^{n \times c}} \quad \mathcal{J}(\mathbf{Y}) = \|\mathbf{Y} - \mathbf{A}\mathbf{Y}\|_{\text{SAV}} = \sum_{l=1}^{c} \|\mathbf{y}_l - \mathbf{A}\mathbf{y}_l\|_1 \tag{18}$$

$$\text{subject to} \quad \mathbf{Y} \text{ is a scaled partition matrix defined in (2)} \tag{19}$$

This is the main objective function we want to optimize. Here, the $L_1$ norm $\|\cdot\|_1$ in equation (18) makes the function not differentiable and difficult to optimize combined with constraint (2) which is not convex. However, the properties of $\mathbf{A}$ in (13) and (14) ensure that the problem can be optimized by eigen-decomposing some sparse matrix. We will derive the algorithm in the next section.

## 3 Algorithm Derivation

In this section, we simplify the optimization problem (18) and obtain an algorithm that's based on exploiting the eigen-structure of a sparse matrix.

### 3.1 Main Theorem

Suppose $\widehat{\mathbf{a}}_i \in \mathbb{R}^{n\times 1}$ denotes the transpose of the $i$-th row vector of $\mathbf{A}$, so $\mathbf{A} = [\widehat{\mathbf{a}}_1, \widehat{\mathbf{a}}_2, \ldots, \widehat{\mathbf{a}}_n]^T$. Then we have

$$\mathcal{J}(\mathbf{Y}) = \|\mathbf{Y} - \mathbf{A}\mathbf{Y}\|_{\text{SAV}} = \sum_{l=1}^{c}\sum_{i=1}^{n} |y_{il} - \widehat{\mathbf{a}}_i^T \mathbf{y}_l| \tag{20}$$

Substituting (2) into (20), we have

$$\mathcal{J}(\mathbf{Y}) = \sum_{l=1}^{c}\left(\sum_{\mathbf{x}_i\in\pi_l} \frac{|1 - \sum_{\mathbf{x}_j\in\pi_l} a_{ij}|}{|\pi_l|} + \sum_{\mathbf{x}_i\notin\pi_l} \frac{|0 - \sum_{\mathbf{x}_j\in\pi_l} a_{ij}|}{|\pi_l|}\right) \tag{21}$$

According to the properties of $\mathbf{A}$ in (13) and (14), we obtain

$$\begin{aligned}
\mathcal{J}(\mathbf{Y}) &= \sum_{l=1}^{c}\left(\sum_{\mathbf{x}_i\in\pi_l} \frac{\sum_{\mathbf{x}_j\notin\pi_l} a_{ij}}{|\pi_l|} + \sum_{\mathbf{x}_i\notin\pi_l} \frac{\sum_{\mathbf{x}_j\in\pi_l} a_{ij}}{|\pi_l|}\right) && (22)\\
&= \sum_{l=1}^{c} \frac{\sum_{\mathbf{x}_i\in\pi_l}\sum_{\mathbf{x}_j\notin\pi_l}(a_{ij}+a_{ji})}{|\pi_l|} && (23)\\
&= \sum_{l=1}^{c} \frac{\mathbf{p}_l^T\left(\text{Deg}(\mathbf{A}+\mathbf{A}^T) - \mathbf{A} - \mathbf{A}^T\right)\mathbf{p}_l}{|\pi_l|} && (24)
\end{aligned}$$

where $\mathbf{p}_l$ is the $l$-th column of partition matrix $\mathbf{P}$ defined in (1).

Define $\mathbf{F} = [\mathbf{f}_1, \ldots, \mathbf{f}_c] = [f_{il}] \in \mathbb{R}^{n\times c}$ as $\mathbf{F} = \mathbf{P}(\mathbf{P}^T\mathbf{P})^{-\frac{1}{2}}$, so $\mathbf{f}_l = \mathbf{p}_l(\mathbf{p}_l^T\mathbf{p}_l)^{-\frac{1}{2}}$. Specifically,

$$f_{il} = \begin{cases} \sqrt{\frac{1}{|\pi_l|}} & \text{if } \mathbf{x}_i \in \pi_l, \\ 0 & \text{if } \mathbf{x}_i \notin \pi_l. \end{cases} \tag{25}$$

Obviously, we have

$$\mathbf{F}^T\mathbf{F} = \left((\mathbf{P}^T\mathbf{P})^{-\frac{1}{2}}\mathbf{P}^T\right)\mathbf{P}(\mathbf{P}^T\mathbf{P})^{-\frac{1}{2}} = \mathbf{I}_c \tag{26}$$

where $\mathbf{I}_c \in \mathbb{R}^{c\times c}$ is the identity matrix of order $c$.

Using $\mathbf{F}$, the clustering objective can be simplified as

$$\begin{aligned}
\mathcal{J}(\mathbf{F}) &= \sum_{l=1}^{c} \mathbf{f}_l^T\left(\text{Deg}(\mathbf{A}+\mathbf{A}^T) - \mathbf{A} - \mathbf{A}^T\right)\mathbf{f}_l && (27)\\
&= \text{Trace}\left(\mathbf{F}^T\left(\text{Deg}(\mathbf{A}+\mathbf{A}^T) - \mathbf{A} - \mathbf{A}^T\right)\mathbf{F}\right) && (28)
\end{aligned}$$

Therefore, we obtain the following theorem:

**Table 2.** Clustering via LOcal Regression (CLOR)

**Input** :
Data set $\{\mathbf{x}_i\}_{i=1}^n$, number of clusters $c$, neighborhood size $k$.
**Output** :
Partition matrix $\mathbf{P}$ as defined in (1).
**Procedure** :
1. Compute the $k$ nearest neighbors ($\mathcal{N}_i$) for each $\mathbf{x}_i$.
2. Construct the matrix $\mathbf{A}$ as defined in (12).
3. Construct the matrix $\mathbf{M} = \text{Deg}(\mathbf{A} + \mathbf{A}^T) - \mathbf{A} - \mathbf{A}^T$.
4. Compute the eigenvectors corresponding to the $c$ smallest eigenvalues of $\mathbf{M}$, thus obtaining $\mathbf{F}^*$.
5. Discretize $\mathbf{F}^*$ to get the partition matrix $\mathbf{P}$.

**Theorem 1.** *The optimization problem in* (18) *is equivalent to the following one:*

$$\min_{\mathbf{F}\in\mathbb{R}^{n\times c}} \quad \text{Trace}\left(\mathbf{F}^T\left(\text{Deg}(\mathbf{A}+\mathbf{A}^T) - \mathbf{A} - \mathbf{A}^T\right)\mathbf{F}\right) \tag{29}$$

$$\text{subject to} \quad \mathbf{F} \textit{ is defined in } (25) \tag{30}$$

### 3.2 Relaxation and Discretization

Following the standard spectral clustering procedure, we relax the $\mathbf{F}$ defined in (25) to be any matrix from $\mathbb{R}^{n\times c}$ that satisfies $\mathbf{F}^T\mathbf{F} = \mathbf{I}_c$. Thus, the optimization problem is as follows:

$$\min_{\mathbf{F}\in\mathbb{R}^{n\times c}} \quad \text{Trace}\left(\mathbf{F}^T\left(\text{Deg}(\mathbf{A}+\mathbf{A}^T) - \mathbf{A} - \mathbf{A}^T\right)\mathbf{F}\right) \tag{31}$$

$$\text{subject to} \quad \mathbf{F}^T\mathbf{F} = \mathbf{I}_c \tag{32}$$

From the *Ky Fan* Theorem [23], the global optimal solution of the above relaxed problem is given by any matrix from the following set

$$\{\mathbf{F}^*\mathbf{Q} \mid \mathbf{Q} \in \mathbb{R}^{c\times c}, \mathbf{Q}^T\mathbf{Q} = \mathbf{I}_c\} \tag{33}$$

where the columns of $\mathbf{F}^* \in \mathbb{R}^{n\times c}$ are the $c$ eigenvectors corresponding to the $c$ smallest eigenvalues of the matrix $\text{Deg}(\mathbf{A}+\mathbf{A}^T) - \mathbf{A} - \mathbf{A}^T$.

After obtaining the relaxed solution, we have to discretize it to get a final solution defined in (1). The discretization approach used in [22] is adopted to obtain the final partition matrix $\mathbf{P}$ since it's previously reported to produce satisfactory results [22,19]. Using an iterative procedure, this discretization method tries to rotate $\mathbf{F}^*$ (after normalizing the rows of $\mathbf{F}^*$ to unit norm) so that it's close to a partition matrix as defined in (1). Details can be found in [22].

The main algorithm is summarized in Table 2. We name our algorithm Clustering via LOcal Regression (CLOR).

### 3.3 Relation to Other Approaches

Although the final optimization problem uses the spectral methods for finding the optimal clustering, our clustering criterion is motivated from a local learning approach. In our method, we adopt sum of absolute error as the discrepancy measure instead of sum of squared error which is used in [19]. In addition, in order for the optimization to be tractable, we use kernel regression as the local label predictor while [19] uses "kernel ridge regression". We will empirically demonstrate that our clustering algorithm usually achieves better performance than the algorithm proposed in [19].

Our algorithm is different from spectral clustering with ratio cut [6] or normalized cut (NCut) [11] on the $k$-nearest neighbor similarity graph. We construct the following $k$-nearest neighbor similarity matrix $\mathbf{G} = [g_{ij}] \in \mathbb{R}^{n \times n}$

$$g_{ij} = \begin{cases} K(\mathbf{x}_i, \mathbf{x}_j) & \text{if } \mathbf{x}_j \in \mathcal{N}_i \text{ or } \mathbf{x}_i \in \mathcal{N}_j \\ 0 & \text{otherwise} \end{cases} \tag{34}$$

Spectral clustering algorithms typically use the affinity matrix $\mathbf{G}$ which is symmetric. In general, the graph laplacian (used by ratio cut) or normalized graph laplacian (used by NCut) derived from this affinity matrix are not equal to $\mathrm{Deg}(\mathbf{A} + \mathbf{A}^T) - \mathbf{A} - \mathbf{A}^T$. Since NCut usually outperforms ratio cut in practice, we compare our algorithm CLOR with NCut in the next section and demonstrate the better performance of our algorithm.

## 4 Experiments

In this section, we present an empirical evaluation of our clustering method in comparison with representative algorithms on a number of data sets.

### 4.1 Data Sets

In this subsection, we introduce the basic information of the data sets used in our experiments. We use 14 document data sets[1] from the CLUTO toolkit [24].

We briefly introduce the basic information of the data sets as follows.

- *cranmed*: This data set comprises the CRANFIELD and MEDLINE abstracts which were previously used to evaluate information retrieval systems.
- *fbis*: The *fbis* data set is derived from the Foreign Broadcast Information Service data of TREC-5 [15].
- *hitech*: This data set is from the San Jose Mercury newspaper articles that are distributed as part of the TREC collection.
- *k1a*, *k1b* and *wap*: These data set are from the WebACE project (WAP).
- *re0* and *re1*: These data sets are from Reuters-21578 text collection [9].
- *tr11*, *tr12*, *tr23*, *tr31*, *tr41*, and *tr45*: These six data sets are derived from the TREC collection [15].

1 `http://glaros.dtc.umn.edu/gkhome/fetch/sw/cluto/datasets.tar.gz`

**Table 3.** Summary of data sets

| Name | Source | Points($n$) | Dims($d$) | Classes($c$) |
|---|---|---|---|---|
| *cranmed* | CRANFIELD/MEDLINE | 2431 | 41681 | 2 |
| *fbis* | FBIS (TREC) | 2463 | 12674 | 17 |
| *k1a* | WebACE | 2340 | 21839 | 20 |
| *k1b* | WebACE | 2340 | 21839 | 6 |
| *hitech* | San Jose Mercury (TREC) | 2301 | 10080 | 6 |
| *re0* | Reuters-21578 | 1504 | 2886 | 13 |
| *re1* | Reuters-21578 | 1657 | 3758 | 25 |
| *tr11* | TREC | 414 | 6429 | 9 |
| *tr12* | TREC | 313 | 5804 | 8 |
| *tr23* | TREC | 204 | 5832 | 6 |
| *tr31* | TREC | 927 | 10128 | 7 |
| *tr41* | TREC | 878 | 7454 | 10 |
| *tr45* | TREC | 690 | 8261 | 10 |
| *wap* | WebACE | 1560 | 8460 | 20 |

Table 3 summarizes the basic properties of the data sets. The smallest of these data sets contains 204 data points and the largest consists of 2463 points. The number of classes ranges from 2 to 25. These data sets are widely used in the literature to evaluate different clustering systems [14,24,25].

## 4.2 Evaluation Criteria

In all the experiments of this paper, the class labels of data points are known to the evaluation process and external validity measures are used to evaluate how much class structure is recovered by a clustering. Besides, the true number of classes $c$ is provided to all the clustering algorithms. We use two popular external validity measures, Normalized Mutual Information (*NMI*) [12] and Clustering Accuracy (*Acc*), as our criteria.

Given a clustering $\mathcal{C}$ and the "true" partitioning $\mathcal{B}$ (class labels). The number of clusters in $\mathcal{C}$ and classes in $\mathcal{B}$ are both $c$. Suppose $n_i$ is the number of objects in the $i$-th cluster, $n_j^{'}$ is the number of objects in the $j$-th class and $n_{ij}$ is the number of objects which are both in the $i$-th cluster and $j$-th class. *NMI* between $\mathcal{C}$ and $\mathcal{B}$ is calculated as follows [12]:

$$NMI(\mathcal{C},\mathcal{B}) = \frac{\sum_{i=1}^{c}\sum_{j=1}^{c} n_{ij}\log\frac{n\cdot n_{ij}}{n_i\cdot n_j^{'}}}{\sqrt{\sum_{i=1}^{c} n_i\log\frac{n_i}{n}\sum_{j=1}^{c} n_j^{'}\log\frac{n_j^{'}}{n}}}. \tag{35}$$

The value of *NMI* equals 1 if and only if $\mathcal{C}$ and $\mathcal{B}$ are identical and is close to 0 if $\mathcal{C}$ is a random partitioning. Larger values of *NMI* indicate better clustering performance.

Suppose $n$ is the total number of objects and other notations are the same as above. Clustering Accuracy (*Acc*) builds a one-to-one correspondence between the clusters and the classes. Suppose the permutation function $\mathrm{Map}(\cdot) : \{i\}_{i=1}^{c} \mapsto \{j\}_{j=1}^{c}$ maps each cluster index to a class index, i.e., $\mathrm{Map}(i)$ is the class index that corresponds to the $i$-th cluster. $Acc$ between $\mathcal{C}$ and $\mathcal{B}$ is calculated as follows:

$$Acc(\mathcal{C}, \mathcal{B}) = \frac{\max\left(\sum_{i=1}^{c} n_{i,\mathrm{Map}(i)}\right)}{n} \tag{36}$$

The value of $Acc$ equals 1 if and only if $\mathcal{C}$ and $\mathcal{B}$ are identical. Larger values of $Acc$ indicate better clustering performance.

### 4.3 Comparison Settings

Each data point (document) is originally represented by a term-frequency vector (Bag-of-Words). We normalize each vector to unit norm so that every point $\mathbf{x}_i$ lies on the nonnegative unit hypersphere, that is, $\mathbf{x}_i \in \mathcal{X} = \{\mathbf{x} \,|\, \mathbf{x}^T\mathbf{x} = 1, \mathbf{x} \geq 0\}$ for all $1 \leq i \leq n$. The cosine kernel defined in (10) is generally considered very suitable for documents. So it is adopted as the kernel function in most experiments unless other kernels are explicitly mentioned to be used. For each point $\mathbf{x}_i$, the neighborhood $\mathcal{N}_i$ consists of $k$-nearest neighbors of $\mathbf{x}_i$ measured by the cosine similarity. Here, $k$ is provided by domain experts.

We test our Clustering via LOcal Regression (CLOR) algorithm in comparison with another two algorithms which are listed as follows:

- Spectral clustering with normalized cut (NCut) [11]. The affinity graph is constructed using the weighted $k$-nearest neighbor graph. The edge weight between two connected data points is calculated with the kernel function.
- Local Learning based Clustering Algorithm[2] (LLCA) proposed in [19]. There is a regularization parameter $\lambda$ in LLCA. As is done in [19], we choose this parameter from: $\lambda \in \{0.1, 1, 1.5\}$. For each data set and $k$, we report the best performance among the results produced when different values of $\lambda$ are used (LLCA1). We also report the result obtained when LLCA automatically select $\lambda$ using the parameter selection method in [19] (LLCA2). Therefore, LLCA1 has an unfair advantage over others.

For each algorithm, we assume that the "true" number of clusters $c$ is given. All the algorithms use the same discretization method whose code is available at `http://www.cis.upenn.edu/~jshi/software/`. Note that the main computational load of all the above algorithms is to eigen-decompose (or singular value decomposition) a sparse $n \times n$ matrix with $O(nk)$ non-zero elements.

### 4.4 Comparison of Clustering Performance

In this section, we compare the clustering performance of the investigated algorithms. We have tested the clustering performance of the algorithms with various neighborhood sizes.

[2] The code is available at `www.kyb.tuebingen.mpg.de/bs/people/mingrui/LLCA.zip` The code for NCut is also included in this package.

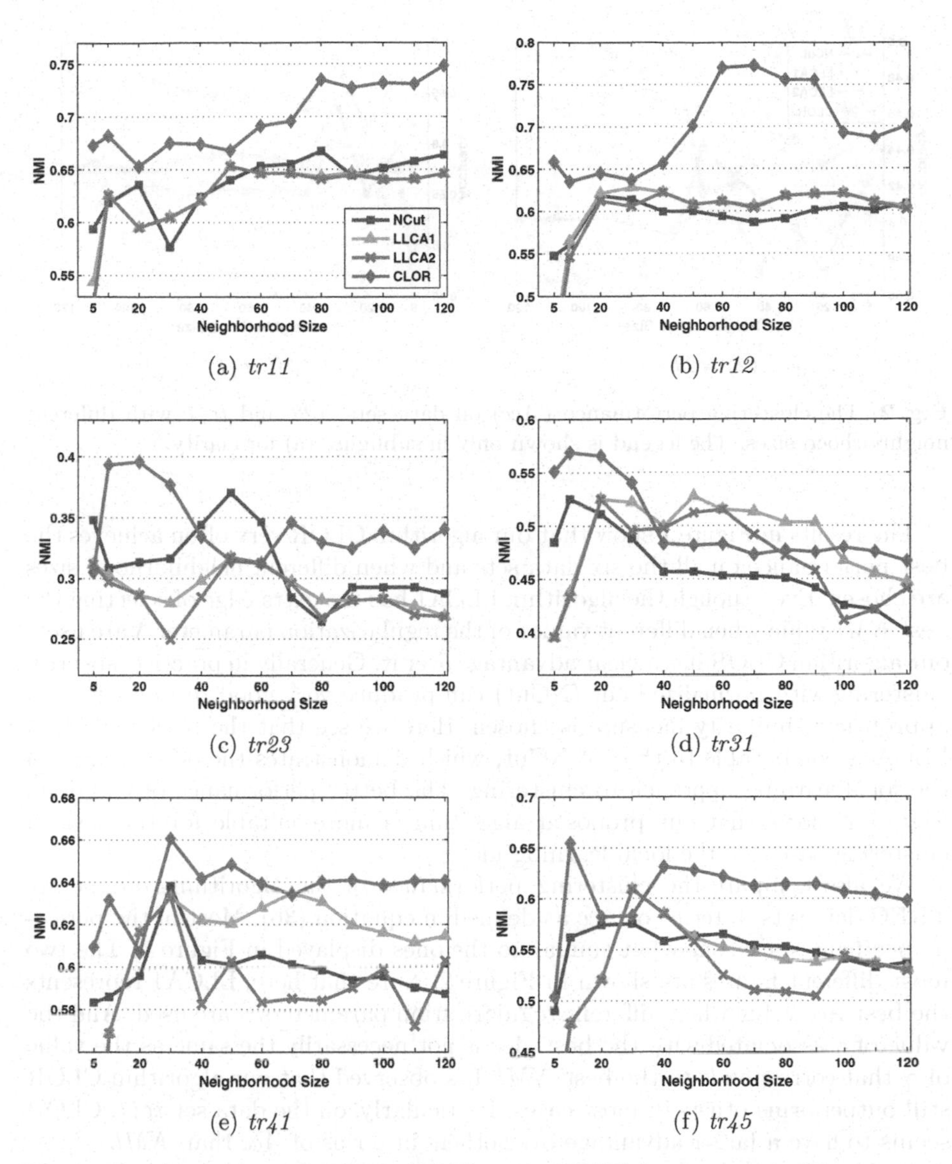


**Fig. 1.** The clustering performance (*NMI*) on six TREC data sets with different number of nearest neighbors ($k$). The legend is shown only in subfigure (a) for clarity.

For the six TREC data sets (*tr11*, *tr12*, *tr23*, *tr31*, *tr41*, and *tr45*), Figure 1 shows the the clustering performance of the algorithms. The $x$-axis denotes the neighborhood size $k$ and the $y$-axis denotes the clustering performance measured by *NMI*. When the neighborhood size is too small ($k = 5$), LLCA1 and LLCA2 produce unsatisfactory clustering results on data sets such as *tr12*, *tr41* and *tr45*. These *NMI* results are not displayed in the subfigures for clarity.

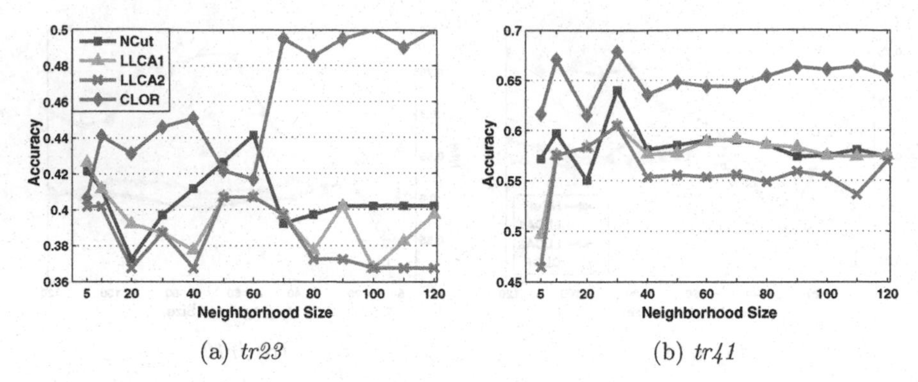


(a) *tr23* (b) *tr41*

**Fig. 2.** The clustering performance (*Acc*) on data sets *tr23* and *tr41* with different neighborhood sizes. The legend is shown only in subfigure (a) for clarity.

The results in Figure 1 show that our algorithm CLOR very often achieves the best performance on all the six data sets and when different neighborhood sizes are chosen. Even though the algorithm LLCA1 has the extra edge of selecting the best *NMI* value when different values of the regularization parameter $\lambda$ are used, our algorithm CLOR has a clear advantage over it. Generally in practice, spectral clustering with normalized cut (NCut) can produce very good results when an appropriate similarity measure is chosen. Here we see that the performance of LLCA is comparable to that of NCut, which demonstrates the effectiveness of the local learning approach to clustering. The better performance of CLOR in Figure 1 shows that our proposed algorithm is more suitable for the task of clustering based on the local learning idea.

We also compare the clustering performance of the algorithms on the six TREC data sets in terms of *Acc* as defined in equation (36). Most of the performance figures are somewhat similar to the ones displayed in Figure 1. The two most different figures are shown in Figure 2. Note that here, LLCA1 represents the best *Acc* value when different regularization parameters $\gamma$ are used. And the value of $\gamma$ associated with the best *Acc* is not necessarily the same as the value of $\gamma$ that corresponds to the best *NMI*. It's observed that our algorithm CLOR still outperforms others in most cases. Particularly, on the data set *tr41*, CLOR seems to have a larger advantage over others in terms of *Acc* than *NMI*.

Besides, we conduct experiments on the six TREC data sets when Gaussian kernel is used instead of cosine kernel. Due to the space limit, we only show the results on data set *tr11* in Figure 3. The left subfigure shows the *NMI* values of the algorithms on *tr11* when different neighborhood sizes $k$ are chosen. Here the algorithms use Gaussian kernel ($\gamma = 1$). The right subfigure displays the performance of CLOR with the two kernels on data set *tr11*. It can be observed from the left subfigure that our algorithm still outperforms the other three. The right subfigure shows that cosine kernel is slightly better than Gaussian kernel for document clustering. However, the result from Gaussian kernel doesn't deviate too much from the result with cosine kernel on *tr11*. Experiments with Gaussian

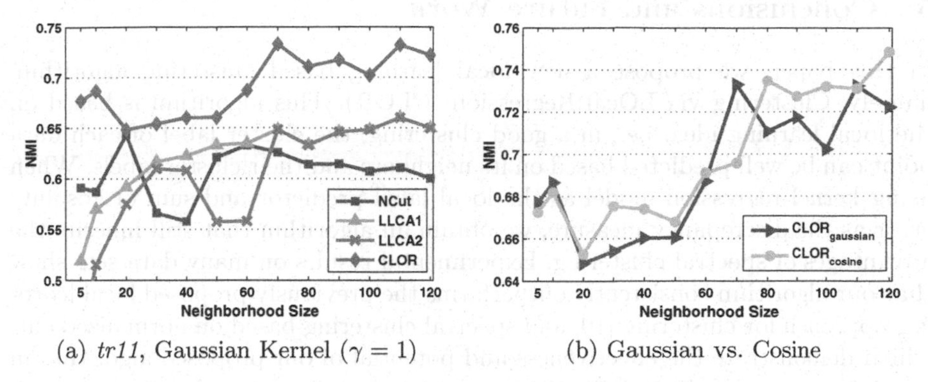


(a) *tr11*, Gaussian Kernel ($\gamma = 1$) (b) Gaussian vs. Cosine

**Fig. 3.** The left subfigure shows the clustering performance (*NMI*) of the considered algorithms on data set *tr11*. Here Gaussian kernel ($\gamma = 1$) is used. The right subfigure displays the performance of CLOR with different kernels on data set *tr11*.

kernel have also been conducted on the other five TREC data sets. From these experiments, we can draw similar conclusions as above. Therefore, the superiority of our algorithm CLOR is not sensitive to what type of kernel is used.

In general, theoretical guidance on how to choose the best neighborhood size $k$ is rare. A rule of thumb is to choose $k$ so that the number of connected components of the $k$-nearest neighbor graph is close to one. An asymptotic result is that if $k$ is chosen on the order of $\log(n)$, then the $k$-nearest neighbor graph will be connected for $n$ points drawn i.i.d. from some probability distribution with a connected support in $\mathbb{R}^d$ [5]. Therefore, for data sets with size around 1000, we can choose $k$ to be a small multiple of 10. We list the experimental results on the remaining 8 data sets when neighborhood size $k = 30$ in Table 4. It can be observed that our algorithm CLOR consistently outperforms the other three. We also note that different performance measures may lead to divergent performance ranks of the clustering algorithms, which is reflected by the results on the data sets *re0* and *re1*.

**Table 4.** Empirical results when neighborhood size $k = 30$. Both *NMI* and *Acc* results are provided. On each data set, the best results of *NMI* and *Acc* are shown in boldface.

| | *NMI* | | | | *Acc* | | | |
|---|---|---|---|---|---|---|---|---|
| | LLCA1 | LLCA2 | NCut | CLOR | LLCA1 | LLCA2 | NCut | CLOR |
| *cranmed* | 0.8479 | 0.8479 | 0.8568 | **0.8927** | 0.9745 | 0.9745 | 0.9770 | **0.9840** |
| *fbis* | 0.5816 | 0.5757 | 0.5832 | **0.5909** | 0.5108 | 0.5108 | 0.5428 | **0.5534** |
| *hitech* | 0.3373 | 0.3373 | 0.2956 | **0.3439** | 0.4920 | 0.4920 | 0.4207 | **0.5502** |
| *k1a* | 0.5267 | 0.5267 | 0.5223 | **0.5557** | 0.4060 | 0.4060 | 0.3987 | **0.4897** |
| *k1b* | 0.6416 | 0.6416 | 0.7180 | **0.7332** | 0.8120 | 0.8120 | 0.8466 | **0.8846** |
| *re0* | 0.3905 | 0.3847 | 0.4030 | **0.4302** | **0.3863** | 0.3138 | 0.3324 | 0.3318 |
| *re1* | 0.4942 | 0.4598 | 0.4967 | **0.5043** | **0.3959** | 0.3693 | 0.3730 | 0.3953 |
| *wap* | 0.5258 | 0.5093 | 0.5173 | **0.5426** | 0.3962 | 0.3333 | 0.3859 | **0.4314** |

## 5 Conclusions and Future Work

In this paper, we propose a new local learning based clustering algorithm, namely, Clustering via LOcal Regression (CLOR). This algorithm is based on the local learning idea, i.e., in a good clustering, the cluster label of each data point can be well predicted based on its neighbors and their cluster labels. When using kernel regression model as the local label predictor and sum of absolute error as the discrepancy measure, we obtain an algorithm that still inherits the advantages of spectral clustering. Experimental results on many data sets show that our algorithm consistently outperforms the previously proposed local learning approach for clustering [19] and spectral clustering based on normalized cut, which demonstrate the effectiveness and potential of our proposed algorithm in obtaining accurate clusterings.

In the future, we want to continue this work on several issues as follows. First, we want to gain a deeper understanding on the underlying reasons for the good performance of our algorithm. We will start this work by investigating the capacity and locality control issues of the local label predictor in our approach and the one in [19]. Second, we will try to derive a good and stable automatic parameter selection procedure for neighborhood size $k$ and parameters in the kernel function. Third, instead of using $k$-nearest neighbors measured by a distance metric which is provided by domain experts, we plan to learn the neighborhood $\mathcal{N}_i$ using semi-supervised information such as instance-level constraints [14]. This problem seems to be easier than learning a distance metric and thus hopefully we'll obtain good results.

**Acknowledgments.** This work is supported in part by NSFC grants 60673103 and 60721061.

## References

1. Benedetti, J.K.: On the Nonparametric Estimation of Regression Functions. Journal of the Royal Statistical Society 39(2), 248–253 (1977)
2. Bishop, C.M.: Pattern Recognition and Machine Learning. Springer, New York (2006)
3. Bottou, L., Vapnik., V.: Local learning algorithms. Neural Computation 4(6), 888–900 (1992)
4. Boyd, S., Vandenberghe, V.: Convex Optimization. Cambridge University Press, Cambridge (2004)
5. Brito, M., Chavez, E., Quiroz, A., Yukich, J.: Connectivity of the Mutual K-nearest-neighbor Graph in Clustering and Outlier Detection. Statistics and Probability Letters 35(1), 33–42 (1997)
6. Chan, P.K., Schlag, M.D.F., Zien, J.Y.: Spectral K-way Ratio-cut Partitioning and Clustering. IEEE Transactions on Computer-Aided Design of Integrated Circuits and Systems 13(9), 1088–1096 (1994)
7. Ding, C., He, X., Zha, H., Gu, M., Simon, H.D.: A Min-max Cut Algorithm for Graph Partitioning and Data Clustering. In: Proceedings of the 2001 IEEE International Conference on Data Mining (2001)

8. Jain, A.K., Dubes, R.C.: Algorithms for Clustering Data. Prentice-Hall, Englewood Cliffs (1988)
9. Lewis, D.D.: Reuters-21578 text categorization test collection, http://www.daviddlewis.com/resources/testcollections/reuters21578/
10. Nadaraya, E.A.: On Estimating Regression. Theory of Probability and its Applications 9(1), 141–142 (1964)
11. Shi, J., Malik, J.: Normalized Cuts and Image Segmentation. IEEE Transactions on Pattern Analysis and Machine Intelligence 22(8), 888–905 (2000)
12. Strehl, A., Ghosh, J.: Cluster Ensembles — A Knowledge Reuse Framework for Combining Multiple Partitions. Journal of Machine Learning Research 3, 583–617 (2002)
13. Takeda, H., Farsiu, S., Milanfar, P.: Kernel Regression for Image Processing and Reconstruction. IEEE Transactions on Image Processing 16(2), 349–366 (2007)
14. Tang, W., Xiong, H., Zhong, S., Wu, J.: Enhancing Semi-Supervised Clustering: A Feature Projection Perspective. In: Proceedings of the 13th ACM SIGKDD International Conference on Knowledge Discovery and Data Mining (2007)
15. TREC: Text REtrieval Conference, http://trec.nist.gov
16. Wang, F., Zhang, C., Li, T.: Clustering with Local and Global Regularization. In: Proceedings of the Twenty-Second AAAI Conference on Artificial Intelligence (2007)
17. Wang, F., Zhang, C., Li, T.: Regularized Clustering for Documents. In: Proceedings of the 30th Annual International ACM SIGIR Conference on Research and Development in Information Retrieval (2007)
18. Weinberger, K.Q., Tesauro, G.: Metric Learning for Kernel Regression. In: Proceedings of the Eleventh International Conference on Artificial Intelligence and Statistics (2007)
19. Wu, M., Schölkopf, B.: A Local Learning Approach for Clustering. In: Advances in Neural Information Processing Systems 19 (2006)
20. Wu, M., Schölkopf, B.: Transductive Classification via Local Learning Regularization. In: Proceedings of the Eleventh International Conference on Artificial Intelligence and Statistics (2007)
21. Wu, M., Yu, K., Yu, S., Schölkopf, B.: Local Learning Projections. In: Proceedings of the Twenty-Fourth International Conference on Machine Learning (2007)
22. Yu, S.X., Shi, J.: Multiclass Spectral Clustering. In: Proceedings of the 9th International Conference On Computer Vision (2003)
23. Zha, H., He, X., Ding, C., Gu, M., Simon, H.D.: Spectral Relaxation for K-means Clustering. In: Advances in Neural Information Processing Systems 14 (2001)
24. Zhao, Y., Karypis, G.: Empirical and Theoretical Comparisons of Selected Criterion Functions for Document Clustering. Machine Learning 55, 311–331 (2004)
25. Zhong, S., Ghosh, J.: A Unified Framework for Model-based Clustering. Journal of Machine Learning Research 4, 1001–1037 (2003)
26. Zhu, X., Goldberg, A.: Kernel Regression with Order Preferences. In: Proceedings of the Twenty-Second AAAI Conference on Artificial Intelligence (2007)

# Decomposable Families of Itemsets

Nikolaj Tatti and Hannes Heikinheimo

HIIT Basic Research Unit, Department of Information and Computer Science,
Helsinki University of Technology, Finland
`ntatti@cc.hut.fi`, `hannes.heikinheimo@tkk.fi`

**Abstract.** The problem of selecting a small, yet high quality subset of patterns from a larger collection of itemsets has recently attracted a lot of research. Here we discuss an approach to this problem using the notion of decomposable families of itemsets. Such itemset families define a probabilistic model for the data from which the original collection of itemsets was derived. Furthermore, they induce a special tree structure, called a junction tree, familiar from the theory of Markov Random Fields. The method has several advantages. The junction trees provide an intuitive representation of the mining results. From the computational point of view, the model provides leverage for problems that could be intractable using the entire collection of itemsets. We provide an efficient algorithm to build decomposable itemset families, and give an application example with frequency bound querying using the model. An empirical study show that our algorithm yields high quality results.

## 1 Introduction

Frequent itemset discovery has been a central research theme in the data mining community ever since the idea was introduced by Agrawal et. al [1]. Over the years, scalability of the problem has been the most studied aspect, and several very efficient algorithms for finding all frequent itemsets have been introduced, Apriori [2] or FP-growth [3] among others. However, it has been argued recently that while efficiency of the mining task is no longer a bottleneck, there is still a strong need for methods that derive compact, yet high quality results with good application properties [4].

In this study we propose the notion of decomposable families of itemsets to address this need. The general idea is to build a probabilistic model of a given dataset $D$ using a small well-chosen subset of itemsets $\mathcal{G}$ from a given candidate family $\mathcal{F}$. The candidate family $\mathcal{F}$ may be generated from $D$ using some frequent itemset mining algorithm. A special aspect of a decomposable family is that it induces a type of tree, called a junction tree, a well-known concept from the theory of Markov Random Fields [5].

As a simple example, consider a dataset $D$ with six attributes $a, \ldots, f$, and a family $\mathcal{G} = \{bcd, bcf, ab, ce, bc, bd, cd, bf, cf, a, b, c, d, e, f\}$. The family $\mathcal{G}$ can be represented as the junction tree shown in Figure 1 such that the nodes in the tree are the maximal itemsets in $\mathcal{G}$. Furthermore, the junction tree defines a decomposable model of the dataset $D$.

W. Daelemans et al. (Eds.): ECML PKDD 2008, Part II, LNAI 5212, pp. 472–487, 2008.

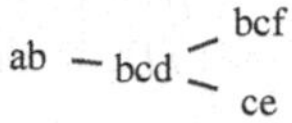

$$p(abcdef) = \frac{p(ab)p(bcd)p(bcf)p(ce)}{p(b)p(bc)p(c)}$$

**Fig. 1.** An example of a junction tree and the corresponding distribution decomposition

Using decomposable itemset families has several notable advantages. First of all, the following junction tree graphs provide an extremely intuitive representation of the mining results. This is a significant advantage over many other itemset selection methods, as even small mining results of, say 50 itemsets, can be hard for humans to grasp as a whole, if just plainly enumerated. Second, from the computational point of view, decomposable families of itemsets provide leverage for accurately solving problems that could be intractable using the entire result set. Such problems include, for instance, querying for frequency bounds of arbitrary attribute combinations. Third, the statistical nature of the overall model enable to incorporated regularization terms, like BIC, AIC, or MDL to select only itemsets that reflect true dependencies between attributes.

In this study we provide an efficient algorithm to build decomposable itemset families while optimizing the likelihood of the model. We also demonstrate how to use decomposable itemset families to execute frequency bound querying, an intractable problem in the general case. We provide empirical results showing that our algorithm works well in practice.

The rest of the paper is organized as follows. Preliminaries are given in Section 2 and the concept of decomposable models are defined in Section 3. A greedy search algorithm is given in Section 4. Section 6 is devoted to experiments. We present the related work in Section 7 and conclude the paper with discussion in Section 8. The proofs for the theorems in this paper are provided in [6].

## 2 Preliminaries and Notations

In this section we describe the notation and the background definitions that are used in the subsequent sections.

A *binary dataset* $D$ is a collection of $N$ *transactions*, binary vectors of length $K$. The dataset can be viewed as a binary matrix of size $N \times K$. We denote the number of transactions by $|D| = N$. The $i$th element of a random transaction is represented by an *attribute* $a_i$, a Bernoulli random variable. We denote the collection of all the attributes by $A = \{a_1, \ldots, a_K\}$. An *itemset* $X = \{x_1, \ldots, x_L\} \subseteq A$ is a subset of attributes. We will often use the dense notation $X = x_1 \cdots x_L$.

Given an itemset $X$ and a binary vector $v$ of length $L$, we use the notation $p(X = v)$ to express the probability of $p(x_1 = v_1, \ldots, x_L = v_L)$. If $v$ contains only 1s, then we will use the notation $p(X = 1)$.

Given a binary dataset $D$ we define $q_D$ to be an *empirical distribution*,

$$q_D(A = v) = |\{t \in D; t = v\}|/|D|.$$

We define the frequency of an itemset to be $fr(X) = q_D(X = 1)$. The *entropy* of an itemset $X = x_1 \cdots x_L$ given $D$, denoted by $H(X; D)$, is defined as

$$H(X; D) = - \sum_{v \in \{0,1\}^L} q_D(X = v) \log q_D(X = v), \tag{1}$$

where the usual convention $0 \log 0 = 0$ is used. We often omit $D$.

We say that a family $\mathcal{F}$ of itemsets is *downward closed* if each subset of a member of $\mathcal{F}$ is also included in $\mathcal{F}$. An itemset $X \in \mathcal{F}$ is maximal if there is no $Y \in \mathcal{F}$ such that $X \subset Y$. We define $m(\mathcal{F}) = \{|X|; X \in \mathcal{F}\}$ to be the maximal number of attributes in a single itemset.

## 3 Decomposable Families of Itemsets

In this section we define the concept of decomposable families. Itemsets of a decomposable family form a junction tree, a concept from the theory of Markov Random Fields [5].

Let $\mathcal{G} = \{G_1, \ldots, G_M\}$ be a downward closed family of itemsets covering the attributes $A$. Let $H$ be a graph containing $M$ nodes where the $i$th node corresponds to the itemset $G_i$. Nodes $G_i$ and $G_j$ are connected if $G_i$ and $G_j$ have a common attribute. The graph $H$ is called the *clique graph* and the nodes of $H$ are called *cliques.*

We are interested in spanning trees of $H$ having a *running intersection property.* To define this property let $\mathcal{T}$ be a spanning tree of $H$. Let $G_i$ and $G_j$ be two sets having a common attribute, say, $a$. These sets are connected in $\mathcal{T}$ by a unique path. Assume that $a$ occurs in every $G_k$ along the path from $G_i$ to $G_j$. If this holds for any $G_i$, $G_j$, and any common attribute $a$, then we say that the tree has a running intersection property. Such a tree is called a *junction tree.*

We should point out that the clique graph can have multiple junction trees and that not all spanning trees are junction trees. In fact, it may be the case that the clique graph does not have junction trees at all. If, however, the clique graph has a junction tree, we call the original family $\mathcal{G}$ *decomposable.*

We label edge $(G_i, G_j)$ of a given junction tree $\mathcal{T}$ with a set of mutual attributes $G_i \cap G_j$. This label set is called a *separator.* We denote the set of all separators by $S(\mathcal{T})$. Furthermore, we denote the cliques of the tree by $V(\mathcal{T})$.

Given a junction tree $\mathcal{T}$ and a binary vector $v$, we define the probability of $A = v$ to be

$$p(A = v; \mathcal{T}) = \prod_{X \in V(\mathcal{T})} q_D(X = v_X) \Big/ \prod_{Y \in S(\mathcal{T})} q_D(Y = v_Y). \tag{2}$$

It is a known fact that the distribution given in Eq. 2 is actually the unique maximum entropy distribution [7,8]. Note that $p(A = v; \mathcal{T})$ can be computed from the frequencies of the itemsets in $\mathcal{G}$ using the inclusion-exclusion principle.

It can be shown that the family $\mathcal{G}$ is decomposable if and only if the maximal sets of $\mathcal{G}$ is decomposable and that Eq. 2 for the maximal sets of $\mathcal{G}$ and the whole $\mathcal{G}$. Hence, we usually construct the tree using only the maximal sets of $\mathcal{G}$. However, in some cases it is convenient to have non-maximal sets as the cliques. We will refer to such cliques as *redundant*. For a tree $\mathcal{T}$ we define a family of itemsets, $s(\mathcal{T})$ to be the downward closure of its cliques, $V(\mathcal{T})$. To summarize, $\mathcal{G}$ is decomposable if and only if there is a junction tree $\mathcal{T}$ such that $\mathcal{G} = s(\mathcal{T})$.

Calculating the entropy of the tree $\mathcal{T}$ directly from Eq. 2 gives us

$$H(\mathcal{T}) = \sum_{X \in V(\mathcal{T})} H(X) - \sum_{Y \in S(\mathcal{T})} H(Y).$$

A direct calculation from Eqs. 1–2 reveals that $\log p(D; \mathcal{T}) = -|D| H(\mathcal{T})$. Hence, maximizing the log-likelihood of the data given $\mathcal{T}$ (whose components are derived from the same data), is equivalent to minimizing the entropy.

We can define the maximum entropy distribution for any cover $\mathcal{F}$ via linear constraints [8]. The downside of this general approach is that solving such a distribution is a **PP**-hard problem [9].

The following definition will prove itself useful in subsequent sections. Given two downward closed covers $\mathcal{G}_1$ and $\mathcal{G}_2$. We say that $\mathcal{G}_1$ *refines* $\mathcal{G}_2$, if $\mathcal{G}_1 \subseteq \mathcal{G}_2$.

**Proposition 1.** *If $\mathcal{G}_1$ refines $\mathcal{G}_2$, then $H(\mathcal{G}_1) \geq H(\mathcal{G}_2)$.*

# 4 Finding Trees with Low Entropy

In this section we describe the algorithm for searching decomposable families. To be more precise, given a candidate set, a downward closed family $\mathcal{F}$ covering the set of attributes $A$, our goal is to find a decomposable downward closed family $\mathcal{G} \subseteq \mathcal{F}$. Hence our goal is to find a junction tree $\mathcal{T}$ such that $s(\mathcal{T}) \subseteq \mathcal{F}$.

## 4.1 Definition of the Algorithm

We search the tree in an iterative fashion. At the beginning of each iteration round we have a junction tree $\mathcal{T}_{n-1}$ whose cliques have at most $n$ attributes, that is $m(s(\mathcal{T})) = n$. The first tree is $\mathcal{T}_0$ containing only single attributes and no edges. During each round the tree is modified so that in the end we will have $\mathcal{T}_n$, a tree with cliques having at most $n+1$ attributes.

In order to fully describe the algorithm we need the following definition: $X$ and $Y$ are said to be $n-1$-*connected* in a junction tree $\mathcal{T}$, if there is a path in $\mathcal{T}$ from $X$ to $Y$ having at least one separator of size $n-1$. We say that $X$ and $Y$ are 0-connected, if $X$ and $Y$ are not connected.

Each round of the algorithm consists of three steps. The pseudo-code of the algorithm is given in Algorithm 1–2.

1. **Generate:** We construct a graph $G_n$ whose nodes are the cliques of size $n$ in $\mathcal{T}_{n-1}$. We add all the edges to $G_n$ having the form $(X, Y)$ such that $|X \cap Y| = n-1$ and $X \cup Y \in \mathcal{F}$. We also set $\mathcal{T}_n = \mathcal{T}_{n-1}$. The weight of the edge is set to

$$w(e) = H(X) + H(Y) - H(X \cap Y) - H(X \cup Y).$$

2. **Augment:** We select the edge, say $e = (X, Y)$, having the largest weight and remove it from $G_n$. If $X$ and $Y$ are $n-1$ -connected in $\mathcal{T}_n$ we add $\mathcal{T}_n$ with a new clique $V = X \cup Y$. Furthermore, for each $v \in V$, we consider $W = V - v$. If $W$ is not in $\mathcal{T}_n$, it is added into $\mathcal{T}_n$. Next, $W$ and $V$ are connected in $\mathcal{T}_n$. At the same time, the node $W$ is also added into $G_n$ and the edges of $G_n$ are added using the same criteria as in Step 1 (Generate). Finally, a possible cycle is removed from $\mathcal{T}_n$ by finding an edge with separator of size $n - 1$. Augmenting is repeated as long as $G_n$ has no edges.
3. **Purge:** The tree $V(\mathcal{T}_n)$ contains redudant cliques after augmentation. We remove these redudant cliques from $\mathcal{T}_n$.

To illustrate the algorithm we provide a toy example.

*Example 1.* Consider that we have a family

$$\mathcal{F} = \{a, b, c, d, e, ab, ac, ad, bc, bd, cd, ce, abc, acd, bcd\}.$$

Assume that we are at the beginning of the second round and we already have the junction tree $\mathcal{T}_1$ given in Figure 2(a). We form $G_2$ by taking the edges $(ab, bc)$ and $(bc, cd)$.

Consider that we pick $ab$ and $bc$ for joining. This will spawn $ac$ and $abc$ in $\mathcal{T}_2$ (Figure 2(c)) and $ac$ in $G_2$ (Figure 2(d)). Note that we also add the edge $(ac, cd)$ into $G_2$. Assume that we continue by picking $(ac, cd)$. This will spawn $acd$ and $cd$ into $\mathcal{T}_2$. Note that $(bc, cd)$ is removed from $\mathcal{T}_2$ in order to break the cycle.

The last edge $(bc, cd)$ is not added into $\mathcal{T}_2$ since $bc$ and $cd$ are not $n - 1$-connected. The final tree (Figure 2(f)) is obtained by keeping only the maximal sets, that is, purging the cliques $bc$, $ab$, $ac$, $ad$, and $cd$. The corresponding decomposable family is $\mathcal{G} = \mathcal{F} - bcd$.

The next theorem states that the **Augment** step does not violate the running intersection property.

**Theorem 1.** *Let $\mathcal{T}$ be a junction tree with cliques of size $n+1$, at maximum, that is, $m(s(\mathcal{T})) = n + 1$. Let $X, Y \in V(\mathcal{T})$ be cliques of size $n$ such that $|X \cap Y| = n - 1$. Set $B = X \cup Y$. Then the family $s(\mathcal{T}) \cup \{C; C \subseteq B\}$ is decomposable if and only if $X$ and $Y$ are $n-1$-connected in $\mathcal{T}$.*

**Theorem 2.** ModifyTree *decreases the entropy of $\mathcal{T}_n$ by $w(e)$.*

Theorems 1–2 imply that SearchTree algorithm is nothing more than a greedy search. However, since we are adding cliques in rounds we can state that under some conditions the algorithm returns an optimal cover for each round.

**Theorem 3.** *Assume that the members of $\mathcal{F}$ of size $n + 1$ are added to $G_n$ at the beginning of the nth round. Let $\mathcal{U}$ be a junction tree such that $s(\mathcal{T}_n) \subseteq s(\mathcal{U})$ and $m(s(\mathcal{U})) = n + 1$. Then $H(\mathcal{T}_{n+1}) \leq H(\mathcal{U})$.*

**Corollary 1.** *The tree $\mathcal{T}_1$ is optimal among the families using the sets of size* 2.

Corollary 1 states that $\mathcal{G}_1$ is the Chow-Liu tree [10].

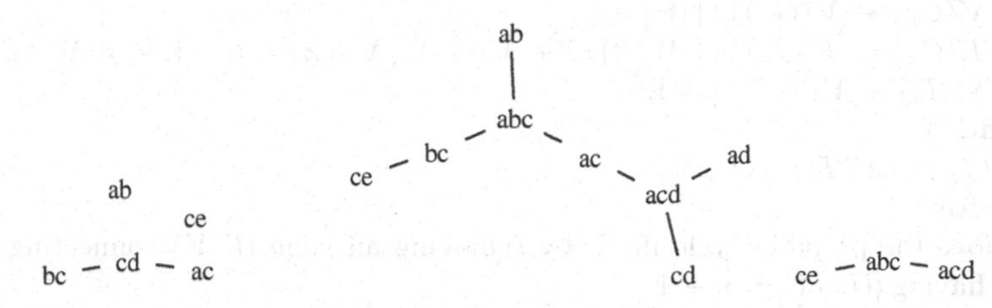


**Fig. 2.** Example of graphs during different stages of SEARCHTREE algorithm

---

**Algorithm 1.** SEARCHTREE algorithm. The input is a downward closed cover $\mathcal{F}$, the output is a junction tree $\mathcal{T}$ such that $V(\mathcal{T}) \subseteq \mathcal{F}$.

---

$V(\mathcal{T}_0) \leftarrow \{x; x \in A\}$ {$\mathcal{T}_0$ contains the single items.}
$n \leftarrow 0$.
**repeat**
  $n \leftarrow n + 1$.
  $\mathcal{T}_n \leftarrow \mathcal{T}_{n-1}$.
  $V(G_n) \leftarrow \{X \in V(\mathcal{T}_n); |X| = n\}$.
  $E(G_n) \leftarrow \{(X, Y); X, Y \in V(G_n), |X \cap Y| = n - 1, X \cup Y \in \mathcal{F}\}$.
  **repeat**
    $e = (X, Y) \leftarrow \arg\max_{x \in E(G_n)} w(x)$.
    $E(G_n) \leftarrow E(G_n) - e$.
    **if** $X$ and $Y$ are $n - 1$-connected in $\mathcal{T}_n$ **then**
      Call MODIFYTREE.
    **end if**
  **until** $E(G_n) = \emptyset$
  Delete nodes marked by MODIFYTREE from $\mathcal{T}_n$, connect the incident nodes.
**until** $G_n$ is empty
**return** $\mathcal{T}$

---

## 4.2 Model Selection

Theorem 1 reveals a drawback in the current approach. Consider that we have two independent items $a$ and $b$ and that $\mathcal{F} = \{a, b, ab\}$. Note that $\mathcal{F}$ is itself decomposable and $\mathcal{G} = \mathcal{F}$. However, a more reasonable family would be $\{a, b\}$ to reflect the fact that $a$ and $b$ are independent. To remedy this problem we will use model selection techniques such as BIC [11], AIC [12], and Refined MDL [13]. All these methods score the model by adding a penalty term to the likelihood.

**Algorithm 2.** ModifyTree algorithm.

$B \leftarrow X \cup Y$.
$V(\mathcal{T}_n) \leftarrow V(\mathcal{T}_n) \cup \{B\}$. {Add new clique $B$ into $\mathcal{T}_n$.}
**for** $v \in B$ **do**
  $W \leftarrow B - v$.
  Mark $W$.
  **if** $W \notin V(G_n)$ **then**
    $V(G_n) \leftarrow V(G_n) \cup \{W\}$.
    $E(G_n) \leftarrow E(G_n) \cup \{(W, Z)\,; Z \in V(G_n)\,, |X \cap Z| = n-1, V \neq X \cup Z \in \mathcal{F}\}$.
    $V(\mathcal{T}_n) \leftarrow V(\mathcal{T}_n) \cup \{W\}$.
  **end if**
  $E(\mathcal{T}_n) \leftarrow E(\mathcal{T}_n) \cup (B, W)$.
**end for**
Remove the possible cycle in $\mathcal{T}_n$ by removing an edge $(U, V)$ connecting $X$ and $Y$ and having $|U \cap V| = n - 1$.

We modify the algorithm by considering only the edges in $G_n$ that improve the score. For BIC this reduces to considering only the edges satisfying

$$|D| w(e) \geq 2^{n-2} \log |D|,$$

where $n$ is the current level of SearchTree algorithm. Using AIC leads to the considering only the edges for which

$$|D| w(e) \geq 2^{n-1}.$$

Refined MDL is more troublesome. The penalty term in MDL is known as *stochastic complexity.* In general, there is no known closed formula for the stochastic complexity, but it can be computed for the multinomial distribution in linear time [14]. However, it is numerically unstable for data with large number of transactions. Hence, we will apply often-used asympotic estimate [15] and define the penalty term

$$C_{\mathrm{MDL}}(k) = \frac{k-1}{2} \log |D| - \frac{1}{2} \log \pi - \log \Gamma\left(k/2\right)$$

for $k$-multinomial distribution.

There are no known exact or approximative solution in a closed form of stochastic complexity for junction trees. Hence we propose the penalty term for the tree to be

$$\sum_{X \in V(\mathcal{T})} C_{\mathrm{MDL}}\left(2^{|X|}\right) - \sum_{Y \in S(\mathcal{T})} C_{\mathrm{MDL}}\left(2^{|Y|}\right).$$

Here we think that a single clique $X$ is a $2^{|X|}$-multinomial distribution and we compensate the possible overlaps of the cliques by subtracting the complexity of the separators. Using this estimate leads to a selection criteria

$$|D| w(e) \geq C_{\mathrm{MDL}}\left(2^{|n+1|}\right) - 2C_{\mathrm{MDL}}\left(2^{|n|}\right) + C_{\mathrm{MDL}}\left(2^{|n-1|}\right).$$

### 4.3 Computing Multiple Decomposable Families

We can use SEARCHTREE algorithm for computing multiple decomposable covers from a single candidate set $\mathcal{F}$. The motivation behind this approach is that we get a sequence of covers, each cover holding partial information of the original cover $\mathcal{F}$. We will show empirically in Section 6.4 that the by exploiting the union information of these covers we are able to improve significantly bounds for boolean queries (see Section 5).

The idea is as follows. Set $\mathcal{F}_1 = \mathcal{F}$ and let $\mathcal{G}_1$ be the first decomposable family constructed from $\mathcal{F}_1$ using SEARCHTREE algorithm. We define

$$\mathcal{F}_2 = \mathcal{F}_1 - \{X \in \mathcal{F}_1;\ \text{there is } Y \in \mathcal{G}_1, |Y| > 1, Y \subseteq X\}.$$

We compute $\mathcal{G}_2$ from $\mathcal{F}_2$ and continue in the iterative fashion until $\mathcal{G}_k$ contains nothing but individual items.

## 5 Boolean Queries with Decomposable Families

One of our motivations for constructing decomposable families is that some computational problems that are hard for general families of itemsets reduce to tractable if the underlying family is decomposable. In this section we will show that the computational burden of a boolean query, a classic optimization problem [16,17], reduces significantly, if we are using decomposable families of itemsets.

Assume that we are given a set of known itemsets $\mathcal{G}$ and a query itemset $Q \notin \mathcal{G}$. We wish to find $fr(Q; \mathcal{G})$, the possible frequencies for $Q$ given the frequencies of $\mathcal{G}$. It is easy to see that the frequencies form an interval, hence it is sufficient to find the maximal and the minimal frequencies. We can express the problem of finding the maximal frequency as a search for the distribution $p$ solving

$$\begin{aligned} &\max p\,(Q = 1) \\ &\text{s.t. } p\,(X = 1) = fr(X)\,, \text{ for each } X \in \mathcal{G}. \\ &\quad p \text{ is a distribution over } A. \end{aligned} \tag{3}$$

We can solve Eq. 3 using Linear Programming [16]. However, the number of variables in the program is $2^{|A|}$ and makes the program tractable only for small datasets. In fact, solving Eq. 3 is an **NP**-hard problem [9].

In the rest of the section we present a method of solving Eq. 3 with a linear program containing only $2^{|Q|}|\mathcal{G}||A|$ variables, assuming that $\mathcal{G}$ is decomposable. This method is an explicit construction of the technique presented in [18]. The idea behind the approach is that instead of solving a joint distribution in Eq. 3, we break the distribution into small component distributions, one for each clique in the junction tree. These components are forced to be consistent by requiring that they are equal at the separators. The details are given in Algorithm 3.

To clarify the process we provide the following simple example.

**Algorithm 3.** QUERYTREE algorithm for solving a query $Q$ from a decomposable cover $\mathcal{G}$. The output is the interval $fr(Q;\mathcal{G})$.

$\{\mathcal{T}_1,\ldots,\mathcal{T}_M\} \leftarrow$ connected components of a junction tree of $\mathcal{G}$.
**for** $i = 1,\ldots,M$ **do**
  $Q_i \leftarrow Q \cap (\bigcup V(\mathcal{T}_i))$. {Items of $Q$ contained in $\mathcal{T}_i$.}
  $\mathcal{U} \leftarrow \arg\min_{\mathcal{S}\subseteq\mathcal{T}_i} \{|V(\mathcal{S})|; Q_i \subseteq \bigcup V(\mathcal{S})\}$. {Smallest subtree containing $Q_i$.}
  **while** there are changes **do**
    Remove the items outside $Q_i$ that occur in only one clique of $\mathcal{U}$.
    Remove redundant cliques.
  **end while**
  Select one clique, say $R$ from $\mathcal{U}$ to be the root.
  $R \leftarrow R \cup Q_i$. {Augment the root with $Q_i$}
  Augment the rest cliques in $\mathcal{U}$ so that the running intersection property holds.
  Let $p_C$ be a distribution over each clique $C \in V(\mathcal{U})$.
  $\alpha_i \leftarrow$ the solution of a linear program

$$\begin{aligned}&\min p_R\left(Q_i = 1\right)\\ &\text{s.t. } p_C\left(X = 1\right) = fr(X)\,, \text{ for each } C \in V(\mathcal{U})\,, X \in \mathcal{G}, X \subseteq C.\\ &\quad\; p_{C_1}\left(C_1 \cap C_2\right) = p_{C_2}\left(C_1 \cap C_2\right), \text{ for each } \left(C_1, C_2\right) \in E(\mathcal{U})\,.\end{aligned}$$

  $\beta_i \leftarrow$ the solution of the maximum version of the linear program.
**end for**
$fr(Q;\mathcal{G}) \leftarrow \left[\max\left(\sum_i^M \alpha_i - (M-1), 0\right), \min_i(\beta_i)\right]$.

*Example 2.* Assume that we have $\mathcal{G}$ whose junction tree is given in Figure 3(a). Let query be $Q = adg$. We begin first by finding the smallest sub-tree containing $Q$. This results in purging $fh$ (Figure 3(b)). We further purge the tree by removing $e$ since it only occurs in one clique (Figure 3(c)). In the next step we pick a root, which in this case is $bc$ and augment the cliques with the members of $Q$ so that the root contains $Q$ (Figure 3(d)). We finally remove the redundant cliques which are $ab$, $cd$, $fg$. The final tree is given in 3(e). Finally, the linear program is formed using two distributions $p_{abcdg}$ and $p_{cfg}$. The number of variables in this program is $2^5 + 2^3 = 40$ opposed to the original $2^8 = 256$.

Note that we did not specify in Algorithm 3 which clique we selected to be the root $R$. The linear program depends on the root $R$ and hence we select the root minimizing the number of variables in the linear program.

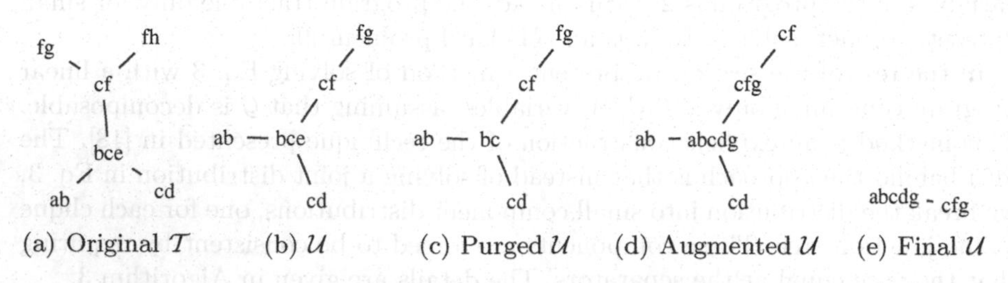

(a) Original $\mathcal{T}$ (b) $\mathcal{U}$ (c) Purged $\mathcal{U}$ (d) Augmented $\mathcal{U}$ (e) Final $\mathcal{U}$

**Fig. 3.** Junction trees during different stages of solving the query problem

**Theorem 4.** QUERYTREE *algorithm solves correctly the boolean query* $fr(q; \mathcal{G})$*. The number of variables occurring in the linear programs is* $2^{|Q|}|\mathcal{G}||A|$*, at maximum.*

# 6 Experiments

In this section we will study empirically the relationship between the decomposable itemset families and the candidate set, the role of the regularization, and the performance of boolean queries using multiple decomposable families.

## 6.1 Datasets

For our experiments we used one synthetic generated dataset, *Path*, and three real-world datasets: *Paleo*, *Courses* and *Mammals*. The synthetic dataset, *Path*, contained 8 items and 100 transactions. Each item was generated from the previous item by flipping it with a 0.3 probability. The first item was generated by a fair coin flip. The dataset *Paleo*[1] contains information of mammal fossils found in specific paleontological sites in Europe [19]. *Courses* describes computer science courses taken by students at the Department of Computer Science of the University of Helsinki. The *Mammals*[2] dataset consists of presence/absence records of current day European mammals [20]. The basic characteristics of the real-world data sets are shown in Table 1.

**Table 1.** The basic properties of the datasets

| Dataset | # of rows | # of items | # of 1s | $\frac{\text{\# of 1s}}{\text{\# of rows}}$ |
|---|---|---|---|---|
| *Paleo* | 501 | 139 | 1980 | 16.0 |
| *Courses* | 3506 | 98 | 16086 | 4.6 |
| *Mammals* | 2183 | 124 | 54155 | 24.8 |

## 6.2 Generating Decomposable Families

In our first experiment we examined the junction trees that were constructed for the *Path* dataset. We calculated a sequence of trees using the technique described in Section 4.3. As input to the algorithm we used an unconstrained candidate collection of itemsets (minimum support = 0) from *Path* and BIC as the regularization method. In Figure 4(a) we see that the first tree corresponds to the model used to generate the dataset. The second tree, given in Figure 4(b), tend to link the items that are one gap away from each other. This is a natural result since close items are the most informative about each other.

[1] NOW public release 030717 available from [19].

[2] The full version of the mammal dataset is available for research purposes upon request from the Societas Europaea Mammalogica (`www.european-mammals.org`)

0, 1 — 1, 2 — 2, 3 — 3, 4 — 4, 5 — 5, 6 — 6, 7

BIC = 522.905958, AIC = 503.367182, MDL = 522.333594

(a) First junction tree of *Path* data.

4, 6 — 2, 4 — 0, 2 — 0, 3 — 3, 5 — 1, 3

7

BIC = 561.499992, AIC = 543.263801, MDL = 560.355263

(b) Second junction tree of *Path* data.

**Fig. 4.** Junction trees for *Path*, a syntetic data in which an item is generated from the previous item by flipping it with 0.3 probability. The junction trees are regularized using BIC. The tree in Figure 4(b) is generated by ignoring the cliques of the tree in Figure 4(a).

With *Courses* data one large junction tree of itemsets is produced with several noticeable components. One distinct component at one end of the tree contains introductory courses like *Introduction to Programming*, *Introduction to Databases*, *Introduction to Application design* and *Java Programming*. Respectively, the other end of the tree features several distinct components with itemsets on more specialized themes in computer science and software engineering. The central node connecting each of these components in the entire tree is the itemset node {*Software Engineering, Models of Programming and Computing, Concurrent systems*}.

Figure 5 shows about two-thirds of the entire *Courses* junction tree, with the component related to introductory courses removed because of the space constraints. We see a concurrent and distributed systems related component in the lower left part of the figure, a more software development oriented component in the lower right quarter and a Robotics/AI component in the upper right corner of the tree. The entire *Courses* junction tree can be found in [6].

We continued our experiments by studying the behavior of the model scores in a sequence of trees induced by a corresponding sequence of decomposable families. For the *Path* data the scores of the two first junction trees are shown in Figure 4, with the first one yielding smaller values. For the real-world datasets, we computed a sequence of trees from each dataset, again, with the unconstrained candidate collection as input and using AIC, BIC, or MDL respectively as the regularization method. Computation took about 1 minute per tree. The corresponding scores are plotted as a function of the order of the corresponding junction tree (Figure 6). The scores are increasing in the sequence, which is expected since the algorithm tries to select the best model and the subsequent trees are constructed from the left-over itemsets. The increase rate slows down towards the end since the last trees tend to have only singleton itemsets as nodes.

### 6.3 Reducing Itemsets

Our next goal was to study the sizes of the generated decomposable families compared to the size of the original candidate set. As input for this experiment, we used several different candidate collections of frequent itemsets resulting from varying the support threshold, and generated the corresponding decomposable itemset families (Table 2).

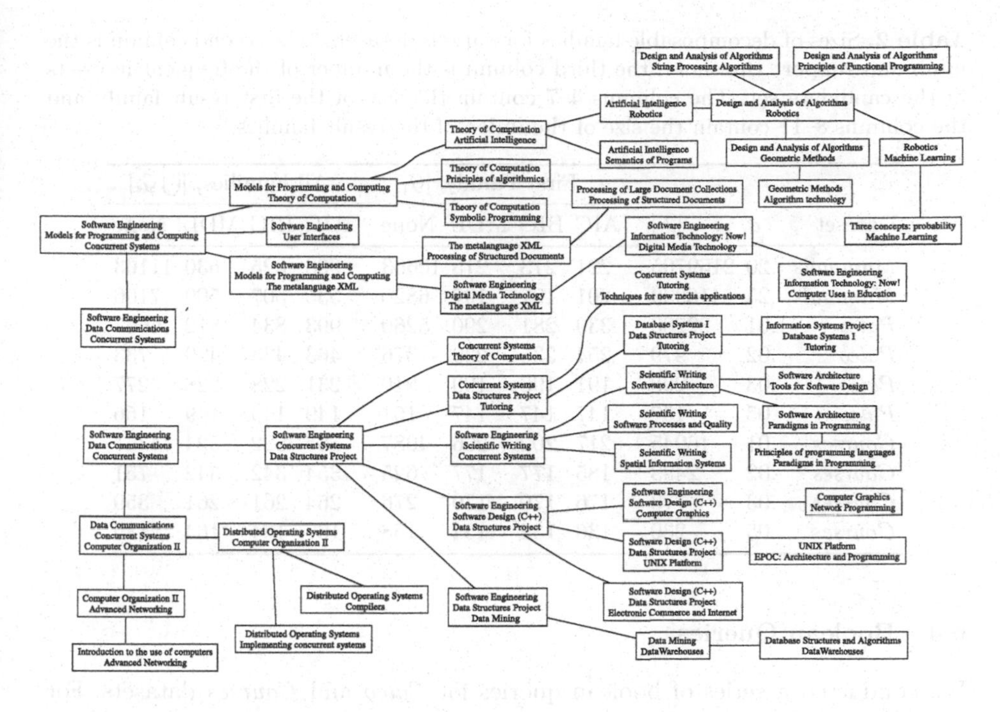


**Fig. 5.** A part of the junction tree constructed from the *Courses* dataset. The tree was constructed using an unconstrained candidate family (min. support = 0) as input and BIC as regularization.

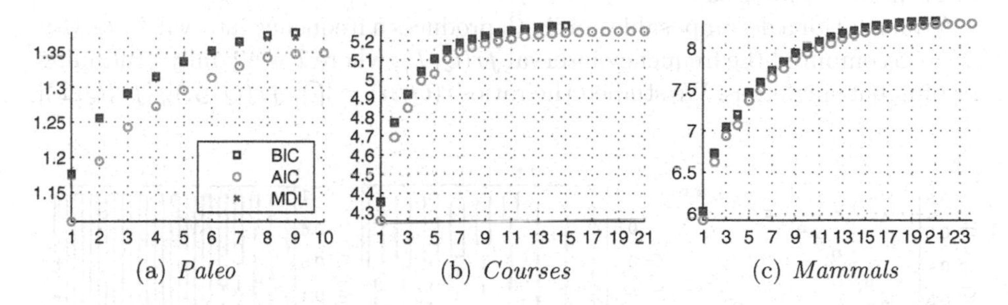


(a) *Paleo* (b) *Courses* (c) *Mammals*

**Fig. 6.** Scores of covers as a function of the order of the cover. Each cover is computed with an unconstrained candidate family (min. support = 0) as input and the corresponding regularization. The $y$-axis is the model score divided by $10^4$.

From the results we see that the decomposable families are much smaller compared to the original candidate set, as a large portion of itemsets are pruned due to the running intersection property. The regularizations AIC, BIC, MDL prune the results further. The pruning is most effective when the candidate set is large.

**Table 2.** Sizes of decomposable families for various datasets. The second column is the minimum support threshold, the third column is the number of the frequent itemsets in the candidate set. The columns 4–7 contain the size of the first result family and the columns 8–11 contain the size of the union of the result families.

| | | | First Family, $\|\mathcal{G}_1\|$ | | | | All Families, $\|\bigcup \mathcal{G}_i\|$ | | | |
|---|---|---|---|---|---|---|---|---|---|---|
| Dataset | $\sigma$ | $\|\mathcal{F}\|$ | AIC | BIC | MDL | None | AIC | BIC | MDL | None |
| *Mammals* | .20 | 2169705 | 221 | 213 | 215 | 10663 | 668 | 625 | 630 | 11103 |
| *Mammals* | .25 | 416939 | 201 | 197 | 197 | 6820 | 535 | 507 | 509 | 7106 |
| *Paleo* | .01 | 22283 | 339 | 281 | 290 | 5260 | 993 | 834 | 812 | 6667 |
| *Paleo* | .02 | 979 | 254 | 235 | 239 | 376 | 463 | 433 | 429 | 733 |
| *Paleo* | .03 | 298 | 191 | 190 | 190 | 210 | 231 | 228 | 228 | 277 |
| *Paleo* | .05 | 157 | 147 | 147 | 147 | 151 | 149 | 149 | 149 | 156 |
| *Courses* | .01 | 16945 | 217 | 202 | 206 | 4087 | 565 | 522 | 524 | 4357 |
| *Courses* | .02 | 2493 | 185 | 177 | 177 | 625 | 354 | 342 | 342 | 751 |
| *Courses* | .03 | 773 | 176 | 170 | 170 | 276 | 264 | 261 | 261 | 359 |
| *Courses* | .05 | 230 | 136 | 132 | 132 | 158 | 167 | 164 | 164 | 186 |

## 6.4 Boolean Queries

We conducted a series of boolean queries for *Paleo* and *Courses* datasets. For each dataset we pick randomly 1000 queries of size 5. We constructed a sequence of trees using BIC and the unconstrained (min. support = 0) candidate set as input. The average computation time for a single query was $0.3s$. A portion (abt. 10%) of queries had to be discarded due to the numerical instability of the linear program solver we used.

A query $Q$ for a decomposable family $\mathcal{G}_i$ produces a frequency interval $fr(Q; \mathcal{G}_i)$. We also computed the frequency interval $fr(Q; \mathcal{I})$, where $\mathcal{I}$ is a family containing nothing but singletons. We studied the ratios $r(Q; n) = |\bigcap_1^n fr(Q; \mathcal{G}_i)| / |fr(Q; \mathcal{I})|$

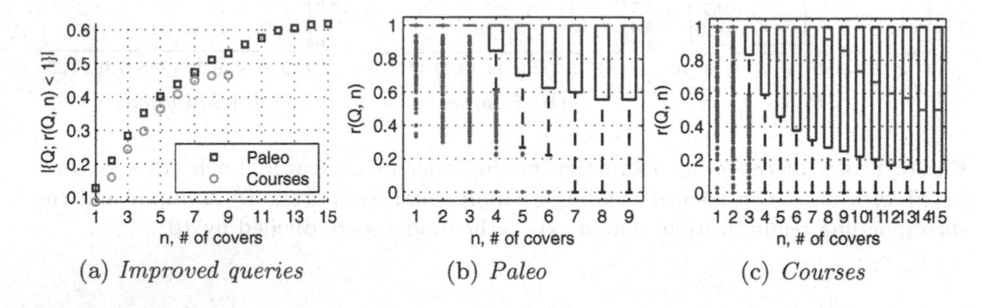


(a) *Improved queries* (b) *Paleo* (c) *Courses*

**Fig. 7.** Boolean query ratios from *Paleo* and *Course* datasets. Figure 7(a) contains the percentage of queries having $r(Q; n) < 1$, that is, the percentage of queries improved over the singleton model as a function of the number of decomposable families. Figures 7(b)–7(c) are box plots of the ratios $r(Q; n)$, where $Q$ is a random query and $n$ is the number of decomposable families.

as a function of $n$, that is, the ratio between the tightness of the bound using $n$ families and the singleton model.

From the results given in Figure 7 we see that the first decomposable family in the sequence yields in about 10 % of the queries an improved bound with respect to the singleton family. As the number of decomposable families increases, the number of queries with tighter bounds goes from 10% up to 60%. Also, in general the absolute bounds become tighter for the queries as we increase the number of decomposable families. For *Courses* the median of the ratio $r(Q; 15)$ is about 0.5.

## 7 Related Work

One of the main uses of our algorithm is in reducing itemset mining results into a smaller and a more manageable group of itemsets. One of the earliest approaches on itemset reduction include close itemsets [21] and maximal frequent itemset [22]. Also more recently, a significant amount of interesting research has been produced on the topic [23,24,25,26]. Yan et al. [24] proposed a statistical model in which $k$ representative patterns are used to summarize the original itemset family as well as possible. This approach has, however, a different goal to that of ours, as our model aims to describe the data itself. From this point of view the work by Siebes et al. [25] is perhaps the most in concordance to ours. Siebes et al. propose an MDL based method where the reduced group of itemsets aim to compress the data as well as possible. Yet, their approach is technically and methodologically quite different and does not provide a probabilistic model of the data as our model does. Furthermore, non of the above approaches provide a naturally following tree based representation of the mining results as our model does.

Traditionally, junction trees are not used as a direct model but rather as a technique for decomposing directed acyclic graph (DAG) models [5]. However, there is a clear difference between the DAG models and our approach. Assume that we have 4 items $a$, $b$, $c$, and $d$. Consider a DAG model $p(a)p(b; a)p(c; a)p(d; bc)$. While we can decompose this model using junction trees we cannot express it exactly. The reason for this is that the DAG model contains the assumption of independence of $b$ and $c$ given $a$. This allows us to break the clique $abc$ into smaller parts. In our approach the cliques are the empirical distributions with no independence assumptions. DAG models and junction tree models are equivalent for Chow-Liu tree models [10].

Our algorithm for constructing junction trees is closely related to EFS algorithm [27,28] in which new cliques are created in a similar fashion. The main difference between the approaches is that we add new cliques in a level-wise fashion. This allows a more straightforward algorithm. Another benefit of our approach is Theorem 3. On the other hand, Corollary 1 implies that our algorithm can be seen also as an extension of Chow-Liu tree model [10].

## 8 Conclusions and Future Work

In this study we applied the concept of junction trees to create decomposable families of itemsets. The approach suits well for the problem of itemset selection, and has several advantages. The naturally following junction trees provide an intuitive representation of the mining results. From the computational point of view, the model provides leverage for problems that could be intractable using generic families of itemsets. We provided an efficient algorithm to build decomposable itemset families, and gave an application example with frequency bound querying using the model. Empirical results showed that our algorithm yields high quality results. Because of the expressiveness and good interpretability of the model, applications such as classification using decomposable families of itemsets could prove an interesting avenue for future research. Even more generally, we anticipate that in the future decomposable models could prove computationally useful with pattern mining applications that otherwise could be hard to tackle.

## References

1. Agrawal, R., Imielinski, T., Swami, A.: Mining association rules between sets of items in large databases. In: ACM SIGMOD international conference on Management of data, pp. 207–216 (1993)
2. Agrawal, R., Mannila, H., Srikant, R., Toivonen, H., Verkamo, A.: Fast discovery of association rules. Advances in knowledge discovery and data mining, 307–328 (1996)
3. Han, J., Pei, J.: Mining frequent patterns by pattern-growth: methodology and implications. SIGKDD Explorations Newsletter 2(2), 14–20 (2000)
4. Han, J., Cheng, H., Xin, D., Yan, X.: Frequent pattern mining: current status and future directions. Data Mining and Knowledge Discovery 15(1) (2007)
5. Cowell, R.G., Dawid, A.P., Lauritzen, S.L., Spiegelhalter, D.J.: Probabilistic Networks and Expert Systems. Statistics for Engineering and Information Science (1999)
6. Tatti, N., Heikinheimo, H.: Decomposable families of itemsets. Technical Report TKK-ICS-R1, Helsinki University of Technology (2008), `http://www.otalib.fi/tkk/edoc/`
7. Jiroušek, R., Přeušil, S.: On the effective implementation of the iterative proportional fitting procedure. Computational Statistics and Data Analysis 19, 177–189 (1995)
8. Csiszár, I.: I-divergence geometry of probability distributions and minimization problems. The Annals of Probability 3(1), 146–158 (1975)
9. Tatti, N.: Computational complexity of queries based on itemsets. Information Processing Letters, 183–187 (June 2006)
10. Chow, C.K., Liu, C.N.: Approximating discrete probability distributions with dependence trees. IEEE Transactions on Information Theory 14(3), 462–467 (1968)
11. Schwarz, G.: Estimating the dimension of a model. Annals of Statistics 6(2), 461–464 (1978)
12. Akaike, H.: A new look at the statistical model identification. IEEE Transactions on Automatic Control 19(6), 716–723 (1974)

13. Grünwald, P.D.: The Minimum Description Length Principle (Adaptive Computation and Machine Learning). MIT Press, Cambridge (2007)
14. Kontkanen, P., Myllymäki, P.: A linear-time algorithm for computing the multinomial stochastic complexity. Information Processing Letters 103(6), 227–233 (2007)
15. Rissanen, J.: Fisher information and stochastic complexity. IEEE Transactions on Information Theory 42(1), 40–47 (1996)
16. Hailperin, T.: Best possible inequalities for the probability of a logical function of events. The American Mathematical Monthly 72(4), 343–359 (1965)
17. Bykowski, A., Seppänen, J.K., Hollmën, J.: Model-independent bounding of the supports of Boolean formulae in binary data. In: Meo, R., Lanzi, P.L., Klemettinen, M. (eds.) Database Support for Data Mining Applications. LNCS (LNAI), vol. 2682, pp. 234–249. Springer, Heidelberg (2004)
18. Tatti, N.: Safe projections of binary data sets. Acta Informatica 42(8-9), 617–638 (2006)
19. Fortelius, M.: Neogene of the old world database of fossil mammals (NOW). University of Helsinki (2005), `http://www.helsinki.fi/science/now/`
20. Mitchell-Jones, A.J., Amori, G., Bogdanowicz, W., Krystufek, B., Reijnders, P.J.H., Spitzenberger, F., Stubbe, M., Thissen, J.B.M., Vohralik, V., Zima, J.: The Atlas of European Mammals. Academic Press, London (1999)
21. Pasquier, N., Bastide, Y., Taouil, R., Lakhal, L.: Discovering frequent closed itemsets for association rules. In: Beeri, C., Bruneman, P. (eds.) ICDT 1999. LNCS, vol. 1540, pp. 398–416. Springer, Heidelberg (1998)
22. Roberto, J., Bayardo, J.: Efficiently mining long patterns from databases. In: ACM SIGMOD international conference on Management of data, pp. 85–93. ACM, New York (1998)
23. Calders, T., Goethals, B.: Mining all non-derivable frequent itemsets. In: European Conference on Principles and Practice of Knowledge Discovery in Databases (2002)
24. Yan, X., Cheng, H., Han, J., Xin, D.: Summarizing itemset patterns: A profilebased approach. In: ACM SIGKDD international conference on Knowledge Discovery and Data Mining (2005)
25. Siebes, A., Vreeken, J., van Leeuwen, M.: Item sets that compress. In: SIAM Conference on Data Mining, pp. 393–404 (2006)
26. Bringmann, B., Zimmermann, A.: The chosen few: On identifying valuable patterns. In: IEEE International Conference on Data Mining (2007)
27. Deshpande, A., Garofalakis, M.N., Jordan, M.I.: Efficient stepwise selection in decomposable models. In: Conference in Uncertainty in Artificial Intelligence, pp. 128–135. Morgan Kaufmann Publishers Inc, San Francisco (2001)
28. Altmueller, S.M., Haralick, R.M.: Practical aspects of efficient forward selection in decomposable graphical models. In: IEEE International Conference on Tools with Artificial Intelligence, pp. 710–715. IEEE Computer Society, Washington (2004)

# Transferring Instances for Model-Based Reinforcement Learning

Matthew E. Taylor, Nicholas K. Jong, and Peter Stone

Department of Computer Sciences, The University of Texas at Austin
{mtaylor,nkj,pstone}@cs.utexas.edu

**Abstract.** *Reinforcement learning* agents typically require a significant amount of data before performing well on complex tasks. *Transfer learning* methods have made progress reducing sample complexity, but they have primarily been applied to model-free learning methods, not more data-efficient model-based learning methods. This paper introduces TIMBREL, a novel method capable of transferring information effectively into a model-based reinforcement learning algorithm. We demonstrate that TIMBREL can significantly improve the sample efficiency and asymptotic performance of a model-based algorithm when learning in a continuous state space. Additionally, we conduct experiments to test the limits of TIMBREL's effectiveness.

## 1 Introduction

In many situations, an agent must learn to execute a series of sequential actions, which is typically framed as a *reinforcement learning* (RL) [1] problem. Although RL approaches have enjoyed past successes (e.g., TDGammon [2], inverted Helicopter control [3], and robot locomotion [4]), they frequently take substantial amounts of data to learn a reasonable control policy. In many domains, collecting such data may be slow, expensive, or infeasible, motivating the need for sample-efficient learning methods.

One recent approach to speeding up RL so that it can be applied to difficult problems with large, continuous state spaces is *transfer learning* (TL). TL is a machine learning paradigm that reuses knowledge gathered in a previous source task to better learn a novel, but related, target task. Recent empirical successes in a variety of RL domains [5,6,7] have shown that transfer can significantly increase an agent's ability to learn quickly, even if agents in the two tasks have different available sensors or actions. Note that TL is related to, but distinct from, the *concept drift* [8] and *multi-task learning* [9,10] paradigms. Concept drift assumes that the environment is non-stationary: at certain points in time, the environment may change arbitrarily and the change is unannounced. Multi-task learning assumes that the agent experiences many problems and that they are all drawn from the same distribution (and thus all tasks have the same actions and state variables). In contrast, TL methods generally assume that the agent is notified when the task changes and generally do not assume that source and target tasks are drawn from the same distribution.

W. Daelemans et al. (Eds.): ECML PKDD 2008, Part II, LNAI 5212, pp. 488–505, 2008.

Model-free algorithms such as Q-Learning [11] and Sarsa [12] learn to predict the utility of each action in different situations, but they do not learn the effects of actions. In contrast, model-based (or model-learning) methods, such as PEGASUS [13], R-MAX [14], and Fitted R-MAX [15], use their experience to learn an internal model of how the actions affect the agent and its environment, an approach empirically shown to often be more sample efficient. Such a model can be used in conjunction with *dynamic programming* [16] to perform off-line planning, often enabling superior action selection without requiring additional environmental samples. Building these models may be computationally intensive, but using CPU cycles to reduce data collection time is a highly favorable tradeoff in many domains (such as physically embodied agents). In order to further reduce sample complexity and ultimately allow RL to be applicable in more complex domains, this paper introduces *Transferring Instances for Model-Based REinforcement Learning* (TIMBREL), a novel approach to combining TL with model-based RL.

The key insight behind TIMBREL is that data gathered in a source task can be used to build beneficial models in a target task. Data is first recorded in a source task, transformed so that it applies to a target task, and then used by the target task learner as it builds its model. In this paper we utilize Fitted R-MAX, an instance-based model-learning algorithm, and show how TIMBREL can help construct a target task model by using source task data. TIMBREL combines the benefits of transfer with those of model-based learning to reduce sample complexity. We fully implement and test our method in a set of mountain car tasks, demonstrating that transfer can significantly reduce the sample complexity of learning.

In principle, the core TIMBREL algorithm (Section 3.1) could be used with other model-learning algorithms, but we leave such extensions to future work. The experiments in this paper use TIMBREL by applying it to Fitted R-MAX (detailed in Section 5), as it can both learn in continuous state spaces and has had significant empirical success [15]. This paper's results thus demonstrate that TIMBREL works in continuous state spaces, as well as between tasks with different state variables and action spaces.

The rest of this paper is organized as follows. Section 2 provides a brief background of RL and Fitted R-MAX, as well as discussing a selection of related TL methods. Section 3 introduces TIMBREL and the experimental domain is detailed in Section 4. Results are presented in Section 6. Section 7 discusses possible future directions and concludes.

## 2 Background and Related Work

In this paper we use the notation of *Markov decision processes* (MDPs) [17]. At every time step the agent observes its state $s \in S$ as a vector of k *state variables* such that $s = \langle x_1, x_2, \ldots, x_k \rangle$. In episodic tasks there is a starting state $s_{initial}$ and often a goal state $s_{goal}$, which terminates the episode if reached by the agent. The agent selects an action from the set of available actions $A$ at every time step.

The start and goal states may be generalized to sets of states. A task also defines the reward function $R : S \times A \mapsto \mathbb{R}$, and the transition function $T : S \times A \mapsto S$ fully describes the dynamics of the system. The agent will attempt to maximize the long-term reward determined by the (initially unknown) reward function $R$ and the (initially unknown) transition function $T$.

A learner chooses which action to take in a state via a policy, $\pi : S \mapsto A$. $\pi$ is modified by the learner over time to improve performance, which is defined as the expected total reward. Instead of learning $\pi$ directly, many RL algorithms instead approximate the action-value function, $Q : S \times A \mapsto \mathbb{R}$, which maps state-action pairs to the expected real-valued return [17]. If the agent has learned the optimal action-value function, it can select the optimal action from any state by executing the action with the highest action-value.

In this paper, we introduce and utilize TIMBREL to improve the performance of Fitted R-MAX [15], an algorithm that approximates the action-value function $Q$ for large or infinite state spaces by constructing an MDP over a small (finite) sample of states $X \subset S$. For each sample state $\mathbf{x} \in X$ and action $a \in A$, Fitted R-MAX estimates the dynamics $T(\mathbf{x}, a)$ using all the available data for action $a$ and for states $s$ near $\mathbf{x}$.[1] Some generalization from nearby states is necessary because we cannot expect the agent to be able to visit $\mathbf{x}$ enough times to try every action. As a result of this generalization process, Fitted R-MAX first approximates $T(\mathbf{x}, a)$ as a probability distribution over predicted successor states in $S$. A value approximation step then approximates this distribution of states in $S$ with a distribution of states in $X$. The result is a stochastic MDP over a finite state space $X$, with transition and reward functions derived from data in $S$. Applying dynamic programming to this MDP yields an action-value function over $X \times A$ that can be used to approximate the desired action-value function $Q$. Past work [15] empirically shows that Fitted R-MAX learns policies using less data than many existing model-free algorithms.

Fitted R-MAX is summarized in Algorithm 1. $s^{\text{opt}}$ is a dummy state that represents unexplored states (where $V(s^{\text{opt}})$ is set to $R_{max}$). $s^{\text{term}}$ is a dummy absorbing state that all discovered terminal states get mapped to. $D$ is a data structure that holds all observed instances. $\phi^{S^a}$ is an averaging object that approximates the effect of action $a$ at state $s$ using nearby sample transitions $d \in D^a$. $\phi^X$ is an averaging object that approximates the value of each predicted successor state using nearby sample states $x \in X$. The reader is referred to [15] for detailed descriptions of the update rules (lines 16 and 17).

Most similar to TIMBREL are methods that transfer between model-free RL algorithms with different state and action spaces. Torrey et. al [5] show how to automatically extract *advice* from a source task by identifying actions which have higher Q-values than other available actions; this advice is then mapped by a human to the target task as initial preferences given to the target task

[1] Fitted R-MAX is an instance-based learning method; our implementation currently retains all observed data to compute the model. It could, in principle, be enhanced to automatically discard instances without significantly decreasing model accuracy.

**Algorithm 1.** Fitted R-MAX ($R_{max}$, $r$, $b$, *minFraction*, *explorationThreshold*)

1: $X \leftarrow \{s^{\text{opt}}, s^{\text{term}}\}$ # Initialize state sample
2: $X.InitializeUniformGrid(r)$
3: **for all** $a \in A$ **do** # Initialize experience sample
4: $D^a \leftarrow \{\langle s^{\text{opt}}, a, V^{\max}, s^{\text{term}}\rangle\}$
5: **loop**
6: $s \leftarrow$ initial state # Begin a trajectory
7: $a \leftarrow \text{argmax}_a \left[R(s) + \sum_{x' \in X} P(x'|s,a)V(x')\right]$
8: **repeat**
9: Execute $a$
10: Observe $r$ and $s'$
11: **if** $s'$ is terminal **then**
12: $s' \leftarrow s^{\text{term}}$
13: **else**
14: $a' \leftarrow \text{argmax}_a \left[R(s) + \sum_{x' \in X} P(x'|s,a)V(x')\right]$
15: $D^a \leftarrow D^a \cup \{\langle s,a,r,s'\rangle\}$ # Update experience sample
16: Update $\phi^X$ and $\phi^{S^a}$ via ⟨experience, *minFraction*, and *explorationThreshold*⟩
17: Update estimates of $R$ and $P$ based on $\phi^X$ and $\phi^{S^a}$
18: Compute $V(x)$ for $x \in X$ via dynamic programming
19: $s \leftarrow s'$
20: $a \leftarrow a'$
21: **until** $s$ is a terminal state # the episode ends

learner. In past work [6], an agent learns an action-value function in a source task, translates the function into a target task via a hand-coded *inter-task mapping*, and then uses the transferred function to initialize the target task agent. Other recent work by Lazaric et. al [7] demonstrates that source task instances (that is, observed $\langle s,a,r,s'\rangle$ tuples) can be usefully transferred between tasks for a batch value-function learning algorithm. In all three cases the transferred knowledge is effectively used to improve learning in the target task, but only for model-free learning methods.

Atkeson and Santamaria [18] show that if only the reward function changed between tasks, a locally weighted regression model can be directly applied from a source task in a novel task. Tanaka and Yamamura [19] consider multi-task learning in a discrete state space. By recording the average and deviation of Q-values for all $(s,a)$ pairs, agents in the $n + 1^{th}$ task can initialize their Q-values to the previously seen average to learn faster. Additionally, agents can order their prioritized sweeping [20] updates based on the average and deviation of each $(s,a)$ pair to gain additional learning speed advantages.

Lastly, two works consider multi-task learning in a Bayesian model-based RL [21] setting. First, Sunmola and Wyatt [22] introduce two methods that use instances from previous tasks to set priors in a Bayesian learner. Initial experiments show that given an accurate estimation of the prior distributions, an agent may learn a novel task faster. Second, Wilson et. al [10] consider learning in a hierarchical Bayesian RL setting over multiple MDP distributions. By setting priors based on previously learned tasks, a new task in a particular distribution can be learned significantly faster. A simple parameterized reward function and the location of an absorbing goal state may change between the different tasks.

TIMBREL differs from previous work along a number of dimensions. Most important, inter-task mappings allow TIMBREL to transfer knowledge suitable for model-learning RL agents, even when transfer is between MDPs with different state variables and actions. Additionally, TIMBREL can run on-line, is not limited to discrete domains, and is designed for transfer (as opposed to multi-task learning).

## 3 Model Transfer

Model-based algorithms learn to estimate the transition model of an MDP, predicting the effects of actions. The goal of transfer for model-based RL algorithms is to allow the agent to build such a model from data gathered both in a previous task, as well as in the current task. To help frame the exposition, we note that transfer methods must typically perform the following three steps:

1. Use the source task agent to record some information during, after, or about, learning. Successful TL approaches include recording learned action-value functions or higher-level advice about high-value policies.
2. Transform the saved source task information so that it applies to the target task. This step is most often necessary if the states and actions in the two tasks are different, as considered in this paper.
3. Utilize the transformed information in the target task. Successful approaches include using source task information to initialize the learner's action-value function, giving advice about actions, and suggesting potentially useful sequences of actions (i.e., *options*).

The following section introduces TIMBREL, a novel transfer method, which accomplishes these steps. Later, in Section 5, we detail how TIMBREL is used in our test domain with Fitted R-MAX, our chosen model-based RL algorithm.

### 3.1 Instance-Based Model Transfer

This section provides an overview of TIMBREL. In order to transfer a model, our method takes the novel approach of transferring observed instances from the source task. The tuples, in the form $(s, a, r, s')$, describe experience the source task agent gathered while interacting with its environment (Step 1). One advantage of this approach as compared to transferring an action-value function or a full environmental model (e.g., the transition function) is that the source task agent is not tied to a particular learning algorithm or representation: whatever RL algorithm that learns will necessarily have to interact with the task and collect experience. This flexibility allows a source task algorithm to be selected based on characteristics of the task, rather than on demands of the transfer algorithm.

To translate a source task tuple into an appropriate target task tuple (Step 2) we utilize *inter-task mappings* [6], which have been successfully used in past transfer learning research to specify how pairs of tasks are related via an action mapping and a state variable mapping. This pair of mappings identifies source task actions which have similar effects as target task actions, and allows a mapping of target task state variables into source task state variables.

When learning in the target task, TIMBREL specifies when to use source task instances to help construct a model of the target task (Step 3). Briefly, when insufficient target task data exists to estimate the effect of a particular $(\mathbf{x}, a)$ pair, instances from the source task are transformed via an action-dependant inter-task mapping, and are then treated as a previously observed transition in the target task model. The TIMBREL method is summarized in Algorithm 2.

Notice that TIMBREL performs the translation of data from the source task to the target task (line 10) on-line while learning the target task. Transfer algorithms more commonly performed such translations off-line, before training in the target task, but this just-in-time approach is justified because of how the source data are utilized. In Section 5, we detail how the current state, x, will affect how the source task sample is translated in our particular task domain. Only transferring instances that will be immediately used in the target task helps to limit computational costs. Furthermore, this method will minimize the number of source instances that must be reasoned over in the target task model by only transferring necessary source task data.

**Algorithm 2.** TIMBREL Overview

```
1: Learn in the source task, recording (s, a, r, s') transitions.
2: Provide recorded transitions to the target task agent.
3: while training in the target task do
4:   if the model-based RL algorithm is unable to accurately estimate some T(x, a)
     or R(x, a) then
5:     while T(x, a) or R(x, a) does not have sufficient data do
6:       Locate 1 or more saved instances that, according to the inter-task mappings,
         are near the current ⟨x, a⟩ to be estimated.
7:       if no such unused source task instances exist then
8:         exit the inner while loop
9:       Use ⟨x, a⟩, the saved source task instance, and the mappings to translate
         the saved instance into one appropriate to the target task.
10:      Add the transformed instance to the current model for ⟨x, a⟩.
```

# 4 Generalized Mountain Car

This section introduces our experimental domain, a generalized version of the standard RL benchmark mountain car task [12]. Mountain car is an appropriate testbed for TIMBREL with Fitted R-MAX because it is among the simplest continuous domains that can benefit from model-based learning, and it is easily generalizable to enable TL experiments.

In mountain car, the agent must generalize across continuous state variables in order to drive an underpowered car up a mountain to a goal state. We also discuss 3D mountain car [23], an extension of the 2D task. In both tasks the transition and reward functions are initially unknown. The agent begins at rest at the bottom of the hill.[2] The reward for each time step is $-1$. The episode ends, and the agent is reset to the start state, after 500 time steps or if it reaches the goal state.

### 4.1 Two Dimensional Mountain Car

In the two dimensional mountain car task, two continuous variables fully describe the agent's state (see Figure 1). The horizontal position ($x$) and velocity ($\dot{x}$) are restricted to the ranges $[-1.2, 0.6]$ and $[-0.07, 0.07]$ respectively. The agent may select one of three actions on every timestep; {`Left`, `Neutral`, `Right`} change the velocity by -0.0007, 0, and 0.0007 respectively.[3] Additionally, $-0.025(\cos(3x))$ is added to $\dot{x}$ on every timestep to account for the x-component of the force of gravity on the car, which depends on the local slope of the mountain. The start state is $(x = -\frac{\pi}{6}, \dot{x} = 0)$, and the goal states are those where $x \geq 0.5$. We use the publicly available[4] version of this code for our experiments.

The transfer experiments in this paper (Section 6) use three variants of the 2D mountain car task. The first, which we will call the *Standard 2D task* is described in the previous paragraph. The *No Goal 2D task* is the same as the standard task, except that goal state has been removed. This task will be used to show how the effectiveness of transfer changes when the reward function changes. The third variant, the *High Power 2D task*, changes the car so that the velocity is changed by $\pm 0.0015$: the car has more than twice the acceleration of the Standard 2D task car. This variant will be used to show how transfer efficacy changes when the source task transition function changes.

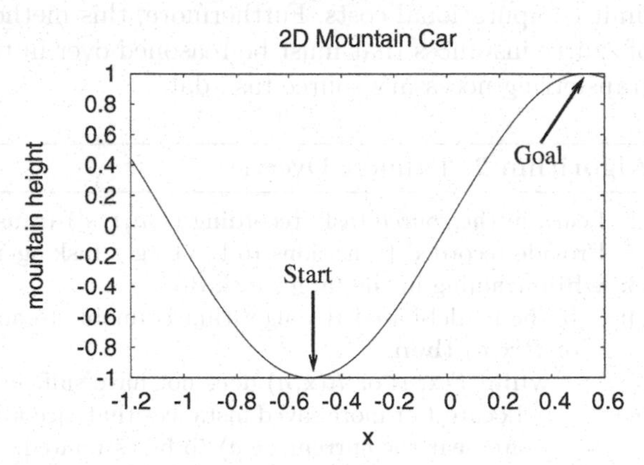


**Fig. 1.** In the standard 2D mountain car the agent must travel along a curve (mountain)

---

[2] Both mountain car tasks are deterministic, and Fitted R-MAX's exploration uses a fixed random seed. To introduce randomness and allow multiple learning trials, when each domain is initialized, $x$ (and $y$ in 3D) in the start state is perturbed by a random number in $[-0.005, 0.005]$.

[3] In the original formulation, the velocity was changed by $\pm 0.001$ due to acceleration. We have reduced the power of the car to make the task more challenging.

[4] Available at `http://rlai.cs.ualberta.ca/RLR/MountainCarBestSeller.html`

### 4.2 Three Dimensional Mountain Car

The 3D task [23] extends the mountain's curve into a surface (see Figure 2).[5] The state is composed of four continuous state variables: $x, \dot{x}, y, \dot{y}$. The positions and velocities have ranges of $[-1.2, 0.6]$ and $[-0.07, 0.07]$, respectively. The agent selects from five actions at each timestep:{`Neutral`, `West`, `East`, `South`, `North`}. West and East modify $\dot{x}$ by -0.0007 and +0.0007 respectively, while South and North modify $\dot{y}$ by -0.0007 and +0.0007 respectively.[6] The force of gravity adds $-0.025(\cos(3x))$ and $-0.025(\cos(3y))$ on each time step to $\dot{x}$ and $\dot{y}$, respectively. The goal region is defined by $x \geq 0.5$ and $y \geq 0.5$.

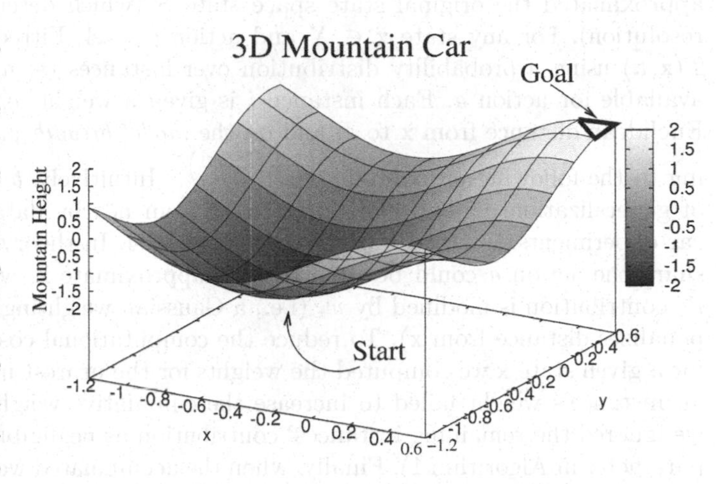


**Fig. 2.** In 3D mountain car the 2D curve becomes a 3D surface. The agent starts at the bottom of the hill with no kinetic energy and attempts to reach the goal area in the Northeast corner.

This task is more difficult than the 2D task because of the increased state space size and additional actions. Furthermore, since the agent can affect its acceleration in only one of the two spatial dimensions at any given time, one cannot simply "factor" this problem into the simpler 2D task. While data gathered from the 2D task should be able to help an agent learn the 3D task, we do expect that some amount of learning will be required after transfer.

### 4.3 Learning Mountain Car

Our experiments used Fitted R-MAX to learn policies in the mountain car tasks. We began by replicating the methods and result of applying Fitted R-MAX to 2D mountain car task as reported in the literature [15]. To apply Fitted R-MAX to 3D mountain car, we first scaled the state space so that each dimension ranges over the unit interval, effectively scaling the state space to a unit hypercube. We sampled a finite state space from this hypercube by applying a grid

[5] An animation of a trajectory from a trained policy can be found at `http://www.cs.utexas.edu/~mtaylor/3dMtnCar.html`.

[6] Although we call the agent's vehicle a "car," it does not turn but simply accelerates in the four cardinal directions.

where each position state variable can be one of 8 values, and each velocity state variable can be one of 9 values. The 3D version of mountain car has 2 of each type of state variable; we obtained a sample $X$ of $8^2 \times 9^2 = 5184$ states that approximated the original state space state $S$ (which determines the model's resolution). For any state $\mathbf{x} \in X$ and action $a \in A$, Fitted R-MAX estimates $T(\mathbf{x}, a)$ using a probability distribution over instances $(s_i, a, r_i, s'_i)$ in the data available for action $a$. Each instance $i$ is given a weight $w_i$ depending on the Euclidean distance from $\mathbf{x}$ to $s_i$ and on the *model breadth* parameter $b$, according to the following formula: $w_i \propto e^{-\left(\frac{|\mathbf{x}-s_i|}{b}\right)^2}$. Intuitively, $b$ controls the degree of generalization used to estimate $T(\mathbf{x}, a)$ from nearby data. In 3D mountain car experiments, we used a parameter of $b = 0.4$. In theory, all instances that share the action $a$ could be used to help approximate $\mathbf{x}$, where each instance $i$'s contribution is modified by $w_i$ (i.e., a Gaussian weighting that exponentially penalizes distance from $\mathbf{x}$). To reduce the computational cost of the algorithm, for a given state $\mathbf{x}$ we computed the weights for the nearest instances first. Once an instance's weight failed to increase the cumulative weight by at least 10%, we ignored the remaining instances' contribution as negligible (the *minFraction* parameter in Algorithm 1). Finally, when the accumulated weight failed to reach a threshold of 1.0, we used Fitted R-MAX's exploration strategy of assuming an optimistic transition to a maximum-reward absorbing state.

Changing the learning parameters for Fitted R-MAX outlined above affects three primary aspects of learning:

- How accurately the optimal policy can be approximated.
- How many samples are needed to accurately approximate the best policy, given the representation.
- How much computation is required when performing dynamic programming.

For this work, it was most important to find settings which allowed the agent to learn a reasonably good policy in relatively few episodes so that we could demonstrate the effectiveness of TIMBREL on sample complexity. We do not argue that the above parameters are optimal. They could be tuned to emphasize any of the above goals, such as achieving higher performance in the limit.

Figure 3 compares the average performance of 12 Fitted R-MAX trials with 12 Sarsa trials in the 3D mountain car task. The $\epsilon$-greedy Sarsa($\lambda$) agent uses a CMAC [24] function approximator with 14 4-dimensional linear tilings, which is analogous to how Singh and Sutton [12] used 14 2-d dimensional linear tile codings for their 2D task.[7] This result demonstrates that Fitted R-MAX can be tuned so that it learns with significantly less data (finding a path to the goal in roughly 50 episodes instead of 10,000 episodes), but does not necessarily achieve optimal performance. Learning with Fitted R-MAX takes substantially more computational resources than Sarsa in this domain; the Fitted R-MAX

[7] We achieved the best performance on this task by setting the learning rate to $\alpha = 0.5$, the exploration rate to $\epsilon = 0.1$, $\lambda = 0.95$, not decaying $\alpha$, and decaying $\epsilon$ at the end of every episode by 0.1%.

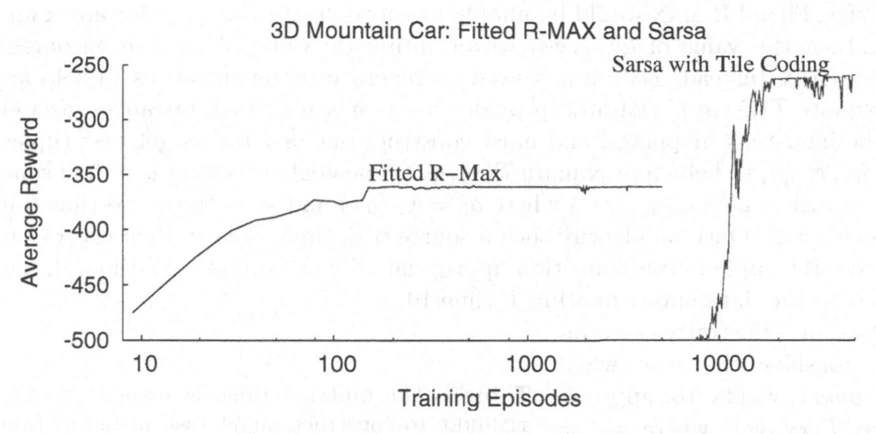


**Fig. 3.** Average learning curves for Fitted R-MAX and Sarsa show the significant speed advantage of model-based RL on the 3D mountain car task (note the log scale). Fitted R-MAX parameters where chosen for relatively low sample and computational complexity requirements at the expense of asymptotic performance.

learning curves were terminated once their performance plateaued (and thus are run for fewer episodes than Sarsa).

## 5 TIMBREL Implementation for Mountain Car

In this section we detail how TIMBREL is used to transfer between tasks in the mountain car domain when using Fitted R-MAX as the underlying RL algorithm. Although TIMBREL is a domain-independent transfer method which is designed to be compatible with multiple model-learning RL algorithms, we will ground our exposition in the context of Fitted R-MAX and mountain car. Throughout this section we use the subscript $\mathbb{S}$ to denote actions, states, and state variables in the source task, and the subscript $\mathbb{T}$ for the target task.

The core result of this paper is to demonstrate transfer between the Standard 2D mountain car task and the 3D mountain car task. After learning the 2D task, TIMBREL must be provided an inter-task mapping between the two tasks. The action mapping, $\chi_A$, maps a target task action into a source task action: $\chi_A(a_\mathbb{T}) = a_\mathbb{S}$, and $\chi_S$ maps a target task state variable into a source task state variable: $\chi_S(s_{(i,\mathbb{T})}) = s_{(j,\mathbb{S})}$. In this work we assume that the inter-task mapping in Table 1 is provided to the agent, but our past work [23] has demonstrated that a mapping between the 2D and 3D mountain car tasks may be learned autonomously. Note that the state variable mapping is defined so that either the target task state variables ($x$ and $\dot{x}$) *or* ($y$ and $\dot{y}$) are mapped into the source task. As we will discuss, the unmapped target task state variables will be set by the state variables' values in the state x that we wish to approximate.

As discussed in Section 2, Fitted R-MAX approximates transitions from a set of sample states $\mathrm{x} \in X$ for all actions. When the agent initially encounters the target task, no target task instances are available to approximate $T$. Without transfer, Fitted R-MAX would be unable to approximate $T(\mathbf{x}_\mathbb{T}, a_\mathbb{T})$ for any $\mathbf{x}$ and would set the value of $Q(s_\mathbb{T}, a_\mathbb{T})$ to an optimistic value ($R_{max}$) to encourage exploration. Instead, TIMBREL is used to generate target instances to help approximate $T(\mathbf{x}_\mathbb{T}, a_\mathbb{T})$. TIMBREL provides a set of source task instances, as well as the inter-task mappings, and must construct one or more target task tuples, $(s_\mathbb{T}, a_\mathbb{T}, r, s'_\mathbb{T})$, to help approximate $T(\mathbf{x}_\mathbb{T}, a_\mathbb{T})$. The goal of transfer is to find some source task tuple $(s_\mathbb{S}, a_\mathbb{S}, r, s'_\mathbb{S})$ where $a_\mathbb{S} = \mathcal{X}_A(a_\mathbb{T})$ and $s_\mathbb{S}$ is "near" $s_\mathbb{T}$ (line 6 in Algorithm 2). Once we identify such a source task tuple, we can then use $\mathcal{X}^{-1}$ to convert the tuple into a transition appropriate for the target task (line 9), and add it to the data approximating $T$ (line 10).

As an illustrative example, consider the case when the agent wants to approximate $T(\mathbf{x}_\mathbb{T}, a_\mathbb{T})$, where $\mathbf{x}_\mathbb{T} = \langle x_\mathbb{T}, y_\mathbb{T}, \dot{x}_\mathbb{T}, \dot{y}_\mathbb{T}\rangle = \langle -0.6, -0.2, 0, 0.1\rangle$ and $a_\mathbb{T} =$ East. TIMBREL considers source task transitions that contain the action Right. $\mathcal{X}_S$ is defined so that either the $x$ or $y$ state variables can be mapped from the target task to the source task, which means that we should consider two transitions selected from the source task instances. The first tuple is selected to minimize the Euclidean distances $D(x_\mathbb{T}, x_\mathbb{S})$ and $D(\dot{x}_\mathbb{T}, \dot{x}_\mathbb{S})$, where each distance is scaled by the range of the state variable. The second tuple is chosen to minimize $D(y_\mathbb{T}, x_\mathbb{S})$ and $D(\dot{y}_\mathbb{T}, \dot{x}_\mathbb{S})$.

**Table 1.** This table describes the mapping used by TIMBREL to construct target task instances from source task data

Inter-task Mapping for Mountain Car

Inter-task Mapping for Mountain Car

| Action Mapping | State Variable Mapping |
|---|---|
| $\mathcal{X}_A(\text{Neutral}) = \text{Neutral}$ | $\mathcal{X}_S(x) = x$ |
| $\mathcal{X}_A(\text{North}) = \text{Right}$ | $\mathcal{X}_S(\dot{x}) = \dot{x}$ |
| $\mathcal{X}_A(\text{East}) = \text{Right}$ | or |
| $\mathcal{X}_A(\text{South}) = \text{Left}$ | $\mathcal{X}_S(y) = x$ |
| $\mathcal{X}_A(\text{West}) = \text{Left}$ | $\mathcal{X}_S(\dot{y}) = \dot{x}$ |

Continuing the example, suppose that the first source task tuple selected was

$$(\langle -0.61, 0.01\rangle, \text{Right}, -1, \langle -0.59, 0.02\rangle).$$

If the inter-task mapping defined mappings for the $x$ and $y$ state variables simultaneously, the inverse inter-task mapping *could* be used to convert the tuple into

$$(\langle -0.61, -0.61, 0.01, 0.01\rangle, \text{East}, -1, \langle -0.59, -0.59, 0.02, 0.02\rangle).$$

However, this point is not near the current $\mathbf{x}_\mathbb{T}$ we wish to approximate. Instead, we recognize that this sample was selected from the source task to be near to $x_\mathbb{T}$ and $\dot{x}_\mathbb{T}$, and transform the tuple, assuming that $y_\mathbb{T}$ and $\dot{y}_\mathbb{T}$ are kept constant. With this assumption, we form the target task tuple

$$(\langle -0.61, y_\mathbb{T}, 0.01, \dot{y}_\mathbb{T}\rangle, \text{East}, -1, \langle -0.59, y_\mathbb{T}, 0.2, \dot{y}_\mathbb{T}\rangle) =$$

$$(\langle -0.61, -0.2, 0.01, 0.1\rangle, \text{East}, -1, \langle -0.59, -0.2, 0.02, 0.1\rangle).$$

The analogous step is then performed for the second selected source task tuple: the source task tuple is transformed with $\mathcal{X}$, and $x_\mathbb{T}$ and $\dot{x}_\mathbb{T}$ are held constant. Finally, both transferred instances are added to the $T(\mathbf{x}, a)$ approximation.

TIMBREL thus transfers pairs of source task instances to help approximate the transition function. Other model-learning methods may need constructed trajectories instead of individual instances, but TIMBREL is able to generate trajectories as well. Over time, the learner will approximate $T(\mathbf{x}_{\mathbb{T}}, a_{\mathbb{T}})$ for different values of $(\mathbf{x}, a)$ in order to construct a model for the target task environment. Any model produced via this transfer may be incorrect, depending on how representative the saved source task instances are of the target task (as modified by $\mathcal{X}$). However, our experiments demonstrate that using transferred data may allow a model learner to produce a model that is more accurate than if the source data were ignored.

As discussed in Section 4.3, Fitted R-MAX uses the distance between instances and $\mathbf{x}$ to calculate instance weights. When an instance is used to approximate $\mathbf{x}$, that instance's weight is added to the total weight of the approximation. If the total weight for an approximation does not reach a threshold value of 1.0, an optimistic value ($R_{max}$) is used because not enough data exists for an accurate approximation. When using TIMBREL, the same calculation is performed, but now instances from both the source task and target task can be used.

As the agent interacts with the target task, more transitions are recorded and the approximations of the transition function at different $(\mathbf{x}, a)$ pairs need to be recalculated based on the new information. Each time an approximation needs to be recomputed, Fitted R-MAX first attempts to use only target task data. If the number of instances available (where instances are weighted by their distance from $\mathbf{x}$) does not exceed the total weight threshold, source task data is transferred to allow an approximation of $T(\mathbf{x}_{\mathbb{T}}, a_{\mathbb{T}})$. This process is equivalent to removing transferred source task data from the model as more target task data is observed and therefore allows the model's accuracy to improve over time. Again, if the total weight from source task and target tasks instances for an approximated $\mathbf{x}$ does not reach 1.0, $R_{max}$ is assigned to the model for $\mathbf{x}$.

As a final implementation note, consider what happens when some $\mathbf{x}$ maps to an $s_{\mathbb{S}}$ that is not near any experienced source task data. If there are no source task transitions near $s_{\mathbb{S}}$, it is possible that using all available source task data will not produce an accurate approximation (recall that instance weights are proportional to the square of the distance from the instance to $\mathbf{x}$). To avoid a significant reduction in performance with limited improvement in approximating $T$, we imposed a limit of 20 source task tuples when approximating a particular point (line 5). This threshold serves a similar purpose as the 10% cumulative weight threshold discussed in Section 4.3.

## 6 Transfer Experiments

To test the efficacy of TIMBREL, we first conducted an experiment to measure the learning speed of Fitted R-MAX in the mountain car domain with and without TIMBREL. Roughly 50 different sets of Fitted R-MAX parameters were used in preliminary experiments to select the best settings for learning the 3D task without transfer (as discussed in Section 4.3). We ran 12 trials for 4,000 episodes and found that 10 out of 12 trials were able to converge to a policy that found the goal area. Recall that Fitted R-MAX is not guaranteed to converge to an optimal policy because it depends on approximation in a continuous state space.

To transfer from the Standard 2D mountain car task into the more complex 3D mountain car, we first allow 12 Fitted R-MAX agents to train for 100 episodes each in the 2D task while recording all observed $(s, a, r, s')$ transitions.[8] We then used TIMBREL to train agents in the target task for 1,000 episodes. 12 out of 12 trials converged to a policy that found the goal area.

After learning, we averaged over all trials for the non-transfer and transfer learning trials. For clarity, we also smoothed the curves by averaging over a 10 episode window. Figure 4(a) shows the first 1000 episodes of training (running the experiments longer than 1,000 episodes did not significantly improve the policy, as suggested by Figure 3). T-tests determined that all the differences in the averages were statistically significant ($p < 0.05$), with the exception of the initial average at episode 9. This result confirms that transfer can significantly improve the performance of agents in the 3D mountain car task.

We hypothesize that the U-shaped transfer learning curve is caused by a group of agents that find an initial path to the goal, spend some number of episodes exploring to find a faster path to the goal, and ultimately return to the original policy (see Figure 4(b)). In addition to improved initial performance, the asymptotic performance is improved, in part because some of the non-transfer tasks failed to successfully locate the goal. The difference in success rates (10 of 12 trials reaching the goal vs. 12 of 12) suggest that transfer may make difficult problems more tractable.

TIMBREL, and its implementation, were designed to minimize sample complexity. However, it is worth noting that there is a significant difference in the computational complexity of the transfer and non-transfer methods. Every time the transfer agent needs to use source task data to estimate $T$, it must locate the most relevant data and then insert it into the model. Additionally, the transfer agent has much more data available initially, and thus its dynamic programming step is significantly slower than the non-transfer agent.[9] These factors cause the transfer learning trials to take roughly twice as much wall clock time as the non-transfer trials. While our code could be better optimized, using the additional transferred data will always slow down the agent, relative to an agent that is not using transfer, but is running for the same number of episodes. However, in many domains a tradeoff of increasing computational requirements and reducing sample complexity is highly advantageous, and is one of the benefits inherent to model-based reinforcement learning.

[8] We experimented with roughly 10 different parameter settings for Fitted R-MAX in the Standard 2D task. Every episode lasts 500 time steps if the goal is not found and the 2D goal state can be reached in roughly 150 time steps. When learning 2D Mountain car, the agent experienced an average of 24,480 source task transitions during the 100 source task episodes.

[9] An analysis of the increase in computational complexity depends on the amount of data transferred into a target task, which in turn depends on the pair of tasks used.

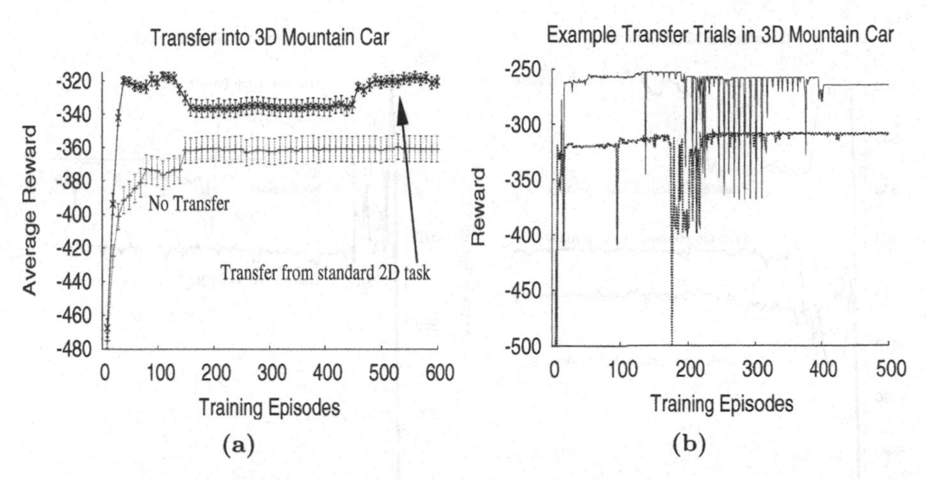


**Fig. 4. (a)** TIMBREL significantly improves the speed of Fitted R-MAX on the 3D mountain car. The average performance is plotted every 10 episodes along with the standard error. **(b)** As Fitted R-MAX explores, the performance can vary significantly, sometimes resulting in a U-shaped learning curve.

Our second experiment examines how the amount of recorded source task data affects transfer. One hypothesis was that more tuples in the source task would equate to higher performance in the target task, because the target task agent would have more data to draw from, and thus would be better able to approximate any given $T(\mathbf{x}, a)$.

We first ran experiments in the Standard 2D task for 5, 10, and 20 episodes, where the average number of steps per episode was 422, 298, 238, respectively.[10] Figure 5(a) shows that transfer from 20 source task episodes is similar to using 100 source task episodes and performs statistically better than no transfer at the 95% level for 98 of the 100 points graphed. While transfer performance degrades for trials that use 10 and 5 source task episodes, both trials do show a statistically significant boost to the agents' *initial* learning performance. This result demonstrates that a significant amount of information can be learned in just a few source task episodes; the source task is less complex than the target and thus a short amount of time spent learning in the source may have a large impact on the target task performance.

Recall the mountain car has a reward of $-1$ on each time step. The agent learns to reach the goal area because transitioning into this area ends the episode and the steady stream of negative reward. The third experiment uses the No Goal 2D task as a source task to examine how changing the reward function in the source task affects transfer. When training in the source task, every episode lasted 500 time steps (the maximum number of steps). After learning for 100 episodes in the source task, we transferred into the target task and found that 9 of the 12 trials successfully discovered policies to reach the goal area. Figure 5(b) suggests that

[10] The average number of steps per episode decreases for longer trials because the agent quickly learns to find the goal.

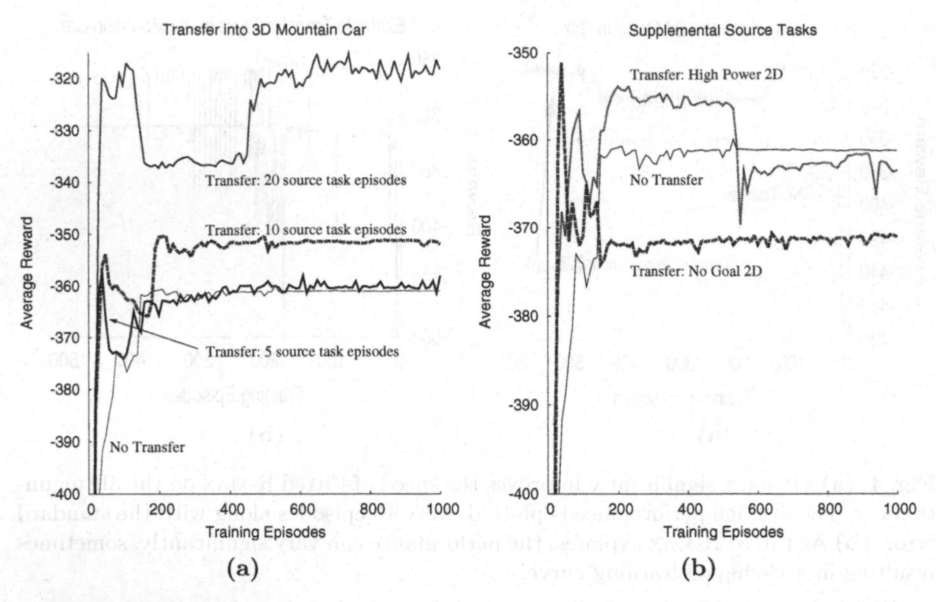


**Fig. 5.** **(a)** This graph shows the effect of different amounts of source task training. Each learning curve is the average of 12 independent trials. **(b)** Transfer from a 2D mountain car task that has no goal state or from a 2D mountain car with significantly stronger acceleration produces statistically significant improvements at the beginning of learning when compare to learning without transfer. However, this relative advantage is lost as the target task agents gain more experience.

transfer from a source task policy with a different reward structure can be initially useful (t-tests confirm that transfer outperforms non-transfer for four of the first five points graphed), but the relative performance of the non-transfer trials soon outperform that of learning with transfer.

Our fourth experiment uses the High Power 2D task as a source task. We again record 100 episodes worth of data for source task learners and use TIMBREL to transfer into 3D mountain car. Because the source task uses a car with a motor more than twice as powerful as in the 3D task, the transition function learned in the source task is less useful to the agent in the target task. 9 of the 12 target task trials successfully converged to a policy that reached the goal. Transferring from the High Power 2D task (Figure 5(b)) is not as useful as transferring from the Standard 2D Task (Figure 5(a)) due to differences in the transition functions. Although t-tests show that there is a statistically significant improvement at the beginning of learning, the transfer and non-transfer curves in Figure 5(b) quickly become statistically indistinct with more target task training.

Figure 5(b) highlights an important drawback of transfer learning. Transfer efficacy is often affected by the similarity of source tasks and target tasks, and in some circumstances transfer may not help the learner. Indeed, other experiments (not shown) confirm that if $T$ or $R$ in the source and target tasks are too dissimilar, transfer may actually cause the learner to learn more slowly than if it had not used transfer. While there is not yet a general solution to avoiding *negative transfer*, our recent results [23] suggest that the "relatedness" of tasks may be possible to measure empirically, and may guide learners when deciding whether or not to transfer.

# 7 Conclusion and Future Work

In this paper we have introduced TIMBREL, a transfer method fully compatible with model-based reinforcement learning. We demonstrate that when learning 3D mountain car with Fitted R-MAX, TIMBREL can significantly reduce the sample complexity and demonstrated how transfer is affected by changes to the source task's reward and transfer functions.

There are a number of future research directions suggested by this work. It would be informative to study how transfer efficacy changes when the amount of exploration is changed in the source task. Put differently, if the agent has 100 episodes to learn the source task, can it intelligently set its parameters to maximize transfer efficacy? In our experiments we used the default threshold value of 1.0 to determine if a particular approximation of $T(\mathbf{x}_{\mathrm{T}}, a_{\mathrm{T}})$ has enough data. This is related to the amount of exploration and its value may impact the efficacy of transfer, but tuning this parameter is left to future work.

All parameters for Fitted R-MAX were tuned when learning without transfer. It is possible that the model breadth parameter, $b$, may change the efficacy of transfer. Section 5 specified that a maximum of 20 source task instances were used to approximate a single target task transition. This parameter was set during initial experimentation, but further tuning could improve transfer performance. Lastly, none of the experiments using Fitted R-MAX attained an asymptotic performance equivalent to Sarsa (Figure 3). It may be worth re-tuning the base learning algorithm's parameters to maximize asymptotic performance (at the expense of computational and sample complexity), and then determine if TIMBREL can compensate so that learning experiments terminate in a reasonable amount of time.

We predict that TIMBREL will work, possibly with minor modifications, in other model-based RL algorithms. For instance, TIMBREL could be directly used in the planning phase of *Dyna-Q* [1] as a source of simulated experience when the agent's model is poor (such as at the beginning of learning a target task). TIMBREL should also be useful, without modification, in R-MAX.

Lastly, we intend to apply TIMBREL in more complex domains with continuous state spaces (which may show relatively more benefit from transfer than the simple tasks discussed in Section 6 [6]). Although this paper focuses on tasks in the mountain car domain, Algorithm 2 is applicable in many settings. Future work to empirically determine how well TIMBREL functions in other domains, and when applied to pairs of tasks with qualitative differences not explored in this paper, will help to better understand and quantify the benefits that instance transfer provides.

## Acknowledgments

We would like to thank Lilyana Mihalkova, Joseph Reisinger, Raymond J. Mooney, and the anonymous reviewers for helpful comments and suggestions. This research was supported in part by DARPA grant HR0011-04-1-0035, NSF CAREER award IIS-0237699, and NSF award EIA-0303609.

## References

1. Sutton, R.S., Barto, A.G.: Introduction to Reinforcement Learning. MIT Press, Cambridge (1998)
2. Tesauro, G.: TD-Gammon, a self-teaching backgammon program, achieves master-level play. Neural Computation 6(2), 215–219 (1994)
3. Ng, A.Y., Coates, A., Diel, M., Ganapathi, V., Schulte, J., Tse, B., Berger, E., Liang, E.: Inverted autonomous helicopter flight via reinforcement learning. In: International Symposium on Experimental Robotics (2004)
4. Kohl, N., Stone, P.: Machine learning for fast quadrupedal locomotion. In: The Nineteenth National Conference on Artificial Intelligence, pp. 611–616 (July 2004)
5. Torrey, L., Walker, T., Shavlik, J., Maclin, R.: Using advice to transfer knowledge acquired in one reinforcement learning task to another. In: Gama, J., Camacho, R., Brazdil, P.B., Jorge, A.M., Torgo, L. (eds.) ECML 2005. LNCS (LNAI), vol. 3720, pp. 412–424. Springer, Heidelberg (2005)
6. Taylor, M.E., Stone, P., Liu, Y.: Transfer learning via inter-task mappings for temporal difference learning. Journal of Machine Learning Research 8(1), 2125–2167 (2007)
7. Lazaric, A., Restelli, M., Bonarini, A.: Transfer of samples in batch reinforcement learning. In: Proceedings of the 25th Annual ICML, pp. 544–551 (2008)
8. Widmer, G., Kubat, M.: Learning in the presence of concept drift and hidden contexts. Machine Learning 23(1), 69–101 (1996)
9. Caruana, R.: Multitask learning. Machine Learning 28, 41–75 (1997)
10. Wilson, A., Fern, A., Ray, S., Tadepalli, P.: Multi-task reinforcement learning: a hierarchical bayesian approach. In: ICML 2007: Proceedings of the 24th international conference on Machine learning, pp. 1015–1022. ACM Press, New York (2007)
11. Watkins, C.J.C.H.: Learning from Delayed Rewards. PhD thesis, King's College, Cambridge, UK (1989)
12. Singh, S., Sutton, R.S.: Reinforcement learning with replacing eligibility traces. Machine Learning 22, 123–158 (1996)
13. Ng, A.Y., Jordan, M.: PEGASUS: A policy search method for large MDPs and POMDPs. In: Proceedings of the 16th Conference on Uncertainty in Artificial Intelligence (2000)
14. Brafman, R.I., Tennenholtz, M.: R-Max – a general polynomial time algorithm for near-optimal reinforcement learning. Journal of Machine Learning Research 3, 213–231 (2002)
15. Jong, N.K., Stone, P.: Model-based exploration in continuous state spaces. In: The Seventh Symposium on Abstraction, Reformulation, and Approximation (July 2007)
16. Bellman, R.E.: Dynamic Programming. Princeton University Press, Princeton (1957)

17. Puterman, M.L.: Markov Decision Processes: Discrete Stochastic Dynamic Programming. John Wiley & Sons, Inc., Chichester (1994)
18. Atkeson, C.G., Santamaria, J.C.: A comparison of direct and model-based reinforcement learning. In: Proceedings of the 1997 International Conference on Robotics and Automation (1997)
19. Tanaka, F., Yamamura, M.: Multitask reinforcement learning on the distribution of MDPs. Transactions of the Institute of Electrical Engineers of Japan. C 123(5), 1004–1011 (2003)
20. Moore, A.W., Atkeson, C.G.: Prioritized sweeping: Reinforcement learning with less data and less time. Machine Learning 13, 103–130 (1993)
21. Dearden, R., Friedman, N., Andre, D.: Model based bayesian exploration. In: Proceedings of the 1999 conference on Uncertainty in AI, pp. 150–159 (1999)
22. Sunmola, F.T., Wyatt, J.L.: Model transfer for Markov decision tasks via parameter matching. In: Proceedings of the 25th Workshop of the UK Planning and Scheduling Special Interest Group (PlanSIG 2006) (December 2006)
23. Taylor, M.E., Kuhlmann, G., Stone, P.: Autonomous transfer for reinforcement learning. In: The Seventh International Joint Conference on Autonomous Agents and Multiagent Systems (May 2008)
24. Albus, J.S.: Brains, Behavior, and Robotics. Byte Books, Peterborough (1981)

# A Simple Model for Sequences of Relational State Descriptions

Ingo Thon, Niels Landwehr, and Luc De Raedt

Department of Computer Science, Katholieke Universiteit Leuven, Celestijnenlaan 200A, 3001 Heverlee, Belgium
{firstname.lastname}@cs.kuleuven.be

**Abstract.** Artificial intelligence aims at developing agents that learn and act in complex environments. Realistic environments typically feature a variable number of objects, relations amongst them, and non-deterministic transition behavior. Standard probabilistic sequence models provide efficient inference and learning techniques, but typically cannot fully capture the relational complexity. On the other hand, statistical relational learning techniques are often too inefficient. In this paper, we present a simple model that occupies an intermediate position in this expressiveness/efficiency trade-off. It is based on CP-logic, an expressive probabilistic logic for modeling causality. However, by specializing CP-logic to represent a probability distribution over sequences of relational state descriptions, and employing a Markov assumption, inference and learning become more tractable and effective. We show that the resulting model is able to handle probabilistic relational domains with a substantial number of objects and relations.

## 1 Introduction

One of the current challenges in artificial intelligence is the modeling of dynamic environments that change due to actions and activities people or other agents take. As one example, consider a model of the activities of a cognitively impaired person [1]. Such a model could be used to assist persons, using common patterns to generate reminders or detect potentially dangerous situations, and thus help to improve living conditions.

As another example and one on which we shall focus in this paper, consider a model of the environment in a *massively multi-player online game* (MMOG). These are computer games that support thousands of players in complex, persistent, and dynamic virtual worlds. They form an ideal and realistic test-bed for developing and evaluating artificial intelligence techniques and are also interesting in their own right (cf. also [2]). One challenge in such games is to build a dynamic probabilistic model of high-level player behavior, such as players joining or leaving alliances and concerted actions by players within one alliance. Such a model of human cooperative behavior in this type of world can be useful in several ways. Analysis of in-game social networks are not only interesting from a sociological point of view but could also be used to visualize aspects of the gaming environment or give advice to inexperienced players (e.g., which alliance to join). More ambitiously, the model could be used to build computer-controlled players that mimic the cooperative behavior of human players, form alliances and jointly pursue

W. Daelemans et al. (Eds.): ECML PKDD 2008, Part II, LNAI 5212, pp. 506–521, 2008.

goals that would be impossible to attain otherwise. Mastering these social aspects of the game will be crucial to building smart and challenging computer-controlled opponents, which are currently lacking in most MMOGs. Finally, the model could also serve to detect non-human players in todays MMOGs — accounts which are played by automatic scripts to give one player an unfair advantage, and are typically against game rules.

From a machine learning perspective, this type of domain poses three main challenges: 1) world state descriptions are inherently relational, as the interaction between (groups of) agents is of central interest, 2) the transition behavior of the world will be strongly stochastic, and 3) a relatively large number of objects and relations is needed to build meaningful models, as the defining element of environments such as MMOGs are interactions among *large* sets of agents. Thus, we need an approach that is both computationally efficient and able to represent complex relational state descriptions and stochastic world dynamics.

Artificial intelligence has already contributed a rich variety of different modeling approaches, for instance, Markov models [3] and decision processes [4], dynamic Bayesian networks [5], STRIPS [6], statistical relational learning representations [7], etc. Most of the existing approaches that support reasoning about uncertainty (and satisfy requirement 2) employ essentially propositional representations (for instance, dynamic Bayesian networks, Markov models, etc.), and are not able to represent complex relational worlds, and hence, do not satisfy requirement 1). A class of models that integrates logical or relational representations with methods for reasoning about uncertainty (for instance, Markov Logic, CP-logic, or BLPs) is considered within statistical relational learning [7] and probabilistic inductive logic programming [8]. However, the inefficiency of inference and learning algorithms causes problems in many realistic applications, and hence, such methods do not satisfy requirement 3).

We aim to alleviate this situation, by contributing a novel representation, called CPT-L (for **CPT**ime-**L**ogic), that occupies an intermediate position in this expressiveness/efficiency trade-off. A CPT-L model essentially defines a probability distribution over sequences of interpretations. Interpretations are relational state descriptions that are typically used in planning and many other applications of artificial intelligence. CPT-L can be considered a variation of CP-logic [9], a recent expressive logic for modeling causality. By focusing on the sequential aspect and deliberately avoiding the complications that arise when dealing with hidden variables, CPT-L is more restricted, but also more efficient to use than its predecessor and alternative formalisms within the artificial intelligence and statistical relational learning literature.

This paper is organized as follows: Section 2 introduces the CPT-L framework; Section 3 addresses inference and parameter estimation, while Section 4 presents some experimental results in both the block's world and the MMOG Travian. Finally, Section 5 discusses related work, before concluding and touching upon future work in Section 6.

## 2 CPT-L

Let us first introduce some terminology. An *atom* is an expression of the form $p(t_1, ..., t_n)$ where $p/n$ is a *predicate symbol* and the $t_i$ are *terms*. Terms are built up from constants, variables and functor symbols. The set of all atoms is called a *language*

$\mathcal{L}$. *Ground* expressions do not contain variables. Ground atoms will be called *facts*. A substitution $\theta$ is a mapping from variables to terms, and $b\theta$ is the atom obtained from $b$ by replacing variables with terms according to $\theta$. We are interested in describing complex world states in terms of *relational interpretations*. A relational interpretation $I$ is a set of ground facts $a_1, ..., a_N$. A *relational stochastic process* defines a distribution $P(I_0, ..., I_T)$ over sequences of interpretations of length $T$, and thereby completely characterizes the transition behavior of the world.

The semantics of CPT-L is based on CP-logic, a probabilistic first-order logic that defines probability distributions over interpretations [9]. CP-logic has a strong focus on causality and constructive processes: an interpretation is incrementally constructed by a process that adds facts which are probabilistic *outcomes* of other already given facts (the *causes*). CPT-L combines the semantics of CP-logic with that of (first-order) Markov processes. Causal influences only stretch from $I_t$ to $I_{t+1}$ (Markov assumption), are identical for all time-steps (stationarity), and all causes and outcomes are observable. Models in CPT-L are also called CP-theories, and can be formally defined as follows:

**Definition 1.** *A* **CPT-theory** *is a set of rules of the form*

$$r = (h_1 : p_1) \vee \ldots \vee (h_n : p_n) \leftarrow b_1, \ldots, b_m$$

*where the $h_i$ are logical atoms, the $b_i$ are literals (i.e., atoms or their negation) and $p_i \in [0, 1]$ probabilities s.t. $\sum_{i=1}^{n} p_i = 1$.*

It will be convenient to refer to $b_1, ..., b_m$ as the body $body(r)$ of the rule and to $(h_1 : p_1) \vee \ldots \vee (h_n : p_n)$ as the head $head(r)$ of the rule. We shall also assume that the rules are range-restricted, that is, that all variables appearing in the head of the rule also appear in its body. Rules define conditional probabilistic events: the intuition behind a rule is that whenever $b_1\theta, ..., b_m\theta$ holds for a substitution $\theta$ in the current state $I_t$, exactly one of the $h_i\theta$ in the head will hold in the next state $I_{t+1}$. In this way, the rule models a (probabilistic) causal process as the condition specified in the body causes one (probabilistically chosen) atom in the head to become true in the next time-step.

*Example 1.* Consider the following CPT-rule:

$$(on(A, table) : 0.9) \vee (on(A, C) : 0.1) \leftarrow free(A), on(A, C), move(A, table).$$

which represents that we try to move a block $A$ from block $C$ to the table. This action succeeds with a probability of 0.9.

We now show how a CPT-theory defines a distribution over sequences $I_0, ..., I_T$ of relational interpretations. Let us first define the concept of the applicable ground rules in an interpretation $I_t$. From a CPT-theory, the rule $(h_1 : p_1\theta) \vee \ldots \vee (h_n : p_n\theta) \leftarrow b_1\theta, \ldots, b_m\theta$ is obtained for a substitution $\theta$. A ground rule $r$ is applicable in $I_t$ if and only if $body(r) = b_1\theta, \ldots, b_m\theta$ is true in $I_t$, denoted $I_t \models b_1\theta, \ldots, b_m\theta$.

One of the main features of CPT-theories is that they are easily extended to include *background knowledge*. The background knowledge $B$ can be any logic program, cf. [10]. In the presence of background knowledge, a ground rule is applicable in an interpretation $I_t$ if $b_1\theta, \ldots, b_m\theta$ can be logically derived from $I_t$ together with the logic program $B$, denoted $I_t \models_B b_1\theta, \ldots, b_m\theta$.

The set of all applicable ground rules in state $I_t$ will be denoted as $\mathbf{R}_t$. Each ground rule applicable in $I_t$ will cause one of its head elements to become true in $I_{t+1}$. More formally, let $\mathbf{R}_t = \{r_1, ..., r_k\}$. A *selection* $\sigma$ is a mapping $\{(r_1, j_1), ..., (r_k, j_k)\}$ from ground rules $r_i$ to indices $j_i$ denoting that head element $h_{ij_i} \in head(r_i)$ is selected. The probability of a selection $\sigma$ is

$$P(\sigma) = \prod_{i=1}^{k} p_{j_i}, \tag{1}$$

where $p_{j_i}$ is the probability associated with head element $h_{ij_i}$ in $r_i$. In the stochastic process to be defined, $I_{t+1}$ is a possible successor for the state $I_t$ if and only if there is a selection $\sigma$ such that $I_{t+1} = \{h_{1\sigma(1)}, ..., h_{k\sigma(k)}\}$. We shall say that $\sigma$ *yields* $I_{t+1}$ in $I_t$, denoted $I_t \xrightarrow{\sigma} I_{t+1}$, and define

$$P(I_{t+1}|I_t) = \sum_{\sigma : I_t \xrightarrow{\sigma} I_{t+1}} P(\sigma). \tag{2}$$

*Example 2.* Consider the theory

$$\begin{aligned} r_1 &= a : 0.2 \lor b : 0.8 \leftarrow \neg a, \neg b \\ r_2 &= a : 0.5 \lor b : 0.5 \leftarrow a \\ r_3 &= a : 0.7 \lor nil : 0.3 \leftarrow a \end{aligned}$$

Starting from $I_t = \{a\}$ only the rules $r_2$ and $r_3$ are applicable, so $\mathbf{R}_t = \{r_2, r_3\}$. The set of possible selections is

$$\{(r_2, j_2), (r_3, j_3) \mid j_2, j_3 \in \{1, 2\}\}.$$

The possible successor states $I_{t+1}$ are therefore

$$\begin{aligned} I^1_{t+1} &= \{a\} \text{ with } P(I^1_{t+1} \mid I_t) = 0.5 \cdot 0.7 + 0.5 \cdot 0.3 = 0.5 \\ I^2_{t+1} &= \{b\} \text{ with } P(I^2_{t+1} \mid I_t) = 0.5 \cdot 0.3 = 0.15 \\ I^3_{t+1} &= \{a, b\} \text{ with } P(I^3_{t+1} \mid I_t) = 0.5 \cdot 0.7 = 0.35 \end{aligned}$$

As for propositional Markov processes, the probability of a sequence $I_0, ..., I_T$ given an initial state $I_0$ is defined by

$$P(I_0, ..., I_T) = P(I_0) \prod_{t=0}^{T} P(I_{t+1} \mid I_t). \tag{3}$$

Intuitively, it is clear that this defines a distribution over all sequences of interpretations of length $T$ much as in the propositional case. More formally:

**Theorem 1 (Semantics of a CPT theory).** *Given an initial state $I_0$, a CPT-theory defines a discrete-time stochastic process, and therefore for $T \in \mathbb{N}$ a distribution $P(I_0, ..., I_T)$ over sequences of interpretations of length $T$.*

# 3 Inference and Parameter Estimation in CPT-L

As for other probabilistic models, we can now ask several questions about the introduced CPT-L model:

- **Sampling:** How to sample sequences of interpretations $I_0, ..., I_T$ from a given CPT-theory $\mathcal{T}$ and initial interpretation $I_0$?
- **Inference:** Given a CPT-theory $\mathcal{T}$ and a sequence of interpretations $I_0, ..., I_T$, what is $P(I_0, ..., I_T \mid \mathcal{T})$?
- **Parameter Estimation:** Given the structure of a CPT-theory $\mathcal{T}$ and a set of sequences of interpretations, what are the maximum-likelihood parameters of $\mathcal{T}$?
- **Prediction:** Let $\mathcal{T}$ be a CPT-theory, $I_0, ..., I_t$ a sequence of interpretations, and $F$ a first-order formula that constitutes a certain property of interest. What is the probability that $F$ holds at time $t+d$, $P(I_{t+d} \models_B F \mid \mathcal{T}, I_0, ..., I_t)$?

Sampling from a CPT-theory $\mathcal{T}$ given an initial interpretation $I_0$ is straightforward due to the causal semantics employed in CP-logic. For $t \geq 0$, $I_{t+1}$ can be constructed from $I_t$ by finding all groundings $r\theta$ of rules $r \in \mathcal{T}$, and sampling for each $r\theta$ a head element to be added to $I_{t+1}$. Algorithmic solutions for solving the inference, parameter estimation, and prediction problem will be presented in turn in the rest of this section.

## 3.1 Inference

Because of the Markov assumption (Equation 3), the crucial task for solving the inference problem is to compute $P(I_{t+1} \mid I_t)$ for given $I_{t+1}$ and $I_t$. According to Equation 2, this involves summing the probabilities of all selections yielding $I_{t+1}$ from $I_t$. However, the number of possible selections $\sigma$ is exponential in the number of ground rules $|\mathbf{R}_t|$ applicable in $I_t$, so a naive generate-and-test approach is infeasible. Instead, we present an efficient approach for computing $P(I_{t+1} \mid I_t)$ without explicitly enumerating all selections yielding $I_{t+1}$, which is strongly related to the inference technique discussed in [11]. The problem is first converted to a DNF formula over boolean variables such that assignments to variables correspond to selections, and satisfying assignments to selections yielding $I_{t+1}$. The formula is then compactly represented as a binary decision diagram (BDD), and $P(I_{t+1} \mid I_t)$ efficiently computed from the BDD using dynamic programming. Although finding satisfying assignments for DNF formulae is a hard problem in general, the key advantage of this approach is that existing, highly optimized BDD software packages can be used.

The conversion of a given inference problem to a DNF formula $f$ is realized as follows:

1. Initialize $f := true$
2. Compute applicable ground rules
$$\mathbf{R}_t = \{r\theta | body(r\theta) \text{ is true in } I_t\}$$
3. For all rules $(r = (p_1 : h_1, ..., p_n : h_n) \leftarrow b_1, ..., b_m)$ in $\mathbf{R}_t$ do:
   (a) $f := f \wedge (r.h_1 \vee ... \vee r.h_n)$, where $r.h$ denotes the proposition obtained by concatenating the name of the rule $r$ with the ground literal $h$ resulting in a new propositional variable $r.h$ (if not $h_i = nil$).
   (b) $f := f \wedge (\neg r.h_i \vee \neg r.h_j)$ for all $i \neq j$

4. For all facts $l \in I_{t+1}$
   (a) Initialize $g := false$
   (b) for all $r \in \mathbf{R}_t$ with $p : l \in head(r)$ do $g := g \vee r.l$
   (c) $f := f \wedge g$

Boolean variables of the form $r.h$ represent that head element $h$ was selected in rule $r$[1]. The second step of the algorithm computes all applicable rules, the third step assures that selections are obtained, and the final step assures that the selection generates the interpretation $I_{t+1}$. It is easily verified that the satisfying assignments for the formula $f$ correspond to the selections yielding $I_{t+1}$.

*Example 3.* The following formula $f$ is obtained for the transition $\{a\} \rightarrow \{a, b\}$ and the CPT-theory given in Example 2.

$$\underbrace{(r2.a \vee r2.b)}_{3.a} \wedge \underbrace{(\neg r2.a \vee \neg r2.b) \wedge (\neg r3.a \vee \neg r3.nil)}_{3.b} \wedge \underbrace{(r2.a \vee r3.a) \wedge r2.b}_{4}$$

The parts of the formula are annotated with the steps in the construction algorithm that generated them.

From the formula $f$, a *reduced ordered binary decision diagram* (BDD) [12] is constructed. Let $x_1, ..., x_n$ denote an ordered set of boolean variables (such as the $r.h$ contained in $f$). A BDD is a rooted, directed acyclic graph, in which nodes are annotated with variables and have out-degree 2, indicating that the variable is either true or false. Furthermore, there are two terminal nodes labeled with 0 and 1. Variables along any path from the root to one of the two terminals are ordered according to the given variable ordering. The graph compactly represents a boolean function $f$ over variables $x_1, ..., x_n$: given an instantiation of the $x_i$, we follow a path from the root to either 1 or 0 (indicating $f$ is true or false). Furthermore, the graph must be reduced, that is, it must not be possible to merge or remove nodes without altering the represented function (cf. [12] for details). Figure 1, left, shows an example BDD.

From the BDD graph, $P(I_{t+1} \mid I_t)$ can be computed in linear time using dynamic programming. This is realized by a straightforward modification of the algorithm for inference in ProbLog theories [11]. The algorithm exploits that paths in the BDD from the root node to the 1-terminal correspond to satisfying assignments for $f$, and thus selections yielding $I_{t+1}$. By sweeping through the BDD from top to bottom contributions from all such selections are summed up (Equation 2) without explicitly enumerating all paths. The efficiency of this method crucially depends on the size of the BDD graph, which in turn depends strongly on the chosen variable ordering $x_1, ..., x_n$. Unfortunately, computing an optimal variable ordering is NP-hard. However, existing implementations of BDD packages contain sophisticated heuristics to find a good ordering for a given function in polynomial time.

Interestingly, it is possible to further reduce complexity for the particular problem we are interested in by adapting a different semantics in the BDD. A *zero-suppressed binary decision* diagrams (or ZDD) is an alternative form of graphical representation

[1] Variables $r.h$ are standardized apart in case head elements coincide after grounding.

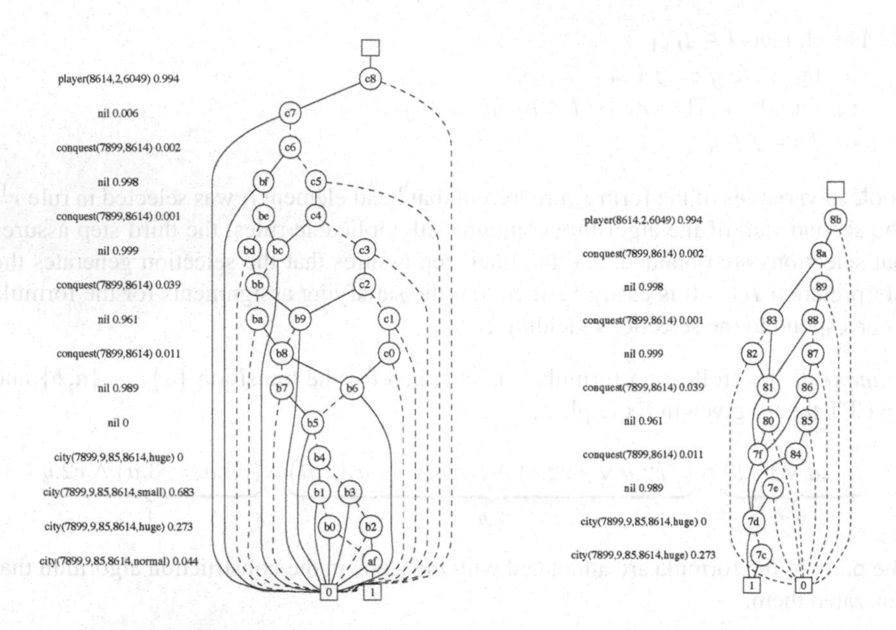


**Fig. 1.** Graphical representation of a formula $f$ resulting from the conversion of a CPT-L inference problem represented as a BDD (left) and ZDD (right)

in which variables appear in a path only if their positive branch is not directly connected to the terminal 0 [13]. Figure 1 shows example BDD and ZDD structures that represent the same function. We will now show that a reduced ZDD representation of $f$ will always be smaller than (or identical to) the corresponding BDD representation for CPT-L:

**Theorem 2.** *Let $f$ be a formula resulting from the conversion of a CPT-L inference problem, $G$ its BDD representation, and $G'$ its ZDD representation (for a fixed variable ordering). Then $size(G') \leq size(G)$.*

*Proof.* We first show that in $G$ every path $Q$ from the root to the 1-terminal contains all variables appearing in $f$. Assume $r.h_1 \notin Q$, and let $r.h_2, ..., r.h_l$ denote the variables corresponding to the other head elements of rule $r$. Because of the constraint added in step 3. of the conversion, $f$ can only be true if exactly one of the $r.h_1, ..., r.h_l$ is true. However, this cannot be verified by looking at any subset of the variables, and therefore they must all be contained in the path. Because all variables appear in every path from the root to 1, the graph structure $G$ is also a faithful representation of $f$ under the ZDD semantics. If $G$ as a ZDD is fully reduced, $G = G'$ because reduced ZDDs, as BDDs, are a canonical representation. Otherwise, $G$ can be further reduced to the ZDD $G'$ with $size(G') < size(G)$.

Typically a ZDD representation of $f$ will be more compact than the BDD representation, as shown in Figure 1.

### 3.2 Parameter Estimation

Assume the structure of a CPT-theory is given, that is, a set $\mathcal{T} = \{r_1, ..., r_k\}$ of rules of the form

$$r_i = (h_{i1} : p_{i1}) \vee \ldots \vee (h_{in} : p_{in}) \leftarrow b_{i1}, \ldots, b_{im},$$

where $\pi = \{p_{ij}\}_{i,j}$ are the unknown parameters to be estimated from a set of training sequences $D$. A standard approach is to find max-likelihood parameters $\pi^* = \arg\max_\pi P(D \mid \pi)$. To determine a model parameter $p_{ij}$, we essentially need to know the number of times head element $h_{ij}$ has been selected in an application of the rule $r_i$ in the training data, which will be denoted by $\kappa_{ij}$. However, the quantity $\kappa_{ij}$ is not directly observable. To see why this is so, first consider a single transition $I_t \rightarrow I_{t+1}$ in one training sequence. We know the set of rules $\mathbf{R}_t$ applied in the transition; however, there are in general many possible selections $\sigma$ of rule head elements yielding $I_{t+1}$. The information which selection was used, that is, which rule has generated which fact in $I_{t+1}$, is hidden. We will now derive an efficient Expectation-Maximization algorithm in which the unobserved variables are the selections used at every transition, and $\kappa_{ij}$ the sufficient statistics. To keep the notation uncluttered, we present the expectation step $\mathbb{E}[\kappa_{ij} \mid \pi, D]$ for a single transition $\tau = I_t \rightarrow I_{t+1}$; contributions from different transitions and different training sequences simply sum up. Let $\Gamma = \{\sigma \mid I_t \xrightarrow{\sigma} I_{t+1}\}$ denote the set of selections yielding $\tau$. The expectation is

$$\begin{aligned} \mathbb{E}[\kappa_{ij} \mid \pi, \tau] &= \sum_{\sigma} P(\delta_{ij} \mid \sigma, \pi, \tau) \\ &= \sum_{\sigma} P(\delta_{ij} \mid \sigma) P(\sigma \mid \pi, \tau) \\ &= \sum_{\sigma \in \Gamma} P(\delta_{ij} \mid \sigma) \frac{P(\sigma \mid \pi)}{\sum_{\sigma' \in \Gamma} P(\sigma' \mid \pi)} \end{aligned} \quad (4)$$

where $\delta_{ij}$ is an indicator variable representing that head $h_{ij}$ was selected in rule $r_i$. Note that $P(\delta_{ij} \mid \sigma)$ is simply 1 if the head is selected in $\sigma$ and 0 otherwise, and $P(\sigma \mid \pi)$ is defined by Equation 1. Given the expectation, the maximization step is

$$p_{ij}^{(new)} = \frac{\mathbb{E}[\kappa_{ij} \mid \pi, D]}{\sum_j \mathbb{E}[\kappa_{ij} \mid \pi, D]}.$$

The key algorithmic challenge is to compute the expectation given by Equation 4 efficiently. As outlined above, the set $\Gamma$ of selections yielding the observed transitions can be compactly represented as the set of paths from the root to the 1-terminal in a (possibly zero-suppressed) decision diagram.

By analogy to the inference problem, the summation given by Equation 4 can be performed in linear time given the BDD (ZDD) structure. This is realized by a dynamic programming algorithm similar to the forward-backward algorithm in hidden Markov models [3] that sweeps through the BDD structure twice to accumulate the sufficient statistics $\kappa_{ij}$. Details of the algorithm are straightforward but somewhat involved, and omitted for lack of space. Note that the presented Expectation-Maximization algorithm, by taking the special structure of our model into account, is significantly more efficient than general-purpose parameter learning techniques employed in CP-logic.

### 3.3 Prediction

Assume we are given a (partial) observation sequence $I_0, ..., I_t$, a CPT-theory $\mathcal{T}$, and a property of interest $F$ (represented as a first-order formula), and would like to compute $P(I_{t+d} \models_B F \mid I_0, ..., I_t, \mathcal{T})$. For instance, a robot might like to know the probability that a certain world state is reached at time $t + d$, given its current world model and observation history. Note that the representation as a first-order formula allows one to express richer world conditions than queries on (sets of) atoms, as they are typically supported in statistical relational learning systems. In CPT-L,

$$P(I_{t+d} \models_B F \mid I_0, ..., I_t, \mathcal{T}) = P(I_{t+d} \models_B F \mid I_t, \mathcal{T})$$

as the world model is Markov. Powerful statistical relational learning systems are in principle able to compute this quantity exactly by "unrolling" the world model into a large dynamic graphical model. However, this is computationally expensive as it requires to marginalize out all (unobserved) intermediate world states $I_{t+1}, ..., I_{t+d-1}$. In contrast, inference in CPT-theories draws its efficiency from the full observability assumption.

As an alternative approach, we propose a straightforward sample-based approximation to $P(I_{t+d} \models_B F \mid I_t, \mathcal{T})$. Given $I_t$, independent samples can be obtained from the conditional distribution $P(I_{t+1}, ..., I_{t+d} \mid I_t, \mathcal{T})$ by simply sampling according to $\mathcal{T}$ from the initial state $I_t$. Ignoring $I_{t+1}, ..., I_{t+d-1}$ and checking $F$ in $I_{t+d}$ yields independent samples of the boolean event $I_{t+d} \models_B F$ from the distribution $P(I_{t+d} \models_B F \mid I_t, \mathcal{T})$. The proportion of positive samples of this variable will thus quickly approach the true probability $P(I_{t+d} \models_B F \mid I_t, \mathcal{T})$.

## 4 Experimental Evaluation

The proposed CPT-L model has been evaluated in two different domains. First, we discuss experiments in a stochastic version of the well-known *blocks world* domain, an artificial domain that allows to perform controlled and systematic experiments e.g. with regard to the scaling behavior of the proposed algorithms. Second, the model is evaluated on real-world data collected from a live server of a massively multi-player online strategy game. Experiments in these two domains will be presented in turn.

### 4.1 Experiments in a Stochastic Blocks World Domain

As an artificial test bed for CPT-L, we performed experiments in a stochastic version of the well-known *blocks world* domain. The domain was chosen because it is truly relational and also serves as a popular artificial world model in agent-based approaches such as planning and reinforcement learning. Application scenarios involving agents that act and learn in an environment are one of the main motivations for CPT-L. In such scenarios world-transition dynamics typically stem from *actions* carried out by the agents according to some *policy*. In the blocks-world domain discussed in this section, we assume that the policy of the agent is known and the task is to probabilistically model transition dynamics given the policy. It is straightforward to represent such conditional world models in CPT-theories by including the policy as part of the background knowledge.

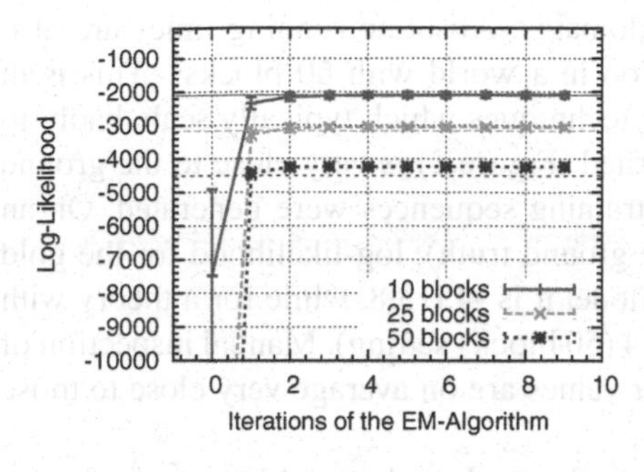


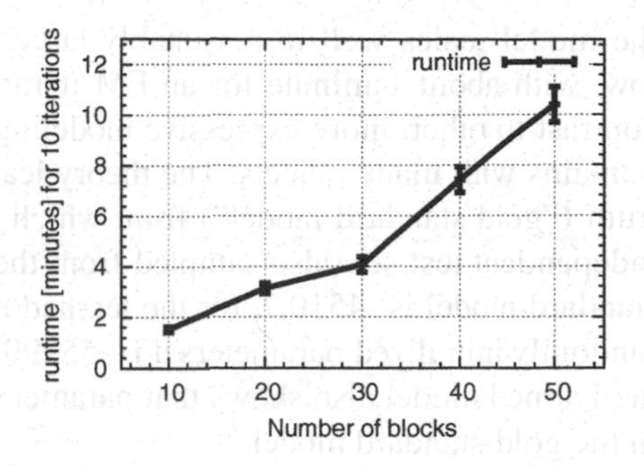


**Fig. 2.** Left graph: per-sequence log-likelihood on the training data as a function of the EM iteration. Right graph: Running time of EM as a function of the number of blocks in the world model.

**World Model.** The blocks world we consider consists of a table and a number of blocks. Every block rests on exactly one other block or the table, denoted by a fact $on(A, B)$. Blocks come in different sizes, denoted by $size\_of(A, N)$ with $N \in \{1, ..., 4\}$. A predicate $free(B) \leftarrow not(on(A, B))$ is defined in the background knowledge. Additionally, a background predicate $stack(A, S)$ defines that block $A$ is part of a stack of blocks, which is represented by its lowest block $S$. Actions derived from the policy are of the form $move(A, B)$. If both $A$ and $B$ are free, the action moves block $A$ on $B$ with probability $1-\epsilon$, with probability $\epsilon$ the world state does not change. Furthermore, a stack $S$ can start to jiggle, represented by $jiggle(S)$. A stack can start to jiggle if its top block is lifted, or a new block is added to it. Furthermore, stacks can start jiggling without interference from the agent, which is more likely if they contain many blocks and large blocks are stacked on top of smaller ones. Stacks that jiggle collapse in the next time-step, and all their blocks fall on the table. Two example rules from this domain are

$$(jiggle(S) : 0.2) \vee (nil : 0.8) \leftarrow move(A, B), stack(A, S)$$

$$(jiggle(S) : 0.2) \vee (nil : 0.8) \leftarrow move(A, B), stack(B, S),$$

they describe that stacks can start to jiggle if blocks are added to or taken from a stack. Furthermore, we consider a simple policy that tries to build a large stack of blocks by repeatedly stacking the free block with second-lowest ID on the free block with lowest ID. This strategy would result in one large stack of blocks if the stack never collapsed.

**Results in the Blocks-World Domain.** In a first experiment, we explore the convergence behavior of the EM algorithm for CPT-L. The world model together with the policy for the agent, which specifies which block to stack next, is implemented by a (gold-standard) CPT-theory $\mathcal{T}$, and a training set of 20 sequences of length 50 each is sampled from $\mathcal{T}$. From this data, the parameters are re-learned using EM. Figure 2, left graph, shows the convergence behavior of the algorithm on the training data for different numbers of blocks in the domain, averaged over 15 runs. It shows rapid and reliable convergence. Figure 2, right graph, shows the running time of EM as a function of the number of blocks. The scaling behavior is roughly linear, indicating that the model scales well to reasonably large domains. Absolute running times are also low, with about 1 minute for an EM iteration in a world with 50 blocks[2]. This is in contrast to other, more expressive modeling techniques which typically scale badly to domains with many objects. The theory learned (Figure 2) is very close to the ground truth ("gold standard model") from which training sequences were generated. On an independent test set (also sampled from the ground truth), log-likelihood for the gold standard model is -4510.7, for the learned model it is -4513.8, while for a theory with randomly initialized parameters it is -55999.4 (50 blocks setting). Manual inspection of the learned model also shows that parameter values are on average very close to those in the gold-standard model.

The experiments presented so far show that relational stochastic domains of substantial size can be represented in CPT-L. The presented algorithms are efficient and scale well in the size of the domain, and show robust convergence behavior.

### 4.2 Experiments in a Massively Multi-player Online Game

As an example for a massively multi-player online game, we consider *Travian*[3], a commercial, large-scale strategy game with a player community of about 3.000.000 players worldwide. In Travian, players are spread over several independent *game worlds*, with approximately 20.000–30.000 players interacting in a single world. Travian game play follows a classical strategy game setup. A game world consists of a large *grid-map*, and each player starts with a single *city* located on a particular tile of the map. During the course of the game, players harvest *resources* from the environment, improve their cities by construction of *buildings* or research of *technologies*, or found new cities on other (free) tiles of the map. Additionally, players can build different military units which can be used to attack and conquer other cities on the map, or trade resources on a global marketplace.

In addition to these low-level game play elements, there are high-level aspects of game play involving *multiple players*, which need to cooperate and coordinate their playing to achieve otherwise unattainable game goals. More specifically, in Travian players dynamically organize themselves into *alliances*, for the purpose of jointly attacking and defending, trading resources or giving advice to inexperienced players. Such alliances constitute social networks for the players involved, where diplomacy is used to settle conflicts of interests and players compete for an influential role in the alliance. In the following, we will take a high-level view of the game and focus on modeling player interaction and cooperation in alliances rather than low-level game elements such as resources, troops and buildings. Figure 3 shows such a high-level view of a (partial) Travian game world, represented as a graph structure relating cities, players and alliances which we will refer to as a *game graph*. It shows that players in one alliance are typically concentrated in one area of the map—traveling over the map takes time, and thus there is little interaction between players far away from each other.

We are interested in the *dynamic* aspect of this world: as players are acting in the game environment (e.g. by conquering other players' cities and joining or leaving

[2] All experiments were run on standard PC hardware, 2.4GHz Intel Core 2 Duo processor, 1GB memory.

[3] www.travian.com;www.traviangames.com

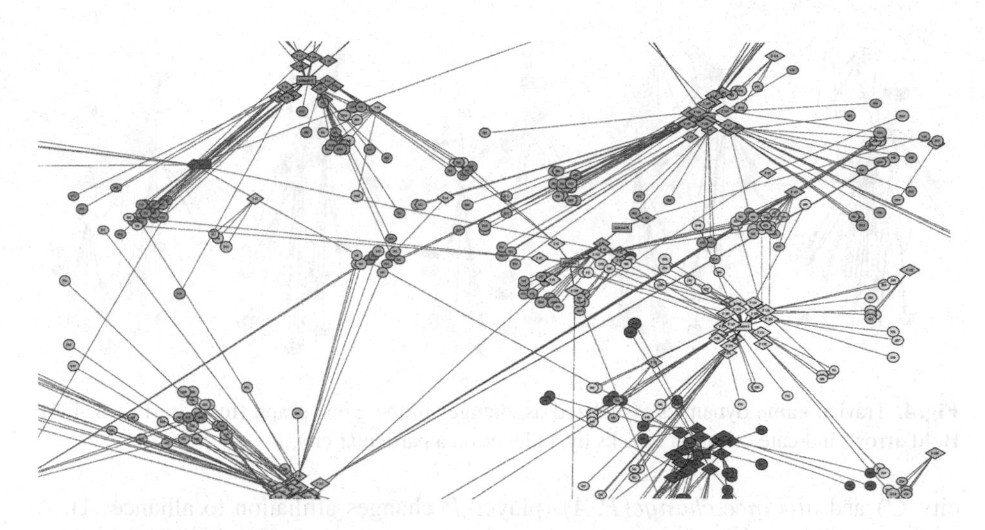

**Fig. 3.** High-level view of a (partial) game world in Travian. Circular nodes indicate cities, shown in their true positions on the game's grid-map. Diamond-shaped nodes indicate players, and are connected to all cities currently owned by the player. Rectangular nodes indicate alliances, and are connected to all players currently members of the alliance. Moreover, players and cities are color-coded according to their alliance affiliation.

alliances), the game graph will continuously change, and thereby reflect changes in the social network structure of the game. As an example for such transition dynamics, consider the sequence of game graphs shown in Figure 4. Here, three players from the pink alliance launch a concerted attack against territory currently held by the green and orange alliances, and partially conquer it.

**Data Collection and Preprocessing.** The data used in the experiments was collected from a "live" Travian server with approximately 25.000 active players. Over a period of three months (December 2007, January 2008, February 2008), high-level data about the current state of the game world was collected once every 24 hours. This included information about all cities, players, and the alliance structure in the game. For cities, their size and position on the map are available; for players, the list of cities they own; and for alliances the list of players currently affiliated with that alliance. From all available data, we extracted 30 sequences of local game world states. Each sequence involves a subset of 10 players, which are tracked over a period of one month (10 sequences each for December, January and February). Player sets are chosen such that there are no interactions between players in different sets, but a high number of interactions between players within one set. Cities that did not take part in any conquest event were removed from the data, leaving approximately 30–40 cities under consideration for every player subset.

**World Model.** The game data was represented using predicates $city(C, X, Y, S, P)$ (city $C$ of size $S$ at coordinates $X, Y$ held by player $P$), $allied(P, A)$ (player $P$ is a member of alliance $A$), $conq(P, C)$ (indicating a conquest attack of player $P$ on

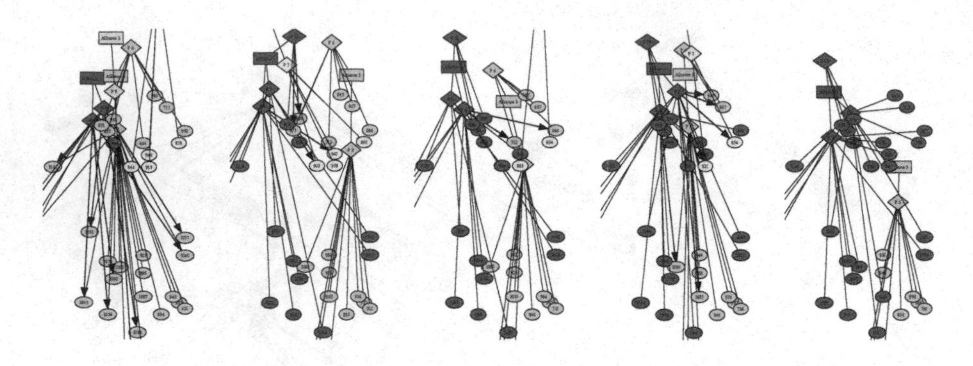

**Fig. 4.** Travian game dynamics visualized as changes in the game graph (for $t = 1, 2, 3, 4, 5$). Bold arrows indicate conquest attacks by a player on a particular city.

city $C$) and $alliance\_change(P, A)$ (player $P$ changes affiliation to alliance $A$). A predicate $distance(C_1, C_2, D)$ with $D \in \{near, medium, far\}$ computing the (discretized) distance between cities was defined in the background knowledge. The final state descriptions (game graphs) on average contain approximately 50 objects (nodes) at every step in time, and relations between them. Sequences consists of between 29 and 31 such state descriptions.

We defined a world model in CPT-L that expresses the probability for player actions such as conquests of cities and changes in alliances affiliation, and updates the world state accordingly. Player actions in Travian—although strongly stochastic—are typically explainable from the social context of the game: different players from the same alliance jointly attack a certain territory on the map, there are retaliation attacks at the alliance level, or players leave alliances that have lost many cities in a short period of time. From a causal perspective, actions are thus triggered by certain (relational) patterns that hold in the game graph, which take into account a player's alliance affiliation together with the actions carried out by other alliance members. Such patterns can be naturally expressed in CPT-L as bodies of rules which trigger actions encoded in the head of the rule. We manually defined a number of simple rules capturing such typical game patterns. As an example, consider the rules

$$conq(P, C){:}0.039 \vee nil{:}0.961 \leftarrow conq(P, C'), city(C', \_, \_, \_, P'), city(C, \_, \_, \_, P')$$
$$conq(P, C){:}0.011 \vee nil{:}0.989 \leftarrow$$
$$city(C, \_, \_, \_, P''), allied(P, A), allied(P', A), conq(P', C'), city(C', \_, \_, \_, P'')$$

The first rule encodes that a player is likely to conquest a city of a player he already attacked in the previous time-step. The second rule generalizes this pattern: a player $P$ is likely to attack a city $C$ of player $P''$ if an allied player has attacked $P''$ in the previous time-step.

Moreover, the world state needs to be updated given the players' actions. After a conquest attack $conq(P, C)$, the city $C$ changes ownership to player $P$ in the next time-step. If several players execute conquest attacks against the same city in one time-step, one of them is chosen as the new owner of the city with uniform probability (note that such simultaneous conquest attacks would not be observed in the training data, as only one

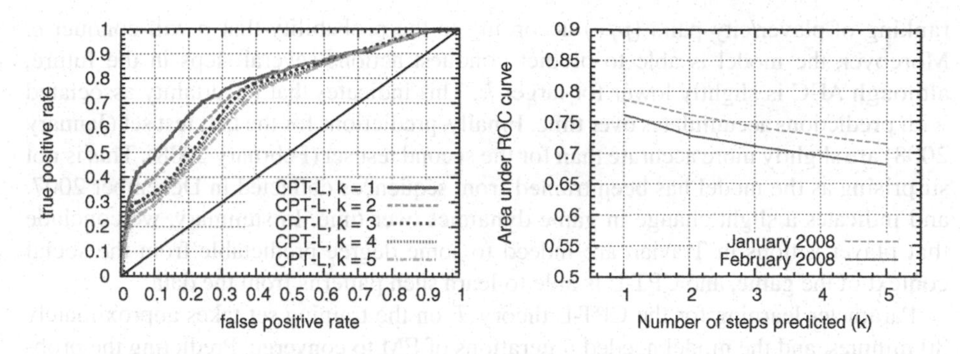


**Fig. 5.** Left figure: ROC curve for predicting that a city $C$ will be conquered by a player $P$ within the next $k$ time-steps, for $k \in \{1, 2, 3, 4, 5\}$. The model was trained on 10 sequences of local game state descriptions from December 2007, and tested on 10 sequences from January 2008. Right figure: AUC as a function of the number $k$ of future time-steps considered in the same experiment. Additionally, AUC as a function of $k$ is shown for 10 test sequences from February 2008.

snapshot of the world is taken every 24 hours). Similarly, an $alliance\_change(P, A)$ event changes the alliance affiliation of player $P$ to alliance $A$ in the next time-step.

**Results in the Travian Domain.** We consider the task of predicting the "conquest" action $conq(P, C)$ based on a learned generative model of world dynamics. The collected sequences of (local) game states were split into one training set (sequences collected in December 2007) and two test sets (sequences collected in January 2008 and sequences collected in February 2008). Maximum-likelihood parameters of a hand-crafted CPT-theory $\mathcal{T}$ as described above were learned on the training set using EM. Afterwards, the learned model was used to predict the player action $conq(P, C)$ on the test data in the following way. Let $S$ denote a test sequence with states $I_0, ..., I_T$. For every $t_0 \in \{0, ..., T-1\}$, and every player $p$ and city $c$ occurring in $S$, the learned model is used to compute the probability that the conquest event $conq(p, c)$ will be observed in the next world state, $P(I_{t_0+1} \models conq(p, c) \mid \mathcal{T}, I_0, ..., I_{t_0})$. This probability is obtained from the sampling-based prediction algorithm described in Section 3. The prediction is compared to the known ground truth (whether the conquest event occurred at that time in the game or not). Instead of predicting whether the player action will be taken in the next step, we can also predict whether it will be taken within the next $k$ steps, by computing

$$P(I_{t_0+1} \models conq(p, c) \vee ... \vee I_{t_0+k} \models conq(p, c) \mid \mathcal{T}, I_0, ..., I_{t_0}).$$

This quantity is also easily obtained from the prediction algorithm for CPT-L. Figure 5, left, shows ROC curves for this experiment with different values $k \in \{1, 2, 3, 4, 5\}$, evaluated on the first test set (January 2008). Figure 5, right, shows the corresponding AUC values as a function of $k$ for both test sets. The achieved area under the ROC curve is substantially above 0.5 (random performance), indicating that the learned CPT-theory $\mathcal{T}$ indeed captures some characteristics of player behavior and obtains a reasonable ranking of player/city pairs $(p/c)$ according to the probability that $p$ will conquer $c$. Moreover, the model is able to predict conquest actions several steps in the future, although AUC is slightly lower for larger $k$. This indicates that uncertainty associated with predictions accumulates over time. Finally, predictions for the first test set (January 2008) are slightly more accurate than for the second test set (February 2008). This is not surprising as the model has been trained from sequences collected in December 2007, and indicates a slight change in game dynamics over time. In summary, we conclude that player actions in Travian are indeed to some degree predictable from the social context of the game, and CPT-L is able to learn such patterns from the data.

Parameter learning for the CPT-L theory $\mathcal{T}$ on the training set takes approximately 30 minutes, and the model needed 5 iterations of EM to converge. Predicting the probability of $conq(p, c)$ for all player/city pairs and the next $k$ time-steps starting from a particular world state takes approximately 1 minute.

## 5 Related Work

There are relatively few existing approaches that can probabilistically model sequences of relational state descriptions. CPT-L can be positioned with respect to them as follows. First, statistical relational learning systems such as Markov Logic [14], CP-logic [9], Probabilistic Relational Models [15] or Bayesian Logic Programs [16] can be used in this setting by adding an extra time argument to predicates (then called *fluents*). However, inference and learning in these systems is computationally expensive: they support very general models including hidden states, and are not optimized for sequential data. A second class of techniques, for instance [17], uses transition models based on (stochastic) STRIPS rules. This somewhat limits the transitions that can be expressed, as only one rule "fires" at every point in time, and it is difficult to model several processes that change the state of the world concurrently (such as an agent's actions and naturally occurring world changes). In contrast, such scenarios are naturally modeled in CP-logic and thus CPT-L. Another approach designed to model sequences of relational state descriptions are relational simple-transition models [18]. In contrast to CPT-L, they focus on domains where the process generating the data is hidden, and inferring these hidden states from observations. This is a harder setting than the fully observable setting discussed in this paper, and typically only approximate inference is possible [18].

## 6 Conclusions and Future Work

We have introduced CPT-L, a probabilistic model for sequences of relational state descriptions. In contrast to other approaches that could be used as a model for such sequences, CPT-L focuses on computational efficiency rather than expressivity. This is essential for many real-life applications. The main direction for future work is to further evaluate the trade-off between representational power and scaling behavior in challenging real-world domains. Furthermore, we want to explore how the model can be extended, for instance to account for hidden data, without sacrificing efficiency.

**Acknowledgments.** The authors would like to thank the anonymous reviewers for helpful comments. This work was supported by FWO-Vlaanderen, and the GOA/08/008 project "Probabilistic Logic Learning".

## References

1. Pollack, M.E.: Intelligent technology for an aging population: The use of AI to assist elders with cognitive impairment. AI Magazine 26(2), 9–24 (2005)
2. Laird, J.E., van Lent, M.: Human-Level AI's Killer Application: Interactive Computer Games. In: Proceedings of the Seventeenth National Conference on Artificial Intelligence and Twelfth Conference on Innovative Applications of Artificial Intelligence (2000)
3. Rabiner, L.R.: A tutorial on hidden Markov models and selected applications in speech recognition. Proceedings of the IEEE 77(2), 257–286 (1989)
4. Puterman, M.: Markov Decision Processes: Discrete Stochastic Dynamic Programming. John Wiley & Sons, Inc., Chichester (1994)
5. Ghahramani, Z.: Learning dynamic bayesian networks. Adaptive Processing of Sequences and Data Structures. In: International Summer School on Neural Networks, pp. 168–197 (1997)
6. Fikes, R.E., Nilsson, N.J.: STRIPS: a new approach to the application of theorem proving to problem solving. In: Computation & intelligence: collected readings, Menlo Park, CA, USA, pp. 429–446. American Association for Artificial Intelligence (1995)
7. Getoor, L., Taskar, B. (eds.): Statistical Relational Learning. MIT Press, Cambridge (2007)
8. De Raedt, L., Frasconi, P., Kersting, K., Muggleton, S. (eds.): Probabilistic Inductive Logic Programming. LNCS (LNAI), vol. 4911. Springer, Heidelberg (2008)
9. Vennekens, J., Denecker, M., Bruynooghe, M.: Representing causal information about a probabilistic process. In: Fisher, M., van der Hoek, W., Konev, B., Lisitsa, A. (eds.) JELIA 2006. LNCS (LNAI), vol. 4160, pp. 452–464. Springer, Heidelberg (2006)
10. Bratko, I.: Prolog Programming for Artificial Intelligence, 2nd edn. Addison-Wesley, Reading (1990)
11. De Raedt, L., Kimmig, A., Toivonen, H.: Problog: A probabilistic Prolog and its application in link discovery. In: Proceedings of the 20th International Joint Conference on Artificial Intelligence, pp. 2462–2467 (2007)
12. Bryant, R.E.: Graph-based algorithms for boolean function manipulation. IEEE Trans. Computers 35(8), 677–691 (1986)
13. Minato, S.: Zero-suppressed BDDs for set manipulation in combinatorial problems. In: DAC 1993: Proceedings of the 30th international conference on Design automation, pp. 272–277. ACM, New York (1993)
14. Richardson, M., Domingos, P.: Markov Logic Networks. Machine Learning 62, 107–136 (2006)
15. Getoor, L., Friedman, N., Koller, D., Pfeffer, A.: Learning probabilistic relational models. In: Relational Data Mining, pp. 307–335. Springer, Heidelberg (2001)
16. Kersting, K., De Raedt, L.: Bayesian Logic Programming: Theory and tool. In: Getoor, L., Taskar, B. (eds.) An introduction to statistical relational learning. MIT Press, Cambridge (2007)
17. Zettlemoyer, L.S., Pasula, H., Kaelbling, L.P.: Learning planning rules in noisy stochastic worlds. In: Proceedings of the 20th National Conference on Artificial Intelligence (AAAI 2005), pp. 911–918 (2005)
18. Fern, A.: A simple-transition model for relational sequences. In: Kaelbling, L., Saffiotti, A. (eds.) IJCAI, pp. 696–701. Professional Book Center (2005)

# Semi-Supervised Boosting for Multi-Class Classification

Hamed Valizadegan, Rong Jin, and Anil K. Jain

Department of Computer Science and Engineering,
Michigan State University
valizade@cse.msu.edu, rongjin@cse.msu.edu, jain@cse.msu.edu

**Abstract.** Most semi-supervised learning algorithms have been designed for binary classification, and are extended to multi-class classification by approaches such as one-against-the-rest. The main shortcoming of these approaches is that they are unable to exploit the fact that each example is only assigned to one class. Additional problems with extending semi-supervised binary classifiers to multi-class problems include imbalanced classification and different output scales of different binary classifiers. We propose a semi-supervised boosting framework, termed **Multi-Class Semi-Supervised Boosting (MCSSB)**, that directly solves the semi-supervised multi-class learning problem. Compared to the existing semi-supervised boosting methods, the proposed framework is advantageous in that it exploits both classification confidence and similarities among examples when deciding the pseudo-labels for unlabeled examples. Empirical study with a number of UCI datasets shows that the proposed MCSSB algorithm performs better than the state-of-the-art boosting algorithms for semi-supervised learning.

**Keywords:** Semi-supervised learning, Multi-Class Classification, Boosting.

## 1 Introduction

Semi-supervised classification combines the hidden structural information in the unlabeled examples with the explicit classification information of labeled examples to improve the classification performance. Many semi-supervised learning algorithms have been studied in the literature. Examples are density based methods [1,2], graph-based algorithms [3,4,5,6], and boosting techniques [7,8]. Most of these methods were originally designed for two class problems. However, many real-world applications, such as speech recognition and object recognition, require multi-class categorization. To adopt a binary (semi-supervised) learning algorithm to problems with more than two classes, the multi-class problems are usually decomposed into a number of independent binary classification problems using techniques such as one-versus-the-rest, one-versus-one, and error-correcting output coding [9]. The main shortcoming with this approach is that the resulting binary classification problems are *independent* binary class problems. As a

W. Daelemans et al. (Eds.): ECML PKDD 2008, Part II, LNAI 5212, pp. 522–537, 2008.

result, it is unable to exploit the fact that each example can only be assigned to one class. This issue was also pointed out in the study with multi-class boosting [10]. In addition, since every binary classifier is trained independently, their outputs may be on different scales, making it difficult to compare them [11]. Though calibration techniques [12] can be used to alleviate this problem in supervised classification, it is rarely used in semi-supervised learning due to the small number of labeled training examples. Moreover, techniques like one-versus-the-rest, where the examples of one class are considered against the examples of all the other classes, could lead to the imbalanced classification problem. Although a number of techniques have been proposed for supervised learning in multi-class problems [10,13,14], they have not addressed semi-supervised multi-class learning problems, which is the focus of this study.

Boosting is a popular learning method because it provides a general framework for improving the performance of any given learner by constructing an ensemble of classifiers. Several boosting algorithms have been proposed for semi-supervised learning [7,8,15]. They essentially operate like self-training where the class labels of unlabeled examples are updated iteratively: a classifier trained by a small number of labeled examples is initially used to predict the pseudo-labels for unlabeled examples; a new classifier is then trained by both labeled and pseudo-labeled examples; the processes of training classifiers and predicting pseudo-labels are altered iteratively till stopping criterion is reached. The main drawback with this approach is that it relies solely on the pseudo-labels predicted by the classifiers learned so far when generating new classifiers. Given the possibility that pseudo-labels predicted in the first few steps of boosting could be inaccurate, the resulting new classifiers may also be unreliable. This problem was addressed in [8] by introduction of a local smoothness regularizer. However, since all the existing semi-supervised boosting algorithms are designed for binary classification, they will still suffer from the aforementioned problems when applied to multi-class problems. In this paper, we develop a semi-supervised boosting framework, termed *Multi-Class Semi-Supervised Boosting (MCSSB)*, that is designed for multi-class semi-supervised learning problems. By directly solving a multi-class problem, we avoid the problems that arise when converting a multi-class classification problem into a number of binary ones. Moreover, unlike the existing semi-supervised boosting methods that only assign pseudo-labels to the unlabeled examples with high classification confidence, the proposed framework decides the pseudo labels for unlabeled examples based on both the classification confidence and the similarities among examples. It therefore effectively explores both the manifold assumption and the clustering assumption for semi-supervised learning. Empirical study with UCI datasets shows the proposed algorithm performs better than the state-of-the-art algorithms for semi-supervised learning.

## 2 Related Work

Most semi-supervised learning algorithms can be classified into three categories: graph-based, density-based, and boosting-based.

Semi-supervised SVMs ($S^3VMs$) or Transductive SVMs (TSVMs) are the semi-supervised extensions to Support Vector Machines (SVM). They are essentially density-based methods and assume that decision boundaries should lie in the sparse regions. Although finding an exact $S^3VM$ is NP-complete [16], there are many approximate solutions for it [1,2,17,18,19]. Except for [19], these methods are designed for binary semi-supervised learning. The main drawback with [19] is its high computational cost due to the semi-definite programming formulation.

Graph-based methods aim to predict class labels that are smooth on the graph of unlabeled examples. These algorithms differ in how to define the smoothness of class labels over a graph. Example graph-based semi-supervised learning approaches include Mincut [3], Harmonic function [4], local and global consistency [5], and manifold regularization [6]. Similar to density based methods, most graph-based methods are mainly designed for binary classification.

Semi-supervised boosting methods such as SSMBoost [15] and Assemble [7] are direct extensions of Adaboost [20]. In [8], a local smoothness regularizer is introduced to improve the reliability of semi-supervised boosting. Unlike the existing approaches for semi-supervised boosting that solve 2-class problems, our study focuses on semi-supervised boosting for multi-class classification.

## 3 Multi-Class Semi-Supervised Learning

### 3.1 Problem Definition

Let $\mathcal{D} = (x_1, .., x_N)$ denote the collection of $N$ examples. Assume that the first $N_l$ examples are labeled by $y_1, ..., y_{N_l}$, where $y_i = (y_i^1, ..., y_i^m) \in \{0, +1\}^m$ is a binary vector and $m$ is the number of classes. $y_i^k = +1$ when $x_i$ is assigned to the $k$th class, and $y_i^k = 0$, otherwise. Since we are dealing with a multi-class problem, we have $\sum_{k=1}^{m} y_i^k = 1$, i.e., each example $x_i$ is only assigned to one and only one class. We denote by $\hat{y}_i = (\hat{y}_i^1, \ldots, \hat{y}_i{}^m) \in \mathbb{R}^m$ the predicted class labels (or confidence) for example $x_i$, and by $\hat{Y} = (\hat{y}_1^\top, \ldots, \hat{y}_N^\top)^\top$ the predicted class labels for all the examples[1]. Let $S = [S_{i,j}]_{N \times N}$ be the similarity matrix where $S_{i,j} = S_{j,i} \geq 0$ is the similarity between $x_i$ and $x_j$. For the convenience of discussion, we set $S_{i,i} = 0$ for any $x_i \in \mathcal{D}$, a convention that is commonly used by many graph-based approaches. Our goal is to compute $\hat{y}_i$ for the unlabeled examples with the assistance of similarity matrix $S$ and $Y = (y_1^\top, \ldots, y_{N_l}^\top)^\top$.

### 3.2 Design of Objective Function

The goal of semi-supervised learning is to combine labeled and unlabeled examples to improve the classification performance. Therefore, we design an objective function that consists of two terms: (a) $F_u$ that measures the consistency between the predicted class labels $\hat{Y}$ of unlabeled examples and the similarity matrix $S$,

[1] $x^\top$ is the transpose of matrix(vector) $x$.

and (b) $F_l$ that measures the consistency between the predicted class labels $\hat{Y}$ and true labels $Y$. Below we discuss these two terms in detail.

Given two examples $x_i$ and $x_j$, we first define the similarity $Z^u_{i,j}$ based on their predicted class labels $\hat{y}_i$ and $\hat{y}_j$:

$$Z^u_{i,j} = \sum_{k=1}^{m} \frac{\exp(\hat{y}_i^k)}{\sum_{k'=1}^{m} \exp(\hat{y}_i^{k'})} \frac{\exp(\hat{y}_j^k)}{\sum_{k'=1}^{m} \exp(\hat{y}_j^{k'})} = \sum_{k=1}^{m} b_i^k b_j^k = b_i^\top b_j \qquad (1)$$

where $b_i^k = \exp(\hat{y}_i^k)/(\sum_{k'=1}^{m} \exp(\hat{y}_i^{k'}))$ and $b_i = (b_i^1, \ldots, b_i^m)$. Note that $b_i^k$ can be interpreted as the probability of assigning $x_i$ to class $k$, and $Z^u_{i,j}$, the cosine similarity between $b_i$ and $b_j$, can be interpreted as the probability of assigning $x_i$ and $x_j$ to the same class. We emphasize it is important to use $b_i^k$, instead of $\exp(\hat{y}_i^k)$, for computing $Z^u_{i,j}$ because normalization in $b_i^k$ allows us to enforce the requirement that each example is assigned to a single class, a key feature of multi-class learning.

Let $Z^u = [Z^u_{i,j}]$ be the similarity matrix based on the predicted labels. To measure the inconsistency between this similarity and the similarity matrix $S$, we define $F_u$ as the distance between the matrices $Z^u$ and $S$ using the Bregman matrix divergence [21], i.e.,

$$F_u = \varphi(Z^u) - \varphi(S) - \mathrm{tr}((Z^u - S)^\top \nabla\varphi(S)), \qquad (2)$$

where $\varphi : \mathbb{R}^{N \times N} \to \mathbb{R}$ is a convex matrix function. By choosing $\varphi(X) = \sum_{i,j=1}^{N} X_{i,j}(\log X_{i,j} - 1)$ [21], $F_u$ is written as [2]

$$F_u = \sum_{i,j=N_l+1}^{N} \left( S_{i,j} \log \frac{S_{i,j}}{Z^u_{i,j}} + Z^u_{i,j} - S_{i,j} \right) \qquad (3)$$

By assuming that $\sum_{i,j=1+N_l}^{N} Z^u_{i,j} \approx \sum_{k=1}^{m} N_k^2$ and $\log x \approx x - 1$, where $N_k$ is the number of examples assigned to class $k$, we simplify the above expression as $F_u \approx \sum_{i,j=N_l+1}^{N} S^2_{i,j}/Z^u_{i,j}$. Since $S^2_{i,j}$ could be viewed as a general similarity measurement, we replace $S^2_{i,j}$ with $S_{i,j}$ and simplify $F_u$ as

$$F_u \approx \sum_{i,j=N_l+1}^{N} \frac{S_{i,j}}{Z^u_{i,j}} = \sum_{i,j=N_l+1}^{N} \frac{S_{i,j}}{\sum_{k=1}^{m} b_i^k b_j^k} \qquad (4)$$

*Remark* We did not use $\varphi(X) = \sum_{i,j=1}^{N} X^2_{i,j}$ [21], which will result in $F_u = \sum_{i,j=N_l+1}^{N} (Z^u_{i,j} - S_{i,j})^2$. This is because the value of $Z^u_{i,j}$ and $S_{i,j}$ may be on different scales.

Similarly, we define the similarity between a labeled example $x_i$ and an unlabeled example $x_j$ based on their class assignments as follows

$$Z^l_{i,j} = \sum_{k=1}^{m} y_i^k b_j^k, \qquad (5)$$

[2] We can only consider the sub-matrices related to unlabeled examples when defining $F_u$.

and the inconsistency measure $F_l$ between the labeled and unlabeled examples as follows:

$$F_l = \sum_{i=1}^{N_l} \sum_{j=N_l+1}^{N} \frac{S_{i,j}}{Z^l_{i,j}} = \sum_{i=1}^{N_l} \sum_{j=N_l+1}^{N} \frac{S_{i,j}}{\sum_{k=1}^m y_i^k b_j^k} \tag{6}$$

Finally, we linearly combine $F_l$ and $F_u$ to form the objective function:

$$F = F_u + CF_l \tag{7}$$

where $C$ weights the importance of $F_l$. It is set to 10,000 in our experiments to emphasize $F_l$ [3]. Given the objective function $F$ in (7), our goal is to find solution $\hat{Y}$ that minimizes $F$.

### 3.3 Multi-Class Boosting Algorithm

In this section, we present a boosting algorithm to solve the optimization problem in (7). Following the architecture of boosting model, we incrementally add weak learners to obtain a better classification model. We denote by $H_i^k$ the solution that is obtained for $\hat{y}_i^k$ so far, and by $h_i^k \in \{0,1\}$ the prediction made by the incremental weak classifier that needs to be learned. Then, our goal is to find $h_i^k, i = N_l+1,\ldots,N, k = 1,\ldots,m$ and a combination weight $\alpha$ such that the new solution $\widetilde{H}_i^k = H_i^k + \alpha h_i^k$ significantly reduces the objective function $F$ in Equation 7. For the convenience of discussion, we use symbol $\sim$ to denote the quantities (e.g., $\widetilde{F}$) associated with the new solution $\widetilde{H}$.

The key challenge in optimizing $F$ with respect to $h_i^k$ and $\alpha$ is that these two quantities are coupled with each other and therefore the solution of one variable depends on the solution of the other. Our strategy to solve the optimization problem is to first upper bound $F$ with a simple convex function in which the optimal solution for $h_i^k$ can be obtained without knowing the solution to $\alpha$. Given the solution to $h_i^k$, we then compute the optimal solution for $\alpha$. Below we give details for these two steps.

First, the following lemma allows us to decouple the interaction between $\alpha$ and $h_i^k$ within $Z^u_{i,j}$ and $Z^l_{i,j}$

**Lemma 1**

$$\frac{1}{\widetilde{Z}^u_{i,j}} \leq \frac{1+e^{6\alpha}+e^{-6\alpha}}{3Z^u_{i,j}} + \frac{e^{6\alpha}-1}{3Z^u_{i,j}} \left( \sum_{k=1}^m (b_i^k - \tau_{i,j}^k) h_i^k \right) \tag{8}$$

$$\frac{1}{\widetilde{Z}^l_{i,j}} \leq \frac{1+e^{6\alpha}+e^{-6\alpha}}{3Z^l_{i,j}} + \frac{e^{6\alpha}-1}{6} \sum_{k=1}^m h_i^k \phi_{i,j}^k \tag{9}$$

*where*

$$\tau_{i,j}^k = \frac{b_i^k b_j^k}{\sum_{k'=1}^m b_i^{k'} b_j^{k'}}, \quad \phi_{i,j}^k = \sum_{k'=1}^m y_j^k \frac{b_i^k}{b_i^{k'}} - \frac{y_i^k}{b_i^k} \tag{10}$$

[3] The algorithm is quite stable with different values of $C$ bigger than 1000 according to our experiment.

The proof of Lemma 1 can be found in Appendix A. Using Lemma 1, we derive an upper bound for $\widetilde{F}$ in the following theorem.

**Theorem 1**

$$\widetilde{F} \leq F\frac{1+\exp(6\alpha)+\exp(-6\alpha)}{3} + \frac{\exp(6\alpha)-1}{3}\sum_{i=N_l+1}^{N}\sum_{k=1}^{m} h_i^k(\alpha_i^k + C\beta_i^k) \quad (11)$$

*where $\alpha_i^k$ and $\beta_i^k$ are defined as follows:*

$$\alpha_i^k = \sum_{j=N_l+1}^{N} \frac{S_{i,j}(b_i^k - \tau_{i,j}^k)}{Z_{i,j}^u}, \quad \beta_i^k = \frac{1}{2}\sum_{j=1}^{N_l} S_{i,j}\phi_{i,j}^k \quad (12)$$

Theorem 1 can be directly verified by replacing $1/\widetilde{Z}_{i,j}^u$ and $1/\widetilde{Z}_{i,j}^l$ in (7) with (8) and (9). Note that the bound in Theorem 1 is tight because by setting $\alpha = 0$, we have $\widetilde{H} = H$ and the inequality in Equation 11 is reduced to an equality. The key feature of the bound in Equation 11 is that the optimal solution for $h_i^k$ can be obtained without knowing the solution for $\alpha$. This is summarized by the following theorem.

**Theorem 2.** *The optimal solution for $h_i^k$ that minimizes the upper bound of $\widetilde{F}$ in Equation 11 is*

$$h_i^k = \begin{cases} 1 & k = \arg\max_{k'}(\alpha_i^{k'} + C\beta_i^{k'}) \\ 0 & otherwise \end{cases} \quad (13)$$

It is straightforward to verify the result in Theorem 2.

We then proceed to find solution for $\alpha$ given the solution for $h_i^k$. The following lemma provides a tighter bound for solving $\alpha$ in $F$ [4].

**Lemma 2**

$$\widetilde{F} - F \leq (e^{2\alpha}-1)(A_u + CA_l) + (e^{-2\alpha}-1)(B_u + CB_l) \quad (14)$$

*where*

$$A_u = \sum_{i,j=N_l+1}^{N} \frac{S_{i,j}}{Z_{i,j}^u}\sum_{k=1}^{m} h_i^k b_i^k \quad (15)$$

$$A_l = \frac{1}{2}\sum_{i=1}^{N_l}\sum_{j=N_l+1}^{N} S_{i,j}\sum_{k,k'=1}^{m}\frac{y_i^k}{b_j^k} b_j^{k'} h_j^{k'} \quad (16)$$

$$B_u = \sum_{i,j=N_l+1}^{N} \frac{S_{i,j}}{Z_{i,j}^u}\sum_{k=1}^{m} h_i^k \tau_{i,j}^k \quad (17)$$

$$B_l = \frac{1}{2}\sum_{i=1}^{N_l}\sum_{j=N_l+1}^{N} S_{i,j}\sum_{k=1}^{m} y_i^k \frac{h_j^k}{b_j^k} \quad (18)$$

[4] Note that this tighter bound can not be used to derive $h_i^k$.

**Algorithm 1.** MCSSB: Multi-Class Semi-Supervised Boosting Algorithm

**Input:**
- $D$: The set of examples; the first $N_l$ examples are labeled.
- $s$: the number of sampled examples from $(N - N_l)$ unlabeled examples
- $T$: the maximum number of iterations

**for** $i = 1$ **to** $T$
- Compute $\alpha_i^k$ and $\beta_i^k$ for every example as given in Equation 12.
- Assign each unlabeled example $x_i$ to class $k_i^* = \arg\max_k(\alpha_i^k + C\beta_i^k)$ and weight $w_i = \alpha_i^{k_i^*} + C\beta_i^{k_i^*}$
- Sample $s$ unlabeled examples using a distribution that is proportional to $w_i$
- Train a multi-class classifier $h(x)$ using the labeled examples and the sampled unlabeled examples with assigned classes
- Predict $h_i^k$ for unlabel examples using $h(x)$, and compute $\alpha$ using Equation 19. Exit the loop if $\alpha \leq 0$.
- $H(x) \leftarrow H(x) + \alpha h(x)$

The proof of Lemma 2 can be found in Appendix B. Using Lemma 2, Theorem 3 gives the optimal solution for $\alpha$.

**Theorem 3.** *The optimal $\alpha$ that minimizes the upper bound of $\widetilde{F}$ in Equation 14 is*

$$\alpha = \frac{1}{4}\log\left(\frac{B_u + CB_l}{A_u + CA_l}\right) \tag{19}$$

Algorithm 1 summarizes the proposed boosting algorithm for multi-class semi-supervised learning. Several issues need to be pointed out: (a) $w_i$, the weight for the $i$th unlabeled example, is guaranteed to be non-negative. This is because $\sum_{k=1}^m \alpha_i^k + C\beta_i^k = 0$ and therefore $w_i = \max_k(\alpha_i^k + C\beta_i^k) \geq 0$; (b) we adopt the sampling approach to train a weak classifier. In our experiments, the number of sampled examples at each iteration is set as $s = \max(20, N/5)$; (c) the maximum number of iteration $T$ is set to be 50 as suggested in [22][5].

Theorem 4 shows that the proposed boosting algorithm reduces the objective function $F$ exponentially.

**Theorem 4.** *The objective function after T iterations, denoted by $F^T$, is bounded as follows:*

$$F^T \leq F^0 \exp\left(-\sum_{t=1}^{T} \frac{(\sqrt{A_u^t + CA_l^t} - \sqrt{B_u^t + CB_l^u})^2}{F^{t-1}}\right) \tag{20}$$

*where $A_u$, $A_l$, $B_u$ and $B_l$ are defined in Lemma 2.*

[5] We run the algorithm with much larger numbers of iterations and find that both the objective function and the classification accuracy remains essentially the same after 50 iterations. We, therefore, set the number of iterations to be 50 to save the computational cost.

*Proof.* Using Lemma 2 and Theorem 3, we have

$$\widetilde{F}-F \leq \sqrt{\frac{B_u+CB_l}{A_u+CA_l}}(A_u+CA_l)+\sqrt{\frac{A_u+CA_l}{B_u+CB_l}}(B_u+CB_l)-(A_u+CA_l+B_u+CB_l)$$
$$= -\left(\sqrt{A_u+CA_l}-\sqrt{B_u+CB_l}\right)^2,$$

which is equivalent to

$$\frac{\widetilde{F}}{F} \leq 1-\frac{\left(\sqrt{A_u+CA_l}-\sqrt{B_u+CB_l}\right)^2}{F}$$
$$\leq \exp\left(-\frac{\left(\sqrt{A_u+CA_l}-\sqrt{B_u+CB_l}\right)^2}{F}\right) \quad (21)$$

The above inequality follows from $\exp(x) \geq 1+x$. We rewrite $F^T$ as

$$F^T = F^0 \prod_{t=1}^{T}(F^t/F^{t-1})$$

By substituting $F^t/F^{t-1}$ with the bound in Equation 21, we have the result in the theorem.

## 4 Experiments

In this section, we present our empirical study on a number of UCI data sets. We refer to the proposed semi-supervised multi-class boosting algorithm as **MC-SSB**. In this study, we aim to show that (1) MCSSB can improve several available multi-class classifiers with unlabeled examples, (2) MCSSB is more effective than the existing semi-supervised boosting algorithms, and (3) MCSSB is robust to the model parameters and the number of labeled examples. It is important to note that it is not our intention to show that the proposed semi-supervised multi-class boosting algorithm always outperforms the other semi-supervised learning algorithms. Instead, our objective is to demonstrate that the proposed semi-supervised boosting algorithm is able to effectively improve the accuracy of different supervised multi-class learning algorithms using the unlabeled examples. Hence, the empirical study is focused on a comparison with the existing semi-supervised boosting algorithms, rather than a wide range of semi-supervised learning algorithms.

We follow [7] and use *Decision Tree* and *Multi-Layer Perceptron (MLP)* as the base multi-class classifiers in our study. In order to create weak classifiers as most boosting algorithms do, we restrict the levels of decision tree to be two, and the structure of MLP to be one hidden layer with two nodes. We create an instance of semi-supervised multi-class learning boosting algorithm for each base classifier, denoted by *MCSSB-Tree* and *MCSSB-MLP*, respectively. We compare the proposed semi-supervised boosting algorithm to *ASSEMBLE*, a state-of-the-art semi-supervised boosting. Similar to MCSSB, two instances of classifiers are

**Table 1.** Description of data sets

| | # samples | # attributes | # Classes |
|---|---|---|---|
| Balance | 625 | 4 | 3 |
| Glass | 214 | 9 | 6 |
| Iris | 150 | 4 | 3 |
| Wine | 178 | 13 | 3 |
| Car | 1728 | 6 | 4 |
| Vowel | 990 | 14 | 11 |
| Contraceptive | 1473 | 9 | 3 |
| Dermatology | 358 | 34 | 6 |
| Ecoli | 336 | 7 | 8 |
| Flag | 194 | 28 | 8 |
| Segmentation | 2310 | 19 | 7 |
| pendigit | 3498 | 16 | 10 |
| Optdigits | 1797 | 64 | 10 |
| Soybean | 686 | 35 | 19 |
| Waves | 5000 | 21 | 3 |
| Yeast | 1484 | 8 | 10 |
| Zoo | 101 | 16 | 7 |

created for ASSEMBLE using decision tree and MLP base classifiers, denoted by *Assemble-Tree* and *Assemble-MLP*, respectively. A Gaussian kernel is used as the measure for similarity in *MCSSB-Tree* and *MCSSB-MLP* with kernel width set to be 15% of the range of the distance between examples [6] for all the experiments, as suggested in [23]. Table 1 summarizes seventeen benchmark data sets from the UCI data repository used in this study.

### 4.1 Evaluation of Classification Performance

Many binary semi-supervised learning studies assume a very small number of labeled examples, e.g. less that 1% of the total number of examples. This setup is difficult to be applied to multi-class cases since it may result in some classes with no labeled examples. As an example, consider Glass data set in Table 1, where 1% of the examples will provide us with only two labeled examples which will cover at most two classes. This motivated us to run two different sets of experiments to evaluate the performance of the proposed algorithm. In the first set of experiments, we assume that 5% of examples are labeled and in the second case we assume that 10% of examples are labeled [7]. We repeat each experiment 20 times and report both the mean and standard deviation of the classification accuracy.

[6] I.e. $0.15 \times (d_{\max} - d_{\min})$, where $d_{\min}$ and $d_{\max}$ are minimum and maximum distance between examples.

[7] Our experience with one labeled example per class shows similar results. We omit the result due to space limitation.

**Table 2.** Classification accuracy with 5% of samples as the labeled set($N_l$)

| | Tree | MLP | Assemble-tree | Assemble-MLP | MCSSB-tree | MCSSB-MLP |
|---|---|---|---|---|---|---|
| Balance | 65.0±0.9 | 82.0±1.4 | 65.0±0.9 | 82.2±1.0 | 72.5±1.0 | **83.2±0.7** |
| Glass | 39.7±1.6 | **40.3±1.8** | 39.7±1.6 | **41.0±1.8** | **40.1±1.2** | **40.4±1.1** |
| Iris | 32.3±0.1 | 71.6±2.5 | 32.4±0.1 | 74.3±3.8 | **77.4±2.6** | 74.0±3.0 |
| Wine | 33.4±1.3 | 70.0±2.7 | 62.4±3.6 | 66.0±2.8 | **78.2±3.5** | 75.0±3.1 |
| Car | 80.5±0.5 | 76.4±1.0 | 80.5±0.5 | 76.8±0.5 | **81.6±0.3** | 77.7±0.5 |
| Vowel | **27.5±1.1** | 17.9±0.6 | 26.0±1.1 | 18.8±0.6 | **28.1±1.1** | 19.3±0.6 |
| Contraceptive | **47.3±0.8** | 45.0±0.7 | **47.2±0.7** | 44.1±0.9 | **47.3±0.8** | 45.4±0.6 |
| Dermatology | 53.6±2.2 | 48.3±2.0 | 53.4±2.3 | 46.2±2.4 | **77.0±1.2** | 68.0±1.8 |
| Ecoli | 57.8±1.7 | **61.5±1.8** | 57.8±1.7 | 59.3±1.7 | 52.0±2.2 | 56.2±1.3 |
| Flag | 23.5±1.3 | 22.1±1.1 | 23.5±1.3 | 25.1±1.2 | **30.3±1.1** | 26.0±1.2 |
| Segmentation | **47.6±2.1** | 43.2±1.3 | 45.9±2.1 | 44.8±1.5 | **47.6±2.1** | 44.5±1.6 |
| pendigit | 33.8±1.5 | 30.0±1.0 | 32.5±1.5 | 29.8±0.8 | **59.3±1.1** | 54.7±1.7 |
| Optdigits | **33.0±1.8** | 23.3±1.0 | 30.6±1.4 | 23.3±0.7 | **33.0±1.8** | 21.9±0.6 |
| Soybean | 37.2±1.3 | 25.0±0.9 | 35.2±1.4 | 25.4±1.1 | **42.4±1.1** | 33.4±0.9 |
| Waves | 65.0±0.3 | 73.3±1.7 | 65.0±0.3 | 73.3±1.8 | 65.4±0.3 | **74.8±0.8** |
| Yeast | **43.6±0.7** | 40.2±0.6 | **43.4±0.6** | 40.4±0.9 | **42.7±0.9** | 39.4±1.2 |
| Zoo | 32.6±2.7 | 39.8±3.0 | 41.7±4.0 | 40.7±3.0 | **59.0±2.7** | 56.9±2.6 |

Table 2 shows the result of different algorithms for the first experiment (5% labeled examples) with the performance of the best approach for each dataset highlighted by bold font. First, notice that MCSSB significantly[8] improves the accuracy of both decision tree and MLP for 10 of the 17 data sets. For six data sets, including 'Glass", "Vowel", "Contraceptive", "Segmentation", "Optdigits", and "Yeast", the classification accuracy remains almost unchanged after applying MCSSB to the base multi-class learning algorithm. Only for data set "Ecoli", MCSSB-MLP performs significantly worse than MLP. Note that for several data sets, the improvement made by the MCSSB is dramatic. For instance, the classification accuracy of decision tree is improved from 32.8% to 77.4% for data set "Iris", and from 33.4% to 78.2% for data set "Wine"; the classification accuracy of MLP is improved from 48.3% to 68.0% for data set "Dermatology", and from 30.0% to 54.7% for data set "pendigit". Second, when compared to ASSEMBLE, we found that the proposed algorithm significantly outperforms ASSEMBLE for 14 of the 16 data sets for both decision tree and MLP. Only for data set "Ecoli", ASSEMBLE performs better than MCSSB when using MLP as the base classifier. The key differences between MCSSB and ASSEMBLE is that MCSSB is not only specially designed for multi-class classification, it does not solely rely on the pseudo-labels obtained in the iterations of boosting algorithm. Thus, the success of MCSSB indicates the importance of designing semi-supervised learning algorithms for multi-class problems.

Table 3 shows the performance of different algorithms when 10% of the examples are labeled. Similar to the previous case, MCSSB outperforms both the base classifiers and the ASSEMBLE method for 8 of the 17 data sets. For the

[8] The variance reported in the table clearly shows the advantage of our method compared to the baseline.

**Table 3.** The accuracy of different methods with 10% labeled examples

| | Tree | MLP | Assemble-tree | Assemble-MLP | MCSSB-tree | MCSSB-MLP |
|---|---|---|---|---|---|---|
| Balance | 67.7±0.7 | **86.0±1.1** | 67.8±0.7 | **87.0±0.5** | 69.5±1.0 | **86.6±0.6** |
| Glass | **46.9±1.7** | 42.7±1.5 | **46.8±1.7** | **45.4±1.6** | **45.3±1.5** | 43.8±1.7 |
| Iris | 68.5±2.4 | 79.2±3.0 | 68.7±2.4 | 77.2±2.6 | 79.7±2.7 | **84.1±2.3** |
| Wine | 73.0±2.7 | 78.2±2.4 | 73.0±2.7 | 74.7±3.0 | 81.8±1.2 | **83.2±1.2** |
| Car | **83.2±0.5** | 77.1±0.6 | 83.1±0.5 | 78.4±0.7 | **83.7±0.5** | 78.0±0.4 |
| Vowel | **27.2±1.1** | 21.2±0.5 | 24.8±1.0 | 21.7±0.9 | **27.6±1.0** | 22.8±1.1 |
| Contraceptive | 42.6±0.0 | 31.9±3.2 | 42.6±0.0 | 28.6±3.5 | **81.4±1.8** | 71.0±2.8 |
| Dermatology | 64.6±1.6 | 48.7±2.0 | 63.9±1.5 | 50.4±2.7 | **78.4±1.4** | 65.6±2.1 |
| Ecoli | **65.1±1.7** | **64.8±1.6** | **65.0±1.7** | **65.2±1.4** | 61.4±1.9 | **64.0±1.6** |
| Flag | **38.7±1.5** | 29.2±1.3 | **38.5±1.4** | 30.3±1.5 | **38.9±1.4** | 28.3±1.1 |
| Segmentation | **48.5±2.3** | 46.3±1.5 | 46.4±2.0 | 42.9±1.3 | **48.5±2.3** | 46.8±1.7 |
| Pendigits | 36.5±1.5 | 31.7±0.8 | 34.0±1.4 | 31.3±0.9 | **57.7±1.2** | 52.2±1.4 |
| Optdigits | **33.9±1.3** | 26.3±0.8 | 31.9±1.1 | 27.5±0.6 | **33.9±1.3** | 27.6±1.1 |
| Soybean | 37.7±1.1 | 33.4±1.0 | 37.2±1.6 | 31.7±1.0 | **42.8±1.1** | 39.9±1.2 |
| Waves | 65.5±0.3 | 76.6±1.4 | 65.5±0.3 | 76.7±2.3 | 65.5±0.3 | **79.3±0.9** |
| Yeast | **47.9±0.7** | 41.0±1.1 | **47.5±0.6** | 41.7±0.9 | **47.8±0.6** | 41.6±1.1 |
| Zoo | 52.0±1.8 | 51.8±2.5 | 52.0±1.8 | 50.8±2.3 | **74.3±1.9** | 71.9±1.5 |

rest of data sets, including "Balance", "Glass","Car", "Vowel", "Ecoli", "Flag", "Segmentation", "Optdigits", "Waves ", and "Yeast", the classification accuracy remains unchanged after applying MCSSB to the base supervised learning algorithms. Noitce that the amount of improvement in this case is less than the case with 5% labeled examples. This is because as the number of labeled examples increases, the improvement gained by a semi-supervised learning algorithm decreases. Moreover, notice that similar to the case of 5% labeled examples, ASSEMBLE is not able to improve the performance of the base classifier. Based on the above observation, we conclude that the proposed semi-supervised boosting algorithm is able to effectively exploit the unlabeled data to improve the performance of supervised multi-class learning algorithms.

## 4.2 Sensitivity to the Number of Labeled Examples

To study the sensitivity of MCSSB to the number of unlabeled examples, we run MCSSB and the baselines by varying the number of labeled examples from 2% to 20% of the total number of examples. Figure 1 shows the result of this experiment on 4 of the datasets when the base classifier is tree[9]. Notice that as the number of labeled examples increases, the performance of difference methods improves. But MCSSB keeps its superiority for almost all the cases when compared to both the base classifier and the ASSEMBLE algorithm. We also observe that overall ASSEMBLE is unable to make significant improvement over the base classifier regardless of the number of labeled examples. More surprisingly, for data set "Soybean", ASSEMBLE performs worse than the base

[9] We omit the result for other data sets and MLP as the base classifier due to space limitation.

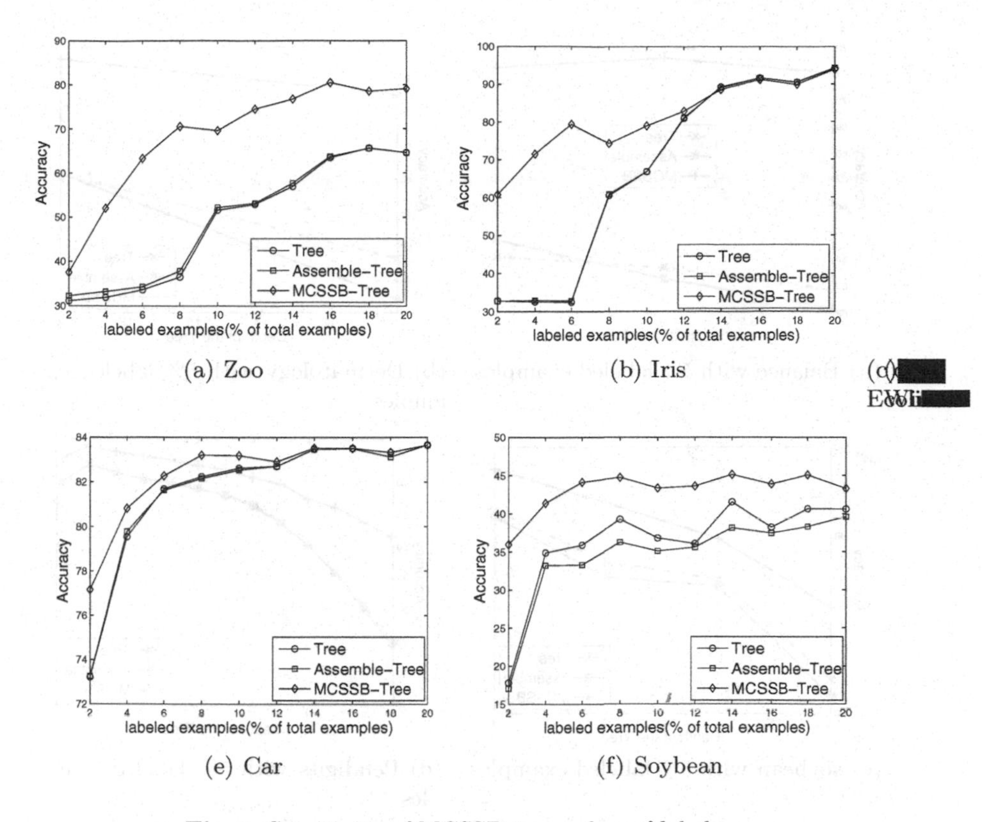


**Fig. 1.** Sensitivity of MCSSB to number of labels

classifier. These results indicate the challenge in developing boosting algorithms for semi-supervised multi-class learning. Compared to ASSEMBLE that relies on the classification confidence to decide the pseudo labels for unlabeled examples, MCSSB is more reliable since it exploits both the classification confidence and similarities among examples when determining the pseudo labels.

### 4.3 Sensitivity to Base Classifier

In this section, we focus on examining the sensitivity of MCSSB to the complexity of base classifiers. This will allow us to understand the behavior of the proposed semi-supervised boosting algorithm for both weak classifiers and strong classifiers. To this end, we use decision tree with varying number of levels as the base classifier. Only the results for datasets Balance, Dermatology, Soybean, and Pendigit are reported in this study because these were the only four data sets for which the fully grown decision tree had more than two levels.

Figure 2 shows the classification accuracy of Tree, ASSEMBLE-tree and MCSSB-tree when we vary the number of levels in decision tree. Notice that in each case, the maximum number of level in the plot for each data set is set to

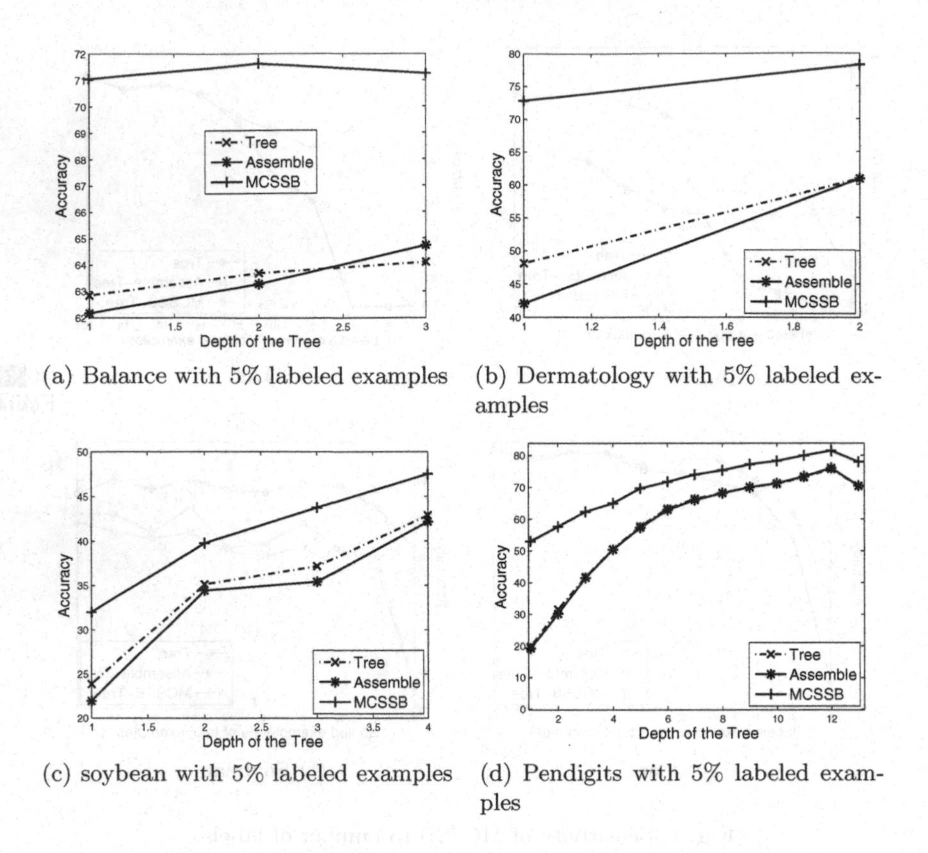


(a) Balance with 5% labeled examples

(b) Dermatology with 5% labeled examples

(c) soybean with 5% labeled examples

(d) Pendigits with 5% labeled examples

**Fig. 2.** Sensitivity of MCSSB to depth of the tree

the tree fully grown for that data set. It is not surprising that overall the classification accuracy is improved with increasing number of levels in decision tree. We also observe that MCSSB is more effective than ASSEMBLE for decision trees with different complexity.

## 5 Conclusion

Unlike many existing semi-supervised learning algorithms that focus on binary classification problems, we address multi-class semi-supervised learning directly. We have proposed a new framework, termed multi-class semi-supervised boosting (MCSSB), that is able to improve the classification accuracy of any given base multi-class classifier. We showed that our proposed framework is able to improve the performance of a given classifier much better than Assemble, a well-known semi-supervised boosting algorithm, on a large set of UCI datasets. We also show that MCSSB is very robust to the choice of base classifiers and the number of labeled examples.

## Acknowledgements

The work was supported in part by the National Science Foundation (IIS-0643494) and Office of Naval Research (N00014-07-1-0255). Any opinions, findings, and conclusions or recommendations expressed in this material are those of the authors and do not necessarily reflect the views of NSF and ONR.


## References


1. Bennett, K.P., Demiriz, A.: Semi-supervised support vector machine. In: NIPS (1999)
2. Chapelle, O., Zien, A.: Semi-supervised classification by low density separation. In: 10th Int. Workshop on AI and Stat. (2005)
3. Blum, A., Chawla, S.: Learning from labeled and unlabeled data using graph mincuts. In: ICML (2001)
4. Zhu, X., Ghahramani, Z., Lafferty, J.: Semi-supervised learning using gaussian fields and harmonic functions. In: ICML (2003)
5. Zhou, D., Bousquet, O., Lal, T., Weston, J., Schölkopf, B.: Learning with local and global consistency. In: NIPS (2003)
6. Belkin, M., Matveeva, I., Niyogi, P.: Regularization and semisupervised learning on large graphs. In: Shawe-Taylor, J., Singer, Y. (eds.) COLT 2004. LNCS (LNAI), vol. 3120. Springer, Heidelberg (2004)
7. Bennett, K.P., Demiriz, A., Maclin, R.: Exploiting unlabeled data in ensemble methods. In: KDD (2002)
8. Chen, K., Wang, S.: Regularized boost for semi-supervised learning. In: NIPS (2007)
9. Dietterich, T.G., Bakiri, G.: Solving multiclass learning problems via error-correcting output codes. J. AI. Res. 2, 263–286 (1995)
10. Jin, R., Zhang, J.: Multi-class learning by smoothed boosting. Mach. Learn. 67(3), 207–227 (2007)
11. Scholkopf, B., Smola, A.J.: Learning with Kernels: Support Vector Machines, Regularization, Optimization, and Beyond. MIT Press, Cambridge (2001)
12. Zadrozny, B., Elkan, C.: Transforming classifier scores into accurate multiclass probability estimates. In: KDD (2002)
13. Eibl, G., Pfeiffer, K.P.: Multiclass boosting for weak classifiers. J. Mach. Learn. Res. 6, 189–210 (2005)
14. Li, L.: Multiclass boosting with repartitioning. In: ICML (2006)
15. d'Alche Buc, F., Grandvalet, Y., Ambroise, C.: Semi-supervised marginboost. In: NIPS (2002)
16. Zhu, X.: Semi-supervised learning literature survey. Technical Report 1530, Computer Science, University of Wisconsin-Madison (2005)
17. Joachims, T.: Transductive inference for text classification using support vector machines. In: ICML (1999)
18. Bie, T.D., Cristianini, N.: Convex methods for transduction. In: NIPS (2004)
19. Xu, L., Schuurmans, D.: Unsupervised and semi-supervised multi-class support vector machines. In: AAAI (2005)
20. Freund, Y., Schapire, R.E.: Experiments with a new boosting algorithm. In: ICML (1996)

21. Higham, N.J.: Matrix nearness problems and applications. In: Gover, M.J.C., Barnett, S. (eds.) Applications of Matrix Theory, pp. 1–27. Oxford University Press, Oxford (1989)
22. Opitz, D., Maclin, R.: Popular ensemble methods: An empirical study. J. AI. Res. 11, 169–198 (1999)
23. Shi, J., Malik, J.: Normalized cuts and image segmentation. IEEE Transactions on Pattern Analysis and Machine Intelligence 22(8), 888–905 (2000)

## Appendix A: Proof of Lemma 1

*Proof.* Bound in Equation (8) can be derived as follows:

$$\frac{1}{\widetilde{Z}^u_{i,j}} = \frac{1}{\sum_{k'=1}^m \widetilde{b}_i^k \widetilde{b}_j^k} = \frac{(\sum_{k'=1}^m b_i^{k'} \exp(\alpha h_i^{k'}))(\sum_{k'=1}^m b_i^{k'} \exp(\alpha h_j^{k'}))}{\sum_{k=1}^m b_i^k b_j^k \exp(\alpha(h_i^k + h_j^k))}$$

$$\leq \left(\sum_{k'=1}^m b_i^{k'} \exp(\alpha h_i^{k'})\right)\left(\sum_{k'=1}^m b_i^{k'} \exp(\alpha h_j^{k'})\right) \times \frac{(\sum_{k=1}^m \tau_{i,j}^k \exp(-\alpha(h_i^k + h_j^k)))}{\sum_{k=1}^m b_i^k b_j^k} \quad (22)$$

$$= \sum_{k_1,k_2,k_3=1}^m \frac{b_i^{k_1} b_j^{k_2} \tau_{i,j}^{k_3}}{Z^u_{i,j}} \exp\left(\alpha(h_i^{k_1} + h_j^{k_2} - h_i^{k_3} - h_j^{k_3})\right)$$

$$\leq \frac{1 + \exp(6\alpha) + \exp(-6\alpha)}{3Z^u_{i,j}} - \frac{\exp(6\alpha) - 1}{3Z^u_{i,j}} \left(\sum_{k=1}^m (\tau_{i,j}^k - b_i^k) h_i^k\right) \quad (23)$$

The inequality in (22) follows the convexity of reciprocal function, i.e.,

$$\frac{1}{\sum_{k=1}^m b_i^k b_j^k \exp(\alpha(h_i^k + h_j^k))} = \frac{1}{\sum_{k=1}^m b_i^k b_j^k} \frac{1}{\sum_{k=1}^m \tau_{i,j}^k \exp(\alpha(h_i^k + h_j^k))}$$

$$\leq \frac{1}{\sum_{k=1}^m b_i^k b_j^k} \sum_{k=1}^m \tau_{i,j}^k \exp(-\alpha(h_i^k + h_j^k))$$

The inequality in (23) follows the convexity of exponential function, i.e.,

$$\exp(\alpha(h_i^{k_1} + h_j^{k_2} - h_i^{k_3} - h_j^{k_3})) = \exp\left(\begin{array}{l} 6\alpha \frac{h_i^{k_1} + h_j^{k_2} - h_i^{k_3} - h_j^{k_3} + 2}{6} + \\ 0 \times \frac{-h_i^{k_1} - h_j^{k_2} + h_i^{k_3} + h_j^{k_3} + 2}{6} + 6\alpha\frac{1}{3} \end{array}\right)$$

$$\leq \frac{h_i^{k_1} + h_j^{k_2} - h_i^{k_3} - h_j^{k_3} + 2}{6} \exp(6\alpha) + \frac{1}{3}\exp(6\alpha) + \frac{-h_i^{k_1} - h_j^{k_2} + h_i^{k_3} + h_j^{k_3} + 2}{6}$$

Bound in Equation 9 can be derived as follows

$$\frac{1}{\widetilde{Z}^l_{i,j}} = \sum_{k',k=1}^m y_i^k \exp(H_j^{k'} - H_j^k + \alpha(h_j^{k'} - h_j^k))$$

$$\leq \frac{1 + \exp(6\alpha) + \exp(-6\alpha)}{3Z^l_{i,j}} + \frac{\exp(6\alpha) - 1}{6} \sum_{k=1}^m h_i^k \left(\sum_{k'=1}^m y_j^{k'} \frac{b_i^k}{b_i^{k'}} - \frac{y_i^k}{b_i^k}\right)$$

The inequality used by the above derivation follows the convexity of exponential function, i.e.,

$$\exp(\alpha(h_i^{k'} - h_j^k)) \leq \exp\left(6\alpha\frac{h_i^{k'} - h_j^k + 2}{6} + 0 \times \frac{-h_i^{k'} + h_j^k + 2}{6} + 6\alpha\frac{1}{3}\right)$$
$$\leq \frac{h_i^{k'} - h_j^k + 2}{6}\exp(6\alpha) + \frac{1}{3}\exp(6\alpha) + \frac{-h_i^{k'} + h_j^k + 2}{6}$$

Using the definition of $\phi_{i,j}^k$, we have the result in Equation 9.

## Appendix B: Proof of Lemma 2

*Proof.* Following the result in (22), we have

$$\frac{1}{\widetilde{Z}_{i,j}^u}\sum_{k_1,k_2,k_3=1}^{m}\frac{b_i^{k_1}b_j^{k_2}\tau_{i,j}^{k3}}{Z_{i,j}^u}\exp(\alpha(h_i^{k_1} + h_j^{k_2} - h_i^{k_3} - h_j^{k_3}))$$
$$\leq \frac{1}{Z_{i,j}^u} + \frac{\exp(2\alpha) - 1}{2Z_{i,j}^u}\left(\sum_{k=1}^{m} h_i^k b_i^k + h_j^k b_j^k\right) + \frac{\exp(-2\alpha) - 1}{2Z_{i,j}^u}\left(\sum_{k=1}^{m} a_{i,j}^k(h_i^k + h_j^k)\right)$$

The inequality in the above derivation follows the convexity of exponential function (similar to the proof of Lemma 1). For $Z_{i,j}^l$, we have

$$\frac{1}{\widetilde{Z}_{i,j}^l} = \sum_{k,k'=1}^{m} y_i^k\frac{b_j^{k'}}{b_j^k}\exp\left(2\alpha\frac{h_j^{k'}}{2} - 2\alpha\frac{h_j^k}{2} + 0\frac{2 - h_j^{k'} - h_j^k}{2}\right)$$
$$\leq \sum_{k,k'=1}^{m} y_i^k\frac{b_j^{k'}}{b_j^k}\left(\frac{h_j^{k'}}{2}\exp(2\alpha) + \frac{h_j^k}{2}\exp(-2\alpha)\right) + \sum_{k,k'=1}^{m} y_i^k\frac{b_j^{k'}}{b_j^k}\frac{2 - h_j^k - h_j^{k'}}{2}$$

Replacing $1/\widetilde{Z}_{i,j}^u$ and $1/\widetilde{Z}_{i,j}^l$ in (8) and (9) with the above bounds, we have the result in Lemma 2.

# A Joint Segmenting and Labeling Approach for Chinese Lexical Analysis

Xinhao Wang, Jiazhong Nie, Dingsheng Luo, and Xihong Wu*

Speech and Hearing Research Center,
Key Laboratory of Machine Perception (Ministry of Education),
School of Electronics Engineering and Computer Science,
Peking University,
100871, Beijing, China
{wangxh,niejz,dsluo,wxh}@cis.pku.edu.cn

**Abstract.** This paper introduces an approach which jointly performs a cascade of segmentation and labeling subtasks for Chinese lexical analysis, including word segmentation, named entity recognition and part-of-speech tagging. Unlike the traditional pipeline manner, the cascaded subtasks are conducted in a single step simultaneously, therefore error propagation could be avoided and the information could be shared among multi-level subtasks. In this approach, Weighted Finite State Transducers (WFSTs) are adopted. Within the unified framework of WFSTs, the models for each subtask are represented and then combined into a single one. Thereby, through one-pass decoding the joint optimal outputs for multi-level processes will be reached. The experimental results show the effectiveness of the presented joint processing approach, which significantly outperforms the traditional method in pipeline style.

**Keywords:** WFSTs, Chinese lexical analysis, joint segmentation and labeling.

## 1 Introduction

The Chinese lexical analysis involves solving a cascade of well-defined segmentation and labeling subtasks, such as word segmentation, named entity recognition and part-of-speech (POS) tagging. Like many problems in natural language processing, the cascade is traditionally processed in a pipeline manner. However, it has the disadvantage that errors introduced by earlier subtasks propagate through the pipeline and will never be recovered in downstream subtasks. Moreover, this manner prevents information sharing among multi-level processes. For example, the POS information is helpful to make better prediction in word segmentation and named entity recognition, while this is prohibited in pipeline processing.

To tackle these problems, several techniques were proposed recently. Reranking method has been widely applied in a number of different natural language

* Corresponding author.

W. Daelemans et al. (Eds.): ECML PKDD 2008, Part II, LNAI 5212, pp. 538–549, 2008.

processing problems, such as parsing [1,2], machine translation [3] and so on. In handling the cascaded tasks, a $k$-best list is preserved at each level firstly, and then reranked in the following subtasks [4,5,6]. Nevertheless, as an approximation of joint processing, reranking may miss the true result, which usually lies out of the $k$-best list. Another intuitive approach is to take multiple subtasks as a single one [7,8]. Such as in [9], the constituent labels of the Penn TreeBank are augmented with semantic role labels (SRL), thus parsing the trees also serves as a SRL system. Similarly in [10,11], word segmentation and POS tagging are performed simultaneously by marking each Chinese character with a word level POS tag. But an obvious obstacle of these label transformations is the requirement of corpus annotated with multi-level information, which is usually unavailable in many situations. Unlike the strategies mentioned above, some unified probabilistic models are proposed to process a cascade jointly. Sutton et al. [12] proposed the Dynamic Conditional Random Fields (DCRFs), which are trained jointly and performs the subtasks in one step, but it is expensive in training and exact inference. Moreover in [13], the Factorial Hidden Markov Model (FHMM) was also introduced to the joint labeling tasks of POS tagging and noun phrase chunking. Compared with DCRFs, FHMM has the computational advantage as a generative model, and the exact inference can be achieved easier. However, both DCRFs and FHMM also suffer from the absence of multi-level annotated corpus.

In this paper, based on Weighted Finite State Transducers (WFSTs), an integrated Chinese lexical analyzer is presented to jointly perform the cascade of segmentation and labeling tasks, including word segmentation, named entity recognition and part-of-speech tagging. Traditionally, WFSTs have already been successfully used in various fields of natural language processing, such as partial parsing [14], named entity recognition [15], semantic interpretation [16], as well as Chinese word segmentation [17,18]. However, being different from those applications, this study employs WFSTs to jointly conduct segmentation and labeling tasks. WFSTs turn to be an ideal choice for our purpose due to two following remarkable features: On one hand, most of the widely used models, like lexical constraints, n-gram language model and Hidden Markov Models (HMMs), can be encoded into WFSTs, and thus a unified transducer representation for these models is able to be achieved. On the other hand, since there exist mathematically well-defined operations to integrate multiple transducers into a single composed one, the optimal candidate can be extracted by one-pass decoding with multi-level knowledge sources represented by each transducer. In contrast to the joint processing techniques mentioned above, the presented approach has the following advantages. Firstly, rather than reranking the $k$-best candidates preserved at each level, it holds the full search space and chooses the optimal results based on the multi-level sources. Secondly, similar to the strategy of [19], the models for each level subtask are trained separately, while the decoding is conducted jointly. Accordingly, it avoids the necessary of corpus annotated with multi-level information. Other than [19], in this study the used generative models bring the benefit in computation, which is important in a joint processing task, especially as the scale of subtasks increasing. Thirdly, in the case when a segmentation task precedes a labeling task, the consistency restriction imposed by the segmentation task must be maintained in the successive labeling task. For instance, the POS tags assigned to each character in a segmented word must be the same. While for the methods taking the smallest characters in the segmentation task as modeling units, such as Chinese Characters in Chinese word segmentation, this restriction is not naturally satisfied. The WFSTs based approach ensures this restriction by the composition operation, i.e., the input sequence of one transducer and the output sequence of the other transducer must be identical. In addition, the unified framework of WFSTs provides the opportunity to easily apply the presented analyzer in other natural language related applications which are also based on WFSTs, such as speech recognition [20] and machine translation [21]. Since more linguistic knowledge in multi-level is modeled by the analyzer, performance improvements possibly can be achieved for those applications.

The remainder of this paper is structured as follows. Section 2 introduces the formal definition and notation of WFST. In section 3, by describing the integrated Chinese lexical analyzer in detail, the joint segmenting and labeling approach is presented. Then simulations are performed to evaluate the new analyzer in section 4. Finally, section 5 draws the conclusion and discusses the future work.

## 2 Weighted Finite State Transducers

The Weighted Finite State Transducer (WFST) is the generalization of the finite state automata. In weighted transducer, besides of an input label, an output label and a weight are also placed on each transition. With these labels, the transducer is capable of realizing a weighted relation between strings. In the most general case, the definition of WFST depends on the algebraic structure of a semiring,$(\mathbb{K}, \oplus, \otimes, \overline{0}, \overline{1})$ [22,23,24]. In a semiring, two operation $\oplus$ and $\otimes$ are associative and closed over the set $\mathbb{K}$. $\overline{0}, \overline{1}$ are their identities respectively. Unlike a ring, a semiring may not have negation. For example, $(\mathbb{N}, +, *, 0, 1)$ is a semiring.

### 2.1 Definition

In this paper, a weighted transducer $W$ over the semiring $\mathbb{K}$ is formally defined as a 6-tuple $W = (Q, i, F, \Sigma, \Delta, T)$ , where

- $Q$ is the set of its states,
- $i \in Q$ is an initial state,
- $F \in Q$ is the set of final states,
- $\Sigma$ is the input alphabet,
- $\Delta$ is the output alphabet,
- $T \subset Q \times Q \times \Sigma \cup \{\epsilon\} \times \Delta \cup \{\epsilon\} \times \mathbb{K}$ is the set of transitions. Each $t \in T$ consists of a source state $src(t) \in Q$, a destination state $des(t) \in Q$, an input label $in(t) \in \Sigma \cup \{\epsilon\}$, an output label $out(t) \in \Delta \cup \{\epsilon\}$ and a weight $wght(t) \in \mathbb{K}$.

A successful path $\pi$ in $W$, is a chain of successive transitions: $p = t_1 t_2 ... t_n$, satisfying $src(t_1) = i$, $des(t_n) \in F$ and $des(t_i) = src(t_{i+1}) \quad 1 \leq i \leq n-1$. The input and output strings mapped by this path are the concatenation of transitions' input and output labels: $in(p) = in(t_1)in(t_2)...in(t_n)$, $out(p) = out(t1)out(t_2)...out(t_n)$, where the symbol $\epsilon$ represents the empty string. The weight of $\pi$ is the $\otimes$-product of its transitions' weights: $wght(\pi) = wght(t_1) \otimes wght(t_2) \otimes ... \otimes wght(t_n)$. For a given input string $s$ and an output string $r$, the set $path_W(s,r)$ consists of all the successful paths in $W$, whose input and output strings match $s$ and $r$ respectively. The weight associated by $W$ to the $(s,r)$ is the $\oplus$-sum of the paths' weights in $path_W(s,r)$ and $\overline{0}$:

$$W(s,r) = (\sum_{p \in path_W(s,r)} wght(p)) \oplus \overline{0}$$

In our system, probabilities are adopted as weights and each string pair is associated with a weight indicating the probability of the mapping between them. Due to the numerical stability, log probabilities are used in implementation instead of probabilities. The appropriate semiring for the finite-state representation of log probabilites and operations is the tropic semiring $(\mathbb{R}_+ \cup \{\infty\}, min, +, \infty, 0)$ [22].

### 2.2 Decoding and Composition

Given an input string $s$ and a WFST $W$, the goal of decoding is to find the best output string $r^*$ maximizing $W(s,r)$. Similarly, when multiple WFSTs are involved, a joint decoding is desired to find the optimal final output string $r^*$, which maximizes the $\otimes$ -product of each mapping $W_1(s,m_1) \otimes W_i(m_{i-1},m_i) \otimes ... \otimes W_n(m_{n-1},r)$, where $m_i$ is an arbitrary string on $W_i$'s output alphabet. An efficient way to implement this desired decoding is using the composition algorithm to combine multiple WFSTs into a single one [22,23,24].

For two WFSTs $E$ and $F$ satisfying the input alphabet of $F$ and output alphabet of $E$ are the same, the composition $G = E \circ F$ represents the composition of the weighted relations realized by $E$ and $F$. As in the classical finite state automata intersection, the states in $G$ are pairs of states in $E$ and $F$. $G$'s initial state is the pair of initial states of $E$ and $F$, and final states are pairs of a final state in $E$ and a final state in $F$. For each pair of transition $t_E$ from $e$ to $e'$ in $E$ and transition $t_F$ from $f$ to $f'$ in $F$, there exists exactly one transition $t$ in $G$ from $(e,f)$ to $(e',f')$. The input label of $t$ is taken from $t_E$ and output label from $t_F$. Wght(t) is the $\otimes$ -product of wght($t_E$) and wght($t_F$), when the weights correspond to probabilities. For transducers have $\epsilon$ transitions, special treatments are needed as in [25]. Figure 1 shows two simple transducer Figure 1(a) and Figure 1(b), and the result of their composition, Figure 1(c). All of them are defined on the tropic semiring.

Apparently, for a path in $E$ mapping $s$ to $v$ and a path in $F$ mapping $v$ to $r$, $G$ has exactly one path mapping $s$ to $r$ directly and its weight is the $\otimes$ -product of the corresponding paths' weights in $E$ and $F$. This property enables us to find

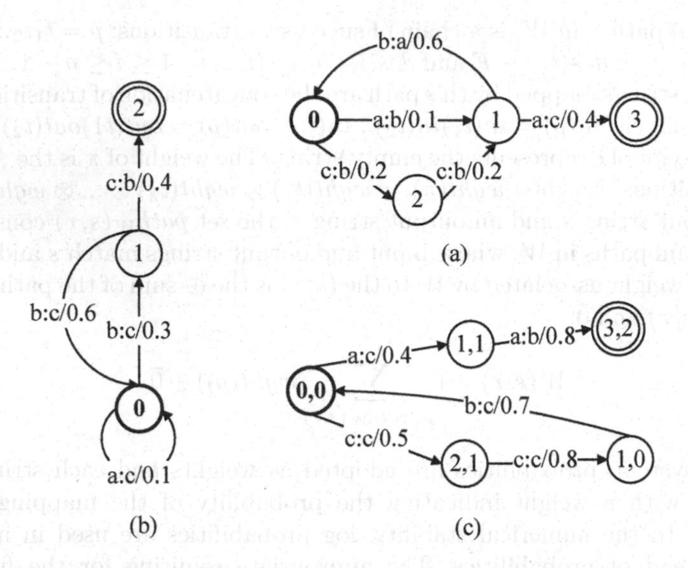


**Fig. 1.** Example of WFSTs composition. Two simple WFSTs are showed in (a) and (b), in which states are represented by circles and labeled with their unique numbers. The bold circles represent initial states and double circles for final states. The input and output labels as well as weight of transition t are marked as in(t):out(t)/wght(t). In (c), the composition of (a) and (b) is illustrated

the optimal final output string among multiple WFSTs by finding the optimal output string in the combination of these WFSTs $W = ((W_1 \circ W_2) ... \circ W_n)$,which can be easily realized by a standard Viterbi search.

# 3 Joint Chinese Lexical Analysis

Word segmentation is the first stage of Chinese text processing since Chinese is typically written without blanks, and POS tagging is to assign the part-of-speech to each segmented word in a sentence. Both tasks face the challenge of tackling unknown words that are out of the dictionary. In this study, two kinds of typical unknown words, person names and location names, are also focused.

In this section, within the unified framework of WFSTs, the models for three level subtasks, i.e. words segmentation, POS tagging and named identity recognition, are presented, and then they are combined to reach an integrated Chinese lexical analyzer.

## 3.1 Multiple Subtasks Modeling

For word segmentation, the class based n-gram technique is adopted. Given an input character sequence, it is encoded by a finite state acceptor $FSA_{input}$. For example, the input “合成分子时”(while synthesizing molecule) is represented as Figure 2(a). Then a dictionary can be represented by a transducer with empty

**Table 1.** Toy dictionary

| Chinese Words | English Words |
|---|---|
| 合 | together |
| 合成 | synthesize |
| 成分 | element |
| 分子 | molecule |
| 子时 | the period of the day from 11 p.m.to 1 a.m. |
| 时 | present |

**Table 2.** Definition of word classes

| Classes | Description |
|---|---|
| $w_i$ | Each word $w_i$ listed in dictionary |
| CNAME | Chinese person names as one class |
| TNAME | Translated person names as one class |
| LOC | Location names |
| NUM | Number expressions |
| LETTER | Letter strings |
| NON | Other non Chinese character strings |
| BEGIN | Beginning of sentence as one class |
| END | End of sentence as one class |

weights, denoted as $FST_{dict}$. Figure 2(b) illustrates a toy dictionary listed in Table 1, in which a successful path encodes a mapping from a Chinese character sequence to some word in dictionary. Afterwards a class based n-gram language model is used to weight the candidate segmentations. In Figure 2(c), a toy bigram with three words is depicted by $WFST_{n-gram}$, and the word classes are defined in Table 2.

For both POS tagging and named entity recognition, the Hidden Markov Model (HMM) is used. Each HMM is represented with two WFSTs. Taking the POS tagging as an example, Figure 3(a) models the generation of words by POS ($P(word/pos)$), and similar to the word n-gram, Figure 3(b) models the transitions between POS tags. For named entity recognition, the HMM states correspond to 30 named entity role tags, such as surname, the first character of a given name with two characters, the first character of a location name, and so on.

Besides the primary WFSTs described above, there are also some other finite state transducers, which are used to represent various rules for recognizing the number strings and letter strings, or to be responsible for the transformation from name roles to word classes.

### 3.2 Integration of Multiple Models

Based on the WFSTs built above, an integrated model is obtained by combining them into a single one using the composition algorithm as describe in section 2.

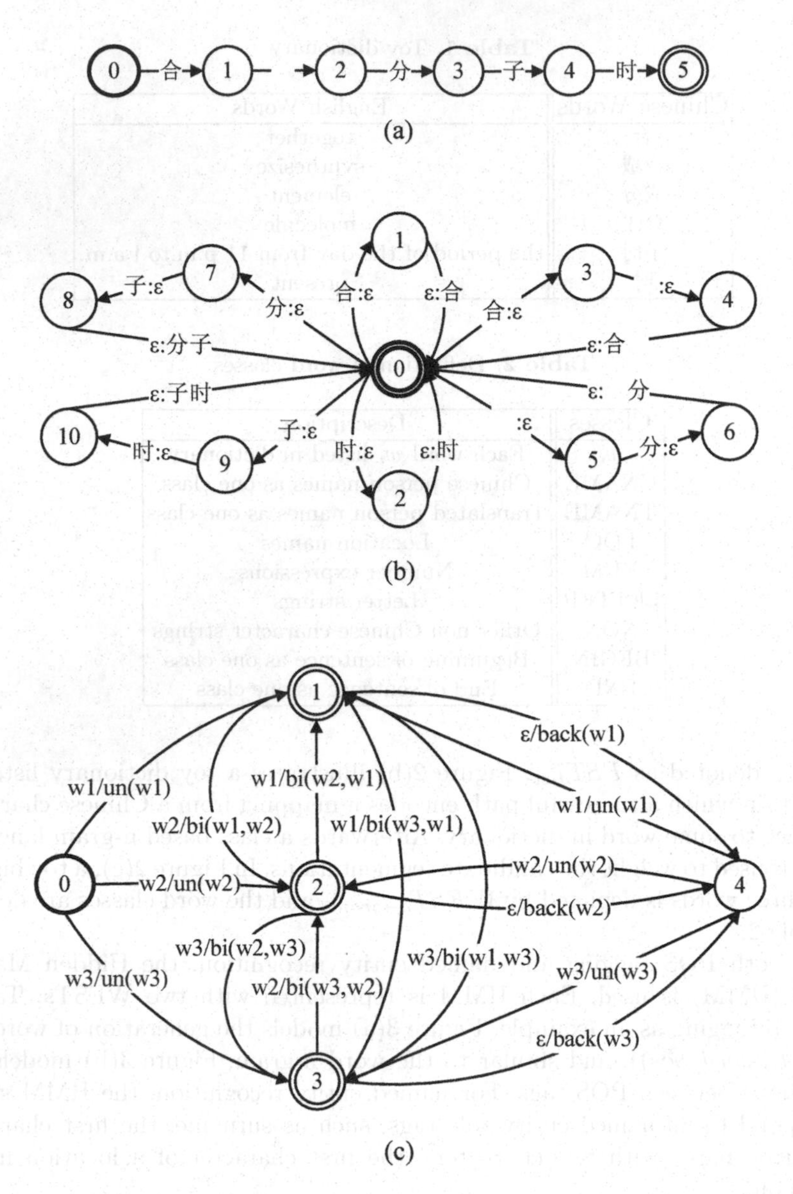


**Fig. 2.** Word WFSTs. (a) is the FSA representing an input example; (b) is the FST representing a toy dictionary; and (c) is the WFSA representing a toy bigram language model, where $un(w1)$ denotes the unigram of $w1$; $bi(w1, w2)$ and $back(w1)$ respectively denotes the bigram of $w2$ and the backoff weight given the word history $w1$.

To perform word segmentation, a WFST embracing all the possible candidates is obtained as below:

$$WFST_{words} = FSA_{input} \circ FST_{dict} \circ WFST_{ne} \circ WFSA_{n-gram} \quad (1)$$

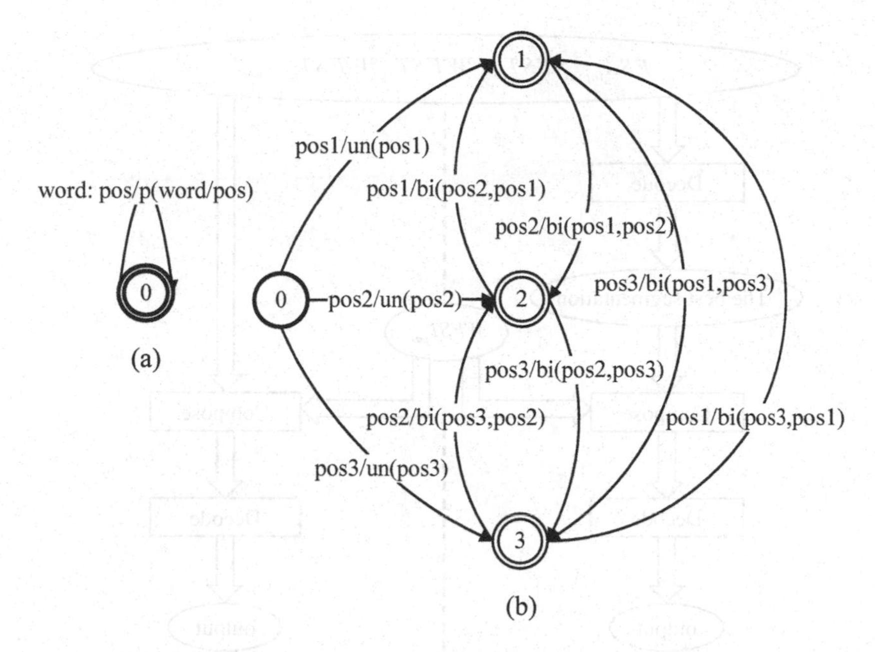


**Fig. 3.** POS WFSTs. (a) is the WFST representing the relationship between the word and the pos; (b) is the WFSA representing a toy bigram of POS.

As the class based n-gram is adopted, the named entity recognition is conducted along with word segmentation. Taking the POS tagging into account, the decoding, which aims to extract the joint optimal results according to multi-level information, is performed on the WFST composed as following, where $\alpha$ is a weight for combining different level subtasks.

$$WFST_{analyzer} = (\alpha * WFST_{words}) \circ WFST_{POS}. \quad (2)$$

# 4 Simulation

To evaluate the presented analyzer, two systems are constructed, as illustrated in Figure 4, where the system based on the pipeline style method is taken as the baseline. The experimental corpus comes from the People's Daily of China in 1998 from January to June, annotated by the Institute of Computational Linguistics of Peking University[1]. The January to May data are taken as the training set. The first two thousand sentences of June data are extracted as the develop set, which is used to fix the composition weight $\alpha$ in equation 2, and the remains are taken as the test set. A dictionary including about 113,000 words is extracted from the training data. The models for different level subtasks are trained separately, where the class based language model is trained with the

[1] http://icl.pku.edu.cn/

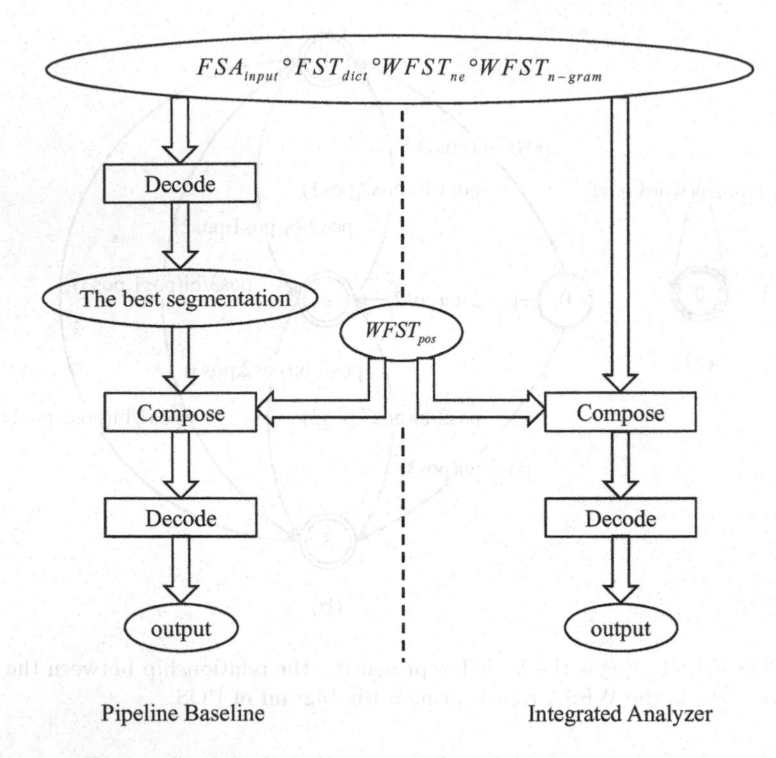


**Fig. 4.** The pipeline system vs The joint processing system

SRI Language Modeling Toolkit[2]. While the decoding is implemented through one-pass Viterbi search on the combined WFST.

In Table 3 the performances of the pipeline baseline and the integrated analyzer are compared. Due to the joint decoding, the integrated analyzer outperforms the pipeline baseline on all the tasks in F1-score metric, especially for the case of person name recognition. It is as expected that person names are greatly improved when incorporating POS information during recognition, where the POS appears effective in preventing to segment a person name into pieces (possibly into characters).

In addition, to further investigate the significance of performance improvement, a statistical test using approximate randomization approach [26] is performed on the word segmentation results. In this approach, the responses for each sentence produced by two systems are shuffled and equally resigned to each system, and then the significance level is computed as to whether shuffles bring differences not smaller than the difference produced when running two systems on the test data. In general, given $n$ test sentences, the shuffle times $s$ is fixed as in equation 3, i.e., when $n$ is small, no larger than 20, the exact randomization

[2] http://www.speech.sri.com/projects/srilm/

**Table 3.** Performance comparison between the pipeline baseline and the integrated analyzer. The system performances are measured with F1-score in the tasks of word segmentation (WS), POS tagging, as well as the person and location name recognition.

| | Pipeline Baseline | Integrated Analyzer |
|---|---|---|
| Word Segmentation | 95.94% | 96.77% |
| POS Tagging | 91.06% | 91.81% |
| Person Name Recognition | 83.31% | 88.51% |
| Location Name Recognition | 89.90% | 90.91% |

is performed, otherwise, only the approximate randomization is performed since the number of different shuffle ways, $2^n$, is too large to be exhaustively evaluated.

$$s = \begin{cases} 2^n, & n \leqslant 20 \\ 2^{20} = 1048576, & n > 20 \end{cases} \tag{3}$$

However, in our test set there are more than 21,000 sentences, the use of $2^{20}$ shuffles to approximate $2^{21000}$ shuffles as in formula 3 turns unreasonable any more. Thus, here ten sets (500 sentences for each) are randomly selected from the test corpus. For each set, we run 1048576 shuffles twice and calculate the significance level $p-value$ according to the shuffled results. Statistical test shows that all $p - values$ are less than 0.001 on the ten sets, which reveals that the performance improvement introduced by the integrated analyzer is statistically significant.

## 5 Conclusion and Future Work

In this research, within the unified framework of WFSTs, a joint processing approach is presented to perform a cascade of segmentation and labeling subtasks. It has been demonstrated that the joint processing is superior to the traditional pipeline manner. The finding suggests two directions for future research: Firstly, more linguistic knowledge will be integrated in the analyzer, such as organization name recognition and shallow parsing. For some tough tasks in related areas, such as large vocabulary continuous speech recognition and machine translation, rich linguistic knowledge will play an important role, thus incorporating our integrated lexical analyzer may lead to a promising performance improvement, and these attempts will be another future work.

## Acknowledgments

The authors would like to thank three anonymous reviewers for their helpful comments. This work was supported in part by the National Natural Science Foundation of China (60435010; 60535030; 60605016), the National High Technology Research and Development Program of China (2006AA01Z196; 2006AA010103), the National Key Basic Research Program of China (2004CB318005), and the

New-Century Training Program Foundation for the Talents by the Ministry of Education of China.

## References

1. Charniak, E., Johnson, M.: Coarse-to-fine n-best parsing and maxent discriminative reranking. In: Proceedings of the 43rd Annual Meeting of the Association for Computational Linguistics, Ann Arbor, Michigan, pp. 173–180 (June 2005)
2. Collins, M., Koo, T.: Discriminative reranking for natural language parsing. Computational Linguistics 31(1), 25–70 (2005)
3. Shen, L., Sarkar, A., Och, F.J.: Discriminative reranking for machine translation. In: Proceedings of the Human Language Technology Conference of the North American Chapter of the Association for Computational Linguistics, Boston, Massachusetts, pp. 177–184 (May 2004)
4. Shi, Y., Wang, M.: A dual-layer CRFs based joint decoding method for cascaded segmentation and labeling tasks. In: Proceedings of the International Joint Conference on Artificial Intelligence, Hyderabad, India, pp. 1707–1712 (January 2007)
5. Sutton, C., McCallum, A.: Joint parsing and semantic role labeling. In: Proceedings of the 9th Conference on Computational Natural Language Learning, Ann Arbor, Michigan, pp. 225–228 (June 2005)
6. Zhang, H., Yu, H., Xiong, D., Liu, Q.: Hhmm-based chinese lexical analyzer ictclas. In: Proceedings of the 2nd SIGHAN Workshop on Chinese Language Processing, Sapporo Japan, pp. 184–187 (July 2003)
7. Luo, X.: A maximum entropy chinese character based parser. In: Proceedings of the Conference on Empirical Methods in Natural Language Processing, Sapporo, Japan, pp. 192–199 (July 2003)
8. Miller, S., Fox, H., Ramshaw, L., Weischedel, R.: A novel use of statistical parsing to extract information from text. In: Proceedings of the 6th Applied Natural Language Processing Conference, Seattle, Washington, pp. 226–233 (April 2000)
9. Yi, S.T., Palmer, M.: The integration of syntactic parsing and semantic role labeling. In: Proceedings of the 9th Conference on Computational Natural Language Learning, Ann Arbor, Michigan, pp. 237–240 (June 2005)
10. Nakagawa, T., Uchimoto, K.: A hybrid approach to word segmentation and pos tagging. In: Proceedings of the 45th Annual Meeting of the Association for Computational Linguistics: Demo and Poster Sessions, Prague, pp. 217–220 (June 2007)
11. Ng, H.T., Low, J.K.: Chinese part-of speech tagging: one-at-a-time or all-at-once? word-based or character-based? In: Proceedings of the Conference on Empirical Methods in Natural Language Processing, Barcelona, Spain, pp. 277–284 (July 2004)
12. Sutton, C., Rohanimanesh, K., McCallum, A.: Dynamic conditional random fields: Factorized probabilistic models for labeling and segmenting sequence data. In: Proceedings of The 21st International Conference on Machine Learning, Banff, Alberta, Canada, pp. 783–790 (July 2004)
13. Duh, K.: Jointly labeling multiple sequences: a factorial hmm approach. In: Proceedings of the 43rd Annual Meeting of the Association for Computational Linguistics: Student Research Workshop, Ann Arbor, Michigan, pp. 19–24 (June 2005)
14. Abney, S.: Partial parsing via finite-state cascades. Natural Language Engineering 2(4), 337–344 (1996)

15. Friburger, N., Maurel, D.: Finite-state transducer cascades to extract named entities in texts. Theoretical Computer Science 313(1), 93–104 (2004)
16. Raymond, C., de ric Bechet, F., Mori, R.D., raldine Damnati, G.: On the use of finite state transducers for semantic interpretation. Speech Communication 48(3-4), 288–304 (2006)
17. Sproat, R., Shih, C., Gale, W., Chang, N.: A stochastic finite-state word-segmentation algorithm for chinese. In: Proceedings of the 32nd Annual Meeting of the Association for Computational Linguistics, Las Cruces, New Mexico, pp. 66–73 (June 1994)
18. Sproat, R., Shih, C., Gale, W., Chang, N.: A stochastic finite-state word-segmentation algorithm for chinese. Computational Linguistics 22(3), 377–404 (1996)
19. Sutton, C., McCallum, A.: Composition of conditional random fields for transfer learning. In: Proceedings of the Joint Conference of Human Language Technology Conference and the Conference on Empirical Methods in Natural Language Processing, Vancouver, pp. 748–754 (October 2005)
20. Mohri, M., Pereira, F., Riley, M.: Weighted finite-state transducers in speech recognition. Computer Speech and Language 16(1), 69–88 (2002)
21. Tsukada, H., Nagata, M.: Efficient decoding for statistical machine translation with a fully expanded wfst model. In: Proceedings of the Conference on Empirical Methods in Natural Language Processing, Barcelona, Spain, pp. 427–433 (July 2004)
22. Mohri, M.: Finite-state transducers in language and speech processing. Computational Linguistics 23(2), 269–311 (1997)
23. Kuich, W., Salomaa, A.: Semirings, Automata, Languages. Monographs in Theoretical Computer Science. An EATCS, vol. 5. Springer, Heidelberg (1986)
24. Berstel, J., Reutenauer, C.: Rational Series and Their Languages. Springer, Berlin (1988)
25. Mohri, M., Pereira, F.C.N., Riley, M.: he design principles of a weighted finite-state transducer library. Theor. Comput. Sci. 231(1), 17–32 (2000)
26. Yeh, A.: More accurate tests for the statistical significance of result differences. In: Proceedings of the 18th International Conference on Computational Linguistics, Saarbrücken, pp. 947–953 (August 2000)

# Transferred Dimensionality Reduction

Zheng Wang, Yangqiu Song, and Changshui Zhang

State Key Laboratory on Intelligent Technology and Systems
Tsinghua National Laboratory for Information Science and Technology (TNList)
Department of Automation, Tsinghua University, Beijing 100084, China

**Abstract.** Dimensionality reduction is one of the widely used techniques for data analysis. However, it is often hard to get a demanded low-dimensional representation with only the unlabeled data, especially for the discriminative task. In this paper, we put forward a novel problem of Transferred Dimensionality Reduction, which is to do unsupervised discriminative dimensionality reduction with the help of related prior knowledge from other classes in the same type of concept. We propose an algorithm named Transferred Discriminative Analysis to tackle this problem. It uses clustering to generate class labels for the target unlabeled data, and use dimensionality reduction for them joint with prior labeled data to do subspace selection. This two steps run adaptively to find a better discriminative subspace, and get better clustering results simultaneously. The experimental results on both constrained and unconstrained face recognition demonstrate significant improvements of our algorithm over the state-of-the-art methods.

**Keywords:** Transfer Learning, Dimensionality Reduction, Clustering.

## 1 Introduction

In many machine learning applications, such as computational biology, appearance-based image recognition and image retrieval, one is confronted with high-dimensional data. However it is considered that the original data naturally reside on lower dimensional manifolds. Finding this compact representation is usually a key step. Using an efficient representation, the subsequent phases, such as clustering or classification, will become much faster and more robust [14]. Thus some dimensionality reduction approaches have been developed. For unsupervised methods, e.g. principle component analysis (PCA) [20] and locality preserving projections (LPP) [14], the compact manifold should preserve the most relevant structure information of the original data point cloud. For supervised case, e.g. linear discriminant analysis (LDA) [1], the low-dimensional representation should find the most discriminative subspace for different classes based on the labeled data. Recently, the semi-supervised method has also been developed [3], which makes use of both labeled and unlabeled data.

In the last few years, several similar works [26,9,23] have been done to couple unsupervised dimensionality reduction with clustering, forming an adaptive dimensionality reduction framework. It performs discriminant analysis and clustering adaptively to select the most discriminative subspace and find a suitable clustering simultaneously. The most recent work [26] uses the method called discriminative k-means (DisKmeans),

W. Daelemans et al. (Eds.): ECML PKDD 2008, Part II, LNAI 5212, pp. 550–565, 2008.

which outgoes the traditional PCA+K-means framework and other similar works in their experiments. However, we observe that this type of methods is efficient only for specific data distributions, which is very limited. For example, we show three cases of a toy problem in Fig. 1.

To alleviate this limitation, additional prior information should be considered. The most straightforward and powerful information is the label, such as the idea of supervised and semi-supervised methods. However, in practice, the label information for these target unknown classes may hardly be obtained. The works from knowledge transfer [22] inspire us to make use of the information from other class domains prior known. Though from different classes, the labeled samples may share some common characteristics with the target task, as they are from the same type of concept.

For example, in face recognition, we want to detect or recognize the face images for a number of persons. When they are all unlabeled, the conventional methods usually cannot get satisfied results, as they cannot use any supervised information. On the other hand, there are already some databases with labeled faces, such as AT&T [18] and Yale [13]. These labeled face data contain some common information for face recognition. So we can use them to improve the original unsupervised learning task. In this situation, though both labeled and unlabeled data appear, the previous semi-supervised methods cannot work, as the labeled and unlabeled data are from different classes. This is a more general problem of learning with both labeled and unlabeled data [15].

This problem brings forward a novel issue which we call transferred dimensionality reduction (TDR). It transfers the task-related information from the classes prior known to the target unlabeled class domains, and finds a better subspace to discriminate them. In this paper, we propose a method called transferred discriminative analysis (TDA) to tackle the TDR problem. This method extracts the discriminative information from the labeled data and transfers it into unsupervised discriminative dimensionality reduction to revise the results iteratively. Finally, using both these labeled and unlabeled data from different classes, we can find the most discriminative subspace and an optimal clustering result simultaneously. The toy problem in Fig. 1 explains this problem more intuitively. It shows that, the labeled samples from known classes can help us to find a much better subspace to discriminate the unknown classes.

The rest of the paper is organized as follows. In section 2, we briefly review the related works. Then we introduce TDA algorithm in section 3. Experiments are given in section 4. Finally, we give our conclusion and suggest some future works based on the novel problem of TDR in section 5.

## 2 Related Work

### 2.1 Discriminative Dimensionality Reduction and Clustering

Over the past few decades, a lot of attention has been paid to dimensionality reduction. Some algorithms have been developed. A large family of them can be explained in a graph view [25]. The low-dimensional vector representation can be obtained from the eigenvectors corresponding to the eigenvalues of the graph Laplacian matrix with certain constraints. It preserves similarities between the pairs of the data, where

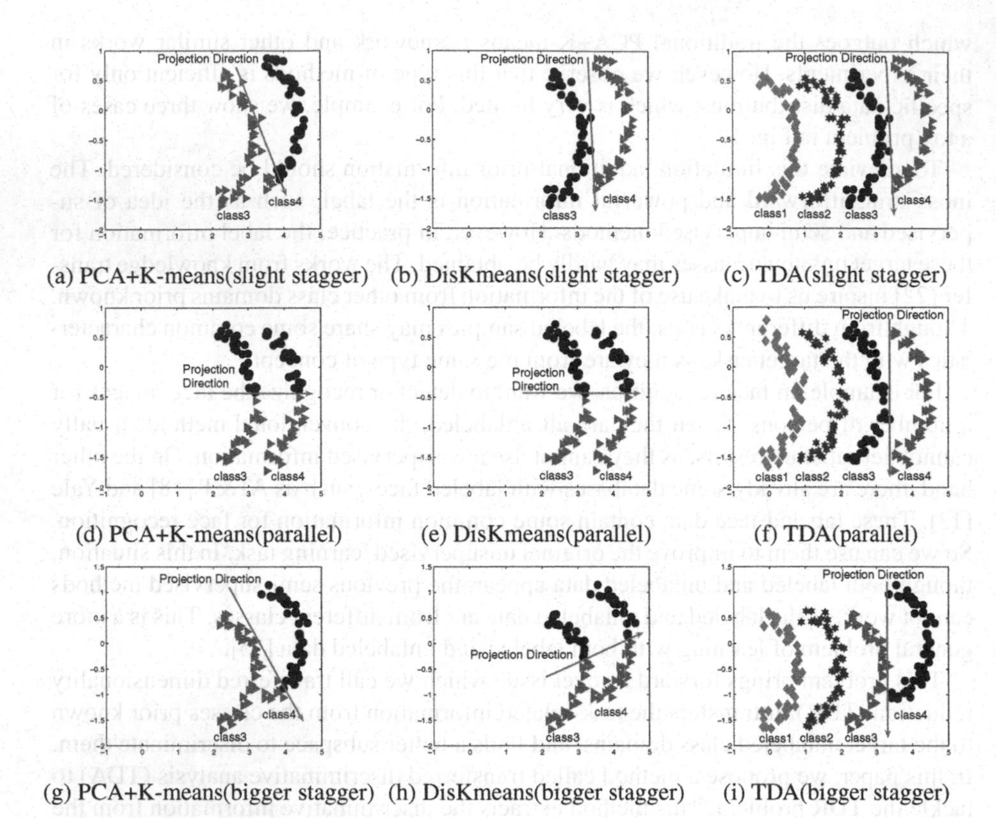


(a) PCA+K-means(slight stagger) (b) DisKmeans(slight stagger) (c) TDA(slight stagger)

(d) PCA+K-means(parallel) (e) DisKmeans(parallel) (f) TDA(parallel)

(g) PCA+K-means(bigger stagger) (h) DisKmeans(bigger stagger) (i) TDA(bigger stagger)

**Fig. 1.** Toy problem: There are four classes of data. Each class contains 50 random samples and forms a moon shape manifold. Suppose the class 3 and 4 are unlabeled, and we want to find the suitable subspace to discriminate them. There are three situations, each one in a row. PCA+K-means framework fails for any case, as in (a), (d) and (g). DisKmeans only works for the case that class 3 and 4 are slightly staggered in (b). When they are paralleled in (e) or staggered too much in (h), it cannot work well either. However, with the help of class 1 and 2, which are labeled beforehand, we can find the suitable subspace for each case as in (c), (f) and (i).

similarity is measured by a graph similarity matrix that characterizes certain statistical or geometric properties of the data set.

To get the discriminative structure of data, supervised methods try to find a transformation that minimizes the within-class scatter and maximizes the between-class scatter simultaneously. Given $l$ labeled samples $\mathbf{X}_L = (\boldsymbol{x}_1, \boldsymbol{x}_2, ..., \boldsymbol{x}_l)$ from C classes, where $\boldsymbol{x}_i \in \mathbb{R}^d$. The within-class scatter matrix $\mathbf{S}_w$, the between-class scatter matrix $\mathbf{S}_b$ and the total-scatter matrix $\mathbf{S}_t$ are defined as:

$$\mathbf{S}_w = \sum_{j=1}^{C} \sum_{i=1}^{l_j} (\boldsymbol{x}_i - \boldsymbol{m}_j)(\boldsymbol{x}_i - \boldsymbol{m}_j)^T = \mathbf{X}\mathbf{L}_w\mathbf{X}^T \quad (1)$$

$$\mathbf{S}_b = \sum_{j=1}^{C} l_j (\boldsymbol{m}_j - \boldsymbol{m})(\boldsymbol{m}_j - \boldsymbol{m})^T = \mathbf{X}\mathbf{L}_b\mathbf{X}^T \quad (2)$$

$$\mathbf{S}_t = \sum_{i=1}^{l} (\boldsymbol{x}_i - \boldsymbol{m})(\boldsymbol{x}_i - \boldsymbol{m})^T = \mathbf{S}_b + \mathbf{S}_w = \mathbf{X}\mathbf{L}_t\mathbf{X}^T, \quad (3)$$

where $\boldsymbol{m}_j = \frac{1}{l_j}\sum_{i=1}^{l_j} \boldsymbol{x}_i$ $(j = 1, 2, ..., C)$ is the mean of the samples in class $j$, $l_j$ is the number of samples in class $j$, and $\boldsymbol{m} = \frac{1}{l}\sum_{i=1}^{l} \boldsymbol{x}_i$ is the mean of all the samples. And the corresponding graph Laplacians are [14]:

$$\mathbf{L}_w = \mathbf{I} - \mathbf{H}(\mathbf{H}^T\mathbf{H})^{-1}\mathbf{H}^T \quad (4)$$

$$\mathbf{L}_b = \mathbf{H}(\mathbf{H}^T\mathbf{H})^{-1}\mathbf{H}^T - \frac{1}{l}\mathbf{1}_l\mathbf{1}_l^T \quad (5)$$

$$\mathbf{L}_t = \mathbf{I} - \frac{1}{l}\mathbf{1}_l\mathbf{1}_l^T, \quad (6)$$

where $\mathbf{H} = \{0, 1\}^{l\times C}$ is an indicator matrix: $H_{ij} = 1$ if $x_i$ belongs to the $j$-$th$ class, and $H_{ij} = 0$ otherwise.

LDA is one of the most popular and representative supervised methods. It is to solve the optimization problem:

$$\max_W \text{trace}((\mathbf{W}^T\mathbf{S}_w\mathbf{W})^{-1}(\mathbf{W}^T\mathbf{S}_b\mathbf{W})). \quad (7)$$

or

$$\max_W \text{trace}((\mathbf{W}^T\mathbf{S}_t\mathbf{W})^{-1}(\mathbf{W}^T\mathbf{S}_b\mathbf{W})). \quad (8)$$

The solution is the eigenvectors corresponding to the $C-1$ largest eigenvalues of $\mathbf{S}_w^{-1}\mathbf{S}_b$ or $\mathbf{S}_t^{-1}\mathbf{S}_b$ [11].

Clustering is another important topic to exploit the discriminative structure of the data. K-means is one of the simplest and most popular algorithms to solve the clustering problem. Given $u$ unlabeled samples $\mathbf{X}_U = (\boldsymbol{x}_1, \boldsymbol{x}_2, ..., \boldsymbol{x}_u)$ from K classes. Standard k-means finds the partition of the data to minimize the energy function:

$$J_K = \sum_{k=1}^{K}\sum_{i\in C_k} \|x_i - m_k\|_2^2 = \text{trace}(\mathbf{S}_w). \quad (9)$$

The clustering state can also be specified by an dummy indicator matrix $\widetilde{\mathbf{H}}^{u\times K}$.

It is clear the k-means clustering is to minimize the within-class scatter matrix $\mathbf{S}_w$, or maximize the between class scatter matrix $\mathbf{S}_b$, since the total scatter $\mathbf{S}_t$ is a constant. It also can be represented in graph form using equation (4). On the other hand, its kernelized version can also be explained under the graph view, which has close connection with other spectral clustering methods [8].

The discriminative analysis and clustering methods all emphasize on pursuing the intrinsic discriminative structure of the data. So [9,23,26] combine them together to get better learning result.

Though the combined method of discriminative k-means does a good job in some situations. It focuses too much on the present unlabeled samples, and sometimes is trapped into a very bad result, even worse than the PCA+K-means method, which is shown in the third case in Fig. 1. To overcome this problem, we consider to introduce more information from outer classes within the same concept domain. As the different classes of data in the same concept domain often lie on similar lower dimensional manifolds in certain extent, they should share some common discriminative structure. We can extract this structure easily from the labeled classes using discriminative analysis. Then, we can transfer the shared information to the unlabeled data, and find their discriminative structure using the clustering method.

### 2.2 Transfer Learning and Semi-supervised Learning

TDA has a similar motivation with knowledge transfer, or transfer learning, which has been recognized as an important topic in machine learning field. It is the ability to apply knowledge and skills learned in previous tasks to novel tasks. Early works raised some significative issues [17,21,4]. There are still more and more attentions paid to this topic recently [16,7]. Most of the previous works focus on transferring the related knowledge for supervised learning tasks. In this work, however, we address on the single-task problem, and transfer the supervised information to unsupervised task. Though it seems like semi-supervised learning [5], they have obvious distinctions. In traditional semi-supervised learning the labeled and unlabeled data come from the same class domains. There should be both labeled and unlabeled data in each class. The unlabeled data should have the same distribution with the labeled ones, then a large number of data points can expose the manifold information and improve the learning result of the labeled data.

In our problem, on the contrary, the labeled and unlabeled data are from different classes, and they have different distributions. We extract the useful discriminative information from the labeled data to improve the subspace selection of the unlabeled data. It is quit different with semi-supervised learning and cannot be solved using existing semi-supervised methods. As a result, we name this problem as transferred dimensionality reduction.

## 3 Transferred Discriminative Analysis

In learning tasks, it is vital to use the prior information. In traditional methods, the prior is often assumed to be given by the designer's experience. We cannot expect this prior to be always right, as it is hard to propose a suitable prior even for an expert. However, in TDR we extract the information directly from the data prior known, and embed this information into the task using the cross similarity part between the source prior known and the target to be dealt with.

In TDR, suppose we have the labeled source data set, contains $l$ points $\mathbf{D_L} = \{\mathbf{X_L}, \mathbf{Y_L}\}$, $\mathbf{X}_L = (\boldsymbol{x}_1, \boldsymbol{x}_2, ..., \boldsymbol{x}_l)$, $\mathbf{Y}_L = (\boldsymbol{y}_1, \boldsymbol{y}_2, ..., \boldsymbol{y}_l)^T$. The label is $\boldsymbol{y} \in \{1, \ldots, C\}$. We want to find the compact subspace of $u$ newly arrived unlabeled points $\mathbf{D_U} = \{\mathbf{X_U}\}$, $\mathbf{X}_U = (\boldsymbol{x}_{l+1}, \boldsymbol{x}_{l+2}, ..., \boldsymbol{x}_{l+u})$, from K classes which are different from the classes in $\mathbf{Y_L}$. Each point $\boldsymbol{x} \in \mathbb{R}^d$ is a $d$-dimensional vector. We denote all data as $\mathbf{D} = \{\mathbf{D_L}, \mathbf{D_U}\}$, and $\mathbf{X} = \{\mathbf{X_L}, \mathbf{X_U}\}$. For simplicity, we assume $n = l + u$, and the sample mean of $\mathbf{D}$ is zero, which is $\boldsymbol{m} = \frac{1}{n}\sum_{i=1}^{n} \boldsymbol{x}_i = 0$.

### 3.1 The Objective Function

The manifold structure is interpreted as that nearby points will have similar embedding. As the labeled and unlabeled data are from different manifolds of the same concept domain. The discriminative structure can be shared to some extent among this manifolds. We can transfer this information from source data $\mathbf{D_L}$ to target data $\mathbf{D_U}$ through the intervention between these two parts.

In our TDR, we measure the between-class information of the data set $\mathbf{D}$ as follows:

$$\mathbf{S}_b = \mathbf{S}_{bl} + \widetilde{\mathbf{S}}_{bu} = \sum_{i=1}^{C} l_i \boldsymbol{m}_i \boldsymbol{m}_i^T + \sum_{j=1}^{K} l_j \widetilde{\boldsymbol{m}}_j \widetilde{\boldsymbol{m}}_j^T. \tag{10}$$

The first part is the between-class scatter of the labled data. However, for the unlabeled data, we can estimate this information using clustering method, which is expressed as the second part, treating each cluster as a class.

In the between-class scatter, the labeled and unlabeled parts are separately presented. To properly describe the structure of all data, we should introduce the relationship between labeled and unlabeled parts.

Under the existence of unlabeled data, Graph Laplacian has been generally used to describe the data structure [5]. We define $G = (V, E)$ as a graph associated with the data. $V$ is the vertex set of graph, which is defined on the observed set, including both labeled and unlabeled data. $E$ is the edge set, which contains the pairs of neighboring vertices $(\boldsymbol{x}_i, \boldsymbol{x}_j)$. A typical adjacency matrix $\mathbf{M}$ of neighborhood graph is defined as:

$$\mathbf{M}_{ij} = \begin{cases} \exp\{-\frac{\|\boldsymbol{x}_i - \boldsymbol{x}_j\|^2}{2\sigma^2}\} & if\ (\boldsymbol{x}_i, \boldsymbol{x}_j) \in E \\ 0 & otherwise \end{cases} \tag{11}$$

then the normalized graph Laplacian [6] is:

$$\mathbf{L} = \mathbf{I} - \mathbf{D}^{-\frac{1}{2}} \mathbf{M} \mathbf{D}^{-\frac{1}{2}}, \tag{12}$$

where the diagonal matrix $\mathbf{D}$ satisfies $\mathbf{D}_{ii} = d_i$, and $d_i = \sum_{j=1}^{l+u} \mathbf{M}_{ij}$ is the degree of vertex $\boldsymbol{x}_i$.

Introducing the graph Laplacian into the total scatter, we can make use of both labled and unlabeled information to describe the structure of the data set $\mathbf{D}$ properly. With the zero sample mean, it becomes

$$(\mathbf{S}_t + \lambda \mathbf{X}\mathbf{L}\mathbf{X}^T) = \mathbf{X}(\mathbf{I} + \lambda \mathbf{L})\mathbf{X}^T \tag{13}$$

It is also can be seen in the regularization of discriminative analysis [10].

As described above, the target of TDA becomes:

$$\max_{\mathbf{W}, \widetilde{\mathbf{H}}_u} \text{trace}((\mathbf{W}^T(\mathbf{S}_t + \lambda \mathbf{X}\mathbf{L}\mathbf{X}^T)\mathbf{W})^{-1}(\mathbf{W}^T(\mathbf{S}_{bl} + \widetilde{\mathbf{S}}_{bu}(\widetilde{\mathbf{H}}_u))\mathbf{W})). \tag{14}$$

It is to optimize the objective function w.r.t two variables. One is the dummy indicator matrix $\widetilde{\mathbf{H}}_u$, representing the clustering structure of the unlabled data, and the other one is the projection direction $\mathbf{W}$ for the dimensionality reduction.

Direct optimizing the objective function is complex and not advisable. Instead, we optimize it alternatively. We can use clustering method to estimate the discriminative structure of the unlabeled data, and project all data into lower dimension by supervised method to revise the clustering result. Using the method in [23], the process will converge to the optimal solution for the objective, while we will using the k-means clustering in our experiment, which gives a local solution but is good enough.

The introduction of the labeled parts in between-class scatter, total scatter and graph Laplcian adds more restriction into the problem. They restrict that in the low-dimensional subspace of unlabeled data, the discriminative structure of labeled data should still be preserved. The labeled data will bring in punishment if the structure is violated. This will force the unlabeled data clustering to form similar discriminative structure with the labeled data, and the information is transferred like this. The alternation process will stop, when the structure consistency of all data in the subspace and the clustering structure within unlabeled data are balanced. Following this process, the knowledge is transferred through the intervention between the labeled and unlabeled structures, and then affects the clustering and projection process.

The above explanation is intuitive. We can also explanation this intervention more explicitly from kernel learning view. [26] analyzes that the clustering step of the adaptive framework is just the kerneled version of k-means clustering, using kernel matrix

$$\mathbf{X}_{\mathbf{U}}^{\mathbf{T}}\mathbf{W}(\mathbf{W}^{\mathbf{T}}(\mathbf{X}_{\mathbf{U}}\mathbf{X}_{\mathbf{U}}^{\mathbf{T}}+\lambda\mathbf{L}_{\mathbf{U}})\mathbf{W})^{-1}\mathbf{W}^{\mathbf{T}}\mathbf{X}_{\mathbf{U}}, \tag{15}$$

which is learned from the unlabeled data. In our method, the kernel matrix becomes

$$\mathbf{X}^{\mathbf{T}}\mathbf{W}(\mathbf{W}^{\mathbf{T}}(\mathbf{X}\mathbf{X}^{\mathbf{T}}+\lambda\mathbf{L})\mathbf{W})^{-1}\mathbf{W}^{\mathbf{T}}\mathbf{X}, \tag{16}$$

which is learned from all available data, both the source and target. So, the prior information from the source is embedded in the kernel matrix and transferred adaptive to the target task. Finally, we can find the most discriminative projection direction, and get a reasonable clustering result of the unlabeled data at the same time.

### 3.2 The Algorithm

Given the labeled data $\mathbf{D}_L = \{\mathbf{X}_L, \mathbf{Y}_L\}$ belong to C classes, and unlabeled data $\mathbf{D}_U = \{\mathbf{X}_U\}$ with their class number K. The TDA algorithm is stated below:

**Step 1. Initialization:** Initially assign the cluster index for the K classes of unlabeled data using k-means. Construct the graph matrix **M** as in equation (11), and calculate the graph Laplacian **L** as in equation (12).

**Step 2. Supervised Dimensionality Reduction:** Find the optimal subspace with dimension $m = C + K - 1$, using eigenvalue decomposition for the objective function (14) w.r.t **W**, which is similar to LDA. Then the optimal solution is given by:

$$(\mathbf{S}_t + \lambda_1 \mathbf{XLX}^T + \lambda_2 \mathbf{I})\boldsymbol{w}_j^* = \eta_j(\mathbf{S}_{bl} + \widetilde{\mathbf{S}}_{bu}(\widetilde{\mathbf{H}}_u))\boldsymbol{w}_j^*,$$

$j = 1, ..., m$, where $\boldsymbol{w}_j^*$ $(j = 1, ..., m)$ are the eigenvectors corresponding to the $m$ largest eigenvalues of $(\mathbf{S}_t + \lambda_1 \mathbf{XLX}^T + \lambda_2 \mathbf{I})^{-1}(\mathbf{S}_{bl} + \widetilde{\mathbf{S}}_{bu}(\widetilde{\mathbf{H}}_u))$, with fixed $\widetilde{\mathbf{H}}_u$. $\lambda_2 \mathbf{I}$ is a regularization term, which ensures the nonsingularity of the matrix $\mathbf{S}_t + \lambda_1 \mathbf{XLX}^T$, and $\lambda_2$ is an arbitrary small real number.

**Step 3. Compact Clustering for Target Data:** Cluster the unlabeled data in the subspace finding in step 2. It is to fix projection direction **W** and use the clustering method to get an optimal indicator matrix $\widetilde{\mathbf{H}}_u$ for the unlabeled data. K-means is used in this step to solve the problem

$$\max_{(\widetilde{\mathbf{H}}_u)} \widetilde{\mathbf{S}}_{bu}(\widetilde{\mathbf{H}}_u)$$

**Step 4. Stop Condition:** Goto step 2 until convergence. It is to stop when the clustering result, the indicator matrix $\widetilde{\mathbf{H}}_u$, for previous two iterations is unchanged.

**Step 5. TDA Embedding:** Let the projection matrix $\mathbf{W}_{tda} = [\boldsymbol{w}_1^*, \ldots, \boldsymbol{w}_m^*]$. The samples can be embedded into $m$ dimensional subspace by: $\mathbf{x} \rightarrow \mathbf{z} = \mathbf{W}_{tda}^T \mathbf{x}$.

### 3.3 Kernelization

In this section we present the generalized version of our algorithm using the kernel trick. We show a simple method under a graph view, using the similar treatment with [2]. It performs TDA in Reproducing Kernel Hilbert Space (RKHS), getting kernel TDA.

Let $\phi : x \rightarrow \mathcal{F}$ be a function mapping the points in the input space to feature space, which is a high-dimensional Hilbert space. We try to replace the explicit mapping with the inner product $K(\boldsymbol{x}_i, \boldsymbol{x}_j) = (\phi(\boldsymbol{x}_i) \cdot \phi(\boldsymbol{x}_j))$. According to Representer Theorem [19], the optimal solution $\boldsymbol{w}_j^*$ can be given by:

$$\boldsymbol{w}_j^{\phi *} = \sum_{i=1}^{l+u} \boldsymbol{\alpha}_{ji}^* \phi(\boldsymbol{x}_i) \quad j = 1, ..., m \tag{17}$$

where $\boldsymbol{\alpha}_{ji}$ is the weight that defines how $\boldsymbol{w}_j^{\phi *}$ is represented in the space spanned by a set of over-complete bases $\{\phi(\boldsymbol{x}_1), \phi(\boldsymbol{x}_2), ..., \phi(\boldsymbol{x}_{l+u})\}$.

For convenience, we rewrite the data matrix in RKHS as $\mathbf{X}_L^\phi = [\phi(\boldsymbol{x}_1), \phi(\boldsymbol{x}_2), ..., \phi(\boldsymbol{x}_l)]$, $\mathbf{X}_U^\phi = [\phi(\boldsymbol{x}_{l+1}), \phi(\boldsymbol{x}_{l+2}), ..., \phi(\boldsymbol{x}_{l+u})]$, and $\mathbf{X}^\phi = (\mathbf{X}_L^\phi, \mathbf{X}_U^\phi)$. Then, $\mathbf{W}^\phi$ can be expressed as $\mathbf{W}^\phi = \mathbf{X}^\phi \boldsymbol{\alpha}$. The kernel matrices are defined as $\mathbf{K} = \mathbf{X}^{\phi T} \mathbf{X}^\phi$. Thus we have

$$\begin{array}{rcl} \mathbf{W}^{\phi T} \mathbf{S}_b^\phi \mathbf{W}^\phi & = & \boldsymbol{\alpha}^T \mathbf{K}^T \mathbf{L}_b \mathbf{K} \boldsymbol{\alpha} \\ \mathbf{W}^{\phi T} \mathbf{S}_t^\phi \mathbf{W}^\phi & = & \boldsymbol{\alpha}^T \mathbf{K}^T \mathbf{I} \mathbf{K} \boldsymbol{\alpha} \\ \mathbf{W}^{\phi T} \mathbf{X}^\phi \mathbf{L} \mathbf{X}^{\phi T} \mathbf{W}^\phi & = & \boldsymbol{\alpha}^T \mathbf{K}^T \mathbf{L} \mathbf{K} \boldsymbol{\alpha} \\ \mathbf{W}^{\phi T} \mathbf{W}^\phi & = & \boldsymbol{\alpha}^T \mathbf{K} \boldsymbol{\alpha} \end{array} \tag{18}$$

Using the graph expression of (1) ~ (6) and the graph Laplacian (12).

As

$$\mathbf{H} = \begin{bmatrix} \mathbf{H}_l^{l \times C} & \mathbf{0} \\ \mathbf{0} & \widetilde{\mathbf{H}}_u^{u \times K} \end{bmatrix}, \mathbf{L}_b = \begin{bmatrix} \mathbf{L}_{bl}^{l \times C} & \mathbf{0} \\ \mathbf{0} & \widetilde{\mathbf{L}}_{bu}^{u \times K}(\widetilde{\mathbf{H}}_u) \end{bmatrix},$$

the indicator matrix composed with two parts for labeled and unlabeled samples individually, and the between-class scatter is also composed by two parts respectively.

We can then give the objective function of kernel TDA (KTDA) as:

$$\max_{\boldsymbol{\alpha}, \widetilde{\mathbf{H}}_u} \operatorname{trace}((\boldsymbol{\alpha}^T \mathbf{K}^T (\mathbf{L}_t + \lambda_1 \mathbf{L} + \lambda_2) \mathbf{K} \boldsymbol{\alpha})^{-1} (\boldsymbol{\alpha}^T \mathbf{K}^T \mathbf{L}_b \mathbf{K} \boldsymbol{\alpha})). \tag{19}$$

The solution is obtained by solving the generalized eigenvalue decomposition problem:

$$\mathbf{K}^T(\mathbf{L}_t + \lambda_1\mathbf{L} + \lambda_2)\mathbf{K}\boldsymbol{\alpha}_j^* = \eta_j\mathbf{K}^T\mathbf{L}_b\mathbf{K}\boldsymbol{\alpha}_j^* \tag{20}$$

where $\boldsymbol{\alpha}^* = (\boldsymbol{\alpha}_1^*, \boldsymbol{\alpha}_2^*, ..., \boldsymbol{\alpha}_m^*)$ corresponds to the $m$ largest eigenvalues. $\boldsymbol{\alpha}_j^*$ should be resized as $\frac{1}{\sqrt{\boldsymbol{\alpha}_j^{*T}\mathbf{K}\boldsymbol{\alpha}_j^*}}\boldsymbol{\alpha}_j^*$ to satisfy the constraint of $\boldsymbol{\alpha}^{*T}\mathbf{K}\boldsymbol{\alpha}^* = \mathbf{I}$.

### 3.4 The Computational Complexity

The TDA contains both dimensionality reduction and clustering. The process may have several iterations. The empirical result shows it converges vary fast. The number of iterations is often less than ten. In supervised dimensionality reduction, it needs to solve a generalized eigenvalue decomposition, which is of order $O(d^2nt)$. $d$ is the dimension of data, $n = l + u$ is the number of total data points, and $t$ is the number of iterations. For clustering method, we use k-means. The computational complexity is $O(dnt)$. As a result, the total computational complexity is of order $O(d^2nt)$, and the complexity is focus on the original dimension of data. As a result, we can use PCA to initially reduce the dimension of the data, and this can accelerate the running speed of our algorithm. The computational complexity of kernel TDA is $O(n^2dt)$ analyzed in the same way.

## 4 Experiments

We have already shown a toy problem in the introduction. Using TDA we can find the true structure with the help of the labeled data using only a few iterations, which is very fast. In this case, the data prior known can exactly express the discriminative information of the unlabeled samples, which is an ideal situation.

In this section, however, we will give the examples of real problems and show that, most of the time, the transferred information is helpful. We perform the comparisons under the problem of face recognition, both constrained and unconstrained. We compare our TDA method with two of the most popular and representative appearance-based methods including *Eigenface* (based on PCA)[20] and *Laplacianface* (based on LPP)[14], and the adaptive dimensionality reduction method with clustering *DisKmeans* [26].

All images in our experiments are preprocessed to the same size of 56 × 46 pixels with 256 gray levels. In each experiment, we randomly choose C classes as labeled data, and K classes unlabeled. TDA runs on all of these data. The comparison methods are operated on the K classes of unlabeled data. We compare their clustering results in the corresponding subspace which they have found. TDA and DisKmeans can cluster the data at the same time of subspace selection. However for the other two methods, we use the k-means for clustering, and run k-means 10 times for each projection then choose the best result in each experiment. For each fixed (C, K), we run the experiment for 50 times, each time on randomly selected labeled and unlabeled classes, then show the average result. For comparison of different methods, we use the clustering result as the measurement of dimensionality reduction performance. We use two standard clustering performance measures, which are Clustering Accuracy (ACC) and Normalized Mutual Information (NMI) [24,26].

The heuristic parameter in TDA and DisKmeans is the Laplacian weight $\lambda_1$. We set it to a fixed value of 1. As a matter of fact, the algorithm is not sensitivity to this parameter for a wide range. For the heuristic parameter of PCA and LPP, the reduced dimensionality, we choose them using cross validation.

### 4.1 What to Transfer

Usually there are several classes of labeled samples in the data set prior known. But not all of them are helpful for a specific unsupervised task. Because each of them has different discriminative structure. Only some of them are the same with the unlabeled samples. The others are not, and using these data is harmful, on the contrary. On the other hand, using all prior data needs much more computational time, which is not practical. As a result, we choose a proper subset of labeled data for our learning task. As the task is to maximize the discriminative ability of the target data, we just use this as the selection criterion. In following experiments, we randomly select C classes from the prior data set, and repeat for R times. Each time we will find an optimal pair of $(\mathbf{W_i}^T, \mathbf{H_i}_u)$, and use the best one. This is,

$$\max_{i \in R} \text{trace}((\mathbf{W_i}^T(\widetilde{\mathbf{S}}_{wu}(\widetilde{\mathbf{H_i}_u}) + \lambda \mathbf{X_U} \mathbf{L} \mathbf{X_U}^T)\mathbf{W_i})^{-1}(\mathbf{W_i}^T \widetilde{\mathbf{S}}_{bu}(\widetilde{\mathbf{H_i}_u})\mathbf{W_i})). \quad (21)$$

As a result, the computational complexity will be multiplied by R to $O(d^2 ntR)$. We fix R = 10. The complexity will not be changed significantly and remain in the same level.

### 4.2 Face Recognition Using Conventional Benchmarks

**Face Data Sets.** In the experiments for this section, we use the face data sets, AT&T [18] and Yale [13]. The typical faces of these data sets are shown in Fig. 2.

**Transferred within the Same Data Set.** In these experiments we use the labeled data and unlabeled data in the same data set.

For AT&T database, we chose each integer C from {2,...,10} and K from {2,...,10}. Table 1 gives a part of the results using ACC measure as the limit of space. However, we show the result of all comparisons in both two measures in Fig. 3, where each point represent an average result of a fixed (C,K). We only show the improvement over DisKmeans in the figures, as it is the second best among all comparison methods. The results tell that TDA is much better than the unsupervised method. For Yale database, we chose C traversing all integers from {2,...,7}, and K traversing from {2,...,7}. The result is also shown in Fig. 3.

(a) AT&T face examples

(b) Yale face examples

**Fig. 2.** Face Data Examples

**Table 1.** Results in AT&T, using ACC measure ($mean \pm std$)

| AT&T | PCA | LPP | DisKmeans | TDA |
|---|---|---|---|---|
| C=2,K=2 | 0.80(0.15) | 0.72(0.12) | 0.91(0.16) | **1.00(0.02)** |
| C=2,K=3 | 0.90(0.11) | 0.78(0.08) | 0.90(0.13) | **0.96(0.10)** |
| C=3,K=2 | 0.84(0.13) | 0.70(0.13) | 0.89(0.16) | **1.00(0.02)** |
| C=3,K=3 | 0.93(0.08) | 0.81(0.08) | 0.89(0.15) | **0.97(0.07)** |
| C=2,K=4 | 0.86(0.10) | 0.80(0.08) | 0.89(0.12) | **0.91(0.11)** |
| C=4,K=2 | 0.84(0.14) | 0.69(0.11) | 0.89(0.20) | **1.00(0.02)** |
| C=3,K=4 | 0.88(0.08) | 0.83(0.06) | 0.88(0.10) | **0.92(0.10)** |
| C=4,K=3 | 0.90(0.11) | 0.80(0.10) | 0.86(0.14) | **0.97(0.07)** |
| C=4,K=4 | 0.88(0.08) | 0.79(0.09) | 0.91(0.11) | **0.92(0.10)** |

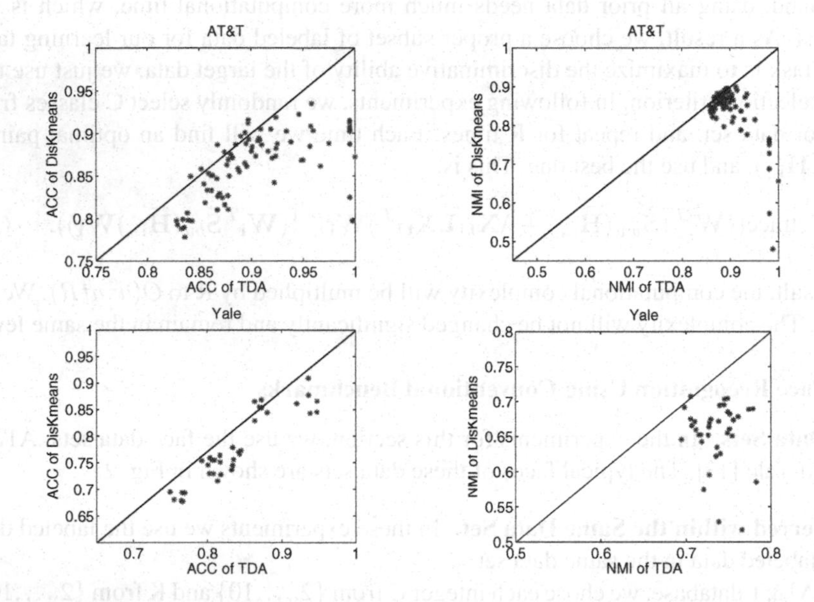


**Fig. 3.** Comparison results of TDA and DisKmeans in ACC and NMI measures, for transfer within either AT&T or Yale, each point represents 50 times average for a fixed (C, K) pair

As the above result cannot show how the change of (C,K) will affect the performance improvement, we give another representation in Table 2. It is the difference matrix between the clustering result of TDA and DisKmeans for each pair of (C,K). In Table 2, we can find that TDA improves significantly over other unsupervised methods for small K, which is the number of unlabeled classes. However, the improvement becomes less significant, with the increase of K. This is because the unknown target discriminative structure becomes more and more complex, and the limited prior cannot describe it properly. However, the increase of the number of labeled classes can not affect the result distinctively. This is because discriminative structure among the labeled data also

**Table 2.** Difference matrix of TDA and DisKmeans for AT&T, each element is calculated as $DM_{kc} = ACC_{kc}^{TDA} - ACC_{kc}^{Dis}$. The bold items show significant improvements of TDA.

| AT&T | C = 2 | C = 3 | C = 4 | C = 5 | C = 6 | C = 7 | C = 8 | C = 9 | C = 10 |
|---|---|---|---|---|---|---|---|---|---|
| K=2 | **0.22** | **0.29** | **0.41** | **0.34** | **0.22** | **0.23** | **0.25** | **0.51** | **0.22** |
| K=3 | **0.08** | **0.14** | **0.16** | 0.05 | 0.05 | **0.11** | **0.10** | **0.09** | **0.13** |
| K=4 | 0.02 | 0.06 | 0.02 | 0.04 | 0.01 | 0.02 | 0.03 | 0.06 | **0.09** |
| K=5 | 0.00 | 0.04 | **0.09** | 0.00 | 0.03 | 0.04 | 0.03 | 0.00 | 0.06 |

becomes more and more complex. On one hand it brings more information, on the other hand it contains some structure not consistent with the unlabeled data and may confuse the unsupervised dimensionality reduction. Another capable reason is the limit of the number of samples in each labeled class. There are only tens of samples in each labeled class, which cannot fully express their class characteristics. The discriminative information should increase exponentially fast in order of the labeled classes number, while the increase of the labeled samples actually in linear order. So the description ability becomes less and less, and the result cannot be much improved. As described above, using limited number of samples in each labeled class, we can only expect significant improvements for not too many classes of unlabeled data.

**Transferred between the Different Data Sets.** It is a more interesting and practical problem to transfer the information from one exiting data set to a newly collected one. We randomly choose the labeled classes from AT&T and unlabeled classes from Yale for every integer C from {2,...,10} and K from {2,...,10}. The result is shown in Table 3. We can get a similar result to transfer Yale into AT&T. Both comparison plots arc shown in Fig. 4.

From these experiments, we can see that though from different data set, the face images still share some common characteristics. This is helpful knowledge to improve the learning result. It suggests that we can use existing labeled data set to handle other unlabeled classes of data, which is a novel and promising learning problem.

**Table 3.** Results for AT&T transferred to Yale, using ACC measure ($mean \pm std$)

| AT&Tto Yale | PCA | LPP | DisKmeans | TDA |
|---|---|---|---|---|
| C=2,K=2 | 0.90(0.03) | 0.68(0.12) | 0.94(0.13) | **0.99(0.02)** |
| C=2,K=3 | 0.84(0.14) | 0.70(0.12) | 0.91(0.12) | **0.95(0.10)** |
| C=3,K=2 | 0.94(0.03) | 0.68(0.08) | 0.93(0.12) | **0.99(0.02)** |
| C=3,K=3 | 0.83(0.15) | 0.71(0.12) | 0.89(0.15) | **0.95(0.10)** |
| C=2,K=4 | 0.90(0.11) | 0.72(0.13) | 0.92(0.10) | **0.97(0.06)** |
| C=4,K=2 | 0.91(0.05) | 0.71(0.08) | 0.96(0.09) | **0.98(0.02)** |
| C=3,K=4 | 0.88(0.12) | 0.73(0.11) | 0.91(0.08) | **0.95(0.08)** |
| C=4,K=3 | 0.83(0.16) | 0.68(0.11) | 0.91(0.12) | **0.97(0.07)** |
| C=4,K=4 | 0.89(0.13) | 0.73(0.10) | 0.90(0.11) | **0.94(0.07)** |

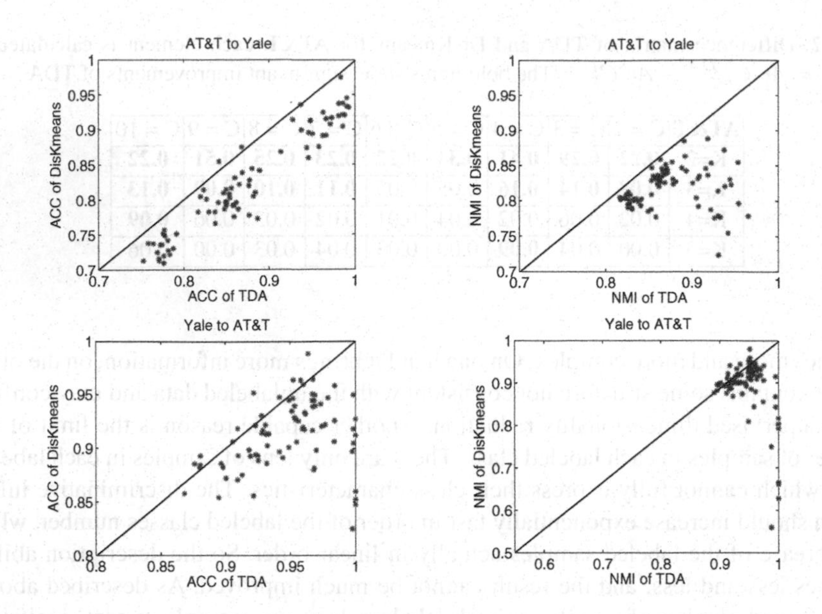


**Fig. 4.** Comparison results of TDA and DisKmeans in ACC and NMI measures, for transfers between different databases. Each point represents 50 times average for a fixed (C, K) pair.

## 4.3 Unconstrained Face Recognition

The databases in the last subsection are created under controlled conditions to facilitate the study of specific parameters on the face recognition problem, such as position, pose, lighting etc. Practically there are also many applications in which the practitioner has little or no control over such parameters. This is provided as a unconstrained face recognition problem. It is much more difficult than the constrained problems and needs novel approaches to solve.

In following experiments, we will use a recently published unconstrained data set and test the performance of our TDA algorithm.

**Unconstrained Face Data Set.** *Labeled Faces in the Wild (LFW):* This is a database of face photographs designed for studying the problem of unconstrained face recognition. The database contains more than 13,000 images of faces collected from the web. 1680 of the people pictured have two or more distinct photos in the database. More details can be found in [12]. To make the data set more balanced and comparable with the constrained data set, we only take the images of persons who have more than 10 and

(a) Original Images (b) Preprocessed Images

**Fig. 5.** LFW Face Data Examples

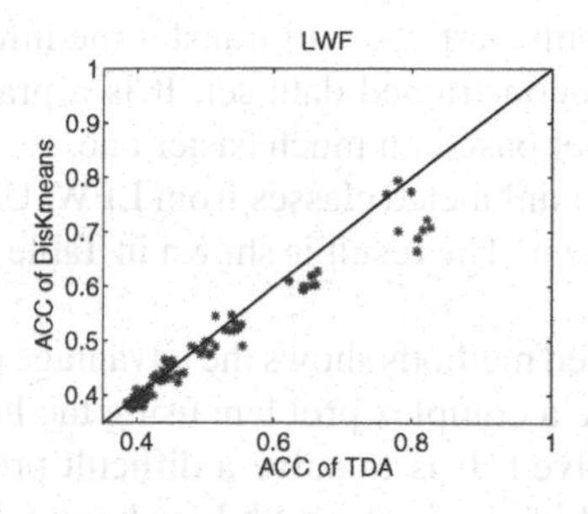


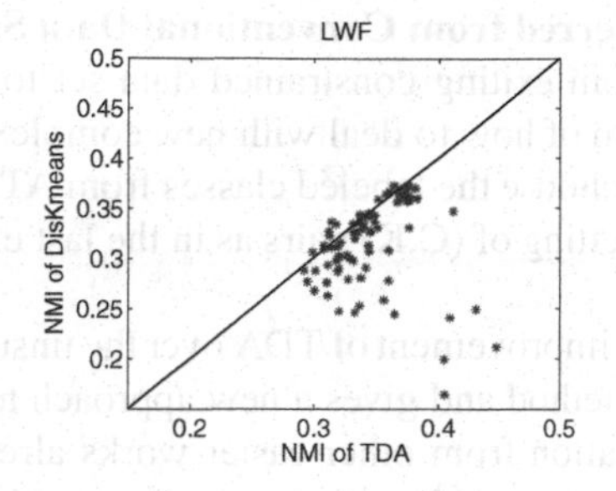


**Fig. 6.** Comparison results of TDA and DisKmeans for LWF, in ACC and NMI measures. Each point represents 50 times average for a fixed (C, K) pair.

**Table 4.** Results for AT&T transferred to LFW, using ACC measure ($mean \pm std$)

| AT&T to LWF | PCA | LPP | DisKmeans | TDA |
|---|---|---|---|---|
| C=2,K=2 | 0.72 (0.14 ) | 0.63 (0.09) | 0.73 (0.15 ) | **0.78 (0.16)** |
| C=3,K=2 | 0.71 (0.14 ) | 0.63 (0.08) | 0.71 (0.17 ) | **0.81 (0.15)** |
| C=4,K=2 | 0.72 (0.15) | 0.63 ( 0.09) | 0.72 (0.17 ) | **0.81 (0.15)** |
| C=5,K=2 | 0.69 (0.12) | 0.61(0.09) | 0.71(0.16) | **0.80(0.16)** |
| C=2,K=3 | 0.60 (0.12 ) | 0.60 (0.11 ) | 0.58 (0.09 ) | **0.61 (0.11)** |

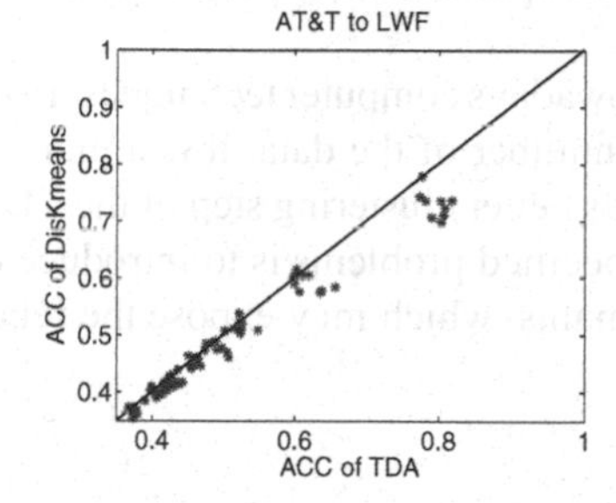


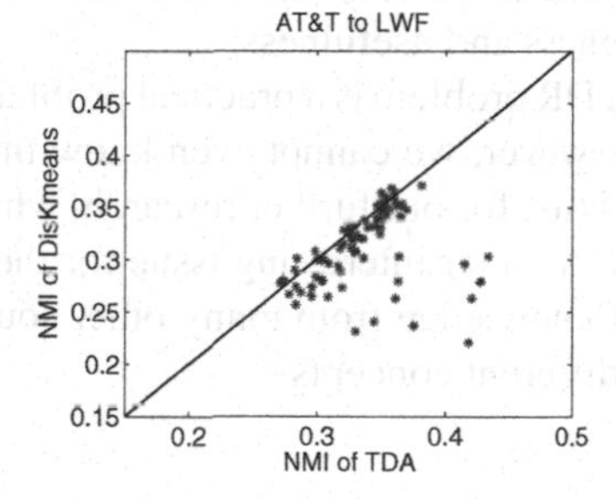


**Fig. 7.** Comparison result of TDA and DisKmeans in ACC and NMI measures, for AT&T transferred to LFW. Each point represents 50 times average for a fixed (C, K) pair.

less than 20 photos in LFW, which are 1401 images for 101 persons. Then take out the head part of the images, resize them to $56 \times 46$ pixels, and turn into gray images. The typical images are shown in Fig. 5.

**Transferred within LFW Data Set.** In this part we use the labeled data and unlabeled data all in the LFW database. We choose C from {2,...,10} and K from {2,...,10}. The results are shown in Fig. 6. Though TDA outperforms other methods, in practice, we cannot always expect that the unconstrained data set is labeled. In this situation, can we use the constrained ones? If yes, it will make the transfer strategy more powerful.

**Transferred from Conventional Data Set.** In this part, we will transfer the information from exiting constrained data set to this unconstrained data set. It is a practical problem of how to deal with new complex data set based on much easier one.

We choose the labeled classes from AT&T and unlabeled classes from LFW. Use the same setting of (C,K) pairs as in the last experiment. The result is shown in Table 4 and Fig. 7.

The improvement of TDA over the unsupervised methods shows the advantage of our TDA method and gives a new approach to tackle a complex problem using the helpful information from other easier works already solved. It is to solve a difficult problem with the knowledge of more easier problems, which is similar with how human learns things.

## 5 Conclusion and Discussion

In this paper, we bring forward a problem of transferred dimensionality reduction. It uses the labeled and unlabeled data from different class domains, which is different from the traditional semi-supervised learning method. And it is more practical for nowadays drastic increase of various sorts of unlabeled information through internet. To solve this problem, we introduce the algorithm, transferred discriminative analysis. It transfers the specific discriminative information from supervised knowledge to the unlabeled samples in other class domains, and finds more suitable subspace for the lower dimensional embedding. It is fast and robust to run. The experimental results demonstrate its effectiveness and usefulness.

The TDR problem is a practical problem for nowadays computer techniques. In many cases, however, we cannot even know the class number of the data. It is a more challenging issue for our further research, which needs better clustering step of the TDA algorithm. Another interesting issue for the task-specified problems is to introduce more types of knowledge from many other source domains, which may expose the relationship of different concepts.

## Acknowledgments

This research was supported by National 863 Project ( 2006AA01Z121 ) and National Science Foundation of China ( 60675009 ).

## References

1. Belhumeur, P., Hespanha, J., Kriegman, D.: Eigenfaces vs. Fisherfaces: Recognition Using Class Specific Linear Projection. IEEE Trans. on PAMI 19(7), 711–720 (1997)
2. Baudat, G., Anouar, F.: Generalized Discriminant Analysis Using a Kernel Approach. Neural Computation 12(10), 2385–2404 (2000)
3. Cai, D., He, X., Han, J.: Semi-Supervised Discriminant Analysis. In: Proceedings of IEEE International Conference on Computer Vision (ICCV), pp. 1–7 (2007)
4. Caruana, R.: Multitask Learning. Machine Learning 28(1), 41–75 (1997)

5. Chapelle, O., Schölkopf, B., Zien, A.: Semi-Supervised Learning. MIT Press, Cambridge (2006)
6. Chung, F. (ed.): Spectral Graph Theory. CBMS Regional Conference Series in Mathematics, vol. 92. American Mathematical Society (1997)
7. Dai, W., Yang, Q., Xue, G., Yu, Y.: Boosting for Transfer Learning. In: Proceedings of International Conference on Machine Learning (ICML), pp. 193–200 (2007)
8. Dhillon, I., Guan, Y., Kulis, B.: A Unified View of Kernel K-means, Spectral Clustering and Graph Partitioning. Technical Report TR-04-25, UTCS (2005)
9. Ding, C., Li, T.: Adaptive Dimension Reduction Using Discriminant Analysis and K-means Clustering. In: Proceedings of International Conference on Machine Learning (ICML) (2007)
10. Friedman, J.: Regularized Discriminant Analysis. Journal of the American Statistical Association 84(405), 165–175 (1989)
11. Fukunaga, K.: Introduction to Statistical Pattern Recognition, 2nd edn. Academic Press (1990)
12. Huang, G.B.: Ramesh, M., Berg, T., Miller, E.: Labeled Faces in the Wild: A Database for Studying Face Recognition in Unconstrained Environments. Technical Report 07-49, University of Massachusetts, Amherst (2007)
13. Georghiades, A., Belhumeur, P., Kriegman, D.: From Few to Many: Illumination Cone Models for Face Recognition under Variable Lighting and Pose. IEEE Trans on PAMI 6(23), 643–660 (2001)
14. He, X., Yan, S., Hu, Y., Niyogi, P., Zhang, H.: Face Recognition Using Laplacianfaces. IEEE Trans. on PAMI 27(3), 328–340 (2005)
15. Miller, D., Browning, J.: A Mixture Model and EM-Based Algorithm for Class Discovery, Robust Classification, and Outlier Rejection in Mixed Labeled/Unlabeled Data Sets. IEEE Trans on PAMI 25(11), 1468–1483 (2003)
16. Raina, R., Battle, A., Honglak, L., Ng, A.: Self-taught Learning: Transfer Learning from Unlabeled Data. In: Proceedings of International Conference on Machine Learning (ICML) (2007)
17. Schmidhuber, J.: On Learning How to Learn Learning Strategies. Technical Report FKI-198-94, Fakultat fur Informatik 28(1), 711–720 (1994)
18. Samaria, F., Harter, A.: Parameterisation of a stochastic model for human face identification. In: IEEE Workshop on Applications of Computer Vision, pp. 138–142 (1994)
19. Schölkopf, B., Herbrich, R., Smola, A.: A Generalized Representer Theorem. In: Helmbold, D.P., Williamson, B. (eds.) COLT 2001 and EuroCOLT 2001. LNCS (LNAI), vol. 2111, pp. 416–426. Springer, Heidelberg (2001)
20. Turk, M., Pentland, A.: Face Recognition Using Eigenfaces. In: Proceedings of IEEE Conference on Computer Vision and Pattern Recognition (CVPR), pp. 586–591 (1991)
21. Thrun, S., Mitchell, T.: Learning One More Thing. In: IJCAI, pp. 1217–1223 (1995)
22. Thrun, S., Pratt, L.: Learning To Learn. Kluwer Academic Publishers, Boston (1998)
23. Torre, F., Kanade, T.: Discriminative cluster analysis. In: Proceedings of International Conference on Machine Learning (ICML), pp. 241–248 (2006)
24. Wu, M., Schölkopf, B.: A Local Learning Approach for Clustering. In: Proceedings of Proceedings of Neural Information Processing Systems (NIPS), pp. 1529–1536 (2007)
25. Yan, S., Xu, D., Zhang, B., Zhang, H., Yang, Q., Lin, S.: Graph Embedding and Extension: A General Framework for Dimensionality Reduction. IEEE Trans. on PAMI 29(1), 40–51 (2007)
26. Ye, J., Zhao, Z., Wu, M.: Discriminative K-Means for Clustering. In: Proceedings of Neural Information Processing Systems (NIPS), pp. 1–8 (2007)

# Multiple Manifolds Learning Framework Based on Hierarchical Mixture Density Model

Xiaoxia Wang, Peter Tiňo, and Mark A. Fardal

School of Computer Science, University of Birmingham, UK
Dept. of Astronomy, University of Massachusetts, USA
{X.Wang.1,P.Tino}@cs.bham.ac.uk, fardal@fcrao1.astro.umass.edu

**Abstract.** Several manifold learning techniques have been developed to learn, given a data, a single lower dimensional manifold providing a compact representation of the original data. However, for complex data sets containing multiple manifolds of possibly of different dimensionalities, it is unlikely that the existing manifold learning approaches can discover all the interesting lower-dimensional structures. We therefore introduce a hierarchical manifolds learning framework to discover a variety of the underlying low dimensional structures. The framework is based on hierarchical mixture latent variable model, in which each submodel is a latent variable model capturing a single manifold. We propose a novel multiple manifold approximation strategy used for the initialization of our hierarchical model. The technique is first verified on artificial data with mixed 1−, 2− and 3−dimensional structures. It is then used to automatically detect lower-dimensional structures in disrupted satellite galaxies.

## 1 Introduction

In the current machine learning literature, the manifold learning has been predominantly understood as learning (a single) underlying low-dimensional manifold embedded in a high dimensional data space, where the data points are assumed to be aligned (up to some noise) along the manifold. Typical representatives of such approaches are principal component analysis (PCA)[1], self-organizing mapping (SOM)[2], locally linear embedding (LLE)[3] and Isomap [4]. In these methods, intrinsic (manifold) dimension $d$ is either treated as a prior knowledge given by user or as a parameter to be estimated. Estimating the intrinsic dimension of a data set (without the structure learning considerations) is discussed e.g. in [5] [6], [7]. To our best knowledge, there has been no systematic work on dealing with situations when the data is aligned along multiple manifolds of various dimensionalities, potentially corrupted by noise.

In this paper we propose a a framework for learning multiple manifolds. With each manifold we associate a probability density (generative model), so that the collection of manifolds can be represented by a mixture of their associated density models. These generative models are formulated as latent variable model along the lines of [8] or [9]. Our proposed approach consists of several steps. First,

W. Daelemans et al. (Eds.): ECML PKDD 2008, Part II, LNAI 5212, pp. 566–581, 2008.

we filter data points according to the intrinsic dimensionality of the local manifold patch they are likely to belong to (modulo some manifold aligned "noise"). Then we detect multiple manifolds in each such dimension-filtered set. Finally, we construct a hierarchical probabilistic model containing density models of the detected noisy manifolds. We illustrate our framework on learning multiple manifolds with dimension $d = 1$ and $d = 2$ embedded in a 3-dimensional space.

The paper is organized as follows: the next section briefly reviews related work on manifold learning, intrinsic dimension estimation and hierarchical latent variable modeling; Section 3 describes our framework of learning multiple manifolds based on probability density modeling; Section 4 contains experimental results on artificial data and a data set produced by realistic galaxy collision models. Finally, section 5 concludes the paper and discusses the directions of our future work.

## 2 Related Work

Some manifold learning algorithms are designed to find a single function $\mathbf{y} = g(\mathbf{x}, \mathbf{U})$ representing the mapping between high dimensional observation $\mathbf{x}$ and its low dimensional representation $\mathbf{y}$. Principal component analysis (PCA) implements the transformation function by linear projection $\mathbf{Ux}$. The $d$ orthonormal principal axes vectors $\mathbf{u}_i$ in the observation (data) space form the matrix $\mathbf{U} = (\mathbf{u}_1, \mathbf{u}_2, \ldots, \mathbf{u}_d)$. In contrast, the Generative Topographic Mapping (GTM) [9] is a probabilistic reformulation of the self-Organizing Map (SOM) [2]. It represents the non-linear mapping from a low-dimensional latent space to the high-dimensional data space as a generalized linear regression model $\mathbf{W}\Phi(\mathbf{y})$, where $\Phi(\mathbf{y})$ consists of $M$ fixed basis functions $\{\phi_j(\mathbf{y}), j = 1, \ldots, M\}$, $\mathbf{W}$ is $D \times M$ weights matrix of outputs of basis functions ($D$ is the dimensionality of the data space). A probabilistic generative model of PCA called probabilistic principal component analysis (PPCA) was also proposed in [8]. Other approaches, like locally linear embedding (LLE) [3], Isomap [4] and Laplacian eigenmaps [10], learn the embedding without formulating an explicit mapping. LLE and Laplacian eigenmaps compute the low dimensional representation preserving the local neighborhood structure in data space. Isomap applies multidimensional scaling (MDS) to estimated geodesic distance between points. To generalise the results of LLE, Saul and Roweis proposed in [11] a probabilistic model for the joint distribution $p(\mathbf{x}, \mathbf{y})$ over the input and embedding spaces, which also provides a way to generalise the results from Isomap or Laplacian eigenmaps.

As in the case of manifold learning, most intrinsic dimensionality estimators assume that all the data points are aligned along a single 'manifold'. In [6], local PCA is applied to each node of the optimal topology preserving map (OPMT), intrinsic dimension is the average over the number of eigenvalues which approximates the intrinsic dimensionality at data clusters. Levina and Bickel [5] also average the estimated dimension over all observation. A point level dimensionality estimator proposed in [7] is appealing due to the ability to deal with manifolds of different dimensionality. The authors first represent data point by a second order, symmetric, non-negative definite tensor, whose eigenvalues and eigenvectors fully describe the local dimensionality and orientation at each point. A voting procedure accumulates votes from its neighbors and provides an estimate of local dimensionality.

Finally, we review some examples of hierarchical model and its structure estimation strategy. To reveal the interesting local structures in a complex data set, a hierarchical visualization algorithm based on a hierarchical mixture of latent variable models is proposed in [12]. The complete data set is visualized at the top level with clusters and subclusters of data points visualized at deeper level. Tino and Nabney [13] extended this visualization by replacing the latent variable model by GTM (generative topographic mapping) so that the non-linear projection manifolds could be visualized. Structures of the hierarchy in these visualization systems are built interactively. A non-interactive hierarchy construction was proposed in [14].

# 3 Multiple Manifolds Learning Framework

In this section, a multiple manifolds learning framework is proposed to learn from the dataset of points aligned along different manifolds of different dimensionalities, as shown in figure 1. Although the methods presented in the previous section could easily learn manifolds from either the first or the second set, no methods have been developed for learning their mixture. To identify these manifolds, we **(1)** cluster them by their intrinsic dimensions $d$; **(2)** use the data with same intrinsic dimension to discover and construct the $d$ dimensional surfaces by the multi-manifolds learning algorithm presented later, then initialize a latent variable model for each manifold, **(3)** build a hierarchical mixture model consisting of the generative model for manifolds.

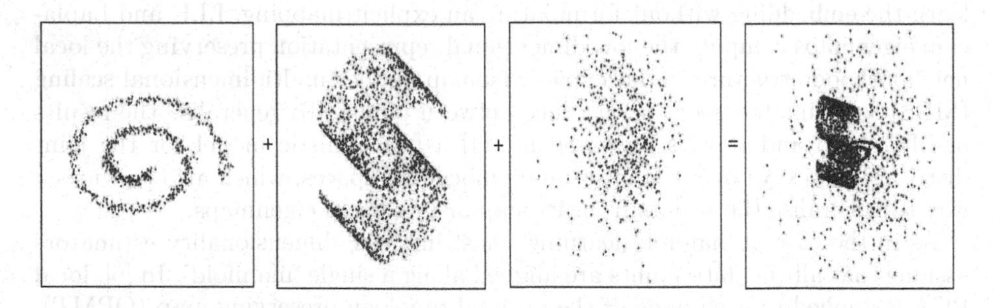


**Fig. 1.** Multiple manifolds example: 700 3D points aligned along a 1D manifold, 2000 3D points from lying on a 2D manifold and 600 3D points generated form a mixture of 3 Gaussians

### 3.1 Intrinsic Dimension Estimation

In the first step, we estimate each point's intrinsic dimension and cluster the entire dataset according to the intrinsic dimensions found. The intrinsic dimension of a point is revealed by the dimensionality of the local manifold patch on which the point is laying. With the assumption that manifold is locally linear, we represent the local patch by the covariance matrix of the points on it. In our implementation, points on $\mathbf{x}_i$'s local patch are $\mathbf{x}_i$'s $K$ nearest neighbours in the dataset (denoted by $\zeta$). The set of $K$ nearest neighbors of $\mathbf{x}_i$ is denoted by $\mathcal{K}(\mathbf{x}_i, \zeta)$. Decomposing the (local) covariance matrix as $\sum_i^D \lambda_i \mathbf{u}_i \mathbf{u}_i^T$, we obtain the orthogonal components as eigenvectors $\mathbf{u}_i$ and their corresponding eigenvalues $\lambda_i$ (we rescale them, so that $\sum_i^D \lambda_i = 1$). The covariance matrix, in the 3D data example, is rewritten as

$$\sum_{i=1}^{3} \lambda_i \mathbf{u}_i \mathbf{u}_i^T = S_1 \mathbf{u}_1 \mathbf{u}_1^T + \frac{1}{2} S_2 (\mathbf{u}_1 \mathbf{u}_1^T + \mathbf{u}_2 \mathbf{u}_2^T) + \frac{1}{3} S_3 (\mathbf{u}_1 \mathbf{u}_1^T + \mathbf{u}_2 \mathbf{u}_2^T + \mathbf{u}_3 \mathbf{u}_3^T),$$

where $\mathbf{u}_1 \mathbf{u}_1^T$, $1/2(\mathbf{u}_1 \mathbf{u}_1^T + \mathbf{u}_2 \mathbf{u}_2^T)$ and $1/3(\mathbf{u}_1 \mathbf{u}_1^T + \mathbf{u}_2 \mathbf{u}_2^T + \mathbf{u}_3 \mathbf{u}_3^T)$ are the covariance matrices of the structures having intrinsic dimension $d = 1$, $d = 2$ and $d = 3$ respectively. Therefore the saliences of these structures are computed as $S_1 = \lambda_1 - \lambda_2$, $S_2 = 2(\lambda_2 - \lambda_3)$ and $S_3 = 3\lambda_3$. Note that $S_1 + S_2 + S_3 = 1$. Intrinsic dimension of the point is then $d = \arg\max_i S_i$.

The intrinsic dimension estimator is a variation of the approach in [7], where in contrast to [7], we operate directly on the tangent spaces of the underlying manifold.

Figure 2 demonstrates the performance of our intrinsic dimension estimator on the multiple manifolds dataset in figure 1. Therefore the whole set $\zeta = \{\mathbf{x}_1, \mathbf{x}_2, \ldots, \mathbf{x}_N\}$ is substituted by a partition $\zeta^1 \cup \ldots \zeta^d \ldots \cup \zeta^D$ with $d$ indicating the intrinsic dimension of the subset.

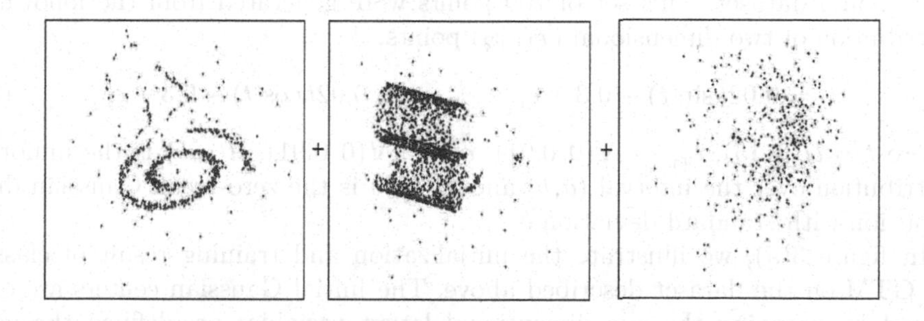


**Fig. 2.** Three intrinsic-dimension-filtered subsets of data from Fig. 1

### 3.2 Multi-manifolds Learning Algorithm

The goal of our multiple manifolds learning algorithm is to represent manifolds contained in each set $\zeta^d (d < D)$ by latent variable models, respectively. We use the notation $\mathcal{M}_d$ to denote the set of manifolds' generative models $p(\mathbf{x}|d,q)$, where $q$ is the index for the manifold found with dimension $d$. In the proposed work, the latent variable model we used is GTM [9]. In the following, we first briefly introduce GTM and then demonstrate the local optima problem in GTM's training. This motivates us to propose a novel robust manifold learning algorithm which provides a better initialization aligned along the non-linear manifold.

**Generative Model for Noisy Manifolds.** Generative topographic mapping (GTM) represents the non-linear transformation by a generalized linear regression model $f^{d,q} = \mathbf{W}^{d,q}\Phi^{d,q}(\mathbf{y}^{d,q})$. The latent variable space is a $d$-dimensional (hyper)cube and is endowed with a (prior) probability distribution $p(\mathbf{y}^{d,q})$ - a set of delta functions. Noise model $p(\mathbf{x}|\mathbf{y}^{d,q})$ is a radially-symmetric Gaussian distribution with mean $f^{d,q}(\mathbf{y}^{d,q})$ and covariance $1/\beta^{d,q}\mathbf{I}$, where $\beta^{d,q} > 0$. The generative model $p(\mathbf{x}|d,q)$ can be obtained by integrating over the latent variables

$$\begin{aligned} p(\mathbf{x}|d,q) &= \frac{1}{Z^{d,q}} \sum_{z=1}^{Z^{d,q}} p(\mathbf{x}|\mathbf{y}_z^{d,q}, \mathbf{W}^{d,q}, \beta^{d,q}) \\ &= \frac{1}{Z^{d,q}} (\frac{\beta^{d,q}}{2\pi})^{(D/2)} \exp\{-\frac{\beta^{d,q}}{2}||f^{d,q}(\mathbf{y}_z^{d,q}) - \mathbf{x}||^2\} \end{aligned} \tag{1}$$

where $Z^{d,q}$ denotes the number of the latent variable $\mathbf{y}^{d,q}$, the map $f^{d,q}$ is defined above and $\Phi^{d,q}(\mathbf{y}_k^{d,q})$ is a column vector

$$\Phi^{d,q}(\mathbf{y}_z^{d,q}) = [\phi_1(\mathbf{y}_z^{d,q}), \phi_2(\mathbf{y}_z^{d,q}), \ldots, \phi_M^{d,q}(\mathbf{y}_z^{d,q})]^T \tag{2}$$

We can in principle determine $\mathbf{W}^{d,q}$ and $\beta^{d,q}$ by maximizing the log likelihood function through E-M. But the optimization procedure cannot guarantee the global optimum in case of a strong non-linear manifold structure, because original GTM is typically initialized through a linear global PCA.

We demonstrate the performance of GTM on a strong non-linear manifold data, spiral dataset. This set of 700 points were generated from the following distribution of two dimensional $(x_1, x_2)$ points:

$$x_1 = 0.02t\sin(t) + 0.3 + \epsilon_{x_1}; \qquad x_2 = 0.02t\cos(t) + 0.3 + \epsilon_{x_2} \tag{3}$$

where $t \sim \mathcal{U}(3, 15)$, $\epsilon_{x_1} \sim \mathcal{N}(0, 0.01)$, $\epsilon_{x_2} \sim \mathcal{N}(0, 0.01)$, $\mathcal{U}(a, b)$ is the uniform distribution over the interval $(a, b)$ and $\mathcal{N}(0, \delta)$ is the zero-mean Gaussian distribution with standard deviation $\delta$.

In figure 3(a), we illustrate the initialization and training result of classical GTM on the dataset described above. The initial Gaussian centers are obtained by mapping the one dimensional latent variables pre-defined through global PCA. The training result with this initialization approximates the non-linear spiral data manifold poorly. In contrast to figure 3(a), figure 3(b) shows an improved fit by the GTM initialized by the method described below.

**Identifying One Manifold.** Here we propose a novel strategy to identify the embedded manifolds and capture their (high) non-linear structure in the learning

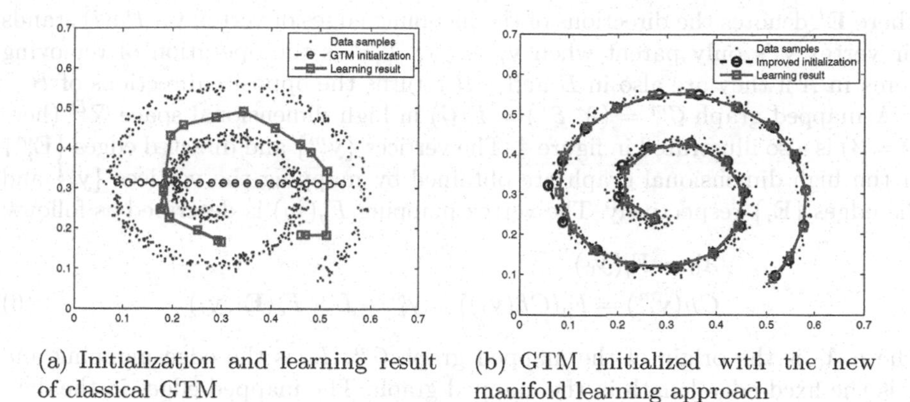


(a) Initialization and learning result of classical GTM

(b) GTM initialized with the new manifold learning approach

**Fig. 3.** Learning results from different GTM initializations

procedure. In our proposed framework, we describe the low dimensional latent manifold by an oriented graph $\mathcal{G} = (\mathcal{Y}, \mathcal{E})$, where $\mathcal{Y}$ represents the graph vertices and $\mathcal{E}$ represents the edges. We label the $i$-th vertex in $\mathcal{Y}$ by a $d$-dimensional point $\mathbf{y}_i \in \mathcal{R}^d$. The directed edges from the vertex $\mathbf{y}_i$ are collected in $\mathbf{E}_i$ that can be also thought of as the set of destination vertices. The coordinates of these destination vertices (children of the source/parent vertex $\mathbf{y}_i$ and denoted by $Ch(\mathbf{y}_i)$) can be calculated as

$$Ch(\mathbf{y}_i) = \mathbf{y}_i + l \times \mathbf{e}_i^{\mathrm{o}} \tag{4}$$

where $\mathbf{e}_i^{\mathrm{o}}$ is a collection of unit directional vectors of the outgoing edges, and $l$ represents the length of the edge (fixed to be 1 in our implementation). We denote the outgoing edges from vertex $\mathbf{y}_i$ as $\mathbf{E}_i = \{\mathbf{e}_i^{\mathrm{o}}\}$ for simplicity.

An example of a graph structure for $d = 2$ is illustrated in figure 4. We describe our manifold learning approach based on this case. Partitioning all the vertices according to the number of their parents, we obtain $\mathcal{Y} = \mathcal{Y}_0 \cup \mathcal{Y}_1 \cup \mathcal{Y}_2$, where $\mathcal{Y}_0$, $\mathcal{Y}_1$, $\mathcal{Y}_2$ represent collections of vertices with 0−, 1− and 2− parents, respectively. The set $\mathcal{Y}_0$ contains the unique parentless vertex in the graph. We refer to this vertex as the origin of the graph and set its coordinates to $\mathbf{0} \in \mathcal{R}^d$. The edges from the origin correspond to $2d = 4$ outgoing edges (orthonormal vectors) denoted by $\mathbf{E}_0$ and specified as $\{(1,0), (-1,0), (0,1), (0,-1)\}$ in the implementation.

Knowing $\mathcal{Y}_0$ and $\mathbf{E}_0$, we could retrieve all the vertices in this graph by eq. (4) from their parents and whose outgoing edges $\mathbf{E}_i$. And we obtain the outgoing edges by the following equation:

$$\mathbf{E}_i = \begin{cases} \mathbf{E}_0 & \mathbf{y}_i \in \mathcal{Y}_0 \\ \mathbf{E}_{Pa(i)} \backslash - \mathbf{E}_i^I & \mathbf{y}_i \in \mathcal{Y}_1 \\ \mathbf{E}_i^I & \mathbf{y}_i \in \mathcal{Y}_2 \end{cases} \tag{5}$$

where $\mathbf{E}_i^I$ denotes the directions of the incoming edges of vertex $\mathbf{y}_i$, $Pa(i)$ stands for vertex $\mathbf{y}_i$'s only parent when $\mathbf{y}_i \in \mathcal{Y}_1$, $A\backslash B$ is the operation of removing items in $A$ if they are also in $B$ and $-B$ returns the opposite directions of $B$.

A mapped graph $\mathcal{G}^m = (\mathcal{X}, \mathcal{E}^x) = F(\mathcal{G})$ in high dimensional space $\mathcal{R}^D$ (here $D = 3$) is also illustrated in figure 4. The vertices $\{\mathbf{y}_i^m\}$ and directed edges $\{\mathbf{E}_i^m\}$ in the high dimensional graph are obtained by mapping the vertices $\{\mathbf{y}_i\}$ and the edges $\{\mathbf{E}_i\}$ respectively. The vertex mapping $F_y(\mathbf{y}_i)$ is described as follows:

$$\begin{aligned} &\mathcal{X}_0 = F_y(\mathcal{Y}_0) \\ &Ch(\mathbf{y}_i^m) = F_y(Ch(\mathbf{y}_i)) = \mathbf{y}_i^m + L \times F_e(\mathbf{E}_i, \mathbf{y}_i) \end{aligned} \tag{6}$$

where $\mathcal{X}_0$ is the origin of the mapped graph $\mathcal{G}^m$. $F_e$ is the edge mapping and $L$ is the fixed edge length in the mapped graph. The mapped edges in the data space are then obtained via the mapping

$$F_e(\mathbf{E}_i, \mathbf{y}_i) = \begin{cases} \mathbf{M}_{\mathbf{y}_i^m} \mathbf{E}_0^m & \mathbf{y}_i \in \mathcal{Y}_0 \\ \mathbf{M}_{\mathbf{y}_i^m} \mathbf{E}_{pa(i)}^m \backslash - \mathbf{E}_i^{mI} & \mathbf{y}_i \in \mathcal{Y}_1 \\ \mathbf{M}_{\mathbf{y}_i^m} \mathbf{E}_i^{mI} & \mathbf{y}_i \in \mathcal{Y}_2 \end{cases} \tag{7}$$

where $\mathbf{M}_{\mathbf{y}_i^m}$ is the projection matrix onto the manifold patch around the point $\mathbf{y}_i^m$ and given by

$$\mathbf{M}_{\mathbf{y}_i^m} = \mathbf{B}_{\mathbf{y}_i^m} \mathbf{B}_{\mathbf{y}_i^m}^T, \tag{8}$$

where $\mathbf{B}_{\mathbf{y}_i^m}$ denotes the matrix of basis vectors which are the first $d$ eigenvectors with largest eigenvalues corresponding to principle directions of the neighborhood of the vertex $\mathbf{y}_i^m$. In eq. (7), $\mathbf{E}_0^m$ denotes the orthonormal outgoing edges of the mapped origin $\mathcal{X}_0$.

We present an algorithm to simultaneously learn the graph $\mathcal{G}$ and the associated mapped graph $\mathcal{G}^m$ from a dataset $\zeta^d$.

- Initialization
  Learning of the mapped graph is initialized by specifying $\mathcal{X}_0$, $L$ in eq. (6), and $\mathbf{E}_0^m$ in eq. (7). In a dataset without outliers (isolated points not lying "close" to any apparent manifold), the origin $\mathcal{X}_0$ of the mapped graph can be associated with any randomly chosen point. Since the presence of outliers cannot be ruled out, we include outlier detection (steps 1, 2) in the algorithm **initialization**.
- Recursive learning
  With the initialization, we set the origins $\mathcal{Y}_0$ and $\mathcal{X}_0$ to be the current generations of graphs $\mathcal{G}$ and $\mathcal{G}^m$, denoted by $CG$ and $CG^m$. The learning procedure forms the graphs together iteratively from the current generation to its next generation ($NG$ and $NG^m$) until the boundary of the manifold is detected.
  - $[NG, NG^m] = Redun\_remove(NG, NG^m)$
    This procedure removes duplicate vertices in $NG$ and $NG^m$. The duplicate vertices are generated in the learning procedure, because vertices may have more than one parent. For example, in the graph $\mathcal{G}$, any vertex $\mathbf{y}_s \in \mathcal{Y}_2$ is a child of two parents. It is then learnt as a set of vertices $\{\mathbf{y}_{s1}, \mathbf{y}_{s2}\} \in NG$ from the current generation. Since these vertices

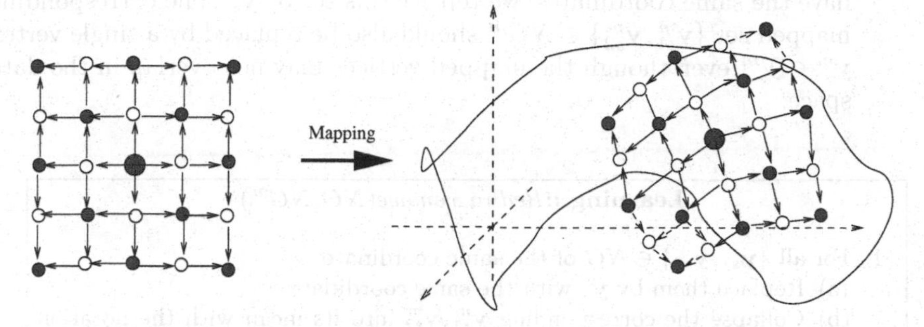


**Fig. 4.** The directed graph representing 2-dimensional latent space and its mapped graph lying on 2 dimensional manifold in the 3 dimensional data space

**Initialization**

1. For all $\mathbf{x}_i \in \zeta^d$ do:
   (a) Count the presence of $\mathcal{K}(\mathbf{x}_i, \zeta)$ in dataset $\zeta^d$, i.e., $k$ =the size of $\{\mathcal{K}(\mathbf{x}_i, \zeta) \cap \zeta^d\}$.
   (b) If $k > \alpha_1 K$ $(0 < \alpha_1 < 1)$, then $\mathcal{X}_0 = \mathbf{x}_i$, break;
2. EndFor
3. If $\mathcal{X}_0! = null$, then
   (a) Find $\mathcal{K}(\mathcal{X}_0, \zeta^d)$ with parameter $K_2$ (neighborhood size).
   (b) $\mathbf{E}_0^m = \{\mathbf{u}_1, \ldots, \mathbf{u}_d, -\mathbf{u}_1, \ldots, -\mathbf{u}_d\}$, where $\{\mathbf{u}_1, \ldots, \mathbf{u}_d\}$ are the first $d$ principal directions of $\mathcal{K}(\mathcal{X}_0, \zeta^d)$.
   (c) $L$ is set to be the average of the distance from the neighbours to $\mathcal{X}_0$.
4. EndIf

**Learning**

1. $NG = \{\}$; $NG^m = \{\}$
2. While $CG! = null$ do:
   (a) i. For all $\mathbf{y}_i \in CG$
         A. $NG = \{NG; Ch(\mathbf{y}_i)\}$;
         B. $NG^m = \{NG^m; F_y(Ch(\mathbf{y}_i)\}$
      ii. EndFor
   (b) $[NG, NG^m] = Redun\_remove(NG, NG^m)$
   (c) $[CG, CG^m] = Boud\_check(NG, NG^m)$
   (d) $\mathcal{Y} = \{\mathcal{Y}; CG\}$;
   (e) $\mathcal{X} = \{\mathcal{X}; CG^m\}$;
3. EndWhile

have the same coordinates, we replace this set by $\mathbf{y}_s$. The corresponding mapped set $\{\mathbf{y}_{s1}^m, \mathbf{y}_{s2}^m\} \in NG^m$ should also be replaced by a single vertex $\mathbf{y}_s^m \in \mathcal{Y}^m$ even though the mapped vertices may not overlap in the data space.

**Learning** $::Redun\_remove(NG, NG^m)$

1. For all $\{\mathbf{y}_{s1}, \mathbf{y}_{s2}\} \in NG$ of the same coordinate
   (a) Replace them by $\mathbf{y}_s$ with the same coordinate.
   (b) Collapse the corresponding $\mathbf{y}_{s1}^m, \mathbf{y}_{s2}^m$ into its mean with the notation $F_y(\mathbf{y}_s)$.
   (c) Collect incoming edges of $\mathbf{y}_s$ and $\mathbf{y}_s^m$ by their directions in $\mathbf{E}_s^I$ and $\mathbf{E}_s^{Im}$
2. EndFor

- $[CG, CG^m] = Boud\_check(NG, NG^m)$
  This procedure checks for the manifold boundary. The learning procedure should stop when the boundary is reached. Close to the boundary, the local density of points around the current point decreases rapidly. In such situations the set of nearest neighbors overlaps significantly (determined by parameter $K_3$) with the set of already visited points on the manifold.

**Learning** $::Boud\_check(NG, NG^m)$

1. $CG = \{\}$; $CG^m = \{\}$;
2. For all $\mathbf{y}_j^m \in NG^m$
   (a) Find $\mathcal{K}(\mathbf{y}_j^m, \zeta^d)$ and $\mathcal{O} =$ the size of $\{\mathcal{K}(\mathbf{y}_j^m, \zeta^d) \cap \mathcal{K}(CG^m, \zeta^d)\}$.
   (b) If $\mathcal{O} < K_3$, then
      i. $CG = \{CG, \mathbf{y}_j\}$;
      ii. $CG^m = \{CG^m, \mathbf{y}_j^m\}$
   (c) EndIf
3. EndFor

**Initializing GTM with Learnt Graphs.** Each detected manifold will be modeled by a single GTM initialized using the graphs $\mathcal{G}$ and $\mathcal{G}^m$. Vertices in graph $\mathcal{G}$ are the low dimensional latent variables and vertices in the mapped graph $\mathcal{G}^m$ are the images the latent variables under the GTM model. Therefore, we determine $\mathbf{W}^{d,q}$ by minimizing the error

$$Err = \frac{1}{2} \sum_{z=1}^{Z^{d,q}} ||\mathbf{W}^{d,q} \Phi(\mathbf{y}_z^{d,q}) - F_y(\mathbf{y}_z^{d,q})||^2. \tag{9}$$

The parameter $\beta^{d,q}$ is initialized to be the larger of the average of the $d+1$ eigenvalues from PCA applied on each vertex in $\mathcal{Y}^m$ and the square of half of the grid spacing of the mapped vertices $\mathcal{Y}^m$ in the data space.

As an example, figure 3(b) shows the result of GTM fitting the spiral data after being initialized with our manifold learning technique. With this initialization, we can see that a little or no further GTM training is required.

**Learn Multiple Manifolds.** Since the proposed manifold learning algorithm explores the dataset starting from a single "seed" point and then "crawls" on the manifold, points in the data space which are not connected in the same manifold will be left unvisited. For datasets having more than one underlying manifolds, we utilize an iterative procedure (see below) to explore the unvisited set $\tilde{\zeta}$ (entire filtered set at the beginning), learn the manifolds and initialize the corresponding generative models one by one.

1. $\tilde{\zeta}^d = \zeta^d (1 \leq d < D)$; $q = 0$
2. While $\tilde{\zeta}^d > K_2$
   (a) Learn $\mathcal{G}$ and $\mathcal{G}^m$ from $\tilde{\zeta}^d$; $q = q + 1$.
   (b) Initialize $\mathbf{W}^{d,q}$, $\beta^{d,q}$ in $p(\mathbf{x}|d,q)$ associating with the $\mathcal{G}$ and $\mathcal{G}^m$.
   (c) $\tilde{\zeta}^d = \tilde{\zeta}^d \backslash \mathcal{K}(\mathcal{X}, \tilde{\zeta}^d)$
3. EndWhile

### 3.3 Hierarchical Mixture Model

In this section, we formulate a two-level hierarchical model $\mathcal{T}$ and the EM algorithm to fit $\mathcal{T}$ to the entire data set $\zeta = \{\mathbf{x}_1, \mathbf{x}_2, \ldots, \mathbf{x}_N\}$.

**Model Formulation.** We formulate the hierarchical model $\mathcal{T}$ by first mixing the models $p(\mathbf{x}|, d, q) \in \mathcal{M}_d$ at the second level hierarchy with $\pi_{q|d}$, for each intrinsic dimensionality $d$, $\sum_q \pi_{q|d} = 1$ $(q > 0)$. If $\mathcal{M}_d = null$, then we set $q = 0$ and $\pi_{0|d} = 0$. We the mix these intrinsic dimensionality groups with $\pi_d$ at the first level hierarchy. Thus we obtain:

$$p(\mathbf{x}|\mathcal{T}) = \sum_{d=1}^{D} \pi_d \sum_{q \in \mathcal{M}_d} \pi_{q|d} p(\mathbf{x}|d,q). \tag{10}$$

The probabilistic models in $\mathcal{M}_d$ $(1 \leq d < D)$ are formulated in eq. (1). We use a Gaussian mixture (11) to model the data points collected in $\zeta^D$.

$$\begin{aligned} p(\mathbf{x}|D,1) &= \sum_{z=1}^{Z^{D,1}} p(z) p(\mathbf{x}|z; D, 1) \\ &= \sum_{z=1}^{Z^{D,1}} p(z) \frac{1}{\sqrt{||2\pi\mathbf{\Sigma}_z||}} \exp\{-\frac{1}{2}(\mathbf{x} - \mu_z)^T \mathbf{\Sigma}_z^{-1} (\mathbf{x} - \mu_z)\} \end{aligned} \tag{11}$$

where $Z^{D,1}$ is the number of Gaussian components, $\mu_z$ and $\mathbf{\Sigma}_z$ are the mean and covariance of the $z$−th component, and $p(z)$'s are the mixing coefficients with $\sum_{z=1}^{Z^{D,1}} p(z) = 1$.

**EM Algorithm.** The mixing coefficients at each hierarchical level and the parameters of each submodel can be determined by maximizing the log likelihood function of the model (10).

$$\mathcal{L} = \sum_{n=1}^{N} \ln p(\mathbf{x}_n|\mathcal{T}) \tag{12}$$

We use binary assignment variables $\upsilon_{n,d}$ to represent that the data $\mathbf{x}_n$ belongs to the group of manifolds having dimension $d$, and $\upsilon_{n,q|d}$ to represent the situation that data $\mathbf{x}_n$ is generated from $q$-th manifold in the group with dimension $d$. Even if this was known, we still need to decide which latent space center $\mathbf{y}_z^{d,q} \in \mathcal{Y}^{d,q}, z = 1, 2, \ldots, Z^{d,q}$ in the latent variable model $p(\mathbf{x}|d,q)$ $(1 \leq d < D)$ corresponds to the Gaussian that generated $\mathbf{x}_n$, We represent this by indicator variables $\upsilon_{n,z}^{d,q}$.

For $d = D$, $\mathcal{M}_D$ contains a single unconstrained Gaussian mixture model. The complete data likelihood function reads

$$\mathcal{L}_c = \sum_{n=1}^{N}\sum_{d=1}^{D} \upsilon_{n,d} \sum_{q\in\mathcal{M}_d} \upsilon_{n,q|d} \sum_{z=1}^{Z^{d,q}} \upsilon_{n,z}^{d,q} \ln\{\pi_{q|d}\pi_d p(\mathbf{x}_n|d,q)\} \tag{13}$$

Taking the expectation (with respect to the posterior given the data) over all types of hidden variables, we arrive at the expected complete-data likelihood

$$< \mathcal{L}_c >= \sum_{n=1}^{N}\sum_{d=1}^{D} R_{d|n} \sum_{q\in\mathcal{M}_d} R_{q|d,n} \sum_{z=1}^{Z^{d,q}} R_{z|q,d,n} \ln\{\pi_d\pi_{q|d} p(\mathbf{x}_n, |d,q)\} \tag{14}$$

The M-step of the EM algorithm involves maximizing (14) w.r.t. the parameters.

We obtain the following updates:

$$\tilde{\pi}_d = \frac{1}{N}\sum_{n=1}^{N} R_{d|n} \tag{15}$$

$$\tilde{\pi}_{q|d} = \frac{\sum_{n=1}^{N} R_{d|n}R_{q|d,n}}{\sum_{n=1}^{N} R_{d|n}} \tag{16}$$

$$p(z) = \frac{\sum_{n=1}^{N} R_{D|n}R_{z|D,n}}{\sum_{n=1}^{N}\sum_{z=1}^{Z^{D,1}} R_{d|n}R_{z|d,n}} \tag{17}$$

As for the manifold models, using eq. (1) and (2), we have

$$\sum_{n=1}^{N} R_{nd} \sum_{z=1}^{Z^{d,q}} R_{z|d,qn}(\mathbf{W}^{d,q}\Phi(\mathbf{y}_z^{d,q}) - \mathbf{x}_n)\Phi^T(\mathbf{y}_z^{d,q}) = 0 \tag{18}$$

The responsibilities $R_{d|n}$ and $R_{z|n,d,q}$ are calculated with the current (old) weight and inverse variance parameters of the probabilistic models $p(\mathbf{x}|d,q)$.

Written in matrix notation, we have to solve

$$(\mathbf{\Phi}^{d,q})^T \mathbf{B}^{d,q} (\mathbf{\Phi}^{d,q})^T (\mathbf{W}^{d,q})^T = (\mathbf{\Phi}^{d,q})^T \mathbf{R}^{d,q} \mathbf{T} \tag{19}$$

for $\mathbf{W}^{d,q}$.

The above system of linear equations involves the following matrices:

- $\mathbf{\Phi}$ is a $Z^{d,q} \times M^{d,q}$ matrix with elements $(\mathbf{\Phi}^{d,q})_{ij} = \phi_j(\mathbf{y}_z^{d,q})$.
- $\mathbf{T}$ is a $N \times D$ matrix storing the data points $\mathbf{x}_1, \mathbf{x}_2, \ldots, \mathbf{x}_N$ as rows.
- $\mathbf{R}^{d,q}$ is a $Z^{d,q} \times N$ matrix containing, for each latent space center $\mathbf{y}_z^{d,q}$, and each data point $\mathbf{x}_n$, scaled responsibilities $(\mathbf{R}^{d,q})_{zn} = R_{d|n} R_{q|n,d} R_{z|q,d,n}$
- $\mathbf{B}^{d,q}$ is a $Z^{d,q} \times Z^{d,q}$ diagonal matrix with diagonal elements corresponding to responsibilities of latent space centers for the whole data sample $\zeta$, where $(\mathbf{B})_{ii} = \sum_{n=1}^{N} R_{d|n} R_{q|d,n} R_{z|q,d,n}$.

The GTM mapping $f^{d,q}$ can be regularized by adding a regularization term to the likelihood (1). Inclusion of the regularizer modifies eq.(19) to

$$[(\mathbf{\Phi}^{d,q})^T \mathbf{B}^{d,q} (\mathbf{\Phi}^{d,q})^T + \frac{\alpha^{d,q}}{\beta^{d,q}} \mathbf{I}] (\mathbf{W}^{d,q})^T = (\mathbf{\Phi}^{d,q})^T \mathbf{R}^{d,q} \mathbf{T} \tag{20}$$

where $\mathbf{I}$ is the $Z^{d,q} \times Z^{d,q}$ identity matrix.

Finally, the re-estimation formulation

$$\frac{1}{\beta^{d,q}} = \frac{\sum_{n=1}^{N} R_{d|n} R_{q|d,n} \sum_{z=1}^{Z^{d,q}} R_{z|n,d,q} ||\mathbf{W}^{d,q} \phi(\mathbf{y}_z^{d,q}) - \mathbf{x}_n||^2}{D \sum_{n=1}^{N} R_{d|n} R_{q|d,n}} \tag{21}$$

where $\mathbf{W}^{d,q}$ is the "new" weight matrix computed by solving (19) in the last step.

We also obtain

$$\mu_z = \frac{\sum_{n=1}^{N} R_{D|n} R_{z|D,n} \mathbf{x}_n}{\sum_{n=1}^{D} R_{D|n}} \tag{22}$$

$$\mathbf{\Sigma}_z^D = \frac{\sum_{n=1}^{N} R_{D|n} R_{z|D,n} (\mathbf{x}_n - \mu_z)(\mathbf{x}_n - \mu_z)^T}{\sum_{n=1}^{N} R_{D|n}} \tag{23}$$

In the E-step of the EM algorithm, we estimate the latent space responsibilities $R_{z|n,d,q}$ in submodels for manifolds and component responsibilities $R_{z|D}$ in Gaussian mixture model. Model responsibilities in each group $R_{q|n,d}$ and the group responsibilities $R_{d|n} (1 \le d \le D)$ are specified as well.

$$R_{z|d,q,n} = \frac{p(\mathbf{x}_n|\mathbf{y}_z, d, q)}{\sum_{z'=1}^{Z^{d,q}} p(\mathbf{x}_n|\mathbf{y}'_z, d, q)} \quad (24)$$

$$R_{z|D,n} = \frac{p(z)p(\mathbf{x}_n|z, D)}{\sum_{z'=1}^{Z^D} p(z')p(\mathbf{x}_n|z, d, q)} \quad (25)$$

$$R_{q|d,n} = \frac{\pi_{q|d} p(\mathbf{x}_n|d, q)}{\sum_{q' \in \mathcal{M}_d} \pi_{q'|d} p(\mathbf{x}_n|d, q')} \quad (26)$$

$$R_{d|n} = \frac{\pi_d \sum_{q \in \mathcal{M}_d} \pi_{q|d} p(\mathbf{x}|d, q)}{\sum_{d'=1}^{D} \pi_{d'} \sum_{q' \in \mathcal{M}_{d'}} \pi_{q'|d'} p(\mathbf{x}_n|d', q')} \quad (27)$$

**Parameter Initialization.** There are two groups of parameters to be initialized before running EM algorithm to fit $\mathcal{T}$ to the data.

An unconstrained GMM is used to model the set $\zeta^D$. The corresponding parameters including Gaussian centers and covariance can be initialized e.g. by a simple K-means algorithm. Parameters in latent variable model $p(\mathbf{x}|d, q)$, where $d < D$ are initialized using our multi manifold learning algorithm described in the previous section.

## 4 Experiments

**Multi-manifolds with Varying Dimensions.** We first present the multiple manifolds learning results on the dataset illustrated in figure 1. Figure 2 shows the filtered subsets of data points with respect to their intrinsic dimensionality. Note that because the manifolds cross, there is a gap splitting the single $1d$ manifold into two parts. The points in the gap were taken to the set of intrinsically 2-dimensional points. Given a strong prior knowledge concerning connectiveness of the data manifolds, we could deal with such situations, however in the absence of such information we would prefer to have several isolated components dictated by the data.

We use our multi manifold initialization alongside with EM algorithm to fit the full hierarchical model to the dataset. The results shown in figure 5. The manifolds of varying dimensionality and shape were correctly identified.

**Identifying Streams and Shells in Disrupted Galaxies.** Recent exciting discoveries of low-dimensional structures such as shells and streams of stars in the vicinity of large galaxies led the astronomers to investigate the possibility that such structures are in fact remnants of disrupted smaller satellite galaxies. One line of investigation models the disruption process using realistic particle models of both the large galaxy (e.g. M31) and the disrupted one (e.g. M32) [15,16]. Realistic set of initial conditions are applied and the results of particle simulations are compared with the observed structures for each initial condition setting. This is of most importance for understanding the disruption process and prediction of the future continuation of the satellite disruption. We will show that

**Fig. 5.** Manifolds learnt from the artificial dataset dataset in figure 1. We use parameters $K = 20, \alpha_1 = 0.8, K_2 = 20, K_3 = 18$.

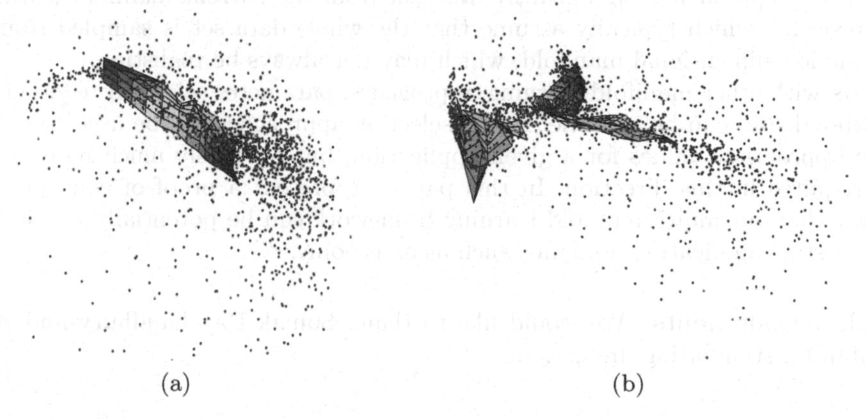

(a) (b)

**Fig. 6.** Identified 2-dimensional manifolds in disrupted satellite galaxy at an early (a) and later (b) stages of disruption by M31

our multi manifold learning methodology can be used for automatic detection of low dimensional streams and shells in such simulations, thus enabling automated inferences in large scale simulations.

Using one realistic setting of the particle model [15,16] for satellite galaxy disruption by M31, we obtained several stages of the disrupted satellite, modeled by approximately 30,000 particles (stars). In each stage we applied the multi manifold learning. In figure 6 we show (along with the stars (particles)) the detected two-dimensional manifold structures ("skeletons of the mixture components) in an early and a later stage of the disruption process. 1- and 3-dimensional structures are not shown. In the early stage a single stream was automatically detected, whereas in the later stage a stream and two shells were correctly identified. Two-dimensional structures are of most importance in such investigations, but our system can be used for investigation of structures across a variety of dimensions. It can also be used to build a hierarchical mixture model for the full system (large galaxy + a satellite) for the purposes of principled comparison of the simulated system with the real observations (in the projected plane). The is a matter for our future work.

## 5 Conclusions and Discussion

We presented a novel hierarchical model based framework to learn multiple manifolds of possibly different intrinsic dimensionalities. We first filter the data points with respect to the intrinsic dimensionality of the manifold patches they lie on. Then our new multi manifold learning algorithm is applied to each such filtered dataset of dimensionality $d$ to detect $d$-dimensional manifolds along which the data are aligned. This is later used to initialize generative latent variable models representing noisy manifolds underlying the data set. The generative models are combine in a hierarchical mixture representing the full data density. The proposed approach is significantly different from the current manifold learning approaches which typically assume that the whole data set is sampled from a single low dimensional manifold, which may not always be realistic.

As with other manifold learning approaches, parameter selection (e.g. neighborhood size) can be an issue. Model selection approaches can be used to select the appropriate values for a given application, but obviously much more work is required in this direction. In this paper we present a proof of concept and show that our multi manifold learning framework can be potentially applied in interesting application domains, such as astronomy.

**Acknowledgments.** We would like to thank Somak Raychaudhury and Arif Babul for stimulating discussions.

## References

1. Moore, B.: Principal component analysis in linear systems: Controllability, observability, and model reduction. IEEE Transactions on Automatic Control, 17–32 (February 1981)
2. Kohonen, T. (ed.): Self-Organization and Associative Memory. Springer, Heidelberg (1997)
3. Roweis, S., Saul, L.: Nonlinear dimensionality reduction by locally linear embedding. Science 290, 2323–2326 (2000)
4. Tenenbaum, J.B., Silva, V.d., Langford, J.C.: A Global Geometric Framework for Nonlinear Dimensionality Reduction. Science 290(5500), 2319–2323 (2000)
5. Levina, E., Bickel, P.: Maximum likelihood estimation of intrinsic dimension. In: Advances in NIPS (2005)
6. Bruske, J., Sommer, G.: Intrinsic dimensionality esitmation with optimally topology preserving maps. IEEE Transactions on Pattern Analysis and Machine Intelligence, 572–575 (May 1998)

7. Mordohai, P., Medioni, G.: Unsupervised dimensionality estimation and manifold learning in high-dimensional spaces by tensor voting. In: International Joint Conference on Artifcial Intelligence, pp. 798–803 (2005)
8. Tipping, M., Bishop, C.: Probabilistic principal component analysis. Journal of the Royal Statistical Society: Series B (Statistical Methodology) (3), 611–622 (1999)
9. Bishop, C.M., Svensen, M., Williams, C.K.I.: GTM: The generative topographic mapping. Neural Computation 10(1), 215–234 (1998)
10. Belkin, M., Niyogi, P.: Laplacian eigenmaps for dimensionality reduction and data representation. In: Neural Computation, pp. 1373–1396 (June 2003)
11. Saul, L.K., Roweis, S.T.: Think globally, fit locally: Unsupervised learning of low dimensional manifolds. Journal of machine learning research, 119–155 (2003)
12. Bishop, C.M., Tipping, M.E.: A hierarchical latent variable model for data visualization. IEEE Transactions on Pattern Analysis and Machine Intelligence 20(3), 281–293 (1998)
13. Tino, P., Nabney, I.: Hierarchical GTM: constructing localized non-linear projection manifolds in a principled way. IEEE Transactions on Pattern Analysis and Machine Intelligence (in print, 2001)
14. Williams, C.: A MCMC approach to hierarchical mixture modelling. In: Advances in Neural Information Processing Systems 12, pp. 680–686. MIT Press, Cambridge (2000)
15. Fardal, M.A., Babul, A., Geehan, J.J., Guhathakurta, P.: Investigating the andromeda stream - ii. orbital fits and properties of the progenitor. Monthly Notices of the Royal Astronomical Society 366, 1012–1028 (2006)
16. Fardal, M., Guhathakurta, P., Babul, A., McConnachie, A.W.: Investigating the andromeda stream - iii. a young shell system in m31. Monthly Notices of the Royal Astronomical Society 380, 15–32 (2007)

# Estimating Sales Opportunity Using Similarity-Based Methods

Sholom M. Weiss[1] and Nitin Indurkhya[2]

[1] IBM Research, Yorktown Heights, NY 10598, USA
[2] School of Comp Sci & Engg, UNSW, Sydney, NSW 2052, Australia

**Abstract.** We describe an application of predicting *sales opportunity* using similarity-based methods. Sales representatives interface with customers based on their potential for product sales. Estimates of this potential are made under optimistic conditions and referred to as the opportunity: How much can be sold if a sale were to be made? Since this can never be verified exactly, the direct use of predictive models is difficult. In building systems for estimating sales opportunity, the key issues are: (a) predictions for targets that cannot be verified, (b) explanatory capabilities (c) capability to incorporate external knowledge (d) parallel computation of multiple targets and other efficiencies (e) capability to calibrate optimism in the predictions. (f) method stability and ease of maintenance for incorporating new examples. Empirical experiments demonstrate excellent predictive accuracy while also meeting these objectives. The methods have been embedded in a widely-used similarity-based system for IBM's worldwide sales force.

## 1 Introduction

Consider the scenario of sales representatives marketing products, often in person, to companies. They prepare for a sales pitch and invest much time and effort. Actual success depends on many factors including the skills of sales representatives, product quality, competition and pricing. The amount of effort can be adjusted by the representatives and their management. For expensive products and services, these efforts are often made based on the magnitude of potential sales under optimistic scenarios, as opposed to the likelihood of an actual sale. Even though a sale is not likely, if the sales opportunity is large enough, a sales representative will do a pitch on the outside chance that the customer may switch from a competitor. In many organizations where sales representatives are involved, estimates of opportunity are taking on an increasing importance as a critical measure relative to sales effort. It gives the organization a rationale for dedicating increased resources to close a sale.

Our objective is to build a system that computes opportunity for sales to a customer. This system will compute sales opportunity both for current customers and for new customers. For current customers, opportunity will suggest possibilities for expanded sales. New customers make up a much larger universe. For these customers, opportunity will suggest direction in a large space of

W. Daelemans et al. (Eds.): ECML PKDD 2008, Part II, LNAI 5212, pp. 582–596, 2008.

possibilities. Representatives cannot call every customer. The opportunity computations can suggest those customers where a potential sale will compensate for the required effort.

In building systems for estimating sales opportunity, the key issues are: (a) predictions for targets that cannot be verified, (b) explanatory capabilities (c) capability to incorporate external knowledge (d) parallel computation of multiple targets and other efficiencies (e) capability to calibrate optimism in the predictions. (f) method stability and ease of maintenance for incorporating new examples. We show how these issues can be successfully addressed by using similarity-based methods to find current customers who are similar to new customers. Our approach results in excellent predictive accuracy in our application while also meeting all the operational requirements.

## 2 Sales Opportunity

Our interest is in sales opportunity, which is a more optimistic target than expected sales. One view of opportunity is the expected magnitude of sales when a sale is made. This relationship is expressed in equation 1, where the distribution of sales to a company is conditional on a company's features like its size and industry.

$$opportunity = expected(sales|sales > 0) \tag{1}$$

Sales opportunity is inherently an optimistic estimate, and optimism can be adjusted by modifying equation 1. Instead of an expected value over the distribution of possible sales to a company, which is an averaged value, a percentile could be used, for example the 75-th percentile. Instead of any positive sale, a constant threshold could be specified to reflect minimum satisfactory sales goals.

The real-world incarnation of this problem is expressed as a sample of company data over a designated feature set. For current customers, the actual sales are known. In our application, information is available for most companies of moderate or large size. For these companies, we know the number of employees, their revenue, and their industry code. We can assemble labeled data about companies, matching company features to actual product sales. To estimate sales to a company, whether expected or potential, a nearest-neighbor method will find similar companies. The sales to these similar companies can be used as a proxy for a sample from the distribution of possible sales to that company. Once similar companies can be identified, estimates for opportunity can readily be computed from recorded sales to these similar companies using equation 1 or an optimistic variation substituting percentiles or thresholds.

## 3 Similarity-Based Methods

Similarity-based methods, nearest neighbors, can offer highly competitive results with other predictive methods. They have strong theoretical bounds on predictive performance. In practice, the similarity or distance function may need extensive tuning to achieve good results. Prediction methods are typically compared and evaluated by measuring their accuracy on sample data. For structured data and very large numbers of examples, alternative prediction methods have often been preferred because they can be applied by invoking a compact function without online access to the original examples. With more powerful computing, similarity-based prediction methods become more enticing for embedding in web-based systems. The similarity-based methods are relatively slow – each new case must be compared to the stored cases. Some new computational approaches can substantially reduce computing time[1]. Alternatively, for some systems, the predictions may be computed and stored prior to invoking the system, thereby reducing the time observed by a system user.

In modern similarity-based systems, for example search or information retrieval systems, users have access to a huge set of examples. These are specialized operating environments, where examples are sparse within the feature space, for example, a word representation within a document, where most words do not occur in a specific document or the search string. For predictive purposes, not all data may be needed, and methods for simplification have evolved from condensed nearest neighbors [2] to conditional random sampling [3]. Still, from a systems perspective, it is important to have access to full data. While this is critical for document retrieval, it is also essential for structured data management systems containing customer information that must be accessible in full detail.

Many commercial applications of similarity methods fall into the category of collaborative filtering methods and recommender systems[4], [5]. From a methodological perspective, our application is closely related to those systems, while maintaining strong explanatory capabilities[6]. There are numerous twists to our application that distinguish it from recommender systems. Here, products are not recommended directly. Rather, an estimate of opportunity for product sales is provided that can range from zero to millions of euros. The sales representative is an intermediary, not a purchaser, who decides whether to pursue an opportunity based on many additional factors including personal knowledge of the customer or industry. Unlike most collaborative filtering applications, like those based on product ratings, the input features are nonuniform and are presented on many different scales from binary to real-valued. Because our target is opportunity, rather than an outright purchase, the exact value of the target variable, i.e. the opportunity, is unknown for all members of the customer database.

## 4 Data Description and Data Transformation

In our application, customers are companies. We obtained data on American and Canadian companies from Dun and Bradstreet. This includes such fields as number of employees, revenue, and an industry code. Extracting data from a proprietary internal database, we consider the sales of 5 IBM brands and the sum of sales for the five brands. These brands constitute most of IBM's software sales. Table 1 summarizes the dimensions of the data we used in all experiments.

**Table 1.** Data Summary

| | |
|---|---|
| total inputs | 16 |
| numericals | 14 |
| codes | 2 |
| number of code values | 3 and 26 |
| outputs | 6 |
| total records | 650,110 |
| non-customers | 612,677 |
| customers | 37,433 |

The feature space is relatively modest, but the number of examples is large. This can be a good environment for a nearest neighbors method because of the absence of large numbers of noisy features that can degrade predictive performance [7] [8]. One of the main challenges of designing a similarity function is to compute similarity on a normalized scale when the features are all measured in different units, for example revenue versus employee counts. One general approach is to weight the features by importance. Both linear [9] and nonlinear approaches have been described [10]. Examples of classification applications using weighted similarity functions are [11] and [12]. A generalized and automated scheme for weighting features is highly advantageous, but the best results may only be achievable by experimentation and application-specific tuning.

We employed a simple, yet highly effective transformation. All features were transformed to binary variables as follows: The codes are mapped into n binary variables, where n is the number of distinct codes. All continuous numerical variables are binned into 4 intervals as determined by k-means clustering. The net result is the transformation of the data from 16 features into 85 binary input features (14*4+3+26). Now all input features are on the same binary scale. The quality of the similarity function remains a question, but the similarity function is completely transparent. Just rank the neighbors by counting the number of matches between the new vector and the stored vectors.

To test hypotheses, we randomly divide the customer data into 70% train and 30% test. In our results, we report on the test data. In our experiments, we stratified the data into 5 groups based on the total sales to make sure that the smaller number of very large sales were represented in both train and test samples. The customer data will have labels in the form of actual sales. The non-customers have zero sales across all products, and estimates will be inferred by comparing to the smaller customer population.

# 5 System Evaluation

## 5.1 Evaluating Predictive Performance

While our major interest is examining factors other than predictive accuracy, we must still consider whether the similarity-based method is competitive with alternatives. We know that our primary target for estimation, the sales opportunity, is not directly known. It is highly related to expected or actual sales which are known for current customers. For a given company's set of similar companies, the computation of expected sales versus sales opportunity differs only in the thresholds or designated percentiles for equation 1. We will use these targets in a sample of companies and their actual sales. Our immediate goal is to see whether a nearest-neighbors method will produce predictions that are close in accuracy to results produced by prominent alternative methods. This goal should not require that nearest neighbors produce the best results for opportunity estimates. But, it does require that the nearest-neighbor approach be competitive to other methods when using the same features.

We compared results for several regression methods using the previously described train and test data sets. Figure 1 summarizes the results on the test examples. The results for prediction were measured by *rmse*, a measure of difference between actual and predicted sales. In the graph, the RMSE values are plotted relative to the baseline RMSE of ten nearest neighbors, 10-NN. For k-NN methods, the average of the k nearest neighbors is taken as the prediction.

The methods that we compared to nearest neighbors, k-nn, were linear least squares, a regression tree, and bagged trees. We also include the results of simply using the mean value of product sales computed from training cases. The linear method was run with all features. The trees were checked at various sizes and k-nn was tested with various values of $k$. Results for k-nn were slightly better with feature selection (FS), when a few features were deleted by greedy backward elimination on the training data.

Clearly, the k-nn method is competitive and effective. The linear method, which is highly popular in industrial circles for predicting sales, was relatively weak. Bagging and k-nn were competitive. Nearest neighbor, and other similarity-based methods, have a rich history of effectiveness in many applications. These results support its use when an appropriate similarity measure can be developed for the application.

## 5.2 Describing and Measuring Optimism

One view of opportunity is the expected sale magnitude given a sale occurs. This was expressed in Equation 1. A company is trying to gauge potential for sales prior to committing additional resources. The hope is that these additional resources, whether additional preparation or a stronger sales team may improve the yield from past experiences, or they may entice a new customer with a better deal.

The degrees of overall optimism can be specified in several ways. In Equation 1, the threshold can be increased to some non-zero threshold, $m$, as in Equation 2.

$$opportunity = Average(sale|sale > m) \tag{2}$$

Increasing the threshold will evaluate customers as if a bigger sale will be closed. It treats the sales below the threshold as zero. Generally, this increases overall opportunity estimates unless raised to the point of non-occurrence of sales.

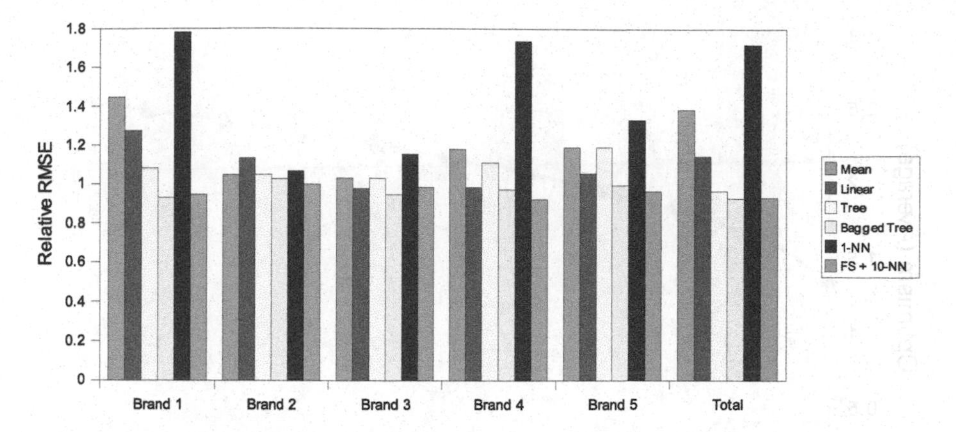


**Fig. 1.** Comparison of Methods for Sales Prediction

An alternative and complementary approach to varying optimism is to replace the "average" with another measure. The median, i.e. the 50th percentile value, is an obvious choice, but the concept can be generalized to any percentile. We have computed values at the 75th and 50th percentile for this application. The average has the highest optimism among these measures. That's because a few extremely large customers, like banks and insurance companies, have much larger purchases than most customers. These very large customers are typically at the tail of the sales distribution. Figure 2 shows the relative ratios to expected sales (i.e. no minimum threshold) for the average-value as a measure of optimism of this application. These are shown for exceeding several threshold values, $m$, of Equation 2 for current customers only. The average is contrasted with the 75th percentile and the 50th percentile in Figure 3 where the optimism at two thresholds is shown. As expected, the average has the highest value of the three. The maximum value could also be used as a measure of opportunity. It does not change with threshold, so the optimism measure would be unchanged over the thresholds. In our application, the optimism as measured by the maximum value was 37.39 – substantially higher than any of the other measures.

The degree of optimism is best decided by knowledge and goals set by the sales executives. Alternatively, the system designers can examine the effects of various measures on global optimism, and then make decisions accordingly. We want optimism, but we do not want to set the bar so high that it is unrealistic. How can we measure the global effect of optimism? Equation 3 describes a simple ratio for yearly sales of a product. The overall optimism is the ratio of total opportunity to total current sales.

$$optimism = opportunity/currentSales \tag{3}$$

All these computations are trivial for a similarity-based method. The k-most similar companies are identified and grouped. This can be a relatively small group, in our case, as small as 10 or 20 similar companies. Once grouped, almost no additional effort is needed to estimate opportunity by any of the measures

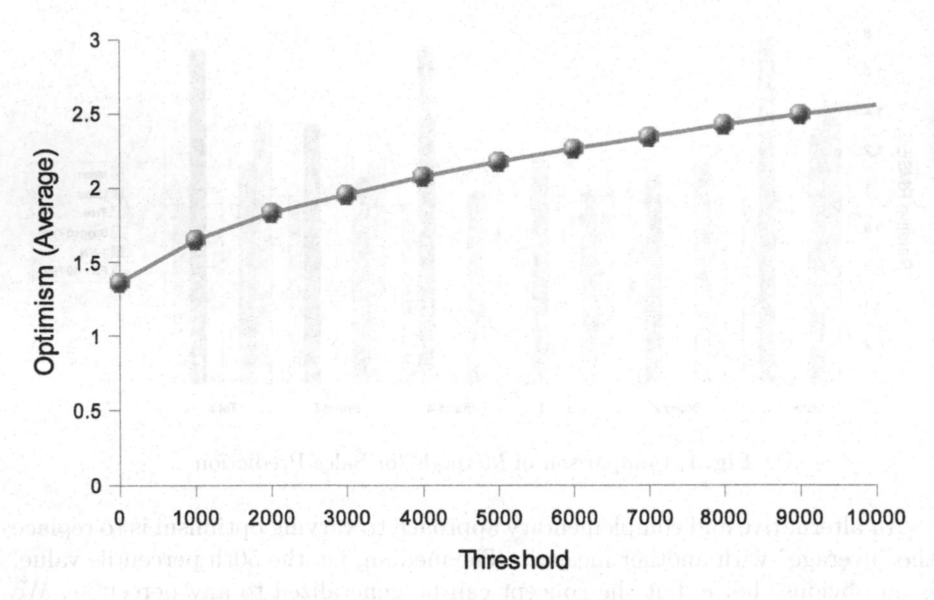


**Fig. 2.** Average as a Measure of Optimism

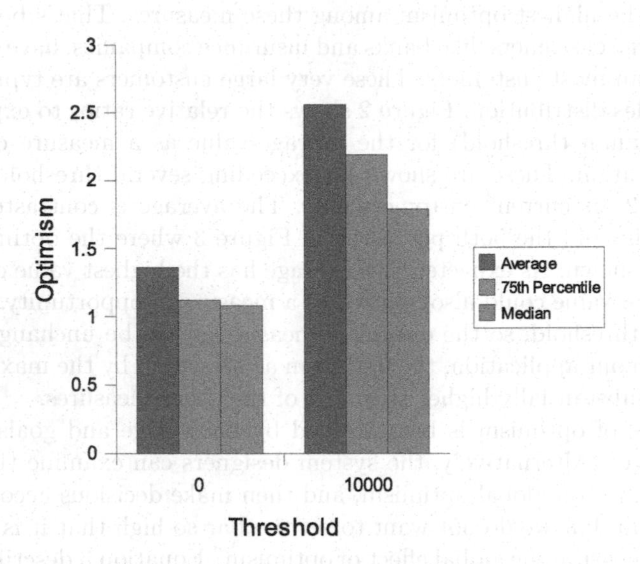


**Fig. 3.** Comparison of Optimism Measures

of optimism. Because our notion of opportunity is based on similar companies, we add a floor to the estimate as in Equation 4. Opportunity for selling to a company can be no less than current sales to that company.

$$finalOpportunity = Max(initialOpportunity, currentSales) \quad (4)$$

## 5.3 Explanatory Capabilities

The ultimate consumers of our opportunity estimates are sales representatives. They will make decisions on sales efforts by examining these numbers in the context of current sales results. An analogy can be made with a search engine and its rankings. There, the rankings are a means to presenting answers to a user. Multiple answers are listed, all of them similar to the query. The objective is to rank based on similarity criteria, but many answers are presented to the user. The user will examine results and make decisions about the usefulness of the retrieved documents. The quality of the answer depends both on the similarity rankings and the actual answers.

We could look narrowly at our sales opportunity task and conclude that the estimates are the sole conveyor of meaning in this application. Yet that would ignore an important component of the estimates. Just like the answers from the search engine, the actual answers, the most similar companies, also have meaning. They explain how the estimate was derived, but just as important, can aid the sales representative in moving forward. For example, when a sale is concluded in a particular industry, it is natural to look for similar companies to make a similar sales pitch. Or even if a sale was not closed and was won by a competitor, it could be expedient to examine similar possibilities.

Because company records are structured with precise values, additional unstructured information may be missing. Once the estimates are derived and similar companies are found, all supporting materials could be examined to formulate a sales plan. The list of similar companies could be considered a form of triage, where these companies are examined carefully to look for additional clues, possibly not recorded in the structured record, that could lead to success in the sales effort. The key concept is that the task need not be just the estimation of sales opportunity. Rather, the component records of companies used to estimate sales opportunity can be most valuable in supporting both the estimate and the planning for future actions.

Using similarity measures, our objectives can be restated as illustrated in Figure 4. Estimating sales opportunity is still a principal goal. To this, we add the presentation of the constituent companies used to formulate the estimate. These can be ordered by similarity as would be done for a search engine. Depending on user queries, full information about these companies and their prior purchases could be made available.

---

- estimate sales opportunity
- order similar companies
- present k most similar companies with complete sales histories

**Fig. 4.** Objectives for Similarity Measures

---

### 5.4 Compatibility with Knowledge

The general approach is to build an automated system. The system will process empirical data about companies and estimate their sales opportunities. After assembling a feature set based on availability and knowledge, we might expect to make this a purely empirical application of similarity-based methods. Yet, we have strong incentives to incorporate knowledge into the system. Two considerations drive our preference for additional knowledge:

1. Sales opportunity is unknown, making our data unlabeled.
2. Presenting a list of similar companies to a user implies consistency with intuitive knowledge of the field.

Enhancing the system with knowledge can constrain the answers to be less variable and more compatible with human expectations, possibly with a reduced accuracy when tested on data sampled from a specific time period. Still, mating knowledge to an empirically-driven system can be satisfying for this application because the consumers are humans who must take action. They will not blindly accept numbers that are incompatible with their experience and might defeat their sales approach.

What kind of knowledge is of interest, and how well does it fit with a similarity-based method? With hundreds of thousands of examples, a case-based reasoning[13] approach to specifying the examples is ill-suited. Our main vehicle for incorporating knowledge is the similarity function. A company may be represented as a vector of data, but the components of that vector also have meaning. When we say that two companies are similar, it must also meet our intuitive notion of similarity. For example, we may not feel that two companies are similar when they engage in completely different industries. Adding conditions that enhance the intuitive notion of similarity will meet the objective of effective explanations to the sales representatives. They will review the matched companies, and various constraints on the matches will increase satisfaction or at least remove some answers that might seem inadmissible. Two simple notions are very sensible, yet readily fit within a function for computing similarity of two companies. If two companies are similar, then

1. they are in the same industry
2. they are approximately the same size.

The industry code for a company is readily obtainable from Dun and Bradstreet or other sources at different resolutions. We used a single letter code with 25 possibilities. For size, both the number of employees and company revenue can be obtained. A window on size can be placed on the compatibility of two companies. We use a heuristic that a company Y must have sales (or employees) no more than twice that of company X for it to be considered a match.

Figure 5 illustrates the difference in predictive performance for predicting actual sales with these heuristics. We show the performance of four simple heuristics and compare with the baseline where no match heuristics are used. For sales

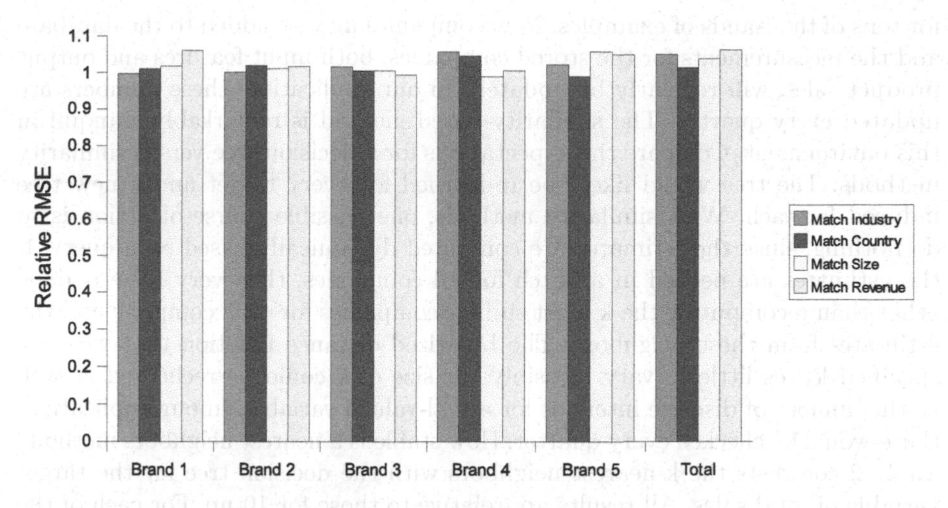


**Fig. 5.** Impact of Match Heuristics

and size matching, we clustered values into four bins (determined by training data) and declared a match if values fell in the same bin. The performance drops slightly on the test data, mostly from the required industry match. However, the matches themselves would be more intuitive and more acceptable to the sales representatives who must take the information and call on customers, while often citing purchases by their competitors.

The records of companies are not extracted from a single data source. Rather, the information is found in different sources and merged. In our application, we consider the sales of five product lines. For each company the proprietary sales numbers for these five products are merged with other information about these companies, such as sales revenue and number of employees. Although companies can be stored with unique identifiers, for example, "duns numbers," errors are unavoidable, both in the record matching process for assembling data or in the data sources. Knowledge of the relationships among variables leads to other heuristic rules for adjusting the numbers used for estimation. For example, given a company has revenue X, we can estimate that no more than 10% of X will be spent on information technology purchases. Although there may be exceptions to this rule, generally it is effective because a higher estimate will not serve as a good template for similar expectations.

## 5.5 Ease of Maintenance

Because a similarity-based method does not directly model labels and does not build any intermediate structure (like a tree), system maintenance is greatly simplified. No need to learn a distinct model for each target. This reduces the problem to computing similarity from the initial set of input features. In our application, the set of features will remain stable. It's very unlikely or at least infrequent that new features will be found that are useful and can be measured for tens of thousands of examples. New companies may be added to the database and the measurements for the stored companies, both input features and output product sales, will regularly be updated. In our application, these numbers are updated every quarter. The similarity-based method is remarkably tranquil in this environment. Compare the expectations for a decision tree versus similarity methods. The tree would likely be re-learned for every target and a new tree induced for each. With similarity methods, one possible course of action is to do nothing, since the estimates are computed dynamically based on a query. If the estimates are needed in a batch for all companies, then very little is done other than recomputing the k most similar companies for each company and the estimates from these neighbors. The binarized distance function that we have specified leaves little to vary. Possibly the size of k could be rechecked as well as the number of discrete intervals for a real-valued variable. In our application, these would be checked every quarter. How stable is a nearest neighbors method? Table 2 contrasts the k-nearest neighbors with the decision tree for the target variable of total sales. All results are relative to those for 10-nn. For each of the n companies in the database, we measure the standard error within the partition for the tree and form the weighted average over all companies as in Equation 5 where $StdDev_i$ for selected tree = standard deviation over all members of partition $i$, and $StdDev_i$ for k-nn = standard deviation across all members of the k neighbors of $i$.

$$stability = \sum_{i=1}^{n} StdDev_i \tag{5}$$

**Table 2.** Stability of tree and k-NN

| Method | Relative Average Mean | Relative Stability |
|---|---|---|
| Tree | 1.5512 | 9.0718 |
| 10-NN | 1.0 | 1.0 |
| 20-NN | 0.8794 | 0.7855 |
| 50-NN | 0.7142 | 0.5531 |

Here we see that the nearest neighbors estimates are far more stable than those for the tree. While the accuracy of results are important, the stability that the similarity-based method is also desirable, since it would need less maintenance and less frequent retraining.

### 5.6 Computational Efficiency

Similarity methods can be quite slow. Given a training set, the simplest variation compares each new example to the full training set, an $O(n)$ process, where $n$ is the size of the training set. To get estimates for all stored companies, the process is $O(n^2)$. Fortunately, when embedded within a system for computing sales opportunity, various factors ameliorate this disadvantage. The fastest algorithms for computing nearest neighbors offer a large improvement over direct comparison. While the high computational cost for finding similar companies is a criticism of nearest neighbors, several critical characteristics of our system, discussed in the previous sections, lessen its importance. Some of these factors may even make nearest neighbors computationally advantageous when compared to other methods.

The results of similarity computations are reusable. For any given company, we need to know its estimated sales opportunity, which is computed from similar companies. However, we need not recompute these number dynamically every time a company is referenced. The possible list of queries is limited to single companies, so we know every possible query, and the estimates can be precomputed and stored. When a user queries a company, the estimates can be directly retrieved. Thus, the more intensive computation is separated from its repeated use, and values can be computed in batch mode with no ramifications for user delays.

The central comparison is "Compare company X to current customers." We want to estimate opportunity for both current customers and potential new customers. The set of potential new customers, sometimes referred to as whitespace, is generally much larger than current customers. In our case, 20 times larger. Given $j$ customers and $k$ new customers, all estimates are made by comparing to the $j$ customers. If sequential comparisons are made, the number of comparisons is $j * (j + k)$. The number of comparisons is substantially less than suggested by the full number of companies in the sample.

A major advantage of the similarity methods over many alternative learning methods is its inherent parallelism in computing estimates for multiple targets. Most learning methods rely on training to a specific target, the label; a separate solution is induced for each target. For example, a separate tree would be induced for each label. In our system, 5 products would require 5 different trees and possibly another for total sales. The estimates computed by similarity can be described by following simple two-step algorithm:

1. Find k most similar companies to company X.
2. From the k companies of step 1, compute all estimates.

For example, if expected sales were to be computed, then the estimates are the average sales for each product to the k companies . To compute the opportunity, an estimate for opportunity in Equation 1 is the average of the positive values for each product. Because of the little work in computing the estimates from the product values of the k companies, the number of products is almost inconsequential to the computations. One thousand products will take just a little more time than one product. In all our results, the same value of k was used over all companies and products.

In our application, we have a mixture of categorical and real-valued features. A similarity or distance function can benefit from re-scaling the variables. Earlier, we described our data transformations, where the original features are all mapped into binary variables. This is straightforward for the categorical variables, and it requires some slight additional computations for the real-valued variables. The transformed data had 85 binary features from which similarity is computed.

This type of binary representation has potential for very efficient computation. Consider the simple similarity function that we actually used:

$$similarity = Count(matchingOnes(X, Y)) \quad (6)$$

We measure similarity as the count of matching ones between the company X and a candidate neighbor, company Y. Because the features are binary-valued, the counts can be computed like words in text for information retrieval (IR). There, for each feature, i.e. a word, token or phrase, an inverted list of records with a positive feature occurrence is computed. When a new search pattern of words is entered, only the inverse lists for these words are accessed to compute similarity. In our application, the search space will not be as sparse, but efficiencies can still be great. When computing the similar companies for company X, only the positive features in the vector for X will be considered. For each of the positive features, the companies appearing in the inverted list will have their matching count incremented by one. This simple algorithm is typical for IR and can be readily adapted to our similarity computations as describe below, where inverse(i,j) is the j-th company on the inverse list of feature i.

1. Initialize counts all training cases to zero.
2. For each positive feature i in company X
   For each j such that inverse(i,j) exists, increment count(j).
3. Select top k counts.

One additional computational efficiency is worth noting. A suitable value for k must be determined. This is determined both empirically and based on knowledge of the domain. Our data support a small value of k, on the order on 10 or 20. All our results are for 10. With such a small value of k, the 10 largest values can be found by dynamically keeping a simple linear table of the 10 current best companies and adjusting when a new insertion is made.

## 6 Discussion

We have described an important application, where opportunities for sales to companies are estimated. The end-user of these estimates is most likely a sales representative who will call on customers. These sales representatives need an estimate that captures the potential for sales under optimistic conditions. The optimism could be a variation of the concept of "how much product could be sold if a sale is made?"

In a situation with human involvement, such as this one with sales representatives, merchants may move away from "expected" value to a "potential" perspective. This allows for optimism and separates the past from the future. A more concerted effort may result in improved sales. However, this effort will only be expended when justified by potential sales.

A similarity-based method is a good fit to this application. All necessary company data is kept online, so no additional resources are devoted to this representation. The method itself has a very strong intuitive appeal for promoting sales initiatives. The process of estimation is not arcane. Home appraisals have long been made by comparing a new home on the market to similar homes that have recently sold. Here, we compare potential product sales to a customer with those sales made to similar customers.

The target of the estimate is sales opportunity which is an unknown quantity. We offered a formulation of the problem that circumvents the need for labeled data and bootstraps from the available prior sales records. We offered several ways to quantify and adjust the optimism in these estimates. Because the training data are unlabeled (technically, the opportunity is not known), but still related to the known and labeled historical sales data, there is room for alternatives to our approach to estimating and predicting opportunity. One related approach for regression is semi-supervised learning using nearest neighbors[14], where both labeled and unlabeled samples are collected. Here our data are labeled, but the actual target, opportunity, is related, but not identical to the label.

From an applications perspective, this project is a success. The system, as described in this paper, has been implemented and deployed. The estimates are embedded in a system that is routinely accessed by hundreds of sales representatives. Starting with just the USA sales, the system has gradually been expanded to cover two dozen countries including most of Europe and the larger countries in the Asia-Pacific region. These estimates are updated worldwide every quarter by geographical region. Besides providing estimates for sales opportunity, the system can function on a query basis, giving details of similar companies and their transactions. Studies of the users' clicks relative to completed sales show strong usage on the part of the sales force, and executives estimate that over $100M of sales are traceable to the system.

From an evaluation perspective, we showed that nearest neighbors was a highly competitive approach. This evaluation could follow the typical scenario of considering machine learning methods as interchangeable building blocks and basing decisions on predictive performance. In our study, that goal remains important, and it is no surprise that k-nearest neighbors do well in this application, even when implemented in a binary fashion.

Our goals for evaluation are more general. The prediction method is embedded in a system that will function in a structured environment, yet can respond to online queries. This leads us to consider an evaluation along dimensions other than empirical and predictive performance. Such factors as compatibility with knowledge, explanatory capabilities, stability, computational parallelism, and quantification of sales optimism can all play important roles in a real-world system's architecture. We have shown that a similarity-based system can perform well along these dimensions and have supplied empirical evidence and practical arguments to support these claims. We have provided the case for similarity-based systems but an alternative approach may also do well along these very same dimensions[15]. In an age of building systems with embedded learners and predictors, one can expect that factors beyond pure empirical performance will play a larger role in the design of applied systems.

## References

1. Beygelzimer, A., Kakade, S., Langford, J.: Cover trees for nearest neighbor. In: ICML 2006: Proceedings of the 23rd International Conference on Machine Learning, pp. 97–104. ACM Press, New York (2006)
2. Hart, P.: The condensed nearest neighbor rule. IEEE Trans. Info. Theory IT-14, 515–516 (1968)
3. Li, P., Church, K., Hastie, T.: Conditional random sampling: A sketch-based sampling technique for sparse data. In: Schölkopf, B., Platt, J., Hoffman, T. (eds.) Advances in Neural Information Processing Systems 19. MIT Press, Cambridge (2007)
4. Herlocker, J.L., Konstan, J.A., Terveen, L.G., Riedl, J.T.: Evaluating collaborative filtering recommender systems. ACM Trans. Inf. Syst. 22(1), 5–53 (2004)
5. Adomavicius, G., Tuzhilin, A.: Toward the next generation of recommender systems: A survey of the state-of-the-art and possible extensions. IEEE Transactions on Knowledge and Data Engineering 17(6), 734–749 (2005)
6. Tintarev, N., Masthoff, J.: A survey of explanations in recommender systems. In: Workshop on Recommender Systems and Intelligent User Interfaces associated with ICDE 2007 (2007)
7. Fukunaga, K., Hummels, D.M.: Bias of nearest neighbor estimates. IEEE Trans. Pattern Anal. Mach. Intell 9(1), 103–112 (1987)
8. Beyer, K., Goldstein, J., Ramakrishnan, R., Shaft, U.: When is "nearest neighbor" meaningful? In: Beeri, C., Bruneman, P. (eds.) ICDT 1999. LNCS, vol. 1540, pp. 217–235. Springer, Heidelberg (1998)
9. Aggarwal, C.C.: Towards systematic design of distance functions for data mining applications. In: KDD 2003: Proceedings of the 9th ACM SIGKDD International Conference on Knowledge Discovery and Data Mining, pp. 9–18. ACM Press, New York (2003)
10. Wu, G., Chang, E.Y., Panda, N.: Formulating distance functions via the kernel trick. In: KDD 2005: Proceeding of the 11th ACM SIGKDD International Conference on Knowledge Discovery and Data Mining, pp. 703–709. ACM Press, New York (2005)
11. Yao, Z., Ruzzo, W.L.: A regression-based K nearest neighbor algorithm for gene function prediction from heterogeneous data. BMC Bioinformatics 7 (2006)
12. Yang, S., Zhang, C.: Regression nearest neighbor in face recognition. In: Proceedings of the 18th International Conference on Pattern Recognition, pp. 515–518 (2006)
13. Kolodner, J.: Case-based Reasoning. Morgan Kaufmann, San Francisco (1993)
14. Zhou, Z., Li, M.: Semi-supervised regression with co-training. In: IJCAI 2005: Proceedings of the 19th International Joint Conference on Artificial Intelligence, pp. 908–913 (2005)
15. Merugu, S., Rosset, S., Perlich, C.: A new multi-view regression approach with an application to customer wallet estimation. In: KDD 2006: Proceeding of the 12th ACM SIGKDD International Conference on Knowledge Discovery and Data Mining, pp. 656–661. ACM Press, New York (2006)

# Learning MDP Action Models Via Discrete Mixture Trees

Michael Wynkoop and Thomas Dietterich

Oregon State University, Corvallis, OR 97370, USA
{wynkoop,tgd}@eecs.oregonstate.edu

**Abstract.** This paper addresses the problem of learning dynamic Bayesian network (DBN) models to support reinforcement learning. It focuses on learning regression tree (context-specific dependence) models of the conditional probability distributions of the DBNs. Existing algorithms rely on standard regression tree learning methods (both propositional and relational). However, such methods presume that the stochasticity in the domain can be modeled as a deterministic function with additive noise. This is inappropriate for many RL domains, where the stochasticity takes the form of stochastic choice over deterministic functions. This paper introduces a regression tree algorithm in which each leaf node is modeled as a finite mixture of deterministic functions. This mixture is approximated via a greedy set cover. Experiments on three challenging RL domains show that this approach finds trees that are more accurate and that are more likely to correctly identify the conditional dependencies in the DBNs based on small samples.

Recent work in model-based reinforcement learning uses dynamic Bayesian network (DBN) models to compactly represent the transition dynamics of the actions and the structure of the reward function. DBN models require much less space than tabular models (Dean & Kanazawa, 1989), and they are able to generalize to novel parts of the state space. Additional compactness can be obtained by representing each conditional probability distribution by a regression tree (Boutilier et al., 1995), a structure we will refer to as a TDBN. Boutilier and colleagues have developed a family of approximate value iteration and policy iteration algorithms that manipulate tree-structured representations of the actions, the rewards, and the value functions (Boutilier et al., 2000).

An additional advantage of DBN representations is that they explicitly identify which state variables at time $t$ influence the state variables at time $t+1$. By analyzing the structure of such dependencies, it is possible to identify legal state abstractions in hierarchical reinforcement methods such as MAXQ (Dietterich, 2000). In recent work, Jonsson and Barto (2006) and Mehta, et al. (2007) have shown how to automatically discover subroutine hierarchies through structural analysis of the action and reward DBNs.

Algorithms for learning TDBNs generally employ the standard set of techniques for learning classification and regression trees (Breiman et al., 1984). Internal nodes split on one or more values of discrete variables or compare continuous values against a threshold. If the target variable is discrete, a classification tree

W. Daelemans et al. (Eds.): ECML PKDD 2008, Part II, LNAI 5212, pp. 597–612, 2008.

is constructed (Quinlan, 1993), and each leaf node contains a multinomial distribution over the values of the target variable. One variation on this is to search for a decision graph (i.e., a DAG, (Chickering et al., 1997)). Search is typically top-down separate-and-conquer with some form of pruning to control overfitting, although Chickering et al. (Chickering et al., 1997) employ a more general search and control overfitting via a Bayesian scoring function. If the target variable is continuous, a regression tree is constructed. Each leaf node contains a Gaussian distribution with a mean and (implicitly) a variance (Breiman et al., 1984).

Many generalizations of the basic methods have been developed. One generalization is to allow the splits at the internal nodes of the tree to be relational (e.g., by evaluating a predicate that takes multiple variables as arguments or by evaluating a function of one or more variables and comparing it against a threshold (threshold Kramer, 1996; Blockeel, 1998)). Another is to allow the leaf nodes of regression trees to contain regression models (so-called *Model Trees*; Quinlan, 1992) or other functions (Torgo, 1997). Gama's (2004) Functional Trees combine functional splits and functional leaves. Vens et al. (2006) combine relational splits with model trees.

It is interesting to note that for discrete random variables, the multinomial distribution in each leaf represents stochasticity as random choice across a fixed set of alternatives. However, in all previous work with regression trees, each leaf represents stochasticity as Gaussian noise added to a deterministic function.

In many reinforcement learning and planning problems, this notion of stochasticity is not appropriate. Consider, for example, the $GOTO(agent, loc)$ action in the real-time strategy game Wargus (2007). If the internal navigation routine can find a path from the agent's current location to the target location $loc$, then the agent will move to the location. Otherwise, the agent will move to the reachable location closest to $loc$. If we treat the reachability condition as unobserved, then this is a stochastic choice between two deterministic outcomes, rather than a deterministic function with additive Gaussian noise. Another case that arises both in benchmark problems and in real applications is where there is some probability that when action $a$ is executed, a different action $a'$ is accidentally executed instead. A third, more mundane, example is the case where an action either succeeds (and has the desired effects) or fails (and has no effect).

The purpose of this paper is to present a new regression tree learning algorithm, DMT (for Discrete Mixture Trees), that is appropriate for learning TDBNs when the stochasticity is best modeled as stochastic choice among deterministic alternatives. Formally, each leaf node in the regression tree is modeled as a multinomial mixture over a finite set of alternative functions. The learning algorithm is given a (potentially large) set of candidate functions, and it must determine which functions to include in the mixture and what mixing probabilities to use. We describe an efficient algorithm for the top-down induction of such TDBNs. Rather than pursuing the standard (but expensive) EM-approach to learning finite mixture models (McLachlan & Krishnan, 1997), we instead apply the greedy set cover algorithm to choose the mixture components to cover the data points in each leaf. The splitting heuristic is a slight variation of the standard mutual information (information gain) heuristic employed in C4.5 (Quinlan, 1993).

We study three variants of DMT. The full DMT algorithm employs relational splits at the internal nodes and mixtures of deterministic functions at the leaves (DMT). DMT-S ("minus splits") is DMT but with standard propositional splits. DMT-F ("minus functions") is DMT but with constant values at the leaves. We compare these algorithms against standard regression trees (CART) and model trees (M5P). All five algorithms are evaluated in three challenging domains. In the evaluation, we compute three metrics: (a) root relative squared error (RRSE; which is most appropriate for Gaussian leaves), (b) Recall over relevant variables (the fraction of relevant variables included in the fitted model), and (c) Precision over relevant variables (the fraction of the included variables that are relevant). The results show that in two of the domains, DMT gives superior results for all three metrics. In the third domain, DMT still has better Recall but produces mixed results for RRSE and Precision.

## 1 Tree Representations of DBNs

Figure 1(a) shows a DBN model involving the action variable $a$, three state variables $x_1, x_2, x_3$, and the reward value $r$. In this model (and the models employed in this paper), there are no probabilistic dependencies within a single time step (no synchronic arcs). Consequently, each random variable at time $t+1$ is conditionally independent given the variables at time $t$. As always in Bayesian networks, each node $x$ stores a representation of the conditional probability distribution $P(x|\mathbf{pa}(x))$, where $\mathbf{pa}(x)$ denotes the parents of $x$.

In this paper, we present a new algorithm for learning functional tree representations of these conditional probability distributions. Figure 1(b) shows an example of this representation. The internal nodes of the tree may contain relational splits (e.g., $x_2(t) < x_3(t)$) instead of simple propositional splits (e.g., $x_2(t) < 1$). The leaves of the tree may contain multinomial distributions over functions. Hence the left leaf in Figure 1(b) increments $x_1$ with probability 0.7 and decrements it with probablity 0.3.

There are many ways in which functional trees provide more compact representations than standard propositional regression trees (Figure 1(c)). First, relational splits are much more compact than propositional splits. To express the condition $x_2(t) < x_3(t)$, a propositional tree must check the conjunction of $x_2(t) < \theta$ and $x_3(t) \geq \theta$ for each value of $\theta$. Second, functional leaves are more compact than constant leaves. To express the leaf condition $x_1(t+1) := x_1(t) + 1$, a standard regression tree must introduce additional splits on $x_1(t) < \theta$ for each value of $\theta$. Finally, standard regression trees approximate the distribution of real values at a leaf by the mean. Hence, the left-most leaf of Figure 1(c) would be approximated by the constant 0.4 with a standard deviation (mean squared error) of 0.92.

This compactness should generally translate into faster learning, because in the functional trees, the data are not subdivided into many "small" leaves. However, if the learning algorithm must consider large numbers of possible splits and leaf functions, this will introduce additional variance into the learning process which could lead to overfitting and poor generalization. Hence, to obtain the

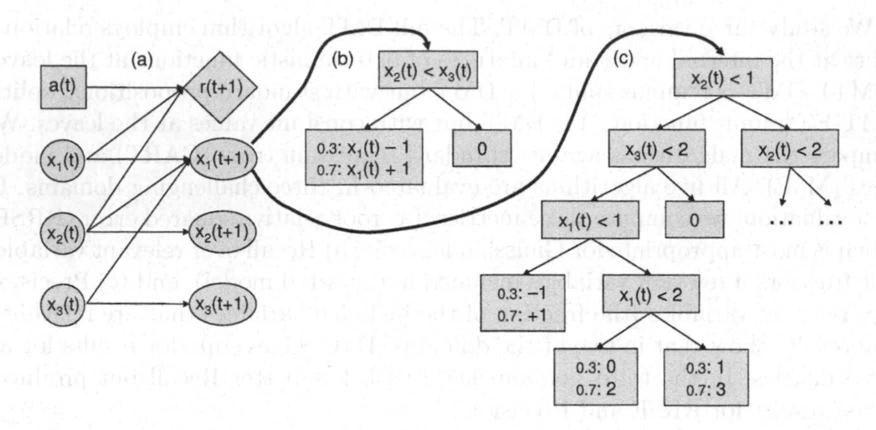


**Fig. 1.** (a) time slice representation of the DBN. The square action node $a(t)$ affects all nodes at time $t+1$, but for readability those arcs have been omitted. Circles represent state variables, and the diamond is the reward node. (b) a tree representation for $P(x_1(t+1)|x_1(t), x_2(t), x_3(t))$ with relational internal nodes and a probability distribution over functions ($x_1(t+1) := x_1(t)+1$ and $x_1(t+1) := x_1(t)-1$) in the left leaf. (c) a tree with propositional nodes and constant leaves must be much more copmlex to represent the same conditional probability distribution.

benefits of functional trees, the engineer must identify a constrained set of candidate relational splits and functional leaves. We adopt a proven approach from inductive logic programming (Lavrac & Dzeroski, 1994) and specify these candidate splits and functions for each domain via a context-free grammar.

Because different actions exhibit different probabilistic dependencies, all work in tree-based DBN learning—including our own—learns a separate set of regression trees for each action.

As mentioned above, other researchers have studied regression trees with relational splits and functional leaves. Our contribution is to extend these to handle multinomial mixtures of functions in the leaves.

## 2 Algorithm

To construct a regression tree for $x_i(t+1)$, we follow the standard recursive top-down divide-and-conquer approach using the values of $x_1(t), \ldots, x_n(t)$ as the input features and $x_i(t+1)$ as the response variable. However, we introduce two modifications. First, given a set of $N$ values for $x_i(t+1)$ (i.e., at a leaf), we fit a mixture of functions by applying the well-known greedy set cover algorithm (Johnson, 1973). That is, we score our candidate leaf functions according to the number of training values that they fit and choose the function that fits the most points. Those points are then removed from consideration, and the process is repeated until all points are covered. The result of the set cover is a list of the form $((f_1, n_1), (f_2, n_2), \ldots, (f_k, n_k))$, where each $f_j$ is a function and $n_j$ is the number of data points covered by $f_j$ that were not covered by functions $f_1, \ldots, f_{j-1}$. We then estimate the multinomial distribution as $P(f_j) = n_j/N$.

This approach introduces two approximations. First, greedy set cover is not optimal set cover (although it does give very good approximations; Slavík, 1996). Second, there may be points that are consistent with more than one of the functions $f_1, \ldots, f_k$. Strictly speaking, the probability mass for such points should be shared equally among the functions, whereas we are assigning it to the first function in the greedy set cover. In our application problems, this second case occurs very rarely and typically only affects one or two data points.

Our second modification concerns the loss function to use for scoring candidate splits. Virtually all regression tree algorithms employ the expected squared error of the children and choose the split that minimizes this squared error. This is equivalent to assuming a Gaussian likelihood function and maximizing the expected log likelihood of the training data. If we followed the same approach here, we would score the expected log likelihood of the training data using the multinomial mixture models. However, this does not work well because we assume that the mixture components (i.e., the individual functions) are themselves deterministic, so if a leaf node contains a single function with assigned probability of 1, the log likelihood of a data point is either 0 (if the function matches a data point) or $-\infty$ (if it does not). This leads to a very non-smooth function that does not work well for scoring splits. Instead, we adopt the approach that has worked well for learning classification trees (Quinlan, 1993): we score each candidate split by the expected entropy of the probability distributions in the leaves and choose the split that minimizes this expected entropy.

To prevent overfitting, we employ a form of "pre-pruning". If no test reduces the expected entropy by more than a constant $\epsilon$, we stop splitting. In future work, we plan to replace this by a more sophisticated technique such as pessimistic pruning or MDL pruning.

Algorithm 1 shows the algorithm. It follows the standard recursive divide-and-conquer schema for top-down induction of decision trees. Ties in split selection are broken in favor of splits that introduce fewer new variables into the tree.

## 2.1 Efficient Splitting Function Search

Algorithm 1 requires performing two greedy set cover computations to evaluate each split. Despite the fact that greedy set cover is very efficient, this is still extremely time-consuming, especially if the set of candidate leaf functions is large. We therefore developed a method based on Uniform Cost Search (UCS) for finding the best set cover without having to evaluate all candidate leaf functions on all candidate splits.

Suppose we define a partial set cover to have the form $((f_1, n_1), (f_2, n_2), \ldots, (f_{k-1}, n_{k-1}), (\text{else}, n_k)))$.This represents the fact that there are $n_k$ data points that have not yet been covered by any leaf function. A node in the Uniform Cost Search consists of the following information:

- the candidate splitting condition $s$, $P_{left}$, and $P_{right}$
- partial set covers $C_{left}$ and $C_{right}$ for the branches

**Algorithm 1.** DMT: Grow a decision tree top-down

1: GROWTREE(examples: $E$, treenode: $T$, setcover: $C$, real: $\epsilon$)
2: $E$ is the set of training examples
3: $T$ is a tree node (initially a leaf)
4: $C$ is the set cover (with associated probability distribution) of the node
5: let $h_{root}$ := ENTROPY($C$)
6: Initialize variables to hold information about the best split:
7: let $h^* := h_{root}$
8: let $E^*_{left} := E^*_{right}$ := empty set
9: let $C^*_{left} := C^*_{right}$ := empty set cover
10: let $s^*$ := null
11: **for all** candidate splits $s$ **do**
12: let $E_{left} := \{e \in E | s(e)\}$ {Examples for which $s$ is true}
13: let $E_{right} := \{e \in E | \neg s(e)\}$ {Examples for which $s$ is false}
14: let $P_{left} := \frac{|E_{left}|}{|E|}; P_{right} := \frac{|E_{right}|}{|E|}$
15: let $C_{left}$ := GREEDYSETCOVER($E_{left}$)
16: let $C_{right}$ := GREEDYSETCOVER($E_{right}$)
17: let $h_s = P_{left} \cdot$ ENTROPY$(C_{left}) + P_{right} \cdot$ ENTROPY$(C_{right})$
18: **if** $h_s < h^*$ **then**
19: let $h^* := h_s; s^* := s$
20: $E^*_{left} := E_{left}; E^*_{right} := E_{right}; C^*_{left} := C_{left}; C^*_{right} := C_{right}$
21: **if** $|h_{root} - h^*| > \epsilon$ **then**
22: set $T.split := s^*$
23: let $T_{left}$ := new treenode($LEAF, C^*_{left}$)
24: let $T_{right}$ := new treenode($LEAF, C^*_{right}$)
25: set $T.left$ := GROWTREE($E^*_{left}, T_{left}, C^*_{left}, \epsilon$)
26: set $T.right$ := GROWTREE($E^*_{right}, T_{right}, C^*_{right}, \epsilon$)

- the entropy of the partial set covers $h_{left}$ and $h_{right}$
- the sets of uncovered response values $V_{left}$ and $V_{right}$
- the current expected entropy $h = P_{left} \cdot h_{left} + P_{right} \cdot h_{right}$

The key observation is that the current expected entropy is a lower bound on the final expected entropy, because any further refinement of either of the partial set covers $C_{left}$ or $C_{right}$ will cause the entropy to increase.

The split selection algorithm starts by creating one UCS node for each candidate split $s$ with empty set covers $C_{left}$ and $C_{right}$ and pushing them on to a priority queue (ordered to mininize $h$). It then considers the best greedy addition of one leaf function to $C_{left}$ and to $C_{right}$ to expand the two set covers, recomputes $V_{left}$, $V_{right}$, and $h$, and pushes this new node onto the priority queue. Note that the size of the priority queue remains fixed, because each candidate split is expanded greedily rather than in all possible ways (which would produce an optimal set cover instead of a greedy set cover).

The algorithm terminates when a node popped off the priority queue has $V_{left} = V_{right}$ = the empty set. In which case, this is the best split $s^*$, because it has the lowest expected entropy and all other items on the priority queue have higher entropy.

## 3 Experiments

To evaluate the effectiveness of our DMT algorithm, we compared it experimentally to four other algorithms: CART (Breiman et al., 1984), Model Trees (Quinlan, 1992), DMT with propositional splits and functional leaves (DMT-S, "minus splits") and DMT with relational splits but constant leaves (DMT-F, "minus functions"). In effect, CART is DMT-SF, DMT without relational splits or functional leaves.

The experiment is structured as follows. We chose three domains: (a) a version of the Traveling Purchase Problem (TPP) adapted from the ICAPS probabilistic planning competition (2006), (b) the Trucks Problem, also adapted from ICAPS, and (c) a resource gathering task that arises in the Wargus real-time strategy game (2007). In each domain, we generated 200 independent trajectories. In TPP and Trucks, each trajectory was generated by choosing at random a legal starting state and applying a uniform random policy to select actions until a goal state was reached. In Wargus, all trajectories started in the same state because they were all generated from the same map, and there is only one legal starting state per map. On average, the trajectories contained 96.9 actions in TPP (standard deviation of 56.7), 600.0 actions in Truck (s.d. of 317.8), and 2065 actions in Wargus (s.d. of 1822). The 200 trajectories were randomly partitioned into a training set of 128 and a test set of 72 trajectories. To generate learning curves and to obtain independent training trials, the 128 training trajectories were further divided into 2 subsets of 64, 4 subsets of 32, 8 subsets of 16, 16 subsets of 8, 32 subsets of 4, 64 subsets of 2, and 128 subsets each containing only one trajectory. For each of these training sets, each of the five algorithms was run. The resulting DBN models were then evaluated according to three criteria:

- Root Relative Squared Error (RRSE). This is the root mean squared error in the predicted value of each state variable divided by the RMS error of simply predicting the mean. Some variables are actually 0-1 variables, in which case the squared error is the 0/1 loss and the RRSE is proportional to the square root of the total 0/1 loss.
- State variable Recall. Because we wish to use the learned trees to guide subroutine discovery algorithms (2006; 2007), we want algorithms that can correctly identify the set of parents $\mathbf{pa}(x)$ of each variable $x$. The recall is the fraction of the true parents $\mathbf{pa}(x)$ that are correctly identified in the DBN model.
- State variable Precision. We also measure precision, which is the fraction of parents in the learned DBN that are parents in the true DBN.

### 3.1 Domains

Here are the detailed specifications of the three domains.

**Traveling Purchase Problem.** (TPP) is a logistics domain where an agent controls a truck that must purchase a number of goods from different markets and then return them to a central depot. Each market has a supply of goods, as well as its own price, both of which are random for each problem instance and are provided as state variables. The state also contains variables that represent the remaining demand. These variables are initialized with the total demand for that product, and they are decremented as the agent buys goods from the markets. Actions in this domain consist of goto actions, actions that buy units of products from markets, and an action to deliver all purchased products to the central depot.

For these experiments, the domain was restricted to two markets, one central depot, and three products. This results in an MDP with 15 state variables and 10 actions. Initial values for product supply and demand can range from zero to 20, which produces an MDP with over $10^{12}$ states. The price variables do not count toward the size of the state space, because their values are constant throughout an instance of the problem.

**The Truck Problem.** is another logistics problem. However in this domain, the focus is on the logistics of picking up packages, placing them in the right order on the truck, dropping them off, and delivering them to their proper destination. Here the agent is in control of two trucks; each truck has two areas (front and rear) in which it can hold packages. As in a real delivery truck, these areas must be loaded and unloaded in the proper order. For example, if there is a package in the front area, an action that attempts to remove the package from the rear of the truck fails.

For our experiments, the two trucks are asked to deliver three packages from one of five locations to another of the five locations. Once a package is dropped off at the correct location, an action must be taken by the agent to deliver the package to the customer. Actions in this domain include loading and unloading a package on a truck, driving a truck to a location, and delivering a package to the customer once it is at its goal location. The domain has 25 actions and 12 state variables, with $4.3 \cdot 10^8$ states.

**Wargus.** is a resource gathering domain where an agent controls one or more peasants and directs them in a grid world that contains gold mines, stands of trees, and town halls. The agent can navigate to anywhere on the map with a set of goto actions. These are temporally extended actions that bring a peasant to specified region of the map. The regions are defined by the "sight radius" of the peasants. Within this radius, they can execute other actions such as mining gold, chopping wood, and depositing their payload in the town hall. Once the peasant has deposited one set of gold and one set of wood to the town hall, the episode ends and reward is received.

For this set of experiments, we use a single peasant on a map with a single gold mine and a single town hall. Trees are distributed randomly around the map. The state contains variables that list the position of the peasant on the map, what the peasant is holding, what objects of interest are within sight radius of the peasant (wood, gold, town hall), and the status of the gold and wood quotas. The domain has 19 actions and $9.8 \cdot 10^4$ states.

This domain differs from the other two domains because it is not fully observable. The map that the peasant is navigating is a hidden variable that determines not only the navigation dynamics but the presence of trees, gold mines, and town halls within the peasant's sight radius.

### 3.2 Results

For each combination of a domain, training set, output state variable (or reward), and action, we measured the three metrics. We performed an analysis of variance on each metric, holding the domain, action, variable and training set size constant and treating the multiple training sets as replications. We treated DMT as the baseline configuration and tested the hypothesis that the metric obtained by each of the other algorithms was significantly worse (a "win"), better (a "loss"), or indistinguishable (a "tie") at the $p < 0.05$ level of significance. The test is a paired-differences t test. Note that the power of this test decreases as the number of training trajectories increases. When the training set contains only 1 trajectory, there are 128 replications, but when the training set contains 64 trajectories, there are only 2 replications. Hence, we generally expect to see the percentage of ties increase with the size of the training set.

Table 1 aggregates the results of these statistical tests over all state variables and all actions in each domain. The large number of ties in each cell is largely an artifact of the loss of power for large training set sizes. Let us first compare DMT with DMT-F (constant leaves). In TTP and Truck, DMT performs much better than DMT-F. In Wargus, the situation is less clear. On Recall, DMT is always at least as good as DMT-F, but on RRSE the algorithms each have a large number of wins, and on precision, DMT-F tends to be better. Next, consider DMT and DMT-S (propositional splits). In this case, DMT is dominant except for Wargus precision, where DMT sometimes includes unnecessary variables that DMT-S avoids. Next, compare DMT with CART (i.e., DMT-SF). Here, DMT is almost always superior on all metrics. Note in particular that for RRSE, it is superior in 1057 out of 1120 cases (94.4%) in TTP, 2085 out of 2100 cases (99.3%) in Truck, and 519 out of 662 cases (78.4%) in Wargus. So the number of ties is actually quite small, despite the low number of replicates at large sample sizes. The only case of less than overwhelming superiority is again Precision in Wargus. Finally, compare DMT with M5P, which is the Weka implementation of model trees. DMT is again dominant in the TPP and Truck domains. In Wargus, DMT is always at least as good as M5P for Recall, but on Precision DMT is more often inferior to M5P than the reverse, and on RRSE, DMT has 247 wins whereas M5P has 207, so there is no overall winner.

Table 1 hides the effect of increasing sample size. To visualize this, Figure 2 shows the win/loss/tie percentages as a function of training set size for Recall comparing DMT versus CART. Each vertical bar is divided into three parts indicating wins, losses, and ties (reading from bottom to top). In virtually all cases, there are no losses, which means that DMT's Recall is almost always better than or equal to CART's Recall. Note also that as the size of the training set gets large (and hence, the number of training sets gets small), we observe

**Table 1.** Statistical wins, losses and ties for DMT all other tested algorithms on each domain. These results are over all non-reward variable models. A win (or loss) is a statistically significant difference between DMT and the indicated algorithm ($p < 0.05$; paired t test).

| TTP | DMT-F | | | DMT-S | | | CART | | | M5P | | |
|---|---|---|---|---|---|---|---|---|---|---|---|---|
| | Win | Loss | Tie | Win | Loss | Tie | Win | Loss | Tie | Win | Loss | Tie |
| Precision | 671 | 3 | 446 | 7 | 3 | 1110 | 787 | 9 | 324 | 358 | 4 | 758 |
| Recall | 404 | 0 | 656 | 20 | 0 | 1040 | 411 | 0 | 649 | 390 | 0 | 670 |
| RRSE | 716 | 40 | 364 | 44 | 21 | 1055 | 1057 | 38 | 25 | 475 | 37 | 608 |

| Truck | DMT-F | | | DMT-S | | | CART | | | M5P | | |
|---|---|---|---|---|---|---|---|---|---|---|---|---|
| | Win | Loss | Tie | Win | Loss | Tie | Win | Loss | Tie | Win | Loss | Tie |
| Precision | 594 | 0 | 1504 | 55 | 0 | 2043 | 1240 | 7 | 851 | 679 | 1 | 1418 |
| Recall | 356 | 22 | 1645 | 117 | 1 | 1905 | 566 | 14 | 1443 | 467 | 16 | 1540 |
| RRSE | 1046 | 83 | 971 | 211 | 56 | 1833 | 2085 | 5 | 10 | 838 | 9 | 1253 |

| Wargus | DMT-F | | | DMT-S | | | CART | | | M5P | | |
|---|---|---|---|---|---|---|---|---|---|---|---|---|
| | Win | Loss | Tie | Win | Loss | Tie | Win | Loss | Tie | Win | Loss | Tie |
| Precision | 14 | 87 | 930 | 15 | 21 | 995 | 291 | 135 | 605 | 120 | 206 | 705 |
| Recall | 118 | 0 | 946 | 15 | 1 | 1048 | 212 | 0 | 852 | 176 | 0 | 888 |
| RRSE | 172 | 182 | 308 | 55 | 16 | 591 | 519 | 87 | 56 | 247 | 207 | 208 |

more ties. This is a consequence of the loss of statistical power of the t test. Space limits prevent us from showing these curves for the other metrics or for the reward TDBNs.

Figure 3 presents learning curves for RRSE. We do not have space to show the learning curve for every combination of action and variable, so we chose one variable-action pair from each domain. For the TPP Supply variable (Purchase action), we see that for training sets of size 8 and above, DMT has the lowest RRSE, DMT-S and M5P come next, the DMT-F and CART are the worst. For the Truck variable Truck Area (Load action), DMT always has the lowest RRSE. At 8 trajectories, it is joined by DMT-F, while the other three algorithms have much higher error levels. This suggests that using relational splits is critical in this domain, and we observed this for several other variable-action-domain combinations. Finally, for the Wargus Reward variable (Navigate action), the three DMT variants have the lowest RRSE (and are indistinguishable). M5P comes next, and CART gives the worst performance. The explanation for this is less clear. Evidentally, good performance requires either relational splits or functional leaves but not both!

Also shown in Figure 3 is the model sizes for the variable-action pairs depicted in the corresponding RRSE plots. For the Supply variable (Purchase action) in the TPP domain, both DMT and DMT-S perform the best, followed by M5P, CART and DMT-F. DMT-S returns a single mixture of functions in this case because it does not have access to the more complex splits of full DMT. In the Truck domain's Load action (Truck Area variable), DMT always produces

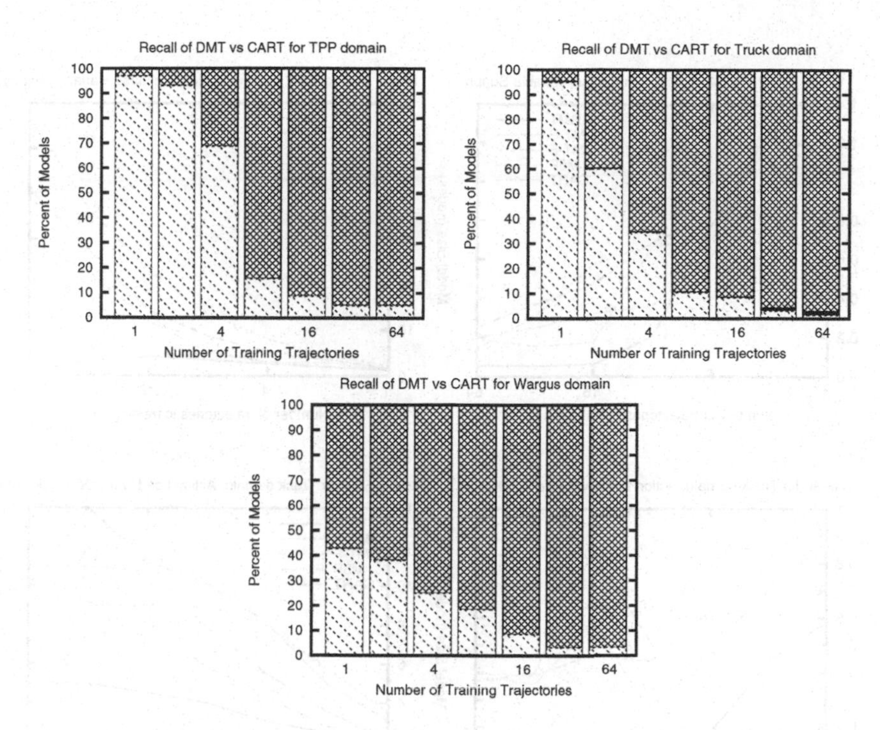


**Fig. 2.** Recall of DMT vs. CART for state variables for TPP, Truck, and Wargus. Each bar is divided into three sections (wins, losses, and ties). The losses are too infrequent to be visible in this plot.

the smallest models, followed by M5P, DMT-F, DMT-S and CART. Finally, in the reward node for a GOTO action in Wargus, we see that DMT and its variants produce the largest models, with the smallest models being produced by M5P and CART. This is consistent with our original hypothesis that the DMT algorithm performs best when the stochasticity is best represented as a mixture over discrete functions. GOTO is a temporally extended action that follows a navigation policy set by Wargus itself, its reward function is represents the distance between the current point and the destination point with noise added in from detours caused by obstacles in the agent's path.

To understand the Precision and Recall behavior of the algorithms, it is not sufficient to plot learning curves of the average Precision and Recall. This is because the distribution of measured Precision and Recall scores is highly skewed, with many perfect scores. Instead, we developed the Precision and Recall profiles shown in Figure 4 as a way of visualizing the distribution of Precision and Recall scores. For each domain, action, result variable, and training set, we computed the Precision and Recall of the fitted TDBN with respect to the set of variables included in the model compared to the variables included in the true DBN. For each domain, we sorted all of the observed scores (either Precision or Recall, depending on the graph) into ascending order and then for each value $\theta$ of the

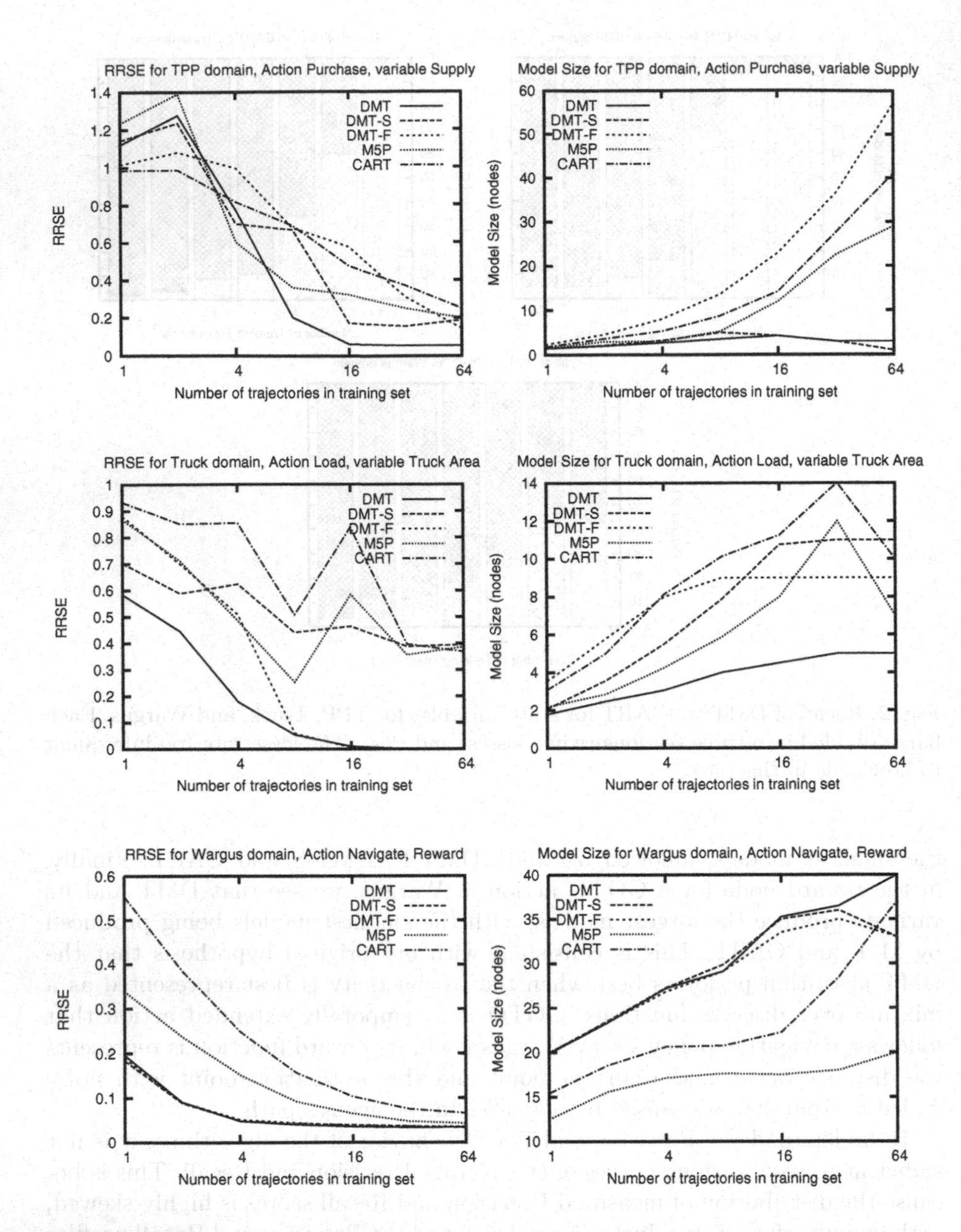


**Fig. 3.** RRSE and Model Size in nodes as a function of the number of trajectories in the training set for one chosen action and variable in each domain. Top: RRSE and Model Size for Market Supply for the Purchase action in TTP; Middle: RRSE and Model Size for Truck Area for the Load action in Truck; Bottom: RRSE and Model Size of the Reward node for a Goto action in Wargus.

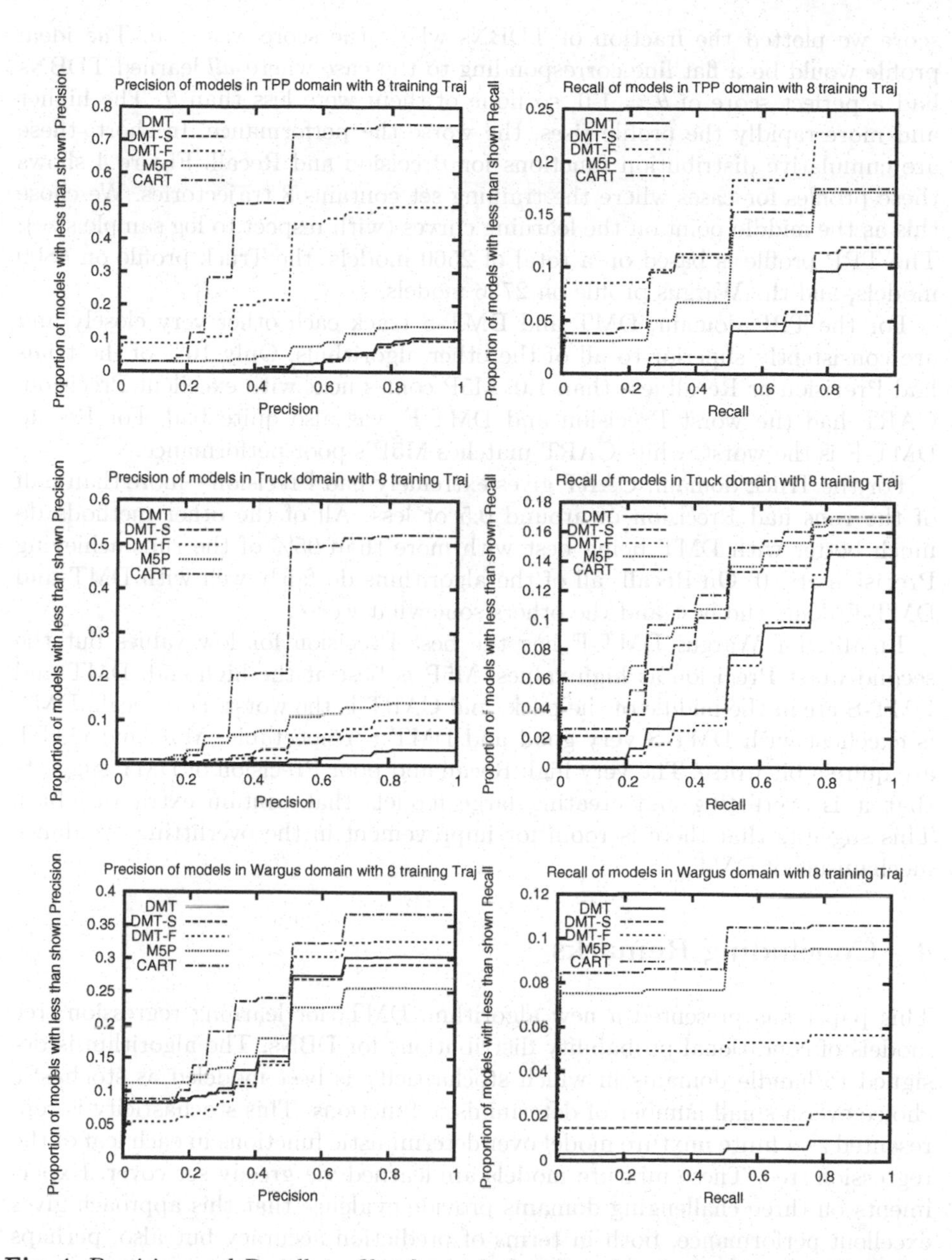


**Fig. 4.** Precision and Recall profiles for each domain when trained on 8 trajectories and compared to the true DBN models. These curves aggregate over all variables, actions, and training sets in each domain. Each plotted point specifies the fraction of learned models with Precision (or Recall, respectively) less than the value specified on the horizontal axis. Hence, the ideal curve would be a flat line at 0, corresponding to the case where all learned models had Precision (or Recall) of 1.0.

score we plotted the fraction of TDBNs where the score was $< \theta$. The ideal profile would be a flat line corresponding to the case where *all* learned TDBNs had a perfect score of $\theta = 1.0$, so none of them were less than $\theta$. The higher and more rapidly the profile rises, the worse the performance. In short, these are cumulative distribution functions for Precision and Recall. Figure 4 shows these profiles for cases where the training set contains 8 trajectories. We chose this as the middle point on the learning curves (with respect to log sample size). The TPP profile is based on a total of 2560 models, the Truck profile on 4800 models, and the Wargus profile on 2736 models.

For the TPP domain, DMT and DMT-S track each other very closely and are consistently superior to all of the other algorithms. Only 10% of the trials had Precision or Recall less than 1.0. M5P comes next with excellent Precision. CART had the worst Precision and DMT-F was also quite bad. For Recall, DMT-F is the worst, while CART matches M5P's poor performance.

For the Truck domain, CART gives extremely bad Precision—more than half of the runs had Precision of around 0.5 or less. All of the other methods do much better with DMT being best with more than 95% of the runs achieving Precision of 1.0. On Recall, all of the algorithms do fairly well with DMT and DMT-F doing the best and the others somewhat worse.

Finally, for Wargus DMT-F has the best Precision for low values but the second-worst Precision at high values. M5P is best at the high end. DMT and DMT-S are in the middle of the pack, and CART is the worst. For Recall, DMT is excellent with DMT-S very good and DMT-F respectable. M5P and CART are quite a bit worse. The very high Recall and poor Precision of DMT suggests that it is overfitting and creating large models that contain extra variables. This suggests that there is room for improvement in the overfitting avoidance mechanisms of DMT.

## 4 Concluding Remarks

This paper has presented a new algorithm, DMT, for learning regression tree models of conditional probability distributions for DBNs. The algorithm is designed to handle domains in which stochasticity is best modeled as stochastic choice over a small number of deterministic functions. This stochasticity is represented as a finite mixture model over deterministic functions in each leaf of the regression tree. These mixture models are learned via greedy set cover. Experiments on three challenging domains provide evidence that this approach gives excellent performance, both in terms of prediction accuracy but also, perhaps more importantly, in terms of the ability to correctly identify the relevant parents of each random variable. In two of the domains, DMT is clearly superior to CART and M5P. In the third domain (Wargus), there are many cases where DMT performs well, but there are also many cases where it gives worse prediction accuracy and precision than CART or M5P. This suggests that DMT may be overfitting in this domain.

It is interesting to note that this problem could also be viewed as a multi-label classification problem with overlapping classes. If each training point were labeled with the set of functions that apply to it, then the optimal mixture of classes for a set of examples represents the optimal discrete mixture of functions for those examples. This leads to some complications when classes partially or fully overlap, such as assigning probability mass of an example that is a member of several classes. These issues can be addressed by the same approximations used to increase efficiency that are described in this paper.

In future work, we plan to incorporate stronger methods for regularizing DMT by controlling both the tree size and the size of the set covers in each leaf. We would also like to extend this approach to allow stochasticity both in the mixture of functions and in the individual functions themselves. In addition, we plan to use the TDBNs learned by DMT as input to the MAXQ discovery algorithm that we have developed (2007).

## Acknowledgement

Thank you Souyma Ray, Prasad Tadepalli and Neville Mehta for all your support. This material is based upon work supported by the Defense Advanced Research Projects Agency (DARPA) under Contract No. FA8750-05-2-0249.

## References

5th International Planning Competition, International conference on automated planning and scheduling (2006)

Blockeel, H.: Top-down induction of first order logical decision trees. Doctoral dissertation, Katholieke Universiteit Leuven (1998)

Boutilier, C., Dearden, R.: Exploiting structure in policy construction. In: IJCAI 1995, pp. 1104–1111 (1995)

Boutilier, C., Dearden, R., Goldszmidt, M.: Stochastic dynamic programming with factored representations. Artificial Intelligence 121, 49–107 (2000)

Breiman, L., Friedman, J., Olshen, R., Stone, C.: Classification and regression trees. Wadsworth Inc. (1984)

Chickering, D.M., Heckerman, D., Meek, C.: A Bayesian approach to learning Bayesian networks with local structure. In: UAI 1997, pp. 80–89 (1997)

Dean, T., Kanazawa, K.: A model for reasoning about persistence and causation. Computational Intelligence 5, 33–58 (1989)

Dietterich, T.G.: Hierarchical reinforcement learning with the MAXQ value function decomposition. JAIR 13, 227–303 (2000)

Gama, J.: Functional trees. Machine Learning 55, 219–250 (2004)

Johnson, D.S.: Approximation algorithms for combinatorial problems. In: STOC 1973, pp. 38–49. ACM, New York (1973)

Jonsson, A., Barto, A.G.: Causal graph based decomposition of factored MDPs. Journal of Machine Learning Research 7, 2259–2301 (2006)

Kramer, S.: Structural regression trees. In: AAAI 1996, pp. 812–819. AAAI Press, Menlo Park (1996)

Lavrac, N., Dzeroski, S.: Inductive logic programming, techniques and applications. Ellis Horwood (1994)

McLachlan, G., Krishnan, T.: The EM algorithm and extensions. Wiley, New York (1997)

Mehta, N., Wynkoop, M., Ray, S., Tadepalli, P., Dietterich, T.: Automatic induction of maxq hierarchies. In: NIPS Workshop: Hierarchical Organization of Behavior (2007)

Quinlan, J.R.: Learning with Continuous Classes. In: 5th Australian Joint Conference on Artificial Intelligence, pp. 343–348 (1992)

Quinlan, J.R.: C4.5: Programs for machine learning. Morgan Kaufmann, San Francisco (1993)

Slavík, P.: A tight analysis of the greedy algorithm for set cover. In: STOC 1996, pp. 435–441. ACM, New York (1996)

The Wargus Team, Wargus sourceforge project (Technical Report) (2007), `wargus.sourceforge.org`

Torgo, L.: Functional models for regression tree leaves. In: Proc. 14th International Conference on Machine Learning, pp. 385–393. Morgan Kaufman, San Francisco (1997)

Vens, C., Ramon, J., Blockeel, H.: Remauve: A relational model tree learner. In: ILP 2006, pp. 424–438 (2006)

# Continuous Time Bayesian Networks for Host Level Network Intrusion Detection

Jing Xu and Christian R. Shelton

Department of Computer Science and Engineering
University of California, Riverside
Riverside, CA 92521, USA
{jingxu,cshelton}@cs.ucr.edu

**Abstract.** We consider the problem of detecting host-level attacks in network traffic using unsupervised learning. We model the normal behavior of a host's traffic from its signature logs, and flag suspicious traces differing from this norm. In particular, we use continuous time Bayesian networks learned from historic non-attack data and flag future event sequences whose likelihood under this normal model is below a threshold. Our method differs from previous approaches in explicitly modeling temporal dependencies in the network traffic. Our model is therefore more sensitive to subtle variations in the sequences of network events. We present two simple extensions that allow for instantaneous events that do not result in state changes, and simultaneous transitions of two variables. Our approach does not require expensive labeling or prior exposure to the attack type. We illustrate the power of our method in detecting attacks with comparisons to other methods on real network traces.

**Keywords:** Unsupervised Machine Learning, CTBNs, Host Based IDS.

## 1 Introduction

Network attacks on computers have become a fact of life for network administrators. Detecting the attacks quickly is important in limiting their scope and destruction. Humans cannot analyze each and every connection or packet to determine if it might be part of an attack, so there is a need for an automated system for performing such checks. In this paper, we look at the problem of detecting such attacks at the host level. Instead of constructing a detector for the network as a whole, we construct a method that can be employed at each computer separately to determine whether a particular host has been compromised. While we lose global information, we gain speed, individual tuning, and robustness. A network under attack may not be able to aggregate information to a central detector.

We focus on an automatic approach to detection. While human intervention can improve security, it comes at the cost of increased time and effort. We would like a method that can adapt to the role and usage patterns of the host, while still detecting attacks automatically.

We approach this problem from the point-of-view of anomaly detection. Attacks vary greatly and new types of attacks are invented frequently. Therefore, a supervised

W. Daelemans et al. (Eds.): ECML PKDD 2008, Part II, LNAI 5212, pp. 613–627, 2008.

learning method that attempts to distinguish good from bad network traffic based on historic labeled data is necessarily limited in its scope. Furthermore, the acquisition of such labeled data is time-consuming and error-prone. By contrast, gathering normal or good network traffic is relatively simple. It is often possible to designate times when we can be reasonably certain that no attacks occurred and use all of the data during that span. For an attack to be successful, it must differ in some way from normal network traffic. Our goal is to detect such abnormalities.

Our approach differs from previous approaches in a number of key ways. It is adaptive and constructed at the host level. It does not treat the packets or connections as *i.i.d.* sequences, but respects the ordering of the network traffic. Finally, it does not model the traffic features as normal or exponential distributions. Many features of network traffic are distinctly non-Gaussian and often multi-modal.

A network flow for a given host machine is a sequence of continuous and asynchronous events. We employ a generative probabilistic model to describe such dynamic processes evolving over continuous time. Furthermore, these events form a complex structured system, where statistical dependencies relate network activities like packets and connections. We apply continuous time Bayesian networks (CTBNs) [1] to the challenge of reasoning about these structured stochastic network processes. CTBNs have been successful in other applications [2,3], but have not previously been used for network traffic modeling.

In Section 2 we relate our method to prior work on this problem. In Section 3 we briefly describe continuous time Bayesian networks. In Section 4 we describe our CTBN model and its use. In Section 5 we describe our experimental results.

## 2 Related Work

As a signature-based detection algorithm, we share many of the assumptions of [4]. In particular, we also assume that we do not have access to the internals of the machines on the networks (which rules out methods like [5] [6] [7] or [8]). However, we differ in that our approach does not rely on preset values, require human intervention and interpretation, nor assume that we have access to network-wide traffic information. Network-wide data and human intervention have advantages, but they can also lead to difficulties (data collation in the face of an attack and increased human effort), so we chose to leave them out of our solution.

Many learning, or adaptive, methods have been proposed for network data. Some of these (for example, [9] and [10]) approach the problem as a classification task which requires labeled data. [11] profiles the statistical characteristics of anomalies by using random projection techniques (sketches) to reduce the data dimensionality and a multi-resolution non-Gaussian marginal distribution to extract anomalies at different aggregation levels. The goal of such papers is usually not to detect attacks but rather to classify non-attacks by traffic type; if applied to attack detection, they would risk missing new types of attacks. Furthermore, they frequently treat each network activity separately, instead of considering their temporal context.

[12] has a nice summary of adaptive (or statistical) methods that look at anomaly detection (instead of classification). They use an entropy-based method for the entire

network traffic. Many of the other methods (such as [13]) use either statistical tests or subspace methods that assume the features of the connections or packets are distributed normally. [14] model the language features like n-grams and words from connection payloads. [15] also uses unsupervised methods, but they concentrate on clustering traffic across a whole network. Similarly, [16] build an anomaly detector based on Markov models, but it is for the network traffic patterns as a whole and does not function at the host level.

[10] is very similar in statistical flavor to our work. They also fit a distribution (in their case, a histogram modeled as a Dirichlet distribution) to network data. However, they model flow-level statistics, whereas we work at the level of individual connections. Additionally, they are attempting network-wide clustering of flows instead of anomaly detection. [17], like our approach, model traffic with graphical models, in particular, Naive Bayes networks. But their goal is to categorize network traffic instead of detecting attacks. [18] present a Bayesian approach to the detecting problem as an event classification task while we only care about whether the host is under attack during an interval. In addition, they also use system monitoring states to build their model, which we do not employ.

[19] is very similar to our work. It is one of the few papers to attempt to find attacks at the host level. They employ nearest neighbor, a Mahalanobis distance approach, and a density-based local outliers method, each using 23 features of the connections. Although their methods make the standard *i.i.d.* assumption about the data (and therefore miss the temporal context of the connection) and use 23 features (compared to our few features), we compare our results to theirs in Section 5, as the closest prior work. We also compare our work with [20]. They present an adaptive detector whose threshold is time-varying. It is similar to our work in that they also rely on model-based algorithms. But besides the usage of network signature data, they look at host internal states like CPU loads which are not available to us.

While there has been a great variety of previous work, our work is novel in that it detects anomalies at the host level using only the timing features of network activities. We do not consider each connection (or packet) in isolation, but rather in a complex context. We capture the statistical dynamic dependencies between packets and connections to find sequences of network traffic that are anomalous *as a group*. CTBNs are a natural modeling method for network traffic, although they have not been previously used for this task.

# 3 Continuous Time Bayesian Networks

We begin by briefly reviewing the definition of Markov processes and continuous time Bayesian networks (CTBNs).

## 3.1 Homogeneous Markov Process

A finite-state, continuous-time, homogeneous Markov process $X_t$ is described by an initial distribution $P_X^0$ and, given a state space $Val(X) = \{x_1, ..., x_n\}$, an $n \times n$ matrix of transition intensities:

$$\mathbf{Q_X} = \begin{bmatrix} -q_{x_1} & q_{x_1x_2} & \cdots & q_{x_1x_n} \\ q_{x_2x_1} & -q_{x_2} & \cdots & q_{x_2x_n} \\ \vdots & \vdots & \ddots & \vdots \\ q_{x_nx_1} & q_{x_nx_2} & \cdots & -q_{x_n} \end{bmatrix}.$$

$q_{x_ix_j}$ is the intensity (or rate) of transition from state $x_i$ to state $x_j$ and $q_{x_i} = \sum_{j \neq i} q_{x_ix_j}$.

The transient behavior of $X_t$ can be described by follows. Variable $X$ stays in state $x$ for time exponentially distributed with parameter $q_x$. The probability density function $f$ for $X_t$ remaining at $x$ is $f(q_x, t) = q_x \exp(-q_x t)$ for $t \geq 0$. The expected time of transitioning is $1/q_x$. Upon transitioning, $X$ shifts to state $x'$ with probability $\theta_{xx'} = q_{xx'}/q_x$.

The distribution over the state of the process $X$ at some future time $t$, $P_x(t)$, can be computed directly from $Q_X$. If $P_X^0$ is the distribution over $X$ at time 0 (represented as a vector), then, letting exp be the matrix exponential,

$$P_X(t) = P_X^0 \exp(Q_X \cdot t).$$

### 3.2 Continuous Time Bayesian Networks

[1] extended this framework to continuous time Bayesian networks, which model the joint dynamics of several local variables by allowing the transition model of each local variable $X$ to be a Markov process whose parametrization depends on some subset of other variables $U$ as follows.

**Definition 1.** A conditional Markov process $X$ *is an inhomogeneous Markov process whose intensity matrix varies as a function of the current values of a set of discrete conditioning variables* $U$*. It is parametrized using a conditional intensity matrix (CIM) —* $Q_{X|U}$ *— a set of homogeneous intensity matrices* $Q_{X|u}$*, one for each instantiation of values* $u$ *to* $U$*.*

**Definition 2.** A continuous time Bayesian network $\mathcal{N}$ *over* $\mathbf{X}$ *consists of two components: an initial distribution* $P_{\mathbf{X}}^0$*, specified as a Bayesian network* $\mathcal{B}$ *over* $\mathbf{X}$*, and a continuous transition model, specified using a directed (possibly cyclic) graph* $\mathcal{G}$ *whose nodes are* $X \in \mathbf{X}$*;* $U_X$ *denotes the parents of* $X$ *in* $\mathcal{G}$*. Each variable* $X \in \mathbf{X}$ *is associated with a conditional intensity matrix,* $\mathbf{Q}_{X|U_X}$*.*

The dynamics of a CTBN are qualitatively defined by a graph. The instantaneous evolution of a variable depends only on the current value of its parents in the graph. The quantitative description of a variable's dynamics are given by a set of intensity matrices, one for each value of its parents.

[21] presented an algorithm based on expectation maximization (EM) to learn parameters of CTBNs from incomplete data. Incomplete data are sets of partially observed trajectories $D = \{\sigma[1], ..., \sigma[w]\}$ that describe the behavior of variables in the CTBNs. A partially observed trajectory $\sigma \in D$ can be specified as a sequence of subsystems $S_i$, each with an associated duration. The transitions inside the subsystems are wholly unobserved. A simplest example of incomplete data is when a variable is hidden. In this case, the observed subsystems are the set of states consistent with the observed variables.

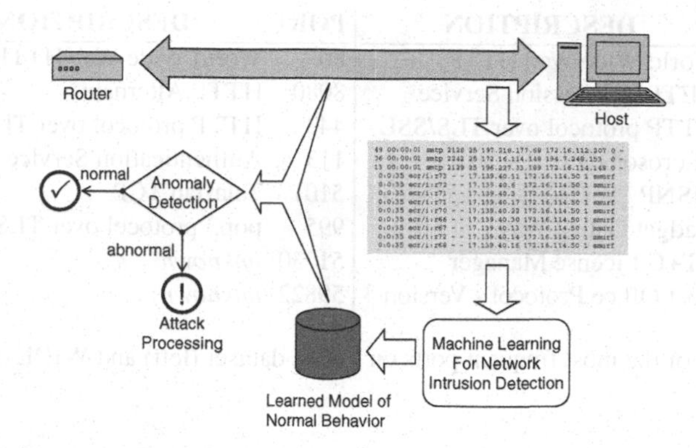


**Fig. 1.** The architecture of the attack detection system. The model is learned from normal traces. Anomaly detection compares the model to current network traffic. The output is the predicted labels for each time window.

# 4 Approach

Figure 4 shows an overview of our host-based network attack detection system. In our work, we only focus on a single computer (a host) on the network. We use unsupervised learning to build a model of the normal behavior of this host-based on its TCP packet header traces, without looking into its system behavioral states like resource usages and file accesses. Those activities that differ from this norm are flagged as possible intrusion attempts.

## 4.1 CTBN Model for Network Traffic

The sequence and timing of events are very important in network traffic flow. It matters not just how many connections were initiated in the past minute, but also their timing: if they were evenly spaced the trace is probably normal, but if they all came in a quick burst it is more suspecious. Similarly, the sequence is important. If the connections were made to sequentially increasing ports it is more likely to be a scanning virus, whereas the same set of ports in random order is more likely to be normal traffic. These are merely simple examples. We would like to detect more complex patterns.

While time-sliced models (like dynamic Bayesian networks [22]) can capture some of these aspects, they require the specification of a time-slice width. This sampling rate must be fast enough to catch these distinctions (essentially fast enough that only one event happens between samples) and yet still long enough to make the algorithm efficient. For network traffic, with long delays between bursts of activity, this is impractical.

As described in Section 3, CTBNs are well suited for describing such structured stochastic processes with finitely many states that evolve over continuous time. By using a CTBN to model the network traffic activities, we can capture the complex context of network behaviors in a meaningful and hierarchical way.

A typical machine in the network may have diverse activities with various service types (*e.g.* HTTP, mail server). Destination port numbers describe the type of service a

| PORT | DESCRIPTION | PORT | DESCRIPTION |
|---|---|---|---|
| 80 | World Wide Wed HTTP | 80 | World Wide Web HTTP |
| 139 | NETBIOS Session Service | 8080 | HTTP Alternate |
| 443 | HTTP protocol over TLS/SSL | 443 | HTTP protocol over TLS/SSL |
| 445 | Microsoft-DS | 113 | Authentication Service |
| 1863 | MSNP | 5101 | Talarian TCP |
| 2678 | Gadget Gate 2 Way | 995 | pop3 protocol over TLS/SSL |
| 1170 | AT+C License Manager | 51730 | *unknown* |
| 110 | Post Office Protocol - Version 3 | 59822 | *unknown* |

**Fig. 2.** Ranking of the most frequent ports on LBNL dataset (left) and WIDE dataset (right)

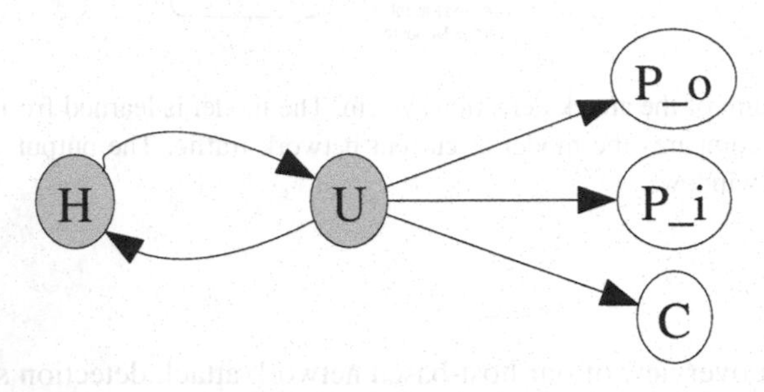


**Fig. 3.** CTBN port level submodel; the whole model contains 9 of such submodels

particular network activity belongs to. Some worms usually propagate malicious traffic towards certain well known ports to affect the quality of the services who own the contact ports. By looking at traffic associated with different ports we are more sensitive to subtle variations that do not appear if we aggregate trace information across ports. Figure 2 shows the most popular ports ranked by their frequencies in the network traffic on the datasets we use (described in more depth later). These services are, to some extent, independent of each other. We therefore model each port's traffic with its own CTBN submodel.

Inside our port-level submodel, we have the fully observed node $C$ — the number of concurrent connections active on the host — and nodes packet-in $P_i$ and packet-out $P_o$ — the transmission of a packet to or from the host. $P_i$ and $P_o$ have no intrinsic state: the transmission of a packet is an essentially instantaneous event. Therefore they have events (or "transitions") without having state. We discuss this further in the next subsection.

To allow complex duration distributions for the states of variables, we introduce another two nodes $H$ and $U$. $U$ is a partially observed variable which we call the mode whose state indicates whether the model can next send a packet, receive a packet, start a new connection, or terminate a connection. It therefore has four states, and we limit the conditional rates of the observed variables ($C$, $P_o$, and $P_i$) to be zero if the current mode differs from the variable's activity type. Therefore, $U$ is observed when an event in $C$, $P_o$ and or $P_i$ occurs, but is unobserved between events.

$H$ is a hidden variable that models the internal state and intrinsic features of the host machine. For the experiments we show later, we let $H$ be a binary process. While the arrival times for events across an entire CTBN are distributed exponentially (as

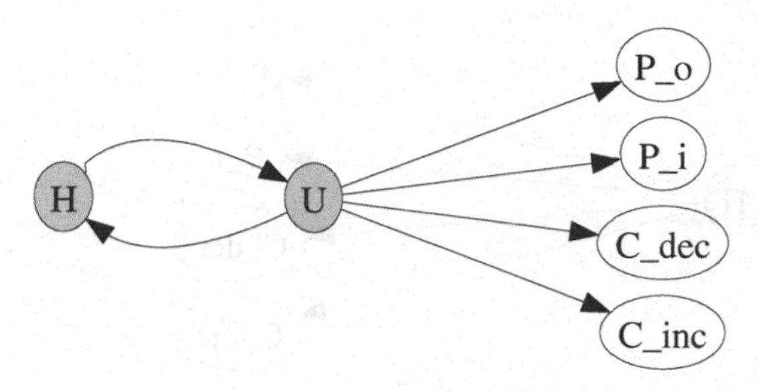


**Fig. 4.** The equivalent CTBN port level submodel

the entire system is Markovian), once certain variables are unobserved (like $H$), the marginal distribution over the remaining variables is no longer Markovian. These nodes make up the structure shown in Figure 3. The results in this paper are for binary hidden variables. We feel this model represents a good balance between descriptive power and tractability.

### 4.2 Adding Toggle Variables to CTBNs

As mentioned above, the variables $P_o$ and $P_i$ do not have state, but rather only events. To describe such transitionless events, we set $P_o$ and $P_i$ to be "toggle variables." That is, they have two indistinguishable states. As a binary variable, they have two parameters for each parent instantiation (the rate of leaving state 0 and the rate of leaving state 1) and we require these two parameters to be the same. A packet event, therefore, consists of flipping the state of the corresponding packet variable.

The concurrent connection count variable, $C$, also poses a slight modeling problem for CTBNs. As originally presented, CTBNs can only deal with finite-domain variables. Although most of the operating systems do have a limit on the number of concurrent connections, this number can potentially be extremely large. Our traffic examples do exhibit a wide range of concurrent connection counts. We tried quantizing $C$ into fixed bins, but the results were fairly poor. We instead note that $C$ can only increase or decrease by one at any given event (the beginning or ending time of a connection). Furthermore, we make the assumption that the arrival of a new connection and the termination of an existing connection are both independent of the number of other connections. This implies that the intensity with which *some* connection starts (stops) is same as any other connections. $C$ is thus a random walk constrained to the non-negative integers.

Let $Q_{inc}$ be the intensity for the arrival of a new connection, and $Q_{dec}$ be the intensity for the termination of a new connection. Let $Q_{change} = Q_{dec} + Q_{inc}$. The resulting intensity matrix has the form:

$$\mathbf{Q_{C|m}} = \begin{pmatrix} & \ddots & & & & & \\ 0 & Q_{dec} & -Q_{change} & Q_{inc} & 0 & \cdots & \\ \cdots & 0 & Q_{dec} & -Q_{change} & Q_{inc} & 0 & \cdots \\ & \cdots & 0 & Q_{dec} & -Q_{change} & Q_{inc} & 0 \\ & & & & & \ddots & \end{pmatrix} .$$

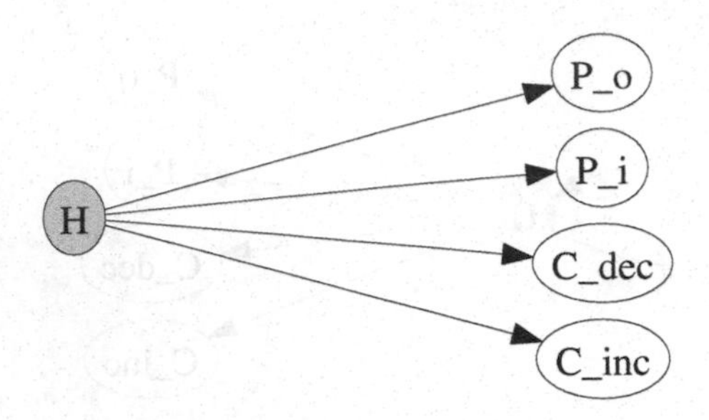


**Fig. 5.** The extended CTBN port level submodel

Note that the only free parameters in the above matrix are $Q_{inc}$ and $Q_{dec}$. Therefore, this model is the same as one in which we replace $C$ with two toggle variables $C_{inc}$ and $C_{dec}$. $C_{inc}$ and $C_{dec}$ operate like $P_o$ and $P_i$ above: their exact state does not matter, it is the transition between states that indicates an event of interest. The model structure is shown in figure 4.

### 4.3 An Extended CTBN Model

CTBN models assume that no two variables change at exactly the same instant. However, we might like $U$ and $H$ to change at the same time. As they both represent abstract concepts, there is no physical reason why they should not change simultaneously (they represent different abstract attributes about the machine's internal state).

The model in Figure 4 has 36 independent parameters.[1] Allowing $U$ and $H$ to change simultaneously requires introducing 24 new parameters: for each of the 8 states for $U$ and $H$ there are 3 new transition possibilities.

Equivalently, we can use the CTBN model shown in Figure 5 where $H$ now has 8 states. However, this diagram does not demonstrate all of the structure. The toggle variables ($P_o$, $P_i$, $C_{inc}$, and $C_{dec}$) are each allowed to change only for 2 of the states of $H$ (the two that corresponded to $U$ from the previous model having the correct mode) and they are required to have the same rate for both of these states.

We have also considered other extensions to the model. For instance, it might be natural to allow the packet rate to have a linear dependence on the number of connections. However, in practice, this extension produced worse results. It seems that the rate is more of a function of the network bandwidth and the currently transmitting application can generate packets than the number of concurrent connections on a port.

### 4.4 Parameter Estimation

A CTBN is a homogeneous Markov process over the joint state space of its constituent variables. Our partially observed trajectory ($H$ is unobserved, $U$ is partially observed) specifies a sequence of subsystems of this process, each with an associated duration.

[1] Each toggle has only 1 because it can only change for one setting of $U$. An intensity matrix for $U$ has 12 independent parameters for each of the two values of $H$ and similarly $H$ has 2 independent parameters for each of the four values of $U$.

We use the expectation maximization (EM) algorithm to estimate the parameters [21]. In short, this algorithm estimates the posterior distribution over the hidden variables' trajectories. The sufficient statistics (summaries of the data necessary for learning) depend on the value of the hidden variables. EM calculates their expected value (using the posterior distribution) and then treats these expected sufficient statistics (ESS) as the true ones for selecting the parameters that maximize the likelihood of the data. This repeats until convergence.

The ESS for any variable $X$ in a CTBN are $\bar{T}_{X|\mathbf{U}}[x|\mathbf{u}]$, the expected amount of time $X$ stays at state $x$ given its parent instantiation $\mathbf{u}$, and $\bar{M}_{X|\mathbf{U}}[x, x'|\mathbf{u}]$, the expected number of transitions from state $x$ to $x'$ given $X$'s parent instantiation $\mathbf{u}$.

For our new types of variables (toggle variables) $P_i, P_o, C_{inc}$ and $C_{dec}$, we need to derive different sufficient statistics. They are quite similar so we just take $P_i$ as an example. Let $\mathbf{U_{P_i}}$ be its parent $U$'s instantiation when event $P_i$ can happen. The ESS we need are $\bar{M}_{P_i}|\mathbf{U_{P_i}}$: the expected number of times $P_i$ changes conditioned on it parent instantiation $\mathbf{U_{P_i}}$. Since event $P_i$ can only occur when $\mathbf{U} = \mathbf{U_{P_i}}$, the ESS $\bar{M}_{P_i}|\mathbf{U_{P_i}}$ is just $M_{P_i}$, the total number of times $P_i$ changes. We also need the total expected amount of time that packet-in occurred while $\mathbf{U} = \mathbf{U_{P_i}}$.

In EM, we use the ESS as if they were the true sufficient statistics to maximize the likelihood with respect to the parameters. For a "regular" CTBN variable $X$ (such as our hidden variable $H$ and $M$), the following equation performs the maximization.

$$Q_X|u(x, x') = \frac{\bar{M}[x, x'|u]}{\bar{T}[x|u]}$$

For our new toggle variable, i.e. $P_i$, the maximization is

$$Q_{P_i|\mathbf{u}} = \frac{M_{P_i}}{T[U = \mathbf{u}]}$$

The above sufficient statistics can be calculated using the exact inference algorithm of [21].

### 4.5 Online Testing Using Forward Pass Likelihood Calculation

Once the CTBN model has been fit to historic data, we detect attacks by computing the likelihood of a window of the data under the model. If the likelihood falls below a threshold, we flag the window as anomalous. Otherwise, we mark it as normal.

In our experiments, we fix the window to be of a fixed time length, $T_w$. Therefore, if the window of interest starts at time $T$, we wish to calculate $p(Y(T, T+T_w) \mid Y(0, T))$ where $Y(s,t)$ represents the joint trajectory of $C_{inc}, C_{dec}, P_i$, and $P_o$ from $s$ to $t$. This involves integrating out the trajectories of $U$ and $H$ and can be calculated in an on-line fashion using the standard forward passing inference algorithm [1].

## 5 Experiment Results

To demonstrate the effectiveness of our methods, we compare our CTBN models to three existing methods in the literature on two different real-world network traffic datasets and with three different attack types.

### 5.1 Data Sets

We verify our approaches on two publicly available real-traffic trace repositories: the MAWI working group backbone traffic [23] and LBNL/ICSI internal enterprise traffic [24].

The MAWI backbone traffic is part of the WIDE project which collects raw daily packet header traces since 2001. It collects the network traffic through the inter-Pacific tunnel between Japan and the USA. The dataset uses `tcpdump` and some IP anonymizing tools to record 15-minute traces everyday, and consists mostly of traffic from or to some Japanese universities. In our experiment, we use the traces from January 1st to 4th of 2008, with 36,592,148 connections over a total time of one hour.

The LBNL traces are recorded from a medium-sized site, with emphasis on characterizing internal enterprise traffic. Publicly released in some anonymized forms, the LBNL data collects more than 100 hours network traces from thousands of internal hosts. From what is publicly released, we take one hour traces from January 7th, 2005 (the latest date available), with 3,665,018 total connections.

### 5.2 Experiment Setup

For each of the datasets, we pick the ten most active hosts. To create a relatively abundant dataset based on the original packet traces, we constructed a training-testing set pair for each IP address. We use half of the traces as a training set to learn the CTBN model for a given IP (host). The other traces we save for testing. Since all the traces contain only normal traffic, we use trace-driven attack simulators to inject abnormal activities into the test traces to form our testing set. The three types of attack simulators are an IP Scanner, W32.Mydoom, and Slammer [25]. We select a point somewhere in the first half of the test trace and insert worm traffic for a duration equal to $\alpha$ times the length of the full testing trace. The shorter $\alpha$ is, the harder it is to detect the anomaly.

We also scaled back the rates of the worms. When running at full speed, a worm is easy to detect for any method. When it slows down (and thus blends into the background traffic better), it becomes more difficult to detect. We let $\beta$ be the scaling rate (*e.g.* 0.1 indicates a worm running at one-tenth its normal speed).

For each algorithm (our CTBN algorithm and those below), we compute its score on consecutive non-overlapping windows of $T_w = 50$ seconds in the testing set. If the score exceeds a threshold, we declare the window a positive example of a threat. We evaluate the algorithm by comparing to the true answer: a window is a positive example if at least one worm connection exists in the window.

### 5.3 Other Methods

We compare against the nearest neighbor algorithms used in [19]. Not all of the features in [19] are available. The features available in our datasets are shown in Figure 6.

Notice that these features are associated with each connection record. To apply the nearest neighbor method to our window based testing framework, we first calculate the nearest distance of each connection inside the window to the training set, and assign the maximum among them as the score for the window.

# packets flowing from source to destination
# packets flowing from destination to source
# connections by the same source in the last 5 seconds
# connections to the same destination in the last 5 seconds
# different services from the same source in the last 5 seconds
# different services to the same destination in the last 5 seconds
# connections by the same source in the last 100 connections
# connections to the same destination in the last 100 connections
# connections with the same port by the same source in the last 100 connections
# connections with the same port to the same destination in the last 100 connections

**Fig. 6.** Features for nearest neighbor approach of [19]

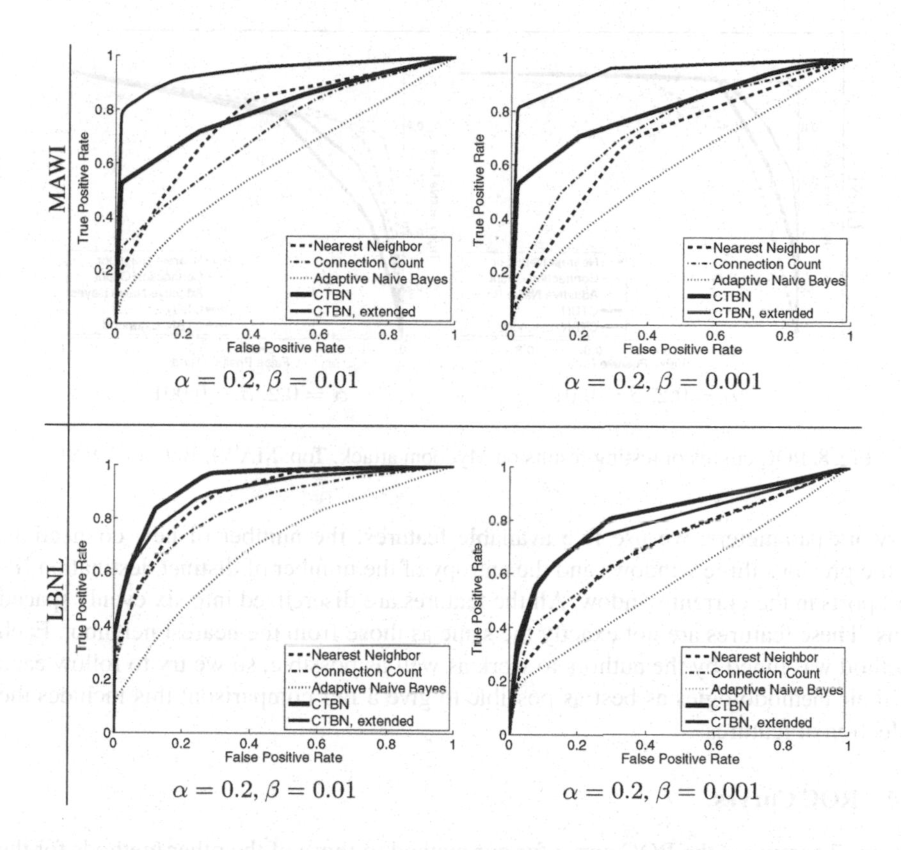


**Fig. 7.** ROC curves of testing results on IP scanning attack. Top: MAWI. Bottom: LBNL.

We also compare with a connection counting method. As most worms aggregate many connections in a short time, this method captures this particular anomaly well. We score a window by the number of initiated connections in the window.

Finally, we employ the adaptive naive Bayes approach of [20], which shows promising results on similar problems. We also follow the feature selection presented in this paper, while not all of them are available in our dataset. To train the Naive Bayes

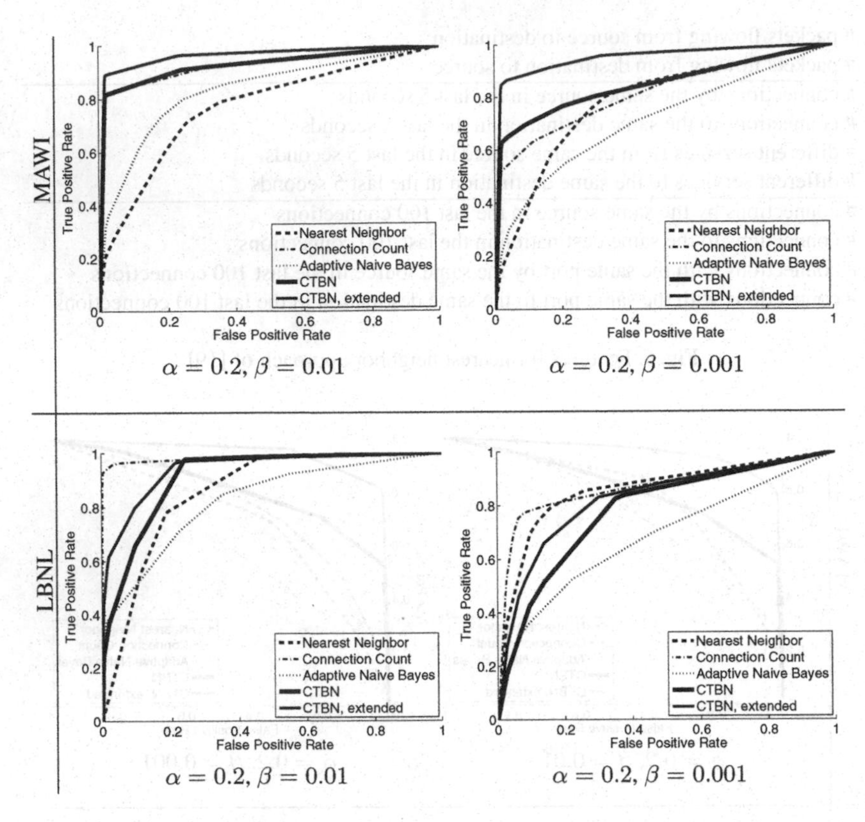


**Fig. 8.** ROC curves of testing results on Mydoom attack. Top: MAWI. Bottom: LBNL.

network parameters, we use five available features: the number of new connections in the previous three windows and the entropy of the number of distinct destination IPs and ports in the current window. All the features are discretized into six evenly spaced bins. These features are not exactly the same as those from the nearest neighbor. Each method was tuned by the authors to work as well as possible, so we try to follow each of their methodologies as best as possible to give a fair comparison; this includes the selection of features.

## 5.4 ROC Curves

Figure 7 compares the ROC curve for our method to those of the other methods for the IP scanning attack. The curves show the overall performance on the 10 hosts we chose for each dataset. $\alpha$ represents the fraction of time during which the attack is present and $\beta$ represents the speed of the attack. The curves demonstrate that as the attack becomes more subtle ($\beta$ is smaller), our method performs relatively better compared with other methods. Figures 8 and 9 show the same curves but for the Mydoom and Slammer attacks.

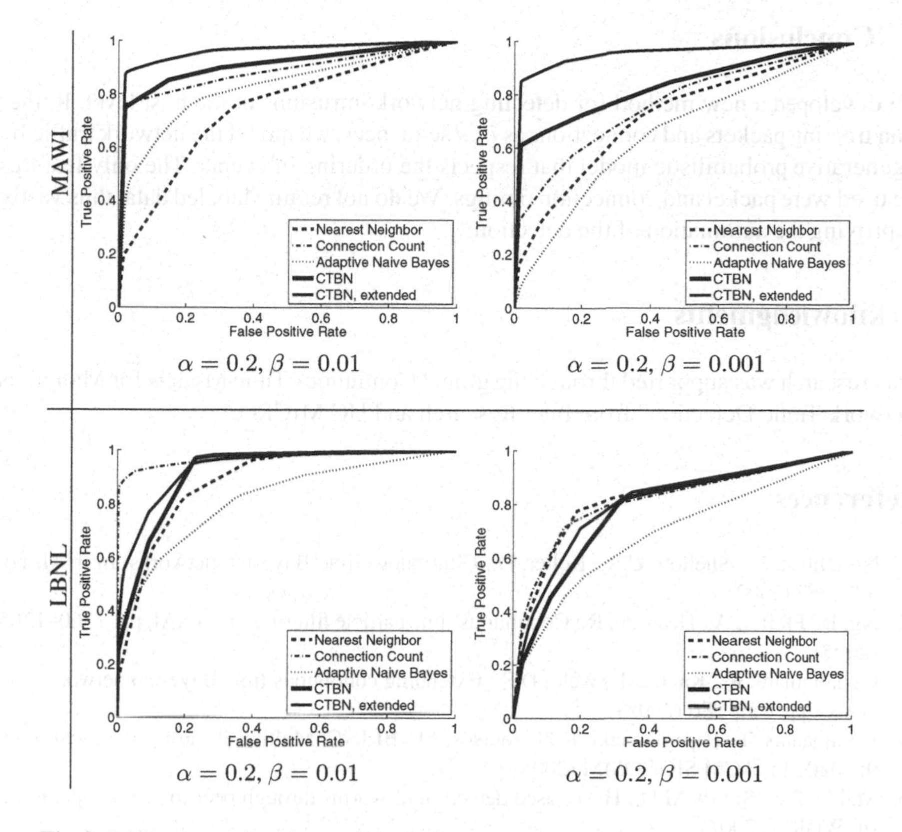


**Fig. 9.** ROC curves of testing results on Slammer attack. Top: MAWI. Bottom: LBNL.

Each point on the curve corresponds to a different threshold of the algorithm. Because attacks are relatively rare, compared to normal traffic, we are most interested in the region of the ROC curve with small false positive rates.

We note that our method out performs the other algorithms consistently for the MAWI dataset. For the LBNL dataset, with the Mydoom and Slammer attacks, some of the other methods have better performance, in particular a simple connection counting method performs the best. We are uncertain of the exact reason, but suspect it may be due to differences in the traffic type (in particular, the LBNL data comes from enterprise traffic). The addition of a hidden variable to our model allows us to model non-exponential durations. However, the complexity of such a phase-type distribution depends on the number of states in the hidden variable. If there are not enough states to model the true duration distribution, the algorithm may end up with an exponential model (at least for some events). Exponential models do not disfavor (in terms of likelihood) many quick transitions, compared to heavier tailed distributions. This might lead to worse performance in exactly the same situations where a connection-count method would work well. We hope to add more hidden states in the future to overcome this limitation.

## 6 Conclusions

We developed a new method for detecting network intrusions in the host level. Rather than treating packets and connections as *i.i.d* sequences, we model the network traffic by a generative probabilistic model that respects the ordering of events. The only features we used were packet and connection timings. We do not require labeled data, thus vastly improving the automation of the detection.

## Acknowledgments

This research was supported through the grant "Continuous Time Models for Malicious Network Trafic Detection" from Intel Research and UC MICRO.

## References

1. Nodelman, U., Shelton, C.R., Koller, D.: Continuous time Bayesian networks. In: UAI, pp. 378–387 (2002)
2. Ng, B., Pfeffer, A., Dearden, R.: Continuous time particle filtering. In: AAAI, pp. 1360–1365 (2005)
3. Gopalratnam, K., Kautz, H., Weld, D.S.: Extending continuous time Bayesian networks. In: AAAI, pp. 981–986 (2005)
4. Karagiannis, T., Papagiannaki, K., Faloutsos, M.: BLINC: Multilevel traffic classification in the dark. In: ACM SIGCOMM (2005)
5. Malan, D.J., Smith, M.D.: Host-based detection of worms through peer to peer cooperation. In: WORM (2005)
6. Cha, B.: Host anomaly detection performance analysis based on system call of neuro-fuzzy using soundex algorithm and n-gram technique. In: Systems Communications (ICW) (2005)
7. Qin, X., Lee, W.: Attack plan recognition and prediction using causal networks. In: Annual Computer Security Application Conference, pp. 370–379 (2004)
8. Eskin, E., Arnold, A., Prerau, M., Portnoy, L., Stolfo, S.: A geometric framework for unsupervised anomaly detection: Detecting intrusions in unlabeled data. In: Barbara, D., Jajodia, S. (eds.) Applications of Data Mining in Computer Security. Kluwer Academic Publishers, Dordrecht (2002)
9. Zuev, D., Moore, A.: Internet traffic classification using Bayesian analysis techniques. In: ACM SIGMETRICS (2005)
10. Soule, A., Salamatian, L., Taft, N., Emilion, R., Papagiannali, K.: Flow classification by histogram. In: ACM SIGMETRICS (2004)
11. Dewaele, G., Fukuda, K., Borgnat, P.: Extracting hidden anomalies using sketch and non Gaussian multiresulotion statistical detection procedures. In: ACM SIGCOMM (2007)
12. Lakhina, A., Crovella, M., Diot, C.: Mining anomalies using traffic feature distributions. In: ACM SIGCOMM, pp. 21–26 (2005)
13. Ye, N., Emran, S.M., Chen, Q., Vilbert, S.: Multivariate statistical analysis of audit trails for host-based intrusion detection. IEEE Transactions of Computers 51(7), 810–820 (2002)
14. Rieck, K., Laskov, P.: Language models for detection of unknown attacks in network traffic. Journal in Computer Virology (2007)
15. Xu, K., Zhang, Z.L., Bhattacharyya, S.: Profiling internet backbone traffic: Behavior models and applications. ACM SIGCOMM (2005)

16. Soule, A., Salamatian, K., Taft, N.: Combining filtering and statistical methods for anomaly detection. In: Internet Measurement Conference, pp. 331–344 (2005)
17. Moore, A.W., Zuev, D.: Internet traffic classification using bayesian analysis techniques. In: ACM SIGMETRICS (2005)
18. Kruegel, C., Mutz, D., Robertson, W., Valeur, F.: Bayesian event classification for intrusion detection. In: Annual Computer Security Applications Conference (2003)
19. Lazarevic, A., Ertoz, L., Kumar, V., Ozgur, A., Srivastava, J.: A compare study of anomaly detection schemes in network intrusion detection. In: SDM (2003)
20. Agosta, J.M., Duik-Wasser, C., Chandrashekar, J., Livadas, C.: An adaptive anomaly detector for worm detection. In: Proceedings of the Second Workshop on Tackling Computer Systems Problems with Machine Learning Techniques (2007)
21. Nodelman, U., Shelton, C.R., Koller, D.: Expectation maximization and complex duration distributions for continuous time Bayesian networks. In: UAI, pp. 421–430 (2005)
22. Dean, T., Kanazawa, K.: A model for reasoning about persistence and causation. Computational Intelligence 5(3), 142–150 (1989)
23. MAWI: MAWI working group traffic archive, `http://mawi.nezu.wide.ad.jp/mawi/`
24. LBNL: LBNL/ICSI enterprise tracing project, `http://www.icir.org/enterprise-tracing/Overview.html`
25. NLANR: National laboratory for applied network research (2006), `http://www.nlanr.net`

# Data Streaming with Affinity Propagation

Xiangliang Zhang, Cyril Furtlehner, and Michèle Sebag

TAO – INRIA CNRS
Université de Paris-Sud 11, F-91405 Orsay Cedex, France

**Abstract.** This paper proposed STRAP (Streaming AP), extending Affinity Propagation (AP) to data steaming. AP, a new clustering algorithm, extracts the data items, or exemplars, that best represent the dataset using a message passing method. Several steps are made to build STRAP. The first one (Weighted AP) extends AP to weighted items with no loss of generality. The second one (Hierarchical WAP) is concerned with reducing the quadratic AP complexity, by applying AP on data subsets and further applying Weighted AP on the exemplars extracted from all subsets. Finally STRAP extends Hierarchical WAP to deal with changes in the data distribution. Experiments on artificial datasets, on the Intrusion Detection benchmark (KDD99) and on a real-world problem, clustering the stream of jobs submitted to the EGEE grid system, provide a comparative validation of the approach.

## 1 Introduction

Data Streaming, one major task of Data Mining [1,2,3,4,5], aims to providing a compact description of the data flows generated by e.g., telecommunications, sensor networks, or Internet traffic. Data Streaming works under severe algorithmic constraints due to the size and dynamics of the data flow, since the underlying distribution of the data flow continuously evolves with the users, the usage and the application context.

Data Streaming is interested in various goals, e.g., computing approximate statistics [6,7], detecting novelties [1], or monitoring the top-K events [8] to name a few. This paper more specifically focuses on the identification of the clusters represented in the data stream, with the modelling of the jobs submitted to the EGEE grid[1] as motivating application.

The challenge is to achieve a good clustering behaviour, specifically enforcing a low distortion (more on this in Section 2) with low computational complexity, while swiftly adapting to the changes in the underlying distribution. An additional requirement is that a cluster should be represented by an actual data item (here, a job), as opposed to an average item, for the application domain hardly enables to consider artefacts. Under this requirement, clustering defines a combinatorial optimization problem, referred to as $K$-centers: i/ find the appropriate

[1] The EGEE grid, established in the EU project *Enabling Grid for e-Science in Europe*, `http://www.eu-egee.org/`) involves 41,000 CPUs and 5 Petabytes storage; it supports 20,000 concurrent jobs on a 24×7 basis.

W. Daelemans et al. (Eds.): ECML PKDD 2008, Part II, LNAI 5212, pp. 628–643, 2008.

number $K$ of clusters; ii/ retain the best $K$ items, referred to as exemplars, which constitute the best representatives of all items.

The proposed approach is based on a new clustering algorithm accommodating the above requirement, called Affinity Propagation (AP) [9,10]. AP (Section 2.1) defines an energy-based formulation of the $K$-centers combinatorial optimization problem, which is solved using a message passing algorithm akin belief propagation. The direct use of AP for Data Streaming however raises two difficulties. Firstly, AP has been designed for batch clustering, and it must be adapted to online clustering to handle data streams. Secondly, and most importantly, AP suffers from its quadratic computational complexity in the number $N$ of items.

The work presented in this paper extends AP to Data Streaming, and proposes three contributions. The first one extends AP to deal with duplicated items with no loss of performance (Weighted AP algorithm, WAP). The second one, called Hierarchical WAP (Hi-WAP), aims at decreasing the computational complexity to $\mathcal{O}(N^{1+\alpha})$ ($\alpha > 0$), taking inspiration from [11]: the idea is to partition the dataset, run AP on each subset, and apply WAP to the collection of exemplars constructed from each subset. The third one, StrAP, achieves online clustering, storing the outliers and occasionally updating the exemplars. Two update criteria have been considered. The first one is based on the number of outliers; when it reaches a predefined threshold, Hi-WAP is applied to the current exemplars and the outliers. The second one is based on the so-called Page-Hinkley change-point detection statistical test [12,13], observing the outlier rate. When a change point is detected, Hi-WAP is likewise applied on the current exemplars and the outliers. The experimental validation on artificial datasets, on the KDD99 benchmark dataset, and on the real-world dataset of the EGEE jobs, demonstrates the relevance of the approach compared to $K$-centers and the *DenStream* algorithm [4].

The paper is organized as follows. After presenting AP for the sake of completeness, Section 2 describes the first two extensions, Weighted and Hierarchical AP. Section 3 gives an overview of StrAP. Section 4 describes the experimental setting and the datasets used for the experimental validation of the approach. Finally, Section 5 and 6 report on the experimental results, using K-centers and *DenStream* as baselines. The approach is discussed w.r.t. related work and the paper concludes with some perspectives for further research.

## 2 Affinity Propagation and Scalable Extensions

For the sake of self-containedness, this section first describes the AP algorithm, referring the reader to [9,10] for a comprehensive introduction. Two AP extensions are thereafter described, respectively handling the case of weighted items, and the merge of partial solutions.

### 2.1 Affinity Propagation

Let $\mathcal{E} = \{e_1, \ldots e_N\}$ be a set of items, and let $d(i,j)$ denote the distance or dissimilarity between items $e_i$ and $e_j$. Letting $K$ denote a positive integer, the

$K$-center problem consists of finding $K$ items in $\mathcal{E}$, referred to as exemplars and denoted $e_{i_1}, \ldots, e_{i_K}$, such that they minimize the sum, over all items $e_j$, of the minimal squared distance between $e_j$ and $e_{i_k}, k = 1 \ldots K$.

The Affinity Propagation approach proposes an equivalent formalization of the $K$-center problem, defined in terms of energy minimization. Let $\sigma(i)$ associate to each item $e_i$ the index of its nearest exemplar, then the goal is to find the mapping $\sigma$ maximizing the functional $E[\sigma]$ defined as:

$$E[\sigma] = \sum_{i=1}^{N} S(e_i, e_{\sigma(i)}) - \sum_{i=1}^{N} \chi_i[\sigma] \quad (1)$$

where $S(e_i, e_j)$ is set to $-d(i,j)^2$ if $i \neq j$, and is set to a small constant $-s^*$, $s^* \geq 0$ called *preference* otherwise. The second term in the energy function expresses that if $e_i$ is selected as an exemplar by some items, it has to be its own exemplar[2], with $\chi_i[\sigma] = \infty$ if $\sigma(\sigma(i)) \neq \sigma(i)$ and 0 otherwise.

Aside from the consistency constraints, the energy function thus enforces a tradeoff between the distortion, i.e. the sum over all items of the squared error $d(i, \sigma(i))^2$ committed by assimilating item $e_i$ to its nearest exemplar $e_{\sigma(i)}$, and the cost of the model, that is $s^* \times |\sigma|$ if $|\sigma|$ denotes the number of exemplars retained. Eq. (1) thus does not directly specify the number of exemplars to be found, as opposed to $K$-centers. Instead, it specifies the penalty $s^*$ for allowing an item to become an exemplar; note that for $s^* = 0$, the best solution is the trivial one, selecting every item as an exemplar.

The resolution of the optimization problem defined by Eq. (1) is achieved by a message passing algorithm, considering two types of messages: availability messages $a(i,k)$ express the accumulated evidence for $e_k$ to be selected as the best exemplar for $e_i$; responsibility messages $r(i,k)$ express the fact that $e_k$ is suitable to be the exemplar of $e_i$.

All availability and responsibility messages $a(i,k)$ and $r(i,k)$ are set to 0 initially. Their values are iteratively adjusted[3] by setting:

$$\begin{aligned} r(i,k) &= S(e_i, e_k) - \max_{k', k' \neq k}\{a(i,k') + S(e_i, e'_k)\} \\ r(k,k) &= S(e_k, e_k) - \max_{k', k' \neq k}\{S(e_k, e'_k)\} \\ a(i,k) &= min\,\{0, r(k,k) + \textstyle\sum_{i', i' \neq i,k} \max\{0, r(i',k)\}\} \\ a(k,k) &= \textstyle\sum_{i', i' \neq k} \max\{0, r(i',k)\} \end{aligned}$$

The index of exemplar $\sigma(e_i)$ associated to $e_i$ is finally defined as:

$$\sigma(i) = argmax\,\{r(i,k) + a(i,k), k = 1 \ldots N\} \quad (2)$$

The algorithm is stopped after a maximal number of iterations or when the exemplars did not change for a given number of iterations.

---

[2] This constraint is relaxed by the soft-constraint AP (SCAP) [14], unveiling the hierarchical cluster structure in the data set while AP is biased toward regularly shaped clusters. The extension of the presented approach to SCAP is left for further study.

[3] Numerical oscillations are avoided by using a relaxation mechanism; empirically, the actual value is set to the half sum of the old and new values [9].

As could have been expected, Affinity Propagation is not to be seen as a universally efficient data clustering approach. Firstly, linear and robust algorithms such as $K$-means should be preferred to AP in domains where artefact items can be constructed[4]. Secondly, and most importantly, AP suffers from a quadratic computational complexity in the number $N$ of items: on the one hand, dissimilarities $d(i, j)$ must be computed; on the other hand, the message passing algorithm converges with complexity $\mathcal{O}(N^2 logN)$, hindering its direct use in large-scale applications. By convention, in the following notation $\widetilde{\mathcal{O}}(P)$ will stand for $\mathcal{O}(P \times polynom(\log P))$.

## 2.2 Weighted and Hierarchical WAP

Two extensions of AP, aiming at a lesser computational complexity, are presented in this section.

**Weighted AP.** A preliminary step is to extend AP in order to deal with multiply-defined items. Let dataset $\mathcal{E}' = \{(e_i, n_i)\}$ involve $n_i$ copies of item $e_i$, for $i = 1 \ldots L$. Let us consider the dissimilarity matrix $S'$ defined as:

$$S'(e_i, e_j) = \begin{cases} -n_i d(i,j)^2 & \text{if } i \neq j \\ s^* + (n_i - 1) \times \varepsilon_i & \text{otherwise}, \varepsilon_i \geq 0 \end{cases}$$

**Proposition.** The combinatorial optimization problem of finding $\sigma : \{1 \ldots L\}$ minimizing

$$E'[\sigma] = \sum_{i=1}^{L} S'(e_i, e_{\sigma(i)}) - \sum_{i=1}^{L} \chi_i[\sigma] \tag{3}$$

is equivalent, for $\varepsilon_i = 0$, to the optimization problem defined by Eq. (1) for $\mathcal{E}$ made of the union of $n_i$ copies of $e_i$, for $i = 1 \ldots L$.

*Proof*
*In the optimization problem defined by Eq. (1), assume that $e_i$ actually represents a set of $n_i$ identical copies; the penalty $S(e_i, e_j)$ of selecting $e_j$ as exemplar of $e_i$ thus is the cost of selecting $e_j$ as exemplar for each one of these copies. Therefore $S'(e_i, e_j) = n_i \times (-d(i, j)^2)$.*

*Likewise, let $e_i$ be unfolded as a set of $n_i$ (almost) identical copies $\{e_{i_1}, \ldots, e_{i_{n_i}}\}$, and let us assume that one of them, say $e_{i_1}$ is selected as exemplar. One thus pays the* preference *penalty $s^*$, plus the sum of the dissimilarities between $e_{i_1}$ and the other copies in $e_i$, modelled as $(n_i - 1)\varepsilon_i$. Constant $\varepsilon_i$ thus models the average dissimilarity among the $n_i$ copies of $e_i$.*

**Hierarchical WAP.** Hierarchical WAP proceeds by launching AP on subsets of the current dataset, and thereafter launching WAP on the dataset made of all exemplars extracted from the subsets.

[4] Selecting the best set of artefacts out of $\tau$ independent runs of $K$-means usually enforce a high-quality distortion, with complexity $\tau \times K \times N$.

Formally, let dataset $\mathcal{E}$ be equally divided into $\sqrt{N}$ subsets noted $\mathcal{E}_i, i = 1 \ldots \sqrt{N}$. Running AP on $\mathcal{E}_i$ outputs $K_i$ exemplars noted $\{e_{i_1}, \ldots e_{i_{K_i}}\}$; let us denote $n_{i_j}$ the number of items in $\mathcal{E}_i$ having $e_{i_j}$ as nearest exemplar.

Define $\mathcal{E}' = \{(e_{i_j}, n_{i_j}), i = 1 \ldots \sqrt{N}, j = 1 \ldots K_i\}$. HI-WAP launches WAP on $\mathcal{E}'$, and returns the produced exemplars.

**Proposition.** The complexity of HI-WAP is $\widetilde{\mathcal{O}}(N^{\frac{3}{2}})$.

*Proof. The construction of $\mathcal{E}'$ is in $\widetilde{\mathcal{O}}(N^{\frac{3}{2}})$, since AP is applied $\sqrt{N}$ times on datasets of size $\sqrt{N}$.*

*Letting $K$ be an upper bound on the number of exemplars learned from every subset $\mathcal{E}_i$, WAP thus achieves the distributed clustering of the exemplars extracted from all $\mathcal{E}_i$ with complexity $\widetilde{\mathcal{O}}(N^{\frac{1}{2}} \times K^2)$. The total complexity then is $\widetilde{\mathcal{O}}(NK^2 + N^{\frac{3}{2}})$, where term $N^{\frac{3}{2}}$ is dominant since $\sqrt{N} > K$.*

Iterating this hierarchical decomposition along the same lines as [11] leads to decrease the complexity to $\widetilde{\mathcal{O}}(N^{1+\alpha})$; on-going research is concerned with bounding the loss of distortion incurred by HI-WAP. Experimentally, the distortion loss is found to be very moderate while the computational cost decreases by one order of magnitude or more. Therefore, in the following we will use indifferently AP or HI-WAP, referred to as *AP, depending on the pressure on the computational resources and the size of the data.

# 3 Data Streaming with AP

This section describes the STRAP algorithm, extending AP to Data Streaming, involving four main steps (Alg. 1):

1. The first bunch of data is used by *AP to compute the first exemplars and initialize the stream model.
2. As the stream flows in, each data item $e_t$ is compared to the exemplars; if too far from the nearest exemplar, $e_t$ is put in the reservoir, otherwise the stream model is updated accordingly (section 3.1).
3. The restart criterion is triggered if the number of outliers exceeds the reservoir size, or upon a change point detection in the data distribution (section 3.2).
4. If it is triggered, the stream model is rebuilt from the current exemplars and the reservoir, using *AP again (section 3.3).

The stream model is available at any time. The performance of the process is measured from the average distortion and the clustering accuracy (section 3.4).

## 3.1 AP-Based Model and Update

The model of the data stream used in STRAP is inspired from *DbScan* [15] and *DenStream* [4]. It consists of 4-tuple $(e_i, n_i, \Sigma_i, t_i)$, where $e_i$ ranges over the exemplars, $n_i$ is the number of items associated to exemplar $e_i$, $\Sigma_i$ is the distortion of $e_i$ (sum of $d(e, e_i)^2$, where $e$ ranges over all items associated to $e_i$), and $t_i$ is the last time stamp when an item was associated to $e_i$.

For each new item $e_t$, its nearest exemplar $e_i$ is computed; if $d(e_t, e_i)$ is less than some threshold $\varepsilon$, heuristically set to the average distance between points and exemplars in the initial model, $e_t$ is affected to the $i$-th cluster and the model is updated accordingly; otherwise, $e_t$ is considered to be an outlier, and put in the reservoir.

In order to avoid the number of exemplars to grow beyond control, one must be able to forget the exemplars that have not been visited for some time. Accordingly, a used-specified window length $\Delta$ is considered; when item $e_t$ is associated to exemplar $e_i$, the model update is thus defined as:

$$n_i := n_i \times \left(\frac{\Delta}{\Delta+(t-t_i)} + \frac{1}{n_i+1}\right) \quad \Sigma_i := \Sigma_i \times \frac{\Delta}{\Delta+(t-t_i)} + \frac{n_i}{n_i+1} d(e_t, e_i)^2 \quad t_i := t$$

Simple calculations show that the above update rules enforce the model stability if exemplar $e_i$ is selected on average by $n_i$ examples during the last $\Delta$ time steps. The sensitivity analysis w.r.t. $\Delta$ is discussed in section 6.2.

### 3.2 Restart Criterion

A key difficulty in Data Streaming is to tell an acceptable ratio of outliers from a change in the generative process underlying the data stream, referred to as drift. In case of drift, the stream model must be updated. In some domains, e.g., continuous spaces, smooth updates can be achieved by gradually moving the centers of the clusters. When artefacts cannot be considered, the centers of the clusters must be redefined. In such domains, the data streaming process thus needs a restart criterion, in order to decide whether to launch the selection of new exemplars.

Two restart criteria have been considered. The first one is most simply based on the number of outliers in the reservoir; when it exceeds the reservoir size, the restart criterion is triggered. The second criterion is based on the distribution of the data items[5]. Let us consider the sequence of items $e_t$; define $p_t$ as $c/(1+o_t)$ where $o_t$ is the fraction of non-outliers and $c$ is 1 (resp. 2) if $e_t$ is an outlier (or not). If a drift occurs in the data distribution, then sequence $p_t$ should display some change; the restart criterion is triggered upon detecting such a change.

Among the many change point detection tests, the so-called Page-Hinkley test (PH) [12,13] has been selected as it minimizes the expected detection time for a prescribed false alarm rate. Formally, the PH test is controlled after a detection threshold $\lambda$ and tolerance $\delta$, as follows:

$$\begin{aligned} \bar{p}_t &= \tfrac{1}{t}\textstyle\sum_{\ell=1}^{t} p_\ell & \qquad m_t &= \textstyle\sum_{\ell=1}^{t} (p_\ell - \bar{p}_\ell + \delta) \\ M_t &= max\{m_\ell, \ell = 1...t\} & \qquad PH_t &= (M_t - m_t) > \lambda \end{aligned}$$

Parameter $\delta$ is set to $10^{-2}$ in all experiments. The sensitivity analysis w.r.t. $\lambda$ is presented in section 6.1.

[5] In case the number of outliers exceeds reservoir size, the new outlier replaces the oldest one in reservoir; a counter keeping track of the removed outliers is incremented.

### 3.3 Model Rebuild

Upon triggering of the restart criterion, Weighted AP is launched on $\mathcal{E} = \{(e_i, n_i)\} \cup \{(e'_j, 1)\}$, where $(e_i, n_i)$ denotes an exemplar of the current stream model together with the associated size $n_i$, and $e'_j$ is an outlier item in the reservoir. Penalties are defined after section 2.2, as follows:

$$S(e_i, e_i) = s^* + \Sigma_i \qquad S(e'_j, e'_j) = s^*$$
$$S(e_i, e_j) = -n_i\, d(e_i, e_j)^2 \qquad S(e_i, e'_j) = -n_i\, d(e_i, e'_j)^2$$
$$S(e'_j, e_i) = -d(e_i, e'_j)^2$$

After the new exemplars have been selected by WAP from $\mathcal{E}$, the stream model is defined as follows. Formally, let $f$ denote a new exemplar and let $e_1, \ldots e_m$ (respectively $e'_1, \ldots, e'_{m'}$) be the previous exemplars (resp. reservoir items) associated to $f$. With no difficulty, the number $n$ of items associated to $f$ is set to $n_1 + \ldots + n_m + m'$. The associated distortion $\Sigma$ is estimated as follows. Let $e$ be an item associated to $e_1$. Indeed $e$ is no longer available; but assuming an Euclidean space, $e$ can be modelled as a random item $e_1 + X\boldsymbol{v}$, where $\boldsymbol{v}$ is a random vector in the unit ball, and $X$ is a scalar random variable with normal distribution. It comes:

$$\begin{aligned} ||f - e||^2 &= ||f - e_1||^2 + ||e_1 - e||^2 - 2\langle f - e_1, X\boldsymbol{v}\rangle \\ &= d(f, e_1)^2 + d(e_1, e)^2 - 2X\langle f - e_1, \boldsymbol{v}\rangle \end{aligned}$$

Therefore, taking the expectation, $\mathbb{E}[d(f, e)^2] = d(f, e_1)^2 + \frac{1}{n_1}\Sigma_1$. Accordingly,

$$\Sigma = \sum_{i=1}^{m} \left(n_i d(f, e_i)^2 + \Sigma_i\right) + \sum_{i=1}^{m'} d(f, e'_i)^2$$

Finally, $t$ is set to the maximal time stamp associated to $e_i$ and $e'_j$, for $e_i$ and $e'_j$ ranging among the exemplars and outliers associated to $f$.

---

**Algorithm 1.** StrAP Algorithm

---

**Datastream** $e_1, \ldots e_t, \ldots$; **fit threshold** $\varepsilon$
**Init**
  $^*\mathrm{AP}(e_1, \ldots, e_T) \rightarrow$ StrAP Model — section 3.1
  Reservoir = {}
**for** $t > T$ **do**
  Compute $e_i$ = nearest exemplar to $e_t$ — section 3.1
  **if** $d(e_t, e_i) < \varepsilon$ **then**
    Update StrAP model — section 3.1
  **else**
    Reservoir $\leftarrow e_t$
  **end if**
  **if** Restart criterion **then** — section 3.2
    Rebuild StrAP model — section 3.3
    Reservoir = {}
  **end if**
**end for**

---

### 3.4 Evaluation Criterion

**Distortion.** The performance of STRAP is first measured after the overall distortion $D$, measured as follows. When a new item $e_t$ is associated to exemplar $e_i$, $D$ is incremented by $d(e_t, e_i)^2$. The distortion due to outliers is estimated a posteriori; after every restart, the average square distance $\bar{d^2}$ of the reservoir items to the new exemplars is computed, and $D$ is incremented by $\bar{d^2}$ times the number of items put in the reservoir since the previous restart (taking into account the outliers removed from reservoir when using PH restart criterion, section 3.2).

**Accuracy of Clustering.** In the case where the items are labeled, the clustering quality is also commonly assessed after the purity of the clusters. An item associated to an exemplar is correctly classified (respectively, misclassified) if its class is same (resp. different) as the exemplar class. The accuracy, the error rate and the percentage of outliers, sum up to 100%.

## 4 Goal of the Experiments and Setting

The algorithms presented in sections 2.2 and 3 raise two main questions.

The first question is whether HI-WAP, designed for reducing the computational effort of AP, does so at the expense of a significant increase of the distortion. The tradeoff between the computational cost and the distortion will be assessed by experimentally comparing HI-WAP with (batch) AP, on the one hand, and with other hierarchical variants (involving $K$-centers and AP, as opposed to WAP) on the other hand (Section 5). The experiments firstly involve benchmark datasets, kindly provided by E. Keogh [16]. As the focus essentially concerns the scalability of HI-WAP, only the largest two datasets (Faces and Swedish leaves, respectively including 2250 and 1125 examples) have been considered. Secondly, a real-world dataset describing the 237,087 jobs submitted to the EGEE grid system, has been considered. Each job is described by five attributes:

1. the duration of waiting time in a queue;
2. the duration of execution;
3. the number of jobs waiting in the queue when the current job arrived;
4. the number of jobs being executed after transiting from this queue;
5. the identifier of queue by which the job was transited.

Note that the behavior might be significantly different from one queue to another. The expert is willing to extract representative actual jobs (as opposed to virtual ones, e.g. executed on queue 1 with weight .3 and on queue 2 with weight .7), which is the main applicative motivation for using AP. The dissimilarity of two jobs $x_i$ and $x_j$ is the sum of the Euclidean distance between the numerical description of $x_i$ and $x_j$, plus a weight $w_q$ if $x_i$ and $x_j$ are not executed on the same queue. Further, the EGEE dataset involves circa 30% duplicated items (different jobs with same description).

The second question regards the performance of StrAP algorithm. As a clustering algorithm, it is meant to enforce a low distortion and high purity; as a streaming algorithm, it is required to efficiently adapt to the changing distribution of data stream. StrAP performances are assessed comparatively to those of *DenStream* [4] (Section 6), using an artificial stream generator, and the Intrusion Detection benchmark data set referred to as KDD Cup 1999 [17,18].

The stream generator is parameterized from the dimension $D$ of the data items, and the number $M$ of target exemplars. Each target exemplar $e_i$ is uniformly selected in $[-1,1]^D$ and its probability $p_i(t)$ evolves along time proportionally to $w_i \times sin(\omega_i t + \varphi_i)$, where weight $w_i$, frequency $\omega_i$ and phase $\varphi_i$ are uniformly selected respectively in $[1, 100]$, $[0, 2\pi]$, and $[-\pi/2, \pi/2]$. At time step $t$, exemplar $e_i$ is selected with probability $p_i(t)$, and item $e_t$ is set to $e_i$ plus gaussian noise.

The Intrusion Detection dataset we used includes 494,021 network connection records (71MB). Records are distributed among 23 classes, the *normal* class and the specific kinds of attack, such as *buffer_overflow*, *ftp_write*, *guess_passwd*, *neptune*. Out of the 41 attributes, only the numeric 34 ones have been used after [4], and cast such as they have same range of variation[6].

All reported computational times have been measured on Intel 2.66GHz Dual-Core PC with 2 GB memory.

## 5 Experimental Validation of Hi-WAP

This section reports on the performances of Hi-WAP, specifically the tradeoff between the computational effort and the distortion, on benchmark datasets and the real-world EGEE dataset.

### 5.1 Experimental Setting

On each dataset $\mathcal{E}$ of size $N$, the experiments were conducted as follows:

- $\mathcal{E}$ is partitioned into $\sqrt{N}$ subsets of equal size noted $\mathcal{E}_i$.
- Hi-WAP (respectively Hi-AP):
  1. On $\mathcal{E}_i$, the preference $s_i^*$ is set to the median of the pair similarities in the subset. WAP (respectively AP) is launched and produces a set of exemplars.
  2. WAP (respectively AP) is launched on the union of the exemplar set, varying preference $s^*$ from the minimum to the median distance of the exemplar pairs.
- Hierarchical $K$-centers:
  1. In parallel, $K$-centers is launched 120 times on each $\mathcal{E}_i$, where $K$ is set to the average number of exemplars extracted from the $\mathcal{E}_i$. The best set of exemplars (w.r.t. distortion) is retained; let $\mathcal{C}$ denote the union of these best sets of exemplars.

[6] Attribute *duration* is changed from seconds into minutes; *src_bytes* and *dst_bytes* are converted from byte to KB; *log* is used on *count*, *srv_count*, *dst_host_count*, *dst_host_srv_count*.

2. For each $K$, varying in the interval defined by the number of clusters obtained by Hi-WAP, 20 independent runs of $K$-centers are launched on $\mathcal{C}$, and the best set of exemplars is returned. The number of independent runs is such that Hierarchical $K$-centers and Hi-WAP have same computational cost for a fair comparison.

– The two approaches are graphically compared, reporting the distortion vs the number of clusters obtained by respectively Hierarchical $K$-Centers, Hi-WAPand Hi-AP.

## 5.2 Experimentation on Benchmark Dataset

As mentioned in Section 4, the benchmark datasets involve the largest two datasets from E. Keogh [16]. Table 1 displays the distortion obtained by $K$-centers and AP in non hierarchical and hierarchical settings. The number $K$ of clusters is set to the number of classes (Table 1.(a)) or defined by AP (Table 1.(b); $s^*$ is set to the median distance). $N$ and $D$ respectively stand for the number of items and the number of dimensions of the dataset.

After Table 1.(a), the loss of distortion incurred by Hi-WAP compared with AP is about 3.3% in the face dataset, and 5.7% in the Swedish leaf case; the distortion is about the same as for (non hierarchical) $K$-centers. The computational time is not relevant here due to the extra-cost of setting $s^*$ in order to enforce the desired number of clusters.

After Table 1.(b), AP significantly improves on $K$-centers (left part) in terms of distortion, by 14% in the face dataset and 27% in the Swedish leaf dataset; Hi-WAP similarly improves on hierarchical $K$-centers (right part), by 7% in the face dataset and 2% in the Swedish leaf dataset. Hierarchical variants improve over batch one by at least one order of magnitude in terms of computational

**Table 1.** Experimental results: Comparative Distortion of $K$-centers, AP, Hierarchical $K$-centers, Hi-AP and Hi-WAP. All results reported for $K$-centers variants are the best ones obtained with same computational effort as for AP variants.

(a). The number $K$ of clusters is set to the number of classes.

| Data | K | N | D | Non Hierarchical | | Hierarchical | | |
|---|---|---|---|---|---|---|---|---|
| | | | | *KC* | AP | *KC* | Hi-AP | Hi-WAP |
| Face (all) | 14 | 2250 | 131 | 189370 | **183265** | 198658 | 190496 | **189383** |
| Swedish Leaf | 15 | 1125 | 128 | 20220 | **19079** | 20731 | 20248 | **20181** |

(b). $K$ is fixed after AP or Hi-WAP, for $s^*$ set to the median distance.

| Data | N | D | Non Hierarchical | | | Hierarchical | | | |
|---|---|---|---|---|---|---|---|---|---|
| | | | K(AP) | *KC* | AP | K(Hi-WAP) | *KC* | Hi-AP | Hi-WAP |
| Face (all) | 2250 | 131 | 168 | 100420 | **88282** | 39 | 172359 | 164175 | **160415** |
| | | | | (128 sec) | | | (3 sec) | | |
| Swedish Leaf | 1125 | 128 | 100 | 12682 | **9965** | 23 | 21525 | **20992** | 21077 |
| | | | | (21 sec) | | | (1.4 sec) | | |

cost; the distortions cannot be directly compared as the number of clusters is different.

### 5.3 Experimentation on the EGEE Dataset

The real-world dataset describing the jobs submitted to the EGEE grid (section 4) is used to compare the distortion incurred by hierarchical clusterings, $K$-centers, HI-AP and HI-WAP, after the same procedure as in section 5.1; AP cannot be used on this dataset as it does not scale up. The number $K$ of clusters for $K$-centers is set to 15; 120 independent $K$-centers runs are launched, and the best distortion is reported, for a fair comparison (same computational cost and overall number of clusters). The first phase (clustering all $\sqrt{N}$ datasets) amounts to 10 minutes for $K$-centers and HI-WAP, and 26 minutes for HI-AP due to the duplications in the dataset). Note that the computational cost of this first phase can trivially be decreased by parallelization.

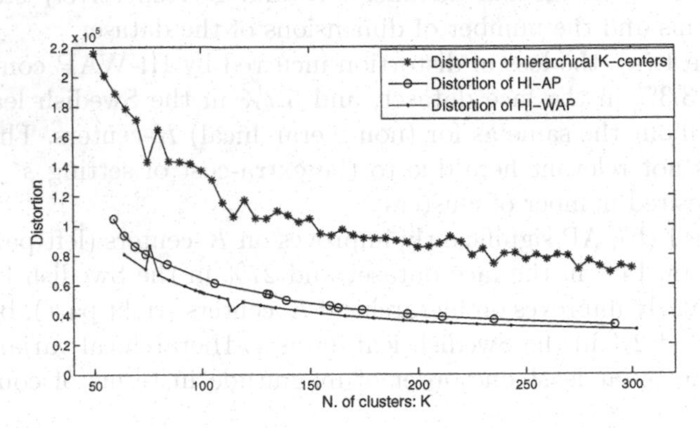


**Fig. 1.** Distortion of hierarchical $K$-centers, HI-AP and HI-WAP on the EGEE job dataset

The various distortions obtained when varying the number $K$ of clusters and the preference $s^*$ are reported in Fig. 1, showing that HI-WAP improves on both HI-AP and $K$-centers; the distortion is decreased by a factor 2 compared to $K$-centers and the computational cost (not shown) is decreased by a factor 3 compared to HI-AP.

## 6 Experimental Validation of StrAP

This section reports on the experimental validation of StrAP on a synthetic dataset and on the Intrusion Detection dataset (KDD99), described in Section 4. Results obtained on the EGEE dataset are omitted due to lack of space.

The initialization of the stream model (Section 3) considers the first 800 (synthetic data) or 1000 (KDD99) items.

### 6.1 Synthetic Data Stream

The dynamics of the synthetic data stream is depicted on Fig. 2; the only clusters represented at the beginning are clusters 2, 7, and 9. Representatives of cluster 0 appear shortly after the initialization; they are first considered to be outliers (legend $*$); using the Page-Hinkley restart criterion ($\lambda$=5, $\delta$=0.01), the first restart indicated by a vertical line occurs soon after. The same pattern is observed when the first representatives of clusters 3, 5 and 8 appear; they are first considered to be outliers, and they respectively trigger the second, third and fourth restarts thereafter. The small number of the "true" clusters makes it unnecessary to use a windowing mechanism (section 3.1).

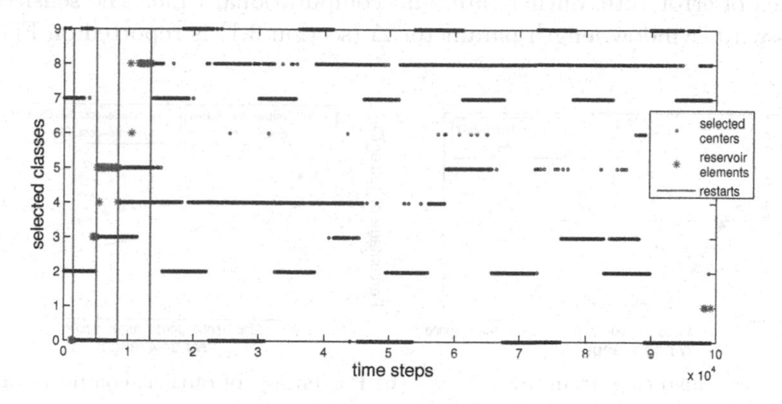


**Fig. 2.** Performance of STRAP on the synthetic data stream

Table 2 displays the performances of STRAP with respect to the percentage of outliers, the error rate[7], and the distortion, depending on the restart criterion used, and the parameters thereof.

**Table 2.** Experimental results of STRAP on the synthetic data stream

| Restart | | Outlier (%) | Error | N. of restart | N. of clusters | Distortion | Runtime |
|---|---|---|---|---|---|---|---|
| PH, $\delta$=0.01, $\lambda$= | 5 | 0.13 | 0 | 4 | 16 | 4369521 | 19 sec |
| | 10 | 0.17 | 0 | 4 | 16 | 4369410 | |
| | 20 | 0.62 | 0 | 4 | 20 | 4159085 | |
| Maximum size of reservoir | 50 | 0.20 | 0 | 4 | 16 | 4334710 | 20 sec |
| | 100 | 0.41 | 0 | 4 | 19 | 4116768 | |
| | 300 | 1.34 | 0 | 4 | 25 | 3896671 | |

All clusters are pure. Interestingly, when the restart criterion becomes less sensitive (increasing parameter $\lambda$ or reservoir size $MaxSizeR$), the outlier rate

[7] The classes of the items in the synthetic dataset correspond to the original exemplars respectively used to generate the items.

increases together with the number of clusters, while the number of restarts remains constant. A tentative interpretation is that, the less sensitive the restart, the more outliers and the more diverse the reservoir becomes; this diversity results in a higher number of clusters, decreasing the distortion. The computational time is circa 20 seconds for 100,000 items. Similar results are obtained for various number of dimensions ($D = 30$ in Table 2).

## 6.2 Intrusion Detection Dataset

The 494,021-transaction Intrusion Detection dataset is handled as a stream after [4]. STRAP performances are measured and compared with those of *Denstream*, in terms of error rate, outlier rate, and computational time. The sensitivity of results w.r.t. window length parameter $\Delta$ (section 3.1) is reported on Fig. 3.

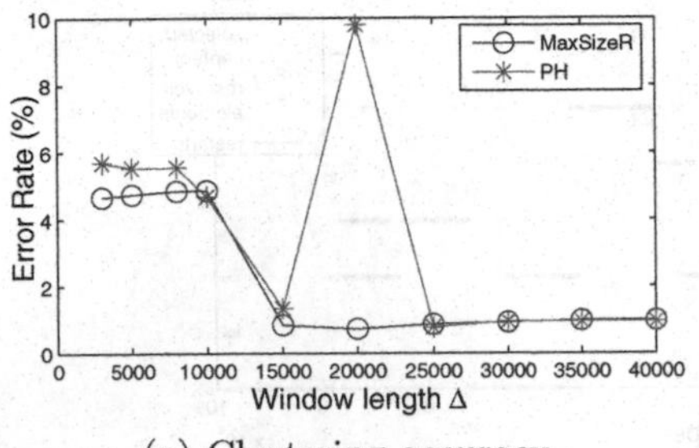


(a) Clustering accuracy

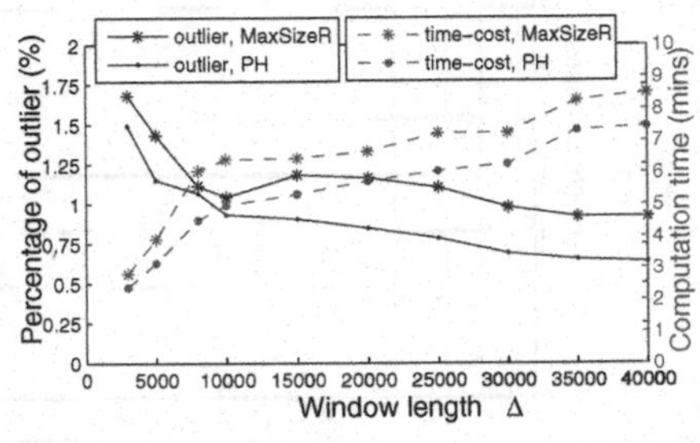


(b) Percentage of outlier, computation time

**Fig. 3.** Performance of STRAP on KDD99 dataset: comparing restart criterion PH ($\lambda = 20$) and Reservoir size ($MaxSizeR = 300$), depending on window length $\Delta$

Fig. 3 (a) shows that STRAP clustered the Intrusion Detection data stream with very low error rate (less than 1% for $\Delta > 15000$; the error peak observed for $\Delta = 20,000$ with the Page-Hinkley criterion is being investigated).

Fig. 3.(b) shows that the PH criterion improves on the Reservoir size, with a lower percentage of outliers and a smaller computational time, though the difference is not significant. The runtime is circa 7 minutes. It is worth noting that STRAP only needs 1% of the data (initial subset plus the outliers) in order to produce an accurate model (less than 1% error rate).

The online performance of STRAP is displayed in Fig. 4, reporting the error rate along time for $\Delta = 15000$ and $MaxSizeR = 300$; restarts are indicated with stars.

Fig. 5 presents a comparative assessment of STRAP and *DenStream* [4], using the same purity measure:

$$Purity = 100 \times (\sum_{i=1}^{K} \frac{|C_i^d|}{|C_i|})/K$$

where $K$ is the number of clusters, $|C_i|$ is the size of cluster $i$ and $|C_i^d|$ is the number of majority class items in cluster $i$. The clustering purity of *DenStream*

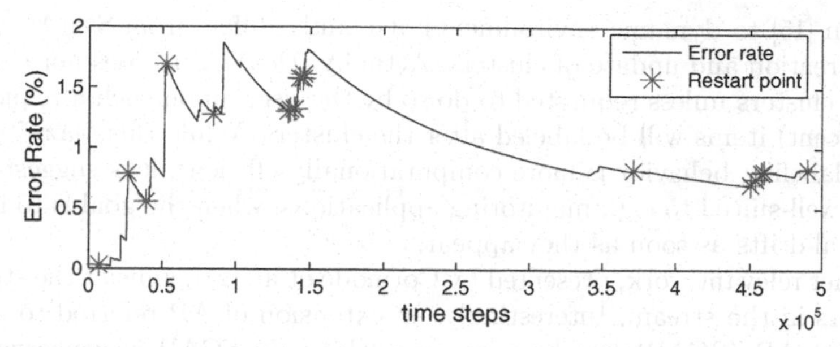


**Fig. 4.** Accuracy of STRAP on KDD99 data when $\Delta = 15000$ and $MaxSizeR = 300$

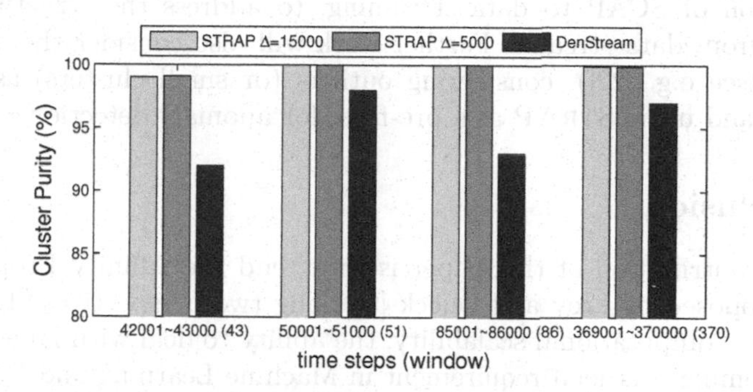


**Fig. 5.** Comparative performances of STRAP and *DenStream* on the Intrusion Detection dataset

on the Intrusion Detection dataset was evaluated during four time windows of length 1000 when some attacks happened. For a fair comparison, the clustering purity of STRAP was computed during the same time windows, considering the same 23 classes.

Fig. 5 respectively reports the results obtained for STRAP with one of the best settings ($\Delta = 15000$ and $MaxSizeR = 300$), an average setting ($\Delta = 5000$ and $MaxSizeR = 300$), and the results of *DenStream* found in [4]. In both STRAP settings, similar number of clusters are obtained (respectively circa 45, 32, 55 and 8 in the four windows).

On the Intrusion dataset, STRAP thus consistently improves on *DenStream*; as a counterpart, the computational time is higher by one or two orders of magnitude ( 7 minutes against 7 seconds for *DenStream*), noting that STRAP is written in Matlab.

### 6.3 Discussion

While many data streaming algorithms actually focus on the extraction of statistical information from data streams [6,7,8], ranging from the approximation of frequent patterns [19] to the construction of decision trees [20], the most related work is that of [4], similarly addressing unsupervised learning and clustering from data streams. The *DenStream* algorithm upgrades the *DbScan* clustering algorithm [15] to dynamic environments; it mainly differs from StrAP regarding the creation and update of clusters. Actually, *DenStream* does not construct the final clusters unless requested to do so by the user; upon such a request, the (most recent) items will be labeled after the clusters. While this "lazy" clustering and labeling behavior is more computationally efficient, it is suggested that it is not well-suited to e.g., monitoring applications, when the goal is to identify behavioral drifts as soon as they appear.

Another relevant work, presented by Cormode et al. [21], aims at the structure of clusters in the stream. Interestingly, an extension of AP referred to as Soft-Constraint AP (SCAP) has been proposed by [14]; SCAP is concerned with identifying the relations between the exemplars. Further research will investigate the extension of SCAP to data streaming, to address the structured cluster extraction from data streams. Further work will also consider the detection of anomalies (see e.g. [22]), considering outliers (or small clusters) as candidate anomalies, and using StrAP as a pre-filter for anomaly detection.

## 7 Conclusion

The main contribution of this paper is to extend the Affinity Propagation algorithm proposed by Frey and Dueck [9] along two perspectives. The first one concerns the computational scalability; the ability to deal with large datasets is indeed becoming a crucial requirement in Machine Learning and Data Mining. The second one, concerned with online clustering, aimed at dealing with evolving data distributions, and seamlessly updating the data model.

These extensions, encapsulated in the StrAP algorithm, have been empirically validated in the data streaming context, on two large sized datasets including the Intrusion Detection dataset used as KDD99 benchmark problem, and compared to the state-of-the art *DenStream* algorithm. While the accuracy of the data model constructed by StrAP was found comparatively satisfactory, the computational time is higher than for *DenStream*.

A first priority for further study is to provide theoretical guarantees for the Hi-WAP algorithm; while it has been shown that the divisive schema can be used to cut down the computational complexity and bring it down to a super-linear one, it is most desirable to provide theoretical guarantees on the loss of distortion incurred along the divisive process. The second priority regards a main limitation of AP, namely the fact that the number of clusters is only indirectly controlled from the preference parameter $s^*$. A strategy for adjusting this parameter, either globally after the desired number of clusters, or locally (depending on the estimated density of the dataset), would alleviate this limitation.

## References

1. Fan, W., Wang, H., Yu, P.: Active mining of data streams. In: SIAM Conference on Data Mining (SDM) (2004)
2. Aggarwal, C., Han, J., Wang, J., Yu, P.: A framework for clustering evolving data streams. In: Int. Conf. on Very Large Data Bases(VLDB), pp. 81–92 (2003)

3. Guha, S., Mishra, N., Motwani, R., O'Callaghan, L.: Clustering data streams. In: IEEE Symposium on Foundations of Computer Science, pp. 359–366 (2000)
4. Cao, F., Ester, M., Qian, W., Zhou, A.: Density-based clustering over an evolving data stream with noise. In: SIAM Conference on Data Mining (SDM) (2006)
5. Muthukrishnan, S.: Data streams: Algorithms and applications. Found. Trends Theor. Comput. Sci. 1, 117–236 (2005)
6. Papadimitriou, S., Brockwell, A., Faloutsos, C.: Adaptive, hands-off stream mining. In: Int. Conf. on Very Large Data Bases(VLDB), pp. 560–571 (2003)
7. Arasu, A., Manku, G.S.: Approximate counts and quantiles over sliding windows. In: ACM Symposium Principles of Database Systems(PODS), pp. 286–296 (2004)
8. Babcock, B., Olston, C.: Distributed topk monitoring. In: ACM International Conference on Management of Data (SIGMOD), pp. 28–39 (2003)
9. Frey, B., Dueck, D.: Clustering by passing messages between data points. Science 315, 972–976 (2007)
10. Frey, B., Dueck, D.: Supporting online material of clustering by passing messages between data points. Science 315 (2007), `http://www.sciencemag.org/cgi/content/full/1136800/DC1`
11. Guha, S., Meyerson, A., Mishra, N., Motwani, R., O'Callaghan, L.: Clustering data streams: Theory and practice. IEEE Transactions on Knowledge and Data Engineering (TKDE) 15, 515–528 (2003)
12. Page, E.: Continuous inspection schemes 41, 100–115 (1954)
13. Hinkley, D.: Inference about the change-point from cumulative sum tests. Biometrika 58, 509–523 (1971)
14. Leone, M., Sumedha, W.M.: Clustering by soft-constraint affinity propagation: Applications to gene-expression data. Bioinformatics 23, 2708 (2007)
15. Ester, M.: A density-based algorithm for discovering clusters in large spatial databases with noisethe uniqueness of a good optimum for k-means. In: International Conference on Knowledge Discovery and Data Mining(KDD) (1996)
16. Keogh, E., Xi, X., Wei, L., Ratanamahatana, C.A.: The UCR time series classification/clustering homepage (2006), `http://www.cs.ucr.edu/~eamonn/time_series_data/`
17. KDD99: Kdd cup 1999 data (computer network intrusion detection) (1999), `http://kdd.ics.uci.edu/databases/kddcup99/kddcup99.html`
18. Lee, W., Stolfo, S., Mok, K.: A data mining framework for building intrusion detection models. In: IEEE Symposium on Security and Privacy, pp. 120–132 (1999)
19. Dang, X.H., Ng, W.K., Ong, K.L.: An error bound guarantee algorithm for online mining frequent sets over data streams. Journal of Knowledge and Information Systems (2007)
20. Gama, J., Rocha, R., Medas, P.: Accurate decision trees for mining highspeed data streams. In: ACM International Conference on Management of Data (SIGMOD), pp. 523–528 (2003)
21. Cormode, G., Korn, F., Muthukrishnan, S., Srivastava, D.: Finding hierarchical heavy hitters in streaming data. ACM Transactions on Knowledge Discovery from Data (TKDD) 1(4) (2008)
22. Agarwal, D.K.: An empirical bayes approach to detect anomalies in dynamic multidimensional arrays. In: International Conference on Data Mining (ICDM) (2005)

# Semi-supervised Discriminant Analysis Via CCCP*

Yu Zhang and Dit-Yan Yeung

Hong Kong University of Science and Technology
{zhangyu,dyyeung}@cse.ust.hk

**Abstract.** Linear discriminant analysis (LDA) is commonly used for dimensionality reduction. In real-world applications where labeled data are scarce, LDA does not work very well. However, unlabeled data are often available in large quantities. We propose a novel semi-supervised discriminant analysis algorithm called $\text{SSDA}_{CCCP}$. We utilize unlabeled data to maximize an optimality criterion of LDA and use the constrained concave-convex procedure to solve the optimization problem. The optimization procedure leads to estimation of the class labels for the unlabeled data. We propose a novel confidence measure for selecting those unlabeled data points with high confidence. The selected unlabeled data can then be used to augment the original labeled data set for performing LDA. We also propose a variant of $\text{SSDA}_{CCCP}$, called $\text{M-SSDA}_{CCCP}$, which adopts the manifold assumption to utilize the unlabeled data. Extensive experiments on many benchmark data sets demonstrate the effectiveness of our proposed methods.

## 1 Introduction

*Linear discriminant analysis* (LDA) [1, 2] is a commonly used method for dimensionality reduction. It seeks a linear projection that simultaneously maximizes the between-class dissimilarity and minimizes the within-class dissimilarity to increase class separability, typically for classification applications. Despite its simplicity, the effectiveness and computational efficiency of LDA make it a popular choice for many applications. Nevertheless, LDA does have its limitations. One of these arises in situations when the sample size is much smaller than the dimensionality of the feature space, leading to the so-called *small sample size* (SSS) problem [3] due to severe undersampling of the underlying data distribution. As a result, the within-class scatter matrix that characterizes the within-class variability is not of full rank and hence it is not invertible. A number of methods have been proposed to overcome this problem, e.g., PseudoLDA [4], PCA+LDA [5], LDA/QR [6], NullLDA [3], and DualLDA [7]. PseudoLDA overcomes the singularity problem by substituting the inverse of the within-class scatter matrix with its pseudo-inverse. PCA+LDA first applies PCA [8] to project the data into a lower-dimensional space so that the within-class scatter matrix computed there is nonsingular, and then applies LDA in the lower-dimensional space. LDA/QR is also a two-stage method which can be divided into two steps: first project the data to the range space of the between-class scatter matrix and then apply LDA in this space. NullLDA first projects the data to the null space of the within-class scatter matrix

* This research has been supported by General Research Fund 621407 from the Research Grants Council of the Hong Kong Special Administrative Region, China.

W. Daelemans et al. (Eds.): ECML PKDD 2008, Part II, LNAI 5212, pp. 644–659, 2008.

and then maximizes the between-class scatter in this space. It is similar to the Discriminative Common Vectors method [9]. DualLDA, which combines the ideas from PCA+LDA and NullLDA, maximizes the between-class scatter matrix in the range space and the null space of the within-class scatter matrix separately and then integrates the two parts together to get the final transformation. There is also another approach to address the SSS problem, with 2DLDA [10] being the representative of this approach. The major difference between 2DLDA and the algorithms above lies in their data representation. Specifically, 2DLDA operates on data represented as (2D) matrices, instead of (1D) vectors, so that the dimensionality of the data representation can be kept small as a way to alleviate the SSS problem. Another limitation of LDA is that it only gives a linear projection of the data points. Fortunately, the kernel approach can be applied easily via the so-called kernel trick to extend LDA to its kernel version, called *kernel discriminant analysis* (KDA), that can project the data points nonlinearly, e.g., [11]. Besides addressing these two limitations of LDA, some interesting recent works also address other issues, e.g., to study the relationships between two variants of LDA [12], to reformulate multi-class LDA as a multivariate linear regression problem [13], and to learn the optimal kernel matrix for KDA using semi-definite programming (SDP) [14, 15].

In many real-world applications, it is impractical to expect the availability of large quantities of labeled data because labeling data requires laborious human effort. On the other hand, unlabeled data are available in large quantities at very low cost. Over the past decade or so, one form of *semi-supervised learning*, which attempts to utilize unlabeled data to aid classification or regression tasks under situations with limited labeled data, has emerged as a hot and promising research topic within the machine learning community. A good survey of semi-supervised learning methods can be found in [16]. Some early semi-supervised learnng methods include Co-Training [17] and transductive SVM (TSVM) [18, 19]. Recently, graph-based semi-supervised learning methods [20, 21, 22] have attracted the interests of many researchers. Unlike earlier methods, these methods model the geometric relationships between all data points in the form of a graph and then propagate the label information from the labeled data points through the graph to the unlabeled data points.

The objective of this paper is to alleviate the SSS problem of LDA by exploiting unlabeled data. We propose a novel *semi-supervised discriminant analysis* algorithm called $\text{SSDA}_{CCCP}$. Although there already exists another semi-supervised LDA algorithm, called SDA [23], which exploits the local neighborhood information of data points in performing dimensionality reduction, our $\text{SSDA}_{CCCP}$ algorithm works in a very different way. Specifically, we utilize unlabeled data to maximize an optimality criterion of LDA and formulate the problem as a constrained optimization problem that can be solved using the *constrained concave-convex procedure* (CCCP) [24, 25]. This procedure essentially estimates the class labels of the unlabeled data points. For those unlabeled data points whose labels are estimated with sufficiently high confidence based on some novel confidence measure proposed by us, we select them to expand the original labeled data set and then perform LDA again. Besides $\text{SSDA}_{CCCP}$, we also propose a variant of $\text{SSDA}_{CCCP}$, called $\text{M-SSDA}_{CCCP}$, which adopts the *manifold assumption* [20] to utilize the unlabeled data. Note that $\text{M-SSDA}_{CCCP}$ shares the spirit of both TSVM and graph-based semi-supervised learning methods.

The remainder of this paper is organized as follows. We first briefly review the traditional LDA algorithm in Section 2. We then present our $\text{SSDA}_{CCCP}$ and $\text{M-SSDA}_{CCCP}$ algorithms in Section 3. Section 4 reports experimental results based on some commonly used data sets. Performance comparison with some representative methods are reported there to demonstrate the effectiveness of our methods. Finally, some concluding remarks are offered in the last section.

## 2 Background

We are given a training set of $n$ data points, $\mathcal{D} = \{x_1, \ldots, x_n\}$, where $x_i \in \mathbb{R}^N$, $i = 1, \ldots, n$. Let $\mathcal{D}$ be partitioned into $C \geq 2$ disjoint classes $\Pi_i$, $i = 1, \ldots, C$, where class $\Pi_i$ contains $n_i$ examples. The between-class scatter matrix $S_b$ and the within-class scatter matrix $S_w$ are defined as

$$S_b = \sum_{k=1}^{C} n_k(\bar{m}_k - \bar{m})(\bar{m}_k - \bar{m})^T$$

$$S_w = \sum_{k=1}^{C} \sum_{x_i \in \Pi_k} (x_i - \bar{m}_k)(x_i - \bar{m}_k)^T,$$

where $\bar{m} = (\sum_{i=1}^{n} x_i)/n$ is the sample mean of the whole data set $\mathcal{D}$ and $\bar{m}_k = (\sum_{x_i \in \Pi_k} x_i)/n_k$ is the class mean of $\Pi_k$. LDA seeks to find a projection matrix $W^*$ that maximizes the trace function of $S_b$ and $S_w$:

$$W^* = \arg\max_{W} \text{trace}((W^T S_w W)^{-1} W^T S_b W), \tag{1}$$

which has an analytically tractable solution. According to [26], the optimal solution $W^*$ for the problem (1) can be computed from the eigenvectors of $S_w^{-1} S_b$, where $S_w^{-1}$ denotes the matrix inverse of $S_w$. Since $W^*$ computed this way is computationally simple yet effective for many applications, the optimality criterion in (1) is often used for many applications. Because the rank of $S_b$ is at most $C - 1$, $W$ contains $C - 1$ columns in most situations.

## 3 Semi-supervised Discriminant Analysis Via CCCP

In this section, we first present a theoretical result on the optimal solution for LDA. We then show how to utilize unlabeled data to solve the optimization problem, leading to the $\text{SSDA}_{CCCP}$ algorithm. Next, we incorporate the manifold assumption into $\text{SSDA}_{CCCP}$ to give $\text{M-SSDA}_{CCCP}$. Finally we give some discussions about our methods.

### 3.1 Optimal Solution for LDA

In our work, we use the following optimality criterion:

$$W^* = \arg\max_W \operatorname{trace}((W^T S_t W)^{-1} W^T S_b W), \tag{2}$$

where $S_t$ is the total scatter matrix with $S_t = S_b + S_w$. It is easy to prove that the optimal solution to the problem (2) is equivalent to that to the problem (1).

We assume that $S_t$ is of full rank, or else we can apply principal component analysis (PCA) [8] first to eliminate the null space of $S_t$ without affecting the performance of LDA since the null space makes no contribution to the discrimination ability of LDA [27].

The following theorem on the optimal solution to the problem (2) is relevant here.

**Theorem 1.** *For* $W \in \mathbb{R}^{N\times(C-1)}$,

$$\max_W \operatorname{trace}((W^T S_t W)^{-1} W^T S_b W) = \operatorname{trace}(S_t^{-1} S_b).$$

The proof of this theorem can be found in [26].

### 3.2 SSDA$_{CCCP}$: Exploiting Unlabeled Data to Maximize the Optimality Criterion

Suppose we have $l$ labeled data points $x_1, \ldots, x_l \in \mathbb{R}^N$ with class labels from $C$ classes $\Pi_i$, $i = 1, \ldots, C$, and $m$ unlabeled data points $x_{l+1}, \ldots, x_{l+m} \in \mathbb{R}^N$ with unknown class labels. So we have totally $n = l + m$ examples available for training. Usually $l \ll m$. When $l$ is too small compared with the input dimensionality, LDA generally does not perform very well. To remedy this problem, we want to incorporate unlabeled data to improve its performance.

Inspired by TSVM [18, 19], which utilizes unlabeled data to maximize the margin, we use unlabeled data here to maximize the optimality criterion of LDA. Since the optimal criterion value is $\operatorname{trace}(S_t^{-1} S_b)$ (from **Theorem 1**), we utilize unlabeled data to maximize $\operatorname{trace}(S_t^{-1} S_b)$ via estimating the class labels of the unlabeled data points.

We first calculate $S_t$ as $S_t = \sum_{i=1}^n (x_i - \bar{m})(x_i - \bar{m})^T$, where $\bar{m} = (\sum_{i=1}^n x_i)/n$ is the sample mean of all the data points. We define the class indicator matrix $A \in \mathbb{R}^{n\times C}$, where the $(i,j)$th element $A_{ij}$ is given by

$$A_{ij} = \begin{cases} 1 & \text{if } x_i \in \Pi_j \\ 0 & \text{otherwise} \end{cases} \tag{3}$$

If $D = (x_1, \ldots, x_l, x_{l+1}, \ldots, x_n)$ is the data matrix and $A_k$ is a vector for the $k$th column of $A$, then the class mean can be expressed as $\bar{m}_k = DA_k/n_k$, where $n_k = A_k^T 1_n$ is the number of data points that belong to the $k$th class and $1_n$ is an

$n$-dimensional column vector of ones. Similarly, we can also express the sample mean as $\bar{m} = D1_n/n$. Then $S_b$ can be calculated as

$$\begin{aligned} S_b &= \sum_{k=1}^{C} n_k(\bar{m}_k - \bar{m})(\bar{m}_k - \bar{m})^T \\ &= \sum_{k=1}^{C} n_k D\left(\frac{A_k}{n_k} - \frac{1_n}{n}\right)\left(\frac{A_k^T}{n_k} - \frac{1_n^T}{n}\right)D^T \\ &= D\left[\sum_{k=1}^{C} n_k\left(\frac{A_k}{n_k} - \frac{1_n}{n}\right)\left(\frac{A_k^T}{n_k} - \frac{1_n^T}{n}\right)\right]D^T. \end{aligned}$$

So $\text{trace}(S_t^{-1}S_b)$ can be calculated as

$$\begin{aligned} \text{trace}(S_t^{-1}S_b) &= \text{trace}\left(S_t^{-1}D\left[\sum_{k=1}^{C} n_k\left(\frac{A_k}{n_k} - \frac{1_n}{n}\right)\left(\frac{A_k^T}{n_k} - \frac{1_n^T}{n}\right)\right]D^T\right) \\ &= \text{trace}\left(\left[\sum_{k=1}^{C} n_k\left(\frac{A_k}{n_k} - \frac{1_n}{n}\right)\left(\frac{A_k^T}{n_k} - \frac{1_n^T}{n}\right)\right]D^T S_t^{-1} D\right) \\ &= \text{trace}\left(\sum_{k=1}^{C} n_k\left(\frac{A_k^T}{n_k} - \frac{1_n^T}{n}\right) S \left(\frac{A_k}{n_k} - \frac{1_n}{n}\right)\right) \\ &= \sum_{k=1}^{C} \frac{1}{n_k}\left(A_k^T - \frac{n_k}{n}1_n^T\right) S \left(A_k - \frac{n_k}{n}1_n\right), \end{aligned}$$

where $S = D^T S_t^{-1} D$ is a positive semi-definite matrix.

Since those entries in $A$ for the unlabeled data points are unknown, we maximize $\text{trace}(S_t^{-1}S_b)$ with respect to $A$. By defining some new variables for the sake of notational simplicity, we formulate the optimization problem as:

$$\begin{aligned} \max_{A,B_k,t_k} \quad & \sum_{k=1}^{C} \frac{B_k^T S B_k}{t_k} \\ \text{s.t.} \quad & t_k = A_k^T 1_n, \; k = 1, \ldots, C \\ & B_k = A_k - \frac{t_k}{n}1_n, \; k = 1, \ldots, C \\ & A_{ij} = \begin{cases} 1 \text{ if } x_i \in \Pi_j \\ 0 \text{ otherwise} \end{cases} \; i = 1, \ldots, l \\ & A_{ij} \in \{0,1\}, \; i = l{+}1, \ldots, n, \; j = 1, \ldots, C \\ & \sum_{j=1}^{C} A_{ij} = 1, \; i = l{+}1, \ldots, n. \end{aligned} \tag{4}$$

Unfortunately this is an integer programming problem which is known to be NP-hard and often has no efficient solution. We seek to make this integer programming problem tractable by relaxing the constraint $A_{ij} \in \{0,1\}$ in (4) to $A_{ij} \geq 0$, giving rise to a modified formulation of the optimization problem:

$$\begin{aligned}
\max_{A,B_k,t_k} & \sum_{k=1}^{C} \frac{B_k^T S B_k}{t_k} \\
s.t.\ & t_k = A_k^T 1_n,\ k = 1,\ldots,C \\
& B_k = A_k - \frac{t_k}{n} 1_n,\ k = 1,\ldots,C \\
& A_{ij} = \begin{cases} 1 \text{ if } x_i \in \Pi_j \\ 0 \text{ otherwise} \end{cases} i = 1,\ldots,l \\
& A_{ij} \geq 0,\ i = l{+}1,\ldots,n,\ j = 1,\ldots,C \\
& \sum_{j=1}^{C} A_{ij} = 1,\ i = l{+}1,\ldots,n.
\end{aligned} \tag{5}$$

With such relaxation, the matrix entries of $A$ for the unlabeled data points may be interpreted as posterior class probabilities. However, even though the constraints in the optimization problem (5) are linear, the problem seeks to maximize a convex function which, unfortunately, does not correspond to a convex optimization problem [28]. If we re-express the optimization problem in (5) as minimizing a concave function, we can adopt the *constrained concave-convex procedure* (CCCP) [24, 25] to solve this non-convex optimization problem. For our case, the convex part of the objective function degenerates to the special case of a constant function which always returns zero.

CCCP is an iterative algorithm. In each iteration, the concave part of the objective function for the optimization problem is replaced by its first-order Taylor series approximation at the point which corresponds to the result obtained in the previous iteration. Specifically, in the $(p{+}1)$th iteration, we solve the following optimization problem:

$$\begin{aligned}
\max_{A,B_k,t_k} & \sum_{k=1}^{C} \left( \frac{2(B_k^{(p)})^T S}{t_k^{(p)}} B_k - \frac{(B_k^{(p)})^T S B_k^{(p)}}{(t_k^{(p)})^2} t_k \right) \\
s.t.\ & t_k = A_k^T 1_n,\ k = 1,\ldots,C \\
& B_k = A_k - \frac{t_k}{n} 1_n,\ k = 1,\ldots,C \\
& A_{ij} = \begin{cases} 1 \text{ if } x_i \in \Pi_j \\ 0 \text{ otherwise} \end{cases} i = 1,\ldots,l \\
& A_{ij} \geq 0,\ i = l{+}1,\ldots,n,\ j = 1,\ldots,C \\
& \sum_{j=1}^{C} A_{ij} = 1,\ i = l{+}1,\ldots,n,
\end{aligned} \tag{6}$$

where $B_k^{(p)}, t_k^{(p)}, k = 1,\ldots,C$ were obtained in the $p$th iteration. The objective function in (6) is just the first-order Taylor series approximation of that in (5) by ignoring some constant terms.

Since the optimization problem (6) is a linear programming (LP) problem, it can be solved efficiently and hence can handle large-scale applications. Because the optimal solution of an LP problem falls on the boundary of its feasible set (or called constraint set), the matrix entries of the optimal $A_{ij}$ computed in each iteration must be in $\{0, 1\}$, which automatically satisfies the constraints in (4).

As the optimization problem is non-convex, the final solution that CCCP obtains generally depends on its initial value. For the labeled data points, the corresponding entries in $A_{ij}$ are held fixed based on their class labels. For the unlabeled data points, we initialize the corresponding entries in $A_{ij}$ with equal prior probabilities for all classes:

$$A_{ij}^{(0)} = \begin{cases} 1 & \text{if } x_i \in \Pi_j \\ 0 & \text{otherwise} \end{cases} \quad i = 1, \ldots, l, \ j = 1, \ldots, C$$

$$A_{ij}^{(0)} = \frac{1}{C}, \quad i = l+1, \ldots, n, \ j = 1, \ldots, C. \tag{7}$$

The initial values for $B_k^{(0)}$ and $t_k^{(0)}$ can be computed based on the equality constraints in (6) which establish the relationships between $A$, $B_k$ and $t_k$.

### 3.3 M-SSDA$_{CCCP}$: Incorporating the Manifold Assumption

The manifold assumption [20] is adopted by many graph-based semi-supervised learning methods. Under this assumption, nearby points are more likely to have the same class label for classification problems and similar low-dimensional representations for dimensionality reduction problems. We adopt this assumption to extend SSDA$_{CCCP}$ to M-SSDA$_{CCCP}$.

Given the data set $\mathcal{D} = \{x_1, \ldots, x_n\}$, we first construct a $K$-nearest neighbor graph $G = (V, E)$, with the vertex set $V = \{1, \ldots, n\}$ corresponding to the labeled and unlabeled data points and the edge set $E \subseteq V \times V$ representing the relationships between data points. Each edge is assigned a weight $w_{ij}$ which reflects the similarity between points $x_i$ and $x_j$:

$$w_{ij} = \begin{cases} \exp\left(-\frac{\|x_i - x_j\|^2}{\sigma_i \sigma_j}\right) & \text{if } x_i \in N_K(x_j) \text{ or } x_j \in N_K(x_i) \\ 0 & \text{otherwise} \end{cases}$$

where $N_K(x_i)$ denotes the neighborhood set of $K$-nearest neighbors of $x_i$, $\sigma_i$ the distance between $x_i$ and its $K$th nearest neighbor, and $\sigma_j$ the distance between $x_j$ and its $K$th nearest neighbor. This way of constructing the nearest neighbor graph is called *local scaling* [29], which is different from that in SDA [23]. In SDA, a constant value of 1 is set for all neighbors. This is unsatisfactory especially when some neighbors are relatively far away.

By incorporating the manifold assumption into our problem, we expect nearby points to be more likely to have the same class label and hence the two corresponding rows in $A$ are more likely to be the same. We thus modify the optimization problem (5) by adding one more term to the objective function:

$$
\begin{aligned}
\max_{A,B_k,t_k} \quad & \sum_{k=1}^{C} \frac{B_k^T S B_k}{t_k} - \lambda \sum_{i=1}^{n} \sum_{j=i+1}^{n} w_{ij} \|A(i) - A(j)\|_1 \\
s.t. \quad & t_k = A_k^T 1_n,\ k = 1, \ldots, C \\
& B_k = A_k - \frac{t_k}{n} 1_n,\ k = 1, \ldots, C \\
& A_{ij} = \begin{cases} 1 \text{ if } x_i \in \Pi_j \\ 0 \text{ otherwise} \end{cases} i = 1, \ldots, l \\
& A_{ij} \geq 0,\ i = l{+}1, \ldots, n,\ j = 1, \ldots, C \\
& \sum_{j=1}^{C} A_{ij} = 1,\ i = l{+}1, \ldots, n,
\end{aligned} \tag{8}
$$

where $\lambda > 0$ is a regularization parameter, $A(i)$ denotes the $i$th row of $A$, and $\|x\|_1$ is the $L_1$-norm of vector $x$.

Since the objective function of the optimization problem (8) is the difference of two convex functions, we can also adopt CCCP to solve it. Similar to SSDA$_{CCCP}$, in each iteration of CCCP, we also need to solve an LP problem:

$$
\begin{aligned}
\max_{A,B_k,t_k} \quad & \sum_{k=1}^{C} \left( \frac{2(B_k^{(p)})^T S}{t_k^{(p)}} B_k - \frac{(B_k^{(p)})^T S B_k^{(p)}}{(t_k^{(p)})^2} t_k \right) - \lambda \sum_{i=1}^{n} \sum_{j=i+1}^{n} w_{ij} \|A(i) - A(j)\|_1 \\
s.t. \quad & t_k = A_k^T 1_n,\ k = 1, \ldots, C \\
& B_k = A_k - \frac{t_k}{n} 1_n,\ k = 1, \ldots, C \\
& A_{ij} = \begin{cases} 1 \text{ if } x_i \in \Pi_j \\ 0 \text{ otherwise} \end{cases} i = 1, \ldots, l \\
& A_{ij} \geq 0,\ i = l{+}1, \ldots, n,\ j = 1, \ldots, C \\
& \sum_{j=1}^{C} A_{ij} = 1,\ i = l{+}1, \ldots, n.
\end{aligned} \tag{9}
$$

One reason for choosing the $L_1$-norm in the problem (8) is to keep the problem (9) as an LP problem which has an efficient and effective solution.

### 3.4 Augmenting the Labeled Data Set with Unlabeled Data

For both SSDA$_{CCCP}$ and M-SSDA$_{CCCP}$, CCCP estimates the class labels of all the unlabeled data points by solving the corresponding optimization problems with respect to $A$. One might then use all these unlabeled data points with estimated class labels to expand the labeled data set and then apply LDA again. However, it should be noted that not all the class labels can be estimated accurately. Thus, including those points with noisy class labels may impair the performance of LDA. Here we propose an effective method for selecting only those unlabeled data points whose labels are estimated with sufficiently high confidence.

Since all matrix entries in $A_{ij}$ obtained by CCCP are either 0 or 1, they cannot serve as posterior class probabilities for defining a measure to characterize the label estimation confidence. Here we propose an alternative scheme. We first use all the unlabeled data points with their estimated labels and the original labeled data set to perform LDA. Then, in the embedding space, we consider the neighborhood of each unlabeled data point by taking into account unlabeled data points only. If an unlabeled point has a sufficiently large proportion (determined by some threshold $\theta$, usually chosen to be larger than 0.5) of neighboring unlabeled points with the same estimated class label as its own, we consider this unlabeled point to have an estimated class label with high confidence and hence select it to augment the labeled data set for performing LDA again.

### 3.5 Discussions

In order to gain some insight into our method, we investigate the dual form of the optimization problem (6). We denote $R_k^{(p)} = \frac{2(B_k^{(p)})^T S}{t_k^{(p)}}$ and $q_k^{(p)} = \frac{(B_k^{(p)})^T S B_k^{(p)}}{(t_k^{(p)})^2}$, for $k = 1, \ldots, C$. We plug the first two equality constraints of the optimization problem (6) into its objective funciton and get the following Lagrangian:

$$L(A, \alpha, \beta) = \sum_{k=1}^{C} \left[ \left( q_k^{(p)} + \frac{R_k^{(p)} 1_n}{n} \right) 1_n^T - R_k^{(p)} \right] A_k - \sum_{k=1}^{C} \sum_{i=1}^{l} \alpha_{ki} (A_{ik} - \delta_k^{c(i)})$$
$$- \sum_{k=1}^{C} \sum_{i=l+1}^{n} \alpha_{ki} A_{ik} - \sum_{i=l+1}^{n} \beta_i \left( \sum_{k=1}^{C} A_{ik} - 1 \right),$$

where $c(i)$ is the class label of labeled data point $i$ and $\delta_k^{c(i)}$ is the delta function whose value is 1 if $c(i) = k$ and 0 otherwise.

So the dual form of the optimization problem (6) is

$$\begin{aligned} \max_{\alpha, \beta} \ & \sum_{k=1}^{C} \sum_{i=1}^{l} \alpha_{ki} \delta_k^{c(i)} + \sum_{i=l+1}^{n} \beta_i \\ s.t. \ & \alpha_{ki} = q_k^{(p)} - R_{ki}^{(p)} + \frac{R_k^{(p)} 1_n}{n}, \ i = 1, \ldots, l, \ k = 1, \ldots, C \\ & \alpha_{ki} + \beta_i = q_k^{(p)} - R_{ki}^{(p)} + \frac{R_k^{(p)} 1_n}{n}, \ i = l+1, \ldots, n, \ k = 1, \ldots, C \\ & \alpha_{ki} \geq 0, \ i = l+1, \ldots, n, \ k = 1, \ldots, C \end{aligned} \quad (10)$$

where $R_{ki}^{(p)}$ is the $i$th element of vector $R_k^{(p)}$.

The *Karush-Kuhn-Tucker* (KKT) condition [28] for the optimization problem (10) is

$$\alpha_{ki} A_{ik} = 0, \ i = l+1, \ldots, n, \ k = 1, \ldots, C. \quad (11)$$

From the first constraint of the optimization problem (10), we can see that each $\alpha_{ki}$ has a constant value for $i = 1, \ldots, l, \ k = 1, \ldots, C$. So we can simplify the optimization problem (10) by eliminating the first summation term in the objective function and the first constraint as

$$\max_{\alpha,\beta} \sum_{i=l+1}^{n} \beta_i$$
$$s.t.\ \alpha_{ki} + \beta_i = q_k^{(p)} - R_{ki}^{(p)} + \frac{R_k^{(p)} 1_n}{n},\ i = l+1, \dots, n,\ k = 1, \dots, C$$
$$\alpha_{ki} \geq 0,\ i = l+1, \dots, n,\ k = 1, \dots, C, \tag{12}$$

which can be further simplified as

$$\max_{\beta} \sum_{i=l+1}^{n} \beta_i$$
$$s.t.\ \beta_i \leq q_k^{(p)} - R_{ki}^{(p)} + \frac{R_k^{(p)} 1_n}{n},\ i = l+1, \dots, n,\ k = 1, \dots, C. \tag{13}$$

So the optimal solution of $\beta_i$ can be obtained as $\beta_i = \min_k \{q_k^{(p)} - R_{ki}^{(p)} + \frac{R_k^{(p)} 1_n}{n}\}$ for $i = l+1, \dots, n$.

For each unlabeled data point, if we assume $A_{ik^\star} > 0$, then from the KKT condition (11) we can get $\alpha_{k^\star i} = 0$ and also $\beta_i = q_{k^\star}^{(p)} - R_{k^\star i}^{(p)} + \frac{R_{k^\star}^{(p)} 1_n}{n}$ according to the first constraint of the optimization problem (12). So

$$q_{k^\star}^{(p)} - R_{k^\star i}^{(p)} + \frac{R_{k^\star}^{(p)} 1_n}{n} = \min_k \left\{ q_k^{(p)} - R_{ki}^{(p)} + \frac{R_k^{(p)} 1_n}{n} \right\}$$

and

$$k^\star = \arg\min_k \left\{ q_k^{(p)} - R_{ki}^{(p)} + \frac{R_k^{(p)} 1_n}{n} \right\}.$$

So $q_k^{(p)} - R_{ki}^{(p)} + \frac{R_k^{(p)} 1_n}{n}$ can be seen as the negative confidence that the $i$th data point belongs to the $k$th class and hence we can classify each data point to the class corresponding to the minimal negative confidence. If there is a unique minimum, then we can get $A_{ik^\star} = 1$ and $A_{ik'} = 0$ for $k' \neq k^\star$; otherwise, we can first find the set of unlabeled data points for which there exist unique minimum and $A_{ik}$ can be easily determined, and then we can solve a smaller LP problem (6) by plugging in the known elements $A_{ij}$. From our experiments, the latter situation seldom occurs and this can speed up the optimization problem (6), which even does not need to solve a LP problem.

[30] proposed a novel clustering method called DisKmeans which also maximize the optimality criterion of LDA to do clustering. However, its purpose is very different. In our work, M-SSDA$_{CCCP}$ and SSDA$_{CCCP}$ utilize unlabeled data to alleviate the SSS problem of LDA and we formulate the learning problem under the semi-supervised setting. On the other hand, DisKmeans aims at clustering high-dimensional data which is an unsupervised learning problem.

The computation cost of SSDA$_{CCCP}$ and M-SSDA$_{CCCP}$ includes performing LDA twice and solving the optimization problem using CCCP. The complexity of LDA

**Table 1.** Algorithm for SSDA$_{CCCP}$ or M-SSDA$_{CCCP}$

| Input: labeled data $x_i$ $(i = 1, \ldots, l)$, unlabeled data $x_i$ $(i = l+1, \ldots, n)$, $K, \theta, \varepsilon$ |
|---|
| Initialize $A^{(0)}$ using Eq. (7);<br>Initialize $B_k^{(0)}$ and $t_k^{(0)}$ based on $A^{(0)}$ for $k = 1, \ldots, C$;<br>Construct the $K$-nearest neighbor graph;<br>$p = 0$;<br>Repeat<br>    $p = p + 1$;<br>    Solve the optimization problem (6) or (9);<br>    Update $A^{(p)}$, $B_k^{(p)}$ and $t_k^{(p)}$ using the result of the optimization problem for $k = 1, \ldots, C$;<br>Until$\lVert A^{(p)} - A^{(p-1)} \rVert_F \leq \varepsilon$<br>Select the unlabeled data points with high confidence based on the threshold $\theta$;<br>Add the selected unlabeled data points with their estimated labels into the labeled data set and perform LDA on the augmented labeled data set to get the transformation $W$. |
| Output: the transformation $W$ |

is $O(N^3)$. The LP problem inside each iteration of CCCP can be solved efficiently. From our experimental results, CCCP converges very fast in less than 10 iterations. So SSDA$_{CCCP}$ and M-SSDA$_{CCCP}$ are efficient under most situations.

Finally, we summary this section by presenting the SSDA$_{CCCP}$ (or M-SSDA$_{CCCP}$) algorithm in Table 1.

## 4 Experiments

In this section, we first study SSDA$_{CCCP}$ and M-SSDA$_{CCCP}$ empirically and compare their performance with several other dimensionality reduction methods, including PCA, LDA [5] and SDA. Note that PCA is unsupervised, LDA is supervised, and SDA is semi-supervised in nature. After dimensionality reduction has been performed, we apply a simple nearest-neighbor classifier to perform classification in the embedding space. We also compare SSDA$_{CCCP}$ and M-SSDA$_{CCCP}$ with two state-of-the-art inductive semi-supervised learning methods, LapSVM and LapRLS [20].

### 4.1 Experimental Setup

We use MATLAB to implement all the algorithms and the CVX toolbox[1] for solving the optimization problems. We use the source code offered by Belkin et al. for LapSVM and LapRLS.[2] We evaluate these algorithms on 11 benchmark data sets, including 8 UCI data sets [31], a brain-computer interface dataset BCI[3] and two image data sets: COIL[3] and PIE [32]. See Table 2 for more details.

[1] http://www.stanford.edu/~boyd/cvx/
[2] http://manifold.cs.uchicago.edu/manifold_regularization/manifold.html
[3] http://www.kyb.tuebingen.mpg.de/ssl-book/

**Table 2.** Summary of data sets used and data partitioning for each data set

| Data set | #Dim ($N$) | #Inst ($n$) | #Class ($C$) | #Labeled ($q$) | #Unlabeled ($r$) |
|---|---|---|---|---|---|
| diabetes | 8 | 768 | 2 | 5 | 100 |
| heart-statlog | 13 | 270 | 2 | 5 | 100 |
| ionosphere | 34 | 351 | 2 | 5 | 50 |
| hayes-roth | 4 | 160 | 3 | 3 | 20 |
| iris | 4 | 150 | 3 | 3 | 20 |
| mfeat-pixel | 240 | 2000 | 10 | 5 | 50 |
| pendigits | 16 | 10992 | 10 | 5 | 95 |
| vehicle | 18 | 864 | 4 | 5 | 100 |
| BCI | 117 | 400 | 2 | 5 | 50 |
| COIL | 241 | 1500 | 6 | 5 | 100 |
| PIE | 1024 | 1470 | 30 | 2 | 20 |

For each data set, we randomly select $q$ data points from each class as labeled data and $r$ points from each class as unlabeled data. The remaining data form the test set. Table 2 shows the data partitioning for each data set. For each partitioning, we perform 20 random splits and report the mean and standard derivation over the 20 trials. For M-SSDA$_{CCCP}$, we choose the number of nearest neighbors $K$ for constructing the $K$-nearest neighbor graph to be the same as that for SDA, LapSVM, and LapRLS.

## 4.2 Experimental Results

We first compare our methods with dimensionality reduction methods and the experimental results are listed in Table 3. There are two rows for each data set: the upper one being the classification error on the unlabeled training data and the lower one being that on the test data. For each data set, the lowest classification error is shown in boldface. From the results, we can see that the performance of SSDA$_{CCCP}$ or M-SSDA$_{CCCP}$ is better than other methods in most situations. For DIABETES, HEART-STATLOG, PENDIGITS, VEHICLE and PIE, the improvement is very significant. Moreover, for the data sets such as DIABETES and HEART-STATLOG which may not contain manifold structure, the performance of SSDA$_{CCCP}$ is better than M-SSDA$_{CCCP}$. For MFEAT-PIXEL, PIE and others which may contain manifold structure, the performance of M-SSDA$_{CCCP}$ is better than SSDA$_{CCCP}$. Thus for data sets such as images which may have manifold structure, we recommend to use M-SSDA$_{CCCP}$. Otherwise SSDA$_{CCCP}$ is preferred. Compared with SDA, SSDA$_{CCCP}$ and M-SSDA$_{CCCP}$ are more stable. Specifically, the performance of SSDA$_{CCCP}$ or M-SSDA$_{CCCP}$ is comparable to or better than that of LDA in most situations. For SDA, however, the performance degradation can sometimes be very severe, especially for MFEAT-PIXEL and PIE.

We also investigate the selection method described in Section 3.4. We record the mean accuracy of label estimation for the unlabeled data over 20 trials before and after applying the selection method. The results in Table 4 show that the estimation accuracy after applying the selection method is almost always higher, sometimes very significantly. This confirms that our selection method for unlabeled data is very effective.

**Table 3.** Average classification errors for each method on each data set. Each number inside brackets shows the corresponding standard derivation. The upper row for each data set is the classification error on the unlabeled training data and the lower row is that on the test data.

| Data set | PCA | LDA | SDA | $\text{SSDA}_{CCCP}$ | $\text{M-SSDA}_{CCCP}$ |
|---|---|---|---|---|---|
| diabetes | 0.4335(0.0775) | 0.4438(0.0878) | 0.4022(0.0638) | **0.3898(0.0674)** | 0.4360(0.0605) |
| | 0.4253(0.1154) | 0.4311(0.0997) | 0.3763(0.0864) | **0.3276(0.0643)** | 0.4125(0.1074) |
| heart-statlog | 0.4288(0.0689) | 0.3978(0.0582) | 0.3680(0.0564) | **0.3293(0.0976)** | 0.3818(0.0662) |
| | 0.3975(0.0669) | 0.3767(0.1055) | 0.3783(0.1076) | **0.3133(0.1174)** | 0.3258(0.1493) |
| ionosphere | 0.2895(0.1032) | 0.2850(0.0876) | **0.2695(0.1056)** | 0.2860(0.1015) | 0.2830(0.1029) |
| | **0.2189(0.0632)** | 0.2365(0.0972) | 0.2241(0.0863) | 0.2351(0.1032) | 0.2399(0.1278) |
| hayes-roth | 0.5175(0.0571) | 0.4942(0.0531) | 0.5058(0.0661) | 0.4867(0.0569) | **0.4758(0.0586)** |
| | 0.5115(0.0605) | 0.5165(0.0690) | 0.5077(0.0752) | 0.5121(0.0770) | **0.5060(0.0627)** |
| iris | 0.0917(0.0417) | 0.0933(0.0613) | 0.0825(0.0506) | 0.0708(0.0445) | **0.0667(0.0493)** |
| | 0.0907(0.0333) | 0.0833(0.0586) | 0.0809(0.0395) | 0.0611(0.0370) | **0.0611(0.0454)** |
| mfeat-pixel | 0.1450(0.0232) | 0.1501(0.0290) | 0.2783(0.0435) | 0.1501(0.0289) | **0.1367(0.0210)** |
| | 0.1429(0.0228) | 0.1486(0.0264) | 0.3428(0.0298) | 0.1485(0.0264) | **0.1329(0.0213)** |
| pendigits | 0.1724(0.0305) | 0.2238(0.0364) | 0.2547(0.0447) | 0.1785(0.0266) | **0.1617(0.0242)** |
| | 0.1761(0.0276) | 0.2192(0.0332) | 0.2544(0.0382) | 0.1779(0.0190) | **0.1650(0.0225)** |
| vehicle | 0.5739(0.0375) | 0.5741(0.0365) | 0.5400(0.0402) | **0.4396(0.0734)** | 0.4838(0.0901) |
| | 0.5808(0.0453) | 0.5879(0.0429) | 0.5462(0.0312) | **0.4329(0.0672)** | 0.4739(0.0791) |
| BCI | 0.4835(0.0460) | 0.4830(0.0557) | 0.4960(0.0476) | **0.4750(0.0432)** | 0.4975(0.0484) |
| | 0.5000(0.0324) | 0.4803(0.0249) | 0.4812(0.0326) | **0.4732(0.0331)** | 0.4741(0.0346) |
| COIL | **0.4443(0.0418)** | 0.5247(0.0371) | 0.5419(0.0607) | 0.5236(0.0374) | 0.5193(0.0401) |
| | **0.4391(0.0364)** | 0.5194(0.0421) | 0.5461(0.04821) | 0.5178(0.0434) | 0.5096(0.0398) |
| PIE | 0.6156(0.0275) | 0.5055(0.1624) | 0.7629(0.0377) | 0.4674(0.1757) | **0.2381(0.0552)** |
| | 0.6207(0.0251) | 0.5126(0.1512) | 0.8277(0.0208) | 0.4777(0.1696) | **0.2424(0.0592)** |

**Table 4.** Accuracy of label estimation for the unlabeled data before and after applying the selection method

| | $\text{SSDA}_{CCCP}$ (%) | | $\text{M-SSDA}_{CCCP}$ (%) | |
|---|---|---|---|---|
| Data set | Before | After | Before | After |
| diabetes | 64.03 | 66.67 | 54.10 | 51.20 |
| heart-statlog | 72.27 | 72.62 | 55.25 | 66.70 |
| ionosphere | 69.05 | 87.51 | 74.10 | 82.07 |
| hayes-roth | 46.75 | 52.73 | 42.00 | 42.64 |
| iris | 75.42 | 93.39 | 91.42 | 95.06 |
| mfeat-pixel | 32.49 | 100.0 | 94.21 | 98.91 |
| pendigits | 75.31 | 86.08 | 88.92 | 94.02 |
| vehicle | 56.30 | 69.88 | 44.80 | 52.26 |
| BCI | 50.75 | 65.42 | 49.00 | 49.15 |
| COIL | 33.57 | 96.07 | 42.64 | 60.03 |
| PIE | 30.48 | 85.00 | 52.64 | 70.41 |

**Table 5.** Average classification errors for each method on each data set. Each number inside brackets shows the corresponding standard derivation. The upper row for each data set is the classification error on the unlabeled training data and the lower row is that on the test data.

| Data set | LapSVM | LapRLS | $\text{SSDA}_{CCCP}$ | $\text{M-SSDA}_{CCCP}$ |
|---|---|---|---|---|
| diabetes | 0.4763(0.0586) | 0.4523(0.0650) | **0.3620(0.0680)** | 0.4015(0.0893) |
| | 0.5643(0.0684) | 0.5009(0.0775) | **0.3488(0.0514)** | 0.4234(0.1107) |
| heart-statlog | 0.3478(0.1059) | 0.3348(0.1070) | **0.3108(0.0901)** | 0.3758(0.0914) |
| | 0.3517(0.1458) | 0.3375(0.1366) | **0.3091(0.0989)** | 0.3442(0.1226) |
| ionosphere | 0.3525(0.0539) | 0.3260(0.0527) | 0.3340(0.0902) | **0.3185(0.0719)** |
| | **0.2245(0.0697)** | 0.2266(0.0732) | 0.2705(0.0969) | 0.2905(0.0933) |
| hayes-roth | 0.6633(0.0149) | 0.6608(0.0261) | **0.4833(0.0824)** | 0.5225(0.0466) |
| | 0.5550(0.0737) | 0.5500(0.0516) | **0.4901(0.0705)** | 0.5104(0.0711) |
| iris | 0.3175(0.1390) | 0.2708(0.1474) | 0.0650(0.0516) | **0.0525(0.0437)** |
| | 0.3049(0.1426) | 0.2741(0.1473) | 0.0772(0.0508) | **0.0593(0.0379)** |
| mfeat-pixel | 0.1488(0.0236) | **0.1359(0.0257)** | 0.1578(0.0268) | 0.1420(0.0249) |
| | 0.2252(0.0187) | 0.2075(0.0181) | 0.1555(0.0263) | **0.1427(0.0183)** |
| pendigits | 0.2571(0.0379) | 0.2368(0.0312) | 0.1856(0.0226) | **0.1697(0.0245)** |
| | 0.2539(0.0334) | 0.2377(0.0283) | 0.1866(0.0244) | **0.1735(0.0217)** |
| vehicle | 0.4713(0.0449) | 0.4921(0.0460) | **0.4219(0.0623)** | 0.4645(0.0770) |
| | 0.4758(0.0477) | 0.5007(0.0452) | **0.4181(0.0600)** | 0.4641(0.0777) |
| BCI | 0.4805(0.0551) | 0.4695(0.0612) | **0.4515(0.0543)** | 0.4665(0.0479) |
| | 0.4631(0.0456) | **0.4562(0.0390)** | 0.4752(0.0362) | 0.4864(0.0372) |
| COIL | 0.5414(0.0496) | 0.5855(0.0617) | **0.5028(0.0576)** | 0.5030(0.0488) |
| | 0.5421(0.0497) | 0.5864(0.0598) | **0.5057(0.0533)** | 0.5062(0.0423) |
| PIE | 0.2561(0.0311) | 0.3405(0.0227) | 0.4096(0.1600) | **0.2497(0.0313)** |
| | 0.2671(0.0235) | 0.3523(0.0151) | 0.4160(0.1575) | **0.2556(0.0235)** |

Next we compare our methods with some representative semi-supervised learning methods. The experimental settings are the same as those in the first experiment. There are many popular semi-supervised learning methods, such as Co-Training [17], TSVM [18, 19], methods in [21, 22], LapSVM and LapRLS [20]. Co-Training requires two independent and sufficient views for the data, but data used in our experiment can not satisfy this requirement. TSVM has high computation cost and hence cannot be used for large-scale problems. Thus it is not included in our experiment. The methods in [21,22] can only work under the transductive setting, in which the test data, in addition to the training data, must be available during model training and the learned model cannot be applied to unseen test data easily. So these methods can not satisfy our experimental settings and are excluded in our experiments. LapSVM and LapRLS, which also adopt the manifold assumption, have efficient solutions and can work under the inductive setting. So we have included them in our experiment for performance comparison. The standard LapSVM and LapRLS algorithms are for two-class problems. For multi-class problems, we adopt the *one vs. rest* strategy as in [20] for LapSVM and LapRLS. Since the methods used here are all linear methods, we use a linear kernel for LapSVM and LapRLS. The experimental results are shown in Table 5. From the experimental results, we can see that the performance of $\text{SSDA}_{CCCP}$ and $\text{M-SSDA}_{CCCP}$ is comparable to or even better than that of LapSVM and LapRLS. Moreover, One advantage of $\text{SSDA}_{CCCP}$

and M-SSDA$_{CCCP}$ is that their formulation and optimization procedure are the same for two-class and multi-class problems. However, this is not the case for LapSVM and LapRLS which require training the models multiple times for multi-class problems.

## 5 Conclusion

In this paper, we have presented a new approach for semi-supervised discriminant analysis. By making use of both labeled and unlabeled data in learning a transformation for dimensionality reduction, this approach overcomes a serious limitation of LDA under situations where labeled data are limited. In our future work, we will investigate kernel extensions to our proposed methods in dealing with nonlinearity. Moreover, we will also apply the ideas here to some other dimensionality reduction methods.

## References


1. Fisher, R.A.: The use of multiple measurements in taxonomic problems. Annals of Eugenics 7, 179–188 (1936)
2. Rao, C.R.: The utilization of multiple measurements in problems of biological classification. Journal of the Royal Statictical Society 10, 159–203 (1948)
3. Chen, L., Liao, H., Ko, M., Lin, J., Yu, G.: A new LDA-based face recognition system which can solve the small sample size problem. Pattern Recognition 33(10), 1713–1726 (2000)
4. Krzanowski, W.J., Jonathan, P., McCarthy, W.V., Thomas, M.R.: Discriminant analysis with singular covariance matrices: methods and applications to spectroscopic data. Applied Statistics 44(1), 101–115 (1995)
5. Belhumeur, P.N., Hespanha, J.P., Kriegman, D.J.: Eigenfaces vs. Fisherfaces: Recognition using class specific linear projection. IEEE Transactions on Pattern Analysis and Machine Intelligence 19(7), 711–720 (1997)
6. Ye, J.P., Li, Q.: A two-stage linear discriminant analysis via QR-Decomposition. IEEE Transactions on Pattern Analysis and Machine Intelligence 27(6), 929–941 (2005)
7. Wang, X., Tang, X.: Dual-space linear discriminant analysis for face recognition. In: Proceedings of the IEEE Computer Society Conference on Computer Vision and Pattern Recognition, Washington, DC, pp. 564–569 (2004)
8. Jolliffe, I.T.: Principal Component Analysis. Springer, New York (1986)
9. Cevikalp, H., Neamtu, M., Wilkes, M., Barkana, A.: Discriminative common vectors for face recognition. IEEE Transactions on Pattern Analysis and Machine Intelligence 27(1), 4–13 (2005)
10. Ye, J.P., Janardan, R., Li, Q.: Two-dimensional linear discriminant analysis. In: Advances in Neural Information Processing Systems 17, Vancouver, British Columbia, Canada, pp. 1529–1536 (2004)
11. Baudat, G., Anouar, F.: Generalized discriminant analysis using a kernel approach. Neural Computation 12(10), 2385–2404 (2000)
12. Ye, J.P., Xiong, T.: Computational and theoretical analysis of null space and orthogonal linear discriminant analysis. Journal of Machine Learning Research 7, 1183–1204 (2006)
13. Ye, J.P.: Least squares linear discriminant analysis. In: Proceedings of the Twenty-Fourth International Conference on Machine Learning, Corvalis, Oregon, USA, pp. 1087–1093 (2007)
14. Kim, S.J., Magnani, A., Boyd, S.: Optimal kernel selection in kernel fisher discriminant analysis. In: Proceedings of the Twenty-Third International Conference on Machine Learning, Pittsburgh, Pennsylvania, USA, pp. 465–472 (2006)

15. Ye, J.P., Chen, J., Ji, S.: Discriminant kernel and regularization parameter learning via semidefinite programming. In: Proceedings of the Twenty-Fourth International Conference on Machine Learning, Corvalis, Oregon, USA, pp. 1095–1102 (2007)
16. Zhu, X.: Semi-supervised learning literature survey. Technical Report 1530, Department of Computer Sciences, University of Wisconsin at Madison, Madison, WI (2006)
17. Blum, A., Mitchell, T.: Combining labeled and unlabeled data with co-training. In: Proceedings of the Workshop on Computational Learning Theory, Madison, Wisconsin, USA, pp. 92–100 (1998)
18. Bennett, K., Demiriz, A.: Semi-supervised support vector machines. In: Advances in Neural Information Processing Systems 11, Vancouver, British Columbia, Canada, pp. 368–374 (1998)
19. Joachims, T.: Transductive inference for text classification using support vector machines. In: Proceedings of the Sixteenth International Conference on Machine Learning, San Francisco, CA, USA, pp. 200–209 (1999)
20. Belkin, M., Niyogi, P., Sindhwani, V.: Manifold regularization: A geometric framework for learning from examples. Journal of Machine Learning Research 7, 2399–2434 (2006)
21. Zhou, D., Bousquet, O., Lal, T., Weston, J., Schölkopf, B.: Learning with local and global consistency. In: Advances in Neural Information Processing Systems 16, Vancouver, British Columbia, Canada (2003)
22. Zhu, X., Ghahramani, Z., Lafferty, J.: Semi-supervised learning using Gaussian fields and harmonic functions. In: Proceedings of the Twentieth International Conference on Machine Learning, Washington, DC, pp. 912–919 (2003)
23. Cai, D., He, X., Han, J.: Semi-supervised discriminant analysis. In: Proceedings of the IEEE International Conference on Computer Vision, Rio de Janeiro, Brazil (2007)
24. Yuille, A., Rangarajan, A.: The concave-convex procedure. Neural Computation 15(4), 915–936 (2003)
25. Smola, A.J., Vishwanathan, S.V.N., Hofmann, T.: Kernel methods for missing variables. In: Proceedings of the Tenth International Workshop on Artificial Intelligence and Statistics, Barbados (2005)
26. Fukunnaga, K.: Introduction to Statistical Pattern Recognition. Academic Press, New York (1991)
27. Yang, J., Yang, J.Y.: Why can LDA be performed in PCA transformed space? Pattern Recognition 36(2), 563–566 (2003)
28. Boyd, S., Vandenberghe, L.: Convex Optimization. Cambridge University Press, New York (2004)
29. Zelnik-Manor, L., Perona, P.: Self-tuning spectral clustering. In: Advances in Neural Information Processing Systems 17, Vancouver, British Columbia, Canada, pp. 1601–1608 (2004)
30. Ye, J.P., Zhao, Z., Wu, M.: Discriminative k-means for clustering. In: Advances in Neural Information Processing Systems 20 (2007)
31. Asuncion, A., Newman, D.: UCI machine learning repository (2007)
32. Sim, T., Baker, S., Bsat, M.: The CMU pose, illumination and expression database. IEEE Transactions on Pattern Analysis and Machine Intelligence 25(12), 1615–1618 (2003)

# A Visualization-Based Exploratory Technique for Classifier Comparison with Respect to Multiple Metrics and Multiple Domains*

Rocío Alaiz-Rodríguez[1], Nathalie Japkowicz[2], and Peter Tischer[3]

[1] Dpto. de Ingeniería Eléctrica y de Sistemas, Universidad de León, Spain
[2] School of Information Technology and Engineering, University of Ottawa, Canada
[3] Clayton School of Information Technology, Monash University, Australia

## 1 Introduction

Classifier performance evaluation typically gives rise to a multitude of results that are difficult to interpret. On the one hand, a variety of different performance metrics can be applied, each adding a little bit more information about the classifiers than the others; and on the other hand, evaluation must be conducted on multiple domains to get a clear view of the classifier's general behaviour.

Caruana et al. [1] studied the issue of selecting appropriate metrics through a visualization method. In their work, the evaluation metrics are categorized into three groups and the relationship between the three groups is visualized. They propose a new metric, SAR, that combines the properties of the three groups.

Japkowicz et al. [2] studied the issue of aggregating the results obtained by different classifiers on several domains. They too use a visualization approach to implement a component-wise aggregation method that allows for a more precise combination of results than the usual averaging or win/loss/tie approaches.

In this demo, we present a visualization tool based on the combination of the above two techniques that allows the study of different classifiers with respect to both a variety of metrics and domains. We, thus, take the view that classifier evaluation should be done on an exploratory basis and provide a technique for doing so. This work is part of a research line that focuses on general issues regarding visualization and its potential benefits to the classifier evaluation process. Our aim is to adapt existing methods to suit our purpose and in this context, this paper extends a work based on MCST (Minimum Cost Spanning Tree) projection [2].

In particular, we assume that classifier evaluation requires two stages. In the first stage, the researcher should compute the results obtained by the various classifiers with respect to several representative metrics on several domains, in order to make the comparison as general as possible. This, of course, will create a considerable amount of data, which, in turn will need to be analyzed, in a second stage, in order to draw valid and useful conclusions about the algorithms

* Supported by the Natural Science and Engineering Council of Canada and the Spanish MEC projects TEC2005-06766-C03-01 and DPI2006-02550.

W. Daelemans et al. (Eds.): ECML PKDD 2008, Part II, LNAI 5212, pp. 660–665, 2008.

under study. We can say that this second stage is a data mining process in and of itself. The tool we are proposing is a visual data mining system for enabling this analysis. It is demonstrated on a study of 15 domains over three representative metrics as per Caruana et al. [1]. In particular, we demonstrate how our tool may allow us to combine information in a way that is more informative than the SAR metric [1].

## 2 Typical Study

Nine classifiers were evaluated by 10-fold cross-validation in the WEKA environment [3] with parameters set as default. Tables 1, 2 and 3 show the Error rate, RMSE and AUC, respectively for the 15 UCI domains assessed here (Sonar, Heart-v, Heart-c, Breast-y, Voting, Breast-w, Credits-g, Heart-s, Sick, Hepatitis, Credits-a, Horse-colic, Heart-h, Labor and Krkp).

**Table 1.** Error rate for different classifiers on several domains

| | ERROR RATE | | | | | | | | | | | | | | |
|---|---|---|---|---|---|---|---|---|---|---|---|---|---|---|---|
| | D1 | D2 | D3 | D4 | D5 | D6 | D7 | D8 | D9 | D10 | D11 | D12 | D13 | D14 | D15 |
| Ideal | 0 | 0 | 0 | 0 | 0 | 0 | 0 | 0 | 0 | 0 | 0 | 0 | 0 | 0 | 0 |
| Ib1 | 0.1342 | 0.2957 | 0.2378 | 0.2757 | 0.0986 | 0.0486 | 0.2800 | 0.2481 | 0.0381 | 0.1937 | 0.1884 | 0.1873 | 0.2317 | 0.1733 | 0.0372 |
| Ib10 | 0.2402 | 0.2160 | 0.1753 | 0.2699 | 0.1077 | 0.0357 | 0.2600 | 0.1851 | 0.0384 | 0.1737 | 0.1405 | 0.1686 | 0.1660 | 0.0833 | 0.0494 |
| NB | 0.3211 | 0.2360 | 0.1652 | 0.2830 | 0.1284 | 0.0400 | 0.2460 | 0.1629 | 0.0739 | 0.1554 | 0.2231 | 0.2200 | 0.1629 | 0.1000 | 0.1210 |
| C4.5 | 0.2883 | 0.2663 | 0.2248 | 0.2445 | 0.0917 | 0.0544 | 0.2950 | 0.2333 | 0.0119 | 0.1620 | 0.1391 | 0.1470 | 0.1893 | 0.2633 | 0.0056 |
| Bagging | 0.2545 | 0.2513 | 0.2080 | 0.2656 | 0.0895 | 0.0415 | 0.2600 | 0.2000 | 0.0127 | 0.1683 | 0.1463 | 0.1442 | 0.2105 | 0.1533 | 0.0056 |
| Boosting | 0.2219 | 0.2965 | 0.1786 | 0.3035 | 0.1010 | 0.0429 | 0.3040 | 0.1963 | 0.0082 | 0.1420 | 0.1579 | 0.1659 | 0.2142 | 0.1000 | 0.0050 |
| RF | 0.1926 | 0.2460 | 0.1850 | 0.3144 | 0.0965 | 0.0372 | 0.2730 | 0.2185 | 0.0188 | 0.2008 | 0.1492 | 0.1524 | 0.2177 | 0.1200 | 0.0122 |
| SVM | 0.2404 | 0.2463 | 0.1588 | 0.3036 | 0.0827 | 0.0300 | 0.2490 | 0.1592 | 0.0615 | 0.1483 | 0.1507 | 0.1740 | 0.1726 | 0.1033 | 0.0456 |
| JRip | 0.2692 | 0.2660 | 0.1854 | 0.2905 | 0.0986 | 0.0457 | 0.2830 | 0.2111 | 0.0177 | 0.2200 | 0.1420 | 0.1306 | 0.2104 | 0.2300 | 0.0081 |

**Table 2.** RMSE for different classifiers on several domains

| | RMSE | | | | | | | | | | | | | | |
|---|---|---|---|---|---|---|---|---|---|---|---|---|---|---|---|
| | D1 | D2 | D3 | D4 | D5 | D6 | D7 | D8 | D9 | D10 | D11 | D12 | D13 | D14 | D15 |
| Ideal | 0 | 0 | 0 | 0 | 0 | 0 | 0 | 0 | 0 | 0 | 0 | 0 | 0 | 0 | 0 |
| Ib1 | 0.3512 | 0.5342 | 0.3045 | 0.5042 | 0.2956 | 0.1860 | 0.5278 | 0.4848 | 0.1936 | 0.4252 | 0.4295 | 0.4261 | 0.2950 | 0.3197 | 0.1936 |
| Ib10 | 0.3931 | 0.4277 | 0.2179 | 0.4305 | 0.2649 | 0.1519 | 0.4193 | 0.3700 | 0.1699 | 0.3406 | 0.3298 | 0.3587 | 0.2192 | 0.3213 | 0.2458 |
| NB | 0.5263 | 0.4164 | 0.2256 | 0.4480 | 0.3310 | 0.1945 | 0.4186 | 0.3542 | 0.2285 | 0.3409 | 0.4346 | 0.4179 | 0.2238 | 0.1997 | 0.3018 |
| C4.5 | 0.5172 | 0.4531 | 0.2689 | 0.4311 | 0.2760 | 0.2105 | 0.4790 | 0.4526 | 0.1035 | 0.3565 | 0.3290 | 0.3521 | 0.2461 | 0.4209 | 0.0638 |
| Bagging | 0.3926 | 0.4177 | 0.2359 | 0.4335 | 0.2564 | 0.1769 | 0.4201 | 0.3768 | 0.0902 | 0.3388 | 0.3186 | 0.3440 | 0.2290 | 0.3412 | 0.0634 |
| Boosting | 0.4366 | 0.4700 | 0.2497 | 0.5105 | 0.2875 | 0.1864 | 0.5054 | 0.4294 | 0.0757 | 0.3507 | 0.3671 | 0.3690 | 0.2579 | 0.2281 | 0.0603 |
| RF | 0.3530 | 0.4166 | 0.2295 | 0.4686 | 0.2607 | 0.1615 | 0.4223 | 0.3912 | 0.1156 | 0.3512 | 0.3323 | 0.3376 | 0.2405 | 0.2962 | 0.1116 |
| SVM | 0.4837 | 0.4942 | 0.2872 | 0.5470 | 0.2667 | 0.1520 | 0.4979 | 0.3934 | 0.2479 | 0.3606 | 0.3837 | 0.4105 | 0.2885 | 0.2249 | 0.2110 |
| JRip | 0.4647 | 0.4360 | 0.2385 | 0.4475 | 0.2828 | 0.1932 | 0.44637 | 0.40846 | 0.1189 | 0.4075 | 0.3419 | 0.336 | 0.2574 | 0.3776 | 0.0782 |

Typical questions we would like to answer after the classifier performance analysis is performed are related to similarities/dissimilarities between classifiers: (a) Which classifiers perform similarly so that they can be considered equivalent? (b) Which classifiers could be worth combining? (c) Does the relative performance of the classifiers change as a function of data dimensionality? (d) Does it change for different domain complexities?

A first attempt at answering these questions could be to analyze directly the data gathered in the three tables. However, it does not seem straightforward given the quantity of results recorded (and there could be worse instances of this).

**Table 3.** AUC* (1-AUC) for different classifiers on several domains

| | AUC* (1-AUC) | | | | | | | | | | | | | | |
|---|---|---|---|---|---|---|---|---|---|---|---|---|---|---|---|
| | D1 | D2 | D3 | D4 | D5 | D6 | D7 | D8 | D9 | D10 | D11 | D12 | D13 | D14 | D15 |
| Ideal | 0 | 0 | 0 | 0 | 0 | 0 | 0 | 0 | 0 | 0 | 0 | 0 | 0 | 0 | 0 |
| Ib1 | 0.1361 | 0.4635 | 0.2403 | 0.3687 | 0.0622 | 0.0256 | 0.3400 | 0.2500 | 0.1912 | 0.3362 | 0.1917 | 0.2035 | 0.2512 | 0.1750 | 0.0105 |
| Ib10 | 0.1373 | 0.4102 | 0.0920 | 0.3201 | 0.0325 | 0.0759 | 0.2553 | 0.1244 | 0.0672 | 0.1890 | 0.0911 | 0.1366 | 0.1138 | 0.0500 | 0.0094 |
| NB | 0.2000 | 0.2826 | 0.0955 | 0.2845 | 0.0483 | 0.0120 | 0.2122 | 0.0994 | 0.0747 | 0.1408 | 0.1040 | 0.1501 | 0.1009 | 0.0125 | 0.0479 |
| C4.5 | 0.2653 | 0.3983 | 0.2032 | 0.3719 | 0.0629 | 0.0515 | 0.3534 | 0.2450 | 0.0505 | 0.3034 | 0.1064 | 0.1507 | 0.2341 | 0.2666 | 0.0012 |
| Bagging | 0.1478 | 0.2869 | 0.1296 | 0.3518 | 0.0362 | 0.0105 | 0.2469 | 0.1291 | 0.0050 | 0.1769 | 0.0771 | 0.1237 | 0.1178 | 0.1583 | 0.0007 |
| Boosting | 0.0938 | 0.3055 | 0.1187 | 0.3569 | 0.0370 | 0.0176 | 0.2770 | 0.1166 | 0.0123 | 0.2003 | 0.0945 | 0.1118 | 0.1389 | 0.0625 | 0.0007 |
| RF | 0.0889 | 0.2914 | 0.1215 | 0.3537 | 0.0376 | 0.0137 | 0.2499 | 0.1386 | 0.0072 | 0.1599 | 0.0886 | 0.1023 | 0.1444 | 0.0916 | 0.0012 |
| SVM | 0.2418 | 0.4335 | 0.1639 | 0.4072 | 0.0869 | 0.0316 | 0.3292 | 0.1633 | 0.5001 | 0.2487 | 0.1434 | 0.1912 | 0.2033 | 0.1250 | 0.0457 |
| JRip | 0.2631 | 0.4366 | 0.1591 | 0.3877 | 0.0839 | 0.0368 | 0.3871 | 0.2041 | 0.0579 | 0.3960 | 0.1285 | 0.1562 | 0.2427 | 0.2416 | 0.0055 |

As an alternative, metrics like SAR try to summarize all the gathered information with a point estimation. Thus, SAR carries out the projection $SAR^* = (1 - SAR) = RMSE + Error + AUC^*$ where $AUC^* = (1 - AUC)$. The closer to zero the SAR values (and all its components) are, the better the classifier performs. Table 4 shows the classifiers' performance values and ranking according to the SAR metric. We consider, however, that combining metrics uniformly may be dangerous. Instead we argue that we should select the information that is relevant to our purpose and concentrate on it to conduct the performance analysis.

**Table 4.** Classifier Ranking according to SAR

| Ideal | RF | Bagging | Ib10 | Boosting | NB | JRip | C4.5 | SVM | Ib1 |
|---|---|---|---|---|---|---|---|---|---|
| 0 | .1958 | .1965 | .2001 | .2037 | .2126 | .2362 | .2365 | .2420 | .2530 |

Visual data mining allows to easily discover data patterns, a task that may be difficult by simply looking at the results organized in tables and inaccurate when summarized by a SAR-like measure. In this work, we demonstrate the use of MDS (MultiDimensional Scaling) to visualize the classifiers in a graph, so that interpoint distances in the high dimensional (metric/domain) space are preserved as much as possible in the 2D space. The technique we propose to conduct the classifier performance analysis has been implemented under Matlab. Performance data, however, can be loaded in standard file formats.

Let us now study what information may be extracted from a graphic where the information provided in Tables 1, 2 and 3 is not simply averaged (over domains and over different metrics) but is projected using MDS. The distance between two points is calculated as the Euclidean distance and the stress criterion (see below) is normalized by the sum of squares of the interpoint distances.

Before starting to explore the graphical representation, it is interesting to assess the stress criterion. It is important to know how much of the original data structure is preserved after projecting the data to two dimensions. We can also get an idea of the information gained when moving from a one dimensional

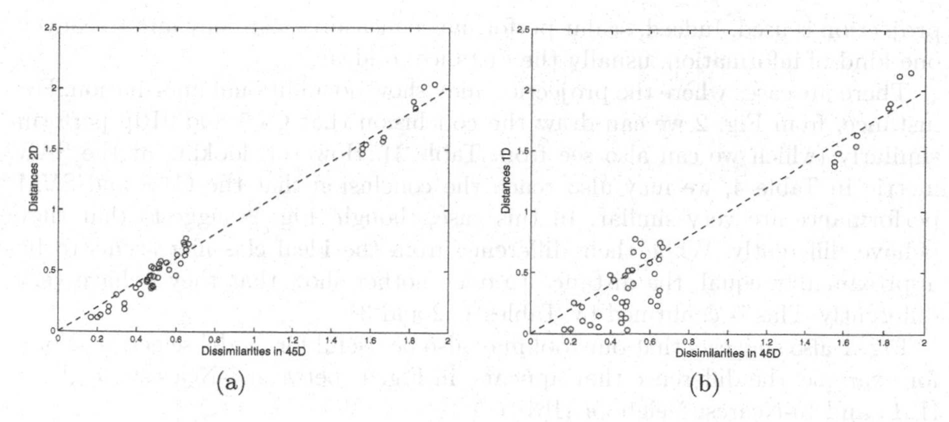


**Fig. 1.** Shepard plot for the metric MDS projection: (a) from 45 to 2 dimensions. (b) from 45 to 1 dimension.

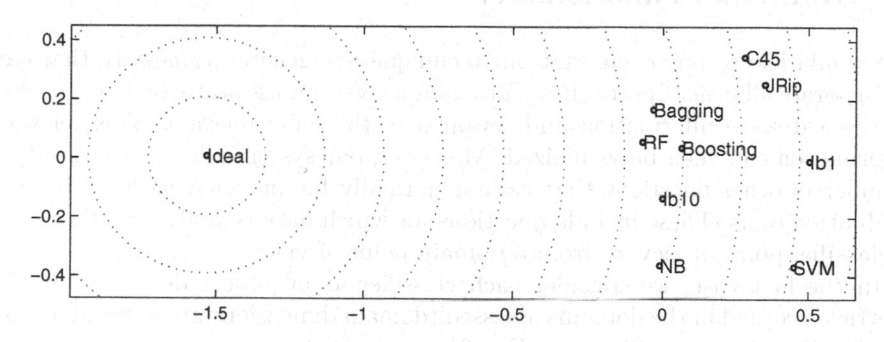


**Fig. 2.** metric MDS projection from 45 dimensions to 2 dimensions based on the RMSE, AUC* and Error rate gathered over 15 domains

representation to a two dimensional one. In our example the stress becomes 0.08 for two dimensions (not much loss of information), but it increases to 0.22 when considering only one dimension. This is supported by the Shepard plot in Fig. 1 that shows the reproduced distances in the new projected space (y axis) versus the dissimilarities in the original space (x axis). It can be seen that a projection to 2D leads to a narrow scatter around the ideal fit, while the scatter with a projection to 1D becomes larger and indicates a higher loss of information.

Now we focus on the whole information (Error rate, RMSE and AUC*) reflected in Fig. 2. In this particular case, we analyze nine classifiers described in the original high dimensional space by 45 dimensions (3 metrics X 15 domains) and then, projected to a 2-dimensional space. The ideal classifier is also introduced, to allow us to compare classifiers by their projected distance to the ideal classifier as well as to their relative position with respect to the other classifiers. Note that this second type of information is lost when a one-dimensional projection is used. Indeed, scalar performance measures, can only aim to convey one kind of information, usually the distance to ideal.[1]

There are cases where the projection may show no additional information. For instance, from Fig. 2 we can draw the conclusion that C4.5 and JRip perform similarly (which we can also see from Table 4). However, looking at the SAR metric in Table 4, we may also reach the conclusion that the C4.5 and SVM performance are very similar. In this case, though, Fig. 2 suggests that they behave differently. While their difference from the ideal classifier seems to be approximately equal, the distance to one another show that they behave very differently. This is confirmed by Tables 1, 2 and 3.

Fig. 1 also suggests that our tool may also be useful for model selection. Note, for example, the difference that appears, in Fig. 1, between 1-Nearest Neighbor (Ib1) and 10-Nearest Neighbor (Ib10).

## 3 Additional Functionality

We would like to point out that our technique is a general framework that can incorporate all other approaches. For example, we can use statistical approaches to discard some information and retain only the most relevant. This relevant information can then be visualized. Moreover, our system allows us to study a number of other questions that cannot normally be answered with traditional evaluation tools. These include questions for which data is analyzed either from a classifier point of view or from a domain point of view.

In the first case, we consider each classifier as an object described by the metrics recorded in the domains assessed (domain dimensions are reduced during the MDS projection as shown in Fig. 2).

In the second case, we can regard each domain as an object with attributes which are a measure of how several classifiers have performed on that domain. Note that the attributes are classifier performance measures and the classifier dimensions are the ones reduced. For example, let us assume that we concentrate on the posterior probability capabilities measured by the RMSE metric. Fig. 3 shows the similarities/dissimilarities among domains in terms of the difficulty for the classifiers to estimate posterior probabilities. The ideal domain D0, for which the estimation is perfect, is included for reference purposes. It is now feasible to identify groups of domains (e.g., {D3, D13, D5} or {D2, D7, D4}) for which the task of estimating posterior probabilities has similar complexity and conduct a further analysis within them.

Some questions our technique allows to address include the following:

- **Classifier-Centric Questions:**
  - Can the classifiers be organized into equivalence classes that perform similarly on a variety of domains?

[1] This is not the only type of information that gets lost, by the way, since, once in two dimensions, a lot more flexibility is possible, especially if we consider colours, motion pictures, and potentially more.

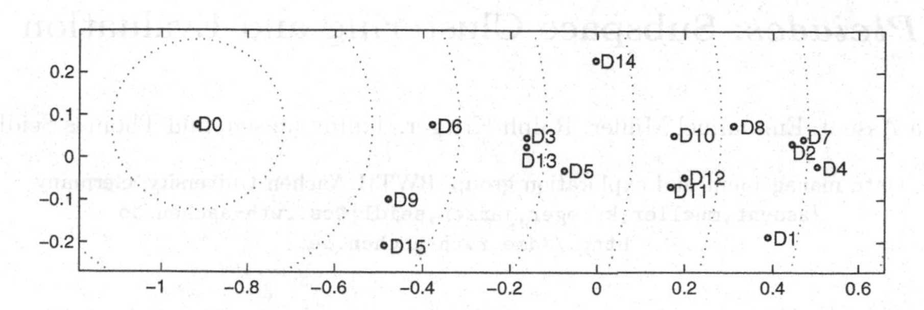


**Fig. 3.** metric MDS projection from 15 dimensions to 2 dimensions based on RMSE metric gathered for nine classifiers

  - In what way are the classifiers similar or different from one another?
  - Which classifiers would it be beneficial to combine? Which combinations would not improve the results?
- **Domain-Centric Questions:**
  - Can domains be organized into equivalence classes within which various classes of classifiers behave predictably?
  - What domain characteristics influence the behaviour of different domains (e.g., domain complexity, dimensionality, etc.)?

## References

1. Caruana, R., Niculescu-Mizil, A.: Data mining in metric space: An empirical analysis of suppervised learning performance criteria. In: Proceedings of the Tenth International Conference on Knowledge Discovery and Data Mining (KDD 2004) (2004)
2. Japkowicz, N., Sanghi, T.P.: A projection-based framework for classifier performance evaluation. In: Daelemans, W., Goethals, B., Morik, K. (eds.) ECML PKDD 2008, Part I. LNCS, vol. 5211. Springer, Heidelberg (2008)
3. Witten, I.H., Frank, E.: Data Mining: Practical Machine Learning Tools and Techniques with Java Implementations. Morgan Kaufmann, San Francisco (1999)

# *Pleiades*: Subspace Clustering and Evaluation

Ira Assent, Emmanuel Müller, Ralph Krieger, Timm Jansen, and Thomas Seidl

Data management and exploration group, RWTH Aachen University, Germany
{assent,mueller,krieger,jansen,seidl}@cs.rwth-aachen.de
http://dme.rwth-aachen.de

**Abstract.** Subspace clustering mines the clusters present in locally relevant subsets of the attributes. In the literature, several approaches have been suggested along with different measures for quality assessment.

*Pleiades* provides the means for easy comparison and evaluation of different subspace clustering approaches, along with several quality measures specific for subspace clustering as well as extensibility to further application areas and algorithms. It extends the popular WEKA mining tools, allowing for contrasting results with existing algorithms and data sets.

## 1 *Pleiades*

In high dimensional data, clustering is hindered through many irrelevant dimensions (cf. "curse of dimensionality" [1]). Subspace clustering identifies locally relevant subspace projections for individual clusters [2]. As these are recent proposals, subspace clustering algorithms and respective quality measures are not available in existing data mining tools. To allow researchers and students alike to explore the strengths and weaknesses of different approaches, our *Pleiades* system provides the means for their comparison and analysis as well as easy extensibility.

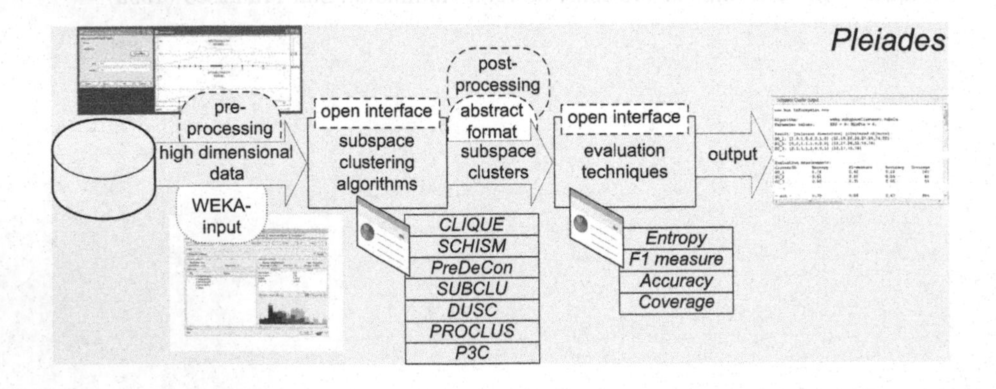


**Fig. 1.** Data processing in *Pleiades*

W. Daelemans et al. (Eds.): ECML PKDD 2008, Part II, LNAI 5212, pp. 666–671, 2008.

Figure 1 gives an overview over *Pleiades*: High-dimensional data is imported using the WEKA input format, possibly using pre-processing tools already available in WEKA and those we specifically added for subspace clustering [3]. The subspace clustering algorithms described in the next section are all part of the *Pleiades* system, and new algorithms can be added using our new subspace clustering interface. The result is presented to the user, and can be subjected to post-processing (e.g. filtering). *Pleiades* offers several evaluation techniques, as described in the next section, to measure the quality of the subspace clustering. Further measures can be plugged into our new evaluation interface.

## 2 *Pleiades* Features

Our *Pleiades* system is a tool that integrates subspace clustering algorithms along with measures designated for the assessment of subspace clustering quality.

### 2.1 Subspace Clustering Algorithms

While subspace clustering is a rather young area that has been researched for only one decade, several distinct paradigms can be observed in the literature. Our *Pleiades* system includes representatives of these paradigms to provide an overview over the techniques available. We provide implementations of the most recent approaches from different paradigms:

**Grid-based subspace clustering** discretizes the data space for efficient detection of dense grid cells in a bottom-up fashion. It was introduced in the *CLIQUE* approach which exploits monotonicity on the density of grid cells for pruning [4]. *SCHISM* [5] extends *CLIQUE* using a variable threshold adapted to the dimensionality of the subspace as well as efficient heuristics for pruning.

**Density-based subspace clustering** defines clusters as dense areas separated by sparsely populated areas. In *SUBCLU*, a density monotonicity property is used to prune subspaces in a bottom-up fashion [6]. *PreDeCon* extends this paradigm by introducing the concept of subspace preference weights to determine axis parallel projections [7]. In *DUSC*, dimensionality bias is removed by normalizing the density with respect to the dimensionality of the subspace [8].

**Projected clustering** methods identify disjoint clusters in subspace projections. *PROCLUS* extends the k-medoid algorithm by iteratively refining a full-space k-medoid clustering in a top-down manner [9]. *P3C* combines one-dimensional cluster cores to higher-dimensional clusters bottom-up [10].

### 2.2 Evaluation Techniques

Quality of clustering or classification is usually measured in terms of accuracy, i.e. the ratio of correctly classified or clustered objects. For clustering, the “ground truth”, i.e. the true clustering structure of the data, is usually not known. In fact, it is the very goal of clustering to detect this structure. As a consequence, clustering algorithms are often evaluated manually, ideally with the help of domain experts. However, domain experts are not always available, and they might not agree on the quality of the result. Their assessment of the clustering result is necessarily only based on the result itself, it cannot be compared to the "optimum" which is not known. Moreover, manual evaluation does not scale to large datasets or clustering result outcomes.

For more realistic evaluation of clustering algorithms, large scale analysis is therefore typically based on pre-labelled data, e.g. from classification applications [10,11]. The underlying assumption is that the clustering structure typically reflects the class label assignment. At least for relative comparisons of clustering algorithms, this provides measures of the quality of the clustering result.

Our *Pleiades* system provides the measures proposed in recent subspace clustering publications. In Figure 2 we present the evaluation output with various measures for comparing subspace clustering results.

Quality can be determined as **entropy and coverage**. Corresponding roughly to the measures of precision and recall, entropy accounts for purity of the clustering (e.g. in [5]), while coverage measures the size of the clustering, i.e. the percentage of objects in any subspace cluster. *Pleiades* provides both coverage and entropy (for readability, inverse entropy as a percentage) [8].

The **F1-value** is commonly used in evaluation of classifiers and recently also for subspace or projected clustering as well [10]. It is computed as the harmonic mean of recall ("are all clusters detected?") and precision ("are the clusters accurately detected?"). The F1-value of the whole clustering is simply the average of all F1-values.

**Accuracy** of classifiers (e.g. C4.5 decision tree) built on the detected patterns compared with the accuracy of the same classifier on the original data is another

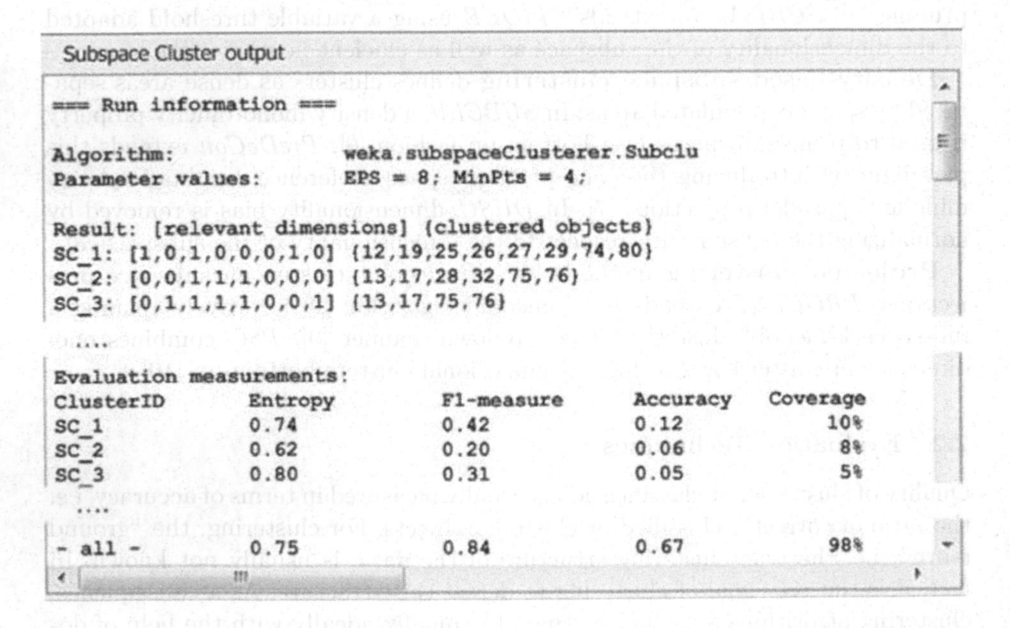
Subspace Cluster output

```
=== Run information ===

Algorithm:          weka.subspaceClusterer.Subclu
Parameter values:   EPS = 8; MinPts = 4;

Result: [relevant dimensions] {clustered objects}
SC_1: [1,0,1,0,0,0,1,0] {12,19,25,26,27,29,74,80}
SC_2: [0,0,1,1,1,0,0,0] {13,17,28,32,75,76}
SC_3: [0,1,1,1,1,0,0,1] {13,17,75,76}
....
Evaluation measurements:
ClusterID   Entropy   F1-measure   Accuracy   Coverage
SC_1        0.74      0.42         0.12       10%
SC_2        0.62      0.20         0.06        8%
SC_3        0.80      0.31         0.05        5%
....
- all -     0.75      0.84         0.67       98%
```

**Fig. 2.** Evaluation screen

quality measure [11]. It indicates to which extend the subspace clustering successfully generalizes the underlying data distribution.

## 2.3 Applicability and Extensibility

Our *Pleiades* system provides the means for researches and students of data mining to use and compare different subspace clustering techniques all in one system. Different evaluation measures allow in-depth analysis of the algorithm properties. The interconnection to the WEKA data mining system allows further comparison with existing full space clustering techniques, as well as pre- and post-processing tools [3]. We provide additional tools for pre- and post-processing for subspace clustering. Figure 3 gives an example of our novel visual assistance for parameter setting in *Pleiades*.

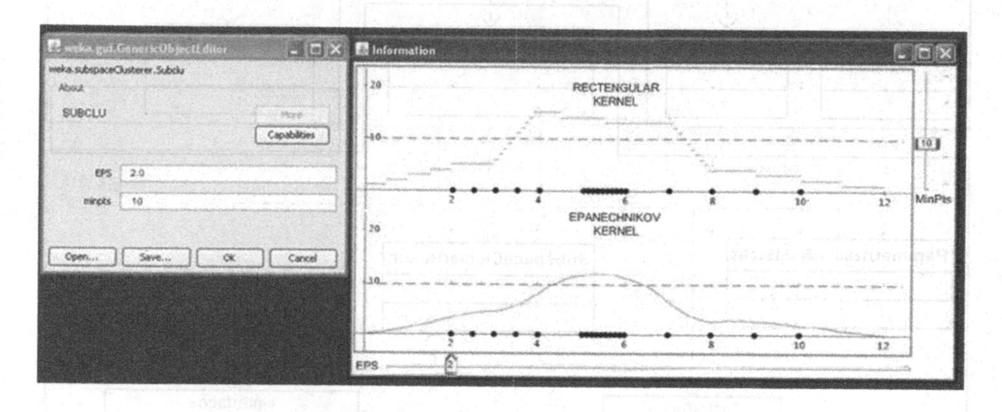


**Fig. 3.** Parametrization screen

*Pleiades* incorporates two open interfaces, which enable extensibility to further subspace clustering algorithms and new evaluation measurements. In Figure 4 we show the main classes of the *Pleiades* system which extends the WEKA framework by a new subspace clustering panel.

Subspace clustering shows major differences compared to traditional clustering; e.g. an object can be part of several subspace clusters in different projections. We therefore do not extend the clustering panel, but provide a separate subspace clustering panel.

Recent subspace clustering algorithms described in Section 2.1 are implemented based on our *Pleiades* system. The abstraction of subspace clustering properties in *Pleiades* allows to easily add new algorithms through our new subspace clustering interface.

Furthermore, *Pleiades* offers several evaluation techniques (cf. Section 2.2) to measure the quality of the subspace clustering. By using these evaluation measures one can easily compare different subspace clustering techniques. Further measures can be added by our new evaluation interface, which allows to define new quality criteria for subspace clustering.

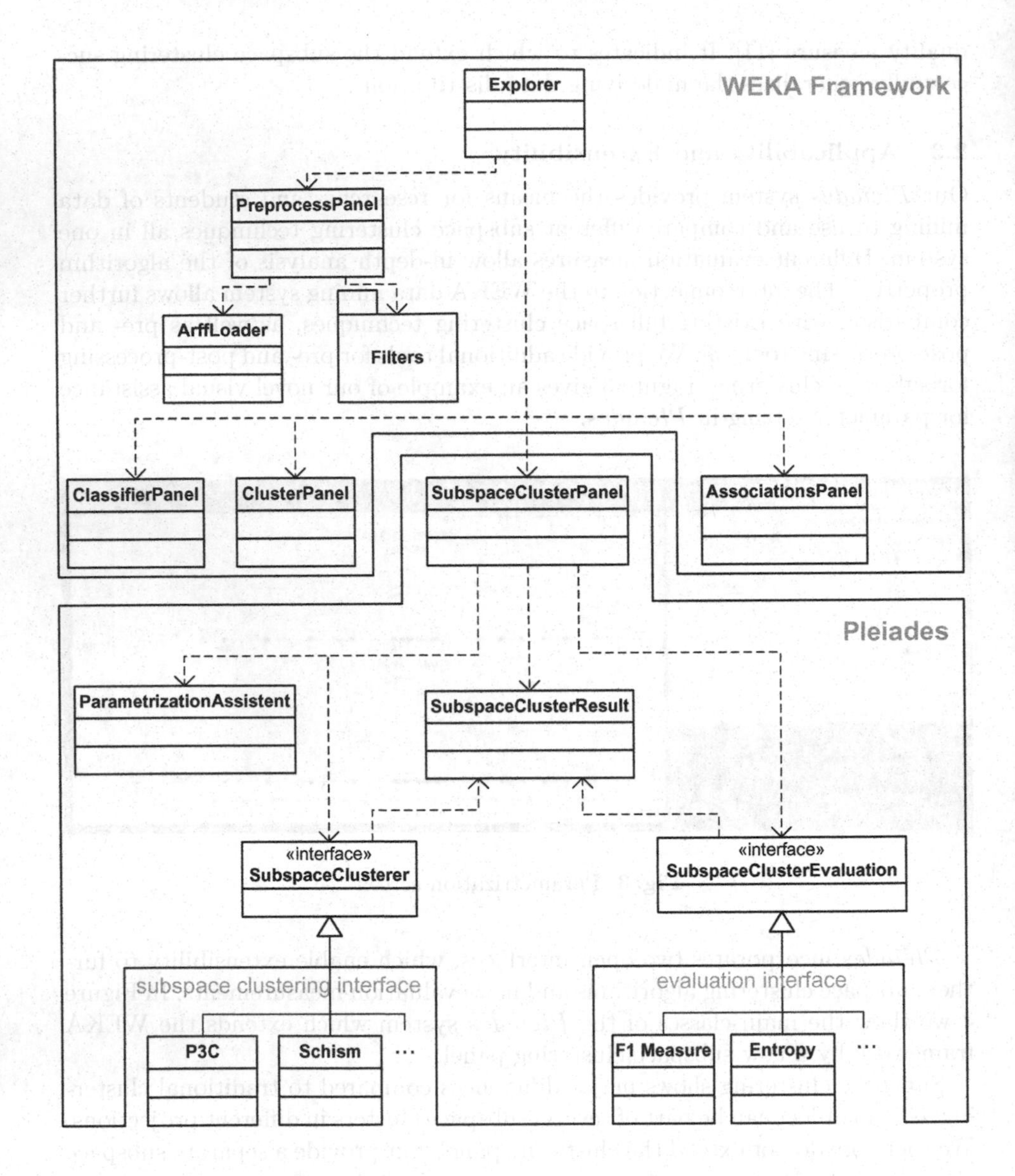


**Fig. 4.** UML class diagram of the *Pleiades* system

# 3 *Pleiades* Demonstrator

The demo will illustrate the above subspace clustering and evaluation techniques for several data sets. It will allow conference attendees to explore the diverse paradigms and measures implemented in the *Pleiades* system, thus raising research interest in the area. Open interfaces will facilitate extension with further subspace clustering algorithms and evaluation measures by other researchers.

## References


1. Beyer, K., Goldstein, J., Ramakrishnan, R., Shaft, U.: When is nearest neighbors meaningful. In: Proceedings International Conference on Database Theory (ICDT), pp. 217–235 (1999)
2. Parsons, L., Haque, E., Liu, H.: Subspace clustering for high dimensional data: a review. ACM SIGKDD Explorations Newsletter 6(1), 90–105 (2004)
3. Witten, I., Frank, E.: Data Mining: Practical Machine Learning Tools and Techniques. Morgan Kaufmann Publishers Inc., San Francisco (2005)
4. Agrawal, R., Gehrke, J., Gunopulos, D., Raghavan, P.: Automatic subspace clustering of high dimensional data for data mining applications. In: Proceedings ACM SIGMOD International Conference on Management of Data, pp. 94–105 (1998)
5. Sequeira, K., Zaki, M.: SCHISM: A new approach for interesting subspace mining. In: Proceedings of the IEEE International Conference on Data Mining (ICDM), pp. 186–193 (2004)
6. Kailing, K., Kriegel, H.P., Kröger, P.: Density-connected subspace clustering for high-dimensional data. In: Proceedings of the IEEE International Conference on Data Mining (ICDM), pp. 246–257 (2004)
7. Böhm, C., Kailing, K., Kriegel, H.P., Kröger, P.: Density connected clustering with local subspace preferences. In: Proceedings of the IEEE International Conference on Data Mining (ICDM), pp. 27–34 (2004)
8. Assent, I., Krieger, R., Müller, E., Seidl, T.: DUSC: Dimensionality unbiased subspace clustering. In: Proceedings of the IEEE International Conference on Data Mining (ICDM), pp. 409–414 (2007)
9. Aggarwal, C., Wolf, J., Yu, P., Procopiuc, C., Park, J.: Fast algorithms for projected clustering. In: Proceedings ACM SIGMOD International Conference on Management of Data, pp. 61–72 (1999)
10. Moise, G., Sander, J., Ester, M.: P3C: A robust projected clustering algorithm. In: Proceedings of the IEEE International Conference on Data Mining (ICDM), pp. 414–425. IEEE Computer Society Press, Los Alamitos (2006)
11. Bringmann, B., Zimmermann, A.: The chosen few: On identifying valuable patterns. In: Proceedings of the IEEE International Conference on Data Mining (ICDM), pp. 63–72 (2007)

# SEDiL: Software for Edit Distance Learning*

Laurent Boyer[1], Yann Esposito[1], Amaury Habrard[2],
Jose Oncina[3], and Marc Sebban[1]

[1] Laboratoire Hubert Curien, Université de Saint-Etienne, France
{laurent.boyer,yann.esposito,marc.sebban}@univ-st-etienne.fr

[2] Laboratoire d'Informatique Fondamentale, CNRS, Aix Marseille Université, France
amaury.habrard@lif.univ-mrs.fr

[3] Dep. de Languajes y Sistemas Informatico, Universidad de Alicante, Spain
oncina@dlsi.ua.es

**Abstract.** In this paper, we present SEDiL, a **S***oftware for* **E***dit* **Di***stance* **L***earning*. SEDiL is an innovative prototype implementation grouping together most of the state of the art methods [1,2,3,4] that aim to automatically learn the parameters of string and tree edit distances.

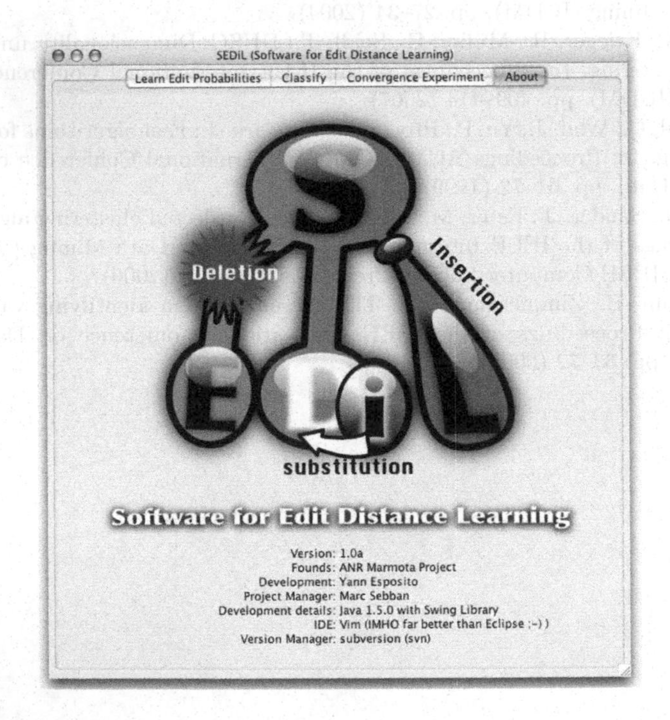


* This work was funded by the French *ANR Marmota project*, the *Pascal Network of Excellence* and the Spanish research programme *Consolider Ingenio-2010 (CSD2007-00018)*. This publication only reflects the authors' views.

W. Daelemans et al. (Eds.): ECML PKDD 2008, Part II, LNAI 5212, pp. 672–677, 2008.

## 1 Introduction

In many machine learning tasks, there is an increasing need for defining similarity measures between objects. In the case of structured data, such as strings or trees, one natural candidate is the well-known edit distance (ED) [5,6]. It is based on edit operations (*insertion, deletion* and *substitution*) required for changing an input data (*e.g.* a misspelled word) into an output one (*e.g.* its correction). In general, an *a priori* fixed cost is associated to each edit operation. Without any specific knowledge, these costs are usually set to 1. However, in many real world applications, such a strategy clearly appears insufficient. To overcome this drawback and to capture background knowledge, supervised learning has been used during the last few years for learning the parameters of edit distances [1,2,3,4,7,8,9], often by maximizing the likelihood of a learning set. The learned models usually take the form of state machines such as stochastic transducers or probabilistic automata.

Since 2004, three laboratories (the LaHC from France, the DLSI from Spain, both members of the RedEx PASCAL, and the LIF from France) have joined their efforts to propose new stochastic models of string and tree EDs in order to outperform not only the standard ED algorithms [5,6] but also the first generative learning methods proposed in [1,2]. This joint work has lead to publications in the previous conferences ECML'06 [7] and ECML'07 [4], and in Pattern Recognition [3,8]. This research has also received funding from the RedEx PASCAL in the form of a pump-priming project in 2007. Since no software platform was available, a part of this financial help has been used to implement these new innovative prototypes in the platform SEDIL [10].

The rest of this paper is organized as follows: in Section 2, we present a brief survey of ED learning methods. Then, Section 3 is devoted to the presentation of our platform SEDIL and its innovative aspects.

## 2 Related Work

### 2.1 Standard ED

The standard ED between two strings (or trees) is the minimal cost to transform by edit operations an input data into an output one. To each operation (*insertion, deletion* and *substitution*) is assigned a so-called *edit cost*. Let us suppose that strings or trees are defined over a finite alphabet $\Sigma$ and that the empty symbol is denoted by $\lambda \notin \Sigma$.

**Definition 1.** *An edit script $e = e_1 \cdots e_n$ is a sequence of edit operations $e_i = (a_i, b_i) \in (\Sigma \cup \{\lambda\}) \times (\Sigma \cup \{\lambda\})$ allowing the transformation of an input data $X$ into an output one $Y$. The cost of an edit script $\pi(e)$ is the sum of the costs of the edit operations involved in the script:*

$$\pi(e) = \sum_{i=1}^{n} c(e_i).$$

*Let $S(X,Y)$ be the set of all the scripts that enable the emission of $Y$ given $X$, the edit distance between $X$ and $Y$ is defined by:*

$$d(X,Y) = min_{e \in S(X,Y)} \pi(e).$$

In order to improve the efficiency of the ED to solve machine learning tasks, learning algorithms have been developed to capture background knowledge from examples in order to learn relevant edit costs.

### 2.2 ED Learning by Memoryless State Machines

When there is no reason that the cost of a given edit operation changes according to the context where the operation occurs, *memoryless* models (see [1,3] in the case of strings and [4] for trees) are very efficient from an accuracy and algorithmic point of view. In practice, many applications satisfy this hypothesis. For instance, the correction of typing errors made with a computer keyboard, the recognition of handwritten digits represented with Freeman codes [3] are some examples of problems that can be solved by these *memoryless* state machines. In such a context, these models learn - by likelihood maximization - one matrix of edit costs not only useful for computing edit similarities between strings or trees, but also to model knowledge of the domain. Actually, by mining the learned matrix it enables us to answer questions such that "*What are the letters of a keyboard at the origin of the main typing errors?*" or "*What are the possible graphical distortions of a given handwritten character?*". To give an illustration of the output of these models, let us suppose that strings are built from an alphabet $\Sigma = \{a, b\}$. Memoryless models return a probability $\delta$ for each possible edit operation in the form of a $3 \times 3$ matrix as shown in Fig.1.

| $\delta$ | $\lambda$ | $a$ | $b$ |
|---|---|---|---|
| $\lambda$ | – | 0.3 | 0.1 |
| $a$ | 0.05 | 0.0 | 0.2 |
| $b$ | 0.05 | 0.1 | 0.2 |

**Fig. 1.** Matrix $\delta$ of edit probabilities

On this toy example, $\delta(a, b) = 0.2$ means that the letter $a$ has a probability of 0.2 to be changed into a $b$. From these edit probabilities, it is then possible, using dynamic programming, to compute in a quadratic complexity the probability $p(X,Y)$ to change an input $X$ into an output $Y$. Note here that, all the possible ways for transforming $X$ into $Y$ are taken into account, in opposition to the standard ED which considers, in general, only the minimal cost. Ristad and Yianilos [1] showed that one can obtain a so-called *stochastic ED* $d_s$ by computing

$$d_s(X,Y) = -\log p(X,Y).$$

### 2.3 ED Learning by Non-memoryless State Machines

In specific applications, the edit operation has an impact more or less important in the data transformation according to its location. For instance, in molecular biology, it is common knowledge that the probability to change a symbol into another one depends on its membership of a transcription factor binding site. To deal with such situations, *non-memoryless* approaches have been proposed [2,9] in the form of probabilistic state machines that are able to take into account the string context[1]. So, the output of the model is not a 2-dimensional matrix anymore, but rather a 3-dimensional matrix where the third dimension represents the state where the operation occurs.

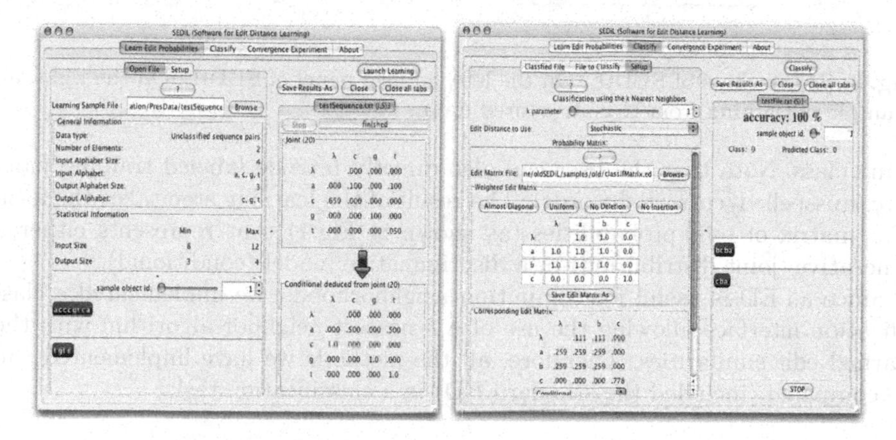

**Fig. 2.** Screen-shots of SEDIL. On the left, an example of learning from strings; On the right, an illustration of the classification step.

## 3 Presentation of the Platform SEDIL

The platform SEDIL groups together the state of the art ED learning algorithms. It is an open source software written in Java and can be used on a local machine using Java Web Start technology or directly from our web with an applet [10]. Some screen-shots of the application are presented on Figures 2 and 3. SEDIL enables us to perform two kinds of tasks: *learning* and *classification.*

### 3.1 Edit Parameters Learning and Classification

All the algorithms we implemented are based on an adaptation of the Expectation-Maximization algorithm to the context of EDs. From a learning sample of labeled pairs of strings or trees, the system automatically generates learning pairs either randomly (by choosing an output data in the same class) or by associating to each example its nearest neighbor from instances of the

[1] Note that so far, no approach has been proposed for trees.

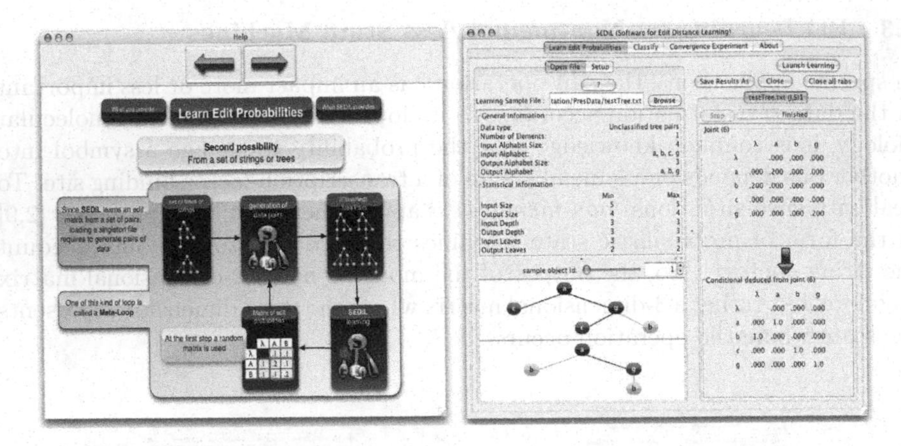


**Fig. 3.** Screen-shots of SEDiL. On the left, an help panel of SEDiL; On the right, an example of learning from tree-structured data.

same class. Note that the user can also directly provide labeled training pairs (*e.g.* misspelled/corrected words). The result of the learning step takes the form of a matrix of edit probabilities (as shown in Fig.1) that represents either a generative (joint distribution) or a discriminative model (conditional).

Since an ED is useful for computing neighborhoods, we implemented a classification interface allowing the use of a $k$-nearest neighbor algorithm with the learned edit similarities. Therefore, all the methods we have implemented can be compared (included the standard ED) in a classification task.

### 3.2 Contribution of SEDiL for the ML and DM Communities

**What makes our piece of software unique and special?**
As far as we know, there does not exist any available software platform to automatically learn the parameters of similarity measures based on the notion of ED. By implementing SEDiL, we filled this gap.

**What are the innovative aspects or in what way/area does it represent the state of the art?**
The two main historical references in ED learning are [1,2]. We implemented not only this state of the art but also all the recent new approaches we published during the past 4 years. In its current version, SEDiL provides five learning methods which are based on the following scientific papers:

- The learning of an edit pair-Hidden Markov Model (Durbin et al. 1998 [2]).
- The learning of a joint stochastic string ED (Ristad & Yianilos [1]).
- The learning of a conditional stochastic string ED (Oncina & Sebban [3]).
- The learning of a stochastic tree ED with deletion and insertion operations on subtrees (Bernard et al. 2006 [7,8]).
- The learning of a stochastic tree ED with deletion and insertion operations on nodes (Boyer et al. 2007 [4]).

**For whom is it most interesting/useful?**
SEDiL has already been used in various situations not only by academic but also by industrial structures. In the context of the ANR BINGO Project (`http://users.info.unicaen.fr/~bruno/bingo`), SEDiL has been used to search regularities in sequences of promoters in molecular biology. It is used at the moment in music recognition by the laboratory DLSI of the University of Alicante. Moreover, it has been recently exploited in a handwritten recognition task by the company A2IA (`http://www.a2ia.com/`). More generally, SEDiL is devoted to researchers who aim to compute distances between structured data.

## 4 Conclusion

SEDiL is the first platform providing several ED learning algorithms. It is devoted to improvements. We plan to implement a new edit model in the form of constrained state machines able to take into account background knowledge and a conditional random field approach recently presented in [9].

## References

1. Ristad, S., Yianilos, P.: Learning string-edit distance. IEEE Transactions on Pattern Analysis and Machine Intelligence 20(5), 522–532 (1998)
2. Durbin, R., Eddy, S., krogh, A., Mitchison, G.: Biological sequence analysis. Probalistic models of protein and nucleic acids. Cambridge University Press, Cambridge (1998)
3. Oncina, J., Sebban, M.: Learning stochastic edit distance: application in handwritten character recognition. Journal of Pattern Recognition 39(9), 1575–1587 (2006)
4. Boyer, L., Habrard, A., Sebban, M.: Learning metrics between tree structured data: Application to image recognition. In: Kok, J.N., Koronacki, J., Lopez de Mantaras, R., Matwin, S., Mladenič, D., Skowron, A. (eds.) ECML 2007. LNCS (LNAI), vol. 4701, pp. 54–66. Springer, Heidelberg (2007)
5. Wagner, R., Fischer, M.: The string-to-string correction problem. Journal of the ACM (JACM) 21, 168–173 (1974)
6. Bille, P.: A survey on tree edit distance and related problem. Theoretical Computer Science 337(1-3), 217–239 (2005)
7. Bernard, M., Habrard, A., Sebban, M.: Learning stochastic tree edit distance. In: Fürnkranz, J., Scheffer, T., Spiliopoulou, M. (eds.) ECML 2006. LNCS (LNAI), vol. 4212, pp. 42–52. Springer, Heidelberg (2006)
8. Bernard, M., Boyer, L., Habrard, A., Sebban, M.: Learning probabilistic models of tree edit distance. Pattern Recognition 41(8), 2611–2629 (2008)
9. McCallum, A., Bellare, K., Pereira, F.: A conditional random field for discriminatively-trained finite-state string edit distance. In: Proceedings of the 21st Conference on Uncertainty in AI, pp. 388–395 (2005)
10. `http://labh-curien.univ-st-etienne.fr/informatique/SEDiL`

# Monitoring Patterns through an Integrated Management and Mining Tool

Evangelos E. Kotsifakos, Irene Ntoutsi, Yannis Vrahoritis, and Yannis Theodoridis

Department of Informatics, University of Piraeus,
80 Karaoli-Dimitriou St, GR-18534 Piraeus, Greece
{ek,ntoutsi,ytheod}@unipi.gr, jb@freemail.gr

**Abstract.** Patterns upon the data of many real applications are affected by changes in these data. We employ PATTERN-MINER tool to detect changes of clusterings extracted from dynamic data and thus, to provide insight on the dataset and to support strategic decisions. PATTERN-MINER, is an integrated environment for pattern (data mining model) management and mining that deals with the whole lifecycle of patterns, from their generation (using data mining techniques) to their storage and querying, putting also emphasis on the comparison between patterns and meta-mining operations over the extracted patterns. In the current version, PATTERN-MINER integrates also an algorithm and technique for monitoring patterns (currently clusters) over time.

**Keywords:** Pattern management, Pattern-base, pattern comparison, pattern monitoring.

## 1 Introduction

Clustering techniques are used to find interesting groups of similar objects in large datasets. As the data in most databases are changing dynamically, clusters upon these data are affected. A lot of research has been devoted in adapting the clusters to the changed dataset. On the other hand research has expanded and it is devoted in tracing of the cluster changes themselves, trying thus to reveal knowledge about the undelying dataset to support strategic decisions. For example, if a business analyst who studies customer profiles, could understand how such profiles change over time he/she could act towards a long-term proactive portfolio design instead of reactive portfolio adaptation.

We demonstrate Pattern-Miner, an integrated environment that deals with pattern modeling, storage and retrieval issues using state-of-the-art approaches in contrast to existing tools that deal with specific aspects of the pattern management problem, mostly storage. Pattern-Miner offers an environment that provides the capability not only to generate and manage the different types of patterns in a unified way, but also to apply more advanced operations over patterns, such as comparison, meta-mining and cluster monitoring without facing interoperability or incompatibility issues as if using different applications for each task. Pattern-Miner follows a modular architecture and integrates the different Data Mining components offering transparency to the end user. A previous version of Pattern-Miner has been demonstrated at KDD2008 conference. In the current

W. Daelemans et al. (Eds.): ECML PKDD 2008, Part II, LNAI 5212, pp. 678–683, 2008.

version, a major addition has been made. A module for monitoring patterns over time. At first we are dealing with cluster monitoring as it has wide application to many scientific or commercial fields.

In order to better understand the theoretical background of PATTERN-MINER we briefly present some basic notions on patterns, following the PBMS approach [5]. A *pattern* is a compact and rich in semantics representation of raw data. Patterns are stored in a *pattern base* for further analysis. The pattern base model consists of three layers: pattern types, patterns, and pattern classes. A *pattern type* is a description of the pattern structure, e.g. decision trees, association rules, etc. A pattern type is a quintuple $pt = (n, ss, ds, ms, f)$, where $n$ is the name of the pattern type, $ss$ (structure schema) describes the structure of the pattern type (e.g. the head and the body of an association rule), $ds$ (source schema) describes the dataset from which patterns are extracted, $ms$ (measure schema) defines the quality of the source data representation achieved by patterns (e.g. the support and the confidence in case of an association rule pattern) and $f$ is the formula that describes the relationship between the source data space and the pattern space. A *pattern* is an instance of the corresponding pattern type and a *class* is a collection of semantically related patterns of the same type.

## 2 Extended Pattern-Miner Architecture

Figure 1 depicts the PATTERN-MINER architecture, including the pattern-monitoring module. *PATTERN-MINER engine* lies in the core of the system arranging the communication between the different peripheral components providing also the end user interface.

In this section, we provide an overview of the funcionality of each module of PATTERN-MINER and we focus on the *Pattern monitoring module* that consists the most recent and advanced module.

**Pattern extraction and representation:** The *Data Mining engine* component is responsible for the extraction of patterns. We employ for this task *WEKA*, since it is an open source tool and offers a variety of algorithms for different Data Mining tasks

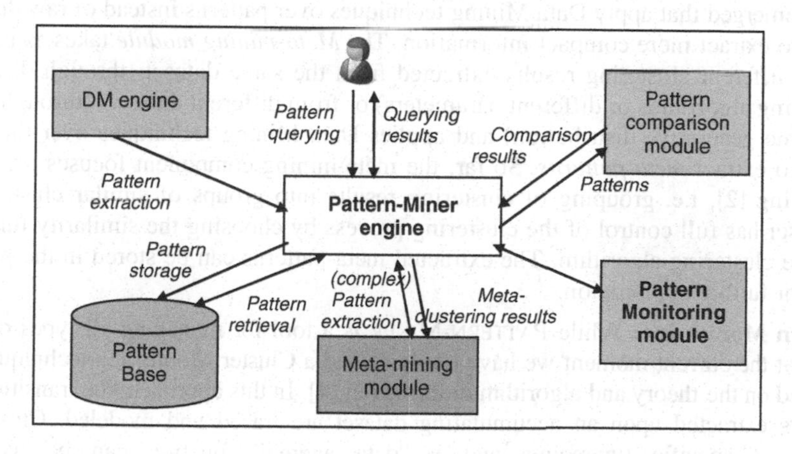


**Fig. 1.** The PATTERN-MINER architecture

as well as preprocessing capabilities over raw data. The output of the Data Mining process is represented with respect to the PBMS approach, described above. Several schemes have been proposed in the literature for the representation of patterns. The most popular choice is *PMML* [3], an XML-based standard that allows the definition of Data Mining and statistical models using a vendor-independent method. Different models are described through different XML schemes. In PATTERN-MINER, we adopt PMML for pattern representation and, thus, we convert the output of the *Data Mining engine* component into PMML format.

**Pattern storage and querying:** Since patterns are represented as XML documents (through PMML), a native XML database system is used for their storage in the *Pattern Base*. In particular, we employ the open source *Berkeley DBXML*, which comprises an extension of the Berkeley DB with the addition of an XML parser, XML indexes and the XQuery data query language. PATTERN-MINER provides a basic environment for querying the pattern base, through the XQuery language. Regarding the supported query types, the user can retrieve the whole pattern or some component of the pattern (measure and/ or structure), as well as to impose constraints over these components. The results are displayed in his/her screen and can be stored in the file system for future analysis.

**Pattern comparison:** One of the most important operations on patterns is that of *pattern comparison* with applications like querying (e.g. k-nearest neighbor queries) and change detection upon dynamic data [4]. Recognizing this fact, we distinguish the comparison process from the querying process and we implement it separately through the *Pattern comparison module*. The comparison is carried out on the basis of *PANDA* [1], a generic and flexible framework for the comparison of patterns defined over raw data and over other patterns as well. Comparison utilizes both structure and measure components of patterns. The user defines the patterns as well as the way that they should be compared, i.e. how the different components of PANDA are instantiated. The output is a dissimilarity score accompanied with a justification, a report actually of how the component patterns have been matched.

**Meta-mining:** Due to the large amount of extracted patterns, several approaches have lately emerged that apply Data Mining techniques over patterns instead of raw data, in order to extract more compact information. The *Meta-mining module* takes as input a set of different clustering results extracted from the same dataset (through different clustering algorithms or different parameters) or from different datasets (through from the same generative distribution) and applies Data Mining techniques over them, in order to extract *meta-patterns*. So far, the meta-mining component focuses on meta-clustering [2], i.e. grouping of clustering results into groups of similar clusterings. The user has full control of the clustering process by choosing the similarity function and the clustering algorithm. The extracted meta-patterns can be stored in the pattern base for further exploitation.

**Pattern Monitoring:** While PATTERN-MINER is a tool for managing all types of patterns, at the current moment we have implemented a Cluster Monitoring technique that is based on the theory and algorithm described in [4]. In this approach, the transitions of clusters extracted upon an accumulating dataset are traced and modeled. Clustering occurs at specific timepoints and a "data ageing" function can be used to assigns lower weights to all or some of the past records. The set of features used for clustering may also change during the period of observation, thus allowing for the inclusion of new features and the removal of obsolete ones. PATTERN-MINER assumes re-clustering rather than cluster adaptation at each timepoint, so that both changes in existing clusters and new clusters can be monitored. Transitions can be detected even when the underlying feature space changes, i.e. when cluster adaptation is not possible. Terms like cluster match, cluster overlap, cluster transition and lifetime of a cluster are core notions of cluster monitoring. This module exploits the clusterings that are stored in the pattern-base and employs the query and comparison capabilities of the system.

## 3 Demo Description

PATTERN-MINER is a tool that can be used in a lot of different areas, scientific or commercial. To point out the major advantages of the integrated environment of PATTERN-MINER we demonstrate a simple senario of a supermarket and its manager as the end-user. The supermarket has a database and everyday transactions are stored in it. The manager is interested in finding useful patterns in the data, like associations in the purchase of the products and clusters of customers with specific profiles and buying habits. Except from these simple patterns, the manager is interested in comparing clusters of customers or products discovered from the same dataset. Moreover, the manager wishes to monitor clusters of customer profiles over time, so he/she can capture any changes in buying preferences or habits. Are there any new clusters that describe a different customer profile? Some clusters may have been disappeared or shrinked while others could have been merged or expanded. PATTERN-MINER can be used to answer these questions supporting the manager on important decisions about strategies, campaigns, supplies etc. In the following paragraphs we describe the steps that the manager as the end-user would follow, to process the data from the dataset of the supermarket in order to extract and manage interesting patterns.

**Pattern extraction and storage:** The manager defines the data source (supermarket database), the Data Mining algorithm and its parameters. He/she would choose the apriori algorithm to find for example associations between products. To find clusters over the customer demographics to create profiles, the K-Means or the EM clustering algorithm would be appropriate. The extraction takes place in WEKA and the results are converted into PMML format before being stored in a user-specified container in the XML pattern base as well as, in a file on the hard disk.

**Pattern query:** The user defines, in the "Query pattetrn base" tab, the pattern set to be queried and the query in the Xquery language. PATTERN-MINER *engine* creates the connection to the pattern base, executes the query and returns the results to the user and also saves them to a file. A sample query is shown in **Figure 2**, described in both natural language and Xquery.

**Pattern comparison:** This module allows the user to define the patterns to be compared, e.g. sets of rules extracted from different months or clusters describing customer profiles. Then, the user chooses the appropriate comparison function from the candidate functions implemented for each pattern type. The results are returned to the user, who can detect any changes in the sales-patterns and decide whether these changes were expected (based on company's strategy) or not (indicating some suspicious or non-predictable behavior).

**Query (natural language):**
*Retrieve the clusters from the super_market dataset that have been extracted using EM algorithm.*
**Query (XQuery):**

```
declare namespace a="http://www.dmg.org/PMML-3_2";
collec-
tion("Clustering.dbxml")[dbxml:metadata("dbxml:dataFileName")="C:\Patter
nMiner\data_files\super_market_data.ARFF"]/a:PMML/a:ClusteringModel[@alg
orithmName="weka.clusterers.EM"]
```

**Fig. 2.** A sample query for the Clustering pattern-model

**Meta-mining:** The user defines the pattern sets to be used as input to the *Meta-mining module* (e.g. sets of rules extracted at each month of 2007), selects the clustering algorithm/ parameters, as well as the similarity measure between sets of rules. The input sets are clustered into groups of similar sets of rules (e.g. March and April are placed in the same group, since they depict similar buying behavior), which can be also stored in the pattern base for future use. The manager can exploit these results in order to decide similar strategies for months belonging to the same cluster.

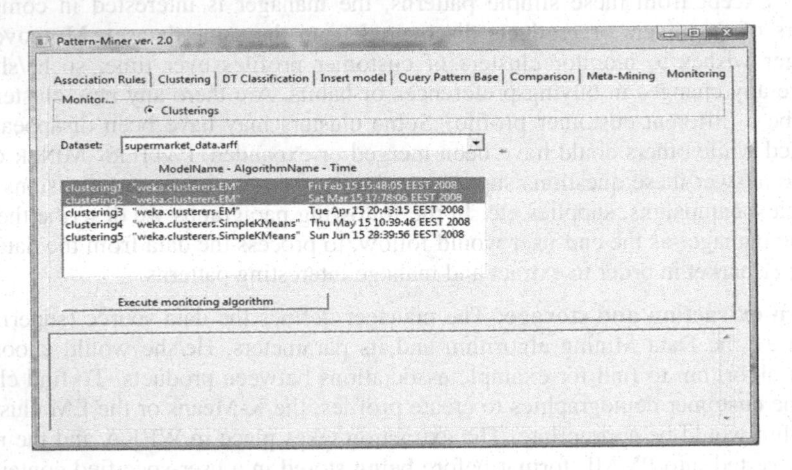


**Fig. 3.** Cluster Monitoring screenshot

**Cluster Monitoring:** User defines the dataset from which the clusters have been extracted. A list of all the clusterings that have been carried out over the spesific dataset is available to the user, sorted by the extraction time. The supermaket manager wants to observe the customer profiles over time. Choosing the apropriate dataset (supermarket.arff), Pattern-Miner returns all the different clusterings that have been created from that dataset, along with the clustering algorithm and the extraction time. The manager

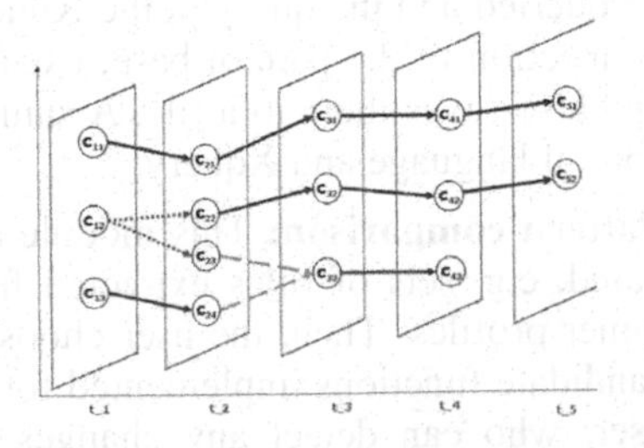

**Fig. 4.** Graphical reprepsentation of cluster monitoring output

chooses two or more clusterings and runs the cluster monitring process. This process results in a matrix showing the clusters of the first clustering and their changes over time (new clusters, clusters that no longer exists, shrinked or expanded clusters etc). Currently the output is in text format, representing the graph depicted in Figure 4.

## 4 Conclusions and Outlook

Advanced operations over patterns, like comparison, meta-mining and cluster monitoring while important for users of a variety of fields, are not supported from data mining or database systems. PATTERN-MINER is an integrated environment for pattern management that supports the whole lifecycle of patterns and also offers sophisticated comparison, meta-mining and cluster monitoring operations over patterns. It follows a modular architecture that employs state-of-the-art approaches at each component. Its advantage lies in the fact that all the operations related to the management of patterns (as data mining results) are integrated into one system in transparent to the user way. It is open source and easily expandable while, because of the use of PMML files, the exhange of data and results with other systems is a simple issue.

Several improvements though, can be carried out: First, existing components can be enhanced, e.g. querying could be improved through appropriate indices and new query types could be supported. Also, the *Meta-mining* and *cluster monitoring modules* can be extended so as to support more pattern types, like decision trees. Second, new components can be added, e.g. some visualization module for better interpretation of the results. Except for the scenario we described, other potential applications include cluster-based image retrieval, pattern validation, comparison of patterns extracted from different sites in a distributed environment setting, etc.

## References

1. Bartolini, I., Ciaccia, P., Ntoutsi, I., Patella, M., Theodoridis, Y.: A Unified and Flexible Framework for Comparing Simple and Complex Patterns. In: Boulicaut, J.-F., Esposito, F., Giannotti, F., Pedreschi, D. (eds.) PKDD 2004. LNCS (LNAI), vol. 3202, pp. 496–499. Springer, Heidelberg (2004)
2. Caruana, R., Elhawary, M., Nguyen, N., Smith, C.: Meta Clustering. In: Proc. ICDM (2006)
3. DMG - PMML, http://www.dmg.org/pmml-v3-1.html
4. Spiliopoulou, M., Ntoutsi, I., Theodoridis, Y., Schult, R.: MONIC: Modelling and monitoring cluster transitions. In: KDD (2006)
5. Terrovitis, P., Skiadopoulos, S., Bertino, E., Catania, B., Maddalena, A., Rizzi, S.: Modeling and language support for the management of pattern-bases. Data Knowl. Eng. 62(2) (August 2007)

# A Knowledge-Based Digital Dashboard for Higher Learning Institutions

[1]Wan Maseri Binti Wan Mohd, [2]Abdullah Embong, and [3]Jasni Mohd Zain

Faculty of Computer System & Software Engineering
University Malaysia Pahang
1maseri@ump.edu.my, aembong@ump.edu.my, asni@ump.edu.my

**Abstract.** We propose a novel approach of knowledge discovery method by adopting dashboard concept and incorporating elements of data clustering, visualization and knowledge codification. The dashboard was designed to help the higher institution to explore the insight of student performance by analyzing significant patterns and tacit knowledge of the experts in order to improve decision making process. The system has been developed using system development life cycle (SDLC) methodology and coded in web-based and open source environment. The dashboard architecture and software are presented in this paper.

**Keywords:** Knowledge Management, Digita Dashboard, Data Mining, Knowledge Discovery, Data Visualization, Knowledge Visualization.

## 1 Introduction

Major application of knowledge discovery identified for higher education institutions is student performance [1]. It is critical for the higher educationist to identify and analyze the relationships among different entities such as students, subjects, lecturers, environment and organizations to ensure the effectiveness of their important processes. Presently, Higher Education Institutions are just utilizing normal report with minimum knowledge exploration to analyze information regarding the performance of the institution such as student performance.

The objective of this project is to design and develop a University Dashboard system to improve the knowledge exploration by combining various techniques; MaxD K-means data clustering technique, graph-based visualization technique, knowledge management elements and dashboard concept. With the new approach of knowledge exploration, it shall helps the Board of Examiner or Senate Members to further explore the findings from the processing of information through the combination of clustering process and visualizing technique in identifying problems and actions to be taken in order to improve student performance.

## 2 A Novel Digital Dashboard for Higher Education Institution

### 2.1 Novelty of the Dashboard

The new digital dashboard for higher education is able to increase the utilization and effectiveness of the generated knowledge to facilitate the decision making process. It

W. Daelemans et al. (Eds.): ECML PKDD 2008, Part II, LNAI 5212, pp. 684–689, 2008.

incorporates Knowledge Codification Technique, Data Clustering (Data Mining), Forecasting Technique, Graph-based Visualization technique and Digital Dashboard concept. The new university dashboard has various advantages and uniqueness compared to the current dashboard as shown in Table 1.

**Table 1.** Comparison of Features and Uniqueness between new University Dashboard and current Dashboard

| No | New University Dashboard | Current Dashboard |
|---|---|---|
| 1 | Illustrate the Overall Summary of Student Performance at a glance using Dashboard Concept and incorporating Key Performance Indicator. | The current dashboard is just a normal reporting without incorporating the Key Performance Indicator and display details without summary for the University. |
| 2 | Incorporating Data Mining engine to cluster the Student's Result for knowledge exploration | The current system does not incorporate Data Mining process. The results are only go through normal information retrieval with predefine requirements |
| 3 | Incorporating Forecasting engine to forecast the Future Student Performance | The current system does not incorporate forecasting engine. It only displays the current status and history. |
| 4 | Incorporating Opinions from Education Experts on the student performance and suggestion for improvement as part of knowledge management mechanism. | The current report does not include expert opinion. Opinions are recorded as part of meeting minutes without proper documentation – no knowledge management mechanism. |
| 5 | Incorporating Related Publications such as article, conference proceeding, paperwork and technical report which relates to the student performance. | No sufficient research being done to improve student performance. The findings are not attached to the academic system for review or to help in decision making. |
| 6 | Incorporating Drill-down Features to zoom into Cluster's Behavior and Analysis by Faculty, year and race | Drill down only for faculty, year, race etc, but not to detail student's behavior. Student behavior is not captured in any university application. |
| 7 | Incorporating Knowledge Management Elements to process the Explicit Knowledge such as Data Clustering of Student's Result, Detail Student Behavior and Analysis of Results. | Very limited knowledge management for explicit knowledge. The current system just display result without exploring the insight of the information. |
| 8 | Incorporating Knowledge Management Elements to process the Tacit Knowledge such as Expert Opinion and Skill and Experience through publications such as knowledge bank in the terms of Technical Report, Article, Journal, Paperwork and etc. | The current systems display only the result of explicit knowledge without tacit knowledge. |

## 2.2 Applicability

Student Performance is one of the benchmark scale which reflect the Institution performance level. To really explore and improve the student performance, knowledge exploration is deemed crucial, thus the need for new KMS is highly demanded. Presently, Higher Education Institutions are just utilizing normal student report with minimum knowledge exploration to analyze student performance. With the new KMS, it shall helps the Board of Examiner or Senate Members to further explore the findings from the processing of information through the combination of knowledge processing and visualizing technique to identify problems and actions to be taken in order to improve student performance.

## 2.3 System Architecture

Fig. 1 shows the system architecture of the dashboard where the process starts from the data capturing from University Database, followed by data clustering process, visualizing process, forecasting process and finally viewing the final results through the dashboard view. Parameter-less K-Means clustering engine was used for data clustering process [2], Multi-layers graph-based technique was used to visualize the generated clusters [3] and fuzzy pattern matching algorithm was used to forecast future performance. At the same time, input from Key Performance Indicator (KPI) System will generate the dashboard for the targeted KPI as a comparison to the current status. For the above processing, interaction with the domain experts shall improve the intelligence of the system.

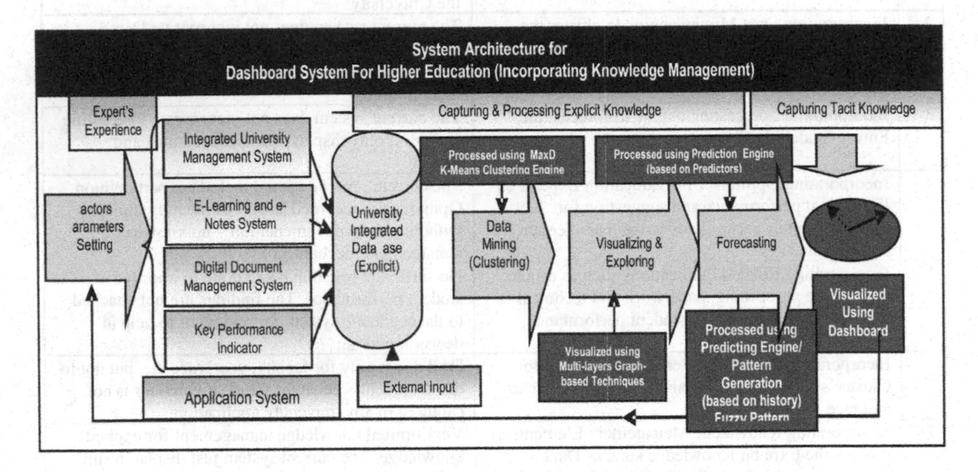


**Fig. 1.** System Architecture of University Dashboard

## 2.4 Snapshots of a Knowledge-Based Digital Dashboard

Fig.2 shows the Max-D K-means Clustering engine [2] used to cluster the student performance data and Fig. 3 shows the forecasting engine which forecast the future academic performance of using Fuzzy Pattern Matching Algorithm. The output of these processing is visualized through the digital dashboard [3] as shown in Fig.3(a) and Fig.3(b) which illustrate the use of dashboard concept with various knowledge exploration techniques; clustering, visualization and knowledge codification [4]. At a glance, the management of the institution can have a feel of their institutions in terms of student performance. In first segment of the dashboard, it shows the current status of the overall performance, followed by the forecast performance and pervious performance. As such, it allows the management to visualize and appreciate the actual performance compared to the previous and future performance. Furthermore, the dashboard shows the student clusters generated from clustering process of the student performance database in the form of graph. Each cluster can be drilled down using multi-layers graph-based technique to visualize the behavior of the students. The clusters allow the management to visualize the group of student in a particular category and further actions can be taken immediately. The dashboard is also comprises of expert opinion which is part of tacit knowledge and related publications for the subject matter as part of explicit knowledge. This segment illustrates the use of knowledge codification technique in the dashboard. Finally, the drill-down features of the dashboard is illustrated through the detail graphs whereby it shows the breakdown of the student performance by various dimensions such as faculty, gender, race etc. The detail information of each graph shall be displayed in the next layer of the dashboard, which becomes the sub-dashboard for the purpose of different group of users.

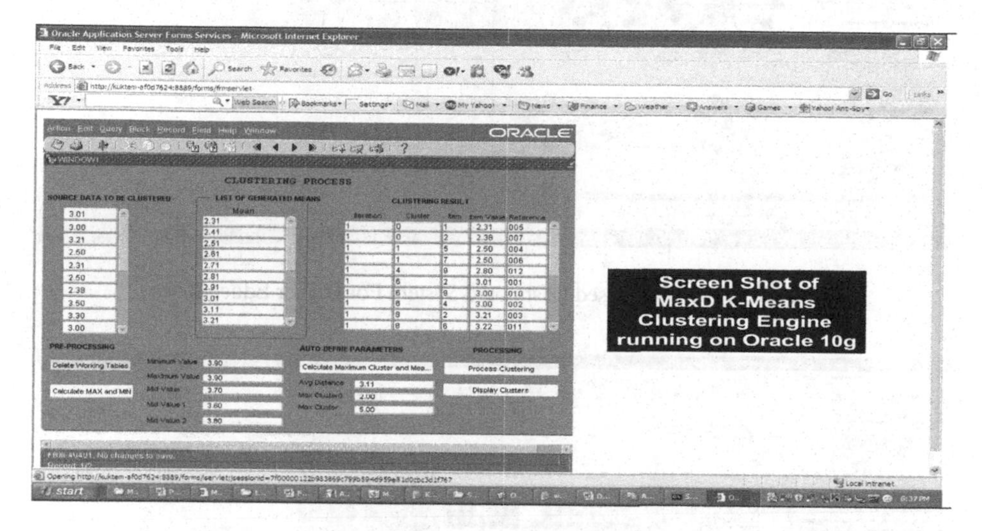


**Fig. 2.** Max-D Clustering Engine

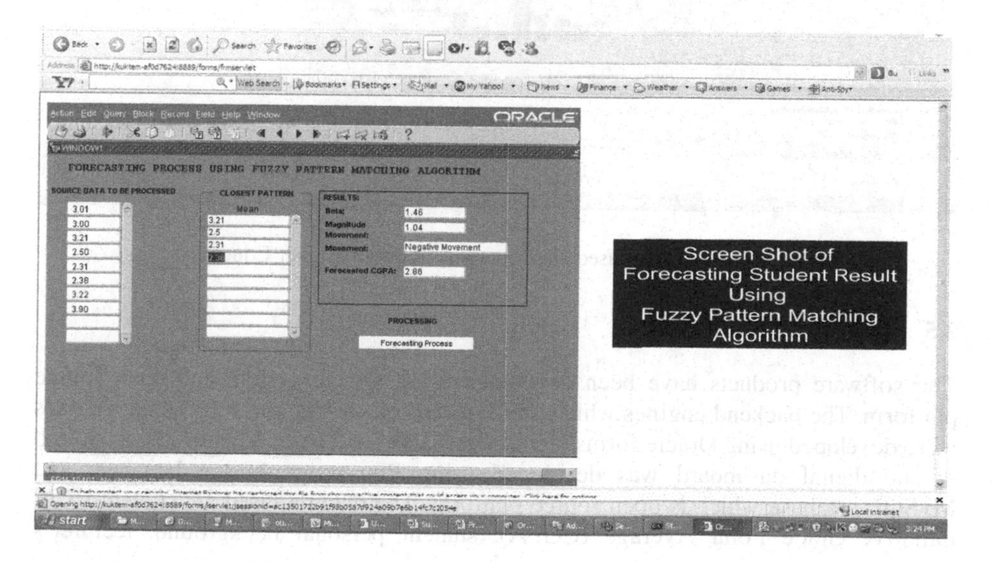


**Fig. 3.** Fuzzy Pattern Forecasting Engine

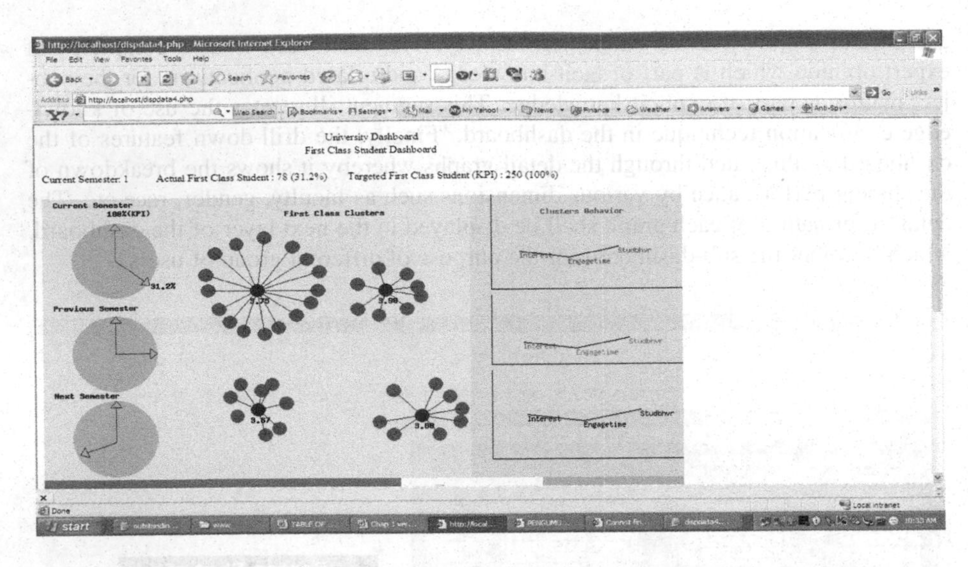


**Fig. 3. (a).** A Web-based Dashboard System For Higher Education

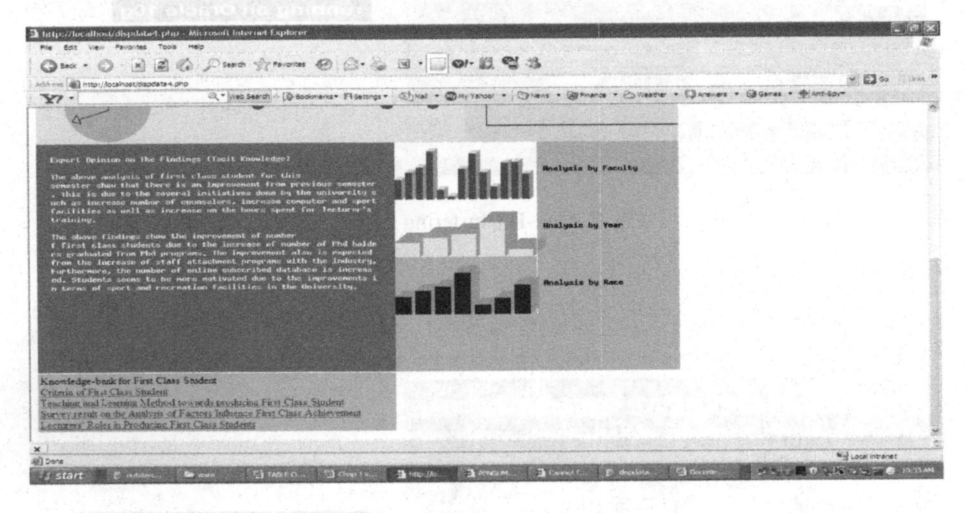


**Fig. 3. (b).** A Web-based Dashboard System For Higher Education

## 2.5 About the Product

The software products have been developed using Oracle system and open source platform. The backend engines which are K-means clustering and forecasting engines were developed using Oracle forms 6i and Oracle 10g database, whereas the front-end of the digital dashboard was developed using Php programming language and MySQL database which is open source platform. The data required are student's Cumulative Grade Point Average (CGPA), student personal background, lecturer's comments and related publications. The system is can be installed in any type of servers which is connected to the internet. The software can be executed by using any browser.

## 3 Conclusion

We propose a holistic design of Knowledge-based Dashboard system which shall improve the understanding of the information through clustering process and enhance the visual impact of the system. By designing the system as a dashboard and embedded with various data retrieval techniques; clustering, visualization and knowledge codification, it shall improve the understanding, visualization and personalization level of the data.

## References

1. Luan, J.: Data Mining and Its Applications in Higher Education. In: Serban, A., Luan, J. (eds.) Knowledge Management: Building a Competitive Advantage for Higher Education. New Directions for Institutional Research, vol. 113. Jossey Bass, San Francisco (2002)
2. Mohd, W.M.W., Embong, A., Zain, J.M.: Parameterless K-Means: Auto-generation of Centroids and Initial Number of Cluster based on Distance of Data Points. In: Malaysian-Japan International Symposium on Advanced Technology 2007 (2007)
3. Mohd, W.M.W., Embong, A., Zain, J.M.: Visual and Interactive Data Mining: A Framework for an Interactive Web-based Dashboard System for Higher Education using Graph-based Visualization Technique. In: Malaysian-Japan International Symposium on Advanced Technology 2007 (2007)
4. Mohd, W.M.W., Embong, A., Zain, J.M.: Enhancing Knowledge Codification Through Integration of Tacit-Explicit Knowledge Codification, Transfer Mechanism and Visualization Dashboard: A Holistic Model for Higher Learning Institutions. In: Malaysian-Japan International Symposium on Advanced Technology (2007)

# SINDBAD and SiQL: An Inductive Database and Query Language in the Relational Model

Jörg Wicker, Lothar Richter, Kristina Kessler, and Stefan Kramer

Technische Universität München, Institut für Informatik I12, Boltzmannstr. 3,
D-85748 Garching b. München, Germany
{joerg.wicker,lothar.richter,stefan.kramer}@in.tum.de

**Abstract.** In this demonstration, we will present the concepts and an implementation of an *inductive database* – as proposed by Imielinski and Mannila – in the relational model. The goal is to support all steps of the knowledge discovery process on the basis of queries to a database system. The query language SiQL (structured inductive query language), an SQL extension, offers query primitives for feature selection, discretization, pattern mining, clustering, instance-based learning and rule induction. A prototype system processing such queries was implemented as part of the SINDBAD (structured inductive database development) project. To support the analysis of multi-relational data, we incorporated multi-relational distance measures based on set distances and recursive descent. The inclusion of rule-based classification models made it necessary to extend the data model and software architecture significantly. The prototype is applied to three different data sets: gene expression analysis, gene regulation prediction and structure-activity relationships (SARs) of small molecules.

## 1 Introduction

Inductive databases are databases handling data, patterns and models, and supporting the complete knowledge discovery process on the basis of inductive query languages. Many of the recent proposals for inductive databases and constraint-based data mining are restricted to single pattern domains (such as itemsets or molecular fragments) or single tasks, such as pattern discovery or decision tree induction. Although the closure property is fulfilled by many of those approaches, the possibilities of combining various techniques in multi-step and compositional data mining are rather limited. In the demonstration, we present a prototype system, SINDBAD (structured inductive database development), supporting the most basic preprocessing and data mining operations such that they can be combined more or less arbitrarily. One explicit goal of the project is to support the complete knowledge discovery process, from pre-processing to post-processing, on the basis of database queries. The research extends ideas discussed at the Dagstuhl perspectives workshop "Data Mining: The Next Generation" [1], where a system of types and signatures of data manipulation and mining operators was proposed to support compositionality in the knowledge

W. Daelemans et al. (Eds.): ECML PKDD 2008, Part II, LNAI 5212, pp. 690–694, 2008.

discovery process. One of the main ideas was to use the simplest possible signature (mapping tables onto tables) as a starting point for the exploration of more complex scenarios. The relational model was chosen, as it possesses several desirable properties, from closure to convenient handling of collections of tuples. Moreover, it is possible to take advantage of mature and optimized database technology. Finally, systems supporting (variants of) SQL are well-known and established, making it easier to get users acquainted with new querying facilities.

## 2 SiQL and SINDBAD: Query Language and Demonstration Overview

SiQL (structured inductive database query language), the query language of the SINDBAD system, is a straightforward extension of SQL. Instead of just adding complicated data mining operators to SQL, we focused on incorporating small, but extensible and adjustable operators that can be combined to build more complex functions. The query language supports the knowledge discovery process by offering features for the successive transformation of data. As each pre-processing and data mining operator returns a table, the queries can be nested arbitrarily, and the kind of compositionality needed in multi-step data mining can be achieved easily. The mining operators were designed in analogy to relational algebra and SQL: For instance, we made heavy use of the extend-add-as operator and devised a feature-select clause in analogy to the select clause.

From each category of preprocessing/mining algorithms, we implemented the most fundamental representatives. For discretization, we included equal frequency and equal width, for feature selection a filter approach based on information gain or variance, for pattern mining, the computation of frequent itemsets using APriori, for clustering k-Medoids, and for classification k-nearest neighbor and rule induction (pFOIL, a propositional variant of FOIL [12]). External tools can be integrated via wrappers.

We adopted the `extend` [4] operator to add the results of various data mining operations as new attributes to a given relation. It computes a function for each tuple and adds the result as the value of a new attribute. In SINDBAD, the `extend` operator adds the results of clustering, instance- or rule-based predictions, and sampling to a table. For clustering/classification, the cluster/class membership is indicated by an additional attribute. In sampling, the sample membership determined by a random number generator is given in the new attribute. In this way, we can split datasets, for instance, into a training set and a test set. For clustering and instance-based learning (k-nearest neighbor), other methods for handling tuples and distances are provided as well.

One of the central concepts of SINDBAD is that of distances between objects. This is not restricted to tuples of a single relation. Using relational distance measures, it is possible to apply clustering and instance-based learning to multi-relational data [13]. Most relational distance measures are based on recursive descent and set distances, i.e., distances between sets of points. In the simplest case, the computation of a distance between two sets of tuples A and B boils down to computing the minimum distance between two elements of each set (single linkage), $d_{SL}(A, B) = \min_{a \in A, b \in B} d(a, b)$.

One of the most recent additions is the inclusion of full-fledged predictive models in the form of rule sets. For simplicity, we chose pFOIL, a propositional variant of the traditional FOIL algorithm [12]. The addition of models required significant extensions of the data model of the system. Models can be composed of component models. The evaluations of *component models* (e.g. class predictions) can be aggregated via *combining functions*. Combining functions can be defined in terms of logical or arithmetic operators. In this way, rule sets, weighted rule sets, trees, linear classifiers, and ensembles can be handled conveniently. For details of the query language and the implementation, we have to refer to more comprehensive publications [9,14].

In the demonstration, we will highlight some of the main features of SINDBAD in three real-world applications. In the first application, we test it on the gene expression data from Golub et al. [6], which contains the expression levels of genes from two different types of leukemia. In the second application (see tables 1,2 and 3), the task is to predict gene regulation dependent on the presence of binding sites and the state of regulators [5]. The third application is to predict anti-HIV activity for more than 40,000 small molecules [10].

**Table 1.** Relational schema of gene regulation data. The relation gene is the main table and connects genes with experimental setups and expression levels. The fun_cat relation gives the functional category membership of a gene according to the FunCat database. The third relation, has_tfbs indicates occurrence of transcription factor binding sites in respective genes whereas in the regulators table experimental conditions and activated regulators are given. The last table p_p_interaction gives the gene product interaction data.

```
gene(gene_id,              fun_cat(gene_id,
     cond_id,                      fun_cat_id)
     level)
has_tfbs(gene_id,          regulators(cond_id,
     yaac3_01,                     ybl005w,
     yacc1_01,                     ycl067c,
     yacs1_07,                     ydl214c,
     yacs2_01,                     ydr277c,
     ...)                          ...)
p_p_interaction(gene1_id,
     gene2_id)
```

## 3 Related Work and Conclusion

In the demonstration, we present a new and comprehensive approach to inductive databases in the relational model. The main contribution is a new inductive query language in the form of a SQL extension, including pre-processing and data mining operators. The approach is similar to the ones by Meo et al. [11], Imielinski and Virmani [8], and Han et al. [7] in extending SQL for

**Table 2.** k-Medoids for gene regulation prediction. The resulting table shows in column 2 the gene identifiers, in column 3 the experimental conditions, followed by the change of expression level and the cluster membership in column 3 and 4.

```
(30)> configure kmedoids_k = 5;
(31)> ...
(32)> extend gene add k medoid membership of gene;
(33)> show table gene;
row|gene_id|cond_id                 |level|cluster|
1  |YAL003W|2.5mM DTT 120 m dtt-1   |-1   |2      |
2  |YAL005C|2.5mM DTT 180 m dtt-1   |-1   |3      |
3  |YAL005C|1.5 mM diamide (20 m)  |+1   |5      |
4  |YAL005C|1.5 mM diamide (60 m)  |+1   |1      |
5  |YAL005C|aa starv 0.5 h         |-1   |2      |
...|...    |...                     |...  |...    |
```

**Table 3.** k-nearest neighbor for gene regulation prediction. This resulting table, column 2 and 3 are the same as in Table 2 followed by the predicted class label in column 4.

```
(40)> configure KNearestNeighbour_K = 10;
(41)> extend gene_test add knn prediction
     > of level from gene_train;
(42)> show table gene_test;
row|gene_id|cond_id                       |class|
1  |YBL064C|aa starv 1 h                  |+1   |
2  |YDL170W|YPD 3 d ypd-2                 |-1   |
3  |YER126C|Heat shock 40 minutes hs -1   |-1   |
4  |YJL109C|dtt 240 min dtt-2             |+1   |
5  |YKL180W|Nitrogen Depletion 1 d        |+1   |
...|...    |...                           |...  |
```

data mining. Another approach of Blockeel et al. [2] focuses on the automatic generation of mining views, in which the relevant parts are materialized on demand when a model is queried. The focus of the work is on the storage model and evaluation logic of data mining results. However, SINDBAD differs from related work [3] in – among other things – the support of pre-processing features. Also, it is a real prototype, useful for exploring concepts and requirements on such systems. Since it is at the moment far from clear what the requirements of a full-fledged inductive database will be, it is our belief that we can only find this out by building such prototype systems. Regarding implementations, the most similar work seems to be MS SQL Server [15]. In contrast to MS SQL Server, SINDBAD focuses on the successive transformation of data. Moreover, the approach presented here seems to be less rigid, especially with respect to feature selection, discretization, and pattern mining. Oracle uses a similar approach to provide data mining functionality in the Oracle Database (see `http://www.oracle.com/technology/documentation/datamining.html`). In future work, we are planning to investigate a more elaborate system of signatures and types. Type signatures would be useful to define admissible inputs and outputs of data manipulation and mining operations. Signatures of operators would enable first steps towards the optimization of inductive queries.

## References


1. Agrawal, R., Bollinger, T., Clifton, C.W., Dzeroski, S., Freytag, J.C., Gehrke, J., Hipp, J.: Data mining: The next generation. In: Agrawal, R., Freytag, J.-C., Ramakrishnan, R. (eds.) Report based on a Dagstuhl perspectives workshop (2005)
2. Blockeel, H., Calders, T., Fromont, E., Goethals, B., Prado, A.: Mining views: Database views for data mining. In: Proc. IEEE ICDE (2008)
3. Boulicaut, J.F., Masson, C.: Data mining query languages. In: Maimon, O., Rokach, L. (eds.) The Data Mining and Knowledge Discovery Handbook, pp. 715–727. Springer, Heidelberg (2005)
4. Date, C.J.: An Introduction to Database Systems, 4th edn. Addison-Wesley, Reading (1986)
5. Fröhler, S., Kramer, S.: Inductive logic programming for gene regulation prediction. Machine Learning 70(2-3), 225–240 (2008)
6. Golub, T.R., Slonim, D.K., Tamayo, P., Huard, P., Gaasenbeek, M., Mesirov, J.P., Coller, H., Loh, M.L., Downing, J.R., Caligiuri, M.A., Bloomfield, C.D., Lander, E.S.: Molecular classification of cancer: class discovery and class prediction by gene expression monitoring. Science 286(5439), 531–537 (1999)
7. Han, J., Fu, Y., Wang, W., Koperski, K., Zaiane, O.: DMQL: A data mining query language for relational databases. In: SIGMOD 1996 Workshop on Research Issues in Data Mining and Knowledge Discovery (DMKD 1996), Montreal, Canada (1996)
8. Imielinski, T., Virmani, A.: MSQL: A query language for database mining. Data Min. Knowl. Discov. 3(4), 373 (1999)
9. Kramer, S., Aufschild, V., Hapfelmeier, A., Jarasch, A., Kessler, K., Reckow, S., Wicker, J., Richter, L.: Inductive databases in the relational model: The data as the bridge. In: Bonchi, F., Boulicaut, J.-F. (eds.) KDID 2005. LNCS, vol. 3933, pp. 124–138. Springer, Heidelberg (2006)
10. Kramer, S., De Raedt, L., Helma, C.: Molecular feature mining in HIV data. In: Proceedings of the Seventh ACM SIGKDD International Conference on Knowledge Discovery and Data Mining (KDD 2001), pp. 136–143 (2001)
11. Meo, R., Psaila, G., Ceri, S.: An extension to sql for mining association rules. Data Mining and Knowledge Discovery 2(2), 195–224 (1998)
12. Quinlan, J.R.: Learning logical definitions from relations. Machine Learning 5, 239 (1990)
13. Ramon, J., Bruynoogh, M.: A polynomial time computable metric between point sets. Acta Informatica 37 (2001)
14. Richter, L., Wicker, J., Kessler, K., Kramer, S.: An inductive database and query language in the relational model. In: Proceedings of the 10th International Conference on Extending Database Technology (EDBT 2008), pp. 740–744. ACM Press, New York (2008)
15. Tang, Z.H., MacLennan, J.: Data mining with SQL Server 2005. Wiley, IN (2005)

# Author Index

# Lecture Notes in Artificial Intelligence (LNAI)

Vol. 4953: N.T. Nguyen, G.S. Jo, R.J. Howlett, L.C. Jain (Eds.), Agent and Multi-Agent Systems: Technologies and Applications. XX, 909 pages. 2008.

Vol. 4946: I. Rahwan, S. Parsons, C. Reed (Eds.), Argumentation in Multi-Agent Systems. X, 235 pages. 2008.

Vol. 4944: Z.W. Raś, S. Tsumoto, D.A. Zighed (Eds.), Mining Complex Data. X, 265 pages. 2008.

Vol. 4938: T. Tokunaga, A. Ortega (Eds.), Large-Scale Knowledge Resources. IX, 367 pages. 2008.

Vol. 4933: R. Medina, S. Obiedkov (Eds.), Formal Concept Analysis. XII, 325 pages. 2008.

Vol. 4930: I. Wachsmuth, G. Knoblich (Eds.), Modeling Communication with Robots and Virtual Humans. X, 337 pages. 2008.

Vol. 4929: M. Helmert, Understanding Planning Tasks. XIV, 270 pages. 2008.

Vol. 4924: D. Riaño (Ed.), Knowledge Management for Health Care Procedures. X, 161 pages. 2008.

Vol. 4923: S.B. Yahia, E.M. Nguifo, R. Belohlavek (Eds.), Concept Lattices and Their Applications. XII, 283 pages. 2008.

Vol. 4914: K. Satoh, A. Inokuchi, K. Nagao, T. Kawamura (Eds.), New Frontiers in Artificial Intelligence. X, 404 pages. 2008.

Vol. 4911: L. De Raedt, P. Frasconi, K. Kersting, S. Muggleton (Eds.), Probabilistic Inductive Logic Programming. VIII, 341 pages. 2008.

Vol. 4908: M. Dastani, A. El Fallah Seghrouchni, A. Ricci, M. Winikoff (Eds.), Programming Multi-Agent Systems. XII, 267 pages. 2008.

Vol. 4898: M. Kolp, B. Henderson-Sellers, H. Mouratidis, A. Garcia, A.K. Ghose, P. Bresciani (Eds.), Agent-Oriented Information Systems IV. X, 292 pages. 2008.

Vol. 4897: M. Baldoni, T.C. Son, M.B. van Riemsdijk, M. Winikoff (Eds.), Declarative Agent Languages and Technologies V. X, 245 pages. 2008.

Vol. 4894: H. Blockeel, J. Ramon, J. Shavlik, P. Tadepalli (Eds.), Inductive Logic Programming. XI, 307 pages. 2008.

Vol. 4885: M. Chetouani, A. Hussain, B. Gas, M. Milgram, J.-L. Zarader (Eds.), Advances in Nonlinear Speech Processing. XI, 284 pages. 2007.

Vol. 4884: P. Casanovas, G. Sartor, N. Casellas, R. Rubino (Eds.), Computable Models of the Law. XI, 341 pages. 2008.

Vol. 4874: J. Neves, M.F. Santos, J.M. Machado (Eds.), Progress in Artificial Intelligence. XVIII, 704 pages. 2007.

Vol. 4870: J.S. Sichman, J. Padget, S. Ossowski, P. Noriega (Eds.), Coordination, Organizations, Institutions, and Norms in Agent Systems III. XII, 331 pages. 2008.

Vol. 4869: F. Botana, T. Recio (Eds.), Automated Deduction in Geometry. X, 213 pages. 2007.

Vol. 4865: K. Tuyls, A. Nowe, Z. Guessoum, D. Kudenko (Eds.), Adaptive Agents and Multi-Agent Systems III. VIII, 255 pages. 2008.

Vol. 4850: M. Lungarella, F. Iida, J.C. Bongard, R. Pfeifer (Eds.), 50 Years of Artificial Intelligence. X, 399 pages. 2007.

Vol. 4845: N. Zhong, J. Liu, Y. Yao, J. Wu, S. Lu, K. Li (Eds.), Web Intelligence Meets Brain Informatics. XI, 516 pages. 2007.

Vol. 4840: L. Paletta, E. Rome (Eds.), Attention in Cognitive Systems. XI, 497 pages. 2007.

Vol. 4830: M.A. Orgun, J. Thornton (Eds.), AI 2007: Advances in Artificial Intelligence. XIX, 841 pages. 2007.

Vol. 4828: M. Randall, H.A. Abbass, J. Wiles (Eds.), Progress in Artificial Life. XII, 402 pages. 2007.

Vol. 4827: A. Gelbukh, Á.F. Kuri Morales (Eds.), MICAI 2007: Advances in Artificial Intelligence. XXIV, 1234 pages. 2007.

Vol. 4826: P. Perner, O. Salvetti (Eds.), Advances in Mass Data Analysis of Signals and Images in Medicine, Biotechnology and Chemistry. X, 183 pages. 2007.

Vol. 4819: T. Washio, Z.-H. Zhou, J.Z. Huang, X. Hu, J. Li, C. Xie, J. He, D. Zou, K.-C. Li, M.M. Freire (Eds.), Emerging Technologies in Knowledge Discovery and Data Mining. XIV, 675 pages. 2007.

Vol. 4811: O. Nasraoui, M. Spiliopoulou, J. Srivastava, B. Mobasher, B. Masand (Eds.), Advances in Web Mining and Web Usage Analysis. XII, 247 pages. 2007.

Vol. 4798: Z. Zhang, J.H. Siekmann (Eds.), Knowledge Science, Engineering and Management. XVI, 669 pages. 2007.

Vol. 4795: F. Schilder, G. Katz, J. Pustejovsky (Eds.), Annotating, Extracting and Reasoning about Time and Events. VII, 141 pages. 2007.

Vol. 4790: N. Dershowitz, A. Voronkov (Eds.), Logic for Programming, Artificial Intelligence, and Reasoning. XIII, 562 pages. 2007.

Vol. 4788: D. Borrajo, L. Castillo, J.M. Corchado (Eds.), Current Topics in Artificial Intelligence. XI, 280 pages. 2007.

Vol. 4775: A. Esposito, M. Faundez-Zanuy, E. Keller, M. Marinaro (Eds.), Verbal and Nonverbal Communication Behaviours. XII, 325 pages. 2007.

Vol. 4772: H. Prade, V.S. Subrahmanian (Eds.), Scalable Uncertainty Management. X, 277 pages. 2007.

Vol. 4766: N. Maudet, S. Parsons, I. Rahwan (Eds.), Argumentation in Multi-Agent Systems. XII, 211 pages. 2007.

Vol. 4760: E. Rome, J. Hertzberg, G. Dorffner (Eds.), Towards Affordance-Based Robot Control. IX, 211 pages. 2008.

Vol. 4755: V. Corruble, M. Takeda, E. Suzuki (Eds.), Discovery Science. XI, 298 pages. 2007.

Vol. 4754: M. Hutter, R.A. Servedio, E. Takimoto (Eds.), Algorithmic Learning Theory. XI, 403 pages. 2007.

Vol. 4737: B. Berendt, A. Hotho, D. Mladenic, G. Semeraro (Eds.), From Web to Social Web: Discovering and Deploying User and Content Profiles. XI, 161 pages. 2007.

Vol. 4733: R. Basili, M.T. Pazienza (Eds.), AI*IA 2007: Artificial Intelligence and Human-Oriented Computing. XVII, 858 pages. 2007.